P9-CLZ-409
10
11
11
13
13
13
14
14
29
29
30
30
30
31
32
33
48
49
51
51
52
54
54
55
62
64
65
66
66
66
67
67
70
71
71
72
73
73
74
75
79
80
81
82
84
84
84
91
92
93
94
95
97
98
105
107
108
109
110
110
111
118
118
118
119
119
120
120

A Field Guide to the BIRDS of South-East Asia

A Field Guide to the BIRDS of South-East Asia

CRAIG ROBSON

Illustrated by Richard Allen, Tim Worfolk, Stephen Message, Jan Wilczur, Clive Byers, Mike Langman, Ian Lewington, Christopher Schmidt, Andrew Mackay, John Cox, Anthony Disley, Hilary Burn, Daniel Cole and Martin Elliott

This edition published in 2011 by New Holland Publishers
London • Cape Town • Sydney • Auckland

www.newhollandpublishers.com

Garfield House, 86-88 Edgware Road, London W2 2EA, United Kingdom
80 McKenzie Street, Cape Town 8001, South Africa
Unit 1, 66 Gibbes Street, Chatswood, NSW 2067, Australia
218 Lake Road, Northcote, Auckland, New Zealand

10 9 8 7 6 5

A catalogue record for this book is available from the British Library.

ISBN 978 1 78009 049 8

Publisher: Simon Papps
Editor: Nigel Collar
Designer: Alan Marshall
Production: Joan Woodroffe
Index: Beth Lucas
Publishing Director: Rosemary Wilkinson

Reproduction by Modern Age Repro Co. Ltd, Hong Kong
Printed and bound by Tien Wah Press (Pte), Ltd in Singapore

CONTENTS

AUTHOR'S PREFACE

Almost ten years have now elapsed since the first edition of this field guide was published. During this time there has been a constant stream of new information on the birds of the region, particularly with regard to taxonomy and distribution. Some corrections and updates to the text and plates have already been published in *A Field Guide to the Birds of Thailand* (Robson 2002) and *New Holland Field Guide to the Birds of South-East Asia* (Robson 2005). However, both of these books had reduced, concise text, so the current work represents the first complete revision.

The past ten years have also seen an amazing change in our understanding of avian taxonomy, with a proliferation of papers on the subject, largely based on the study of DNA. In fact, so many papers are now being produced that the situation is being updated almost monthly. Combine this with an upsurge in home-grown birders and some very active locally based organisations in the region, and it is easy to see how information soon becomes out of date. More than 100 new references have been included.

This new edition covers 1,327 species, 76 more than the first edition. It includes at least 172 new illustrations, and more than 170 existing figures have been improved. There are now 120 plates, instead of the original 104. I have implemented widespread changes to the taxonomy and species order (see Taxonomy and nomenclature), and there are also many changes to the common and scientific names in general. Since the deadline for text completion, there has been news of a further new species for South-East Asia, Slaty-backed Gull *Larus schistisagus*. There have now been at least two records of first-winter birds from the coast of Central Thailand that have been photographed and reliably identified, in November 2002 and December 2007-January 2008.

Unfortunately, deadlines and time restraints are placed on authors when producing such a book and although much time has been spent trawling literature and corresponding with visitors to the region, some readers might find occasional errors and omissions. The author (c/o the publishers, New Holland) would be pleased to receive any information which updates or corrects that presented herein, in the hope that a further updated edition may appear in the future.

ACKNOWLEDGEMENTS

First and foremost, I would like to extend my sincere thanks to all of the artists for the great effort that they put into the very complex colour plates.

Special thanks must also go to Graeme Green and his co-worker Colin Bushell for their painstaking work checking a large proportion of the original colour plates against the bird collections at the Natural History Museum, Tring, in order to provide accurate correction guidelines for the artists. Carole and Graeme Green also went out of their way to accommodate me during my many stints at Tring. At the museum itself, I am indebted to the staff (Mark Adams, Robert Prys-Jones, Frank Steinheimer, Cyril Walker, Michael Walters and F.E 'Effie' Warr) for their unbridled cooperation and assistance, despite their own overloaded schedules, during my numerous visits over many years.

I am also very grateful to Nigel Collar and Jo Hemmings for their considerable support and efforts, often under difficult circumstances, above and beyond their respective duties. Lorna Sharrock, Jane Morrow, Charlotte Judet, Marianne Taylor, Beth Lucas and Simon Papps at New Holland all showed a high level of commitment to the project for which I am very grateful. I also owe my gratitude to the designer Alan Marshall, at Heron Recreations, for doing such a great job.

The following people were kind enough to provide me with comments on certain species groups: Per Alström (pipits, wagtails, larks, warblers), Dave Bakewell (plovers), Bill Clark (raptors), Peter Clement (thrushes), Martin Elliott (gulls), Johannes Erritzoe (cuckoos), David Gibbs (pigeons), Roy Hargreaves (accentors, pipits, wagtails, larks), Peter Kennerley (plovers, *Locustella* warblers), Killian Mullarney (waders, terns, gulls) and Barry Taylor (rails).

A number of people also allowed me valuable access to significant works either in preparation, in press or unpublished: Per Alström (bushlarks), Per Alström and Urban Olsson (*Seicercus* warblers), Peter Davidson (status and distribution of birds in Cambodia and Laos), Pamela Rasmussen (South Asian birds), Jonathan Eames (new subspecies from Vietnam), Tim Inskipp (annotated checklists and bibliographies for Myanmar and Indochina), Steve Madge (*Arborophila* partridges), John Pilgrim (Vietnamese birds), Colin Poole (Cambodian birds), Subaraj Rajathurai (Singapore birds), and David Wells (Thai-Malay Peninsula birds).

Tim Inskipp also assisted me greatly by supplying certain important reference material throughout the project.

Many people provided other kinds of assistance. For the use of tape recordings, I am indebted to Des Allen, Hem Baral, David Bishop, Peter Davidson, Nick Dymond, Jonathan Eames, Dave Farrow, Simon Harrap, Ben King and Edward Vercruysse; for help with photographic references, I would like to thank Krys Kazmierczak, Pamela Rasmussen, Philip Round, Paul Sweet, Uthai Treesucon and Tim Worfolk; and for supplying me with valuable information, Des Allen, Per Alström, Dave Bakewell, Bird Conservation Society of Thailand Records Committee, Seb Buckton, Nigel Collar, Peter Davidson, Edward Dickinson, Will Duckworth, Nick Dymond, Jonathan Eames, James Eaton, Martin Elliott, Tom Evans, Tony Htin Hla, Tim Inskipp, Mikhail Kalyakin, Peter Kennerley, Robert Kirk, Guy Kirwan, Frank Lambert, Paul Leader, Lim Kim Seng, Steve Madge, Killian Mullarney, Nivesh Nadee, Kiyoaki Ozaki, John Pilgrim, Colin Poole, Le Hai Quang, Subaraj Rajathurai, Pamela Rasmussen, Roger Safford, Yoshimitsu Shigeta, Tony Stones, Rob Timmins, Joost van der Ven, David Wells, and James Wolstencroft. I would like to extend a special thankyou to Philip Round, the world authority on Thai birds, who has constantly helped me by supplying important information during this and previous projects.

SOUTH-EAST ASIA

In this work, South-East Asia is primarily a political area, consisting of Myanmar (Burma), Thailand, Peninsular Malaysia, Singapore, Cambodia, Laos and Vietnam and, in the absence of distribution maps, is further divided into ornithological regions (see endpapers). The ornithological regions largely follow those which appeared in Ben King's *Birds of South-East Asia* (King *et al.* 1975) but have been modified in a number of ways. In Myanmar, SW ('Arakan') is treated as a separate unit, W has been expanded to include the southern part of 'Upper Chindwin' and N replaces 'NE' and includes the northern part of 'Upper Chindwin'. In Thailand, the exact borders of regions are modified slightly, following Boonsong Lekagul and Round (1991), W replaces 'SW' and S replaces 'Peninsular'. In Vietnam, 'Tonkin' is divided into W Tonkin and E Tonkin, following Vo Quy (1983) and the extent of N Annam is slightly reduced.

The term 'Indochina' is used to describe Cambodia, Laos and Vietnam collectively.

TAXONOMY AND NOMENCLATURE

In the first edition, the taxonomy, sequence and nomenclature followed *An Annotated Checklist of the Birds of the Oriental Region* (Inskipp *et al.* 1996), with a few exceptions. Unfortunately, this excellent work has never been updated and, for this edition, I had to look elsewhere for guidance. I have followed three primary references in order to arrive at the taxonomic sequence used in this edition. The main reference followed was *The Howard and Moore Complete Checklist of the Birds of the World*, third edition (Dickinson 2003). The sequence follows this reference closely, from the beginning through to Eupetidae, apart from some rearrangements among the shorebirds, terns and gulls etc. For the oscine passerines, I have largely used a combination of 'A phylogenetic supertree of oscine passerine birds (Aves: Passeri)' by Jønsson & Fjeldså (2006), and 'Phylogeny and classification of the avian superfamily Sylvioidea' (Alström *et al.* 2006). I am aware that there is a certain amount of resistance to the on-going changes in taxonomy that are taking place, but personally I find it impossible to ignore them. All of the taxonomic references that I have used appear in the Selected references section. Table 1 lists all of the new species to the region, omitted species, and the changes in nomenclature since the first edition.

For subspecies, the primary reference used is still Peters's outstanding *Check-list of Birds of the World* (Peters and others 1931-1986), but there are many modifications, particularly following the recent proliferation of books on various bird families, and publication of the *Handbook of the Birds of the World* (Barcelona: Lynx Edicions).

Most of the changes to common names are the result of my employment of hyphenation, while changes to the scientific names are primarily the result of recent taxonomic changes and the resolution of gender issues.

Table 1: New species, name changes, and taxonomic changes since the first edition in 2000 (excluding hyphenation changes)

COMPLETELY NEW TAXA TO THE REGION (REGARDLESS OF TAXONOMIC CHANGES)

2 **Snow Partridge** *Lerwa lerwa*
78 **Long-tailed Duck** *Clangula hyemalis*
83 **Red-breasted Merganser** *Mergus serrator*
84 **Yellow-billed Loon** *Gavia adamsii*
93 **Horned Grebe** *Podiceps auritus*
202 **Common Buzzard** *Buteo buteo* (the 'Common Buzzard' in the first edition is now treated as Himalayan Buzzard)
251 **White-headed Stilt** *Himantopus leucocephalus*
281 **Red Phalarope** *Phalaropus fulicarius*
338 **Arctic Tern** *Sterna paradisaea*
345 **Mongolian Gull** *Larus mongolicus*
346 **Lesser Black-backed Gull** *Larus fuscus*
349 **Laughing Gull** *Larus atricilla*
355 **Little Gull** *Hydrocoloeus minutus*
356 **Black-legged Kittiwake** *Rissa tridactyla*
357 **Ancient Murrelet** *Synthliboramphus antiquus*
481 **Alpine Swift** *Tachymarptis melba*
645 **Indian Golden Oriole** *Oriolus kundoo*
781 **Mekong Wagtail** *Motacilla samveasnae*
804 **Japanese Grosbeak** *Eophona personata*
813 **Red-headed Bunting** *Emberiza bruniceps*
817 **Grey-necked Bunting** *Emberiza buchanani*
821 **Rustic Bunting** *Emberiza rustica*
841 **Wallcreeper** *Tichodroma muraria*
878 **Tickell's Thrush** *Turdus unicolor*
886 **Naumann's Thrush** *Turdus naumanni*
894 **Rusty-bellied Shortwing** *Brachypteryx hyperythra*
1084 **Martens's Warbler** *Seicercus omeiensis*
1120 **Common Chiffchaff** *Phylloscopus collybita*
1142 **Ludlow's Fulvetta** *Fulvetta ludlowi*
1263 **Chestnut-eared Laughingthrush** *Ianthocincla konkakinhensis*
1298 **Large-billed Reed-warbler** *Acrocephalus orinus*
1305 **Pleske's Warbler** *Locustella pleskei*

SPECIES THAT WERE PREVIOUSLY THOUGHT TO OCCUR, OR WERE UNCONFIRMED IN THE REGION BUT HAVE NOT ACTUALLY BEEN RECORDED, AND ARE NOW OMITTED

CURRENT NOMENCLATURE	DIFFERING NOMENCLATURE IN 2000 EDITION
Vega Gull *Larus vegae*	
Saunders's Tern *Sternula saundersi*	*Sterna saundersi*

SPECIES THAT HAVE BEEN OMITTED DUE TO TAXONOMIC CHANGE

	TAXONOMY AND NOMENCLATURE IN THIS EDITION	TAXONOMY AND NOMENCLATURE IN 2000 EDITION
16	**Chestnut-headed Partridge** *Arborophila cambodiana diversa*	**Siamese Partridge** *Arborophila diversa*
30a	**'Imperial Pheasant'** *Lophura* x *imperialis* - Hybrid origin	**Imperial Pheasant** *Lophura imperialis*
31	**Edwards's Pheasant** *Lophura edwardsi hatinhensis*	**Vietnamese Pheasant** *Lophura hatinhensis*

ADDITIONAL SPECIES FOR THE REGION DUE TO TAXONOMIC SPLITS

	TAXONOMY AND NOMENCLATURE IN THIS EDITION	TAXONOMY AND NOMENCLATURE IN 2000 EDITION
64	**Chinese Spot-billed Duck** *Anas zonorhyncha*	**Spot-billed Duck** *Anas poecilorhyncha*
234	**Black-backed Swamphen** *Porphyrio indicus*	**Purple Swamphen** *Porphyrio porphyrio*
377	**Andaman Green-pigeon** *Treron chloropterus*	**Pompadour Green Pigeon** *Treron pompadora*
401	**Dark Hawk-cuckoo** *Hierococcyx bocki*	**Large Hawk Cuckoo** *Hierococcyx sparverioides*
405	**Hodgson's Hawk-cuckoo** *Hierococcyx nisicolor*	**Hodgson's Hawk Cuckoo** *Hierococcyx fugax*
408	**Oriental Cuckoo** *Cuculus horsfieldi*	**Oriental Cuckoo** *Cuculus saturatus*
410	**Sunda Cuckoo** *Cuculus lepidus*	**Oriental Cuckoo** *Cuculus saturatus*
457	**Northern Boobook** *Ninox japonica*	**Brown Hawk Owl** *Ninox scutulata*
522	**Southern Brown Hornbill** *Ptilolaemus tickelli*	**Brown Hornbill** *Anorrhinus tickelli*
544	**Annam Barbet** *Megalaima annamensis*	**Black-browed Barbet** *Megalaima oorti*
561	**Spot-breasted Woodpecker** *Dendrocopos analis*	**Fulvous-breasted Woodpecker** *Dendrocopos macei*
682	**Large-billed Crow** *Corvus japonensis*	**Large-billed Crow** *Corvus macrorhynchos*
683	**Eastern Jungle Crow** *Corvus levaillantii*	**Large-billed Crow** *Corvus macrorhynchos*
762a	**White-capped Munia** *Lonchura ferruginosa*	**Black-headed Munia** *Lonchura malacca*
784	**Eastern Yellow Wagtail** *Motacilla tschutschensis*	**Yellow Wagtail** *Motacilla flava*
837	**Neglected Nuthatch** *Sitta neglecta*	**Chestnut-bellied Nuthatch** *Sitta castanea*
870	**White's Thrush** *Zoothera aurea*	**Scaly Thrush** *Zoothera dauma*
884	**Black-throated Thrush** *Turdus atrogularis*	**Dark-throated Thrush** *Turdus ruficollis*
928	**Himalayan Bluetail** *Tarsiger rufilatus*	**Orange-flanked Bush Robin** *Tarsiger cyanurus*
943	**Large Blue Flycatcher** *Cyornis magnirostris*	**Hill Blue Flycatcher** *Cyornis banyumas*
945	**Chinese Blue Flycatcher** *Cyornis glaucicomans*	**Blue-throated Flycatcher** *Cyornis rubeculoides*
960	**Green-backed Flycatcher** *Ficedula elisae*	**Narcissus Flycatcher** *Ficedula narcissina*
992	**Japanese Tit** *Parus minor*	**Great Tit** *Parus major*
1059	**Rufous-bellied Swallow** *Cecropis badia*	**Striated Swallow** *Hirundo striolata*
1071	**Sunda Bush-warbler** *Cettia vulcania*	**Aberrant Bush Warbler** *Cettia flavolivacea* (part)
1079	**Grey-crowned Tit** *Aegithalos annamensis*	**Black-throated Tit** *Aegithalos concinnus*
1081	**Burmese Tit** *Aegithalos sharpei*	**Black-browed Tit** *Aegithalos bonvaloti*
1094	**Limestone Warbler** *Phylloscopus* sp.	**Sulphur-breasted Warbler** *Phylloscopus ricketti*
1096	**White-tailed Leaf-warbler** *Phylloscopus ogilviegranti*	**White-tailed Leaf Warbler** *Phylloscopus davisoni*
1099	**Claudia's Warbler** *Phylloscopus claudiae*	**Blyth's Leaf Warbler** *Phylloscopus reguloides*
1101	**Hartert's Warbler** *Phylloscopus goodsoni*	**Blyth's Leaf Warbler** *Phylloscopus reguloides*
1125	**Black-crowned Parrotbill** *Psittiparus margaritae*	**Grey-headed Parrotbill** *Paradoxornis gularis*
1134	**Buff-breasted Parrotbill** *Suthora ripponi*	**Black-throated Parrotbill** *Paradoxornis nipalensis*
1135	**Black-eared Parrotbill** *Suthora beaulieui*	**Black-throated Parrotbill** *Paradoxornis nipalensis*

1157 **Chestnut-collared Yuhina** *Staphida torqueola*	**Striated Yuhina** *Yuhina castaniceps*
1159 **Schaeffer's Fulvetta** *Alcippe schaefferi*	**Grey-cheeked Fulvetta** *Alcippe morrisonia*
1180 **Black-streaked Scimitar-babbler** *Pomatorhinus gravivox*	**Spot-breasted Scimitar Babbler** *Pomatorhinus erythrocnemis*
1189 **Pale-throated Wren-babbler** *Spelaeornis kinneari*	**Long-tailed Wren Babbler** *Spelaeornis chocolatinus*
1190 **Chin Hills Wren-babbler** *Spelaeornis oatesi*	**Long-tailed Wren Babbler** *Spelaeornis chocolatinus*
1223 **White-throated Wren-babbler** *Rimator pasquieri*	**Long-billed Wren Babbler** *Rimator malacoptilus*
1226 **Collared Babbler** *Gampsorhynchus torquatus*	**White-hooded Babbler** *Gampsorhynchus rufulus*
1229 **Black-crowned Fulvetta** *Pseudominla klossi*	**Rufous-winged Fulvetta** *Alcippe castaneceps*
1234 **Vietnamese Cutia** *Cutia legalleni*	**Cutia** *Cutia nipalensis*
1270 **Silver-eared Laughingthrush** *Trochalopteron melanostigma*	**Chestnut-crowned Laughingthrush** *Garrulax erythrocephalus*
1271 **Malayan Laughingthrush** *Trochalopteron peninsulae*	**Chestnut-crowned Laughingthrush** *Garrulax erythrocephalus*
1277 **Scarlet-faced Liocichla** *Liocichla ripponi*	**Red-faced Liocichla** *Liocichla phoenicea*
1309 **Baikal Bush-warbler** *Bradypterus davidi*	**Spotted Bush Warbler** *Bradypterus thoracicus*
1327 **Hill Prinia** *Prinia superciliaris*	**Hill Prinia** *Prinia atrogularis*

CHANGES IN NOMENCLATURE DUE TO TAXONOMIC SPLITS

TAXONOMY AND NOMENCLATURE IN THIS EDITION	DIFFERING TAXOMONY AND NOMENCLATURE IN 2000 EDITION
48 **Taiga Bean-goose** *Anser fabalis*	**Bean Goose**
63 **Indian Spot-billed Duck** *Anas poecilorhyncha*	**Spot-billed Duck**
66 **Andaman Teal** *Anas albogularis*	**Sunda Teal** *Anas gibberifrons*
70 **Eurasian Teal** *Anas crecca*	**Common Teal**
125 **Eastern Cattle Egret** *Bubulcus coromandus*	**Cattle Egret** *Bubulcus ibis*
150 **Oriental Darter** *Anhinga melanogaster*	**Darter**
179 **Slender-billed Vulture** *Gyps tenuirostris*	**Long-billed Vulture** *Gyps indicus*
201 **Himalayan Buzzard** *Buteo burmanicus*	**Common Buzzard** *Buteo buteo*
204 **Indian Spotted Eagle** *Aquila hastata*	**Lesser Spotted Eagle** *Aquila pomarina*
208 **Eastern Imperial Eagle** *Aquila heliaca*	**Imperial Eagle**
215 **Changeable Hawk-eagle** *Nisaetus limnaeetus*	*Spizaetus cirrhatus*
222 **Eastern Water Rail** *Rallus indicus*	**Water Rail** *Rallus aquaticus*
233 **Grey-headed Swamphen** *Porphyrio poliocephalus*	**Purple Swamphen** *Porphyrio porphyrio*
245 **Indian Thick-knee** *Burhinus indicus*	**Eurasian Thick-knee** *Burhinus oedicnemus*
376 **Ashy-headed Green-pigeon** *Treron phayrei*	**Pompadour Green Pigeon** *Treron pompadora*
404 **Malaysian Hawk-cuckoo** *Hierococcyx fugax*	**Hodgson's Hawk Cuckoo**
409 **Himalayan Cuckoo** *Cuculus saturatus*	**Oriental Cuckoo**
434 **Eastern Grass-owl** *Tyto longimembris*	**Grass Owl** *Tyto capensis*
439 **Collared Scops-owl** *Otus lettia*	*Otus bakkamoena*
441 **Indian Eagle-owl** *Bubo bengalensis*	**Eurasian Eagle Owl** *Bubo bubo*
451 **Himalayan Wood-owl** *Strix nivicola*	**Tawny Owl** *Strix aluco*
463 **Blyth's Frogmouth** *Batrachostomus affinis*	**Javan Frogmouth** *Batrachostomus javensis*
466 **Grey Nightjar** *Caprimulgus jotaka*	*Caprimulgus indicus*
521 **Northern Brown Hornbill** *Ptilolaemus austeni*	**Brown Hornbill** *Anorrhinus tickelli*
638 **Jerdon's Minivet** *Pericrocotus albifrons*	**White-bellied Minivet** *Pericrocotus erythropygius*
642 **Scarlet Minivet** *Pericrocotus speciosus*	*Pericrocotus flammeus*
684 **Southern Jungle Crow** *Corvus macrorhynchos*	**Large-billed Crow**
713 **Van Hasselt's Sunbird** *Leptocoma brasiliana*	**Purple-throated Sunbird** *Nectarinia sperata*
742 **Plain Flowerpecker** *Dicaeum minullum*	*Dicaeum concolor*
762 **Chestnut Munia** *Lonchura atricapilla*	**Black-headed Munia** *Lonchura malacca*
783 **Western Yellow Wagtail** *Motacilla flava*	**Yellow Wagtail**
796 **Sharpe's Rosefinch** *Carpodacus verreauxii*	**Spot-winged Rosefinch** *Carpodacus rodopeplus*
827 **Hodgson's Treecreeper** *Certhia hodgsoni*	**Eurasian Treecreeper** *Certhia familiaris*
830 **Hume's Treecreeper** *Certhia manipurensis*	**Brown-throated Treecreeper** *Certhia discolor*

CHANGES IN NOMENCLATURE DUE TO TAXONOMIC SPLITS (CONTINUED FROM P.11)

	TAXONOMY AND NOMENCLATURE IN THIS EDITION	DIFFERING TAXOMONY AND NOMENCLATURE IN 2000 EDITION
836	**Chestnut-bellied Nuthatch** *Sitta cinnamoventris*	**Chestnut-bellied Nuthatch** *Sitta castanea*
877	**Chinese Blackbird** *Turdus mandarinus*	**Eurasian Blackbird** *Turdus merula*
885	**Red-throated Thrush** *Turdus ruficollis*	**Dark-throated Thrush**
922	**Eastern Stonechat** *Saxicola maurus*	**Common Stonechat** *Saxicola torquata*
929	**Red-flanked Bluetail** *Tarsiger cyanurus*	**Orange-flanked Bush Robin**
970	**Taiga Flycatcher** *Ficedula albicilla*	**Red-throated Flycatcher** *Ficedula parva*
991	**Grey Tit** *Parus cinereus*	**Great Tit** *Parus major*
1014	**Black-crested Bulbul** *Pycnonotus flaviventris*	*Pycnonotus melanicterus*
1045	**Himalayan Black Bulbul** *Hypsipetes leucocephalus*	**Black Bulbul**
1052	**Grey-throated Sand-martin** *Riparia chinensis*	**Plain Martin** *Riparia paludicola*
1056	**House Swallow** *Hirundo tahitica*	**Pacific Swallow**
1066	**Hume's Bush-warbler** *Cettia brunnescens*	**Yellowish-bellied Bush Warbler** *Cettia acanthizoides*
1097	**Davison's Warbler** *Phylloscopus davisoni*	**White-tailed Leaf Warbler**
1126	**Greater Rufous-headed Parrotbill** *Psittiparus bakeri*	*Paradoxornis ruficeps*
1133	**Grey-breasted Parrotbill** *Suthora poliotis*	**Black-throated Parrotbill** *Paradoxornis nipalensis*
1143	**Streak-throated Fulvetta** *Fulvetta manipurensis*	*Alcippe cinereiceps*
1158	**Grey-cheeked Fulvetta** *Alcippe fratercula*	*Alcippe morrisonia*
1179	**Spot-breasted Scimitar-babbler** *Pomatorhinus mcclellandi*	*Pomatorhinus erythrocnemis*
1186	**Chevron-breasted Babbler** *Sphenocichla roberti*	**Wedge-billed Wren Babbler** *Sphenocichla humei*
1188	**Grey-bellied Wren-babbler** *Spelaeornis reptatus*	**Long-tailed Wren Babbler** *Spelaeornis chocolatinus*
1197	**Pin-striped Tit-babbler** *Macronus gularis*	**Striped Tit Babbler** *Macronous gularis*
1233	**Himalayan Cutia** *Cutia nipalensis*	**Cutia**
1253	**Spectacled Laughingthrush** *Rhinocichla mitrata*	**Chestnut-capped Laughingthrush** *Garrulax mitratus*
1258	**Chinese Hwamei** *Leucodioptron canorum*	**Hwamei** *Garrulax canorus*
1269	**Assam Laughingthrush** *Trochalopteron chrysopterum*	**Chestnut-crowned Laughingthrush** *Garrulax erythrocephalus*
1276	**Crimson-faced Liocichla** *Liocichla phoenicea*	**Red-faced Liocichla**
1302	**Indian Reed-warbler** *Acrocephalus brunnescens*	**Clamorous Reed Warbler** *Acrocephalus stentoreus*
1326	**Black-throated Prinia** *Prinia atrogularis*	**Hill Prinia**

OTHER CHANGES IN NOMENCLATURE

	NOMENCLATURE IN THIS EDITION	DIFFERING NOMENCLATURE IN 2000 EDITION
56	**White-winged Duck** *Asarcornis scutulata*	*Cairina scutulata*
72	**Red-crested Pochard** *Netta rufina*	*Rhodonessa rufina*
119	**Malaysian Night-heron** *Gorsachius melanolophus*	**Malayan Night Heron**
121	**Little Heron** *Butorides striata*	*Butorides striatus*
131	**Great Egret** *Ardea alba*	*Casmerodius albus*
151	**White-rumped Pygmy-falcon** *Polihierax insignis*	**White-rumped Falcon**
166	**Oriental Honey-buzzard** *Pernis ptilorhynchus*	*Pernis ptilorhyncus*
182	**Red-headed Vulture** *Aegypius calvus*	*Sarcogyps calvus*
209	**Bonelli's Eagle** *Aquila fasciata*	*Hieraaetus fasciatus*
210	**Booted Eagle** *Aquila pennata*	*Hieraaetus pennatus*
212	**Rufous-bellied Eagle** *Lophotriorchis kienerii*	*Hieraaetus kienerii*
213	**Blyth's Hawk-eagle** *Nisaetus alboniger*	*Spizaetus alboniger*
214	**Mountain Hawk-eagle** *Nisaetus nipalensis*	*Spizaetus nipalensis*
216	**Wallace's Hawk-eagle** *Nisaetus nanus*	*Spizaetus nanus*
218	**Bengal Florican** *Houbaropsis bengalensis*	*Eupodotis bengalensis*
223	**Corncrake** *Crex crex*	**Corn Crake**
242	**Small Buttonquail** *Turnix sylvaticus*	*Turnix sylvatica*

289	**Far Eastern Curlew** *Numenius madagascariensis*	**Eastern Curlew**
293	**Grey-tailed Tattler** *Tringa brevipes*	*Heteroscelus brevipes*
325	**Sooty Tern** *Onychoprion fuscatus*	*Sterna fuscata*
326	**Bridled Tern** *Onychoprion anaethetus*	*Sterna anaethetus*
327	**Aleutian Tern** *Onychoprion aleuticus*	*Sterna aleutica*
328	**Little Tern** *Sternula albifrons*	*Sterna albifrons*
330	**Caspian Tern** *Hydroprogne caspia*	*Sterna caspia*
333	**Whiskered Tern** *Chlidonias hybrida*	*Chlidonias hybridus*
340	**Lesser Crested Tern** *Thalasseus bengalensis*	*Sterna bengalensis*
341	**Great Crested Tern** *Thalasseus bergii*	*Sterna bergii*
342	**Chinese Crested Tern** *Thalasseus bernsteini*	*Sterna bernsteini*
350	**Relict Gull** *Chroicocephalus relictus*	*Larus relictus*
351	**Brown-headed Gull** *Chroicocephalus brunnicephalus*	*Larus brunnicephalus*
352	**Black-headed Gull** *Chroicocephalus ridibundus*	*Larus ridibundus*
353	**Slender-billed Gull** *Chroicocephalus genei*	*Larus genei*
354	**Saunders's Gull** *Chroicocephalus saundersi*	*Larus saundersi*
360	**Speckled Woodpigeon** *Columba hodgsonii*	**Speckled Wood Pigeon**
361	**Ashy Woodpigeon** *Columba pulchricollis*	**Ashy Wood Pigeon**
380	**Yellow-footed Green-pigeon** *Treron phoenicopterus*	*Treron phoenicoptera*
421	**Asian Koel** *Eudynamys scolopaceus*	*Eudynamys scolopacea*
423	**Black-bellied Malkoha** *Rhopodytes diardi*	*Phaenicophaeus diardi*
424	**Chestnut-bellied Malkoha** *Rhopodytes sumatranus*	*Phaenicophaeus sumatranus*
425	**Green-billed Malkoha** *Rhopodytes tristis*	*Phaenicophaeus tristis*
426	**Raffles's Malkoha** *Rhinortha chlorophaeus*	*Phaenicophaeus chlorophaeus*
427	**Red-billed Malkoha** *Zanclostomus javanicus*	*Phaenicophaeus javanicus*
428	**Chestnut-breasted Malkoha** *Zanclostomus curvirostris*	*Phaenicophaeus curvirostris*
431	**Andaman Coucal** *Centropus andamanensis*	**Brown Coucal**
433	**Common Barn-owl** *Tyto alba*	**Barn Owl**
456	**Brown Boobook** *Ninox scutulata*	**Brown Hawk Owl**
472	**Himalayan Swiftlet** *Aerodramus brevirostris*	*Collocalia brevirostris*
473	**Black-nest Swiftlet** *Aerodramus maximus*	*Collocalia maxima*
474	**Edible-nest Swiftlet** *Aerodramus fuciphaga*	*Collocalia fuciphaga*
475	**Germain's Swiftlet** *Aerodramus germani*	*Collocalia germani*
499	**Stork-billed Kingfisher** *Pelargopsis capensis*	*Halcyon capensis*
500	**Brown-winged Kingfisher** *Pelargopsis amauroptera*	*Halcyon amauroptera*
506	**Black-backed Kingfisher** *Ceyx erithaca*	*Ceyx erithacus*
512	**Crested Kingfisher** *Ceryle lugubris*	*Megaceryle lugubris*
516	**Little Green Bee-eater** *Merops orientalis*	**Green Bee-eater**
528	**Helmeted Hornbill** *Rhinoplax vigil*	*Buceros vigil*
529	**White-crowned Hornbill** *Berenicornis comatus*	*Aceros comatus*
557	**Rufous-bellied Woodpecker** *Hypopicus hyperythrus*	*Dendrocopos hyperythrus*
567	**Rufous Woodpecker** *Micropternus brachyurus*	*Celeus brachyurus*
569	**Banded Woodpecker** *Chrysophlegma mineaceus*	*Picus mineaceus*
570	**Greater Yellownape** *Chrysophlegma flavinucha*	*Picus flavinucha*
571	**Checker-throated Woodpecker** *Chrysophlegma mentalis*	*Picus mentalis*
616	**Eared Pitta** *Anthocincla phayrei*	*Pitta phayrei*
624	**White-bellied Erpornis** *Erpornis zantholeuca*	**White-bellied Yuhina** *Yuhina zantholeuca*
643	**Mangrove Whistler** *Pachycephala cinerea*	*Pachycephala grisola*
658	**Rufous-winged Philentoma** *Philentoma pyrhoptera*	*Philentoma pyrhopterum*
659	**Maroon-breasted Philentoma** *Philentoma velata*	*Philentoma velatum*
663	**Yellow-bellied Fantail** *Chelidorhynx hypoxantha*	*Rhipidura hypoxantha*
674	**Hair-crested Drongo** *Dicrurus hottentottus*	**Spangled Drongo**
712	**Purple-rumped Sunbird** *Leptocoma zeylonica*	*Nectarinia zeylonica*
714	**Copper-throated Sunbird** *Leptocoma calcostetha*	*Nectarinia calcostetha*
715	**Purple Sunbird** *Cinnyris asiaticus*	*Nectarinia asiatica*
716	**Olive-backed Sunbird** *Cinnyris jugularis*	*Nectarinia jugularis*
724	**Ruby-cheeked Sunbird** *Chalcoparia singalensis*	*Anthreptes singalensis*
736	**Yellow-breasted Flowerpecker** *Dicaeum maculatus*	*Prionochilus maculatus*

OTHER CHANGES IN NOMENCLATURE (CONTINUED FROM P.13)

NOMENCLATURE IN THIS EDITION	DIFFERING NOMENCLATURE IN 2000 EDITION
737 **Crimson-breasted Flowerpecker** *Dicaeum percussus*	*Prionochilus percussus*
738 **Scarlet-breasted Flowerpecker** *Dicaeum thoracicus*	*Prionochilus thoracicus*
786 **Eurasian Siskin** *Spinus spinus*	*Carduelis spinus*
787 **Tibetan Serin** *Serinus thibetana*	**Tibetan Siskin** *Carduelis thibetana*
789 **Grey-capped Greenfinch** *Chloris sinica*	*Carduelis sinica*
790 **Black-headed Greenfinch** *Chloris ambigua*	*Carduelis ambigua*
791 **Vietnamese Greenfinch** *Chloris monguilloti*	*Carduelis monguilloti*
792 **Yellow-breasted Greenfinch** *Chloris spinoides*	*Carduelis spinoides*
812 **Crested Bunting** *Emberiza lathami*	*Melophus lathami*
850a **Black-winged Myna** *Acridotheres melanopterus*	**Black-winged Starling** *Sturnus melanopterus*
851 **Vinous-breasted Myna** *Acridotheres burmannicus*	**Vinous-breasted Starling** *Sturnus burmannicus*
852 **Black-collared Starling** *Gracupica nigricollis*	*Sturnus nigricollis*
853 **Asian Pied Starling** *Gracupica contra*	*Sturnus contra*
865 **Common Hill-myna** *Gracula religiosa*	**Hill Myna**
887 **Dusky Thrush** *Turdus eunomus*	*Turdus naumanni*
897 **Japanese Robin** *Luscinia akahige*	*Erithacus akahige*
909 **Plumbeous Water-redstart** *Rhyacornis fuliginosa*	*Rhyacornis fuliginosus*
921 **Grey Bushchat** *Saxicola ferreus*	*Saxicola ferrea*
951 **Verditer Flycatcher** *Eumyias thalassinus*	*Eumyias thalassina*
974 **Rufous-gorgeted Flycatcher** *Muscicapa strophiata*	*Ficedula strophiata*
987 **Black-bibbed Tit** *Poecile hypermelaena*	*Parus hypermelaena*
988 **Grey-crested Tit** *Lophophanes dichrous*	*Parus dichrous*
989 **Coal Tit** *Periparus ater*	*Parus ater*
990 **Rufous-vented Tit** *Periparus rubidiventris*	*Parus rubidiventris*
1003 **Indochinese Bushlark** *Mirafra erythrocephala*	*Mirafra marionae*
1044 **Mountain Bulbul** *Ixos mcclellandii*	*Hypsipetes mcclellandii*
1046 **White-headed Bulbul** *Cerasophila thompsoni*	*Hypsipetes thompsoni*
1047 **Northern House-martin** *Delichon urbicum*	*Delichon urbica*
1049 **Nepal House-martin** *Delichon nipalense*	*Delichon nipalensis*
1050 **Common Sand-martin** *Riparia riparia*	**Sand Martin**
1051 **Pale Sand-martin** *Riparia diluta*	**Pale Martin**
1053 **Dusky Crag-martin** *Ptyonoprogne concolor*	*Hirundo concolor*
1057 **Red-rumped Swallow** *Cecropis daurica*	*Hirundo daurica*
1058 **Striated Swallow** *Cecropis striolata*	*Hirundo striolata*
1064 **Mountain Tailorbird** *Phyllergates cucullatus*	*Orthotomus cuculatus*
1098 **Grey-hooded Warbler** *Phylloscopus xanthoschistos*	*Seicercus xanthoschistos*
1112 **Chinese Leaf-warbler** *Phylloscopus yunnanensis*	*Phylloscopus sichuanensis*
1123 **Brown Parrotbill** *Cholornis unicolor*	*Paradoxornis unicolor*
1124 **Grey-headed Parrotbill** *Psittiparus gularis*	*Paradoxornis gularis*
1128 **Lesser Rufous-headed Parrotbill** *Chleuasicus atrosuperciliaris*	*Paradoxornis atrosuperciliaris*
1129 **Brown-winged Parrotbill** *Suthora brunneus*	*Paradoxornis brunneus*
1130 **Vinous-throated Parrotbill** *Suthora webbianus*	*Paradoxornis webbianus*
1131 **Ashy-throated Parrotbill** *Suthora alphonsianus*	*Paradoxornis alphonsianus*
1132 **Fulvous Parrotbill** *Suthora fulvifrons*	*Paradoxornis fulvifrons*
1136 **Golden Parrotbill** *Suthora verreauxi*	*Paradoxornis verreauxi*
1137 **Short-tailed Parrotbill** *Neosuthora davidiana*	*Paradoxornis davidianus*
1140 **Golden-breasted Fulvetta** *Lioparus chrysotis*	*Alcippe chrysotis*
1141 **White-browed Fulvetta** *Fulvetta vinipectus*	*Alcippe vinipectus*
1144 **Indochinese Fulvetta** *Fulvetta danisi*	*Alcippe danisi*
1156 **Striated Yuhina** *Staphida castaniceps*	*Yuhina castaniceps*
1165 **Rufous-throated Fulvetta** *Schoeniparus rufogularis*	*Alcippe rufogularis*
1166 **Rusty-capped Fulvetta** *Schoeniparus dubius*	*Alcippe dubia*
1171 **Spot-necked Babbler** *Stachyris strialata*	*Stachyris striolata*
1176 **Sickle-billed Scimitar-babbler** *Xiphirhynchus superciliaris*	**Slender-billed Scimitar Babbler**
1184 **Orange-billed Scimitar-babbler** *Pomatorhinus ochraceiceps*	**Red-billed Scimitar Babbler**

1191 **Spotted Wren-babbler** *Elachura formosa*	*Spelaeornis formosus*
1194 **Golden Babbler** *Stachyridopsis chrysaea*	*Stachyris chrysaea*
1195 **Rufous-capped Babbler** *Stachyridopsis ruficeps*	*Stachyris ruficeps*
1196 **Rufous-fronted Babbler** *Stachyridopsis rufifrons*	*Stachyris rufifrons*
1198 **Grey-faced Tit-babbler** *Macronus kelleyi*	*Macronous kelleyi*
1199 **Fluffy-backed Tit-babbler** *Macronus ptilosus*	*Macronous ptilosus*
1201 **Rufous-rumped Grass-babbler** *Graminicola bengalensis*	**Rufous-rumped Grassbird**
1210 **Grey-breasted Babbler** *Ophrydornis albogularis*	*Malacopteron albogulare*
1212 **Horsfield's Babbler** *Malacocincla sepiaria*	*Malacocincla sepiarium*
1217 **Marbled Wren-babbler** *Turdinus marmoratus*	*Napothera marmorata*
1218 **Large Wren-babbler** *Turdinus macrodactylus*	*Napothera macrodactyla*
1219 **Limestone Wren-babbler** *Gypsophila crispifrons*	*Napothera crispifrons*
1224 **Indochinese Wren-babbler** *Rimator danjoui*	**Short-tailed Scimitar Babbler** *Jabouilleia danjoui*
1227 **Yellow-throated Fulvetta** *Pseudominla cinerea*	*Alcippe cinerea*
1228 **Rufous-winged Fulvetta** *Pseudominla castaneceps*	*Alcippe castaneceps*
1235 **Grey-sided Laughingthrush** *Dryonastes caerulatus*	*Garrulax caerulatus*
1236 **Black-throated Laughingthrush** *Dryonastes chinensis*	*Garrulax chinensis*
1237 **Chestnut-backed Laughingthrush** *Dryonastes nuchalis*	*Garrulax nuchalis*
1238 **Rufous-vented Laughingthrush** *Dryonastes gularis*	*Garrulax gularis*
1239 **Yellow-throated Laughingthrush** *Dryonastes galbanus*	*Garrulax galbanus*
1240 **White-cheeked Laughingthrush** *Dryonastes vassali*	*Garrulax vassali*
1241 **Rufous-necked Laughingthrush** *Dryonastes ruficollis*	*Garrulax ruficollis*
1252 **Black Laughingthrush** *Melanocichla lugubris*	*Garrulax lugubris*
1255 **Striated Laughingthrush** *Grammatoptila striata*	*Garrulax striatus*
1256 **Spot-breasted Laughingthrush** *Stactocichla merulina*	*Garrulax merulinus*
1257 **Orange-breasted Laughingthrush** *Stactocichla annamensis*	*Garrulax annamensis*
1259 **Striped Laughingthrush** *Strophocincla virgata*	*Garrulax virgatus*
1260 **White-browed Laughingthrush** *Pterorhinus sannio*	*Garrulax sannio*
1261 **Moustached Laughingthrush** *Ianthocincla cineracea*	*Garrulax cineraceus*
1262 **Rufous-chinned Laughingthrush** *Ianthocincla rufogularis*	*Garrulax rufogularis*
1264 **Spotted Laughingthrush** *Ianthocincla ocellata*	*Garrulax ocellatus*
1265 **Scaly Laughingthrush** *Trochalopteron subunicolor*	*Garrulax subunicolor*
1266 **Brown-capped Laughingthrush** *Trochalopteron austeni*	*Garrulax austeni*
1267 **Blue-winged Laughingthrush** *Trochalopteron squamatum*	*Garrulax squamatus*
1268 **Black-faced Laughingthrush** *Trochalopteron affine*	*Garrulax affinis*
1272 **Golden-winged Laughingthrush** *Trochalopteron ngoclinhense*	*Garrulax ngoclinhensis*
1273 **Collared Laughingthrush** *Trochalopteron yersini*	*Garrulax yersini*
1274 **Red-winged Laughingthrush** *Trochalopteron formosum*	*Garrulax formosus*
1275 **Red-tailed Laughingthrush** *Trochalopteron milnei*	*Garrulax milnei*
1278 **Bar-throated Minla** *Chrysominla strigula*	**Chestnut-tailed Minla** *Minla strigula*
1280 **Blue-winged Siva** *Siva cyanouroptera*	**Blue-winged Minla** *Minla cyanouroptera*
1281 **Silver-eared Mesia** *Mesia argentauris*	*Leiothrix argentauris*
1285 **Grey Sibia** *Malacias gracilis*	*Heterophasia gracilis*
1286 **Black-headed Sibia** *Malacias desgodinsi*	*Heterophasia desgodinsi*
1287 **Dark-backed Sibia** *Malacias melanoleucus*	*Heterophasia melanoleuca*
1288 **Beautiful Sibia** *Malacias pulchellus*	*Heterophasia pulchella*
1289 **Rufous-backed Sibia** *Leioptila annectens*	*Heterophasia annectens*
1324 **Striated Prinia** *Prinia crinigera*	*Prinia criniger*

COLOUR PLATES

All species are illustrated on the colour plates, with the exception of five birds that are currently thought to be unrecognisable on plumage in the field: **Oriental Cuckoo** *Cuculus horsfieldi*, **Northern Boobook** *Ninox japonica*, **Martens's Warbler** *Seicercus omeiensis*, **Hartert's Warbler** *Phylloscopus goodsoni* and **Claudia's Warbler** *P. claudiae*. An attempt has also been made, despite restraints on the number of plates, to illustrate the majority of distinctive plumage variations (particularly sex/age) and subspecies. The layout mostly follows the systematic order, although some species or blocks of species have been moved in order to balance out the average number of figures on each plate or to enable more useful comparisons to be made. Species depicted on any one plate have been illustrated to the same scale (smaller in the case of flight figures) unless stated on the plate.

The caption page facing each plate is intended to provide a summary of important identification features for quick reference. For more detailed information, it is essential to consult the main text. Where any discrepancy exists between caption and illustration, the reader should always be guided by the caption.

FAMILY AND SUBFAMILY INFORMATION

A brief summary is given of the distinctive characteristics and diet. Worldwide totals are given for comparison with South-East Asia totals and, in many cases, are only approximate.

SPECIES ACCOUNT INFORMATION

Identification

The total length of each species appears at the beginning of each account and an attempt has been made to improve on measurements existing in other works covering the region's birds, many of which are highly inaccurate. A range was found for most species and, as far as time allowed, this information was gathered from museum specimen labels.

A comparative approach has been adopted with species descriptions, where scarcer species are generally compared to commoner or more widespread species. In general, those species considered to be easily identifiable have been afforded less coverage than the more difficult species.

Males are described first (except in polyandrous species) and female plumage compared directly to the male plumage.

Comparisons between similar species are dealt with directly and separately under the various sex/age or other headings.

In cases where more than one subspecies occurs in the region, a 'primary subspecies' is described and other distinctive subspecies compared directly to it (not to each other, unless stated). The primary subspecies (given after length) is usually that most likely to be encountered by birdwatchers visiting Thailand. For example, the primary subspecies of **Silver-eared Mesia** is *Mesia argentauris galbana*, the form occurring in NW Thailand. Apart from a few exceptions, the subspecies listed under 'Other subspecies in SE Asia' are not considered to differ markedly from the primary subspecies. Where a primary subspecies is question-marked, this indicates doubt over its occurrence (i.e no specimen) rather than the plumage description. In most cases, ranges have been given for all subspecies except the primary one, the range of which can be deduced by subtracting the ranges of other subspecies from the entire SE Asia distribution.

Voice

Transcription of bird vocalisations is very subjective and different authors tend to prefer the use of different letters to represent certain sounds. An attempt has been made to describe structure, loudness and tone etc. of most vocalisations before the transcriptions themselves.

Spacing between sounds indicates elapsed time. For example, slowly repeated notes appear as *tit tit..* or *tit...tit..* etc., whereas more quickly repeated sounds are transcribed *tit-tit..* and very quickly repeated sounds *tit'tit..* or *tittit..* etc.

Habitat & Behaviour

Along with range, these are important identification tools, which should be used in conjunction with identification material; many species only occur in a certain habitat type and/or in a certain altitudinal range. Altitude ranges refer to South-East Asia only.

Range & Status

To create a broader perspective, extralimital distribution is summarised for all species and breeding and wintering ranges dealt with separately. The status given for each species is necessarily general and subjective but is intended to give a comparative overview.

Use of the word 'except' in SE Asian ranges refers to distribution only. For example, a species which is an 'uncommon resident (except C Thailand)', occurs in all regions of SE Asia apart from C Thailand and is considered to be uncommon in all of the regions where it occurs.

Breeding

Space restraints allow only a brief summary, without room for such detail as nest materials etc. Season spans the period from nest construction to fledged but dependent young.

Arboreal: tree-dwelling.
Axillaries: the feathers at the base of the underwing.
Bare-parts: collective term for bill, legs and feet, eyering, exposed facial skin, etc.
Cap: well-defined patch of colour or bare skin on top of the head
Casque: an enlargement of the upper mandible, as in many hornbill species.
Cere: a fleshy structure at the base of the bill which contains the nostrils.
Colonial: nesting or roosting in tight colonies.
Comb: erect unfeathered fleshy growth, situated lengthwise on crown.
Crest: tuft of feathers on crown of head, sometimes erectile.
Distal: (of the part) farther from the body.
Eclipse: a dull short-term post-nuptial plumage.
Endemic: restricted or confined to a specific country or region.
Face: informal term for the front part of the head, usually including the forehead, lores, cheeks and often the chin.
Flight feathers: In this work, a space-saving collective term for primaries and secondaries.
Fringe: complete feather margin.
Frugivorous: fruit-eating.
Galliform: belonging to the order Galliformes, the typical game birds, including pheasants and partridges.
Graduated tail: tail on which each feather, starting outermost, is shorter than the adjacent inner feather.
Gregarious: living in flocks or communities.
Gular: pertaining to the throat.
Gunung: Malay word for mountain.
Hackles: long, pointed neck feathers.
Hepatic: brownish-red (applied to the rufous morph of some cuckoos).
Knob: a fleshy protrusion on the upper mandible of the bill.
Lappet: a fold of skin (wattle) hanging or protruding from the head.
Lateral: on or along the side.
Leading edge: the front edge (usually of the forewing in flight).
Local: occurring or relatively common within a small or restricted area.
Mask: informal term for the area of the head around the eye, often extending back from the bill and covering (part of) the ear-coverts.
Mesial: down the middle (applied to streak on chin/throat, mostly of raptors); interchangeable with gular.
Morph: a permanent alternative plumage exhibited by a species, having no taxonomic standing and usually involving base colour, not pattern.
Nomadic: prone to wandering, or occurring erratically, with no fixed territory outside breeding season.
Nuchal: pertaining to the nape and hindneck.
Ocelli: eye-like spots, often iridescent.
Orbital: surrounding the eye.
Pelagic: of the open sea.
Polyandrous: mating with more than one male (usually associated with sex-role reversal).
Post-ocular: behind the eye.
Pre-ocular: in front of the eye.
Race: see subspecies.
Rami: barbs of feathers.
Shaft-streak: a pale or dark line in the plumage produced by the feather shaft.
Subspecies: a geographical population whose members collectively show constant differences, in plumage and/or size etc., from those of other populations of the same species.
Subterminal: immediately before the tip.
Terminal: at the tip.
Terrestrial: living or occurring mainly on the ground.
Tibia: upper half of often visible avian leg (above the reverse "knee").
Trailing edge: the rear edge (usually of the wing in flight).
Underparts: the lower parts of the body (loosely applied).
Underside: the entire lower surface of the body.
Upperparts: the upper parts of the body, usually excluding the head, tail and wings (loosely applied).
Upperside: the entire upper surface of the body, tail and wings.
Vagrant: a status for a species nationally or regionally when it is accidental (rare and irregular) in occurrence.
Vermiculated: marked with narrow wavy lines, often only visible at close range.
Web: a vane (to one side of the shaft) of a feather.
Wing-bar: a line across a closed wing formed by different-coloured tips to the greater or median coverts, or both.
Wing-panel: a lengthwise strip on closed wing formed by coloured fringes (usually on flight feathers).
Zygodactyl: arrangement of feet in which two toes point forward, two backward.

AVIAN TOPOGRAPHY

The figures below illustrate the main plumage tracts and bare-part features. This terminology for bird topography has been used extensively in the species descriptions, and a full understanding of these terms is important if the reader is to make full use of this book; they are a starting point in putting together a description.

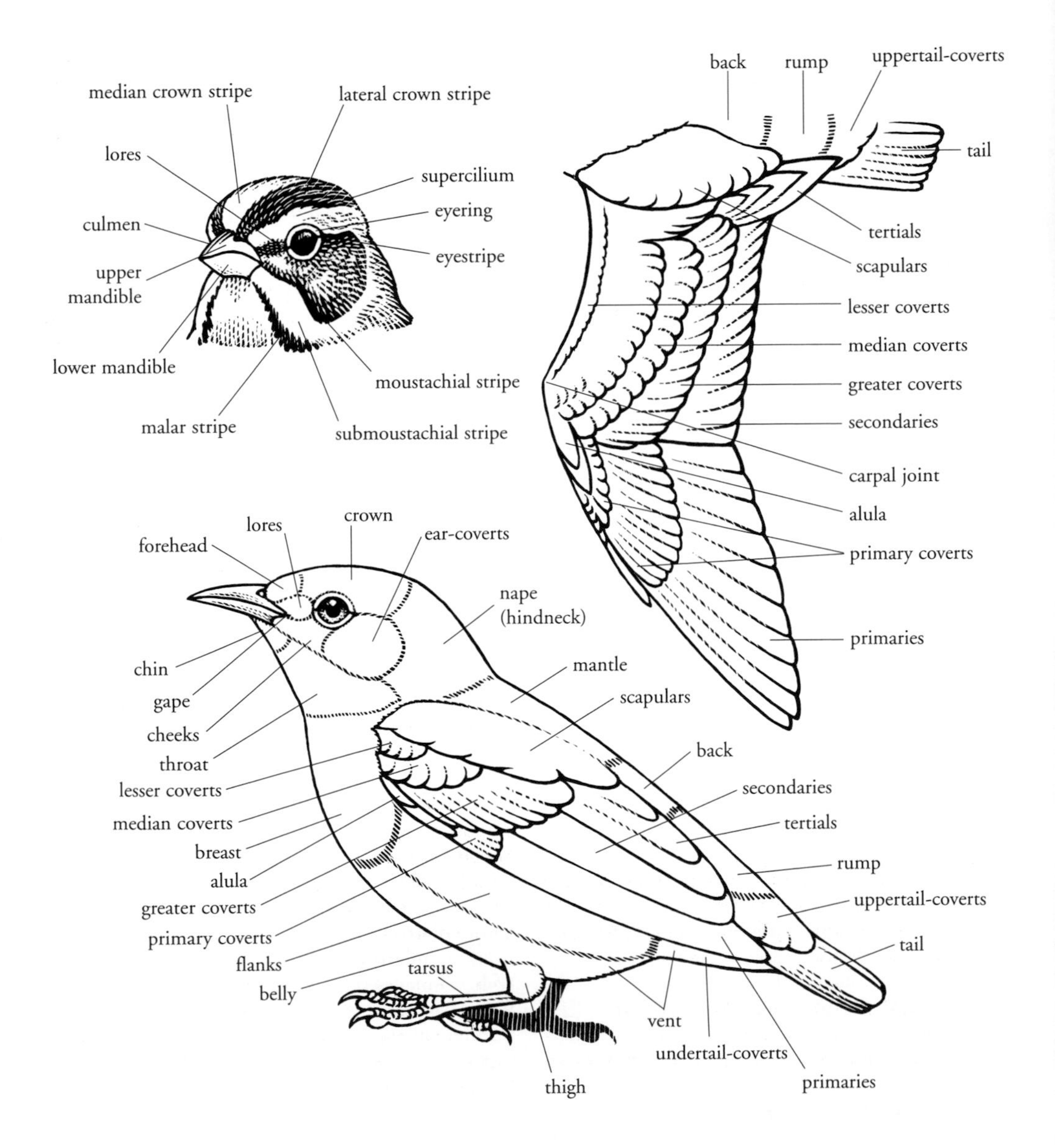

OUTLINE OF MAIN NATURAL HABITAT CATEGORIES IN SOUTH-EAST ASIA

Vegetation cover is determined by three main factors: climate, landform and human disturbance. The distribution of forest types in the region is rather complex, and two or three different types with intergradations between them often occur locally in a mosaic pattern, even in a relatively small area. No fully standardised habitat classification exists for the region, and the habitats described below largely follow those described for the region by MacKinnon and MacKinnon (1986) and for Thailand by Round (1988) and Boonsong Lekagul and Round (1991).

DECIDUOUS FOREST

Originally predominant in lowlands of continental Myanmar, Thailand and Cambodia. Very widely exploited and often replanted with monocultures of Teak and other species. Few completely untouched tracts remain.

Dry dipterocarp forest

Tends to be relatively uniform, low and open, with a grassy understorey. Occurs on the poorest, stoniest soils. Often quite rich in medium-sized arboreal birds, including parakeets and woodpeckers; smaller birds are scarcer due to the lack of middle storey and understorey foraging niches. Other species typical of this habitat include White-rumped Pygmy-falcon, Yellow-crowned Woodpecker, Common Woodshrike and White-browed Fantail. Bird diversity is typically lower than in other broadleaved forest types in the region.

Mixed deciduous forest

Richer and more diverse than dry dipterocarp forest, with a greater variety of trees in mixed association. Trees are generally taller and the forest more layered, with three main strata, including a middle storey. Bamboo is frequent, though usually occurring as a result of human disturbance (mainly through the use of fire). The structure of bird communities is similar to that found in broadleaved evergreen forests, and representatives of most landbird families occur.

BROADLEAVED EVERGREEN FOREST

Relatively dense with pronounced stratification, particularly in the case of lowland evergreen rain forest.

Lowland evergreen rain forest

Originally predominant from extreme S Thailand southwards and also in parts of south-west, S, E Myanmar, west Tenasserim, SE Thailand and south-west Cambodia. Probably the most highly threatened forest type in the region. Bird communities are very rich and diverse, with a high proportion of Sundaic species in the south. Key families include cuckoos, trogons, hornbills, barbets, woodpeckers, spiderhunters, flowerpeckers, bulbuls and babblers.

Lowland semi-evergreen rain forest

In the current work, this and the next category are included under the umbrella of broadleaved evergreen forest (see SEMI-EVERGREEN FOREST below). Subtly different to the last category and somewhat more seasonal, although undoubtedly largely evergreen and generally very moist in aspect. Varies considerably across the region. For example, such forest on the east side of the Annamite mountain chain in N and S Annam is predominantly very wet, while that to the west (in Laos) is relatively dry. Originally predominant in east and south Tenasserim, S Thailand and much of Indochina, as well as parts of W and NE Thailand. Bird richness is very similar to lowland evergreen forest but somewhat more species-poor. Specialities include important endemic or near-endemic species, such as Gurney's Pitta in south Tenasserim and S Thailand, Ratchet-tailed Treepie in parts of Tenasserim, W Thailand and Indochina, Blue-rumped Pitta in SE Thailand and Indochina, Red-collared Woodpecker and Grey-faced Tit-babbler in parts of Indochina and several galliforms in Vietnam.

Montane evergreen forest

While the lowland forests in South-east Asia are dominated by the tree family Dipterocarpaceae, at higher elevations, particularly above 900-1,000 m, families such as oaks and chestnuts (Fagaceae) predominate, forming a distinct forest type. At the highest levels, rhododendrons and other trees or shrubs of the families Ericaceae and Magnoliaceae are frequent. The largest areas of this habitat are found in northern parts of the region, while smaller pockets exist on montane isolates throughout; they are generally less threatened than other forest types in the region (except the next category). Bird communities are very rich, with a high proportion of Sino-Himalayan species (more so in the north) and marked endemicity on montane isolates. Typical of this category are such species as Rufous-throated Partridge, Wedge-tailed Green-pigeon, Rufous-necked Hornbill, Golden-throated Barbet, Maroon Oriole, Yellow-billed Nuthatch and a wide variety of flycatchers, warblers and babblers.

Forest on limestone

A sub-type of broadleaved evergreen forest, found exclusively in association with rocky limestone areas and outcrops. In some parts of the region Limestone

Wren-babbler is confined to wooded rocky limestone outcrops. More specialised subtypes occur in some areas, including C Laos and C Annam, where Sooty Babbler is endemic. Probably the least threatened forest type in the region.

Forest on islands

An ill-defined subtype, normally with a low number of tree species of relatively low stature, usually no marked stratification and little or no understorey. Generally very poor in bird diversity, supporting only a small number of more tolerant species. Two pigeons however, are specifically adapted to live on small forested islands: Pied Imperial-pigeon and Nicobar Pigeon.

SEMI-EVERGREEN FOREST

In this work, refers to evergreen forest with a varying deciduous component of up to 50%. Dense and stratified with dipterocarps predominating. Often represents an intergradation zone between lowland deciduous forests and montane evergreen forest.

SWAMP FOREST

Best regarded as a special sub-type of lowland broadleaved evergreen forest. Supports a rich variety of arboreal forest birds but galliforms and other ground-feeding birds are scarce, depending on state of inundation. Originally an important habitat for a number of globally threatened species, including Giant and White-shouldered Ibises. Swamp forests are severely threatened throughout the region and only small isolated pockets remain.

Freshwater swamp forest

Mixed to fairly species-poor and of variable stature and structure. Occurs on regularly to almost permanently flooded, typically non-acid sites. Formerly predominant in major delta areas in the region. Now much depleted, only occurring in tiny isolated patches.

Peatswamp forest

Similar to the last category but grows on deep, rain-fed, permanently water-logged acid peat-muck. More localised historically and now reduced to tiny remnants in S Thailand, Peninsular Malaysia, SE Cambodia and Cochinchina.

MANGROVES

A distinct forest type, composed of trees with a number of specialised adaptations enabling them to survive inundation by saline or brackish waters. Occurs in silt-rich sheltered inlets and near the mouths of rivers, often in association with areas of extensive intertidal mudflats. Over-exploitation and extensive replanting of monocultures is widespread and large areas of mature mangrove forest are now well isolated. A small number of bird species, including Chestnut-bellied Malkoha, Brown-winged Kingfisher, Mangrove Pitta, Mangrove Whistler, Copper-throated Sunbird and Mangrove Blue Flycatcher are restricted or largely restricted to this habitat type in South-East Asia. Mangroves also hold important large waterbird colonies, notably of Milky Stork in Peninsular Malaysia and a variety of species in Cochinchina.

CONIFEROUS FOREST

Pine forest

Occurs primarily on dry plateaux and ridges, with a very patchy distribution in continental parts of the region. Often mixed with broadleaved evergreen forest at higher elevations and with dry dipterocarp forest at lower to moderate elevations. Pine forest usually lacks any marked stratification and has a species-poor, often grassy understorey. Native pine forests are heavily exploited and replanted, with mature stands relatively scarce. The number of bird species is low throughout much of the region, with few species restricted to this habitat; Giant Nuthatch being a notable example. In S Annam more distinctive bird communities occur, including key species such as Slender-billed Oriole, Eurasian Jay, Red Crossbill and the endemic Vietnamese Greenfinch.

Other types of coniferous forest

Firs (*Abies*) occur in pure stands or in mixed associations with rhododendrons at higher elevations in N Myanmar and locally W Tonkin. Hemlocks and other conifers also occur in such areas. Spotted Nutcracker, Goldcrest and several species of tit are typical of such habitats in Myanmar.

BAMBOO

Usually more frequent in occurrence where forest has been destroyed or disturbed, especially through use of fire. Uniform stands of bamboo are generally much less rich in bird species than broadleaved forest but a number of species are associated with bamboo where it occurs as a component of deciduous or broadleaved evergreen forest. These include piculets, Pale-headed and Bamboo Woodpeckers, Pin-tailed Parrotfinch, Blue-fronted Robin, Yellow-bellied Warbler, a number of parrotbill species and White-hooded and Collared Babblers.

NATURAL SAVANNA AND THORN SCRUB

A localised 'semi-desert' habitat occurring in C and north S Myanmar, often referred to as the 'dry zone'. Four species endemic to Myanmar, Jerdon's Minivet, Hooded Treepie, Burmese Bushlark and White-throated Babbler, are largely confined here. Much of this habitat has been modified or fallen under cultivation and it is probable that little of the original habitat remains in a natural state.

GRASSLAND

Extensive natural and semi-natural grasslands were formerly widespread in the region, particularly bordering freshwater wetlands and along seasonally inundated watercourses but also in drier areas. They have been severely depleted throughout and are now one of the most threatened habitats in the region. Many were at least formerly maintained, primarily for cattle fodder, by annual burning. Key species include certain quails and buttonquails, Bengal Florican in Cambodia and Cochinchina, Jerdon's, Striated and Slender-billed Babblers in Myanmar, bushlarks, weavers, a number of estrildine finches, Jerdon's Bushchat and Rufous-rumped Grass-babbler.

SUBALPINE/ALPINE

At very high elevations in northern Myanmar, areas of rhododendron and juniper scrub and alpine meadows occur above the 'tree-line' (c.3,650–3,950 m), before giving way to bare rock. Alpine meadows and scrub provide breeding habitat for such species as Himalayan and Sclater's Monals, Alpine Accentor, a number of finches and Grandala.

WETLANDS

Wetland habitats are severely threatened throughout and support a high proportion of globally and regionally threatened species.

Freshwater wetlands

Most of the region's freshwater lakes have been altered or over-exploited while most marshes have been drained, dammed to create lakes, reclaimed for building sites or canalised and turned into rice-paddy or other cultivation. Additionally, many reservoirs have been created, few of which provide a valuable habitat for birds. Freshwater wetlands originally supported a wide variety of breeding large waterbirds, as well as large numbers of wintering waterfowl. However, most large waterbirds, particularly storks, herons, ibises, egrets, pelicans, cormorants and Oriental Darter, are now severely threatened throughout the region. Smaller waterbirds such as bitterns, crakes, rails and jacanas, and wintering flocks of ducks, are still quite widespread and many make use of certain man-made wetlands.

The sandbanks and shores of the region's large, open rivers additionally support a distinctive avifauna, with key species including Great Thick-knee, River Lapwing, Small Pratincole, Indian Skimmer, River and Black-bellied Terns and Grey-throated Sand-martin. This habitat is also under threat, particularly along the Mekong River, where populations of the above-mentioned species are now much depleted or in some cases virtually extinct.

Slow-moving rivers through forest also provide valuable habitat for some rare and local species, including Green Peafowl, White-winged Duck and Blyth's Kingfisher. See swamp forest (page 20).

Out of South-East Asia's original avifauna, only two species are now thought to have become globally extinct, Pink-headed Duck and White-eyed River-martin; both occurred in freshwater wetlands.

Coastal wetlands

Intertidal mudflats, brackish marshes and man-made habitats such as salt-pans and prawn- and fish-ponds are of very great importance for a host of migrant waders, terns and gulls, many of which winter in South-East Asia. Key species include Black-faced Spoonbill, Spoon-billed Sandpiper and Saunders's Gull, which all winter in the Red River Delta in Vietnam.

Beaches

Undisturbed sandy beaches, dunes and coastal flats provide valuable breeding areas for certain species, some of which are shared with the larger rivers. These include Great and Beach Thick-knees, Kentish and Malaysian Plovers, Little Tern and bee-eaters. Extensive stretches of undisturbed beach are now largely restricted to Myanmar and Vietnam.

Offshore islands

Small islets are important as nesting areas for terns and locally some larger seabirds such as boobies. Disturbance and collection of eggs and young by fishermen have led to widespread local extinctions.

BIRD STUDY AND CONSERVATION ORGANISATIONS

NATIONAL

CAMBODIA

Wildlife Protection Office
40 Norodom Boulevard, Phnom Penh, Cambodia
Email: wildlifedfw@online.com.kh

Cambodia Bird News
c/o Wildlife Conservation Society (Cambodia Programme), PO Box 1620, Phnom Penh, Cambodia
Email: cambodia@wcs.org

LAOS

Wildlife Conservation Society (Lao programme)
P.O. Box 6712, Vientiane, Lao PDR
Email: wcslao@laonet.net

WWF Lao Programme
c/o Department of Forestry, Vientiane, Lao PDR
Email: wwflao@laonet.net / wwflao@loxinfo.co.th

MYANMAR

Myanmar Bird and Nature Society
69 Myaynigone Zay Street, Sanchaung Township, Yangon 11111, Myanmar
Email: SST@mptmail.net.mm

PENINSULAR MALAYSIA

Malaysian Nature Society
P.O.Box 10750, 50724 Kuala Lumpur, Malaysia
Email: mns@streamyx.com

WWF Malaysia
49 Jalan SS23/15, Taman SEA, 47400 Petaling Jaya, Selangor DE, Malaysia.
Email: wwfmal.org.my

SINGAPORE

Nature Society (Singapore)
510 Geylang Road, # 02-05, The Sunflower, Singapore 389466
Email: contact@nss.org.sg

THAILAND

Bird Conservation Society of Thailand
69/12 Ramintra 24, Jarakheebua, Lardprao, Bangkok 10230, Thailand
Email: bcst@box1.a-net.net.th

VIETNAM (also Myanmar)

BirdLife International *in Indochina*
4, Lane 209, Doi Can Street, Hanoi, Vietnam
Email: birdlife@birdlife.netnam.vn

INTERNATIONAL

BirdLife International
Wellbrook Court, Girton Road, Cambridge CB3 0NA, United Kingdom.
Email: birdlife@birdlife.org.uk

Oriental Bird Club
c/o The Lodge, Sandy, Bedfordshire SG19 2DL, United Kingdom.
Email: mail@orientalbirdclub.org

TRAFFIC Southeast Asia
Unit 9-3A, 3rd Floor, Jalan SS23/11, Taman SEA, 47400 Petaling Jaya, Selangor, Malaysia
Email: tsea@po.jaring.my

Wetlands International–Asia Pacific
3A39, Block A, Kelana Centre Point, Jalan SS7/19, 47400 Petaling Jaya, Selangor, Malaysia
Email: malaysia@wetlands.org.my

SELECTED REFERENCES

Abdulali, H. (1964) The birds of the Andaman and Nicobar Islands. *J. Bombay Nat. Hist. Soc.* 61: 483-571.

Ali, S. and Ripley, S. D. (1968-1974) *Handbook of the Birds of India and Pakistan.* 10 vols. First edition. Bombay: Oxford University Press.

Alström, P. (1998) Taxonomy of the *Mirafra assamica* complex. *Forktail* 13: 97-107.

Alström, P. and Mild, K. (2003) *Pipits and Wagtails of Europe, Asia and North America.* London: Christopher Helm.

Alström, P. and Olsson, U. (1999) The Golden-spectacled Warbler: a complex of sibling species, including a previously undescribed species. *Ibis* 141: 545-568.

Alström, P. Davidson, P., Duckworth, J. W., Eames, J., Nguyen Cu, Olsson, U., Robson, C. and Timmins, R.J. (in prep.) Description of a new species of *Phylloscopus* from Vietnam and Laos.

Alström, P, Ericson, P. G. P., Olsson, U. and Sundberg, P. (2006) Phylogeny and classification of the avian superfamily Sylvioidea. *Molec. Phylogen. Evol.* 38: 381-397.

Alström, P., Olsson, U., Lei Fumin, Wang Hai-tao, Gao Wei and Sundberg, P. (2008) Phylogeny and classification of the Old World Emberizini (Aves, Passeriformes). *Molec. Phylogen. Evol.* 47: 960-973.

Alström, P., Olsson, U., Rasmussen, P. C., Yao, C.-t., Ericson, P. G. P. and Sundberg, P. (2007) Morphological, vocal and genetic divergence in the *Cettia acanthizoides* complex Aves: Cettiidae). *Zool. J. Linn. Soc.* 149: 437-452.

Alström, P., Rasmussen, P. C., Olsson, U. and Sundberg, P. (in press) Species delimitation based on multiple criteria: the Spotted Bush Warbler *Bradypterus thoracicus* complex (Aves: Megaluridae). *Zool. J. Linn. Soc.*

Anderson, J. (1887) List of birds, chiefly from the Mergui Archipelago, collected for the Trustees of the Indian Museum, Calcutta. *J. Linn. Soc.* 21: 136-153.

Anon. (1994) From the field. *Oriental Bird Club Bull.* 19: 65-67.

Arnaiz-Villena, A., Álvarez-Tejado, M., Ruiz-del-Valle, V., García-de-la-Torre, C., Varela, P., Recio, M. J., Ferre, S. and Martínez-Laso, J. (1998) Phylogeny and rapid Northern and Southern Hemisphere speciation of goldfinches during the Miocene and Pliocene Epochs. *Cell. Mol. Life Sci.* 54: 1031-1041.

Arnaiz-Villena, A., Guillén, J., Ruíz-del-Valle, V., Lowy, E., Zamora, J., Varela, P., Stefani, D. and Allende L. M. (2001) Phylogeography of crossbills, bullfinches, grosbeaks, and rosefinches. *Cell. Mol. Life Sci.* 58: 1-8.

Arnaiz-Villena, A., Moscoso, J., Ruiz-del-Valle, V., Gonzalez, J., Reguera, R., Ferri, A., Wink, M. and Serrano-Vela, J. I. (2008) Mitochondrial DNA phylogenetic definition of a group of 'arid-zone' cardueline finches. *Open Ornithology J.* 2008(1): 1-7.

Bairlein, F., Alström, P., Aymí, R., Clement, P., Dyrcz, A., Gargallo, G., Hawkins, F., de Bridge, E. S., Jones A. W. and Baker, A. J. (2005) A phylogenetic framework for the terns (Sternini) inferred from mtDNA sequences: implications for taxonomy and plumage evolution. *Molec. Phylogen. Evol.* 35: 459-469.

Baker, E. C. S. (1922-1930) *The fauna of British India: Birds.* Second edition. 8 vols. Taylor and Francis, London.

Bakewell, D. N. and Kennerley, P. R. (2008) Field characteristics and distribution of an overlooked *Charadrius* plover from South-East Asia. *BirdingASIA* 9: 46–57.

Bangs, O. (1921) The birds of the American Museum of Natural History's Asiatic zoological expedition of 1916–1917. *Bull. Amer. Mus. Nat. Hist.* 44: 575-612.

Banks, R. C., Chesser, R. T., Cicero, C., Dunn, J. L., Kratter, A. W., Lovette, I. J., Rasmussen, P. C., Remsen, J. V. Jr, Rising, J. D. and Stotz, D. F. (2007) Forty-eighth supplement to the American Ornithologists' Union check-list of North American birds. *Auk* 124: 1109-1115.

Banks, R. C., Cicero, C., Dunn, J. L., Kratter, A. W., Rasmussen, P. C., Remsen, J. V. Jr, Rising, J. D. and Stotz, D. F. (2006) Forty-seventh supplement to the American Ornithologists' Union check-list of North American birds. *Auk* 123: 926-936.

Barker, F. K. (2004) Monophyly and relationship of wrens (Aves: Troglodytidae): a congruence analysis of heterogeneous mitochondrial and nuclear DNA sequence data. *Molec. Phylogen. Evol.* 31: 486-504.

Beaman, M. and Madge, S. (1999) *The Handbook of Bird Identification for Europe and the Western Palearctic.* London: Christopher Helm.

Becking, J. H. (1981) Notes on the breeding of Indian cuckoos. *J. Bombay Nat. Hist. Soc.* 78: 201-231.

Beresford, P., Barker, F. K., Ryan, P. G. and Crowe, T. M. (2005) African endemics span the tree of songbirds (Passeri): molecular systematics of several evolutionary 'enigmas'. *Proc. R. Soc. B* 272: 849-858.

Bingham, C. T. (1878) After the adjutants. *Stray Feathers* 7: 25-33.

Bingham, C. T. (1900) On the birds collected and observed in the Southern Shan States of Upper Burma. *J. Asiatic Soc. Bengal* 69: 102-143.

Bingham, C. T. (1903) A contribution to our knowledge of the birds occurring in the Southern Shan States, Upper Burma. *Ibis* (8)3: 584-606.

BirdLife International (2001) *Threatened birds of Asia: the BirdLife International Red Data Book.* Cambridge, UK: BirdLife International.

Boonsong Lekagul and Round, P. D. (1991) *A Guide to the Birds of Thailand.* Bangkok: Saha Karn Bhaet.

Bourret, R. (1943) Liste des oiseaux récemment entrés dans les collections du Laboratoire de Zoologie. *Notes et travaux de l'Ecole supérieure des Sciences de l'Université Indochinoise, Hanoi* 2: 18-37.

Bourret, R. (1944) Liste des oiseaux dans la collection du Laboratoire de Zoologie, troisième liste, 1943. *Notes et travaux de l'Ecole supérieure des Sciences de l'Université Indochinoise, Hanoi* 3: 19-36.

Brazil, M. A. (1991) *The Birds of Japan.* London: Christopher Helm.

Brown, L. and Amadon, D. (1968) *Eagles, Hawks and Falcons of the World.* Feltham, Middlesex: Country Life Books.

Byers, C., Olsson, U. and Curson, J. (1995) *Buntings and Sparrows.* Robertsbridge, UK: Pica Press.

Caldwell, H. R. & Caldwell, J. C. (1931) *South China Birds.* Shanghai: Hester May Vanderburgh.

Chantler, P. and Driessens, G. (1995) *Swifts.* Robertsbridge, UK: Pica Press.

Cheng Tso-hsin (1987) *A Synopsis of the Avifauna of China.* Beijing: Science Press, and Hamburg: Paul Parey.

Christison, P., Buxton, A., Emmet, A. M. and Ripley, D. (1946) Field notes on the birds of coastal Arakan and the foothills of the Yomas. *J. Bombay Nat. Hist. Soc.* 46: 13-32.

Cleere, N. and Nurney, D. (1998) *Nightjars.* Robertsbridge, UK: Pica Press.

Clement, P., Harris, A. and Davis, J. (1993) *Finches and Sparrows.* London: Christopher Helm.

Coates, B. J. and Bishop, K. D. (1997) *A Guide to the Birds of Wallacea.* Alderley, Australia: Dove Publications.

Collar, N. J. (2006) A taxonomic reappraisal of the Black-browed Barbet *Megalaima oorti. Forktail* 22: 170-173.

Collar, N. J. and Pilgrim, J. D. (2007) Species-level changes proposed for Asian birds, 2005-2006. *BirdingASIA* 8: 14-30.

Collar, N. J. and Robson, C. (2007) Family Timaliidae (Babblers). Pp. 70-291 in del Hoyo, J., Elliott, A. and Christie, D. A. (eds.) *Handbook of Birds of the World,* 12. Barcelona: Lynx Edicions.

Collar, N. J., Crosby, M. J. and Stattersfield, A. J. (1994) *Birds to Watch 2: the world list of threatened birds.* Cambridge, U.K.: BirdLife International (BirdLife Conservation Series 4).

Colston, P. R. (1978) A new bird for Burma – Pallas's Reed Bunting *Emberiza pallasi. Bull. Brit. Orn. Club* 98: 21-22.

Cook, J. P. (1912) Notes on some of the bird life at Thandoung. *J. Bombay Nat. Hist. Soc.* 21: 668-675.

Cook, J. P. (1913) A list of Kalaw birds, with bird-nesting notes. *J. Bombay Nat. Hist. Soc.* 22: 260-270.

Coomans de Ruiter, L. (1946) Oölogische en biologische aanteekeningen over eenige hoendervogels in de Westerafdeeling van Borneo. *Limosa* 19: 129-140.

Crosby, M. (1995) From the field. *Oriental Bird Club Bull.* 21: 68-73.

David-Beaulieu, A. (1932a) Les oiseaux de la région de Honquan (Province de Thudaumot, Cochinchine). *Oiseau et R.F.O.* 2: 133-154.

David-Beaulieu, A. (1932b) Supplément à la liste des oiseaux de Honquan (Province de Thudaumot, Cochinchine). *Oiseau et R.F.O.* 2: 619-620.

David-Beaulieu, A. (1939) Les oiseaux de la région de Pleiku (hauts plateaux de Sud-Annam). *Oiseau et R.F.O.* 9: 13-32, 163-182.

David-Beaulieu, A. (1944) *Les oiseaux du Tranninh.* Hanoi: Université Indochinoise.

David-Beaulieu, A. (1949) Les oiseaux de la province de Savannakhet (Bas-Laos). *Oiseau et R.F.O.* 19: 41-84, 153-194.

David-Beaulieu, A. (1950) Les oiseaux de la province de Savannakhet (Bas-Laos) (continuation). *Oiseau et R.F.O.* 20: 9-50.

Deignan, H. G. (1945) Birds of northern Thailand. *U.S. Natn. Mus. Bull.* 186.

Deignan, H. G. (1963) Checklist of the Birds of Thailand. *U.S. Natn. Mus. Bull.* 226.

Delacour, J. (1929) On the birds collected during the fourth expedition to French Indo-China. *Ibis* (12)5: 193-220, 403-429.

Delacour, J. (1930) On the birds collected during the fifth expedition to French Indo-China. *Ibis* (12)6: 564-599.

Delacour, J. (1951) Commentaires, modifications et additions à la liste des oiseaux de l'Indochine française (II). *Oiseau et R.F.O.* 21: 1-32, 81-119.

Delacour, J. (1977) *The Pheasants of the World.* Second edition. Hindhead, UK: Spur Publications and World Pheasant Association.

Delacour, J. and Greenway, J. C. (1940) VIIe expédition ornithologique en Indochine française. *Oiseau et R.F.O.* 10: 1-24.

Delacour, J. and Greenway, J. C. (1941) Commentaires, additions et modifications à la liste des oiseaux de l'Indo-chine française. *Oiseau et R.F.O.* 11 (Suppl.): i-xxi.

Delacour, J. and Jabouille, P. (1925) On the birds of Quang-tri, Central Annam; with notes on others from other parts of French Indo-China. *Ibis* (12)1: 209-260.

Delacour, J. and Jabouille, P. (1927) *Recherches ornithologiques dans les provinces du Tranninh (Laos), de Thua-Thien et de Kontoum (Annam) et quelques autres régions de l'Indochine française.* Paris: Société Nationale d'Acclimatation de France (Archives d'Histoire Naturelle).

Delacour, J. and Jabouille, P. (1931) *Les oiseaux de l'Indochine française,* 1-4. Paris: Exposition Coloniale Internationale.

Delacour, J., Jabouille, P. and Lowe, W. P. (1928) On the birds collected during the third expedition to French Indo-China. *Ibis* (12)4: 23-51, 285-317.

Dickinson, E. C. (1970a) Birds of the Legendre Indochina expedition 1931-1932. *Amer. Mus. Novit.* 2423.

Dickinson, E. C. (1970b) Notes upon a collection of birds from Indochina. *Ibis* 112: 481-487.

Dickinson, E. C., ed. (2003) *The Howard and Moore Complete Checklist of the Birds of the World.* Third edition. London: Christopher Helm.

Dickinson, E. C. and Dekker, R. W. R. J. (2001) Systematic notes on Asian birds. 13. A preliminary review of the Hirundinidae. *Zool. Verh. Leiden* 335: 127-144.

Dickinson, E. C., Eck, S. and Martens. J. (2004) Systematic notes on Asian birds 44. A preliminary review of the Corvidae. *Zoologische Verhandelingen* 350: 85-109.

Dickinson, E. C., Kennedy, R. S. and Parkes, K. C. (1991) *The birds of the Philippines: an annotated check-list.* Tring, UK: British Ornithologists' Union.

Dickinson, E. C., Rasmussen, P. C., Round, P. D. and Rozendaal, F. G. (1998) Reinstatement of *Bradypterus seebohmi* to the Indian avifauna, and revalidation of an earlier name. *Ostrich* 69: 399.

Duckworth, J. W. and Hedges, S. (1998) Bird records from Cambodia in 1997, including records of sixteen species new for the country. *Forktail* 14: 29-36.

Duckworth, J. W., Alström, P., Davidson, P., Evans, T. D., Poole, C. M., Setha, T. and Timmins, R. J. (2001) A new species of wagtail from the lower Mekong basin. *Bull. Brit. Orn. Club* 121: 152-182.

Duckworth, J. W., Davidson, P., Evans, T. D., Round, P. D. and Timmins, R. J. (2002) Bird records from Laos, principally the Upper Lao/Thai Mekong and Xiangkhouang Province, in 1998–2000. *Forktail* 18: 11-44.

Duckworth, J. W., Davidson, P and Timmins, R. J. (1999 in press) Birds. In Duckworth, J. W., Salter, R. E. and Khounboline, K. (compilers) *Wildlife in Lao PDR: 1999 status report.* Vientiane: IUCN/WCS/CPAWM.

Duckworth, J. W., Timmins, R., Thewlis, R. M., Robichaud, W. G. and Evans, T. D. (1998) Ornithological records from Laos, October 1994-August 1995. *Forktail* 13: 33-68.

Dymond, J. N. (1998) Birds in Vietnam in December 1993 and December 1994. *Forktail* 13: 7-12.

Eames, J. C. (1997) Some additions to the list of birds of Vietnam. *Forktail* 12: 163-166.

Eames, J.C. (2002) Eleven new sub-species of babbler (Passeriformes: Timaliinae) from Kon Tum Province, Vietnam. *Bull. Brit. Orn. Club* 122: 109-141.

Eames, J.C. and Eames, C. (2001) A new species of laughingthrush (Passeriformes: Garrulacinae) from the Central Highlands of Vietnam. *Bull. Brit. Orn. Club* 121: 10-23.

Eames, J. C. and Ericson, P. G. P. (1996) The Björkegren expedition to French Indochina: a collection of birds from Vietnam and Cambodia. *Nat. Hist. Bull. Siam Soc.* 44: 75-111.

Eames, J. C., Le Trong Trai and Nguyen Cu (1995) Rediscovery of the Sooty Babbler *Stachyris herberti* in central Vietnam. *Bird Conserv. Internatn.* 5: 129-135.

Eames, J. C., Le Trong Trai and Nguyen Cu (1995) Rediscovery of the Grey-crowned Crocias *Crocias langbianis. Bird Conserv. Internatn.* 5: 525-535.

Eames, J. C., Le Trong Trai and Nguyen Cu (1999) A new species of laughingthrush (Passeriformes: Garrulacinae) from the Western Highlands of Vietnam. *Bull. Brit. Orn. Club* 119: 4-15.

Eames, J. C., Le Trong Trai, Nguyen Cu and Eve, R. (1999) New species of barwing *Actinodura* (Passeriformes: Sylviinae: Timaliini) from the Western Highlands of Vietnam. *Ibis* 141: 1-10.

Eames, J. C., Robson, C. R. and Nguyen Cu (1995) A new subspecies of Spectacled Fulvetta *Alcippe ruficapilla* from Vietnam. *Forktail* 10: 141-157.

Eames, J.C., Steinheimer, F.D. and Bansok, R. (2002) A collection of birds from the Cardamom Mountains, Cambodia, including a new subspecies of *Arborophila cambodiana. Forktail* 18: 67-86.

Engelbach, P. (1927) Une collection d'oiseaux du Bas Laos. *Bull. Soc. Zool. France* 52: 239-250.

Engelbach, P. (1932) Les oiseaux de Laos méridional. *Oiseau et R.F.O.* 2: 439-498.

Ericson, P. G. P., Envall, I., Irestedt, M. and Norman J. A. (2003) Inter-familial relationships of the shorebirds (Aves: Charadriiformes) based on nuclear DNA sequence data. *BMC Evolutionary Biology* 3: 16-29.

Ericson, P. G. P., Jansén, A.-L., Johansson, U. S. and Ekman, J. (2005) Inter-generic relationships of the crows, jays and allied groups (Aves: Corvidae) based on nucleotide sequence data. *J. Avian Biol.* 36: 222–234.

Evans, T. D. (2001) Ornithological records from Savannakhet Province, Lao PDR, January-July 1997. *Forktail* 17: 21-28.

Evans, T. D. and Timmins, R. J. (1998) Records of birds from Laos during January-July 1994. *Forktail* 13: 69-96.

Evans, T. D., Towll, H. C., Timmins, R. J., Thewlis, R. M., Stones, A. J., Robichaud, W. G. and Barzen, J. (2000) Ornithological records from the lowlands of southern Laos during December

1995-September 1996, including areas on the Thai and Cambodian borders. *Forktail* 16: 29-52.

Fain M. G. and Houde, P. (2004) Parallel radiations in the primary clades of birds. *Evolutionary Biol.* 3: 16-29.

Feare, C. and Craig, A. (1999) *Starlings and Mynas.* Princeton, New Jersey: Princeton University Press.

Feinstein, J., Yang Xiaojun and Li Shou-Hsien (2008) Molecular systematics and historical biogeography of the Black-browed Barbet species complex (*Megalaima oorti*). *Ibis* 150: 40-49.

Fischer, W. (1961) Kleine Beiträge zur Vogelkunde Vietnams. *Beitr. Vogelk.* 7: 285-317.

Fischer, W. (1963) Weitere Beiträge zur Vogelkunde Vietnams. *Beitr. Vogelk.* 9: 102-123.

Fischer, W. (1965) Neue Beiträge zur Vogelkunde Vietnams. *Beitr. Vogelk.* 10: 361-379.

Fischer, W. (1974) Vorläufiger Abschlussbericht über Vogelbeobachtungen in Vietnam. *Beitr. Vogelk.* 20: 249-300.

Fischer, W. (1983) Ein Beiträg zum Vorkommen, Durchzug und zur Ubersommerung von Limikolen (Charadriiformes) in Vietnam. *Beitr. Vogelk.* 29: 297-305.

Flint, V. E., Boehme, R. L., Kostin, Y. V. and Kuznetsov, A. A. (1984) *A Field Guide to Birds of the USSR.* Princeton, New Jersey: Princeton University Press.

Fry, C. H., Fry, K. and Harris, A. (1992) *Kingfishers, Bee-eaters and Rollers.* London: Christopher Helm.

Fuchs, J., Fjeldså, J. and Pasquet, E. (2006). An ancient African radiation of corvoid birds (Aves: Passeriformes) detected by mitochondrial and nuclear sequence data. *Zoologica Scripta* 35: 375–385.

Fuchs, J., Pons, J.-M., Ericson, P. G. P , Bonillo, C., Couloux, A. and Pasquet, E. (2008) Molecular support for a rapid cladogenesis of the woodpecker clade Malarpicini, with further insights into the genus *Picus* (Piciformes: Picinae). *Molec. Phylogen. Evol.* 48: 34–46.

Gill, F. B., Slikas, B. and Sheldon, F. H. (2005) Phylogeny of titmice (Paridae): II. Species relationships based on sequences of mitochondrial cytochrome-*B* Gene. *Auk* 122: 121-143.

Goroshko, O. A. (1993) [Taxonomic status of the Pale (Sand?) Martin *Riparia (riparia) diluta* (Sharpe et Wyatt 1893).] *Russ. J. Orn.* 2: 303-323. (In Russian.)

Gretton, A. (1990) Recent reports. *Oriental Bird Club Bull.* 11: 40-48.

Grimmett, R., Inskipp, C. & Inskipp, T. (1998) *Birds of the Indian Subcontinent.* London: Christopher Helm.

Hancock, J. and Kushlan, J. (1984) *The Herons Handbook.* London & Sydney: Croom Helm.

Hancock, J. A., Kushlan, J. A. and Kahl, M. P. (1992) *Storks, Ibises and Spoonbills of the World.* London: Academic Press.

Haring, E., Kvaløy, K., Gjershaug, J.-O., Røv, N. and Gamauf, A. (2007) Convergent evolution and paraphyly of the hawk-eagles of the genus *Spizaetus* (Aves, Accipitridae) – phylogenetic analyses based on mitochondrial markers. *J. Zool. Syst. Evol. Res.* 45: 353-365.

Harington, H. H. (1909-1910) A list of the birds of the Bhamo District of Upper Burma. *J. Bombay Nat. Hist. Soc.* 19: 107-128, 299-313.

Harington, H. H. (1911a) A further list of birds of the Bhamo District, Upper Burma. *J. Bombay Nat. Hist. Soc.* 20: 373-379.

Harington, H. H. (1911b) Some Maymyo birds. *J. Bombay Nat. Hist. Soc.* 20: 1002-1011.

Harington, H. H. (1911c) Some Maymyo birds. Part II. *J. Bombay Nat. Hist. Soc.* 21: 585-587.

Harrap, S. (2008) Family Aegithalidae (long-tailed tits). Pp.76-101 in del Hoyo, J., Elliott, A. and Christie, D.A. (eds.) *Handbook of Birds of the World,* 13. Lynx Edicions. Barcelona.

Harrap, S. (2008) Family Sittidae (nuthatches). Pp.102-145 in del Hoyo, J., Elliott, A. and Christie, D.A. (eds.) *Handbook of Birds of the World,* 13. Lynx Edicions. Barcelona.

Harrap, S. and Quinn, D. (1996) *Tits, Nuthatches and Treecreepers.* London: Christopher Helm.

Harrison, P. (1985) *Seabirds: an identification guide.* London: Croom Helm.

Hayman, P., Marchant, J. and Prater, T. (1986) *Shorebirds: an identification guide to the waders of the world.* London: Croom Helm.

Hennache, A., Rasmussen, P., Lucchini, V., Rimondi, S. and Randi, R. (2003) Hybrid origin of the imperial pheasant *Lophura imperialis* (Delacour and Jabouille, 1924) demonstrated by morphology, hybrid experiments, and DNA analyses. *Biol. J. Linn. Soc.* 80: 573-600.

Herbert, E. G. (1923–1926) Nests and eggs of birds in Central Siam. *J. Nat. Hist. Soc. Siam* 6: 81-123, 215-222, 293-311, 323-356.

Hopwood, C. (1908) A further list of birds from the Chindwin, Upper Burma. *J. Bombay Nat. Hist. Soc.* 18: 432-433.

Hopwood, C. (1912a) Notes on some birds from the Chindwin Valley. *J. Bombay Nat. Hist. Soc.* 21: 1089-1090.

Hopwood, C. (1912b) A list of birds from Arakan. *J. Bombay Nat. Hist. Soc.* 21: 1196-1221.

Hopwood, J. C. and Mackenzie, J. M. D. (1917) A list of birds from the North Chin Hills. *J. Bombay Nat. Hist. Soc.* 25: 72-91.

del Hoyo, J., Elliott, A. and Sargatal, J. eds. (1992–1997) *Handbook of Birds of the World,* 1-4. Barcelona: Lynx Edicions.

Htin Hla, T., Eames, J. C., Sein Ohn Maung, Ko Pan, Saw Moses and Saw Nyunt Tin (in prep.). [Records from BANCA/BirdLife International surveys of S. Tanintharyi forests in May 2003 and April–May 2004.]

Hume, A. O. (1874) A first list of the birds of the Tenasserim Provinces. *Stray Feathers* 2: 467-484.

Hume, A. O. (1875a) A first list of the birds of upper Pegu. *Stray Feathers* 3: 1-194.

Hume, A. O. (1875b) A second list of the birds of Tenasserim. *Stray Feathers* 3: 317-326.

Hume, A. O. (1876) A third list of the birds of the Tenasserim Provinces. *Stray Feathers* 4: 223-225.

Hume, A. O. and Davison, W. R. (1878) A revised list of the birds of Tenasserim. *Stray Feathers* 6: i-viii, 1-524.

Inskipp, T. P. (in prep.) *Annotated bibliography and checklist of the birds of Myanmar (Burma).*

Inskipp, T. P. and Mlikovsky, J. (in prep.) *Annotated checklist and bibliography of the birds of Indochina.*

Inskipp, T., Lindsey, N. and Duckworth, W. (1996) *An annotated checklist of the birds of the Oriental region.* Sandy, UK: Oriental Bird Club.

Irwin, D., Alström, P., Olsson, U. and Benowitz-Fredericks, Z. M. (2001) Cryptic species in the genus *Phylloscopus* (Old World leaf warblers). *Ibis* 143: 233–247.

Jepson, P. (1987) Recent reports. *Oriental Bird Club Bull.* 6: 36-40.

Jeyarajasingam, A. and Pearson, A. (1999) *A Field Guide to the Birds of West Malaysia and Singapore.* Oxford, New York, Tokyo: Oxford University Press.

Johansson, U. S., Alström, P., Olsson, U., Ericson, P. G. P., Sundberg, P. and Price, T. D. (2007) Build-up of the Himalayan avifauna through immigration: a biogeographical analysis of the *Phylloscopus* and *Seicercus* warblers. *Evolution* 61: 324-333.

Jones, D. N., Dekker, R. W. R. J. and Roselaar, C. S. (1995) *The Megapodes.* Oxford: Oxford University Press.

Jønsson, K. A. and Fjeldså, J. (2006) A phylogenetic supertree of oscine passerine birds (Aves: Passeri). *Zoologica Scripta* 35: 149–186.

Juana, E., Suárez, F., Ryan, P., Alström, P. and Donald, P. (2004) Family Alaudidae (larks). Pp.496–601 in del Hoyo, J., Elliott, A. and Christie, D.A. (Eds.). *Handbook of the Birds of the World,* 9. Barcelona: Lynx Edicions.

Juniper, T. and Parr, M. (1998) *Parrots.* Robertsbridge, UK: Pica Press.

Kemp, A. C. (1995) *The Hornbills* (Bird Families of the World, 1). Oxford: Oxford University Press.

King, B. (1983) New bird distribution data for Burma. *Nat. Hist. Bull. Siam Soc.* 31: 55-62.

King, B. (2004) The taxonomic status of the three subspecies of *Cuculus saturatus. Bull. Brit. Orn. Club* 125: 48-55.

King, B. (2007) Some 1960s additions to the list of Thailand's birds. *Nat. Hist. Bull. Siam Soc.* 55: 105-119.

King, B. and Robson, C. (2008) The taxonomic status of the three subspecies of Greater Rufous-headed Parrotbill *Paradoxornis ruficeps. Forktail* 24: 120-122.

King, B., Abramson, I. J., Keith, A. R., Weiss, W. J., Jr. and Carrott, J. (1973) Some new bird records for Burma and Thailand. *Nat. Hist. Bull. Siam Soc.* 25: 157-160.

King, B., Buck, H., Ferguson, R., Fisher, T., Goblet, C., Nickel, H. and Suter, W. (2001) Birds recorded during two expeditions to north Myanmar (Burma). *Forktail* 17: 29-40.

King, B. F. (1997) *Checklist of the Birds of Eurasia.* Vista, California: Ibis Publishing Company.

King, B. F. (2002) The *Hierococcyx fugax,* Hodgson's Hawk Cuckoo, complex. *Bull. Brit. Orn. Club* 122: 74-80.

King, B. F., Dickinson, E. C. and Woodcock, M. W. (1975) *A Field Guide to the Birds of South-East Asia.* London: Collins.

Kinnear, N. B. (1929) On the birds collected by Mr H. Stevens in northern Tonkin in 1923–1924, with notes by the collector. *Ibis* (12)5: 107-150, 292-344.

Kinnear, N. B. (1934) On the birds of the Adung Valley, north-east Burma. *J. Bombay Nat. Hist. Soc.* 37: 347-368.

Klicka, J., Voelker, G. and Spellman, G. M. (2005). A molecular phylogenetic analysis of the 'true thrushes' (Aves: Turdinae). *Molec. Phylogen. Evol.* 34: 486–500.

Kyaw Nyunt Lwin and Khin Ma Ma Thwin (2003) *Birds of Myanmar.* Yangon: Swiftwinds.

Lambert, F. and Woodcock, M. (1996) *Pittas, Broadbills and Asities.* Robertsbridge, UK: Pica Press.

Lambert, F. R., Eames, J. C. and Nguyen Cu (1994) *Surveys for endemic pheasants in the Annamese lowlands of Vietnam, June–July 1994: status of and conservation recommendations for Vietnamese Pheasant Lophura hatinhensis and Imperial Pheasant L. imperialis.* Gland, Switzerland and Cambridge, UK: IUCN Species Survival Commission.

La Touche, J. D. D. (1931) *A handbook of the birds of eastern China,* 1-2. London: Taylor and Francis.

Laurie, A., Ha Dinh Duc and Pham Trong Anh (1989) Survey for Kouprey (*Bos sauveli*) in western Daklak Province, Vietnam. The Kouprey Conservation Trust. Unpublished.

Lefranc, N. and Worfolk, T. (1997) *Shrikes.* Robertsbridge, UK: Pica Press.

Lovette, I. J. and Rubenstein, D. R. (2007) A comprehensive molecular phylogeny of the starlings (Aves: Sturnidae) and mockingbirds (Aves: Mimidae): congruent mtDNA and nuclear trees for a cosmopolitan avian radiation. *Molec. Phylogen. Evol.* 44: 1031-1056.

Lowe, W. P. (1933) A report on the birds collected by the Vernay Expedition to Tenasserim and Siam. *Ibis* (13)3: 259-283, 473-491.

Macdonald, K. C. (1906) A list of birds found in the Myingyan district of Burma. *J. Bombay Nat. Hist. Soc.* 17: 184-194, 492-504.

Macdonald, K. C. (1908) Notes on birds in the Amherst district, Lower Burma. *J. Bombay Nat. Hist. Soc.* 18: 912-913.

MacKinnon, J. (1988) *Field Guide to the Birds of Java and Bali.* Yogyakarta, Indonesia: Gadjah Mada University Press.

MacKinnon, J. and MacKinnon, K. (1986) *Review of the protected areas system in the Indo-Malayan realm.* Gland, Switzerland and Cambridge, UK: IUCN.

MacKinnon, J. and Phillipps, K. (1993) *A Field Guide to the Birds of Borneo, Sumatra, Java and Bali.* Oxford, New York, Tokyo: Oxford University Press.

MacKinnon, J. and Vu Van Dung (1992) Draft management plan for Vu Quang Nature Reserve, Huong Khe District, Ha Tinh Province, Vietnam. Unpublished.

Madge, S. and Burn, H. (1988) *Wildfowl.* London: Christopher Helm.

Madge, S. and Burn, H. (1993) *Crows and Jays.* London: Christopher Helm.

Madge, S., Pearson, D. and Svensson, L. (2006) Family Sylviidae (Warblers). Pp. 492-709 in del Hoyo, J., Elliott, A. and Christie, D.A. (Eds.). *Handbook of the Birds of the World,* 12. Barcelona: Lynx Edicions.

Madoc, G. C. (1956) *An introduction to Malayan birds.* Revised edition. Kuala Lumpur: Malayan Nature Society.

Martens, J. and Tietze, D. T. (2006) Systematic notes on Asian birds. 65. A preliminary review of the Certhiidae. *Zool. Med. Leiden* 80-5(17): 273-286.

Martens J., Eck, S., Packert, M. and Sun, Y.-H. (1999) The Golden-spectacled Warbler *Seicercus burkii* – a species swarm (Aves: Passeriformes: Sylviidae), part 1. *Zool. Abhandl. Mus. Dresden* 50: 281-327.

Martins, R. (1986) Recent reports. *Oriental Bird Club Bull.* 3: 33-36.

McGowan, P. J. K. and Panchen, A. L. (1994) Plumage variation and geographical distribution in the Kalij and Silver Pheasants. *Bull. Brit. Orn. Club* 114: 113-123.

van Marle, J. G. and Voous, K. H. (1988) *The Birds of Sumatra: an annotated check-list.* Tring, UK: British Ornithologists' Union.

Mayr, E. (1938) The birds of the Vernay-Hopwood Chindwin Expedition. *Ibis* (14)2: 277-320.

Mears, A. and Oates, E. W. (1907) On the birds of the Chindwin, Upper Burma. *J. Bombay Nat. Hist. Soc.* 18: 78-87.

Medway, Lord and Wells, D. R. (1976) *The birds of the Malay Peninsula,* 5. London: H. F. & G. Witherby in association with Penerbit Universiti Malaya.

Meyer de Schauensee, R. (1934) Zoological results of the third de Schauensee Siamese Expedition. Part II, birds from Siam and the Southern Shan States. *Proc. Acad. Nat. Sci. Philadelphia* 86: 165-280.

Meyer de Schauensee, R. (1946) On two collections of birds from the Southern Shan States, Burma. *Proc. Acad. Nat. Sci. Philadelphia* 98: 99-122.

Meyer de Schauensee, R. (1984) *The Birds of China.* Oxford, New York, Tokyo: Oxford University Press.

Morris, G. E. (1988) Recent sight records of birds at Nam Cat Tien. *Garrulax* 4: 11-13.

Moulin, S., Randi, E., Tabarroni, C. and Hennache, A. (2003) Mitochondrial DNA diversification among the subspecies of the Silver and Kalij Pheasants, *Lophura nycthemera* and *L. leucomelanos,* Phasianidae. *Ibis* 145 (online): E1-E11.

Moyle, R. G., Chesser, R. T., Prum, R. O., Schikler, P. and Cracraft, J. (2006) Phylogeny and evolutionary history of Old World suboscine birds (Aves: Eurylaimides). *Amer. Mus. Novit.* 3544.

Mullarney, K., Svensson, L., Zetterström, D. and Grant, P. J. (1999) *Collins Bird Guide.* London: HarperCollins.

Oates, E. W. (1894) On some birds collected on Byingyi Mountain, Shan States, Burma. *Ibis* (6)6: 478-484.

Oates, E. W. (1895) Notes on some birds from the Ruby Mines District, Burma. *J. Bombay Nat. Hist. Soc.* 10: 108-112.

Olsen, K. M. and Larsson, H. (1995) *Terns of Europe and North America.* London: Christopher Helm.

Olsen, K. M. and Larsson, H. (1997) *Skuas and Jaegers.* Robertsbridge, UK: Pica Press.

Olsen, K. M. and Larsson, H. (2004) *Gulls of Europe, Asia and North America.* London: Christopher Helm.

Olsson, U., Alström, P., Ericson, P. G. P. and Sundberg, P. (2005) Non-monophyletic taxa and cryptic species-evidence from a molecular phylogeny of leaf-warblers (*Phylloscopus,* Aves). *Molec. Phylogen. Evol.* 36: 261-276.

Olsson, U., Alström, P. and Sundberg, P. (2004) Non-monophyly of the avian genus *Seicercus* (Aves: Sylviidae) revealed by mitochondrial DNA. *Zoologica Scripta* 33: 501-510.

Olsson, U., Alström, P., Gelang, M., Ericson, P. G. P. and Sundberg, P. (2006) Phylogeography of Indonesian and Sino-Himalayan region bush warblers (*Cettia,* Aves). *Molec. Phylogen. Evol.* 41: 556-565.

Parry, S. J., Clark, W. S. and Prakash, V. (2002) On the taxonomic status of the Indian Spotted Eagle *Aquila hastata. Ibis* 144: 665-675.

Paton, T. A., Baker, A. J., Groth, J. G. and Barrowclough G. F. (2003) RAG-1 sequences resolve phylogenetic relationships within charadriiform birds. *Molec. Phylogen. Evol.* 29: 268-278.

Pavlova, A., Zink, R. M., Rohwer, S., Koblik, E. A., Red'kin, Y. A., Fadeev, I. V. and Nesterov, E. V. (2005) Mitochondrial DNA and plumage evolution in the white wagtail *Motacilla alba. J. Avian Biol.* 36: 322-336.

Payne, R. B. (1997) Family Cuculidae (cuckoos). Pp. 508-607 in del Hoyo, J., Elliott, A. and Sargatal, J. (eds.) *Handbook of Birds of the World*, 4. Barcelona: Lynx Edicions.

Payne, R. B. (2005) *The Cuckoos*. Oxford: Oxford University Press.

Penhallurick, J. and Robson, C. (in prep.) The taxonomy of parrotbills (Aves, Paradoxornithidae).

Pereira, S. L. and Baker, A. J. (2005) Multiple gene evidence for parallel evolution and retention of ancestral morphological states in the shanks (Charadriiformes: Scolopacidae). *Condor* 107: 514–526.

Peters, J. L. and others (1931–1986) *Check-list of Birds of the World*, 1–15. Cambridge, Mass.: Harvard University Press and Museum of Comparative Zoology.

Pilgrim, J. D., Bijlmakers, P., Crutchley, A., Crutchly, G., de Bruyn, T. and Tordoff, A. W. (in prep.). Common Ringed Plover *Charadrius hiaticula* and Black-headed Bunting *Emberiza melanocephala*: two species of bird new to Vietnam.

Pilgrim, J. D., Bijlmakers, P., de Bruyn, T., Doppagne, S., Mahood, S. and Tordoff, A. W. (in prep.). Updates to the distribution and status of birds in Vietnam.

Qiao-Wa Pan, Fu-Min Lei, Zuo-Hua Yin, Kristín, A. and Kanuch, P. (2007) Phylogenetic relationships between *Turdus* species: mitochondrial cytochrome *b* gene analysis. *Ornis Fennica* 84: 1-11.

Rajathurai, S. (1998) Checklist of birds of Singapore. Unpublished.

Rasmussen, P. C. and Anderton, J. C. (2005). *Birds of South Asia: the Ripley Guide*, 1-2. Washington D.C. and Barcelona: Smithsonian Institution and Lynx Edicions.

Rasmussen, P.C. and Parry, S.J. (2001) The taxonomic status of the 'Long-billed' Vulture *Gyps indicus*. *Vulture News* 44: 18–21.

Restall, R. (1996) *Munias and Mannikins*. Robertsbridge, UK: Pica Press.

Rheindt, F. E. (2006) Splits galore: the revolution in Asian leaf warbler systematics. *BirdingASIA* 5: 25-39.

Riley, J. H. (1938) Birds from Siam and the Malay Peninsula. *U.S. Natn. Mus. Bull.* 172.

Ripley, S. D. (1961) Some bird records from northern Burma with a description of a new subspecies. *J. Bombay Nat. Hist. Soc.* 58: 279-283.

Ripley, S. D. (1982) *A synopsis of the birds of India and Pakistan.* Second edition. Bombay: Bombay Natural History Society.

Rippon, G. (1896) Notes on some birds obtained at Kalaw in the Southern Shan States. *Ibis* (7)2: 357-362.

Rippon, G. (1897) An additional list of birds obtained at Kalaw, Southern Shan States, during April and May 1896. *Ibis* (7)3: 1-5.

Rippon, G. (1901) On the birds of the Southern Shan States, Burma. *Ibis* (8)1: 525-561.

Roberts, T. J. (1991-1992) *The Birds of Pakistan*, 1–2. Karachi: Oxford University Press.

Robinson, H. C. and Kloss, C. B. (1919) On birds from South Annam and Cochin China. *Ibis* (11)1: 392-453, 565-625.

Robson, C. (1985-1999) Recent reports/From the field. *Oriental Bird Club Bull.* 1: 24-28; 2: 36-40; 4: 29-31; 5: 33-36; 7: 35-40; 8: 32-36; 9(10): 38-44; 10: 41-44; 12: 40-44; 13: 47-52; 14: 48-52; 15: 43-47; 16: 50-52; 17: 49-53; 18: 67-70; 20: 55-61; 22: 57-62; 23: 49-53; 24: 59-65; 25: 61-69; 26: 60-66; 27: 61-66; 28: 44-48; 29: 51-56; 30: 52-56; 31: 49-57; 32: 66-76; 33: 68-78; 34: 83-93; 35: 83-93; 36: 61-71; 37: 77-87; 38: 72-79.

Robson, C. (2004-2008) From the field. *BirdingASIA* 1: 78-87; 2: 98-106; 3: 77-85; 4: 84-92; 5: 88-93; 6: 92-96; 7: 92-96; 8: 90-95; 9: 107-112.

Robson, C. (2007) Family Paradoxornithidae (parrotbills). Pp. 292-320 in del Hoyo, J., Elliott, A. and Christie, D.A. (eds.) *Handbook of Birds of the World.* 12. Barcelona: Lynx Edicions.

Robson, C. R., Buck, H., Farrow, D. S., Fisher, T. & King, B. F. (1998) A birdwatching visit to the Chin Hills, West Burma (Myanmar), with notes from nearby areas. *Forktail* 13: 109-120.

Robson, C. R., Eames, J. C., Nguyen Cu and Truong Van La (1993) Further recent records of birds from Viet Nam. *Forktail* 8: 25-52.

Robson, C. R., Eames, J. C., Nguyen Cu and Truong Van La (1993) Birds recorded during the third BirdLife/Forest Birds Working Group expedition in Viet Nam. *Forktail* 9: 89-119.

Robson, C. R., Eames, J. C., Wolstencroft, J. A., Nguyen Cu and Truong Van La (1989) Recent records of birds from Viet Nam. *Forktail* 5: 71-97.

Roseveare, W. L. (1949) Notes on birds of the irrigated area of Shwebo District, Burma. *J. Bombay Nat. Hist. Soc.* 48: 515-534, 729-749.

Roseveare, W. L. (1950) Notes on birds of the irrigated area of Minbu District, Burma. *J. Bombay Nat. Hist. Soc.* 49: 244-287.

Round, P. D. (1988) *Resident forest birds in Thailand.* Cambridge, UK: International Council for Bird Preservation (Monogr. 2).

Round, P. D. (2000) *Field check-list of Thai birds.* Bangkok: Bird Conservation Society of Thailand.

Round, P. D. (2003) BCST Records Committee: Review of additions to the list of Thai birds since June 2000. *Bird Conserv. Soc. of Thailand Bull.* 20: 16-19.

Round, P. D. and Robson, C. (2001) Provenance and affinities of the Cambodian Laughingthrush *Garrulax ferrarius*. *Forktail* 17: 41-44.

Round, P. D., Bengt, H., Pearson D. J., Kennerley P. R. and Bensch, F. (2007) Lost and found: the enigmatic large-billed reed warbler *Acrocephalus orinus* rediscovered after 139 years. *J. Avian Biol.* 38: 133-138.

Rozendaal, F. (1990) Report on surveys in Hoang Lien Son, Lai Chau and Nghe Tinh Provinces, Vietnam. Unpublished.

Rozendaal, F., Nguyen Cu, Truong Van La and Vo Quy (1991) Notes on Vietnamese pheasants, with description of female plumage of *Lophura hatinhensis. Dutch Birding* 13: 12-15.

Sangster G., van den Berg A. B., van Loon, A. J. and Roselaar C. S. (2003) Dutch avifaunal list: taxonomic changes in 1999-2003. *Ardea* 91: 279-286.

Sangster, G., Hazevoet, C. J., van den Berg, A. B. and Roselaar, C. S. (1998) Dutch avifaunal list: species concepts, taxonomic instability and taxonomic changes in 1998. *Dutch Birding* 20: 22-32.

Sangster, G., Collinson, J. M., Helbig A. J., Knox, A. G. and Parkin D. T. (2005) Taxonomic recommendations for British birds: third report. *Ibis* 147: 821-826.

Sangster, G., Collinson, J. M., Knox, A. G., Parkin, D. T. and Svensson, L. (2007) Taxonomic recommendations for British birds: fourth report. *Ibis* 149: 853-857.

Scott, D. A. (1988) Bird observations in Vietnam: 7–21 March 1988. Appendix XII of Wetland projects in Vietnam, report on a visit to the Red River and Mekong deltas during March 1988. Unpublished.

Scott, D. A. (1989) *A directory of Asian wetlands.* Gland, Switzerland, and Cambridge, UK: IUCN.

Scott, D. A., Howes, J. and Le Dien Duc (1989) *Recommendations for management of Xuan Thuy Reserve, Red River Delta, Vietnam.* Appendix III: birds recorded in Xuan Thuy District, Red River Delta, 10–12 March 1988 & 17–31 March 1989. Kuala Lumpur: Asian Wetland Bureau Publication No. 44.

Sewell, J. H. (1899) Birds of the Kyaukse District. *J. Bombay Nat. Hist. Soc.* 12: 758-760.

Sheldon, F. H., Whittingham, L. A., Moyle, R. G., Slikas, B. and Winkler, D. W. (2005) Phylogeny of swallows (Aves: Hirundinidae) estimated from nuclear and mitochondrial DNA sequences. *Molec. Phylogen. Evol.* 35: 254-270.

Sibley, C. G. and Monroe, B. L. (1990) *Distribution and taxonomy of birds of the world.* New Haven: Yale University Press.

Sibley, C. G. and Monroe, B. L. (1993) *A supplement to distribution and taxonomy of birds of the world.* New Haven: Yale University Press.

Smith, H. C., Garthwaite, P. F., Smythies, B. E. and Ticehurst, C. B. (1940) Notes on the birds of Nattaung, Karenni. *J. Bombay Nat. Hist. Soc.* 41: 557-593.

Smith, H. C., Garthwaite, P. F., Smythies, B. E. and Ticehurst, C. B. (1943–1944) On the birds of the Karen Hills and Karenni found over 3,000 feet. *J. Bombay Nat. Hist. Soc.* 43: 455-474; 44: 60-72, 221-232.

Smythies, B. E. (1940) *The birds of Burma.* Rangoon: American Baptist Mission Press.

Smythies, B. E. (1949) A reconnaissance of the N'Mai Hka drainage, northern Burma. *Ibis* 91: 627-648.

Smythies, B. E. (1953) *The Birds of Burma.* Second edition. Edinburgh: Oliver and Boyd.

Smythies, B. E. (1981) *The Birds of Borneo.* Third edition. Kota Kinabalu: The Sabah Society; and Kuala Lumpur: Malayan Nature Society.

Smythies, B. E. (1986) *The Birds of Burma,* Third edition. Liss: Nimrod.

Stanford, J. K. and Mayr, E. (1940–1941) The Vernay-Cutting Expedition to northern Burma. *Ibis* (14)4: 679-711; (14)5: 56-105, 213-245, 353-378, 479-518.

Stanford, J. K. and Ticehurst, C. B. (1931) The birds of the Prome district of lower Burma. *J. Bombay Nat. Hist. Soc.* 34: 665-672, 901-915; 35: 32-50.

Stanford, J. K. and Ticehurst, C. B. (1935) Notes on the birds of the Sittang-Irrawaddy Plain, Lower Burma. *J. Bombay Nat. Hist. Soc.* 37: 859-889.

Stanford, J. K. and Ticehurst, C. B. (1938–1939) On the birds of northern Burma. *Ibis* (14)2: 65-102, 197-229, 391-428, 599-638; (14)3: 1-45, 211-258.

Stepanyan, L. S., Võ Quy, Nguyen Cu and Truong Van La (1983) [Results of the research on composition and distribution of avifauna in region Kon Ha Nung (Tay Nguyen Plateau).] *Thong bao khoa hoc* 1: 15-33. (In Vietnamese.)

Stresemann, E. and Heinrich, G. (1940) Die Vögel des Mount Victoria. *Mitt. Zool. Mus. Berlin* 24: 151-264.

Sun Hean and Poole, C. (1998) Checklist to the birds of Cambodia. Unpublished draft.

Taylor, B. and van Perlo, B. (1998) *Rails.* Robertsbridge, UK: Pica Press.

Thewlis, R. M., Duckworth, J. W., Anderson, G. Q. A., Dvorak, M., Evans, T. D., Nemeth, E., Timmins, R. J. and Wilkinson, R. J. (1996) Ornithological records from Laos, 1992–1993. *Forktail* 11: 47-100.

Thewlis, R. M., Timmins, R. J., Evans, T. D. & Duckworth, J. W. (1998) The conservation status of birds in Laos: a review of key species. *Bird Conserv. Internatn.* 8 (supplement).

Tietze, D. T., Martens, J. and Sun Yue-Hua (2006) Molecular phylogeny of treecreepers (*Certhia*) detects hidden diversity. *Ibis* 148: 477-488.

Thompson, H. N. and Craddock, W. H. (1902) Notes on the occurrence of certain birds in the Southern Shan States of Burma. *J. Bombay Nat. Hist. Soc.* 14: 600.

Ticehurst, C. B. (1933) Notes on some birds from southern Arakan. *J. Bombay Nat. Hist. Soc.* 36: 920-937.

Thomas, W. W. (1964) A preliminary list of the birds of Cambodia. Unpublished.

Timmins, R. J. and Vongkhamheng, C. (1996) A

preliminary wildlife and habitat survey of Xe Sap National Biodiversity Conservation Area and mountains to the south, Saravan Province, Lao PDR. Final report to CPAWM, Department of Forestry. Vientiane: Wildlife Conservation Society.

Tiraut, G. (1873) *Les oiseaux de la Basse Cochin Chine.* Saigon.

Treesucon, U. (2000) *Birds of Kaeng Krachan: checklist and guide to birds finding.* Bangkok: Bird Conservation Society of Thailand.

Turner, A. and Rose, C. (1989) *A Handbook to the Swallows and Martins of the World.* London: Christopher Helm.

'Vagrant' (1868) *Random notes on Indian and Burman ornithology.* Bangalore: privately published.

van der Ven, J. (2000) Myanmar Expedition 1999-2000. Report, visit January 2000. Unpublished.

van der Ven, J. (2001) Myanmar Expedition 1999-2000. Report. Second expedition, December 2000/January 2001. Unpublished.

van der Ven, J. (2002) Myanmar Expedition 2001. Report. Third expedition, December 2001. Unpublished.

van der Ven, J. (2003) Myanmar Expedition 4 2001. Report. Fourth expedition, January 2003. Unpublished.

van der Ven, J. (2004) Myanmar Expedition. 5th Expedition to North Myanmar. Report. December 2003-March 2004. Unpublished.

van der Ven, J. (2005) 6th Expedition to North Myanmar. Report. December 2004-March 2005. Unpublished.

van der Ven, J. and Thet Zaw Naing (2005) 6th Expedition to North Myanmar. Report. December 2004-March 2005. Unpublished.

van der Ven, J. and Thet Zaw Naing (2005) Spring visit to Indawgyi Lake, Kachin State, Myanmar. 22nd-28th April 2005. Unpublished.

van der Ven, J. and Thet Zaw Naing (2005) Summer visit to Indawgyi Lake, Kachin State and Ayeyarwady River (Myitkyina–U Laut), Myanmar. 16th-21st June 2005. Unpublished.

Venning, F. E. W. (1912) Some birds and birds' nests from Haka, Chin Hills. *J. Bombay Nat. Hist. Soc.* 21: 621-633.

Võ Quy (1971) [Biology of common bird species in Vietnam. Part I: The preliminary history of bird studies in Indochina and Vietnam. Part II: The fundamental characteristics of avifauna in north Vietnam. Biology of common birds in north Vietnam.] Hanoi: Science and technology. (In Vietnamese.)

Võ Quy (1975) [*Birds of Vietnam*] 1. Hanoi: Nha Xuat Ban Khoa Hoc Va Ky Thuat. (In Vietnamese.)

Võ Quy (1981) [*Birds of Vietnam*] 2. Hanoi: Nha Xuat Ban Khoa Hoc Va Ky Thuat. (In Vietnamese.)

Võ Quy (1983) [A catalogue of the birds of Vietnam.] Pp. 12-43 in L.N. Medvedev, ed. [*Fauna and ecology of the animals of Vietnam.*] Moskva: Nauka. (In Russian.)

Võ Quy, Stepanyan, L. S., Nguyen Cu and Truong Van La (1983) [Materials on the avifauna of the Tay Nguyen Plateau.] Pp. 44-61 in L.N. Medvedev, ed. [*Fauna and ecology of the animals of Vietnam.*] Moskva: Nauka. (In Russian.)

Voelker, G. and Spellman, G. M. (2004). Nuclear and mitochondrial DNA evidence of polyphyly in the avian superfamily Muscicapoidea. *Molec. Phylogen. Evol.* 30: 386-394.

Wells, D. R. (1976–1981) Bird report. *Malayan Nat. J.* 36: 197-218; 38: 113-150; 39: 279-298; 43: 116-147; 43: 148-171; 43: 172-210.

Wells, D. R. (1999) *The birds of the Thai-Malay Peninsula,* 1. San Diego and London: Academic Press.

Wells, D. R. (2007) *The birds of the Thai-Malay Peninsula,* 2. London: Christopher Helm.

Wells, D. R. and Medway, Lord (1976) Taxonomic and faunistic notes on birds of the Malay Peninsula. *Bull. Brit. Orn. Club* 96: 20-34.

White, C. M. N. and Bruce, M. D. (1986) *The birds of Wallacea (Sulawesi, the Moluccas and Lesser Sunda Islands, Indonesia): an annotated check-list.* London: British Ornithologists' Union (Check-list 7).

Wickham, P. F. (1929) Notes on the birds of the upper Burma hills. *J. Bombay Nat. Hist. Soc.* 33: 799-827; 34: 46-63, 337-349.

Wink, M., Sauer-Gürth, H. and Gwinner, E. (2002) Evolutionary relationships of stonechats and related species inferred from mitochondrial-DNA sequences and genomic fingerprinting. *Brit. Birds* 95: 349-355.

Wink, M., Seibold, I., Lotfikhah, F. and Bednarek, W. (1998) Molecular systematics of Holarctic raptors (Order Falconiformes). Pp. 29-48 in Chancellor, R. D., Meyburg, B.-U. and Ferrero, J. J. (eds.) *Holarctic birds of prey: proceedings of an international conference.* Mérida & Berlin: Adenex & WWGBP.

Winkler, H., Christie, D. A. and Nurney, D. (1995) *Woodpeckers.* Robertsbridge, UK: Pica Press.

Wood, H. and Finn, F. (1902) On a collection of birds from Upper Burmah. *J. Asiatic Soc. Bengal* 71: 121-131.

Yésou, P. (2001) Phenotypic variation and systematics of Mongolian Gull. *Dutch Birding* 23: 65-82.

Yésou, P. (2002) Systematics of *Larus argentatus-cachinnans-fuscus* complex revisited. *Dutch Birding* 24: 271-298.

Zou, F., Haw Chuan Lim, Marks, B. D., Moyle, R. G. and Sheldon, F. H. (2007) Molecular phylogenetic analysis of the Grey-cheeked Fulvetta (*Alcippe morrisonia*) of China and Indochina: a case of remarkable genetic divergence in a 'species'. *Molec. Phylogen. Evol.* 44: 165-174.

PLATE 1. NICOBAR SCRUBFOWL, PARTRIDGES, CHINESE FRANCOLIN, QUAILS & BUTTONQUAILS

1 [1] **NICOBAR SCRUBFOWL** *Megapodius nicobariensis*
Adult *nicobariensis*: Plain brown above, grey-brown below; red skin round eye, face grey.

2 [2] **SNOW PARTRIDGE** *Lerwa lerwa*
(a,b) **Adult:** Head, neck and upperside intensely barred blackish and whitish, lower breast to belly chestnut with bold white streaks. Whitish trailing edge to secondaries in flight.

3 [3] **CHINESE FRANCOLIN** *Francolinus pintadeanus*
(a) **Male** *phayrei*: Black body boldly spotted and barred whitish to buffy-white, bold black, chestnut and white head pattern, chestnut scapulars.
(b) **Female**: Duller than male, chestnut areas washed brown, facial pattern softer, underparts barred.
(c) **Juvenile**: Duller than female, less rufous on crown-sides, pale-streaked upperparts.

4 [4] **LONG-BILLED PARTRIDGE** *Rhizothera longirostris*
(a) **Male** *longirostris*: Large bill, light chestnut head-sides and underparts, grey foreneck and upper breast, buffish wing-coverts, yellow legs.
(b) **Female**: No grey on neck and breast.
(c) **Juvenile**: Like female but paler-faced, buff streaks on throat, breast and mantle.

5 [5] **BLACK PARTRIDGE** *Melanoperdix nigra*
(a) **Male** *nigra*: Glossy black, slightly browner wings.
(b) **Female**: Rather uniform dark chestnut, buffier on face and vent, black spots on scapulars.
(c) **Juvenile**: Like female but upperparts finely vermiculated, less black on scapulars, large whitish spots and dark bars on sides of breast and flanks.

6 [6] **COMMON QUAIL** *Coturnix coturnix*
(a) **Male** *coturnix*: From non-breeding Japanese Quail by less chestnut on breast and flanks, browner upperparts. Possibly not separable in field.
(b) **Female**: From Japanese Quail as male.

7 [7] **JAPANESE QUAIL** *Coturnix japonica*
(a) **Male non-breeding**: Like female but throat and foreneck white with dark centre and transverse throat-band, richer buff-and-chestnut breast.
(b) **Male breeding** *japonica*: Uniform pale pinkish-chestnut head-sides and throat.
(c,d) **Female**: Small; greyish-brown upperparts with dark markings and pale streaks, pale ear-coverts, whitish throat with double dark bar at sides, rufescent to chestnut breast and flanks with dark markings and whitish streaks; barred primaries. Plain upperwing in flight with barring on primaries.

8 [8] **RAIN QUAIL** *Coturnix coromandelica*
(a) **Male**: Large black breast-patch, black flank-streaks.
(b) **Female**: Washed-out (often greyish-tinged) breast with irregular dark spots; unbarred primaries.

9 [9] **BLUE-BREASTED QUAIL** *Coturnix chinensis*
(a) **Male** *chinensis*: Relatively unstreaked upperparts; slaty-blue face, breast and flanks, black-and-white markings on throat and upper breast, chestnut lower breast to vent.
(b,c) **Female**: Smaller than other quail with less streaked upperparts, long buff supercilium, barring on breast and flanks, yellowish legs; uniform brown upperwing.

10 [21] **CRESTED PARTRIDGE** *Rollulus rouloul*
(a) **Male**: Glossy blue-black above and below (usually bluer on mantle, greener towards tail), large fan-shaped chestnut-maroon crest, long wire-like forehead-plumes, red orbital skin, bill-base and legs, dark brownish wings.
(b) **Female**: Deep green with red eyering, dark grey hood, blackish nape, chestnut scapulars, rusty-brown wings with darker vermiculations, shortish forehead plumes, red legs.
(c) **Juvenile**: Like female but warm brown crown-sides, greyer-tinged mantle, duller breast, dull greyish belly and vent; pale spots on wing-coverts.

11 [242] **SMALL BUTTONQUAIL** *Turnix sylvatica*
(a,b) **Female** *mikado*: Like Yellow-legged Buttonquail but smaller, with pale chestnut wing-coverts (contrast less with flight feathers than in other buttonquail), strong buff stripes on mantle and tertials, paler buff and sharply defined breast-patch, slaty-blue to blackish bill, fleshy-greyish legs.
(c) **Juvenile**: Less distinct buff breast-patch, blackish spots across breast.

12 [243] **YELLOW-LEGGED BUTTONQUAIL** *Turnix tanki*
(a,b) **Female** *blanfordii*: Sandy-buff wing-coverts (more contrasting with flight feathers than in other buttonquail) with large black spots, deep buff breast-band, round black spots on upper flanks (otherwise rather plain below), rufous nuchal collar (lacking on male), extensive yellowish bill, yellow legs.
(c) **Juvenile**: Duller with less distinct breast-patch, faint narrow bars on throat/breast, less obvious wing-covert spots.

13 [244] **BARRED BUTTONQUAIL** *Turnix suscitator*
(a,b) **Female** *thai*: Small; head, breast, flanks and wing-coverts boldly but densely patterned: whitish-buff speckles on head, black throat and upper breast, black-and-whitish bars on lower throat to flanks, black-and-buff bars on wing-coverts, rufous vent.
(c) **Male**: Like female but no black patch on throat and breast.
(d) **Female** *blakistoni*: More rufous-chestnut above, buffier below (both sexes).

1
2b
2a
3c
3a
3b
4c
4b
4a
5a
5b
5c
6b
6a
8b
8a
7a
7b
7c
7d
9c
9b
9a
10a
10b
10c
11b
11c
11a
13b
13d
12b
12c
12a
13c
13a

PLATE 2. PARTRIDGES, BLOOD PHEASANT & TRAGOPANS

1 [10] **HILL PARTRIDGE** *Arborophila torqueola*
(a) **Adult** *batemani*: Chestnut crown, rufous-chestnut ear-coverts, black face and eyestripe, black-and-white throat, wavy blackish bars on upperparts, white gorget.
(b) **Female**: Like Rufous-throated but buffy-rufous on head-sides to throat, dark-barred upperparts, browner breast.
(c) **Juvenile**: Like female but has buff to whitish spots on breast, reduced flank markings.

2 [11] **RUFOUS-THROATED PARTRIDGE** *Arborophila rufogularis*
(a) **Adult** *tickelli*: Dark-streaked brown crown, plain olive-brown mantle, whitish lores, supercilium and ear-coverts with dark streaks, orange-rufous throat, grey breast to belly, chestnut flank markings.
(b) **Juvenile**: Plain buffish supercilium and throat, underparts spotted and streaked white.
(c) **Adult** *intermedia*: Mostly black throat.
(d) **Adult** *annamensis*: Whitish throat.

3 [12] **WHITE-CHEEKED PARTRIDGE** *Arborophila atrogularis*
Adult: Like Bar-backed Partridge but throat black, breast and upper belly slaty-greyish.

4 [13] **BAR-BACKED PARTRIDGE** *Arborophila brunneopectus*
(a) **Adult** *brunneopectus*: Pale buff face, dark eyestripe and crown, black bars on mantle, black foreneck with buff streaks, warm brown breast to upper belly, black-and-white flank markings.
(b) **Adult** *henrici*: Richer buff head markings.
(c) **Adult** *albigula*: Tends to have whiter head markings.

5 [14] **MALAYAN PARTRIDGE** *Arborophila campbelli*
Adult: Like Bar-backed but largely black head and neck, slaty-greyish breast/upper belly, pale rufous and black flanks.

6 [15] **ORANGE-NECKED PARTRIDGE** *Arborophila davidi*
Adult: Like Bar-backed but broad black eyestripe extends down neck and across lower foreneck, neck orange, broader whitish supercilium behind eye, black flanks with white bars.

7 [16] **CHESTNUT-HEADED PARTRIDGE** *Arborophila cambodiana*
(a) **Adult** *cambodiana*: Dull chestnut head/breast, black crown and post-ocular stripe, black bars above, black-and-white flanks.
(b) **Adult** *diversa*: Narrower black upperpart bars, more extensive flank markings; head pattern similar to Bar-backed.
(c) **Adult** *diversa*: Less well-marked individual.

8 [17] **CHESTNUT-NECKLACED PARTRIDGE** *Arborophila charltonii*
(a) **Adult** *charltonii*: Like Annam but chestnut breast-band, pale chestnut ear-covert patch, more orange-buff flanks.
(b) **Adult** *tonkinensis*: Chestnut areas much reduced.

9 [18] **ANNAM PARTRIDGE** *Arborophila merlini*
Adult: Very like Scaly-breasted but blackish heart-spots on lower breast and flanks, yellow legs.

10 [19] **SCALY-BREASTED PARTRIDGE** *Arborophila chloropus*
(a) **Adult** *chloropus*: Rather uniform; upperparts/breast olive-brown with blackish vermiculations, foreneck, lower breast/upper belly orange-buff; blackish flank markings, greenish legs, reddish bill with dull greenish-yellow tip.
(b) **Adult** *cognacqi*: Colder with whiter foreneck.

11 [20] **FERRUGINOUS PARTRIDGE** *Caloperdix oculea*
Male *oculea*: Chestnut head and breast, black-and-whitish scaled mantle and flanks, black-and-rufous back to tail, black spots on pale wing-coverts.

12 [22] **MOUNTAIN BAMBOO-PARTRIDGE** *Bambusicola fytchii*
(a) **Male** *fytchii*: Rather long neck and tail; buffish head-sides and throat with blackish stripe behind eye, greyish-brown above with distinct dark spots, chestnut streaks on neck and breast and bold black markings on flanks.
(b) **Female**: Like male but eyestripe brown.

13 [23] **BLOOD PHEASANT** *Ithaginis cruentus*
(a) **Male** *marionae*: Crimson and black head, slaty-grey crest, white-streaked grey upperside, green on wing-coverts/tertials; crimson breast, grey belly (both streaked greenish-white), crimson undertail-coverts/tail fringes.
(b) **Female**: Uniform dark brown with more rufous head, grey hindcrown and nape, short crest.

14 [24] **BLYTH'S TRAGOPAN** *Tragopan blythii*
(a,b) **Male** *blythii*: Short-legged and tailed, yellow face/throat, orange-red and black on rest of head/neck, orange-red breast, white-spotted chestnut-red and brown upperparts and lower belly, brownish-grey upper belly.
(c) **Female**: Nondescript, greyish-brown above with subtle mottling and speckling, paler below, yellowish eyering.

15 [25] **TEMMINCK'S TRAGOPAN** *Tragopan temminckii*
(a,b) **Male**: Recalls Blyth's but face/throat blue, underparts crimson, bold greyish-white spots from mid-breast to vent.
(c) **Female**: Very like female Blyth's but eyering blue, more distinct pale streaks and spots on underparts, warmer throat and neck, blacker base colour of crown.

1a
1b
1c
2a
2b
2c
2d
3
4a
4b
4c
5
6
7a
7b
7c
8a
8b
9
10a
10b
11
12a
12b
13a
13b
13-15 to different scale
14a
14b
14c
15a
15b
15c

PLATE 3. RED JUNGLEFOWL & *LOPHURA* PHEASANTS

1 [28] **RED JUNGLEFOWL** *Gallus gallus*
(a) **Male** *spadiceus*: Long rufous to golden-yellow hackles, maroon scapulars and lesser coverts, blackish underparts, glossy dark green high-arched tail, red comb, facial skin and lappets.
(b) **Male eclipse**: No hackles, leaving all-blackish crown and neck; smaller comb and lappets.
(c) **Female**: Smaller; short blackish and golden-buff hackles, drab brown upperside with fine blackish vermiculations and buffy-white shaft-streaks, paler below, plain greyish-brown vent, rather short and blunt dark tail, bare pinkish face.
(d) **Male** *gallus*: Conspicuous white 'ear-patch' (smaller on female).

2 [29] **KALIJ PHEASANT** *Lophura leucomelanos*
(a) **Male** *lathami*: Glossy blue-black including long crest, broad white scaling on lower back to uppertail-coverts, dark legs, red facial skin.
(b) **Female**: Like female Silver Pheasant (subspecies *lineata*) but darker above with distinctive pale scaling, blackish tail with chestnut-brown central feathers, underparts brown with pale scaling and shaft-streaks, dark legs.
(c) **Male** *williamsi*: Dense, fine whitish vermiculations on upperparts and tail, indistinct whitish scaling on lower back to uppertail-coverts.
(d) **Female**: A shade paler than *lathami*, notably central tail feathers, rest of tail vermiculated white.
(e) **Male** *lineata*: Much denser black markings above, creating very grey appearance; often has dark grey or greyish-brown to pinkish-brown legs.
(f) **Female**: Light scaling on upperparts, sharp black-and-white V-shapes on hindneck, largely dull chestnut breast and belly with white streaks, paler creamier central tail feathers.

3 [30] **SILVER PHEASANT** *Lophura nycthemera*
(a) **Male** *nycthemera*: White upperparts, wings and tail with black chevrons and lines, glossy blue-black crest and underparts, red facial skin and legs.
(b) **Female**: Plain mid-brown crown and upperside (sometimes warm-tinged) with faint vermiculations, short blackish crest, broad white and blackish scales on underparts, blackish-and-whitish barring on outertail feathers.
(c) **Male** *lewisi*: Very like Kalij (subspecies *williamsi*) but stronger markings above, no white scales on back to uppertail-coverts, white on tail, red legs.
(d) **Female**: Chestnut-tinged upperside with greyish scaling, greyer and relatively plain below.

4 [30A] **'IMPERIAL PHEASANT'** *Lophura x imperialis*
(a) **Male**: Like Edwards's Pheasant but larger, with longer, pointed glossy dark blue crest, longer tail, less prominent, only slightly greener-blue fringes to upperwing-coverts; overall less brilliant.
(b) **Female**: Very like some female Silver (see text) but tail uniform blackish-chestnut, plumage more chestnut-tinged overall.

5 [31] **EDWARDS'S PHEASANT** *Lophura edwardsi*
(a) **Male** *edwardsi*: Mostly glossy dark purplish-blue, blackish vent, tail and flight feathers, prominent greenish-blue fringes to wing-coverts, short tufted white crest, relatively short and blunt tail, red facial skin and legs.
(b) **Female**: Head, neck, mantle and breast rather plain cold greyish-brown, underparts not paler than upperparts, scapulars and wing-coverts warmer-tinged, blackish tail with chestnut-brown central feathers.
(c) **Male** *hatinhensis*: Like Edwards's but with white central tail feathers.

6 [32] **CRESTLESS FIREBACK** *Lophura erythrophthalma*
(a) **Male** *erythrophthalma*: Purplish-blue-black with fine whitish vermiculations on mantle, wings and sides of breast, rufous-chestnut shading to dark maroon from lower back to uppertail-coverts, shortish warm buff tail (base blackish), red facial skin, greyish legs.
(b) **Female**: Blackish overall, extensively glossed dark purplish- to greenish-blue.

7 [33] **CRESTED FIREBACK** *Lophura ignita*
(a) **Male** *rufa*: Blue facial skin, purplish-blue-black crest and body, golden-rufous upper back grading to maroon shorter uppertail-coverts, white streaks on flanks, strongly arched white central tail feathers, reddish legs.
(b) **Female**: Blue facial skin, dull rufous-chestnut crest, head, upperside and tail, breast similar with white streaks, blackish belly to vent with bold white scales.

8 [34] **SIAMESE FIREBACK** *Lophura diardi*
(a) **Male**: Extensive red facial skin, thin pendant-tipped crest, grey body with black-and-white bars on scapulars and wing-coverts, golden-buff patch on back, bluish barring on maroon rump to uppertail-coverts and on glossy purplish-black belly to vent, long down-curled glossy blackish-green tail.
(b) **Female**: Red facial skin, rufous-chestnut mantle, underparts (white-scaled on belly and flanks) and outertail, bold blackish-and-whitish bars on wings and central tail feathers.

1d
1b
1c
2b
1a
2a
2d
3d
2c
3c
3a
4b
3b
2f
4a
2e
5b
6b
5a
5c
6a
8b
7a
7b
8a

PLATE 4. PHEASANTS

1 [26] **HIMALAYAN MONAL** *Lophophorus impejanus*
(a) **Male**: Like Sclater's Monal but with long upright crest, glossy purple and turquoise on upperparts, all-rufous tail.
(b) **Female**: Like Sclater's but throat all white, streaked below, less obvious pale tail-tip, slight crest, darker head-sides.

2 [27] **SCLATER'S MONAL** *Lophophorus sclateri*
(a) **Male** *sclateri*: Large; metallic green curly-crested crown, greenish to purplish upperparts with white back to uppertail-coverts, chestnut tail with white tip. (White uppertail-coverts, not tail-base as illustrated.)
(b) **Female**: Blackish-brown above with small buffish streaks, paler back to uppertail-coverts, tail barred blackish and whitish with a clear whitish tip, off-white throat-centre, speckled head-sides, dark brownish below with narrow wavy buffish barring.

3 [35] **MRS HUME'S PHEASANT** *Syrmaticus humiae*
(a) **Male** *burmanicus*: Red facial skin, mostly glossy dark greyish-purple head to upper breast and inner wing-coverts (too green on plate), rich chestnut body, two white wing-bands, long greyish dark-barred tail.
(b) **Female**: Smaller than male with shorter white-tipped tail; generally warm brown, with blackish markings above and whitish wing-bars, warmer below with whitish scales on lower breast to vent.

4 [36] **COMMON PHEASANT** *Phasianus colchicus*
(a) **Male** *elegans*: Red facial skin, purplish-green head to breast, rufous and chestnut body with black streaks and bars, long brown dark-barred tail, mostly greenish-grey wing-coverts, back and rump.
(b) **Female**: Smaller and shorter-tailed than male. Rufous to buffish-brown upperside with blackish bars and mottling, buffish underparts with blackish scales on breast and flanks.
(c) **Male** *takatsukasae*: White ring at base of neck (usually broken at front), lighter mantle and rump, coppery pinkish-maroon breast.

5 [37] **LADY AMHERST'S PHEASANT** *Chrysolophus amherstiae*
(a) **Male**: Black-scaled white neck-ruff, green mantle and breast, white belly, buffy-yellow rump, very long white tail with black bars.
(b) **Female**: Like female Common Pheasant but with distinct dark barring above and below.

6 [38] **MOUNTAIN PEACOCK-PHEASANT** *Polyplectron inopinatum*
(a) **Male**: From Malayan by lack of crest or pale facial skin; dark greyish head and neck with whitish speckles, blackish underparts, strong chestnut tinge to upperparts, with small bluish ocelli.
(b) **Female**: Ocelli smaller and black, tail shorter.

7 [39] **GERMAIN'S PEACOCK -PHEASANT** *Polyplectron germaini*
(a) **Male**: Like Grey but smaller and darker with finer pale markings, ocelli more greenish-blue, dull blood-red facial skin, no crest, dark lower throat.
(b) **Female**: Like Grey but darker; smaller, more defined ocelli above, no pale scaling below, dark lower throat.

8 [40] **GREY PEACOCK-PHEASANT** *Polyplectron bicalcaratum*
(a) **Male** *bicalcaratum*: Dark greyish-brown with fine whitish-buff bars and speckles, whitish throat, dark green and purplish light-bordered ocelli above, short bushy crest, flesh-coloured facial skin.
(b) **Female**: Smaller, darker and plainer with less distinct ocelli, duller facial skin.

9 [41] **MALAYAN PEACOCK-PHEASANT** *Polyplectron malacense*
(a) **Male**: Like Grey Peacock-pheasant but warmer brown, greener ocelli, long dark crest, pale orange facial skin, darker ear-coverts, plainer underparts.
(b) **Female**: Smaller and shorter-tailed with less distinct, more pointed ocelli, little crest, indistinct paler scales above.

10 [42] **CRESTED ARGUS** *Rheinardia ocellata*
(a) **Male** *nigrescens*: Blackish-brown peppered whitish all over; extremely long, broad tail, head with drooping brown and white crest extending back from buff supercilium, buff throat.
(b) **Female**: Smaller and much shorter-tailed, with same head pattern but shorter crest; appears plain warm brown below, barred black above.
(c) **Male** *ocellata*: Shorter, mostly brownish crest, white supercilium and upper throat, chestnut-tinged lower throat and foreneck, more dark chestnut and grey (less blackish) on tail.

11 [43] **GREAT ARGUS** *Argusianus argus*
(a) **Male** *argus*: Naked blue head and foreneck with short-crested black crown, warm brown above with fine pale speckles and mottling, mostly dark chestnut below, very long secondaries and very long white-spotted tail.
(b) **Female**: Head and neck like male, has complete rufous-chestnut collar, less distinct markings above, duller and plainer below, much shorter barred tail and much shorter secondaries.

12 [44] **GREEN PEAFOWL** *Pavo muticus*
(a) **Male** *imperator*: Huge, long-necked, glossy green with blackish scales, long upright crest, extremely long broad train with large colourful ocelli.
(b) **Female**: Like male but duller, lacks train.

1b
1a
2b
2a
4c
3a
4b
4a
5a
5b
3b
6b
7a
7b
6a
8a
9b
8b
9a
10c
10a
10b
11a
10-12 to different scale
11b
12a
12b

PLATE 5. GEESE, COMB DUCK, SHELDUCKS, WHITE-WINGED DUCK & PINK-HEADED DUCK

1 [47] **SWAN GOOSE** *Anser cygnoides*
(a,b) **Adult**: Thick blackish bill, dark crown and hindneck contrasts with pale creamy-brownish lower head-side to foreneck; narrow whitish frontal band (borders bill-base). Wing pattern recalls Greater White-fronted.
(c) **Juvenile**: Crown/hindneck duller, no frontal band.

2 [48] **TAIGA BEAN-GOOSE** *Anser fabalis*
(a,b) **Adult** *middendorffii*: Black bill with pale orange subterminal band, orangey legs, dark head and neck, rather uniform belly; rather uniformly dark upperwing and all-dark underwing.

3 [49] **GREYLAG GOOSE** *Anser anser*
(a,b) **Adult** *rubrirostris*: Relatively uniform pale greyish plumage with pinkish bill and legs; pale wing-coverts contrast with dark flight feathers above and below.

4 [50] **GREATER WHITE-FRONTED GOOSE** *Anser albifrons*
(a,b) **Adult** *albifrons*: Pinkish bill, white frontal patch, orange legs, irregular transverse black belly-patches.
(c) **Juvenile**: All-dark head, rather uniform belly.

5 [51] **LESSER WHITE-FRONTED GOOSE** *Anser erythropus*
(a,b) **Adult**: Small size and bill; white frontal patch extending to above yellow-ringed eye, relatively short, thick neck, small amount of black markings on belly.
(c) **Juvenile**: All-dark head, rather uniform belly, fainter yellow orbital ring.

6 [52] **BAR-HEADED GOOSE** *Anser indicus*
(a,b) **Adult**: Striking black-and-white pattern on head and neck, yellow bill and legs; pale wing-coverts contrast with dark flight feathers above and below.
(c) **Juvenile**: Rather uniform brownish-grey hindcrown to hindneck, with greyish lores and upper foreneck.

7 [53] **COMB DUCK** *Sarkidiornis melanotos*
(a) **Male non-breeding** *melanotis*: Black-speckled head and neck, whitish remainder of underparts with grey-washed flanks, dark bill with broad knob (comb) on upper base; wings appear all blackish, contrasting with pale underparts.
(b,c) **Female**: Lacks knob on bill.
(d) **Juvenile**: Browner overall; dark crown to hindneck and eyestripe offsetting long buffy supercilium; distinctive all-dark wings.

8 [54] **COMMON SHELDUCK** *Tadorna tadorna*
(a,b) **Male**: White and black appearance, with red knobbed bill, dark green head and upper neck, chestnut breast-band.
(c) **Female**: Lacks knob on bill, face marked with whitish.

9 [55] **RUDDY SHELDUCK** *Tadorna ferruginea*
(a,b) **Male breeding**: Striking orange-rufous plumage, black bill, narrow black collar, restricted whitish facial markings; white wing-coverts contrast with dark flight feathers above and below.
(c) **Female**: Lacks black collar, face extensively white, head buffier.

10 [56] **WHITE-WINGED DUCK** *Asarcornis scutulata*
(a,b) **Male**: All-dark body with contrasting black-speckled whitish head and upper neck, yellowish bill; white wing-coverts contrast with black primaries above and below.

11 [71] **PINK-HEADED DUCK** *Rhodonessa caryophyllacea*
(a,b) **Male**: Blackish-brown with mostly deep pink head, neck and bill; leading edge of upperwing-coverts whitish, secondaries buffish, underwing extensively pinkish.
(c) **Female**: Dull pinkish bill, head and sides of neck, brown wash on crown and hindneck, body browner than male.

1b
1a
3b
3a
2b
1c
5b
4c
4b
2a
4a
5a
6c
7c
7d
5c
7b
7a
6a
6b
9c
10b
9b
8b
10a
8c
9a
11c
11a
8a
11b

PLATE 6. WHISTLING-DUCKS, COTTON PYGMY-GOOSE & DABBLING DUCKS

1 [45] **FULVOUS WHISTLING-DUCK** *Dendrocygna bicolor*
(a,b) **Adult**: Rich rufous head and underbody, prominent streaked patch on neck, bold white flank-streaks, white uppertail-coverts.

2 [45A] **WANDERING WHISTLING-DUCK** *Dendrocygna arcuata*
(a,b) **Adult** *arcuata*: Blackish-brown of forecrown reaches eye, prominent blackish line down hindneck, richer flanks with large black-and-white markings, white outer uppertail-coverts.

3 [46] **LESSER WHISTLING-DUCK** *Dendrocygna javanica*
(a,b) **Adult**: Brown head with dark cap, brownish-rufous underparts, reddish-chestnut lesser wing-coverts and uppertail-coverts.

4 [57] **COTTON PYGMY-GOOSE** *Nettapus coromandelianus*
(a,b) **Male** *coromandelianus*: White head and neck with blackish cap, dark glossy green upperparts and collar; broad white wing-band.
(c,d) **Female**: Whitish head and neck, black crown and eyestripe; all-dark wings with white-tipped secondaries; indistinct collar.

5 [58] **MANDARIN DUCK** *Aix galericulata*
(a) **Male**: Bulky head, red bill, long pale supercilium, erect orange-rufous wing-sails.
(b,c) **Female**: Greyish head with white spectacles, extensive heavy whitish streaks/spotting on breast and flanks.
(d) **Juvenile**: Browner than female, with less distinct spectacles.

6 [63] **INDIAN SPOT-BILLED DUCK** *Anas poecilorhyncha*
(a,b) **Male** *haringtoni*: Yellow-tipped black bill, red spot on lores at bill-base, pale head with blackish crown and eyestripe, breast and flanks rather spotted; secondaries dark green, bordered white, extensive white on tertials, white underwing-coverts contrasting with rest of wing.

7 [64] **CHINESE SPOT-BILLED DUCK** *Anas zonorhyncha*
(a,b) **Adult**: Lacks red spot on lores (see text for female Indian Spot-billed), has band across head-side, relatively uniform dark body; mostly dark tertials, secondaries typically dark bluish with indistinct white border.

8 [66] **ANDAMAN TEAL** *Anas albogularis*
(a,b) **Adult**: Dark brown plumage, white patch around eye and on throat; wing pattern like female Eurasian Teal, but underwing mostly dark.
(c) **Adult variant**: Irregular white patches on head and neck.

9 [68] **GARGANEY** *Anas querquedula*
(a,b) **Male**: Dark brown head with broad white supercilium; upperwing-coverts pale grey.
(c,d) **Female**: Dark eyestripe, bordered above and below by pale line, short dark line below white lower lores; upperwing-coverts greyish with white tips to greater coverts.
(e) **Juvenile**: Darker than female with less defined head pattern.

10 [69] **BAIKAL TEAL** *Anas formosa*
(a) **Male**: Complex buff, green, white and black head pattern, pinkish breast.
(b) **Male eclipse**: Similar to female but less distinct loral spot.
(c,d) **Female**: Circular whitish loral spot encircled by dark brown, pale vertical stripe below eye, buffish supercilium broken above eye; upperwing-coverts dark with buffish-brown tips to greater coverts, broad dark leading edge to underwing-coverts.
(e) **Juvenile**: Sides of head duller than female with slightly larger loral spot.

11 [70] **EURASIAN TEAL** *Anas crecca*
(a) **Male** *crecca*: Dark chestnut head, broad dark green band from lores to nape.
(b,c) **Female**: Small size and bill, head and neck rather uniform with darker crown, nape and eyestripe; upperwing-coverts dark, with whitish tips to greater coverts, narrow dark leading edge to underwing-coverts.

1a
1b
2a
2b
3a
3b
4a
4b
4c
4d
5a
5b
5c
5d
6a
6b
7a
7b
8a
8b
8c
9a
9b
9c
9d
9e
10a
10b
10c
10d
10e
11a
11b
11c

PLATE 7. DUCKS

1 [59] **GADWALL** *Anas strepera*
(a) **Male**: Relatively uniform greyish plumage, blackish bill and vent; square white patch on inner secondaries.
(b,c) **Female**: Rather uniform squarish head, defined orange bill-sides, square white patch on inner secondaries.

2 [60] **FALCATED DUCK** *Anas falcata*
(a) **Male**: Greyish plumage, glossy green and purplish head, white throat, black foreneck-band.
(b,c) **Female**: Relatively small and compact, plain greyish-brown head, dark bill; rich brown breast and flanks with dark brown scales, secondaries blackish with white border.

3 [61] **EURASIAN WIGEON** *Anas penelope*
(a,b) **Male**: Rufescent-chestnut head, yellowish forehead-patch, pinkish breast, grey flanks, whitish belly; extensively white upperwing-coverts.
(c,d) **Female**: Relatively small and compact, with blackish eye-patch, shortish pale grey bill; overall rather plainer than other female *Anas* ducks, with sharply contrasting white belly, upperwing-coverts paler and greyer than rest of wing.

4 [62] **MALLARD** *Anas platyrhynchos*
(a) **Male** *platyrhynchos*: Yellowish bill, glossy purplish-green head, white neck-collar, purplish-brown breast; greyish upperwing-coverts.
(b) **Male eclipse**: Like female but breast more chestnut, bill dull yellowish.
(c,d) **Female**: Bill dull orange with irregular dark markings, dark eyestripe contrasts with pale brown supercilium and head-sides; secondaries dark blue bordered white, underwing relatively pale.

5 [65] **NORTHERN SHOVELER** *Anas clypeata*
(a,b) **Male**: Huge bill, dark glossy green head, white underparts with chestnut sides; largely blue upperwing-coverts, bold white underwing-coverts.
(c) **Male eclipse**: Like female but flanks and belly more rufous, body markings blacker and upperwing-coverts bluer.
(d,e) **Female**: Huge bill, grey with orange edges; darker crown and eyestripe not sharply contrasting; bluish-grey on upperwing-coverts.

6 [67] **NORTHERN PINTAIL** *Anas acuta*
(a,b) **Male**: Dark chocolate-brown head, white of breast extending in line up and behind ear-coverts, long pointed tail-streamers; extensively dark underwing-coverts.
(c,d) **Female**: Slender grey bill, plain, brown head and noticeably long neck; extensively dark underwing-coverts.

7 [78] **LONG-TAILED DUCK** *Clangula hyemalis*
(a) **Male non-breeding**: White head and neck, grey-brown and black face-/neck-patches; narrow, elongated black tail-streamers, large pink bill-patch.
(b) **Male breeding** Head and neck mostly blackish, with grey-brown to white facial markings, upperparts blackish with warm brown fringing.
(c,d) **Female non-breeding** Whitish head and upper neck, with contrasting blackish crown and large patch on lower head-side; dark upperside, breast-band and head markings contrast with white collar and belly.
(e) **Female breeding** More brown and less white on face and neck, with white markings mainly behind eye and on neck-side; greyer breast-band.
(f) **Juvenile** Dull head/neck markings.

8 [79] **COMMON GOLDENEYE** *Bucephala clangula*
(a,b) **Male** *clangula*: Glossy greenish-black head, white loral spot, breast and underparts; large white patch across secondaries and greater upperwing-coverts.
(c,d) **Female**: Greyish with dark brown head, yellow eye and subterminal patch on bill, white collar; white patch on upperwing bisected by two dark lines.

9 [80] **SMEW** *Mergellus albellus*
(a) **Male**: Mostly white; black patches on face and nape, black lines on sides of breast and upperparts.
(b,c) **Female**: Small size and bill, greyish plumage, contrasting chestnut crown and nape, blackish lores, white throat and lower head-sides; broad white band across upperwing-coverts, narrow white tips to greater coverts and secondaries.

1a
1b
1c
2a
2b
2c
3a
3b
3c
3d
4a
4b
4c
4d
5a
5b
5c
5d
5e
6a
6b
6c
6d
7a
7b
7c
7d
7e
7f
8a
8b
8c
8d
9a
9b
9c

PLATE 8. DIVING DUCKS

1 [72] **RED-CRESTED POCHARD** *Netta rufina*
(a) **Male**: Red bill, bulky orange-rufous head, black breast, tail-coverts and vent, broad white flank-patch.
(b,c) **Female**: Plain brownish with dark crown extending round eye and contrasting whitish sides of head to upper foreneck, pink-tipped bill; relatively pale brown upperwing-coverts, broad whitish band across flight feathers, largely whitish underwing.

2 [73] **COMMON POCHARD** *Aythya ferina*
(a) **Male**: Chestnut head, black breast and tail-coverts, grey body.
(b) **Male eclipse**: Duller and browner overall.
(c,d) **Female non-breeding**: Pale head markings, mottled greyish-brown body, dark bill with grey subterminal band; upperwing appears all greyish.

3 [74] **BAER'S POCHARD** *Aythya baeri*
(a) **Male**: Glossy blackish-green head, whitish eye, chestnut-brown breast, flanks mixed chestnut-brown and white, white undertail-coverts.
(b,c) **Female**: Dark brown head, slightly paler dark chestnut loral area, dark eye, brown and white flanks, white undertail-coverts; dark upperwing with broad, sharply contrasting white band across flight feathers.

4 [75] **FERRUGINOUS POCHARD** *Aythya nyroca*
(a) **Male**: Rich dark chestnut plumage, white eye, blackish upperparts, sharply demarcated white undertail-coverts.
(b,c) **Female**: Duller than male, dark eye, peaked crown, dark flanks, white undertail-coverts; dark upperwing with broad, sharply contrasting white band across flight feathers.

5 [76] **TUFTED DUCK** *Aythya fuligula*
(a) **Male**: Blackish plumage with white flanks, long nuchal tuft.
(b) **Male eclipse**: Duller, greyish flanks, short nuchal tuft.
(c-e) **Female**: Dark brownish plumage, lighter on flanks, slight nuchal tuft or bump, yellow eye, sometimes shows some white on face at sides of bill-base; dark upperwing with relatively narrow white band across flight feathers.

6 [77] **GREATER SCAUP** *Aythya marila*
(a) **Male** *marila*: Glossy greenish-black head, pale greyish upperparts, white flanks, black breast and tail-coverts.
(b,c) **Female**: White face-patch encircles bill-base, squarish head-shape; greyish upperwing-coverts, broad white band across upperside of flight feathers.
(d) **Female (worn)**: Pale patch on ear-coverts.

7 [81] **COMMON MERGANSER** *Mergus merganser*
(a,b) **Male** *comatus*: Large size, long narrow red bill, glossy dark green head, unmarked white breast and flanks (variably washed salmon-pink); large unmarked white upperwing-patch.
(c,d) **Female**: White throat sharply demarcated from dark rufous-chestnut hood, rather uniform grey sides of breast and flanks with paler scaling; white on upperwing restricted to secondaries and unmarked.

8 [82] **SCALY-SIDED MERGANSER** *Mergus squamatus*
(a,b) **Male**: Spiky, uneven nuchal crest, dark grey scaling on flanks; large white upperwing-patch bisected by two dark lines.
(c,d) **Female**: Spiky nuchal crest, ill-defined whitish throat, dark grey scaling on white sides of breast and flanks; white upperwing-patch bisected by single dark line.

9 [83] **RED-BREASTED MERGANSER** *Mergus serrator*
(a,b) **Male**: Thin-based slender bill, red eyes, shaggy crest, white collar, black-streaked rufescent lower neck/breast, white-spotted black breast-sides. Large white upperwing-patch bisected by two black lines.
(c,d) **Female**: Rufescent hood with untidy crest, paler throat/foreneck (not demarcated), pale and dark loral lines, variable pale eyering; reddish-brown eyes, brownish-grey body with vaguely pale-scaled flanks. Smaller white upperwing-patch bisected by single line.

1c
1a
1b
2a
2b
2c
2d
3a
3b
3c
4a
4b
4c
5a
5b
5c
5d
5e
6a
6b
6c
6d
7a
7c
7b
7d
8a
8c
8b
8d
9a
9c
9b
9d

PLATE 9. YELLOW-BILLED LOON, SHEARWATERS, PETRELS, GREBES & ANCIENT MURRELET

1 [84] **YELLOW-BILLED LOON** *Gavia adamsii*
(a,b) **Adult non-breeding**: Thick pointed pale bill (usually held upward), thick head/neck, steep forehead, blackish-brown crown/hindneck, shadowy half-collar, white below.
(c) **Adult breeding**: Black head/neck with black-striped white patches, black above with white chequers and spots; yellower-tinged more uniformly pale bill.
(d) **Juvenile**: Paler/browner than adult non-breeding, neatly scaled above/along flanks.

2 [85] **STREAKED SHEARWATER** *Calonectris leucomelas*
(a,b) **Adult**: Relatively large; white head with variable dark streaking on crown, nape and ear-coverts, white underwing-coverts with dark patches on primary coverts.

3 [86] **WEDGE-TAILED SHEARWATER** *Puffinus pacificus*
(a,b) **Adult pale morph**: All-dark crown/face, dark above, mostly white underwing-coverts, tail rather pointed.
(c) **Adult dark morph**: Broad-winged, longish pointed tail, rather uniformly dark underwing, pinkish feet.

4 [87] **SHORT-TAILED SHEARWATER** *Puffinus tenuirostris*
(a,b) **Adult**: Relatively small; short squarish tail (toes extend beyond tail-tip), dark overall, with pale chin, paler breast and belly and silvery underwing with dark base and surround; dark feet.

5 [88] **BULWER'S PETREL** *Bulweria bulwerii*
(a,b) **Adult**: Larger than Swinhoe's Storm-petrel, with longer wings and distinctive long, graduated tail (usually closed in flight), indistinct paler band across upperwing-coverts, otherwise all dark.

6 [89] **WILSON'S STORM-PETREL** *Oceanites oceanicus*
(a,b) **Adult** *oceanicus*: Small and blackish; pale band across upperwing-coverts, white rump and uppertail-coverts to vent sides, fairly short square-cut tail, paler band along underwing-coverts.

7 [90] **SWINHOE'S STORM-PETREL** *Oceanodroma monorhis*
(a,b) **Adult**: Relatively small, blackish with paler band across upperwing-coverts, prominently forked tail, white shafts on base of primaries (above), all-dark underwing.

8 [91] **LITTLE GREBE** *Tachybaptus ruficollis*
(a,b) **Adult non-breeding** *poggei*: Pale brownish head and sides of neck, mostly pale bill, dark eye; all-dark upperwing with narrow whitish tips to secondaries.
(c) **Adult breeding**: Dark rufous-chestnut throat and sides of head and neck, blackish crown and hindneck, dark flanks; yellow eye and gape-skin.

9 [92] **GREAT CRESTED GREBE** *Podiceps cristatus*
(a,b) **Adult non-breeding** *cristatus*: Large size. White head-sides and neck, black crown, hindneck and loral stripe, long pinkish bill; two broad white bands on upperwing.
(c) **Adult breeding**: Black crown-tuft, rufous-chestnut and blackish frills on rear head-sides.

10 [94] **HORNED GREBE** *Podiceps auritus*
(a,b) **Adult non-breeding**: Flatter crown than Black-necked, thicker/straighter bill (tip often pale), black cap demarcated at eye-level, pale loral spot; upperwing with white at shoulder and white secondaries.
(c) **Adult breeding** Outstanding black and gold 'head-frills', reddish-chestnut foreneck/underparts

11 [94] **BLACK-NECKED GREBE** *Podiceps nigricollis*
(a,b) **Adult non-breeding** *nigricollis*: Blackish crown extending down round eye, blackish hindneck, white throat, sides of nape and foreneck/breast, red eye; broad white trailing edge to upperwing.
(c) **Adult breeding** Black head and neck with orange-yellow flash on rear head-sides, red eye, rich chestnut flanks.

12 [357] **ANCIENT MURRELET** *Synthliboramphus antiquus*
(a) **Adult non-breeding** *antiquus*: Dark-based pale pinkish to yellowish bill, blackish face, crown and nuchal-collar, slaty-grey upperside, white breast to neck.
(b) **Adult breeding** Streaky white supercilium, white streaks on nuchal collar, black throat to ear-coverts.

1a
1c
1d
1b
3c
3b
2a
3a
5b
5a
2b
4b
4a
2-7 to different scale
6b
6a
8b
8a
8c
7a
7b
10b
9b
10a
10c
9a
9c
11b
11a
11c
12a
12b

PLATE 10. GREATER FLAMINGO & STORKS

1 [95] **GREATER FLAMINGO** *Phoenicopterus ruber*
(a) **Adult** *roseus*: Mostly pinkish-white, extremely long neck and legs, broad downcurved pink bill with black tip.
(b) **Juvenile**: Largely brownish-grey, pale greyish bill with black tip, dark brownish legs.

2 [96] **MILKY STORK** *Mycteria cinerea*
(a,b) **Adult non-breeding**: All white with blackish flight feathers, pale pinkish-yellow bill, dark red naked head skin.
(c) **Adult breeding**: Bright yellow to orange-yellow bill, brighter red head skin, plumage suffused pale creamy-buffish.
(d,e) **Juvenile**: From Painted by browner and more uniform head and neck, paler lesser and median upper wing-coverts (hardly any contrast), no breast-band.

3 [97] **PAINTED STORK** *Mycteria leucocephala*
(a,b) **Adult non-breeding**: White with blackish flight feathers, black-and-white patterned upper- and under wing-coverts and breast-band, pinkish tertials and inner greater coverts, pinkish-yellow bill, naked orange-red head.
(c) **Adult breeding**: Bright pinkish-peach bill, redder head.
(d,e) **Juvenile**: Head and neck pale greyish-brown with whitish streaks, naked head skin dull yellowish and less extensive, mostly pale greyish-brown mantle and wing-coverts, obviously darker lesser and median coverts, indistinct dusky breast-band, dark underwing-coverts.

4 [98] **ASIAN OPENBILL** *Anastomus oscitans*
(a,b) **Adult non-breeding**: Dull bill with gap between mandibles, dirty greyish-white plumage with black tail, lower scapulars and flight feathers.

5 [99] **BLACK STORK** *Ciconia nigra*
(a,b) **Adult**: Glossy blackish plumage with white lower breast to vent, red bill and orbital skin; white patch on inner underwing-coverts.

6 [100] **WOOLLY-NECKED STORK** *Ciconia episcopus*
(a,b) **Adult** *episcopus*: Glossy blackish plumage with white neck, vent and tail-coverts, dark bill.

7 [101] **STORM'S STORK** *Ciconia stormi*
(a,b) **Adult**: Like Woolly-necked but bill red, facial skin dull orange with broad golden-yellow area around eye, lower foreneck black.

8 [102] **WHITE STORK** *Ciconia ciconia*
(a,b) **Adult** *asiatica*: White with black lower scapulars and flight feathers, white tail, red bill.

9 [103] **BLACK-NECKED STORK** *Ephippiorhynchus asiaticus*
(a) **Female** *asiaticus*: Huge, owing partly to very long bill and legs; white and glossy blackish plumage, white flight feathers and leading edge to wing, black bill; bright yellow eyes.
(b) **Male**: Brown eyes.
(c,d) **Juvenile**: Dull brown head, neck and upperside, dark bill; all-dark wings.

10 [104] **LESSER ADJUTANT** *Leptoptilos javanicus*
(a) **Male non-breeding**: Very large and bulky; broad dull-coloured bill, naked pinkish head and yellowish neck skin, all-blackish upperparts and wings, white underparts and patch on inner underwing-coverts.
(b) **Male breeding**: Redder head-sides.

11 [105] **GREATER ADJUTANT** *Leptoptilos dubius*
(a) **Adult non-breeding**: Huge and bulky; very large deep-based bill, pinkish naked head, neck and pro nounced drooping neck-pouch, white neck ruff, paler, greyer upperside, contrasting paler grey greater coverts and tertials; underwing-coverts paler than flight feathers.
(b) **Adult breeding**: Blacker on face, yellower neck-pouch.

1a
1b
2a
2b
2c
2d
2e
3a
3b
3c
3d
3e
4a
4b
5a
5b
6a
6b
7a
7b
8a
8b
9a
9b
9c
9d
10a
10b
11a
11b
9-11 to different scale

PLATE 11. IBISES, SPOONBILLS, CORMORANTS & ORIENTAL DARTER

1 [106] **BLACK-HEADED IBIS** *Threskiornis melanocephalus*
(a,b) **Adult non-breeding**: White with naked blackish head and upper neck; naked reddish skin on underwing.

2 [107] **RED-NAPED IBIS** *Pseudibis papillosa*
(a,b) **Adult**: Dark overall; blackish naked head, with red patch on hindcrown and nape, dull red legs; white patch on inner forewing.
(c) **Juvenile**: Feathered head, without red patch.

3 [108] **WHITE-SHOULDERED IBIS** *Pseudibis davisoni*
(a,b) **Adult**: Dark overall; blackish naked head, whitish collar, dull red legs; white patch on inner forewing.

4 [109] **GIANT IBIS** *Pseudibis gigantea*
(a,b) **Adult**: Very large; naked greyish head and neck with dark bands at rear, reddish legs, pale wings with dark bars, dark outer upperwing and underwing.

5 [110] **GLOSSY IBIS** *Plegadis falcinellus*
(a,b) **Adult non-breeding** *falcinellus*: Relatively small, all dark but head and neck with pale streaking, indistinct white facial lines; all-dark upperwing.
(c) **Adult breeding**: Head, neck and body mostly deep chestnut, forecrown glossed green, pronounced white facial lines.

6 [111] **EURASIAN SPOONBILL** *Platalea leucorodia*
(a) **Adult non-breeding** *major*: Like Black-faced but larger, has all-white forehead and cheeks, pale fleshy-yellow patch on upperside of bill 'spoon'.
(b) **Adult breeding**: Differs from Black-faced as non-breeding adult; also shows yellow-orange gular skin.
(c,d) **Juvenile**: Dull pinkish bill and loral skin; similar wing markings to Black-faced.

7 [112] **BLACK-FACED SPOONBILL** *Platalea minor*
(a) **Adult non-breeding**: All-blackish bill, blackish face encircling bill-base.
(b) **Adult breeding**: Yellowish to buffish nuchal crest and breast-patch.
(c) **Juvenile**: Similar to adult non-breeding but has blackish edges to outer primaries and small blackish tips to primaries, primary coverts and secondaries.

8 [147] **LITTLE CORMORANT** *Phalacrocorax niger*
(a,b) **Adult non-breeding**: Relatively small; short, stubby, mostly pale bill, short neck, all dark with whitish chin.
(c) **Adult breeding**: Head and underparts glossy black, dense silvery-white streaks on head, blackish bill.
(d) **Juvenile**: Browner than non-breeding adult with head and neck paler, throat whitish, underparts scaled pale brownish (shows paler crown and hindneck and darker belly than other juvenile cormorants).

9 [148] **INDIAN CORMORANT** *Phalacrocorax fuscicollis*
(a,b) **Adult non-breeding**: Similar to Little but larger with long slender bill, relatively long tail, more extensive white on throat, paler lower head-sides.
(c) **Adult breeding**: Blackish head and neck with silvery peppering over eye and white tuft on rear head-side; browner above than Little.
(d) **Juvenile**: Upperside browner than non-breeding adult, underparts mostly whitish with dark brown marks on foreneck and breast and dark brown flanks.

10 [149] **GREAT CORMORANT** *Phalacrocorax carbo*
(a,b) **Adult non-breeding** *sinensis*: Much bigger and larger-billed than other cormorants. Prominent defined white area from head-side to upper throat, prominent yellow facial and gular skin, strong brown cast to upperparts.
(c) **Adult breeding**: Dense white streaks on head and neck, orange-yellow facial and darker gular skin, more white on head-side to throat.
(d) **Juvenile**: Whitish head-sides and underparts, with dark streaks on foreneck and breast, and dark brown flanks and thighs.

11 [150] **ORIENTAL DARTER** *Anhinga melanogaster*
(a,b) **Adult non-breeding**: Long, slender bill and neck, long tail, white streaks on upperparts.

1a
1b
2a
2b
2c
3a
3b
4a
4b
5a
5b
5c
6a
6b
6c
6d
7a
7b
7c
8a
8b
8c
8d
9a
9b
9c
9d
10a
10b
10c
10d
11a
11b

PLATE 12. BITTERNS, NIGHT-HERONS & LITTLE HERON

1 [113] **GREAT BITTERN** *Botaurus stellaris*
(a,b) **Adult** *stellaris*: Large size, buffish, cryptically patterned with blackish streaks and vermiculations, thick yellowish bill, plain rufous-buff head-sides; flight feathers browner, barred black.

2 [114] **YELLOW BITTERN** *Ixobrychus sinensis*
(a,b) **Male**: Light buffish-brown with darker sandy-brown mantle, blackish crown, tail and flight feathers.
(c,d) **Juvenile**: Heavy dark streaking above and below; rather plain whitish underwing-coverts.

3 [115] **VON SCHRENCK'S BITTERN** *Ixobrychus eurhythmus*
(a,b) **Male**: Dark chestnut head-sides and upperparts with blackish median crown-stripe; mostly buffish upperwing-coverts contrast with upperparts, flight feathers dark slaty-grey, chestnut patch at wing-bend.
(c,d) **Female/juvenile**: Bold white to buff speckling and spotting on upperparts and wing-coverts, blackish-grey flight feathers and tail.

4 [116] **CINNAMON BITTERN** *Ixobrychus cinnamomeus*
(a,b) **Male non-breeding**: Almost uniform rich cinnamon-rufous upperside.
(c) **Female**: Duller above with vague buffish speckling, brown streaking on underparts; upperwing rather uniform warm brown in flight.
(d,e) **Juvenile**: Duller above than female, buffish-streaked head-sides, dense buffish markings above, darker underpart streaking.

5 [117] **BLACK BITTERN** *Dupetor flavicollis*
(a,b) **Male** *flavicollis*: Blackish plumage, whitish throat and breast with broad dark streaks, yellowish-buff neck-patch; all-dark upperwing.
(c) **Female**: Browner than male.
(d) **Juvenile**: Head and upperparts brown with paler fringing, breast washed buffish-brown.

6 [118] **WHITE-EARED NIGHT-HERON** *Gorsachius magnificus*
Male: Black head with white markings, uniform dark brownish upperparts and wings, rufescent neck-sides, dark streaks/scales below, green legs.

7 [119] **MALAYSIAN NIGHT-HERON** *Gorsachius melanolophus*
(a) **Adult** *melanolophus*: Chestnut-tinged brown above with blackish vermiculations, black crown and long crest, blackish streaks on centre of throat and foreneck, dark-marked belly, short bill.
(b) **Juvenile**: Irregular white markings on crown and nape, upper- and underparts heavily vermiculated whitish to buffish and greyish, whitish throat with broken dark mesial streak.

8 [120] **BLACK-CROWNED NIGHT-HERON** *Nycticorax nycticorax*
(a,b) **Adult non-breeding** *nycticorax*: Quite large, grey with black crown, mantle and scapulars, whitish nape-plumes, yellow legs; wings broad and uniform.
(c) **Juvenile**: Brown streaks on head, neck and breast, dark brown upperparts and wings with buffish to whitish spots/streaks.

9 [121] **LITTLE HERON** *Butorides striata*
(a,b) **Adult** *javanicus*: Small and greyish with black crown, nape-plumes and streak on head-side, whitish to buffish-white streaks on scapulars and upperwing-coverts; rather uniform in flight. Dull yellowish-orange legs.
(c,d) **Juvenile**: Dull brown with darker crown and nape, underparts all streaked, greenish to yellowish-green legs; rather uniform in flight.

1a
1b
2a
2b
2c
2d
3a
3b
3c
3d
4a
4b
4c
4d
4e
5a
5b
5c
5d
6
7a
7b
8a
8b
8c
9a
9b
9c
9d

PLATE 13. HERONS, CRANES & PELICANS

1 [126] **GREY HERON** *Ardea cinerea*
(a,b) **Adult non-breeding** *jouyi*: Grey upperside, mostly white head and neck with black markings, black nape-plumes, yellowish bill; grey wing-coverts contrast with blackish flight feathers.
(c) **Adult breeding**: Deep orange to reddish bill.
(d) **Juvenile**: Dark crown, grey neck-sides, short nape-plumes, duller bill.

2 [127] **WHITE-BELLIED HERON** *Ardea insignis*
(a,b) **Adult**: Large size, very long neck, mostly greyish plumage with clean white throat and whitish belly and vent; whitish underwing-coverts contrast with dark flight feathers.

3 [128] **GREAT-BILLED HERON** *Ardea sumatrana*
(a,b) **Adult breeding** *sumatrana*: Like White-bellied but shorter-necked, browner overall, less defined pale throat, dull greyish belly and vent, dark underwing-coverts.

4 [129] **GOLIATH HERON** *Ardea goliath*
(a,b) **Adult**: Huge size, black bill, plain rufous-chestnut crown and hindneck, dark chestnut belly, blackish legs; rather plain upperwing, chestnut underwing-coverts.

5 [130] **PURPLE HERON** *Ardea purpurea*
(a,b) **Adult** *manilensis*: Black crown and nape-plumes, neck mostly rufous-chestnut with black lines down side and front, dark chestnut-maroon belly, flanks and vent, yellow bill; upperwing pattern not sharply con trasting but underwing-coverts mostly chestnut-maroon.
(c,d) **Juvenile**: Brownish upperparts and upperwing-coverts, neck duller with less distinct markings.

6 [238] **DEMOISELLE CRANE** *Grus virgo*
(a,b) **Adult**: Like Common Crane but smaller, with smaller bill and shorter neck, mostly slaty-black head to breast, long white tuft of feathers extending from behind eye.
(c) **First winter**: Duller and browner, few or no elongated feathers.

7 [239] **SARUS CRANE** *Grus antigone*
(a,b) **Adult** *sharpii*: Huge; grey with naked red head and upper neck, long bill, reddish legs. Blackish flight feathers and primary coverts with paler secondaries.
(c) **Juvenile**: Feathered buffish head and upper neck, duller overall with brownish-grey feather fringes, those of upperparts cinnamon-brown.

8 [240] **COMMON CRANE** *Grus grus*
(a,b) **Adult** *lilfordi*: Huge, generally greyish; blackish head and upper neck with broad white band from ear-coverts down upper neck, red midcrown, long drooping tertials mixed with black; grey wing-coverts, blackish flight feathers.
(c) **First winter**: Warm buffish to grey head and upper neck, rest of plumage often mixed with brown.

9 [241] **BLACK-NECKED CRANE** *Grus nigricollis*
(a,b) **Adult**: Like Common Crane but even larger, paler and whiter with contrasting all-blackish drooping tertials, all-black head and upper neck, apart from red midcrown and white patch behind eye.
(c) **Juvenile/immature**: Poorly documented. Buff-washed/-scaled plumage, buffish on crown, blackish neck.

10 [142] **GREAT WHITE PELICAN** *Pelecanus onocrotalus*
(a,b) **Adult non-breeding**: Huge; mostly whitish plumage, yellowish pouch, pinkish legs; black underside of flight feathers contrast with whitish underwing-coverts.
(c) **Adult breeding**: White plumage tinged pinkish, bright deep yellow pouch, tufted nuchal crest, yellowish-buff patch on lower foreneck/breast.
(d,e) **Juvenile**: Head, neck and upperside predominantly greyish-brown; dark brownish leading edge to underwing, dark underside of flight feathers.

11 [143] **SPOT-BILLED PELICAN** *Pelecanus philippensis*
(a,b) **Adult breeding**: Huge; mostly whitish plumage, dark spots on upper mandible, pinkish pouch with heavy purplish-grey mottling, tufted dusky nape and hindneck, dark legs; dark greyish flight feathers (above and below).
(c) **Juvenile**: Sides of head, hindneck, upperparts and wing-coverts browner. Plain dull pinkish pouch.

1c
1b
1d
1a
2b
4a
3b
5a
4b
2a
5c
3a
5b
5d
8b
6-9 to different scale
7c
7b
6c
7a
8c
8a
9a
6a
6b
9c
10e
10b
11b
9b
10d
10c
11c
11a
10a

PLATE 14. POND-HERONS, EGRETS & BUSTARDS

1 [122] **INDIAN POND-HERON** *Ardeola grayii*
Adult breeding: Brownish-buff head, neck and breast, rich brownish-maroon mantle and scapulars. Long white head-plumes.

2 [123] **CHINESE POND-HERON** *Ardeola bacchus*
(a,b) **Adult non-breeding**: Brown-streaked head, neck and breast, brown upperparts; white wings, often with dusky tips to primaries.
(c) **Adult breeding**: Chestnut-maroon head, neck and breast, blackish-slate mantle and scapulars.

3 [124] **JAVAN POND-HERON** *Ardeola speciosa*
(a) **Adult non-breeding**: Possibly shows whiter primary tips than Chinese Pond-heron.
(b) **Adult breeding**: Buffish head and neck, deep cinnamon-rufous breast, blackish-slate mantle and scapulars, white head-plumes.

4 [125] **EASTERN CATTLE EGRET** *Bubulcus coromandus*
(a,b) **Adult non-breeding**: Relatively small size, all-white plumage, short yellow bill, short neck, heavy-jowled appearance, relatively short dark legs.
(c) **Adult breeding**: Variable amount of rufous-buff on head, neck, back and breast. Short nape and breast-plumes, long back-plumes.

5 [131] **GREAT EGRET** *Ardea alba*
(a,b) **Adult non-breeding** *modestus*: Large size, long snake-like neck, sharply kinked when retracted, long sharply pointed yellow bill, blackish legs.
(c) **Adult breeding**: Blackish bill, legs reddish, long back-plumes, short coarse breast-plumes. Cobalt-blue facial skin.

6 [132] **INTERMEDIATE EGRET** *Mesophoyx intermedia*
(a,b) **Adult non-breeding**: Size intermediate between Little and Great (nearer former), bill shorter and somewhat blunter than Great, neck shorter, less kinked, appears somewhat rounder-crowned and heavier-jowled.
(c) **Adult breeding**: Bill often shows dark on tip and ridge of upper mandible (blacker during courtship), long plumes on breast and back, blackish legs.

7 [133] **LITTLE EGRET** *Egretta garzetta*
(a,b) **Adult non-breeding** *garzetta*: All-white plumage, mostly blackish bill, blackish legs with contrasting yellow feet.
(c) **Adult breeding**: Long nape-, back- and breast-plumes, reddish facial skin, blackish bill, black legs, yellowish to redder feet.

8 [134] **PACIFIC REEF-EGRET** *Egretta sacra*
(a) **Adult dark morph non-breeding** *sacra*: Overall dark slaty-grey plumage.
(b) **Adult dark morph breeding**: Plumes on nape, back and breast.
(c,d) **Adult white morph non-breeding**: All-white plumage, bill relatively thick and blunt-tipped, mostly pale greenish to yellowish, upper mandible usually darker, legs relatively short, greenish to yellowish.

9 [135] **CHINESE EGRET** *Egretta eulophotes*
(a,b) **Adult non-breeding**: Dull flesh-coloured to yellowish basal two-thirds of lower mandible, greenish to greenish-yellow legs.
(c) **Adult breeding**: Yellow bill, shaggy nuchal crest, blue facial skin, legs blackish, feet greenish-yellowish to yellow.

10 [217] **GREAT BUSTARD** *Otis tarda*
(a,b) **Male non-breeding** *dybowskii*: Very large, with short thick bill and medium legs, bluish-grey head and neck, golden-buffish upperparts with strong black barring, white below with some rufous-chestnut on sides of upper breast and lower hindneck; extensive white to pale grey on upperwing, all-whitish underwing with broad blackish trailing edge.
(c) **Male breeding**: White moustachial whiskers, thicker-looking neck with more rufous-chestnut on front and sides.
(d) **Female non-breeding**: Much smaller, with narrower bill, thinner neck, less white on upperwing.

11 [218] **BENGAL FLORICAN** *Houbaropsis bengalensis*
(a,b) **Male** *blandini*: Largely black with mostly white wings, fine golden-buff markings from mantle to uppertail.
(c,d) **Female**: Thin neck, short bill, longish legs; spangled buffish-brown and black with striped head, whitish belly; in flight shows buff wing-coverts, blackish outer primaries.

1
2a
2b
2c
3a
3b
4a
4b
4c
5a
5b
5c
6a
6b
6c
7a
7b
7c
8a
8b
8c
8d
9a
9b
9c
10-11 to different scale
10a
10b
10c
10d
11a
11b
11c
11d

PLATE 15. TROPICBIRDS, FRIGATEBIRDS & BOOBIES

1 [136] **RED-BILLED TROPICBIRD** *Phaethon aethereus*
(a,b) **Adult** *indicus*: Upperparts and wing-coverts barred black, largely black primary coverts, orange-red bill.
(c) **Juvenile**: Upperparts and wing-coverts densely barred blackish, largely black primary coverts, diffuse blackish band across hindcrown/nape, small black spots on tail-tip, bill yellowish-cream with dark tip.

2 [137] **RED-TAILED TROPICBIRD** *Phaethon rubricauda*
(a,b) **Adult** *westralis?*: Red bill and tail-streamers, almost completely white plumage. Often flushed pink.
(c) **Juvenile**: Greyish to blackish bill, no obvious black on primaries/primary coverts.

3 [138] **WHITE-TAILED TROPICBIRD** *Phaethon lepturus*
(a,b) **Adult** *lepturus*: White upperparts and wing-coverts with broad black bar across latter, white primary coverts.
(c) **Juvenile**: Upperparts and wing-coverts sparsely barred/scaled blackish, white primary coverts, rather faint blackish spotting/barring on crown, small black spots on tail-tip, bill yellowish-cream with indistinct dark tip.

4 [139] **CHRISTMAS ISLAND FRIGATEBIRD** *Fregata andrewsi*
(a) **Male**: All blackish with white belly-patch.
(b) **Female**: Underparts predominantly white, black bar from shoulder to breast-side, white axillary spur.
(c) **Juvenile**: Similar to Lesser but larger, hexagonal white belly-patch, more parallel-sided and more forward-pointing (towards forewing) white axillary spur (rarely absent), originating from behind line of breast-band; upperwing-band whitish and very prominent.
(d) **Immature (second year)**: Differs from juvenile as Lesser.

5 [140] **GREAT FRIGATEBIRD** *Fregata minor*
(a) **Male** *minor*: Overall blackish plumage.
(b) **Female**: Pale throat, white breast and upper flanks, all-black belly and underwing-coverts.
(c) **Juvenile**: All-black inner underwing-coverts (30% show short, outward-angled white axillary spur that originates well behind line of breast-band), elliptical white belly-patch. Upperwing-band similar to Lesser, bill intermediate between Lesser and Christmas Island.
(d) **Immature (second year)**: As for Lesser.

6 [141] **LESSER FRIGATEBIRD** *Fregata ariel*
(a,b) **Male** *ariel*: Overall blackish with whitish patches extending from sides of body to inner underwing-coverts.
(c) **Female**: Black hood, belly and lower flanks, white remainder of underparts, extending in axillary spur onto inner underwing-coverts.
(d) **Juvenile**: Rufous to brownish-white head, black breast-band, triangular white belly-patch. White axillary spur always present, originating from line of breast-band and angled outwards (towards wing-tip). Band across upperwing-coverts buffish, moderately prominent.
(e) **Immature (second year)**: Gradually loses black breast-band and acquires blackish plumage-parts of respective adults. Third and fourth year birds are sexually dimorphic.

7 [144] **MASKED BOOBY** *Sula dactylatra*
(a) **Adult** *personata*: White, with blackish face, flight feathers and tail, yellowish bill, greyish feet.
(b) **Juvenile**: Head, neck and upperside warmish brown, flight feathers browner, white hind-collar, white underwing-coverts with defined dark central band.

8 [145] **RED-FOOTED BOOBY** *Sula sula*
(a) **Adult white morph** *rubripes*: Light blue-grey and pinkish bill and facial skin, red feet, white tail and tertials.
(b) **Adult intermediate morph**: Mantle, back and wing-coverts brown (above and below).
(c) **Juvenile**: Dark greyish-brown overall, dark grey bill, purplish facial skin, yellowish-grey to pinkish feet.
(d) **Immature white morph**: Untidy whitish areas on head, body and underwing-coverts, darker breast-band.

9 [146] **BROWN BOOBY** *Sula leucogaster*
(a,b) **Adult** *plotus*: Dark brown with contrasting white lower breast to vent, yellowish bill, pale yellowish feet, white underwing-coverts with dark leading edge and incomplete diagonal bar.
(c) **Juvenile**: Duller than adult, with browner belly, vent and underwing-coverts, bill pale bluish-grey, feet pinker.

1a
1b
1c
2a
2b
2c
3a
3b
3c
1-3 to different scale
4a
4b
4c
4d
5a
5b
5c
5d
6a
6b
6c
6d
6e
7-9 to different scale
7a
7b
8a
8b
8c
8d
9a
9b
9c

PLATE 16. WHITE-RUMPED PYGMY-FALCON, FALCONETS, FALCONS & BAT HAWK

1 [151] **WHITE-RUMPED PYGMY-FALCON** *Polihierax insignis*
(a,b) **Male** *cinereiceps*: Small size, long tail, pale greyish ear-coverts and forehead to upper mantle with blackish streaks, dark slate-grey remainder of upperparts, white rump and uppertail-coverts, whitish underparts.
(c) **Female**: Deep rufous crown to upper mantle.
(d) **Juvenile**: Both sexes resemble male, but with rufous nuchal collar, rest of upperparts brown-washed.
(e) **Male** *insignis*: Paler grey mantle and scapulars, blackish streaks on lower throat, breast and flanks.

2 [152] **COLLARED FALCONET** *Microhierax caerulescens*
(a,b) **Adult** *burmanicus*: Very small, with black crown, ear-covert patch and upperside, white forehead and supercilium meeting broad white nuchal collar, chestnut throat; barred underwing and -tail.
(c) **Juvenile**: Forehead and supercilium washed pale chestnut, throat whitish.

3 [153] **BLACK-THIGHED FALCONET** *Microhierax fringillarius*
Adult: Very small, little white on forehead, narrow white supercilium, large black ear-covert patch, no white nuchal collar, mostly white throat, black lower flanks and thighs.

4 [154] **PIED FALCONET** *Microhierax melanoleucus*
Adult: Slightly larger than other falconets, with all-whitish underparts, black lower flanks, little white on forehead, narrow white supercilium, broad black ear-covert patch, and no white nuchal collar.

5 [155] **LESSER KESTREL** *Falco naumanni*
(a-c) **Male**: From Common Kestrel by plain bluish-grey head, plain rufous-chestnut on upperparts, mostly bluish-grey tertials, greater coverts and median covert-tips, relatively plain (some flank spots) vinous-tinged warm buff underparts; cleaner, whiter underwing, contrasting with dark tip, tail-tip usually more wedge-shaped.
(d-f) **Female**: From Common by slightly smaller and slimmer build, proportionately shorter tail, weaker facial markings, finer crown-streaks, narrower, more V-shaped upperpart markings, finer streaking below; usually slightly cleaner, whiter underwing.

6 [156] **COMMON KESTREL** *Falco tinnunculus*
(a-c) **Male** *interstinctus*: Grey, lightly streaked head, dark moustachial/cheek-stripe, grey rump to uppertail, broad black subterminal tail-band, rufous-chestnut dark-spotted upperparts, pale buffish underparts with dark streaks; largely whitish underwing with dark markings. Two-tone upperwing in flight.
(d-f) **Female**: Warm brown crown and nape with dark streaks, rufescent uppertail with fairly narrow blackish bars and broad black subterminal band, rather dull pale rufous upperparts with numerous distinct dark triangular or bar-shaped markings, evenly dark-streaked underparts, dark moustachial/cheek-stripe and post-ocular stripe, greyish-brown ear-coverts; largely whitish dark-marked underwing.

7 [158] **MERLIN** *Falco columbarius*
(a-c) **Male** *insignis*: Fairly small and compact, with uniform bluish-grey upperside and blackish flight feathers, broad blackish subterminal tail-band, narrow dark streaks on warm buffish underparts, faint darker moustachial/cheek-stripe, buffish-white supercilium, rufous nuchal collar; wings short and broad but pointed, closely barred pale-and-dark on underwing, broad whitish-and-blackish bands on undertail.
(d-f) **Female**: Drab brownish crown and upperside with pale buffish-brown markings, faint moustachial/cheek-stripe, pale supercilium, underparts and underwing similar to male, but tail broadly barred buff-and-blackish above.

8 [167] **BAT HAWK** *Macheiramphus alcinus*
(a,b) **Adult** *alcinus*: Falcon-like; blackish-brown, dark mesial streak, contrasting with whitish throat and centre of upper breast.
(c) **Juvenile**: Browner, with more extensive whitish areas on underparts.

1a
1b
1c
1d
1e
2a
2b
2c
3
4
5a
5b
5c
5d
5e
5f
6a
6b
6c
6d
6e
6f
7a
7b
7c
7d
7e
7f
8a
8b
8c
8 to different scale

PLATE 17. FALCONS

1 [157] **AMUR FALCON** *Falco amurensis*
(a,b) **Male**: Slaty-grey overall with paler grey underparts, rufous-chestnut thighs and vent, and red eyering, cere and legs; white underwing-coverts contrast sharply with black remainder of underwing.
(c,d) **Female**: Recalls adult Eurasian Hobby but has dark-barred upperparts and uppertail, buffy-white thighs and vent, different bare-part colours (similar to male); whiter base colour to the underwing, more pronounced bands on undertail.
(e,f) **Juvenile**: Like juvenile Eurasian Hobby but broader buff fringing above, dark-barred upperparts and uppertail, whiter base colour on head, underparts and underwing.
(g) **First-summer male**: Variable, showing mixed characters of adult and juvenile.

2 [159] **EURASIAN HOBBY** *Falco subbuteo*
(a-c) **Adult** *streichi*: Like small Peregrine, but upperparts more uniform, has narrower moustachial stripe, uppertail unbarred, breast and belly heavily streaked blackish, thighs and vent reddish-rufous; slenderer wings.
(d,e) **Juvenile**: Head and upperparts duller with narrow buffish fringes, sides of head and vent buffish.

3 [160] **ORIENTAL HOBBY** *Falco severus*
(a,b) **Adult** *severus*: All-blackish head-sides, buffish-white throat and forecollar, slate-grey upperside and reddish-rufous underparts; reddish-rufous underwing-coverts.
(c,d) **Juvenile**: Resembles adult but darker and browner above with narrow pale feather fringes, dark-streaked rufous below and on underwing-coverts, barred outertail feathers.

4 [161] **LAGGAR FALCON** *Falco jugger*
(a-c) **Adult**: Recalls Peregrine but crown rufous with dark streaks, head whiter and less boldly marked, unbarred uppertail, unmarked whitish breast, broad dark greyish-brown patch on lower flanks and thighs, narrow dark streaks on lower belly; wings less broad-based, tail proportionately longer, mostly dark underwing-coverts and axillaries.
(d,e) **Juvenile**: Like Peregrine but with more uniformly dark brown underparts and underwing-coverts with a few whitish streaks.

5 [162] **PEREGRINE FALCON** *Falco peregrinus*
(a-c) **Adult** *japonensis*: Large, with slate-grey upperside, broad blackish moustachial streak and whitish lower head-sides and underparts with dark bars on flanks, belly and undertail-coverts; wings appear broad-based and pointed, tail rather shortish, underwing very densely dark-barred.
(d,e) **Juvenile**: Upperparts and wing-coverts duller with narrow warm brown to buffish fringes, forehead and supercilium whitish with indistinct dark streaks, lower sides of head and underparts buffish-white with dark streaks, some broken buffish bars on uppertail; underwing like adult but coverts more boldly marked.
(f,g) **Adult** *peregrinator*: Unbarred, strongly rufous-washed underparts and (barred) underwing-coverts, buffish tail-tip.
(h) **Juvenile**: Darker brown upperside than *japonensis*, no obvious whitish areas on head, rufous-washed lower underparts and underwing-coverts.
(i,j) **Adult** *ernesti*: Smaller, with darker upperside, solid blackish sides of head, denser barring on duller underparts.

1a
1c
1g
1e
1b
1d
1f
2a
2d
2b
2c
2e
3a
3c
3b
3d
4a
4c
4b
4d
4e
5a
5c
5b
5f
5h
5g
5d
5e
5i
5j

PLATE 18. BAZAS, KITES & CRESTED SERPENT-EAGLE

1 [164] **JERDON'S BAZA** *Aviceda jerdoni*
(a-c) **Adult** *jerdoni*: Relatively small. Long white-tipped blackish crest, warm brown sides of head and nape, paler area on upperwing-coverts, dark mesial streak, indistinct rufous breast-streaks and broad rufous bars on belly and vent; cinnamon-rufous and white bars on underwing-coverts, few dark bars on flight feathers, three unevenly spaced blackish tail-bands.
(d,e) **Juvenile**: Head mostly buffish-white with blackish streaks, four evenly spaced dark tail-bands.

2 [165] **BLACK BAZA** *Aviceda leuphotes*
(a-c) **Adult** *syama*: Relatively small, head black with long crest, upperside mostly black, bold white on scapulars, white and chestnut on greater coverts and secondaries, whitish underparts with black and chestnut bars, and black vent; underwing shows black coverts and primary tips, white remainder of primaries, grey secondaries.
(d) **Juvenile**: Upperside with more white markings; underside with small streaks.

3 [168] **BLACK-SHOULDERED KITE** *Elanus caeruleus*
(a,b) **Adult** *vociferus*: Smallish, with short tail; pale grey above with black lesser and median coverts, most of head and underparts whitish; black underside of primaries.
(c) **Juvenile**: Upperparts tinged browner and pale-fringed, breast initially washed warm buff.

4 [169] **BLACK KITE** *Milvus migrans*
(a-c) **Adult** *govinda*: Dull brownish with long shallow-forked tail; outer wing broadly fingered and angled back, pale diagonal band across coverts, underwing showing small whitish patch at base of primaries.
(d,e) **Juvenile**: Head to mantle and belly streaked whitish-buff, rest of upperside and underwing-coverts with whitish-buff tips.

5 [170] **BLACK-EARED KITE** *Milvus lineatus*
(a,b) **Adult**: Larger and more rufescent than Black, typically whiter face and throat, extensive whitish area on underside of primaries.
(c,d) **Juvenile**: Similar to Black, but with broader, whiter streaks and feather-tips, and more prominent blackish mask. Also differs in similar underwing features to adult.

6 [171] **BRAHMINY KITE** *Haliastur indus*
(a,b) **Adult** *indus*: Bright cinnamon-rufous with whitish, narrowly streaked head to breast; outer primaries blackish.
(c,d) **Juvenile**: Recalls juvenile Black Kite but smaller, with shorter, rounded tail, warmer plumage and less obvious streaking; shorter, broader wings with unbarred buffish-white area across underside of primaries.

7 [184] **CRESTED SERPENT-EAGLE** *Spilornis cheela*
(a,b) **Adult** *burmanicus*: Medium-sized, with large full-crested head, yellow cere and facial skin, mostly rather dark brownish (white-spotted) plumage, paler below, and black tail with broad pale central band; broad wings, underwing has distinctive white and black band on flight feathers, tail black with broad white central band.
(c,d) **Juvenile**: Paler with head scaled black-and-whitish, blackish ear-coverts, upperparts fringed whitish, underparts buffy-white with broad dark streaks, tail whitish with three blackish bands; mostly whitish underwing with fine dark bars and indistinct trailing edge.
(e) **Adult** *malayensis*: Distinctly smaller and darker below.

1a
1d
1c
2a
3a
3c
3b
2b
2c
2d
1e
1b
4a
4b
5a
4c
4e
4d
6a
5c
5b
5d
6b
6d
6c
7 to different scale
7a
7c
7b
7d
7e

PLATE 19. OSPREY, SEA-EAGLES & FISH-EAGLES

1 [163] **OSPREY** *Pandion haliaetus*
(a) **Male** *haliaetus*: Dark brown above, white head and underparts with dark eyestripe. Narrow dark breast-streaking (often forms complete band).
(b) **Female**: Breast-band broader; adults have long, fairly slender wings with white underwing-coverts and contrasting blackish carpal patch, shortish, evenly barred tail.
(c) **Juvenile**: Upperparts broadly pale-fringed.

2 [172] **WHITE-BELLIED SEA-EAGLE** *Haliaeetus leucogaster*
(a,b) **Adult**: Very large, with grey upperparts and wing-coverts, white head, neck and underparts, white tail with blackish base; bulging secondaries and relatively narrow outer wing, white upperwing-coverts contrasting sharply with blackish flight feathers.
(c,d) **Juvenile**: Upperparts and wings mostly dark brownish, head and underparts dingy brownish, tail off-white with brownish subterminal band; underwing has warm buffish coverts, dark secondaries and large whitish patch on primaries.
(e) **Third year**: Resembles adult but has duller breast and underwing-coverts and largely pale underside of flight feathers with contrasting blackish primary tips.

3 [173] **PALLAS'S FISH-EAGLE** *Haliaeetus leucoryphus*
(a,b) **Adult**: Recalls Grey-headed Fish-eagle but larger, hood warm whitish, has darker brown upperside, thighs and vent, blackish base of tail; longer, straighter wings and longer blackish tail with broad white central band.
(c,d) **Juvenile**: Like White-bellied Sea-eagle but has blackish mask, tail blackish-brown; underwing mostly dark with whitish band across median and greater coverts, whitish primary flash.
(e) **Second/third year**: Dark flight feathers, pale tail-band. (Darker lesser and greater underwing-coverts and more contrasting pale band along median coverts than illustrated.)

4 [174] **WHITE-TAILED EAGLE** *Haliaeetus albicilla*
(a,b) **Adult** *albicilla*: Very large and brownish, paler hood, big yellow bill, short white tail.
(c-e) **Juvenile**: Upperparts and wing-coverts mostly darker, head and underparts blackish-brown with pale streaks on neck and breast, bill mostly dusky-greyish, blackish-bordered tail feathers; very broad, parallel-edged wings, dark below with pale bands across wing-coverts, whitish spikes on tail feathers. (Darker vent and darker breast markings than illustrated on flight figure.)
(f) **Second/third year**: Darker body and underwing-coverts, whiter tail with narrow black terminal band. (Slightly paler hood and whiter tail than illustrated in flight figure.)

5 [175] **LESSER FISH-EAGLE** *Ichthyophaga humilis*
(a-c) **Adult** *humilis*: Like Grey-headed Fish-eagle but smaller, with paler upperparts and dull greyish tail; underwing similar but may show whitish bases to outer primaries (less pronounced than illustrated), undertail dark brownish with only indistinct terminal band.
(d,e) **Juvenile**: From juvenile Grey-headed by plainer head and body with only vague streaking, plainer, darker tail, paler upperparts.

6 [176] **GREY-HEADED FISH-EAGLE** *Ichthyophaga ichthyaetus*
(a-c) **Adult**: Rather large and long-necked, with plain greyish hood, white thighs and vent, and rounded white tail with broad black terminal band. Upperparts greyish-brown, breast mostly warm brown to brownish-grey; wings rather broad, rounded and all dark, contrasting with white tail-base and vent.
(d,e) **Juvenile**: Head to upper belly mostly warm brownish with whitish streaks; upperside browner than adult, tail dark with whitish mottling (showing as faint pale bands); underwing mostly whitish with darker tips.
(f) **Second/third year**: Whitish patches on primaries, white thighs and vent, dark end to tail.

1a
2a
2b
2d
2e
1b
1c
2c
4e
4f
3b
3d
3e
4b
4d
3a
4a
3c
5b
5c
5e
4c
5a
5d
6a
6d
6b
6c
6e
6f

PLATE 20. VULTURES

1 [177] **EGYPTIAN VULTURE** *Neophron percnopterus*
(a-c) **Adult** *ginginianus*: Mostly dirty-whitish, with shaggy nape and neck, naked yellowish face, long yellowish bill and pointed, diamond-shaped whitish tail; rather long wings, black below with contrasting whitish coverts.
(d,e) **Juvenile**: Dark brown with dull greyish tail, grey facial skin; fairly uniform dark underwing.
(f) Second/third year: Paler on body and underwing-coverts, shadowing adult pattern.

2 [178] **WHITE-RUMPED VULTURE** *Gyps bengalensis*
(a-c) **Adult**: Blackish with contrasting white neck-ruff and white lower back and rump, mostly greyish-brown naked head and longish neck; very broad wings with well-spaced fingers, short tail, greyish secondaries and inner primaries (above and below) and white underwing-coverts with black leading edge.
(d,e) **Juvenile**: Browner and nondescript, underparts narrowly streaked whitish; in flight shows uniform dark plumage with short narrow whitish bands across underwing-coverts, which have darker leading edge (see Himalayan Griffon).

3 [179] **SLENDER-BILLED VULTURE** *Gyps tenuirostris*
(a-c) **Adult**: Pale sandy-brown body and wing-coverts, naked blackish head and longish neck, with relatively long dark bill, white neck-ruff, lower back and rump; pale wing-coverts and underbody contrast with blackish head, neck and flight feathers.
(d,e) **Juvenile**: Neck-ruff browner, upperwing-coverts duller and browner, underwing-coverts and (unstreaked) underparts duller and browner.

4 [180] **HIMALAYAN GRIFFON** *Gyps himalayensis*
(a-c) **Adult**: Huge and bulky with thick pale bill, pale bare head, thickish neck and mostly sandy-buffy body and wing-coverts; from flying Long-billed Vulture by pale head, paler upper body and wing-coverts, whitish underwing-coverts.
(d,e) **Juvenile**: Like juvenile White-rumped but larger and more heavily built, underwing-coverts have longer, more clearly separated buffish bands and lack contrasting darker leading edge.
(f) **Subadult**: Develops paler underbody, aiding separation from White-rumped.

5 [181] **CINEREOUS VULTURE** *Aegypius monachus*
(a,b) **Adult**: Huge, with uniform blackish-brown plumage, large bill, pale crown and nape, black face and foreneck; very broad, rather straight-edged wings.
(c) **Juvenile**: Plumage even blacker, head and neck largely blackish.

6 [182] **RED-HEADED VULTURE** *Aegypius calvus*
(a,b) **Male**: Blackish plumage, red head and neck, red legs, white front of neck-ruff and lateral body-patches; in flight these white areas and pale bases of flight feathers contrast sharply.
(c,d) **Juvenile**: Plumage somewhat browner, head and neck pinkish with some whitish down, legs pinkish, white vent; flight feathers more uniform (above and below).

1d
1a
1c
1b
1f
1e
2d
2a
2b
2c
2e
3d
3a
3b
3c
3e
4d
4a
4b
4c
4e
4f
5c
5a
5b
6c
6a
6d
6b

PLATE 21. HARRIERS

1 [187] **HEN HARRIER** *Circus cyaneus*
(a-c) **Male** *cyaneus*: Grey with white lower breast to undertail-coverts; wing-tips fall well short of tail-tip when perched; all-black 'five-fingered' outer primaries, darker trailing edge to wing, white band across uppertail-coverts.
(d-f) **Female**: Heavy arrowhead or drop-shaped markings on thighs and undertail-coverts; rounder-winged than similar female harriers, with clearly defined broad white uppertail-covert band, broad dark bands across flight feathers (above and below).
(g,h) **Juvenile**: Like female (underparts streaked) but rustier; boldly streaked duller underparts distinguish it from juvenile Pallid and Montagu's.

2 [188] **PALLID HARRIER** *Circus macrourus*
(a-c) **Male**: Like Hen but smaller and slimmer, with paler grey plumage, no obvious cut-off between breast and belly; wings narrower and more pointed (almost reach tail-tip when perched), narrower black wedge on outer primaries, less white on uppertail-coverts, no obvious darker trailing edge to wing.
(d-f) **Female**: Very like female Hen and Montagu's, more variable than latter; upperwing usually plainer, underwing shows darker secondaries than primaries, and white uppertail-covert band usually narrower than Hen.
(g,h) **Juvenile**: Unstreaked warm buff underparts, very similar to Montagu's but typically with less white above and below eye, more dark brown on sides of head, wider, paler, unstreaked collar; underwing has paler-tipped inner primaries than Montagu's.

3 [189] **PIED HARRIER** *Circus melanoleucos*
(a-c) **Male**: Black head, mantle, back, upper breast and median coverts, large whitish patch on lesser coverts; black outer primaries (above and below).
(d-f) **Female**: Whitish leading edge of lesser upperwing-coverts, almost unmarked whitish thighs and vent; grey dark-barred flight feathers, whitish uppertail-covert band, uniformly barred tail.
(g,h) **Juvenile**: Like Pallid and Montagu's but more uniform dark rufous-brown wing-coverts, dark rufous-brown underparts. Warmer than Eastern Marsh.

4 [190] **MONTAGU'S HARRIER** *Circus pygargus*
(a-c) **Male**: Like Hen and Pallid but belly streaked rufous-chestnut, closed wing-tips extend to tail-tip; upperwing has blackish bar on secondaries, underwing has blackish bars on secondaries and coverts.
(d-f) **Female**: Very like Hen and Pallid but smaller, slimmer and narrower-winged than Hen with narrower white uppertail-covert band, underwing with inner primaries darker-tipped than Pallid.
(g) **Adult** dark morph: Sooty-brown with extensive silvery patch on underside of primaries.
(h,i) **Juvenile**: From juvenile Hen by unstreaked buffish-rufous underparts, from Pallid by more white above and below eye, less dark on sides of head, no clear collar, streaked sides of breast; underwing has all-blackish primary tips.

1a
1b
1c
1d
1e
1f
1g
1h

2a
2b
2c
2d
2e
2f
2g
2h

3a
3b
3c
3d
3e
3f
3g
3h

4a
4b
4c
4d
4e
4f
4g
4h
4i

PLATE 22. MARSH-HARRIERS & *BUTASTUR* BUZZARDS

1 [185] **WESTERN MARSH-HARRIER** *Circus aeruginosus*
(a-c) **Male** *aeruginosus*: Recalls Eastern Marsh-harrier but head to breast browner, ear-coverts paler, upperparts browner and more uniform, lower underparts rufous-chestnut; buff leading edge to inner wing.
(d-f) **Female**: Similar to juvenile Eastern but has dark eye-line extending down neck-side, no neck-streaks, smaller pale flash on underside of primaries, creamy-buff leading edge to inner wing and breast-band.
(g,h) **Juvenile**: Resembles female but initially without creamy-buff on wings or breast.

2 [186] **EASTERN MARSH-HARRIER** *Circus spilonotus*
(a-c) **Male** *spilonotus*: From Pied by streaked neck and breast and lack of whitish patch on wing-coverts.
(d-f) **Female**: Resembles female Pied but belly and thighs dull rufous; no pale leading edge to inner upperwing, thinner bars on uppertail (plain centrally), extensively dull chestnut-brown underwing-coverts.
(g,h) **Juvenile** (dark individual illustrated): Dark brown with paler hood, rather plain head-sides, large pale flash on underside of primaries.

3 [198] **WHITE-EYED BUZZARD** *Butastur teesa*
(a-c) **Adult**: Like Grey-faced Buzzard but eyes whitish, head browner, broad pale area across wing-coverts, rufescent rump to uppertail; in flight distinguished by paler upperwing-coverts (more pronounced than illustrated) and rufescent tail (above), narrower tail-bands and darker body, lesser underwing-coverts and axillaries.
(d,e) **Juvenile**: Head, neck and underparts deep rich buff with narrow streaks, pale band across upperwing-coverts, rufescent uppertail, rich buff underwing-coverts.

4 [199] **RUFOUS-WINGED BUZZARD** *Butastur liventer*
(a-c) **Adult**: Recalls Grey-faced Buzzard but greyish, indistinctly streaked head and underside, mostly rufous-brown to chestnut upperside, narrower tail-bands; mostly rufous-chestnut upperside of flight feathers, rump and tail, plain whitish underwing-coverts and indistinctly patterned undertail.
(d) **Juvenile**: Duller and browner above and below.

5 [200] **GREY-FACED BUZZARD** *Butastur indicus*
(a-c) **Male**: Rather slim and large-headed, recalling *Accipiter* and *Buteo*. Upperside and breast mostly rather plain greyish-brown, white throat, blackish submoustachial and mesial stripes, greyish-brown and white bars on belly, three dark bands across tail; underwing rather pale with darker trailing edge (primary tips blackish).
(d) **Juvenile**: Crown and neck brown, narrowly streaked whitish, broad white supercilium framing blackish-brown 'mask', upperparts and wing-coverts darker, underparts dull whitish with pronounced dark streaks.

1g
1a
1d
1e
1h
1b
1c
1f
2g
2d
2a
2e
2b
2h
2c
2f
3a
3d
3e
3b
3c
4b
4c
5d
5a
5b
5c
4a
4d

PLATE 23. SPARROWHAWKS

1 [191] **CRESTED GOSHAWK** *Accipiter trivirgatus*
(a,b) **Male** *indicus*: Relatively large, with short crest, slaty crown and sides of head, brownish-grey upperparts, dark mesial streak, streaked breast and barred belly; uppertail greyish with equal-width bands; in flight, wings appear broad and rounded.
(c) **Juvenile**: Like adult but head and upperparts browner (latter fringed buff), underparts mostly streaked brown.

2 [192] **SHIKRA** *Accipiter badius*
(a-c) **Male** *poliopsis*: Pale grey sides of head, upperparts and tail, dark primary tips, white throat with faint grey mesial streak, dense narrow orange-rufous bars on breast and belly; underwing whitish with some dark bars across outer primaries, undertail with four dark bands in centre. Eyes reddish.
(d,e) **Female**: Larger, eyes yellow, washed brownish above; more dark tail-bands.
(f,g) **Juvenile**: Crown and upperparts brown with paler fringing, darker mesial streak, tear-drop breast-streaks (bars on flanks and spots on thighs), five dark bands on brown uppertail; even dark bars across whitish underside of flight feathers.

3 [193] **CHINESE SPARROWHAWK** *Accipiter soloensis*
(a) **Male**: Like Shikra but darker grey; breast indistinctly barred pale pinkish-rufous, underwing whitish with black-tipped primaries, grey trailing edge and unmarked, pinkish-buff washed coverts. Eyes reddish.
(b) **Female** More rufous on underparts, underwing-coverts more rufous-tinged, eyes yellow.
(c,d) **Juvenile**: Slate-greyish crown and head-sides, dark brown upperparts, chestnut tinge to neck-streaks, upperpart fringing and underpart markings, four dark bands on uppertail, barred thighs; underwing-coverts unbarred, two dark bands across inner secondaries (apart from trailing edge); undertail has three dark bands, five narrow ones on outertail.

4 [194] **JAPANESE SPARROWHAWK** *Accipiter gularis*
(a-c) **Male**: Like Besra but somewhat paler above, underparts diffusely barred pale pinkish-rufous, uppertail with narrow dark bands (usually four visible); wings appear more pointed.
(d,e) **Female**: Larger and browner than male, more prominent mesial streak, more obviously barred below; darker and plainer above than female Besra, no breast-streaks and narrower tail-bands.
(f) **Juvenile female**: Like juvenile Besra but underpart markings less pronounced, mesial streak narrower, uppertail with narrower dark bands.

5 [195] **BESRA** *Accipiter virgatus*
(a,b) **Male** *affinis*: Very dark slate-greyish upperside, prominent dark mesial streak, blackish and rufous-chestnut breast-streaks, broad rufous-chestnut bars on underparts, broad dark tail-bands; wings short and rounded. Eyes reddish.
(c,d) **Female**: Larger, with yellow eyes, browner-tinged above with blackish crown and nape.
(e) **Juvenile female**: From juvenile Shikra and Japanese Sparrowhawk by more prominent dark mesial streak, broader tail-bands, blunter wings. Also from latter by contrast between crown and mantle, heavier-marked underparts.

6 [196] **EURASIAN SPARROWHAWK** *Accipiter nisus*
(a-c) **Male** *nisosimilis*: Slaty-grey upperside, orange-rufous wash on cheeks, faint orange-rufous bars on underparts and faint darker bands on uppertail, no mesial streak. Relatively long-winged and -tailed.
(d,e) **Female**: Larger, with whitish supercilium, browner-tinged upperside.
(f) **Juvenile**: Told by size, shape and heavy rufous-chestnut to blackish bars below.

1a
1c
1b
2a
2b
2c
2d
2e
2f
2g
3a
3b
3c
3d
4a
4b
4c
4d
4e
4f
5a
5b
5c
5d
5e
6f
6a
6b
6c
6d
6e

PLATE 24. NORTHERN GOSHAWK & BUZZARDS

1 [197] **NORTHERN GOSHAWK** *Accipiter gentilis*
(a) **Male** *schvedowi*: Rather large, with bold white supercilium, head darker than mantle, throat whitish.
(b,c) **Female**: Larger. Underparts finely barred brownish-grey; proportionately long-winged and short-tailed.
(d,e) **Juvenile female**: Upperside darker and browner than adult, with buffish fringing, underparts buff with strong dark brown streaks; heavier markings on underwing, more bulging secondaries.

2 [201] **HIMALAYAN BUZZARD** *Buteo burmanicus*
(a-c) **Adult pale morph** *burmanicus*: Combination of size, robust build, rather large head, mostly dark brown upperside and mostly whitish underparts with variable, large dark brown patch across belly distinctive. Typically shows dark brown and whitish streaks on crown and neck, more whitish sides of head with dark eyestripe, heavy brown throat-streaks (particularly at sides), sparse brown breast-streaks and greyish-brown uppertail with numerous faint narrow darker bars. Often has pale thighs, unlike Common and Long-legged Buzzards, and also shows distinctive, nearly unbarred uppertail. Tarsi more than half feathered (mostly unfeathered in Common and Long-legged). In flight shows broad rounded wings and shortish rounded tail; upperwing usually has paler area across primaries, underwing with contrasting blackish outer primary tips and blackish carpal patches, rest of underwing rather pale with whiter primaries and rather narrow dark trailing edge; has numerous indistinct narrow dark bars across underside of flight-feathers and tail.
(d,e) **Adult dark morph**: Head, body and wing-coverts (above and below) rather uniform blackish-brown, tail with broader terminal band; rest of underwing similar to pale morph.
(f-g) **Juvenile pale morph**: Similar to adult but has pale eyes, initially a strong buffy wash below, narrower, paler and more diffuse trailing edge to underwing and evenly barred tail, without broader subterminal band.
(h) **Juvenile dark morph**: Eyes and tail differ similarly to pale morph.

3 [202] **COMMON BUZZARD** *Buteo buteo*
(a-c) **Adult pale morph** *vulpinus*: Head to underparts more heavily dark-marked than Long-legged; less rufescent overall. A *Buteo* with heavily barred underparts will be this species. Smaller than Himalayan.
(d) **Adult rufous morph**: The commonest morph. May appear almost identical in plumage to Long-legged, but smaller and weaker-billed, lacking long-necked appearance at rest; upperside more solidly brown, with narrow rufous edgings, and tail usually crisply and narrowly barred (at least lightly).
(e-f) **Juvenile pale morph**: Differs from adult in similar way to Himalayan. Apparently told from Himalayan by streakier head to breast, barring on belly and thighs, less distinct carpal patches on underwing.

4 [203] **LONG-LEGGED BUZZARD** *Buteo rufinus*
(a-c) **Adult pale morph** *rufinus*: Similar to Himalayan and common but larger and larger-headed, with long-necked appearance at rest, and rufous on upperparts, underparts and tail. Typically has pale cream-coloured head, neck and breast, with indistinct darker streaks on crown, hindneck and breast, slightly darker eyestripe and malar line, rufous thighs and belly-patch, and rather plain pale rufous to cinnamon-coloured tail (above and below). Rufous on underparts almost always heavier on belly than breast. In flight shows longer, more eagle-like wings, and rather long square-tipped tail; underwing-coverts often tinged rufescent.
(d) **Adult dark morph**: Very like dark morph Himalayan , but broader dark bands on underwing and typically broader tail-bands.
(e) **Adult rufous morph**: Head, body and wing-coverts (above and below) rufous-chestnut, underwing and tail pattern like pale morph.
(f,g) **Juvenile pale morph**: Upperparts browner, underparts less rufous, more brownish; tail has faint narrow dark bars towards tip. In flight shows narrower, paler and more diffuse trailing edge to underwing.

1b
1a
1c
1d
1e
3a
3c
3b
3d
3f
3e
2a
2c
2b
2d
2h
2f
2e
2g
4c
4e
4g
4a
4b
4d
4f

PLATE 25. *AQUILA* EAGLES

1 [204] **INDIAN SPOTTED EAGLE** *Aquila hastata*

(a-c) **Adult**: Smaller than Greater Spotted; proportionately smaller bill with larger gape (pronounced wide flange or 'thick lips'; extending further under eye than on Greater Spotted). Head, upperparts and upper-wing-coverts very dark brown with black shaft-streaks. More prominent whitish flash at base of primaries on upperwing than Greater Spotted, double whitish crescent in vicinity of primary underwing-coverts, under-wing-coverts uniform apart from paler brown lesser coverts.
(d-f) **Juvenile**: Head, body and wing-coverts paler than Greater Spotted; less pronounced white spotting on tertials, median and lesser upperwing-coverts, buffish-brown below with darker streaks. Uppertail-coverts are very pale brown with white barring. In flight, from Greater Spotted by same features as adult.
(g) **Subadult**: Mixture of adult and juvenile characters.

2 [205] **GREATER SPOTTED EAGLE** *Aquila clanga*

(a-c) **Adult**: From similar large dark eagles by relatively short wings, smallish bill, shortish tail, narrow pale U on uppertail-coverts; upperwing dark with indistinct pale primary bases, underwing dark with whitish patch or crescent at base of outer primaries.
(d-f) **Juvenile**: Blackish with whitish tips to most of wing feathers and tail, broad pale buffish streaks on belly and thighs, whitish bands across upperwing-coverts and another across uppertail-coverts, underwing with bases of primaries paler than adult, and coverts darker than flight feathers.
(g) **Subadult**: Mixture of adult and juvenile features.
(h,i) **Juvenile pale morph**: Mostly buffy body and wing-coverts (as adult pale morph), pale-tipped wing feathers and tail.
(j) **Juvenile pale morph**: Younger individual with rufescent-brown body and wing-coverts.

3 [206] **TAWNY EAGLE** *Aquila rapax*

(a-c) **Adult pale morph** *vindhiana*: From pale morph Greater Spotted by somewhat larger bill, longer neck, fuller 'trousers', yellower eyes; somewhat paler greater and primary underwing-coverts, paler inner primaries and bases of outer primaries, no defined whitish patch or crescent.
(d,e) **Adult dark morph**: Somewhat smaller and more weakly built than Steppe, often more upright posture, darker nape and throat, gape ending under eye (behind eye in Steppe); in flight from below, greyish inner primaries and bases of outer primaries, uniformly dark secondaries.
(f,g) **Juvenile pale morph**: Whitish tips to upperwing-coverts, secondaries, inner primaries and tail.

4 [207] **STEPPE EAGLE** *Aquila nipalensis*

(a-c) **Adult** *nipalensis*: Largest and largest-billed of region's uniformly dark eagles. Key features include full 'trousers', rufous-buff nape-patch, somewhat paler throat and indistinct dark bars on secondaries; wings and tail distinctly longer than spotted eagles, underwing with mostly blackish primary coverts, darker trailing edge and faint dark bars on flight feathers.
(d-f) **Juvenile**: Head, body and wing-coverts paler and more grey-brown, with whitish tips to greater and primary upperwing-coverts, secondaries and tail feathers, and broader, whiter band across uppertail-coverts; diagnostic broad whitish underwing-band.
(g) **Subadult**: Underwing shows vestigial whitish band and signs of adult-like pattern.

1a
1d
1c
1b
1f
1e
1g
2a
2c
2b
2j
2h
2e
2g
2f
2i
2d
3c
3d
3a
3b
3e
3g
3f
4b
4c
4e
4a
4d
4f
4g

PLATE 26. EAGLES

1 [183] **SHORT-TOED SNAKE-EAGLE** *Circaetus gallicus*
(a,b) **Adult pale individual**: Medium-sized, big pale head, yellow eyes, and longish unfeathered legs. Pale wing-coverts, underparts whitish with indistinct markings; wings distinctly broad (pinched-in at base), underwing white with faint darker bands and dark trailing edge, tail pale with narrow dark bands.
(c) **Adult dark individual**: Distinctive dark hood, extensive dark barring below and darker bands across underwing.

2 [208] **EASTERN IMPERIAL EAGLE** *Aquila heliaca*
(a-c) **Adult**: Large blackish-brown eagle with golden-buff nape and hindneck, white markings on upper scapulars and broadly black-tipped greyish tail; almost uniformly blackish underwing, pale undertail-coverts.
(d-f) **Juvenile**: Recalls pale morph Spotted and Tawny Eagles but larger, with diagnostic distinct dark streaks on neck, breast and wing-coverts (above and below); pale greyish wedge on inner primaries.

3 [209] **BONELLI'S EAGLE** *Aquila fasciata*
(a,b) **Adult** *fasciata*: Whitish dark-streaked underparts and barred thighs recall hawk-eagles but smaller and uncrested with single broad subterminal band on greyish tail; underwing greyish with whitish leading edge to mostly black coverts.
(c-e) **Juvenile**: Head and upperparts browner, underparts warm buffish with dark streaks; in flight tail shows even, narrow, darker bars, warm buffish underwing-coverts with dark tips.

4 [210] **BOOTED EAGLE** *Aquila pennata*
(a-c) **Adult pale morph**: Smallish with pale brownish hood, dark sides of head and mostly whitish underparts; on (relatively narrow) upperwing, buff-brown band across coverts contrasts with otherwise dark feathers, on underwing whitish coverts and pale wedge on inner primaries (concolorous with underparts) contrast with otherwise blackish feathers, (fairly long) undertail greyish with indistinct dark terminal band and central feathers.
(d,e) **Adult dark morph**: Head and body rather uniform blackish-brown, scapulars and wing-coverts browner; undertail as in pale morph.
(f) **Adult rufous morph**: Strongly rufescent head, body and underwing-coverts.

5 [211] **BLACK EAGLE** *Ictinaetus malayensis*
(a,b) **Adult** *malayensis*: Overall blackish, yellow cere and feet, longish tail with slightly paler bands; long broad wings with narrow, pinched-in bases and well-spread 'fingers'.
(c,d) **Juvenile**: Head to underparts pale buffish with heavy blackish streaks, some pale tips on upperparts; pale buffish underwing-coverts with blackish streaks, and darker, pale-barred underside of flight feathers and tail.

6 [212] **RUFOUS-BELLIED EAGLE** *Lophotriorchis kienerii*
(a,b) **Adult** *formosae*: Sides of slightly crested head and upperside blackish, throat and upper breast white, underparts rufous-chestnut (breast to undertail-coverts dark-streaked); rufous-chestnut underwing-coverts, pale greyish flight feathers with dark trailing edge, tail greyish with blackish subterminal band.
(c,d) **Juvenile**: Upperside brown, sides of head whitish with dark eyestripe, underparts whitish with dark flank-patch; whitish underwing-coverts with broken dark border, narrower trailing edge to flight feathers and tail.

1a
1b
1c
2a
2d
2c
2b
2e
2f
3a
3c
3b
3e
3d
4a
4b
4c
4e
4f
6a
4d
6c
5c
5a
5d
5b
6b
6d

PLATE 27. ORIENTAL HONEY-BUZZARD & HAWK-EAGLES

1 [166] **ORIENTAL HONEY-BUZZARD** *Pernis ptilorhynchus*
(a-c) **Male pale morph** *ruficollis*: Relatively large with small head, longish neck, slight crest. Highly variable. Typically has greyish sides of head and pale underparts with dark throat-border and mesial streak, gorget of dark streaks, warm brown bars below; narrow tail with two complete, well-spaced blackish bands, underwing typically whitish with blackish trailing edge, narrow bars on coverts and usually three blackish bands across primaries and outer secondaries.
(d) **Female pale morph**: Narrower tail-bands and narrower trailing edge to underwing, which shows three bands across secondaries and inner primaries.
(e,f) **Adult dark morph**: Mostly dark chocolate-brown head, body and wing-coverts.
(g) **Adult dark morph variant**: Whitish bars/scales on underbody and underwing-coverts.
(h,i) **Juvenile pale morph**: Equally variable. Typically shows paler head, neck, underparts and underwing-coverts than adult. Has less distinct dark bands across underside of flight feathers and three or more dark tail-bands.

2 [213] **BLYTH'S HAWK-EAGLE** *Nisaetus alboniger*
(a-c) **Adult**: Erectile crest blackish (sometimes narrowly white-tipped), sides of head and crown to mantle blacker than other hawk-eagles, underparts whitish with prominent black mesial streak, bold blackish breast-streaks and bold blackish bars on belly, thighs and vent, tail blackish with pale greyish broad central band and narrow tip; whitish underwing with heavy blackish bars on coverts, four blackish bands across flight feathers, distinctive tail pattern.
(d,e) **Juvenile**: Crown, hindneck and ear-coverts sandy-rufous, upperside browner than adult with whitish fringes, underparts plain pale buff to whitish with buffish breast and flanks, tail whitish with two or three medium-width dark bands and broader terminal dark band, underwing-coverts plain creamy-whitish, crest as adult.

3 [214] **MOUNTAIN HAWK-EAGLE** *Nisaetus nipalensis*
(a-c) **Adult** *nipalensis*: Like pale morph Changeable but has long erectile white-tipped blackish crest, whitish bars on rump, broader dark mesial streak, broader dark bars on lower underparts (including belly) and broader dark tail-bands; from below, wings broader and more rounded, tail relatively shorter, wing-coverts with heavy dark markings, bases of outer primaries barred, and dark tail-bands broader.
(d,e) **Juvenile**: From Changeable by distinctive crest, pale to warm buff head and underparts, darker-streaked crown, hindneck and sides of head, and buff-barred rump; in flight by shape, barred outer primaries and broader, less numerous dark tail-bands.

4 [215] **CHANGEABLE HAWK-EAGLE** *Nisaetus limnaeetus*
(a-c) **Adult pale morph** *limnaeetus*: Rather nondescript brown above, whitish below with dark mesial streak, short crest, dark streaks on breast and belly, faint narrow rufous barring on thighs and undertail-coverts, four dark bands on tail (terminal one broader), broad, rather parallel-edged wings, rather plain underwing-coverts.
(d,e) **Adult dark morph**: Blackish with greyer, broadly dark-tipped tail; underwing shows greyer bases to flight feathers.
(f-h) **Juvenile pale morph**: Head, neck and underparts almost unmarked whitish, prominent whitish fringing on upperparts (particularly wing-coverts), narrower, more numerous dark tail-bands (lacks wider terminal band); underside like adult apart from paler body and wing-coverts and tail pattern.

5 [216] **WALLACE'S HAWK-EAGLE** *Nisaetus nanus*
(a-c) **Adult** *nanus*: Like Blyth's but smaller, sides of head and hindneck rufescent-brown with blackish streaks, crest broadly white-tipped, tail greyish with three dark bands (terminal one slightly broader); buffish-white base colour to underside of flight feathers, warm buffish coverts with narrow dark barring.
(d,e) **Juvenile**: Very like Blyth's but smaller, with broader white tip to crest.

1a
1e
1c
1b
1f
1g
1i
1h
1d
2c
2e
2d
2a
2b
3d
3a
3e
3c
3b
4a
4d
4e
4h
4c
4b
4f
4g
5d
5a
5c
5e
5b

PLATE 28. CRAKES & RAILS

1 [219] **RED-LEGGED CRAKE** *Rallina fasciata*
(a) **Adult**: Chestnut-tinged brown upperside, dull chestnut head and breast, black-and-whitish bars on wing-coverts and flight feathers, bold blackish and whitish ventral bars, red legs.
(b) **Juvenile**: Like adult but duller, buffier, less bold wing markings and ventral barring, brownish-yellow legs.

2 [220] **SLATY-LEGGED CRAKE** *Rallina eurizonoides*
(a) **Adult** *telmatophila*: Like Red-legged Crake but larger, with bigger bill, duller and colder brown upper side, no barring/spotting on wings, greyish to black legs.
(b) **Juvenile/first winter**: Like Red-legged but larger, bigger-billed, no bars/spots on wing-coverts, paler-faced.

3 [221] **SLATY-BREASTED RAIL** *Gallirallus striatus*
(a) **Adult** *albiventer*: Chestnut crown, dull olive-brown above with black-and-white mottling and barring, grey head-sides and breast, blackish-and-whitish barring ventrally; longish reddish bill, dark legs.
(b) **Juvenile**: Duller and paler, streaked blackish above, paler and browner below with duller bars, shorter bill.

4 [222] **EASTERN WATER RAIL** *Rallus indicus*
Adult: Like Slaty-breasted Rail but no chestnut on crown, grey supercilium, broad dark brownish eyestripe, buffy olive-brown upperparts with broad blackish streaks, broader underpart bars, paler legs.

5 [223] **CORNCRAKE** *Crex crex*
(a,b) **Male non-breeding**: Like female Watercock but smaller, with greyish supercilium, unbarred breast, rufous-and-whitish bars on flanks; largely rufous upperwing.
(c) **Male breeding**: More defined blue-grey supercilium and foreneck, blue-grey tinge to breast.

6 [224] **WHITE-BREASTED WATERHEN** *Amaurornis phoenicurus*
(a) **Adult** *phoenicurus*: Dark slaty olive-brown crown, sides of body and upperparts, white face to upper belly, rufous-chestnut rear flanks and vent, yellowish-green bill and legs.
(b) **Juvenile**: Slightly browner above, face clouded, broken dark bars on white of underparts, bare parts duller.

7 [225] **BROWN CRAKE** *Porzana akool*
Female *akool*: Rather dark olive-brown upperside, grey sides of head and underparts, greenish bill, red eyes, fleshy-brown to reddish legs.

8 [226] **BLACK-TAILED CRAKE** *Porzana bicolor*
(a) **Adult**: Like Brown Crake but smaller, with rufescent upperside, blackish tail and slaty-grey head, neck and underparts, brick-red legs.
(b) **Juvenile**: Upperparts mixed with blackish-brown, brownish tinge below.

9 [227] **BAILLON'S CRAKE** *Porzana pusilla*
(a) **Adult** *pusilla*: Blackish and white streaks and speckles above, blue-grey supercilium and underparts, black and white bars on rear flanks/vent, greenish, yellowish or brownish to pinkish legs.
(b) **Juvenile**: Lacks blue-grey; has whitish throat to belly-centre and rufous-buff on upper breast and upper flanks.

10 [228] **SPOTTED CRAKE** *Porzana porzana*
(a) **Male non-breeding**: Larger than Baillon's, darker above with more white markings, extensive white speckling and barring on browner underparts, plain buff undertail-coverts.
(b) **Male breeding**: More bluish-grey underlying head, neck and breast markings, brighter red bill-base.
(c) **First winter**: Whitish throat, denser white speckling on head and sides of neck, duller bill.

11 [229] **RUDDY-BREASTED CRAKE** *Porzana fusca*
(a) **Adult** *fusca*: Uniform, rather cold dark olive-brown above, reddish-chestnut below with indistinct narrow whitish bars on greyish-black rear flanks and vent, blackish bill with greenish base, reddish legs.
(b) **Juvenile**: Whitish underside with dense dull brownish-grey vermiculations/mottling, dull legs.

12 [230] **BAND-BELLIED CRAKE** *Porzana paykullii*
(a) **Adult**: Like Slaty-legged Crake but has darker, colder brown crown and upperside, paler, more extensive chestnut below, some whitish and dark bars on wing-coverts, salmon-red legs, greenish-slate bill with pea-green base.
(b) **First winter**: Paler, washed-out head-sides and breast (vaguely barred) to upper belly with little pale chestnut, purplish-brown legs.

13 [231] **WHITE-BROWED CRAKE** *Porzana cinerea*
Adult: Small; blackish and buffish streaks above, white eyebrow and line from chin to ear-coverts, blackish around eye, greenish bare parts.

14 [232] **WATERCOCK** *Gallicrex cinerea*
(a) **Male breeding**: Fairly large; blackish with grey to buff feather fringing on mantle, scapulars and wing-coverts, yellow bill with red base, red frontal shield and legs.
(b) **Male non-breeding**: Like non-breeding female but larger, with more distinct underpart bars.
(c,d) **Female breeding**: Smaller; dark brown upperside with buff feather fringes, buff below with wavy greyish-brown bars, greenish bill and legs.
(e) **Juvenile**: Like female non-breeding with warmer, broader streaks above and narrow, less distinct bars below.

1b
2b
3b
1a
3a
5a
2a
5b
5c
4
6a
7
6b
8a
8b
9b
9a
10b
11b
11a
10a
12a
10c
12b
13
14d
14c
14b
14a
14e

PLATE 29. GALLINULES, MASKED FINFOOT, THICK-KNEES, STILTS, AVOCETS, NORTHERN LAPWING

1 [233] **GREY-HEADED SWAMPHEN** *Porphyrio poliocephalus*
(a) **Adult** *poliocephalus*: Large; deep purple-blue with big red bill and frontal shield, medium-length red legs, silvery wash on face, dark turquoise throat, foreneck, upper breast and wings, white undertail-coverts.
(b) **Juvenile**: Duller, with brown on hindneck to rump, smaller, blackish bill and shield.

2 [234] **BLACK-BACKED SWAMPHEN** *Porphyrio indicus*
Adult *viridis*: Smaller than Grey-headed; mostly blackish-brown upperside, turquoise shoulder-patch.

3 [235] **COMMON MOORHEN** *Gallinula chloropus*
(a) **Adult** *chloropus*: Dark slaty-brown above, greyer below, with white lateral undertail-coverts and line along flanks, yellow-tipped red bill, red frontal shield, yellow-green legs.
(b) **Juvenile**: Paler and browner above, much paler sides of head and underparts with whitish throat and centre of abdomen, greenish-brown bill.

4 [236] **COMMON COOT** *Fulica atra*
(a) **Adult** *atra*: Slaty-black plumage, white bill and frontal shield.
(b) **Juvenile**: Browner-tinged plumage, whitish throat and centre of foreneck, paler belly, smaller frontal shield.

5 [237] **MASKED FINFOOT** *Heliopais personata*
(a) **Male**: Grebe-like; grey hindcrown and -neck, black face and upper foreneck with white border, thick yellow bill; body mostly brown above, whitish below, dark eye.
(b) **Female**: Whitish on lores, throat and foreneck, pale eye.

6 [245] **INDIAN THICK-KNEE** *Burhinus indicus*
(a,b) **Adult**: From other thick-knees by smaller size, much smaller bill, streaked upperparts, neck and breast, and lack of prominent black head markings. Upperparts pale sandy-brown with blackish and white bands along folded wing. Flight feathers blackish, primaries with relatively small white patches.
(c) **Juvenile**: Scapulars, inner coverts and tertials fringed rufous-buff, indistinct bars on closed wings, whiter tips to greater coverts.

7 [246] **GREAT THICK-KNEE** *Esacus recurvirostris*
(a,b) **Adult**: Fairly large with long legs and longish, very thick, slightly upturned black bill with yellow base. Complex black, white and sandy-greyish head pattern, upperparts sandy-grey with narrow blackish and whitish bands along lesser coverts, underparts whitish with brownish wash on foreneck, upper breast. In flight, upperwing shows mostly grey greater coverts, mostly black flight feathers, primaries with large white patches.

8 [247] **BEACH THICK-KNEE** *Esacus neglectus*
(a,b) **Adult**: Similar to Great Thick-knee but bill bulkier, head largely blackish with white supercilium. In flight, upperwing shows mostly grey secondaries and mostly white inner primaries.

9 [250] **BLACK-WINGED STILT** *Himantopus himantopus*
(a,b) **Male non-breeding** *himantopus*: Slimly built with a medium-long needle-like blackish bill and very long pinkish-red legs. Mostly white with grey cap and hindneck, black ear-coverts and black mantle, scapulars and wings.
(c) **Male breeding**: Head and neck typically all white (may have variable grey and black on head and hindneck).
(d) **Female breeding**: Mantle and scapulars browner than male.
(e) **Juvenile**: Crown and hindneck brownish-grey, upperparts and wing-coverts greyish-brown with buffish fringes.

10 [251] **WHITE-HEADED STILT** *Himantopus leucocephalus*
(a) **Adult**: Like male breeding Black-winged, but shows long black 'mane' on back of neck and an otherwise white head; slightly smaller, but wing and bill longer relative to body size.
(b) **First immature**: Possibly not distinguishable from adult non-breeding Black-winged.

11 [252] **PIED AVOCET** *Recurvirostra avosetta*
(a,b) **Male**: Long, strongly upturned blackish bill, long bluish-grey legs, mainly white plumage with black crown to hindneck, scapulars, median and lesser coverts, and primaries.
(c) **Juvenile**: Dark parts of plumage tinged dull brown, white of mantle and scapulars mottled pale greyish-brown.

12 [256] **NORTHERN LAPWING** *Vanellus vanellus*
(a,b) **Adult non-breeding**: Long black crest and crown, sides of head buffish with black facial patch, upperparts dark green, underparts white with broad blackish breast-band and orange-rufous undertail-coverts; very broad wings with mostly black flight feathers, white underwing-coverts.
(c) **Male breeding**: Throat and foreneck black, sides of head white.
(d) **Female breeding**: Like breeding male but lores and throat marked with white.
(e) **Juvenile**: Like non-breeding adult but crest short, upperparts prominently fringed buff.

1-5 and 12 to different scale
1a
1b
2
3a
3b
4a
4b
5a
5b
6a
6b
6c
7a
7b
8a
8b
9a
9b
9c
9d
9e
10a
10b
11a
11b
11c
12a
12b
12c
12d
12e

PLATE 30. EURASIAN OYSTERCATCHER, CRAB-PLOVER, IBISBILL, JACANAS, GREATER PAINTED-SNIPE & PRATINCOLES

1 [253] **EURASIAN OYSTERCATCHER** *Haematopus ostralegus*
(a) **Adult non-breeding** *osculans*: Robust shape, medium-short pink legs, long, thick orange-red bill with duller tip, black plumage with white back to uppertail-coverts and lower breast to vent, usually also a prominent white band across lower throat.
(b) **Adult breeding**: Throat all black, bill uniform orange-red; broad white band across upperwing at all seasons.
(c) **First winter**: Like adult non-breeding but black of plumage tinged brown, bill much duller and narrower.

2 [254] **CRAB-PLOVER** *Dromas ardeola*
(a,b) **Adult**: Bill very thick, pointed and blackish, legs long and bluish-grey, plumage mainly white with black mantle, scapulars, upperside to flight feathers, primary coverts and outer greater coverts. Sometimes has blackish speckles on hindcrown and nape.
(c) **Juvenile**: Pronounced speckling on hindcrown and nape, upperparts mostly greyish with paler brownish-grey scapulars and wing-coverts, dark of upperwing greyer.

3 [255] **IBISBILL** *Ibidorhyncha struthersii*
(a,b) **Adult non-breeding**: Fairly large, with mainly grey head, neck and upperparts, blackish face with white mottling, black crown, narrow white and broad black breast-band, strongly downcurved reddish bill. Dark flight feathers with slight white bar visible in flight.
(c) **Juvenile**: Upperparts browner with buff fringes, face whitish to dark brown with numerous white feather tips, no white breast-band, lower breast-band dark brown.

4 [269] **PHEASANT-TAILED JACANA** *Hydrophasianus chirurgus*
(a) **Adult breeding**: Blackish-brown with white head and foreneck, yellow-buff hindneck, long pointed blackish tail, mostly white wings.
(b,c) **Adult non-breeding**: Drab brown upperparts, pale buff supercilium extends down neck bordered by black eyestripe that leads to breast-band, all-white underside, shorter tail.
(d) **Juvenile**: Like non-breeding adult but has rufous-chestnut crown, rufous-buff fringes to upperparts, weaker breast-band.

5 [270] **BRONZE-WINGED JACANA** *Metopidius indicus*
(a) **Adult**: Green-glossed black head, neck and underparts with white supercilium, purplish gloss on lower hindneck, bronze-olive lower mantle to wing-coverts, chestnut-maroon back to uppertail.
(b) **Juvenile**: Dull chestnut crown, blacker on hindneck, whitish face and underparts with rufous-buff on neck, black and white tail.

6 [271] **GREATER PAINTED-SNIPE** *Rostratula benghalensis*
(a-c) **Female** *benghalensis*: Slightly drooping bill, mostly plain dark head, neck, upper breast and upperside, broad whitish spectacles, buffish median crown-stripe and mantle-lines, unmarked white lower breast to vent. Barred tail and flight feathers, underwing with largely clear white coverts.
(d,e) **Male**: Mostly greyish-brown head, neck, upper breast and upperside, buffish spectacles. Large rich buff markings on wing-coverts.

7 [316] **ORIENTAL PRATINCOLE** *Glareola maldivarum*
(a-c) **Adult breeding**: Rather tern-like with short bill, long pointed wings, short forked tail. Warmish grey-brown, throat and upper foreneck buff with narrow black 'necklace', red bill-base; white rump, uppertail-coverts and chestnut underwing-coverts in flight.
(d) **Adult non-breeding**: Duller, necklace frayed.
(e) **Juvenile**: Greyish-brown above, fringed whitish to buff and blackish, breast streaked/mottled greyish-brown.

8 [317] **SMALL PRATINCOLE** *Glareola lactea*
(a-c) **Adult breeding**: Smaller and much paler than Oriental, no dark necklace; both wing surfaces show broad white band and black trailing edge, black band on squarish tail.
(d) **Adult non-breeding**: Throat faintly streaked, paler lores.
(e) **Juvenile**: Lower throat bordered by brownish spots/streaks, upperparts fringed buffish-and-brownish.

1b
1a
1c
2a
2c
2b

3b
3c
4d
4c
3a
4a
4b
5a
5b
6e
6d
6b
6a
6c
7a
7c
7b
8c
8b
8d
7d
7e
8a
8e

PLATE 31. LAPWINGS & PLOVERS

1 [257] **RIVER LAPWING** *Vanellus duvaucelii*
(a,b) **Adult**: Black crest, crown, face and stripe down throat to breast, whitish head-sides, white underparts with sandy breast-band and small black belly-patch, legs black; upperwing shows broad white band narrowly bordered black on inner wing.
(c) **Juvenile**: Black of head fringed brownish, upperparts fringed dark-and-buff.

2 [258] **YELLOW-WATTLED LAPWING** *Vanellus malabaricus*
(a,b) **Adult non-breeding**: All-black crown, long yellow wattle in front of eye, long white post-ocular band and uniform sandy greyish-brown remainder of head, neck and upper breast (latter with blackish lower border); more white on underwing than Red-wattled.
(c) **Juvenile**: Crown brown with buffish speckling, upperparts fringed buff and dark, small wattles, no dark breast-border.

3 [259] **GREY-HEADED LAPWING** *Vanellus cinereus*
(a,b) **Adult non-breeding**: Relatively large, with long yellowish legs and bill (latter tipped black), plain brownish-grey head to breast, with broad blackish breast-band, white belly, otherwise sandy upperparts; white greater coverts and secondaries, black outer wing.
(c) **Juvenile**: Head to breast and upperparts sandy-brownish, scaled buffish.

4 [260] **RED-WATTLED LAPWING** *Vanellus indicus*
(a,b) **Adult** *atronuchalis*: Largely black head and breast, white ear-coverts, red facial skin and black-tipped bill, cold sandy greyish-brown upperside, whitish underparts, long yellow legs; in flight recalls Yellow-wattled and River but breast black.
(c) **Juvenile**: Duller, throat whitish.

5 [266] **LESSER SAND-PLOVER** *Charadrius mongolus*
(a,b) **Adult non-breeding** *schaeferi*: From other small plovers (except Greater Sand-) by broad breast-patches and no white nuchal collar. Upperside sandy greyish-brown with whitish forehead and supercilium; prominent white upperwing-bar.
(c) **Male breeding**: Black forehead and mask, white throat encircled by deep orange-rufous on neck and breast.
(d) **Female breeding**: Duller with less orange-rufous on neck and breast and browner mask.
(e) **Juvenile**: Like non-breeding adult but upperparts fringed buffish, buff wash to supercilium and breast-patches.
(f) **Adult non-breeding** *mongolus*: Bill smaller.
(g) **Male breeding**: White sides to forehead (as in Greater), clearer black upper border to breast-band.

6 [267] **GREATER SAND-PLOVER** *Charadrius leschenaultii*
(a,b) **Adult non-breeding** *leschenaultii*: From Lesser Sand- by larger size, longer appearance, longer, more tapered bill, longer tibia, often paler legs, toes distinctly projecting beyond tail-tip in flight.
(c) **Male breeding**: Plumage like Lesser (subspecies *mongolus*) but breast-band narrower with no black border.
(d) **Female breeding**: Like non-breeding adult but with narrow orange-brown breast-band.
(e) **Juvenile**: Differs in same way as Lesser.

7 [268] **ORIENTAL PLOVER** *Charadrius veredus*
(a-c) **Adult non-breeding**: Larger than sand-plovers with longer neck, legs and wings and slenderer bill, longer, broader supercilium; buffish-brown breast, no wing-bar, dark underwing.
(d) **Male breeding**: Rufous-chestnut breast with broad black lower border.
(e) **Female breeding**: Like non-breeding adult but upper breast with rufescent wash.
(f) **Juvenile**: Like non-breeding adult with stronger buff fringing above.

1-4 to different scale
1c
1b
1a
2c
2b
2a
3b
3a
3c
4b
4a
4c
5g
5f
5d
5c
5a
5e
6e
6d
6a
6c
5b
6b
7f
7e
7a
7d
7b
7c

PLATE 32. PLOVERS

1 [248] **PACIFIC GOLDEN PLOVER** *Pluvialis fulva*
(a,b) **Adult non-breeding**: Medium-sized, short-billed and rather nondescript with distinctive golden spangling on upperside, neck and breast with dusky-grey streaks and mottling, dark patch on rear ear-coverts; dull underwing.
(c) **Male breeding**: Face to belly black, bordered white from supercilium to flanks.
(d) **Juvenile**: Like non-breeding adult but upperparts more boldly patterned, neck and breast strongly washed golden and spotted/streaked darker.

2 [249] **GREY PLOVER** *Pluvialis squatarola*
(a,b) **Adult non-breeding**: Larger, stockier and bigger-headed than Pacific Golden Plover, with stouter bill, upperside distinctly greyish with whitish speckling, whitish supercilium and base colour to neck and breast; strong white bar across upperwing, largely whitish underwing with black axillaries.
(c) **Male breeding**: Pattern similar to breeding Pacific Golden but more white on head and sides of breast, upperparts black with silvery-white spangling.
(d) **Juvenile**: Like non-breeding adult but upperside blacker with more distinct yellowish-buff to whitish speckles and spangling, neck and breast washed yellowish-buff and more distinctly dark-streaked.

3 [261] **COMMON RINGED PLOVER** *Charadrius hiaticula*
(a,b) **Adult non-breeding** *tundrae*: Like Little Ringed but slightly larger and more robust, with broader dark breast-band, orange legs; white wing-bar.
(c) **Male breeding**: Bold black pattern on white and pale brown head and neck; from Little by orange bill with black tip, orange legs, no yellow eyering or white band across midcrown.
(d) **Juvenile**: Like Adult non-breeding but breast-band narrower, upperparts fringed buffish-and-dark, bill all dark, legs duller.

4 [262] **LONG-BILLED PLOVER** *Charadrius placidus*
(a,b) **Adult non-breeding**: Larger but more attenuated than Little Ringed, with longer bill, broader dark band across forecrown, broader supercilium, narrower breast-band; upperwing subtly different.
(c) **Adult breeding**: Bands on crown and breast black, no black on lores and ear-coverts, weaker eyering.
(d) **Juvenile**: Neat buff fringes to upperparts, no dark band across midcrown, buffier supercilium, greyish-brown breast-band.

5 [263] **LITTLE RINGED PLOVER** *Charadrius dubius*
(a,b) **Adult non-breeding** *jerdoni*: Rather dainty, small-headed and attenuated, mainly greyish-brown; slender dark bill, narrow pale yellowish eyering, pinkish to yellowish legs, white collar, (almost) complete greyish-brown breast-band; uniform upperwing.
(c) **Male breeding**: Lores, ear-coverts and breast-band black, forehead and supercilium white; prominent black band across forecrown, backed by distinctive narrow white band, pronounced broad yellow eyering.
(d) **Juvenile**: Like non-breeding adult but with buff-and-dark fringing above, breast-band browner and broken, supercilium buffier.

6 [264] **KENTISH PLOVER** *Charadrius alexandrinus*
(a,b) **Adult non-breeding** *alexandrinus*: Plain upperside, white collar, narrow lateral breast-patches, blackish bill, dark greyish legs; white upperwing-bar and outertail.
(c,d) **Adult non-breeding**: Worn individuals can be very pale ('bleached').
(e) **Male breeding**: Strongly rufous-washed midcrown to nape, blackish mask and forecrown-band, narrow black lateral breast-patches.
(f) **Male breeding variant**: Less well marked.
(g) **Female breeding**: Like faded breeding male (more like non-breeding adult).
(h) **Juvenile**: Like non-breeding adult but head paler and washed buff, upperparts narrowly fringed buff, breast-patches paler and more diffuse.

7 [265] **MALAYSIAN PLOVER** *Charadrius peronii*
(a) **Male**: Like breeding male Kentish but slightly smaller and shorter-billed, upperparts appear scaly or mottled, breast-patches narrower and forming complete border to white nuchal collar.
(b,c) **Female**: Lacks black markings. From Kentish by rufous-washed ear-coverts and lateral breast-patches, scaly upperparts.
(d) **Juvenile**: Slightly duller than female.

1c
1d
2d
2c
1b
1a
2b
2a
3b
3a
4a
3c
4b
4c
3d
4d
5d
5b
7c
7b
5a
7d
7a
5c
6g
6h
6a
6e
6d
6b
6c
6f

PLATE 33. EURASIAN WOODCOCK & SNIPES

1 [272] **EURASIAN WOODCOCK** *Scolopax rusticola*
(a,b) **Adult**: Larger and more robust than snipes, broad blackish bars on hindcrown, strongly rufescent and densely patterned upperside, buffish-white dark-barred underparts. Heavy and broad-winged in flight, upperwing rather uniform and strongly rufescent.

2 [273] **JACK SNIPE** *Lymnocryptes minimus*
(a-c) **Adult**: Small size, relatively short bill, dark crown-centre, split supercilium, dark purple- and green-glossed mantle and scapulars with broad buff lines, unbarred underparts, streaked foreneck, upper breast and flanks.

3 [274] **SOLITARY SNIPE** *Gallinago solitaria*
(a-c) **Adult** *solitaria*: Relatively large; whitish face and lengthwise lines on mantle, extensive rufous vermiculations on upperparts, gingery breast-patch. In flight, toes do not extend beyond tail-tip, underwing-coverts heavily barred.

4 [275] **WOOD SNIPE** *Gallinago nemoricola*
(a-c) **Adult**: Relatively large, bulky and short-billed; pronounced buff stripes and scaling on distinctly dark upperparts, lower breast to vent entirely barred, no gingery breast-patch. In flight, wings appear relatively broad and rounded, with densely dark-barred underwing-coverts.
(d) **Juvenile**: Finer, whiter fringing on mantle and scapulars, paler fringes to median coverts.

5 [276] **PINTAIL SNIPE** *Gallinago stenura*
(a-c) **Adult**: From Common by relatively short bill, supercilium always broader than dark eyestripe at bill-base, thin pale brown to whitish scapular edges, tail only projects slightly beyond closed wing-tips, primaries only project slightly beyond tertials. In flight toes extend almost entirely beyond tail-tip, upperwing-coverts show contrasting pale panel, whitish trailing edge to secondaries very indistinct, underwing-coverts densely dark-barred, very little white on uppertail corners.
(d) **Juvenile**: Narrower pale fringing on upperparts and wing-coverts.

6 [277] **SWINHOE'S SNIPE** *Gallinago megala*
(a-c) **Adult**: Like Pintail Snipe but slightly larger, bill longer, head somewhat larger and squarer, eye set further back, crown peak more obviously behind eye (not apparent on plate), primaries may clearly extend beyond tertials, tail projects more beyond primaries, legs often rather yellowish. In flight heavier, more barrel-chested, wings slightly longer and more pointed, toes extending less beyond tail, more white on tail corners but less than Common (not apparent on plate).
(d) **Juvenile**: Mantle and scapulars more narrowly fringed; tertials and wing-coverts sharply fringed whitish-buff.

7 [278] **GREAT SNIPE** *Gallinago media*
(a-c) **Adult**: Relatively large, bulky and short-billed; bold white tips to wing-coverts, relatively indistinct pale lines on upperparts (not apparent on plate), lower breast to vent mostly heavily dark-barred. In flight, greater coverts distinctly dark and narrowly bordered white at front and rear, narrow white trailing edge to secondaries, underwing-coverts more extensively dark-barred than Common, extensive unmarked white on tail corners.
(d) **Juvenile**: Narrow upperpart streaks, less white on wing-coverts, white of tail has some brown barring.

8 [279] **COMMON SNIPE** *Gallinago gallinago*
(a-c) **Adult** *gallinago*: Relatively long bill, dark eyestripe broader than pale supercilium at bill-base, pronounced pale buff to buffish-white lengthwise stripes on upperparts, largely white belly and vent, no obvious white tips to wing-coverts, tail clearly projects beyond closed wings. In flight, prominent white trailing edge to secondaries, panel of unbarred white on underwing-coverts, toes project only slightly beyond tail-tip, little white on outertail, but more than Pintail and Swinhoe's (not apparent on plate).
(d) **Juvenile**: Mantle and scapulars with fine whitish lines, wing-coverts more narrowly fringed buffish-white.

1b
2c
2a
2b
1a
3c
4c
4b
3b
4a
3a
4d
5b
5c
6c
5d
6d
6b
5a
6a
8b
7d
7c
8c
8d
7b
7a
8a

PLATE 34. GODWITS, CURLEWS & RUFF

1 [282] **BLACK-TAILED GODWIT** *Limosa limosa*
(a,b) **Adult non-breeding** *melanuroides*: Fairly large, with long black legs, long bicoloured bill, long neck, greyish plumage, pale supercilium; white upperwing-bar and rump-band, mostly black tail in flight.
(c) **Male breeding**: Head-sides, neck and upper breast reddish-rufous, close dark barring on rest of underparts, bold blackish and chestnut markings on upperparts.
(d) **Female breeding**: Larger, longer-billed and duller.
(e) **Juvenile**: Like breeding female but more heavily marked above, wing-coverts fringed cinnamon-buff.

2 [283] **BAR-TAILED GODWIT** *Limosa lapponica*
(a,b) **Adult non-breeding** *lapponica*: Like non-breeding Black-tailed but more streaked and buffier above and below, slightly upturned bill, slightly shorter-legged. In flight lacks upperwing-bar, has white base to uppertail-coverts, white tail with close dark barring; underwing whitish.
(c) **Male breeding**: Mostly reddish-chestnut with dark-marked upperparts.
(d) **Female breeding**: Larger and longer-billed than male. Differs relatively little from non-breeding plumage but darker-marked above and with deep apricot flush on head-sides, neck and breast.
(e) **Juvenile**: Differs from non-breeding adult by brown-centred upperpart feathers with broad buff edges, fine dark streaks on neck and breast.
(f) **Adult non-breeding** *baueri*: Dark back and rump, dark-barred underwing-coverts.

3 [286] **LITTLE CURLEW** *Numenius minutus*
(a,b) **Adult**: Like Whimbrel but much smaller (not much larger than Pacific Golden Plover) and finer-billed, sharper head pattern, buffier below. Underwing-coverts mostly buffish-brown. Legs yellowish to bluish-grey.

4 [287] **WHIMBREL** *Numenius phaeopus*
(a,b) **Adult** *phaeopus*: Fairly large; markedly down-kinked bill, cold greyish-brown above with whitish to pale buff mottling, prominent blackish lateral crown-stripes and eyestripe, broad whitish supercilium, buffy-white below with heavy dark streaks. In flight, dark upperside with clean white back and rump, mostly plain whitish underwing-coverts.
(c) **Juvenile**: Clear buff on scapulars and tertials, buffish breast.
(d) **Adult** *variegatus*: Back and rump concolorous with mantle, heavy dark barring on underwing-coverts.

5 [288] **EURASIAN CURLEW** *Numenius arquata*
(a,b) **Adult non-breeding** *orientalis*: Like Whimbrel but larger; longer and more downcurved bill, more uniform head, coarsely marked upperparts, more pronounced streaks below. In flight, strong contrast between outer and inner wing, contrasting white back and rump, largely white underwing-coverts.
(c) **Juvenile**: Buffier than adult, with shorter bill.

6 [289] **FAR EASTERN CURLEW** *Numenius madagascariensis*
(a,b) **Adult non-breeding**: Like Eurasian Curlew but larger and longer-billed, more uniform, and browner/buffier below. In flight shows rufescent-tinged greyish-brown back and rump and densely barred underwing.
(c) **Adult breeding**: Upperparts distinctly washed rufous, underparts warmer-tinged.

7 [314] **RUFF** *Philomachus pugnax*
(a) **Male non-breeding**: Mid-sized, longish neck (shown retracted), small head, shortish, slightly drooping bill. Upperside greyish-brown with pale buff to whitish fringing, underside mainly whitish with greyish wash and mottling on foreneck and upper breast. Longish, usually orange to yellowish legs.
(b) **Male breeding**: Broad loose 'ruff' varying from black or white to rufous-chestnut, with or without bars and streaks; naked face, pink bill.
(c) **Male breeding/non-breeding (transitional)**: Face feathered, bill bicoloured, patchy plumage.
(d) **Female non-breeding**: Notably smaller than male. In flight both sexes show broad-based wings, broad white sides to rump, narrowish white wing-bar, toes projecting beyond short tail.
(e) **Female breeding**: Greyish-brown with blackish markings on upperparts and wing-coverts, variable bold blackish markings on neck and breast.
(f) **Juvenile male**: Upperside like breeding female but with simpler warm buff to whitish fringing; head-sides to breast rather uniform buff.
(g) **Juvenile female**: Usually darker buff on neck and breast than juvenile male, and smaller.

1e
1d
1a
1c
1b
2e
2c
2a
2d
2b
2f
3b
3a
5c
4c
6c
4d
4b
5a
6a
4a
5b
6b
7b
7a
7g
7e
7d
7c
7f

PLATE 35. DOWITCHERS & SANDPIPERS

1 [284] **LONG-BILLED DOWITCHER** *Limnodromus scolopaceus*
(a,b) **Adult non-breeding**: Similar to Asian but smaller, basal half of bill greenish, upperside somewhat plainer, neck and upper breast rather plain brownish-grey, legs shorter and paler. In flight shows unmarked white back, somewhat darker inner primaries and secondaries with narrow, defined white trailing edge and finely dark-barred underwing-coverts.
(c) **Adult breeding**: Similar to Asian but upperparts boldly mottled rather than streaked, supercilium and sides of head noticeably paler, underparts paler with dark speckles and bars. Bill and leg colour similar to adult non-breeding.
(d) **Juvenile**: Resembles adult non-breeding but mantle and scapulars dark brown with fine chestnut fringes, sides of head and breast washed buff.

2 [285] **ASIAN DOWITCHER** *Limnodromus semipalmatus*
(a,b) **Adult non-breeding**: Similar to Bar-tailed Godwit (subspecies *baueri*) but smaller, shorter-necked and -legged, flattish forehead and straight all-black bill slightly swollen at tip. In flight, upperside of secondaries, greater coverts and inner primaries somewhat paler than rest of wing, underwing-coverts white.
(c) **Male breeding**: From Bar-tailed Godwit by largely white vent, size, shape and bill.
(d) **Female breeding**: Duller than male.
(e) **Juvenile**: Like non-breeding adult but upperparts blacker with neat pale buff fringes, neck and breast washed buff and lightly dark-streaked.

3 [290] **TEREK SANDPIPER** *Xenus cinereus*
(a,b) **Adult non-breeding**: Smallish, with upturned yellow-based bill, shortish orange-yellow legs. In flight rather uniform above, with white trailing edge to secondaries.
(c) **Adult breeding**: Much clearer grey above with black lines on scapulars, bill all dark.
(d) **Juvenile**: Like non-breeding adult but browner above with buffish fringes, short blackish lines on scapulars.

4 [291] **COMMON SANDPIPER** *Actitis hypoleucos*
(a,b) **Adult non-breeding**: Fairly small, plain brownish upperside and lateral breast-patches, otherwise white below. Tail extends beyond closed wing-tips, legs greyish- to yellowish. In flight shows all-dark upperparts and white wing-bar.
(c) **Adult breeding**: Faint dark streaks and bars on upperparts, lateral breast-patches browner with dark streaks.
(d) **Juvenile**: Like non-breeding adult but upperparts narrowly fringed buff with some darker subterminal markings, wing-coverts with prominent buff tips and subdued dark barring.

5 [292] **GREEN SANDPIPER** *Tringa ochropus*
(a,b) **Adult non-breeding**: Smallish and quite short-legged; blackish olive-brown above and on breast-sides with buff speckling, white rump, uppertail-coverts and underparts. Plainer and darker than Wood, with greener legs, no post-ocular supercilium, distinct tail-bars. Larger and darker than Common, with more extensively dark breast, different pattern to upperside in flight.
(c) **Adult breeding**: Bold streaks on crown, neck and breast, more distinct white speckles on upperparts.

6 [293] **GREY-TAILED TATTLER** *Tringa brevipes*
(a,b) **Adult non-breeding**: Plain grey above and on upper breast, with white supercilium, rather stout straight yellowish-based bill and shortish yellow legs. In flight, grey above with darker outer wing.
(c) **Adult breeding**: Sides of throat and neck streaked grey, breast and flanks scaled grey, bill-base duller.

7 [298] **WOOD SANDPIPER** *Tringa glareola*
(a,b) **Adult non-breeding**: Smallish, with dark brown, buffish-speckled upperparts, whitish supercilium, lightly streaked breast, pale yellowish legs. In flight quite uniform above, with darker flight feathers, white rump, weakly marked tail.
(c) **Adult breeding**: Upperparts more blackish with much bolder whitish speckles and fringes, more clearly streaked on head, neck and breast.

8 [315] **RUDDY TURNSTONE** *Arenaria interpres*
(a,b) **Adult non-breeding** *interpres*: Brownish above, short stout bill, orange-red legs, complex blackish pattern on lower head-sides and breast; complex white markings on upperside in flight.
(c) **Male breeding**: Head patterned black and white, upperparts boldly patterned blackish and orange-chestnut.
(d) **Female breeding**: Crown washed brown.
(e) **Juvenile**: Like non-breeding adult but head and breast markings duller, upperparts fringed buffish, legs duller.

1-2 to different scale
1b
1c
1a
1d
2d
2a
2b
2c
2e
3d
3b
4d
3a
4c
4a
4b
3c
6a
5c
5a
6c
6b
5b
7b
8b
8d
8c
7c
7a
8e
8a

PLATE 36. LARGER *TRINGA* SANDPIPERS & PHALAROPES

1 [294] **SPOTTED REDSHANK** *Tringa erythropus*
(a,b) **Adult non-breeding**: Resembles Common Redshank but paler overall, bill longer and slenderer, more distinct supercilium, unstreaked underparts, longer legs. No white in wing.
(c) **Adult breeding**: Blackish overall, almost uniform on head and underparts.
(d) **Juvenile**: Recalls non-breeding adult but brownish-grey overall, finely white-speckled upperparts, dark-streaked neck, closely dark-barred breast to undertail-coverts.

2 [295] **COMMON GREENSHANK** *Tringa nebularia*
(a,b) **Adult non-breeding**: Fairly large with stoutish, slightly upturned dark-tipped bill, long greenish legs. Prominent streaks on crown, hindneck, mantle and breast-sides. In flight dark bars conspicuous on white uppertail-coverts and tail; toes extend beyond tail-tip.
(c) **Adult breeding**: Some scapulars have prominent blackish centres; crown, neck and breast heavily streaked and spotted blackish.

3 [296] **NORDMANN'S GREENSHANK** *Tringa guttifer*
(a,b) **Adult non-breeding**: Like Common Greenshank but legs shorter and yellower, neck shorter, bill distinctly bicoloured, crown to breast only faintly streaked, upperside plainer, more white above eye, paler lores. In flight shows all-white uppertail-coverts and greyish tail; toes do not extend beyond tail-tip.
(c) **Adult breeding**: Very boldly marked: upperside blackish with whitish spangling, head and upper neck dark-streaked, with broad spots on lower neck and breast.
(d) **Juvenile**: Like non-breeding adult but crown and upperparts tinged pale brown, whitish notching on scapulars, wing-coverts fringed buff, breast washed brown, sides faintly streaked.

4 [297] **MARSH SANDPIPER** *Tringa stagnatilis*
(a,b) **Adult non-breeding**: Resembles Common Greenshank but smaller, slimmer, with thin straight blackish bill and lankier legs, broad white supercilium. Upperside greyish, underside white, legs greenish.
(c) **Adult breeding**: Upperside boldly patterned black, with dark speckles and streaks on crown, neck and breast, arrow-shapes along flanks; legs often yellowish.

5 [299] **COMMON REDSHANK** *Tringa totanus*
(a,b) **Adult non-breeding** *eurhinus*: Plain brownish-grey above, whitish below with fine dark breast-streaks, straight red-based bill, bright red legs. In flight shows white secondaries and white-tipped inner primaries.
(c) **Adult breeding**: Browner above with small black markings, more heavily marked below.
(d) **Juvenile**: Like breeding adult but has neat pale buffish spotting and spangling on upperparts and wing-coverts, often more yellowish-orange legs.

6 [280] **RED-NECKED PHALAROPE** *Phalaropus lobatus*
(a,b) **Adult non-breeding**: Small size, needle-like bill, blackish mask, grey above, white below; upperwing blackish with narrow white wing-bar.
(c) **Female breeding**: Head to breast slate-grey with rufous-chestnut band from ear-coverts to foreneck.
(d) **Male breeding**: Like breeding female but duller.
(e) **Juvenile**: Like non-breeding adult but blacker mask and upperside, latter with rufous-buff markings.

7 [281] **RED PHALAROPE** *Phalaropus fulicarius*
(a,b) **Adult non-breeding**: Somewhat larger/heavier than Red-necked; shorter/thicker bill, whiter crown, paler and plainer upperside. Bill dark with yellowish to yellowish-brown tinge. Legs and feet proportionately very short, greyish to brownish. Wings proportionately larger than Red-necked, with more pronounced white bar.
(c) **Female breeding**: Blackish crown/face, white head-sides, chestnut-red neck/underparts. Otherwise blackish-brown above with mainly buff to rufous edgings. Bill rich yellow with black tip, legs and feet yellowish-brown.
(d) **Male breeding**: Duller than female; pale-streaked crown, buff-washed head-sides, often much white on belly.
(e) **Juvenile/first winter**: Less pronounced ochre-buff 'mantle-V' than Red-necked, lacks obvious scapular-V, has weaker pinkish-buff wash on face and neck. Acquires first pale grey scapulars of first winter plumage rather quickly.

1c
5d
1d
1a
5a
1b
2c
5c
5b
2a
2b
3a
4c
4b
3b
3c
4a
3d
7d
7c
6c
6d
6b
7b
7a
6e
6a
7e

PLATE 37. SMALL SANDPIPERS

1 [302] **SANDERLING** *Calidris alba*
(a,b) **Adult non-breeding**: Small; shortish black bill, grey above (often a dark area at bend of wing) and white below, legs blackish. Prominent white upperwing-bar.
(c) **Adult breeding**: Like large Red-necked Stint with dark streaks on duller chestnut sides of head, throat and breast, mostly chestnut centres of scapulars, broader wing-bar.
(d) **Adult breeding variant**: Fresh individual has faint chestnut on head and breast, hence sharper breast markings.
(e) **Juvenile**: Like non-breeding adult but has darker streaks and mottling above, bold black and buffish-white patterned scapulars.

2 [303] **SPOON-BILLED SANDPIPER** *Calidris pygmeus*
(a,b) **Adult non-breeding**: Like Red-necked and Little Stints but with spatulate bill, bigger head, whiter forehead and breast, broader supercilium.
(c) **Adult breeding**: Like Red-necked Stint but bill distinctive.
(d) **Juvenile**: Like Red-necked and Little Stints but bill distinctive, whiter forehead and face, darker lores and ear-coverts, more uniform upperpart fringing.

3 [304] **LITTLE STINT** *Calidris minuta*
(a,b) **Adult non-breeding**: Like Red-necked Stint but slightly slimmer, longer-legged, somewhat finer bill, broader dark centres to upperpart feathers, more streaked on head and breast, sometimes a more diffuse greyish breast-band. Legs blackish.
(c) **Adult breeding**: Like Red-necked but chin and throat whitish, more extensive but less intense rufous on breast.
(d) **Juvenile**: Like Red-necked but head usually more contrastingly patterned, darker-centred and more rufous-edged upperpart feathers, more coarsely streaked breast-sides, usually prominent mantle-lines and often split supercilium.

4 [305] **RED-NECKED STINT** *Calidris ruficollis*
(a,b) **Adult non-breeding**: Small, with shortish black bill, greyish above, white below with greyish breast-patches. Supercilium whitish, legs blackish.
(c) **Adult breeding**: Variable. Ear-coverts to upper breast unstreaked rufous to brick-red.
(d) **Adult pre-breeding**: Fresh-plumaged bird appears like very washed-out breeding adult.
(e) **Juvenile**: Crown and upperparts darker than non-breeding adult with whitish to warm fringing, breast-sides washed pinkish-grey and faintly streaked; sometimes faint mantle-lines.

5 [306] **TEMMINCK'S STINT** *Calidris temminckii*
(a,b) **Adult non-breeding**: Recalls miniature Common Sandpiper: uniform greyish-brown upperside, drab breast and greenish-yellow legs, tail usually projecting beyond wing-tips, and with white sides.
(c) **Adult breeding**: Upperparts olive-brown patterned with blackish patches and rufous edges, sides of head and breast washed brown and indistinctly streaked.
(d) **Juvenile**: Like non-breeding adult but browner above with buff fringes and blackish subterminal markings.

6 [307] **LONG-TOED STINT** *Calidris subminuta*
(a,b) **Adult non-breeding**: Like Red-necked and Little but neck longer, bill finer (lower mandible pale-based), upperparts browner with larger dark markings, neck and breast washed brown and darker-streaked, legs yellowish-brown to greenish.
(c) **Adult breeding**: Dark-streaked rufous crown, buffy neck-sides and breast distinctly streaked; upperparts and tertials broadly fringed rufous.
(d) **Juvenile**: Like breeding adult but slight 'split supercilium', prominent white mantle-lines, greyer lower scapulars.

7 [313] **BROAD-BILLED SANDPIPER** *Limicola falcinellus*
(a,b) **Adult non-breeding** *sibirica*: Stint-like but larger, bill longer, tip kinked down. Grey above, white 'split supercilium', white below with lateral breast-streaks; dark leading edge to upperwing-coverts.
(c) **Adult breeding**: Fresh plumage (May). Crown blackish, white 'split supercilium', upperparts blackish with rufous and whitish fringing, white lines on mantle and scapulars, boldly streaked neck and breast.
(d) **Juvenile**: Like breeding adult but mantle-lines more prominent, wing-coverts broadly fringed buff.

1c
2b
2c
1e
1a
2a
1d
2d
1b
4b
4e
3b
3c
4c
4a
3a
3d
4d
5b
5a
5c
6b
5d
7c
6a
6d
7b
6c
7d
7a

PLATE 38. MEDIUM-SIZED SANDPIPERS

1 [300] **GREAT KNOT** *Calidris tenuirostris*

(a,b) **Adult non-breeding**: Medium size, attenuated shape, broad-based, slightly downward-tapering blackish bill, grey upperside, streaked greyish head to upper breast, blackish spots on sides of breast and upper flanks. In flight shows blackish primary coverts, white uppertail-coverts contrast with dark tail.
(c) **Adult breeding**: Scapulars bright chestnut and black, head and neck with bold black streaks, breast and flanks with dense black spots.
(d) **Juvenile**: Like adult non-breeding but more strongly streaked blackish above, scapulars blackish-brown with whitish to buffish fringes, breast washed buffish and more distinctly dark-spotted/streaked.

2 [301] **RED KNOT** *Calidris canutus*

(a,b) **Adult non-breeding** *canutus*: Smaller and more compact than Great, shorter neck and bill, no black spots on breast; uniform dark scales/bars on rump, greyer tail.
(c) **Adult breeding**: Blackish and chestnut above, face and underparts deep reddish-chestnut.
(d) **Adult post-breeding**: Head and neck turn greyish.
(e) **Juvenile**: Like adult non-breeding but slightly buffier with scaly pattern above, finely dark-streaked below.

3 [308] **PECTORAL SANDPIPER** *Calidris melanotos*

(a,b) **Adult non-breeding**: Like Sharp-tailed but head pattern less striking, dark streaks on foreneck and breast sharply demarcated from white belly.
(c) **Male breeding**: Foreneck and breast may be blackish-brown with whitish mottling.
(d) **Female breeding**: Upperparts blackish-brown with rufous-buff fringes, foreneck and breast washed buffish.
(e) **Juvenile**: Similar to female breeding but usually has clear white lengthwise lines on mantle and scapulars, wing-coverts neatly fringed buffish to whitish-buff. From Sharp-tailed primarily by less distinct supercilium, duller sides of head and neck, duller breast with much more extensive and bolder dark streaks.

4 [309] **SHARP-TAILED SANDPIPER** *Calidris acuminata*

(a,b) **Adult non-breeding**: Medium-length, slightly down-tapering dark bill, rich brown crown, whitish supercilium, diffusely streaked neck and breast. Recalls much smaller Long-toed Stint.
(c) **Adult breeding**: Crown rufous with dark streaks, upperparts blackish-brown with mostly bright rufous fringes, distinct streaks on neck and upper breast become bold arrow-head markings on lower underparts.
(d) **Juvenile**: Like breeding adult but supercilium and throat plainer and whiter, breast lightly streaked rich buff, rest of underparts white.

5 [310] **DUNLIN** *Calidris alpina*

(a,b) **Adult non-breeding** *sakhalina*: Shorted-billed, -necked and -legged than Curlew Sandpiper, bill also less decurved, supercilium less distinct, foreneck and upper breast duller with fine streaking; dark centre to rump and uppertail-coverts. From Broad-billed by larger size, longer legs, no distinct kink in bill or 'split supercilium'.
(c) **Male breeding**: Large black belly-patch, mantle and scapulars mostly bright rufous-chestnut.
(d) **Juvenile**: Mantle and scapulars blackish with rufous, buff and whitish fringing, extensive blackish streaks on hindneck, breast and belly.

6 [311] **CURLEW SANDPIPER** *Calidris ferruginea*

(a,b) **Adult non-breeding**: Smallish to mid-sized, rather long downcurved blackish bill, fairly long blackish legs, rather plain greyish upperside, white supercilium and underparts (greyish wash on breast), narrow white wing-bar and white lower rump and uppertail-coverts.
(c) **Female breeding**: Head and underparts deep reddish-chestnut, mantle and scapulars boldly patterned chestnut, black and whitish.
(d) **Juvenile**: Like non-breeding adult but upperparts browner with buff-and-dark scaly pattern, head and breast washed peachy-buff and faintly dark-streaked.

7 [312] **STILT SANDPIPER** *Micropalama himantopus*

(a,b) **Adult non-breeding**: From Curlew Sandpiper by longer, straighter, less pointed bill (slightly downturned at tip), longer dull yellowish to greenish legs, greyish-streaked foreneck, breast and flanks. White lower rump to uppertail-coverts but almost no wing-bar, feet project prominently beyond tail-tip.
(c) **Adult breeding**: Bright chestnut on crown/head-sides, bold blackish streaks/spots on neck and upper breast, blackish-barred lower breast to vent, buffish-pink wash below, blackish mantle/scapulars with bold rufous, pinkish and white fringes; dark markings partly obscure white of rump/uppertail-coverts.
(d) **Juvenile**: From Curlew Sandpiper by structure, darker feather-centres above, rufescent upper scapular fringes, lightly streaked flanks, more contrasting head pattern.

1c
1d
2c
2b
1b
2d
1a
2e
3b
3c
3e
2a
3d
3a
5b
4b
5c
5d
4a
4c
5a
4d
7a
7b
6b
6a
6d
6c
7c
7d

PLATE 39. NODDIES & TERNS

1 [322] **BROWN NODDY** *Anous stolidus*
(a,b) **Adult** *pileatus*: Dark chocolate-brown plumage with whitish forehead, grey crown, broken white eyering, long wedge-shaped tail; paler brownish band across upperwing-coverts.
(c) **Juvenile**: Crown, upperparts and wing-coverts mostly with indistinct pale buffish fringing, forecrown browner.

2 [323] **BLACK NODDY** *Anous minutus*
(a,b) **Adult** *worcesteri*: Smaller and slimmer than Brown Noddy with narrower, longer bill (longer than head), blacker plumage with whiter crown, shorter tail; uniform dark upperwing.
(c) **Juvenile**: Smaller and longer-billed than juvenile Brown with narrower brownish-buff fringing above, whiter forehead and crown.

3 [324] **WHITE TERN** *Gygis alba*
(a,b) **Adult** *monte*: Small and white with blackish bill, eyering and legs.
(c) **Juvenile**: Like adult but with dark flecks on crown and nape, brown and buff bars/scales on upperparts.

4 [325] **SOOTY TERN** *Onychoprion fuscatus*
(a) **Adult non-breeding** *nubilosa*: Like Bridled Tern but larger, blacker and more uniform above, no whitish eyebrow.
(b-d) **Adult breeding**: From Bridled by concolorous (and continuous) blackish crown, hindneck and upperside, broader square-cut white forehead-patch, blackish tail with more contrasting white edge, darker underside of flight feathers.
(e) **Juvenile/first winter**: Mostly sooty-blackish, with contrasting whitish vent, feather tips on upperside.

5 [326] **BRIDLED TERN** *Onychoprion anaethetus*
(a) **Adult non-breeding** *anaethetus*: Brownish-grey upperside with uneven paler feather tips, whitish forehead-patch and blackish crown, nape and mask.
(b-d) **Adult breeding**: Clean white forehead and short eyebrow, black eyestripe, crown and nape, plain brownish-grey upperside; mostly whitish underwing with darker trailing edge.
(e) **Juvenile**: Whitish tips and dark subterminal bars on upperparts, head as non-breeding adult.

6 [328] **LITTLE TERN** *Sternula albifrons*
(a,b) **Adult non-breeding** *sinensis*: Small with longish dark bill and short dark legs, hindcrown, rear eyestripe and nape blackish; grey above with narrow dark band on leading edge of lesser coverts and blackish outermost primaries.
(c,d) **Adult breeding**: Bill yellow with black tip, legs orange to yellow, defined white forehead-patch.
(e) **Juvenile**: Like non-breeding adult but with dark subterminal markings on upperparts and wing-coverts. Dark leading edge to upperwing.

7 [331] **BLACK TERN** *Chlidonias niger*
(a,b) **Adult non-breeding/first winter** *niger*: Like White-winged but more solidly black hindcrown, larger ear-patch, dark smudge on breast-side, darker grey above, including rump and uppertail-coverts.
(c) **Adult breeding**: Rather uniform greyish; darker grey body, blackish hood, white vent, blackish legs; pale grey underwing-coverts.
(d) **Juvenile**: Differs from White-winged by greyer, more clearly barred/scaled 'saddle' (contrasts less with darker grey upperwing), grey rump and uppertail-coverts, prominent dark smudge on breast-side.

8 [332] **WHITE-WINGED TERN** *Chlidonias leucopterus*
(a,b) **Adult non-breeding**: Like Whiskered but smaller, finer bill and roundish ear-patch ('headphones'); white rump and uppertail-coverts, darker outer primaries, dark bands across lesser coverts and secondaries.
(c) **Adult breeding**: Black head, body and underwing-coverts, whitish upperwing-coverts, rump and tail-coverts.
(d) **Adult non-breeding/breeding (transitional)**: Patchy plumage.
(e) **Juvenile**: Like Whiskered but with white rump and uppertail-coverts, darker and more uniform 'saddle', darker lesser coverts and secondaries.

9 [333] **WHISKERED TERN** *Chlidonias hybrida*
(a,b) **Adult non-breeding** *javanicus*: Rather small, compact, with blackish bill, mask and hindcrown/nape, white crown (dark-streaked at rear), dark reddish legs (less dark than illustrated), shortish shallow-forked tail; relatively short, broad wings, grey above with darker secondaries and outer primaries.
(c) **Adult breeding**: Dark red bill and legs, black crown and nape, white throat and lower head-sides, dark grey underparts with white vent.
(d) **Juvenile**: Like non-breeding adult but 'saddle' fawn brown barred/scaled blackish-and-buff, forecrown and face washed brownish-buff.

1-5 to different scale
1a
1b
1c
2a
2b
2c
3a
3b
3c
4a
4b
4c
4d
4e
5a
5b
5c
5d
5e
6a
6b
6c
6d
6e
7a
7b
7c
7d
8a
8b
8c
8d
8e
9a
9b
9c
9d

PLATE 40. ALEUTIAN & *STERNA* TERNS

1 [327] **ALEUTIAN TERN** *Onychoprion aleuticus*
(a,b) **Adult non-breeding**: Whitish crown, dark band along underside of secondaries.
(c,d) **Adult breeding**: Blackish bill and legs, white forehead and cheeks, grey below; broad white edge to inner wing, pale inner primaries, dark secondary band on underwing.
(e) **Juvenile**: More uniform warm brown forehead than Common, mostly dark brown above with buff tips.

2 [334] **RIVER TERN** *Sterna aurantia*
(a) **Adult non-breeding**: Medium-sized with dark-tipped yellow bill, greyish dark-streaked crown, black mask and nape, reddish legs; greyish-white below.
(b) **Adult breeding**: Bill orange-yellow, forehead to nape and mask black, upperwing all grey, long streamer-like outertail feathers.
(c) **Juvenile**: Dark-tipped yellow bill, blackish mask and streaks on head, whitish supercilium, blackish-brown fringing above, blackish-tipped primaries.

3 [335] **ROSEATE TERN** *Sterna dougallii*
(a,b) **Adult non-breeding** *bangsi*: Very like Common but slimmer, paler grey above, tail longer, bill longer and slenderer, defined white line along inner edge of closed wing; in flight shows paler upperparts, whiter secondaries and less dark primaries, underwing only faintly dark on primary tips.
(c,d) **Adult breeding**: Bill black with red on basal half only, underparts often pinkish, tail-streamers very long; from Common by paler, plainer upperwing, darker greyish outermost primaries, no dark wedge on mid-primaries.
(e) **Juvenile**: From Common by all-blackish bill, darker forecrown, bolder subterminal markings on upperparts.
(f) **First winter/summer**: Like non-breeding adult, but in flight upperwing shows dark bands across lesser coverts and secondaries, darker leading edge of outer wing.
(g) **Adult breeding** *korustes*: Less black on bill.

4 [336] **BLACK-NAPED TERN** *Sterna sumatrana*
(a) **Adult breeding** *sumatrana*: Crown all white, sharply defined nape-band; blackish bill and legs.
(b) **Adult non-breeding**: Some dark streaks on hindcrown, uniform very pale grey upperwing with blackish outer edge of outermost primary.
(c) **Juvenile**: White forehead, dark streaks on crown, black band from eye to nape, blackish subterminal markings on upperparts, dark-centred tail feathers, darker grey flight feathers.

5 [337] **COMMON TERN** *Sterna hirundo*
(a,b) **Adult non-breeding** *tibetana*: Medium-sized with fairly slender blackish bill, dark red legs (too black on plate), white forehead and lores, blackish mask and nape and medium grey upperparts; darkish grey of upperwing contrasts with white rump and uppertail-coverts, underwing shows relatively extensive dark tips to outer primaries.
(c,d) **Adult breeding**: Bill orange-red with black tip, forehead to nape black, longer tail-streamers; dark wedge on mid-primaries, underwing with dark trailing edge to outer primaries.
(e) **Juvenile (late)**: Head as non-breeding adult, upperparts with dark brown subterminal markings and, initially, buffish fringes (mostly lacking on illustrated bird); upperwing shows pronounced blackish band at leading edge of lesser coverts.
(f) **First winter/summer**: Like non-breeding adult but upperwing has somewhat bolder dark bands across lesser coverts and secondaries.
(g) **Adult breeding** *longipennis*: Mostly black bill, greyer below.

6 [338] **ARCTIC TERN** *Sterna paradisaea*
(a) **Adult breeding**: Shortish dark red bill, shorter neck than Common, very short legs; flight feathers near-white (apart from black-tipped outer primaries), translucent from below; tail-streamers longer than Common (extend a little beyond wing-tips at rest, rather than roughly equal).
(b,c) **Juvenile/first winter**: From Common by less patterned upperwing, whitish secondaries/inner primaries; bill all-black from Aug/Sep.
(d) **First summer**: Darker leading edge to upperwing than non-breeding adult.

7 [339] **BLACK-BELLIED TERN** *Sterna acuticauda*
(a) **Adult non-breeding**: Overlaps habitat with River Tern, differing by smaller size and slenderer, dark-tipped orange bill, sometimes mottled blackish towards vent.
(b) **Adult breeding**: Black crown and nape, orange bill, grey breast, blackish belly and vent, whitish head-sides and throat, deeply forked tail.
(c) **Juvenile**: From River by size, bill (similar to non-breeding adult), lack of whitish supercilium, whiter sides of head and neck.

1c
1a
2b
1b
1d
2c
1e
2a
3a
3c
4c
4a
3g
3e
4b
3f
3b
3d
5a
5f
5e
5c
5d
5b
5g
6b
6c
7a
7c
6d
6a
7b

PLATE 41. INDIAN SKIMMER & LARGER TERNS

1 [321] **INDIAN SKIMMER** *Rynchops albicollis*
(a,b) **Adult breeding**: Black crown, nape and upperside, white forehead, collar and underparts, long, yellow-tipped deep orange bill with longer lower mandible; white trailing edge to black upperwing.
(c) **Juvenile**: Mostly greyish-brown above, whitish to pale buff fringing on upperparts and wing-coverts.

2 [329] **GULL-BILLED TERN** *Gelochelidon nilotica*
(a-c) **Adult non-breeding** *affinis*: Rather large; white head with dark mask, heavy blackish bill, slender wings, silvery-grey rump and uppertail and shallow tail-fork.
(d) **Adult breeding**: Forehead to nape black.
(e,f) **First winter**: Like non-breeding adult but secondaries and primary upperwing-coverts darker, tail dark-tipped.

3 [330] **CASPIAN TERN** *Hydroprogne caspia*
(a-c) **Adult non-breeding**: Huge with thick red bill.
(d) **Adult breeding**: Forehead to nape and mask all black.
(e) **First winter**: Like non-breeding adult but secondaries, primary upperwing-coverts and tail somewhat darker.

4 [340] **LESSER CRESTED TERN** *Thalasseus bengalensis*
(a-c) **Adult non-breeding** *bengalensis*: Like Great Crested but smaller, bill narrower and yellowish-orange, more solid black hindcrown and 'mane' on nape, paler grey on upperside; when worn, shows darker outer primaries and bar on secondaries.
(d) **Adult breeding**: Crown to nape black, including extreme forehead, bill more orange.
(e) **Juvenile**: From Great by size, bill, darker upperside of inner primaries.
(f) **First winter**: Greater/median upperwing-coverts plainer grey than Great (not illustrated), head as adult non-breeding.

5 [341] **GREAT CRESTED TERN** *Thalasseus bergii*
(a-c) **Adult non-breeding** *velox*: Largish, stocky; thick cold yellowish bill, white forecrown, blackish mask, blackish hindcrown and nape streaked with whitish, grey upperside; uniform wings with darker-tipped outer primaries (may appear patchy when worn).
(d) **Adult breeding**: Bill brighter yellow, extreme forehead white, crown to shaggy nape black.
(e) **Juvenile**: Bill duller than non-breeding adult, upperparts brownish-grey with whitish fringing, darker tail; four dark bands across upperwing-coverts and secondaries.
(f) **Second winter**: As non-breeding adult but retains juvenile outer primaries, primary coverts and secondaries.

6 [342] **CHINESE CRESTED TERN** *Thalasseus bernsteini*
(a-c) **Adult non-breeding**: Like Great Crested but smaller with blackish-tipped yellow bill, paler grey upperside; pale grey upperwing with blackish outer primaries.
(d) **Adult breeding**: Forehead to shaggy nape black; black-tipped yellow bill diagnostic.

1a
1b
1c
2a
2b
2c
2d
2e
2f
3a
3b
3c
3d
3e
3 not to scale
4a
4b
4c
4d
4e
4f
5a
5b
5c
5d
5e
5f
6a
6b
6c
6d

PLATE 42. JAEGERS & PELAGIC GULLS

1 [318] **POMARINE JAEGER** *Stercorarius pomarinus*
(a) **Adult pale morph non-breeding**: Similar to breeding but throat/neck dark-mottled, tail-coverts barred whitish.
(b) **Adult pale morph breeding**: Rather large, heavily built; blackish-brown with broad whitish collar, neck washed yellowish, large whitish patch on underparts, long broad twisted central tail feathers; upperwing with whitish shaft-streaks at base of primaries, underwing with white crescent at base of primaries.
(c) **Juvenile pale morph**: Dark face, prominently pale-barred uppertail-coverts, distinctive double pale patch on underside of primaries.
(d) **Second-winter pale morph**: Similar to non-breeding adult but underwing like juvenile.
(e) **Juvenile dark morph**: Rather uniform blackish-brown, pale-barred tail-coverts; distinctive double pale patch on underside of primaries.

2 [319] **PARASITIC JAEGER** *Stercorarius parasiticus*
(a) **Adult pale morph non-breeding**: Cap less neat than breeding adult, mottled breast-band, barred tail-coverts.
(b) **Adult pale morph breeding**: Similar to Pomarine but smaller, slimmer, with pointed tail-streamers.
(c) **Juvenile pale morph**: From Pomarine by warmer plumage; rarely shows double pale patch on underside of primaries.
(d) **Second-winter pale morph**: From non-breeding adult as Pomarine but lacks double pale underwing-patch.
(e) **Juvenile dark morph**: From Pomarine by all-dark underwing-coverts (including primary coverts).

3 [320] **LONG-TAILED JAEGER** *Stercorarius longicaudus*
(a) **Adult non-breeding** *pallescens*: Tail-streamers short or lacking; from Pomarine and Parasitic by all-dark underwing, little white on upperside of primaries.
(b) **Adult breeding**: Slighter and more tern-like than other jaegers, very long tail-streamers, two-tone upperwing, little white on upperside of primaries, all-dark underwing, underparts greyer towards vent.
(c) **Juvenile pale morph**: From Parasitic by more attenuated rear end, generally greyer and more contrasting pattern on underparts and underwing, less white on primaries.
(d) **Juvenile dark morph**: From Parasitic by boldly barred tail-coverts and axillaries, less white on primaries.
(e) **Second winter**: Approaching non-breeding adult but underwing somewhat barred.

4 [343] **BLACK-TAILED GULL** *Larus crassirostris*
(a,b) **Adult non-breeding**: Medium-sized, dark grey upperside, broad black subterminal tail-band, greyish-streaked hindcrown and nape.
(c) **Adult breeding**: All-white head and neck.
(d,e) **First winter**: Dark-tipped pinkish bill, grey-brown body, blackish tail with narrow white terminal band.
(f) **Second winter**: Like non-breeding adult but brown-tinged on mantle and upperwing.

5 [349] **LAUGHING GULL** *Larus atricilla*
(a,b) **Adult non-breeding** *megalopterus?*: Told by size, relatively dark grey upperside, longish dark bill (tip often reddish), reddish-black legs, dark greyish smudged ear-coverts and hindcrown (no defined ear-spot). Uniform grey upperwing with black tips and bold white trailing edge to secondaries and inner primaries.
(c) **Adult breeding**: Black hood (including nape) with broken white eyering, dark red bill and legs (former usually with black subterminal band).
(d,e) **First winter**: Head, mantle, scapulars and underparts similar to adult non-breeding, but dusky nape, breast-band and flanks; blackish bill and legs. Mostly dark greyish-brown centres to wing-coverts and tertials. Mostly blackish flight feathers, with similar trailing edge to adult, dark markings across mid-underwing; tail with broad black subterminal band and dark greyish sides.
(f) **Second winter**: Like adult non-breeding but greyer nape to breast-sides/flanks, some dark markings on primary coverts, may show faint suggestion of tail-band.

6 [356] **BLACK-LEGGED KITTIWAKE** *Rissa tridactyla*
(a,b) **Adult non-breeding** *pollicaris*: Relatively dark grey upperparts and upperwing, grey nape, vertical blackish bar behind eye, yellowish bill, shortish dark brown to blackish legs (rarely tinged pinkish to reddish). Tail slightly notched. Rather narrow outer wing turns whitish before neat black tip.
(c) **Adult breeding**: All-white head.
(d,e) **First winter**: Differs from non-breeding adult by upperwing pattern, with broadly black outer primaries and black diagonal band across coverts, contrasting sharply with largely whitish secondaries and inner primaries. Black-tipped tail, black bill (may be slightly paler at base); head as non-breeding adult but may show black band across hindneck.

1a
1b
1c
1d
1e
2a
2b
2c
2d
2e
3a
3b
3c
3d
3e
1-3 to different scale
4a
4b
4c
4d
4e
4f
5a
5b
5c
5d
5e
5f
6a
6b
6c
6d
6e

PLATE 43. LARGER GULLS

1 [344] **MEW GULL** *Larus canus*
(a,b) **Adult non-breeding** *kamtschatschensis*: Recalls Heuglin's but smaller and smaller-billed, head rounder and heavily streaked, upperparts paler, wing-tip with larger white 'mirrors'.
(c) **Adult breeding**: Head all white.
(d,e) **First winter**: Black-tipped pinkish bill, pinkish legs, mostly plain grey upperparts, mostly plain brownish-grey greater upperwing-coverts, broad blackish subterminal tail-band.
(f) **Second winter**: Like non-breeding adult but upper primary coverts marked black, 'mirrors' smaller.

2 [345] **MONGOLIAN GULL** *Larus mongolicus*
(a,b) **Adult non-breeding**: Averages larger than Heuglin's, distinctly paler grey above; typically whiter-headed. Eyes usually look yellowish in field, legs yellowish-flesh to flesh/pink (rarely yellow). In February/March, already has white head of breeding plumage (while Heuglin's still shows head-streaking).
(c) **Adult breeding**: Head and neck all white.
(d,e) **First winter**: Best separated from Heuglin's by much paler overall coloration, and distinctly 'frosty' upperside; paler greater coverts. In flight, note paler inner primaries.
(f) **Second winter**: From Heuglin's by same features as first winter; additionally has paler grey on upperparts.

3 [346] **LESSER BLACK-BACKED GULL** *Larus fuscus*
(a,b) **Adult non-breeding** *fuscus*: Similar to Heuglin's Gull, but smaller, slimmer and slightly smaller-billed, with proportionately longer wings, upperparts almost jet-black, head and neck only weakly streaked; legs yellow.
(c) **Adult breeding**: White head and neck, bright yellow legs.
(d,e) **First winter**: Apart from size and build, somewhat darker overall than Heuglin's, mantle and scapulars with typically less contrasting dark 'anchor-marks' on centres; perhaps shows less clean white on head and underparts.
(f) **Second winter**: Upperpart coloration close to that of adult, but head and neck more heavily streaked than adult non-breeding.

4 [347] **HEUGLIN'S GULL** *Larus heuglini*
(a,b) **Adult non-breeding** *tamyrensis*: Similar to Mongolian but legs yellow, slightly darker grey above, more extensive black on primaries. Fairly light head/neck-streaking.
(c) **Adult breeding**: Head and neck all white.
(d,e) **First winter**: Best separated from Mongolian by darker greater upperwing-coverts, somewhat darker inner primaries.
(f) **Second winter**: Like first winter, but mantle, back and scapulars largely grey.

5 [348] **PALLAS'S GULL** *Larus ichthyaetus*
(a,b) **Adult non-breeding**: Large; yellowish bill with blackish subterminal band, mask of dark streaks, pale grey upperparts and little black on outer primaries.
(c) **Adult breeding**: Head black with broken white eyering.
(d,e) **First winter**: Like second-winter Heuglin's and Mongolian but has dark mask and hindcrown streaking, densely dark-marked lower hindneck and breast-sides, paler grey above, mostly grey, unbarred greater coverts, white rump to tail-base, broad blackish subterminal tail-band.
(f) **Second winter**: Like non-breeding adult, with remnants of dark markings on coverts, mostly black outer primaries, narrow subterminal tail-band.

1e
1c
2b
2d
1f
1b
1d
2a
2c
2f
2e
1a
3d
3c
3e
3f
3b
3a
5e
4f
4b
5c
4e
5f
4d
5b
5d
5a
4a
4c

PLATE 44. SMALLER GULLS

1 [350] **RELICT GULL** *Chroicocephalus relictus*
(a,b) **Adult non-breeding**: Recalls Brown-headed and Black-headed Gulls but larger, thicker-billed and longer-legged, with more uniformly dark-smudged ear-coverts and hindcrown, white tips to primaries; upperwing shows white-tipped primaries with prominent black subterminal markings, no white leading edge to outer wing.
(c) **Adult breeding**: Blackish hood (including nape), broad broken white eyering.
(d,e) **First winter**: Blackish bill (greyish base to lower mandible) and legs, mostly rather pale grey upperside with dark markings on nape, neck, breast-sides and wing-coverts; upperwing has solid black outer primaries, narrow markings on remaining flight feathers, no white on outer wing, narrow black subterminal tail-band.
(f) **Second winter**: Like non-breeding adult but bill blacker, tertials dark with white edges, legs darker.

2 [351] **BROWN-HEADED GULL** *Chroicocephalus brunnicephalus*
(a,b) **Adult non-breeding**: Like Black-headed but pale-eyed, slightly bulkier, thicker-billed and -necked; broader, more rounded, broadly black-tipped wings with enclosed white mirrors.
(c) **Adult breeding**: From Black-headed by shape, pale eye, paler face.
(d,e) **First winter**: Black outer and broad-tipped inner primaries (above and below), whitish patch extending across upper primary coverts and inner primaries.

3 [352] **BLACK-HEADED GULL** *Chroicocephalus ridibundus*
(a,b) **Adult non-breeding**: Small, slim, with narrow black-tipped red bill, dark red legs, pale grey above, white head with dark eyes, ear-spot and smudges on crown; white leading edge to outer wing and black-tipped outer primaries.
(c) **Adult breeding**: Dark brown hood, broken white eyering, uniform dark red bill.
(d,e) **First winter**: Resembles non-breeding adult but bill paler, legs duller and pinker, greyish-brown mottling on wing-coverts; broad greyish-brown band across upperwing-coverts, broad blackish band along flight feathers, blackish subterminal tail-band.

4 [353] **SLENDER-BILLED GULL** *Chroicocephalus genei*
(a,b) **Adult non-breeding**: Like Black-headed but longer-necked and -billed, with all-white head (sometimes with faint ear-spot), pale eyes.
(c) **Adult breeding**: Like non-breeding adult but often washed pink below, head always white.
(d,e) **First winter**: From Black-headed by shape, more orange bill with only slight dark tip, fainter head markings, pale eyes, longer, paler legs; less contrasting wing pattern.

5 [354] **SAUNDERS'S GULL** *Chroicocephalus saundersi*
(a,b) **Adult non-breeding**: Like Black-headed Gull but smaller with shorter, thicker, blackish bill, white tips to primaries.
(c) **Adult breeding**: Full black hood, broad broken white eyering.
(d,e) **First winter**: From Black-headed by size and bill; subtly different upperwing pattern with less black on tips of flight feathers, no white leading edge, narrower black tail-band.

6 [355] **LITTLE GULL** *Hydrocoloeus minutus*
(a,b) **Adult non-breeding**: Told by size and neat proportions, weak blackish bill, pinkish legs and feet. Dark head markings recall several other small gulls but has more capped appearance. Rounded white wing-tip and extensively blackish underwing.
(c) **Adult breeding**: Black hood, reddish-brown bill (looks black), scarlet legs and feet, pink-flushed underparts.
(d,e) **First winter**: Smaller than similar Black-legged Kittiwake; no black nuchal band, has faint darker subterminal band along secondaries and inner primaries (also visible on underwing, along with dark tertial mark); grey hindneck, slender blackish bill, dull flesh legs and feet.

1c
2c
2a
1e
2b
1b
1a
2d
1f
1d
2e
3b
4e
3e
4b
3a
4a
4d
3d
3c
4c
5c
6b
6e
5a
6c
5d
6d
5b
5e
6a

PLATE 45. *COLUMBA* PIGEONS, JAMBU FRUIT-DOVE, IMPERIAL-PIGEONS & NICOBAR PIGEON

1 [358] **ROCK PIGEON** *Columba livia*
(a,b) **Adult** *intermedia*: Mainly grey with darker hood and breast, paler upperwing (primaries dark) with two broad blackish bars, blackish tail-tip, green and purple neck gloss. Silvery-whitish underwing-coverts.
(c) **Juvenile**: Duller and browner; browner head, neck and breast, initially lacks neck gloss, wings mostly pale greyish-brown.
(d-f) **Adult feral variants**: Highly variable with patches of white and brown in plumage.

2 [359] **SNOW PIGEON** *Columba leuconota*
(a,b) **Adult** *gradaria*: Slaty-grey head contrasts sharply with white collar and underparts; white patch on lower back, blackish rump and uppertail-coverts, blackish tail with broad whitish central band, greyish wing-coverts with three dark bands.

3 [360] **SPECKLED WOODPIGEON** *Columba hodgsonii*
(a,b) **Male**: Pale grey head, neck and upper breast, dark maroon mantle, scapulars and lesser coverts, slaty remaining wing-coverts, bold whitish speckles on scapulars and wing-coverts; dark vent.
(c) **Female**: Darker grey head and breast, lacks maroon on mantle and scapulars; base colour of underparts dark brownish-grey; spotting on coverts duller.

4[361] **ASHY WOODPIGEON** *Columba pulchricollis*
(a,b) **Adult**: Grey head and broad buffish, black-barred neck-collar, dark slaty upperside and breast with greenish gloss on upper mantle and upper breast, pale belly and vent, all-dark tail, red legs.
(c) **Juvenile**: Buff of neck replaced by pale grey, with reduced black spotting/barring, no green gloss on upper breast and upper mantle, browner breast with thin dull rufous bars, rufous-tinged lower breast and belly, browner wings.

5 [362] **PALE-CAPPED PIGEON** *Columba punicea*
(a,b) **Male**: Relatively plain overall with whitish crown, purplish-maroon upperparts with greenish gloss on sides and back of neck, dark slate-coloured rump and uppertail-coverts, blackish tail and flight feathers, vinous-brown lower head-sides and underparts, slaty-grey undertail-coverts, red orbital skin and base to pale bill.
(c) **Female**: Greyish crown.
(d) **Juvenile**: Crown initially concolorous with mantle, duller wing-coverts and scapulars with rufous fringes, no neck gloss, greyer underparts.

6 [371] **NICOBAR PIGEON** *Caloenas nicobarica*
(a,b) **Adult** *nicobarica*: Robust, blackish-slate head, neck (including long hackles), breast and flight feathers, mostly coppery-green remainder of plumage, short white tail, blackish bill with short 'horn' near base.
(c) **Juvenile**: Duller and browner, with rather uniform dark greenish-brown head, mantle and underparts, no Neck-hackles and dark brownish-green tail.

7 [385] **JAMBU FRUIT-DOVE** *Ptilinopus jambu*
(a) **Male**: Crimson face and forecrown, white eyering, greenish upperparts, white underparts with pink flush on foreneck and upper breast, chestnut undertail-coverts, orange-yellow bill.
(b) **Female**: Mostly uniform green with purplish-grey face, white eyering and buffish undertail-coverts.
(c) **Juvenile**: Like female but has green crown, brownish face, whitish central throat (washed dull rufous); upperparts initially fringed warm brown.

8 [386] **GREEN IMPERIAL-PIGEON** *Ducula aenea*
(a,b) **Adult** *sylvatica*: Large; mostly dark metallic green upperparts with rufous-chestnut gloss, rather uniform vinous-tinged pale grey head, neck and underparts, dark chestnut undertail-coverts, all-dark tail.

9 [387] **MOUNTAIN IMPERIAL-PIGEON** *Ducula badia*
(a,b) **Adult** *griseicapilla*: Very large; extensively purplish-maroon mantle and wing-coverts, bluish-grey crown and face, white throat, vinous-tinged pale grey neck and underparts with paler, whitish (buffish) undertail-coverts, dark tail with broad greyish terminal band, reddish orbital skin, red bill with pale tip.
(c) **Juvenile**: Less pink on hindneck, rusty-brown fringes to mantle, wing-coverts and flight feathers.
(d) **Adult** *badia*: More extensively purplish-maroon upperparts, duller grey crown and face (contrasting less with hindneck), darker and more strongly pink-tinged underparts.

10 [388] **PIED IMPERIAL-PIGEON** *Ducula bicolor*
(a,b) **Adult** *bicolor*: All white with sharply contrasting black flight feathers and black tail with much white on outer feathers.

1a
1b
1c
1d
1e
1f
2a
2b
3a
3b
3c
4a
4b
4c
5a
5b
5c
5d
6a
6b
6c
7a
7b
7c
8a
8b
9a
9b
9c
9d
10a
10b

PLATE 46. DOVES

1 [363] **ORIENTAL TURTLE-DOVE** *Streptopelia orientalis*
(a,b) **Adult** *agricola*: Like Spotted Dove but larger, darker and shorter-tailed; broad rufous scales on mantle and wing-coverts, barred sides of neck, bluish-slate rump; in flight lacks pale bar across upperwing, outertail feathers dark.
(c) **Juvenile**: Narrower, paler fringing on mantle, scapulars and wing-coverts, pale-fringed breast feathers, indistinct neck-patch.
(d) **Adult** *orientalis*: Larger and greyer, less vinous.

2 [364] **EURASIAN COLLARED-DOVE** *Streptopelia decaocto*
(a,b) **Adult** *xanthocyclus*: Like female Red Collared-dove but larger and longer-tailed, with paler mantle, and wing-coverts, no slate-grey on rump and uppertail-coverts, paler, pinker-grey breast.
(c) **Juvenile**: Duller, browner crown, mantle and underparts, no neck-bar, indistinct buff fringes above.

3 [365] **RED COLLARED-DOVE** *Streptopelia tranquebarica*
(a) **Male** *humilis*: Relatively small and compact; pale bluish-grey head, black hindneck-band, brownish vinous-red body and wing-coverts, grey rump and uppertail-coverts.
(b,c) **Female**: Similar pattern to male but mostly brownish instead of reddish, less grey on head; in flight, square-cut tail has white corners (also male).
(d) **Juvenile**: Like female but crown tinged rufous, no neck-bar, distinct buffish fringes above.

4 [366] **SPOTTED DOVE** *Streptopelia chinensis*
(a,b) **Adult** *tigrina*: Greyish crown and head-sides, greyish-brown upperparts with dark streaks and light fringing, broad black half-collar with white spots, pale pinkish-brown below; in flight, long graduated tail shows broad white tips, broad pale bar across wing contrasts with blackish flight feathers.
(c) **Juvenile**: Much browner, almost no grey on crown and wing-coverts, vague brown collar with buffish-brown bars, buff fringes on breast.
(d) **Adult** *chinensis*: Bluer-grey crown, pinker breast and unstreaked upperparts.

5 [367] **BARRED CUCKOO-DOVE** *Macropygia unchall*
(a,b) **Male** *tusalia*: Slender and long-tailed, dark brown-and-rufous barring above (head plainer), buffish-brown
underparts with darker, vinous-tinged breast, violet and green gloss on nape and upper mantle.
(c) **Female**: Paler buffish below with dense blackish bars.
(d) **Juvenile**: Like female but darker, head and neck barred.

6 [368] **LITTLE CUCKOO-DOVE** *Macropygia ruficeps*
(a,b) **Male** *assimilis*: Smaller than Barred Cuckoo-dove with rufous-chestnut crown, no bars, rufous-buff underparts (extending onto underwing-coverts) with whitish breast-scales, chestnut scaling on wing-coverts.
(c) **Female**: Blackish mottling on breast, stronger chestnut fringes on wing-coverts.
(d) **Male** *malayana*: Darker, with some blackish mottling on breast and broader chestnut fringing above.
(e) **Female**: Heavier mottling on throat and breast.

7 [369] **EMERALD DOVE** *Chalcophaps indica*
(a,b) **Male** *indica*: Blue-grey crown and nape, white forehead and eyebrow, red bill, metallic green wings with white shoulder, vinous-pinkish face and underparts. Whitish double band on back/upper rump.
(c) **Female**: Reduced grey on head, no white in wing.
(d) **Juvenile**: Darker and extensively barred rufous-buff, almost no green above.

8 [370] **ZEBRA DOVE** *Geopelia striata*
(a,b) **Male**: Recalls Spotted Dove but much smaller, with black-fringed greyer upperparts, black-and-white bars on hindneck to flanks, blue-grey face.
(c) **Juvenile**: Brownish and buffish bars on crown, upperparts and wing-coverts, duller bars on hindneck and underparts.

1a
1b
1c
1d
2a
2b
2c
3a
3b
3c
3d
4a
4b
4c
4d
5a
5b
5c
5d
6a
6b
6c
6d
6e
7a
7b
7c
7d
8a
8b
8c

PLATE 47. GREEN-PIGEONS

1 [372] **CINNAMON-HEADED GREEN-PIGEON** *Treron fulvicollis*
(a) **Male** *fulvicollis*: Rufous-chestnut head and neck.
(b) **Female**: Like female Thick-billed Pigeon but has narrower bill and eyering, greener crown, yellowish thighs, streaked undertail-coverts.

2 [373] **LITTLE GREEN-PIGEON** *Treron olax*
(a) **Male**: Bluish-grey hood, maroon mantle to lesser coverts, broad orange breast-patch, dark slaty tail with paler grey terminal band.
(b) **Female**: Dark grey crown, dark green upperparts, dull green underparts, buffish green-streaked undertail-coverts, blackish-slate outertail, no red on bill.

3 [374] **PINK-NECKED GREEN-PIGEON** *Treron vernans*
(a) **Male** *griseicapilla*: Like Orange-breasted but has grey head, vinous-pink nape and neck, blackish subterminal tail-band.
(b) **Female**: Like female Orange-breasted but has grey-green crown and nape, blackish subterminal tail-band.

4 [375] **ORANGE-BREASTED GREEN-PIGEON** *Treron bicincta*
(a) **Male** *bicincta*: Green head with grey nape, brownish-tinged green upperparts, vinous-pink and orange breast-patches, plain grey central tail feathers.
(b) **Female**: Lacks breast-patches.

5 [376] **ASHY-HEADED GREEN-PIGEON** *Treron phayrei*
(a) **Male**: Like Thick-billed but with thinner and greyish bill, yellower-tinged throat, orange wash on breast, no broad eyering.
(b) **Female**: Like female Thick-billed but with thinner, greyish bill, no broad eyering.

6 [377] **ANDAMAN GREEN-PIGEON** *Treron chloropterus*
(a) **Male**: Quite large; green lesser coverts, bright green rump, yellow-tipped grey-green undertail-coverts.
(b) **Female**: Bright lime-green rump and undertail-coverts (as male). Note range.

7 [378] **THICK-BILLED GREEN-PIGEON** *Treron curvirostra*
(a) **Male** *nipalensis*: Thick pale bill with red base, broad bluish-green eyering, grey crown, maroon mantle to lesser coverts, green underparts.
(b,c) **Female**: Bill, head and eyering as male; all-green upperparts.

8 [379] **LARGE GREEN-PIGEON** *Treron capellei*
(a) **Male**: Relatively large with very stout bill, green upperparts, all-green head (greyish on face) and underparts with yellow-orange breast-patch.
(b) **Female**: Yellowish breast-patch.

9 [380] **YELLOW-FOOTED GREEN-PIGEON** *Treron phoenicopterus*
(a) **Male** *annamensis*: Largely grey head, yellowish-green neck and upper breast, pale grey-green upperparts, grey lower breast and belly, yellow legs, grey and yellow-green tail.
(b) **Male** *viridifrons*: Yellower collar, throat and breast, greener upperparts.

10 [381] **PIN-TAILED GREEN-PIGEON** *Treron apicauda*
(a) **Male** *apicauda*: Mostly rather bright green, apricot flush on breast, wedge-shaped grey tail with long pointed central feathers, blue round eye to base of slender bill.
(b,c) **Female**: All-green breast, rather shorter central tail feathers.
(d) **Male** *lowei*: Duller green head with feathered lores, light brownish-grey wash on mantle and wing-coverts, contrasting greenish-yellow rump and uppertail-coverts, uniform green underparts.
(e) **Female**: Differs as male.

11 [382] **YELLOW-VENTED GREEN-PIGEON** *Treron seimundi*
(a,b) **Male** *seimundi*: Like Pin-tailed Green-pigeon but darker with maroon shoulder-patch, whitish belly, mostly yellow undertail-coverts, pinkish-orange wash on breast, shorter central tail feathers.
(c) **Female**: No maroon shoulder-patch, greener breast.

12 [383] **WEDGE-TAILED GREEN-PIGEON** *Treron sphenura*
(a) **Male** *sphenura*: Maroon from upper mantle to lesser coverts, yellow forehead and strong apricot wash on crown and breast, broad wedge-shaped tail, blue base of bill.
(b,c) **Female**: No apricot or maroon in plumage.
(d) **Male** *robinsoni*: Smaller and darker, maroon restricted to shoulder, little or no apricot colour.

13 [384] **WHITE-BELLIED GREEN-PIGEON** *Treron sieboldii*
(a) **Male** *murielae*: Very like Wedge-tailed Green-pigeon but has mostly greyish-white belly, creamy-whitish base colour to undertail-coverts.
(b) **Female**: From Wedge-tailed by same features as male.

1a
1b
2a
2b
3a
3b
4a
4b
5a
5b
6a
7a
7c
7b
8a
8b
6b
9a
9b
10a
10c
10b
10d
10e
11a
11b
11c
12c
13a
12a
12b
12d
13b

PLATE 48. PARROTS & PARAKEETS

1 [389] **VERNAL HANGING-PARROT** *Loriculus vernalis*
(a,b) **Male**: Very small, short tail; bright green, red back to uppertail-coverts, light blue flush on throat/upper breast, red bill, dull orange legs; turquoise underwing with green coverts.
(c) **Female**: All colours duller, little blue on breast.

2 [390] **BLUE-CROWNED HANGING-PARROT** *Loriculus galgulus*
(a) **Male**: Like Vernal but with dark blue crown-patch, golden mantle-patch, golden-yellow band across lower back, red patch on throat, upper breast, black bill, duller legs.
(b) **Female**: Duller, no red on throat or breast, less distinct colours throughout.
(c) **Juvenile**: Like female but duller still; dusky bill.

3 [391] **BLUE-RUMPED PARROT** *Psittinus cyanurus*
(a,b) **Male** *cyanurus*: Small, stocky; greyish-blue head, blackish mantle, yellowish-green wing-covert fringes, pale breast and flanks, blackish underwing with largely red coverts, red and dark bill.
(c) **Female**: Brown head, yellower throat, dark green above, paler below, dark brown bill.
(d) **Juvenile female**: Green crown and sides of head.

4 [392] **ALEXANDRINE PARAKEET** *Psittacula eupatria*
(a) **Male** *siamensis*: Relatively large; big red bill, largely green with narrow black collar turning deep pink at ear, yellowish-green head with blue wash on hindcrown/nape, maroon-red shoulder-patch.
(b) **Female**: Smaller with no collar, only faint blue tinge on head.
(c) **Male** *avensis*: More uniformly green head.

5 [393] **ROSE-RINGED PARAKEET** *Psittacula krameri*
(a) **Male** *borealis*: Like Alexandrine Parakeet but smaller, smaller bill, no shoulder-patch, blackish lower mandible.
(b) **Female**: Lacks collar.

6 [394] **GREY-HEADED PARAKEET** *Psittacula finschii*
(a) **Male**: Slaty-grey head, narrow black collar, green body, mostly red bill with yellow lower mandible, maroon shoulder-patch (lacking in female), tail-streamers purplish-blue basally, yellowish distally.
(b) **Juvenile**: Green head with darker crown, no shoulder-patch.
(c) **First summer**: Pale slaty head, no collar.

7 [395] **BLOSSOM-HEADED PARAKEET** *Psittacula roseata*
(a) **Male** *juneae*: Relatively small; yellow upper, black lower mandible, rosy-pink face with greyer hindcrown, narrow black collar, green body, maroon shoulder-patch, deep turquoise tail-streamers shading yellow distally.
(b) **Female**: Violet-grey head, blackish malar patch, no black collar.
(c) **Juvenile**: Like female but head green with grey on malar area, no shoulder-patch, all-yellowish bill.

8 [396] **RED-BREASTED PARAKEET** *Psittacula alexandri*
(a) **Male** *fasciata*: Red upper mandible, blackish lower mandible, lilac-grey crown and head-sides, black line before eye, broad black malar band, yellowish wash on wing-coverts, deep pink breast, largely turquoise tail.
(b) **Female**: All-black bill, richer pink breast.
(c) **Juvenile**: Dull vinous-grey face, duller dark head markings, otherwise green head, breast and upper belly.

9 [397] **LONG-TAILED PARAKEET** *Psittacula longicauda*
(a) **Male** *longicauda*: Red upper and black lower mandible, reddish-pink head-sides and nape, deep green crown, broad black malar band, pale blue-green mantle, pale turquoise back; long purplish-blue tail-streamers.
(b) **Female**: Green nape, darker green crown/upperparts, dull brown bill, dark green malar band, shorter tail.
(c) **Juvenile**: Pinkish face, duller malar band.
(d) **Male** *tytleri*: Brighter green crown, flame-red head-sides, nape as mantle, darker green underparts.
(e) **Female**: Like female *longicauda* but crown brighter green, no reddish-pink on supercilium.

1a
1b
1c
3a
3b
3c
3d
1-3 to different scale
2a
2b
2c
4a
4b
4c
5a
5b
6a
6b
6c
9a
9b
9c
9d
9e
7a
7b
7c
8a
8b
8c

PLATE 49. HAWK-CUCKOOS & TYPICAL CUCKOOS

1 [400] **LARGE HAWK-CUCKOO** *Hierococcyx sparverioides*
(a) **Adult** *sparverioides*: Relatively large; slaty-grey crown and sides of head, brownish-grey mantle and wings, extensively dark chin, dark rufous breast-patch with prominent darker streaks, dark banding across whitish underparts, greyish tail with broad dark bands and buffish tip.
(b) **Juvenile**: Dark brown upperparts with rufous bars, buffy-white underparts with bold drop-like streaks.
(c) **Immature/subadult**: Greyish-brown crown, dull pale rufous bars on upperparts and wings; underparts similar to juvenile but with pale rufous breast-patch.

2 [401] **DARK HAWK-CUCKOO** *Hierococcyx bocki*
(a) **Adult**: Smaller, darker and richer-coloured than Large Hawk-cuckoo; dark streaks below restricted to breast, broader richer orange-rufous breast-band, less dark on chin.
(b) **Immature/subadult**: Apart from size, like Large, but richer rufescent barring above, whiter throat with less dark on chin.

3 [402] **COMMON HAWK-CUCKOO** *Hierococcyx varius*
(a) **Adult**: Like Large but smaller, upperparts usually ashier-grey (no contrast with crown), less blackish chin, rufous of underparts often paler/more extensive, restricted dark underpart markings, narrower dark tail-bands.
(b) **Juvenile**: Best separated from Large by smaller size, less distinct underpart markings and narrower tail-bands.
(c) **Immature/subadult**: Best separated from Large by same characters as juvenile.

4 [403] **MOUSTACHED HAWK-CUCKOO** *Hierococcyx vagans*
Adult: Relatively small and long-tailed, slaty crown and nape, dark moustachial/cheek-bar separating whitish upper throat from whitish ear-coverts, creamy underparts with blackish-brown streaks on lower throat to belly and flanks, white tail-tip.

5 [404] **MALAYSIAN HAWK-CUCKOO** *Hierococcyx fugax*
(a) **Adult**: All-dark head-side, dark chin, mostly pale chestnut tail-tip (very tip narrowly whitish); even tail-bars.
(b) **Juvenile**: Dark brown above with faint pale fringing, white on nape-side, spot-streaks below.

6 [405] **HODGSON'S HAWK-CUCKOO** *Hierococcyx nisicolor*
(a) **Adult**: Greyer above than Malaysian, no rufous bars on wing, inner tertial usually whiter, pinkish-rufous on breast (quite uniform if fine-streaked); penultimate dark tail-band is narrowest.
(b) **Juvenile**: Warmer/buffier fringes above than Malaysian, usually paler innermost tertial.

7 [406] **INDIAN CUCKOO** *Cuculus micropterus*
(a) **Male** *micropterus*: Brownish-tinged mantle, wings and tail, grey head, dark subterminal tail-band, broad dark underpart bars.
(b) **Female**: Rufescent wash across breast.
(c) **Juvenile**: Browner crown and sides of head with very heavy buffish-white bars/blotches, prominent rufous or buffish to whitish feather tips on upperparts and wings, buffish underparts with broken dark bars; may be washed rufous on throat and breast.

8 [407] **EURASIAN CUCKOO** *Cuculus canorus*
(a) **Male** *bakeri*: Like Himalayan but usually cleaner white below, dark bars often fainter and sometimes narrower, typically has whitish undertail-coverts with distinct blackish bars.
(b) **Female hepatic morph**: Rufescent-brown head and upperparts, buffy-rufous throat and breast, white lower underparts; strong blackish-brown bars overall. More narrowly barred than Himalayan.
(c) **Juvenile**: Dark head and upperside with whitish fringing, white nuchal patch (fresh plumage). Usually more narrowly barred below than Himalayan.
(d) **Juvenile female hepatic morph**: Whitish fringing on upperside, white nuchal patch (fresh plumage), more narrowly barred than Himalayan.

9 [409] **HIMALAYAN CUCKOO** *Cuculus saturatus*
(a) **Male**: Like Eurasian but usually slightly smaller, underparts typically buff-tinged, often with somewhat bolder and sometimes slightly broader dark bars, typically has less obvious or no blackish bars on buffish undertail-coverts.
(b) **Female hepatic morph**: Like Eurasian but has broader, blackish bars, notably on back to uppertail and breast.
(c) **Juvenile**: Like Eurasian but usually has broader dark bars below.
(d) **Juvenile female hepatic morph**: Like Eurasian but has warmer buffish underparts and broader dark bars overall.

10 [410] **SUNDA CUCKOO** *Cuculus lepidus*
(a) **Male**: Smaller and somewhat darker overall than Oriental and Himalayan Cuckoos, with broader dark bands below, and more rusty-buff undertail-coverts; eyes brown to dark reddish. See Lesser and Indian Cuckoos. Note voice and range.
(b) **Female hepatic morph**: Apart from size, differs from Himalayan by darker rufous upperparts with broader black barring, and broader black barring below.

11 [411] **LESSER CUCKOO** *Cuculus poliocephalus*
(a) **Male**: Like Himalayan but usually with finer bill, usually darker rump and uppertail-coverts which contrast less with blackish tail, usually even buffier underparts with wider-spaced, bolder black bars.
(b) **Female hepatic morph**: Like Himalayan but usually more rufous, sometimes with almost unbarred crown, nape, rump and uppertail-coverts.
(c) **Juvenile**: Dark grey-brown upperparts with narrow whitish to rufous bars and a few whitish spots on nape. Upperside somewhat darker and more uniform than Himalayan, underpart barring more contrasting.
(d) **Juvenile female hepatic morph**: More barring on crown and mantle, whitish nuchal patch.

1a
1c
1b
2a
4
3b
2b
5a
6a
3a
3c
5b
6b
7a
7b
8b
8d
7c
8c
9a
8a
9b
9c
11b
9d
10b
11d
10a
11a
11c

PLATE 50. CUCKOOS

1 [398] **PIED CUCKOO** *Clamator jacobinus*
(a) **Adult** *jacobinus*: Recalls Chestnut-winged Cuckoo but somewhat smaller, with black hindneck and wings, all-white underparts, short white bar on primaries, white-tipped outertail.
(b) **Juvenile**: Duller above, throat and breast initially dull greyish, otherwise buff-tinged below, shorter crest.

2 [399] **CHESTNUT-WINGED CUCKOO** *Clamator coromandus*
(a) **Adult**: Slender, long-tailed; blackish upperparts and crest, white hindcollar, rufous-chestnut wings, whitish below with buffy-rufous throat and blackish vent, white tail corners (not apparent on plate).
(b) **Juvenile**: Dark greenish-brown above with pale chestnut to buff feather tips, buffish hindcollar, all whitish below, paler bill, reduced crest.

3 [412] **BANDED BAY CUCKOO** *Cacomantis sonneratii*
(a) **Adult** *sonneratii*: Like hepatic female Plaintive but dark mask isolated by broad whitish supercilium; whiter underparts with narrower dark bars.
(b) **Juvenile**: Whitish to pale buff bars on sides of head and upperparts (particularly crown and mantle), weaker mask, coarser and more spaced dark barring.

4 [413] **GREY-BELLIED CUCKOO** *Cacomantis passerinus*
(a) **Male**: Grey with white vent.
(b) **Female hepatic morph**: Like Plaintive but has mostly plain deep rufous to rufous-chestnut crown, mantle, rump to uppertail, throat and upper breast.
(c) **Juvenile male**: Upperside, throat and breast more sooty-brown with fine pale to buffish fringing.
(d) **Juvenile female hepatic morph**: Like brown plumages of Plaintive but lacks distinct tail-barring; mostly plain rump and uppertail-coverts, darker breast.

5 [414] **PLAINTIVE CUCKOO** *Cacomantis merulinus*
(a) **Male** *querulus*: Relatively small; grey head, throat and upper breast, peachy-rufous remainder of underparts.
(b) **Female hepatic morph**: Rufescent upperside with even blackish barring, pale underside with less pronounced barring, slightly paler supercilium, completely barred tail.
(c) **Female hepatic morph variant**: Whiter below and more evenly barred.
(d) **Juvenile**: Like hepatic female but has broad blackish streaks on head and upper breast.

6 [415] **RUSTY-BREASTED CUCKOO** *Cacomantis sepulcralis*
(a) **Adult** *sepulcralis*: Like male Plaintive but slightly slatier-grey, with entire underparts peachy-rufous, yellow eyering.
(b) **Adult hepatic morph**: Difficult to separate from hepatic female Plaintive but slightly larger and proportionately longer-tailed, much broader dark barring on upperside, throat and breast, rufous tail-barring restricted to notches along outer fringes of feathers.
(c) **Juvenile**: Like hepatic adult but darker above, with only narrow buffish to rufous barring.

7 [416] **HORSFIELD'S BRONZE CUCKOO** *Chrysococcyx basalis*
Adult: Like Little Bronze but browner above, no pale forehead, plainer supercilium, pale brownish-white fringing on scapulars and wing-coverts, brownish throat-streaks, unbarred belly-centre, more rufous-chestnut on outertail.

8 [417] **LITTLE BRONZE CUCKOO** *Chrysococcyx minutillus*
(a) **Male** *peninsularis*: Like female Violet but has black bill, red eyering, pale forehead and dark ear-covert patch, more uniform upperparts, glossy bottle-green crown and bronzy-green mantle.
(b) **Female**: Paler, less glossy crown, less bronzy mantle, reduced pale forehead, duller eyering.
(c) **Juvenile**: Like female but duller and browner above, almost uniform greyish- to brownish-white underparts.

9 [418] **ASIAN EMERALD CUCKOO** *Chrysococcyx maculatus*
(a) **Male**: Small; glossy gold-tinged emerald-green; white bars on lower breast to vent, dark-tipped rich yellowish bill.
(b) **Female**: Plain rufous crown and nape, coppery-green upperparts, entirely barred underparts.
(c) **Juvenile**: Like female but less green above with dark and buff bars on crown to mantle, rufous-tipped upperpart feathers and wing-coverts, darker tail with more rufous (no white bars), darker bill.

10 [419] **VIOLET CUCKOO** *Chrysococcyx xanthorhynchus*
(a) **Male** *xanthorhynchus*: Small; glossy violet-purple with white bars on lower breast to vent, orange bill with red base.
(b) **Female**: Like Asian Emerald Cuckoo but has mostly dark bronzy-brown crown to upper mantle, rest of upperside browner, bill yellowish with some red at base.
(c) **Juvenile**: Like juvenile Asian Emerald but with more rufous-chestnut crown to upper mantle, bold rufous-chestnut and dull dark greenish bars on upperparts and wing-coverts, rufous-chestnut fringes to flight feathers, no rufous wash below.

1b
1a
2b
2a
1-2 to different scale
3a
3b
4c
4d
4a
4b
5a
5d
5b
5c
6b
6a
6c
8a
8b
8c
10c
10b
10a
7
9a
9b
9c

PLATE 51. DRONGO CUCKOO, ASIAN KOEL, CORAL-BILLED GROUND-CUCKOO, MALKOHAS, COUCALS

1 [420] **DRONGO CUCKOO** *Surniculus lugubris*
(a) **Adult** *dicruroides*: Glossy greenish-black with longish, slightly forked tail (resembling some drongos), slender, slightly decurved bill, white bars on undertail-coverts and undertail; some white feathers on nape (often concealed).
(b) **Juvenile**: Browner, less glossy, with white spots on body and wing-coverts.

2 [421] **ASIAN KOEL** *Eudynamys scolopaceus*
(a) **Male** *chinensis*: Fairly large, long-tailed; glossy blue-black, stout greenish bill, red eye.
(b) **Female**: Predominantly streaked, spotted and barred white.
(c) **Juvenile**: Brownish-black above with whitish feather tips, dull rufous bars on tail, variable broad whitish to pale buff bars on breast to vent.
(d) **Female** *malayana*: Predominantly streaked, spotted and barred buff to rufous.

3 [422] **CORAL-BILLED GROUND-CUCKOO** *Carpococcyx renauldi*
(a) **Adult**: Large; greyish with glossy blackish head, neck, upper breast, primaries and tail, mostly greyish remainder of plumage, violet and red facial skin, stout red bill and legs.
(b) **Juvenile**: Blackish-brown crown and nape, brownish, green- and purple-tinged dull upperparts, dull rufous-chestnut edging on wings, dull drab rufous-chestnut throat and breast.

4 [423] **BLACK-BELLIED MALKOHA** *Rhopodytes diardi*
Adult *diardi*: Like Green-billed Malkoha but smaller, shorter-tailed, throat and breast darker, no whitish border to facial skin, narrower white tips to tail feathers.

5 [424] **CHESTNUT-BELLIED MALKOHA** *Rhopodytes sumatranus*
Adult *sumatranus*: Chestnut belly and undertail-coverts, thick bill, orange facial skin.

6 [425] **GREEN-BILLED MALKOHA** *Rhopodytes tristis*
(a) **Adult** *longicaudatus*: Large, and much longer-tailed than other malkohas, greyish head and underparts with dark shaft-streaks and dark vent, white-bordered red facial skin, green bill.
(b) **Juvenile**: Browner-tinged crown, mantle and wings, browner vent, blacker bill, duller facial skin with less clearly defined whitish surround.

7 [426] **RAFFLES'S MALKOHA** *Rhinortha chlorophaeus*
(a) **Male** *chlorophaeus*: Relatively small; rufescent head and breast, rufous-chestnut mantle and wings, greyish belly, blackish white-tipped tail, greenish bill, turquoise facial skin.
(b) **Female**: Greyish head and breast, buffish belly, rufous-chestnut tail with black subterminal band and white tip.

8 [427] **RED-BILLED MALKOHA** *Zanclostomus javanicus*
Adult *pallidus*: Red bill, blue orbital skin, rusty-rufous throat and upper breast, chestnut vent, pale greyish remainder of underparts.

9 [428] **CHESTNUT-BREASTED MALKOHA** *Zanclostomus curvirostris*
Male *singularis*: Dark oily-green upperside, dark chestnut underparts and distal quarter of uppertail, blackish vent, pale yellowish upper and red lower mandible, red facial skin.

10 [429] **SHORT-TOED COUCAL** *Centropus rectunguis*
(a) **Adult**: Like Greater Coucal but considerably smaller.
(b) **Juvenile**: Chestnut-brown crown and mantle with blackish bars, narrow blackish bars on tertials and wing-coverts, dark brown underparts with whitish to buff bars and shaft-streaks, browner bill.

11 [430] **GREATER COUCAL** *Centropus sinensis*
(a) **Adult** *intermedius*: Relatively large; glossy purplish blue-black with uniform chestnut back and wings.
(b) **Juvenile**: Back and wings duller with heavy blackish bars; rest of plumage blackish with small whitish streaks and flecks turning to buffier bars on lower body, tail barred brownish- to greyish-white, browner bill.

12 [431] **ANDAMAN COUCAL** *Centropus andamanensis*
Adult: Buffy greyish-brown head, mantle and underparts (duller on vent), dark reddish-chestnut back and wings, uppertail uniform dark brownish or shading paler basally.

13 [432] **LESSER COUCAL** *Centropus bengalensis*
(a) **Adult breeding** *bengalensis*: Like Short-toed Coucal but with duller back and darker wing-tips.
(b) **Adult non-breeding**: Dark brown head, mantle and scapulars with whitish-buff streaks, dull buff below with some blackish-brown bars and whitish-buff shaft-streaks; pale bill.
(c) **Juvenile**: Like adult non-breeding but strongly rufescent above with broad blackish-brown streaks on crown and broad blackish bars on rest of upperparts and wings, more rufescent below with broader dark bars. Has unusual long uppertail-coverts (also adult non-breeding).

1a
1b
2d
2b
2a
3b
2c
3a
4
5
1-2 & 3 to different scale
6b
6a
7b
7a
8
11b
11a
9
12
13c
13b
10b
10a
13a

PLATE 52. SMALLER OWLS

1 [435] **ORIENTAL BAY OWL** *Phodilus badius*
Adult *badius*: Like small Common Barn-owl but with distinctive facial shape involving dark-outlined buffy facial discs rising to triangular rudimentary ear-tufts and downturning through large dark eyes; sparsely spotted pinkish-buff underparts, dull rufous-chestnut upperside with black speckles and sparse white markings, largely golden-buff nuchal band and scapulars.

2 [436] **WHITE-FRONTED SCOPS-OWL** *Otus sagittatus*
Adult: Relatively large and long-tailed, resembling a giant Reddish Scops-owl but with brighter dark rufous to rufous-chestnut upperside (sometimes also breast), broad whitish forehead-patch and eyebrows (extends to ear-tufts), whiter throat, bluish-white bill, brown eyes, no dark bars on primaries.

3 [437] **REDDISH SCOPS-OWL** *Otus rufescens*
Adult *malayensis*: Small and dark with dull rufescent-tinged plumage, recalling Mountain Scops-owl but with plainer upperparts, less distinct scapular markings, plain rufescent-buff underparts with distinct blackish spots (slightly highlit above with pale buff), brown eyes, broad dark and buff barring on primaries.

4 [438] **MOUNTAIN SCOPS-OWL** *Otus spilocephalus*
(a) **Adult** *siamensis*: Small with relatively short, rounded ear-tufts, usually warm buff eyebrows, rufous-chestnut upperparts with large white scapular markings, buffish to rufous underparts with white markings and dark vermiculations but no streaks, yellow eyes, yellowish-brown bill.
(b) **Adult** *spilocephalus*: Dull variant. More greyish-brown overall.
(c) **Adult** *vulpes*: More deeply rufescent, particularly below, coarser blackish markings on upperparts with broad blackish crown-streaks.

5 [439] **COLLARED SCOPS-OWL** *Otus lettia*
(a,b) **Adult** *lettia*: Greyish variant. Larger than most scops-owls, densely vermiculated and streaked plumage, prominent broad whitish eyebrows and nuchal collar, pronounced ear-tufts, dark eyes, greyish-brown buff-marked upperparts, whitish underparts with fine vermiculations and sparse streaks.
(c) **Adult buff variant**: Buffish eyebrows and nuchal collar, deep buff base colour to underparts.
(d) **Juvenile**: Paler (often warmer) head and body, darker bars overall.

6 [440] **ORIENTAL SCOPS-OWL** *Otus sunia*
(a) **Adult greyish morph** *distans*: Recalls Collared Scops-owl but smaller and slimmer, usually distinctly greyer, with yellow eyes, no pale nuchal collar, prominent white scapular markings, usually bolder dark streaks on underparts.
(b) **Adult rufous morph**: Like Mountain Scops-owl but with distinct narrow blackish streaks on crown, more pronounced ear-tufts, contrasting white belly and vent with blackish to rufous streaks and vermiculations.
(c) **Adult rufous morph** *stictonotus*: Paler and more rufous (less chestnut).

7 [452] **COLLARED OWLET** *Glaucidium brodiei*
(a,b) **Adult** *brodiei*: Very small; like Asian Barred but much smaller, has buff and blackish imitation face pattern on nape, typically greyer, more speckled crown, more tear-shaped streaks on belly and lower flanks, no prominent white markings on wing-coverts.

8 [453] **ASIAN BARRED OWLET** *Glaucidium cuculoides*
(a) **Adult** *bruegeli*: Quite small, robust and 'neck-less'; broadly rounded head without ear-tufts, dull brown with pale buffish bars, narrow whitish eyebrows, white ventral line, whitish belly and lower flanks with broad brown streaks, yellow eyes.
(b) **Juvenile**: Pale bars on upperparts more diffuse and spot-like, dark underpart bars broken and more diffuse, streaking more diffuse, crown more speckled.
(c) **Adult** *deignani*: Rustier belly-streaks, grey head with whiter bars.

9 [454] **JUNGLE OWLET** *Glaucidium radiatum*
Adult *radiatum*: Like Asian Barred Owlet but smaller, somewhat greyer and more densely barred, with prominent dull rufous bars on flight feathers, smaller white scapular markings, all-barred underparts.

10 [455] **SPOTTED OWLET** *Athene brama*
Adult *mayri*: Fairly small; brownish-grey upperparts with white spots, broad white eyebrows, broken dark foreneck collar, whitish underparts with broken dark bars, yellow eyes.

11 [456] **BROWN BOOBOOK** *Ninox scutulata*
Adult *burmanica*: Slim, small-headed and relatively long-tailed, dark slaty-brown above, whitish patch between eyes, whitish to buffish-white underparts with broad drab chestnut-brown streaks (denser on breast), pale brownish-grey tail with broad blackish bars, no obvious facial discs, golden-yellow eyes.

1
2
3
4a
4b
6a
6b
6c
4c
5b
5d
7b
7a
5a
8a
5c
11
8c
9
10
8b

PLATE 53. LARGER OWLS

1 [433] **COMMON BARN-OWL** *Tyto alba*
(a,b) **Adult** *stertens*: Pale buffy-grey above with golden-buff markings, pale facial discs forming heart-shape, buffy-white underparts with sparse blackish speckles; relatively pale, uniform upperside in flight.

2 [434] **EASTERN GRASS-OWL** *Tyto longimembris*
(a,b) **Female** *longimembris*: Like Common Barn-owl but mostly dark brown above with deep golden-buff markings, rufous wash on face, neck and breast, dark area around eyes; in flight shows dark above and light below, with golden-buff patch at base of outer primaries (upperwing) contrasting with broadly dark-tipped primary coverts (both surfaces).

3 [441] **INDIAN EAGLE-OWL** *Bubo bengalensis*
Adult: Huge; streaked crown, orange eyes, dark bill, black border to facial discs, dark breast-streaks.

4 [442] **SPOT-BELLIED EAGLE-OWL** *Bubo nipalensis*
(a) **Adult** *nipalensis*: Huge; long part-barred ear-tufts, whitish underparts with distinctive blackish-brown heart-shaped bars/spots, dark eyes, pale yellow bill.
(b) **Juvenile**: Whitish to buffy-white head and body with blackish-brown bars (more prominent above).

5 [443] **BARRED EAGLE-OWL** *Bubo sumatranus*
Adult *sumatranus*: Like Spot-bellied Eagle-owl but much smaller, with brown-and-buff barring above and dense narrow blackish-and-whitish barring below (denser on breast, where base colour browner).

6 [444] **DUSKY EAGLE-OWL** *Bubo coromandus*
(a) **Adult** *klossii*: Like Brown Fish-owl but has rather uniform drab dark greyish-brown upperparts (without obvious markings), duller and greyer below without white gorget, primaries unbarred, ear-tufts more erect.
(b) **Juvenile**: Creamy-whitish with very faint darker markings.

7 [445] **BROWN FISH-OWL** *Ketupa zeylonensis*
(a) **Adult** *leschenault*: Rather warm buffish-brown, bold dark markings above. Floppy streaked ear-tufts, strong blackish-brown and whitish/buff wing markings, whitish scapular markings and gorget, long blackish-brown streaks and narrow dark cross-bars below, golden-yellow eyes.
(b) **Juvenile**: Pale creamy-buffish with long dark brown streaks, paler wings.

8 [446] **TAWNY FISH-OWL** *Ketupa flavipes*
Adult: Huge; very like large Buffy but more orange-buff, typically more unmarked buff on scapulars and greater coverts, richer buff wing- and tail-bars, broader dark streaks below (particularly on breast).

9 [447] **BUFFY FISH-OWL** *Ketupa ketupu*
(a) **Adult** *aagaardi*: Like Brown Fish-owl but richer buff, with broad blackish streaks/markings above (contrasting with bar-free underparts), white forehead/eyebrows, less obvious white gorget, broader streaks on ear-tufts.
(b) **Adult** *ketupu*: Warmer, richer buff above and below.

10 [448] **SPOTTED WOOD-OWL** *Strix seloputo*
Adult *seloputo*: Like Brown Wood-owl but with plain rufous-buff facial discs, white speckles and spots on upperparts, white to buffy-white underparts (mixed rich buff) with well-spaced dark bars.

11 [449] **MOTTLED WOOD-OWL** *Strix ocellata*
Adult *grisescens?*: Like Spotted Wood-owl but facial discs much less distinct, narrower barring below, overall plumage pattern less contrasting, with greyish vermiculations and patches of rufous.

12 [450] **BROWN WOOD-OWL** *Strix leptogrammica*
(a) **Adult** *laotiana*: Large, with rounded head, no ear-tufts; buffy-brown facial discs with blackish-brown border, dark eyes surrounded by dark brown, dark brown upperparts with whitish to buff markings on scapulars and upper mantle, buff underparts (breast often dark brown) with dense dark brown bars.
(b) **Juvenile**: Whitish-buff with narrow dark bars, contrasting dark-and-rufous bars on wings.

13 [451] **HIMALAYAN WOOD-OWL** *Strix nivicola*
(a) **Adult pale morph** *nivicola*: Blackish-brown, pale buff and pale greyish mottling, vermiculations and streaks on upperparts, buffy-white markings on scapulars and wing-coverts, pale greyish facial discs, buffy-white underparts with heavy dark streaks and vermiculations.
(b) **Adult dark morph**: Blacker-brown and richer rufous above, rufous-buff discs, rufous and white mixed below.
(c) **Juvenile pale morph**: Paler, with even dark bars.

14 [458] **LONG-EARED OWL** *Asio otus*
(a,b) **Female** *otus*: Like Short-eared Owl but ear-tufts longer, upperparts greyer, facial discs rufescent with less dark around (orange) eyes, more evenly streaked underparts; less boldly barred upperwing and -tail, more rufous primary bases, less black wing-tips, wings somewhat shorter.

15 [459] **SHORT-EARED OWL** *Asio flammeus*
(a,b) **Female** *flammeus*: Rather robust and round-headed, short ear-tufts, dark around (yellow) eyes with pale cross-like mid-pattern, uneven broad dark breast-streaks; broad dark bars on flight feathers and tail, buff primary bases abut blackish primary coverts, whitish underwing with pronounced black bands.

1a
1b
2a
2b
3
4a
4b
5
6a
6b
7a
7b
8
9a
9b
10
11
12a
12b
13a
13b
13c
14a
14b
15a
15b

PLATE 54. FROGMOUTHS & NIGHTJARS

1 [460] **LARGE FROGMOUTH** *Batrachostomus auritus*
Adult: Large; bold white tips to wing-coverts, buffy-white nuchal collar with blackish bars, warm brown throat and breast with a few small white markings.

2 [461] **GOULD'S FROGMOUTH** *Batrachostomus stellatus*
(a) **Adult**: White spots on wing-coverts, rufescent-brown scales and no blackish markings on underparts, whiter belly and undertail-coverts, protruding bill.
(b) **Adult**: Darker individual: colder, browner upperparts and scales on underparts.
(c) **Juvenile**: More uniform, with dark-barred upperparts.

3 [462] **HODGSON'S FROGMOUTH** *Batrachostomus hodgsoni*
(a) **Male** *indochinae*: Like Blyth's (subspecies *affinis*) but has heavier black vermiculations and markings on upperparts and breast, no rufous on breast, smaller, less protruding bill.
(b) **Female**: Very like Blyth's but paler, with more prominent white markings on underparts, smaller bill.
(c) **Juvenile**: Warm-tinged upperparts with blackish and pale brown bars, no nuchal collar, underparts similar to upperparts, grading to plainer and whiter on vent.

4 [463] **BLYTH'S FROGMOUTH** *Batrachostomus affinis*
(a) **Male** *continentalis*: Complex buff, white and blackish markings on warm brown upperside, narrow white and black nuchal collar, white scapular spots, buffy-white below with more rufous throat and breast, dark vermiculations and large white markings.
(b) **Female**: Rather plain rufous-chestnut, white and black nuchal band, large white spots on scapulars, black-tipped white markings on breast; yellow eyes (both sexes).
(c) **Male** *affinis*: Less rufous-tinged, particularly on throat and breast, usually has more whitish vermiculations on crown and more black markings on upperparts and underparts, usually more distinctly barred tail.

5 [464] **MALAYSIAN EARED-NIGHTJAR** *Eurostopodus temminckii*
(a,b) **Adult**: Like Great Eared- but smaller and darker, crown darker, 'ear-tufts' smaller (often not visible), dark tail with less contrasting bands.

6 [465] **GREAT EARED-NIGHTJAR** *Eurostopodus macrotis*
(a,b) **Adult** *cerviniceps*: Much larger than other nightjars; pale brownish crown with 'ear-tufts', pale collar (white at front, buff at rear), distinct chestnut tinge around shoulders, blackish and buffish bars on uppertail, long wings and tail with no whitish or pale markings.

7 [466] **GREY NIGHTJAR** *Caprimulgus jotaka*
(a-c) **Male** *hazarae*: Smaller than Large-tailed with darker crown, no rufous on nape, heavier black vermiculations on upperparts, stronger, more even bars on uppertail, less contrastingly patterned scapulars, fainter pale bars across upperwing-coverts, duller ear-coverts and throat (latter usually with smaller whitish area, sometimes two distinct patches). Smaller white wing- and tail-patches.
(d) **Female**: Smaller, buff wing-patches, no obvious pale tail-patches.

8 [467] **LARGE-TAILED NIGHTJAR** *Caprimulgus macrurus*
(a-c) **Male** *bimaculatus*: Fairly large, strongly variegated; relatively pale crown with dark centre, large area of white across lower throat, prominent row of black scapulars with broad whitish-buff fringes, whitish to buff bars across upperwing-coverts, brownish-grey tail with uneven dark bars; large white patches on wings and tail.
(d) **Female**: Wing-patches small and buff, tail-patches much smaller.

9 [468] **INDIAN NIGHTJAR** *Caprimulgus asiaticus*
(a-c) **Adult** *asiaticus*: Like Large-tailed Nightjar but smaller, shorter-tailed and paler, distinct buff nuchal collar with darker markings, more extensive whitish-buff fringes to scapulars, normally two large white throat-patches; slightly smaller white to buffy-white wing- and tail-patches.

10 [469] **SAVANNA NIGHTJAR** *Caprimulgus affinis*
(a-c) **Male** *monticolus*: From other nightjars by uniform, heavily vermiculated brownish-grey upperparts, lack of defined dark median crown-stripe, two roundish buffish-white patches on lower throat, almost all-white outertail feathers.
(d) **Female**: Slightly smaller, buff wing-patches, no obvious pale or whitish tail markings.

2a
3a
4c
1
2b
3b
2c
3c
4b
4a
5a
5b
6a
6b
7a
7b
7c
7d
8a
8d
8c
8b
9a
10d
9c
9b
10a
10c
10b

PLATE 55. SWIFTS

1 [470] **WATERFALL SWIFT** *Hydrochous gigas*
(a,b) **Adult**: Relatively large size, blackish-brown overall, obvious tail-notch.

2 [471] **GLOSSY SWIFTLET** *Collocalia esculenta*
(a,b) **Adult** *cyanoptila*: Small size, glossy bluish-black on upperparts, whitish belly.

3 [472] **HIMALAYAN SWIFTLET** *Aerodramus brevirostris*
(a,b) **Adult** *brevirostris*: Blackish-brown above with paler greyish rump-band, throat and breast rather uniform and slightly paler than ear-coverts and chin, brownish-grey belly and vent, obvious tail-notch.
(c) **Adult** *inopina*: Slightly longer-winged, rump-band almost same colour as rest of upperparts.
(d) **Adult** *rogersi*: Slightly shorter-winged, rump-band slightly darker than *brevirostris*.

4 [473] **BLACK-NEST SWIFTLET** *Aerodramus maximus*
(a,b) **Adult** *maximus*: Like Germain's but longer-winged, bulkier, bigger-headed, with little or no tail-notch.
(c) **Adult** *lowi*: Rump-band concolorous with rest of upperparts.

5 [474] **EDIBLE-NEST SWIFTLET** *Aerodramus fuciphaga*
(a,b) **Adult** *inexpectata*: Like Germain's but slightly shorter-winged, slightly narrower and duller rump-band, slightly darker underparts.

6 [475] **GERMAIN'S SWIFTLET** *Aerodramus germani*
(a,b) **Adult** *germani*: Blackish-brown above with obvious whitish-grey rump-band, throat and breast paler than chin and obviously paler than ear-coverts, pale brownish-grey belly and vent, slight tail-notch.

7 [476] **SILVER-RUMPED NEEDLETAIL** *Rhaphidura leucopygialis*
(a,b) **Adult**: Small, robust, blackish with contrasting silvery-white lower back to uppertail-coverts, short squarish tail, very broad paddle-shaped wings.

8 [477] **WHITE-THROATED NEEDLETAIL** *Hirundapus caudacutus*
(a,b) **Adult** *caudacutus*: Relatively large; clear white throat, vent and patch on forehead/upper lores, extensive whitish saddle.
(c) **Adult** *nudipes*: Blackish forehead and lores.

9 [478] **SILVER-BACKED NEEDLETAIL** *Hirundapus cochinchinensis*
(a,b) **Adult** *cochinchinensis*: Relatively large; brownish-grey throat, white vent, blackish forehead and lores, dark brown saddle with brownish-white centre.

10 [479] **BROWN-BACKED NEEDLETAIL** *Hirundapus giganteus*
(a,b) **Adult** *indicus*: Relatively large; mostly dark brown throat, white vent, white loral spot, dark brown saddle.

11 [480] **ASIAN PALM-SWIFT** *Cypsiurus balasiensis*
(a,b) **Adult** *infumatus*: Small and slim; dark greyish-brown with slightly paler sides of head, rump, breast and belly, throat paler still, slender sickle-shaped wings, long, pointed and deeply forked tail.

12 [482] **FORK-TAILED SWIFT** *Apus pacificus*
(a,b) **Adult** *cooki*: Fairly large; appears all blackish with narrow white rump-band and paler central throat (whitish scales on underparts may be visible), long sickle-shaped wings, deeply forked tail.
(c) **Adult** *pacificus*: Slightly browner upperparts, broader white rump-band, whiter throat.

13 [483] **DARK-RUMPED SWIFT** *Apus acuticauda*
(a,b) **Adult**: Like Fork-tailed Swift but has all-dark rump, usually slightly darker throat.

14 [484] **HOUSE SWIFT** *Apus affinis*
(a,b) **Adult** *subfurcatus*: Quite broad-winged, blackish with sharply contrasting broad white rump-band and distinctly whitish throat.

1a
2b
3a
3d
4a
1b
2a
3b
3c
4b
4c
5a
6a
7a
7b
5b
6b
8c
8a
9a
9b
10b
8b
10a
12a
11b
12c
14b
13a
12b
13b
11a
14a

PLATE 56. ALPINE SWIFT, TREESWIFTS & TROGONS

1 [481] **ALPINE SWIFT** *Tachymarptis melba*
(a,b) **Adult** *nubifuga*: Large, long-winged; dark brown, with white throat (often hard to see at distance) and breast/belly-patch.

2 [485] **CRESTED TREESWIFT** *Hemiprocne coronata*
(a) **Male**: Slim, with long, slender, dark wings and (very deeply forked) tail; (slightly bluish-) grey above with dark erectile crest, pale rufous ear-coverts, throat-sides and upper throat, paler grey below with whitish belly and vent.
(b,c) **Female**: No rufous on head, blackish lores, blackish-slate ear-coverts, dusky-whitish line above eye and along edge of all-grey throat, plain underwing.

3 [486] **GREY-RUMPED TREESWIFT** *Hemiprocne longipennis*
(a) **Male** *harterti*: Like Crested Treeswift but has dark glossy green crown and mantle (less grey than illustrated), all-grey throat, tail-tip falls short of primary tips, mostly whitish-grey tertials.
(b,c) **Female**: Blackish ear-coverts; grey back and rump and whitish tertials contrast with dark mantle; contrasting blackish underwing-coverts.

4 [487] **WHISKERED TREESWIFT** *Hemiprocne comata*
(a) **Male** *comata*: Small, mostly olive-bronze body, glossy blue-black crown and wings, white head-stripes, tertials and vent, dull rufous-chestnut ear-coverts, no erectile crest.
(b,c) **Female**: Blackish ear-coverts. Dark underwing with white-tipped secondaries and inner primaries (also male), demarcated white vent.

5 [488] **RED-NAPED TROGON** *Harpactes kasumba*
(a) **Male** *kasumba*: Like Diard's but has broad cobalt-blue facial skin, broad red nuchal patch meeting facial skin, narrow white breast-band.
(b) **Female**: Like female Diard's but brownish-buff lower breast to vent, unmarked white on undertail (both sexes).

6 [489] **DIARD'S TROGON** *Harpactes diardii*
(a) **Male** *sumatranus*: Black forehead and throat, maroon-washed blackish hindcrown, pale pink nuchal band, black upper breast, reddish-pink underparts, pale pink breast-band, violet orbital skin.
(b) **Female**: Rather uniform brown head, upperparts and upper breast, reddish-pink belly, dark vermiculations on white of undertail (both sexes).

7 [490] **CINNAMON-RUMPED TROGON** *Harpactes orrhophaeus*
(a) **Male** *orrhophaeus*: Like Scarlet-rumped but larger and thicker-billed, no red on rump or uppertail-coverts, much pinker underparts.
(b) **Female**: Recalls Scarlet-rumped but lacks pink in plumage, brown of head darker and richer with deep rufous lores and around eye, mostly blackish-brown throat with dull rufous centre.

8 [491] **SCARLET-RUMPED TROGON** *Harpactes duvaucelii*
(a) **Male**: Black head, blue 'eye-brow', bill and gape skin, bright deep pinkish-red rump, uppertail-coverts and underparts.
(b) **Female**: Drab brown head, paler lores, around eye and throat, mostly buffy-brown rump and uppertail-coverts (mixed with pink), deep buffy-brown breast, pinkish belly and vent.
(c) **Juvenile female**: Lacks obvious pink in plumage, rump and uppertail-coverts rufescent, underparts rufous-buff with pale buff centrally.

9 [492] **ORANGE-BREASTED TROGON** *Harpactes oreskios*
(a) **Male** *stellae*: Greenish-olive head and upper breast, chestnut-maroon upperparts, yellowish-orange underparts becoming yellower on lower belly.
(b) **Female**: Drab olive-brownish head, mantle and back, duller rump and uppertail-coverts, grey-washed throat and upper breast, paler remainder of underparts.

10 [493] **RED-HEADED TROGON** *Harpactes erythrocephalus*
(a) **Male** *erythrocephalus*: Dark red head and upper breast, narrow white breast-band, pinkish-red belly, whitish wing vermiculations.
(b) **Female**: Brown head and upper breast, pale warmish brown wing vermiculations. White undertail with black border (both sexes).

11 [494] **WARD'S TROGON** *Harpactes wardi*
(a) **Male**: Slaty head, upperparts and breast with pinkish-maroon wash, dark pinkish-red forehead, red bill (purplish at gape), pale blue orbital skin, reddish-pink underparts, blackish uppertail.
(b) **Female**: Head, upperparts and breast mostly olive-brownish, bright yellow forehead, pale yellow belly and vent, mostly pale yellow undertail (pink on male).

1b
2b
3b
4b
3c
1a
2c
4c
2a
1-4 to different scale
5b
5a
4a
3a
6b
8b
8c
7b
7a
6a
8a
11a
10a
9b
9a
10b
11b

PLATE 57. KINGFISHERS

1 [497] **RUFOUS-COLLARED KINGFISHER** *Actenoides concretus*
(a) **Male**: *concretus*: Greenish crown, black eyestripe, dark blue submoustachial stripe, rufous collar/breast.
(b) **Female**: Mostly dull green above, pale buffish speckling on scapulars and wing-coverts.

2 [498] **BANDED KINGFISHER** *Lacedo pulchella*
(a) **Male** *amabilis*: Long tail, red bill, chestnut forehead and mask, blue hindcrown and nape, blue-barred upperparts, warm buffish breast-band on whitish underparts.
(b) **Female**: Rufous-and-black bars on head and upperside, white below with blackish scales on breast/flanks.

3 [499] **STORK-BILLED KINGFISHER** *Pelargopsis capensis*
(a) **Male** *burmanica*: Relatively large; huge red bill, pale dull brownish crown and head-sides, greenish-blue mantle, wings and tail, pale turquoise back and rump, warm buffish neck and underparts.
(b) **Male**: *malaccensis*: Darker crown and head-sides, bluer upperparts, wings and tail.

4 [500] **BROWN-WINGED KINGFISHER** *Pelargopsis amauroptera*
Adult: Like Stork-billed Kingfisher but with orange-buff head and underparts, dark brown mantle/wings//tail.

5 [501] **RUDDY KINGFISHER** *Halcyon coromanda*
(a) **Adult** *coromanda*: Bright rufous, violet-tinged above, bluish-white lower back and rump-patch, red bill.
(b) **Juvenile**: Browner; narrow dark bars below, duller bill.

6 [502] **WHITE-THROATED KINGFISHER** *Halcyon smyrnensis*
(a,b) **Adult** *perpulchra*: Dark chestnut head and belly, white throat and breast; whitish patch on primaries.

7 [503] **BLACK-CAPPED KINGFISHER** *Halcyon pileata*
(a,b) **Adult**: Black cap, white collar to breast, deep blue above; whitish patch on primaries.

8 [504] **COLLARED KINGFISHER** *Todiramphus chloris*
Adult *armstrongi*: Blue upperside with variable turquoise wash, white collar and underparts, faint creamy-buffish wash on lower flanks, mostly dark bill.

9 [505] **SACRED KINGFISHER** *Todiramphus sanctus*
Adult *sanctus*: Smaller than Collared, blackish-green head-sides, distinctly buff supercilium, clearly buff-washed nape and flanks.

10 [506] **BLACK-BACKED KINGFISHER** *Ceyx erithaca*
Adult *erithaca*: Very small; red bill, blackish-blue mantle and scapulars, dark wings, blue head-patches.

11 [507] **RUFOUS-BACKED KINGFISHER** *Ceyx rufidorsa*
Adult *rufidorsa*: Like Black-backed but mantle and scapulars all rufous, no blue head-patches.

12 [508] **BLUE-BANDED KINGFISHER** *Alcedo euryzona*
(a) **Male** *peninsulae*: Mostly dull dark brownish wings, blue breast-band, blackish bill.
(b) **Female**: Like Common but larger and bulkier, with much duller, browner crown, scapulars and wings, dark ear-coverts, paler blue stripe down upperparts. Mostly dull reddish lower mandible.

13 [509] **BLUE-EARED KINGFISHER** *Alcedo meninting*
(a) **Female** *verreauxii*: Like Common but slightly smaller, with blue ear-coverts, deeper blue crown, upperparts and wings (without turquoise), deeper orange-rufous below (male with mostly blackish bill).
(b) **Juvenile**: Rufous on cheeks and ear-coverts, dark scales on duller breast, initially mostly reddish bill.
(c) **Female** *coltarti*: Pale turquoise crown-bars, spots on wing-coverts and stripe down upperparts.

14 [510] **COMMON KINGFISHER** *Alcedo atthis*
(a) **Male** *bengalensis*: Small; rufous ear-coverts, turquoise stripe down upperparts, pale rufous below with whitish throat, mostly blackish bill (orange-reddish on lower mandible of female).
(b) **Juvenile**: Duller, paler below with dusky wash across breast.

15 [511] **BLYTH'S KINGFISHER** *Alcedo hercules*
Male: Like Blue-eared Kingfisher but much larger with longer, heavier, all-black bill (female has reddish base of lower mandible), darker lores, darker crown, ear-coverts and wings.

16 [512] **CRESTED KINGFISHER** *Ceryle lugubris*
(a) **Male** *guttulata*: Very large, black-and-white, uneven spiky crest, densely barred and spotted above, white below with blackish speckling and washed-out pale chestnut flecking on malar area and breast.
(b) **Female**: Usually no chestnut flecking but pale rufous underwing-coverts (white on male).

17 [513] **PIED KINGFISHER** *Ceryle rudis*
(a,b) **Male** *leucomelanura*: More distinct areas of black and white than on larger Crested; flattish crest, two black bands on side of breast, white supercilium; contrasting upperwing and tail pattern, whitish underwing.
(c) **Female**: Single breast-band.

1b
1a
2a
3a
3b
4
2b
6b
6a
5b
5a
7b
8
7a
10
9
11
12a
14b
13a
12b
14a
16b
15
13c
16a
17a
13b
17c
17b

PLATE 58. COCKATOOS, ROLLERS, BEE-EATERS & COMMON HOOPOE

1 [390A] **YELLOW-CRESTED COCKATOO** *Cacatua sulphurea*
(a,b) **Male** *sulphurea*: All white; upcurved yellow crest, black bill; ear-coverts, underwing and -tail suffused yellow.

2 [390B] **TANIMBAR CORELLA** *Cacatua goffini*
(a,b) **Male**: All white; salmon-pinkish lores, short deep pale bill, short crest; underwing and undertail washed yellow.

3 [495] **INDIAN ROLLER** *Coracias benghalensis*
(a,b) **Adult** *affinis*: Turquoise crown, greenish-olive upperparts, purplish-blue rump, light turquoise uppertail-coverts and vent, vinous-brown head-sides to belly; brilliant purplish-blue and light turquoise wings and outertail.
(c) **Juvenile**: Browner-bodied, reduced turquoise on crown, paler, browner head-sides.

4 [496] **DOLLARBIRD** *Eurystomus orientalis*
(a,b) **Adult** *orientalis*: Thick red bill; brownish/greenish above, dark turquoise below; silvery patch on primaries.
(c) **Juvenile**: Browner upperparts, all-brown head, mostly dark bill.

5 [514] **RED-BEARDED BEE-EATER** *Nyctyornis amictus*
(a) **Male**: Like Blue-bearded but with shaggy red feathers on throat and breast, purplish-pink forecrown, broad black terminal band on undertail.
(b) **Female**: Usually shows red forehead.
(c) **Juvenile**: Mostly green head and breast.

6 [515] **BLUE-BEARDED BEE-EATER** *Nyctyornis athertoni*
Adult *athertoni*: Green upperparts with light blue wash, blue forecrown and shaggy throat and breast feathers, pale buffish-yellow belly with broad green streaks, golden-yellowish undertail with indistinct terminal band.

7 [516] **LITTLE GREEN BEE-EATER** *Merops orientalis*
(a) **Adult** *ferrugeiceps*: Small; coppery-rufous crown to mantle, green throat, light blue chin and cheeks, black breast-patch, mostly green uppertail.
(b) **Juvenile**: Mostly green crown and mantle, mostly yellowish throat, plain green breast.

8 [517] **BLUE-TAILED BEE-EATER** *Merops philippinus*
(a,b) **Adult**: Bronze-green crown to back, mid-blue rump and uppertail-coverts, pale yellowish upper throat, dull chestnut wash on lower throat and upper breast.
(c) **Juvenile**: Washed-out chestnut on lower throat and upper breast, more bluish-green crown and mantle.

9 [518] **BLUE-THROATED BEE-EATER** *Merops viridis*
(a,b) **Adult** *viridis*: Dark chestnut crown to mantle, blue throat, distinctly pale blue rump and uppertail-coverts.
(c) **Juvenile**: Green crown and mantle, pale chin.

10 [519] **CHESTNUT-HEADED BEE-EATER** *Merops leschenaulti*
(a) **Adult** *leschenaulti*: Chestnut crown to mantle, pale blue rump and uppertail-coverts, tail without elongated central feathers, pale yellow throat with chestnut and black gorget.
(b) **Juvenile**: Crown and mantle mostly green, some dull chestnut on hindcrown/nape, washed-out gorget.

11 [520] **COMMON HOOPOE** *Upupa epops*
(a,b) **Adult** *longirostris*: Pinkish-buff with long bill, black-marked crest, broad black-and-whitish bands above.

1b
1a
2a
2b
3c
3a
3b
4a
4b
4c
1-4 & 11 to different scale
5b
5a
5c
6
7a
9b
7b
9a
9c
10b
10a
8c
8a
8b
11b
11a

PLATE 59. HORNBILLS

1 [521] **NORTHERN BROWN HORNBILL** *Ptilolaemus austeni*
(a,b) **Male**: Relatively small; mostly brownish, whitish throat and upper breast, brownish-rufous lower breast and belly, pale yellowish bill with small casque; white tips to outer primaries and outertail feathers.
(c) **Female**: Dark brownish throat and underparts, smaller casque.

2 [522] **SOUTHERN BROWN HORNBILL** *Ptilolaemus tickelli*
(a) **Male**: Bright brownish-rufous throat and underparts.
(b) **Female**: From Northern Brown mainly by horn-coloured to blackish bill.

3 [523] **BUSHY-CRESTED HORNBILL** *Anorrhinus galeritus*
(a,b) **Male**: Thick blackish drooping crest, dirty brownish-grey basal two-thirds of tail, blackish bill with relatively small casque, pale orbital and gular skin.
(c) **Female**: Often has extensively dull yellowish to ivory-coloured bill.

4 [524] **ORIENTAL PIED HORNBILL** *Anthracoceros albirostris*
(a,b) **Male** *albirostris*: Relatively small; white belly and vent, white facial markings, pale yellowish bill and casque with dark markings; black wings with broad white trailing edge, black tail with broadly white-tipped outer feathers.
(c) **Female**: Smaller casque with more extensive dark areas.
(d) **Male** *convexus*: Mostly white outertail feathers (tail often appears white from below).

5 [525] **BLACK HORNBILL** *Anthracoceros malayanus*
(a,b) **Male**: Relatively small with all-black plumage, yellowish-white bill and casque, broadly white-tipped outertail feathers, blackish orbital skin.
(c) **Male** variant: Broad white supercilium.
(d) **Female**: Blackish bill and casque, pinkish orbital skin and submoustachial patch.

6 [526] **GREAT HORNBILL** *Buceros bicornis*
(a) **Male** *homrai*: Very large; black and yellowish-white head pattern, large yellowish bill and casque, white tail with broad black central band, reddish eye; double broad yellowish-white and white bands across wings.
(b) **Female**: Whitish eye; shows less black on casque than male.

7 [527] **RHINOCEROS HORNBILL** *Buceros rhinoceros*
(a) **Male** *rhinoceros*: Very large; yellowish bill with red at base, bright red and yellow upwardly curved casque, all-black head and neck, reddish eyes; all-black wings.
(b) **Female**: Whitish eyes.

8 [528] **HELMETED HORNBILL** *Rhinoplax vigil*
(a,b) **Male**: Very large; white tail with black central band and elongated central feathers, bare red skin on sides of head, throat and neck, short straight bill, short, rounded reddish casque with yellow tip; broad white trailing edge to wing.

9 [529] **WHITE-CROWNED HORNBILL** *Berenicornis comatus*
(a) **Male**: White head and throat to belly, long shaggy white crest, all-white tail, dark bill without prominent casque, pale blue facial skin; black wings with white trailing edge.
(b,c) **Female**: Black neck and underparts.

10 [530] **RUFOUS-NECKED HORNBILL** *Aceros nipalensis*
(a) **Male**: Very large, bright rufous head, neck and underparts, white tail with black basal third/half, pale yellowish bill with virtually no casque, bright red gular skin; black wings with white-tipped outer primaries.
(b,c) **Female**: Black head, neck and underparts.

11 [531] **WRINKLED HORNBILL** *Aceros corrugatus*
(a,b) **Male**: Recalls Wreathed Hornbill but smaller, has smaller red-based yellow bill with squarer red casque, blue orbital skin, plain gular pouch (less markedly bulging) and black tail-base.
(c) **Female**: Black head-sides and neck, yellowish casque, blue gular pouch.

12 [532] **PLAIN-POUCHED HORNBILL** *Aceros subruficollis*
(a,b) **Male**: Very like Wreathed Hornbill but somewhat smaller, bill relatively shorter and deeper with more peaked casque and no lateral corrugations or blackish band on gular pouch.
(c) **Female**: Very similar to Wreathed, differing as male.

13 [533] **WREATHED HORNBILL** *Aceros undulatus*
(a,b) **Male**: Large; brownish-white sides of crown, neck and upper breast, dark brown nape and hindneck, white tail, dull yellowish bill with lateral corrugations and small casque, bulging yellow gular pouch with blackish band, reddish orbital skin, all-black wings.
(c) **Female**: Blue gular pouch, all-black head, neck and underparts.

4c
4a
4b
4d
1b
1c
1a
2b
2a
7a
7b
5c
5a
5b
5d
8a
8b
6a
6b
3a
3c
9a
9c
3b
10a
9b
10b
11a
10c
11b
13c
13a
11c
12b
13b
12c
12a

PLATE 60. BARBETS

1 [534] **FIRE-TUFTED BARBET** *Psilopogon pyrolophus*
(a) **Adult**: Green with brownish-maroon hindcrown and nape, whitish band across forecrown, pale yellowish-green bill with vertical dark band, short green supercilium, grey ear-coverts, yellow and blackish breast-bands.
(b) **Juvenile**: Dull olive-brown hindcrown and nape, dull supercilium.

2 [535] **GREAT BARBET** *Megalaima virens*
Adult *virens*: Relatively large and long-tailed, large yellowish bill, dark bluish head, brownish mantle and breast, dark greenish-blue streaks on yellow belly, red undertail-coverts.

3 [536] **RED-VENTED BARBET** *Megalaima lagrandieri*
(a) **Adult** *lagrandieri*: Relatively large; large, mostly dark bill, head brownish with paler greyish sides and throat, narrow blue eyebrow, green upper- and underside, red undertail-coverts.
(b) **Juvenile**: Uniform brownish head.

4 [537] **LINEATED BARBET** *Megalaima lineata*
Adult *hodgsoni*: Thick pale bill, dark brown head and breast with broad whitish streaks, yellow orbital skin.

5 [538] **GREEN-EARED BARBET** *Megalaima faiostricta*
Adult *faiostricta*: Like Lineated Barbet but smaller, bill mostly dark, dark orbital skin, green ear-coverts, all-green mantle, red spot on breast-side.

6 [539] **GOLD-WHISKERED BARBET** *Megalaima chrysopogon*
Adult *laeta*: Dark bill, broad eyestripe above large yellow patch on lower head-side, pale greyish-buff throat.

7 [540] **RED-CROWNED BARBET** *Megalaima rafflesii*
(a) **Adult** *malayensis*: Red crown, blue throat and supercilium, black eyestripe bordered yellow and red below.
(b) **Juvenile**: Much duller.

8 [541] **RED-THROATED BARBET** *Megalaima mystacophanos*
(a) **Male** *mystacophanos*: Yellow forehead, blue-bordered red throat, yellow malar stripe, blue cheeks.
(b) **Female**: Mostly greenish head without black, red patches on lores, hindcrown and upper breast-side.
(c) **Juvenile**: Head all green with yellower-tinged forehead and throat.

9 [542] **GOLDEN-THROATED BARBET** *Megalaima franklinii*
(a) **Adult** *ramsayi*: Red and yellow crown, yellow then grey on throat, grey-streaked blackish ear-coverts.
(b) **Juvenile**: Duller, with less distinct head pattern.
(c) **Adult** *auricularis*: Violet ear-coverts, black eyestripe, blue band from eyestripe to mostly yellow throat.
(d) **Adult** *franklinii*: Plain black eyestripe, deeper yellow upper throat.

10 [543] **BLACK-BROWED BARBET** *Megalaima oorti*
(a) **Adult**: Blue ear-coverts, black brow and whisker, blue border to yellow throat, with red spot.
(b) **Juvenile**: Duller, less red and black on head.

11 [544] **ANNAM BARBET** *Megalaima annamensis*
Adult: Broader black eyestripe than Black-browed, more blue on lower throat.

12 [545] **BLUE-THROATED BARBET** *Megalaima asiatica*
(a) **Adult** *davisoni*: All-blue head-sides and throat, red crown with blue mid-band, red spot on breast-side.
(b) **Juvenile**: Duller.
(c) **Adult** *asiatica*: Black and yellow bands across centre of crown.
(d) **Adult** *chersonesus*: More extensively blue crown.

13 [546] **MOUSTACHED BARBET** *Megalaima incognita*
(a) **Adult** *elbeli*: Small red patch on rear crown, blue cheeks and throat, black eye- and submoustachial stripe.
(b) **Juvenile**: Duller, greener sides of head and throat, narrower submoustachial stripe.

14 [547] **YELLOW-CROWNED BARBET** *Megalaima henricii*
(a) **Adult** *henricii*: Yellow crown with blue at rear, blue throat, black eyestripe, green head-sides.
(b) **Juvenile**: Duller, with more washed-out head pattern.

15 [548] **BLUE-EARED BARBET** *Megalaima australis*
(a) **Male** *cyanotis*: Small; black front of head, blue midcrown, ear-coverts and throat, red patches above and below ear-coverts, orange-red cheek-patch.
(b) **Female**: Head pattern duller.
(c) **Juvenile**: Rather uniform dull green with blue-tinged ear-coverts and throat.
(d) **Male** *duvaucelii*: Black head-side with larger red patches, broader breast-band.

16 [549] **COPPERSMITH BARBET** *Megalaima haemacephala*
(a) **Adult** *indica*: Small; red forecrown, yellow head-sides and throat, black eyestripe and submoustachial stripe, pale greenish underparts with broad dark green streaks, red band across upper breast.
(b) **Juvenile**: Duller; lacks red on crown and breast.

17 [550] **BROWN BARBET** *Calorhamphus fuliginosus*
Adult *hayii*: Smallish; dark brown with paler underparts, pinkish-orange legs.

1b
1a
2
3a
3b
4
6
7a
7b
5
9d
8c
9b
8b
10a
10b
11
9a
9c
8a
12b
12a
12c
12d
13b
13a
14b
14a
15c
15a
15b
15d
16a
16b
17

PLATE 61. HONEYGUIDES, EURASIAN WRYNECK, PICULETS & PIED WOODPECKERS

1 [551] **MALAYSIAN HONEYGUIDE** *Indicator archipelagicus*
Male: Plain; lemon-yellow shoulder-patch (often obscured; lacking on female), thick bill with paler lower mandible, deep red eye, dirty whitish underparts, greyish wash across breast, dark-streaked lower flanks.

2 [552] **YELLOW-RUMPED HONEYGUIDE** *Indicator xanthonotus*
(a) **Male**: Orange-yellow forehead, cheeks and band down back, tertials fringed white, streaked underparts.
(b) **Female**: Duller than male, yellow areas on head much duller and less extensive.

3 [553] **EURASIAN WRYNECK** *Jynx torquilla*
Adult *chinensis*: Cryptically patterned; upperparts greyish-brown with broad dark stripe down centre, underparts buffish-white, barred blackish, wings and tail heavily barred and vermiculated.

4 [554] **SPECKLED PICULET** *Picumnus innominatus*
(a) **Male** *malayorum*: Tiny and short-tailed, with bold olive-slate and white pattern on head, rufous-buff forehead with black barring, olive-green upperparts, whitish underparts with bold blackish spots and bars.
(b) **Female**: Forehead concolorous with crown.
(c) **Male** *chinensis*: Cinnamon-brown crown, ear-coverts and submoustachial stripe.

5 [555] **RUFOUS PICULET** *Sasia abnormis*
(a) **Male** *abnormis*: Yellowish forehead, pinkish eyering, greenish-olive above, deep rufous head/underparts.
(b) **Female**: Forehead rufous.
(c) **Juvenile**: Dull olive above, brownish-slate below.

6 [556] **WHITE-BROWED PICULET** *Sasia ochracea*
(a) **Male** *reichenowi*: White supercilium behind eye, rufescent-olive mantle/scapulars, buffish-rufous below.
(b) **Female**: Forehead rufous.
(c) **Male** *hasbroucki*: Blackish eyering.

7 [557] **RUFOUS-BELLIED WOODPECKER** *Hypopicus hyperythrus*
(a) **Male** *hyperythrus*: Red crown and nape, whitish face, deep rufous underparts, red undertail-coverts.
(b) **Female**: Crown and nape black with white speckling.

8 [558] **SUNDA PYGMY WOODPECKER** *Dendrocopos moluccensis*
Male *moluccensis*: Like Grey-capped but browner above, dull brownish crown, broad dark greyish-brown ear-covert band and defined submoustachial stripe, broader, more diffuse underpart streaks.

9 [559] **GREY-CAPPED PYGMY WOODPECKER** *Dendrocopos canicapillus*
(a) **Male** *canicapillus*: Small; brownish-grey crown, ear-covert band and washed-out submoustachial stripe; white-barred blackish upperside, dark-streaked underparts, short red streak on rear crown-side.
(b) **Female**: Lacks red on rear crown.
(c) **Male** *kaleensis*: Larger, mantle more extensively black, underparts buffier.

10 [560] **FULVOUS-BREASTED WOODPECKER** *Dendrocopos macei*
(a) **Male** *macei*: Red crown/nape, buffish below, red vent, darker-barred flanks, black central tail.
(b) **Female**: Like male but with black crown and nape.

11 [561] **SPOT-BREASTED WOODPECKER** *Dendrocopos analis*
(a) **Male** *longipennis*: Red crown, black nape, whitish below with spot-streaks then bars, pinkish vent.
(b) **Female**: Black crown.

12 [562] **STRIPE-BREASTED WOODPECKER** *Dendrocopos atratus*
(a) **Male**: Like Fulvous-breasted but larger, upper mantle unbarred, uniform distinct streaks below.
(b) **Female**: Like male but crown and nape black.

13 [563] **YELLOW-CROWNED WOODPECKER** *Dendrocopos mahrattensis*
(a) **Male** *aurocristatus*: Brownish-yellow forecrown grading to red hindcrown, densely white barred/spotted upperparts, no blackish head markings, brown streaking below, red belly-centre, no red on undertail-coverts.
(b) **Female**: Yellowish-brown hindcrown.

14 [564] **CRIMSON-BREASTED WOODPECKER** *Dendrocopos cathpharius*
(a) **Male** *tenebrosus*: All-black mantle and scapulars, large white wing-patch, red hindcrown and nape, heavy dark underpart-streaks, red breast-patch and undertail-coverts.
(b) **Female**: Black hindcrown and nape, red breast-patch duller and smaller.
(c) **Male** *pyrrhothorax*: Entirely red nape, more black on neck-side.
(d) **Female**: Like female *tenebrosus* but more black on neck-side.

15 [565] **DARJEELING WOODPECKER** *Dendrocopos darjellensis*
(a) **Male**: Relatively large and long-billed, small red patch on hindcrown, all-black upperparts, large white wing-patch, rich buff breast and belly with bold blackish streaks.
(b) **Female**: All-black crown.

16 [566] **GREAT SPOTTED WOODPECKER** *Dendrocopos major*
(a) **Male** *cabanisi*: All black above, white wing-patch, brownish-white below, red vent and hindcrown-patch.
(b) **Female**: All-black crown.

1
2a
2b
3
4c
4b
4a
5a
5b
5c
6a
6b
6c
8
9b
9c
9a
11a
11b
10a
10b
12b
12a
13a
13b
7a
7b
14b
14a
14c
14d
15a
15b
16a
16b

PLATE 62. LARGER TYPICAL WOODPECKERS

1 [568] **WHITE-BELLIED WOODPECKER** *Dryocopus javensis*
(a) **Male** *feddeni*: Large size, black and white plumage, red crown, crest and submoustachial band.
(b) **Female**: Black forecrown, no red submoustachial band.

2 [569] **BANDED WOODPECKER** *Chrysophlegma mineaceus*
(a) **Male** *malaccensis*: Reddish-rufous ear-coverts, red of crown extends to nape-sides, distinctly scaled mantle and scapulars, dull rufescent wash on neck and breast, closely barred underparts, barred primaries.
(b) **Female**: No reddish tinge on ear-coverts; whitish speckled forehead and face.

3 [570] **GREATER YELLOWNAPE** *Chrysophlegma flavinucha*
(a) **Male** *lylei*: Pronounced yellow nuchal crest, dull rufous-brown crown, plain olive sides of head, yellow throat, broad blackish streaks on white lower throat and uppermost breast, barred primaries.
(b) **Female**: Hindcrown olive-tinged, dull rufous-chestnut submoustachial stripe, upper throat striped blackish and whitish.
(c) **Male** *wrayi*: Smaller and darker, defined yellow chin and submoustachial stripe, narrow whitish streaks on dark lower throat, crown indistinctly washed dull rufous-chestnut, dark olive at rear, crest shorter with less yellow.
(d) **Female**: Differs partly as male, forecrown indistinctly washed chestnut, hindcrown dark olive.
(e) **Male** *flavinucha*: Olive hindcrown, darker sides of neck and breast.

4 [571] **CHECKER-THROATED WOODPECKER** *Chrysophlegma mentalis*
(a) **Male** *humii*: Yellow nuchal crest, no red head markings, olive crown and ear-coverts, chestnut sides of neck and upper breast, unbarred belly and flanks, largely dull rufous-red wings, barred primaries.
(b) **Female**: Neck-sides, upper breast and submoustachial stripe dull chestnut.

5 [572] **LESSER YELLOWNAPE** *Picus chlorolophus*
(a) **Male** *chlorolophus*: Narrowly red crown-sides, red submoustachial stripe, narrow white moustachial stripe, olive and whitish barring on lower breast to undertail-coverts, yellow nuchal crest, unbarred primaries.
(b) **Female**: Only shows red on side of rear crown, lacks red submoustachial stripe.
(c) **Male** *rodgeri*: Red crown with blackish feather tips and centre, darker sides of neck and underparts, narrow pale underpart barring.
(d) **Male** *krempfi*: Mostly red crown.

6 [573] **CRIMSON-WINGED WOODPECKER** *Picus puniceus*
(a) **Male** *observandus*: Deep red crown, submoustachial stripe and wings, olive-greenish head-sides, mantle and breast, yellow nuchal crest.
(b) **Female**: Lacks red submoustachial stripe.

7 [576] **STREAK-BREASTED WOODPECKER** *Picus viridanus*
(a) **Male** *viridanus*: Red crown and nape. Like Laced but neck-sides and throat duller olive, throat and upper breast streaked.
(b) **Female**: Black crown and nape.

8 [577] **LACED WOODPECKER** *Picus vittatus*
(a) **Male**: Red crown and nape, greyish ear-coverts, blackish submoustachial stripe, unmarked buffy yellowish-olive neck-sides, throat and upper breast, olive-whitish belly with dark streaks/loops, barred primaries and outertail feathers.
(b) **Female**: Black crown and nape.

9 [578] **STREAK-THROATED WOODPECKER** *Picus xanthopygaeus*
(a) **Male**: Like Streak-breasted and Laced but with pale eye, white supercilium, little black on submoustachial stripe, prominently streaked ear-coverts, neck-side, throat and breast.
(b) **Female**: Greyish streaks on black crown, white supercilium.

10 [579] **RED-COLLARED WOODPECKER** *Picus rabieri*
(a) **Male**: Red head and upper breast with olive-greyish head-sides and throat, plain olive underparts with whitish belly-bars.
(b) **Female**: Blackish forehead and crown, less red on submoustachial stripe.

11 [575] **BLACK-HEADED WOODPECKER** *Picus erythropygius*
(a) **Male** *nigrigenis*: Black head and band down nape-centre, red patch on crown-centre, yellow throat and sides of nape, red rump.
(b) **Female**: All-black crown.

12 [574] **GREY-HEADED WOODPECKER** *Picus canus*
(a) **Male** *hessei*: Red forecrown, black hindcrown and nape-centre, plain grey head-sides, black loral and submoustachial stripes, unmarked greyish-olive underparts.
(b) **Female**: Crown all black with grey streaks.

13 [595] **GREAT SLATY WOODPECKER** *Mulleripicus pulverulentus*
(a) **Male** *harterti*: Large size, long thin neck, grey plumage, red submoustachial patch, buff throat and foreneck.
(b) **Female**: Lacks red submoustachial patch.

1a
1b
2a
2b
3a
3b
3c
3d
3e
4a
4b
5a
5b
5c
5d
6a
6b
7a
7b
8a
8b
9a
9b
10a
10b
11a
11b
12a
12b
13a
13b

PLATE 63. SMALLER TYPICAL WOODPECKERS & FLAMEBACKS

1 [567] **RUFOUS WOODPECKER** *Micropternus brachyurus*
(a) **Male** *phaioceps*: Rufescent, blackish bars above, red cheek-patch; shortish black bill, speckled throat.
(b) **Female**: Lacks red cheek-patch.
(c) **Male** *squamigularis*: More heavily marked throat, barred belly.
(d) **Male** *fokiensis*: Creamy head, heavy dark streaks on crown and throat; otherwise much darker.
(e) **Male** *annamensis*: Similar to *fokiensis* but head darker.

2 [580] **OLIVE-BACKED WOODPECKER** *Dinopium rafflesii*
(a) **Male** *rafflesii*: Like flamebacks but olive-green above, breast to vent plain dull olive. Red crown and crest.
(b) **Female**: Black crown and crest.

3 [581] **HIMALAYAN FLAMEBACK** *Dinopium shorii*
(a) **Male** *anguste*: Like Common but mantle redder, brownish submoustachial stripe, less heavily marked breast.
(b) **Female**: From Common by submoustachial loop; from Greater by streaked crown and black hindneck.

4 [582] **COMMON FLAMEBACK** *Dinopium javanense*
(a,b) **Male** *intermedium*: Red crown and crest. Like Greater Flameback but has single black submoustachial stripe, all-black hindneck, no black on lores or at sides of crown.
(c) **Female**: Black crown and crest with white streaks.

5 [583] **BLACK-RUMPED FLAMEBACK** *Dinopium benghalense*
(a) **Male** *benghalense*: From other flamebacks by black throat, rump and streaks on red forehead.
(b) **Female**: Forecrown black with white streaks.

6 [584] **GREATER FLAMEBACK** *Chrysocolaptes lucidus*
(a,b) **Male** *guttacristatus*: Rather large and long-billed; red crown and crest, white head with black lores, crown-edge, mask and looped submoustachial stripe extending in line down neck, olive-golden above, red rump and uppertail-coverts, blackish-scaled white underparts and upper mantle, white centre of hindneck.
(c) **Female**: Black crown and crest with white spots.

7 [585] **PALE-HEADED WOODPECKER** *Gecinulus grantia*
(a) **Male** *indochinensis*: Like Bamboo but mainly maroon-chestnut above, pinkish-red crown-centre, barred wings.
(b) **Female**: No red crown-patch.
(c) **Male** *grantia*: More crimson crown-patch, redder upperside, yellower head-sides.

8 [586] **BAMBOO WOODPECKER** *Gecinulus viridis*
(a) **Male** *viridis*: Quite small; greyish-olive above with red midcrown to nape, red tips on rump, olive-brown below.
(b) **Female**: Pale yellowish-green head.

9 [587] **MAROON WOODPECKER** *Blythipicus rubiginosus*
(a) **Male**: Like Bay but smaller, upperparts and wings unbarred maroon-chestnut (except flight feathers and tertials), tail blackish with faint pale bars. Prominent red neck-patch, often some red on submoustachial area.
(b) **Female**: Lacks red on head.

10 [588] **BAY WOODPECKER** *Blythipicus pyrrhotis*
(a) **Male** *pyrrhotis*: Long pale bill; dull rufescent-brown with paler head, black bars above, red neck-patch.
(b) **Female**: Lacks red on neck.
(c) **Male** *cameroni*: Darker with more heavily barred flight feathers, small red neck-patch.

11 [589] **ORANGE-BACKED WOODPECKER** *Reinwardtipicus validus*
(a) **Male** *xanthopygius*: Long-neck; red crown/underparts, broad whitish stripe down centre of upperparts.
(b) **Female**: Blackish crown and crest, whitish back and rump, dull greyish-brown underparts.

12 [590] **BUFF-RUMPED WOODPECKER** *Meiglyptes tristis*
(a) **Male** *grammithorax*: Small; dense pale barring on blackish plumage, unmarked whitish-buff lower back/rump.
(b) **Female**: Lacks red submoustachial stripe.

13 [591] **BLACK-AND-BUFF WOODPECKER** *Meiglyptes jugularis*
(a) **Male**: Like Heart-spotted but nape and hindneck white, throat mostly blackish, reddish submoustachial stripe.
(b) **Female**: Lacks red submoustachial stripe.

14 [592] **BUFF-NECKED WOODPECKER** *Meiglyptes tukki*
(a) **Male** *tukki*: Recalls Buff-rumped but larger, with narrower pale barring, plain head, pale buff neck-patch.
(b) **Female**: Lacks red submoustachial stripe.

15 [593] **GREY-AND-BUFF WOODPECKER** *Hemicircus concretus*
(a) **Male** *sordidus*: Very small, short-tailed; red crown, pointed crest, sooty-greyish with whitish-buff scales above.
(b) **Female**: Greyish crown.

16 [594] **HEART-SPOTTED WOODPECKER** *Hemicircus canente*
(a) **Male**: Very small, short-tailed, pointed crest, white lower scapulars/tertials with black heart-shaped markings.
(b) **Female**: Whitish forecrown.

1b
1d
2b
3b
4c
1a
1c
2a
3a
4a
5b
4b
1e
5a
6a
7c
6b
7b
8b
6c
7a
9b
10b
8a
9a
11a
10c
11b
12a
12b
10a
13a
13b
14b
14a
15b
16b
15a
16a

PLATE 64. BROADBILLS & PITTAS

1 [596] **GREEN BROADBILL** *Calyptomena viridis*
(a) **Male** *continentis*: Chunky, short-tailed, green with black ear-patch and black bars on wing-coverts.
(b) **Female**: Duller, with no black markings.

2 [597] **LONG-TAILED BROADBILL** *Psarisomus dalhousiae*
(a) **Adult** *dalhousiae*: Mostly green body, long blue tail (black below), black crown, nape and ear-coverts, yellow lores, throat, narrow collar and patch on rear crown-side.
(b) **Juvenile**: Crown, nape and ear-coverts dark green, underparts more uniformly green.

3 [598] **DUSKY BROADBILL** *Corydon sumatranus*
(a) **Adult** *laoensis*: Blackish-brown, buffish throat and upper breast, massive dark reddish bill with greyish tip.
(b) **Juvenile**: Browner, with darker throat and upper breast, pinker bill.

4 [599] **SILVER-BREASTED BROADBILL** *Serilophus lunatus*
(a) **Male** *lunatus*: Brownish upperparts, greyish-white forehead and underparts, black supercilium.
(b) **Female**: Thin whitish necklace across upper breast.
(c) **Male** *rubropygius*: All-grey crown to upper mantle, dark grey supercilium, less blue on wings.

5 [600] **BLACK-AND-RED BROADBILL** *Cymbirhynchus macrorhynchos*
(a) **Adult** *malaccensis*: Upperparts, chin and cheeks black, rump and underparts dark maroon-red with narrow black breast-band, white streak along scapulars to tertials, bright bill.
(b) **Juvenile**: Browner overall, spotted wing-coverts.

6 [601] **BANDED BROADBILL** *Eurylaimus javanicus*
(a) **Male** *harterti*: Uniform dull vinous-reddish head and underparts, dark upperparts with prominent yellow markings, blackish-brown wings with broad yellow band on flight feathers, blackish breast-band.
(b) **Female**: Lacks breast-band.
(c) **Juvenile**: Underparts paler and streaked, browner crown and head-sides, yellow eyebrow, spots/streaks on mantle and tips to wing-coverts.

7 [602] **BLACK-AND-YELLOW BROADBILL** *Eurylaimus ochromalus*
(a) **Male** *ochromalus*: Small with black head and breast-band and diagnostic white collar.
(b) **Female**: Breast-band broken in centre.
(c) **Juvenile**: Duller than female, throat whitish with dark chin, eyebrow yellowish; breast with streaks, no band.

8 [603] **HOODED PITTA** *Pitta sordida*
(a,b) **Adult** *cucullata*: Black head, dark brown crown to nape-centre, green body and greater coverts.
(c) **Juvenile**: Head-sides and upperparts duller and browner, throat mostly white, underparts dull brownish.

9 [604] **FAIRY PITTA** *Pitta nympha*
(a,b) **Adult**: Like Blue-winged but smaller, crown-sides rufous, narrow whitish-buff supercilium.

10 [605] **BLUE-WINGED PITTA** *Pitta moluccensis*
(a,b) **Adult**: Blackish head with buff crown-sides/supercilium and white throat, green mantle and scapulars, plain buff underparts with red vent.
(c) **Juvenile**: Head and upperparts duller, crown-sides/supercilium dark-scaled, blue of plumage duller, chin whiter.

11 [606] **MANGROVE PITTA** *Pitta megarhyncha*
(a) **Adult**: Like Blue-winged but bill much longer, crown drabber, underparts slightly duller, chin whitish.
(b) **Juvenile**: Duller. From Blue-winged by same characters as adult.

1a
2b
3b
1b
2a
3a
4b
6c
6b
4a
4c
5a
5b
6a
7b
8-11 to different scale
8c
9b
7c
7a
8a
8b
9a
10c
10a
10b
11b
11a

PLATE 65. PITTAS

1 [607] **GIANT PITTA** *Pitta caerulea*
(a) **Male** *caerulea*: Large; pale greyish head, black centre of crown and nape, black eyestripe, blue upperparts.
(b) **Female**: Like male but base colour of head warm buffish-brown, black nuchal collar, rufescent-brown upperside, blue lower rump/tail.
(c) **Juvenile**: Resembles female but upperparts dark brown, underparts whitish with smudgy breast-band, head and upper breast indistinctly scaled.

2 [608] **BLUE PITTA** *Pitta cyanea*
(a) **Male** *cyanea*: Blue upperside, orange-scarlet rear crown-sides and nape, bluish-whitish underparts, spotted and barred black.
(b) **Female**: Duller overall and browner above.
(c) **Juvenile**: Crown and nape buffish-brown with dark scales, darker median stripe and eyestripe, upperparts dark brown with warm buff streaks, breast and belly mostly dark brown with warm buff streaks.

3 [609] **RUSTY-NAPED PITTA** *Pitta oatesi*
(a) **Male** *oatesi*: Deep rufous head and underparts, well-defined narrow blackish post-ocular stripe.
(b) **Juvenile**: Upperparts, breast and wing-coverts dark brown with whitish-buff spots, crown streaked whitish.

4 [610] **BLUE-RUMPED PITTA** *Pitta soror*
(a) **Male** *petersi*: Blue lower back/rump, pale blue crown and nape, mostly lilac-pinkish head-sides.
(b) **Female**: Duller; greenish crown and nape, upperparts browner-tinged.
(c) **Juvenile**: Like Rusty-naped but head-sides and throat buffier, richer buff marks above, lighter breast marks.
(d) **Male** *tonkinensis*: Green crown and nape with faint blue tinge.

5 [611] **BLUE-NAPED PITTA** *Pitta nipalensis*
(a) **Male** *nipalensis*: Like Blue-rumped but forehead and head-sides rufous, no blue on lower back/rump.
(b) **Female**: Hindcrown to upper mantle green; more rufescent forehead and head-sides than Blue-rumped.
(c) **Juvenile**: From Blue-rumped by paler, buff spots on wing-coverts, weaker-marked crown. From Rusty-naped by buff ear-coverts, buffier crown and wing-covert markings.

6 [612] **BAR-BELLIED PITTA** *Pitta elliotii*
(a) **Male**: Green crown and upperparts, largely yellow underparts with narrow dark bars.
(b) **Female**: Crown and breast buffy-brown, head-sides blackish-brown with buffish streaks.
(c) **Juvenile**: Dark brown with darker head-sides, paler throat, pale buff spots on upperparts and breast.

7 [613] **BANDED PITTA** *Pitta guajana*
(a) **Male** *irena*: Crown-sides and supercilium bright yellow turning vivid orange-red on nape, breast to vent blue-black, white band along inner wing-coverts and secondaries.
(b) **Female**: Underparts whitish with narrow dark barring; less orange-red on nape.
(c) **Juvenile**: Duller than female, dark brown breast with buff spots/streaks.

8 [614] **GURNEY'S PITTA** *Pitta gurneyi*
(a) **Male**: Blue crown/nape, black forecrown/head-sides, yellow underparts with black centrally, black-barred flanks.
(b) **Female**: Crown and nape buffy-rufous, head-sides blackish-brown, underparts mostly buffy-whitish with dark bars.
(c) **Juvenile**: Much like female but forehead to nape, breast and upper belly dark brown with buff streaks.

9 [615] **GARNET PITTA** *Pitta granatina*
(a) **Adult** *coccinea*: Black head-sides and throat, scarlet crown and nape and crimson belly and vent.
(b) **Juvenile**: Mostly dark brown with paler throat and some red on nape and vent, wing-coverts and tail duller blue.

10 [616] **EARED PITTA** *Anthocincla phayrei*
(a) **Male**: Blackish crown-centre, head-sides and nape, buffy-whitish elongated supercilium; rather plain above, buffish below with dark-scaled flanks.
(b) **Female**: Head browner, heavier markings below.

1c
2b
2c
2a
1b
1a
5b
3b
4b
4c
5a
3a
4a
4d
5c
7c
7a
7b
6c
8c
6b
6a
8b
8a
9a
10a
10b
9b

PLATE 66. GOLDEN-BELLIED GERYGONE, RAIL-BABBLER, SHRIKE-BABBLERS & WHITE-BELLIED ERPORNIS

1 [617] **GOLDEN-BELLIED GERYGONE** *Gerygone sulphurea*
(a) **Adult** *sulphurea*: Shortish black bill, greyish-brown upperparts, pale yellow underparts, darkish head-sides and whitish lores and spots on tail.
(b) **Juvenile**: Complete narrow whitish eyering, greyish head-sides, slightly paler yellow underparts.

2 [618] **RAIL-BABBLER** *Eupetes macrocerus*
(a) **Adult** *macrocerus*: Slender, with long neck, bill and tail. Warm brown, with buffy-rufous forehead, chestnut-red crown, hindneck and foreneck, black band through ear-coverts to neck-side and long white supercilium/neck-stripe.
(b) **Juvenile**: Resembles adult but dull chestnut crown and hindneck, warmer upperparts, duller head-/neck-stripes, whitish throat and greyer belly.

3 [619] **BLACK-HEADED SHRIKE-BABBLER** *Pteruthius rufiventer*
(a) **Male** *rufiventer*: Black head, wings and tail, rufous-chestnut upperparts, pale grey throat and breast, and deep pinkish belly and vent.
(b) **Female**: Crown scaled grey, head-sides grey, upperparts, wing-fringes and uppertail mostly bright olive-green, underparts darker than male.
(c) **Immature male**: Crown and nape duller and browner than adult, mantle yellowish-green with blackish bars.

4 [620] **WHITE-BROWED SHRIKE-BABBLER** *Pteruthius flaviscapis*
(a) **Male** *aeralatus*: Black head with white supercilium, grey rest of upperparts, black tail and wings, pale grey throat and underparts, chestnut tertials with black tips.
(b) **Female**: Head greyish, supercilium duller, upperparts greyish-brown, wings and tail mostly golden-olive, underparts creamy-buffish, throat paler.
(c) **Juvenile male**: Crown to uppertail-coverts pale brown, lores and ear-coverts black, white below.
(d) **Male** *ricketti*: Greyer ear-coverts, grey below.
(e) **Female**: Grey throat and upper breast.
(f) **Male** *validirostris*: Whiter underparts, all-chestnut tertials.
(g) **Female**: All-chestnut tertials, dark grey head, whiter throat and upper breast.

5 [621] **GREEN SHRIKE-BABBLER** *Pteruthius xanthochlorus*
(a) **Male** *hybrida*: Short stout bill, greyish crown and head-sides, white eyering, greyish-white throat and breast, yellowish-buff flanks and vent, narrow pale bar on greenish-olive wings.
(b) **Female**: Paler grey on head, duller upperparts.
(c) **Male** *pallidus*: Slaty-grey head to upper mantle, broad white eyering, brighter lower flanks and vent, largely grey fringes to wing feathers.

6 [622] **BLACK-EARED SHRIKE-BABBLER** *Pteruthius melanotis*
(a) **Male** *melanotis*: Like Chestnut-fronted but forehead yellowish, throat paler, nape slate-grey, rear ear-coverts with broad black border, flight feathers fringed slaty-grey.
(b) **Female**: Rufous-buff wing-bars, all-yellowish underparts, pale chestnut malar area.
(c) **Juvenile**: Like female but has less distinct head pattern, brown nape, much paler underparts.
(d) **Male** *tahanensis*: Chestnut restricted to throat.
(e) **Female**: Whitish-yellow below.

7 [623] **CHESTNUT-FRONTED SHRIKE-BABBLER** *Pteruthius aenobarbus*
(a) **Male** *intermedius*: Small, robust and stout-billed with chestnut then yellow forehead, whitish-grey supercilium, white eyering, deep yellow underparts with dark chestnut centre to throat and breast-wash, two white wing-bars.
(b) **Female**: Pale rufous-chestnut forehead, no yellow band on forecrown, creamy-whitish underparts, wing-bars dull pale rufous-buff, rufous-buff primary fringes.
(c) **Male** *indochinensis*: Reduced chestnut on breast.

8 [624] **WHITE-BELLIED ERPORNIS** *Erpornis zantholeuca*
(a) **Adult** *zantholeuca*: Yellowish green above, short crest, pale greyish below with whiter throat, yellow vent.
(b) **Juvenile**: Crest shorter, unscaled.
(c) **Adult** *tyrannula*: Greener upperparts, darker grey underparts.
(d) **Adult** *canescens*: Pale grey wash above.

2b
1a
1b
2a
3c
3a
4f
4g
3b
4a
5b
5c
5a
4c
4d
4e
4b
7a
7c
7b
6a
8a
6c
8b
6b
6e
6d
8c
8d

PLATE 67. CUCKOOSHRIKES, PIED TRILLER & FLYCATCHER-SHRIKES

1 [625] **LARGE CUCKOOSHRIKE** *Coracina macei*
(a) **Male** *siamensis*: Large; thick bill, pale grey with blackish lores, dark grey ear-coverts, whitish vent.
(b) **Female**: Paler, with paler lores and ear-coverts, whiter below with grey barring, pale barring on rump.
(c) **Juvenile**: Head and body heavily barred/scaled; broad whitish fringes to wing-coverts, tertials and secondaries.
(d) **Male** *rexpineti*: Darker forehead, head-sides and throat.

2 [626] **JAVAN CUCKOOSHRIKE** *Coracina javensis*
(a) **Male** *larutensis*: Like Large but darker, primaries and primary coverts blacker, less white on tail-tip.
(b) **Female**: Differs from Large in same way as male.

3 [627] **BAR-BELLIED CUCKOOSHRIKE** *Coracina striata*
(a) **Male** *sumatrensis*: Pale eyes, rather pale lores, pale and dark bars on rump, faint grey bars on vent.
(b) **Female**: Rump and lower breast to undertail-coverts broadly barred blackish and whitish.
(c) **Juvenile**: Head and body heavily scaled, has whiter fringes to wing feathers.

4 [628] **INDOCHINESE CUCKOOSHRIKE** *Coracina polioptera*
(a) **Male** *indochinensis*: From Black-winged by paler wing feather fringes, tail less graduated, greyer above and with broader white tips below.
(b,c) **Female**: Head-sides and throat to belly paler than male, with darker/paler scales, undertail-coverts whitish with blackish scales; large pale patch on underwing.
(d) **Male** *jabouillei*: Much darker, with more uniform wings.

5 [629] **BLACK-WINGED CUCKOOSHRIKE** *Coracina melaschistos*
(a) **Male** *avensis*: Very like Indochinese but wings blackish (also shown by subspecies *jabouillei* of Indochinese), tail more graduated, blackish above and with narrower white tips below.
(b,c) **Female**: Typically plainer below than Indochinese, different tail shape and pattern (as male), usually plainer-winged; little or no pale patch on underwing.
(d) **Juvenile**: Head and body paler, heavily barred/scaled buffish to whitish and dark sooty-brownish, whitish to buffish-white tips to wing feathers.
(e) **Male** *melaschistos*: Darker.
(f) **Male** *saturata*: Much darker, showing little contrast between mantle and wings.

6 [630] **LESSER CUCKOOSHRIKE** *Coracina fimbriata*
(a) **Male** *neglecta*: Small; squarish tail with small white tips below; otherwise like Black-winged.
(b) **Female**: Like Indochinese but more uniformly barred below, smaller white tips to undertail, usually clearer supercilium, dark eyestripe and pale ear-covert streaking.

7 [631] **BLACK-HEADED CUCKOOSHRIKE** *Coracina melanoptera*
(a) **Male** *sykesi*: Black hood.
(b) **Female**: Like Indochinese but with stronger pale wing-covert fringes, stronger black bars below, white vent.

8 [632] **PIED TRILLER** *Lalage nigra*
(a) **Male** *nigra*: Black upperside, whitish supercilium, black eyestripe, grey back to uppertail-coverts, large white wing markings.
(b) **Female**: Dark of plumage greyish-brown, underparts tinged buffish with indistinct dark scales.
(c) **Juvenile**: Crown and upperparts with pale buff scales, underparts with dark streaks on throat and breast.

9 [656] **BAR-WINGED FLYCATCHER-SHRIKE** *Hemipus picatus*
(a) **Male** *picatus*: Black upperside, whitish underside and prominent long white wing-patch.
(b) **Female**: Like male but black of upperside replaced by brown, underparts paler.
(c) **Juvenile**: Barred/scaled buff above, wing-patch buffish with dark bars/scales.
(d) **Male** *capitalis*: Dark brown mantle, back and scapulars.

10 [657] **BLACK-WINGED FLYCATCHER-SHRIKE** *Hemipus hirundinaceus*
(a) **Male**: Like Bar-winged but lacks white wing-patch.
(b) **Female**: Like male but upperside browner.
(c) **Juvenile**: Buff bars/scales above, wing-coverts mostly buffish with dark markings.

1d
1a
1c
2a
3c
3a
1b
2b
3b
4a
4d
5c
5a
5f
5e
4b
4c
5b
5d
6a
6b
7b
7a
8b
9d
9a
10c
9c
9b
10a
10b
8c
8a

PLATE 68. MINIVETS

1 [633] **ROSY MINIVET** *Pericrocotus roseus*
(a) **Male** *roseus*: Brownish-grey above, greyer crown, red rump and wing-patch, rosy-pink below, whitish throat.
(b) **Female:** Red replaced by pale yellow; yellow wing markings, pale yellowish-white throat, washed-out yellow underparts.
(c) **Male** '*stanfordi*': Like Swinhoe's but has pale pink forecrown, often pinkish on outertail-feathers and wing-patch, reddish on rump, pink tinge below.
(d) **Female:** Like Swinhoe's but yellow markings on wing and tail, tinge to uppertail-coverts.

2 [634] **SWINHOE'S MINIVET** *Pericrocotus cantonensis*
(a) **Male:** Like Ashy but hindcrown dark grey, white behind eye, upperparts tinged brown, breast and belly washed vinous-brownish, rump pale drab brownish; wing-patch (if present) pale yellowish-buff.
(b) **Female:** Paler above, rump less sharply contrasting, wing-patch (if present) yellower-tinged. From Ashy by paler rump, browner upperparts, no dark forehead-band, underparts less clean.

3 [635] **ASHY MINIVET** *Pericrocotus divaricatus*
(a) **Male** *divaricatus*: Black lores, crown and nape, all-grey mantle to uppertail-coverts, creamy-white forecrown and underparts.
(b) **Female:** Crown and upperparts pale grey; whitish forehead-band, blackish loral stripe narrowly across extreme forehead.

4 [636] **SMALL MINIVET** *Pericrocotus cinnamomeus*
(a) **Male** *vividus*: Small; grey above with darker ear-coverts, reddish-orange rump, breast and flanks, orange-yellow vent, orange and yellow wing-patch.
(b) **Female:** Pale dull grey above, bright reddish-orange rump, greyish-white throat and upper breast, pale yellow belly and vent.

5 [637] **FIERY MINIVET** *Pericrocotus igneus*
(a) **Male** *igneus*: Like miniature Scarlet but more orange, wing-patch red-orange, no markings on tertials and secondaries.
(b) **Female:** Orange-yellow forehead-band, bright red-orange rump, yellow below.
(c) **Juvenile:** Like female but browner above with dark and whitish markings, dark mottling on breast.

6 [638] **JERDON'S MINIVET** *Pericrocotus albifrons*
(a) **Male:** Glossy black above with white supercilium/forehead-band and wing markings, orange-centred rump; white below with orange breast-patch.
(b) **Female:** Crown to back mousy grey-brown; grey-washed breast.

7 [639] **GREY-CHINNED MINIVET** *Pericrocotus solaris*
(a) **Male** *rubrolimbatus*: Recalls Long-tailed but head and mantle dark grey, whitish chin, orange-yellow wash on throat; more orange underparts, rump and wing-patch.
(b) **Female:** Olive-yellow with grey head and mantle, greyish-white chin, dark wings with yellow patch.
(c) **Juvenile:** Like female but slightly darker above with yellow scales/bars and blackish subterminal markings.
(d) **Male** *deignani*: More uniform lower head-sides and throat, yellowish-olive tint on throat and redder orange parts of plumage.
(e) **Female:** Greyish-white throat, olive cast to yellow underparts, slightly more olive rump.
(f) **Male** *montanus*: Much redder, with black upperparts and ear-coverts, dark grey throat.
(g) **Female:** Blackish-slate crown and mantle, orange-yellow on underparts, wing-patch and rump.

8 [640] **LONG-TAILED MINIVET** *Pericrocotus ethologus*
(a) **Male** *ethologus*: Glossy black with bright red lower back to uppertail-coverts, breast to vent, wing-patch and outertail-feathers. Slimmer than Scarlet, less extensively red tail.
(b) **Female:** Upperparts greyish-olive, yellowish olive-green lower back to uppertail-coverts; bright yellow below with paler lower and whitish upper throat.
(c) **Juvenile:** Like female but scaly above, wing-coverts tipped yellowish-white, breast and flanks spotted/barred and washed drab olive-greyish.
(d) **First-summer male:** Orange extreme forehead, darker crown to mantle than female, deep orange to red-orange rump and wing-patch, deep orange below.
(e) **Female** *annamensis*: Grey crown to mantle, brighter yellow plumage, yellower extreme forehead, cheeks and upper throat.

9 [641] **SHORT-BILLED MINIVET** *Pericrocotus brevirostris*
(a) **Male** *neglectus*: Like Long-tailed but single red line along primary fringes, more red on outertail feathers, black throat extends onto upper breast.
(b) **Female:** Like Long-tailed (subspecies *ethologus*) but forehead golden-yellow, forecrown washed yellow, mantle greyer, throat yellow, more yellow on outertail.

10 [642] **SCARLET MINIVET** *Pericrocotus speciosus*
(a) **Male** *semiruber*: Larger, larger-billed and shorter-tailed than Long-tailed, red marks on tertials and secondaries.
(b) **Female:** From other minivets by size, build, yellow forehead and forecrown, slaty-grey hindcrown to mantle, bright yellow throat and underparts, isolated yellow markings on tertials and secondaries.
(c) **Juvenile:** Like Long-tailed and Short-billed but has yellower underparts.

1a
1b
1c
1d
2a
2b
3a
3b
4a
4b
5b
5c
6a
5a
6b
7d
7e
7a
7b
7c
7f
7g
8a
8d
8b
8c
8e
9a
9b
10a
10b
10c

PLATE 69. ORIOLES & WOODSWALLOWS

1 [644] **DARK-THROATED ORIOLE** *Oriolus xanthonotus*
(a) **Male** *xanthonotus*: Small; black hood and wings and white lower breast to belly with prominent black streaks.
(b) **Female**: Upperside olive-green with brighter lower rump and uppertail-coverts and darker, greyer crown and head-sides, underparts whitish with bold blackish streaks and yellow undertail-coverts.
(c) **Immature male**: Like female but upperparts yellower-green, sooty crown and nape, greyish-washed throat.

2 [645] **INDIAN GOLDEN ORIOLE** *Oriolus kundoo*
(a) **Male**: Yellow nape, restricted black eye-patch, black shoulders diagnostic. Wings black with bold yellow markings.
(b) **Female**: From similar Slender-billed and Black-naped Orioles by lack of blackish head-band and blacker wing-coverts.
(c) **Juvenile**: Similar to Slender-billed and Black-naped (which both often show shadowy head-band); smaller, with thinner streaking below, which tends to be more lacking on throat/vent; wing-coverts with darker centres?

3 [646] **SLENDER-BILLED ORIOLE** *Oriolus tenuirostris*
(a) **Male** *tenuirostris*: Very like female Black-naped but bill longer and thinner, black nape-band narrower.
(b) **Female**: Like male but yellow of plumage tinged greener, shows some indistinct, narrow darker streaks on underparts.
(c) **Juvenile**: Difficult to separate from Black-naped, except by longer, thinner bill.
(d) **Immature**: As juvenile.

4 [647] **BLACK-NAPED ORIOLE** *Oriolus chinensis*
(a) **Male** *diffusus*: Relatively large; golden-yellow body and wing-coverts, and broad black band from lores to hindcrown and nape.
(b) **Female**: Upperparts and wing-coverts mostly olive-yellow.
(c) **Juvenile**: Much duller than female, lacking black head-band. Possibly not separable in field from Slender-billed, except by thicker, slightly shorter bill.
(d) **Immature**: Gradually attains dark head-band.

5 [648] **BLACK-HOODED ORIOLE** *Oriolus xanthornus*
(a) **Male** *xanthornus*: Uniform golden-yellow body and wing-coverts and contrasting black hood.
(b) **Female**: Lower mantle to rump washed olive.
(c) **Juvenile**: Like female but crown often streaked olive, yellowish-white eyering, whitish throat with blackish streaks, and blackish bill.

6 [649] **BLACK-AND-CRIMSON ORIOLE** *Oriolus cruentus*
(a) **Male** *malayanus*: Glossy bluish-black plumage, large dark crimson patch on lower breast and upper belly, dark crimson primary coverts and pale bluish bill, legs and feet.
(b) **Female**: Lacks crimson in plumage, has pale greyish lower breast and belly.

7 [650] **MAROON ORIOLE** *Oriolus traillii*
(a) **Male** *traillii*: Dark maroon body, black hood and wings, pale dull maroon tail, bluish-grey bill and pale yellowish eyes.
(b) **Female**: Crown, nape and head-sides blackish-brown, upperparts dark brown with maroon tinge and dark reddish-chestnut rump, tail brownish-maroon, underparts whitish with heavy dark streaks and pale reddish-maroon undertail-coverts.
(c) **Female variant**: Throat and breast may be darker.
(d) **Juvenile**: Like paler-throated female but has pale streaks on forehead, rufescent tips to wing-coverts and scales on mantle, back and scapulars, narrower dark streaks on underparts.
(e) **Immature male**: Attains black hood and variable maroon wash on body while underparts still streaked.
(f) **Male** *nigellicauda*: Reddish wash on mantle and back, distinctly deep reddish rump, tail-coverts and tail.

8 [651] **SILVER ORIOLE** *Oriolus mellianus*
(a) **Male**: Silvery-whitish body plumage, black hood and wings and dull maroon tail.
(b) **Female**: Like Maroon but mostly greyish mantle to rump and paler undertail-coverts with pale fringes; tends to have narrower dark streaks on underparts.

9 [652] **ASHY WOODSWALLOW** *Artamus fuscus*
(a,b) **Adult**: Bill pale bluish, hood and upperparts brownish-grey, underparts paler and browner. In flight wings pale below, broad-based and pointed.
(c) **Juvenile**: Bill duller; browner above with pale fringing, paler below with vague darker barring.

10 [653] **WHITE-BREASTED WOODSWALLOW** *Artamus leucorynchus*
(a,b) **Adult** *leucorynchus*: Like Ashy but clean white underparts and underwing-coverts.
(c) **Juvenile**: Like Ashy but underparts whiter, whiter feather tips on upperside.

1b
1a
1c
2a
2b
3d
3b
3a
2c
4a
3c
4d
4c
4b
5b
5a
7a
7f
5c
7b
6a
8a
7c
6b
7e
7d
8b
9c
10c
9a
10a
9b
10b

PLATE 70. MANGROVE WHISTLER, WOODSHRIKES, PHILENTOMAS, FANTAILS, MONARCHS & PARADISE-FLYCATCHERS

1 [643] **MANGROVE WHISTLER** *Pachycephala cinerea*
Adult *cinerea*: Drab brown upperparts, rather slaty-grey crown, white underparts with duller throat and greyish-washed breast. From some flycatchers by thick black bill, no obvious head/wing markings or rufous tones in plumage.

2 [654] **LARGE WOODSHRIKE** *Tephrodornis gularis*
(a) **Male** *jugans*: Greyish crown and nape, pale greyish-brown rest of upperside, blackish mask, white rump-band.
(b) **Female**: Crown duller, mask and bill browner, wash on throat and breast buffier.
(c) **Juvenile**: Whitish shafts and tips on crown to scapulars, buffish-and-whitish bars on wing-coverts.
(d) **Male** *fretensis*: Smaller; fairly uniform pale bluish-slate crown, mantle and scapulars.

3 [655] **COMMON WOODSHRIKE** *Tephrodornis pondicerianus*
(a) **Adult** *pondicerianus*: Like Large but smaller, with whitish supercilium and outertail feathers.
(b) **Juvenile**: Spotted whitish-buff above, wing-coverts tipped whitish-buff, dusky mottling below.

4 [658] **RUFOUS-WINGED PHILENTOMA** *Philentoma pyrhoptera*
(a) **Male typical morph** *pyrhoptera*: Rather stocky and stout-billed; mostly dull blue head and body, buffy-whitish belly and vent, reddish-chestnut greater coverts, tertials, secondaries and tail.
(b) **Male blue morph**: All dull blue, with greyish vent (mixed whitish).
(c) **Female**: No blue; crown and head-sides cold dark greyish-brown (sometimes tinged blue), rest of upperparts mid-brown, underparts buffy-whitish, breast-sides and flanks mid-brown.

5 [659] **MAROON-BREASTED PHILENTOMA** *Philentoma velata*
(a) **Male** *caesia*: Dull blue plumage with black forehead, upper throat and head-sides and dark maroon lower throat and breast (often appears blackish).
(b) **Female**: Duller above; has dull blackish lores, cheeks, throat and upper breast, and duller and darker lower breast and upper belly.

6 [663] **YELLOW-BELLIED FANTAIL** *Chelidorhynx hypoxantha*
(a) **Male**: Dull greenish upperparts, long dark white-tipped graduated tail, blackish mask and bright deep yellow forehead, supercilium and underparts; whitish tips to greater coverts.
(b) **Female**: Mask same colour as crown.

7 [664] **WHITE-THROATED FANTAIL** *Rhipidura albicollis*
(a) **Adult** *celsa*: Mostly dark greyish to blackish-slate plumage (crown blacker) with contrasting white supercilium and throat and fan-shaped white-tipped tail.
(b) **Juvenile**: Resembles adult but upperparts browner-tinged, scapulars and back to uppertail-coverts scaled/barred and wing-coverts indistinctly tipped warm brown; indistinct paler barring on underparts and almost no white on throat, supercilium often tinged buffish.

8 [665] **WHITE-BROWED FANTAIL** *Rhipidura aureola*
(a) **Adult** *burmanica*: From other fantails by long broad white supercilium and mostly whitish underparts.
(b) **Juvenile**: Like adult but throat darker, upperparts browner with pale warm brown scales, mainly on scapulars, back and rump, broad whitish to dull pale rufous tips to wing-coverts.

9 [666] **PIED FANTAIL** *Rhipidura javanica*
Adult *longicauda*: From other fantails by pale underparts with contrasting blackish breast-band.

10 [667] **SPOTTED FANTAIL** *Rhipidura perlata*
Adult: From other fantails by uniform blackish-slate plumage, white vent and prominent white spots/streaks on throat and breast.

11 [676] **BLACK-NAPED MONARCH** *Hypothymis azurea*
(a) **Male** *styani*: Predominantly blue plumage with black nuchal tuft/bump, narrow black band across uppermost breast, and whitish belly and vent.
(b) **Female**: Like male but blue duller and restricted mainly to head, no black nuchal patch or breast-band, upperparts warm-tinged greyish-brown, breast greyish (tinged blue).

12 [677] **ASIAN PARADISE-FLYCATCHER** *Terpsiphone paradisi*
(a) **Male** *indochinensis*: Bright rufous-chestnut upperside and extremely long rufous-chestnut tail.
(b) **Male white 'morph'**: Almost all white, with glossy black head.
(c) **Female**: Like rufous male but lacks tail streamers, crest shorter, eyering usually duller.
(d) **Juvenile**: Upperparts paler and more rufous (including crown and head-sides), underparts whitish, initially scaled/mottled dull rufous on breast, bill dark brownish with flesh-coloured base, no pronounced eyering.
(e) **Male** *incei*: Glossy black hood, grey breast, darker, deeper chestnut above.
(f) **Female**: Throat obviously darker than breast.

13 [678] **JAPANESE PARADISE-FLYCATCHER** *Terpsiphone atrocaudata*
(a) **Male** *atrocaudata*: Shape recalls Asian but plumage all black apart from purple mantle to rump and whitish belly and vent.
(b) **Female**: Very like Asian but duller crown contrasts less with head-sides and throat, breast usually appears more sharply cut off from whitish belly, upperparts duller, tail darker and browner.

2d
1
2a
2b
3a
3b
2c
6b
6a
4c
4a
4b
5a
5b
8b
8a
7b
13b
12a
12e
7a
12b
9
13a
10
12c
12f
11a
12d
11b

PLATE 71. DRONGOS, CROWS, BLACK MAGPIE & CRESTED JAY

1 [668] **BLACK DRONGO** *Dicrurus macrocercus*
(a) **Adult** *thai*: All-blackish plumage (slightly glossed bluish) and long, deeply forked tail.
(b) **First winter**: Whitish scales on uppertail-coverts, breast, belly and undertail-coverts.

2 [669] **ASHY DRONGO** *Dicrurus leucophaeus*
(a) **Adult** *mouhoti*: Dark steely-grey with slightly bluish gloss, paler below, shallowish tail-fork.
(b) **Adult** *nigrescens*: Blacker above and darker below, particularly on throat and breast.
(c) **Adult** *leucogenis*: Very pale ashy-grey, black forehead, whitish lores, ear-coverts and vent.
(d) **Adult** *salangensis*: Like *leucogenis* but a shade darker with mostly grey ear-coverts.

3 [670] **CROW-BILLED DRONGO** *Dicrurus annectans*
(a) **Adult**: Tail broad with shallow fork and strongly upcurled outer tips, bill broad-based and longish.
(b) **First winter**: White spots on underparts.

4 [671] **BRONZED DRONGO** *Dicrurus aeneus*
Adult *aeneus*: Small; small bill, very strong blue to greenish-blue gloss on upperparts, throat and breast.

5 [672] **LESSER RACKET-TAILED DRONGO** *Dicrurus remifer*
Adult *tectirostris*: Relatively small; square tail-tip with very long bare outer shafts ending in longish pendants.

6 [673] **GREATER RACKET-TAILED DRONGO** *Dicrurus paradiseus*
(a) **Adult** *rangoonensis*: Large; tufted forehead, forked tail with long bare outer shafts ending in twisted pendants.
(b) **Adult** *platurus*: Much shorter crest.

7 [674] **HAIR-CRESTED DRONGO** *Dicrurus hottentottus*
Adult *hottentottus*: Large; rather long, slender, slightly downcurved pointed bill, and triangular square-ended tail with strongly upcurled outer tips.

8 [675] **ANDAMAN DRONGO** *Dicrurus andamanensis*
Adult *dicruriformis*: Like Black but larger, tail longer with shallower fork and upcurling outer tips.

9 [679] **HOUSE CROW** *Corvus splendens*
(a) **Adult** *insolens*: All-blackish plumage with broad dull greyish collar encompassing nape, upper mantle, rear ear-coverts, neck and breast.
(b) **Adult** *protegatus*: Paler, more contrasting brownish-grey collar.
(c) **Adult** *splendens*: Much paler, sharply contrasting grey collar.

10 [680] **SLENDER-BILLED CROW** *Corvus enca*
Adult *compilator*: Very similar to Large-billed but slimmer with shorter, squarer-ended tail, relatively long slender bill with less strongly arched upper mandible.

11 [681] **CARRION CROW** *Corvus corone*
Adult *orientalis*: Like Large-billed Crow but bill obviously shorter, slenderer and more pointed; shorter, squarer-ended tail.

12 [682] **LARGE-BILLED CROW** *Corvus japonensis*
Adult *tibetosinensis*: Large, with strong purplish/bluish gloss; rather long, high-arched bill, steep forehead, peaked crown, distinctly wedge-shaped tail, heavy feet.

13 [683] **EASTERN JUNGLE CROW** *Corvus levaillantii*
Adult *levaillantii*: Smaller than Large-billed, with squarer-ended and relatively longer tail, less steep forehead, less high-arched bill, less heavy feet.

14 [684] **SOUTHERN JUNGLE CROW** *Corvus macrorhynchos*
(a) **Adult** *macrorhynchos*: Hard to separate from very similar Eastern Jungle. Apparently somewhat larger, with relatively shorter tail that has more rounded tip, plumage somewhat glossier below.
(b) **Juvenile**: Duller and less glossy than adult.

15 [685] **COLLARED CROW** *Corvus torquatus*
(a) **Adult**: Like Large-billed and Carrion but has white collar encompassing nape, hindneck and lower breast.
(b) **Juvenile**: Body browner and glossless, collar duller with dark feather tips.

16 [699] **BLACK MAGPIE** *Platysmurus leucopterus*
Adult *leucopterus*: All black with white band along greater coverts and tertials and fairly long, broad tail.

17 [701] **CRESTED JAY** *Platylophus galericulatus*
(a) **Adult** *ardesiacus*: Blackish plumage, white patch on neck-side and slightly forward-pointing crest.
(b) **Juvenile**: Upperparts tinged warm brown, underparts duller and barred whitish; warm buff spots on wing-coverts, short crest with buffish tips.
(c) **Immature**: Like adult but paler bars on underparts, pale shaft-streaks on throat and breast.

1-8 to different scale
1b
1a
2a
2d
2b
2c
5
6a
6b
3b
3a
4
7
8
9c
9b
9a
10
11
12
14a
14b
13
15b
15a
16
17a
17b
17c

PLATE 72. SPOTTED NUTCRACKER, MAGPIES, JAYS & TREEPIES

1 [686] **SPOTTED NUTCRACKER** *Nucifraga caryocatactes*
(a) **Adult** *macella*: Dark brown with large white spots on mantle, breast and upper belly, broad white streaks on upper throat and head-sides, white vent and mostly white outertail feathers.
(b) **Juvenile**: Paler; narrower and buffier spots and streaks on upperparts, and paler, mottled underparts.

2 [687] **BLACK-BILLED MAGPIE** *Pica pica*
Adult *sericea*: Slim build, long tail and mostly black plumage with white scapulars and belly.

3 [688] **EURASIAN JAY** *Garrulus glandarius*
(a) **Adult** *leucotis*: Black cap and broad submoustachial band, white head-sides and throat and buffish-grey upperparts.
(b) **Adult** *sinensis*: Warm pinkish-buff crown, nape and head-sides, more pinkish-buff mantle, smaller submoustachial band, pinker underparts.
(c) **Adult** *oatesi*: Warm pinkish-buff crown with indistinct blackish streaks, slightly duller head-sides and throat.

4 [689] **COMMON GREEN MAGPIE** *Cissa chinensis*
(a) **Adult** *chinensis*: Mostly bright green plumage, largely reddish-chestnut wings with black and whitish tertial markings, black band from lores to nape-sides and bright red bill, eyering, legs and feet.
(b) **Adult** (worn): Worn birds may have strongly bluish plumage and browner wings.
(c) **Juvenile**: Duller with paler lower breast to vent, smaller dark subterminal markings on tertials and inner secondaries, browner bill and duller eyering, legs and feet.
(d) **Adult** *margaritae*: Bright golden-yellow crown.

5 [690] **INDOCHINESE GREEN MAGPIE** *Cissa hypoleuca*
(a) **Adult** *hypoleuca*: Very like Common Green but tertials and inner secondaries appear mostly pale greenish, strong lemon-yellow wash on underparts; tail shorter.
(b) **Adult** *concolor*: Greener underparts, darker upperparts and buffish tinge to tail feather tips.

6 [691] **YELLOW-BILLED BLUE MAGPIE** *Urocissa flavirostris*
(a) **Adult** *flavirostris*: Like Red-billed Blue Magpie but has yellow bill and all-blackish hood with small white nuchal patch.
(b) **Juvenile**: Like adult but initially browner on head, mantle and breast; duller nuchal patch, dull yellowish-olive bill, duller legs and feet.

7 [692] **RED-BILLED BLUE MAGPIE** *Urocissa erythrorhyncha*
(a) **Adult** *magnirostris*: Red bill, black hood, white band down hindneck, blue upperparts and very long white-tipped blue tail.
(b) **Juvenile**: Dark areas of head and upper breast much duller and browner, upperparts and wing-coverts browner (latter pale-tipped).
(c) **Adult** *erythrorhyncha*: Rather more extensive, very pale bluish-grey hindcrown/hindneck-band, greyer upperparts, smaller bill.

8 [693] **WHITE-WINGED MAGPIE** *Urocissa whiteheadi*
(a) **Adult** *xanthomelana*: Dark brown head, upper breast and upperside, orange bill and large white markings on wing-coverts, tertials and tail-tips.
(b) **Juvenile**: Paler greyer head, nape and breast, and greyish to brownish bill and eyes.

9 [694] **RUFOUS TREEPIE** *Dendrocitta vagabunda*
(a) **Adult** *kinneari*: Dark rufescent-brown upperparts, deep buff underparts and contrasting blackish-grey hood and upper breast.
(b) **Juvenile**: Hood paler and browner, underparts paler buff, greater coverts, tertials and tail feathers tipped buffish.

10 [695] **GREY TREEPIE** *Dendrocitta formosae*
(a) **Adult** *assimilis*: Recalls Rufous but has paler grey hindcrown and nape, blackish forecrown and face, all-blackish wings with white patch at base of primaries, and dull greyish underparts with deep rufous undertail-coverts.
(b) **Juvenile**: Less black on forecrown, paler hindcrown, nape, head-sides, lower throat and breast with warm buffish infusion and whiter belly.
(c) **Adult** *sinica*: Darker brown mantle and scapulars, whiter rump and uppertail-coverts and all-blackish uppertail.

11 [696] **COLLARED TREEPIE** *Dendrocitta frontalis*
Adult Very pale grey hindcrown, collar, breast and upper belly, rufous remainder of body (paler below) and sharply demarcated black forecrown, ear-coverts and throat.

12 [697] **RACKET-TAILED TREEPIE** *Crypsirina temia*
(a) **Adult**: Relatively small and slim with distinctive all-blackish plumage and long, straight, spatulate-tipped tail.
(b) **Juvenile**: Head and body duller and browner, has less contrasting dark face, brown eyes and narrower tail-tip.

13 [698] **HOODED TREEPIE** *Crypsirina cucullata*
(a) **Adult**: Smallest treepie in region with distinctive pale greyish body and wing-coverts and contrasting black head, primaries and central tail feathers.
(b) **Juvenile**: Head much paler and browner, body browner-tinged, no whitish collar.

14 [700] **RATCHET-TAILED TREEPIE** *Temnurus temnurus*
Adult: Resembles Racket-tailed but has broad tail with long spikes projecting from tips of outer webs of feathers.

1a
1b
2
3b
3a
3c
4b
4a
4c
4d
5a
5b
6a
6b
8b
8a
7b
7c
7a
9b
9a
10b
10a
10c
11
12a
12b
13b
13a
14

PLATE 73. IORAS & SHRIKES

1 [660] **COMMON IORA** *Aegithina tiphia*
(a) **Male non-breeding** *philipi*: Yellow head-sides and underparts, olive-washed flanks, rather deep olive-green upperparts and mostly black wings and tail with two pronounced white to yellowish-white wing-bars.
(b) **Male breeding variant**: May show some black on mantle to rump; bright yellow head sides and underparts.
(c) **Female non-breeding**: Like male non-breeding but upperparts paler green, wings greyer with less distinct bars, yellow of underparts duller/paler.
(d) **Male breeding** *deignani*: Frequently has black crown to back with pale yellow admixed on mantle.

2 [661] **GREEN IORA** *Aegithina viridissima*
(a) **Male** *viridissima*: Like Common but body dark olive-green with paler belly, yellow vent; dark lores and distinctive broad broken yellow eyering.
(b) **Female**: Like Common but upperparts deeper green, wing-bars always yellow, breast and flanks darker and greener, rest of underparts more greenish-yellow.

3 [662] **GREAT IORA** *Aegithina lafresnayei*
(a) **Male** *innotata*: From other ioras by size, thicker, longer bill and lack of wing-bars.
(b) **Juvenile**: Underparts dull and washed olive, particularly on flanks.
(c) **Male** *lafresnayei*: Upperside mostly black.
(d) **Male variant**: Individual with less black on upperparts.

4 [702] **TIGER SHRIKE** *Lanius tigrinus*
(a) **Male**: Grey crown and nape, black forehead and mask and all-whitish underparts, blackish bars/scales above.
(b) **Female**: Duller, buff-tinged flanks with blackish scales; whitish patch on lores, narrow white supercilium.
(c) **First winter**: Crown and head-sides scaly brown, heavily scaled overall, pale bill-base.

5 [703] **BULL-HEADED SHRIKE** *Lanius bucephalus*
(a) **Male** *bucephalus*: Rufescent-brown crown and nape, greyish mantle to uppertail-coverts with brown wash, white wing-patch.
(b) **Female**: Head more uniform, narrower supercilium, upperparts duller, no wing-patch.
(c) **First winter**: Like female but upperparts and sides of body more rufous, tail mid-brown, almost no mask.

6 [704] **BROWN SHRIKE** *Lanius cristatus*
(a) **Male** *confusus*: Warmish brown above, black mask, whitish supercilium and underparts, buff wash along sides.
(b) **Female**: Slightly duller; cream-tinged supercilium, dusky vermiculations on breast and flanks.
(c) **Juvenile**: Duller, but with blackish mask, whitish supercilium and all-brown base colour to upperside.
(d) **Male** *superciliosus*: Rich chestnut upperparts, demarcated white forehead and supercilium.
(e) **Male** *lucionensis*: Pale grey crown, less distinct supercilium, pale grey wash on mantle and scapulars.

7 [705] **BURMESE SHRIKE** *Lanius collurioides*
(a) **Male** *collurioides*: Crown and nape slaty, upperparts unbarred, white wing-patch, white-edged blackish tail.
(b) **Female**: Duller above, whitish lores.
(c) **Juvenile**: Like female but head with blackish-and-buff bars, upperparts with subterminal scales/bars, wing feathers fringed paler, breast, flanks and thighs with wavy bars.
(d) **Male** *nigricapillus*: Much darker, blackish-slate crown and nape and darker chestnut upperparts.

8 [706] **LONG-TAILED SHRIKE** *Lanius schach*
(a) **Adult** *longicaudatus*: Black head with white throat.
(b) **Juvenile**: Black bars/scales on crown, mantle and scapulars, wavy blackish-brown bars on breast, flanks and vent.
(c) **Adult** *schach*: Grey midcrown to upper mantle, smaller wing-patch.
(d) **Adult '*fuscatus*' morph**: Brownish-black throat; brownish-grey crown, nape and body (paler below); mostly blackish scapulars.
(e) **Juvenile**: Relatively plain, blackish mask and throat.

9 [707] **GREY-BACKED SHRIKE** *Lanius tephronotus*
(a) **Adult** *tephronotus*: Mantle and scapulars uniform grey, little black on forehead, no white wing-patch.
(b) **Juvenile**: Scaly brownish-grey crown, mantle and scapulars, bars/scales on rump, uppertail-coverts and flanks.

1a
2a
3a
3c
1c
1b
2b
3b
3d
1d
4a
4b
4c
5b
5a
5c
8a
8b
8c
8d
8e
6a
6b
6c
6d
6e
7a
7b
7c
7d
9a
9b

PLATE 74. SUNBIRDS & SPIDERHUNTERS

1 [717] **CRIMSON SUNBIRD** *Aethopyga siparaja*
(a) **Male** *seheriae*: Red head, mantle and upper breast, iridescent dark green crown and tail, greyish belly.
(b) **Male eclipse**: Like female but throat and breast red.
(c) **Female**: Overall dull olive, slightly yellower below.

2 [718] **TEMMINCK'S SUNBIRD** *Aethopyga temminckii*
(a) **Male** *temminckii*: Like Crimson but head-sides, nape, mantle, wing-coverts, throat, breast and tail all scarlet.
(b) **Female**: Like Crimson but underparts yellowish-olive, wings and tail fringed reddish-rufous.

3 [719] **MRS GOULD'S SUNBIRD** *Aethopyga gouldiae*
(a) **Male** *dabryii*: Red upperparts, uniform red breast, and yellow rump-band and belly.
(b) **Female**: Yellow rump-band, yellow belly (both brighter than in Black-throated).
(c) **Male** *annamensis*: Yellow breast and iridescent blue rump (no yellow).
(d) **Female**: Greyish hood and upper breast, rump yellowish but not as prominent band.

4 [720] **GREEN-TAILED SUNBIRD** *Aethopyga nipalensis*
(a) **Male** *koelzi*: Like Mrs Gould's but crown, throat and tail iridescent dark green, mantle dark red, breast yellow with orange-scarlet streaks.
(b) **Female**: From Mrs Gould's and Black-throated by white-tipped graduated undertail, no clear-cut rump-band. From Crimson by yellower belly and tail.
(c) **Male** *karensis*: Yellow breast, red on upper mantle-sides only.
(d) **Female**: Very like Mrs Gould's (*A.g.annamensis*) but bill longer, upperparts greener, rump less yellow, lower breast more solidly olive-green, tail more graduated.

5 [721] **BLACK-THROATED SUNBIRD** *Aethopyga saturata*
(a) **Male** *petersi*: From Mrs Gould's and Green-tailed by black throat and upper breast, pale yellow lower breast, and dull whitish-yellow belly and rump-band.
(b) **Female**: Greyer overall than Mrs Gould's with paler, narrower yellow rump-band, duller underparts. From other sunbirds by clear-cut pale yellow rump-band.
(c) **Male** *johnsi*: Brighter red mantle, yellow-streaked orange-scarlet breast and entirely iridescent purple-blue throat.
(d) **Male** *assamensis*: All-black breast and darker, greyish belly.

6 [722] **FIRE-TAILED SUNBIRD** *Aethopyga ignicauda*
(a) **Male** *ignicauda*: Like Mrs Gould's but uppertail-coverts and tail red, tail-streamers much longer.
(b) **Female**: Like Green-tailed but rump yellower, tail squarer with no white.

7 [723] **FORK-TAILED SUNBIRD** *Aethopyga christinae*
(a) **Male** *latouchii*: Small; iridescent green crown and tail, dark olive mantle, yellow rump-band, dark crimson throat and upper breast.
(b) **Female**: Like female Crimson but crown greyish, belly and undertail-coverts yellowish, bold white undertail-tips. Smaller than Green-tailed with shorter bill, yellow-olive throat/breast, less graduated tail.

8 [725] **PURPLE-NAPED SUNBIRD** *Hypogramma hypogrammicum*
(a) **Male** *lisettae*: Large; plain greenish-olive above with iridescent purple-blue nuchal band, rump and uppertail-coverts, boldly streaked below.
(b) **Female**: Lacks purple-blue markings.

9 [726] **GREY-BREASTED SPIDERHUNTER** *Arachnothera affinis*
Adult *modesta*: Bright olive-green above, paler olive-greyish below, narrowly dark-streaked on throat and breast.

10 [727] **STREAKED SPIDERHUNTER** *Arachnothera magna*
Adult *musarum*: Yellowish-olive above, whitish below, heavily streaked head and body, orange legs.

11 [728] **LITTLE SPIDERHUNTER** *Arachnothera longirostra*
Adult *longirostra*: Long downcurved bill, slaty-grey head with whitish lores/cheeks and broken eyering, whitish throat, yellow rest of underparts, white tail-tip.

12 [729] **THICK-BILLED SPIDERHUNTER** *Arachnothera crassirostris*
Adult: Like Little but bill thicker, throat/breast greyish-olive, broad broken eyering yellowish, no white tail-tip.

13 [730] **LONG-BILLED SPIDERHUNTER** *Arachnothera robusta*
Adult *robusta*: Very long, strongly downcurved bill, uniform head-sides, faintly dark-streaked yellowish-olive throat and breast, and yellow belly; outertail tipped white.

14 [731] **SPECTACLED SPIDERHUNTER** *Arachnothera flavigaster*
Adult: Relatively large and robust; thick broad-based bill, broad yellow eyering, yellow patch on ear-coverts.

15 [732] **YELLOW-EARED SPIDERHUNTER** *Arachnothera chrysogenys*
Adult *chrysogenys*: Like Spectacled but smaller with relatively thinner bill, narrower eyering, slightly larger yellow patch on ear-coverts.

1a
1b
1c
2a
2b
3a
3b
3c
3d
4a
4b
4c
4d
5a
5b
5c
5d
6a
6b
7a
7b
8a
8b
9
10
11
12
13
14
15

PLATE 75. SUNBIRDS & FLOWERPECKERS

1 [709] **PLAIN SUNBIRD** *Anthreptes simplex*
(a) **Male**: Olive-green above, pale greyish-olive below (greyer on throat and breast), browner tail, relatively short straight bill, no obvious yellow in plumage, dark iridescent blue-green forehead-patch.
(b) **Female**: No forehead-patch.

2 [710] **BROWN-THROATED SUNBIRD** *Anthreptes malacensis*
(a) **Male** *malacensis*: Relatively robust, with glossy dark green and purple upperparts, dull brownish head-sides and throat, and yellow lower breast to vent.
(b) **Female**: Relatively straight bill, all yellow below, broad yellowish eyering, no white on tail (but beware Red-throated; see below).

3 [711] **RED-THROATED SUNBIRD** *Anthreptes rhodolaema*
(a) **Male**: From Brown-throated by pale brick-red throat, maroon-red head-sides and mostly chestnut-maroon wing-coverts, with only small metallic purple shoulder-patch.
(b) **Female**: Plainer-faced than Brown-throated with greyish-tinged eyering, breast-sides tinged greenish, underparts less bright, sometimes orange-tinged on throat and breast.

4 [712] **PURPLE-RUMPED SUNBIRD** *Leptocoma zeylonica*
(a) **Male** *flaviventris*: Maroon head, upper breast and upperparts, iridescent green crown and shoulder-patch, purple throat and rump, yellow lower breast and upper belly.
(b) **Female**: Like Olive-backed and Purple but throat, upper breast and flanks pale greyish-white. Browner above than Copper-throated with shorter all-dark tail.

5 [713] **VAN HASSELT'S SUNBIRD** *Leptocoma brasiliana*
(a) **Male** *brasiliana*: Small and dark: iridescent green crown and purple throat, deep red belly.
(b) **Female**: Small; dull olive-coloured above, plain head, dull yellow below with olive-washed throat and upper breast.

6 [714] **COPPER-THROATED SUNBIRD** *Leptocoma calcostetha*
(a) **Male**: All dark and relatively long-tailed; iridescent green crown, shoulder-patch and uppertail-coverts, iridescent copper-red throat and upper breast.
(b) **Female**: Greyish crown and head-sides, white throat, mostly yellow underparts, and longish tail with white outer tips.

7 [715] **PURPLE SUNBIRD** *Cinnyris asiaticus*
(a) **Male** *intermedia*: Uniformly dark: mostly iridescent dark bluish to purplish.
(b) **Male eclipse**: Like Olive-backed but with mostly iridescent dark blue wing-coverts and darker wings.
(c) **Female**: Like Olive-backed but pale yellow below, less white on tail-tips.

8 [716] **OLIVE-BACKED SUNBIRD** *Cinnyris jugularis*
(a) **Male** *flamaxillaris*: Plain olive-green above, iridescent blue-black forehead, throat and breast (latter with reddish band), yellow belly and vent, white on tail.
(b) **Male eclipse**: Blue-black stripe on central throat and breast.
(c) **Female**: Downcurved bill, all-yellow underside and extensive white on tail.

9 [724] **RUBY-CHEEKED SUNBIRD** *Chalcoparia singalensis*
(a) **Male** *assamensis*: Iridescent dark green above, copper-red ear-coverts, pale orange-rufous throat and upper breast, and yellow belly and vent.
(b) **Female**: Greenish-olive above, as male below.

10 [735] **YELLOW-BELLIED FLOWERPECKER** *Dicaeum melanoxanthum*
(a) **Male**: Large; upper body black with white throat and centre of breast, rest of underparts yellow.
(b) **Female**: Pattern like male but head and breast-sides dull olive-greyish, upperparts dull greyish-brown, yellow of underparts duller and paler.

11 [737] **CRIMSON-BREASTED FLOWERPECKER** *Dicaeum percussus*
(a) **Male** *ignicapilla*: Slaty-blue above, bright yellow below, with red crown- and breast-patches and white submoustachial stripe.
(b) **Female**: Nondescript, with thick bill, dull orange crown-patch, greyish-olive yellow-centred underparts, indistinct whitish submoustachial stripe.
(c) **Juvenile**: Duller and more uniformly olive-coloured than female, bill mostly pinkish.

12 [738] **SCARLET-BREASTED FLOWERPECKER** *Dicaeum thoracicus*
(a) **Male**: Black head and breast with inset red crown-patch and large breast-patch, yellowish body and black wings and tail.
(b) **Female**: Thick bill, greyish head, whitish throat and submoustachial stripe, and yellowish underparts with greyish-mottled reddish-orange wash on breast.
(c) **Juvenile**: Duller and more greyish-olive below than female.

13 [740] **ORANGE-BELLIED FLOWERPECKER** *Dicaeum trigonostigma*
(a) **Male** *rubropygium*: Slaty-blue crown, nape, wings and tail, grey throat and upper breast, and orange lower breast, mantle and back, yellower on uppertail-coverts and vent.
(b) **Female**: Slender, slightly downcurved bill, plain olive head, orange-yellow rump, greyish throat and upper breast, dull yellow belly-centre and vent.

1b
1a
3b
4a
4b
2a
3a
5a
2b
6a
5b
6b
8a
7a
8b
8c
7c
7b
9a
9b
10a
10b
12c
12b
11a
12a
11c
11b
13a
13b

PLATE 76. FLOWERPECKERS, LEAFBIRDS & ASIAN FAIRY-BLUEBIRD

1 [733] **THICK-BILLED FLOWERPECKER** *Dicaeum agile*
Adult *modestum*: From Brown-backed by reddish to orange eyes, somewhat narrower bill, thin malar line, olive-green above, paler below with darker streaks, whitish tail-tips.

2 [734] **BROWN-BACKED FLOWERPECKER** *Dicaeum everetti*
Adult *sordidum*: Plain earth-brown above, pale buffy-brown below with faint brown streaks, thick bill, pale yellowish eye.

3 [736] **YELLOW-BREASTED FLOWERPECKER** *Dicaeum maculatus*
Adult *septentrionalis*: Distinctly greenish above, with orange crown-patch, thick dark bill; yellowish below with broad olive-green streaks.

4 [739] **YELLOW-VENTED FLOWERPECKER** *Dicaeum chrysorrheum*
Adult *chrysochlore*: Narrow, slightly downcurved bill, whitish loral stripe, white underparts with bold blackish streaks and yellow to orange-yellow undertail-coverts; yellowish-olive upperparts.

5 [741] **PALE-BILLED FLOWERPECKER** *Dicaeum erythrorynchos*
Adult *erythrorynchos*: Like Plain but bill pinkish.

6 [742] **PLAIN FLOWERPECKER** *Dicaeum minullum*
(a) **Adult** *olivaceum*: Greenish-olive above and pale olive-greyish below, with cream belly-centre (sometimes throat), dark bill and relatively pale head-sides.
(b) **Juvenile**: Bill pinkish with darker culmen.

7 [743] **FIRE-BREASTED FLOWERPECKER** *Dicaeum ignipectus*
(a) **Male** *ignipectus*: Glossy dark greenish-blue above, buffish below with red breast-patch and black line on belly-centre, black sides of head and breast.
(b) **Female**: Dark bill, greenish-olive upperparts and head-sides, uniform buffish underside with olive-tinged flanks.
(c) **Male** *cambodianum*: No red breast-patch.

8 [744] **SCARLET-BACKED FLOWERPECKER** *Dicaeum cruentatum*
(a) **Male** *cruentatum*: Red crown and upperparts, blackish head- and breast-sides, glossy blue-blackish wings.
(b) **Female**: Bright red rump and uppertail-coverts, pale underparts.
(c) **Juvenile**: Like female but bill reddish-pink basally, upperparts uniform with orange-tinged uppertail-coverts.

9 [745] **GREATER GREEN LEAFBIRD** *Chloropsis sonnerati*
(a) **Male** *zosterops*: Stout bill, all-green plumage apart from black face and bib and purple-blue malar band.
(b) **Female**: Stout bill, all-green plumage with demarcated yellow throat and eyering. Blue malar band faint.
(c) **Juvenile**: Like female but blue malar band faint or absent; has yellowish submoustachial band.

10 [746] **LESSER GREEN LEAFBIRD** *Chloropsis cyanopogon*
(a) **Male** *septentrionalis*: Very like Greater Green but smaller, smaller-billed; yellowish forehead and border to black bib, no blue shoulder-patch.
(b) **Female**: All-green with golden-green forehead and blue to purplish-blue malar band.
(c) **Juvenile**: Very like Greater Green but smaller, smaller-billed, less defined yellow throat, no obvious eyering.
(d) **Male** *cyanopogon*: Much less yellow on forehead and bordering bib; more like Greater Green.

11 [747] **BLUE-WINGED LEAFBIRD** *Chloropsis cochinchinensis*
(a) **Male** *chlorocephala*: Turquoise-blue flight-feather fringes, leading edge of wing and tail, blue shoulder-patch.
(b) **Female**: Mostly greenish head with diffuse blue malar band, less blue on wings and tail.
(c) **Juvenile**: Like female but almost lacks any blue on face and malar area, crown and nape greener.
(d) **Male** *moluccensis*: Even yellower forecrown and border to bib, more defined yellowish or golden nape.

12 [748] **GOLDEN-FRONTED LEAFBIRD** *Chloropsis aurifrons*
(a) **Adult** *pridii*: Forecrown golden-orange, throat/malar area purple-blue, bill slender, slightly downcurved.
(b) **Juvenile**: All green, with purple-blue and black malar band and blue on shoulder, golden forehead.

13 [749] **ORANGE-BELLIED LEAFBIRD** *Chloropsis hardwickii*
(a) **Male** *hardwickii*: From other leafbirds by dull yellowish-orange lower breast to vent.
(b) **Female**: Dull yellowish-orange centre to abdomen and undertail-coverts, broad purplish-blue malar band.
(c) **Juvenile**: Underparts uniform light green, little blue on malar, no blue shoulder-patch. Like Blue-winged but darker above, little blue on wings and tail, bill longer.
(d) **Female** *melliana*: From other leafbirds by bluish colour on crown and ear-coverts, broader, purpler malar band, darker green upperside.

14 [750] **ASIAN FAIRY-BLUEBIRD** *Irena puella*
(a) **Male** *puella*: Shining deep blue upperparts and undertail-coverts, otherwise mostly blackish. Eyes red.
(b) **Female**: Overall dull turquoise-blue with blackish tail and flight feathers. Eyes reddish.

1
2
3
4
5
6a
6b
8c
8a
7a
7b
8b
10a
9c
9a
7c
9-14 to different scale
10b
9b
10c
11d
11c
12a
12b
10d
11a
14a
14b
11b
13a
13b
13c
13d

PLATE 77. RED AVADAVAT, JAVA SPARROW, MUNIAS, PARROTFINCHES & SPARROWS

1 [758] **RED AVADAVAT** *Amandava amandava*
(a) **Male breeding** *punicea*: Bright red with white spotting on scapulars, uppertail-coverts and underparts.
(b) **Female**: Greyish-brown, paler below, with bright red rump, uppertail-coverts and bill, white spots on coverts.
(c) **Juvenile**: Like female but lacks red, wing-bars buff.

2 [758A] **JAVA SPARROW** *Padda oryzivora*
(a) **Adult**: Grey with black head and rump to tail, white ear-coverts, red bill and legs, pinkish belly.
(b) **Juvenile**: Washed-out brownish; crown and tail darkest, ear-coverts lightest, dull buffish below, breast-streaking.

3 [759] **WHITE-RUMPED MUNIA** *Lonchura striata*
(a) **Adult** *subsquamicollis*: Dark brownish plumage with contrasting whitish rump and belly.
(b) **Juvenile**: Dark parts of plumage paler and browner, rump and belly tinged buffish.

4 [759A] **JAVAN MUNIA** *Lonchura leucogastroides*
(a) **Adult**: Rich dark brown above, face, throat, upper breast and tail blackish, belly white.
(b) **Juvenile**: Brownish throat and upper breast, dark-barred undertail-coverts and pale buffish belly.

5 [760] **SCALY-BREASTED MUNIA** *Lonchura punctulata*
(a) **Adult** *topela*: Drab brown above, yellowish-olive on rump/tail, chestnut-tinged head-sides, brown-scaled below.
(b) **Juvenile**: Paler and plain above, slightly buffish drab brown below, lower mandible paler than upper.
(c) **Adult** *subundulata*: Bolder, blacker scaling below.

6 [761] **WHITE-BELLIED MUNIA** *Lonchura leucogastra*
Adult *leucogastra*: Very dark brown, olive-yellow tail-fringes, whitish belly, no whitish rump.

7 [762] **CHESTNUT MUNIA** *Lonchura atricapilla*
(a) **Adult** *deignani*: Rufous-chestnut with black hood, blackish central belly and vent, blue-grey bill.
(b) **Juvenile**: Plain brown above, buff below, bluish bill, crown darker than White-headed, underparts buffer.
(c) **Adult** *atricapilla*: Orange-yellow on tail, blacker on vent.

8 [762A] **WHITE-CAPPED MUNIA** *Lonchura ferruginosa*
Adult: Whitish head, black throat and patch from lower mid-breast to vent.

9 [763] **WHITE-HEADED MUNIA** *Lonchura maja*
(a) **Adult** *maja*: White head, broad pale vinous-brown collar.
(b) **Juvenile**: Like Black-headed but crown duller and paler, head-sides paler, underparts duller buff.

10 [764] **TAWNY-BREASTED PARROTFINCH** *Erythrura hyperythra*
(a) **Female** *malayana*: Short green tail, warm buff head-sides and uppertail-coverts, dark brown then blue on forehead.
(b) **Juvenile**: Paler; all-green crown.

11 [765] **PIN-TAILED PARROTFINCH** *Erythrura prasina*
(a) **Male** *prasina*: Green above, blue face and throat, bright red lower rump, uppertail-coverts and long tail.
(b) **Female**: Much shorter-tailed, powder-blue on head-sides, washed-out buffish underside, shorter red tail.
(c) **Juvenile**: Like female but duller, lower mandible yellowish.

12 [766] **HOUSE SPARROW** *Passer domesticus*
(a) **Male breeding** *indicus*: Grey crown, whitish head-sides, broad black bib, dark-streaked mantle, greyish rump.
(b) **Male non-breeding**: Paler bill, browner crown, greyer mantle.
(c) **Female**: Pale brown, with buffy supercilium and underparts, dark streaks on mantle and scapulars.

13 [767] **PLAIN-BACKED SPARROW** *Passer flaveolus*
(a) **Male**: Unstreaked above, rufous-chestnut from eye to nape-side and on scapulars, yellowish head-sides to vent.
(b) **Female**: Unstreaked above and pale yellowish-tinged below with pale greyish breast-wash.

14 [768] **EURASIAN TREE-SPARROW** *Passer montanus*
(a) **Adult** *malaccensis*: Whitish head-sides with isolated blackish patch, crown and nape dull chestnut, small bib.
(b) **Juvenile**: Duller; crown paler with dark forecrown, black head marks less defined, bill pale-based.

15 [769] **RUSSET SPARROW** *Passer rutilans*
(a) **Male** *intensior*: Rufous-chestnut above, dingy yellowish on head-sides and abdomen, narrow black bib.
(b) **Female**: Resembles House but eyestripe darker, upperside with rufous-chestnut and black admixed.

1a
2b
1b
2a
3b
3a
1c
5c
6
4b
4a
5b
8
7c
5a
9b
9a
7b
7a
12b
11b
10a
11c
12a
12c
11a
10b
13a
14b
15a
13b
15b
14a

PLATE 78. ACCENTORS, PIPITS & FOREST WAGTAIL

1 [751] **ALPINE ACCENTOR** *Prunella collaris*
(a) **Adult** *nipalensis*: Robust and rounded; brownish-grey head and breast, chestnut-streaked flanks and black-and-white barred throat.
(b) **Juvenile**: Like adult but browner, flank-streaks dark brownish.

2 [752] **RUFOUS-BREASTED ACCENTOR** *Prunella strophiata*
(a) **Adult** *strophiata*: Pipit-like but with orange-rufous supercilium, dark lores and ear-coverts, orange-rufous breast, rufous-buff flanks.
(b) **Juvenile**: No orange-rufous. From small pipits by small size, streaked throat, rufescent wing-fringing.

3 [753] **MAROON-BACKED ACCENTOR** *Prunella immaculata*
(a) **Adult**: Dark grey head, whitish eyes and dark chestnut scapulars and wings with grey greater coverts, brownish-olive above, dull grey below with rufescent flanks/vent.
(b) **Juvenile**: Head olive-greyish, upperparts dark-streaked warm brownish, underparts whitish-buff with dark brown streaks.

4 [770] **BUFF-BELLIED PIPIT** *Anthus rubescens*
(a) **Adult non-breeding** *japonicus*: Resembles Rosy but only faint streaks on cold greyish-brown upperparts.
(b) **Adult breeding**: Greyer above, plainer and buffier below with fine streaks confined to warmer breast.

5 [771] **ROSY PIPIT** *Anthus roseatus*
(a) **Adult non-breeding**: Like first-winter Red-throated but darker head-sides, bold whitish supercilium, blacker bill.
(b) **Adult breeding**: Vinous-pinkish flush on supercilium, throat and breast, no malar and breast-streaks.
(c) **Juvenile**: Browner above, less heavily streaked below than non-breeding adult.

6 [772] **RED-THROATED PIPIT** *Anthus cervinus*
(a) **Adult**: Brick-red head-sides to upper breast.
(b) **Adult variant (dull female)**: No dark markings on head-sides, throat and upper breast.
(c) **First winter**: Like Tree but more boldly streaked with whiter wing-bars and underparts.

7 [773] **OLIVE-BACKED PIPIT** *Anthus hodgsoni*
(a) **Adult** *yunnanensis*: Greenish above, buff to white supercilium, whitish/black ear-marks, strong streaking below.
(b) **Juvenile**: Browner, more boldly streaked upperparts.
(c) **Adult** *hodgsoni*: Heavier and more extensive streaking above and below.

8 [774] **TREE PIPIT** *Anthus trivialis*
(a) **Adult** *trivialis?*: Like Olive-backed but buffier/greyer above with heavy streaks, plainer head-sides.
(b) **Adult (worn)**: Browner above, whiter below.

9 [775] **BLYTH'S PIPIT** *Anthus godlewskii*
(a) **Adult**: See under Richard's for differences.
(b) **First winter**: Differs as Richard's.

10 [776] **RICHARD'S PIPIT** *Anthus richardi*
(a) **Adult**: From Blyth's by longer bill, legs, tail, less contrasting pattern above, more pointed, less clear-cut centres to median coverts; hindclaw longer. Larger than Paddyfield; longer bill and tail, heavier breast-streaking.
(b) **Adult (worn)**: Greyer above with stronger streaking, paler below.
(c) **First winter**: Retains whiter-fringed juvenile wing-coverts; median coverts very like Blyth's (until moulted).

11 [777] **PADDYFIELD PIPIT** *Anthus rufulus*
(a) **Adult** *rufulus*: See under Richard's for differences.
(b) **Adult (worn)**: Differs as Richard's.
(c) **Juvenile**: More scalloped above, dark spotting on breast.

12 [778] **LONG-BILLED PIPIT** *Anthus similis*
(a) **Adult** *yamethini*: Large; long bill and tail; faint streaks above, mostly plain buff below, outertail fringed buffish.
(b) **Juvenile**: Pale scalloping above, stronger streaking below.

13 [779] **FOREST WAGTAIL** *Dendronanthus indicus*
Adult: Brownish-olive crown and upperparts, whitish underparts with double dark breast-band, blackish and whitish wing pattern.

1b
1a
2b
2a
3b
3a
4b
4a
5c
5b
5a
6b
6a
6c
7b
7a
7c
8b
8a
10c
10a
10b
9b
9a
11b
11a
11c
12b
12a
13

PLATE 79. WAGTAILS, BRAMBLING & CHAFFINCH

1 [780] **WHITE WAGTAIL** *Motacilla alba*
(a) **Male non-breeding** *leucopsis*: White head and underparts, extensively white wings and outertail, isolated black breast-patch, black hindcrown, and upperparts.
(b) **Male breeding**: Black breast-patch reaches lower throat and joins mantle.
(c) **Female non-breeding**: Slaty-grey above, narrower breast-patch.
(d) **Juvenile**: Crown and nape grey, breast-patch more diffuse.
(e) **Male non-breeding** *alboides*: Black ear-coverts and neck-sides joining broader breast-patch.
(f) **Male breeding**: All-black throat.
(g) **Female non-breeding**: Greyer above than male.
(h) **Juvenile**: Recalls adult more than *leucopsis*.
(i) **Male non-breeding** *personata*: Black hood, white forecrown, eyering and upper throat, grey upperparts.
(j) **Male breeding**: All-black throat.
(k) **Female non-breeding**: Hindcrown and nape greyer than male.
(l) **Male non-breeding** *baicalensis*: Grey upperparts, large black breast-patch.
(m) **Female non-breeding**: Crown duller than male, upperparts much paler grey than *leucopsis*.
(n) **Male non-breeding** *ocularis*: Black eyestripe.
(o) **Male breeding**: Black lower throat and upper breast.
(p) **Female non-breeding**: Black of crown/nape, greyer than male.
(q) **Male non-breeding** *lugens*: Black on upperparts, white forehead and supercilium, black eyestripe, white cheeks and ear-coverts, smudgy black breast-patch, extensive white on wing. Upperparts most often with large black patches, rather than solidly black.
(r) **Male breeding**: Black lower throat /upper breast and upperparts, white head-side connected to neck-patch.
(s) **Female non-breeding**: Slate-grey nape to upperparts, though noticeably darker above than similar *ocularis*.

2 [781] **MEKONG WAGTAIL** *Motacilla samveasnae*
(a) **Male**: Black forehead and head-sides, broad white supercilium, white throat and enclosed neck-patch.
(b) **Female**: Distinctly paler and greyer above, but still showing distinctive combination of features as male.
(c) **Juvenile**: Like washed-out female; less distinct head pattern, dark malar stripe, smudgy greyish breast.

3 [782] **GREY WAGTAIL** *Motacilla cinerea*
(a) **Female** *cinerea*: Slaty-grey crown, ear-coverts and upperparts, narrow whitish supercilium, blackish wing-coverts and bright yellow vent, rump and tail-coverts.
(b) **Male breeding**: Underparts all yellow with diagnostic black throat and upper breast.

4 [783] **WESTERN YELLOW WAGTAIL** *Motacilla flava*
(a) **Male breeding** *thunbergi*: Grey-tinged olive-green upperparts, yellow underparts (throat whiter), mid-grey crown/nape/rear head-sides, blackish-grey face to ear-coverts . Blackish-grey of lores typically extends up onto forehead, lacks supercilium (or short whitish one behind eye), clay-cream edging/tipping on wing-coverts, sometimes a necklace of darkish flecks across upper breast.
(b) **Female breeding**: Dull greyish- to brownish-olive above, duller and less extensive yellow below.
(c) **First winter**: Like female but greyer above, whiter below.

5 [784] **EASTERN YELLOW WAGTAIL** *Motacilla tschutschensis*
(a) **Male breeding** *macronyx*: Hard to separate from Western Yellow. Averages slightly paler/cleaner grey on forehead to nape (blackish-grey of lores usually not extending to forehead as most Western *thunbergi*), slightly cleaner grey (less extensively blackish-grey) ear-coverts; tends to brighter/greener above, slightly cleaner yellow underparts appear to rarely show necklace, slightly wider/clearer/yellower wing-bars. Typically no supercilium.
(b) **Male breeding** *tschutschensis*: Pronounced full whitish supercilium, face to ear-coverts grey to blackish, crown/nape bluish-grey, upperparts purer olive-green.
(c) **Male breeding** *taivana*: Olive-green crown and head-sides, broad yellow supercilium.
(d) **Female**: Broad, clear-cut, yellowish supercilium.

6 [785] **CITRINE WAGTAIL** *Motacilla citreola*
(a) **Male breeding** *citreola*: All-yellow head and underparts, black nuchal band and grey mantle, back and rump.
(b) **Female**: Like female Yellow but supercilium, throat and breast yellowish, upperparts grey and undertail-coverts whitish. Non-breeding male similar.
(c) **First winter**: Like Yellow but purer grey above, whitish supercilium extends down behind ear-coverts, buff wash on forehead, all-dark bill.
(d) **Male breeding** *calcarata*: Black mantle to uppertail-coverts.

7 [810] **BRAMBLING** *Fringilla montifringilla*
(a) **Male non-breeding**: Blackish head and mantle with heavy grey to brown scaling, pale orange throat, breast, flanks and scapulars, white rump.
(b) **Male breeding**: Black and orange colours more solid, blackish bill.
(c) **Female**: Like non-breeding male but crown and ear-coverts plainer greyish-brown.
(d) **Juvenile**: Like female but head buffier.

8 [811] **CHAFFINCH** *Fringilla coelebs*
(a) **Male non-breeding** *coelebs*: Mostly blue-grey crown, vinous-pinkish face/underparts, grey-green rump.
(b) **Male breeding**: Crown/nape smooth blue-grey, forehead black, more rufescent face and underparts.
(c) **Female**: Duller than Brambling; no orange, grey-buff to whitish below, vague head-bands; green-grey rump.

1d
1c
1e
1g
1h
1a
1b
1f
1i
1j
1k
1m
1l
1o
1n
1p
2a
1q
1r
2b
2c
1s
3a
3b
4a
4c
4b
5a
6b
6a
6d
5d
5c
5b
6c
8c
7a
7b
7c
7d
8a
8b

PLATE 80. EURASIAN SISKIN, TIBETAN SERIN, RED CROSSBILL & FINCHES

1 [786] **EURASIAN SISKIN** *Spinus spinus*
(a) **Male**: Small; black forehead and chin, yellow breast, ear-covert surround, rump, wing-bars and fringing to wings/tail.
(b) **Female**: Lacks black head markings; narrowly streaked above and below, wing-bars duller and whiter. From Tibetan and juvenile greenfinches by whiter underside and broad wing-bars.
(c) **Juvenile**: Like female but browner above, more streaked.

2 [787] **TIBETAN SERIN** *Serinus thibetana*
(a) **Male**: Small; unstreaked greenish-yellow, darker on crown, ear-coverts and upperparts.
(b) **Female**: Upperparts darker greyish-green with blackish streaks, belly and vent whiter, lower breast to vent streaked.
(c) **Juvenile**: Duller than female, wing-coverts fringed buffish, underparts paler, upper breast streaked.

3 [788] **RED CROSSBILL** *Loxia curvirostra*
(a) **Male** *meridionalis*: Heavy crossed mandibles, largely red body, whitish vent, dark brownish wings and tail.
(b) **Female**: Dull dark-streaked greenish-grey, with paler, yellower rump, breast and belly.
(c) **Juvenile**: Paler and darker-streaked than female.
(d) **Male** *himalayensis*: Much smaller and smaller-billed; mandible tips overlap more.

4 [789] **GREY-CAPPED GREENFINCH** *Chloris sinica*
(a) **Male** *sinica*: Yellow on wing, vent and basal tail-sides, greyish crown and nape, rich brown mantle and scapulars, warm brownish breast and flanks, secondaries fringed greyish-white to buffish-white.
(b) **Female**: Head more uniform greyish-brown, upperparts duller, underparts more washed out.
(c) **Juvenile**: From Yellow-breasted and Black-headed by paler wing-coverts and tertials and pale greyish fringing on secondaries; paler overall.

5 [790] **BLACK-HEADED GREENFINCH** *Chloris ambigua*
(a) **Male** *ambigua*: Dull olive-green upperparts, mottled dull olive-green throat to flanks, blackish crown and ear-coverts, greyish-white greater covert bar.
(b) **Female**: From Yellow-breasted by lack of obvious supercilium, plainer upperparts and greyish wing-bars.
(c) **Juvenile**: Darker and greener than Yellow-breasted, yellow wing-slash broader.

6 [791] **VIETNAMESE GREENFINCH** *Chloris monguilloti*
(a) **Male**: Black head, blackish-green nape to back, yellow throat and underparts, mottling on breast.
(b) **Female**: Like duller, paler male, with streaked nape to back and more mottling on breast and flanks.
(c) **Juvenile**: Like female but paler and duller.

7 [792] **YELLOW-BREASTED GREENFINCH** *Chloris spinoides*
(a) **Male** *heinrichi*: Blackish crown and head-sides, yellow supercilium, throat and underparts, yellow wing-bars.
(b) **Female**: Duller, paler; nape to back streaked.
(c) **Juvenile**: Paler and browner than Black-headed, yellow wing-slash narrower.

8 [799] **SCARLET FINCH** *Haematospiza sipahi*
(a) **Male**: Bright red head and body, pale yellowish bill.
(b) **Female**: Scaly brownish-olive, with pale bill, yellow rump.
(c) **First-year male**: Like female but has rufous tinge above and below, orange rump.

9 [802] **CRIMSON-BROWED FINCH** *Propyrrhula subhimachala*
(a) **Male**: Warm to reddish-brown head and upperparts, greenish wing-fringing, red rump and forehead, greyish underparts with pinkish-speckled red throat and breast.
(b) **Female**: Dull green and grey with yellowish rump, forehead and upper breast, grey-streaked whitish throat.
(c) **First-year male**: Rufous and orange tinges where red on adult.

10 [809] **GOLD-NAPED FINCH** *Pyrrhoplectes epauletta*
(a) **Male**: Small and blackish, with golden-orange crown/nape-patch and shoulder-spot, white tertial fringes.
(b) **Female**: Drab rufescent-chestnut with olive-green hindcrown and nape, grey face and upper mantle, dark flight feathers and tail, and white tertial fringes.
(c) **Juvenile**: Like female but richer brown, nape greyer.

1c
1b
1a
2a
2b
2c
3d
3a
3b
3c
4b
4a
4c
5b
5c
5a
6a
6c
6b
7b
7c
7a
8b
8a
8c
9c
9b
9a
10a
10c
10b

PLATE 81. ROSEFINCHES, BULLFINCHES & GROSBEAKS

1 [793] **PINK-RUMPED ROSEFINCH** *Carpodacus eos*
(a) **Male**: Pink-tinged greyish above with blackish streaking, reddish-pink rump, supercilium and underparts.
(b) **Female**: Like juvenile Common but has defined supercilium, more extensive streaking, only vague wing-bars.

2 [794] **VINACEOUS ROSEFINCH** *Carpodacus vinaceus*
(a) **Male** *vinaceus*: Body almost uniform dark red, pink supercilium, paler red rump, whitish tertial spots.
(b) **Female**: Dull brown above and dull buffy-brown below with vague streaking, has tertial marks but no wing-bars.
(c) **Juvenile**: Like female but more boldly streaked.

3 [795] **DARK-RUMPED ROSEFINCH** *Carpodacus edwardsii*
(a) **Male** *rubicunda*: Paler than Dark-breasted, more streaky above, complete pale pink supercilium, pinker belly.
(b) **Female**: Like Spot-winged but larger and bulkier, mantle lacks whitish-buff streaks, deeper buffish-brown below.

4 [796] **SHARPE'S ROSEFINCH** *Carpodacus verreauxii*
(a) **Male**: Plain dark reddish crown, narrow pinkish streaks on mantle and tertials, pink rump, supercilium and darker-mottled underparts.
(b) **Female**: Smaller than Dark-rumped, has some thin whitish-buff streaks on mantle, paler below with broader streaks on lower throat and breast.

5 [797] **COMMON ROSEFINCH** *Carpodacus erythrinus*
(a) **Male breeding** *roseatus*: Red head, wing-bars and underparts, darker line through eye, darker upperparts.
(b) **Female**: Greyish-brown above, whitish below, finely streaked; two buffy-whitish wing-bars, plain head-sides.
(c) **Juvenile**: Browner and darker-streaked above and below than female, with broader, buffier wing-bars.
(d) **Male breeding** *erythrinus*: Greyer above, paler lower breast, whiter belly to vent.

6 [798] **DARK-BREASTED ROSEFINCH** *Carpodacus nipalensis*
(a) **Male** *intensicolor*: Dark brownish-red above and on breast, pinkish-red supercilium, throat and belly.
(b) **Female**: Unstreaked drab brown, vague mantle streaks, brown wing-bars and tertial fringes.

7 [800] **GREY-HEADED BULLFINCH** *Pyrrhula erythaca*
(a) **Male** *erythaca*: Mid-grey head and upperparts, black white-bordered face, deep orange breast and belly, glossy black wings and tail, broad greyish wing-bar, black band above white rump.
(b) **Female**: Resembles male but no orange, body mostly suffused pinkish-buff.
(c) **Juvenile**: Like female but head more uniform buffish-brown, no black on forehead.

8 [801] **BROWN BULLFINCH** *Pyrrhula nipalensis*
(a) **Adult** *waterstradti*: Greyish-brown; blackish face, whitish ear-coverts and rump-band, blackish wings and tail.
(b) **Juvenile**: No dark face, body tinged buffish, buffish bar on median coverts.
(c) **Adult** *ricketti*: Darker crown with pale scaling, little whitish on ear-coverts.

9 [804] **JAPANESE GROSBEAK** *Eophona personata*
(a) **Male** *magnirostris?*: Larger than Yellow-billed; heavier, yellow-tipped bill, black of head restricted more to crown/face, body greyer, wings with white on mid-primaries only. Tertials as brownish-grey mantle, or washed browner.
(b) **Juvenile**: Browner overall (not as brown as Yellow-billed), wing-coverts tipped pale buffish, black of head reduced to mask; flight feathers similar.

10 [805] **YELLOW-BILLED GROSBEAK** *Eophona migratoria*
(a) **Male** *migratoria*: Dark-tipped yellow bill, greyish-brown body, bluish-black hood, tail and wings with white markings on wing-tips and primary coverts.
(b) **Female**: Duller, with brownish-grey head, less white on wings.

11 [806] **WHITE-WINGED GROSBEAK** *Mycerobas carnipes*
(a) **Male** *carnipes*: Sooty-black with greenish-yellow back, rump, belly and vent, white patch on base of primaries.
(b) **Female**: Paler and greyer than male.

12 [807] **COLLARED GROSBEAK** *Mycerobas affinis*
(a) **Male**: Large; black hood, wings and tail, yellow nape, back, rump and underparts, nape and rump flushed orange-rufous.
(b) **Female**: Hood grey, upperparts pale greyish-green, primaries and tail blackish, underparts yellowish-olive.
(c) **Juvenile male**: Like adult but yellow has olive tinge, head duller, greyish-brown throat-mottling.

13 [808] **SPOT-WINGED GROSBEAK** *Mycerobas melanozanthos*
(a) **Male**: Black with yellow breast to vent and whitish tips to greater coverts, secondaries and tertials.
(b) **Female**: Paler, with yellow streaks on crown to mantle, streaky yellow supercilium and lower head-sides, and blackish streaking on breast and belly.
(c) **Juvenile**: Like female but paler and buffier.

1b
1a
2c
2a
3b
2b
4b
5a
4a
3a
4a
6a
5c
5b
7c
8c
6b
5d
7a
8a
10b
8b
7b
10a
9b
9a
11b
11a
13b
12b
12c
13a
12a
13c

PLATE 82. WEAVERS, PLAIN MOUNTAIN-FINCH & BUNTINGS

1 [754] **BLACK-BREASTED WEAVER** *Ploceus benghalensis*
(a) **Male breeding**: Yellow crown, blackish upperside and broad breast-band, whitish throat and belly/vent.
(b) **Male non-breeding**: Like female but with buffish-scaled blackish breast-band.
(c) **Female**: Like Streaked but crown/nape plainer and greyer, breast unstreaked, rump plainer.

2 [755] **STREAKED WEAVER** *Ploceus manyar*
(a) **Male breeding** *williamsoni*: Yellow crown, blackish head-sides, breast-streaks, buff-streaked upperparts.
(b) **Female**: Yellowish-white supercilium, submoustachial stripe and neck-patch; streaked above and below.
(c) **Male breeding** *peguensis*: Broader dark streaking above and below.

3 [756] **BAYA WEAVER** *Ploceus philippinus*
(a) **Male breeding** *angelorum*: Yellow crown and unstreaked warm buffish-brown breast.
(b) **Female**: Only lightly mottled on buffish breast, no blackish and yellowish markings on head-sides.
(c) **Juvenile**: Like female but breast and flanks deeper buff, crown-streaking more broken.
(d) **Male breeding** *infortunatus*: More rufescent base colour above and below.

4 [757] **ASIAN GOLDEN WEAVER** *Ploceus hypoxanthus*
(a) **Male breeding** *hymenaicus*: Yellow head and body, black head-sides, throat and upperpart-streaking.
(b) **Female**: From Baya by more conical bill, no breast-mottling, heavier-streaked (darker-looking) crown.

5 [803] **PLAIN MOUNTAIN-FINCH** *Leucosticte nemoricola*
(a) **Adult** *nemoricola*: Brownish above with broad streaks, black-marked wing-coverts with whitish wing-bars, pale supercilium, plain greyish-brown below with whiter vent, whitish outertail fringes.
(b) **Juvenile**: Warmer brown overall, head more uniform.

6 [813] **RED-HEADED BUNTING** *Emberiza bruniceps*
(a) **Male non-breeding**: Bright colours obscured by pale buffish to greyish feather tips.
(b) **Male breeding**: Chestnut-red crown, head-sides, throat and upper breast, yellow rump and lower underparts. Mantle and scapulars yellowish-olive, with blackish streaks.
(c) **Female non-breeding**: Buffier head-sides than Black-headed, contrasting less with throat, more greenish-yellow lower back and uppertail-coverts, warm buffish breast, less yellow remainder of underparts; throat typically whiter than belly.
(d) **First winter**: Extremely similar to Black-headed, which see for differences.

7 [814] **BLACK-HEADED BUNTING** *Emberiza melanocephala*
(a) **Male non-breeding**: Duller than male breeding; pale-fringed head and upperparts, duller underparts.
(b) **Male breeding**: Black crown and head-sides, rufous-chestnut nape and upperparts, all-yellow underparts.
(c) **Female non-breeding**: Relatively large size and long bill, plain washed-out appearance, no white on outertail. Pale sandy-brown above, with indistinct fine darker streaks, plain pale buffish (often tinged yellow) below, pale yellow undertail-coverts; wings darker with whitish feather fringes.
(d) **First winter**: Features that can confirm identification from Red-headed are: five (rarely six) primary tips clearly showing beyond longest tertial (four/five in Red-headed), darker/more numerous crown-streaks, weaker dark streaking on darker, warmer scapulars and/or mantle, strongly chestnut-tinged rump/uppertail-coverts, presence of extensive yellow on underparts, somewhat longer/narrower bill than Red-headed.

8 [823] **BLACK-FACED BUNTING** *Emberiza spodocephala*
(a) **Male** *sordida*: Plain greenish-olive head and breast, blackish face and yellow belly and vent.
(b) **Female breeding**: Like male but lacks blackish face and shows yellow throat/breast, darker malar and streaking below.
(c) **First winter**: No rufous in plumage or yellow below; greyish ear-coverts, neck-sides and lesser coverts, greyish-brown rump, clear underpart-streaking.

9 [824] **CHESTNUT BUNTING** *Emberiza rutila*
(a) **Male non-breeding**: Chestnut has pale fringing.
(b) **Male breeding**: Bright chestnut with yellow breast to vent.
(c) **Female**: Like Yellow-breasted but smaller, with plain rufous-chestnut rump, rather plain unbordered ear-coverts, little white on outertail.

10 [825] **YELLOW-BREASTED BUNTING** *Emberiza aureola*
(a) **Male non-breeding** *ornata*: Yellow throat and underparts with warm brown breast-band and white median and lesser coverts.
(b) **Male breeding**: Chestnut upperparts and breast-band, blackish face and throat, yellow underparts.
(c) **Female**: Like non-breeding male but no breast-band, paler yellow below, paler and less rufescent above.
(d) **Juvenile**: Like female but paler below, breast finely dark-streaked.

1a
2c
3a
1c
2a
3b
1b
2b
3c
3d
5a
4a
5b
4b
7d
7c
6a
7b
6c
7a
6b
9a
6d
9c
10b
9b
10c
8c
8b
8a
10a
10d

PLATE 83. BUNTINGS

1 [812] **CRESTED BUNTING** *Emberiza lathami*
(a) **Male non-breeding**: Blackish with chestnut wings and tail and long crest.
(b) **Female non-breeding**: Olive-brown above, streaked darker, with chestnut wings and tail (no white) and short crest; paler below with faint dark breast-streaking.
(c) **Juvenile**: Like female but buffier below with more extensive but fainter breast-streaking.

2 [815] **CHESTNUT-EARED BUNTING** *Emberiza fucata*
(a) **Male breeding** *fucata*: Grey, dark-streaked crown and nape, chestnut ear-coverts, black and chestnut breast-bands.
(b) **Female**: Like breeding male but head pattern less distinct and buffier, underparts buffier.
(c) **Juvenile**: Duller than female, ear-coverts dull greyish-brown, with pale centre and broad dark border.
(d) **First-winter female (dull individual)**: Crown, breast and flanks washed buffish, breast-streaking and malar line darker than on juvenile, submoustachial stripe whitish.

3 [816] **GODLEWSKI'S BUNTING** *Emberiza godlewskii*
(a) **Male** *yunnanensis*: Grey hood, with dull chestnut lateral crown-stripes, eyestripe and rear border of ear-coverts.
(b) **Female**: Like male but less rufous-chestnut on scapulars, underparts paler, streaky on flanks.
(c) **Juvenile**: Head and breast buffish-brown with darker streaks; lacks prominent head markings.

4 [817] **GREY-NECKED BUNTING** *Emberiza buchanani*
(a) **Male breeding** *neobscura?*: Bluish-grey head/breast-sides, whitish/buffish-white eyering and submoustachial stripe, extensively pinkish-rufous below; pinkish bill, largely rufous-chestnut scapulars.
(b) **Female breeding**: Usually duller than male, with head and nape buffier (showing little contrast with mantle); crown and nape often with some streaking, scapulars duller, breast slightly paler.
(c) **First-winter female**: Brownish crown, head-side and nape, dark-streaked crown, blackish spots/streaks on breast to flanks. Note pinkish bill.

5 [818] **YELLOW-THROATED BUNTING** *Emberiza elegans*
(a) **Male non-breeding** *elegantula*: Pattern as breeding but duller, with black replaced by dark greyish.
(b) **Male breeding**: Black (crested) crown, ear-coverts, chin and breast-band, yellow supercilium and throat.
(c) **Female breeding**: Similar to male non-breeding but crown, ear-coverts and breast-band dark warmish brown, supercilium and throat light yellowish-buff; lacks black on chin, breast-band less solid.
(d) **Juvenile**: Broad pale supercilium, rufescent wing-fringing, pale below.

6 [819] **PALLAS'S BUNTING** *Emberiza pallasi*
(a) **Male non-breeding** *pallasi?*: Like non-breeding female but some black mottling on throat and ear-coverts.
(b) **Male breeding**: Black hood, white submoustachial stripe and mostly white collar and underparts; bold dark streaking above, white rump and uppertail-coverts.
(c) **Female non-breeding**: Pale rufescent-tinged buffish-brown crown and ear-coverts, white submoustachial stripe and throat, dark malar line, only vaguely streaked on buffy underparts.
(d) **Juvenile**: Like non-breeding female but more rufescent ear-coverts and heavier body-streaking.

7 [820] **REED BUNTING** *Emberiza schoeniclus*
(a) **Male non-breeding** *pyrrhulina*: Much thicker bill than Pallas's, rufous wing fringing (including lesser coverts), streakier flanks, usually warmer above.
(b) **Male breeding**: From Pallas's as non-breeding.
(c) **Female non-breeding**: Best told from Pallas's by thick dark bill, rufescent lesser/median covert tips, plainer crown and particularly hindneck.
(d) **Female breeding/first winter**: From Pallas's as female non-breeding.

8 [821] **RUSTIC BUNTING** *Emberiza rustica*
(a) **Male non-breeding** *rustica*: Less black on head than breeding, less solid breast-band and collar.
(b) **Male breeding**: Bold black and white head, dark reddish-chestnut collar, rump and uppertail-coverts, lesser coverts and breast/flank streaks.
(c) **Female non-breeding**: Less black on crown and ear-coverts than breeding, crown more uniformly streaked.
(d) **First-winter female**: Buffier than female non-breeding, browner neck and rump, some dark throat-/breast-streaks. Pale hindcrown-/ear-spot.

9 [822] **LITTLE BUNTING** *Emberiza pusilla*
(a) **Adult non-breeding**: Small; finely dark-streaked breast and flanks, bold broad blackish (streaky) lateral crown-stripes, eyestripe and border to rufous-chestnut ear-coverts.
(b) **Adult breeding**: Chestnut flush over most of head; lateral crown-stripes solid black.
(c) **Juvenile**: Like adult non-breeding but weaker lateral crown-stripes, plumage browner-tinged.

10 [826] **TRISTRAM'S BUNTING** *Emberiza tristrami*
(a) **Male non-breeding**: Pattern as breeding but duller and more diffuse.
(b) **Male breeding**: Black head with white supercilium, median crown- and submoustachial stripe.
(c) **Female**: Like non-breeding male but throat buffish-white and centre of ear-coverts, cheeks and lores pale brownish; breast and flanks streaked brown.

1a
1b
1c
2a
2b
2c
2d
3a
3b
3c
4a
4b
4c
5a
5b
5c
5d
6a
6b
6c
6d
7a
7b
7c
7d
8a
8b
8c
8d
9a
9b
9c
10a
10b
10c

PLATE 84. TREECREEPERS, NUTHATCHES, WALLCREEPER, WINTER WREN & DIPPERS

1 [827] **HODGSON'S TREECREEPER** *Certhia hodgsoni*
Adult *khamensis*: From other treecreepers by unbarred tail, relatively short bill, broad white supercilium and mostly whitish underparts.

2 [828] **BAR-TAILED TREECREEPER** *Certhia himalayana*
Adult *ripponi*: From other treecreepers by rather pale greyish-brown uppertail with contrasting narrow dark cross-bars, and relatively long and strongly downcurved bill.

3 [829] **RUSTY-FLANKED TREECREEPER** *Certhia nipalensis*
(a) **Adult**: Cinnamon-coloured breast-sides, flanks, belly and vent; broad whitish supercilium encircles dark ear-coverts.
(b) **Juvenile**: Underparts washed dull buffish, with some faint dark feather fringing.

4 [830] **HUME'S TREECREEPER** *Certhia manipurensis*
(a) **Adult** *shanensis*: Drab greyish throat and underparts, buffy vent and indistinct supercilium.
(b) **Adult** *manipurensis*: Deep buffish throat to belly.
(c) **Adult** *meridionalis*: Darker, greyer throat to belly and warmer, darker upperparts.

5 [831] **VELVET-FRONTED NUTHATCH** *Sitta frontalis*
(a) **Male** *frontalis*: Red bill, violet-blue upperparts, black forehead and post-ocular line, whitish throat and pale dull beige underparts (washed lavender).
(b) **Female**: No black post-ocular line, cinnamon-tinged underparts.
(c) **Juvenile**: Like adult but bill blackish, upperparts greyer, underparts washed cinnamon-orange.

6 [832] **YELLOW-BILLED NUTHATCH** *Sitta solangiae*
(a) **Male** *fortior*: Like Velvet-fronted but bill yellow, underparts pale; vague pale nuchal collar.
(b) **Female**: No black post-ocular line.
(c) **Male** *solangiae*: Paler crown, slightly paler upperparts and drabber underparts.

7 [833] **BLUE NUTHATCH** *Sitta azurea*
Adult *expectata*: Blackish, with whitish throat and breast, broad bluish-white eyering and silver-blue and black wing markings.

8 [834] **BEAUTIFUL NUTHATCH** *Sitta formosa*
Adult: Large; black above, streaked and edged blue and white, rufous-buff below with paler throat and head-sides.

9 [835] **CHESTNUT-VENTED NUTHATCH** *Sitta nagaensis*
Male *montium*: From Chestnut-bellied and White-tailed by pale greyish-buff underparts, reddish-chestnut lower flanks and vent (latter marked white).

10 [836] **CHESTNUT-BELLIED NUTHATCH** *Sitta cinnamoventris*
(a) **Male** *tonkinensis*: Darker above than Neglected, underparts deep dark reddish-chestnut, contrasting sharply with white cheeks (finely barred with blackish), undertail-coverts blackish, marked with white.
(b) **Female**: From male Neglected by deeper pale chestnut underparts with more contrasting white cheeks; somewhat darker overall, notably on bases/centres of undertail-coverts.

11 [837] **NEGLECTED NUTHATCH** *Sitta neglecta*
(a) **Male**: Medium bluish-grey upperside, pale buffish-chestnut underparts (darker on flanks), rather contrasting white cheeks, mostly dark grey undertail-coverts with large white markings.
(b) **Female**: Like male but underparts pale, drab orange-buff, with whiter cheeks.

12 [838] **GIANT NUTHATCH** *Sitta magna*
(a) **Male** *magna*: Like Chestnut-vented but much larger and bigger-billed, broader black head-band, pale grey crown-centre, no chestnut on lower flanks or buff wash on underparts.
(b) **Female**: Washed buff below, head-band duller.

13 [839] **WHITE-TAILED NUTHATCH** *Sitta himalayensis*
Male: Very like female Chestnut-bellied but smaller with plain cinnamon-orange undertail-coverts; white at tail-base.

14 [840] **WHITE-BROWED NUTHATCH** *Sitta victoriae*
Adult: Smaller than Chestnut-bellied and Neglected; plain cinnamon-orange undertail-coverts, white at tail-base.

15 [841] **WALLCREEPER** *Tichodroma muraria*
(a,b) **Adult non-breeding** *nepalensis*: Grey; whiter throat and upper breast, thin curved dark bill, pale-tipped undertail-coverts and tail, crimson and black on wings, white primary spots in flight.
(c) **Male breeding**: Black face to breast.

16 [842] **WINTER WREN** *Troglodytes troglodytes*
Adult *talifuensis*: Small, short-tailed; warm dark brown, paler below, barred blackish on lower body, wings and tail; indistinct buffish supercilium and slender bill.

17 [843] **WHITE-THROATED DIPPER** *Cinclus cinclus*
Adult *przewalskii*: Like Brown Dipper but has white throat and breast.

18 [844] **BROWN DIPPER** *Cinclus pallasii*
(a) **Adult** *dorjei*: Robust shape, stout bill, short tail and uniform dark brown plumage.
(b) **Juvenile**: Paler and greyer-brown, with fine blackish scales on body plumage.

1-4 to different scale
1
2
3a
3b
4a
4b
4c
5b
5c
5a
6a
6b
6c
7
7-8 to different scale
8
9
11a
11b
13
10b
10a
14
15c
12a
12b
15a
15b
18b
17
18a
16
15 and 17-18 to different scale

PLATE 85. MYNAS & STARLINGS

1 [846] **CRESTED MYNA** *Acridotheres cristatellus*
(a,b) **Adult** *brevipennis*: Like White-vented but bill ivory, eyes pale orange, crest shorter, undertail-coverts black with narrow white fringes; very large white wing-patches and narrow white outertail-tips.
(c) **Juvenile**: Like White-vented (see above for differences).

2 [847] **WHITE-VENTED MYNA** *Acridotheres grandis*
(a,b) **Adult**: Yellow bill, tufted crest, reddish eyes and uniform slaty-black plumage with white undertail-coverts; large white wing-patch, mostly blackish underwing-coverts and white-tipped outertail.
(c) **Juvenile**: Browner with no obvious crest, undertail-coverts dark brown with pale scaling, little or no white on tail-tip, bill duller. Like Crested but bare parts yellower, undertail-coverts paler, wing-patch smaller.

3 [848] **JUNGLE MYNA** *Acridotheres fuscus*
(a,b) **Adult** *fuscus*: Like White-vented but bill orange with bluish base, eyes yellow, crest short; smaller white wing-patch and mostly greyish underwing-coverts.
(c) **Juvenile**: Browner overall, no obvious crest, bill yellowish.

4 [848A] **JAVAN MYNA** *Acridotheres javanicus*
(a,b) **Adult**: Like Jungle but body darker with white undertail-coverts, bill yellow; blacker underwing-coverts and broader white tail-tip.
(c) **Juvenile**: Like Jungle but body greyer.

5 [849] **COLLARED MYNA** *Acridotheres albocinctus*
(a,b) **Adult**: Like White-vented but with broad white collar on neck-side (buff-tinged in winter), white-tipped dark undertail-coverts, short crest and pale blue eyes; similar wing pattern to Jungle Myna.
(c) **Juvenile**: Browner than adult with weaker neck-patch.

6 [850] **COMMON MYNA** *Acridotheres tristis*
(a,b) **Adult** *tristis*: Brown with greyish-black hood, whitish vent and yellow bill and facial skin; large white patch on primary coverts and bases of primaries, white underwing-coverts.
(c) **Juvenile**: Hood paler. Very like Jungle Myna but some yellow facial skin, larger wing-patch, white underwing-coverts.

7 [863] **ASIAN GLOSSY STARLING** *Aplonis panayensis*
(a) **Adult** *strigata*: Glossy blackish-green; red eyes.
(b) **Juvenile**: Greyish-brown above, buffish-white below with bold dark streaks; eyes often paler red.

8 [864] **GOLDEN-CRESTED MYNA** *Ampeliceps coronatus*
(a,b) **Male**: Glossy blackish with yellow crown, cheeks and throat, and patch at base of primaries. From Hill Myna in flight by smaller size, yellow wing-patch.
(c) **Female**: Yellow on head greatly reduced.
(d) **Juvenile**: Duller and browner than adult with no yellow on crown, yellowish-white lores, throat and wing-patch, faint streaking below.

9 [865] **COMMON HILL-MYNA** *Gracula religiosa*
(a,b) **Adult** *intermedia*: Large; glossy black, heavy deep orange bill (often tipped yellowish), connected yellow wattles on ear-coverts and nape, and white wing-patch.
(c) **Adult** *religiosa*: Larger and thicker-billed, with separated head wattles.

4c
4b
2b
2c
4a
1b
2a
1c
6a
6c
3c
1a
6b
3a
3b
5c
5a
5b
7a
9c
8b
8c
7b
9a
8a
8d
9b

PLATE 86. STARLINGS

1 [845] **SPOT-WINGED STARLING** *Saroglossa spiloptera*
(a,b) **Male**: Scaly greyish above, rufescent below, blackish ear-coverts, pale eyes and dark chestnut throat.
(c,d) **Female**: Scaly brown above, pale below with darker streaking and scaling, pale eyes. In flight, small white wing-patch as in male.

2 [854] **WHITE-CHEEKED STARLING** *Sturnus cineraceus*
(a,b) **Male**: Black-tipped orange bill, blackish head and breast with mostly white forehead and ear-coverts, body mostly dark, whitish uppertail-covert band.
(c) **Female**: Paler than male, throat mixed with whitish.
(d,e) **Juvenile**: From other starlings by bare-part colours, rather uniform plumage and contrasting whitish ear-coverts, uppertail-covert band and tail-border.

3 [855] **RED-BILLED STARLING** *Sturnus sericeus*
(a) **Male**: Greyish body, whitish hood and dark-tipped red bill, white on primary coverts and bases of primaries, legs and feet orange.
(b) **Female**: Similar to male but body brown-tinged.
(c) **Juvenile**: Browner than female, less distinct whitish wing-patch, bill yellower.

4 [856] **BRAHMINY STARLING** *Sturnus pagodarum*
(a) **Adult**: Blackish crown and nape, greyish upperparts and salmon-pinkish head-sides, breast and belly.
(b) **Juvenile**: Duller above and below.

5 [857] **CHESTNUT-TAILED STARLING** *Sturnus malabaricus*
(a,b) **Adult** *nemoricola*: Blue-based yellowish bill, greyish-white hood and rufous-chestnut outertail feathers; small white area on wing-bend.
(c) **Adult** *malabaricus*: Pale chestnut below.

6 [858] **WHITE-SHOULDERED STARLING** *Sturnus sinensis*
(a,b) **Male**: Mostly grey with black wings and wholly white upperwing-coverts and scapulars, white-bordered black tail.
(c) **Female**: Wings have smaller white patch.
(d,e) **Juvenile**: Like female but lacks white wing-patch. Bare-part colour, all-dark upperwing and pale-bordered dark tail exclude similar starlings.

7 [859] **PURPLE-BACKED STARLING** *Sturnus sturninus*
(a,b) **Male**: Greyish head and underparts, purplish upperparts and dark green upperwing, with whitish scapular
band, tips of median and greater coverts and tertials; buff wing-panels.
(c,d) **Female**: Pattern as male but purple and green replaced with brown, crown browner.

8 [860] **CHESTNUT-CHEEKED STARLING** *Sturnus philippensis*
(a,b) **Male**: Like Purple-backed but head pale with chestnut patch on ear-coverts and neck-side; darker grey below, less marked wings.
(c,d) **Female**: Like Purple-backed but has less marked wings.

2c
2d
1d
1c
1b
1a
2a
2b
2e
5b
5c
3b
5a
3c
3a
8b
8c
8a
8d
6e
4b
6b
6a
4a
6d
7a
7b
6c
7c
7d

PLATE 87. MYNAS, STARLINGS & LARKS

1 [850A] **BLACK-WINGED MYNA** *Acridotheres melanopterus*
(a) **Adult** *melanopterus*: White with black wings and tail, yellowish bill, facial skin and legs.
(b) **Juvenile**: Crown to scapulars brownish-streaked grey, wings and tail duller.

2 [851] **VINOUS-BREASTED MYNA** *Acridotheres burmannicus*
(a) **Adult** *leucocephalus*: Whitish head, narrow naked black mask, yellow to orange-yellow bill and pale vinous-brownish underparts; buff rump/uppertail-coverts and tail-tip.
(b,c) **Juvenile**: Browner overall with dull mask and bill and buffish-fringed wing feathers; large wing-patch.
(d) **Adult** *burmannicus*: Red bill with blackish base, paler above, darker below, whiter tail-tip.

3 [852] **BLACK-COLLARED STARLING** *Gracupica nigricollis*
(a) **Adult**: Relatively large; whitish head and underparts with broad blackish collar.
(b,c) **Juvenile**: Lacks black collar, head and breast dull brownish, white plumage parts duller.

4 [853] **ASIAN PIED STARLING** *Gracupica contra*
(a,b) **Adult** *floweri*: Black and white with longish, red-based yellow bill.
(c) **Juvenile**: Black of plumage replaced by dark brown, including crown; throat paler to whitish, pale plumage parts duller, bill brownish.
(d) **Adult** *superciliaris*: Pale grey underparts, less white streaking on forecrown.

5 [861] **ROSY STARLING** *Sturnus roseus*
(a) **Adult non-breeding**: Dull buffish-pink with blackish hood, wings and tail. Bill brownish-pink, black vent scaled paler, shaggy crest.
(b) **Adult breeding**: Body pinker, hood glossy purplish-black, bill pink with black base.
(c) **Juvenile**: Overall pale sandy greyish-brown with darker wings and tail. Like White-shouldered but has paler bill, legs, wings and tail, slight streaking on crown and breast.

6 [862] **COMMON STARLING** *Sturnus vulgaris*
(a) **Adult non-breeding** *poltaratskyi*: Blackish with heavy white to buff speckling, bill blackish.
(b) **Adult breeding**: More uniform glossy purplish- and greenish-black, sparse speckling, bill yellow.
(c) **Juvenile/first winter (transitional)**: Like non-breeding adult but with dusky-brown hood.

7 [1000] **AUSTRALASIAN BUSHLARK** *Mirafra javanica*
(a) **Adult** *williamsoni*: Weakish streaking on brown-washed breast, all-whitish outertail.
(b) **Juvenile**: Crown, mantle and ear-coverts less clearly streaked, crown darker, breast paler with more diffuse streaking.

8 [1001] **BENGAL BUSHLARK** *Mirafra assamica*
(a) **Adult**: Larger, larger-billed than Burmese, much greyer above, breast less boldly streaked, belly richer buff.
(b) **Juvenile**: Greyer, plainer above, breast less boldly marked than Burmese (less bold than illustrated).

9 [1002] **BURMESE BUSHLARK** *Mirafra microptera*
Adult: Like Indochinese but warmer and more boldly streaked above, small breast-streaks more spot-shaped.

10 [1003] **INDOCHINESE BUSHLARK** *Mirafra erythrocephala*
(a) **Adult**: Heavier-marked breast than Australasian, no white on outertail. For differences from Burmese see above.
(b) **Juvenile**: Scaled buff above, breast-streaks more diffuse.

11 [1004] **ORIENTAL SKYLARK** *Alauda gulgula*
(a) **Adult** *herberti*: Slender bill, crest, dark-streaked above, pale buffy-whitish below, clear breast-streaking.
(b) **Juvenile**: Crown to scapulars paler with whitish fringes, wing-coverts tipped whitish, breast-streaking more diffuse.
(c) **Adult** *inopinata*: Whiter throat, belly and vent, more strongly rufescent outer wing-fringes.

12 [1005] **GREATER SHORT-TOED LARK** *Calandrella brachydactyla*
Adult *dukhunensis*: Patch on breast-side. From Asian also by browner, streakier upperparts, longer tertials.

13 [1006] **SAND LARK** *Calandrella raytal*
(a) **Adult** *raytal*: Smallish and short-tailed, very pale greyish above, white below, slender bill, fine breast-streaks.
(b) **Juvenile**: Vague pale/dark scaling on upperparts.

14 [1007] **ASIAN SHORT-TOED LARK** *Calandrella cheleensis*
Adult *kukunoorensis*: Fine breast-streaking, head fainter-patterned than Greater (which see for differences).

7-14 to different scale

PLATE 88. TYPICAL THRUSHES & COCHOAS

1 [868] **PLAIN-BACKED THRUSH** *Zoothera mollissima*
Adult *mollissima*: Like Long-tailed but wing-bars faint, flight feathers plainer, warmer above.

2 [869] **LONG-TAILED THRUSH** *Zoothera dixoni*
(a) **Adult**: Head pattern and underparts recall Scaly but plain above, two prominent buffy-whitish wing-bars.
(b) **Juvenile**: Like adult but buff streaks on crown, mantle and scapulars, dark bars from mantle to uppertail-coverts.

3 [870] **WHITE'S THRUSH** *Zoothera aurea*
Adult *aurea*: Larger and somewhat paler and greyer above than Scaly; typically with longer bill, more prominent whitish eyering, well-mottled ear-coverts with less contrasting blackish patch at rear, stronger spots on malar area; often bolder, broader and more rounded scales on upperparts.

4 [871] **SCALY THRUSH** *Zoothera dauma*
Adult *dauma*: Mostly warm olive-brown to buffish upperbody and whitish underbody with heavy blackish scales. Two buffy-white bands on underwing, white-tipped outertail-feathers.

5 [872] **LONG-BILLED THRUSH** *Zoothera monticola*
Adult *monticola*: Like Dark-sided but larger, bill bigger, greyer and dark-scaled above, spotted below.

6 [873] **DARK-SIDED THRUSH** *Zoothera marginata*
Adult: Long bill, relatively short tail, warm olive-brown above, dark olive-brown flanks, scaled body-sides.

7 [875] **CHINESE THRUSH** *Turdus mupinensis*
Adult: Smallish size, shortish bill, spotted below; two buff wing-bars, no white on outertail feathers.

8 [876] **WHITE-COLLARED BLACKBIRD** *Turdus albocinctus*
(a) **Male**: Blackish plumage with broad white collar. Bill, legs and feet yellow.
(b) **Female**: Pattern as male but warm-tinged dark brown with greyish- to dusky-white collar.
(c) **Juvenile female**: Buffish streaks above, buff tipped wing-coverts, brownish-buff below with blackish scales.

9 [877] **CHINESE BLACKBIRD** *Turdus mandarinus*
(a) **Male** *mandarinus*: Sooty-black with yellow bill.
(b) **Female**: Browner, notably below, throat paler. Like Grey-winged but darker, darker bill, uniform wings.

10 [882] **GREY-WINGED BLACKBIRD** *Turdus boulboul*
(a) **Male**: Blackish with large pale greyish wing-patch. Bill orange.
(b) **Female**: Warm olive-brown with yellowish bill; wing-patch only slightly paler than rest of wing.
(c) **Juvenile female**: Upperparts and wings with buff streaks/spots; buffish below with dark mottling.

11 [883] **CHESTNUT THRUSH** *Turdus rubrocanus*
(a) **Male** *gouldi*: Largely chestnut with dark brownish-grey hood, blackish wings and tail.
(b) **Female**: Like male but body somewhat paler and more rufous, wings and tail browner.
(c) **Male** *rubrocanus*: Pale grey hood.
(d) **Female**: Duller and paler with more uniform crown, nape and upper mantle.

12 [890] **PURPLE COCHOA** *Cochoa purpurea*
(a) **Male**: Brownish-purple with pale lavender-blue crown.
(b) **Female**: Dark rufescent-brown above, deep buffish-rufous below.
(c) **Juvenile male**: White-marked blackish crown, upperparts streaked buff, underparts buff, barred blackish.

13 [891] **GREEN COCHOA** *Cochoa viridis*
(a) **Male**: Green; blue crown and nape, silvery-blue markings on black wings, black-tipped blue tail.
(b) **Female**: Like male but wings washed brownish.
(c) **Juvenile**: White-marked blackish crown, spotted buff above, scaled below.
(d) **First-summer male**: Whitish strip from chin to lower ear-coverts and neck-side, washed golden-buff below.

1
2a
2b
3
4
5
6
7
8a
8b
8c
9a
9b
10a
10b
10c
11a
11b
11c
11d
12a
12b
12c
13a
13b
13c
13d

PLATE 89. TYPICAL THRUSHES, ROCK-THRUSHES & WHISTLING-THRUSHES

1 [866] **CHESTNUT-CAPPED THRUSH** *Zoothera interpres*
(a) **Adult** *interpres*: Chestnut crown and nape, black throat and breast, white on wings, white vent.
(b) **Juvenile**: Streaked dull chestnut crown and mantle; head-sides/breast buff, blackish ear-covert bars and malar line.

2 [867] **ORANGE-HEADED THRUSH** *Zoothera citrina*
(a) **Male** *innotata*: Orange-rufous head and underparts (vent whitish) and bluish-grey upperparts and wings.
(b) **Female**: Greyish-olive above; head and underparts duller orange-rufous.
(c) **Juvenile**: Like female but crown and mantle darker; two ear-covert bars, scaled below.
(d) **Male** *aurimacula*: White-tipped median coverts, two ear-covert bars, whiter throat.

3 [892] **GRANDALA** *Grandala coelicolor*
(a) **Male**: Brilliant purplish-blue plumage with black lores, wings and tail.
(b) **Female**: Brownish with whitish streaks, greyish-blue rump, white tips and patch on wings.

4 [916] **BLUE-CAPPED ROCK-THRUSH** *Monticola cinclorhynchus*
(a) **Male non-breeding**: Like White-throated but blue on throat, deep rufous on underparts, blackish lores.
(b) **Male breeding**: Differs from White-throated in similar way to non-breeding.
(c) **Female**: Like White-throated but upperparts largely plain.

5 [917] **WHITE-THROATED ROCK-THRUSH** *Monticola gularis*
(a) **Male non-breeding**: Blue crown and nape, blue-black mantle, chestnut lores, rump and underparts, white wing-patch, white patch on throat and upper breast; greyish-white fringing above.
(b) **Male breeding**: No fringing on head and upperparts.
(c) **Female**: Greyish-brown above with black bars, heavy blackish scales below, white throat-patch.

6 [918] **BLUE ROCK-THRUSH** *Monticola solitarius*
(a) **Male breeding** *pandoo*: Lacks scales on body, bluer overall than non-breeding.
(b) **Male non-breeding**: Rather uniform dull grey-blue, whitish-and-blackish scales, unbarred crown.
(c) **Female non-breeding**: More uniform than other rock-thrush females.
(d) **Juvenile**: Like female but crown and mantle speckled dull pale buffish, underparts paler.
(e) **Male non-breeding** *philippensis*: From Chestnut-bellied by dense scales, duller head and upperparts.
(f) **Male breeding**: Cleaner; closer to Chestnut-bellied.

7 [919] **CHESTNUT-BELLIED ROCK-THRUSH** *Monticola rufiventris*
(a) **Male breeding**: Upperparts blue, face and throat dark blue, rest of underparts chestnut.
(b) **Female**: Strong dark/buff malar/submoustachial pattern, buffish-white behind ear-coverts.
(c) **Juvenile female**: Broad pale buff to whitish spots above and below.

8 [937] **MALAYAN WHISTLING-THRUSH** *Myophonus robinsoni*
Adult: Small; blue on wing-bend but no whitish covert tips or bluish spots above, on neck-sides or throat.

9 [938] **BLUE WHISTLING-THRUSH** *Myophonus caeruleus*
(a) **Adult** *eugenei*: Relatively large; dark purplish-blue with blue spangling. Bill yellow.
(b) **Juvenile**: Body mostly dark brown, bill duller.
(c) **Adult** *caeruleus*: Smaller all-blackish bill.
(d) **Adult** *dicrorhynchus*: Duller, whitish median covert tips, bluish spots above and on throat and breast.

1a
2d
2b
2a
1b
2c
3a
3b
4a
4b
4c
5a
5b
5c
6a
6b
6e
6c
7a
7b
7c
6d
6f
8
9b
9a
9c
9d

PLATE 90. TYPICAL THRUSHES

1 [874] **SIBERIAN THRUSH** *Zoothera sibirica*
(a) **Male** *sibirica*: Dark slaty; broad white supercilium and white vent with dark-scaled undertail-coverts.
(b) **Female**: Buffish supercilium, dark eyestripe, plain above, buffy-whitish below, scales on breast and flanks.
(c) **First-winter male**: Mixture of male and female features.

2 [878] **TICKELL'S THRUSH** *Turdus unicolor*
(a) **Male**: Smallish; ashy-grey, white belly and vent, orange-yellow to yellow bill. Eyering yellow, legs and feet brownish-yellow.
(b) **Female**: Upperside olive-brown, breast and flanks clay-brown with warm buff wash; blackish streaks/spots on malar to upper breast but mostly whitish throat. Relatively dull flanks with no markings.

3 [879] **GREY-BACKED THRUSH** *Turdus hortulorum*
(a) **Male**: Bluish-grey head, breast and upperside, orange-rufous flanks and whitish underparts.
(b) **Female**: Like Black-breasted but bill brownish, paler and more olive-tinged above, thinner breast streaks.

4 [880] **BLACK-BREASTED THRUSH** *Turdus dissimilis*
(a) **Male**: Black hood and upper breast, dark slaty upperside, orange-rufous lower breast and flanks.
(b) **Female**: Yellowish bill, plain brown upperside, blackish spots on throat and breast, orange-rufous flanks.
(c) **Juvenile**: Like female but upperparts tipped darker and streaked buff.

5 [881] **JAPANESE THRUSH** *Turdus cardis*
(a) **Male**: Yellow bill, blackish head, upper breast and upperside, white (dark-spotted) underparts.
(b) **Female**: Recalls Black-breasted but has blackish spots on flanks.
(c) **First-winter male**: Paler above than adult; whitish throat-centre, grey breast with blackish markings.

6 [884] **BLACK-THROATED THRUSH** *Turdus atrogularis*
(a) **Male breeding**: Black base colour to supercilium, throat and breast.
(b) **Female**: Whiter throat and submoustachial area.
(c) **First-winter female**: No rufous on tail or on head and breast.

7 [885] **RED-THROATED THRUSH** *Turdus ruficollis*
(a) **Male breeding**: Brownish-grey above and white below; rufous-red supercilium, throat and breast.
(b) **Female**: Black streaks on throat-sides and upper breast, less rufous-red on supercilium, throat and breast.
(c) **First-winter female**: Little rufous-red; supercilium whitish, throat/breast whitish with blackish streaks.

8 [886] **NAUMANN'S THRUSH** *Turdus naumanni*
(a) **Male**: Similar to Dusky, but much greyer and almost unmarked above, rump to tail strongly rufescent, pale parts of head-side washed reddish-rufous, breast and body-sides heavily mottled reddish-rufous. Reddish-rufous edgings on scapulars to median coverts.
(b) **Female**: Duller; lacks reddish-rufous edgings above, tail less rufescent, rufous of head-side duller and more pinkish to buffish-tinged, upper throat whitish, with stronger dark malar marking.
(c) **First-winter female**: Generally duller than female. Dullest females still show rufescent tinges on tail, head-side and underparts that are lacking on Dusky.

9 [887] **DUSKY THRUSH** *Turdus eunomus*
(a) **Male**: Whitish supercilium, largely blackish white-scaled breast and flanks, rufous-chestnut wings.
(b) **First-winter female**: Like female or duller.

10 [888] **GREY-SIDED THRUSH** *Turdus feae*
(a) **Male**: Like Eyebrowed but warmer/browner above; grey underparts with white chin and belly.
(b) **First winter**: Warmer breast and flanks, darker streaks on throat, buffish tips to greater coverts.

11 [889] **EYEBROWED THRUSH** *Turdus obscurus*
(a) **Male**: Grey hood, white supercilium, olive-brown above, pale orange-rufous lower breast and flanks.
(b) **Female**: Head much browner than male, flanks more washed out.

1c
1b
1a
2a
2b
4c
4b
4a
3b
3a
5c
6c
6a
6b
5a
5b
7b
8a
8c
7a
7c
8b
9b
9a
10b
10a
11b
11a

PLATE 91. SHORTWINGS, ROBINS, REDSTARTS & SHAMAS

1 [893] **GOULD'S SHORTWING** *Brachypteryx stellata*
(a) **Adult** *stellata*: Chestnut above, grey-and-black vermiculations below and small white star-shapes on lower breast to vent.
(b) **Juvenile**: Generally blackish-brown with rufous to rufous-chestnut streaks, greyer below with arrow-shapes (no vermiculations) on belly.

2 [894] **RUSTY-BELLIED SHORTWING** *Brachypteryx hyperythra*
(a) **Male**: Short white eyebrow, dark slaty-blue above, orange-rufous below.
(b) **Female**: Duller below than male, dark brown above, no eyebrow.

3 [895] **LESSER SHORTWING** *Brachypteryx leucophrys*
(a) **Adult** *carolinae*: Brown above, paler and buffier below with whiter throat, belly-centre and vent. Short white eyebrow may be concealed.
(b) **Juvenile**: Darker, rufous streaks above, buff-and-blackish scaling below, wing-coverts tipped rufous-chestnut.
(c) **Male** *langbianensis*: Grey on cheeks, rear supercilium, breast-band and flanks.
(d) **Male** *wrayi*: Slaty-blue head-sides and upperside, bluish-grey on breast and flanks.

4 [896] **WHITE-BROWED SHORTWING** *Brachypteryx montana*
(a) **Male** *cruralis*: Dull dark blue with clear white supercilium.
(b) **Female**: Brown with rufous forehead, lores, orbital area and supercilium.
(c) **Juvenile**: From Lesser by more uniformly dark throat and breast.
(d) **First-winter male**: Like female but has white supercilium, darker lores.

5 [907] **WHITE-TAILED ROBIN** *Myiomela leucura*
(a) **Male** *leucura*: Blackish, blue forehead, supercilium and shoulder-patch, and long white line on outertail feathers.
(b) **Female**: Cold olive-brown above, paler, buffish-tinged below with paler buffish throat-patch; white tail-lines.
(c) **Juvenile male**: Like female but darker (particularly below) with warm buff streaks/spots.

6 [908] **BLUE-FRONTED ROBIN** *Cinclidium frontale*
(a) **Male** *orientale*: Like White-tailed but tail longer with no white, rather ashier-blue overall.
(b) **Female**: Like White-tailed but tail longer with no white, darker above, more russet-brown below.

7 [909] **PLUMBEOUS WATER-REDSTART** *Rhyacornis fuliginosa*
(a) **Male** *fuliginosa*: Slaty-blue with chestnut tail-coverts and tail.
(b) **Female**: Blue-grey above, grey-and-whitish scales below, whitish wing-bars, dark tail with white at base.

8 [910] **WHITE-CAPPED WATER-REDSTART** *Chaimarrornis leucocephalus*
(a) **Adult**: White crown, black upper body and wings, chestnut-red lower body and tail-base.
(b) **Juvenile**: Crown fringed dark, dark of head, mantle and breast browner.

9 [939] **WHITE-BELLIED REDSTART** *Hodgsonius phaenicuroides*
(a) **Male** *ichangensis*: Slaty-blue with white belly-centre, long blackish tail with orange-rufous basal outer half.
(b) **Female**: Brown, buffier below with tail patterned as male but browner and duller.

10 [984] **ORIENTAL MAGPIE-ROBIN** *Copsychus saularis*
(a) **Male** *erimelas*: Glossy blackish head, upperside and upper breast, white belly, wing-stripe and outertail.
(b) **Female**: Pattern like male but dark grey not black.

11 [985] **WHITE-RUMPED SHAMA** *Copsychus malabaricus*
(a) **Male** *interpositus*: Head, breast and upperparts glossy blue-black, underparts deep orange-rufous, white rump and uppertail-coverts, long blackish tail with white outer feathers (not apparent on plate).
(b) **Female**: Pattern as male but dark greyish not blue-black, duller, paler rufous below.
(c) **Juvenile**: Buff speckles above, buff tipped wing-coverts, broad buff wing-fringes; throat/breast dark-scaled buffish.

12 [986] **RUFOUS-TAILED SHAMA** *Trichixos pyrropyga*
(a) **Male**: Resembles female White-rumped Shama but bright rufous rump, uppertail-coverts and most of shorter tail; white pre-ocular mark.
(b) **Female**: Head-sides and upperparts grey-brown, throat and breast buffy-rufous and belly whitish.
(c) **Juvenile**: Like female but heavily marked rich buff, throat and breast buff with broad dark streaks.

1a
1b
2b
2a
3c
3a
3b
3d
4a
4d
4b
4c
5c
5b
5a
6b
6a
7a
7b
8a
8b
9b
9a
10b
10a
12a
11a
12c
11c
12b
11b

PLATE 92. *LUSCINIA* ROBINS & ISABELLINE WHEATEAR

1 [897] **JAPANESE ROBIN** *Luscinia akahige*
(a) **Male** *akahige*: Rufous-orange head and breast, dark breast-band, grey belly, rufous-brown above.
(b) **Female**: Like male but browner below, rufous-orange duller, no breast-band.

2 [898] **SIBERIAN RUBYTHROAT** *Luscinia calliope*
(a) **Male non-breeding**: Brilliant red throat, short white supercilium and white submoustachial stripe.
(b) **Female**: Resembles male but throat white, supercilium and submoustachial stripe less distinct.

3 [899] **WHITE-TAILED RUBYTHROAT** *Luscinia pectoralis*
(a) **Male breeding** *tschebaiewi*: Like Siberian but black breast, slatier above, blackish tail with white at base and tip.
(b) **Male non-breeding**: Browner upperparts, whitish scaling on breast.
(c) **Female**: Like Siberian but colder and greyer with white spots on outertail-tips.
(d) **Juvenile**: Upperparts streaked/tipped pale buffish, underparts buffish-white with brownish-grey scales/streaks on throat/breast, outertail-tips buffish.

4 [900] **BLUETHROAT** *Luscinia svecica*
(a) **Male non-breeding** *svecica*: Pale below with scaly blue, black and rufous-red breast-bands, broad whitish supercilium, rusty-rufous at base of outertail.
(b) **Male breeding**: Blue, black and red solid, unscaled.
(c) **Female**: Blackish malar stripe and breast-band.

5 [901] **RUFOUS-HEADED ROBIN** *Luscinia ruficeps*
(a) **Male**: Slaty-grey above, whitish below, orange-rufous crown, ear-coverts and nape, black mask, clean white throat bordered black.
(b) **Female**: Like Siberian Blue but all-dark bill, no blue on rump to uppertail, outertail edged warm brown.

6 [902] **BLACKTHROAT** *Luscinia obscura*
(a) **Male**: Like Siberian Blue but has all-black throat and upper breast and largely white base of outertail.
(b) **Female**: Like Siberian Blue but tail warm, legs dark, no obvious scales below.

7 [903] **FIRETHROAT** *Luscinia pectardens*
(a) **Male breeding**: Like Blackthroat but has bright orange-red throat to upper breast, white neck-patch.
(b) **Male non-breeding**: Head-sides and underparts similar to female; hint of neck-patch.
(c) **Female**: Like Indian Blue but more uniform warm buff and unscaled below, legs and feet dark.
(d) **First-winter male**: Like adult female but scapulars and back slaty-blue.

8 [904] **INDIAN BLUE ROBIN** *Luscinia brunnea*
(a) **Male** *wickhami*: Dark blue above, white supercilium, bright orange-rufous below, white vent.
(b) **Female**: Like Siberian Blue but has unscaled rich buff breast and flanks, brown rump to uppertail.
(c) **Juvenile**: Darker above than female and streaked buffish, breast and flanks heavily mottled brown.

9 [905] **SIBERIAN BLUE ROBIN** *Luscinia cyane*
(a) **Male** *cyane*: Dull dark blue above, white underparts and broad black line from lores to breast-side.
(b) **Female**: Upperparts greyish-brown, often with dull blue on rump and uppertail-coverts (sometimes tail), throat-sides and breast variably mottled/scaled. Legs distinctly pinkish.
(c) **First-winter female**: Like adult but outer greater coverts tipped rufous-buff, often no blue on rump to tail.

10 [906] **RUFOUS-TAILED ROBIN** *Luscinia sibilans*
Adult: Like female Siberian Blue Robin but rump and tail rufescent, no buff below but distinct scaling.

11 [920] **ISABELLINE WHEATEAR** *Oenanthe isabellina*
Adult: Robust, short-tailed, longish-legged, sandy-brown above, white uppertail-coverts, blackish tail with white base.

1a
1b
2a
2b
3a
3b
3c
3d
4a
4b
4c
5a
5b
6a
6b
7a
7b
7c
7d
8a
8b
8c
9a
9b
9c
10
11

PLATE 93. REDSTARTS & CHATS

1 [911] **HODGSON'S REDSTART** *Phoenicurus hodgsoni*
(a) **Male:** Resembles Daurian but mantle to wings grey, upper breast black, white wing-patch narrower.
(b) **Female:** Like Black but paler above, stronger pale eyering and whiter underparts.

2 [912] **WHITE-THROATED REDSTART** *Phoenicurus schisticeps*
(a) **Male non-breeding**: Crown, mantle and breast edged pale brownish.
(b) **Male breeding**: Light blue crown, black head-sides, white throat-patch, long white wing-patch.
(c) **Female:** White throat-patch and long white wing-patch.

3 [913] **DAURIAN REDSTART** *Phoenicurus auroreus*
(a) **Male non-breeding** *leucopterus*: Crown/mantle brownish-grey, pale greyish fringing on throat and upper breast.
(b) **Male breeding**: Grey crown to upper mantle, black lower mantle and throat, broad white wing-patch.
(c) **Female**: Brown with paler underparts, broad white wing-patch and pale eyering.

4 [914] **BLUE-FRONTED REDSTART** *Phoenicurus frontalis*
(a) **Male non-breeding**: Blue feathers fringed pale brown.
(b) **Male breeding**: Dark blue head, breast and mantle, largely orange-rufous tail with blackish central feathers and tip.
(c) **Female**: Tail pattern as male, no white wing-patch; buffy greater covert tips and tertial fringes.
(d) **Juvenile**: Like female but upperparts and throat to upper belly blackish speckled buff.

5 [915] **BLACK REDSTART** *Phoenicurus ochruros*
(a) **Male non-breeding** *rufiventris*: Blackish head/breast with greyish scales, buff fringed-wings, no wing-patch.
(b) **Male breeding**: Head, back, breast and wings blacker.
(c) **Female**: Like female Daurian but lacks white wing-patch, less distinct pale eyering.

6 [921] **GREY BUSHCHAT** *Saxicola ferreus*
(a) **Male non-breeding**: Upperparts tipped brown, head and underparts browner-washed than breeding male.
(b) **Male breeding**: Slaty-grey above with dark streaks, black head-sides, white supercilium and throat.
(c) **Female non-breeding**: Autumn bird. Ear-coverts, crown, mantle and breast-band browner, upperparts less distinctly streaked.
(d) **Female breeding**: Supercilium buffish-white, ear-coverts dark brown, whitish throat, greyish-brown breast.

7 [922] **EASTERN STONECHAT** *Saxicola maurus*
(a) **Male non-breeding** *stejnegeri*: Like female but black-flecked head-sides and throat.
(b) **Male breeding**: Black hood, blackish upperparts and wings, white neck-side and wing-patch, whitish rump, orange-rufous breast.
(c) **Female non-breeding**: Fresh autumn bird. Rump and tail-tip warm buffish to rufous-buff.
(d) **Male breeding** *przewalskii*: Slightly larger; rufous-chestnut extends to belly.
(e) **Juvenile**: Blackish-brown above, spotted buff, uniform buffish below with brown streaks on breast.

8 [923] **WHITE-TAILED STONECHAT** *Saxicola leucura*
(a) **Male non-breeding**: Very like Common but tail pattern still distinctive.
(b) **Male breeding**: Very like Common but outertail mostly whitish, belly whiter and clear-cut from breast.
(c) **Female non-breeding**: Like Common but paler and plainer, tail paler-fringed.

9 [924] **PIED BUSHCHAT** *Saxicola caprata*
(a) **Male non-breeding** *burmanica*: Blackish fringed brownish, white vent and wing-streak, white rump and uppertail-coverts tipped rufous.
(b) **Male breeding**: Solid black and white.
(c) **Female non-breeding**: Much darker than Common Stonechat, dark-streaked above and below, rusty uppertail-coverts and tinge to belly, no white wing-streak.
(d) **Female breeding**: Plainer, less distinct streaks.

10 [925] **JERDON'S BUSHCHAT** *Saxicola jerdoni*
(a) **Male**: Glossy blackish above, all-white below.
(b) **Female**: No supercilium, longish, rather plain tail, white throat and belly-centre.

1a
1b
2b
2c
2a
3b
3a
3c
4d
4c
4b
4a
5b
5a
5c
6a
6d
6b
6c
7d
7a
7b
7c
7e
8a
8b
8c
9b
9c
9d
9a
10b
10a

PLATE 94. BUSH-ROBINS, BLUETAILS & FORKTAILS

1 [926] **WHITE-BROWED BUSH-ROBIN** *Tarsiger indicus*
(a) **Male** *yunnanensis*: Recalls Indian Blue but longer-tailed, darker-legged; slatier above, supercilium longer.
(b) **Female**: Narrow white supercilium, no blue above, buffy-brownish below. Bill blackish, legs and feet dark.

2 [927] **RUFOUS-BREASTED BUSH-ROBIN** *Tarsiger hyperythrus*
(a) **Male**: Like bluetails, but orange-rufous throat to belly.
(b) **Female**: Like bluetails, but much darker above, lacks white throat-patch, shows little rufous-orange on flanks.
(c) **Juvenile**: Like Himalayan Bluetail but much darker above and buffier below.

3 [928] **HIMALAYAN BLUETAIL** *Tarsiger rufilatus*
(a) **Male** *rufilatus*: Dark blue upperside, throat- and breast-side, white below with rufous-orange flanks.
(b) **Female**: Grey-brown above, with contrasting blue rump to uppertail; narrow white throat-patch enclosed by grey-brown breast-band, rufous-orange flanks.
(c) **Juvenile**: Marked with buff specks and streaks, flanks richer buff, uppertail-coverts and tail mostly blue.

4 [929] **RED-FLANKED BLUETAIL** *Tarsiger cyanurus*
(a) **Male**: Like Himalayan, but duller, lighter, almost turquoise-tinged upperside, paler supercilium (whitish in front of eye), less pure white underparts.
(b) **Female**: Like Himalayan, but breast paler, contrasting less with white throat-patch; greyer belly?

5 [930] **GOLDEN BUSH-ROBIN** *Tarsiger chrysaeus*
(a) **Male** *chrysaeus*: Rufous-yellow supercilium, rump, uppertail-coverts and underparts, blackish head-sides.
(b) **Female**: Greenish-olive above, tail as male but browner, yellowish eyebrow, underparts, and eyering.
(c) **Juvenile**: Like Himalayan Bluetail but tail similar to respective adults.

6 [931] **LITTLE FORKTAIL** *Enicurus scouleri*
(a) **Adult** *scouleri*: Recalls White-crowned but much smaller, tail much shorter with white restricted to sides.
(b) **Juvenile**: Upperparts browner, lacks white forehead, throat/breast white with sooty scales.

7 [932] **CHESTNUT-NAPED FORKTAIL** *Enicurus ruficapillus*
(a) **Male**: Only forktail with chestnut crown to upper mantle, dark scales on white breast.
(b) **Female**: Chestnut extends onto back.
(c) **Juvenile**: Duller above than female, weaker-marked below, with white throat, black malar line.

8 [933] **BLACK-BACKED FORKTAIL** *Enicurus immaculatus*
(a) **Adult**: Like Slaty-backed but crown and mantle black. From White-crowned by white breast, black midcrown.
(b) **Juvenile**: Ear-coverts, crown and upperparts duller and browner; no white forehead, white throat.

9 [934] **SLATY-BACKED FORKTAIL** *Enicurus schistaceus*
(a) **Adult**: Black and white with long forked tail and slaty-grey crown, nape and mantle.
(b) **Juvenile**: Upperparts tinged brown, lacks white forehead; white throat with greyish flecks, dull streaks below.

10 [935] **WHITE-CROWNED FORKTAIL** *Enicurus leschenaulti*
(a) **Adult** *indicus*: Large size, steep white forehead, black mantle and breast.
(b) **Juvenile**: Black plumage parts tinged brown; head all brown, indistinct streaks on throat and breast.

11 [936] **SPOTTED FORKTAIL** *Enicurus maculatus*
(a) **Adult** *robinsoni*: Like White-crowned but mantle heavily spotted white.
(b) **Juvenile**: Very like White-crowned but broad white tips on outer webs of tertials and secondaries.

1a
2a
2b
1b
2c
3a
3c
3b
5c
5a
4a
5b
4b
7c
7b
6a
6b
7a
8a
8b
9a
9b
10b
10a
11a
11b

PLATE 95. BLUE FLYCATCHERS

1 [940] **HAINAN BLUE FLYCATCHER** *Cyornis hainanus*
(a) **Male**: Dark blue upperside, throat and breast, pale bluish-grey belly and flanks grading to whiter vent.
(b) **Male variant**: White triangle on throat-centre.
(c) **Female**: Very like Blue-throated but duller and darker above (Blue-throated subspecies *klossi* similar), throat-and breast-sides coloured as mantle (Blue-throated nominate similar), throat and breast duller buffy-rufous.

2 [941] **PALE BLUE FLYCATCHER** *Cyornis unicolor*
(a) **Male** *unicolor*: Recalls Verditer Flycatcher but wings duller, tail-fringes brighter, belly and vent paler and greyer.
(b) **Female**: Grey-tinged crown/nape, rufous uppertail-coverts/tail, brownish-grey underparts, whiter belly-centre.

3 [942] **HILL BLUE FLYCATCHER** *Cyornis banyumas*
(a) **Male** *whitei*: Very like Tickell's Blue but orange-rufous grades into white belly.
(b) **Female**: Same gradation of orange-rufous to white on underparts as male, upperparts warm brown, tail and wing-fringes rufescent.
(c) **Juvenile female**: Dark tips and warm buff spotting above; buff-and-dark scaling below.

4 [943] **LARGE BLUE FLYCATCHER** *Cyornis magnirostris*
(a) **Male**: Very similar to Hill subspecies *whitei*, but much bigger, hook-tipped bill, longer wings (primary projection), slightly deeper blue above, orange-rufous of throat slightly paler than breast, very little visible darkness on uppermost chin at base of bill.
(b) **Female**: Much bigger bill than Hill, longer wings, more strongly rufescent tail; possibly paler throat (throat slightly paler than breast).

5 [944] **TICKELL'S BLUE FLYCATCHER** *Cyornis tickelliae*
(a) **Male** *indochina*: Sharp demarcation of orange-rufous breast and white belly.
(b) **Female**: From similar *Cyornis* by greyish to bluish-grey tinge to upperparts, demarcated underpart colours.
(c) **Female** *sumatrensis*: Often bluer above.

6 [945] **CHINESE BLUE FLYCATCHER** *Cyornis glaucicomans*
(a) **Male**: From Blue-throated subspecies *dialilaema* by blue of throat restricted to sides/chin (visible at close range), deeper/duller orange-rufous throat and breast, extensive brown wash on flanks, somewhat darker blue. Much deeper blue above than Hill, with contrasting shining azure shoulder-patch and uppertail-coverts, more azure-blue forehead/eyebrow. Always has dark blue chin.
(b) **Female**: Darker and warmer above than Blue-throated subspecies *dialilaema*, darker and less rufescent tail, deeper orange-rufous breast, contrasting pale buff throat, brown-washed flanks.

7 [946] **BLUE-THROATED FLYCATCHER** *Cyornis rubeculoides*
(a) **Male** *dialilaema*: Dark blue throat with orange-rufous triangle extending up its centre.
(b) **Female**: Like Hill Blue but throat and breast paler, more rufous-buff, lores paler, crown and mantle paler.
(c) **Juvenile female**: Like Hill Blue but lighter breast-scaling.
(d) **Male** *rubeculoides*: Uniform dark blue throat and darker upperparts.
(e) **Male** *klossi*: Paler rufous-orange breast, whitish apex to throat-triangle.
(f) **Male variant**: Whitish throat-triangle, warm buff wash/smudges on breast.

8 [947] **MALAYSIAN BLUE FLYCATCHER** *Cyornis turcosus*
(a) **Male** *rupatensis*: Like Blue-throated (nominate) but throat bright deep blue, breast paler rufous-orange, upperparts brighter blue, paler rump.
(b) **Female**: Like male but throat warm buff (whiter on chin and sides), upperparts less deep blue.

9 [948] **MANGROVE BLUE FLYCATCHER** *Cyornis rufigastra*
(a) **Male** *rufigastra*: Resembles Tickell's and Hill but somewhat duller blue above, less obvious lighter forehead and eyebrow, dull deep orange-rufous below, with buffier vent and whiter belly-centre.
(b) **Female**: Like male but shows distinctive whitish lores, cheek-spot and chin.

10 [949] **WHITE-TAILED FLYCATCHER** *Cyornis concretus*
(a) **Male** *cyanea*: Relatively large; blue head and body, white belly and vent, white lines down outertail.
(b) **Female**: White crescent on upper breast, white lines on outertail.
(c) **Female** *concretus*: Brighter rufous-brown plumage, buff throat, whiter, more demarcated belly.

11 [972] **PYGMY BLUE FLYCATCHER** *Muscicapella hodgsoni*
(a) **Male** *hodgsoni*: Tiny, flowerpecker-like. Dark blue above, brighter forecrown, buffy rufous-orange below.
(b) **Female**: Like Sapphire Flycatcher but smaller, shorter-tailed, warmer above, pale rufescent-buff below.

1a
1c
1b
2b
2a
3c
3b
3a
4b
4a
5c
5b
5a
7e
7d
7a
7f
6a
6b
7c
7b
8b
8a
9b
9a
10c
10a
10b
11a
11b

PLATE 96. NILTAVAS & FLYCATCHERS

1 [952] **FUJIAN NILTAVA** *Niltava davidi*
(a) **Male**: Like Rufous-bellied but only front and sides of crown shining blue; breast more rufous, belly and vent buffier.
(b) **Female**: Like Rufous-bellied but upperparts a shade darker, less rufous on wings and tail; slightly darker below.

2 [953] **RUFOUS-BELLIED NILTAVA** *Niltava sundara*
(a) **Male** *sundara*: Dark blue above, shining blue crown, neck, shoulder-patch and rump, blackish face/throat, dark orange-rufous underparts.
(b) **Female**: Strongly rufescent wings and tail (no white), blue neck-patch, whitish breast-patch.

3 [954] **RUFOUS-VENTED NILTAVA** *Niltava sumatrana*
(a) **Male**: Like Rufous-bellied but smaller, shining blue duller, no shoulder-patch, less obvious neck-patch.
(b) **Female**: Recalls Rufous-bellied but crown, nape and underparts greyer, undertail-coverts pale rufous.

4 [955] **VIVID NILTAVA** *Niltava vivida*
(a) **Male** *oatesi*: Like Rufous-bellied but orange-rufous forms wedge on centre of lower throat.
(b) **Female**: Lacks blue neck-patch: otherwise similar to Large.

5 [956] **LARGE NILTAVA** *Niltava grandis*
(a) **Male** *grandis*: Relatively large; dark blue above, lighter crown, neck and rump, mostly blue-black below.
(b) **Female**: Large; light blue neck-patch, brown below with buffish throat-patch.
(c) **Juvenile female**: Tipped black and spotted buff above, rich buff below and narrowly scaled blackish.
(d) **Female** *decorata*: Darker blue neck-patch and shining blue crown and nape.

6 [957] **SMALL NILTAVA** *Niltava macgrigoriae*
(a) **Male** *signata*: Like miniature Large but much paler belly, whitish vent; lighter blue forehead and neck-patch.
(b) **Female**: Like miniature Large but darker throat-centre, generally greyer underparts.
(c) **Juvenile female**: Resembles miniature Large.

7 [980] **BLUE-AND-WHITE FLYCATCHER** *Cyanoptila cyanomelana*
(a) **Male** *cyanomelana*: Large; azure to cobalt-blue above, blackish head-sides, throat and breast, otherwise white.
(b) **Female**: Clear-cut white belly and vent, no blue, no white on tail.
(c) **First-winter male**: Like female but wings and scapulars to uppertail-coverts similar to adult male.

8 [967] **LITTLE PIED FLYCATCHER** *Ficedula westermanni*
(a) **Male** *australorientis*: Only black and white flycatcher in region.
(b) **Female**: Greyish above, rufescent uppertail-coverts, white below with greyish wash on breast-sides/flanks.
(c) **Juvenile female**: Darker above than adult with blackish fringes and buff spots, whitish below, lightly scaled.

9 [968] **ULTRAMARINE FLYCATCHER** *Ficedula superciliaris*
(a) **Male** *aestigma*: Deep blue above, broad dark blue lateral breast-patches, white underparts.
(b) **Female**: Very like Little Pied but larger, no rufous on uppertail-coverts, brownish-grey sides of throat and breast.
(c) **Juvenile female**: Darker and browner above than adult, with blackish tips and buff spotting, buffy-white below with blackish scaling.
(d) **First-winter male**: Like female but blue on wings, tail and mantle, broad buff wing-bar.

10 [971] **SAPPHIRE FLYCATCHER** *Ficedula sapphira*
(a) **Male breeding** *sapphira*: Like Ultramarine but blue brighter, orange-rufous centre of throat and breast.
(b) **Male non-breeding**: Like female but scapulars to uppertail-coverts and wings as breeding male.
(c) **Female**: Small; shortish tail, warm brown above, rufescent uppertail-coverts, deep buffish-orange throat and upper breast.
(d) **Juvenile**: Blackish-brown above spotted rich buff, throat and upper breast buffier than female with fine blackish scaling.

1a
1b
2a
2b
3a
3b
4a
4b
5c
5b
5a
6b
6a
5d
6c
8c
7a
8a
7c
8b
7b
10d
9d
10a
10b
9a
9b
9c
10c

PLATE 97. FLYCATCHERS

1 [958] **YELLOW-RUMPED FLYCATCHER** *Ficedula zanthopygia*
(a) **Male (winter)**: Black and yellow above, yellow below, white supercilium, wing-patch and undertail-coverts.
(b) **Male (spring)**: Orange flush on throat and breast.
(c) **Female**: Dull greyish-olive upperparts, but yellow rump and white wing-patch.

2 [959] **NARCISSUS FLYCATCHER** *Ficedula narcissina*
(a) **Male** *narcissina*: Like Yellow-rumped but supercilium yellow, belly yellowish-white, tertials without white line.
(b) **Female**: Browner above (olive-tinged back and rump), whitish below, mottling on throat-sides and breast.

3 [960] **GREEN-BACKED FLYCATCHER** *Ficedula elisae*
(a) **Male**: Greyish olive-green above, dark wings and tail, bright yellow loral stripe, eyering, rump and underparts, broad white wing-patch.
(b) **Female**: Duller above and below with no yellow on rump or white wing-patch, two pale wing-bars.

4 [961] **MUGIMAKI FLYCATCHER** *Ficedula mugimaki*
(a) **Male**: Blackish-slate above, white supercilium, wing-patch and tertial edges, rufous-orange below.
(b) **Female**: Greyish-brown above, throat and breast buffish-orange, two narrow wing-bars.
(c) **First-winter male**: Like female but head-sides greyer, supercilium stronger, throat/breast brighter.

5 [962] **SLATY-BACKED FLYCATCHER** *Ficedula hodgsonii*
(a) **Male**: Dark, dull bluish-slate above, blackish tail with white at base, orange-rufous below.
(b) **Female**: Dull olive-brown above, greyish throat/breast, rufescent uppertail-coverts, whitish eyering.

6 [963] **WHITE-GORGETED FLYCATCHER** *Ficedula monileger*
Adult *leucops*: Small; mostly brown with clean white throat bordered black.

7 [964] **RUFOUS-BROWED FLYCATCHER** *Ficedula solitaris*
(a) **Adult** *submonileger*: Like White-gorgeted but more rufescent, rufous spectacles; faint throat border.
(b) **Adult** *malayana*: Richer rufous-chestnut.

8 [965] **SNOWY-BROWED FLYCATCHER** *Ficedula hyperythra*
(a) **Male** *hyperythra*: Small; dark slaty-blue above, orange-rufous throat and breast, white eyebrow.
(b) **Female**: Shortish tail, uniform plumage tone, buffy loral stripe, eyering and underparts, rufescent wings.
(c) **Juvenile female**: Darker above than adult with blackish fringes and warm buff streaks/speckles; heavily marked blackish below.
(d) **Female** *annamensis*: Crown washed bluish-grey, more orange-buff below.

9 [966] **RUFOUS-CHESTED FLYCATCHER** *Ficedula dumetoria*
(a) **Male** *muelleri*: Like Mugimaki but smaller, long supercilium, white wing-covert streak, all-dark tertials.
(b) **Female**: From Mugimaki by size and proportions, paler throat, warm buff eyering, rusty-buffish covert tips.

10 [969] **SLATY-BLUE FLYCATCHER** *Ficedula tricolor*
(a) **Male** *diversa*: Slaty-blue above, blackish tail with white at base, buffy-white throat and buffy blue-grey breast-band.
(b) **Female**: Like Slaty-backed but darker, warmer upperparts, rufous-chestnut tail, buffish eyering and underparts.
(c) **Male** *cerviniventris*: Underparts warm buff (apart from breast-band).
(d) **Female**: Rich buff underparts.

11 [970] **TAIGA FLYCATCHER** *Ficedula albicilla*
(a) **Male non-breeding/female**: Buffish-grey on breast, blackish tail with white at side; pale eyering, blackish bill.
(b) **Male breeding**: Rufous-orange throat, bordered grey.

12 [999] **GREY-HEADED CANARY-FLYCATCHER** *Culicicapa ceylonensis*
(a) **Adult** *calochrysea*: Head and breast grey, rest of underparts bright yellowish, flanks washed olive.
(b) **Juvenile**: Like adult but duller.

1b
1a
1c
2a
2b
3a
3b
4c
4a
4b
5b
5a
6
7b
7a
8b
8c
8a
8d
10d
10a
9b
10c
10b
9a
11a
12a
12b
11b

PLATE 98. FLYCATCHERS

1 [950] **PALE-CHINNED FLYCATCHER** *Cyornis poliogenys*
(a) **Adult** *poliogenys*: Head greyish, throat whitish, breast rufescent. Pale buff eyering, rufescent tail.
(b) **Juvenile**: Upperparts darker-tipped and speckled buff, breast dark-scaled.
(c) **Adult** *cachariensis*: Deeper, purer rufous-buff on underparts.

2 [951] **VERDITER FLYCATCHER** *Eumyias thalassinus*
(a) **Male breeding** *thalassinus*: Turquoise-tinged pale blue with black lores.
(b) **Male non-breeding**: Duller and more turquoise.
(c) **Female breeding**: Like male but duller and slightly greyish-tinged, with dusky lores.
(d) **Female non-breeding**: Underparts duller and greyer.

3 [973] **FERRUGINOUS FLYCATCHER** *Muscicapa ferruginea*
(a) **Adult (fresh)**: Body richly rufescent, distinctly slaty-grey on head.
(b) **Juvenile**: Crown blackish with buff streaks, upperparts blackish, mottled buff, breast scaled blackish.

4 [974] **RUFOUS-GORGETED FLYCATCHER** *Muscicapa strophiata*
(a) **Male** *strophiata*: Blackish-grey face, paler breast with orange-rufous centre, whitish eyebrow.
(b) **Female**: Duller, with paler breast, less distinct eyebrow, smaller gorget, buffish chin.
(c) **Juvenile**: Upperparts with blackish tips and rich buff spots, buff below, scaled blackish.

5 [975] **GREY-STREAKED FLYCATCHER** *Muscicapa griseisticta*
Adult: Like Dark-sided but larger, cleaner below with greyish streaks across entire breast and flanks.

6 [976] **DARK-SIDED FLYCATCHER** *Muscicapa sibirica*
(a) **Adult** *rothschildi*: Small bill, smudgy, darker-streaked breast/flanks, dark-centred undertail-coverts.
(b) **Juvenile**: Blacker above with pale buff spots, blackish-marked throat to flanks, coverts tipped buff.
(c) **Adult** *sibirica*: More white on abdomen, clearer breast-streaks.
(d) **Adult (worn)**: Less breast-streaking.

7 [977] **ASIAN BROWN FLYCATCHER** *Muscicapa dauurica*
(a) **Adult** *dauurica*: Plain brownish-grey above, whitish below, breast brownish-grey; pale covert fringes.
(b) **Adult (worn)**: Greyer above and paler below.
(c) **Adult** *siamensis*: Browner above, duller, more uniform below, mostly pale lower mandible, less distinct eyering.

8 [978] **BROWN-STREAKED FLYCATCHER** *Muscicapa williamsoni*
(a) **Adult**: Warmer above than Asian Brown nominate *dauurica*; yellowish lower mandible, brownish streaks below.
(b) **Adult (worn)**: Vaguer streaking, plainer wings.

9 [979] **BROWN-BREASTED FLYCATCHER** *Muscicapa muttui*
Adult (typical winter bird): Like Asian Brown but larger, pale yellowish lower mandible, broad whitish eyering, rufescent upperparts, greyish-brown crown and ear-coverts, warm greyish-brown breast/flanks.

10 [981] **BROWN-CHESTED JUNGLE-FLYCATCHER** *Rhinomyias brunneata*
Adult *brunneata*: Rather long bill with pale yellow lower mandible, faint dark mottling on throat.

11 [982] **FULVOUS-CHESTED JUNGLE-FLYCATCHER** *Rhinomyias olivacea*
(a) **Adult** *olivacea*: All-dark bill, pinkish legs, whitish throat and warm buffish-brown breast and flanks.
(b) **Juvenile**: Warmer above with brown tips and buff speckles, breast-band and flanks mottled.

12 [983] **GREY-CHESTED JUNGLE-FLYCATCHER** *Rhinomyias umbratilis*
(a) **Adult**: From other *Rhinomyias* by grey upper breast, clean, gleaming white throat and dark malar area.
(b) **Juvenile**: Like Fulvous-chested but darker above, with stronger buff spots, breast washed greyish.

2a
2c
2b
1b
1c
1a
2d
3b
3a
4a
4b
4c
5
6d
6c
6b
6a
7b
7a
7c
8b
8a
9
12b
10
11b
11a
12a

PLATE 99. TITS

1 [987] **BLACK-BIBBED TIT** *Poecile hypermelaena*
(a) **Adult**: Greyish-brown above, pale below, no crest or wing-bars.
(b) **Juvenile**: Dark of head browner, upperparts warmer, whitish head-side tinged buff, bib smaller.

2 [988] **GREY-CRESTED TIT** *Lophophanes dichrous*
Adult *wellsi*: Greyish upright crest, pale buffish throat, half-collar and underparts, mid-greyish upperparts.

3 [989] **COAL TIT** *Periparus ater*
(a) **Adult** *aemodius*: Like Black-bibbed and Rufous-vented but greyish above, pinkish to buffish-white below, two pronounced wing-bars. Paler below than Rufous-vented.
(b) **Juvenile**: Dark plumage parts duller and browner with fainter bib, no obvious crest or nape-patch, whitish of head-sides washed yellowish.

4 [990] **RUFOUS-VENTED TIT** *Periparus rubidiventris*
(a) **Adult** *beavani*: Greyish; black head with upright crest and whitish patches on side and nape, rufous vent.
(b) **Juvenile**: Black of head duller, crest shorter, duller above, drabber below, whitish head-patches washed buff.

5 [991] **GREY TIT** *Parus cinereus*
(a) **Male** *ambiguus*: Relatively large; grey upperparts, black head and ventral stripe, large white patch on head-side, small patch on nape and single wing-bar.
(b) **Juvenile**: Dark parts of head duller and browner, upperparts tinged olive, ventral stripe reduced.

6 [992] **JAPANESE TIT** *Parus minor*
Male *nubicolus*: Larger than Grey, yellowish-green wash on upper mantle, blue-grey fringes to flight-feathers.

7 [993] **GREEN-BACKED TIT** *Parus monticolus*
(a) **Male** *legendrei*: Greenish above, pale yellow below, very broad black ventral stripe, two wing-bars.
(b) **Juvenile**: Black parts browner, head-sides and underparts duller, nuchal patch and wing-bars yellowish.
(c) **Male** *yunnanensis*: Greener above, yellower below with narrower black ventral stripe, broader wing-bars.

8 [994] **YELLOW-CHEEKED TIT** *Parus spilonotus*
(a) **Male** *subviridis*: Black crest (yellow at rear), yellow sides of head and underparts, black post-ocular stripe, bib and ventral stripe, and broad whitish wing-bars.
(b) **Female**: Less black above (more uniform), bib and ventral stripe olive-yellow (sometimes absent).
(c) **Juvenile male**: Crown and bib duller than adult male, crest shorter, wing-bars washed yellow.
(d) **Male** *rex*: Blue-grey upperparts with black streaks, broader bib (joining post-ocular stripe) and ventral stripe, greyish rest of underparts.
(e) **Female**: Greyer throat and ventral band.

9 [995] **SULTAN TIT** *Melanochlora sultanea*
(a) **Male** *sultanea*: Very large; longish tail, black plumage with yellow crest and lower breast to vent.
(b) **Female**: Browner above, head-sides and wings browner, throat and upper breast dull yellowish olive-green.
(c) **Male** *gayeti*: Glossy blue-black crest.

10 [996] **YELLOW-BROWED TIT** *Sylviparus modestus*
Adult *modestus*: Small and short-tailed with greyish-olive upperparts and paler underparts, slight tufted crest, short pale yellowish eyebrow, and narrow pale bar on greater coverts.

11 [997] **CHINESE PENDULINE TIT** *Remiz consobrinus*
(a) **Male non-breeding**: Greyish crown, black mask, dull chestnut upper mantle band, buffish wing fringing.
(b) **Female non-breeding**: Duller than male; browner mask and upperparts.
(c) **First-year female**: Pale mask (hardly contrasts with crown).

12 [998] **FIRE-CAPPED TIT** *Cephalopyrus flammiceps*
(a) **Male breeding** *olivaceus*: Tiny and warbler-like, yellowish-green above (yellower on rump), yellowish below, reddish-orange forehead-patch and faint reddish wash on throat.
(b) **Female breeding**: Like breeding male but forehead-patch golden-olive, throat dull olive-yellow.
(c) **Adult non-breeding**: Like breeding female but throat whitish. Recalls Yellow-browed Tit but lacks crest, greener above with yellowish fringes on wing-coverts, yellower below.
(d) **Juvenile**: Whitish below. Daintier and thinner-billed than Green Shrike-babbler.

13 [1078] **BLACK-THROATED TIT** *Aegithalos concinnus*
(a) **Adult** *pulchellus*: Very small with long tail, long black mask, white bib and neck-side with large isolated black throat-patch, rufous-chestnut breast-band extending along flanks; otherwise grey above, whitish below.
(b) **Juvenile**: No black throat-patch, dark smudges on border of upper breast.
(c) **Adult** *talifuensis*: Orange-rufous crown.

14 [1079] **GREY-CROWNED TIT** *Aegithalos annamensis*
Adult: Dull grey crown, narrow drab greyish breast-band and greyish-pink wash on flanks.

15 [1080] **BLACK-BROWED TIT** *Aegithalos bonvaloti*
(a) **Adult** *bonvaloti*: Like Black-throated but black head-band broader, lower head-side pale cinnamon-rufous, throat white with central triangle of dense black speckles, breast-band broader.
(b) **Juvenile**: Duller black head-band and throat, paler below with greyish mottling.

16 [1081] **BURMESE TIT** *Aegithalos sharpei*
Adult: Slightly narrower black head-bands than Black-browed, white lower head-sides and lower throat, dark brownish breast-band, uniformly buffish remainder of underparts.

1b
1a
2
3a
3b
4a
4b
5a
5b
6
7a
7b
7c
8a
8b
8c
8d
8e
9a
9b
9c
10
11a
11b
11c
12a
12b
12c
12d
13a
13b
13c
14
15a
15b
16

PLATE 100. BULBULS

1 [1008] **CRESTED FINCHBILL** *Spizixos canifrons*
(a) **Adult** *ingrami*: Relatively large, greenish above, yellowish-green below, with thick pale bill and erect pointed crest.
(b) **Juvenile**: Crown and crest paler, throat greener.

2 [1009] **COLLARED FINCHBILL** *Spizixos semitorques*
(a) **Adult** *semitorques*: Like Crested but uncrested, white patch/streaks on head, whitish half-collar.
(b) **Juvenile**: Browner on head.

3 [1011] **STRIATED BULBUL** *Pycnonotus striatus*
Adult *paulus*: Yellowish-white streaking on head and body, prominent crest, yellow undertail-coverts.

4 [1026] **RED-WHISKERED BULBUL** *Pycnonotus jocosus*
(a) **Adult** *pattani*: Tall black crest, whitish ear-coverts and underparts, red ear-patch and undertail-coverts.
(b) **Juvenile**: Browner-tinged overall, crest shorter, no red ear-patch.
(c) **Adult** *monticola*: Larger; darker red ear-patch.

5 [1027] **BROWN-BREASTED BULBUL** *Pycnonotus xanthorrhous*
(a) **Adult** *xanthorrhous*: Like Sooty-headed but browner above, brown ear-coverts and breast, little white on tail.
(b) **Juvenile**: Browner and less distinctly marked.

6 [1028] **LIGHT-VENTED BULBUL** *Pycnonotus sinensis*
(a) **Adult** *hainanus*: Undertail-coverts whitish, breast-band yellowish-grey, dark ear-coverts with whitish patch.
(b) **Juvenile**: Paler above, largely greyish-brown head, breast-band less distinct.
(c) **Adult** *sinensis*: Broad white patch from eye to nape.

7 [1029] **RED-VENTED BULBUL** *Pycnonotus cafer*
Adult *melanchimus*: Mostly dark with whitish scales, white rump and red undertail-coverts.

8 [1030] **SOOTY-HEADED BULBUL** *Pycnonotus aurigaster*
(a) **Adult** *klossi*: Black cap and cheeks, whitish rump, greyish underparts and red undertail-coverts.
(b) **Adult** *thais*: Yellow undertail-coverts.

9 [1042] **ASHY BULBUL** *Hemixos flavala*
(a) **Adult** *hildebrandi*: Black crown/face, brown ear-coverts, grey above, yellowish wing-panel, grey-washed breast.
(b) **Adult** *flavala*: Dark grey crown.
(c) **Adult** *remotus*: Paler, browner head, mid-brown upperparts and brownish-tinged breast.
(d) **Adult** *cinereus*: Plainer and browner above, grey-fringed crown, black patch on lores and cheeks, plain wings.

10 [1043] **CHESTNUT BULBUL** *Hemixos castanonotus*
(a) **Adult** *canipennis*: Black crown, chestnut head-sides and upperparts, whitish throat.
(b) **Juvenile**: Duller overall.

11 [1045] **HIMALAYAN BLACK BULBUL** *Hypsipetes leucocephalus*
(a) **Adult** *concolor*: Relatively large, all blackish (greyer below), red bill and legs.
(b) **Juvenile**: Browner overall.
(c) **Adult** *stresemanni*: All-white head.
(d) **Adult** *leucothorax*: All-white head and breast.

12 [1046] **WHITE-HEADED BULBUL** *Cerasophila thompsoni*
(a) **Adult**: Like white-headed subspecies of Black but body paler grey, undertail-coverts rufous-chestnut, lores black.
(b) **Juvenile**: Browner overall, head brownish-grey.

1b
2b
4b
1a
3
2a
4a
4c
6c
5a
5b
6b
6a
7
9d
9a
8b
8a
9b
9c
10a
11a
11d
12a
10b
11c
11b
12b

PLATE 101. BULBULS

1 [1013] **BLACK-HEADED BULBUL** *Pycnonotus atriceps*
(a) **Adult** *atriceps*: Mostly yellowish-green, with glossy black head, blackish primaries and broad black tail-band.
(b) **Adult**: Greener variant.
(c) **Adult grey morph**: Neck, breast and belly grey.
(d) **Juvenile**: Generally duller, head largely dull greenish.

2 [1014] **BLACK-CRESTED BULBUL** *Pycnonotus flaviventris*
(a) **Adult** *caecilli*: Bright yellow below, glossy black head and tall crest; greenish-olive above.
(b) **Juvenile**: Head duller and browner, throat mixed with olive-yellow, crest short.
(c) **Adult** *johnsoni*: Deeper yellow below with red throat.

3 [1015] **SCALY-BREASTED BULBUL** *Pycnonotus squamatus*
Adult *weberi:* Black head, white throat, white-scaled black breast and flanks, yellow undertail-coverts.

4 [1016] **GREY-BELLIED BULBUL** *Pycnonotus cyaniventris*
Adult *cyaniventris*: Grey head and underparts, yellow undertail-coverts, upperparts and tail olive-green.

5 [1017] **PUFF-BACKED BULBUL** *Pycnonotus eutilotus*
Adult: Brown above, whitish throat and underparts; short crest.

6 [1018] **STRIPE-THROATED BULBUL** *Pycnonotus finlaysoni*
(a) **Adult** *eous*: Yellow streaks on forecrown, ear-coverts, throat and upper breast, yellow vent.
(b) **Juvenile**: Crown and upperparts browner, little yellow streaking.
(c) **Adult** *davisoni*: No yellow streaks on forecrown and ear-coverts, greener crown and rump, pale eyes.

7 [1019] **FLAVESCENT BULBUL** *Pycnonotus flavescens*
(a) **Adult** *vividus*: Yellowish below, greyish head, blackish crown and lores, whitish pre-ocular supercilium.
(b) **Juvenile**: Browner and plainer on head and upperparts, bill paler.

8 [1020] **YELLOW-VENTED BULBUL** *Pycnonotus goiavier*
(a) **Adult** *personatus*: White supercilium and throat, dark crown, lores and bill, yellow vent.
(b) **Juvenile**: Supercilium weaker, crown and bill paler.

9 [1021] **OLIVE-WINGED BULBUL** *Pycnonotus plumosus*
(a) **Adult** *plumosus*: Like Streak-eared but ear-streaks weaker, red eyes, dark bill; yellowish-green wing fringes.
(b) **Juvenile**: Browner overall.

10 [1022] **STREAK-EARED BULBUL** *Pycnonotus blanfordi*
(a) **Adult** *conradi*: Brownish; paler throat and belly, yellowish vent, whitish-streaked ear-coverts, pale eyes.
(b) **Juvenile**: Less distinctively marked.

11 [1023] **CREAM-VENTED BULBUL** *Pycnonotus simplex*
(a) **Adult** *simplex*: Nondescript with whitish eyes.
(b) **Juvenile**: Eyes pale brown, crown and upperparts warmer than adult.

12 [1024] **RED-EYED BULBUL** *Pycnonotus brunneus*
(a) **Adult** *brunneus*: Nondescript with red eyes.
(b) **Juvenile**: Eyes brownish, upperparts warmer than adult, bill paler.

13 [1025] **SPECTACLED BULBUL** *Pycnonotus erythropthalmos*
(a) **Adult** *erythropthalmos*: Very like Red-eyed but has orange to yellow-orange eyering, paler throat and vent.
(b) **Juvenile**: Eyering duller, upperparts warmer brown.

1d
1b
2b
3
1a
2c
2a
1c
4
5
6b
6a
7a
6c
7b
8b
12b
9a
8a
12a
11b
10a
9b
11a
10b
13a
13b

PLATE 102. BULBULS

1 [1010] **STRAW-HEADED BULBUL** *Pycnonotus zeylanicus*
Adult: Large; golden-yellowish crown and cheeks, blackish eye- and submoustachial stripes.

2 [1012] **BLACK-AND-WHITE BULBUL** *Pycnonotus melanoleucus*
(a) **Male**: Blackish-brown with mostly white wing-coverts.
(b) **Juvenile**: Cold brownish plumage, darker-centred upperpart feathers and dark streaks on breast.

3 [1031] **OLIVE BULBUL** *Iole virescens*
Adult *virescens*: From Grey-eyed by dark reddish to brown eyes, typically stronger olive above, yellower below.

4 [1032] **GREY-EYED BULBUL** *Iole propinqua*
(a) **Adult** *propinqua*: Small and slim; pale supercilium, yellowish below with rufous-buffish undertail-coverts.
(b) **Adult** *cinnamomeoventris*: Less yellow below than Olive and darker-vented than Buff-vented.

5 [1033] **BUFF-VENTED BULBUL** *Iole olivacea*
Adult *cryptus*: From Grey-eyed by browner wings, greyer underparts and buffish undertail-coverts.

6 [1034] **HAIRY-BACKED BULBUL** *Tricholestes criniger*
Adult *criniger*: Relatively small; broad yellowish orbital area, breast mottled greyish-olive, vent yellow.

7 [1035] **FINSCH'S BULBUL** *Alophoixus finschii*
Adult: Small; brownish-olive with shortish dark bill and yellow throat and vent.

8 [1036] **YELLOW-BELLIED BULBUL** *Alophoixus phaeocephalus*
Adult *phaeocephalus*: Bluish-grey head, whitish-grey lores, white throat and bright yellow underparts.

9 [1037] **GREY-CHEEKED BULBUL** *Alophoixus bres*
Adult *tephrogenys*: No obvious crest, grey head-sides, olive wash across breast.

10 [1038] **WHITE-THROATED BULBUL** *Alophoixus flaveolus*
(a) **Adult** *burmanicus*: Like Puff-throated but breast to vent all yellow, lores and ear-coverts whitish-grey, crest longer.
(b) **Juvenile**: Crown and upperparts brown, whitish throat, rest of underparts suffused brown.

11 [1039] **PUFF-THROATED BULBUL** *Alophoixus pallidus*
(a) **Adult** *henrici*: Browner below than White-throated, buffier vent, darker head-sides; darker below than Ochraceous, strong greenish-olive tinge above.
(b) **Adult** *griseiceps*: Whiter ear-coverts, yellower underparts (to separate from White-throated see that text).
(c) **Adult** *annamensis*: Darker breast and flanks, slatier-grey head-sides.

12 [1040] **OCHRACEOUS BULBUL** *Alophoixus ochraceus*
(a) **Adult** *sacculatus*: Like Puff-throated but slightly smaller and shorter-crested, browner above, no yellow below.
(b) **Adult** *hallae*: Yellowish-tinged below (to separate from Puff-throated see that text).

13 [1041] **STREAKED BULBUL** *Ixos malaccensis*
(a) **Adult**: Olive upperparts, greyish throat and breast with narrow whitish streaks, white vent.
(b) **Juvenile**: Upperparts warm rufescent-brown, breast less distinctly streaked.

14 [1044] **MOUNTAIN BULBUL** *Ixos mcclellandii*
(a) **Adult** *tickelli*: Greenish-olive above, whitish streaks on shaggy crown, throat and upper breast, yellow vent.
(b) **Adult** *similis*: Greyish-brown upperparts, pinkish-chestnut ear-coverts and breast.

1
2b
2a
3
4a
4b
5
6
7
8
9
10a
10b
11a
11b
12a
12b
11c
13b
13a
14a
14b

PLATE 103. MARTINS & SWALLOWS

1 [1047] **NORTHERN HOUSE-MARTIN** *Delichon urbicum*
(a,b) **Adult** *lagopoda*: Like Asian but with whiter and larger rump-patch, whiter underparts, greyish-white underwing-coverts, deeper-forked tail.
(c) **Juvenile**: Browner above, greyish-washed below often with duller breast-sides; some dark scaling on vent and tail-coverts.

2 [1048] **ASIAN HOUSE-MARTIN** *Delichon dasypus*
(a,b) **Adult** *dasypus*: Smallish; shallow tail-fork, blue-glossed black upperparts and greyish-white rump (faintly streaked) and underside, darkish underwing-coverts.
(c) **Juvenile**: Upperparts browner, tail less deeply forked.

3 [1049] **NEPAL HOUSE-MARTIN** *Delichon nipalense*
(a,b) **Adult** *nipalense*: Like Asian but tail almost square-cut, throat mostly dark, undertail-coverts black, rump-band narrower.
(c) **Juvenile**: Browner above; throat and undertail-coverts mixed with whitish, breast greyish.

4 [1050] **COMMON SAND-MARTIN** *Riparia riparia*
(a) **Adult** *ijimae*: Larger than Grey-throated, whitish below with broad brown breast-band, tail-fork somewhat deeper.
(b) **Juvenile**: Upperparts fringed buffish, throat tinged buff, rest of underparts often less white.

5 [1051] **PALE SAND-MARTIN** *Riparia diluta*
Adult *fokiensis*: From Common by slightly smaller size, bill and tail-fork, slightly paler and greyer above, paler breast-band, creamy wash on belly and vent, no dark centre of breast/upper belly.

6 [1052] **GREY-THROATED SAND-MARTIN** *Riparia chinensis*
(a,b) **Adult** *chinensis*: Similar to Common and Pale but smaller and daintier, with shallower tail-fork, no breast-band, throat and breast greyish-brown.

7 [1053] **DUSKY CRAG-MARTIN** *Ptyonoprogne concolor*
(a,b) **Adult**: Uniform dark brown, tail barely forked showing row of whitish spots when spread.

8 [1054] **BARN SWALLOW** *Hirundo rustica*
(a,b) **Adult breeding** *gutturalis*: Blue-black upperside and breast-band, chestnut-red forehead and throat, all-whitish rest of underparts and deeply forked tail with outertail-streamers.
(c) **Juvenile**: Browner above, forehead and throat duller, breast-band browner; short tail-fork.
(d) **Adult non-breeding** *tytleri*: Pale rufous below (seasonally lacking tail-streamers).

9 [1055] **WIRE-TAILED SWALLOW** *Hirundo smithii*
(a) **Adult** *filifera*: Recalls Barn but crown chestnut and entire underparts and underwing-coverts snowy-white; bluer above, tail square-cut with much longer streamers.
(b) **Juvenile**: Browner above, crown paler, throat vaguely buffish, no streamers. Dark thigh-patch (also adult).

10 [1056] **HOUSE SWALLOW** *Hirundo tahitica*
(a,b) **Adult** *abbotti*: Like Barn but upper breast chestnut-red without blue-black band, belly tinged greyish-brown, dark mottling on vent, underwing-coverts dusky, tail less forked (no streamers).
(c) **Juvenile**: Browner above than adult, throat and upper breast paler.

11 [1057] **RED-RUMPED SWALLOW** *Cecropis daurica*
(a,b) **Adult** *japonica*: From Barn by orange-rufous neck-sides and rump and dark-streaked whitish underparts. From Striated by smaller size, almost complete nuchal collar and narrower streaks.

12 [1058] **STRIATED SWALLOW** *Cecropis striolata*
(a) **Adult** *stanfordi*: Slightly larger than Red-rumped, lacks (near complete) orange-rufous nuchal collar, has much broader streaks on rump and underparts. Only shows a little reddish-rufous behind ear-coverts.
(b) **Juvenile**: Duller above, rump paler, tertials browner and pale-tipped, tail-streamers shorter.
(c) **Adult** *mayri*: Narrower streaks (only slightly broader than on Red-rumped).

13 [1059] **RUFOUS-BELLIED SWALLOW** *Cecropis badia*
Adult: Larger; deep rufous-chestnut below.

14 [1060] **WHITE-EYED RIVER-MARTIN** *Pseudochelidon sirintarae*
(a) **Adult**: Big-headed with stout yellow bill, white eyes and broad eyering, all-dark underparts, white rump-band and long central tail-streamers.
(b) **Juvenile**: Head and underparts browner with paler throat; no tail-streamers.

1a
2a
3a
3b
3c
1b
2b
4a
4b
6a
1c
2c
5
6b
8a
10a
7a
7b
10b
8c
8d
8b
10c
9a
11a
12c
9b
11b
12a
14b
14a
12b
13

PLATE 104. *CETTIA* BUSH-WARBLERS, ASIAN STUBTAIL & *ACROCEPHALUS* WARBLERS

1 [1066] **HUME'S BUSH-WARBLER** *Cettia brunnescens*
Adult: Small, short-tailed; rufescent upperparts, yellowish belly, dull throat and breast.

2 [1067] **GREY-SIDED BUSH-WARBLER** *Cettia brunnifrons*
(a) **Adult** *umbraticus*: Like Chestnut-crowned but smaller, supercilium whiter, breast and upper flanks greyer.
(b) **Juvenile**: Upperparts as adult but crown as mantle, drab brownish-olive below, buffish supercilium.

3 [1068] **CHESTNUT-CROWNED BUSH-WARBLER** *Cettia major*
Adult: Rufescent-chestnut crown and nape, rufescent lores, buffy supercilium, brownish flanks.

4 [1069] **BROWNISH-FLANKED BUSH-WARBLER** *Cettia fortipes*
(a) **Adult** *fortipes*: Rufescent-brown above, supercilium and underparts paler, dark eyestripe, buffy-brownish flanks, vent and wash across breast.
(b) **Adult** *davidiana*: Whiter on throat to belly-centre, with less buff on breast and flanks.

5 [1070] **MANCHURIAN BUSH-WARBLER** *Cettia canturians*
(a) **Male**: From larger reed-warblers by head pattern.
(b) **Female**: Smaller; similar to Brownish-flanked but bulkier and stouter-billed with rufescent forecrown.

6 [1071] **SUNDA BUSH-WARBLER** *Cettia vulcania*
Adult *intricata*: Slimmer and longer-tailed than Brownish-flanked, with paler, more olive upperside, yellowish-washed underparts and paler flanks.

7 [1072] **ABERRANT BUSH-WARBLER** *Cettia flavolivacea*
Adult *weberi*: Somewhat more rufescent above than Sunda (never as dark rufescent as Brownish-flanked), slightly yellower below and buffier on flanks (also lightly across breast).

8 [1073] **PALE-FOOTED BUSH-WARBLER** *Cettia pallidipes*
Adult *laurentei*: Striking pale supercilium and blackish eyestripe, cold olive-tinged upperparts, square-ended tail, whitish underparts, pale pinkish legs.

9 [1074] **ASIAN STUBTAIL** *Urosphena squameiceps*
Adult: Very small, short-tailed; long buffy-whitish supercilium, black eyestripe, pale pinkish legs.

10 [1296] **BLACK-BROWED REED-WARBLER** *Acrocephalus bistrigiceps*
(a) **Adult (worn)**: Paler and more greyish-olive than fresh adult.
(b) **Adult (fresh)**: Long broad buffy-white supercilium, blackish lateral crown-stripe, crown and upperparts warm olive-brown, whitish below with warm buff breast-sides and flanks.

11 [1297] **BLYTH'S REED-WARBLER** *Acrocephalus dumetorum*
(a) **Adult (worn)**: Greyer-olive and plainer above than Blunt-winged.
(b) **Adult (fresh)**: From Blunt-winged by slightly longer bill and shorter tail, slightly longer primary projection, colder, less rufescent upperside and duller flanks.

12 [1298] **LARGE-BILLED REED-WARBLER** *Acrocephalus orinus*
Adult (fresh): Difficult to separate from Blunt-winged and Blyth's, but bill longer, more wedge-shaped when seen from below, and with entirely flesh-pink lower mandible. Supercilium shorter than Blunt-winged; brown parts tend to be very slightly darker/warmer than Blyth's. In hand, legs and claws average longer than Blyth's.

13 [1299] **MANCHURIAN REED-WARBLER** *Acrocephalus tangorum*
(a) **Adult (worn)**: Longer bill than Black-browed, darker greyish-brown above, thin dark line above supercilium.
(b) **Adult (fresh)**: Like Paddyfield but longer bill usually has completely pale lower mandible, stronger blackish line above supercilium.

14 [1300] **PADDYFIELD WARBLER** *Acrocephalus agricola*
(a) **Adult (worn)**: Slightly greyer above and less rufescent on rump than fresh adult.
(b) **Adult (fresh)**: Like Blunt-winged but supercilium extends further behind eye, bill shorter with dark tip to lower mandible, longer primary projection.

15 [1301] **BLUNT-WINGED WARBLER** *Acrocephalus concinens*
(a) **Adult (worn)** *concinens*: Greyer above than fresh adult (rump warm-tinged).
(b) **Adult (fresh)**: Longish bill (lower mandible pale with darker shadow near tip), short whitish supercilium (ending just behind eye), no dark line above supercilium, short primary projection (beyond tertial tips).

16 [1302] **INDIAN REED-WARBLER** *Acrocephalus brunnescens*
(a) **Adult (worn)** *amyae*: Very like Oriental but has narrower bill, shorter supercilium, slightly shorter primary projection and longer tail without whitish tips.
(b) **First winter (fresh)**: Differs from Oriental as worn adult.

17 [1303] **ORIENTAL REED-WARBLER** *Acrocephalus orientalis*
(a) **Adult (worn)**: Greyish streaks on lower throat/upper breast more obvious than fresh adult, colder above.
(b) **Adult (fresh)**: Prominent whitish supercilium, dark eyestripe, warm olive-brown above and whitish below with buffish wash on flanks, vent. From Indian by stouter bill, bolder supercilium (particularly behind eye), slightly longer primary projection and shorter tail with whitish feather tips; usually less buffish below.

18 [1304] **THICK-BILLED-WARBLER** *Acrocephalus aedon*
(a) **Adult (worn)** *stegmanni*: Like Oriental and Indian but short, stout bill and no supercilium or eyestripe.
(b) **First winter (fresh)**: Rufescent tinge to upperparts, wings and flanks.

1
2a
2b
3
4a
4b
5a
5b
6
7
8
9
10a
10b
11a
11b
12
13a
13b
14a
14b
15a
15b
16a
16b
17a
17b
18a
18b
16-18 to different scale

PLATE 105. TESIAS, *ABROSCOPUS* & *SEICERCUS* WARBLERS & ALLIES

1 [1061] **YELLOW-BELLIED WARBLER** *Abroscopus superciliaris*
Adult *superciliaris*: Greyish head, whitish supercilium, greenish upperparts, whitish throat/breast, yellow belly.

2 [1062] **RUFOUS-FACED WARBLER** *Abroscopus albogularis*
(a) **Adult** *hugonis*: Dull crown, rufous head-sides, dark throat, whitish below, yellow breast-band and vent.
(b) **Juvenile**: Head-sides washed out, no lateral crown-stripes, throat-streaking less obvious.
(c) **Adult** *fulvifacies*: Very faint yellow breast-band.

3 [1063] **BLACK-FACED WARBLER** *Abroscopus schisticeps*
(a) **Adult** *ripponi*: Slaty crown, nape, upper breast, dark mask, yellow supercilium, sooty-centred yellow throat.
(b) **Adult** *flavimentalis*: All-yellow throat.

4 [1064] **MOUNTAIN TAILORBIRD** *Phyllergates cucullatus*
(a) **Adult** *coronatus*: Rufous forecrown, yellowish-white supercilium, dark eyestripe, yellow belly and vent.
(b) **Juvenile**: Crown and mantle uniform dull green, rest of plumage duller and plainer.

5 [1065] **BROAD-BILLED WARBLER** *Tickellia hodgsoni*
Adult *hodgsoni*: Dark rufous crown, dull green upperparts, grey throat and breast, yellow belly.

6 [1075] **GREY-BELLIED TESIA** *Tesia cyaniventer*
(a) **Adult**: Almost tail-less, dark olive-green above, black eyestripe, yellowish supercilium, grey underparts.
(b) **Juvenile**: Upperparts dark brown, supercilium and eyestripe duller, underparts drab olive.

7 [1076] **SLATY-BELLIED TESIA** *Tesia olivea*
Adult: Like Grey-bellied but crown washed golden-yellow, underparts dark slaty-grey.

8 [1077] **CHESTNUT-HEADED TESIA** *Tesia castaneocoronata*
(a) **Adult** *castaneocoronata*: Almost tail-less, chestnut head, mostly yellow below, whitish patch behind eye.
(b) **Juvenile**: Darker and browner above, dark rufous below.

9 [1082] **BIANCHI'S WARBLER** *Seicercus valentini*
Adult *valentini*: As Plain-tailed but crown-stripes blacker/more extensive, more white on outertail, wing-bar usually distinct.

10 [1083] **WHISTLER'S WARBLER** *Seicercus whistleri*
Adult *nemoralis*: Green median crown-stripe with greyish streaks, wing-bar typically distinct, much white on outertail.

11 [1085] **PLAIN-TAILED WARBLER** *Seicercus soror*
Adult: Little white on outertail, forehead greenish, relatively short greyish-black lateral crown-stripes.

12 [1086] **GREY-CROWNED WARBLER** *Seicercus tephrocephalus*
(a) **Adult**: Greyish crown with black lateral stripes, yellow eyering, no wing-bar, much white on outertail.
(b) **Juvenile**: Lateral crown-stripes weaker, upperparts slightly darker, underparts duller.

13 [1087] **WHITE-SPECTACLED WARBLER** *Seicercus affinis*
(a) **Adult** *affinis*: Crown and ear-coverts grey, eyering white, prominent wing-bar, yellow lores.
(b) **Juvenile**: Faded crown, duller above, paler below.
(c) **Adult** *intermedius*: Crown less pure grey, broad yellow eyering, sometimes second wing-bar.

14 [1088] **GREY-CHEEKED WARBLER** *Seicercus poliogenys*
Adult: Like White-spectacled but darker grey head, greyish-white chin and lores, broader eyering.

15 [1089] **CHESTNUT-CROWNED WARBLER** *Seicercus castaniceps*
(a) **Adult** *collinsi*: Rufous crown with blackish lateral stripes, yellow rump, vent and wing-bars, grey throat/breast.
(b) **Adult** *butleri*: Darker crown, dark grey below and on mantle, no yellow rump-patch.

16 [1090] **YELLOW-BREASTED WARBLER** *Seicercus montis*
(a) **Adult** *davisoni*: Like Chestnut-crowned but head-sides rufous, mantle greenish, underparts all yellow.
(b) **Juvenile**: Duller head, less intense underparts.

1
2c
2b
3b
3a
4a
2a
5
4b
6-8 to different scale
6a
6b
7
8b
8a
9
10
11
12a
12b
13c
13a
13b
14
15b
15a
16b
16a

PLATE 106. *PHYLLOSCOPUS* WARBLERS

1 [1091] **EASTERN CROWNED WARBLER** *Phylloscopus coronatus*
Adult: Like Arctic but crown darker, with well-defined pale median stripe, upperparts yellower-green, ear-coverts and underparts whiter, undertail-coverts pale yellow; lower mandible pale.

2 [1092] **YELLOW-VENTED WARBLER** *Phylloscopus cantator*
Adult *cantator*: Bright yellow on head, upper breast and undertail-coverts, dark eye- and crown-stripes, whitish lower breast and belly.

3 [1093] **SULPHUR-BREASTED WARBLER** *Phylloscopus ricketti*
Adult: Bright yellow on head and underparts, blackish eye- and median crown-stripes, narrow wing-bars.

4 [1094] **LIMESTONE WARBLER** *Phylloscopus* sp.
Adult: Smaller and proportionately longer-billed than Sulphur-breasted. Possible differences (slight and to be confirmed) are greyer-tinged upperparts, paler, less intense yellow of plumage. See voice in main text.

5 [1095] **MOUNTAIN LEAF-WARBLER** *Phylloscopus trivirgatus*
(a) **Adult** *parvirostris*: Greenish above, yellowish below, no wing-bars.
(b) **Juvenile**: Head-stripes duller than adult, underparts washed olive and less yellow.

6 [1096] **WHITE-TAILED LEAF-WARBLER** *Phylloscopus ogilviegranti*
Adult *klossi*: Recalls smaller *Phylloscopus* warblers, but lacks pale rump-patch and whitish tertial markings; larger and longer-billed with mostly pale lower mandible. From Davison's by less white on tail (which is yellow-tinged), much brighter yellow underparts (including throat), much yellower-green upperparts; note status and range. Also very similar to Blyth's Leaf-, Claudia's and Hartert's (latter two not illustrated), but has much more white on two outermost tail-feathers and yellower supercilium and median crown-stripe (less marked when worn).

7 [1097] **DAVISON'S LEAF-WARBLER** *Phylloscopus davisoni*
Adult: Difficult to separate from White-tailed, which see for differences.

8 [1098] **GREY-HOODED WARBLER** *Phylloscopus xanthoschistos*
Adult *tephrodiras*: Grey crown, nape and head-sides, dark eye- and lateral crown-stripes, whitish supercilium.

9 [1100] **BLYTH'S LEAF-WARBLER** *Phylloscopus reguloides*
(a) **Adult** *assamensis*: From White-tailed by less yellow on supercilium and underparts, less white on tail.
(b) **Juvenile**: Like adult but crown plainer (barely discernible median stripe), underparts duller.
(c) **Adult** *ticehursti*: Pale areas yellower, slightly greener upperparts.

10 [1102] **ARCTIC WARBLER** *Phylloscopus borealis*
(a) **Adult (fresh)** *borealis*: From Greenish and Two-barred by slightly larger size, slightly longer, heavier bill, supercilium falls short of bill-base, ear-coverts more prominently mottled, legs browner or yellower, greyer and slightly streaked breast-sides. From Eastern Crowned by plain crown, no yellow on undertail-coverts, paler legs.
(b) **Adult (worn)**: Duller, with narrower wing-bars.
(c) **Adult** *xanthodryas*: Yellower-tinged pale parts of plumage.

11 [1103] **LARGE-BILLED LEAF-WARBLER** *Phylloscopus magnirostris*
(a) **Adult (fresh)**: Like Greenish and Two-barred but stockier and larger- and darker-billed, greener-olive above, eyestripe broader, supercilium longer and crown darker, ear-coverts mottled greyish, underparts rather dirty and streaky. From Arctic by somewhat darker upperparts, more dark on lower mandible, duller legs.
(b) **Adult (worn)**: Duller above, wing-bars narrower.

12 [1104] **PALE-LEGGED LEAF-WARBLER** *Phylloscopus tenellipes*
Adult: Resembles Arctic and Greenish Warblers but crown uniform dark greyish, rump paler olive-brown, legs pale greyish-pink; usually two narrow wing-bars.

13 [1105] **GREENISH WARBLER** *Phylloscopus trochiloides*
(a) **Adult (fresh)** *trochiloides*: Very like Two-barred but usually shows only one narrow wing-bar on greater coverts.
(b) **Adult (worn)**: Greyer above, wing-bar/s narrower.

14 [1106] **TWO-BARRED WARBLER** *Phylloscopus plumbeitarsus*
(a) **Adult (fresh)**: Very like Greenish but shows two broader yellowish-white wing-bars. Can resemble Yellow-browed but larger and longer-billed, no whitish fringes to tertials. See Arctic for differences.
(b) **Adult (worn)**: Often worn in winter, with narrower wing-bars (may lack shorter upper wing-bar).

15 [1107] **EMEI LEAF-WARBLER** *Phylloscopus emeiensis*
Adult: Very like Blyth's but crown pattern duller, inner webs of two outertail feathers have narrower whitish margins. From White-tailed by less distinct crown pattern, less greenish upperparts and less yellow head markings, throat, belly-centre and undertail-coverts.

1
2
3
4
5a
5b
6
7
8
9a
9b
9c
10a
10b
10c
11a
11b
12
13a
13b
14a
14b
15

PLATE 107. GOLDCREST, *PHYLLOSCOPUS* WARBLERS & LESSER WHITETHROAT

1 [708] **GOLDCREST** *Regulus regulus*
Male *yunnanensis*: From small *Phylloscopus* by bright yellowish-orange median crown-stripe with blackish border and broad pale orbital area, blackish and whitish markings at base of flight feathers.

2 [1108] **ASHY-THROATED WARBLER** *Phylloscopus maculipennis*
Adult *maculipennis*: Tiny; rump pale yellow, head-stripes and wing-bars recall Lemon-rumped, but supercilium white, throat/breast greyish, outertail largely white.

3 [1109] **BUFF-BARRED WARBLER** *Phylloscopus pulcher*
(a) **Adult (fresh)** *pulcher*: Broad orange-buff wing-bar, yellowish rump and extensively white outertail.
(b) **Adult (worn)**: Wing-bar narrower and buffier.

4 [1110] **YELLOW-BROWED WARBLER** *Phylloscopus inornatus*
(a) **Adult (fresh)**: Broad yellowish-white supercilium and wing-bars, plain olive-green upperside with no median crown-stripe, pale rump-patch or white on tail. Like Two-barred but smaller and shorter-tailed, with whitish tips to tertials and broader, more contrasting wing-bars, bill weaker, usually darker-tipped.
(b) **Adult (worn)**: Narrower wing-bars.

5 [1111] **HUME'S WARBLER** *Phylloscopus humei*
(a) **Adult (fresh)** *mandellii*: From Yellow-browed by grey wash on upperparts, rather darker lower mandible and legs; median covert bar usually slightly less distinct than greater covert bar and slightly duller throat and breast.
(b) **Adult (worn)**: Greyer above, wing-bars weaker. Otherwise differs from Yellow-browed as fresh adult.

6 [1112] **CHINESE LEAF-WARBLER** *Phylloscopus yunnanensis*
Adult: Like Lemon-rumped but crown-sides slightly paler, median crown-stripe fainter, eyestripe slightly paler and fairly straight, no pale spot on ear-coverts; slightly larger, more elongated and longer-billed.

7 [1113] **LEMON-RUMPED WARBLER** *Phylloscopus chloronotus*
Adult *chloronotus*: Tiny; small dark bill, pale yellow rump, supercilium, median crown-stripe and double wing-bar, no white on tail. From Pallas's by slightly duller head-stripes, wing-bars and mantle.

8 [1114] **PALLAS'S LEAF-WARBLER** *Phylloscopus proregulus*
(a) **Adult (fresh)**: Very like Lemon-rumped, but supercilium and median crown-stripe much yellower, mantle greener.
(b) **Adult (worn)**: Duller above with narrower wing-bars, less strongly marked yellow.

9 [1115] **TICKELL'S LEAF-WARBLER** *Phylloscopus affinis*
(a) **Adult (fresh)**: Like Buff-throated but more greenish-tinged above, supercilium, ear-coverts and underparts washed lemon-yellow.
(b) **Adult (worn)**: More washed out, less distinctly yellow; similar to Buff-throated above.

10 [1116] **RADDE'S WARBLER** *Phylloscopus schwarzi*
(a) **Adult (fresh)**: Thicker-billed and -legged than other *Phylloscopus*. Larger than Yellow-streaked, pre-ocular supercilium less defined. Larger than Dusky, more olive-tinged above, supercilium broader, often bordered darkish above and ill-defined and buffish in front of eye, rusty-buff undertail-coverts.
(b) **Adult (worn)**: Lacks olive-tinge on upperparts.

11 [1117] **YELLOW-STREAKED WARBLER** *Phylloscopus armandii*
(a) **Adult (fresh)**: From Radde's at close range by fine yellow streaking on breast and belly; slightly smaller, smaller-headed, with thinner bill and legs.
(b) **Adult (worn)**: Yellow streaking less obvious.

12 [1118] **DUSKY WARBLER** *Phylloscopus fuscatus*
(a) **Adult (fresh)** *fuscatus*: Rather dark brown above, dirty whitish below, often with distinct buffish wash on breast-sides to vent. Smaller/slimmer than Radde's, with finer bill, thinner legs, whiter supercilium in front of eye.
(b) **Adult (worn)**: Upperparts paler greyish-brown.
(c) **Adult variant**: More olive-tinged individual.

13 [1119] **BUFF-THROATED WARBLER** *Phylloscopus subaffinis*
Adult (fresh): Yellowish-buff supercilium and underparts (including throat), no crown-stripes or wing-bars. Browner above than fresh Tickell's, lower mandible typically tipped dark (not apparent on plate).

14 [1120] **COMMON CHIFFCHAFF** *Phylloscopus collybita*
(a) **Adult (fresh)** *tristis*: Recalls Dusky and Buff-throated but grey-brown above, whitish below with buffy-washed breast and flanks, olive-green wing-fringing, slight pale wing-bar, blackish bill/legs.
(b) **Adult (worn)**: Greyer and whiter, duller wings, no wing-bar.

15 [1121] **LESSER WHITETHROAT** *Sylvia curruca*
(a) **Adult** *blythi*: Dark grey crown, contrasting dark lores and ear-coverts, rather square-ended tail with white on outer feathers.
(b) **First winter**: Crown sullied brownish, less blackish lores and ear-coverts, narrow pale supercilium.

1
2
3b
3a
4a
5b
4b
5a
6
8a
8b
7
9a
10a
10b
9b
11a
11b
13
12b
14b
15b
12c
12a
14a
15a

PLATE 108. PARROTBILLS, *CHRYSOMMA* BABBLERS & CHESTNUT-CAPPED BABBLER

1 [1122] **GREAT PARROTBILL** *Conostoma oemodium*
Adult: Large; bill relatively long and yellowish, plumage brown with greyish-white forehead, blackish pre-ocular patch, and greyer underparts and fringes to primaries and tail.

2 [1123] **BROWN PARROTBILL** *Cholornis unicolor*
Adult: Greyish-brown, browner upperparts, narrow blackish lateral crown-stripes and short deep yellowish bill.

3 [1124] **GREY-HEADED PARROTBILL** *Psittiparus gularis*
Adult *transfluvialis*: Greyish head with white lores and eyering, long black lateral crown-stripes, black throat, pale buff underparts.

4 [1125] **BLACK-CROWNED PARROTBILL** *Psittiparus margaritae*
Adult: All-black crown, mottled ear-coverts, darker above, all-white below.

5 [1126] **GREATER RUFOUS-HEADED PARROTBILL** *Psittiparus bakeri*
Adult *bakeri*: Like Lesser but larger, rounder-headed, more uniform rufous face, no black eyebrow.

6 [1127] **SPOT-BREASTED PARROTBILL** *Paradoxornis guttaticollis*
Adult: Rufous crown and nape, white head-sides with broad black patch on rear ear-coverts, white throat and upper breast with pointed blackish spots.

7 [1128] **LESSER RUFOUS-HEADED PARROTBILL** *Chleuasicus atrosuperciliaris*
Adult *atrosuperciliaris*: Buffy-rufous head, peaked crown, black eyebrow, buffish-white below.

8 [1129] **BROWN-WINGED PARROTBILL** *Suthora brunneus*
Adult *brunneus*: Dull brown wings, much more vinous throat and upper breast with darker chestnut streaks.

9 [1130] **VINOUS-THROATED PARROTBILL** *Suthora webbianus*
(a) **Adult** *suffusus*: Warm above, rufous-chestnut crown/nape and wing-fringes, brown-streaked throat/upper breast.
(b) **Adult** *elisabethae*: Slightly duller with fainter streaks on throat/upper breast.

10 [1131] **ASHY-THROATED PARROTBILL** *Suthora alphonsianus*
Adult *yunnanensis*: Brownish-grey head- and neck-sides, whitish throat and breast with faint greyish streaks.

11 [1132] **FULVOUS PARROTBILL** *Suthora fulvifrons*
Adult *albifacies*: Rather uniform warm buff head and breast, dark brownish-grey lateral crown-stripes.

12 [1133] **GREY-BREASTED PARROTBILL** *Suthora poliotis*
Adult *feae*: Black lateral crown-stripes and throat, grey head-sides and breast, thin rufous supercilium.

13 [1134] **BUFF-BREASTED PARROTBILL** *Suthora ripponi*
Adult *ripponi*: Broader lateral crown-stripes than Grey-breasted, mostly white narrow supercilium, rufous-buff breast.

14 [1135] **BLACK-EARED PARROTBILL** *Suthora beaulieui*
(a) **Adult** *beaulieui*: Shorter, broader lateral crown-stripes than Grey-breasted, black ear-coverts, greyish and buff breast, white supercilium.
(b) **Adult** *kamoli*: Rear supercilium rufous, rear ear-coverts grey, lateral crown-stripes narrow.

15 [1136] **GOLDEN PARROTBILL** *Suthora verreauxi*
Adult *craddocki*: Like Black-throated but has rufous-buff ear-coverts, no crown-stripes or grey on head/breast.

16 [1137] **SHORT-TAILED PARROTBILL** *Neosuthora davidiana*
Adult *thompsoni*: Short-tail, thick bill, chestnut head and upper mantle, mostly black throat.

17 [1138] **JERDON'S BABBLER** *Chrysomma altirostre*
(a) **Adult** *altirostre*: Lores, supercilium, throat and breast grey, bill pale, eyes brown, eyering greenish-yellow.
(b) **Adult** *griseigularis*: Throat and upper breast greyer, lower breast to vent deeper, richer buff.

18 [1139] **YELLOW-EYED BABBLER** *Chrysomma sinense*
Adult *sinense*: Long tail, short black bill, white face to breast, orange-yellow eyes, orange eyering.

19 [1200] **CHESTNUT-CAPPED BABBLER** *Timalia pileata*
(a) **Adult** *smithi*: Thick black bill, chestnut cap, black mask, white supercilium, cheeks, throat and upper breast.
(b) **Juvenile**: Warmer brown above, head duller, lower mandible paler.

1
2
3
4
5
6
7
8
9a
9b
10
11
16
17b
17a
13
12
15
14a
14b
19b
19a
18

PLATE 109. FULVETTAS

1 [1140] **GOLDEN-BREASTED FULVETTA** *Lioparus chrysotis*
(a) **Adult** *forresti*: White median crown-stripe, silvery ear-coverts, black and grey throat, orange-yellow below.
(b) **Adult** *amoena*: All-blackish throat, yellow eyering.

2 [1141] **WHITE-BROWED FULVETTA** *Fulvetta vinipectus*
(a) **Adult** *perstriata*: Warm crown, white supercilium, blackish head-sides, dark-streaked white throat and breast.
(b) **Adult** *ripponi*: Head-sides and throat/breast-streaks warm brown.
(c) **Adult** *austeni*: Like *ripponi* but supercilium post-ocular, ear-coverts darker.
(d) **Adult** *valentinae*: Greyish crown to mantle, lower breast and upper belly.

3 [1142] **LUDLOW'S FULVETTA** *Fulvetta ludlowi*
Adult: Dark above, plain-looking crown and head-sides, bold throat streaks.

4 [1143] **STREAK-THROATED FULVETTA** *Fulvetta manipurensis*
Adult *manipurensis*: Greyish-brown above, dark lateral crown-stripes, streaked throat, greyish head-sides.

5 [1144] **INDOCHINESE FULVETTA** *Fulvetta danisi*
Adult *danisi*: Recalls Streak-throated but darker above, streaked head-sides and upper breast, plainer wings.

6 [1158] **GREY-CHEEKED FULVETTA** *Alcippe fratercula*
Adult *fratercula*: Distinctly buff underparts, black lateral crown-stripes, white eyering.

7 [1159] **SCHAEFFER'S FULVETTA** *Alcippe schaefferi*
(a) **Adult** *schaefferi*: Faint lateral crown-stripe, less buffy below than Western, greyish-white throat with very faint streaks.
(b) **Adult** *laotiana*: Somewhat more pronounced crown-stripes, less warm-tinged above, slightly buffier below.

8 [1160] **NEPAL FULVETTA** *Alcippe nipalensis*
Adult *commoda*: Brown-tinged crown, rufescent upperparts, whitish centre to underparts, broad eyering.

9 [1161] **MOUNTAIN FULVETTA** *Alcippe peracensis*
(a) **Adult** *peracensis*: Small; dull slate-grey crown to upper mantle, blackish lateral crown-stripes, mid-grey head-sides, white eyering, whitish centre to throat and abdomen.
(b) **Adult** *annamensis*: Paler grey on head, more olive upperside and paler underparts.

10 [1162] **BLACK-BROWED FULVETTA** *Alcippe grotei*
(a) **Adult** *grotei*: Like Mountain subspecies *annamensis* but larger, crown slaty-grey, upperside rufescent dark brown, head-sides brown-washed, indistinct eyering, whitish underparts with faint brown wash on sides.
(b) **Juvenile**: Warmer above, grey duller and restricted to crown, browner head.

11 [1163] **BROWN-CHEEKED FULVETTA** *Alcippe poioicephala*
(a) **Adult** *haringtoniae*: Plain greyish-buff head-sides, buff below, narrow lateral crown-stripes, olive-brown above.
(b) **Adult** *phayrei*: Paler plain crown, paler upperparts and underparts with whitish centre.
(c) **Adult** *karenni*: Roughly intermediate between two preceding, with indistinct lateral crown-stripes.

12 [1164] **BROWN FULVETTA** *Alcippe brunneicauda*
Adult *brunneicauda*: Rather plain greyish head, whiter throat, dull below.

13 [1165] **RUFOUS-THROATED FULVETTA** *Schoeniparus rufogularis*
(a) **Adult** *major*: Rufous-brown crown, white supercilium, broad rufous-chestnut breast-band.
(b) **Adult** *collaris*: Broader breast/throat-band.
(c) **Adult** *stevensi*: Duller and buffier on head, more rufous breast/throat-band.

14 [1166] **RUSTY-CAPPED FULVETTA** *Schoeniparus dubius*
(a) **Adult** *genestieri*: Warm brown crown, black over white supercilium, white throat and upper breast, plain wings.
(b) **Adult** *mandellii*: Blackish ear-coverts and bold black-and-whitish streaks on neck-sides.

15 [1227] **YELLOW-THROATED FULVETTA** *Pseudominla cinerea*
Adult: Tiny; greyish-olive above, yellow supercilium and throat, breast and centre of abdomen.

16 [1228] **RUFOUS-WINGED FULVETTA** *Pseudominla castaneceps*
Adult *castaneceps*: Chestnut crown, black and white head-sides, blackish coverts, rufous wing-panel.

17 [1229] **BLACK-CROWNED FULVETTA** *Pseudominla klossi*
Adult: Blackish crown, brownish coverts, no rufous wing-panel.

1a
1b
2b
2d
2a
2c
3
5
7a
4
6
7b
8
9b
10b
9a
10a
11b
11a
12
11c
13b
13a
14a
14b
15
13c
17
16

PLATE 110. WHITE-EYES, YUHINAS & FIRE-TAILED MYZORNIS

1 [1145] **CHESTNUT-FLANKED WHITE-EYE** *Zosterops erythropleurus*
Adult: Chestnut flanks, green forehead.

2 [1146] **ORIENTAL WHITE-EYE** *Zosterops palpebrosus*
(a) **Adult typical morph** *siamensis*: Yellowish-white ventral stripe, extensive yellow forehead.
(b) **Adult yellow morph**: Underparts completely yellow.
(c) **Adult** *williamsoni*: Reduced ventral stripe, paler yellow and grey below, duller green above.
(d) **Adult** *auriventer*: Clear yellow mid-ventral stripe, less yellow on forehead.

3 [1147] **JAPANESE WHITE-EYE** *Zosterops japonicus*
(a) **Adult** *simplex*: From Oriental by darker upperparts, no ventral stripe; defined yellow loral band.
(b) **Juvenile**: Like adult but eyering greyer at first.

4 [1148] **EVERETT'S WHITE-EYE** *Zosterops everetti*
Adult *wetmorei*: Green forehead, deep grey sides, deep yellow throat and ventral stripe.

5 [1149] **BLACK-CHINNED YUHINA** *Yuhina nigrimenta*
Adult: Black and red bill, black face and chin, grey head with short black-streaked crest, short tail.

6 [1150] **WHITE-COLLARED YUHINA** *Yuhina diademata*
Adult: Mostly brownish-grey with darker face, tall crest, broad white nuchal patch and undertail-coverts.

7 [1151] **STRIPE-THROATED YUHINA** *Yuhina gularis*
(a) **Adult** *gularis*: Robust; tall crest, blackish-streaked throat, pale orange-buff wing-panel, mostly blackish primaries.
(b) **Adult** *uthaii*: Broader black throat-streaks, colder upperside and greyer head-sides.

8 [1152] **RUFOUS-VENTED YUHINA** *Yuhina occipitalis*
Adult *obscurior*: Tall greyish crest, white eyering, black malar streak, rufous nuchal patch and rufous-buff vent.

9 [1153] **WHITE-NAPED YUHINA** *Yuhina bakeri*
Adult: Dark rufous head, broad white streaks on ear-coverts, white nape-patch, shortish crest.

10 [1154] **WHISKERED YUHINA** *Yuhina flavicollis*
Adult *rouxi*: Erect crest, white eyering, blackish moustachial stripe, golden-yellow nuchal collar and white-streaked olive-brown flanks.

11 [1155] **BURMESE YUHINA** *Yuhina humilis*
Adult *clarki*: Like Whiskered but has greyish-brown crown and ear-coverts, grey nuchal collar, grey base to flanks.

12 [1156] **STRIATED YUHINA** *Staphida castaniceps*
(a) **Adult** *striata*: Slim build, short crest and graduated white-edged tail; cold brownish above with narrow greyish streaks, pale brown ear-coverts, narrow whitish eyebrow, buffish-white below.
(b) **Adult** *castaniceps*: Pale-fringed warmer brown forehead and crown, rufous-chestnut nape.
(c) **Adult** *plumbeiceps*: Greyer upperside (notably crown) and plainer, brighter rufous ear-coverts.

13 [1157] **CHESTNUT-COLLARED YUHINA** *Staphida torqueola*
Adult: Broad chestnut nuchal collar and ear-coverts with strong whitish streaks, greyer crown.

14 [1295] **FIRE-TAILED MYZORNIS** *Myzornis pyrrhoura*
(a) **Male**: Largely green, black mask, red and black on tail, reddish throat/breast-patch, red, yellow, black and white on wings.
(b) **Female**: Bluer-tinged below (and often above), throat/breast and vent duller.
(c) **Immature male**: Like female but underparts greener with more orange centre of throat/breast.

1
2a
2b
2c
2d
1–4 to different scale
3a
3b
4
5
6
7a
7b
8
9
10
11
12a
12b
12c
13
14a
14b
14c

PLATE 111. *STACHYRIS* & *STACHYRIDOPSIS* BABBLERS, RUFOUS-RUMPED GRASS-BABBLER & TIT-BABBLERS

1 [1167] **SOOTY BABBLER** *Stachyris herberti*
Adult: Sooty dark brown with whiter throat and pale bill and orbital ring.

2 [1168] **BLACK-THROATED BABBLER** *Stachyris nigricollis*
(a) **Adult**: Rufescent above, black face, throat and upper breast, broad white cheek-patch, white forehead-streaking, short white eyebrow and white line of scales on lower breast.
(b) **Juvenile**: Head and underside sooty-brownish but with white eyebrow and cheek-patch.

3 [1169] **WHITE-NECKED BABBLER** *Stachyris leucotis*
(a) **Adult** *leucotis*: Like Black-throated but ear-coverts greyer bordered with bold white spots, wing-coverts and tertials tipped pale, no white cheek-patch or breast-line.
(b) **Juvenile**: Dull chestnut above, neck-spots buffish, ear-coverts browner, throat to belly dark brown.

4 [1170] **SNOWY-THROATED BABBLER** *Stachyris oglei*
Adult: Recalls Spot-necked but bold white supercilium, broad black mask, clean white throat, grey breast.

5 [1171] **SPOT-NECKED BABBLER** *Stachyris strialata*
Adult *guttata*: White throat, rufous-chestnut breast and belly, blackish malar streak and white-flecked supercilium, neck and mantle-sides.

6 [1172] **GREY-HEADED BABBLER** *Stachyris poliocephala*
(a) **Adult**: Dark rufescent-brown above, rufous-chestnut below, greyish head with whitish streaks on forehead and throat, whitish eyes.
(b) **Juvenile**: Duller above, paler below, head-streaking less distinct, eyes browner.

7 [1173] **GREY-THROATED BABBLER** *Stachyris nigriceps*
(a) **Adult** *spadix*: Crown streaked black-and-silver, black over white supercilium, grey throat with white sub-moustachial patch, warm buffish underparts.
(b) **Juvenile**: Upperparts washed chestnut, head-sides and underparts rufous-buff, hindcrown unstreaked.
(c) **Adult** *coltarti*: Darker throat, warmer ear-coverts and underparts.
(d) **Adult** *rileyi*: Uniform pale grey throat, paler underparts, paler crown with restricted streaking.
(e) **Adult** *davisoni*: Much like *spadix*, crown like *rileyi*.

8 [1174] **CHESTNUT-WINGED BABBLER** *Stachyris erythroptera*
(a) **Adult** *erythroptera*: Drab brown above with rufous-chestnut on wings, mostly greyish head to upper belly with buffy vent, blue orbital skin.
(b) **Juvenile**: Crown and upperparts more rufescent, grey much paler, orbital skin duller.

9 [1175] **CHESTNUT-RUMPED BABBLER** *Stachyris maculata*
Adult *maculata*: Large, with black throat, broad blackish streaks on whitish lower throat to upper belly, and rufous-chestnut lower back to uppertail-coverts; whitish eyes, blue orbital skin.

10 [1194] **GOLDEN BABBLER** *Stachyridopsis chrysaea*
(a) **Adult** *assimilis*: Bright yellow forehead and underparts, black face and dark crown-streaking. Ear-coverts and nape yellowish-olive, upperparts and flanks greyish-olive.
(b) **Adult** *binghami*: Grey ear-coverts, stronger crown-streaks.

11 [1195] **RUFOUS-CAPPED BABBLER** *Stachyridopsis ruficeps*
(a) **Adult** *pagana*: From Rufous-fronted by paler, more olive upperparts, yellowish lores, head-sides and underparts, no eyebrow, orange-rufous crown; usually pinker lower mandible.
(b) **Juvenile**: Crown paler, underparts more washed out.
(c) **Adult** *bhamoensis*: Greyer flanks, darker mantle.

12 [1196] **RUFOUS-FRONTED BABBLER** *Stachyridopsis rufifrons*
(a) **Adult** *rufifrons*: Rufous forehead to midcrown and buffish underparts. From Rufous-capped by somewhat darker, warmer upperparts, pale greyish lores, eyebrow and eyering, duller rufous crown, and buffier underparts.
(b) **Adult** *poliogaster*: Darker, duller crown and upperparts, paler lower breast to vent.

13 [1197] **PIN-STRIPED TIT-BABBLER** *Macronus gularis*
(a) **Adult** *sulphureus*: Olive-brown above, rufous crown and largely yellowish head-sides and underparts. From Rufous-fronted by yellow supercilium and narrow streaks on lower throat and breast.
(b) **Juvenile**: More uniform above, paler below, supercilium narrower.
(c) **Adult** *connectens*: Darker above and more chestnut-tinged; stronger streaks below.
(d) **Adult** *gularis*: Like *connectens* with darker upperparts.

14 [1198] **GREY-FACED TIT-BABBLER** *Macronus kelleyi*
Adult: Very like Pin-striped but forehead, crown and upperparts more uniform rufescent-brown, supercilium and head-sides grey, underparts paler yellow and barely streaked.

15 [1199] **FLUFFY-BACKED TIT-BABBLER** *Macronus ptilosus*
Adult *ptilosus*: Dark brown with rufous crown, black cheeks and throat and blue spectacles.

16 [1201] **RUFOUS-RUMPED GRASS-BABBLER** *Graminicola bengalensis*
(a) **Adult** *striata*: Resembles Rusty-rumped Warbler but larger, tail blackish with broad white crescent-shaped tips, undertail-coverts shorter, mantle-streaking broader, rump plain rufous, bill shorter, thicker.
(b) **Juvenile**: Warmer streaks on crown and mantle, duller dark streaking on upperparts.

1
2b
3b
2a
3a
4
6a
5
7b
6b
7c
7a
8b
9
7d
7e
8a
11b
10a
10b
11a
11c
13b
12b
12a
13a
13c
16a
13d
16b
14
15

PLATE 112. CHEVRON-BREASTED BABBLER & WREN-BABBLERS

1 [1186] **CHEVRON-BREASTED BABBLER** *Sphenocichla roberti*
Adult: Heavy white chevron-scaling, dark-barred wings and tail, conical, pointed bill.

2 [1187] **BAR-WINGED WREN-BABBLER** *Spelaeornis troglodytoides*
(a) **Male** *souliei*: Buffish-grey wings and tail with prominent dark bars, rufescent mantle to uppertail-coverts, white throat and rufous flanks and vent, black-tipped pale streaks above and on belly.
(b) **Juvenile**: Crown, head-sides and upperparts indistinctly mottled dark brown and dull rufescent-brown, underparts plain deep rufous.

3 [1188] **GREY-BELLIED WREN-BABBLER** *Spelaeornis reptatus*
(a) **Male**: Relatively long, unmarked tail and plain wings, dark scaling above, grey head-sides, mostly white throat, black-and-white scales on brownish underparts.
(b) **Female**: Washed rufescent-buff below.

4 [1189] **PALE-THROATED WREN-BABBLER** *Spelaeornis kinneari*
(a) **Male**: Darker moustachial line than Grey-bellied, stronger scales below.
(b) **Female**: Buffier throat than Grey-bellied, browner base colour of breast and flanks, stronger scales on breast.

5 [1190] **CHIN HILLS WREN-BABBLER** *Spelaeornis oatesi*
Adult: Reduced grey on head-sides, white throat to belly-centre with black spots and brown wash on throat-sides, breast and flanks.

6 [1191] **SPOTTED WREN-BABBLER** *Elachura formosa*
Adult: Brown peppered white, black-barred rufous wings and tail.

7 [1192] **SCALY-BREASTED WREN-BABBLER** *Pnoepyga albiventer*
(a) **Adult dark morph** *albiventer*: Small, tail-less, dark brown with buff scales below (throat sometimes white). From Pygmy by slightly larger size, pale speckling on head.
(b) **Adult pale morph**: Mostly white-scaled below.
(c) **Juvenile (dark morph)**: Rather plain brown above; underparts mottled dark brown and dark buff.
(d) **Juvenile (pale morph)**: Underparts mottled pale and dark.

8 [1193] **PYGMY WREN-BABBLER** *Pnoepyga pusilla*
(a) **Adult dark morph** *pusilla*: Slightly smaller and slimmer than Scaly-breasted, unspeckled head.
(b) **Adult pale morph**: From Scaly-breasted as dark morph.

9 [1217] **MARBLED WREN-BABBLER** *Turdinus marmoratus*
Adult *grandior*: Large with relatively long tail, rufous ear-coverts and blackish underparts with white throat and scales.

10 [1218] **LARGE WREN-BABBLER** *Turdinus macrodactylus*
(a) **Adult** *macrodactylus*: Large with black lores and ear-coverts, white loral supercilium and throat, and light blue orbital skin; faintly patterned below.
(b) **Juvenile**: Upperparts paler, plainer and more rufous, with pale buff shaft-streaks; underparts pale rufescent-brown, with whiter throat and belly-centre.

11 [1219] **LIMESTONE WREN-BABBLER** *Gypsophila crispifrons*
(a) **Adult** *crispifrons*: Like Streaked but larger and longer-tailed, unspotted coverts, bolder throat-streaking and colder greyish-brown underparts, streaked white on centre of belly.
(b) **Adult white-throated morph**: White face and throat.
(c) **Adult** *calcicola*: Rufescent-brown below.
(d) **Adult** *annamensis*: Greyer crown, mantle, lower breast and belly.

12 [1220] **STREAKED WREN-BABBLER** *Napothera brevicaudata*
(a) **Adult** *brevicaudata*: Fairly small, short-tailed; blackish-scaled brown above, small whitish spots on tips of tertials, secondaries and greater coverts, greyish head-sides, rufescent-brown below, dark throat/breast-streaks.
(b) **Juvenile**: Plain dark brown, with paler chin and throat-centre, small dull wing-spots, pale shaft-streaks on crown to upper back.
(c) **Adult** *stevensi*: Bigger, colder olive above, colder brown below.
(d) **Adult** *leucosticta*: Very broad sooty-brown streaks on throat and breast.

13 [1221] **EYEBROWED WREN-BABBLER** *Napothera epilepidota*
(a) **Adult** *davisoni*: Resembles Streaked but smaller and shorter-tailed, with long, pale buff supercilium, broad dark eyestripe and large whitish spots on tips of greater and median coverts.
(b) **Juvenile**: Plain warm dark brown above and dark rufous below, with indistinct pale supercilium, buff spots on wing-coverts.
(c) **Adult** *roberti*: Duller above, blackish streaks below.
(d) **Adult** *amyae*: Larger; colder above, whiter throat and supercilium.

1
3a
4b
3b
2a
2b
4a
5
6
7a
7b
7c
7d
8a
8b
10b
10a
9
11b
11d
12d
12a
11a
11c
12c
12b
13c
13b
13a
13d

PLATE 113. JUNGLE BABBLERS

1 [1202] **BUFF-BREASTED BABBLER** *Pellorneum tickelli*
(a) **Adult** *fulvum*: Longer-tailed and thinner-billed than Abbott's, upperparts paler and more olive-brown, head-sides tinged buffish, breast buff with faint streaks, flanks and vent paler.
(b) **Juvenile**: Upperparts strongly rufescent.
(c) **Adult** *assamense*: Darker, more rufescent above, pale shaft-streaks on crown.

2 [1203] **BLACK-CAPPED BABBLER** *Pellorneum capistratum*
(a) **Adult** *nigrocapitatum*: Warm dark brown above, deep rufous below, black crown, nape and moustachial stripe, greyish-white supercilium and white throat.
(b) **Juvenile**: Like adult but head pattern much less distinct, upperparts more rufescent.

3 [1204] **PUFF-THROATED BABBLER** *Pellorneum ruficeps*
(a) **Adult** *chthonium*: Rufescent crown, buffy-whitish supercilium and whitish underparts with dark streaking on breast and flanks.
(b) **Juvenile**: Upperparts (not crown) more rufescent. streaking below indistinct.
(c) **Adult** *stageri*: Strong streaking on upper mantle and neck.
(d) **Adult** *minus*: Unstreaked upper mantle, narrow breast-streaks.
(e) **Adult** *acrum*: Unstreaked upper mantle, neat light breast-streaks.

4 [1205] **SPOT-THROATED BABBLER** *Pellorneum albiventre*
(a) **Adult** *cinnamomeum*: Dark-spotted whitish throat. Resembles Buff-breasted but smaller, with shorter, rounded tail, shorter bill and greyish head-sides.
(b) **Adult** *albiventre*: Broadly white centre to abdomen, much duller flanks.

5 [1206] **MOUSTACHED BABBLER** *Malacopteron magnirostre*
(a) **Adult** *magnirostre*: Olive-brown crown and mantle, reddish eyes, greyish head-sides and dark moustachial stripe, underparts whitish with light greyish wash/streaks, tail strongly rufescent.
(b) **Juvenile**: Moustachial stripe less distinct, lower mandible flesh-coloured, eyes greyish/brownish.

6 [1207] **SOOTY-CAPPED BABBLER** *Malacopteron affine*
(a) **Adult** *affine*: Like Moustached but smaller and more slender-billed, crown sooty, no moustachial stripe.
(b) **Juvenile**: Crown paler, lower mandible dull flesh.

7 [1208] **SCALY-CROWNED BABBLER** *Malacopteron cinereum*
(a) **Adult** *cinereum*: From Rufous-crowned by smaller size, slenderer bill, lack of greyish streaks on breast, pinkish legs.
(b) **Adult** *indochinense*: No blackish nape-patch, slightly paler above and buffier-tinged below.

8 [1209] **RUFOUS-CROWNED BABBLER** *Malacopteron magnum*
Adult *magnum*: Like Scaly-crowned but larger and bigger-billed with greyish breast-streaks and greyish legs.

9 [1210] **GREY-BREASTED BABBLER** *Ophrydornis albogularis*
Adult *albogularis*: Small and relatively short-tailed with slaty-grey head, white supercilium and throat and grey breast-band.

10 [1211] **ABBOTT'S BABBLER** *Malacocincla abbotti*
(a) **Adult** *abbotti*: Short-tailed and rather large-billed with rufous-buff flanks and vent. Like Horsfield's but crown and upperparts concolorous and paler, crown with shaft-streaks, less contrasting grey supercilium, lacks breast-streaking. Shorter-tailed and bigger-billed than Buff-breasted, no obvious buffy breast-wash.
(b) **Juvenile**: Crown and upperparts dark rufescent-brown (similar to adult Ferruginous Babbler).

11 [1212] **HORSFIELD'S BABBLER** *Malacocincla sepiaria*
Adult *tardinata*: See under Abbott's Babbler above, crown darker than mantle, breast vaguely streaked.

12 [1213] **SHORT-TAILED BABBLER** *Malacocincla malaccensis*
Adult *malaccensis*: From Abbott's and Horsfield's by smaller size, short tail, thinner bill, blackish moustachial line, grey ear-coverts, white throat.

13 [1214] **WHITE-CHESTED BABBLER** *Trichastoma rostratum*
Adult *rostratum*: Like Abbott's and Horsfield's but clean white below, breast-sides lightly washed grey, rather slender bill, cold dark brown above.

14 [1215] **FERRUGINOUS BABBLER** *Trichastoma bicolor*
Adult: Bright rufescent above and rather clean creamy or buffy-whitish below.

1a
1b
1c
2a
2b
3a
3b
3c
3d
3e
4a
4b
5a
5b
6a
6b
7a
7b
8
9
10a
10b
11
12
13
14

PLATE 114. SCIMITAR-BABBLERS, *TURDOIDES* BABBLERS & CHINESE BABAX

1 [1176] **SICKLE-BILLED SCIMITAR-BABBLER** *Xiphirhynchus superciliaris*
Adult *forresti*: Very long bill, blackish head with whitish supercilium and throat; deep rufous below.

2 [1177] **LARGE SCIMITAR-BABBLER** *Pomatorhinus hypoleucos*
(a) **Adult** *tickelli*: Large; chestnut neck-patch, pale supercilium, grey sides with broad white streaks.
(b) **Adult** *hypoleucos*: Whiter breast-sides, rufous rear supercilium and neck-patch, darker above.
(c) **Adult** *wrayi*: Colder and darker above, patterned supercilium.

3 [1178] **RUSTY-CHEEKED SCIMITAR-BABBLER** *Pomatorhinus erythrogenys*
Adult *celatus*: Orange-rufous head-sides, flanks and vent, white throat, breast and belly.

4 [1179] **SPOT-BREASTED SCIMITAR-BABBLER** *Pomatorhinus mcclellandi*
Adult: Drab brown flanks and breast markings (latter more spot-shaped than on Black-streaked).

5 [1180] **BLACK-STREAKED SCIMITAR-BABBLER** *Pomatorhinus gravivox*
Adult *odicus*: Like Rusty-cheeked but with blackish spots/streaks on breast.

6 [1181] **WHITE-BROWED SCIMITAR-BABBLER** *Pomatorhinus schisticeps*
(a) **Adult** *olivaceus*: Yellowish bill, drab olive-brown upperparts and unmarked white throat, breast and belly-centre.
(b) **Adult** *ripponi*: Duller, greyer above, broader rufous-chestnut nuchal collar and rufous-chestnut flanks.
(c) **Adult** *mearsi*: More blackish-brown crown, chestnut flanks with white streaks.

7 [1182] **CHESTNUT-BACKED SCIMITAR-BABBLER** *Pomatorhinus montanus*
Adult *occidentalis*: Like White-browed but has black crown and dark chestnut mantle to rump and flanks.

8 [1183] **STREAK-BREASTED SCIMITAR-BABBLER** *Pomatorhinus ruficollis*
(a) **Adult** *reconditus*: Like White-browed but smaller and shorter-billed, heavy chestnut streaking below.
(b) **Juvenile**: Mask warm dark brown, underparts uniform buffish-brown with whitish throat.
(c) **Adult** *bakeri*: Dull streaking below.
(d) **Adult** *beaulieui*: Duller streaking, with centre of breast and abdomen whiter.

9 [1184] **ORANGE-BILLED SCIMITAR-BABBLER** *Pomatorhinus ochraceiceps*
(a) **Adult** *ochraceiceps*: Long narrow red bill, rufescent crown and upperparts, white breast and belly-centre.
(b) **Adult** *stenorhynchus*: Warm buff breast and belly.

10 [1185] **CORAL-BILLED SCIMITAR-BABBLER** *Pomatorhinus ferruginosus*
(a) **Adult** *albogularis*: Like Red-billed but thicker bill, black line above supercilium, broadly black head-sides.
(b) **Adult** *phayrei*: Duller above and orange-buff lower throat to belly.

11 [1230] **WHITE-THROATED BABBLER** *Turdoides gularis*
Adult: Largish; very long tail, pale eyes, white throat and upper breast, rufous-buff below.

12 [1231] **STRIATED BABBLER** *Turdoides earlei*
Adult *earlei*: Long tail, brown above with blackish streaks, pale brownish below, thinly streaked throat/breast.

13 [1232] **SLENDER-BILLED BABBLER** *Turdoides longirostris*
(a) **Adult**: Plain; slightly downcurved bill, pale eyes, whitish throat and upper breast, buff below.
(b) **Juvenile**: Paler and more rufescent above, more rufescent-buff below.

14 [1254] **CHINESE BABAX** *Babax lanceolatus*
(a) **Adult** *lanceolatus*: Large and pale with bold blackish, brown and chestnut streaks.
(b) **Juvenile**: Buffier overall with less prominent streaking, plainer crown, paler ear-coverts.

1
3
2b
5
4
2a
6b
2c
6a
7
6c
8b
8a
9a
8c
8d
9b
10b
10a
12
13b
13a
14b
11
14a

PLATE 115. LAUGHINGTHRUSHES

1 [1235] **GREY-SIDED LAUGHINGTHRUSH** *Dryonastes caerulatus*
Adult *latifrons*: Warm brown above, white below with broadly grey flanks, black face, white ear-patch.

2 [1236] **BLACK-THROATED LAUGHINGTHRUSH** *Dryonastes chinensis*
(a) **Adult** *lochmius*: Olive-brown, slate-grey crown, black face to upper breast, white ear-coverts and throat-sides.
(b) **Adult '*lugens*' morph**: Ear-coverts and throat-sides grey to blackish (various intergrades occur).
(c) **Adult** *germaini*: Deep rufescent-brown body.

3 [1237] **CHESTNUT-BACKED LAUGHINGTHRUSH** *Dryonastes nuchalis*
Adult: Resembles Black-throated but smaller, nape and upper mantle rufous-chestnut, breast whiter.

4 [1238] **RUFOUS-VENTED LAUGHINGTHRUSH** *Dryonastes gularis*
Adult: Resembles Yellow-throated but larger, heavier, with plain brown tail, yellow chin and rufous vent.

5 [1239] **YELLOW-THROATED LAUGHINGTHRUSH** *Dryonastes galbanus*
Adult *galbanus*: Warm brown above, greyish crown and nape, black mask and chin, pale yellow underparts and greyish tail with broad dark terminal band and white outer tips.

6 [1240] **WHITE-CHEEKED LAUGHINGTHRUSH** *Dryonastes vassali*
Adult: Recalls Black-throated but upper ear-coverts black, white mostly on throat-sides, narrow black stripe on throat-centre (none on breast), paler below, tail with broad black subterminal band and white outer tips.

7 [1241] **RUFOUS-NECKED LAUGHINGTHRUSH** *Dryonastes ruficollis*
Adult: Smallish, grey crown, blackish face to upper breast, light rufous-chestnut neck-patch and vent.

8 [1242] **RUFOUS-CHEEKED LAUGHINGTHRUSH** *Garrulax castanotis*
Adult *varennei*: Recalls White-necked but shows large broad orange-rufous ear-covert patch, greyish crown, white rear supercilium.

9 [1243] **GREY LAUGHINGTHRUSH** *Garrulax maesi*
Adult *maesi*: Greyish with black face and chin, whitish ear-coverts and neck-patch.

10 [1244] **BLACK-HOODED LAUGHINGTHRUSH** *Garrulax milleti*
Adult *milleti*: Brownish-black head and upper breast bordered by whitish band from mantle to breast.

11 [1245] **CAMBODIAN LAUGHINGTHRUSH** *Garrulax ferrarius*
Adult: Like Black-hooded but hood browner, upper mantle and belly darker, isolated white neck-patch.

12 [1246] **WHITE-NECKED LAUGHINGTHRUSH** *Garrulax strepitans*
(a) **Adult**: Brown crown, blackish-brown face to upper breast, rufous ear-coverts, white neck-patch; mostly brown.
(b) **Adult variant**: Individual with browner upper breast.

13 [1247] **WHITE-CRESTED LAUGHINGTHRUSH** *Garrulax leucolophus*
(a) **Adult** *diardi*: Broad whitish crest, black mask, white below with rufescent flanks and undertail-coverts.
(b) **Adult** *patkaicus*: Chestnut upper mantle, lower breast and centre of abdomen.

14 [1248] **LESSER NECKLACED LAUGHINGTHRUSH** *Garrulax monileger*
(a) **Adult** *mouhoti*: Smaller than Greater; black line on lores, orange-yellow to yellowish-brown eyes, dark eyering, paler primary coverts, often lacks black line below cheeks, narrow black necklace.
(b) **Adult variant**: Individual with more black on ear-coverts.
(c) **Adult** *pasquieri*: Smaller; darker above, deep rufous breast, dark line above supercilium.

15 [1249] **GREATER NECKLACED LAUGHINGTHRUSH** *Garrulax pectoralis*
(a) **Adult** *subfusus*: Larger than Lesser; pale lores, brown to crimson eyes, golden-yellow eyering, blackish-brown primary coverts; complete black line below ear-coverts/cheeks, mostly buff throat/breast, with broader necklace.
(b) **Adult** *pectoralis*: Usually more complete, solid black necklace.
(c) **Adult variant**: Occasionally all-black ear-coverts.
(d) **Adult** *robini*: Rufous-chestnut nuchal collar, grey and black behind ear-coverts, broken necklace.

16 [1250] **WHITE-THROATED LAUGHINGTHRUSH** *Garrulax albogularis*
Adult *eous*: Brown above, white throat and upper breast, greyish-brown breast-band, buffish below.

17 [1251] **MASKED LAUGHINGTHRUSH** *Garrulax perspicillatus*
(a) **Adult**: Greyish-brown with broad blackish-brown mask and orange-buff vent.
(b) **Juvenile**: Mask fainter, rest of head and breast browner, upperparts warmer-tinged.

18 [1252] **BLACK LAUGHINGTHRUSH** *Melanocichla lugubris*
Adult: Blackish with naked bluish-white post-ocular patch, orange-red bill.

19 [1255] **STRIATED LAUGHINGTHRUSH** *Grammatoptila striata*
(a) **Adult** *cranbrooki*: Broad rounded crest, rich brown pale-streaked plumage, broad blackish supercilium.
(b) **Adult** *brahmaputra*: Narrower supercilium, whitish shaft-streaks on forehead and above eye.

1
3
2a
2c
4
5
2b
6
7
8
9
10
11
13b
12b
12a
13a
15c
14a
15b
15a
14c
14b
15d
17a
19a
18
16
17b
19b

PLATE 116. LAUGHINGTHRUSHES

1 [1253] **SPECTACLED LAUGHINGTHRUSH** *Rhinocichla mitrata*
Adult *major*: Largely greyish; chestnut crown, broad white eyering, rufous-chestnut undertail-coverts.

2 [1256] **SPOT-BREASTED LAUGHINGTHRUSH** *Stactocichla merulina*
(a) **Adult** *merulina*: Plain brown; blackish spots on buffish-white throat and breast, thin buff supercilium.
(b) **Adult** *obscurus*: Richer buff below, with much heavier blackish spotting on throat and breast.

3 [1257] **ORANGE-BREASTED LAUGHINGTHRUSH** *Stactocichla annamensis*
Adult: Black throat, black streaks on deep orange-rufous breast, narrow pale orange-rufous supercilium.

4 [1258] **CHINESE HWAMEI** *Leucodioptron canorum*
Adult *canorum*: Warm brown; dark streaks on crown, nape, mantle, throat and breast, white spectacles.

5 [1259] **STRIPED LAUGHINGTHRUSH** *Strophocincla virgata*
Adult: Rufescent, with narrow whitish streaks and broad whitish supercilium and submoustachial stripe.

6 [1260] **WHITE-BROWED LAUGHINGTHRUSH** *Pterorhinus sannio*
Adult *comis*: Brownish; buffish-white loop round front of eye (supercilium to cheek-patch), rufescent vent.

7 [1261] **MOUSTACHED LAUGHINGTHRUSH** *Ianthocincla cineracea*
(a) **Adult** *cineracea*: Sandy-brown above, with black crown and subterminal tail-band, greyish-white head-sides with frayed submoustachial stripe, black, grey and white wings.
(b) **Adult** *strenua*: Buffier-brown overall.

8 [1262] **RUFOUS-CHINNED LAUGHINGTHRUSH** *Ianthocincla rufogularis*
(a) **Adult** *rufiberbis*: Warm olive-brown above with black crown, scaling, wing-bands and subterminal tail-band; whitish below with pale rufous-chestnut chin and vent, and dark-spotted breast.
(b) **Juvenile**: Reduced dark markings and less rufous on chin.
(c) **Adult** *intensior*: Deeper rufous above, less black on crown but more on head-sides, larger spots below and paler rufous chin.

9 [1263] **CHESTNUT-EARED LAUGHINGTHRUSH** *Ianthocincla konkakinhensis*
Adult: Chestnut ear-coverts, whitish chin, unbanded flight-feathers, pale tail-tip, grey on forehead/supercilium.

10 [1264] **SPOTTED LAUGHINGTHRUSH** *Ianthocincla ocellata*
Adult *maculipectus*: Relatively large; black-and-white spotting on mostly chestnut upperparts, black crown and lower throat, rufous-buff underparts with short black-and-white bars on breast.

11 [1265] **SCALY LAUGHINGTHRUSH** *Trochalopteron subunicolor*
(a) **Adult** *griseatus*: Dark-scaled body like Blue-winged, but with dark eyes, no supercilium, yellowish-olive wings, golden-brown tail with white outer tips.
(b) **Adult** *fooksi*: Slatier crown, extensively dark throat.

12 [1266] **BROWN-CAPPED LAUGHINGTHRUSH** *Trochalopteron austeni*
Adult *victoriae*: Warm brown above, white streaks on neck, brown-and-whitish scaled underparts.

13 [1267] **BLUE-WINGED LAUGHINGTHRUSH** *Trochalopteron squamatum*
Adult: White eyes, black supercilium, scaled body, largely rufous-chestnut wings, black tail with reddish tip.

14 [1268] **BLACK-FACED LAUGHINGTHRUSH** *Trochalopteron affine*
Adult *oustaleti*: Blackish-brown head, whitish submoustachial and neck-patches, body rufescent-brown, wings and tail grey and yellowish-olive.

15 [1269] **ASSAM LAUGHINGTHRUSH** *Trochalopteron chrysopterum*
(a) **Adult** *erythrolaemum*: Mostly rufous head-sides/underparts, grey supercilium, dark-spotted mantle and breast.
(b) **Adult** *woodi*: Black-streaked silvery-grey forehead and ear-coverts, dark-spotted mantle and breast.

16 [1270] **SILVER-EARED LAUGHINGTHRUSH** *Trochalopteron melanostigma*
Adult *schistaceum*: Rufous-chestnut crown, blackish face, rufescent greater coverts and yellowish-olive fringing on wing and tail, mostly plain olive-greyish body, mostly silvery-grey ear-coverts.

17 [1271] **MALAYAN LAUGHINGTHRUSH** *Trochalopteron peninsulae*
Adult: Plain rufescent-brown; dark grey head-side, mostly brownish ear-coverts, indistinct whitish eyering.

18 [1272] **GOLDEN-WINGED LAUGHINGTHRUSH** *Trochalopteron ngoclinhense*
Adult: Dark head and body, golden to orange fringing on wing and tail feathers.

19 [1273] **COLLARED LAUGHINGTHRUSH** *Trochalopteron yersini*
Adult: Black hood with silvery-grey ear-coverts, deep orange-rufous collar and breast.

20 [1274] **RED-WINGED LAUGHINGTHRUSH** *Trochalopteron formosum*
Adult *greenwayi*: From Red-tailed by grey forecrown and ear-coverts, brown upperparts, browner underparts.

21 [1275] **RED-TAILED LAUGHINGTHRUSH** *Trochalopteron milnei*
Adult *sharpei*: Bright rufous crown and nape and extensively red wings and tail.

22 [1276] **CRIMSON-FACED LIOCICHLA** *Liocichla phoenicea*
Adult *bakeri*: Darker red on head than Scarlet-faced, which does not extend to throat-side, browner-tinged crown, deeper/warmer brown body-plumage.

23 [1277] **SCARLET-FACED LIOCICHLA** *Liocichla ripponi*
Adult *ripponi*: Relatively small, rather plain brown body and striking red head and throat-sides.

1
2b
3
5
4
2a
6
8c
7a
7b
8a
8b
11b
10
9
11a
13
14
12
16
15a
17
18
15b
23
20
19
21
22

PLATE 117. *KENOPIA* & *RIMATOR* WREN-BABBLERS, CUTIAS, MINLAS, MESIAS, LEIOTHRIXES & ALLIES

1 [1216] **STRIPED WREN-BABBLER** *Kenopia striata*
(a) **Adult**: Bold white streaking on dark crown, mantle, wing-coverts and breast-sides, whitish head-sides, throat and underparts with buff lores and flanks.
(b) **Juvenile**: Crown and breast-sides browner, streaks buffier, breast mottled, bill paler.

2 [1222] **LONG-BILLED WREN-BABBLER** *Rimator malacoptilus*
Adult: Small; long, slightly downcurved bill, very short tail, heavily streaked plumage.

3 [1223] **WHITE-THROATED WREN-BABBLER** *Rimator pasquieri*
Adult: Brown of plumage darker and colder than Long-billed, throat white, body-streaking whiter.

4 [1224] **INDOCHINESE WREN-BABBLER** *Rimator danjoui*
(a) **Adult** *parvirostris*: Long, slightly downcurved bill and shortish tail; dark brown above with light shaft-streaks, whitish below, rufescent breast with smudgy blackish streaks.
(b) **Adult** *danjoui*: Browner breast-streaking, paler rufous on neck and breast, longer bill.

5 [1233] **HIMALAYAN CUTIA** *Cutia nipalensis*
(a) **Male** *melanchima*: Bluish-slate crown, black head-sides, rufous-chestnut upperparts, whitish underparts with bold black bars on buff-washed sides; wings black with bluish-grey fringing.
(b) **Female**: Like male but mantle, back and scapulars black-streaked, head-sides dark brown.

6 [1234] **VIETNAMESE CUTIA** *Cutia legalleni*
(a) **Male** *legalleni*: Underparts entirely barred black.
(b) **Female**: Underparts as male, drab brown crown, black-streaked greyish-brown mantle, back and scapulars.
(c) **Male** *hoae*: Much narrower underpart-barring, broken by white centre to abdomen.

7 [1278] **BAR-THROATED MINLA** *Chrysominla strigula*
(a) **Adult** *castanicauda*: Golden-rufous crown, blackish supercilium and submoustachial streak, scaly black-and-white throat; olive-greyish above, yellowish below, brownish-chestnut, black and white on wings and tail.
(b) **Adult** *malayana*: Duller overall, with broader black throat-scales.
(c) **Adult** *traii*: Mostly whitish lores and upper ear-coverts, broader black moustachial stripe, brighter crown, greyer upperparts.

8 [1279] **RED-TAILED MINLA** *Minla ignotincta*
(a) **Male** *mariae*: Black crown and head-sides, broad white supercilium, pale yellow throat and underparts, olive-brownish upperparts, black, white and red on wings, black and much red on tail.
(b) **Female**: Little or no red on wings, pinker tail-fringes and tip.
(c) **Male** *ignotincta*: Deeper brown upperparts.

9 [1280] **BLUE-WINGED SIVA** *Siva cyanouroptera*
(a) **Adult** *sordida*: Slim and long-tailed with violet-blue fringing on primaries and tail, indistinct whitish supercilium, warm brown upperparts, greyish-white underparts, bluish crown.
(b) **Adult** *aglae*: Bluer on crown, warmer mantle, greyer throat and breast.
(c) **Adult** *orientalis*: Brownish crown to mantle, duller wings and tail, and whiter underparts.

10 [1281] **SILVER-EARED MESIA** *Mesia argentauris*
(a) **Male** *galbana*: Yellow bill and forehead-patch, black head, silver-grey ear-coverts, orange-yellow throat and breast, reddish wing-patch and tail-coverts.
(b) **Female**: Nape and uppertail-coverts dull golden-olive; generally less intense yellows.
(c) **Male** *cunhaci*: Larger yellow forehead-patch.
(d) **Male** *ricketti*: Orange-red throat and upper breast and slightly darker remainder of underparts.

11 [1282] **RED-BILLED LEIOTHRIX** *Leiothrix lutea*
(a) **Male** *kwangtungensis*: Red bill, yellowish face, golden-olive crown and nape, dark submoustachial stripe, deep yellow throat, orange-rufous upper breast and complex coloured wing pattern.
(b) **Female**: More greenish crown, greyer ear-coverts, weaker submoustachial stripe, paler below.
(c) **Juvenile**: Crown concolorous with mantle, underparts olive-grey with whitish centre, bill paler.
(d) **Male** *yunnanensis*: Dull crown, greyer above, redder breast, more black and less chestnut on wings.

1b
2
3
1a
4b
4a
7b
5a
7c
5b
7a
6b
9a
6a
9b
8a
9c
6c
8c
8b
10d
10c
11a
11b
10b
10a
11c
11d

PLATE 118. WHITE-HOODED BABBLERS, GREY-CROWNED CROCIAS, SIBIAS & BARWINGS

1 [1225] **WHITE-HOODED BABBLER** *Gampsorhynchus rufulus*
(a) **Adult**: More olive-tinged above than Collared, paler below with whiter breast and neck-side; distinctive white shoulder-slash formed by white median and some lesser coverts.
(b) **Juvenile**: Head rufous, throat buff-tinged white.

2 [1226] **COLLARED BABBLER** *Gampsorhynchus torquatus*
(a) **Adult** *torquatus*: White hood, rufescent-brown upperside, warm buff below, dark on neck/breast-side.
(b) **Adult** *saturatior*: More pronounced blackish collar, indistinct at rear, broken at front, warmer buff below.
(c) **Adult** *luciae*: Blackish hindcrown, rufescent nape, full blackish necklace, warmer buff below.

3 [1283] **GREY-CROWNED CROCIAS** *Crocias langbianis*
(a) **Adult**: Slaty crown and nape, blackish mask, dark streaks above, blackish flank-streaks on pale underparts.
(b) **Juvenile**: Crown browner with buffish streaks, head-sides duller, flank-streaks smaller.

4 [1284] **LONG-TAILED SIBIA** *Heterophasia picaoides*
(a) **Adult** *cana*: All grey with very long whitish-tipped tail and long broad white wing-patch.
(b) **Adult** *wrayi*: Browner with smaller wing-patch.

5 [1285] **GREY SIBIA** *Malacias gracilis*
(a) **Adult** *dorsalis*: Like Black-headed but crown colour merges into mantle, uppertail and tertials grey, vent buff.
(b) **Adult** (N Myanmar): Greyer nape to scapulars.

6 [1286] **BLACK-HEADED SIBIA** *Malacias desgodinsi*
(a) **Adult** *desgodinsi*: Black head, wings and tail, greyish above, white below with mauvish-grey flank-wash.
(b) **Adult** *engelbachi*: Dark brown lower mantle, scapulars and back, broken white eyering.
(c) **Adult** *robinsoni*: Broad broken white eyering and white-streaked ear-coverts.

7 [1287] **DARK-BACKED SIBIA** *Malacias melanoleucus*
(a) **Adult** *radcliffei*: Blackish above, white below/on tail-tips. Mantle/scapulars to uppertail-coverts brownish-black.
(b) **Adult** *melanoleucus*: Distinctly brown mantle and scapulars to uppertail-coverts.
(c) **Adult** *castanoptera*: Rufous on inner greater coverts and outer webs of tertials.

8 [1288] **BEAUTIFUL SIBIA** *Malacias pulchellus*
Adult *pulchellus*: Bluish-grey with black mask and wing-coverts, brown tertials and basal two-thirds of tail.

9 [1289] **RUFOUS-BACKED SIBIA** *Leioptila annectens*
(a) **Adult** *mixta*: Black crown and head-sides, rufous-chestnut back to uppertail-coverts, white below, buff vent.
(b) **Adult** *saturata*: More black on mantle and scapulars, chestnut on rest of upperparts.
(c) **Adult** *davisoni*: Almost completely black mantle and scapulars.

10 [1290] **STREAK-THROATED BARWING** *Actinodura waldeni*
(a) **Adult** *poliotis*: Bulky, relatively short-tailed; no pale eyering, streaked dark on crown and light on underparts.
(b) **Adult** *waldeni*: Paler on head and darker rufescent below with less distinct streaking.
(c) **Adult** *saturatior*: More strongly contrasting streaks on head and underparts.

11 [1291] **STREAKED BARWING** *Actinodura souliei*
Adult *griseinucha*: Like Streak-throated but ear-coverts and nape plainer, buffish body with broad blackish streaks.

12 [1292] **RUSTY-FRONTED BARWING** *Actinodura egertoni*
Adult *ripponi*: Slim, long-tailed; brownish-grey hood with chestnut face, and mostly rufous-buff below.

13 [1293] **SPECTACLED BARWING** *Actinodura ramsayi*
(a) **Adult** *ramsayi*: Greyish-olive upperside, buffy-rufous forehead, white eyering, blackish lores, deep buff below.
(b) **Adult** *yunnanensis*: Rufous crown, narrow throat-streaks.

14 [1294] **BLACK-CROWNED BARWING** *Actinodura sodangorum*
Adult: Black crown, olive-brown above, mostly black wings. (Eyering broader than illustrated.)

1a
1b
2c
2b
2a
3a
3b
4a
4b
5a
5b
6a
6b
6c
7a
7b
7c
8
9a
9b
9c
10a
10b
10c
11
12
13a
13b
14

PLATE 119. *LOCUSTELLA* & *BRADYPTERUS* BUSH-WARBLERS & CISTICOLAS

1 [1305] **PLESKE'S WARBLER** *Locustella pleskei*
Adult: Larger/bulkier than Rusty-rumped. Plain-looking, greyish-brown above (no rufous); whitish below with greyish-brown flanks (and to lesser extent breast), relatively indistinct supercilium, but obvious pale cream eyering (broken at front/rear).

2 [1306] **RUSTY-RUMPED WARBLER** *Locustella certhiola*
(a) **Adult** *certhiola*: Largish; dark rufescent above, whitish-tipped tail, rufescent rump, unstreaked below.
(b) **Juvenile**: Like adult but washed yellowish below, with faint breast-streaks.
(c) **Adult** *rubescens*: Deeper rufescent-brown above.

3 [1307] **LANCEOLATED WARBLER** *Locustella lanceolata*
(a) **Adult (worn)**: Small, heavily streaked; streaked on rump, throat, breast, flanks and undertail-coverts.
(b) **Adult (fresh)**: Upperparts and flanks rather warmer.

4 [1308] **SPOTTED BUSH-WARBLER** *Bradypterus thoracicus*
(a) **Adult** *thoracicus*: Told from Baikal by more rufescent-tinged upperparts, greyer supercilium and breast-sides, heavier dark throat and breast-spotting, all-blackish bill.
(b) **Adult variant**: Weakly spotted individual.
(c) **Juvenile**: Yellowish wash below, browner breast and flanks, vague throat/breast spots.

5 [1309] **BAIKAL BUSH-WARBLER** *Bradypterus davidi*
(a) **Adult** *suschkini*: Cold, dark brown above, speckling on throat, contrasting white tips to undertail-coverts.
(b) **Adult variant**: Weakly spotted individual.

6 [1310] **CHINESE BUSH-WARBLER** *Bradypterus tacsanowskius*
(a) **Adult**: Relatively long-tailed; greyish-olive upperparts, indistinct darker centres to undertail-coverts, pale lower mandible, no speckling on throat/breast.
(b) **First winter**: Supercilium and underparts washed yellow, often light speckling on throat/breast.

7 [1311] **BROWN BUSH-WARBLER** *Bradypterus luteoventris*
(a) **Adult**: Dark rufescent upperparts, buffy-rufous flanks, plain throat/breast and undertail-coverts, pale lower mandible.
(b) **Juvenile**: Upperparts tinged rufous-chestnut, pale areas of underparts washed yellow.

8 [1312] **RUSSET BUSH-WARBLER** *Bradypterus mandelli*
(a) **Adult** *mandelli*: Similar to Brown but dark brown undertail-coverts with broad whitish tips, all-blackish bill, usually speckled on throat/breast. More rufescent above and on flanks than Spotted.
(b) **Adult variant**: Little or no throat/breast-spotting.
(c) **Juvenile**: Like Spotted but more rufescent above, broader dark flanks, rufous on breast-band.

9 [1313] **STRIATED GRASSBIRD** *Megalurus palustris*
(a) **Adult** (female) *toklao*: Very large, with long graduated pointed tail, dark-streaked buffish-brown upperside, white supercilium and largely whitish underparts with fine dark breast- and flank-streaks.
(b) **Juvenile**: Supercilium and underparts washed yellow, streaking below fainter, bill paler.

10 [1314] **ZITTING CISTICOLA** *Cisticola juncidis*
(a) **Adult non-breeding** *malaya*: Buffish-brown with bold blackish streaks above, rump rufescent, tail tipped white; whitish below with buff wash on flanks. Very like non-breeding male and female Bright-headed Cisticola, but has whiter supercilium, duller nape and whiter tail-tip.
(b) **Adult breeding**: Broader dark streaks above.
(c) **Juvenile**: Underparts washed light yellow.

11 [1315] **BRIGHT-HEADED CISTICOLA** *Cisticola exilis*
(a) **Male breeding** *equicaudata*: Unstreaked golden-rufous crown, rich buff breast.
(b) **Female breeding**: Broad dark streaks on crown and mantle, plain warm supercilium and nape.
(c) **Juvenile**: Browner above, pale yellow below, washed buff on flanks.
(d) **Male non-breeding** *tytleri*: Rather heavier streaking above.
(e) **Male breeding**: Creamy-buffish crown, bolder streaks on mantle.

1
2c
2a
2b
3b
3a
4b
5a
5b
4a
6a
4c
6b
7a
7b
9 to different scale
8b
9a
8c
9b
8a
11a
11b
11d
10b
10a
10c
11c
11e

PLATE 120. TAILORBIRDS & PRINIAS

1 [1316] **ASHY TAILORBIRD** *Orthotomus ruficeps*
(a) **Male** *cineraceus*: Rufous 'face', dark grey throat, breast and flanks.
(b) **Female**: Mostly whitish on centre of underparts.
(c) **Juvenile**: Browner above without rufous, like female below but more washed out.

2 [1317] **RUFOUS-TAILED TAILORBIRD** *Orthotomus sericeus*
(a) **Adult** *hesperius*: Rufous-chestnut crown, dull rufous tail, grey upperparts, mostly whitish underside.
(b) **Juvenile**: Little rufous-chestnut on crown, browner than adult above, tail paler (slightly too plain on plate).

3 [1318] **DARK-NECKED TAILORBIRD** *Orthotomus atrogularis*
(a) **Male breeding** *nitidus*: Like Common but shows solid blackish-grey lower throat and upper breast, lacks pale supercilium, tail much shorter, vent yellow.
(b) **Female**: Relatively weak dark streaking on throat/breast-sides.
(c) **Juvenile**: Like female but duller above, rufous on crown initially lacking.

4 [1319] **COMMON TAILORBIRD** *Orthotomus sutorius*
(a) **Male breeding** *inexpectatus*: Rufescent forecrown, greenish upperparts, pale underparts, long bill and tail.
(b) **Female**: Shorter tail.
(c) **Adult** *maculicollis*: Darker overall, dark streaks on breast. From Dark-necked by duller, less extensive rufous on crown, pale supercilium, duller upperparts and lack of yellow on vent.

5 [1320] **RUFESCENT PRINIA** *Prinia rufescens*
(a) **Adult non-breeding** *beavani*: Small; plain rufescent mantle, strongly graduated tail with pale-tipped feathers. Bill slightly thicker than Grey-breasted, has buffier flanks and vent, longer supercilium.
(b) **Adult breeding**: Head mostly slaty-grey.

6 [1321] **GREY-BREASTED PRINIA** *Prinia hodgsonii*
(a) **Adult non-breeding** *erro*: Like Rufescent (which see for differences); often greyish on neck-sides and breast.
(b) **Adult breeding**: Dark grey head and broad breast-band, whitish throat and belly.
(c) **Juvenile**: Like non-breeding adult but more rufescent above, bill pale.

7 [1322] **YELLOW-BELLIED PRINIA** *Prinia flaviventris*
(a) **Adult** *delacouri*: Greyish head, greenish mantle, whitish throat and breast, yellow belly and vent.
(b) **Juvenile**: Plain rufescent olive-brown above, pale yellow below and on supercilium, buffier flanks.
(c) **Adult** *sonitans*: More rufescent above, light buffish throat and breast, and deep buff belly and vent.
(d) **Juvenile**: Whiter throat to centre of abdomen and duller head.

8 [1323] **PLAIN PRINIA** *Prinia inornata*
(a) **Adult** *herberti*: Long broad whitish supercilium. Larger and longer-tailed than Rufescent and Grey-breasted, smaller and plainer above than Striated and Brown.
(b) **Juvenile**: Like adult but warmer above, washed yellowish below.
(c) **Adult non-breeding** *extensicauda*: Deep buff below, warmer above, long tail.

9 [1324] **STRIATED PRINIA** *Prinia crinigera*
(a) **Adult non-breeding** *catharia*: Relatively large, long-tailed; heavily streaked upperparts, ear-coverts and neck-sides, speckling on throat and breast.
(b) **Adult breeding**: Darker above, streaks less defined, breast more mottled, bill black (male only?).
(c) **Juvenile**: Like non-breeding adult but warmer and less well marked.

10 [1325] **BROWN PRINIA** *Prinia polychroa*
(a) **Adult non-breeding** *cooki*: Like Striated but less distinctly streaked. Larger than Plain with weak supercilium and streaks above.
(b) **Adult breeding**: Greyer and generally less distinctly streaked than non-breeding. Bill black (male only?).
(c) **Juvenile**: Plainer above than adult breeding; warmer than adult non-breeding; slight yellow tinge below.

11 [1326] **BLACK-THROATED PRINIA** *Prinia atrogularis*
(a) **Adult non-breeding** *khasiana*: Rufescent upperparts, paler, browner head-sides, almost no breast-streaks.
(b) **Adult breeding**: Black throat merges to scales on breast, white submoustachial stripe.

12 [1327] **HILL PRINIA** *Prinia superciliaris*
(a) **Adult breeding** *erythropleura*: Large, long-tailed; plain above, white supercilium, greyish head-sides, sparse dark spots/streaks on breast.
(b) **Adult non-breeding**: Breast-streaking longer and more extensive.
(c) **Adult breeding** *waterstradti*: Duller crown and upperparts, narrow supercilium, smudgy breast-streaks.

1a
1b
1c
2a
2b
3a
3b
3c
4a
4b
4c
5a
5b
6a
6b
6c
7a
7b
7c
7d
8a
8b
8c
9a
9b
9c
10a
10b
10c
11a
11b
12a
12b
12c

MEGAPODIIDAE: Scrubfowl

Worldwide c.22 species. SE Asia 1 species. Stocky with rather short rounded wings, short tails, straight bills and large strong legs and feet with long straight claws. Terrestrial, feeding on snails, insects and vegetable matter etc.

1 NICOBAR SCRUBFOWL
Megapodius nicobariensis Plate **1**

IDENTIFICATION 37–40.5 cm. *M.n.nicobariensis*: **Adult** Resembles female Red Junglefowl but tail much shorter, bill longer and straighter, plumage plainer with brown upperparts including slightly tufted crown, grey-brown underparts, and grey face and throat with prominent red facial skin but no wattle. Note range and habitat. **Juvenile** Has smaller and paler area of red facial skin, brown underparts and initially more buffish forehead, ear-coverts and upper throat. **VOICE** Male territorial call is ***kyouououou-kyou-kou-koukoukouk-oukou***, the first note rising in pitch, remainder gradually decreasing in a staccato series. Contact call is a cackling ***kuk-a-kuk-kuk***. **HABITAT & BEHAVIOUR** Understorey of broadleaved evergreen forest on islands, adjacent sandy beaches. Usually found singly or in pairs, sometimes in fairly large family parties. Runs from danger, flight weak. Partly nocturnal or crepuscular. **RANGE & STATUS** Resident Nicobar Is. **SE Asia** Former resident (current status unknown) Coco Is, off S Myanmar. **BREEDING** All year? Colonial (1–3 females per mound). **Nest** Very large mound of sand and leaves etc., usually just inside forest next to sandy beach. **Eggs** Individual clutch size unknown; pinkish, turning dull buffish to ochre-brown, then dull whitish; 76.4–85.5 x 46.2–57.1 mm.

PHASIANIDAE: PERDICINAE: Partridges, francolins, quails

Worldwide c.112 species. SE Asia 21 species. Terrestrial, generally feeding on invertebrates, grain, buds, seeds, fallen fruit and other vegetable matter. **Chinese Francolin & partridges** (*Lerwa, Francolinus, Rhizothera, Melanoperdix, Arborophila, Caloperdix, Rollulus, Bambusicola*) Medium-sized, very robust and short-tailed with rounded wings and strong legs and feet, largely in forest or wooded areas. **Quails** (*Coturnix*) Smaller, inhabiting more open areas and cultivation.

2 SNOW PARTRIDGE *Lerwa lerwa* Plate **1**

IDENTIFICATION 38–40 cm. Monotypic. **Adult** Appears dark greyish above and chestnut below, with red bill, legs and feet. At close range, head, neck and upperside intensely barred blackish and whitish, with some dull chestnut on scapulars and wing-coverts. Darkness of upperside varies, with crown sometimes almost blackish. Lower breast to belly chestnut with bold white streaks, merging to whiter vent with chestnut-vermiculated undertail-coverts. Whitish trailing edge to secondaries can be striking in flight. **Juvenile** Like adult, but finely spotted, rather than barred, on head and neck; upper breast finely barred rufous. **VOICE** Male territorial call is a loud, discordant, screechy, whiplashing whistled ***skLEE'Eer, skLEE'Eer...***, with each note quickly dropping away. Notes given singly or at regular or irregular intervals. **HABITAT & BEHAVIOUR** Open mountainsides above tree-line, alpine pastures, grassy clearings with rhododendron scrub; 3,000–5,500 m (rarely down to 2,000 m) in extralimital areas. In groups of up to 30 outside breeding season. Can be tame, but extremely well camouflaged. **RANGE & STATUS** Resident N Pakistan, NW,N,NE Indian subcontinent, S,SE,E Tibet, W China. **SE Asia** Presumed resident N Myanmar (specimen in a Putao museum thought to have been collected locally). **BREEDING** Indian subcontinent: May–July. **Nest** Lined or unlined scrape on ground, under rock, grass-tuft or bush. **Eggs** 3–5; pale buff or yellowish-buff to grey-buff, with fine reddish freckles and blotches; 54.6 x 35.4 mm (av.).

3 CHINESE FRANCOLIN
Francolinus pintadeanu Plate **1**

IDENTIFICATION 30.5–33.5 cm. *F.p.phayrei*: **Male** Blackish plumage with whitish to buffy-white spots and bars, black, grey and whitish bars on rump and uppertail-coverts, white ear-coverts surrounded by black, and white throat distinctive. Sides of crown rufous, scapulars chestnut, vent rufous-buff. See Bar-backed Partridge but note habitat. **Female** Much browner, head pattern duller with less black on crown-centre, upperparts greyish-brown and less distinctly marked, little chestnut on scapulars, underparts more barred than spotted. From *Arborophila* partridges by rufescent sides of crown, whitish sides of head and throat bisected by blackish malar line and rather uniformly barred underparts. **Juvenile** Duller than female with less rufous on crown-sides, pale streaks on upperparts, hindneck more barred than spotted, eyestripe and moustachial streak almost lacking or latter reduced to spots. **VOICE** Male territorial call is a loud harsh metallic ***wi-ta-tak-takaa***, repeated after rather long intervals. **HABITAT** Open forest and woodland, grass and scrub; up to 1,800 m. **RANGE & STATUS** Resident NE India, southern China. Introduced Philippines (Luzon), Mauritius. **SE Asia** Fairly common to common resident (except Tenasserim, S Thailand, Peninsular Malaysia, Singapore). **BREEDING** March–September. **Nest** Scrape on ground. **Eggs** 3–7; pale buff or cream to warm coffee; 35.3 x 28.7 mm (av.).

4 LONG-BILLED PARTRIDGE
Rhizothera longirostris Plate **1**

IDENTIFICATION 36–40.5 cm. *R.l.longirostris*: **Male** Told by combination of relatively large size, long stout bill, dark chestnut crown, blackish line above supercilium and through eye, light chestnut supercilium, sides of head and throat, and grey foreneck, upper breast and upper mantle (broken in centre). Rest of upperparts mottled dark chestnut-brown and black with buff streaks, back to uppertail-coverts finely vermiculated buff and grey, rest of underparts light chestnut (belly-centre paler), wing-coverts mostly buffish with some brown, grey and black vermiculations, legs yellow. See Ferruginous Partridge. **Female** Lacks grey on breast and mantle; throat and breast light chestnut; back to uppertail-coverts warmer buffish. From Ferruginous by size and bill structure, lack of dark eyestripe or black-and-white scales on mantle and flanks, relatively plain rump and uppertail-coverts (noticeably paler than mantle) and plainer brown wings. See Black Partridge. **Juvenile** Initially like female (both sexes) but upperparts more chestnut, has some dark spots and bars on breast, and buff streaks on throat, breast, neck and mantle. Males gradually attain patches of adult plumage. **VOICE** Territorial call is a far-carrying double whistle with distinctly higher second note; usually given in duet, producing a repetitive, rising, four-note sequence. **HABITAT** Broadleaved evergreen forest, bamboo; up to 1,500 m. **RANGE & STATUS** Resident Sumatra, Borneo. **SE Asia** Scarce to uncommon resident south Tenasserim, W(south),S Thailand, Peninsular Malaysia. **BREEDING** Borneo: February–March. **Nest** Slight lined scrape on ground. **Eggs** 2–5; slightly glossy, whitish to pinkish-white, sparingly spotted, blotched and smeared chestnut over more scattered violet underspots; 36.6–36.7 x 26.8–26.9 mm.

5 BLACK PARTRIDGE
Melanoperdix nigra Plate **1**

IDENTIFICATION 24–27 cm. *M.n.nigra*: **Male** Glossy black plumage with slightly browner wings diagnostic. See Crested Partridge. **Female** Distinctive, with rather uniform dark chestnut plumage, more buffish sides of head, throat and vent, and broad black spots on scapulars. Has chestnut scales/bars on sides of head and throat (sometimes blacker on cheeks). See Long-billed and Ferruginous Partridges. **Juvenile** Similar to female but upperparts finely vermiculated a little paler and darker with some pale spots; has less black on scapulars, large whitish spots and dark bars broadly down sides of breast and flanks, whiter vent. See Ferruginous Partridge. **VOICE** Undocumented. **HABITAT & BEHAVIOUR** Broadleaved evergreen forest; up to 610 m. Very shy. Said to sometimes cock tail like a rail. **RANGE & STATUS** Resident Sumatra, Borneo. **SE Asia** Scarce resident Peninsular Malaysia. **BREEDING** July–August. **Nest** Lined scrape on ground. **Eggs** Borneo: 3–5; dull white; 36.4–40.9 x 31.3–32.7 mm.

6 COMMON QUAIL *Coturnix coturnix* Plate 1

IDENTIFICATION 20–20.5 cm. *C.c.coturnix*: **Male** Like non-breeding male Japanese Quail but slightly larger, possibly tends to have less chestnut base colour to breast and flanks, and browner upperparts. In hand, wing over 105 mm. See Rain Quail. **Female** Lacks dark gular stripe of male. Possibly inseparable in field (either sex), on current knowledge, from Japanese. **First winter** Similar to female. **VOICE** Male territorial call is a fast, rhythmic, dripping whistle: ***pit pil-it***. Sometimes gives slightly ringing ***pik-kreee*** when flushed. **HABITAT** Grassy areas, cultivation; lowlands. **RANGE & STATUS** Resident Atlantic Is, NE,E and southern Africa, Madagascar, Comoros. Breeds N Africa, Palearctic east to Baikal region, northern Indian subcontinent, NW China; winters Africa, Indian subcontinent (except islands). **SE Asia** Rare winter visitor/vagrant W,S Myanmar.

7 JAPANESE QUAIL *Coturnix japonica* Plate 1

IDENTIFICATION 19 cm. *C.j.japonica*: **Male non-breeding** Similar to female but throat and foreneck white with blackish-brown to pale chestnut line down centre and transverse throat-band, breast richer buff, often with more chestnut and fewer blackish streaks. **Male breeding** Uniform pale pinkish-chestnut sides of head and throat diagnostic. Often shows dark throat-bands. **Female** Greyish-brown above with narrow whitish to buff streaks and speckles and small blackish markings, whitish below with rufescent to chestnut breast and flanks, marked with black and long whitish streaks. Has narrow whitish median crown-stripe, throat pale buff to whitish with double dark bar at sides, often with short moustachial line almost joining first bar. Difficult to separate from Common Quail but slightly smaller (wing under 105 mm), base colour of breast and flanks tends to be more chestnut, upperparts typically somewhat greyer. Sharply pointed feathers on sides of throat (beard) distinctive if visible; broad chestnut breast-streaking may be distinctive if present. From Rain Quail by warmer, more distinctly dark-streaked breast, paler ear-coverts and barring on primaries. See Blue-breasted Quail and buttonquail. **First winter** Similar to female. **VOICE** Male territorial call is a loud ***choo-peet-trrr*** or ***guku kr-r-r-r-r***. **HABITAT** Grassy areas, cultivation; up to 500 m. **RANGE & STATUS** Breeds E Palearctic, NE China, N Korea, Japan; winters Bhutan and NE India (probably breeds), C and southern China (probably breeds SW,S), S Korea, southern Japan. Introduced Réunion, Hawaii. **SE Asia** Scarce to uncommon winter visitor Myanmar (may breed in N), NW Thailand, N Laos, W,E(may breed) Tonkin. Also recorded on passage E Tonkin (sight records; extralimital Common Quail *C. coturnix* not definitely eliminated). Vagrant Cambodia, C Annam. **BREEDING** Likely to be March–July at least. **Nest** Lightly lined scrape on ground. **Eggs** 6–11; yellowish-buff to yellowish- or reddish-brown, speckled and blotched dark brown to chestnut-brown.

8 RAIN QUAIL *Coturnix coromandelica* Plate 1

IDENTIFICATION 16.5–18.5 cm. Monotypic. **Male** Similar to Common and Japanese Quails but has diagnostic large black breast-patch and large black markings on flanks. **Female** Similar to Common and Japanese but has duller, more uniform, often greyish-tinged breast with more irregular dark spots (rather than blackish streaks) and more pronounced whitish shaft-streaks and unbarred primaries. See Blue-breasted Quail and buttonquail. **Juvenile** Similar to female (both sexes) but perhaps initially more heavily speckled on breast. **VOICE** Male territorial call is a loud sharp metallic ***whit-whit*** or ***which-which***, repeated every 0.5–1 s, in series of 3–5. **HABITAT** Dry grassland and scrub, dry cultivation; up to 1,525 m (mainly lowlands). **RANGE & STATUS** Resident Pakistan, India, Bangladesh. **SE Asia** Scarce to locally common resident (subject to some local movements) Myanmar (except Tenasserim), W,NW,NE(south-west),C,SE,S(Phuket) Thailand, Cambodia, S Annam. **BREEDING** June–October. **Nest** Sparsely lined scrape on ground. **Eggs** 4–11; quite glossy, pale buff to rich brownish-buff, finely and evenly speckled, boldly blotched and freckled or marbled blackish or purplish to olive-brown; 27.9 x 21.3 mm (av.).

9 BLUE-BREASTED QUAIL *Coturnix chinensis* Plate 1

IDENTIFICATION 13–15 cm. *C.c.chinensis*: **Male** Slaty-blue forehead, upper sides of head, breast and flanks and chestnut lower breast to vent diagnostic. Smaller than other quails with darker upperparts (useful in flight), black throat with enclosed white malar patch and broad white crescent on upper breast bordered below by black band. **Female** From other quails by smaller size, duller, more uniform upperparts with much less distinct buff to whitish streaks, buff supercilium, sides of throat and foreneck, distinct blackish bars on breast and flanks; lacks dark eyestripe but has narrow dark cheek-stripe. From buttonquail by combination of face pattern, lack of dark spots on sides of body, stubby bill, uniform brown upperwing (useful in flight) and yellowish legs. **Juvenile** Similar to female but initially (at least when very small) lacks rufous on head and neck and has dense blackish-brown mottling and whitish streaks on breast and flanks. Males soon attain patches of adult plumage. **VOICE** Male territorial call is a sweet whistled ***ti-yu ti-yu*** (sometimes three notes). **HABITAT** Dry to slightly marshy grassland, scrub, cultivation; up to 1,300 m. **RANGE & STATUS** Resident (subject to local movements) India, Nepal, Bangladesh, Sri Lanka, S China, Taiwan, Greater Sundas, Philippines, Wallacea, New Guinea, Bismarck Archipelago, Australia. Introduced Réunion. **SE Asia** Uncommon to common resident (except W Myanmar, S Laos, N,S Annam). Subject to local movements in Peninsular Malaysia at least. **BREEDING** All year. **Nest** Lightly lined scrape on ground. **Eggs** 4–8; pale buffish or buffish-olive to rich sienna, either plain, freckled darker or finely speckled black; 24.5 x 19 mm (av.).

10 HILL PARTRIDGE *Arborophila torqueola* Plate 2

IDENTIFICATION 27.5–30.5 cm. *A.t.batemani*: **Male** Distinguished by chestnut crown and nape, paler and more rufous ear-coverts, black face and eyestripe, black throat and foreneck with white streaks, and narrow white band on upper breast. Upperside similar to Rufous-throated Partridge but with wavy blackish bars. *A.t.griseata* (W Tonkin) is greyer above with more rufous-chestnut crown and nape and slatier-grey breast. **Female** Similar to Rufous-throated but base colour of head-sides, throat and foreneck buffy-rufous (deeper and richer on foreneck), upperparts barred blackish, breast scaled rufous-brown and grey, has more dull chestnut and less grey on flanks. Note habitat and range. **Juvenile** Similar (both sexes) to female but breast spotted buff to whitish, upperparts may be warmer and less olive, has some blackish-brown and chestnut markings on flanks. **VOICE** Territorial call is a loud, drawn-out, mournful whistle, lasting c.1.5 s and repeated every 3 s or so. **HABITAT** Broadleaved evergreen forest; 2,135–3,005 m. **RANGE & STATUS** Resident N,NE Indian subcontinent, S,SE Tibet, SW China. **SE Asia** Locally common resident W,N Myanmar, W Tonkin. **BREEDING** India: April–May. **Nest** Sparsely or well-lined depression in ground. **Eggs** 3–9; glossy, white; 40.6 x 31.9 mm (av.).

11 RUFOUS-THROATED PARTRIDGE *Arborophila rufogularis* Plate 2

IDENTIFICATION 25.5–29 cm. *A.r.tickelli*: **Adult** Told by combination of brown crown with darker streaks, orange-rufous throat, foreneck and sides of neck with black streaks (centre of lower throat/upper foreneck plainer), and plain olive-brown mantle. Lores whitish, supercilium and ear-coverts with black streaks, breast to upper belly greyish, flanks grey with broad chestnut and narrow white streaks, scapulars, wing-coverts and tertials with buffish-grey, chestnut and black markings. *A.r.intermedia* (SW,W,N,S Myanmar) has mostly black throat; *annamensis* (S Annam) has whitish throat and often a black demarcation line across upper breast. See Hill, White-cheeked and Bar-backed Partridges. **Juvenile** Throat plainer and buffier, underparts more uniformly dark grey with white spots and streaks (may have some chestnut and dark flank markings), supercilium buffish; has small black markings but no broad pale grey to buffish-grey markings on scapulars and wing-coverts. Lack of clear dark bars on mantle rules out Hill. **Other subspecies in SE Asia** *A.r.euroa* (N Indochina), *guttata* (C Annam). **VOICE** Territorial call is a long clear plaintive whistle leading into repeated series of double whistles: ***whu-whu whu-whu whu-whu...***, gradually ascending scale and increasing in pitch. Partner may join in with more rapid, monotonous ***kew-kew-kew-kew...*** Continuous, nervous ***whiRR-whiRR-whiRR-whiRR-whiRR-whiRR-whiRR....*** when agitated. **HABITAT** Broadleaved evergreen forest; 1,000–2,590 m. **RANGE & STATUS** Resident N,NE Indian subcontinent, SE Tibet, SW China. **SE Asia** Common resident SW,W,N,S,E Myanmar, northern Tenasserim, W,NW,NE Thailand, Laos, W,E(north-west) Tonkin, N,C,S Annam. **BREEDING** February–August. **Nest** Scrape with roof or leaf pile, on ground. **Eggs** 3–6; glossy, white; 39.2 x 29.8 mm (av.).

12 WHITE-CHEEKED PARTRIDGE

Arborophila atrogularis Plate **2**

IDENTIFICATION 25.5–27 cm. Monotypic. **Adult** Similar to Bar-backed Partridge but has black throat, black foreneck and upper breast with white streaks, slaty-greyish remainder of breast and upper belly, much smaller black-and-white flank markings, usually less chestnut and rufous-chestnut on scapulars and wing-coverts, and narrower, whiter supercilium. See Hill and Rufous-throated Partridges. **Juvenile** Undocumented. **VOICE** Territorial call recalls Bar-backed. Accelerating and ascending series of 8–18 throaty, quavering ***prrrer*** notes, usually followed by similar number of emphatic ***wi-hu*** or ***wa-hu*** couplets, with sharply stressed first syllable. Partner often gives simultaneous stressed, monotonous ***chew-chew-chew-chew...*** **HABITAT** Broadleaved evergreen forest, secondary growth; 610–1,220 m. **RANGE & STATUS** Resident NE India, E Bangladesh, SW China. **SE Asia** Uncommon resident SW,W,N,E(north) Myanmar. **BREEDING** April–June. **Nest** Well-padded scrape on ground. **Eggs** 3–7; fairly glossy, white; 37.6 x 28.4 mm (av.).

13 BAR-BACKED PARTRIDGE

Arborophila brunneopectus Plate **2**

IDENTIFICATION 26.5–29.5 cm. *A.b.brunneopectus*: **Adult** Shows diagnostic combination of broad pale buffish supercilium and upper throat (sometimes with some dark streaks on centre) to ear-coverts, broad black band through eye to sides of neck, black lower throat to uppermost breast with buff streaks, heavy black barring on mantle, warm brown breast (variably tinged rufous) to upper belly, and large black-and-white flank markings. *A.b.henrici* (N,C Indochina) tends to have pale head markings richer buff; *albigula* (S Annam) tends to have whiter head markings. From Rufous-throated Partridge by barring on mantle, warm brown breast, and head and flank pattern; from Scaly-breasted, Annam and Chestnut-necklaced Partridges by heavy black barring on mantle, sharply contrasting wing and flank markings, plain warm brown breast, broad red orbital skin and all-blackish bill. See White-cheeked and Orange-necked Partridges. **Juvenile** Uncertain; may show broken blackish barring across breast. **VOICE** Male territorial call is a series of loud ***brr*** notes, increasing in volume and leading up to separate series of ***wi-wu*** couplets (with stressed first note), which also gradually become louder before ending abruptly: ***brr-brr-brr-brr-brr-brr WI-wu WI-wu WI-wu WI-wu...*** Partner often gives a rapid short ***chew-chew-chew-chew-chew...*** at same time. Alarm call is a quiet, sibilant ***wu-wirr wu-wirr wu-wirr...*** **HABITAT & BEHAVIOUR** Broadleaved evergreen forest; 500–1,525 m; rarely to 1,850 m in S Annam. **RANGE & STATUS** Resident SW China. **SE Asia** Uncommon to locally common resident C(east),S,E Myanmar, Tenasserim, W,NW,NE Thailand, north-east Cambodia, Laos, Vietnam (except Cochinchina). **BREEDING** June–July. **Nest** Well-padded scrape or dome on ground. **Eggs** 4; white; 37.2 x 28.4 mm (av.).

14 MALAYAN PARTRIDGE

Arborophila campbelli Plate **2**

IDENTIFICATION 26–27 cm. Monotypic. **Adult** Resembles Bar-backed Partridge but has largely black head and neck, white sides of forehead (sometimes with an indistinct broken supercilium) and band from base of bill to ear-coverts, some white streaks on sides of nape and foreneck, dark slaty-greyish upper mantle, breast and upper belly, and pale rufous and black flank markings (no white); upperparts darker and browner with less distinct bars. Note range. **Juvenile** Upperparts more chestnut-tinged, breast darker and barred blackish, grey and dull rufous, flanks more heavily marked chestnut, blackish and buffish. **VOICE** Territorial call recalls Bar-backed: a single whistled ***oii*** note repeated monotonously c.13 times every 10 s, usually followed by loud, shrill, whistled ***pi-hor*** couplets with second note nearly an octave lower. Alarm call is a subdued, rapidly repeated ***wut-wit, wut-wit...*** **HABITAT** Broadleaved evergreen forest; 1,000–1,600 m. **RANGE & STATUS** Endemic. Uncommon to fairly common resident extreme S Thailand, Peninsular Malaysia. **BREEDING** March–April. **Nest** Loose pad or leaf dome on ground, beneath palm etc. **Eggs** 2–4; white; 42 x 31.5–32 mm. **NOTE** Recently treated as a subspecies of **Grey-breasted Partridge** *Arborophila orientalis* but shows strong morphological differences consistent with other closely related species in this genus.

15 ORANGE-NECKED PARTRIDGE

Arborophila davidi Plate **2**

IDENTIFICATION 27 cm. Monotypic. **Adult** Like Bar-backed Partridge but has more solid broad black band through eye, continuing down side of neck and across lower foreneck (forming necklace); rest of neck orange with narrow black line down hindneck, has broader whitish supercilium behind eye (sides of forehead greyish), mostly grey lower breast and upper belly, and black flanks with white bars. See Rufous-throated and Scaly-breasted Partridges. Note habitat and range. **Juvenile** Undocumented. **VOICE** Territorial call is an accelerating series of ***prruu*** notes, soon running into a rapid, gradually higher-pitched, plaintive series of up to 70 ***pwi*** notes (lasting up to 23 s); sometimes leading to series of ***wi-wu*** couplets recalling Bar-backed. Also a very rapid series of up to 60 plaintive piping notes (lasting up to 14 s): ***tututututututututututu...*** (recalls some cuckoos). Partner often accompanies with slower, stressed ***tchew-tchew-tchew-tchew...*** Weak, airy ***pher*** or ***phu*** notes when agitated. **HABITAT** Broadleaved evergreen and semi-evergreen forest, bamboo, secondary growth; 140–600 m. **RANGE & STATUS** Endemic. Locally fairly common to common resident border regions of north-east Cambodia to north-east Cochinchina. **BREEDING** Undocumented.

16 CHESTNUT-HEADED PARTRIDGE

Arborophila cambodiana Plate **2**

IDENTIFICATION 28 cm. *A.c.cambodiana*: (south-west Cambodia): **Adult** Dull chestnut head, neck and breast with black crown, nape and post-ocular stripe and broadly black-and-white scalloped flanks diagnostic. Has variable white and black markings across lower breast/upper belly. Upperside similar to Bar-backed Partridge but usually more heavily marked black (sometimes largely black). *A.c.diversa* (SE Thailand) lacks chestnut on head and neck (restricted to patch on breast), black bars on upperparts narrower; black upper foreneck with some white streaks; head pattern more like Bar-backed Partridge. From Bar-backed by chestnut patch on breast, more extensive black-and-white markings on flanks and belly, and less contrasting head pattern. *A.c.chandamonyi* (west Cambodia) is roughly intermediate between other races. See Scaly-breasted Partridge. Note range. **Juvenile** *A.c.diversa* has chestnut of breast paler than adult and washed out, head pattern less contrasting. **VOICE** Said to recall Bar-backed Partridge. **HABITAT** Broadleaved evergreen forest; 200–1,500 m (above 700 m in Thailand, below 1,000 m in Cambodia). **RANGE & STATUS** Endemic. Common to uncommon local resident SE Thailand, west and south-west Cambodia. **BREEDING** May–June. Otherwise undocumented. **NOTE** Subspecies *diversa* and *chandamonyi* included in this species, following Eames *et al.* (2002).

17 CHESTNUT-NECKLACED PARTRIDGE

Arborophila charltonii Plate **2**

IDENTIFICATION 26–32 cm. *A.c.charltonii*: **Adult** Similar to Scaly-breasted and Annam Partridges but has diagnostic chestnut band across upper breast and pale chestnut patch on ear-coverts. Rest of underparts similar to Annam but lower breast and flanks more orange-buff. *A.c.tonkinensis* (E Tonkin, N Annam) has narrower chestnut breast-band, much smaller pale chestnut patch on ear-coverts and less bold blackish markings on lower breast and upper flanks. **Juvenile** Undescribed. **VOICE** Possibly not distinguishable from Scaly-breasted. **HABITAT** Broadleaved evergreen forest; up to 500 m. **RANGE & STATUS** Resident Sumatra, Borneo. **SE Asia** Rare resident south Tenasserim, S Thailand (almost extinct), Peninsular Malaysia. Locally common resident E Tonkin, N Annam. **BREEDING** Undocumented.

18 ANNAM PARTRIDGE

Arborophila merlini Plate **2**

IDENTIFICATION 29 cm. Monotypic (syn. *vivida*). **Adult** Very similar to Scaly-breasted Partridge (particularly *A.c.cognacqi*) but has yellow legs and feet, warm buff base colour to foreneck, and distinct large blackish heart-shaped spots and bar-like markings on lower breast and flanks. From Chestnut-necklaced Partridge by lack of chestnut on upper breast and ear-coverts, warm buff base colour to sides and front of neck, and less orange-buff lower breast and upper belly. See Rufous-throated and Bar-backed Partridges. Note range and habitat. **Juvenile** Undescribed. **VOICE** Possibly not distinguishable from Scaly-breasted. **HABITAT** Broadleaved evergreen forest, secondary growth; up to 600 m. **RANGE & STATUS** Endemic. Local resident C Annam. **BREEDING** Undocumented. **NOTE** Variously lumped with Chestnut-necklaced and Scaly-breasted Partridge by some authors but morphological differences (somewhat intermediate between these two species) suggest it is best kept separate at present.

19 SCALY-BREASTED PARTRIDGE

Arborophila chloropus Plate **2**

IDENTIFICATION 27–31.5 cm. *A.c.chloropus*: **Adult** Rather nondescript, with dull green to pale green legs and feet, and reddish bill with dull greenish-yellow tip. Upperparts, wings and breast-band rather uniform warmish olive-tinged brown with indistinct blackish vermiculations; lower foreneck, lower breast and upper belly orange-buff; rest of underparts pale buffish (whiter in centre) with indistinct blackish and buffish flank markings. *A.c.cognacqi* (S Laos, S Annam, Cochinchina) has distinctly colder, less warm-tinged plumage, whiter foreneck and much duller buff on underparts; *peninsularis* (south W Thailand) has underparts (only) similar to *cognacqi*. See Bar-backed, Orange-necked, Siamese, Annam and Chestnut-necklaced Partridges. Note habitat and range. **Juvenile** Like adult but breast and flank feathers have whitish shafts and tips. **Other subspecies in SE Asia** *A.c.olivacea* (Cambodia, N,C Laos, W Tonkin). **VOICE** Territorial call is a series of c.20–90 variously spaced plaintive notes (often doubled), gradually accelerating and leading to very loud, harsh, shrill series of 5–7 quickly repeated undulating couplets: ***tu-tu....tu-tu....tu-tu-tu..tu-tu.tu-tu-tu-tutututututututu TCHIRRA-TCHWIU-TCHIRRA-TCHWIU-TCHIRRA-TCHWIU-TCHIRRA-TCHWIU-TCHIRRA-TCHWIU...*** Whole series lasts c.15–70 s. May be followed by more spaced encore: ***TCHRA-TWI-TCHRA.*** **HABITAT** Broadleaved evergreen, semi-evergreen and mixed deciduous forest, secondary growth, bamboo; up to 1,000 m. **RANGE & STATUS** Resident SW China. **SE Asia** Common resident N,C,S Myanmar, north Tenasserim, W,NW,NE,SE Thailand, Indochina (except E Tonkin, N,C Annam). **BREEDING** April–June. **Eggs** 3. Otherwise undocumented. **NOTE** Lumped with Chestnut-necklaced Partridge by some authors but exhibits marked morphological differences consistent with other closely related species in this genus.

20 FERRUGINOUS PARTRIDGE

Caloperdix oculea Plate **2**

IDENTIFICATION 27.5–32 cm. *C.o.oculea*: **Male** Combination of chestnut head (sides and throat paler) and breast, blackish post-ocular stripe, and blackish mantle and sides of body with whitish to warm buff scales diagnostic. Rest of upperparts black, with pointed bright rufous markings, centre of lower belly whitish, has large black spots on olive-brown wing-coverts and tertials. **Female** Lacks leg spurs or has single short one. See Long-billed and Black Partridges. **Juvenile** Has black bars on nape and irregular blackish spots and bars on breast. **VOICE** Territorial call is an ascending, gradually accelerating series of high-pitched notes, terminating abruptly with harsher couplets: ***p-pi-pi-pipipipipipi dit-duit dit-duit.*** **HABITAT** Broadleaved evergreen forest, bamboo, freshwater swamp forest; up to 915 m. **RANGE & STATUS** Resident Sumatra, Borneo. **SE Asia** Scarce to uncommon resident south Tenasserim, W,S Thailand, Peninsular Malaysia. **BREEDING** August–September. Otherwise not reliably recorded.

21 CRESTED PARTRIDGE

Rollulus rouloul Plate **1**

IDENTIFICATION 24–29.5 cm. Monotypic. **Male** Unmistakable: glossy blackish plumage (usually bluer on mantle and greener on back to uppertail-coverts), blue-black underparts, large fan-shaped chestnut-maroon crest and red orbital skin, base of bill, legs and feet. Has white patch on centre of crown, long wire-like plumes on forehead and dark brownish wings with warm buffish edges to primaries. See Black Partridge. **Female** Distinctive with deep green plumage, dark grey hood, blackish nape, chestnut scapulars and rusty-brown wings with darker vermiculations. Lacks crest but orbital skin, legs and forehead plumes like male. **Juvenile** Similar (both sexes) to female but has warm brown sides of crown, greyer-tinged mantle, dull greyish belly and vent, less green breast and some pale buffish spots on wing-coverts. **VOICE** Territorial call is a melancholy, upslurring, persistently uttered whistle: ***su-il.*** **HABITAT** Broadleaved evergreen forest; up to 1,220 m. **RANGE & STATUS** Resident Sumatra, Borneo. **SE Asia** Scarce to locally common resident south Tenasserim, W(south),S Thailand, Peninsular Malaysia. **BREEDING** All year. **Nest** Lined hollow or simple scrape on ground. **Eggs** 5–6; white to greyish-white; 35.3–41 x 30.2–32.5 mm.

22 MOUNTAIN BAMBOO-PARTRIDGE

Bambusicola fytchii Plate **2**

IDENTIFICATION 32–37 cm. *B.f.fytchii*: **Male** Distinguished by relatively large size, rather long neck and tail, broad blackish post-ocular stripe, broad chestnut streaks on neck and breast, and pale buff lower breast to vent with large black markings. Has whitish-buff supercilium, rich buff lower sides of head and throat, and greyish olive-brown upperside with bold black and dark chestnut spot-like markings on mantle, scapulars and wing-coverts. Flight-feathers distinctly chestnut (useful in flight). Note habitat and range. **Female** Like male but post-ocular stripe brown. **Juvenile** Similar (both sexes) to female but crown more rufescent, has less rufous-chestnut on hindneck, more buffy-grey upperparts with dark greyish vermiculations and larger, darker markings, and greyer breast with less chestnut and some darker bars. **Other subspecies in SE Asia** *B.f.hopkinsoni* (SW,W Myanmar). **VOICE** Territorial call consists of very loud bouts of explosive shrill chattering (two notes rapidly repeated), starting more loudly and quickly, then slowing and dying away. Often interspersed with harsh, hoarse ***tch-hherrrr*** call-notes. **HABITAT & BEHAVIOUR** Grass and scrub, bamboo, secondary growth, overgrown clearings; 1,200–2,135 m. Usually in small flocks, easily flushed. **RANGE & STATUS** Resident NE India, E Bangladesh, SW China. **SE Asia** Common resident W,N,E Myanmar, NW Thailand, N Laos, W Tonkin. **BREEDING** May–September. **Nest** Roughly lined scrape on ground, amongst grass or bamboo. **Eggs** 3–7; pale buff to deep warm buff; 40.2 x 29.6 mm (av.).

PHASIANIDAE: PHASIANINAE: Pheasants & junglefowl

Worldwide c.48 species. SE Asia 22 species. Terrestrial, generally feeding on invertebrates, grain, buds, seeds, fallen fruit and other vegetable matter. Medium-sized to very large, chicken-like, mostly with long to very long tails (longer on males) and colourful facial skin; show marked sexual dimorphism; shy and unobtrusive, mostly in forest and wooded habitats.

23 BLOOD PHEASANT *Ithaginis cruentus* Plate 2

IDENTIFICATION Male 44–48 cm, female 39.5–42 cm. *I.c.marionae*: **Male** Unmistakable: black sides of head and neck, deep crimson forehead, supercilium and throat, warm buff crown with slaty-grey crest (streaked pale buff), slaty-greyish upperside with whitish streaks and much green on wing-coverts and tertials, crimson upper breast with black streaks, crimson lower breast with pale green streaks, grey upper belly and flanks with green to pale green streaks, crimson undertail-coverts with greyish-white markings, and pinkish-crimson fringes to tail-feathers. **Female** Rather uniform dark brown with extremely fine vermiculations; dull rufous head with darker, pale-streaked ear-coverts and slaty-grey hindcrown and nape and only short crest. **Juvenile** Similar (both sexes) to female but may have some buff speckles on crown, upperparts and wing-coverts, and buff streaks on underparts. Male acquires patches of adult plumage before fully grown. **VOICE** Shrill, high-pitched hissing ***huewerrrr...hieu-hieu-hieu-hieu***. Short sharp ***tchwik*** notes. **HABITAT** Broadleaved and mixed broadleaved/coniferous forest, forest edge, bamboo; 2,590–3,355 m. **RANGE & STATUS** Resident Nepal, NE Indian subcontinent, S,E Tibet, SW,W,NC China. **SE Asia** Uncommon to locally fairly common resident N Myanmar. **BREEDING** NE India: April–July. **Nest** Depression on ground. **Eggs** 5–12; pinkish-buff, profusely speckled and blotched rich brown.

24 BLYTH'S TRAGOPAN *Tragopan blythii* Plate 2

IDENTIFICATION Male 65–70 cm, female 58–59 cm. *T.b.blythii*: **Male** Robust, short-legged and short-tailed with distinctive orange-red crown-sides, neck, upper breast and uppermost mantle, black crown, nape, ear-coverts and upper foreneck, and naked yellow face and throat (can become strongly reddish-tinged). Rest of upperparts, lower belly and (narrowly along) flanks dark brown with round white and chestnut-red spots (larger on wing-coverts) and buff and black vermiculations; lower breast and belly dull brownish-grey with faint dark scales, has yellow extendible throat lappet with blue border, used in display. See Temminck's Tragopan; note range. **Female** Rather nondescript. Upperside slightly greyish-brown (warmer on wing-coverts and tail) with heavy pale buff, warm buff and blackish mottling, speckles and vermiculations, underparts paler greyish-brown with indistinct whitish spot-like markings on belly and flanks, and brown vermiculations on breast and vent (both also with whitish spots/streaks). Eyering yellowish. See Temminck's. **Juvenile** Initially like female (both sexes). Males gradually attain orange-red on neck during first year. **VOICE** Male territorial call is a loud, moaning ***ohh.. ohhah...ohaah...ohaaah...ohaaaha...ohaaaha...ohaaaha...*** **HABITAT & BEHAVIOUR** Oak and mixed oak/rhododendron forest, bamboo; 1,830–2,600 m, locally down to 1,525 m. Very shy. **RANGE & STATUS** Resident E Bhutan, NE India, S Tibet. **SE Asia** Scarce to uncommon resident W,N Myanmar. **BREEDING** April–May. **Nest** Rough bulky structure on cliff ledge or amongst orchids in tree, 7.6 m above ground. **Eggs** 4; pale buff, finely and lightly speckled brown; 58.5 x 44 mm (av.).

25 TEMMINCK'S TRAGOPAN *Tragopan temminckii* Plate 2

IDENTIFICATION Male 62–64 cm, female 55–58 cm. Monotypic. **Male** Unmistakable: overall crimson plumage, marked with pale grey spots. Plain flame-red neck; orange-red crown-sides, upper foreneck and area behind ear-coverts; black-ringed spots on upperside, larger and more broadly black-ringed spots on wing-coverts, and much larger spots on underparts which almost form broad streaks. Head pattern very like Blyth's Tragopan but black (not red) above eye and blue naked skin on face and throat. Extendible throat lappet is dark blue with pale blue spots and light blue at sides, with row of eight red bars at each side. **Female** Very like Blyth's but eyering blue, has more distinct blackish-brown and whitish to buff streaks on lower throat and breast, more distinct whitish spots/streaks on rest of underparts, and usually distinctly rufescent-tinged throat and neck; crown may tend to be blacker with more distinct pale buff streaks. Note range. **Juvenile** Initially like female (both sexes) but may appear warmer with buff throat and buff streaks on body. **Immature** (first-year) males may appear intermediate with adult. See male Blyth's. **VOICE** Male territorial call is a series of 6–9 eerie moaning notes, gradually increasing in length and volume, terminated by a curious nasal grumbling note: ***woh...woah...woaah....waaah....waaah....waaah....waaah.... griiiik***. **HABITAT & BEHAVIOUR** Broadleaved evergreen forest, secondary growth; 2,135–3,050 m. Rather shy. **RANGE & STATUS** Resident NE India, SE Tibet, SW,W,C China. **SE Asia** Scarce to uncommon resident N Myanmar, W Tonkin. **BREEDING Nest** Bulky structure 0.5–8 m above ground. **Eggs** 3–5; buff, finely freckled light brown; 54 x 40 mm (av.).

26 HIMALAYAN MONAL *Lophophorus impejanus* Plate 4

IDENTIFICATION 63–72 cm. Monotypic. **Male** Resembles Sclater's Monal but has long upright crest, uniform dull rufous tail, much greener mantle and glossy purple and turquoise scapulars, wing-coverts, lower rump and uppertail-coverts; white of upperparts restricted to large patch on back. **Female** Like Sclater's but has short erectile crest (often lies flat over nape), much plainer sides of head, all-white throat (contrasts sharply with sides of head and breast), variable whitish streaks on rest of underparts, and narrow whitish band across tips of uppertail-coverts; lower back to uppertail-coverts paler than rest of upperparts but less sharply contrasting, pale tail-tip narrower and buff-tinged (much less conspicuous). **Juvenile** Initially (both sexes) like female but darker above, crown almost uniform, rest of upperparts streaked buff, underparts (including lower throat) more barred and less streaked. **First-winter male** shows similar pattern of adult features to Sclater's. **VOICE** Male territorial call is a series of loud upward-inflected whistles, recalling Eurasian Curlew *Numenius arquata*: ***kur-leiu***; ***kleeh-vick*** etc., alternating with higher-pitched ***kleeh*** notes. Female may give similar calls. Alarm call is a similar ***kleeh-wick-kleeh-wick...***, alternating with shorter, more urgent ***kwick-kwick...*** **HABITAT** Open broadleaved and mixed broadleaved/coniferous forest, forest edge, alpine meadows; recorded at 3,050 m (occurs higher elsewhere). **RANGE & STATUS** Resident NE Afghanistan, N Pakistan, NW,N,NE

Indian subcontinent, S,SE Tibet. **SE Asia** Rare resident extreme N Myanmar. **BREEDING** May–July. **Nest** Scrape on ground, often sheltered by rock or fallen tree. **Eggs** 4–6; pale buffish to reddish-buff with reddish-brown freckles and spots; 63.5 x 44.9 mm (av.).

27 SCLATER'S MONAL
Lophophorus sclateri Plate **4**

IDENTIFICATION 63–68 cm. *L.s.sclateri*. **Male** Large size, robust build, white back to uppertail-coverts, relatively short white-tipped chestnut tail, short curly crest and blue facial skin diagnostic. Rest of upperside glossy, with metallic green crown, reddish-copper sides and rear of neck and shoulders, and greenish to purplish mantle, scapulars, greater coverts and tertials; blackish foreneck and underparts, pale yellowish bill. See Himalayan Monal. **Female** Upperside blackish-brown with warmer brown to buff streaks and other markings, and contrasting blackish and whitish vermiculations on back to uppertail-coverts (often washed buffish); tail blackish with wavy whitish bars and rather broad whitish tip; crown and sides of head distinctly speckled/streaked whitish to buffish; centre of throat whitish, rest of underparts dark brownish, barred and vermiculated pale buffish to whitish. See Himalayan. **Juvenile** Initially (both sexes) like female but upperparts darker with buff streaks. Female soon resembles adult. **First-winter male** shows some black on throat and undertail-coverts and paler back to uppertail-coverts, sometimes a few odd glossy feathers on upperparts. **VOICE** In India, a low, prolonged, slightly quavering whistled ***fuuuuuuuuWHIK***, repeated every few seconds, and a similar but more miaowing, rising then falling whistle. Decelerating series of hard yapping ***yark*** notes when alarmed. **HABITAT** Open broadleaved and mixed broadleaved/coniferous forest, forest edge, scrub and cliffs above treeline, alpine meadows; 2,630–3,960 m. **RANGE & STATUS** Resident NE India, SE Tibet, SW China. **SE Asia** Uncommon local resident east N Myanmar. **BREEDING** April–July (SE Tibet, SW China). **Eggs** 3–5 in captivity. No other information.

28 RED JUNGLEFOWL *Gallus gallus* Plate **3**

IDENTIFICATION Male 65–78 cm (including tail up to 28 cm), female 41–46 cm. *G.g.spadiceus*: **Male** Striking, with long pointed maroon and rufous to golden-yellow neck feathers (hackles), glossy dark green to bluish-green wing-coverts, maroon scapulars and lesser coverts, silvery-maroon back to uppertail-coverts (mixed orange-rufous and green on latter), glossy dark green tail (central feathers long and strongly arched) and blackish underparts. Has red comb, facial skin and lappets. *G.g.gallus* (NE[eastern], SE Thailand, S Indochina) has conspicuous white 'ear-patch' (smaller on female). **Male eclipse** After breeding. Crown and neck all blackish, lacks hackles, tail shorter, comb and lappets may appear shrivelled. **Female** Drab brown above, with fine blackish vermiculations and buffy-white shaft-streaks, short blackish and golden-buff neck-hackles (sides and back), pale chestnut-tinged brown breast and upper belly with whitish-buff shaft-streaks and plain greyish-brown vent; bare pinkish face. From *Lophura* pheasants by size, rather plain plumage with contrasting neck-hackles, shorter, less pointed, unbarred tail, and lack of crest. **Juvenile** Both sexes resemble female. Male has more blackish underparts and darker, plainer upperparts and wings. Female may be slightly richer brown above than adult and show some dark vermiculations on breast. **Other subspecies in SE Asia** *G.g.jabouillei* (east N Laos, W,E Tonkin, N Annam). **VOICE** Territorial call is similar to domestic fowl but higher-pitched, with last syllable cut short. **HABITAT** Forest edge, open woodland, overgrown clearings, scrub and grass; up to 1,830 m. **RANGE & STATUS** Resident N,NE,E Indian subcontinent, SW,S China. **SE Asia** Common resident (except C Thailand, Singapore). Recorded (status uncertain, perhaps introduced) Singapore. **BREEDING** All year. **Nest** Lined scrape on ground, sometimes no nest or construction in low tree fork. **Eggs** 4–9; fairly glossy, whitish to deep creamy-buff; 45.3 x 34.4 mm (av.).

29 KALIJ PHEASANT
Lophura leucomelanos Plate **3**

IDENTIFICATION Male 63–74 cm (including tail up to 35 cm), female 50–60 cm. *L.l.lathami*: **Male** Glossy blue-black plumage (tinted purplish to greenish) with broad white scales on lower back to uppertail-coverts distinctive. Wings and vent mostly glossless blackish-brown, legs and feet greyish to greenish. *L.l.williamsi* (W,C,S[north-west] Myanmar) has dense fine whitish vermiculations on hindneck, upperparts, wings and tail, and less contrasting white scales on rump and uppertail-coverts; *oatesi* (south-east SW and S Myanmar west of Irrawaddy R), *lineata* (S Myanmar east of Irrawaddy R, E[southern] Myanmar, W,NW[western] Thailand) and *crawfurdii* (Tenasserim, south W Thailand) lack white scales on rump and uppertail-coverts, and resemble dark southern forms of Silver Pheasant, but are more densely marked above, and typically have dark greyish to pinkish-brown legs and feet. Intergrades between various subspecies are widespread. Note range. **Female** Resembles Silver but darker above (often more chestnut-tinged) with distinctive greyish-olive scales which become whiter and more defined on wing-coverts; tail black with chestnut-tinged brown central feathers (faintly vermiculated darker), underparts like upperparts but with warm buff to whitish shaft-streaks, legs and feet brown to greenish-grey. *L.l.williamsi* is a shade paler overall, has pale brownish central tail-feathers with indistinct dark vermiculations and black remainder of tail with neat white vermiculations; *oatesi*, *lineata* and *crawfurdii* have less distinct greyish-olive scales/mottling above, paler central tail-feathers with heavier markings, mid-brown to chestnut-brown remainder of tail with varied dark brown to blackish vermiculations, varied blackish-and-whitish streaks to V shapes on hindneck, and largely dull chestnut-brown to blackish (*crawfurdii*) breast and belly with narrow to broad whitish streaks. Note range. **Juvenile** Like female (both sexes) but may show some black bars/spots on scapulars and wing-coverts. Assumes adult plumage during first year. **VOICE** When alarmed gives a high-pitched, watery ***WHiiii***, repeated quite quickly, and a more subdued but quite sharp ***whit-whit-whit...*** **HABITAT & BEHAVIOUR** Mixed deciduous and broadleaved evergreen forest, secondary growth, bamboo; up to 2,590 m. Often encountered in small parties. May be shy. **RANGE & STATUS** Resident NE Pakistan, NW,N,NE Indian subcontinent, adjacent S,SE Tibet. **SE Asia** Common resident Myanmar (except those areas given for Silver), W,NW(western) Thailand. **BREEDING** February–October. **Nest** Scrape on ground, under bush. **Eggs** 4–10; whitish to buff; 47.2 x 36.3 mm (av.). **NOTE** Taxonomy follows provisional findings of McGowan & Panchen (1994) and Moulin *et al.* (2003).

30 SILVER PHEASANT
Lophura nycthemera Plate **3**

IDENTIFICATION Male c.80–127 cm (including tail up to 76 cm; longest in north-east), female 56–71 cm. *L.n.nycthemera* (including *rufipes*, *occidentalis*, *ripponi*, *jonesi*, *beaulieui*; also possibly *berliozi*): **Male** Distinguished by white upperparts, wings and tail with black chevrons and lines, glossy blue-black crest and underparts (tinted purple to greenish), and red facial skin, legs and feet. Shows cline from whitest in north-east to more heavily marked with black in west/south-west. Also, a southern subspecies group comprising *engelbachi* (S Laos), *beli* (C Annam), *annamensis* (C[southern],S Annam), and *lewisi* (SE Thailand, south-west Cambodia) have much broader black chevrons and lines on upperside (dominant over white markings), giving very grey appearance. Note range. **Female** Crown and upperside plain mid-brown (sometimes warm-tinged) with very faint vermiculations (mainly on wings and central tail-feathers), crest blackish (shorter than male), outertail-feathers barred/vermiculated blackish and whitish, underparts broadly scaled white and blackish (throat plainer and whiter). Underparts grade to duller and buffier with browner scales to west/south-west. Birds in E Tonkin (syn. *nycthemera*) and those in southern subspecies group have underparts similar to upperparts (tinged greyer on former to warmer on *engelbachi*) with whitish shaft-streaks and paler throat; tail may be rather plain with finer vermiculations on outer feathers, crest brown (except birds in E Tonkin). *L.n.beli* and *engelbachi* may have blackish and whitish vermiculations on lower breast to vent; *lewisi* has strongly chestnut-tinged upperside with greyish scaling and whitish shaft-streaks on upperparts and wing-coverts, underparts and neck slightly paler and greyer with prominent shaft-streaks. **Juvenile** Both sexes like female but may have some black spots/bars on scapulars and wing-coverts. Males are soon distinguishable and assume adult plumage during first year. **VOICE** When alarmed, throaty grunting ***WWERK*** and ***WWICK*** notes, running to ***WWERK wuk-uk-uk-uk-uk***, combined with very sharp high-pitched ***HSSiik*** or ***SSSiik***. Also similar, rising ***hwiiieeik*** or ***swiiieeik*** and ***hswiiiii***. Contact and foraging calls include throaty ***wutch-wutch-wutch...*** and short throaty ***UWH*** or ***ORH*** notes. **HABITAT & BEHAVIOUR** Broadleaved evergreen and mixed deciduous forest; up to 2,020 m, mainly above 500 m. Often in small flocks; usually shy but locally more confiding. **RANGE & STATUS** Resident southern China. **SE Asia** Uncommon to locally common resident N(south-east),C(east),E(north-eastern) Myanmar, NW(eastern),NE,SE Thailand, Indochina (except Cochinchina?). **BREEDING** February–June. **Nest** Scrape on ground. **Eggs** 4–10; pale to dark rosy-buff; 51 x 39 mm (av.; *nycthemera*). **NOTE** Taxonomy follows provisional findings of McGowan & Panchen (1994) and Moulin *et al.* (2003).

30A 'IMPERIAL PHEASANT'
Lophura x imperialis Plate **3**

IDENTIFICATION Male 75 cm (including tail up to 38 cm), female c.60 cm. **Male** Like Edwards's Pheasant but larger with longer, more pointed tail, short pointed glossy dark blue crest, and somewhat less brilliant purplish-blue tinge to plumage with only slightly greener blue on wing-coverts. **Female** Very similar to Silver Pheasant (*L.n.annamensis*). **NOTE** It has now been demonstrated that Imperial Pheasant is the result of occasional hybridisation between Edwards's and Silver Pheasants (Hennache *et al.* 2003). Only 4 specimens are known from N,C(north) Annam.

31 EDWARDS'S PHEASANT
Lophura edwardsi Plate **3**

IDENTIFICATION 58–65 cm (including male tail up to 26 cm). *L.e.edwardsi* (north-central C Annam): **Male** Told by glossy dark blue (slightly purplish) plumage, short tufted white crest, rather short blunt tail, greenish-blue fringes to wing-coverts and red facial skin, legs and feet. Has blackish vent, primaries and secondaries. *L.e.hatinhensis* (north C and N Annam) differs by having 2–4 white pairs of central tail-feathers. See 'Imperial Pheasant'. **Female** Head, neck, mantle and breast rather plain cold greyish-brown, scapulars and wings warmer-tinged, tail blackish with dark brown, slightly warm-tinged central feathers. Has red facial skin, legs and feet. Smaller, duller and plainer than 'Imperial Pheasant'. See Silver Pheasant. **Juvenile male** Similar to female but with patches of adult plumage; assumes complete adult plumage during first year. **Juvenile female** Like adult but may have black spots/bars on mantle, scapulars and wing-coverts. **VOICE** Alarm call is a low guttural ***uk uk uk uk uk...*** **HABITAT** Broadleaved evergreen forest; below 300 m. **RANGE & STATUS** Endemic. Rare local resident N,C Annam. **BREEDING** In captivity: March–June. **Eggs** 4–7; rosy-buff to creamy-buff; 45 x 36 mm (av.). **NOTE** There is currently no strong evidence to suggest that *hatinhensis* is a distinct species, as previously treated by some authors.

32 CRESTLESS FIREBACK
Lophura erythrophthalma Plate **3**

IDENTIFICATION Male 47–51 cm, female 42–44 cm. *L.e.erythrophthalma*: **Male** Distinguished by blackish plumage (glossed purplish-blue) with fine whitish vermiculations on mantle, upper back, scapulars, wings and sides of breast, and relatively short buffish to warm caramel-coloured tail with blackish base. Has rufous-chestnut lower back, grading to dark maroon uppertail-coverts, red facial skin and greyish legs and feet. See Crested Fireback; note range. **Female** Distinctive. Blackish overall (glossed dark purplish- to greenish-blue) with browner head and paler throat. Tail, centre of abdomen, vent, primaries and secondaries largely glossless. **Juvenile** Both sexes like female but body feathers have rusty-brown tips. **VOICE** Repeated low-pitched ***tak-takrau***. Vibrating throaty purr and loud ***kak*** when alarmed. Low clucking notes when foraging. **HABITAT & BEHAVIOUR** Broadleaved evergreen forest; lowlands. Shy. Wing-whirrs loudly during display. **RANGE & STATUS** Resident Sumatra, Borneo. **SE Asia** Uncommon to locally fairly common resident Peninsular Malaysia. **BREEDING** March–September. **Nest** Scrape on ground, between tree buttresses. **Eggs** 4–6; brownish-white; 46.5–49 x 35–36.5 mm.

33 CRESTED FIREBACK *Lophura ignita* Plate **3**

IDENTIFICATION Male 65–73.5 cm (including tail up to 26.5 cm), female 56–59 cm. *L.i.rufa*: **Male** Unmistakable: blackish with dark purplish- to greenish-blue gloss (except flight-feathers and vent), golden-rufous upper back, grading to maroon rump and shorter uppertail-coverts, white-streaked flanks and strongly arched white central tail-feathers. Has upright crest of tufted shafts, blue facial skin and reddish legs and feet. See Crestless Fireback; note range. **Female** Also distinctive, with dull rufous-chestnut upperside and tail, dark chestnut breast streaked white and mottled blackish, and blackish remainder of underparts with bold white scaling (centre of belly mostly buffish-white). Has fine blackish vermiculations on scapulars, wings and tail; crest shorter and fuller than male, facial skin blue. **Juvenile female** Like adult but lacks crest, may show some black bars on nape, mantle, scapulars and wing-coverts. **Subadult male** Similar to adult but has chestnut central tail-feathers and rufous streaks on upper flanks. **VOICE** Alarm call is continuous guttural ***UKHH-UKHH-UKHH...*** (end of notes more metallic), interspersed with lower ***uur*** notes. **HABITAT & BEHAVIOUR** Broadleaved evergreen forest; up to 200 m, locally to 1,200 m. Display includes loud wing-whirring. **RANGE & STATUS** Resident Sumatra, Borneo. **SE Asia** Uncommon to locally common resident south Tenasserim, S Thailand (rare), Peninsular Malaysia. **BREEDING** July–September. **Nest** Lined scrape on ground, under bush. **Eggs** Up to 5 (captivity); stone-white; 50.6–56.8 x 39.3–43.6 mm.

34 SIAMESE FIREBACK *Lophura diardi* Plate **3**

IDENTIFICATION Male c.70–80 cm (including tail up to 39 cm), female 53–60 cm. Monotypic. **Male** Grey hindneck, mantle and wings and glossy purplish blue-black belly and vent recalls dark subspecies of Silver Pheasant, but has diagnostic crest of pendant-tipped erect bare shafts, golden-buff patch on back, maroon and dark purplish-blue bars on rump and uppertail-coverts, long down-curled glossy blackish blue-green tail, and grey foreneck and breast. Shows fine blackish vermiculations on wings and broad black and narrow white bars on wing-coverts. Note range and habitat. **Female** Told by rufous-chestnut mantle, bold blackish-and-white bars on wings, finer blackish-and-buff bars on rump and uppertail-coverts, rufous-chestnut underparts with white scales on belly and flanks, and mostly dark chestnut outertail-feathers. Central tail-feathers blackish with pale buffish and brown vermiculations, lacks crest. **Juvenile male** Like juvenile female but lacks rufous/chestnut plumage tones; soon attains patches of adult plumage. **Juvenile female** Like adult but may have duller mantle with dark vermiculations and duller base colour to underparts. **VOICE** Alarm calls include short metallic ***tsik tik-tik..tik tik tik...*** etc. and combination of low guttural grunts with more metallic ending and thinner, slightly rising notes: ***UKHT...UKHT...UKHT...***; ***UKHT..hewer...UKHT***. Low ***yurk-yurk*** when flushed. **HABITAT** Broadleaved evergreen and semi-evergreen forest, forest edge, secondary growth; up to 800 m. **RANGE &**

STATUS Endemic. Uncommon to locally common resident NW(eastern),NE,SE Thailand, Indochina (except W,E Tonkin). **BREEDING** April–June. **Nest** Simple affair on ground, sometimes in hollow fallen tree. **Eggs** 4–8; pale buff; 46.7–47.5 x 36.6–38 mm.

35 MRS HUME'S PHEASANT

Syrmaticus humiae Plate **4**

IDENTIFICATION Male 90–92 cm (including tail up to 53.5 cm), female c.60–61 cm. *S.h.burmanicus*: **Male** Unmistakable: red facial skin, glossy dark greyish-purple head, neck, upper mantle, upper breast and inner wing-coverts (crown and ear-coverts browner), mostly dark chestnut lower mantle, underparts and wings (tinged pinkish on mantle and scapulars), white scapular band and tips to greater coverts and tertials, white back to uppertail-coverts with black bars, and long, straight, pointed, pale greyish tail with brown and black bars. **Female** Somewhat smaller; tail relatively short and blunt with mostly chestnut outer feathers, blackish subterminal band and distinctive white tip; upperside brown (tinged greyish) with bold blackish markings (scapulars mostly plain), triangular white streaks on mantle and whitish tips to wing-coverts (forming bars), has pale warm brownish underparts with whiter throat and broad whitish scales from lower breast down; foreneck and upper breast variably spotted/streaked black. See Common, Kalij and Silver Pheasants; note habitat and range. **Subadult male** Differs from adult by showing some female-like wing-feathers (April). **Other subspecies in SE Asia** *S.h.humiae* (Myanmar range west of Irrawaddy R). **VOICE** Male territorial call is a crowing ***cher-a-per, cher-a-per, cher cher, cheria, cheria***. Repeated cackling ***waaak*** notes. Sharp ***tuk tuk*** when alarmed. **HABITAT** More open broadleaved evergreen and mixed broadleaved evergreen and coniferous forest (mainly oak and mixed oak/pine), forest edge, grass and scrub; 1,200–2,285 m. **RANGE & STATUS** Resident NE India, SW China. **SE Asia** Uncommon to locally fairly common resident W,N,C,S(east),E Myanmar, NW Thailand. **BREEDING** March–May. **Nest** Natural hollow in ground, usually under bush, rock or grass clump. **Eggs** 6–10; cream-coloured to rosy-white; 48.7 x 35.3 mm (av.).

36 COMMON PHEASANT

Phasianus colchicus Plate **4**

IDENTIFICATION Male 75–91 cm (including tail up to 59 cm), female 53–62 cm. *P.c.elegans*: **Male** Distinguished by metallic dark green head, neck (tinged purple on sides and front) and breast, generally rufous and chestnut rest of body (vent dull blackish), black streaks on mantle, black scales/bars on sides, and long straight pointed warm brown tail (tinged grey) with black and maroon bars. Has mostly pale greenish-grey wing-coverts, back and rump, the latter two with metallic green scales, and red facial skin. *P.c.rothschildi* (north-west E Tonkin) has more purple on sides and front of neck and sides of breast; *takatsukasae* (north-east E Tonkin) has white ring at base of neck (usually broken at front), variable slight white supercilium, more golden-buff base colour to mantle, paler and greener rump and coppery pinkish-maroon breast with black streaks/scales. **Female** Upperside buffish-brown with distinctly rufous mantle and broad blackish bars and mottling overall, tail similar to male but lacks maroon, central feathers also barred whitish and marked with warm brown, throat whitish, rest of underparts buffish (sides of breast sometimes rufous-chestnut) with blackish scales on breast and flanks. *P.c.takatsukasae* has plainer, more distinctly buff underparts with fewer scales on breast and flanks. See Mrs Hume's and Silver Pheasants. **Juvenile** Both sexes like female but a little duller and less evenly patterned. Male starts to attain adult plumage from first autumn. **VOICE** Male territorial call is a sudden, very loud, resonant ***KO-or OK***, ***korrk-kok*** or ***kok-ok-ok*** etc. Alarm call is a loud ***gogOK gogOK gogOK***... **HABITAT & BEHAVIOUR** Forest edge, secondary growth, grass and scrub, clearings; 1,220–1,830 m. Brief loud wing-whirring may follow male's territorial call. Flies strongly when flushed. **RANGE & STATUS** Resident south-east W Palearctic, C Asia, SE Siberia, Ussuriland, China, Taiwan, N,S Korea, Japan. Introduced Europe, Australia, New Zealand, Hawaii, N America. **SE Asia** Local resident N(east),E(north-east) Myanmar, E Tonkin. **BREEDING** No information on season. **Nest** Lined depression on ground. **Eggs** 7–14; olive-brown; c.45 x 36 mm.

37 LADY AMHERST'S PHEASANT

Chrysolophus amherstiae Plate **4**

IDENTIFICATION Male 130–173 cm (including tail up to 117 cm), female 63.5–68 cm. Monotypic. **Male** Unmistakable: black-scaled white neck-ruff, buffy-yellow rump with dark bluish-green bars and very long, arched and pointed white tail with black bars and broad vermiculations. Has glossy dark green crown, fine pointed red crest extending from hindcrown, blue-tinted dark green mantle, back, scapulars and breast with black scales, black-and-white uppertail-coverts mixed with orange to red (long, pointed orange tips to longer feathers) and mostly blackish throat and foreneck. **Female** Resembles Common Pheasant but has blackish-brown and warm buff bars on upperside (rump and uppertail-coverts plainer), white chin/upper throat, rufous-buff foreneck and upper breast with narrow blackish-brown bars, and buffish-white remainder of underparts with broad blackish-brown bars on flanks. Has distinctly chestnut sides of crown (above eye) and buff bars on primaries and secondaries. See Mrs Hume's Pheasant. **Juvenile** Like female (both sexes) but lacks chestnut on crown, has duller dark bars on crown, upperparts and flanks, buff and dark grey bars on tail and narrower buff bars on primaries and secondaries. Male starts to attain adult plumage from first autumn. **VOICE** Territorial call of male is a loud, hoarse, slightly metallic and grating ***hirk hik-ik***, repeated after rather lengthy intervals and interspersed with 1–3 quick ***hwik*** notes (possibly used as introduction or encore). **HABITAT & BEHAVIOUR** Open broadleaved evergreen forest, secondary growth, bracken-covered hillsides; 1,525–2,440 m. Shy and retiring but males call from trees in early morning. **RANGE & STATUS** Resident extreme SE Tibet, SW,W China. Introduced British Isles. **SE Asia** Uncommon local resident N(east),E(north-east) Myanmar. **BREEDING** No information on season. **Nest** Scantily lined scrape on ground, often under bush. **Eggs** 6–11; glossy, buff to creamy-white; 42–52 x 32–37 mm.

38 MOUNTAIN PEACOCK-PHEASANT

Polyplectron inopinatum Plate **4**

IDENTIFICATION Male c.65 cm (including tail up to 40 cm), female c.46 cm. Monotypic. **Male** Only likely to be confused with Malayan Peacock-pheasant but note habitat. Differs by lack of obvious crest or pale facial skin, much smaller dark bluish ocelli on upperparts and wings, which are bordered at base by whitish to buff spots, strongly chestnut-tinged lower mantle and scapulars to uppertail-coverts, darker tail with no obvious rings around dark green ocelli, dark greyish head and neck with whitish speckles (particularly on throat), and blackish underparts with some fine grey vermiculations (browner on undertail-coverts). See female Crested Argus. **Female** Similar to male but ocelli smaller and black, tail shorter and less graduated with almost no ocelli. **Juvenile** Similar to female (both sexes). **VOICE** Male territorial call is a series of 1–4 fairly loud harsh clucks or squawks (about 0.5 s apart), repeated every 5–6 s. **HABITAT & BEHAVIOUR** Broadleaved evergreen forest; 800–1,600 m, reported down to 600 m (1,200–1,400 m in Thailand). Very shy. **RANGE & STATUS** Endemic. Uncommon resident Extreme S Thailand, Peninsular Malaysia. **BREEDING** January–June. **Nest** Scrape on ground. **Eggs** 2; pale cream-terracotta; 51.3–53 x 43.8–44 mm.

39 GERMAIN'S PEACOCK-PHEASANT

Polyplectron germaini Plate **4**

IDENTIFICATION Male 56–60 cm (including tail up to 33.5 cm), female c.48 cm. Monotypic. **Male** Like Grey Peacock-pheasant but somewhat smaller and darker with finer, denser pale markings and darker ocelli with more greenish-blue (often appear darker purple), has much darker head and neck with more distinct pale bars, dull blood-red facial skin, no crest, and white of head restricted to upper throat. Note range and habitat. **Female** Like Grey but distinctly darker and more uniform overall, ocelli on upperparts and wings smaller, rather pointed and much more defined, no obvious pale scales on upperparts, more distinct pale bars/speckles on head, white of head restricted to upper throat. **Juvenile** Both sexes like female but ocelli blacker and less defined; indistinct pale scales on scapulars and wing-coverts, underparts more uniform. **VOICE** Male calls with a series of low purring or growling rattles (2–5 s long): ***erraar-rrrrrr***, repeated after 3–11 s intervals. These change to repeated series of 4–7 much louder and harsher cackling notes, gradually becoming louder and angrier, when responding to rival: ***erraarrrrrakak... aarrrr-akh-akh-akh-akh... AKH-AKH-AKH-AKH...*** Generally similar to Grey but somewhat faster and higher-pitched. Rarely, a distressed rising then falling, shrill yelping ***hwii hwer***. **HABITAT** Broadleaved evergreen and semi-evergreen forest, bamboo; up to 1,400 m. **RANGE & STATUS** Endemic. Locally common resident north-east Cambodia, C(south),S Annam, Cochinchina. **BREEDING** March–April. **Eggs** 1–2; creamy-white; 43.8–45.4 x 33.4–34.6 mm (2 in wild).

40 GREY PEACOCK-PHEASANT
Polyplectron bicalcaratum Plate **4**

IDENTIFICATION Male 56–76 cm (including tail up to 42 cm), female 48.5–53 cm. *P.b.bicalcaratum*: **Male** Told by dark greyish-brown plumage with fine whitish to buff bars and speckles, whitish throat (faintly speckled/barred darker) and glossy dark green and pinkish-purple ocelli on mantle, wings and tail (greener on latter). Has bushy-crested crown and pale yellowish to pinkish facial skin; ocelli are variably ringed whitish to buff (more grey-buff on tail). See other peacock-pheasants; note habitat and range. **Female** Somewhat smaller, darker and plainer with less distinct blackish ocelli (much reduced on tail), whitish-buff scales on mantle, scapulars and wing-coverts, and duller facial skin. See Crested Argus. **Juvenile** Both sexes like female. Male develops more defined and colourful ocelli during first winter and assumes adult plumage by second winter. **Other subspecies in SE Asia** *P.b.ghigii* (W,E Tonkin, N,C[northern] Annam). **VOICE** Male territorial call is a very loud, airy ***PU PWOI*** or ***POI PWOI*** (second note more drawn and rising), repeated up to 5 times. More commonly gives a long series of low rolling or growling rattles (c.2 s long): ***uhrrrrr***, starting very quietly and repeated after 3–4 s intervals. These gradually get louder, then run into series of 3–6 much louder and harsher cackling notes, when responding to rival: ***uhrrrrr uhrrrrruk orrokhokhokhokh OKH-OKH-OKH-OKH-OKH ORKH-ORKH-ORKH ORKH-ORKH-ORKH...*** **HABITAT** Broadleaved evergreen forest; up to 2,320 m. **RANGE & STATUS** Resident NE Indian subcontinent, SW China. **SE Asia** Uncommon to locally fairly common resident Myanmar, W,NW,NE,S(north) Thailand, north-east Cambodia, Laos, W,E Tonkin, N,C(northern) Annam. **BREEDING** March–April. **Nest** Crudely lined scrape or natural depression on ground. **Eggs** 2–6; pale cream-coloured to rich dark buff; 46.5 x 35.8 mm (av.).

41 MALAYAN PEACOCK-PHEASANT
Polyplectron malacense Plate **4**

IDENTIFICATION Male 50–53.5 cm (including tail up to 25.5 cm), female 40–45 cm. Monotypic. **Male** Similar to Grey Peacock-pheasant but warmer brown (particularly above) with greener ocelli, long dark green-glossed crest, blacker crown and hindneck, darker ear-coverts (contrasting with pale surrounding area) and plainer underparts with variable black and buffish vermiculations on upper breast. Has orange-pink facial skin. See Mountain Peacock-pheasant; note habitat and range. **Female** Somewhat smaller and shorter-tailed, very short crest, blacker and more pointed ocelli, indistinct paler to buff scales on upperparts, more uniform underparts. **Juvenile** Both sexes like female. Male as adult by first summer but with darker ocelli, plainer breast and no gloss on crest. **VOICE** Male territorial call (dawn and dusk) is a repeated loud, clear, slow melancholy ***PUU PWOII*** or ***PUU PWORR*** (second note more drawn-out and rising). Typically gives sudden harsh explosive cackle, running into long series of throaty clucks: ***TCHI-TCHI-TCHAO-TCHAO wuk-wuk-wuk-wuk-wuk...*** Also, repeated loud harsh grating ***TCHOW*** or ***KAAOW***. **HABITAT & BEHAVIOUR** Broadleaved evergreen forest; up to 305 m. Holds crest forwards over bill during display. **RANGE & STATUS** Endemic. Uncommon to locally common resident south Tenasserim, S Thailand (possibly extinct), Peninsular Malaysia. **BREEDING** March–September. Nest Slight scrape among leaves on ground or weathered old termite mound. Eggs 1; off-white; 49 x 40 mm.

42 CRESTED ARGUS *Rheinardia ocellata* Plate 4

IDENTIFICATION Male 190–239 cm (including tail up to 175 cm), female 74–75 cm. *R.o.nigrescens*: **Male** Unmistakable, with blackish-brown crown and sides of head, buff supercilium and throat, long drooping blackish-brown and white crest, blackish-brown upperside peppered with small white spots, and incredibly long, broad (laterally compressed) and pointed tail; base colour of neck, underparts and tail browner, less blackish than above. *R.o.ocellata* (Indochina) has shorter, fuller, mostly brownish crest (some white), white supercilium and upper throat, dull chestnut-tinged lower throat and foreneck, more numerous, smaller and buffier markings on upperside, generally paler and more uniform upperparts and wings, and more dark chestnut and grey (less blackish) on tail. See Great Argus. **Female** Smaller and much shorter-tailed. Head similar to male but lacks long white feathers in crest; overall plumage warm-tinged dark brown with blackish and warm buff bars, speckles and vermiculations (mantle plainer, wings and tail more heavily barred); underparts somewhat paler and very finely vermiculated darker, warmer on foreneck and upper breast. *R.o.ocellata* has white supercilium and upper throat, plumage somewhat greyer-tinged with finer, more intricate markings. See *Lophura* pheasants, peacock-pheasants and Great Argus. **Juvenile** Both sexes initially similar to female. Male shows some adult features at early age and assumes adult plumage after a year but with shorter tail; full adult plumage in third year. **VOICE** In Indochina, territorial call at dancing ground is a very loud, resonant, trisyllabic ***WOO'O-WAO*** or ***WUUUA-WAO*** (first part rising, last part louder and more resonant), usually given singly but sometimes twice in quick succession. More commonly heard (male only?) giving series of 4–7 loud, far-carrying, disyllabic ***oowaaaa*** or ***oowaaau*** calls, with variations including ***woyawaa***, ***woyaaa*** and ***waaaaauu*** (rising at end). Distress or alarm call is a repeated, yelping ***pook*** or ***puwoo***. Populations in Peninsular Malaysia are thought to have similar vocalisations but have not been compared directly. **HABITAT & BEHAVIOUR** Broadleaved evergreen forest; 790–1,100 m in Peninsular Malaysia; up to 1,500 m in Indochina but 1,700–1,900 m only in S Annam. Extremely shy and adept at avoiding detection. Creates long bare dancing grounds on forest floor. Male has simple display involving ruffling of head feathers and spreading of crest. **RANGE & STATUS** Endemic. Scarce and very local resident Peninsular Malaysia. Scarce to locally common resident C,S(east) Laos, N,C,S Annam. **BREEDING** April–August. **Nest** Scrape on ground. **Eggs** 2; deep pinkish-buff with fine purplish-brown spots; 63–66 x 45–47 mm (*ocellata*).

43 GREAT ARGUS *Argusianus argus* Plate 4

IDENTIFICATION Male 160–203 cm (including tail up to 145 cm, secondaries up to 102 cm), female 72–76 cm. *A.a.argus*: **Male** Told by warm brown plumage, naked blue sides of head, throat and foreneck, black crown with short dark crest at rear, very long laterally compressed tail (greyer along top and spotted whitish) and extremely long secondaries with large round ocelli. Upperparts and wings finely speckled and mottled whitish, buffish and blackish, back and rump pale warm brown with dark brown spots and streaks; underparts dark chestnut (lower foreneck plainer and more rufous, vent duller and plainer) with wavy blackish and whitish bars. See

Crested Argus; note habitat and range. **Female** Head and neck similar to male, upperpart markings less distinct and buff, back and rump as mantle and more vermiculated than speckled, has complete rufous-chestnut collar, rest of underparts duller and plainer with fine blackish vermiculations, tail much shorter and blackish-brown with paler warm brown to buff markings, no elongated secondaries; primaries dark chestnut with indistinct dark brown markings. See Crested Argus. **Juvenile** Both sexes like female. Male soon develops longer tail with numerous tiny white spots. **VOICE** Male territorial call is a very loud, far-carrying ***KWAH-WAU*** with louder and longer second note. Female gives a series of 25–35 loud ***WAU*** notes, the last few becoming progressively longer and more upward-inflected. **HABITAT & BEHAVIOUR** Broadleaved evergreen forest; up to 950 m. Very shy. Makes similar dancing grounds to Crested but male has elaborate display during which the secondaries are raised and fanned towards female. **RANGE & STATUS** Resident Sumatra, Borneo. **SE Asia** Uncommon to locally common resident south Tenasserim, S Thailand, Peninsular Malaysia. **BREEDING** March–August. **Nest** On ground between tree buttresses, sometimes leaf-filled concave top of tree stump 1.5 m above ground. **Eggs** 2; pale buff to creamy-white; 61.5–66.3 x 45–45.7 mm.

44 GREEN PEAFOWL *Pavo muticus* Plate 4

IDENTIFICATION Male 180–250 cm (including 'train' up to 162 cm), female 100–110 cm. *P.m.imperator*: **Male** Unmistakable, very large: long neck, long upright crest and extremely long broad train (elongated uppertail-coverts) covered with large colourful ocelli. Plumage largely brilliant glossy green, scaled blackish on upperparts and wings, crown and crest darker and bluer-green, shoulders blue, throat largely blackish, remainder of underparts dark brown, tinged green on lower breast and flanks, wings mostly blackish-brown (tinged green) with contrasting caramel-coloured primaries; facial skin yellow and blue, with black loral stripe. Train is lacking during post-breeding moult. **Female** Head, neck, breast and wings similar to male but duller, upperparts and tail blackish-brown with pale buffish bars and vermiculations, long uppertail-coverts (not reaching tail-tip, i.e. no train) mixed green and bronze, rest of underparts like male; facial skin duller, with reddish-brown loral stripe. **Juvenile** Similar to female (both sexes) but duller. During early stages whole head feathered, lores black, has large whitish patch around eye to throat. Immature males develop more golden-green colour on back to uppertail-coverts during first year. **Second-year male** Similar to adult but train much shorter and lacks ocelli; maximum length may not be attained until about fifth year. **Other subspecies in SE Asia** *P.m.spicifer* (Myanmar west of Irrawaddy R), *muticus* (Tenasserim and S Thailand southward). **VOICE** Male territorial call is a very loud, far-carrying ***KI-WAO*** or ***YEE-OW***, often repeated: ***YEE-OW..KI-WAO KI-WAO KI-WAO...*** Female gives loud ***AOW-AA***, with emphasis on first syllable, often repeated after short intervals. Most vocal at dawn and dusk. **HABITAT & BEHAVIOUR** Open mixed deciduous, broadleaved evergreen and semi-evergreen forest, particularly along rivers and bordering wetlands, forest edge, secondary growth, bamboo; up to 915 m. Rather shy but males often call from trees in early morning. During display, male's train is raised to vertical and fanned. **RANGE & STATUS** Resident SW China, Java; formerly reported NE India and Bangladesh (but no specimens). **SE Asia** Very local resident W,N,C,S Myanmar, W,NW Thailand, Cambodia, N,S Laos, N,C,S Annam, Cochinchina. Formerly resident (current status uncertain) SW,E Myanmar, Tenasserim, C Laos. Formerly resident (now extinct) NE,S Thailand, Peninsular Malaysia. **BREEDING** March–October. **Nest** Lined scrape on ground. **Eggs** 3–8; glossy, pale fawn-coloured to warm buff, sometimes freckled darker buff or reddish-brown; 72.7 x 53.5 mm (av.; *spicifer*).

ANATIDAE: DENDROCYGNINAE: Whistling-ducks

Worldwide 9 species. SE Asia 3 species. Similar to typical ducks but relatively long-necked, long-legged and short-bodied with broad rounded wings. Feed on aquatic vegetation, grain, small fish, amphibians, snails, worms etc.

45 FULVOUS WHISTLING-DUCK *Dendrocygna bicolor* Plate 6

IDENTIFICATION 45–53 cm. Monotypic. **Adult** Similar to Lesser Whistling-duck but larger, head and underparts richer dark rufous, crown only slightly darker than sides of head, has pronounced blackish line down centre of nape and hindneck, white neck-patch with fine blackish streaks, white uppertail-coverts, more prominent whitish flank markings, and more uniform upperwing with darker coverts. **Juvenile** Somewhat duller and greyer with greyish uppertail-coverts and less pronounced neck and flank markings. **VOICE** Thin whistled ***k-weeoo*** (often repeated); mostly given in flight. Harsh ***kee***. **HABITAT & BEHAVIOUR** Lowland lakes, large rivers and marshes. Gregarious but generally in smaller flocks than Lesser. **RANGE & STATUS** Resident (subject to relatively local movements) south N America, S America, sub-Saharan Africa, Madagascar, S Pakistan, N,NE and peninsular India (mainly NE). Winter visitor Bangladesh. **SE Asia** Rare resident (subject to local movements) S Myanmar. Formerly resident (current status unknown) C Myanmar. Recorded (status uncertain) SW,N,E Myanmar, Tenasserim, Cochinchina. **BREEDING** June–September. **Nest** Simple structure in tree-cavity or old nest of other bird, sometimes on ground amongst vegetation. **Eggs** 6–8; ivory-white; 56.6 x 42.9 mm (av.).

45A WANDERING WHISTLING-DUCK *Dendrocygna arcuata* Plate 6

IDENTIFICATION 40–45 cm. *D.a.arcuata*: **Adult** Similar to Lesser Whistling-duck but a little larger, blackish-brown of forecrown extends down to eye-level, has pronounced blackish line down centre of nape and hindneck, richer chestnut flanks with prominent large black and white markings, white outer uppertail-coverts and duller upperwing-coverts. **Juvenile** Duller overall with paler belly and less distinct flank pattern. **VOICE** Cheerful, high-pitched, multisyllabic twittering ***pwit-wit-ti-t-t-t...*** and high-pitched whistles; mainly when flying. **HABITAT & BEHAVIOUR** Freshwater lakes and marshes, various wetlands. Gregarious, often in large flocks. **RANGE & STATUS** Resident (subject to relatively local movements) Sundas, Philippines, Sulawesi, Sula Is, New Guinea, N Australia, New Britain; formerly New Caledonia, Fiji. **SE Asia** Scarce feral resident Singapore. **BREEDING** No information on season. **Nest** Well-lined depression on ground, amongst vegetation, sometimes in tree-hole. **Eggs** 6–15; cream-coloured.

46 LESSER WHISTLING-DUCK *Dendrocygna javanica* Plate 6

IDENTIFICATION 38–41 cm. Monotypic. **Adult** Common and widespread resident. Combination of rather long neck, relatively dull, plain plumage and broad, rounded, dark wings with reddish-chestnut lesser and median upperwing-coverts distinctive. Has dark cap, prominent rufous fringes to mantle feathers and scapulars, and reddish-chestnut uppertail-coverts. Centre of nape and hindneck only slightly darker than sides of neck; only indistinct thin whitish flank-streaks. **Juvenile** Somewhat duller overall, crown often paler and more greyish-brown. **VOICE** Incessantly repeated, clear, low whistled ***whi-whee***, usually when flying. Wings also make whistling sound in flight. **HABITAT & BEHAVIOUR** Lakes, marshes, sometimes mangroves, various wetlands; up to 1,450 m. Very gregarious, often in large flocks. **RANGE & STATUS** Resident (subject to relatively local movements) Indian subcontinent, Greater Sundas, W Lesser Sundas. Breeds southern China, formerly Taiwan, Japan (Nansei Is); some populations winter to south/south-west. **SE Asia** Locally common resident, subject to some movements (except W Tonkin, N Annam); scarce in Singapore. **BREEDING** All year. **Nest** Simple structure in tree-cavity or old nest of other bird, up to 6 m above ground, sometimes on ground amongst vegetation. **Eggs** 7–12; ivory-white, becoming brownish-stained; 46.9 x 36.8 mm (av.).

ANATIDAE: ANSERINAE: Geese & allies

Worldwide c.28 species. SE Asia 6 species. Characterised by broad, flattened bills, fairly long necks, bulky bodies, short tails and short strong legs with webbed feet; most have rather pointed wings. Sexes do not differ markedly. Flight strong and direct with neck outstretched. Large and gregarious with relatively dull plumage, generally associated with wetland habitats, mostly feeding on aquatic plants and animals.

47 **SWAN GOOSE** *Anser cygnoides* Plate **5**
IDENTIFICATION 81–94 cm. Monotypic. **Adult** Resembles Greylag Goose but bill thicker-based, more triangular and blackish; uniform dark brown crown, nape and hindneck contrasts strongly with very pale creamy-brownish lower sides of head, throat and foreneck. Has whitish band from lores across forehead, bordering base of bill. In flight, wing pattern very similar to Greater White-fronted Goose. See juvenile Bar-headed Goose. **Juvenile** Crown, nape and hindneck duller, no whitish face-band. **VOICE** Prolonged resounding honks, ending at higher pitch. Short harsh note repeated 2–3 times in alarm. **HABITAT** Banks of large rivers, marshy edges of freshwater wetlands; recorded at 450 m. **RANGE & STATUS** Breeds NE Kazakhstan, Mongolia, S,SE Siberia, NE China; winters C,E China, Taiwan, N,S Korea. **SE Asia** Vagrant NW Thailand, N Laos (same record).

48 **TAIGA BEAN-GOOSE** *Anser fabalis* Plate **5**
IDENTIFICATION 75–90 cm. *A.f.middendorffii*: **Adult** Combination of black bill with pale orange subterminal band, rather dark brown head and neck, and orange to yellow-orange legs and feet diagnostic. Rest of plumage similar to Greylag Goose, but in flight shows darker upperwing-coverts and uniform dark underwing. See Greater and Lesser White-fronted Geese. **Juvenile** Has duller orange on bill and duller legs and feet. **VOICE** Relatively quiet. May give fairly deep ***hank-hank*** or ***wink-wink*** in flight. **HABITAT & BEHAVIOUR** Large rivers, lakes; recorded in lowlands. Usually gregarious; may join other geese. **RANGE & STATUS** Breeds northern Palearctic; winters Europe, Iran, NW,C,E China, N,S Korea, Japan. **SE Asia** Vagrant N Myanmar.

49 **GREYLAG GOOSE** *Anser anser* Plate **5**
IDENTIFICATION 78–90 cm. *A.a.rubrirostris*: **Adult** The most likely goose to be seen in the region. Distinguished by combination of pink bill, legs and feet and relatively pale brownish-grey plumage with plain head and neck and no black belly-patches (has some dark speckles). In flight, shows contrastingly pale upperwing-coverts and paler underwing-coverts. See juvenile Greater and Lesser White-fronted Geese and Taiga Bean-goose. **Juvenile** Less prominent pale fringing on upperparts, flanks and belly, lacks dark speckling on belly, bill and legs somewhat duller. **VOICE** Relatively noisy. In flight, utters a loud series of clanging honking notes: ***aahng-ahng-ung*** or ***ank-ang-ang***, which are deeper than other geese. **HABITAT & BEHAVIOUR** Lakes, rivers, estuaries, arable fields, grassy areas; lowlands. Gregarious, usually found in flocks, may associate with other geese. **RANGE & STATUS** Breeds W,C Palearctic, S,SE Siberia, Ussuriland, Mongolia, northern China; winters N Africa, W,S Europe, Turkey, Middle East, Indian subcontinent (except south), southern China, N,S Korea; rarely Japan. **SE Asia** Rare to local winter visitor SW,W(north),N,C Myanmar, E Tonkin; formerly N,C Annam. Vagrant NW Thailand, C Laos.

50 **GREATER WHITE-FRONTED GOOSE**
Anser albifrons Plate **5**
IDENTIFICATION 65–75 cm. *A.a.albifrons*: **Adult** Shows distinctive combination of pinkish bill, broad white patch on forehead and lores (around base of bill), irregular black patches on belly, and orange to yellow-orange legs and feet. In flight from Greylag Goose by darker upperwing-coverts and uniform dark underwing. See very similar Lesser White-fronted Goose. **Juvenile** Lacks white on forehead/lores and black on belly. From Greylag by smaller size and bill, darker overall plumage, particularly head and neck, and more yellow or orange legs and feet. See Lesser White-fronted and Taiga Bean-goose. **VOICE** Fairly noisy: in flight a repeated musical ***lyo-lyok***, variable in pitch but higher than Greylag and Taiga Bean-. **HABITAT & BEHAVIOUR** Lakes, rivers, grain fields, grassy areas; lowlands. Gregarious in its usual range, may flock with other geese. **RANGE & STATUS** Breeds N Holarctic; winters W,S Europe, south-east W Palearctic, Middle East, NE,C,E China, N,S Korea, Japan, N(south),C America; rarely northern Indian subcontinent. **SE Asia** Vagrant W,E Myanmar.

51 **LESSER WHITE-FRONTED GOOSE**
Anser erythropus Plate **5**
IDENTIFICATION 53–66 cm. Monotypic. **Adult** Very similar to Greater White-fronted Goose, but somewhat smaller in size, with rather shorter body, neck and legs, smaller, shorter and brighter pink bill, distinct yellow eyering (visible at fairly close range) and more rounded head (forecrown steeper and higher), the white frontal patch extending slightly further onto forecrown and ending in more of a point. Longer wings project noticeably beyond tail-tip at rest (not or only slightly projecting in Greater); generally darker overall (particularly head and neck) with smaller black belly-patches (usually) and clearer white line along inner edge of flanks. **Juvenile** Best told from Greater by size, proportions, primary projection and darker overall coloration. **VOICE** Flight calls are quicker, higher and squeakier than in Greater, typically including repeated ***kyu-yu-yu***. **HABITAT & BEHAVIOUR** Freshwater lakes, marshes; probably recorded at 800 m. Feeds and walks faster than Greater.**RANGE & STATUS** Breeds N Palearctic; winters south-eastern W Palearctic, Iran, Turkmenistan, C,E China; rarely northern Indian subcontinent, N,S Korea, Japan (formerly common). **SE Asia** Vagrant E Myanmar.

52 **BAR-HEADED GOOSE** *Anser indicus* Plate **5**
IDENTIFICATION 71–76 cm. Monotypic. **Adult** Unmistakable, with two black bands on back of white head, broad white line down side of neck, yellow bill with dark tip and orange-yellow legs and feet. Upper foreneck and upper hindneck are blacker than lower neck. In flight, wing pattern similar to Greylag Goose but more uniform pale grey upperwing-coverts contrast sharply with blacker flight-feathers. **Juvenile** Hindcrown to back of neck dark greyish-brown, has dusky line on lores, rest of neck more uniform pale greyish. See Swan Goose. **VOICE** Soft, nasal, repeated honking: ***oh-wa***, ***aah-aah*** and ***ooh-ah*** etc. Notes somewhat lower, more nasal and wider-spaced than other geese. **HABITAT & BEHAVIOUR** Large rivers, lakes, arable fields, grassy areas; up to 400 m. Usually in small flocks. **RANGE & STATUS** Breeds C Asia, NW India (Ladakh), Tibet, Mongolia, NW,W,N China; winters Indian subcontinent (except islands), S Tibet. **SE Asia** Rare to scarce winter visitor SW,W,N(locally common),C,S Myanmar. Vagrant north Tenasserim, NW Thailand and N Laos (same record), E Tonkin.

ANATIDAE: TADORNINAE: Comb Duck, shelducks & allies

Worldwide 23 species. SE Asia 3 species. As for Anserinae species. Fairly large with generally striking plumage; feed on land and in water.

53 COMB DUCK *Sarkidiornis melanotos* Plate 5

IDENTIFICATION 56–76 cm. *S.m.melanotos*: **Male non-breeding** Relatively large size, mostly whitish head, neck and underparts, and contrasting blackish upperside distinctive. Has prominent knob ('comb') on top of bill, blackish crown and hindneck, black speckles on rest of head and neck, green, bluish-purple and bronze gloss on upperparts, wing-coverts and secondaries, narrow black bar on sides of breast and grey-washed flanks. In flight, all-dark wings contrast sharply with pale underside. See White-winged Duck and Common Shelduck. **Male breeding** Much larger knob on top of bill, rich buffish wash to sides of head and neck. **Female** Much smaller, upperside duller and less glossy, lacks knob on bill. **Juvenile** Crown and eyestripe dark brown, upperparts dark brown with warm buff feather-fringes, rest of head, neck and underparts strongly washed brownish-buff with some dark brown markings on sides of breast and flanks. **VOICE** Occasionally utters low croaking sounds. Also wheezy whistles and grunts during breeding season. **HABITAT & BEHAVIOUR** Freshwater lakes and marshes; lowlands. Found singly, in pairs or small flocks, often associating with other ducks. Feeds on land and in water. **RANGE & STATUS** Resident (subject to some movements) sub-Saharan Africa, Madagascar, India, S Nepal, Bangladesh, S America; formerly S Pakistan, Sri Lanka. **SE Asia** Rare to scarce resident (subject to some movements) N,C,S Myanmar, Cambodia. Former resident (current status unknown) SW,E Myanmar). Rare winter visitor NW,NE,C Thailand, Cochinchina. Vagrant C Laos, E Tonkin. **BREEDING** June–September. **Nest** Cavity in tree, bank or wall, rarely old nest of other bird. **Eggs** 7–15; ivory white to pale cream, becoming stained; 61.8 x 43.3 mm (av.).

54 COMMON SHELDUCK
Tadorna tadorna Plate 5

IDENTIFICATION 58–67 cm. Monotypic. **Male** Unmistakable, with white body and contrasting black head, neck, scapular and ventral stripes and broad chestnut breast-band, extending to upper mantle. Bill red with prominent knob at base, head and upper neck glossed bottle-green. In flight, wing pattern very similar to Ruddy Shelduck. **Male eclipse** Has smaller knob on bill, whitish mottling on face, less sharply defined breast-band and fine greyish bars on body. **Female** Somewhat smaller, bill duller with no knob, head and neck duller, face marked with white near base of bill, breast-band narrower and duller. **Female eclipse** Somewhat duller and greyer with more white face markings and even less distinct breast-band. May closely resemble juvenile. **Juvenile** Head, neck and upperparts predominantly sooty-brownish with face, eyering and foreneck whitish, breast-band absent, underparts all whitish, secondaries and inner primaries tipped white, bill dull pinkish. **VOICE** Female utters a rapid chattering ***gag-ag-ag-ag-ag...***; male calls with thin low whistles. **HABITAT** Large rivers, lakes, coastal mudflats and other wetlands; up to 400 m. **RANGE & STATUS** Breeds W,C Palearctic, S Siberia, Iran, W Pakistan (occasional), Mongolia, NW,N,NE China; most populations winter to south, N Africa, southern W Palearctic, Middle East, Pakistan, W and northern India, S,E China, S Korea, S Japan, rarely Nepal, Bangladesh. **SE Asia** Rare to uncommon winter visitor W,N Myanmar. Vagrant SW,C,S Myanmar, W(coastal),NW,C Thailand, N Laos, E Tonkin, C Annam.

55 RUDDY SHELDUCK
Tadorna ferruginea Plate 5

IDENTIFICATION 61–67 cm. Monotypic. **Male non-breeding** Relatively large size and predominantly orange-rufous plumage with mostly creamy-buff head diagnostic. May show faint black neck-collar. In flight, shows distinctive blackish wings with contrasting whitish coverts (above and below); bottle-green gloss on secondaries. **Male breeding** Has well-defined narrow black neck-collar. **Female** Lacks black neck-collar, head more buff with contrasting whiter face. **Juvenile** Like female but head and upperparts strongly washed greyish-brown, underparts duller. **VOICE** Rather vocal. Typically utters a rolling, honking ***aakh*** and trumpeted ***pok-pok-pok-pok...*** **HABITAT & BEHAVIOUR** Large rivers, lakes; up to 900 m. Normally in flocks. **RANGE & STATUS** Breeds NW Africa, Ethiopia, south-east W Palearctic, C Asia, S,SE Siberia, NW India (Ladakh), N Nepal, Tibet, Mongolia, W,NW,N,NE China; most populations winter to south, NW Africa, Middle East, Indian subcontinent, S Tibet, southern and E China, S Korea, S Japan. **SE Asia** Locally common winter visitor Myanmar. Vagrant W,NW,C,NE Thailand, N Laos, E Tonkin, N Annam.

ANATIDAE: ANATINAE: Typical ducks and pygmy-geese

Worldwide c.107 species. SE Asia 28 species. Similar to Anserinae species but generally smaller and more exclusively aquatic. Exhibit strong sexual dimorphism. **Dabbling ducks** (*Asarcornis, Aix, Anas*) Prefer shallower water, feeding on surface or by up-ending; typically take off direct from water surface; *Asarcornis* more secretive in wooded habitats. **Cotton Pygmy-goose** (*Nettapus*) Small and short-billed, wings very rounded, always feeds on water. **Diving ducks** (*Rhodonessa, Netta, Aythya, Clangula, Bucephala, Mergellus, Mergus*) Prefer deeper waters and mostly feed by diving; typically patter along surface of water before taking off; *Rhodonessa* and *Netta* also feed by dabbling, up-ending or head-dipping.

56 WHITE-WINGED DUCK
Asarcornis scutulata Plate 5

IDENTIFICATION 66–81 cm. Monotypic. **Male** All-dark body, contrasting whitish head and upper neck and mostly dull yellowish to orange-yellowish bill distinctive. Head and upper neck variably mottled blackish (can be mainly white), lesser and median upperwing-coverts and inner edges of tertials white, secondaries bluish-grey. In flight, white wing-coverts (above and below) contrast sharply with rest of wings. See Comb Duck. **Female** Smaller and slightly duller, usually with more densely mottled head and upper neck. **Juvenile** Duller and browner, initially with pale brownish head and neck. **VOICE** Flight call is a prolonged, vibrant series of honks, often ending with a nasal whistle; mainly at dawn and dusk. Also single short harsh honks. **HABITAT & BEHAVIOUR** Pools and rivers in forest, freshwater swamp forest; up to 800 m, rarely 1,500 m. Usually encountered singly or in pairs. Feeds mostly at night, flying to and from roosting sites at dawn and dusk. **RANGE & STATUS** Local resident NE India, Sumatra; formerly SE Bangladesh, W Java. **SE Asia** Rare local resident N Myanmar, W,NE,S Thailand, Cambodia, C,S Laos, N,S Annam, Cochinchina; formerly NW Thailand. Formerly resident (current status uncertain) SW,W,C,E,S Myanmar, Tenasserim, Peninsular Malaysia. **BREEDING** March–September. **Nest** Tree-cavity; 2–6 m above ground. **Eggs** 6–13; glossy, creamy-white; 65 x 48.1 mm (av.).

57 COTTON PYGMY-GOOSE
Nettapus coromandelianus Plate 6

IDENTIFICATION 33–38 cm. *N.c.coromandelianus*: **Male** Small size, blackish cap, breast-band/collar and upperparts, and contrasting white sides of head, neck and underparts diagnostic. Upperparts and wings strongly glossed green, flanks washed grey, undertail-coverts blackish. In flight, shows diagnostic white band across primaries and along tips of secondaries; wings rather rounded. **Male eclipse** Sides of head and neck washed greyish, has darker eyestripe, greyish mottling on breast and flanks and no obvious breast-band/collar. Retains distinctive wing pattern. **Female** Narrow black eyestripe, duller (faintly glossed) and browner upperparts; neck and underparts washed greyish-brown and mottled darker, particularly on breast; undertail-coverts pale; lacks defined breast-band/collar. In flight, wings dark with narrow white trailing edge to secondaries. See Mandarin Duck and Little Grebe. **Juvenile** Like female but sides of head less white, eyestripe broader, lacks any obvious gloss on upperparts. **VOICE** Male has a staccato cackling ***WUK-wirrarrakWUK-wirrarrakWUK-wirrarrak...***, usually in flight; female gives a weak ***quack***. **HABITAT** Lakes, marshes and other freshwater wetlands; up to 800 m. **RANGE & STATUS** Breeds Indian subcontinent, southern China, N New Guinea, N Australia; some northern populations winter to south, Sumatra, Borneo, N Philippines, rarely Java, N Sulawesi. **SE Asia** Scarce to locally common resident, subject to local movements (except SE Thailand, Singapore, W Tonkin, N Annam). Scarce non-breeding visitor Singapore. **BREEDING** June–August. **Nest** Cavity in tree, sometimes building; 2–21 m above ground. **Eggs** 8–14; glossy, ivory-white; 43.1 x 32.9 mm (av.).

58 **MANDARIN DUCK** *Aix galericulata* Plate 6

IDENTIFICATION 41–49 cm. Monotypic. **Male** Very distinctive, with big-headed appearance, pinkish-red bill (tip pale), very broad whitish supercilium, broad orange-rufous fan of pale-streaked 'hackles' on lower sides of head, erect pale-tipped dull orange-rufous wing 'sails' and two white bands on sides of black breast. **Male eclipse** Similar to female but bill reddish, 'spectacles' less pronounced, neck feathers shaggier, upperparts glossier. **Female** Distinctive. Has greyish head with white 'spectacles', rather full nape/hindneck feathers, dark greenish-brown upperparts, white throat, belly-centre and vent, dark brown breast and flanks with heavy whitish streaks and mottling, and pink-tinged dark greyish bill with pale tip. In flight, upperwing quite uniform with white-tipped greenish secondaries. Like Cotton Pygmy-goose but larger, head distinctly bulky and more uniform, breast and flanks much darker. See Eurasian Wigeon. **Juvenile** Duller and browner overall than female, particularly head, with less pronounced 'spectacles' (sometimes lacking) and more diffuse markings on breast and flanks. **VOICE** Usually silent. **HABITAT & BEHAVIOUR** Freshwater lakes and pools; up to 400 m. Often found with other ducks. **RANGE & STATUS** Breeds SE Siberia, Ussuriland, S Sakhalin, Kuril Is, NE China, Taiwan, N,S Korea, Japan; some populations winter to south, E,S China. Introduced W Europe. **SE Asia** Vagrant C Myanmar, NW,C Thailand, N Laos, W Tonkin.

59 **GADWALL** *Anas strepera* Plate 7

IDENTIFICATION 46–56 cm. Monotypic. **Male** Overall greyish plumage with contrasting black vent and blackish bill diagnostic. In flight, from above, square white patch on inner secondaries contrasts sharply with black outer secondaries and inner greater coverts; has maroon patch on median upperwing-coverts. **Male eclipse** Like female but upperparts greyer and more uniform; retains tertial and upperwing colour and pattern. Bill can be all dark (see female Falcated Duck). **Female** Very similar to Mallard but smaller and more compact with squarer head, head pattern less contrasting, bill finer and blackish with orange sides. Easily identified in flight by square white patch on inner secondaries (often visible at rest). See Eurasian Teal, Northern Pintail and Falcated Duck. **Juvenile** Like female but underparts richer brown with more distinctly streaked breast; contrasts more with grey head and neck. White patch on secondaries may be very indistinct on females. **VOICE** Usually silent. Courting males utter a short ***nheck*** and low whistle; females a repeated ***gag-ag-ag-ag-ag...*** **HABITAT** Freshwater lakes and marshes; up to 800 m. **RANGE & STATUS** Breeds Europe, W(east) and C Palearctic, S,SE Siberia, Mongolia, NW,NE China, Japan (Hokkaido), N America; winters N Africa, W and southern Europe, Middle East, Indian subcontinent, S Tibet, southern China, Taiwan, S Korea, southern Japan, southern N and C America. **SE Asia** Scarce to uncommon winter visitor SW,W,N(locally common to common),C,E,S Myanmar, north Tenasserim, NW Thailand, E Tonkin. Vagrant C,W(coastal) Thailand, Singapore.

60 **FALCATED DUCK** *Anas falcata* Plate 7

IDENTIFICATION 48–54 cm. Monotypic. **Male** Distinctive. Mostly greyish; head dark glossy green with long 'mane' and purple crown and cheeks, throat and upper foreneck white, bisected by black band, has long curved black and whitish tertials and black-bordered yellowish-white patch on sides of vent. In flight, rather pale grey upperwing-coverts contrast sharply with green-glossed black secondaries. **Male eclipse** Like female but crown, hindneck and upperparts darker, breast and flanks richer brown; wing pattern retained but tertials shorter. **Female** From similar Gadwall and Eurasian Wigeon by combination of longish, narrow, dark grey bill, rather plain greyish-brown head with rather full nape feathers, rich brown breast and flanks with dark brown scales and, in flight, all-dark secondaries, white bar across tips of greater upperwing-coverts, and rather dark greyish underwing with contrasting white coverts. **Juvenile** Like female but buffier, with greyer tips to greater upperwing-coverts. **VOICE** Deep, nasal ***bep*** noted from males at least. In flight may give distinctive short low whistle followed by wavering ***uit-trr***. **HABITAT** Lowland lakes and marshes. **RANGE & STATUS** Breeds Siberia, E Palearctic, Mongolia, NE China, Japan (Hokkaido); winters N,NE Indian subcontinent, E and southern China, Taiwan, Japan. **SE Asia** Rare to uncommon winter visitor SW, N,C,E Myanmar, NW,NE,C Thailand, N Laos, E Tonkin, C Annam.

61 **EURASIAN WIGEON** *Anas penelope* Plate 7

IDENTIFICATION 45–51 cm. Monotypic. **Male** Bright chestnut head with broad yellowish median stripe (ends in point on crown), pinkish breast and mostly greyish remainder of body with black vent diagnostic. Bill pale grey with black tip, centre of abdomen and rear flanks white. In flight shows large white patch on upperwing-coverts, green-glossed secondaries and mostly greyish underwing with whiter greater and primary coverts. See Common Pochard. **Male eclipse** Similar to female but head and breast richer brown; retains white patch on upperwing-coverts (often visible at rest). **Female** Combination of fairly small size, rounded head, shortish pale grey bill with black tip, and rather plain dark brownish head, neck, breast and flanks diagnostic. Breast and flanks are more chestnut-tinged. In flight, underside appears uniform brownish with sharply contrasting white belly and vent; upperwing-coverts paler and greyer than rest of wing. See Falcated Duck. **Juvenile** Like female but has almost glossless secondaries and some brown mottling on belly. **VOICE** Male utters clear, piercing, whistled ***wheeooo*** and more subdued ***whut-whittoo***; female gives low growling ***krrr*** or ***karr***. **HABITAT** Lakes, large rivers, various wetlands; up to 800 m. **RANGE & STATUS** Breeds northern Palearctic, NE China; winters equatorial Africa, southern Palearctic, Indian subcontinent, S Tibet, C and southern China, Taiwan, S Korea, Japan, N Philippines. **SE Asia** Scarce to uncommon winter visitor Myanmar (except Tenasserim), Thailand (except W,SE,S), N Laos, E Tonkin. Vagrant W,S Thailand, Peninsular Malaysia, Singapore, Cambodia, C Annam, Cochinchina.

62 **MALLARD** *Anas platyrhynchos* Plate 7

IDENTIFICATION 50–65 cm. *A.p.platyrhynchos*: **Male** Unmistakable. Large and mostly pale brownish-grey with yellowish bill, glossy dark green head with purple sheen, white neck-collar and purplish-brown breast. Tail-coverts and lower scapulars black. In flight, shows glossy dark bluish secondaries, bordered at front and rear by defined white band and dark underwing with contrasting white coverts and bases of secondaries. See Falcated Duck. **Male eclipse** Like female but breast more chestnut, tertials grey, bill dull yellowish, crown and eyestripe may be glossed dark green, upperwing-coverts greyer. **Female** Similar to several other female *Anas* ducks. Distinguished by combination of relatively large size, elongated shape, dull orange to dull reddish bill with dark brown markings (not distinctly bicoloured), contrasting dark crown and eyestripe and, in flight, wing colour and pattern (like male but upperwing-coverts browner). See Gadwall, Northern Pintail, Northern Shoveler and much smaller Eurasian Teal. **Juvenile** Like female but crown and eyestripe blackish, breast neatly streaked, flanks more streaked (less scaled), bill initially mostly dull reddish to dull orange. **VOICE** Male utters a rasping ***kreep***; female gives series of mocking quacks, descending towards end: ***QUACK-QUACK-QUACK-quack-quack-quack...*** **HABITAT** Lakes, large rivers, various wetlands; up to 420 m. **RANGE & STATUS** Breeds Holarctic, S Tibet, NW,N,NE China, N Korea, Japan, locally NE Pakistan, NW India, Nepal; more northerly populations winter to south, N Africa, Middle East, Indian subcontinent (except S), S Tibet, C and southern China, Taiwan, N,S Korea, Japan, south N and north C America. **SE Asia** Rare to locally common winter visitor W,N,C,E Myanmar, NW Thailand. Vagrant E Tonkin.

63 INDIAN SPOT-BILLED DUCK

Anas poecilorhyncha Plate **6**

IDENTIFICATION 55–63 cm. *A.p.haringtoni*: **Male** Recalls some female *Anas* ducks but has pale, rather plain head and neck, sharply contrasting blackish crown and eyestripe, distinctive blackish bill with broad yellow tip, red spot on lores at bill-base and almost completely whitish outer webs of longest two tertials. Upperparts dark brown with narrow buffish fringes; breast and flanks rather pale with dark brown streaks and mottling, vent darker. In flight, shows glossy green secondaries, bordered at front and rear by black and white band and sharply contrasting white on tertials and underwing-coverts. *A.p.poecilorhyncha* (SW,W Myanmar?) has larger red spot on lores, paler base colour to sides of head, neck and breast, and more distinct dark streaks and spots on breast and flanks. See Chinese Spot-billed Duck. **Female** Somewhat smaller, no obvious red spot on lores (may be absent), usually has smaller and less distinct markings on breast and flanks. *A.p.poecilorhyncha* normally shows small red spot on lores (otherwise differs in same way as male). **Juvenile** Like female but browner below with less distinct markings, no red on lores. **VOICE** Descending series of ***quark*** notes. **HABITAT & BEHAVIOUR** Lakes, large rivers, marshes; up to 800 m. Usually found in pairs or small flocks. **RANGE & STATUS** Resident (subject to relatively local movements) Indian subcontinent, SW,S China. **SE Asia** Uncommon resident (subject to some movements) N(commoner in winter),C,E,S Myanmar, Cambodia, S Laos, E Tonkin, Cochinchina. Scarce to uncommon winter visitor NW,NE,C,SE Thailand, N Laos. Recorded (status uncertain) W Myanmar(common), N,C Annam. **BREEDING** All year. **Nest** Well-made pad, on ground amongst vegetation (usually near water). **Eggs** 6–14; greyish-buff to olive-white; 52.7 x 39.6 mm (av.).

64 CHINESE SPOT-BILLED DUCK

Anas zonorhyncha Plate **6**

IDENTIFICATION 55–60 cm. Monotypic. **Adult** Like Indian Spot-billed Duck but lacks red on lores, has dark band from below eye to base of lower mandible, darker and more uniform body and dark green to dark purplish-blue secondaries with almost no white borders; longest two tertials have white restricted to tip and narrow outer fringe. **Juvenile** Not clearly documented, but similar to adult in most respects (including head pattern). **VOICE** Undocumented. **HABITAT** Lakes, large rivers, and marshes. **RANGE & STATUS** Breeds China (except SW,NW), Taiwan, S,SE Siberia, Ussuriland, Sakhalin, Kuril Is, N,S Korea, Japan; winters south to E,S China, S Korea. **SE Asia** Doubtfully recorded N Myanmar, but likely to occur in north-east of region. **NOTE** Formerly lumped in Indian Spot-billed Duck by most authors, but best treated as a separate species due to its highly distinctive morphology, and sympatric breeding with *A. poecilorhyncha* in S China.

65 NORTHERN SHOVELER

Anas clypeata Plate **7**

IDENTIFICATION 43–52 cm. Monotypic. **Male** Long, wide, spatula-shaped bill, glossy dark green head and upper neck, white breast and mostly chestnut sides diagnostic. Has yellow eyes, white patch on sides of vent and black tail-coverts. In flight, shows distinctive blue median and lesser upperwing-coverts, broadly white-tipped secondaries, glossy green secondaries and contrasting white underwing-coverts. **Male eclipse** Similar to female but more rufous, particularly on flanks and belly; retains eye and upperwing colour and pattern. Sub-eclipse birds have whitish crescent on face. **Female** From other scaly brown *Anas* ducks by distinctive bill, greyish-blue median and lesser upperwing-coverts and broadly white-tipped greater coverts. Eyes usually brown. See Mallard and Garganey. **Juvenile male** Like juvenile female but upperwing similar to adult. Immatures can resemble sub-eclipse adults. **Juvenile female** Like adult female but crown and nape darker, underparts paler and more spotted, greater covert bar indistinct, lacks obvious gloss on secondaries. **VOICE** Courting male utters a repeated, liquid, hollow ***sluk-uk*** or ***g'dunk***; female gives a descending ***gak-gak-gak-ga-ga***. **HABITAT** Lakes, large rivers, marshes, various wetlands; up to 800 m. **RANGE & STATUS** Breeds Holarctic, NW,NE China, Japan (Hokkaido); mostly winters N and equatorial Africa, southern Palearctic, Indian subcontinent, S Tibet, C and southern China, Taiwan, S Korea, Japan, Philippines, south N,C and north S America. **SE Asia** Uncommon winter visitor SW,W,N,C,E,S Myanmar, north Tenasserim, Thailand (except SE,S), E Tonkin. Vagrant Peninsular Malaysia, Singapore, Cambodia, C Annam, Cochinchina.

66 ANDAMAN TEAL *Anas albogularis* Plate 6

IDENTIFICATION 36–43 cm. Monotypic. **Adult** Small size and dark grey-brown head with white patches around eye and on upper throat diagnostic. May have much more white on head and neck or only show white eyering. Otherwise similar to Eurasian Teal but head squarer, forehead steep, bill dark grey (sometimes mixed with pink). In flight very like Common but underwing darker with only a narrow white band along coverts. **Juvenile** Similar to adult but with less white on head, and mottled rather than scaled below. **VOICE** Said to include a low, soft whistle, and low quacking notes. **HABITAT & BEHAVIOUR** Lakes, marshes, coastal wetlands, rice paddies; lowlands. Typically feeds in open habitats at night, roosting in mangroves and on coastal rocks etc. **RANGE & STATUS** Resident (subject to local movements) Andaman Is. **SE Asia** Formerly occurred (status unknown) Coco Is (off S Myanmar). Vagrant mainland S Myanmar. **BREEDING** July–September. **Nest** Situated in natural tree-cavity. **Eggs** 10; cream-coloured; 49 x 36.3 mm (av.)

67 NORTHERN PINTAIL *Anas acuta* Plate 7

IDENTIFICATION 51–56 cm (male's tail up to 10 cm more). Monotypic. **Male** Unmistakable. Slender and long-necked, predominantly grey with dark chocolate-brown head and upper neck, white line from side of head down neck to whitish lower foreneck and upper breast, yellowish-white patch on rear flanks and black tail-coverts. Has relatively long slender grey bill with blackish median stripe and distinctive elongated central tail-feathers (streamers). In flight, shows grey upperwing-coverts, rufous-buff tips to greater coverts, glossy blackish-green secondaries with broad white tips and mostly greyish underwing with blackish median and lesser coverts. **Male eclipse** Like female but greyer above, has grey tertials; retains bill and wing colour and pattern. **Female** Slender proportions, longish neck, long darkish grey bill, rather plain brown head and distinctly pointed tail diagnostic. In flight shows distinctive greyish underwing with dark median and lesser coverts and white-tipped secondaries. Upperwing-coverts duller than male, secondaries much duller and browner, tips of greater coverts whiter. See Gadwall and Mallard. **Juvenile** Like female but upperparts darker, flanks more boldly patterned. **VOICE** Male gives a low ***preep-preep***; female utters weak descending quacks and low growling croaks when flushed. **HABITAT** Lakes, large rivers, marshes, various wetlands; up to 800 m. **RANGE & STATUS** Breeds Holarctic, NW China; winters N and equatorial Africa, southern Palearctic, Indian subcontinent, S Tibet, southern China, Taiwan, S Korea, Japan, Philippines, south N,C and north S America. **SE Asia** Uncommon to locally common winter visitor SW,W,N,E,C,S Myanmar, north Tenasserim, W,NW,NE,C Thailand, Cambodia, N,C Laos, E Tonkin, N,C Annam, Cochinchina. Vagrant S Thailand, Peninsular Malaysia, Singapore.

68 **GARGANEY** *Anas querquedula* Plate **6**
IDENTIFICATION 36–41 cm. Monotypic. **Male** Relatively small size and mostly dark brownish head and neck with blacker crown and pronounced long white supercilium diagnostic. Rest of plumage brownish-grey with distinctly pale grey, dark-vermiculated flanks and elongated grey scapulars with long black and white streaks. In flight shows mostly bluish-grey upperwing-coverts and glossy blackish-green secondaries, bordered at front and rear by broad white band. See Northern Shoveler. **Male eclipse** Like female but lacks defined white line below blackish eyestripe, throat whiter; retains wing colour and pattern. **Female** Relatively small size and bold head pattern distinctive. Has dark crown, narrow whitish supercilium, bold dark eyestripe, large whitish loral patch continuing in narrow line below blackish eyestripe, dark cheek-bar and whitish throat. Centre of belly extensively whitish. In flight, shows grey tinge to upperwing-coverts, mostly dark brownish secondaries (little green gloss), bordered at front and rear by narrow white band and distinctly dark leading edge to underwing-coverts. See Eurasian and Baikal Teals and Northern Shoveler. **Juvenile** Like female but somewhat darker with dark markings on belly; less defined head pattern. **VOICE** Male utters a dry rattling ***knerek***, female a short high ***quack***. **HABITAT** Lakes, marshes, various wetlands; up to 800 m. **RANGE & STATUS** Breeds Palearctic, NW,NE China; winters equatorial Africa, Middle East, Indian subcontinent, southern China, Taiwan, Greater Sundas, Philippines, Sulawesi, locally elsewhere Wallacea, New Guinea, rarely Japan, Bismarck Archipelago, Australia. **SE Asia** Scarce to locally common winter visitor (except W Tonkin).

69 **BAIKAL TEAL** *Anas formosa* Plate **6**
IDENTIFICATION 39–43 cm. Monotypic. **Male** Unmistakable. Relatively small with striking complex buff, green, white and black head pattern, dark-spotted pinkish breast and grey flanks bordered at front and rear by single vertical white bands. Has black undertail-coverts and elongated chestnut, black and whitish scapulars. In flight, wing pattern like Eurasian Teal but with rufous-chestnut tips to greater coverts and blacker leading edge to underwing-coverts. See Garganey. **Male eclipse** Like female but mantle has darker, more rufous fringes, breast and flanks warmer brown, loral spot less distinct. **Female** Like Garganey but with isolated round white spot (encircled darker) on lower lores, broad vertical whitish stripe from below/behind eye down to throat, supercilium broken above eye and more buffish behind eye, narrow buffish-white line along outer edge of undertail-coverts. In flight, wing pattern like Eurasian Teal but has rufescent tips to greater coverts (may be hard to see), broader white tips to secondaries and blacker leading edge to underwing-coverts. Centre of belly white. **Juvenile** Like female but has buffier and less distinct, slightly larger loral spot, and dark mottling on whitish belly. See Common. **VOICE** Male sustains a deep chuckling ***wot-wot-wot...***; female has a low ***quack***. **HABITAT** Freshwater lakes; lowlands. **RANGE & STATUS** Breeds north-eastern Palearctic; winters south-eastern China, S Korea, southern Japan. **SE Asia** Vagrant N Myanmar, NW,C Thailand.

70 **EURASIAN TEAL** *Anas crecca* Plate **6**
IDENTIFICATION 34–38 cm. *A.c.crecca*: **Male** Easily identified by small size and dark chestnut head with buff-bordered, broad glossy dark green patch from eye to nape. Rest of plumage mostly greyish with black-bordered buffish sides of vent and long white line along outer edge of scapulars. In flight, shows broad white band along tips of greater coverts, glossy dark green secondaries with rather narrow white tips, and greyish underwing with somewhat darker leading edge and white axillaries and band across coverts. See Baikal Teal and Garganey. **Male eclipse** Like female but upperparts darker and more uniform, underparts with larger dark markings, eyestripe faint or absent. **Female** Like Garganey but smaller, bill smaller, often with some dull flesh-colour or orange at base, head rather plain greyish-brown apart from darker crown and eyestripe; has narrow buffish-white line along outer edge of undertail-coverts, and more restricted whitish belly-centre. Wings like male but upperwing-coverts browner. See Baikal and Andaman Teals, Gadwall and Mallard. **Juvenile** Like female but upperparts somewhat plainer, belly speckled dark; may show darker area on ear-coverts. **VOICE** Male utters a soft, liquid ***preep-preep...***; female may give a sharp high ***quack*** when flushed. **HABITAT** Lakes, large rivers, marshes, various wetlands; up to 1,830 m. **RANGE & STATUS** Breeds northern Palearctic, NW,NE China, northern Japan, Aleutian Islands; mostly winters equatorial Africa, southern Palearctic, Indian subcontinent, C and southern China, Taiwan, S Korea, Japan, Philippines, N Borneo. **SE Asia** Uncommon to locally common winter visitor (except Tenasserim, SE,S Thailand, Peninsular Malaysia, Singapore, S Laos, S Annam). Vagrant Peninsular Malaysia, Singapore.

71 **PINK-HEADED DUCK**
Rhodonessa caryophyllacea Plate **5**
IDENTIFICATION 60 cm. Monotypic. **Male** Very striking. Relatively large, long-necked and long-billed, crown distinctly peaked, head and neck mostly deep pink, centre of throat, foreneck and most of remaining plumage blackish-brown, bill rosy-pinkish. In flight shows contrasting pale brownish-buff secondaries, narrow whitish leading edge to wing-coverts and extensively pale pink underwing. **Female** Body duller and browner, head and upper neck pale greyish-pink with brownish wash on crown and hindneck, bill duller. See Red-crested Pochard. **Juvenile** Body duller brown than female with fine whitish feather-fringes. **VOICE** Male utters a weak whizzy whistle, female a low ***quack***. **HABITAT & BEHAVIOUR** Pools and swamps in open forest and areas of elephant grass; lowlands. Shy and wary; sometimes in flocks outside breeding season. Feeds by dabbling as well as diving. **RANGE & STATUS** Formerly resident (subject to some movements) N,NE India, Bangladesh; scattered records peninsular India, Nepal; probably extinct. **SE Asia** Recorded (status uncertain) SW,N,C Myanmar. Last confirmed sighting 1910.

72 **RED-CRESTED POCHARD**
Netta rufina Plate **8**
IDENTIFICATION 53–57 cm. Monotypic. **Male** Relatively large size, red bill, bulky dull orange-rufous head and upper neck, black lower neck and underparts, and broad white flank-patch diagnostic. Eyes reddish. In flight, wings show distinctive combination of brown upperwing-coverts, broad white band across primaries and secondaries, and almost completely whitish underwing. See Common Pochard. **Male eclipse** Like female but bill red, eyes reddish. **Female** Overall plain brownish plumage with contrasting whitish sides of head, throat and upper foreneck and pink-tipped blackish bill distinctive. Forecrown distinctly darker, eyes dark brownish, wing pattern like male but upperwing-coverts paler brown. See Pink-headed Duck and Smew. **Juvenile** Like female but bill all dark. **VOICE** Usually silent. Courting male gives a rasping wheeze; female has a grating chatter. **HABITAT & BEHAVIOUR** Freshwater lakes and marshes, large rivers; up to 800 m. Primarily feeds by diving, sometimes by up-ending and head-

dipping. **RANGE & STATUS** Breeds C,SW Europe, southeast W Palearctic, C Asia, Mongolia, NW,N China; mostly winters to south, NE Africa, south W Palearctic, Middle East, Indian subcontinent (except islands), S Tibet, SW China. **SE Asia** Scarce to uncommon winter visitor W(north),N,C,E Myanmar. Vagrant C Thailand, E Tonkin.

73 **COMMON POCHARD** *Aythya ferina* Plate **8**
IDENTIFICATION 42–49 cm. Monotypic. **Male** Plain chestnut head and upper neck and rather plain grey remainder of plumage with contrasting black lower neck, breast and tail-coverts diagnostic. Bill blackish with broad pale bluish-grey central band. In flight, shows distinctive greyish upperwing, lacking white band shown by other *Aythya* ducks. **Male eclipse** Duller and browner-tinged overall. **Female non-breeding** From other *Aythya* ducks by combination of peaked head with pale spectacles and variable pale markings on lores, cheeks and throat, dark eyes, long dark grey bill with pale bluish-grey subterminal band and black tip, mottled greyish-brown body and dark undertail-coverts. In flight, wing colour and pattern distinctive (like male but upperwing duller). **Female breeding** Body somewhat plainer and browner, sides of head somewhat plainer. **Juvenile** Generally duller than female with more uniform upperparts, all-dark bill and much plainer head (may lack obvious spectacles). **VOICE** Female sometimes utters a harsh ***krrr*** or ***krrah***; courting male may give a repeated soft wheezy whistled ***pee***. **HABITAT** Freshwater lakes; up to 800 m. **RANGE & STATUS** Breeds W,C Palearctic (east to C Siberia), Afghanistan, Mongolia, NE China; mostly winters to south, equatorial Africa, west and southern W Palearctic, Middle East, Indian subcontinent (except islands), south-eastern China, N,S Korea, Japan. **SE Asia** Rare to scarce winter visitor N(locally common),C,E Myanmar, NW Thailand, E Tonkin. Vagrant C Thailand.

74 **BAER'S POCHARD** *Aythya baeri* Plate **8**
IDENTIFICATION 41–46 cm. Monotypic. **Male** From other *Aythya* ducks by combination of glossy greenish-black head and upper neck, whitish eyes, blackish upperparts, rich chestnut-brown lower neck and breast, chestnut-brown sides with large white patch on foreflanks, and white undertail-coverts. In flight, wing pattern like Ferruginous Pochard but white upperwing-band does not extend quite as far onto outer primaries. **Male eclipse** Similar to female but eyes whitish. **Female** Similar to male but head and upper neck dark brown, typically with a large diffuse dark chestnut loral patch, eyes dark, often has some whitish mottling on throat, duller breast and flanks, and smaller white patch on foreflanks (may not be visible above water when swimming). Combination of domed head without nuchal tuft, contrast between dark head and warm brown breast, and presence of white on foreflanks rule out Ferruginous and Tufted Duck. **Juvenile** Similar to female but head tinged more chestnut with darker crown and hindneck; no defined loral patch. **VOICE** Usually silent. **HABITAT** Lakes, large rivers and their deltas; up to 800 m. **RANGE & STATUS** Breeds SE Siberia, Ussuriland, NE China; winters southern China, locally north-eastern Indian subcontinent, Japan. **SE Asia** Scarce to local winter visitor Myanmar (except Tenasserim), NW,C Thailand, E Tonkin. Vagrant S Thailand.

75 **FERRUGINOUS POCHARD**
Aythya nyroca Plate **8**
IDENTIFICATION 38–42 cm. Monotypic. **Male** White eyes, rich chestnut (domed) head, neck, breast and flanks, blackish upperparts and sharply demarcated white undertail-coverts diagnostic. In flight, upperwing has most extensive white bar across flight-feathers of any *Aythya* duck in region, white belly-centre sharply defined. See Baer's Pochard and female Tufted Duck. **Male eclipse** Like female but head, neck and breast somewhat brighter; retains white eyes. **Female** Duller, more chestnut-brown, eyes dark. See Tufted Duck. **Juvenile** Like female but sides of head, foreneck, flanks and upperparts somewhat paler, belly and sides of undertail-coverts mottled brown. **VOICE** Usually silent. Courting male has a short ***chuk*** and soft ***wheeoo***; female utters a snoring ***err err err...*** and harsh ***gaaa***. **HABITAT** Freshwater lakes and marshes, large rivers; up to 800 m. **RANGE & STATUS** Breeds C,S Europe, east W Palearctic, C Asia, Iran, NW India, S Tibet, NW,N China, rarely W Pakistan; mainly winters to south, equatorial Africa, southern W Palearctic, Middle East, Indian subcontinent (except islands), SW China. **SE Asia** Rare to uncommon winter visitor SW,W,N(locally common),C,S,E Myanmar, W,NW,NE,C Thailand, E Tonkin.

76 **TUFTED DUCK** *Aythya fuligula* Plate **8**
IDENTIFICATION 40–47 cm. Monotypic. **Male** Blackish plumage with contrasting white flanks and drooping crest diagnostic. Bill grey with whitish subterminal band and black tip, eyes yellow, head glossed dark purplish. In flight, upperwing has least extensive white bar across flight-feathers of any *Aythya* duck in region. See Greater Scaup. **Male eclipse** Crest reduced to tuft; head, breast and upperparts more brownish-black, flanks greyish, bill duller. **Female** From other *Aythya* ducks by rather uniform dull dark brownish plumage with paler lower neck, breast and particularly flanks, squarish head, usually with suggestion of crest (short tuft or bump), yellow eyes and upperwing pattern (like male). May have white undertail-coverts, recalling Ferruginous and Baer's Pochards, or white patches on face, recalling Greater Scaup. Bill duller than male. **Juvenile** Similar to female but head and upperparts somewhat lighter brown (crown dark), has pale area on lores, little or no sign of crest and browner-tinged eyes (particularly female). **VOICE** Female sometimes gives a low, gruff growling ***err err err...***; courting male utters a low vibrant whistled ***wheep-wee-whew***. **HABITAT** Lakes, large rivers; up to 1,300 m. **RANGE & STATUS** Breeds northern Palearctic, NE China; mostly winters to south, equatorial Africa, west and southern W Palearctic, Middle East, Indian subcontinent, southern China, Taiwan, N,S Korea, Japan, Philippines. **SE Asia** Scarce to uncommon winter visitor W,N(locally fairly common),C,S,E Myanmar, NW,NE,C Thailand, E Tonkin. Vagrant S Thailand, Peninsular Malaysia, Singapore, C Annam.

77 **GREATER SCAUP** *Aythya marila* Plate **8**
IDENTIFICATION 42–51 cm. *A.m.marila*: **Male** Similar to Tufted Duck but larger; has longer, plain bluish-grey bill with almost no black at tip, rounder head with dark green gloss and no crest, and pale grey lower mantle and scapulars with fine blackish vermiculations; in flight shows greyish upperwing-coverts and broader, longer white band across flight-feathers. **Male eclipse** Head, neck and breast duller and brownish-tinged, upperpart vermiculations darker and browner, flanks faintly vermiculated grey and brown; may show whitish patch on lores and/or ear-coverts. **Female** From Tufted by structure (as male), broad distinct white face-patch (encircling base of bill), dark (never white) undertail-coverts and usually some grey vermiculations on upperparts and flanks. Worn (summer) birds show distinctive pale patch on ear-coverts. Wings as male but upperwing-coverts browner. **Juvenile** Like female but initially has less white on face but usually a whitish patch on ear-coverts, no grey vermiculations on upperparts and flanks, flanks buffier. **VOICE** Usually silent. Courting male utters soft cooing and whistling; female gives a harsh gruff ***arr arr arr...*** **HABITAT** Lowland lakes, coastal waters. **RANGE & STATUS** Breeds N Holarctic; mostly winters to south (particularly along coasts), north-western and SC Europe, Black and Caspian Seas, south-eastern China, Taiwan, S Korea, Japan, USA, rarely northern Indian subcontinent. **SE Asia** Vagrant N Myanmar, NW Thailand, E Tonkin.

78 **LONG-TAILED DUCK**
Clangula hyemalis Plate **7**
IDENTIFICATION 39–47 cm. Monotypic. **Male non-breeding** Combination of white head and neck with grey-brown and black face- and neck-patches, and mostly black, white and grey remainder of plumage diagnostic. Has narrow,

elongated black tail-streamers, and large pink bill-patch. In flight, white scapulars contrast with blackish wings. **Male breeding** Head and neck mostly blackish, with grey-brown to white facial markings, upperparts blackish with warm brown fringing; less white on scapulars than non-breeding. **Female non-breeding** Whitish head and upper neck, with contrasting blackish crown and large patch on lower head-side, blackish upperparts with brown fringing, warm brownish breast-band; lacks tail-streamers of male. In flight, dark upperside, breast-band and head markings contrast with white collar and belly. **Female breeding** More brown and less white on face and neck, with white markings mainly behind eye and on neck-side; greyer breast-band. **Juvenile** Similar to female breeding, but pale head and neck markings duller and less contrasting, large patch on lower head-side less black. **VOICE** Nasal ***gak*** from both sexes. Male utters far-carrying, nasal, yodelling ***ow ow-owDELEE, ow-owDELEE***, during migration, as well as in display. Shorter ***a-GLUU-ah*** (***GLUU*** high, slurred). **HABITAT & BEHAVIOUR** Lowland lakes, large rivers; recorded in plains. **RANGE & STATUS** Breeds N Holarctic; winters mainly at sea and offshore south to NW Europe, NE China, N,S Korea, N Japan, N Mexico; rarely N Indian subcontinent. **SE Asia** Vagrant NW Thailand.

79 COMMON GOLDENEYE

Bucephala clangula Plate **7**

IDENTIFICATION 42–50 cm. *B.c.clangula*: **Male** Black head, upperparts and rear end, oval white loral spot, white neck, breast and belly, and black and white outer scapulars distinctive. Shows broad, almost triangular head and bill profile, green gloss on head, blackish bill and yellow eyes. In flight shows blackish wings with large white patch on secondaries and upperwing-coverts and small white patch on underside of secondaries. **Male eclipse** Like female but head a little darker, usually with trace of loral patch; retains bill and wing colour and pattern. **Female** Combination of rather dark greyish body, whitish collar and wing markings and uniform dark brown head with yellow eyes diagnostic. Bill blackish with pale yellowish subterminal band. Wings like male but large white patch on upperwing bisected by two narrow black bands. See Smew and Common Merganser. **Juvenile** Like female but duller, body browner, lacks whitish collar, bill all dark. Males attain white on lores during first winter. **VOICE** Usually silent. Courting male utters strange whistles and grating notes, including ***be-beeezh*** (when head-tossing), followed by low rattle; female has a harsh ***berr*** or ***graa***. **HABITAT & BEHAVIOUR** Lowland lakes, large rivers; recorded in plains. Male has whistling wingbeats; performs head-tossing display in spring. **RANGE & STATUS** Breeds northern Holarctic, NE China; mostly winters to south (particularly along coasts), C,SE and north-western Europe, Black and Caspian Sea regions, C Asia, C and northern China, N,S Korea, Japan, USA, irregularly northern Indian subcontinent, S Tibet. **SE Asia** Rare to scarce winter visitor N Myanmar. Vagrant C Myanmar.

80 SMEW *Mergellus albellus* Plate 7

IDENTIFICATION 38–44 cm. Monotypic. **Male** Relatively small size and white head with floppy black and white crest and black face-patch diagnostic. Neck and underparts white with two fine black lines on sides of breast, flanks greyer, scapulars black and white. In flight shows black upperwing with white median coverts and narrow bands along tips of greater coverts and secondaries, and dark grey underwing with similar white markings, which extend to lesser coverts. **Male eclipse** Like female but retains wing pattern. **Female** Size, small bill, greyish plumage, chestnut crown and nape and sharply contrasting white lower sides of head, throat and upper foreneck distinctive. Has blackish lores (browner in summer), and all-dark bill; black and white wing markings may be visible at rest. Wings like male but has smaller white patch on median upperwing-coverts and less white on median and lesser underwing-coverts. See Red-crested Pochard. **Juvenile** Very similar to female. **VOICE** Usually silent. Courting male utters occasional low croaks and whistles; female sometimes gives low growling notes. **HABITAT** Lakes and large rivers; up to 450 m. **RANGE & STATUS** Breeds northern Palearctic, NE China; mostly winters to south, C and north-western Europe, south-eastern W Palearctic, C Asia, Iran, China, N,S Korea, Japan, irregularly northern Indian subcontinent. **SE Asia** Vagrant W,N Myanmar.

81 COMMON MERGANSER

Mergus merganser Plate **8**

IDENTIFICATION 61–72 cm. *M.m.comatus*: **Male** Distinctive. Relatively large and elongated with long slender red bill, black hood with full nape feathers, white lower neck, underparts and outer scapulars, and black mantle and inner scapulars. Hood glossed dark green, underparts variably flushed salmon-pink. In flight, white basal half of wing with grey leading edge contrasts with black primaries; underwing-coverts and secondaries mostly whitish. See Scaly-sided and Red-breasted Mergansers. **Male eclipse** Similar to female but mantle darker, flanks whiter; retains wing pattern. **Female** Distinguished by size and structure (as male), rufous-chestnut hood with well-defined white throat, and greyish remainder of plumage with white centre of underparts. Nape slightly shaggier than male. Wings similar to male but median and lesser upperwing-coverts grey. See Scaly-sided and Red-breasted Mergansers, Common Goldeneye and Smew. **Juvenile** Duller than female with pale loral stripe and slightly less sharply defined white throat. **VOICE** Usually silent. Courting male repeats a soft frog-like ***kuoorrp kuoorrp...*** and similar ***drruu-drro***; female gives harsh calls, including ***skrrak skrrak***. **HABITAT** Large rivers, sometimes lakes; up to 1,135 m. **RANGE & STATUS** Breeds Holarctic, NW India, Tibet, NW,N,NE China, Japan (Hokkaido); mostly winters to south, C,SC and north-western Europe, Black and Caspian Sea regions, C Asia, northern Indian subcontinent, C and southern China, N,S Korea, Japan, USA. **SE Asia** Scarce to fairly common winter visitor W(north),N Myanmar.

82 SCALY-SIDED MERGANSER

Mergus squamatus Plate **8**

IDENTIFICATION 52–58 cm. Monotypic. **Male** Similar to Common Merganser but head and neck more slender with long spiky uneven crest, flanks white with pointed dark grey scales, greater coverts tipped black; in flight, large white upperwing-patch is bisected by two black lines. See Red-breasted Merganser. **Male eclipse** Similar to female but darker above; retains wing pattern. **Female** From Common by slender head and neck, shaggy crest, ill-defined whitish throat-centre, grey scales on whitish sides of lower breast and flanks and, in flight, black band along tips of greater upperwing-coverts. See Red-breasted Merganser. **Juvenile** Like female but flanks may be more uniformly grey. **VOICE** Usually silent. **HABITAT** Large rivers, lakes; up to 500 m. **RANGE & STATUS** Breeds SE Siberia, Ussuriland, NE China, N Korea; mainly winters C,E China, S Korea, southern Japan. **SE Asia** Vagrant NW Thailand, W,E Tonkin.

83 RED-BREASTED MERGANSER

Mergus serrator Plate **8**

IDENTIFICATION 52–58 cm. Monotypic. **Male** Told by combination of thin-based, slender reddish bill, red eyes, glossy greenish-black head with shaggy crest, white collar, black-streaked rufescent lower neck and breast, and white-spotted black breast-sides. In flight, large white upperwing-patch is bisected by two black lines. See Common and Scaly-sided Mergansers. **Male eclipse** Mantle blacker than female; retains wing pattern and red eyes. **Female** From other mergansers by combination of dull rufescent hood with untidy crested appearance, undemarcated paler throat and foreneck, pale and dark loral lines, variable pale eyering; reddish-brown eyes, and brownish-grey body with only vaguely pale-scaled flanks. In flight, shows smaller white upperwing-patch than male, bisected by single line. **Juvenile** As female but bill duller, crest shorter, breast and central underparts greyer.

VOICE Male's display-call is weak ***chika...pitchee***; female gives grating ***prrak prrak prrak...*** **HABITAT** Sea coasts. **RANGE & STATUS** Breeds northern Holarctic, NE China; winters mainly along coasts south to southern Palearctic, S China, N,S Korea, Taiwan, Japan, northern C America. **SE Asia** Vagrant E Tonkin.

GAVIIDAE: Loons

Worldwide 5 species. SE Asia 1 species. Aquatic and relatively large (larger than grebes), longish necks, longish pointed bills, almost no tails, strong legs near rear of body adapted for foot-propelled diving. Fly reluctantly and with relatively quick wing-beats. Feed mainly on fish and aquatic invertebrates.

84 YELLOW-BILLED LOON
Gavia adamsii Plate **9**

IDENTIFICATION 77–90 cm. Monotypic. **Adult non-breeding** Could be confused with swimming juvenile cormorant. Combination of thick pointed ivory-coloured to yellowish-white bill (usually held upward) with dark on culmen, thick head and neck, very steep forehead, blackish-brown crown and hindneck, shadowy half-collar, and whitish underside distinctive. Greyish-brown above and along flanks, with blackish mottling. **Adult breeding** Black head and neck with black-striped white patches, black upperparts with white chequers and spots; yellower-tinged and more uniformly pale bill. **Juvenile** Paler and browner than adult non-breeding, neatly scaled above and along flanks. **VOICE** Usually silent in winter. **HABITAT** Sea coasts, rarely inland on lowland lakes and large rivers. **RANGE & STATUS** Breeds northern Holarctic, NE China, winters mainly along coasts south to north-west Europe, E China, N Korea, northern Japan, west USA. **SE Asia** Vagrant N Myanmar (one sight record).

PROCELLARIIDAE: Petrels & shearwaters

Worldwide c.80 species. SE Asia 4 species. Marine seabirds, with slender bodies, short necks, long narrow wings, webbed feet and rather slender bills with long tubular external nostrils (used to excrete excess salt). Flight strong and direct, with lengthy glides. Feed on zooplankton, cephalopods, fish and offal, caught on or below water surface.

85 STREAKED SHEARWATER
Calonectris leucomelas Plate **9**

IDENTIFICATION 48–49 cm. Monotypic. **Adult** Larger than other shearwaters in region, with mostly greyish-brown upperside and upperwing, white underparts, white underwing-coverts with dark patches on primary coverts, and distinctive white head with variable dark streaking on crown, nape and ear-coverts. Crown and ear-coverts may be mostly dark, offsetting white eyering, bill pale grey to pinkish-grey with darker tip. See similar pale morph Wedge-tailed Shearwater. **HABITAT & BEHAVIOUR** Open seas; very rarely inland. During calm conditions, flight direct but rather languid, with outer wing slightly angled backward; often glides on bowed wings. **RANGE & STATUS** Breeds offshore islands E China, Taiwan, N,S Korea, Japan, W Pacific. Outside breeding season, wanders to coastal waters of Sundas, Philippines, Wallacea, New Guinea, N,E Australasia, rarely S India, Sri Lanka. **SE Asia** Vagrant NE Thailand; offshore west S Thailand, west Peninsular Malaysia, C Annam, Cochinchina.

86 WEDGE-TAILED SHEARWATER
Puffinus pacificus Plate **9**

IDENTIFICATION 41–46 cm. Monotypic. **Adult pale morph** Similar to Streaked Shearwater but upperside darker and browner (including crown and sides of head), primary underwing-coverts mostly white, tail longer and pointed (wedge-shaped when fanned). **Adult dark morph** Most likely to be confused with Short-tailed Shearwater but larger and broader-winged, with much longer, pointed tail (toes do not extend beyond tail-tip), with more uniformly dark underwing, pinkish feet, head not contrastingly dark. See Bulwer's Petrel. **HABITAT & BEHAVIOUR** Open seas. In calm weather, flight rather lazy, with much gliding and banking, wings held forward and bowed. **RANGE & STATUS** Breeds islands in Indian and Pacific Oceans, including offshore W,E Australia; wanders widely outside breeding season, including coastal waters of S Indian subcontinent, S Korea, Japan, Sumatra, Java, Philippines, Wallacea. **SE Asia** Offshore/coastal vagrant C Thailand, west Peninsular Malaysia, Singapore.

87 SHORT-TAILED SHEARWATER
Puffinus tenuirostris Plate **9**

IDENTIFICATION 41–43 cm. Monotypic. **Adult** Smaller and slimmer than other shearwaters in region. Overall dark sooty-brown with pale chin, slightly paler breast and belly, and prominent whitish area along centre of underwing, contrasting with dark base and border; toes extend beyond tip of short tail. Bill, legs and feet dark. See Wedge-tailed Shearwater (dark morph) and Bulwer's Petrel. **HABITAT & BEHAVIOUR** Open seas. Flight direct, consisting of a flapping rise followed by a long downward glide. **RANGE & STATUS** Breeds SE Australia; wanders north into Pacific Ocean outside breeding season. **SE Asia** Offshore vagrant C, west S Thailand, Singapore (inland).

88 BULWER'S PETREL *Bulweria bulwerii* Plate **9**

IDENTIFICATION 26–27 cm. Monotypic. **Adult** Similar to Swinhoe's Storm-petrel but larger, with longer wings and distinctive long, graduated tail (usually held closed in flight); upperwing shows less obvious paler band across coverts and lacks whitish shaft-streaks at base of primaries. See Wedge-tailed (dark morph) and Short-tailed Shearwaters. **HABITAT & BEHAVIOUR** Open seas. Flight buoyant and erratic with wings usually held forward and slightly bowed, usually close to water surface; in windy conditions, flies with faster wingbeats and glides in shallow arcs. **RANGE & STATUS** Breeds Pacific and Atlantic Oceans, including islands off W Africa, SE China, Taiwan; wanders widely outside breeding season, including Indian Ocean, coastal waters of Sundas, N Sulawesi. **SE Asia** Offshore vagrant south-east Peninsular Malaysia.

HYDROBATIDAE: OCEANITINAE: *Oceanites* storm-petrels & allies

Worldwide 7 species. SE Asia 1 species. Small; shorter-winged than petrels and shearwaters. Flight buoyant and erratic, low over water surface. Feed on tiny marine animals picked from surface of sea.

89 WILSON'S STORM-PETREL
Oceanites oceanicus Plate **9**

IDENTIFICATION 15–19 cm. *O.o.oceanicus*: **Adult** Similar to Swinhoe's Storm-petrel but shows distinctive white patch extending from rump and uppertail-coverts onto sides of vent. Also has shorter wings, slightly pale band across underwing-coverts, square-cut tail and no white shaft-streaks at base of primaries on upperwing. At close range, shows yellowish webs to feet (difficult to see in field). **VOICE** Occasional soft, rapid squeaking or chattering when feeding. **HABITAT & BEHAVIOUR** Open seas. Flight quite direct and purposeful but skips and bounds close to water surface when feeding.

RANGE & STATUS Breeds Antarctica, islands in S Atlantic and S Indian Oceans; migrates north throughout all oceans outside breeding season, including coastal waters of W,S Indian subcontinent, Sumatra, Java, Bali, Wallacea, New Guinea, Australia. **SE Asia** Offshore/coastal vagrant Tenasserim, west Peninsular Malaysia.

HYDROBATIDAE: HYDROBATINAE: *Oceanodroma* storm-petrels & allies

Worldwide 14 species. SE Asia 1 species. Small; shorter-winged than petrels and shearwaters. Flight buoyant and erratic, low over water surface. Feed on tiny marine animals picked from surface of sea.

90 SWINHOE'S STORM-PETREL

Oceanodroma monorhis Plate **9**

IDENTIFICATION 20 cm. Monotypic. **Adult** Combination of small size, forked tail and all-dark body diagnostic. Upperwing has contrasting paler diagonal bar across coverts and indistinct whitish shaft-streaks on base of primaries, underwing uniformly dark. See Bulwer's Petrel and Wilson's Storm-petrel. **HABITAT & BEHAVIOUR** Open seas, sometimes inshore. Has erratic swooping, bounding flight pattern. **RANGE & STATUS** Breeds offshore islands E China, Taiwan, N,S Korea, S Ussuriland, Japan, probably off NW Africa (Salvage Is); outside breeding season, eastern populations wander to Indian Ocean, offshore S India, Sri Lanka, Sunda Straits. **SE Asia** Scarce to locally common non-breeding offshore visitor (spring and autumn) Peninsular Malaysia, Singapore. Vagrant S Thailand.

PODICIPEDIDAE: Grebes

Worldwide 21 species. SE Asia 4 species. Aquatic and duck-like, longish necks, almost no tails, strong legs near rear of body and lobed feet. Fly reluctantly and with rapid wingbeats. Feed mainly on fish and aquatic invertebrates.

91 LITTLE GREBE *Tachybaptus ruficollis* Plate 9

IDENTIFICATION 25–29 cm. *T.r.poggei*: **Adult non-breeding** Small, stocky and duck-like, with puffed-up rear end and rather narrow, mostly pale bill. Brownish-buff sides of head and underparts (throat and vent whiter) contrast with dark brown crown, hindneck and upperside. Eyes dark. In flight, upperwing all dark with narrow whitish trailing edge to secondaries. *T.r.capensis* (Myanmar) has much more white on secondaries. See female Cotton Pygmy-goose. **Adult breeding** Sides of head, throat and foreneck dark rufous-chestnut, flanks rich dark brown, eyes yellow, bill blackish with prominent yellow gape-skin. **Juvenile** Similar to adult non-breeding but sides of head have dark brown stripes, neck and breast often tinged rufous. **VOICE** Territorial call is a shrill whinnying trill, recalling some *Porzana* crakes. Sharp ***bee-eep*** and ***wit*** or ***bit*** notes when alarmed. **HABITAT & BEHAVIOUR** Lakes, pools, well-watered marshes; up to 1,450 m. Swims buoyantly, dives frequently. **RANGE & STATUS** Resident (subject to local movements) Africa, Madagascar, W Palearctic, C Asia, Middle East, Indian subcontinent, China, Taiwan, N,S Korea, Japan, Java, Bali, Philippines, Wallacea, New Guinea, Solomon Is (Bougainville). **SE Asia** Uncommon to common resident (subject to local movements) throughout. **BREEDING** All year. **Nest** Floating or anchored mound of aquatic vegetation. **Eggs** 3–7; white, staining to earthy-brown; 33–39.3 x 24.1–29.8 mm (*poggei*).

92 GREAT CRESTED GREBE

Podiceps cristatus Plate **9**

IDENTIFICATION 46–51 cm. *P.c.cristatus*: **Adult non-breeding** Relatively large and long-necked, bill rather long, slender and pinkish, crown, hindneck and upperside blackish-brown, sides of head, neck and underparts white with isolated black loral stripe and greyish-brown flanks. In flight, long neck extends forwards, legs and feet protrude beyond short rear end, wings rather long and narrow with white leading edge to upperwing-coverts, white scapular band and white secondaries, contrasting with dark remainder of upperwing. See mergansers. **Adult breeding** Develops short blackish crest and rufous-chestnut and blackish 'frills' on rear sides of head, flanks washed rufous. **Juvenile** Similar to adult non-breeding but has brown stripes across sides of head. **VOICE** Usually silent. Gives harsh, rolling ***aooorrr*** and chattering ***kek-kek-kek...*** on breeding grounds. **HABITAT** Lakes, large rivers, coastal waters; lowlands. **RANGE & STATUS** Breeds Africa, W,C Palearctic, Iran, W Pakistan, W,NW India, Tibet, northern China, S Ussuriland, N Japan, Australia, New Zealand; some northern breeders winter south to N Africa, Middle East, northern Indian subcontinent, S Tibet, southern China, S Korea, Japan. **SE Asia** Rare to uncommon winter visitor W(north),N,C,E Myanmar. Vagrant NW Thailand.

93 HORNED GREBE *Podiceps auritus* Plate 9

IDENTIFICATION 31–38 cm. Monotypic. **Adult non-breeding** Flatter-crowned than Black-necked, with thicker and straighter bill (tip often pale), black cap demarcated from white head-sides at eye-level, thinner line down hindneck, and pale loral spot. In flight, upperwing shows small white shoulder-patch and white secondaries. **Adult breeding** Exhibits outstanding black and gold 'head-frills', reddish-chestnut foreneck and underparts. **Juvenile** Browner than adult non-breeding, with dusky facial band. **VOICE** Feeble trembling ***hii-arrr*** or nasal rattling ***joarrrh***. Whinnying trill in diminishing pulses, during display. **HABITAT** Large rivers; lowlands. **RANGE & STATUS** Breeds northern Holarctic, NW China; mostly winters south to southern Palearctic, E China, Taiwan, N,S Korea, Japan, north C America. **SE Asia** Vagrant N Myanmar.

94 BLACK-NECKED GREBE

Podiceps nigricollis Plate **9**

IDENTIFICATION 28–34 cm. *P.n.nigricollis*: **Adult non-breeding** Similar to Little but somewhat larger and longer-necked, bill dark, more pointed and slightly upturned, crown more peaked, with broad blackish area around red eye, white nape-sides and throat, whiter underparts with greyish-washed sides of neck. In flight, mostly white secondaries and inner primaries contrast with dark remainder of upperwing. **Adult breeding** Unmistakable. Head, neck and upperside black with contrasting large orange-yellow patch on rear side of head, flanks chestnut. **Juvenile** Like adult non-breeding but may have buffish wash on side of head and foreneck. **VOICE** Fluty rising ***poo-eeet*** and short whistled ***wit***. Shrill trilled ***tsssrrroooeep*** in display. **HABITAT** Lakes, pools, well-watered marshes, coastal waters; lowlands. **RANGE & STATUS** Breeds S,E Africa, Palearctic, Iran, W Pakistan, W,NW India, NW,NE China, N America, north-west S America; some populations winter south to N Africa, Middle East, Pakistan, W and northern India, Nepal, southern China, Taiwan, S Korea, Japan, C America. **SE Asia** Rare winter visitor N Myanmar. Vagrant C Thailand, E Tonkin.

PHOENICOPTERIDAE: Flamingos

Worldwide 5 species. SE Asia 1 species. Large, with extremely long neck and legs, webbed feet and downward-kinked bill, adapted for filter-feeding. Flight strong and steady with outstretched neck and legs. Feeds with bill inverted and immersed in water; mainly eats small larvae, crustaceans and molluscs.

95 **GREATER FLAMINGO**
Phoenicopterus roseus Plate **10**
IDENTIFICATION 125–145 cm. *P.r.roseus*: **Adult** Unmistakable large, mostly pinkish-white wading bird with very long neck, very long pinkish legs and broad, downward-kinked, deep pink bill with black tip. In flight shows pinkish-red upper- and underwing-coverts and blackish flight-feathers. **Juvenile** Mostly brownish-greyish with dark streaks on scapulars and upperwing-coverts, browner flight-feathers, pale greyish bill with blackish tip and dark brownish legs and feet. Has shorter neck and legs than adult, pink in plumage restricted to flush on underwing-coverts. **VOICE** May give a repeated goose-like honking ***ka-ha*** in flight; a softer ***kuk-kuk ke-kuk kuk-kuk...*** when foraging. **HABITAT** Shallow, mainly brackish lowland lakes and lagoons, sometimes mudflats and saltpans. **RANGE & STATUS** Resident (subject to relatively local movements) Africa, SW Europe, Turkey, Iran, Kazakhstan, SE Pakistan, W India, C America, Caribbean, northern S America; winter/non-breeding visitor N Africa, Madagascar, S Europe, Middle East, Pakistan, W,C,S India. **SE Asia** Formerly occurred (status unclear) Cambodia (Tonle Sap). Recent records from C Thailand thought to relate to escaped captives.

CICONIIDAE: Storks

Worldwide 20 species. SE Asia 10 species. Large to very large with long necks, long, thick, mostly straight and pointed bills, long legs, short tails and long broad wings. Flight strong, with slowish wingbeats, neck and legs extended (neck retracted in adjutants); often spend time soaring on thermals. Feed mainly on fish, amphibians, reptiles, large insects, crustaceans and molluscs; Greater Adjutant also feeds on carrion.

96 **MILKY STORK** *Mycteria cinerea* Plate **10**
IDENTIFICATION 92–97 cm. Monotypic. **Adult non-breeding** Resembles Painted Stork but plumage all white, apart from blackish primaries, secondaries and tail, has limited dark red head-skin. Bill pale pinkish-yellow, legs and feet dull pinkish-red. See Asian Openbill and White Stork. **Adult breeding** White parts of plumage suffused very pale creamy-buffish, bill bright yellow to orange-yellow, bare head brighter red, legs and feet deep magenta. **Juvenile** Similar to Painted but has browner, more uniform head and neck, paler lesser and median upperwing-coverts (hardly contrasting with mantle), no defined darker breast-band and slightly less extensive naked head-skin. In flight shows off-whitish tips to underwing-coverts, creating overall paler appearance. **HABITAT & BEHAVIOUR** Tidal mudflats, mangroves. Gregarious. Pair greeting displays at nest include bowing, upward-stretching and bill raising, touching and crossing. Bill-clattering also recorded at nest sites. **RANGE & STATUS** Resident S,E Sumatra, Java, probably S Sulawesi; non-breeding visitor Bali. **SE Asia** Rare to local resident west Peninsular Malaysia, Cambodia. Rare non-breeding visitor NE,C Thailand, Cochinchina (formerly bred). Vagrant (or escapee?) south Peninsular Malaysia, Singapore. Formerly occurred (status unclear) extreme S(west) Thailand. **BREEDING** August–December. Usually colonial. **Nest** Large flimsy platform, in tree; 8–10 m above ground. **Eggs** 3–4; white (usually very dirty); 67.9 x 46.9 (mean of 6).

97 **PAINTED STORK**
Mycteria leucocephala Plate **10**
IDENTIFICATION 93–102 cm. Monotypic. **Adult non-breeding** Very distinctive. Mostly white with black and white median and lesser coverts and breast-band, pink-washed inner greater coverts and tertials, and blackish flight-feathers and tail. Has long thick pinkish-yellow bill which droops slightly at tip, naked orange-red head and pinkish-red to brownish-red legs and feet. In flight shows white-barred black underwing-coverts. See Milky Stork. **Adult breeding** Head-skin redder, bill pinkish-peach, legs and feet brighter reddish-magenta, brighter pink on tertials etc. **Juvenile** Head and neck pale greyish-brown with whitish streaks, naked head-skin dull yellowish and reduced to patch around eye to throat, mantle and greater coverts pale greyish-brown with whitish fringes, lesser and median coverts obviously darker, with whitish fringes, back to uppertail-coverts creamy-whitish; rest of underparts dull whitish with indistinct but defined dusky breast-band. Often has slight pinkish suffusion on tertials. In flight shows uniformly dark underwing-coverts. See Milky and Black-necked Storks. **HABITAT & BEHAVIOUR** Marshes, lakes, freshwater swamp forest, sometimes wet rice paddies; up to 1,000 m (mainly lowlands). Often in flocks. Bill-clatters at nest. **RANGE & STATUS** Breeds S Pakistan, India (except NE), S Nepal, Sri Lanka; formerly Bangladesh; southern China? Some populations disperse outside breeding season, reaching Pakistan, NE India, Bangladesh. **SE Asia** Rare, local resident (subject to some movements) C,S Thailand (almost extinct in latter), Cambodia; formerly C,S Annam, Cochinchina. Former resident (current status unknown) C Myanmar. Rare to scarce non-breeding (mainly winter) visitor SW,S,C Myanmar, Tenasserim, W(coastal),C,NE,SE Thailand, Laos, E Tonkin, C,S Annam, Cochinchina. Rare to scarce passage migrant W,C Thailand. Vagrant Peninsular Malaysia. Formerly occurred (status uncertain) E Myanmar. **BREEDING** November–May. Colonial. **Nest** Large flimsy platform, in tree or tall bush; up to 20 m above ground. **Eggs** 2–5; dull white, sometimes sparsely spotted and streaked brown; 69.5 x 49 mm (av.).

98 **ASIAN OPENBILL** *Anastomus oscitans* Plate **10**
IDENTIFICATION 68–81 cm. Monotypic. **Adult non-breeding** Broad, dull horn-coloured to greyish bill with open space between mandibles diagnostic. Relatively small, with mostly dull greyish-white plumage (including head) and contrasting glossy black lower scapulars, tertials, primaries, secondaries and tail. Legs and feet pinkish to greyish-pink. See White and Milky Storks. **Adult breeding** Appears whiter with redder legs at onset of breeding season. **Juvenile** Like adult non-breeding but head, neck, mantle, scapulars and breast brownish-grey, bill brownish and initially shorter with no space between mandibles, legs and feet duller. **HABITAT & BEHAVIOUR** Freshwater marshes, rice paddies, ditches through cultivation. Gregarious. Bill-clatters at nest. **RANGE & STATUS** Resident (subject to relatively local movements) India, S Nepal, Bangladesh, Sri Lanka; formerly S Pakistan. **SE Asia** Scarce to local resident (subject to some movements) C Thailand, Cambodia, Cochinchina. Scarce to locally fairly common non-breeding visitor Myanmar, W,NW,C,NE Thailand, S Laos. Also recorded on passage SW Myanmar. Vagrant S Thailand. **BREEDING** October–April. Colonial. **Nest** Bulky platform with slight depression, in tree. **Eggs** 2–5; creamy-white (becoming soiled); 57.9 x 41.2 mm (av.).

99 **BLACK STORK** *Ciconia nigra* Plate **10**
IDENTIFICATION 95–100 cm. Monotypic. **Adult** Overall glossy greenish- to purplish-black, with white lower breast to undertail-coverts and inner underwing-coverts diagnostic. Bill, orbital skin, legs and feet red. **Juvenile** Patterned like adult but dark parts of plumage mostly dark brown, white parts somewhat duller. Has pale brown flecks on neck and breast, pale brown tips to scapulars and upperwing-coverts, and mostly dull greyish-olive bill, orbital skin, legs and feet. **HABITAT** Freshwater marshes, pools and ditches, rivers, cultivation, open areas; up to 1,525 m. **RANGE & STATUS** Breeds southern Africa, SW,C Europe, W(east),C,E Palearctic, northern China, Mongolia; mostly winters sub-Saharan Africa, Indian subcontinent, southern China, N,S Korea. **SE Asia** Rare to uncommon winter visitor W,N(locally common),C,S,E Myanmar, NW Thailand, N Laos, E Tonkin. Vagrant NE Thailand, Cambodia, C Laos.

100 WOOLLY-NECKED STORK

Ciconia episcopus Plate **10**

IDENTIFICATION 75–91 cm. *C.e.episcopus*: **Adult** Glossy purplish- to greenish-black, with black cap and contrasting white neck and vent distinctive. Bill blackish with some dark red at tip and along ridge of upper mandible; facial skin dark grey, legs and feet dull red, bronzy area along inner upperwing-coverts and short, forked black tail (appears white due to extended undertail-coverts). See Storm's Stork. **Juvenile** Patterned like adult but dark parts of plumage mostly dull brown, has feathered forehead and duller bill, legs and feet. **HABITAT & BEHAVIOUR** Marshes, freshwater swamp forest, pools and streams in open forest. Not very gregarious. Bill-clatters at nest, with head resting back on mantle. **RANGE & STATUS** Resident (subject to some movements) sub-Saharan Africa, India, Nepal, Bhutan, Sri Lanka, Java, Bali, W Lesser Sundas, Philippines, Sulawesi; formerly N Pakistan. Recorded (status unclear) N Sumatra. **SE Asia** Rare to locally fairly common resident SW,W,N,C,S Myanmar, W,SE,S(north) Thailand, Cambodia, S Laos, C(south-west),S Annam, Cochinchina; formerly NW,C,NE,S(southern) Thailand. Recorded (status uncertain) W Tonkin. Former resident (current status uncertain) E Myanmar, Tenasserim, N,C Laos. Formerly occurred (status unknown) north-west Peninsular Malaysia (Langkawi I). **BREEDING** February–May and August–November (probably all year). **Nest** Large bulky platform, in tree; 20–30 m above ground. **Eggs** 3–5; white (becoming brown-stained); 62.9 x 47.4 mm (av.).

101 STORM'S STORK *Ciconia stormi* Plate **10**

IDENTIFICATION 75–91 cm. Monotypic. **Adult** Similar to Woolly-necked Stork but bill bright red, facial skin dull orange with broad golden-yellow area around eye, lower foreneck glossy black, no bronzy area along inner wing-coverts. Note range. **Juvenile** Blackish parts of plumage browner, bill dark-tipped, facial skin, legs and feet duller. **HABITAT & BEHAVIOUR** Freshwater swamp forest, rivers, streams and pools in broadleaved evergreen forest; lowlands. Solitary or in pairs; rarely in small loose groups. Performs bill-clattering displays at nest sites. **RANGE & STATUS** Resident Sumatra, Borneo; formerly W Java. **SE Asia** Rare to scarce and local resident south Tenasserim, S Thailand, Peninsular Malaysia. **BREEDING** September–November. **Nest** Large bulky platform, in tree; 19–30 m above ground. **Eggs** 2–3, otherwise undocumented.

102 WHITE STORK *Ciconia ciconia* Plate **10**

IDENTIFICATION 100–115 cm. *C.c.asiatica*: **Adult** Similar to Asian Openbill and Milky Stork but shows straight, pointed red bill and red legs and feet. Plumage all white (including head and tail), apart from contrasting black lower scapulars, tertials, greater coverts, primaries and secondaries. **Juvenile** Has browner greater upperwing-coverts and brownish-red bill, legs and feet. **HABITAT** Marshes, rice paddies, open areas; recorded in lowlands. **RANGE & STATUS** Breeds W Palearctic, north-western Iran, C Asia, NW China; mostly winters sub-Saharan Africa, S Pakistan, India. **SE Asia** Vagrant C Thailand.

103 BLACK-NECKED STORK

Ephippiorhynchus asiaticus Plate **10**

IDENTIFICATION 121–135 cm. *E.a.asiaticus*. **Male** Unmistakable. Huge and mostly white, with glossy black head, neck, back to tail, greater and median coverts, tertials and lower scapulars. Has distinctly long black bill, very long red legs, strong blue to greenish and purplish gloss on black of plumage (particularly head and neck) and brown eyes. In flight shows diagnostic white wings with broad black central band (above and below). **Female** Eyes bright yellow. **Juvenile** Head, neck and upperparts dull brown with whitish lower back to base of tail, flight-feathers blackish-brown, rest of underparts whitish, bill dark olive-brown, legs and feet dull olive. In flight shows all-dark wings. See Painted Stork. Gradually attains adult plumage and soon shows suggestion of distinctive adult wing pattern. **HABITAT & BEHAVIOUR** Freshwater marshes, marshy areas and pools in open forest, rarely also mud- and sandflats; up to 1,200 m. Pair display involves extended, fluttering wings and simultaneous bill-clattering. **RANGE & STATUS** Resident India, Nepal, Sri Lanka, New Guinea, Australia; formerly Pakistan, Bangladesh, Java. **SE Asia** Rare to scarce resident N Myanmar, Cambodia, S Laos, north-west S Annam; formerly NW,C Thailand, C Annam, Cochinchina. Former resident (current status unknown) C,E,S Myanmar, west S Thailand. Formerly occurred (status unclear) SW Myanmar, N,C Laos. Vagrant Cochinchina. **BREEDING** October–March. **Nest** Large, bulky platform, in tree 6–25 m above ground or on cliff ledge. **Eggs** 3–5; white (becoming stained); 72.1 x 53.4 mm (av.).

104 LESSER ADJUTANT

Leptoptilos javanicus Plate **10**

IDENTIFICATION 122.5–129 cm. Monotypic. **Male non-breeding** Distinctive. Very large and deep-billed with naked head and neck, glossy black above and white below. Bill mostly horn-coloured with straight-edged culmen, head- and neck-skin mostly yellowish with more vinous-tinged sides of head and contrasting pale forehead, legs and feet dark greyish. In flight shows all-blackish wings, apart from prominent white patch on inner underwing-coverts. See Greater Adjutant and Black Stork. Note range and status. **Male breeding** Has oval coppery spots near tips of median upperwing-coverts and narrow whitish edges to lower scapulars, tertials and inner greater coverts; sides of head redder. **Female** As male non-breeding but shorter, with somewhat less massive bill. **Juvenile** Duller above, with downier head and neck. **VOICE** Usually silent away from nest, where may give deep guttural sounds. **HABITAT & BEHAVIOUR** Freshwater marshes and pools in or near open forest, freshwater swamp forest, mangroves, sometimes rice paddies and open areas including mudflats; up to 550 m. Flies with neck retracted (like Grey Heron). Performs similar display to Black-necked Stork. **RANGE & STATUS** Resident (subject to wandering movements) India (except W,NW), S Nepal, Bangladesh, Sri Lanka, S China (at least formerly), Greater Sundas. **SE Asia** Rare resident S Myanmar, Tenasserim, S Thailand, Peninsular Malaysia, Cambodia, S Laos, C(south-west),S Annam, Cochinchina; formerly C Laos. Rare to scarce non-breeding visitor C,SE Thailand. Vagrant/rare non-breeding visitor Singapore. Scarce (status uncertain) SW,N Myanmar. Formerly occurred (current status unknown) C Myanmar. **BREEDING** October–June. Sometimes colonial. **Nest** Large, bulky platform, in tree 12–46 m above ground or on rock ledge. **Eggs** 2–4; white (becoming stained); 76.4 x 55.3 mm (av.).

105 GREATER ADJUTANT

Leptoptilos dubius Plate **10**

IDENTIFICATION 145–150 cm. Monotypic. **Adult non-breeding** Similar to Lesser Adjutant but larger and much deeper-billed (culmen convex), no obvious hair-like feathers on nape, more uniformly pinkish head and neck, drooping pouch on lower foreneck, pronounced white neck-ruff (broken at rear) and paler, more bluish-grey upperside with contrasting grey greater coverts and tertials. In flight differs by greyer underwing with slightly less contrasting white area on inner coverts and sooty-grey undertail-coverts. Iris pale greyish-blue to whitish or yellowish-white. **Adult breeding** Forehead and face mostly blackish, head and neck redder, upperparts paler bluish-grey, with more silvery greater coverts and tertials, neck-pouch bright saffron-yellow. **Juvenile** Bill narrower (similar to adult Lesser), has denser pale brownish to grey down/hair-like feathering on head and neck, wings initially all dark but soon with paler underwing-coverts and brown band across greater coverts and tertials. Iris brownish to blue-brown. **VOICE** Usually silent away from nest, where may give loud grunting, croaking and roaring sounds. **HABITAT & BEHAVIOUR** Freshwater marshes and pools in or near open drier forests, freshwater swamp forest, sometimes rice paddies and open areas; lowlands. Performs similar display to Black-necked Stork. **RANGE & STATUS** Resident (subject to wandering movements) NE India; formerly occurred much more widely India, Bangladesh (may still breed). Non-breeding visitor Nepal. **SE Asia** Rare resident Cambodia; formerly NW,C,SE Thailand, S Annam, Cochinchina. Formerly bred (current status unknown) S Myanmar, Tenasserim. Formerly occurred (current/former status unclear) SW,N,C,E Myanmar. Rare non-breeding visitor W,C,NE Thailand, S Laos, Cochinchina. Vagrant S Thailand. Formerly occurred (status unclear) C Annam. **BREEDING** October–April. Sometimes colonial. **Nest** Large, bulky, deeply cupped structure, in tree 12–30 m above ground, or on rock ledge. **Eggs** 2–5; white (becoming stained); 77.3 x 57.5 mm (av.).

THRESKIORNITHIDAE: THRESKIORNITHINAE: Ibises

Worldwide 26 species. SE Asia 5 species. Appearance and habits similar to herons and storks. Flight strong, with neck and legs extended, flaps interspersed with glides. Ibises (*Threskiornis, Pseudibis, Plegadis*) have long downcurved bills and relatively short legs. Feed on invertebrates, fish and frogs etc.

106 BLACK-HEADED IBIS
Threskiornis melanocephalus Plate **11**

IDENTIFICATION 75–76 cm. Monotypic. **Adult non-breeding** White plumage and blackish downcurved bill, (naked) head, upper neck, legs and feet diagnostic. In flight, bare reddish skin shows through underwing-coverts. **Adult breeding** Has variable yellowish wash on mantle and breast, greyish wash on scapulars and tertials, white plumes extending from lower neck and elongated tertials. **Juvenile** Has brownish to greyish-white feathering on head and neck, black edges and tips to outer primaries, and blackish bare skin showing through underwing-coverts. **VOICE** Peculiar vibrant grunting sounds at nesting colonies. **HABITAT** Marshy wetlands, mudflats, mangroves; up to 800 m. **RANGE & STATUS** Resident (subject to erratic movements) SE Pakistan, western and S India, SE Nepal, Sri Lanka, Sumatra, Java, probably NE China. Non-breeding/winter visitor NE India, Bangladesh, southern China, Japan, Sumatra, N Borneo, Philippines. **SE Asia** Scarce to local resident Cambodia, Cochinchina. Former resident C Thailand. Rare to uncommon non-breeding/winter visitor Myanmar (locally fairly common in S), W,NE,C,S,SE Thailand, E Tonkin (may breed?). Vagrant west Peninsular Malaysia (formerly occurred more frequently), C,S Laos, C Annam. **BREEDING** October–April. Colonial. **Nest** Bulky cupped structure, in tree or bush. **Eggs** 2–4; chalky white with faint bluish tinge (becoming stained), sometimes lightly spotted or blotched brown (mostly towards broader end); 63.5 x 43.1 mm (av.).

107 RED-NAPED IBIS *Pseudibis papillosa* Plate **11**

IDENTIFICATION 62–72 cm. Monotypic. **Adult** Like White-shouldered Ibis but smaller, bill shorter, lacks whitish collar, has diagnostic red patch on hindcrown and nape. **Juvenile** Like White-shouldered but head and neck uniform dark brown (feathered). **VOICE** A loud, nasal scream, repeated several times. **HABITAT** Lakes, large rivers, open areas, cultivation; lowlands. **RANGE & STATUS** Resident SE Pakistan, India, S Nepal. **SE Asia** Former resident (current status unknown) SW Myanmar. Records perhaps doubtful. **BREEDING** India: March–October. **Nest** Large rough platform or old raptor nest, in tree, 6–12 m above ground. **Eggs** 2–4; pale bluish-green, usually sparsely spotted and blotched pale reddish; 63 x 43.8 mm (av.).

108 WHITE-SHOULDERED IBIS
Pseudibis davisoni Plate **11**

IDENTIFICATION 75–85 cm. Monotypic. **Adult** Large, thickset and mostly dark brownish with long, downcurved greyish bill, dull red legs and feet and distinctive white collar (faintly tinged bluish). Has naked blackish head and dark greenish-blue gloss on upperwing. In flight, shows distinctive white patch on inner lesser upperwing-coverts. See Red-naped, Glossy and Giant Ibises. **Juvenile** Duller and browner than adult with dirty white collar, less gloss on wings, shorter bill and dull legs and feet. Has similar white patch on inner upperwing-coverts. **VOICE** Group territorial calls include long, loud, unearthly, hoarse screams: ***ERRRRRRH*** or ***ERRRRRRRROH*** (repeated after longish intervals), accompanied antiphonally by monotonous, subdued, moaning, rhythmic ***errh errh errh errh...*** Also, screams mixed with honking sounds: ***errrrh OWK OWK OWK OWK...*** and more subdued ***ohhaaa ohhaaa...*** and ***errrr-ah***. **HABITAT** Pools, streams and marshy areas in open lowland forest. **RANGE & STATUS** Resident SE Borneo; formerly SW China. **SE Asia** Rare resident Cambodia, S Laos, Cochinchina; formerly NW,C,S Thailand, N,C Laos. Former resident (current status unknown) SW,S,C Myanmar, Tenasserim. Recorded (status uncertain) C Annam. **BREEDING** December–May. **Nest** Large rough structure or old raptor nest, in tree; 5–10 m above ground. **Eggs** 3–4; sea-green to pale bluish, sometimes spotted or streaked brown to yellowish-brown; 61.7 x 43.2 mm (av.).

109 GIANT IBIS *Pseudibis gigantea* Plate **11**

IDENTIFICATION 102–106.5 cm. Monotypic. **Adult** Very distinctive. Huge, with long downcurved pale horn bill, naked greyish head and neck with black bars at rear, mostly blackish-slate body and contrasting grey wings with black feather-tips. Has faint greenish gloss on body and deep red eyes, legs and feet. In flight shows strongly contrasting upperwing pattern and all-dark underwing. See much smaller White-shouldered Ibis. **Juvenile** Has short black feathers on back of head and neck, shorter bill and brown eyes. **HABITAT** Pools, streams and marshy areas in open lowland forest. **RANGE & STATUS** Endemic. Rare resident Cambodia, S Laos, north-west S Annam; formerly C,W Thailand, Cochinchina. Formerly occurred (status uncertain) S Thailand, C Laos. **BREEDING** July–February. **Nest** Simple stick structure situated high up in leafy tree. **Eggs** 2+. No other information.

110 GLOSSY IBIS *Plegadis falcinellus* Plate **11**

IDENTIFICATION 55–65 cm. *P.f.falcinellus*: **Adult non-breeding** Smaller than other ibises. Mostly uniform dark brownish (slightly purplish-tinged) with white streaks on head and neck and green gloss on scapulars and upperwing-coverts. Bill pale brownish, legs and feet dark brownish, dark facial skin bordered above and below by distinctive narrow white lines. In flight shows rather bulbous head, thin neck and all-dark wings. See juvenile Red-naped and White-shouldered Ibises and much larger Giant Ibis. **Adult breeding** Head, neck and body mostly deep chestnut, forecrown glossed green, much of plumage with purplish tinge, no streaks on head and neck but more pronounced white border to lores; bill mostly flesh-coloured. **Juvenile** Duller than adult non-breeding, head and neck densely mottled whitish, throat whitish, no white border to lores and little greenish gloss on upperside. **VOICE** May give a low harsh ***graa*** and subdued grunting sounds in flight. **HABITAT & BEHAVIOUR** Marshy wetlands; up to 800 m. Flies with rapid wingbeats, interspersed with glides. **RANGE & STATUS** Breeds sub-Saharan Africa, Madagascar, south-eastern W and SC Palearctic, Iran, Java, New Guinea, Australia, south USA, C,S(north) America; possibly southern China, Sulawesi; occasionally/formerly SE Pakistan, northern India, Sri Lanka, Philippines. Northern populations winter northern Africa, Middle East, Indian subcontinent. Recorded (status uncertain) Lesser Sundas, Moluccas. **SE Asia** Rare to locally common resident Cambodia, Cochinchina. Former resident (current status unknown) C Myanmar. Rare to scarce winter/non-breeding visitor SW,N,S(locally fairly common),E Myanmar. Vagrant C,S Thailand, Singapore, E Tonkin. **BREEDING** October–February. Colonial. **Nest** Bulky platform, situated in tree. **Eggs** 2–3; deep blue-green; 52.2 x 36.9 mm (av.).

THRESKIORNITHIDAE: PLATALEINAE: Spoonbills

Worldwide 6 species. SE Asia 2 species. Appearance and habits similar to herons and storks. Flight strong, with neck and legs extended, flaps interspersed with glides. Long straight bills with spatulate tips, adapted for sifting floating and swimming aquatic invertebrates and small fish.

111 EURASIAN SPOONBILL
Platalea leucorodia Plate **11**

IDENTIFICATION 82.5–89 cm. *P.l.major*: **Adult non-breeding** Very similar to Black-faced Spoonbill but larger, has all-white forehead and cheeks and pale fleshy-yellow patch on upperside of bill-tip ('spoon'). **Adult breeding** Separated from Black-faced by same features as adult non-breeding. Additionally, shows yellow-orange throat skin; patch on upperside of bill-tip is yellower than on adult non-breeding. **Juvenile** Easily separated from Black-faced by dull pinkish bill and loral skin. **HABITAT** Marshes, lakes, tidal mudflats; lowlands. **RANGE & STATUS** Breeds W,N Africa, W(southern),C Palearctic, S Siberia, Middle East, S Pakistan, India (except NE), S Nepal, Sri Lanka, northern China, S Ussuriland; mostly winters to south, N Africa, Middle East, Pakistan, India (except NE), Nepal, S China. **SE Asia** Rare winter visitor SW,S Myanmar. Vagrant NW Thailand, Cambodia, E Tonkin.

112 BLACK-FACED SPOONBILL
Platalea minor Plate **11**

IDENTIFICATION 76 cm. Monotypic. **Adult non-breeding** All-white plumage, long spatulate all-black bill and black facial skin (encircling base of bill) diagnostic. Eyes dark. See Eurasian Spoonbill. **Adult breeding** Has yellowish to buffish nuchal crest and breast-patch and reddish eyes. At close range, shows narrow yellow crescent over eye. See Eurasian. **Juvenile** Similar to adult non-breeding but has blackish edges to outer primaries and small blackish tips to primaries, primary coverts and secondaries (particularly former). See Eurasian. **HABITAT & BEHAVIOUR** Tidal mudflats, coastal pools. Likely to be encountered in groups. **RANGE & STATUS** Breeds islets of N,S Korea, E China; winters locally S China, Taiwan, rarely E China, southern Japan, Borneo (Brunei); formerly Philippines (Luzon). **SE Asia** Local winter visitor E Tonkin, rarely C Annam, Cochinchina. Vagrant NE,C,S Thailand.

ARDEIDAE: BOTAURINAE: Bitterns

Worldwide 14 species. SE Asia 5 species. Smallish to medium-sized with dagger-like bills; more squat with shorter necks and legs than herons. Feed on various aquatic animals, including fish, amphibians, crustaceans, invertebrates.

113 GREAT BITTERN *Botaurus stellaris* Plate **12**

IDENTIFICATION 70–80 cm. *B.s.stellaris*: **Adult** Large size, overall buffish plumage and cryptic pattern with black streaks and vermiculations diagnostic. Has thick yellowish bill, blackish crown and submoustachial stripe, plain rufous-buff sides of head, more golden-buff sides of neck, and greenish legs and feet; base colour below whiter than above. In flight, appears heavy and broad-winged, upperside of flight-feathers are blackish-brown with broad rufous-brown bars and contrast somewhat with coverts. See juvenile Black-crowned Night and Purple Herons. **VOICE** May give harsh, nasal ***kau*** or ***krau*** in flight. Territorial call (unlikely to be heard in region) is a slow deep resonant far-carrying boom, ***up-RUMBH***. **HABITAT** Well-vegetated freshwater marshes, reedy ditches. **RANGE & STATUS** Breeds N Africa, Palearctic, NW,NE China, N Japan; many northerly populations winter Africa, south-eastern W Palearctic, Middle East, Pakistan, India, southern China, southern Japan, Philippines. **SE Asia** Rare to scarce winter visitor N,C,S,E Myanmar, W(coastal),NW,NE,C Thailand, Cambodia, N,S Laos, E Tonkin. Vagrant south-west Peninsular Malaysia, Singapore, C Annam, Cochinchina.

114 YELLOW BITTERN
Ixobrychus sinensis Plate **12**

IDENTIFICATION 36–38 cm. Monotypic. **Male** Small, skinny and long-necked with long dagger-like bill. Told by overall light brown to buffy coloration with contrasting blackish crown, rather cold olive-brown mantle to uppertail-coverts and scapulars, and white underside with vague darker lines down foreneck and upper breast. Often shows strong vinous wash on sides of head and neck and upperparts (particularly when breeding). In flight, buff greater upperwing-coverts and plain whitish underwing-coverts contrast strongly with blackish flight-feathers and tail. Rest of upperwing-coverts gradually become darker towards leading edge. Bill yellow-horn with dark ridge of upper mandible and often tip, legs and feet yellowish-green (yellower when breeding). See Von Schrenck's Bittern. **Female** More uniform above, with mostly warm brown crown (virtually no dark coloration), upperparts and scapulars, and obvious warm brown lines down foreneck to upper breast. Crown and upperparts may appear vaguely streaked. Lacks vinous wash of male. **Juvenile** Similar to female but has bold dark streaks on crown, upperparts and upperwing-coverts, and bolder dark streaking on underparts. Easily separated from other small bitterns, at rest, by buffish upperside with dark streaks, and in flight by contrasting buffish base colour to upperwing-coverts and plain whitish underwing-coverts. **VOICE** Territorial call is a series of low-pitched ***ou*** notes. In flight may give a staccato ***kak-kak-kak***. Also a rhythmic harsh, stressed ***ik..ik'RR.ikh*** (with rasping ***RR***), and strange thin rasping creaking notes. **HABITAT & BEHAVIOUR** Densely vegetated freshwater wetlands, reedbeds, sometimes rice paddies; up to 1,000 m. May freeze, with neck upstretched, when approached. **RANGE & STATUS** Resident (subject to some movements) Seychelles, Indian subcontinent, Sumatra, Philippines, Sulawesi, N Melanesia, W Micronesia. Breeds China (except NW,N), Taiwan, N,S Korea, Japan, Sakhalin, Kuril Is; mostly winters to south, S China, Borneo, Java, Bali, Wallacea, New Guinea. **SE Asia** Scarce to locally common resident Myanmar (except W,N), Thailand, Peninsular Malaysia, Singapore, Cambodia, E Tonkin, Cochinchina. Uncommon to locally common winter visitor Thailand, Peninsular Malaysia, Singapore (at least). Also recorded on passage Peninsular Malaysia. Recorded (status uncertain) N Myanmar, N,S Laos, C,S Annam. **BREEDING** March–November. **Nest** Simple platform, in small tree or amongst grass or rice. **Eggs** 3–6; milky-blue to greenish-blue; 31.2 x 23.9 mm (av.).

115 VON SCHRENCK'S BITTERN
Ixobrychus eurhythmus Plate **12**

IDENTIFICATION 39–42 cm. Monotypic. **Male** Resembles Cinnamon Bittern but crown (apart from blackish median stripe), ear-coverts, mantle to back and scapulars rich dark chestnut, contrasting strongly with mostly buffish underparts and wing-coverts. Has single dark chestnut line from chin to centre of upper breast. In flight resembles Yellow Bittern but differs by chestnut upperparts and leading edge of upperwing-coverts (particularly patch at wing-bend) and greyer, less contrasting upperside of flight-feathers and more silvery-grey underwing with dark markings on coverts. Has similar plain whitish underwing-coverts. **Female/juvenile** Similar to Cinnamon but has much bolder white to buff speckling and spotting on upperparts and upperwing-coverts. In flight, easily separated by mostly blackish-grey flight-feathers, tail and underwing colour (similar to adult male). **HABITAT & BEHAVIOUR** Swampy areas or pools in or near forest or secondary growth, sometimes in more open, well-vegetated wetlands; lowlands. Secretive, usually seen in flight. **RANGE & STATUS** Breeds eastern China, N,S Korea, SE Siberia, Ussuriland, Sakhalin, Japan; winters Sumatra, Java, Borneo, Philippines, Sulawesi. **SE Asia** Rare to scarce passage migrant Tenasserim, W,NW,C,S Thailand, Peninsular Malaysia, Singapore, Cambodia, Laos, E Tonkin, C Annam. Also recorded in winter Singapore, S Laos.

116 CINNAMON BITTERN
Ixobrychus cinnamomeus Plate **12**

IDENTIFICATION 38–41 cm. Monotypic. **Male** Almost uniform rich cinnamon-rufous upperside diagnostic. Underparts mostly warm buffish with indistinct dark chestnut line from centre of throat to centre of upper breast. In flight shows almost uniform rich cinnamon-rufous wings, paler below. Yellowish eyes and facial skin (like other small bitterns) turn red when breeding and yellow of bill becomes more orange. **Female** Upperside slightly duller and darker with vague buffish speckling on scapulars and upperwing-coverts; underparts have dark brown lines down sides of neck and upper breast, and darker central line. See Von Schrenck's Bittern. **Juvenile** Duller and darker above than female, with narrow buffish streaks on sides of head and dense buffish speckling and feather-fringing on upperparts and upperwing-coverts, has darker streaking below, wings a shade duller and darker. See Von Schrenck's. **VOICE** Territorial call is a throaty, 9–18 note ***ukh-ukh-ukh-ukh-ukh-ukh-ukh...*** (tailing off towards end), repeated after lengthy intervals. Flight call is a low, clicky ***ikh*** or ***ikh-ikh***. **HABITAT & BEHAVIOUR** Rice paddies, marshes, various freshwater wetland habitats; up to 1,830 m. Usually seen in flight. **RANGE & STATUS** Resident (subject to some movements) Indian subcontinent, C,E and southern China, Taiwan, Greater Sundas, W Lesser Sundas, Philippines, Sulawesi. **SE Asia** Fairly common to common resident, subject to some movements (except W Myanmar). Also recorded on passage Peninsular Malaysia, E Tonkin, N Annam. **BREEDING** All year. **Nest** Sturdy platform (sometimes slightly canopied), in tall grass or bush. **Eggs** 3–5; whitish, sometimes faintly bluish-tinged (becoming stained); 33.1–35.6 x 25.4–27.9 mm.

117 BLACK BITTERN
Dupetor flavicollis Plate **12**

IDENTIFICATION 54–61 cm. *D.f.flavicollis*: **Male** Combination of size, all-blackish ear-coverts and upperside, long dagger-like bill and blackish legs and feet diagnostic. Lacks head-plumes/crest. Throat and breast whitish with broad dark chestnut to blackish streaks and plain yellowish-buff patch on sides of neck and upper breast, lower underparts sooty-greyish, bill horn to yellowish-horn with dark tip and ridge of upper mandible. In flight shows rather broad, all-dark wings. See Little Heron and White-eared Night-heron. Note habitat and range. **Female** Like male but dark parts of plumage browner, breast-streaking more rufescent. **Juvenile** Similar to female but feathers of crown and upperside have narrow rufous fringes, breast washed buffish-brown. **VOICE** Territorial call is described as a loud booming. In flight, utters a rather throaty ***ukh-ukh-ukh-ukh*** and ***ukh...ukh.ukh.ukh*** etc. **HABITAT & BEHAVIOUR** Marshy freshwater wetlands, rice-paddy margins, freshwater swamp forest, mangroves; up to 1,100 m (usually lowlands). Usually seen in flight. **RANGE & STATUS** Breeds Indian subcontinent, C,S,SE China, Sumatra, Philippines, New Guinea, Australia, N Melanesia; probably Sulawesi, Moluccas. Some northern populations winter to south, Borneo, Java, Bali, probably Sulawesi. Recorded (status unclear) Lesser Sundas. **SE Asia** Uncommon resident S Thailand, Cambodia, Cochinchina. Uncommon to fairly common breeding visitor Myanmar, W(coastal),NW,C,NE Thailand, E Tonkin. Uncommon to fairly common passage migrant S Thailand, Peninsular Malaysia, Singapore. Recorded (status uncertain) N,S Laos, N,C Annam. **BREEDING** June–October. **Nest** Simple platform, in bush or amongst grass or reeds. **Eggs** 3–5; bluish- to greenish-white (becoming stained); 41.6 x 31.4 mm (av.).

ARDEIDAE: ARDEINAE: Herons & egrets

Worldwide c.45 species. SE Asia 18 species. Smallish to very large with dagger-like bills. Feed on various aquatic animals, including fish, amphibians, crustaceans, invertebrates, small mammals and young birds; *Gorsachius* feed mainly on earthworms and frogs. **Night-herons** (*Gorsachius, Nycticorax*) Similarly shaped to bitterns but nocturnal, usually roosting in trees during daytime. **Pond-herons** (*Ardeola*) and **Little Heron** (*Butorides*) Smallish to medium-sized, more squat with shorter necks and legs; bittern-like. **Typical herons** (*Ardea*) Large to very large with long necks and legs and powerful bills. **Egrets** (*Mesophoyx, Egretta, Bubulcus*) Medium-sized to fairly large, relatively slim and long-legged, generally with relatively slender bills; plumages mostly white.

118 WHITE-EARED NIGHT-HERON
Gorsachius magnificus Plate **12**

IDENTIFICATION 54–56 cm. Monotypic. **Male** Shows unmistakable combination of blackish head with white post-ocular and cheek-stripes, uniform dark brown upperparts (tinged purple) and whitish underparts with dark brown streaks/scales. Rear sides of neck orange-rufous to rufous-chestnut, bordered black at front and rear, throat white with dark mesial streak, long black nape-plumes, mostly dark bill, yellowish eyes and pea-green legs and feet; may show white spotting on lower mantle. See Black Bittern. Note habitat. **Female** Head and neck pattern less pronounced, has whitish streaks and spots on mantle and upperwing-coverts and shorter nape-plumes. **Juvenile** Similar to female but blackish parts of plumage browner, with heavier whitish to buff markings on upperparts and wing-coverts. **HABITAT & BEHAVIOUR** Streams in broadleaved evergreen forest. Probably feeds at night. **RANGE & STATUS** Resident (probably subject to some movements) S,SE China. **SE Asia** Rare resident E Tonkin. **BREEDING** Undocumented.

119 MALAYSIAN NIGHT-HERON
Gorsachius melanolophus Plate **12**

IDENTIFICATION 48–51 cm. *G.m.melanolophus*: **Adult** Stocky and thick-necked with shortish bill and legs. Shows distinctive combination of deep rufous sides of head and neck, black crown and long crest, black streaks down foreneck and breast, and chestnut-tinged brown upperparts and wing-coverts with fine blackish vermiculations. Belly and vent densely marked blackish and whitish, tail blackish, facial skin blue to greenish-blue, legs and feet greenish. In flight, from above, shows rufous secondaries with broad blackish subterminal band, white-tipped rufous primary coverts and buff-tipped blackish primaries. Facial skin may turn reddish when breeding. **Juvenile** Duller than adult with dense irregular whitish to buffish and greyish bars and vermiculations overall. Crown and nape blacker with more pronounced white markings, throat whitish with broken dark mesial streak. In flight, shows more contrast between upperwing-coverts and flight-feathers. Gradually attains adult plumage. **VOICE** Territorial call is a series of 10–11 deep ***oo*** notes, given at c.1.5 s intervals; heard between dusk and dawn. **HABITAT & BEHAVIOUR** Swampy areas and streams in broadleaved evergreen and mixed deciduous forests, freshwater swamp forest, secondary forest, bamboo; up to 1,600 m. Very secretive, feeding mostly at night. **RANGE & STATUS** Breeds SW,NE India, Nicobar Is, SW,S China, Philippines; some populations winter to south, Greater Sundas. Recorded Sulawesi subregion (Talaud Is, Peleng). **SE Asia** Scarce to uncommon resident or breeding visitor W,NW,NE,SE Thailand, Cambodia, Laos, C Annam, Cochinchina. Scarce breeding visitor E Tonkin. Scarce winter visitor Singapore; probably also extreme S Thailand,

Peninsular Malaysia. Scarce passage migrant C,S Thailand, Peninsular Malaysia. Recorded (status uncertain) SW,C,S,E Myanmar, Tenasserim, W Tonkin, N Annam. **BREEDING** May–July. Sometimes multi-brooded. **Nest** Flimsy platform, in tree; 5–12 m above ground. **Eggs** 3–5; white, tinged bluish; 46.2 x 37.2 mm (av.).

120 BLACK-CROWNED NIGHT-HERON

Nycticorax nycticorax Plate **12**

IDENTIFICATION 58–65 cm. *N.n.nycticorax*: **Adult non-breeding** Unmistakable. Thickset and mostly grey with contrasting black crown, mantle and scapulars. Has blackish bill and lores, yellow legs and feet and long whitish nape-plumes. In flight appears bulky and big-headed with broad, rounded wings. **Adult breeding** Black of plumage glossed bluish-green, lores and legs turning red during courtship. **Juvenile** Similar to non-breeding pond-herons but larger, stockier, shorter-necked and thicker-billed, wings and tail dark brown, with prominent buffish to whitish drop-like spots on mantle, scapulars and upperwing-coverts. Gradually attains adult plumage; mature at 3–4 years. See much smaller Little Heron and Von Schrenck's, Cinnamon and Great Bitterns. **VOICE** A deep hollow croaking ***kwok***, ***quark*** or more sudden ***guk***, particularly in flight. **HABITAT & BEHAVIOUR** Marshes, swamps, rice paddies, mangroves. Gregarious. Mainly feeds at night, roosting in thick cover during day. Mostly seen at dusk and dawn. **RANGE & STATUS** Breeds Africa, Madagascar, west and southern W and C Palearctic, Middle East, Indian subcontinent, China (except NW,N), Taiwan, S Korea, Japan, Java, Borneo, N,C,S America; most northern populations winter to south, Africa, S China, Sumatra, N Borneo, Philippines, Sulawesi, C America, Caribbean. **SE Asia** Uncommon to locally common resident (subject to some movements) C,S Myanmar, W(coastal),C,NE,SE Thailand, Peninsular Malaysia, Singapore, Cambodia, E Tonkin, S Annam, Cochinchina. Former resident (current status unknown) SW,W,C,S,E Myanmar, Tenasserim, C Annam. Uncommon to common non-breeding/winter visitor N Myanmar (formerly bred), NW,S(south-east) Thailand, Peninsular Malaysia, Singapore. Recorded (status uncertain) Laos. **BREEDING** March–January. Usually colonial. **Nest** Rough platform, in tree. **Eggs** 3–5; pale blue-green; 49 x 35.1 mm (av.).

121 LITTLE HERON *Butorides striata* Plate 12

IDENTIFICATION 40–48 cm. *B.s.javanicus*: **Adult** Distinctive. Small and mostly slaty-greyish with black crown, long black nape-plumes (often raised in untidy crest), green-tinged upperside, whiter sides of head, throat and central breast, prominent black streak along lower ear-coverts, and narrow whitish to buffish-white fringes to scapulars and upperwing-coverts. Bill blackish with yellower base to lower mandible, facial skin olive-yellow, legs and feet dull yellowish-orange. See Black Bittern. **Juvenile** Browner above than adult with less uniform dark crown and nape, indistinct buffish markings on crown, mantle and lesser coverts, underparts streaked dark brown and whitish, legs and feet dull greenish to yellowish-green. See Von Schrenck's, Cinnamon and Black Bitterns. **Other subspecies in SE Asia** *B.s.spodiogaster* (Coco Is, S Myanmar); *actophilus* (northern Myanmar, northern Thailand, northern Indochina; exact distribution and status unclear); *amurensis* (visitor, recorded S Thailand southwards; E Myanmar?). **VOICE** May give a distinctive short harsh ***skeow***, ***k-yow*** or ***k-yek*** when flushed, and a high-pitched, raspy ***kitch-itch itch*** when alarmed. **HABITAT** Mangroves, tidal mudflats, offshore islands, rivers and streams in or near forest, lakes; up to 1,400 m. **RANGE & STATUS** Resident sub-Saharan Africa, Madagascar, Red Sea coast, Indian subcontinent, Greater Sundas, Philippines, Wallacea, New Guinea, Australia, Melanesia, Polynesia, S America. Breeds China (except NW,N), Taiwan, N,S Korea, SE Siberia, Ussuriland, Japan; some northern populations winter to south, N Sumatra, N Borneo, Philippines, N Sulawesi. **SE Asia** Scarce to locally common resident (mostly coastal) Myanmar, W,C,SE,S Thailand, Peninsular Malaysia, Singapore, Cambodia, S Laos, E Tonkin, Cochinchina. Fairly common to common winter visitor throughout. Also recorded on passage Peninsular Malaysia, N Laos. **BREEDING** All year. Loosely colonial. **Nest** Crude platform, in tree or bush; 2–12 m above ground. **Eggs** 3–5; pale bluish-green; 35.6–41.4 x 27.7–29 mm (av.; *javanicus*).

122 INDIAN POND-HERON

Ardeola grayii Plate **14**

IDENTIFICATION 45 cm. Monotypic. **Adult non-breeding** Possibly indistinguishable from other pond-herons. **Adult breeding** Unmistakable. Head, neck and breast brownish-buff with long white head-plumes, throat whitish, mantle and scapulars dark brownish-maroon, greater upperwing-coverts washed buff. **Juvenile** Possibly indistinguishable from other pond-herons. **VOICE** May give a gruff rolling ***urrh urrh urrh...*** and abrupt hollow ***okh*** in flight. Repeated conversational ***wa-koo*** at breeding sites. **HABITAT** Various freshwater wetlands, sometimes coastal pools; lowlands. **RANGE & STATUS** Resident (subject to some movements) Iran, Afghanistan, Indian subcontinent. **SE Asia** Uncommon to locally common resident Myanmar. Rare to scarce passage migrant/non-breeding visitor C,W(coastal),S Thailand, north-west Peninsular Malaysia. **BREEDING** May–September. Colonial. **Nest** Untidy platform, in tree; 2–4 m above ground or water. **Eggs** 3–5; pale sea-green; 38 x 28.5 mm (av.).

123 CHINESE POND-HERON

Ardeola bacchus Plate **14**

IDENTIFICATION 45–52 cm. Monotypic. **Adult non-breeding** At rest appears rather small, stocky and nondescript, but in flight shows distinctive white wings and tail. Head, neck and breast mostly buffish with bold dark brown streaks, mantle and scapulars darkish olive-brown, rest of plumage white. Bill yellowish with dark tip, sometimes also with darker upper mandible, legs and feet greenish-yellow to yellow. Possibly indistinguishable from other pond-herons but may tend to show more obvious dusky tips to outermost primaries than Javan. **Adult breeding** Chestnut-maroon head (including plumes), neck and breast, white throat and blackish-slate mantle and scapulars diagnostic. **Juvenile** Similar to adult non-breeding but rather more spotted below than streaked (markings fainter), has brownish markings on tail, brown inner primaries and grey-washed upperwing-coverts. **VOICE** May give a high-pitched harsh squawk when flushed. **HABITAT** Various freshwater wetlands, also mangroves and tidal pools; up to 1,600 m. **RANGE & STATUS** Breeds NE India, China (except NW), S Ussuriland; more northerly populations winter south to NE India, Andaman Is, S China, N Sumatra, N Borneo. **SE Asia** Scarce to local resident (subject to some movements) E Tonkin; possibly N Laos. Uncommon to common winter visitor (except SW Myanmar). Also recorded on passage Laos, E Tonkin, C Annam. **BREEDING** March–May. Colonial. **Nest** Simple platform, in tree. **Eggs** 3–6; greenish-blue; 37.7 x 28.4 mm (av.).

124 JAVAN POND-HERON

Ardeola speciosa Plate **14**

IDENTIFICATION 45 cm. Monotypic. **Adult non-breeding** Possibly indistinguishable from other pond-herons but may show less obvious dusky tips to outermost primaries than Chinese. **Adult breeding** Pale brownish-buff to creamy-whitish head and neck, white head-plumes, deep cinnamon-rufous breast and blackish-slate mantle and scapulars diagnostic. **Juvenile** Possibly indistinguishable from other pond-herons. **VOICE** Similar to other pond-herons. **HABITAT** Various wetlands, particularly along coast; lowlands. **RANGE & STATUS** Resident Java, Bali, SE Borneo, W Lesser Sundas, Philippines (Mindanao), Sulawesi. Recorded (status unclear) S Sumatra. **SE Asia** Common resident C Thailand, Cambodia, Cochinchina. Recorded (status uncertain) Tenasserim. Vagrant S Thailand, Peninsular Malaysia,

Singapore. **BREEDING** June–September. Colonial. **Nest** Untidy platform, in tree. **Eggs** 3–5; dark greenish-blue; 37.7 x 28.7 mm (av.).

125 EASTERN CATTLE EGRET
Bubulcus coromandus Plate **14**

IDENTIFICATION 48–53 cm. Monotypic. **Adult non-breeding** Smaller and stockier than other egrets, often appearing rather hunched. Has shortish neck, rather rounded head with pronounced 'jowl', relatively short thick yellow bill and blackish legs and feet (often tinged brownish to greenish). Facial skin yellowish to greenish-yellow. Only likely to be confused with Intermediate Egret but considerably smaller, with notably shorter bill and legs (legs and feet somewhat paler) and much shorter neck without obvious kink. In flight shows shorter, rounder wings and much less downward-bulging neck; legs and feet extend less beyond tail-tip. See white morph Pacific Reef-egret. **Adult breeding** Head and neck extensively rufous-buff, with short rufous-buff nape and breast-plumes, long rufous-buff back-plumes and more yellowish legs and feet. During courtship, bill is reddish with yellower tip and (briefly) face is violet to purple, eyes red, and legs dusky-reddish. **Juvenile** Like adult non-breeding but may show blackish legs and feet and grey tinge to plumage. **VOICE** Sometimes gives quiet croaking ***ruk*** or ***RIK-rak*** in flight. Low conversational rattling at roost sites. **HABITAT & BEHAVIOUR** Various wetlands, cultivation (usually avoids saline wetlands); up to 800 m. Gregarious. Often feeds near cattle, on disturbed insects. **RANGE & STATUS** Resident (subject to dispersive movements) Pakistan, Indian subcontinent, C and southern China, Taiwan, S Korea, Japan, Sundas, Philippines, Wallacea; self-introduced Australasia. **SE Asia** Scarce to local resident (subject to some movements) N,S Myanmar, W(coastal),C,SE,S Thailand, north-west Peninsular Malaysia, Cambodia, Cochinchina. Former resident (current status uncertain) SW,W,C,E Myanmar, Tenasserim. Fairly common winter visitor (except W Tonkin). Also recorded on passage Peninsular Malaysia, Laos. **BREEDING** April–July. Colonial. **Nest** Rough platform, in tree. **Eggs** 3–5; whitish, tinged blue or green; 44.1 x 36.5 mm (av.). **NOTE** The treatment of *coromandus* as a full species follows Rasmussen & Anderton (2005).

126 GREY HERON *Ardea cinerea* Plate **13**

IDENTIFICATION 90–98 cm. *A.c.jouyi*: **Adult non-breeding** Distinctive. Large, long-necked and long-legged, upperparts and wing-coverts greyish, head, foreneck and underparts mostly whitish with prominent broad black head-stripes, extending to long plumes and blackish streaks down centre of foreneck to upper breast. Has thick dull yellowish bill and black shoulder-patch. In flight, mostly greyish upperwing-coverts contrast with blackish primaries and secondaries, underwing similar but with darker leading edge to coverts; shows black band down side of body and white leading edge to wing (when viewed head-on). See other *Ardea* herons. **Adult breeding** Bill bright orange-yellow, has elongated white scapular plumes; during courtship, bill, legs and feet become deeper orange to vermilion. **Juvenile** Duller than adult non-breeding, head and neck pattern less contrasting, crown rather uniform dark greyish, much shorter nape-plumes, grey sides of neck, duller dark band down side of body and duller bill. See much larger and darker White-bellied and Great-billed Herons. **VOICE** Loud, harsh, abrupt ***krahnk***, particularly in flight. Also, deep, grating ***raark*** when flushed. **HABITAT** Various inland and coastal wetlands, rice paddies, mangroves; up to 1,000 m. **RANGE & STATUS** Breeds W and sub-Saharan Africa, Madagascar, Palearctic, Iran, S Pakistan, India, Bangladesh, Sri Lanka, China, Taiwan, N,S Korea, Japan, Sumatra, Java, Sumbawa, Sumba; most northerly populations winter south to Africa, Madagascar, Middle East, Indian subcontinent, Greater Sundas, Philippines. **SE Asia** Scarce to local resident (subject to some movements) western Peninsular Malaysia, Singapore, Cambodia, Cochinchina. Former resident (current status uncertain) C,S Myanmar. Occurs in summer C Thailand (formerly bred), C Annam (breeds?). Uncommon to fairly common winter visitor (except Peninsular Malaysia, Singapore). **BREEDING** All year. Usually colonial. **Nest** Large untidy cupped structure, in tree. **Eggs** 3–5; blue to greenish-blue (soon fading); 58.6 x 43.5 mm (av.).

127 WHITE-BELLIED HERON
Ardea insignis Plate **13**

IDENTIFICATION 127 cm. Monotypic. **Adult** Very large and long-necked; dark greyish plumage, sharply defined white throat, and whitish belly and vent distinctive. Has greyish-white nape-plumes, narrow white scapular plumes, white streaks on lower foreneck and breast, mostly blackish bill with dull yellowish-green base to lower mandible and dark greyish legs and feet. In flight, wings similar to Grey Heron but upperwing-coverts uniformly darker grey, underwing-coverts uniformly whitish. See very similar Great-billed Heron and juvenile Grey. Note range. **Juvenile** Grey of plumage browner than adult. **VOICE** Loud, deep, braying, croaking ***ock ock ock ock urrrrrr***. **HABITAT & BEHAVIOUR** Shingle banks and shores of larger rivers and nearby wetlands; up to 1,135 m. Often occurs singly, and can be quite secretive. **RANGE & STATUS** Resident NE India, Bhutan; formerly Nepal. **SE Asia** Rare to scarce resident N Myanmar. Former resident (current status unknown) SW,W,C,S Myanmar. **BREEDING** April–June. **Nest** Large cupped platform, high in tree. **Eggs** 4; pale sea-green or bluish-green (soon fading); c.70 x 50 mm.

128 GREAT-BILLED HERON
Ardea sumatrana Plate **13**

IDENTIFICATION 114–115 cm. *A.s.sumatrana*: **Adult non-breeding** Similar to White-bellied Heron but shorter-necked, somewhat browner overall with less defined pale streaks on scapulars, foreneck and breast, belly and vent dull greyish, pale ear-coverts and less sharply contrasting pale throat. In flight shows distinctive, uniformly dark wings. Note range. See juvenile Grey Heron. **Adult breeding** Has more distinct white-tipped greyish plumes on scapulars and breast. **Juvenile** Warmer-tinged than adult non-breeding, upperparts tipped buff to rufous-buff, neck and underparts tinged vinaceous, lower foreneck and breast more broadly streaked whitish. **VOICE** Occasional loud harsh croaks. Also a series of unnerving loud deep guttural roars during breeding season (mainly at night). **HABITAT & BEHAVIOUR** Mangroves, islets, undisturbed beaches, sometimes venturing upstream along large rivers. Solitary or in pairs. Display by pairs includes upward-stretching of neck with throat puffed, fluffing and twisting of neck, and bill-clasping. **RANGE & STATUS** Coastal resident Greater Sundas, Philippines, Wallacea, New Guinea, N Australia. **SE Asia** Rare and local coastal resident SW Myanmar, Tenasserim, S Thailand, Peninsular Malaysia, Singapore, Cambodia. Former resident (current status unknown) SE Thailand, Cochinchina. **BREEDING** March–September. **Nest** Large platform, in tree; 4.5–6 m above ground. **Eggs** 2; greyish to bluish-green; 68.6 x 47.2 mm.

129 GOLIATH HERON *Ardea goliath* Plate **13**

IDENTIFICATION 135–150 cm. Monotypic. **Adult** Reminiscent of Purple Heron but almost twice as large; bill, legs and feet blackish, crown, ear-coverts and hindneck uniform bright rufous-chestnut with indistinct bushy crest and no black markings. In flight shows more uniformly grey upperwing with less contrasting primaries and secondaries, underwing-coverts slightly paler chestnut than that of belly. **Juvenile** Forehead and crown blackish, hindneck paler rufous-chestnut, mantle feathers and upperwing-coverts fringed rufous, black markings on foreneck less clearly defined, belly and vent streaked grey and pale chestnut, underwing-coverts dark grey. See much smaller Grey Heron. **VOICE** Flight call is a loud harsh deep ***kowoorrk-kowoorrk-woorrk-work-worrk***, likened to sound made by bellowing calf. **HABITAT & BEHAVIOUR** Various inland and coastal wetlands, particularly large lakes and rivers. Usually solitary. **RANGE & STATUS** Resident sub-Saharan Africa, SW

Arabian Peninsula, Iraq, S Iran. Recorded (status uncertain) C,NE India, Bangladesh, Sri Lanka. **SE Asia** Vagrant S Myanmar (one old record, possibly dubious).

130 **PURPLE HERON** *Ardea purpurea* Plate **13**
IDENTIFICATION 78–90 cm. *A.p.manilensis*: **Adult** Combination of black crown and nape-plumes, mostly rufous-chestnut neck with black lines down sides and front, and dark chestnut-maroon belly, flanks and vent diagnostic. Bill, legs and feet mostly yellowish. Smaller, slimmer and longer-necked than Grey Heron, upperparts darker with chestnut-maroon shoulder-patch and wash on scapulars. In flight, recalls Grey but upperwing shows darker coverts and buff (rather than white) markings near wing-bend and mostly chestnut-maroon underwing-coverts; at distance, feet look bigger and extend slightly further beyond tail-tip, retracted neck is more deeply bulging. See much larger Goliath Heron. **Juvenile** Sides of head and hindneck duller and more buffish, with less defined dark markings, no nape-plumes, distinctive dark brown upperparts and wing-coverts with warm buffish-brown fringes, buffier-brown lower underparts and browner underwing-coverts. See Great Bittern. Gradually attains adult plumage; mature at 3–5 years. **VOICE** Flight call similar to Grey but thinner and higher-pitched. Loud, hoarse ***raanka*** and ***raank*** calls noted at roosting sites. **HABITAT & BEHAVIOUR** Well-vegetated freshwater wetlands, marshes, lakes, reedbeds, large rivers, occasionally coastal wetlands; up to 1,000 m. More secretive than Grey Heron. **RANGE & STATUS** Breeds sub-Saharan Africa, Madagascar, southern W Palearctic, C Asia, Middle East, Indian subcontinent, C,S and eastern China, Taiwan, Ussuriland, Sundas, Philippines, Sulawesi, S Moluccas; some northern populations winter to south, Africa. **SE Asia** Scarce to local resident (subject to some movements) W,NE,C,SE,S Thailand, Peninsular Malaysia, Singapore, Cambodia, S Laos, Cochinchina, probably E Tonkin. Former resident (currently?) S Myanmar, N Laos. Scarce to locally fairly common winter visitor (except W Myanmar, W Tonkin, N Annam). Also recorded on passage N Laos. **BREEDING** All year. Colonial. **Nest** Bulky structure, in tree or on ground amongst reeds. **Eggs** 3–6; pale green or greenish-blue; 54.6 x 39.7 mm (av.).

131 **GREAT EGRET** *Ardea alba* Plate **14**
IDENTIFICATION 85–102 cm. *A.a.modestus*: **Adult non-breeding** Considerably larger than other egrets, with very long neck, dagger-like yellow bill (extreme tip sometimes black) and blackish legs and feet (upper tibia often yellowish). Facial skin olive-yellow. Only likely to be confused with smaller Intermediate Egret but more elegant, neck longer and more strongly kinked (though distinctly long and thin when held outstretched), head thinner and bill much longer. At close range, shows diagnostic pointed extension of facial skin below/behind eye. In flight, retracted neck is more deeply bulging. **Adult breeding** Has very long back-plumes and very short coarse breast-plumes. During courtship, bill turns black, facial skin cobalt-blue, legs reddish. See Intermediate and Little Egrets. **Juvenile** Similar to adult non-breeding. **VOICE** May utter a harsh but high-pitched rolling ***krr'rr'rr'rra*** when flushed. Various guttural calls at breeding colonies. **HABITAT & BEHAVIOUR** Various inland and coastal wetlands, rice paddies, mangroves; up to 800 m. Gregarious. Flies with slower wingbeats than Intermediate. **RANGE & STATUS** Breeds sub-Saharan Africa, Madagascar, C,SE Europe, east W Palearctic, C Asia, S,SE Siberia, Ussuriland, Indian subcontinent, NW,NE,S,SE China, Taiwan, N,S Korea, Japan, Java, S Sulawesi, Australia, N(south),C,S America. More northerly populations winter Africa, S Europe, south-east W Palearctic, Greater Sundas, Philippines, C America; Australian populations disperse to New Guinea and New Zealand. Non-breeding visitor Wallacea. **SE Asia** Rare to local resident (subject to some movements) S Myanmar, W(coastal),C,SE Thailand, west Peninsular Malaysia, Cambodia, Cochinchina. Former resident (currently?) elsewhere Myanmar. Fairly common to common winter visitor throughout. Also recorded on passage Laos, Cochinchina. **BREEDING** May–October. Colonial. **Nest** Rough platform, in tree. **Eggs** 3–4; pale greenish-blue; 48.8–58.9 x 36.8–43.2 mm.

132 **INTERMEDIATE EGRET**
Mesophoyx intermedia Plate **14**
IDENTIFICATION 65–72 cm. Monotypic. **Adult non-breeding** Easily confused with Great Egret. Differs primarily by smaller size, considerably shorter bill, more rounded head, shorter and less distinctly kinked neck and more hunched appearance. At close range, facial skin does not extend in point below/behind eye. In flight, retracted neck has less pronounced downward bulge. Recalls Eastern Cattle Egret but much larger and more graceful, with longer neck and bill, less rounded head and longer, darker legs. Facial skin pale yellow to greenish-yellow, legs all blackish. **Adult breeding** Has long back-plumes and longish breast-plumes, bill yellow, sometimes with black or dark brown on tip and ridge of upper mandible (blacker with yellow base during courtship); facial skin, legs and feet as adult non-breeding. See Great and Chinese Egrets. **Juvenile** Similar to adult non-breeding. **VOICE** May give a harsh croaking ***kwark*** or ***kuwark*** when flushed. Distinctive buzzing sounds during display at breeding colonies. **HABITAT & BEHAVIOUR** Various wetlands; up to 800 m. Gregarious. Wingbeats more rapid than Great Egret, slower and more graceful than Cattle. Often raises crown-feathers. **RANGE & STATUS** Resident (subject to local movements) sub-Saharan Africa, Indian subcontinent, Sumatra, Java, Bali, N Sulawesi, Australia, probably New Guinea. Breeds southern China, Taiwan, S Korea, Japan; mostly winters to south, Borneo, Philippines, probably Sulawesi. Non-breeding visitor Lesser Sundas, Moluccas (probably from Australasia). **SE Asia** Scarce to local resident (subject to local movements) S Myanmar, Cambodia, Cochinchina; ? SE Thailand. Uncommon to fairly common winter visitor (except W Tonkin). Also recorded on passage Laos. **BREEDING** June–September. **Nest** Rough platform, in tree. **Eggs** 3–5; pale sea-green; 47.6 x 35.8 mm (av.).

133 **LITTLE EGRET** *Egretta garzetta* Plate **14**
IDENTIFICATION 55–65 cm. *E.g.garzetta*: **Adult non-breeding** From other white egrets by combination of size, mostly dark bill and blackish legs with yellow to greenish-yellow feet (can be hard to see when muddy). Bill may have small (variable) amount of pale yellowish to flesh-colour at base of lower mandible (often difficult to see), facial skin dull greenish to yellowish-grey. In flight, legs and feet extend well beyond tail-tip. *E.g.nigripes* (probable scarce non-breeding visitor Peninsular Malaysia, Singapore) has blackish feet. See very similar Chinese Egret and adult breeding Great Egret. **Adult breeding** Has pronounced nape-, back- and breast-plumes, blackish bill. During courtship, feet often more reddish, facial skin more reddish to pinkish-purple. **Juvenile** Similar to adult non-breeding. **VOICE** Hoarse, grating ***kgarrk*** or longer ***aaahk*** when flushed. Also, various guttural calls at breeding colonies. **HABITAT & BEHAVIOUR** Various open freshwater and coastal wetlands, cultivation; up to 800 m. Gregarious. **RANGE & STATUS** Breeds W and sub-Saharan Africa, southern W Palearctic, C Asia, Middle East, Indian subcontinent, C and southern China, Taiwan, S Korea, Japan, Java, Sulawesi, Australia. Most northerly populations winter Africa, S Europe, Middle East, Indian subcontinent, S China, Sumatra, Borneo, Philippines, N Sulawesi, Moluccas; Australian population disperses northwards to New Guinea, W Micronesia and N Melanesia at least. **SE Asia** Scarce to locally common resident (subject to local movements) N Myanmar, W(coastal),C,SE,S Thailand, Peninsular Malaysia, Cambodia, Cochinchina. Former resident (current status uncertain) elsewhere Myanmar. Uncommon to common winter visitor throughout. Also recorded on passage Laos. Recorded in summer (with no breeding evidence) Singapore. **BREEDING** June–April. Usually colonial. **Nest** Large rough shallow cup, in tree; 2–6 m above ground. **Eggs** 3–5; pale blue-green (soon fading); 44.4 x 31.7 mm (av.).

134 **PACIFIC REEF-EGRET** *Egretta sacra* Plate **14**
IDENTIFICATION 58 cm. *E.s.sacra*: **Adult dark morph non-breeding** Overall dark greyish plumage diagnostic. Often has whitish chin and throat line, bill as in white morph but facial skin may be greyer. **Adult white morph non-breeding** Shorter-legged and somewhat less elegant than other white egrets (except Cattle), differing by combination of size, bill length and colour, and leg colour. Facial skin greenish; upper mandible greenish-horn with variable amount of blackish to dark brownish, lower mandible horn- to greenish-yellow with slightly darker tip; legs and feet mostly rather uniform olive-green to yellowish. Very similar to Chinese Egret but legs shorter (tarsus always shorter than bill), bill somewhat thicker and less pointed, typically with paler upper mandible (darker on culmen) and less contrasting dark tip to lower mandible. In flight, only feet and small amount of legs project behind tail-tip. See Little Egret. **Adult breeding** Both morphs acquire shortish, tufted plumes on nape, back and breast; bill, legs and feet often yellower. **Juvenile** Similar to adult non-breeding. Dark morph is a somewhat paler smoky-grey. **VOICE** An occasional grunting ***ork*** when foraging and harsh ***squak*** when flushed. **HABITAT & BEHAVIOUR** Rocky shores, islets, beaches, sometimes mudflats. Usually found singly or in pairs. **RANGE & STATUS** Coastal resident Andaman and Nicobar Is, S,SE China, Taiwan, S Korea, Japan, Sundas, Philippines, Wallacea, New Guinea, Australia, New Zealand, Melanesia, Polynesia, Micronesia. **SE Asia** Uncommon to common coastal resident (except N Annam). **BREEDING** May–December. Usually colonial. **Nest** Rough, cupped platform, in small tree or bush or on rock ledge. **Eggs** 3–5; pale bluish-green; 44.8 x 33.3 mm (av.).

135 **CHINESE EGRET** *Egretta eulophotes* Plate **14**
IDENTIFICATION 68 cm. Monotypic. **Adult non-breeding** Very similar to white morph Pacific Reef-egret but legs longer (tarsus longer than bill), bill somewhat more slender and pointed, upper mandible black to brownish-black, lower mandible yellowish-flesh to yellowish with contrasting blackish terminal third. Facial skin pale greenish to greenish-yellow. In flight, very similar to Little Egret, with legs and feet extending well beyond tail-tip. At rest, differs from Little Egret by mostly dull greenish legs and feet and more extensive pale area on lower mandible. **Adult breeding** Recalls Little but has yellow to orange-yellow bill and distinctive shaggy nape-plumes (crest). Facial skin light blue to grey-blue, legs blackish with greenish-yellow to yellow feet. **Juvenile** Similar to adult non-breeding. **HABITAT & BEHAVIOUR** Tidal mudflats, mangroves. Gregarious. **RANGE & STATUS** Breeds islands off N,S Korea; formerly coastal S,E China; winters locally Greater Sundas, Philippines, Sulawesi. **SE Asia** Local winter visitor west S Thailand, west (and probably south-east) Peninsular Malaysia, Singapore; occasionally C Thailand, E Tonkin, probably Cochinchina. Rare to scarce passage migrant C,S Thailand, Peninsular Malaysia, Singapore, E Tonkin, S Annam, Cochinchina.

PHAETHONTIDAE: Tropicbirds

Worldwide 3 species. SE Asia 3 species. Graceful aerial seabirds, resembling terns but with relatively shortish, thick bills and greatly elongated central tail-feathers. Flight rather pigeon-like with flapping and circling interspersed with long glides. Feed mainly on fish and squid by hovering and then plunge-diving.

136 **RED-BILLED TROPICBIRD**
Phaethon aethereus Plate **15**
IDENTIFICATION 46–51 cm (tail-streamers up to 56 cm or more). *P.a.indicus*: **Adult** Similar to White-tailed Tropicbird but bill thicker and bright orange-red with black cutting edge, black barring on mantle to uppertail-coverts and upperwing-coverts, mostly black primary upperwing-coverts, no broad black band across upperwing-coverts. See Red-tailed Tropicbird. **Juvenile** From White-tailed by mostly white crown, darker nape, narrower and denser blackish barring on upperparts and upperwing-coverts, and mostly blackish primary upperwing-coverts. Bill yellowish-cream with dark tip and black cutting edge (greyish pre-fledging). See Red-tailed. **HABITAT** Open seas, islets. **RANGE & STATUS** Breeds offshore islands W,NE Africa, Arabian Peninsula, Iran, C America, Caribbean, northern S America, SC Atlantic Ocean; wanders outside breeding season, including offshore Indian subcontinent. **SE Asia** Offshore/coastal vagrant S Myanmar (including Coco Is), south Tenasserim; all old sight records and perhaps doubtful. Recorded (said to breed) Cochinchina (Con Dao Is). **BREEDING** Often colonial. **Nest** None, eggs laid on rock ledge or in crevice. **Eggs** 1, white to fawn, mottled darker purplish-brown.

137 **RED-TAILED TROPICBIRD**
Phaethon rubricauda Plate **15**
IDENTIFICATION 46–48 cm (tail-streamers up to 35 cm more). *P.r.westralis?*: **Adult** Combination of red bill and tail-streamers and almost completely white plumage diagnostic. Has narrow black mask and blackish shafts-streaks on primaries, primary coverts, innermost wing-coverts, inner secondaries and outertail-feathers, and black chevrons on tertials. Body and wings often flushed pink. See Red-billed and White-tailed Tropicbirds. **Juvenile** Easily distinguished from other tropicbirds by greyish to blackish bill and lack of obvious black markings on upperside of primaries and primary coverts. **HABITAT** Open seas, islets. **RANGE & STATUS** Breeds widely tropical and subtropical Indian and Pacific Oceans, including Nicobar Is, E Wallacea, offshore SW,NE Australia; wanders quite widely outside breeding season, including coastal waters of Sumatra, Java. **SE Asia** Vagrant west S Thailand.

138 **WHITE-TAILED TROPICBIRD**
Phaethon lepturus Plate **15**
IDENTIFICATION 38–41 cm (tail-streamers up to 40 cm). *P.l.lepturus*: **Adult** Very distinctive. White with narrow black mask, diagonal black band across upperwing-coverts, mostly black upperside of outer primaries and very long white tail-streamers with black shafts. Bill yellowish to orange. See Red-billed and Red-tailed Tropicbirds. **Juvenile** Lacks black band across upperwing, has black bars on crown, coarse, well-spaced black bars/scales on mantle to uppertail- and upperwing-coverts, and black-tipped tail-feathers with no streamers. Bill yellowish-cream with indistinct dark tip (greyish pre-fledging). See Red-billed and Red-tailed. **HABITAT** Open seas, islets. **RANGE & STATUS** Breeds widely tropical and subtropical Indian (including Maldive Is, India), Pacific and Atlantic Oceans, Caribbean; wanders widely outside breeding season, including coastal waters of Sri Lanka, Sumatra, Java, southern and eastern Wallacea. **SE Asia** Offshore/coastal vagrant south-east S Thailand; Myanmar?

FREGATIDAE: Frigatebirds

Worldwide 5 species. SE Asia 3 species. Large aerial seabirds, with long hooked bills, short necks, long pointed wings and deeply forked tails. Flight strong and agile, often spending long periods soaring. Feed by diving down to snatch prey at water surface and by chasing other seabirds (particularly boobies) to make them regurgitate food. Feed mainly on fish and squid, also offal, plus eggs and young of seabirds and turtles.

139 **CHRISTMAS ISLAND FRIGATEBIRD**
Fregata andrewsi Plate **15**

IDENTIFICATION 92–102 cm. Monotypic. **Male** Overall blackish plumage with contrasting white belly-patch diagnostic. **Female** Similar to Lesser Frigatebird but larger, with white belly and black bar extending from base of forewing to side of breast. **Juvenile** Apart from size and much longer bill, told from Lesser by hexagonal white belly-patch, generally more parallel-sided white axillary spur (rarely absent), which originates from behind line of breast-band and is more forward-pointing (towards forewing); upperwing-band whitish and very prominent. See Great Frigatebird. **Immature (second-fourth year)** As for Lesser. **HABITAT** Open seas, islets. **RANGE & STATUS** Breeds Christmas I. Non-breeders wander to offshore waters of Greater Sundas; rarely Sri Lanka, Andaman Is. **SE Asia** Scarce to locally fairly common non-breeding offshore visitor S Thailand, Peninsular Malaysia, Cambodia (April–July at least); occasionally strays inland within these regions. Coastal/offshore vagrant C Thailand, Singapore, E Tonkin, Cochinchina.

140 **GREAT FRIGATEBIRD**
Fregata minor Plate **15**

IDENTIFICATION 86–100 cm. *F.m.minor*. **Male** Overall blackish plumage diagnostic, but Lesser Frigatebirds with dull whitish axillaries can cause confusion. **Female** From other frigatebirds by combination of pale greyish throat, black belly and black inner underwing-coverts. **Juvenile** From other frigatebirds by all-black inner underwing-coverts (30% of birds show a short, outward-angled white axillary spur that originates well behind line of breast-band), elliptical white belly-patch. Upperwing-band similar to Lesser, bill intermediate between Lesser and Christmas Island. **Immature (second-fourth year)** As for Lesser. **HABITAT** Open seas, islets. **RANGE & STATUS** Breeds tropical and subtropical Indian, Pacific and Atlantic Oceans, including Philippines, S,E Wallacea, islands off NE Australia; non-breeders wander widely, including coastal waters of S Indian subcontinent, Greater Sundas, Philippines, Wallacea. **SE Asia** Scarce non-breeding offshore visitor west S Thailand. Coastal/offshore vagrant W Thailand, south-east Peninsular Malaysia, Singapore, C Annam. Inland vagrant N Laos.

141 **LESSER FRIGATEBIRD**
Fregata ariel Plate **15**

IDENTIFICATION 71–81 cm. *F.a.ariel*. **Male** Overall blackish plumage with prominent whitish patches extending from sides of body to inner underwing-coverts diagnostic. Some individuals have much duller patches and may be mistaken for Great at long distance. Has red gular pouch. **Female** Combination of black hood, belly and lower flanks and white remainder of underparts, extending in axillary spur onto inner underwing-coverts, diagnostic. **Juvenile** Shows rufous to brownish-white head, black breast-band, triangular white belly-patch. White axillary spur always present, originating from line of breast-band and angled outwards (towards wing-tip). Band across upperwing-coverts buffish, moderately prominent. See Christmas Island and Great Frigatebirds. **Immature (second-fourth year)** Gradually loses black breast-band and acquires blackish plumage-parts of respective adults. Third and fourth year birds are sexually dimorphic. Attains adult plumage during fifth year. **HABITAT** Open seas, islets. **RANGE & STATUS** Breeds tropical and subtropical Indian, Pacific and Atlantic Oceans, including E Wallacea, islands off New Guinea, northern Australia; non-breeders wander widely, including coastal waters of southern Indian subcontinent, Greater Sundas, Philippines, Wallacea. **SE Asia** Scarce to locally common non-breeding offshore visitor S Thailand, Peninsular Malaysia, Cambodia. Coastal/offshore vagrant Singapore, Cochinchina.

PELECANIDAE: Pelicans

Worldwide 8 species. SE Asia 2 species. Large, short-tailed and short-legged with webbed feet and long straight flattened bills supporting huge expandable pouches. Gregarious; spend much time swimming, bodies high in water. Feed on fish, often cooperatively. Flight strong, with steady wingbeats interspersed with glides, on long broad wings.

142 **GREAT WHITE PELICAN**
Pelecanus onocrotalus Plate **13**

IDENTIFICATION 140–175 cm. Monotypic. **Adult non-breeding** Similar to Spot-billed Pelican but larger, with whiter, less tufted head and neck, grey bill with pinkish cutting edge, plain yellowish pouch, pale lores and pale pinkish legs and feet. In flight shows mostly black flight-feathers (above and below) offsetting whitish wing-coverts. **Adult breeding** Pale parts of plumage whiter with variable pinkish flush on body and wing-coverts, bill bluer-grey with redder cutting edge, pouch deeper yellow, legs and feet redder, with distinctive floppy nuchal crest and yellowish-buff patch on lower foreneck and upper breast. **Juvenile** Plumage similar to Spot-billed but head and neck more uniform greyish-brown, upperparts and wing-coverts darker; bill, pouch and lores differ as adult non-breeding (though pouch initially duller). In flight, shows distinctive dark brownish leading edge to underwing-coverts as well as blackish-brown underside of flight-feathers. **VOICE** May give quiet deep croaking notes in flight. At breeding colonies, utters low-pitched grunting and growling sounds. **HABITAT** Various inland and coastal wetland habitats, including large lakes, lagoons and tidal creeks; lowlands. **RANGE & STATUS** Breeds sub-Saharan Africa, SE Europe, Turkey, N Black Sea region, C Asia, W India; some populations winter to south, NE Africa, Middle East, S Pakistan, northern India, Bangladesh, possibly NW China. **SE Asia** Former resident (no recent records) Cochinchina. Rare to scarce winter visitor (at least formerly) S Myanmar. Vagrant Tenasserim, west Peninsular Malaysia, Cambodia (perhaps more frequent historically), C Annam. **BREEDING** Colonial. **Nest** Skimpy to fairly substantial, on ground. **Eggs** 2–4; ivory-white; 95.6 x 61.6 mm (av.).

143 **SPOT-BILLED PELICAN**
Pelecanus philippensis Plate **13**

IDENTIFICATION 127–140 cm. Monotypic. **Adult non-breeding** Only likely to be confused with Great White Pelican. Differs by smaller size, dirtier plumage, tufted dusky nape and hindneck, yellowish-pink bill with dark spots along upper mandible, pinkish pouch with heavy purplish-grey mottling, dark bluish to purplish lores and blackish legs and feet. In flight shows much greyer flight-feathers (above and below) contrasting less with wing-coverts, and duller underwing-coverts with obvious whitish band along greater coverts. **Adult breeding** Rump, tail-coverts and underwing-coverts variably washed cinnamon-pinkish, underparts variably flushed pink, with faint yellowish-buff patch on lower foreneck and upper breast. **Juvenile** Similar to adult non-breeding but sides of head, nape, hindneck, mantle and upperwing-coverts browner, with paler fringes to mantle feathers and upperwing-coverts, unmarked yellower bill, plain dull pinkish pouch, duller lores and pinkish-grey legs and feet. In flight shows browner flight-feathers but similar underwing-pattern. See Great White. **HABITAT** Lakes, lagoons, large rivers, estuaries, mudflats; lowlands. **RANGE & STATUS** Breeds S,SE,NE India, Sri Lanka; formerly Bangladesh, Philippines; E China? Some populations disperse outside breeding season, reaching northern India, Nepal, Java. Recorded (status unclear) Sumatra. **SE Asia** Rare to locally common resident Cambodia, Cochinchina; formerly S Myanmar. Rare to scarce non-breeding/winter visitor W,N,S Myanmar, north Tenasserim, W,C,NE,S Thailand, C,S Laos, E Tonkin, N Annam. Former non-breeding visitor (current status uncertain) SW,C Myanmar. Vagrant Peninsular Malaysia (formerly more abundant). **BREEDING** August–May. Usually colonial. **Nest** Huge structure, in tree; up to 30 m above ground. **Eggs** 3–4; chalky-white (becoming stained); 78.8 x 53.4 mm (av.).

SULIDAE: Boobies

Worldwide 9 species. SE Asia 3 species. Large, slender-bodied seabirds with longish, pointed conical bills, long, narrowish pointed wings and longish, wedge-shaped tails. Flight strong and direct with flapping interspersed by long glides. Feed mainly on fish and squid, by plunge-diving.

144 MASKED BOOBY *Sula dactylatra* Plate 15

IDENTIFICATION 74–86 cm. *S.d.personata*: **Adult** White plumage, with contrasting black primaries, secondaries, tertials and tail distinctive. Has blackish facial skin ('mask'), pale yellowish bill, dark greyish legs and feet, and white underwing-coverts with only a small dark area on primary coverts. See white morph Red-footed Booby. **Juvenile** Head, neck, upperparts and upperwing-coverts warmish brown, with dark brown band across underwing-coverts and browner primaries and secondaries than adult. Similar to Brown Booby but upperparts and upperwing-coverts somewhat paler and warmer brown, with narrow whitish band across upper mantle, white underwing-coverts with clearly defined dark central band, whiter underparts and dark legs and feet. Gradually attains adult plumage. See Red-footed Booby. **HABITAT** Open seas, islets. **RANGE & STATUS** Breeds widely tropical and subtropical Indian, Pacific and Atlantic Oceans, Caribbean, Philippines, E Wallacea, offshore northern Australia; wanders widely outside breeding season, including offshore western Indian subcontinent, Java, southern and eastern Wallacea. **SE Asia** Rare offshore/coastal visitor Peninsular Malaysia (formerly bred Pulau Perak). Offshore/coastal vagrant W,S Thailand. Reported to breed Cochinchina (Con Dao Is). **BREEDING** January–February. **Nest** Simple arrangement of fragments on ground or rock ledge. **Eggs** 1–2; whitish, often covered with chalky deposits; c.66 x 41 mm.

145 RED-FOOTED BOOBY *Sula sula* Plate 15

IDENTIFICATION 68–72.5 cm. *S.s.rubripes*: **Adult white morph** Like Masked Booby but smaller and slimmer, tail and tertials white, bill and facial skin light blue-grey and pinkish, legs and feet rose-red, large blackish patch on primary underwing-coverts. Plumage may show apricot flush, particularly on crown and hindneck. **Adult brown morph** Similar to juvenile but bill, facial skin, legs and feet as adult white morph. May show apricot flush on crown and hindneck. **Adult intermediate morph** Similar to white morph but mantle, back and wing-coverts brown (above and below). See juvenile Masked and Brown Boobies. **Juvenile** Rather uniform dark greyish-brown with darker primaries, secondaries and tail, dark grey bill, purplish facial skin and yellowish-grey to flesh-coloured legs and feet. Gradually attains adult features. **Immature** often shows diffuse darker breast-band; white morph bird soon has untidy whitish areas on wing-coverts. See Masked and Brown Boobies. **HABITAT** Open seas, oceanic islets; sometimes coastal waters. **RANGE & STATUS** Breeds tropical and subtropical Indian, Pacific and Atlantic Oceans, Caribbean, Philippines, southern and eastern Wallacea, offshore N Australia; wanders widely outside breeding season. **SE Asia** Coastal/offshore vagrant west S Thailand. **BREEDING** Colonial. **Nest** Simple construction on ground. **Eggs** 1–2; white.

146 BROWN BOOBY *Sula leucogaster* Plate 15

IDENTIFICATION 73–83 cm. *S.l.plotus*: **Adult** Dark chocolate-brown plumage with contrasting white lower breast to vent diagnostic. In flight, underwing-coverts also white apart from dark leading edge, incomplete dark diagonal bar and dark primary coverts. Bill pale yellowish, facial skin blue to blue-grey, legs and feet pale yellow to greenish-yellow. See juvenile Masked Booby. **Juvenile** Brown of plumage somewhat paler, white of plumage washed dusky-brownish, bill and facial skin mostly pale bluish-grey, legs and feet dull pink to pale yellowish. Gradually attains adult plumage. See Masked. **VOICE** At breeding colonies, male utters a goose-like whistling hiss and ***koe-el*** calls; female gives honking quacks and crow-like growling sounds. **HABITAT** Open seas, islets. **RANGE & STATUS** Breeds tropical and subtropical Indian, Pacific and Atlantic Oceans, including offshore islands W,NE Africa, W,S Arabian Peninsula, Taiwan, southern Japan, Philippines, southern and eastern Wallacea, northern Australia; wanders widely outside breeding season, including offshore western Indian subcontinent, Greater Sundas. **SE Asia** Scarce offshore resident west Peninsular Malaysia (Pulau Perak), Cochinchina (Con Dao Is). Offshore/coastal vagrant Tenasserim, SE,C,S Thailand, Peninsular Malaysia, Singapore, C,S Annam; formerly more frequent. **BREEDING** All year. Colonial. **Nest** Simple arrangement of fragments on ground or rock ledge. **Eggs** 2; whitish; 52.3–62.7 x 39.1–40.8 mm.

PHALACROCORACIDAE: Cormorants

Worldwide c.38 species. SE Asia 3 species. Medium-sized to large waterbirds; mostly blackish with strong hook-tipped bills, longish necks, rather long, broad stiff tails and webbed feet. Legs are set well back on body. Swim low in water with upstretched neck and head tilted upward. Flight strong and direct on outstretched wings. Feed mainly on fish, by diving.

147 LITTLE CORMORANT *Phalacrocorax niger* Plate 11

IDENTIFICATION 51–54.5 cm. Monotypic. **Adult non-breeding** Relatively small and short-necked with distinctive shortish, stubby bill. Mostly blackish-brown, scapulars and wings greyer with black feather-edges, chin whitish, bill mostly dull greyish flesh-colour. In flight, appears relatively short-necked, small-winged and long-tailed. See Indian Cormorant. **Adult breeding** Head, neck and underparts black with bluish to greenish gloss and dense silvery-white streaks on crown, ear-coverts and nape, bill blackish. See Indian. **Juvenile** Similar to adult non-breeding but browner overall, head and neck paler and browner, throat whitish, underparts streaked/scaled pale brownish. Has paler crown and hindneck and darker belly than other cormorants. **HABITAT & BEHAVIOUR** Various wetlands, mainly fresh water but also estuaries and mangroves; up to 1,450 m (breeds in lowlands). Flies with regular, rather fast wingbeats. **RANGE & STATUS** Resident (subject to local movements) Indian subcontinent, SW China, Java. **SE Asia** Scarce to locally fairly common resident (subject to local movements) S,E Myanmar, W(coastal),C,SE Thailand, Cambodia, Cochinchina. Former resident (current status uncertain) C Myanmar, Tenasserim, N,C Laos, E Tonkin, C,S Annam. Scarce to locally common non-breeding visitor W,NW,S Thailand, S Laos (formerly bred). Vagrant western Peninsular Malaysia, Singapore. Scarce to locally common (status unclear) SW,W(north),N Myanmar (formerly bred in these areas and may still), N Annam (rare). **BREEDING** October–June. Colonial. **Nest** Fairly large, rough platform, in tree, bamboo or reeds. **Eggs** 3–5; pale bluish-green with chalky coating (becoming stained); 44.8 x 29 mm (av.).

148 INDIAN CORMORANT *Phalacrocorax fuscicollis* Plate 11

IDENTIFICATION 61–68 cm. Monotypic. **Adult non-breeding** Similar to Little Cormorant but larger, bill relatively long and slender, head, neck and underparts blacker with whitish lower sides of head and throat, and uneven whitish to pale brown streak-like markings on foreneck and breast, base colour of scapulars and wings browner. In flight appears longer-necked, larger-winged and shorter-tailed; more closely resembles Great Cormorant but smaller, slimmer and thinner-necked with smaller, more oval-shaped head and proportionately longer tail. **Adult breeding** Apart from size and shape, similar to Little but has silvery peppering over eye, distinctive white tuft on rear side of head, and browner base colour of scapulars and wings. **Juvenile** Upperside browner than adult non-breeding, underparts mostly whitish with dark brown smudging and streaks on foreneck and breast and dark brown flanks. Best separated from Great by size, shape, much thinner bill, dark ear-coverts, less extensive and duller gular skin and darker breast. **HABITAT** Various wetlands in both fresh and salt water; lowlands. **RANGE & STATUS** Resident (subject to local movements) S Pakistan, India, Bangladesh, Sri Lanka. **SE Asia** Scarce to locally common resident (subject to some movements) S Myanmar, C,SE Thailand, Cambodia, Cochinchina, Vagrant S Thailand. Scarce (status uncertain) N Myanmar, southern NE Thailand. **BREEDING** January–August. Colonial. **Nest** Fairly large, rough platform, in tree. **Eggs** 3–6; pale bluish-green with chalky coating (becoming stained); 51.3 x 33.2 mm (av.).

149 **GREAT CORMORANT**

Phalacrocorax carbo Plate **11**

IDENTIFICATION 80–100 cm. *P.c.sinensis*: **Adult non-breeding** Much bigger and larger-billed than other cormorants. Head, neck and underparts black with prominent white area on lower sides of head and upper throat, has browner scapulars and wings with black feather-edges and extensive yellow facial and gular skin. In flight, apart from large size, appears relatively thick-necked, large-winged and short-tailed, with squarish head and slightly kinked neck. See Indian Cormorant. **Adult breeding** Has dense white streaks forming sheen across sides of crown and neck, more orange facial skin and darker gular skin, larger and more defined white area on ear-coverts and throat, greenish-glossed scapulars and wings, and large white thigh-patch. **Juvenile** Upperside much browner than adult non-breeding, sides of head and underparts mostly whitish, with dark brown streaks on foreneck and upper breast, and dark brown flanks and thighs. Apart from size and bill, differs from similar Indian by more extensive, yellower facial and gular skin, whitish ear-coverts and sparser, more contrasting breast markings. Gradually attains adult plumage. **VOICE** Usually silent but utters a variety of deep, guttural calls at breeding colonies. **HABITAT & BEHAVIOUR** Various wetlands in both fresh and salt water; up to 1,830 m. Flies with slower wingbeats than other cormorants. **RANGE & STATUS** Breeds Africa, W,C Palearctic, S,SE Siberia, Ussuriland, Iran, Indian subcontinent, S and northern China, S Korea, Japan, Australia, New Zealand, east N America, S Greenland; formerly Sumatra. Some northern populations winter south to Africa, Middle East, northern Indian subcontinent, southern China, Philippines, N Borneo; some Australian populations disperse northwards, reaching New Guinea. **SE Asia** Scarce to local resident Cambodia. Former resident (current status uncertain) S Myanmar, SE Thailand, Cochinchina. Rare to scarce winter/non-breeding visitor W(north),N(locally common),C,S Myanmar, Thailand (except S), E Tonkin, S Annam, Cochinchina; formerly S Thailand. Former winter/non-breeding visitor (current status uncertain) SW,E Myanmar, Tenasserim, N Laos, C Annam. Recorded in summer (with no evidence of breeding) NE Thailand. Winter vagrant north-west Peninsular Malaysia (formerly commoner), Singapore (possibly escapees). **BREEDING** India: September–February. Colonial. **Nest** Large, rough, shallow cup, in tree (particularly where partly submerged in water). **Eggs** 3–6; pale bluish-green with chalky coating (becoming stained); 60.6 x 39.3 mm (av.).

ANHINGIDAE: Darters

Worldwide 4 species. SE Asia 1 species. Similar to cormorants but chiefly in fresh water, with long and slender bills and necks, and long, broad tails. Swim sometimes with only head and neck visible. Flight strong with neck extended and wings outstretched. Feed mainly on fish.

150 **ORIENTAL DARTER**

Anhinga melanogaster Plate **11**

IDENTIFICATION 85–97 cm. Monotypic. **Adult non-breeding** Similar to cormorants but easily distinguished by longish slender pointed bill, long thin neck and relatively long tail. Head, neck and mantle mostly dark brown with pale throat and long whitish stripe extending from eye down sides of neck, prominent white streaks on upper mantle, scapulars and upperwing-coverts. **Adult breeding** Crown, hindneck and base colour of upper mantle blackish, foreneck more chestnut. **Juvenile** Somewhat paler and browner, underside buffish-white with contrasting dark flanks and vent, buff-fringed upperwing-coverts and narrow pale tail-tip. **VOICE** Usually silent. On breeding grounds utters unusual gruff, slightly nasal ***uk ukukukuk-errr uk-uk*** or ***ok ok ok ok ukukukukuk-err rerr-rerr-rer-ruh***. **HABITAT & BEHAVIOUR** Lakes, marshes, large rivers; up to 1,200 m. Swims on open water; sometimes submerges, with only snake-like neck visible. In flight, often spends time soaring. **RANGE & STATUS** Resident (subject to some movements) Indian subcontinent, Greater Sundas, Philippines, Wallacea. **SE Asia** Rare to locally fairly common resident (subject to local movements) N,S Myanmar, C,NE,SE Thailand, Cambodia, Cochinchina. Scarce non-breeding visitor SW,W(north),N Myanmar (formerly bred SW,W), W,NW,S Thailand (formerly bred). Former resident (current status uncertain) C,E Myanmar, Tenasserim, C Annam. Rare to scarce (status uncertain) Laos (formerly bred throughout), S Annam (formerly bred), E Tonkin. Vagrant western Peninsular Malaysia (possibly former resident). **BREEDING** June–March. Colonial. **Nest** Fairly large rough platform, in tree. **Eggs** 3–6; pale greenish-blue with chalky coating (becoming stained); 52.9 x 33.5 mm

FALCONIDAE: FALCONINAE: Falcons

Worldwide c.61 species. SE Asia 12 species. Females generally larger than males. Have tooth-like projection near tip of upper mandible. Feed on birds and insects (often caught in mid-air), as well as small mammals and reptiles etc. **Typical falcons** (*Falco*) Wings pointed and often rather long and slender, tails generally rather long and narrow, flight strong and rapid. **White-rumped Pygmy-falcon & falconets** (*Polihierax*, *Microhierax*) Diminutive with shorter wings; spend long periods perched.

151 **WHITE-RUMPED PYGMY-FALCON**

Polihierax insignis Plate **16**

IDENTIFICATION 25–26.5 cm. *P.i.cinereiceps*: **Male** Told by smallish size, long tail, pale greyish ear-coverts and forehead to upper mantle with blackish streaks, white rump and uppertail-coverts, and unmarked whitish underparts. Rest of upperparts dark slate-grey. Often shows whiter nuchal collar. *P.i.insignis* (W,C,S Myanmar) has much paler grey mantle and scapulars (more concolorous with crown) and shows blackish streaks on lower throat, breast and flanks; *P.i.harmandi* (Indochina) has whiter crown to upper mantle, with less distinct dark streaks. **Female** Resembles male but crown to upper mantle deep rufous. *P.i.insignis* is paler grey above and shows prominent blackish underpart-streaking (mainly on lower throat, breast and flanks), often mixed with some dark rufous. **Juvenile** Both sexes similar to male but lower nape/upper mantle broadly rufous, rest of upperparts washed brown. *P.i.insignis* also has more streaked underparts than adult male. **VOICE** Rapid, high, modulated, almost gibbon-like yelping cries: ***WIHY'ah***; ***WIHY'AH-ha***; ***WIHY'A'YA-ha***; ***WIHYUYUYU*** etc. (more stressed at beginning). **HABITAT & BEHAVIOUR** Open deciduous woodland, clearings in deciduous forest; up to 915 m. Often perches (in concealed or exposed position) for long periods. Flight direct, with rapid wingbeats. **RANGE & STATUS** Endemic. Scarce to uncommon resident SW,W,C,S Myanmar, north Tenasserim, W,NW,NE Thailand, Cambodia, C,S Laos, S Annam, Cochinchina. **BREEDING** February–May. **Nest** Hole in tree. **Eggs** 2; white.

152 COLLARED FALCONET

Microhierax caerulescens Plate **16**

IDENTIFICATION 15.5–18 cm. *M.c.burmanicus*: **Adult** Told by very small size, black ear-covert patch and upperside, broad white forehead and supercilium (latter meeting white nuchal collar), and chestnut throat. Thighs and vent chestnut, breast-centre and belly variable, white to chestnut. See Black-thighed and Pied Falconets. **Juvenile** Pale areas of forehead, sides of head and supercilium washed pale chestnut, throat whitish, chestnut of underparts paler and restricted mainly to undertail-coverts. **VOICE** A high-pitched ***kli-kli-kli*** or ***killi-killi-killi***. **HABITAT & BEHAVIOUR** Deciduous forest, clearings in broadleaved evergreen and mixed forests; up to 1,830 m (locally 2,310 m). Perches in exposed places for long periods. Flight direct and rapid, shrike-like. **RANGE & STATUS** Resident N and north-eastern India, Nepal, Bhutan. **SE Asia** Fairly common resident Myanmar, Thailand (except S), Cambodia, Laos, S Annam, Cochinchina. **BREEDING** February–May. **Nest** Old woodpecker or barbet hole; 6–9 m above ground (or higher). **Eggs** 4–5; dirty white (may become stained or spotted reddish); 27.9 x 21.7 mm (av.).

153 BLACK-THIGHED FALCONET

Microhierax fringillarius Plate **16**

IDENTIFICATION 15–17 cm. Monotypic. **Adult** Similar to Collared but lacks white nuchal collar, has much less white on forehead and supercilium, more black on ear-coverts, whiter throat and black lower flanks (extending to thighs). See Pied Falconet. Note range. **Juvenile** Similar to adult but pale areas of forehead, sides of head and supercilium washed very pale chestnut, throat white, chestnut of underparts much paler. **VOICE** A shrill, squealing ***kweer WEEK***. **HABITAT** Clearings in broadleaved evergreen forest, forest edge, partly wooded cultivation and parkland; up to 1,700 m. **RANGE & STATUS** Resident Greater Sundas. **SE Asia** Fairly common to common resident south Tenasserim, S Thailand, Peninsular Malaysia, Singapore (scarce). **BREEDING** February–August. **Nest** Old woodpecker or barbet hole, sometimes hole under eaves of building; 10–20 m above ground. **Eggs** 2–5; white (perhaps becoming stained or spotted reddish); 25.2–29.4 x 20.2–23.2 mm.

154 PIED FALCONET

Microhierax melanoleucus Plate **16**

IDENTIFICATION 19–20 cm. Monotypic. **Adult** Similar to Black-thighed Falconet but larger, lacks chestnut on underparts. Note range. See Collared Falconet. **Juvenile** Very similar to adult. **VOICE** A shrill whistle, low chattering and, when agitated, prolonged hissing sounds. **HABITAT** Clearings in broadleaved evergreen forest, forest edge; up to 1,080 m. **RANGE & STATUS** Resident SE Bhutan, NE India, SW,S,SE China; formerly E Bangladesh. **SE Asia** Scarce resident N,C Laos, W,E Tonkin, N,C Annam. **BREEDING** India: March–May. **Nest** Old woodpecker or barbet hole; 13–30 m above ground. **Eggs** 3–4; white; 23.9–31.2 x 20.5–25.4 mm.

155 LESSER KESTREL *Falco naumanni* Plate **16**

IDENTIFICATION 29–32 cm. Monotypic. **Male** Similar to Common Kestrel but slightly smaller and slimmer, with proportionately shorter tail; unmarked bluish-grey crown, nape and sides of head (lacks dark moustachial stripe); unmarked rufous-chestnut mantle, scapulars, median and lesser coverts; mostly bluish-grey tertials, greater coverts and median-covert tips; and plainer vinous-tinged warm buff underparts with fewer, rounder dark markings. At rest, primary tips reach subterminal tail-band (all plumages). Claws whitish. In flight, wing-tips somewhat narrower and more rounded than Common, underwing cleaner and whiter with contrasting dark tip (and often diffuse trailing edge), tail somewhat narrower and usually more wedge-shaped (central tail-feathers project further). **Female** Very similar to Common but differs by structure (as in male); typically has finer, more diffuse moustachial/cheek-stripe and no dark post-ocular stripe; plainer, greyish-white cheek/ear-covert patch, often extending widely to throat; crown-streaks usually finer; dark upperpart bars usually more V-shaped, protruding in sharper point towards feather-tip (more triangular or bar-like on Common); uppertail-coverts more often washed pale greyish; underparts usually more neatly and less densely marked, with somewhat sparser, finer (more drip-like) streaks, which usually do not extend along flanks as far as tips of secondaries and tend to be denser on upper breast (more evenly spread on Common). In flight, underwing usually somewhat cleaner and whiter than Common, less heavily dark-marked (particularly primary coverts) with fainter, more diffuse darker bars across flight-feathers (relatively dark primary tips contrast more). **Juvenile** Like female (both sexes) but upperparts, wing-coverts, flight and tail-feathers tend to be more heavily marked, uppertail-coverts rufescent grey-brown. **First-summer male** Similar to adult but lacks grey on wings and shows some dark and pale brown bars on upperside of flight-feathers and outertail-feathers. See male Common. **VOICE** Weaker and less piercing than Common; a rapid, rather rasping ***kik-kik-kik-kik-kik...*** or ***keh-chet-chet-kick***, hoarse ***kye-kye*** or ***kye-kiki*** and rasping ***kee-chee-chee*** or ***chet-che-che*** and ***kihik*** or ***kichit***. **HABITAT & BEHAVIOUR** Open areas, cultivation; up to 1,065 m. Flight is more buoyant than Common's, with slower, deeper, softer wingbeats. Hovers as often as Common, but less persistently in one place. Likely to be encountered in loose flocks. **RANGE & STATUS** Resident NW Africa. Breeds southern Europe, W(south-east),SC Palearctic, Middle East, S Siberia, NW,N,NE China, Mongolia; mostly winters sub-Saharan Africa, rarely C,S India. **SE Asia** Rare passage migrant/winter visitor N Myanmar; formerly S Myanmar, N Laos. Vagrant Singapore.

156 COMMON KESTREL

Falco tinnunculus Plate **16**

IDENTIFICATION 30–34 cm. *F.t.interstinctus* (throughout): **Male** Medium-size, slaty-grey crown, nape and rump to uppertail, broad black subterminal tail-band and rufous remainder of upperparts and wing-coverts with blackish markings distinctive. Has contrasting dark moustachial/cheek-stripe and pale buffish underparts with dark streaks/spots on breast, belly and flanks. Wing-tips fall short of dark subterminal tail-band on perched birds. In flight, underwing appears largely whitish with dark markings (mainly on coverts), tail quite strongly graduated. *F.t.tinnunculus* (recorded east to NW and C Thailand) is somewhat paler-toned and less heavily marked with paler grey parts of plumage. See Lesser Kestrel. **Female** Typically lacks grey in plumage (may show grey on uppertail-coverts and uppertail); crown and nape warm brown with dark streaks, has dark line behind eye and long, dark moustachial stripe, uppertail rufous with narrow blackish bars and broad black subterminal band, rufous of upperparts duller and paler than male with more numerous and distinct dark bars (less spotted), underparts more heavily dark-streaked. See very similar Lesser Kestrel and Merlin. **Juvenile** Like female (both sexes) but upperparts, wing-coverts, flight and tail-feathers tend to be more strongly and heavily marked, uppertail-coverts rufescent grey-brown. **VOICE** A sharp, piercing ***keee-keee-keee...***, in a rapid series or singly. Also ***kik*** notes given singly or repeated, and trilling ***kreeeee*** or ***wrreeee*** when agitated.

HABITAT & BEHAVIOUR Various open habitats, cultivation, urban areas, cliffs; up to 2,000 m. Often hovers. Flies with shallow, winnowy wingbeats. **RANGE & STATUS** Resident Africa, west and southern W Palearctic, Middle East, C and northern Pakistan, NW,S India, Nepal, Bhutan, Sri Lanka, W,S,E Tibet, southern China, N,S Korea, C Japan. Breeds W(northern),C,E Palearctic, northern China, N Japan; many populations winter south to Africa, Middle East, Indian subcontinent, southern China, Taiwan, N,S Korea, Japan, Borneo, Philippines. **SE Asia** Local resident or breeding visitor E Tonkin. Recorded in summer (with no evidence of breeding) N Myanmar. Scarce to fairly common winter visitor throughout. Also recorded on passage SW Myanmar, Cambodia, E Tonkin. **BREEDING** April–May. **Nest** Simple structure in cavity in tree, cliff or building, on rock ledge, or in disused tree nest of other species. **Eggs** 3–6; pale pinkish- to yellowish-stone-colour, densely speckled and blotched shades of reddish; 39.7 x 31.8 mm (av.; *tinnunculus*, UK)

157 **AMUR FALCON** *Falco amurensis* Plate **17**

IDENTIFICATION 28–31 cm. Monotypic. **Male** Unmistakable. Slaty-grey overall with paler grey underparts, rufous-chestnut thighs and vent and red eyering, cere, legs and feet. In flight, diagnostic white underwing-coverts contrast sharply with blackish remainder of underwing. **Female** Similar to adult Eurasian Hobby but with dark-barred upperparts and uppertail, buffy-white thighs and vent, and different bare-part colours (similar to male). In flight, differs by whiter base colour of underwing (particularly coverts) and more pronounced dark and pale bars on undertail. Thighs and undertail-coverts are buffy. See much larger Peregrine Falcon. **Juvenile** Similar to Eurasian Hobby but upperparts and wing-coverts more broadly and prominently tipped/fringed buff, has dark-barred upperparts and uppertail, pale parts of head, underparts and underwing somewhat whiter, legs and feet reddish, eyering and cere pale yellow. **First-summer male** Variable, showing mixed characters of adult and juvenile. **VOICE** A shrill screaming ***kew-kew-kew...*** at roost sites. **HABITAT & BEHAVIOUR** Various open habitats, wooded areas; up to 1,900 m. Likely to be encountered in flocks. Often hovers. **RANGE & STATUS** Breeds SE Siberia, Ussuriland, northern and north-eastern China, N Korea, rarely NE Indian subcontinent; winters sub-Saharan Africa, rarely S,NE India, Nepal. **SE Asia** Scarce to locally common, erratic passage migrant W,N,C,S,E Myanmar, W,NW Thailand, N Laos, W,E Tonkin. Vagrant SE,S Thailand, Peninsular Malaysia.

158 **MERLIN** *Falco columbarius* Plate **16**

IDENTIFICATION 25–30 cm. *F.c.insignis*: **Male** Told by fairly small size and compact shape, rather uniform, bluish-grey upperside, blackish flight-feathers, broad blackish subterminal tail-band and warm buffish to rufescent-buff underparts with rather narrow dark streaks on breast, belly and flanks. Has only indistinct darker moustachial/cheek-stripe, fairly pronounced buffish-white supercilium, rufous nuchal collar and a few broken narrower dark bars on uppertail. Legs and feet orange to yellowish. In flight shows rather short, broad but pointed wings, shortish tail, heavy pale-and-dark markings on underwing, and broad whitish and blackish bands on undertail. See female Amur Falcon. **Female** Resembles kestrels, juvenile Amur Falcon and Eurasian Hobby. Differs from all by combination of small size, compact shape, faint moustachial/cheek-stripe, fairly prominent pale supercilium, rather drab brown crown and upperside with paler buffish-brown markings (including uppertail) but no blackish bars and heavier brown underpart-streaking. Lacks bicoloured upperwing of kestrels. See much larger, darker Peregrine Falcon. **Juvenile** Like female but upperparts tend to be darker brown. **VOICE** A shrill chattering ***quik-ik-ik-ik*** or ***kek-kek-kek*** when alarmed; a coarse lower-pitched ***zek-zek-zek*** given by females. **HABITAT** Open areas, cultivation; up to 1,065 m. Flight is rapid and direct; does not hover. **RANGE & STATUS** Breeds northern Holarctic, NW China; mostly winters south to N Africa, southern Palearctic, Pakistan, northern Indian subcontinent, southern Tibet, southern China, S Korea, Japan, north S America. **SE Asia** Vagrant NW Thailand, N Laos, E Tonkin, C Annam.

159 **EURASIAN HOBBY** *Falco subbuteo* Plate **17**

IDENTIFICATION 30–36 cm. *F.s.streichi*: **Adult** Similar to Peregrine Falcon but smaller, moustachial stripe narrower, upperparts more uniform, uppertail unbarred, breast and belly heavily streaked blackish, thighs and vent reddish-rufous. In flight shows slenderer wings. See Oriental Hobby and Amur Falcon. **Juvenile** Crown, upperparts and wing-coverts duller with narrow pale buffish feather-fringes, darkly streaked underparts, and buffish vent. From Peregrine Falcon by size, shape, lack of obvious pale supercilium, and plainer, deeper buff or rufous vent. See similar Amur Falcon and larger Laggar Falcon. **VOICE** May give a rapid, sharp scolding ***kew-kew-kew-kew...*** **HABITAT & BEHAVIOUR** Wooded and open areas; up to 2,000 m. Flight swift and dashing. Does not hover. **RANGE & STATUS** Breeds Palearctic, N Pakistan, NW India, Nepal, SW Tibet, China, N Korea, N Japan; winters sub-Saharan Africa, Indian subcontinent (except Sri Lanka); rarely S China, S Japan, Java. **SE Asia** Rare to local passage migrant W,N Myanmar (occasionally winters), NW Thailand, N Laos, W,E Tonkin. Vagrant S,E Myanmar, W(coastal),C,NE,S Thailand, Peninsular Malaysia, Singapore.

160 **ORIENTAL HOBBY** *Falco severus* Plate **17**

IDENTIFICATION 27–30 cm. *F.s.severus*: **Adult** Unmistakable, with all-blackish head-sides, buffish-white throat and forecollar and reddish-rufous remainder of underparts. Upperside slate-grey with darker flight-feathers. In flight, also shows distinctive reddish-rufous underwing-coverts. Size and shape recalls Eurasian Hobby. **Juvenile** Resembles adult but upperparts and wing-coverts darker and browner with narrow pale feather-fringes, breast to vent rufous with blackish drop-like streaks on breast and belly, outertail-feathers barred. In flight shows rufous underwing-coverts with indistinct darker markings. See Eurasian and larger Peregrine Falcon (particularly subspecies *peregrinator*). **VOICE** A high-pitched rapid ***ki-ki-ki-ki-ki-ki...*** or ***hiu-hiu-hiu-hiu-hiu-hiu...***, repeated at intervals. **HABITAT & BEHAVIOUR** Open areas in broadleaved evergreen and deciduous forest, secondary growth, cultivation, mangroves, vicinity of limestone cliffs; up to 1,525 m (locally 1,915 m). Flight swift and dashing; does not hover. **RANGE & STATUS** Resident (subject to some movements) northern and NE India, Nepal, Bhutan, SW,S China, Greater Sundas, Philippines, Sulawesi, Sula Is, Moluccas, New Guinea region, N Melanesia; formerly E Bangladesh. **SE Asia** Scarce to uncommon resident (except SE Thailand, Peninsular Malaysia, Singapore, W Tonkin). Rare (status uncertain) Peninsular Malaysia. Vagrant Singapore. **BREEDING** April–June. **Nest** Old stick nest of other bird, in tree or on cliff ledge. **Eggs** 2–4; buff to cream, speckled reddish-brown overall; 35–41.4 x 30–34 mm.

161 **LAGGAR FALCON** *Falco jugger* Plate **17**

IDENTIFICATION 41–46 cm. Monotypic. **Adult** Similar to Peregrine Falcon but crown rufous with dark streaks, has whitish forehead and supercilium, much narrower moustachial streak, unbarred uppertail, unmarked whitish breast, broad dark greyish-brown patch on lower flanks/thighs and narrow dark streaks on lower belly. In flight, wings less broad-based, tail proportionately longer; from below, shows diagnostic dark area on axillaries and flanks/thighs and mostly solid dark brownish-grey greater and primary coverts. **Juvenile** Has more uniformly dark brown underparts and underwing-coverts, with a few whitish streaks; upperside similar to Peregrine Falcon. See Eurasian Hobby, Amur Falcon and Merlin. **VOICE** Occasionally utters a shrill ***whi-ee-ee***, particularly on breeding grounds. **HABITAT &**

BEHAVIOUR Semi-desert, dry cultivation; lowlands. Flight direct, not as heavy as Peregrine Falcon's, with relatively shallow but strong wingbeats. **RANGE & STATUS** Resident Indian subcontinent (except Bhutan, Sri Lanka). **SE Asia** Uncommon resident C,S(north),E Myanmar. **BREEDING** January–April. **Nest** Lacking or slight, in old nest of other bird in tree, or on cliff ledge or building; 9–15 m above ground (when in tree). **Eggs** 3–5; pale reddish to buffish, thickly spotted red to reddish-brown overall (sometimes capped); 46.5–54.7 x 36.2–41.7 mm.

162 PEREGRINE FALCON

Falco peregrinus Plate **17**

IDENTIFICATION 38–48 cm. *F.p.japonensis* (*calidus*?): **Adult** Told by large size, slate-grey upperside, broad blackish moustachial streak and whitish lower sides of head and underparts with dark bars on flanks and belly to undertail-coverts. Has somewhat paler back and rump and indistinct bands on uppertail (terminal one broader). In flight, wings appear broad-based and pointed, tail shortish, underwing uniformly darkish (due to dense dark barring), with darker-tipped primaries and darker trailing edge. *F.p.ernesti* (resident N Laos, E Tonkin at least) smaller, with darker upperside (can appear blackish), solid blackish head-sides, duller breast to vent with denser dark barring. Resident birds further south in region (S Thailand, Pen Malaysia; ? W,NE Thailand) have plainer breast, varying from whitish to strongly reddish, lower underparts more densely barred and grey-washed, appearing blackish in field: perhaps an undescribed taxon? *F.p.peregrinator* (resident S,E Myanmar, recorded C Thailand; ? NW Thailand) has strongly rufous-washed underparts (usually barred) with whiter throat and sides of neck, buffish tail-tip and rufous-washed underwing-coverts. See Oriental Hobby. **Juvenile** Upperparts and wing-coverts duller with narrow warm brown to buffish fringes, forehead and supercilium whitish with indistinct dark streaks, lower head-sides and underparts buffy with dark streaks (turning to chevrons on vent), some broken buffish bars on uppertail. In flight, underwing like adult but coverts more boldly marked. See Laggar Falcon and smaller Eurasian Hobby and Amur Falcon. *F.p.peregrinator* has darker, browner upperside, no obvious whitish forehead-patch or supercilium, strongly rufous-washed underparts and underwing-coverts. See much smaller hobbies. **VOICE** May give shrill ***kek-kek-kek...*** when alarmed. **HABITAT & BEHAVIOUR** Various open habitats: wetlands, coastal habitats, offshore islets, cliffs; up to 2,900 m. Large, powerful falcon that captures birds in mid-air, usually after a spectacular stoop. **RANGE & STATUS** Resident Africa, southern W Palearctic, Iran, Indian subcontinent, southern China, S Korea, Japan, Sundas, Philippines, Wallacea, New Guinea, Australia, Melanesia, south S America, Falkland Is. Breeds northern Holarctic; some populations winter south to S Africa, Middle East, Indian subcontinent, S China, S Korea, S Japan, Sundas, N Moluccas, S America (New Guinea?). **SE Asia** Uncommon resident S,E Myanmar, north Tenasserim, W,NE,S Thailand, Peninsular Malaysia, N Laos, E Tonkin. Scarce to uncommon winter visitor throughout. Also recorded on passage C Myanmar, S Thailand. **BREEDING** January–May. **Nest** Lacking or well-made, on rock ledge. **Eggs** 3–4; pale buffish to pale brick-red, variably blotched brick-red or reddish-brown; 48.9–58.5 x 38.8–44 mm (*peregrinator*).

FALCONIDAE: PANDIONINAE: Osprey

Monotypic subfamily. Highly adapted for catching relatively large fish, with specialised feet and oily plumage.

163 OSPREY *Pandion haliaetus* Plate **19**

IDENTIFICATION 55–63 cm. *P.h.haliaetus*: **Male** Fairly large size, uniform dark brown upperside and contrasting white head and underparts with dark line through eye and dark streaks on breast (often in complete band, but can be lacking) distinctive. In flight shows long, rather slender wings (typically angled back from carpal joints and bowed), relatively shortish tail, white underwing-coverts with contrasting blackish primary coverts and tips to greater coverts, and evenly barred tail. See buzzards. **Female** Breast-band (when present) typically heavier. **Juvenile** Upperparts and wing-coverts broadly fringed white to buffish, breast markings less defined. **VOICE** May give series of hoarse, falling whistles: ***piu-piu-piu-piu...*** **HABITAT & BEHAVIOUR** Lakes, large rivers, sea coasts; up to 1,400 m. Catches fish in feet in plunge-dive onto surface of water; often hovers over water. **RANGE & STATUS** Resident northern Africa, S Europe, Java, Bali, Wallacea, New Guinea, Australia, Melanesia, Caribbean. Breeds Holarctic, NW,N,NE India, S,SE and northern China, Taiwan, N Korea, Japan; mostly winters to south, Africa, Indian subcontinent, S China, Greater Sundas, Philippines, Sulawesi, S America. **SE Asia** Scarce to fairly common winter visitor (except W Tonkin). Recorded on passage S Thailand, Peninsular Malaysia, Cambodia. Also recorded in summer N,C,S Myanmar, C,NE,SE,S Thailand, Peninsular Malaysia, Singapore, Cambodia, Cochinchina.

FALCONIDAE: ACCIPITRINAE: Hawks, eagles & allies

Worldwide c.239 species. SE Asia 53 species. Highly variable, with pointed hooked bills, very acute sight and powerful feet with long, curved claws (except vultures). Mostly predatory with some scavengers, feeding mainly on mammals, birds, reptiles, amphibians, fish, crabs, molluscs and insects. **Bazas** (*Aviceda*) Small to medium-sized, thickset with long crest, broad rounded wings and square-cut tails. **Oriental Honey-buzzard** Rather small-headed and long-necked, with relatively long, broad wings and longish, narrow tail. **Bat Hawk** Wings broad-based and pointed, tail shortish and square-cut. Mostly crepuscular. **Kites** (*Elanus, Milvus, Haliastur*) Wings mostly long and angular (broader and round on *Haliastur*), tails forked (unforked on *Haliastur*). Often seen soaring and gliding (*Elanus* also hovers), mostly in open habitats. **Fish-eagles** (*Haliaeetus, Ichthyophaga*) Medium to large, heavily built with large bills, broad wings and short rounded to wedge-shaped tails. Distal half of tarsus unfeathered. Often seen soaring and gliding; normally near water. Feed mostly on fish. **Vultures** (*Neophron, Gyps, Aegypius*). Large to very large with very broad wings and short rounded or wedge-shaped tails. Heads featherless or partly down-covered; *Gyps* have longish necks. Masterful flyers, often seen soaring and gliding. Scavenge on dead animals. **Short-toed Snake-eagle** Relatively large with big head and broad wings and tail. Legs unfeathered. Often hovers. **Crested Serpent-eagle** Relatively large, thickset and short-necked with very broad wings and broad rounded tail. Often seen soaring and gliding. Legs unfeathered. Feeds mainly on snakes. **Harriers** (*Circus*) Medium-sized and slim with long wings and tail. Face somewhat owl-like. Legs unfeathered. Glide low over open habitats with wings held in shallow V. **Sparrowhawks** (*Accipiter*) Small to medium-sized with relatively short rounded wings and long tails. Flight often quite fast and direct. Mostly in or near wooded areas. **Buzzards** (*Butastur, Buteo*) Medium-sized with broad rounded wings and tails and smaller heads and bills than eagles. Often seen soaring and gliding, occasionally hovering. *Butastur* species are smaller with narrower wings and proportionately longer, narrower tails. **Larger eagles** (*Aquila, Ictinaetus*) Large and broad-winged with big bills. Plumage relatively unbarred. Legs feathered. Often seen soaring and gliding. *Ictinaetus* inhabits forested areas, *Aquila* prefer open landscapes. **Rufous-bellied Eagle** Medium-sized with relatively longish wings and tail. **Hawk-eagles** (*Nisaetus*) Medium-sized to large with short to long crests and broad wings and tails.

164 **JERDON'S BAZA** *Aviceda jerdoni* Plate **18**
IDENTIFICATION 46 cm. *A.j.jerdoni*: **Adult** Relatively small, with long, erectile, white-tipped blackish crest, warm brown sides of head and nape, paler and warmer area on upperwing-coverts, dark mesial streak, indistinct rufous breast-streaks and broad rufous bars on belly and vent. Resembles some hawk-eagles (particularly juveniles) but smaller, tips of primaries fall closer to tail-tip (at rest). In flight, broad wings have strongly 'pinched-in' bases; shows slightly contrasting paler and warmer band across upperwing-coverts, cinnamon-rufous and white bars on underwing-coverts, fewer dark bars on flight-feathers and three unevenly spaced blackish tail-bands. See Crested Goshawk. **Juvenile** Head mostly buffish-white with blackish streaks, breast plainer, bars on belly and vent often broken; has four evenly spaced dark tail-bands (subterminal one broadest). **VOICE** A rather high-pitched, airy ***pee-weeeow*** or ***fiweeoo*** and shorter ***ti-wuet***, repeated 3–4 times. When agitated, a very high-pitched ***chi chichitchit chit-chit*** and ***chu chit-chit chu-chit chu-chit*** etc., interspersed with ***he-he-hew*** or descending ***he-he-wi-wiwi***. **HABITAT & BEHAVIOUR** Broadleaved evergreen forest, freshwater swamp forest; up to 1,900 m. Rarely soars high over forest; often travels and hunts below canopy. **RANGE & STATUS** Resident S,E,NE India, NE Bangladesh, Sri Lanka, SW,S China, Sumatra, Borneo, Sulawesi, Philippines, Sula Is. **SE Asia** Scarce to uncommon resident S Myanmar, Tenasserim, W,NW,NE,SE,S Thailand, Peninsular Malaysia (rare), N Annam, Cochinchina, Cambodia. Scarce to uncommon passage migrant S Thailand, E Tonkin (also recorded in winter). Vagrant Singapore. Recorded (status uncertain) Laos (probably at least passage migrant N), S Annam. **BREEDING** April–June (NE India); January–March (Sumatra). **Nest** Fairly simple (Sumatra) to quite bulky structure with fairly deep cup (NE India), in tree; 7–20 m above ground. **Eggs** 2; white (becoming stained); 44.7 x 36.6 mm (av.; NE India).

165 **BLACK BAZA** *Aviceda leuphotes* Plate **18**
IDENTIFICATION 31.5–33 cm. *A.l.syama*: **Male** Distinctive. Relatively small, head black with long erectile crest, upperside mostly black, prominent white markings on scapulars, variable white and chestnut markings on greater coverts, tertials and secondaries; whitish remainder of underparts with black band on breast, c.3 chestnut bars on lower breast/upper belly, and black vent. In flight, underwing shows black coverts and primary tips and grey secondaries, contrasting with whitish remainder of primaries; has rather broad wings and shortish tail. **Female** Shows c.7 thicker chestnut bars below (upper 4 complete, rest broken by black of vent) , and smaller white markings on secondaries and tertials. *A.l.leuphotes* (breeds SW,S,E Myanmar, W,NW Thailand) apparently tends to have more chestnut and white on upperparts, mostly chestnut breast-band and less barring on more rufous-buff lower breast and belly. **Juvenile** Duller overall, upperside more brownish-black with more white markings; narrow whitish streaks on throat and brown streaks on white of upper breast. **VOICE** A plaintive, high-pitched rather shrill ***chi-aah***, ***tchi'euuah*** or ***tcheeoua***, with stressed first syllable; often repeated. **HABITAT & BEHAVIOUR** Broadleaved evergreen and deciduous forest, freshwater swamp forest, other habitats on migration; up to 1,500 m. Often in small flocks, particularly during migration. Flight rather crow-like; glides and soars on level wings. **RANGE & STATUS** Breeds SW,N,NE India, Nepal, Bangladesh, SW,S China; some northern (mainly Chinese) populations winter to south, S India, Sri Lanka, Sumatra, W Java. **SE Asia** Uncommon to common resident Myanmar (except Tenasserim), W,NW,NE,SE Thailand, Cambodia, Laos, N,C,S Annam, Cochinchina. Uncommon to fairly common winter visitor Tenasserim, Thailand (except much of NE), Peninsular Malaysia, Singapore. Uncommon to locally common passage migrant S Myanmar, Tenasserim, Thailand (except much of NE), Peninsular Malaysia, Singapore, Cambodia, C Laos, E Tonkin, N Annam. **BREEDING** February–May. **Nest** Bulky structure, in tree; 18–21 m above ground. **Eggs** 2–3; greyish-white (usually stained); 37.3 x 31 mm (av.; Myanmar).

166 **ORIENTAL HONEY-BUZZARD**
Pernis ptilorhynchus Plate **27**
IDENTIFICATION 55–65 cm. *P.p.ruficollis*: **Male pale morph** Relatively large with rather small head, longish neck and short crest or tuft on hindcrown. Highly variable. Typically has greyish sides of head and pale underparts with dark throat-border and mesial streak, gorget of dark streaks on lower throat/upper breast and warm brown to rufous bars on lower breast to vent. Eyes dark brown, cere grey. May have mostly cinnamon-rufous or whitish head and underparts, latter with streaks and/or bars or unmarked. In flight shows relatively long, rather broad wings and rather long narrow tail with two complete, well-spaced blackish bands (often appears blackish with pale central band); underwing typically whitish with complete blackish trailing edge, narrow dark bars on coverts and (usually) three blackish bands across primaries and outer secondaries. Dark markings on flight-feathers are also visible from above. *P.p.torquatus* (resident S Tenasserim and south W Thailand southwards) has a relatively long crest and at least some (perhaps all) have yellow eyes; *orientalis* (only visiting subspecies) is large and long-winged, with little or no crest, often gorgeted. **Female pale morph** Eyes yellow. In flight, from below, shows narrower tail-bands, four narrower blackish bands across primaries and outer secondaries, and narrower dark trailing edge to wing. Neither of latter two features is visible from above. Typically has browner sides of head and upperparts than male. See hawk-eagles and *Buteo* buzzards. **Adult dark morph** Mostly dark chocolate-brown head, body and wing-coverts. Some (*P.p.torquatus* only?) have whitish barring/scaling on underparts and underwing-coverts. **Juvenile** Equally variable. Has less distinct dark bands across underside of flight-feathers and three or more dark tail-bands. Typically shows paler head, neck, underparts and underwing-coverts than adult. May have dark mask and/or streaked underparts. Eyes typically dark. **VOICE** A high-pitched screaming whistled ***wheeew***. **HABITAT & BEHAVIOUR** Broadleaved evergreen and deciduous forest, open wooded country; up to 2,000 m (breeds mostly below 1,220 m). Holds wings level when soaring, often slightly arched when gliding. Flight-display involves shallow upward swoop and almost vertical upstretching of wings, which are briefly winnowed. **RANGE & STATUS** Breeds C,E Palearctic, Indian subcontinent (N Pakistan only), SW,NE China, N,S Korea, Japan, Greater Sundas, Philippines; most northern populations winter to south, E Arabian Peninsula, Pakistan, southern Indian subcontinent, Sundas, Philippines, Sulawesi. **SE Asia** Fairly common resident (except C,SE Thailand, Singapore, W,E Tonkin). Uncommon to fairly common passage migrant and winter visitor W,NE,SE,S Thailand, Peninsular Malaysia, Singapore. Uncommon to fairly common passage migrant NW,C Thailand, N Laos, W,E Tonkin. **BREEDING** All year. **Nest** Large, bulky cupped structure in tree; 6–20 m above ground. **Eggs** 2; pale cream or yellowish-buff to pale reddish-buff, variably freckled, mottled or blotched reddish- or chestnut-brown; 52.8 x 42.8 mm (av.; *ruficollis*).

167 **BAT HAWK** *Macheiramphus alcinus* Plate **16**
IDENTIFICATION 46 cm. *M.a.alcinus*: **Adult** Falcon-like appearance, blackish-brown plumage and dark mesial streak on contrasting whitish throat and centre of upper breast diagnostic. Has broken white eyering and variable amount of white on breast. In flight, most likely to be confused with Peregrine Falcon but, apart from distinctive plumage, wings appear longer and broader-based. **Juvenile** Browner with paler base to uppertail and more extensive whitish areas on underparts. **VOICE** A high, yelping ***kwik kwik kwik kwik***... **HABITAT & BEHAVIOUR** Open areas in or near broadleaved evergreen forest, vicinity of bat caves; up to 1,220 m. Mostly hunts at night, with rapid flight on shallow stiff wingbeats. **RANGE & STATUS** Resident sub-Saharan Africa, Sumatra, Borneo, Sulawesi, New Guinea. **SE Asia** Scarce to uncommon resident south Tenasserim, S Thailand, Peninsular Malaysia. Vagrant Singapore. **BREEDING** March–October. **Nest** Bulky struc-

ture with broad shallow cup, in large tree; 9–60 m above ground. **Eggs** 1–2; white with a few faint grey markings; 59.8–46.5 mm (av.).

168 BLACK-SHOULDERED KITE
Elanus caeruleus Plate **18**

IDENTIFICATION 31–35 cm. *E.c.vociferus*: **Adult** Relatively small size, rather pale grey upperside with contrasting black lesser and median coverts diagnostic. Sides of head and underparts whitish, black eyebrow, shortish white tail with grey central feathers. In flight, wings appear rather pale with contrasting black median and lesser upperwing-coverts and underside of primaries. See male Hen and Pallid Harriers. **Juvenile** Crown streaked darker, grey of upperparts tinged browner with whitish-buff feather-tips, tail darker-tipped, ear-coverts and breast initially washed warm buff, breast and flanks sparsely and narrowly streaked brown. **VOICE** Soft piping ***pii-uu*** or ***pieu*** (particularly when displaying). Sharp ***gree-ah*** or harsh, screaming ***ku-eekk*** when alarmed. **HABITAT & BEHAVIOUR** Open country, semi-desert, cultivation; up to 1,500 m. Glides and soars with wings raised; often hovers. Male courtship display includes mock dive-attacks at female. **RANGE & STATUS** Resident Africa, SW Europe, Middle East, SE Uzbekistan, Indian subcontinent, southern China, Sundas, Philippines, Sulawesi, New Guinea. **SE Asia** Uncommon to common resident throughout. Spreading in many areas, following deforestation. **BREEDING** All year. Multi-brooded. **Nest** Untidy, bulky structure in tree. **Eggs** 3–5; whitish to pale buff, heavily blotched, smeared and streaked brown to dark red; 39.3 x 30.9 mm (av.).

169 BLACK KITE *Milvus migrans* Plate **18**

IDENTIFICATION 55–60 cm. *M.m.govinda*: **Adult** Rather nondescript dull brownish plumage and longish tail with shallow fork distinctive. Head, mantle and underparts indistinctly streaked darker, tail with indistinct narrow dark barring, cere, legs and feet yellow. In flight, outer wing is broadly fingered and angled backwards, paler diagonal band across coverts contrasts with mostly blackish remainder of upperwing, inner primaries somewhat paler; underwing similar but with uniform coverts and variably sized (usually small) whitish patch at base of outer primaries. See very similar Black-eared Kite, as well as juvenile Brahminy Kite and harriers. **Juvenile** Crown, upper mantle, breast and belly streaked whitish to pale buff; rest of upperside and wings (including underwing-coverts) have whitish to pale buff tips; shows rather contrasting dark mask. **VOICE** A high-pitched whinnying ***pee-errrr*** or ***ewe-wirrrrr***. Also transcribed as a steeply downslurred, hoarse, squealing whistle, followed by lower, slightly descending, rapid whinnying: ***fEEYA'A'A'A'O'O'OW***. **HABITAT & BEHAVIOUR** Open areas, coastal habitats, large rivers, cities; up to 1,525 m. Spends much time soaring and gliding, with wings slightly arched; twists tail. **RANGE & STATUS** Breeds NW Africa, W Palearctic, west Siberia, west C Asia, Middle East; mainly winters sub-Saharan Africa. Largely resident Indian subcontinent, SW,S China, Taiwan, Lesser Sundas, Sulawesi, S Moluccas, E New Guinea, Australia; most northern populations winter to south, southern Africa, Indian subcontinent (except S), S China. **SE Asia** Scarce to locally common resident (subject to some movements) Myanmar, C,NE Thailand, Cambodia, E Tonkin; formerly Cochinchina. Rare to scarce winter visitor and passage migrant S Thailand, Peninsular Malaysia, Singapore. **BREEDING** October–February. **Nest** Large untidy cupped structure, in tree; 7–14 m above ground. **Eggs** 2–4; pale greenish- to greyish-white, blotched, smeared or freckled blackish- or reddish-brown, purplish or blood-red; 52.7 x 42.7 mm (av.).

170 BLACK-EARED KITE
Milvus lineatus Plate **18**

IDENTIFICATION 61–66 cm. Monotypic. **Adult** Larger than Black Kite, with more rufescent-tinged body and tail, typically paler forehead, face and throat, more pronounced dark area on ear-coverts. In flight, shows broader wings, with deeply splayed six-fingered primaries (five-fingered and less splayed in Black) resulting in a squarish-looking wing-tip. When splayed, the exposed parts of the longest primary fingers (third and fourth counting outwards to inwards) are typically longer than the remaining visible part of the same feathers, whereas on Black Kite, they are typically equal or even shorter. The sixth (innermost) primary 'finger' is usually two-thirds to three-quarters the length of the fifth 'finger', whereas in Black it is only about half the length. Underwing shows extensive unmarked whitish area across base of outer inner primaries (much bigger than in Black), and prominent pale and dark barring across inner primaries (only vaguely barred on Black). See juvenile Brahminy Kite and harriers. **Juvenile** Similar to Black Kite, but with broader, whiter streaks and feather-tips, and more prominent blackish mask. Also differs in similar underwing features to adult. **VOICE** Squealing, hissing whistle, starting explosively, decreasing, rising, then falling in pitch/volume, followed by lower, slightly descending, crescendoing, rapid, musical whinnying: ***FEEeeeya'A'A'A'O'O'OW***. **HABITAT & BEHAVIOUR** As for Black Kite. **RANGE & STATUS** Breeds EC,E Palearctic, Mongolia, NW Indian subcontinent (perhaps further east in Himalayas), Tibet, China, N,S Korea, Japan; winters to Indian subcontinent (except S), S China; rarely Sumatra, Borneo. **SE Asia** Rare to uncommon winter visitor Myanmar, Thailand, Peninsular Malaysia, Singapore, Cambodia, N Laos, E Tonkin, N,S Annam, Cochinchina. Former winter visitor (current status unclear) C,S Laos, C Annam. Scarce to uncommon passage migrant N,C Myanmar, Tenasserim, NW,W,S Thailand, Peninsular Malaysia, Singapore, N,C Laos, W,E Tonkin.

171 BRAHMINY KITE *Haliastur indus* Plate **18**

IDENTIFICATION 44–52 cm. *H.i.indus*: **Adult** Bright cinnamon-rufous plumage and contrasting whitish head, neck and breast with dark shaft-streaks diagnostic. Outer primaries largely blackish. *H.i.intermedius* (south Tenasserim and south Thailand southwards) tends to have narrower dark shaft-streaks on whitish parts of plumage. **Juvenile** Recalls Black Kite but smaller, with shorter, rounded tail, overall warmer-tinged plumage and less obviously streaked crown and nape. In flight shows shorter, broader wings, some rufous markings on upperwing-coverts, rounded tail-tip and distinctive, unbarred buffish-white area across underside of primaries, contrasting with darker secondaries and black tips of outer primaries. **VOICE** A thin, high, stressed note followed by a hoarse gasping: ***tsss, herhehhehhehhehheh...*** A drawn-out, mewing ***kyeeeer*** or ***kyerrh***. **HABITAT & BEHAVIOUR** Coastal areas, large lakes and rivers; mostly lowlands. Often scavenges around harbours. **RANGE & STATUS** Resident Indian subcontinent, southern China, Sundas, Philippines, Wallacea, New Guinea, N,E Australia, Solomon Is. **SE Asia** Locally common (mostly coastal) resident (except NW Thailand, W Tonkin, N,C Laos). Former resident (current status uncertain) C Laos. Recorded (status unknown) N Laos. **BREEDING** October–August. **Nest** Bulky structure with fairly shallow cup, in tree; 5–25 m above ground. **Eggs** 2–4; greyish-white, plain or faintly speckled, blotched or squiggled reddish-brown to brown; 50.7 x 40.2 mm (av.; *indus*).

172 WHITE-BELLIED SEA-EAGLE
Haliaeetus leucogaster Plate **19**

IDENTIFICATION 70–85 cm. Monotypic. **Adult** Unmistakable. Very large, with grey upperparts and wings, white head, neck and underparts and short, diamond-shaped white tail with blackish base. In flight shows bulging secondaries and relatively narrow outer wing, white coverts contrasting sharply with blackish remainder of underwing. **Juvenile** Upperparts and wings mostly dark brownish, head, neck and underparts dull cream to buffish with dingy brownish wash across breast, tail off-white with broad dark brownish subterminal band. In flight, upperwing shows paler band across median coverts and paler area on inner primaries; underwing has warm buffish coverts and large whitish patch on primaries, contrasting with blackish secondaries and tips of primaries. See Pallas's Fish, Tawny and Imperial Eagles. **Third year**

Resembles adult but has duller breast and underwing-coverts, paler underside of secondaries and black-tipped whitish underside of primaries. **VOICE** A loud, honking ***kank kank kank...*** or ***blank blank blank blank...***; shorter, ***quicker ken-ken-ken-ken...*** and ***ka ka kaa...*** **HABITAT & BEHAVIOUR** Rocky coasts, islets, sometimes larger inland waterbodies; up to 1,400 m (breeds in lowlands). Glides and soars with wings in V. Aerial courtship display includes acrobatic somersaults, side-slipping and stoops. **RANGE & STATUS** Resident India, Bangladesh, Sri Lanka, Andaman and Nicobar Is, S,SE China, Sundas, Philippines, Wallacea, New Guinea region, Bismarck Archipelago, Australia. **SE Asia** Scarce to uncommon coastal resident (except N,C Annam). Often travels some distance inland to feed, formerly as far as S Laos. **BREEDING** September–July. **Nest** Very large and bulky with fairly deep cup, in tree or on rock ledge or pylon; 10–50 m above ground. **Eggs** 2; white; 77.7 x 53.4 mm (av.).

173 PALLAS'S FISH-EAGLE

Haliaeetus leucoryphus Plate **19**

IDENTIFICATION 76–84 cm. Monotypic. **Adult** Recalls Grey-headed Fish-eagle but larger, head, neck and upper mantle warm buffish to whitish, has darker brown upperside, dark brown thighs and vent and blackish base of tail. In flight shows longer, straighter wings and longer blackish tail with broad white central band. **Juvenile** Similar to White-bellied Sea-eagle but underparts more uniformly dark, narrow pale supercilium contrasts with blackish mask, tail blackish-brown. In flight shows similar underwing pattern but coverts mostly dark brown with whitish band across median coverts to axillaries, outer primaries all dark. Whitish primary-flashes rule out White-tailed Eagle. **Second/third year** Similar to juvenile but underside paler and more uniform. In flight, underwing shows broader whitish band across median coverts, and almost all-dark flight-feathers. **VOICE** A series of loud, guttural notes, repeated up to 14 times or more: ***kha-kha-kha-kha...***; ***gho-gho-gho-gho...***; ***gao-gao-gao-gao...*** etc. Calling may speed up and run to a higher-pitched excited yelping. **HABITAT** Large lakes and rivers, marshes; lowlands. **RANGE & STATUS** Breeds C Asia, S Siberia, Pakistan, northern Indian subcontinent, S,E Tibet, northern China, Mongolia; some populations winter to south, Iraq, northern Indian subcontinent. **SE Asia** Rare to scarce resident (subject to some movements) S Myanmar. Formerly resident (current status unknown) SW,N,C Myanmar, north Tenasserim. Vagrant NW Thailand, Cambodia, Cochinchina. **BREEDING** November–February. **Nest** Huge rough platform, in tree; 15–35 m above ground. **Eggs** 2–4; white; 69.7 x 55.1 mm (av.).

174 WHITE-TAILED EAGLE

Haliaeetus albicilla Plate **19**

IDENTIFICATION 70–90 cm. *H.a.albicilla*: **Adult** Very large size, nondescript brownish plumage, big yellow bill and short, diamond-shaped white tail diagnostic. Has paler brown to creamy-whitish hood and upper breast, mostly blackish secondaries and primaries, indistinct darker streaks on head and body, and mottled wing-coverts. **Juvenile** Upperparts and wing-coverts mostly darker, warmish brown with blackish tips, head and underparts blackish-brown with pale streaks on neck and breast, bill dusky-greyish, tail-feathers broadly bordered blackish. In flight shows distinctive combination of very broad, parallel-edged wings, dark underwing with whitish axillaries and narrow whitish bands across coverts, and whitish spikes on tail-feathers. See Egyptian, White-rumped and Cinereous Vultures, Himalayan Griffon and *Aquila* eagles. **Second/third year** Broad buffish-white line on underwing-coverts, brown-buff belly to vent. **Third/fourth year** Similar to juvenile but body darker. In flight shows less prominent whitish bands across underwing-coverts and white tail with narrow blackish terminal band. **VOICE** A loud rapid yelping ***klee-klee-klee-klee-klee-klee...*** **HABITAT & BEHAVIOUR** Large lakes and rivers, open country; recorded in lowlands. Soars with wings held level or only slightly raised; glides on level wings or with outer wing slightly depressed. **RANGE & STATUS** Breeds Palearctic, NE China, N Japan, Greenland; some northern populations winter south to S Europe, Middle East, Pakistan, northern Indian subcontinent, S Tibet, southern China, N,S Korea, Japan. **SE Asia** Scarce to local winter visitor W(north),N Myanmar. Vagrant C Thailand, E Tonkin.

175 LESSER FISH-EAGLE

Ichthyophaga humilis Plate **19**

IDENTIFICATION 51–68 cm. *I.h.humilis*: **Adult** Like Grey-headed Fish-eagle but smaller, with somewhat paler upperparts, wing-coverts and breast, and dull greyish tail with darker terminal band (often appears rather uniform). In flight, underwing similar but may show whitish bases to outer primaries, undertail dark brownish with only slightly contrasting dark terminal band. *I.h.plumbea* (N,E,S Myanmar, W,NW Thailand, Indochina) averages 20% larger (within size range given). See juvenile Pallas's Fish-eagle. **Juvenile** From Grey-headed by paler upperparts, plainer head, neck and breast, with only vague paler streaks, no whitish supercilium, more contrasting white vent, darker tail. **VOICE** Various deep, yelping and whining gull-like sounds: ***yow***; ***yow-ow***; ***ow-ow-ow-ow***; ***yow-yow-yow-yow***; ***yaa'aaah***; ***eeyaauuah***; ***yow-eee-aaa...yow-aaa*** etc. **HABITAT & BEHAVIOUR** Larger rivers in forest; up to 900 m. Spends long periods perched in waterside trees. Glides and soars on level wings. **RANGE & STATUS** Resident NW,N,NE India, Nepal, Bhutan, S China (Hainan), Sumatra, Borneo, Sulawesi, Sula Is, S Moluccas. **SE Asia** Scarce to locally fairly common resident N,E,S Myanmar, Tenasserim, W,NW,S Thailand, Peninsular Malaysia, Cambodia, Laos, W Tonkin, C,S Annam, Cochinchina. **BREEDING** November–April. **Nest** Large and bulky with fairly deep cup, c.20 m up in large tree. **Eggs** 2–3; white (often stained); 55.1–68 x 50.2–53.1 (*humilis*).

176 GREY-HEADED FISH-EAGLE

Ichthyophaga ichthyaetus Plate **19**

IDENTIFICATION 69–74 cm. *I.i.ichthyaetus*: **Adult** Rather large and long-necked, with plain greyish hood, sharply contrasting white thighs and vent, and rounded white tail with broad black terminal band. Upperparts and wing-coverts greyish-brown, breast mostly warm brown to brownish-grey. In flight, wings appear rather broad and rounded, white tail-base and vent contrasting sharply with all-dark wings. See Lesser and Pallas's Fish-eagles. **Juvenile** Head, neck, breast and upper belly mostly warm brownish, with narrow white supercilium and whitish streaks on crown, sides of head, foreneck and breast; upperside browner-tinged than adult, tail dark with whitish mottling showing as faint pale bands. In flight, underwing-coverts mostly whitish, flight-feathers mostly whitish with darker tips and some faint narrow darker bars towards their tips. **Immature** Plumages poorly documented. Gradually develops darker underwing and adult tail pattern. **Second/third year** Whitish patches on primaries, white thighs and vent, dark end to tail. **VOICE** During courtship display utters a powerful barking ***kroi-ork*** and repeated loud eerie ***tiu-weeeu***. **HABITAT & BEHAVIOUR** Lakes, swamps, large rivers; up to 1,525 m. Soars with wings held level or in shallow V. **RANGE & STATUS** Resident India, Nepal, Bangladesh, Sri Lanka, Greater Sundas, Philippines, Sulawesi. **SE Asia** Rare to locally fairly common resident Myanmar, W,NE(south-west),S Thailand, Peninsular Malaysia, Singapore, Cambodia, N,S Laos, Vietnam (except W Tonkin); ? SE Thailand (possibly escaped bird). Former resident (current status unclear) NW,C Thailand, W Tonkin. **BREEDING** August–March. **Nest** Very large and bulky with deep cup, in large tree; 7.5–30 m above ground. **Eggs** 2–4; white (often stained); 59.8–72 x 50–54.5 mm.

177 EGYPTIAN VULTURE

Neophron percnopterus Plate **20**

IDENTIFICATION 60–70 cm. *N.p.ginginianus*: **Adult** Unmistakable. Rather slim and mostly dirty-whitish, with shaggy nape and neck, naked yellowish forehead, face and throat, long slender yellowish bill and pointed, diamond-shaped whitish tail. In flight shows mostly whitish upperside with black-tipped greater coverts and secondaries, and mostly black primaries and primary coverts; underwing black with contrasting whitish coverts. See White-bellied Sea-eagle. **Juvenile** Generally dark brownish with paler vent and mostly dull greyish tail with darker base; facial skin grey, bill pale, often with dark tip. In flight shows mostly blackish-brown upperwing with narrow whitish bands across coverts and fairly uniform dark underwing. Note size, shape and distinctive head and bill. Gradually attains white in plumage with age. **Third/fourth year** Similar to adult but has dark brownish collar, some darker mottling on underparts and duller tail. In flight shows dark mottling on underwing-coverts. **VOICE** Typically silent but may utter mewing, hissing and low grunting sounds when alarmed. **HABITAT & BEHAVIOUR** Open country; recorded at c.1,000 m. Soars with wings held level or slightly bowed. **RANGE & STATUS** Breeds Africa, southern W Palearctic, Middle East, C Asia, Pakistan, India, Nepal; some northern populations winter to south, N Africa, Pakistan, NW India. **SE Asia** Vagrant E Myanmar.

178 **WHITE-RUMPED VULTURE**
Gyps bengalensis Plate **20**

IDENTIFICATION 75–85 cm. Monotypic. **Adult** Blackish plumage with contrasting white neck-ruff, lower back and rump diagnostic. Has mostly greyish-brown naked head and longish neck and rather short, thick dark bill with pale bluish-grey upper mandible. In flight shows very broad wings with well-spaced fingers, short tail, distinctive white patch on back and rump, mostly greyish secondaries and inner primaries (above and below) and diagnostic white underwing-coverts with black leading edge. See Egyptian Vulture and White-bellied Sea-eagle. **Juvenile** Browner overall, neck-ruff dark brownish, back and rump dark brownish, head and neck mostly covered with whitish down, bill blackish, lesser and median wing-coverts vaguely streaked paler, underparts narrowly streaked whitish. In flight shows rather uniform dark plumage with short narrow whitish bands across underwing-coverts; leading edge of underwing-coverts is darker than greater underwing-coverts. See Himalayan Griffon, Cinereous Vulture and White-tailed Eagle. Gradually attains adult features with age. **Second/third year** Like juvenile but primary and greater underwing-coverts mostly whitish (shows some whitish feathers on lower back and beginnings of short buffy neck-ruff). **VOICE** Often silent but gives occasional grunts, croaks, hisses and squeals at nests sites, roosts and when feeding. **HABITAT** Open country, vicinity of abattoirs, cliffs; mostly lowlands but up to 2,600 m. **RANGE & STATUS** Resident (subject to local movements) Indian subcontinent (except Sri Lanka), SW China (SW,S Yunnan). **SE Asia** Rare to scarce resident (subject to some movements) SW,W,N,E Myanmar, Cambodia, S Laos, S Annam; formerly C Thailand, northern Peninsular Malaysia, N,C Laos, Cochinchina. Former resident (current status unknown) C,S Myanmar, Tenasserim. Rare winter visitor W(formerly resident),NW,NE Thailand; ? SE Thailand (perhaps escaped bird). Vagrant S Thailand (formerly resident). **BREEDING** October–February. Often loosely colonial. **Nest** Large and bulky with shallow cup, in tree; 4.5–30 m above ground. **Eggs** 1; white, usually faintly marked (sometimes boldly) with red and red-brown over grey or lavender undermarkings (rarely unmarked); 80.5–107 x 61–69 mm.

179 **SLENDER-BILLED VULTURE**
Gyps tenuirostris Plate **20**

IDENTIFICATION 81–103 cm. Monotypic. **Adult** Told by mostly rather pale sandy-brown body and wing-coverts and contrasting naked blackish head and longish neck. Narrow head profile, relatively long, slender dark bill, relatively small whitish neck-ruff, whitish lower back and rump, dark-centred greater coverts (above and below), dark greyish legs. In flight, relatively uniform pale upper- and underwing-coverts and underbody contrast with dark head and neck and blackish secondaries and primaries. **Juvenile** Neck-ruff browner, upperwing-coverts duller and browner and vaguely streaked, underwing-coverts and underparts duller and browner. Gradually becomes paler with age. **VOICE** Occasional hissing and cackling sounds. **HABITAT** Open country; lowlands. **RANGE & STATUS** Resident (subject to local movements) N,NE India, Nepal; ? visits NE Bangladesh. **SE Asia** Rare to scarce resident (subject to local movements) N,E(south-west) Myanmar, Cambodia, S Laos; formerly NW,NE,C,SE Thailand, N,C Laos, Cochinchina. Former resident (current status unknown) Myanmar (except N,E), S Annam. Former rare non-breeding visitor S Thailand, Pen Malaysia. **BREEDING** October–March. May be loosely colonial. **Nest** Large and bulky with shallow cup, on rock ledge or in tree. **Eggs** 1; white, sometimes flecked and blotched light reddish (rarely heavily marked); 84.7 x 63.6 mm. **NOTE** Formerly united with extralimital Indian Vulture *G. indicus* as Long-billed Vulture *G. indicus*, but see Rasmussen & Parry (2001).

180 **HIMALAYAN GRIFFON**
Gyps himalayensis Plate **20**

IDENTIFICATION 115–125 cm. Monotypic. **Adult** Huge and bulky with shortish, thick pale bill, pale head, thickish neck and mostly sandy-buffy body and wing-coverts. Neck-ruff brownish-buff, legs and feet pinkish. In flight, plumage resembles Slender-billed Vulture but head pale, upper body and wing-coverts much paler and more contrasting, underwing-coverts (including greater coverts) uniformly whitish with narrow dark leading edge. **Juvenile** Similar to White-rumped Vulture but considerably larger and more heavily built, with whitish streaks on mantle to uppertail-coverts and scapulars, broader, more prominent whitish streaks on upper- and underwing-coverts and underparts, and paler legs and feet. In flight, plumage very similar to White-rumped but underwing-coverts typically have longer, more clearly separated buffish bands, leading edge of underwing-coverts same colour as greater underwing-coverts (not darker). **Subadult** Develops paler underbody and upperwing-coverts, aiding separation from White-rumped. **VOICE** Occasional grunts and hissing sounds. **HABITAT** Open country; recorded in lowlands. **RANGE & STATUS** Resident (subject to some movements) SE Kazakhstan, Kirghizstan, Tajikistan, NE Afghanistan, N Pakistan, NW,N India, Nepal, Bhutan, S,E Tibet, W,N,NW China; vagrant Riau Archipelago (Indonesia). **SE Asia** Vagrant N Myanmar, W(coastal),NW,C,NE,S Thailand, Peninsular Malaysia, Singapore, north Cambodia.

181 **CINEREOUS VULTURE**
Aegypius monachus Plate **20**

IDENTIFICATION 100–110 cm. Monotypic. **Adult** Huge size and uniform blackish-brown plumage distinctive. Bill larger than other vultures, pale crown and nape contrast with black face and foreneck, legs and feet greyish-white. In flight shows very broad, relatively straight-edged wings, appears all dark except for narrow pale line at base of flight-feathers. See juvenile White-rumped Vulture and Himalayan Griffon. **Juvenile** Plumage even blacker, head and neck mostly blackish. **HABITAT** Open country; lowlands. **RANGE & STATUS** Breeds S Europe east to Caspian Sea, Turkey, C Asia, Iran, Afghanistan, Pakistan, NW India, E Tibet, northern China, Mongolia; some non-breeders wander to south, S India, S China. **SE Asia** Vagrant/rare winter visitor W,N,S,E Myanmar, W,NE,SE,S Thailand, Peninsular Malaysia, Cambodia, E Tonkin.

182 **RED-HEADED VULTURE**
Aegypius calvus Plate **20**

IDENTIFICATION 76–86 cm. Monotypic. **Male** Unmistakable, with blackish plumage and red head, neck, legs and feet. Has yellowish eyes, paler dark-tipped tertials and secondaries, white frontal part of neck-ruff and white lateral body-patches. In flight, these white areas and pale band across bases of flight-feathers (less pronounced on upperwing) contrast sharply with black remainder of wings and body. **Female** Eyes dark brown to dark red, lower scapulars white. **Juvenile** Plumage somewhat browner, head and neck pinkish with some whitish down, eyes brown, legs and feet pinkish, vent whitish, flight-feathers uniform (above and below). In flight shows similar underside pattern to adult, apart from all-dark flight-feathers and whitish vent. **VOICE** Occasional squeaks, hisses and grunts. **HABITAT** Open country and wooded areas, dry deciduous forest with rivers; up to 1,525 m (mostly lowlands). **RANGE & STATUS** India, Nepal, NW Bangladesh, SW China (W,S Yunnan); formerly Pakistan. **SE Asia** Rare resident W Thailand, S Laos, north-east Cambodia, N,S Annam; formerly NW,C,S Thailand, northern Peninsular Malaysia, N,C Laos, Cochinchina. Former resident (current status unknown) Myanmar, C Annam. Formerly occurred (status unclear) Singapore. **BREEDING** October–April. **Nest** Bulky with deep cup (after some years) or old nest of other large raptor, in tree or bush; 1–30 m above ground. **Eggs** 1; greenish-white or white (rarely marked reddish-brown); 79–90 x 61.5–71.1 mm.

183 SHORT-TOED SNAKE-EAGLE

Circaetus gallicus Plate **26**

IDENTIFICATION 62–67 cm. Monotypic. **Adult** Medium-sized with rather big-headed, top-heavy appearance, yellow to orange-yellow eyes, longish unfeathered legs and relatively long wings, which reach tail-tip at rest. Plumage very variable but often shows distinctive darker hood, dark barring on underparts and in flight, long broad wings (distinctly pinched-in at base), narrow dark barring and trailing edge on otherwise pale and rather featureless underwing, and three bands across undertail (terminal one broadest). Head may be mostly pale and underparts and underwing can be very pale and rather featureless, but always shows tail-banding and some dark markings on underwing. Upperside normally shows contrastingly pale lesser and median coverts. See juvenile Crested Serpent-eagle and Rufous-bellied Eagle. **Juvenile** Shows uniform narrow pale tips to wing-feathers, paler and more contrasting median and lesser upperwing-coverts, pale-tipped greater coverts and poorly marked underwing, without dark trailing edge. **VOICE** May give plaintive musical whistled ***weeo*** or longer ***weeooo***, sometimes followed by gull-like ***woh-woh-woh*** or ***quo-quo-quo***. **HABITAT & BEHAVIOUR** Open and coastal areas; lowlands. Flies with rather slow, heavy wingbeats; soars with wings held flat or slightly raised, with gently drooping primaries; frequently hovers. **RANGE & STATUS** Breeds Africa, W Palearctic, SW C Asia, Middle East, Pakistan, India, Nepal, N China, Lesser Sundas; some northern populations winter to south, northern Africa, northern Indian subcontinent. Non-breeding visitor E Java, Bali. **SE Asia** Rare passage migrant Singapore. Vagrant/rare migrant SW,C,S Myanmar, north Tenasserim, Thailand, Peninsular Malaysia, Cambodia, N,S Laos, C Annam.

184 CRESTED SERPENT-EAGLE

Spilornis cheela Plate **18**

IDENTIFICATION 56–74 cm. *S.c.burmanicus*: **Adult** Distinctive. Medium-sized to largish, with large head, mostly rather dark brownish plumage and black tail with broad white central band. Has prominent yellow cere, facial skin and eyes, short, full, white-marked blackish crest on nape and hindneck, small white spots on median and lesser coverts, paler, warm-tinged sides of head, neck and underparts, with black-bordered whitish speckles turning to short bars on lower breast to undertail-coverts. In flight appears very broad-winged; from below shows distinctive combination of darkish body and underwing-coverts, broad white band across underwing contrasting with broad black trailing edge, and black tail with broad white central band. *S.c.malayensis* (south Tenasserim and S Thailand southwards) is distinctly smaller, with darker sides of head and underparts and more pronounced markings below; *ricketti* (E Tonkin) averages larger and has somewhat paler upperparts, thin whitish bars on breast and fewer spots and bars on rest of underparts. See Oriental Honey-buzzard and hawk-eagles. **Juvenile** Much paler. Head scaled black and whitish with blackish ear-coverts; upperparts and wing-coverts have whitish tips/fringes, underparts whitish usually with faint buffish wash and dark streaks on throat-centre, breast and belly; has whitish tail with three complete blackish bands. In flight shows mostly whitish underwing, with buff-washed and dark-streaked wing-coverts, many fine dark bars across underside of flight-feathers and indistinct broad darker trailing edge. See hawk-eagles, Rufous-bellied Eagle and Short-toed Snake-eagle. **VOICE** Rather vocal, particularly when soaring. Utters a loud plaintive high-pitched 2–4 note ***hwiii-hwi***; ***h'wi-hwi***; ***hwi-hwi-hwi***; ***h'wee hew-hew***; ***hii-hwi-hwi*** etc., sometimes introduced by spaced ***hu-hu-hu-hu...*** 'Song' is a sustained crescendo: ***ha-ha-ha-ha hu-hu-hu-hu-h'weeooleeoo***. **HABITAT & BEHAVIOUR** Broadleaved evergreen, deciduous and peatswamp forest, secondary forest; up to 2,470 m. Soars with wings in shallow V. Displaying pairs soar with head and tail raised, dive at one another and undertake short flights with wings winnowed shallowly below horizontal. **RANGE & STATUS** Resident Indian subcontinent (except Pakistan), southern S China, Taiwan, Greater Sundas, Philippines (Palawan). **SE Asia** Fairly common resident (subject to minor movements) throughout; rare (may no longer breed) Singapore. Rare to scarce passage migrant NW,S Thailand. **BREEDING** January–October. **Nest** Large, loosely built cup, in tree; 6–30 m above ground. **Eggs** 1; bluish- to greenish-white with pale purple, purplish- or reddish-brown spots and clouds; 66.1–73.1 x 54–58.2 mm (*burmanicus*).

185 WESTERN MARSH-HARRIER

Circus aeruginosus Plate **22**

IDENTIFICATION 48–56 cm. *C.a.aeruginosus*: **Male** Recalls Eastern Marsh-harrier but markings on head, neck and breast browner, ear-coverts paler, dark of upperparts and wing-coverts brown and more uniform, lower underparts and thighs rufous-chestnut. In flight also shows buff leading edge to lesser upperwing-coverts. **Female** Similar to juvenile Eastern but shows dark eye-line that extends down side of neck, lacks neck-streaks. In flight shows smaller pale flash on underside of primaries; normally shows broad creamy-buff leading edge to inner wing and creamy-buff breast-band. **Juvenile** Resembles female but initially lacks creamy-buff on wing-coverts and breast. Male gradually acquires grey parts of plumage with age. See female Pied Harrier. **VOICE** May give weak ***kyik*** or cackling ***chek-ek-ek-ek-ek...*** **HABITAT & BEHAVIOUR** Marshes, rice paddies, open areas; mostly in lowlands. Typical harrier flight. **RANGE & STATUS** Breeds NW Africa, W,C Palearctic (east to Baikal region and W Mongolia), NW China; most populations winter to south, Africa, Middle East, Indian subcontinent. **SE Asia** Uncommon to fairly common winter visitor Myanmar, C Thailand (rare), E Tonkin (rare). Vagrant W(coastal),NW,S Thailand, Peninsular Malaysia, Cambodia.

186 EASTERN MARSH-HARRIER

Circus spilonotus Plate **22**

IDENTIFICATION 48–56 cm. *C.s.spilonotus*: **Male** Similar to Pied Harrier but shows black streaks on neck and breast and lacks large white patches on upperwing-coverts. Has pale scaling on upperparts and wing-coverts. In flight, darker lesser upperwing-coverts are apparent. **Female** Resembles Pied but belly and thighs dull rufous. In flight further differs by lack of pale leading edge to lesser upperwing-coverts, thinner bars on uppertail with plain central feathers, extensively dull chestnut-brown underwing-coverts and less boldly marked underside of secondaries. Other harriers ruled out by combination of grey on wings and tail, streaked head and breast and rufescent belly and thighs. **Juvenile** Rather uniform dark brown (faint paler area across breast) with pale hood, with or without dark crown and band across lower throat; head-sides rather uniform, variable darker streaking often on neck, pronounced creamy-buff streaks sometimes on mantle. In flight, blackish-tipped creamy-greyish underside of primaries, contrasting with dark remainder of underwing, distinctive. See Western Marsh-harrier. Male gradually acquires grey plumage-parts with age. **VOICE** May utter a kite-like mewing ***keeau***, particularly at roosts. **HABITAT & BEHAVIOUR** Marshes, rice paddies, open areas; up to at least 1,000

m. Large numbers may gather at roosts. Typical harrier flight. **RANGE & STATUS** Breeds C?,E Palearctic (west to Baikal region), NE China, N Japan; winters south to NE Indian subcontinent, southern China, Taiwan, S Korea, southern Japan, Borneo, Philippines. **SE Asia** Uncommon to fairly common winter visitor (except SW Myanmar, W Tonkin). Also recorded on passage C Myanmar, W,S Thailand, Cambodia, N Laos, E Tonkin.

187 **HEN HARRIER** *Circus cyaneus* Plate **21**
IDENTIFICATION 44–52 cm. *C.c.cyaneus*: **Male** Grey plumage with contrasting white lower breast to undertail-coverts distinctive. When perched, wing-tips fall well short of tail-tip. In flight shows distinctive combination of all-black, 'five-fingered' outer primaries (above and below), darker trailing edge to rest of wing (more prominent below) and obvious unmarked white band across uppertail-coverts. See Pallid and Montagu's Harriers and Black-shouldered Kite. **Female** Rather nondescript brownish, with dark-barred wings and tail and dark-streaked pale neck and underparts. Difficult to separate from Pallid and Montagu's. Best separated by heavy arrowhead or drop-shaped markings on thighs and undertail-coverts (wide arrowhead shapes on undertail-coverts are diagnostic), and by wing-tips, which fall well short of tail-tip at rest. In flight appears rounder-winged with clearly defined broad white uppertail-covert band; has two well-marked dark bands on upperside of secondaries (terminal one broadest and most obvious); underwing has broadest dark band along trailing edge, the other two differing in width and variably spaced, but with pale bands between them reaching body; dark tips to inner primaries and dark, unpatterned axillaries. See Pied Harrier. **Juvenile** Similar to female (underparts streaked) but generally rustier, with darker underside of secondaries. From Pallid and Montagu's by boldly streaked, duller underparts, less solid and distinct pale band across upperwing-coverts; always shows narrow whitish collar. See Pied. **VOICE** May give a rapid, quacking chatter, ***quek-ek-ek-ek***, ***quik-ik-ak-ik-uk-ik*** etc. **HABITAT & BEHAVIOUR** Open areas; up to 1,500 m. Glides with wings in shallow V. **RANGE & STATUS** Breeds Holarctic, NW,NE China; winters south to N Africa, Middle East, N Pakistan, northern Indian subcontinent, southern China, N,S Korea, Japan, C America. **SE Asia** Rare to uncommon winter visitor W,N,C Myanmar, NW Thailand, N Laos. Rare passage migrant S Thailand. Vagrant S Myanmar, NE Thailand, Peninsular Malaysia, Singapore, Cambodia, C Laos, E Tonkin, Cochinchina.

188 **PALLID HARRIER** *Circus macrourus* Plate **21**
IDENTIFICATION 40–48 cm. Monotypic. **Male** Similar to Hen Harrier but somewhat smaller and slimmer, with paler and more uniform grey plumage, lacking obvious contrast between breast and belly. In flight, wings narrower and more pointed, with narrower black wedge on outer primaries, less white on uppertail-coverts and no darker trailing edge to wing (subadult male shows dusky trailing edge to secondaries). See Montagu's Harrier. **Female** Very similar to Hen and Montagu's and more variable than latter. In flight, upperwing usually lacks paler bar through dark secondaries, and white uppertail-covert band usually narrower than Hen; on the underwing primaries are paler than secondaries, with dark bars sometimes absent from bases (may show distinctive pale crescent at base of primaries), inner primaries are relatively pale-tipped; broadest and darkest underwing-band shows along trailing edge of secondaries; pale bands become distinctly narrower and darker towards body (see juvenile); underwing-coverts and axillaries lack bold bars. See juvenile Hen. **Juvenile** Easily separated from Hen by unstreaked warm buff underparts. Very similar to Montagu's but typically has less white above and below eye, more dark brown on sides of head (extending under base of lower mandible), wider, paler, unstreaked collar bordered behind/below by uniform dark brown. Also lacks faint narrow streaking on sides of breast and flanks shown by most juvenile Montagu's. In flight, underwing has outer primaries either evenly barred or pale with narrow dark tips, inner primaries have paler tips (may show distinctive pale crescent on primary bases like female); also has paler buff coverts and axillaries, without obvious dark bars. First-spring males are often almost unmoulted, though more faded on head and breast (juvenile head pattern still clear); during moult, thin rusty streaks (similar to Montagu's) appear on body. See Pied Harrier. **VOICE** Sometimes gives a high-pitched, plaintive, single or repeated ***siehrrr***. **HABITAT** Open country, cultivation; up to 1,320 m. **RANGE & STATUS** Breeds E Europe, C Palearctic, NW China; winters Africa, SE Europe, Middle East, Indian subcontinent, possibly southern China. **SE Asia** Rare to scarce winter visitor SW,N,C,S,E Myanmar. Vagrant S Thailand, Peninsular Malaysia, E Tonkin (old sight record).

189 **PIED HARRIER** *Circus melanoleucos* Plate **21**
IDENTIFICATION 43–46 cm. Monotypic. **Male** Unmarked black head, mantle, back, upper breast and median coverts and large whitish patch on lesser coverts diagnostic. In flight, black outer primaries (above and below) and black band across upperwing-coverts contrasts sharply with mostly pale remainder of wings. See Eastern Marsh Harrier and Black-shouldered Kite. **Female** Separated from other female harriers by whitish area along upper edge of lesser coverts, grey outer edge to wing-coverts and secondaries, and almost unmarked whitish thighs and vent. In flight, from above, shows distinctive combination of whitish leading edge to lesser coverts, whitish uppertail-covert band, uniformly barred tail and grey primary coverts and flight-feathers with dark bars; also has broadly dark-tipped primaries and darker trailing edge to wing. Underwing pattern most similar to Montagu's but overall paler with narrower dark bands across secondaries. **Juvenile** Similar to Pallid and Montagu's but has more uniform dark rufous-brown wing-coverts and dark rufous-brown underparts. Also resembles Eastern Marsh Harrier but more rufous-brown. In flight shows narrow whitish uppertail-covert band and, from below, distinctive combination of darkish rufous-brown body and wing-coverts, indistinct paler bands across blackish secondaries (tapering towards body) and pale greyish primaries with dark bars restricted to tips and inner feathers (leading edge also darker). **VOICE** Displaying male gives ***keee-veeee*** calls and female utters a rapid ***kee-kee-kee*** or rapid, chattering ***chak-chak-chak-chak...*** Otherwise usually silent but occasionally utters a rapid ***wek-wek-wek***. **HABITAT** Marshes, grassland, open areas, cultivation; up to 2,130 m. **RANGE & STATUS** Breeds E Palearctic, NE China, N Korea; has bred NE India, Philippines (Luzon); winters to S and eastern Indian subcontinent, southern China, Borneo, Philippines. **SE Asia** Has bred N Myanmar. Uncommon to fairly common winter visitor (except W Tonkin, N Annam). Uncommon to fairly common passage migrant NW,S Thailand, Cambodia, N Laos, W,E Tonkin. Vagrant Singapore. **BREEDING** April–June. **Nest** Lined hollow on dry ground. **Eggs** 4–6; white or greenish-white, sometimes with a few brown to reddish-brown spots or blotches; 40.5–47.7 x 32–37 mm.

190 **MONTAGU'S HARRIER**
Circus pygargus Plate **21**
IDENTIFICATION 43–47 cm. Monotypic. **Male** Similar to Hen and Pallid Harriers at rest but belly streaked rufous-chestnut, closed wing-tips extend to tail-tip (all plumages). In flight, easily distinguished by blackish bar across upperside of secondaries, two blackish bars on underside of secondaries, and dark bands across underwing-coverts and axillaries. **Female** Very similar to Hen and Pallid but somewhat smaller and slimmer than former, underparts quite narrowly and evenly marked, with narrow dark streaks on thighs and undertail-coverts. In flight, narrower-winged than Hen, narrower white uppertail-covert band, upperside of flight-feathers usually obviously barred; underwing has inner primaries darker-tipped than Pallid, broadest dark band on secondaries at trailing edge but central band darkest, pale intervening bands reaching body; also shows diagnostic boldly barred axillaries

and chequered barring on underwing-coverts. **Adult dark morph** Rare (may not occur in region). Sooty to blackish overall with paler tail and dark-tipped silvery underside of primaries with darker bars on outer feathers. **Juvenile** Easily separated from Hen by unstreaked buffish-rufous underparts. Very similar to Pallid but typically shows narrow dark streaks on sides of breast and flanks, has more white above and below eye, less dark brown on sides of head, no clear pale collar, sometimes with darker sides of neck but usually mottled or streaked and not uniform (may have diffuse paler area behind ear-coverts but never forms collar). In flight, underwing has blackish primary tips (broader on outer feathers) and usually more rufescent coverts and axillaries with prominent blackish barring. First-spring male usually has head and body similar to adult and some strong rusty streaks on belly. See Pied and female Eastern Marsh Harrier. **VOICE** May give an occasional ***chock-chock, chock-ok-ok***; other calls similar to Hen Harrier. **HABITAT** Open areas, cultivation; lowlands. **RANGE & STATUS** Breeds NW Africa, W,C Palearctic, probably NW China; winters sub-Saharan Africa, Indian subcontinent (rare NE). **SE Asia** ? vagrant Myanmar (doubtful old records from S, south E, and Tenasserim).

191 CRESTED GOSHAWK

Accipiter trivirgatus Plate **23**

IDENTIFICATION 40–46 cm. Populations of S Thailand southwards are somewhat smaller. *A.t.indicus*: **Male** Combination of relatively large size, short crest, slaty crown and sides of head, brownish-grey upperparts, dark mesial streak, streaked breast and barred belly distinctive. Mesial streak blackish, sides of breast rufous-chestnut, streaks on central breast and broad bars on belly rufous-chestnut; uppertail greyish with three complete broad blackish bands (pale bands of equal width), eyes yellow to orange-yellow, legs relatively short and stout, closed wing-tips fall close to base of tail. In flight wings appear broad, blunt-tipped and rounded with bulging secondaries (pinched in at base), underside shows four dark bands across flight-feathers, three broad dark bands across central tail-feathers and four narrow bands across outermost tail-feathers. See juvenile Shikra, female and juvenile Besra and Jerdon's Baza. **Female** Larger, with browner-tinged crown and sides of head and browner breast-streaks and bars on belly. **Juvenile** Generally similar to adult but eyes yellow, sides of head and upperparts browner, has rufous to buffish feather-fringes on crown and nape, pale buffish fringing on mantle and wing-coverts, streaked ear-coverts, underparts washed buffish and mostly streaked brown, bars restricted to lower flanks and thighs. **VOICE** A shrill, screaming ***he he hehehehe*** and high-pitched squeaking ***chiup*** when alarmed. **HABITAT & BEHAVIOUR** Broadleaved evergreen, deciduous and mixed broadleaved and coniferous forest; up to 1,950 m (below 245 m in Peninsular Malaysia). Performs display-flights with shallow, winnowing, bowed wings and fluffed-out undertail-coverts. **RANGE & STATUS** Resident SW,N,NE India, Nepal, Bhutan, Sri Lanka, south-western and S China, Taiwan, Greater Sundas, Philippines; recorded E Bangladesh. **SE Asia** Fairly common to common resident (except SW Myanmar, C Thailand); rare Singapore. **BREEDING** All year. **Nest** Large bulky cup in tree; 12–45 m above ground. **Eggs** 2–3; white to bluish-white; 48.4 x 39.6 mm (av.).

192 SHIKRA *Accipiter badius* Plate **23**

IDENTIFICATION 30–36 cm. *A.b.poliopsis*: **Male** Pale grey sides of head and upperparts, white throat with faint grey mesial streak and dense narrow orange-rufous bars on breast and belly diagnostic. Tail has central and outer pairs of feathers plain grey, the rest with 3–4 dark cross-bars, outer primary tips blackish, eyes orange-red to scarlet. Barring below sometimes more pinkish and less marked. In flight, underwing appears rather whitish with some dark bars across dark-tipped outer primaries and indistinct warm buffish markings on coverts. See Chinese and Eurasian Sparrowhawks. **Female** Larger, eyes yellow, upperside washed brownish (particularly nape and mantle), underparts more coarsely barred rufous. In flight undertail shows 7–8 narrow dark bands on outer feathers. **Juvenile** Crown, upperparts and wing-coverts brown with paler feather-fringes. Has warm-tinged dark brown mesial streak, tear-drop breast-streaks turning to barring on flanks and spots on thighs, five complete narrow dark bands across uppertail (subterminal one broadest) and greenish-yellow to yellow eyes. In flight shows even dark bars across whole of whitish underside of flight-feathers. Best separated from similar Besra by larger size, spots on thighs, paler upperside and narrower dark tail-bands. See that species, Crested Goshawk and Chinese Sparrowhawk. **VOICE** A loud harsh thin ***titu-titu*** and long, drawn-out screaming ***iheeya iheeya...*** **HABITAT & BEHAVIOUR** Deciduous, open broadleaved evergreen, mixed broadleaved and coniferous forest, open areas, cultivation; up to 1,600 m. Often seen hunting in open. **RANGE & STATUS** Resident sub-Saharan Africa, C Asia, Iran, Afghanistan, Indian subcontinent, NW,SW,S China, Sumatra. **SE Asia** Common resident, subject to some movements (except southern S Thailand, Peninsular Malaysia, Singapore, W Tonkin). Rare to scarce winter visitor S Thailand, northern Peninsular Malaysia. Vagrant southern Peninsular Malaysia, Singapore. **BREEDING** February–May. **Nest** Loosely built bulky cup, in tree (sometimes on epiphytic fern clump); 6–20 m above ground. **Eggs** 3–5; pale bluish-white, sometimes lightly speckled grey; 39.4 x 31 mm (av.).

193 CHINESE SPARROWHAWK

Accipiter soloensis Plate **23**

IDENTIFICATION 29–35 cm. Monotypic. **Male** Similar to Shikra but grey of plumage considerably darker, breast pale pinkish to pinkish-rufous (either unbarred or indistinctly barred), belly whitish, closed wing-tips fall more than half-way along visible part of uppertail. Eyes dark brownish-red (cere prominent and orange). In flight shows diagnostic, mostly whitish underwing with contrasting broadly black-tipped primaries, dark grey trailing edge to secondaries and pinkish-buff tinge to unmarked coverts. **Female** Usually has somewhat darker, more rufous breast and upper belly, often with some faint greyish barring on latter, and underwing-coverts typically more rufous-tinged; eyes yellow to orange-yellow. **Juvenile** Eyes like female. Difficult to separate from Shikra, Japanese Sparrowhawk and Besra. Important features at rest include: slate-greyish crown and sides of head contrasting with dark brown upperparts, short pale supercilium (not extending obviously behind eye), barred thighs, relatively long primary projection, distinct chestnut tinge to neck-streaks, upperpart fringing and underpart markings, and four visible complete dark bands on uppertail. In flight, pointed wing-tips and unmarked underwing-coverts are distinctive; shows only two dark bands across inner secondaries (apart from darker trailing edge); undertail has three visible complete dark bands but five narrower ones on outer feathers. **VOICE** Usually silent. A rapid, shrill, nasal, accelerating ***kee pe-pe-pe-petu-petu*** (descending in pitch) on breeding grounds. **HABITAT & BEHAVIOUR** Open country, wooded areas. Often seen in migrating flocks over forest; up to 2,135 m. Often in flocks during migration. **RANGE & STATUS** Breeds W,C,E,S China, Taiwan, N,S Korea, S Ussuriland; winters Sundas, Philippines, Sulawesi, N Moluccas, rarely New Guinea (W Papuan Is). **SE Asia** Scarce to uncommon passage migrant Tenasserim, Thailand, Peninsular Malaysia, Singapore, Indochina (except N,C Annam).

194 JAPANESE SPARROWHAWK

Accipiter gularis Plate **23**

IDENTIFICATION 25–31 cm. *A.g.gularis*: **Male** Similar to Besra but upperside tends to be paler, mesial streak faint or lacking, has rather diffuse pale pinkish-rufous barring on breast, belly and flanks, uppertail has narrow dark bands (usually four complete ones visible), closed wing-tips fall about halfway along tail. Eyes deep red. In flight, wings appear more pointed. See Eurasian and Chinese Sparrowhawks. **Female** Larger than male, with orange-yellow to yellow eyes, distinctly browner upperparts, more prominent mesial streak and more obviously barred underparts. From Besra by noticeably darker upperside with less contrasting crown, much narrower mesial streak, lack of breast-streaks, dense but even greyish-brown bars on breast, belly and flanks, and uppertail-bands (see male). See much larger Eurasian. **Juvenile** Difficult to separate from Besra but mesial streak normally much thinner, breast-streaks less pronounced (never black), often less heavily marked belly and flanks and narrower dark bands on uppertail (as adult); may show slaty-grey hindcrown contrasting somewhat with mantle. In flight, from above, shows narrower, less distinct dark bands across flight-feathers and tail than Besra; underside very similar. Eyes yellow. See Chinese and Shikra. **HABITAT & BEHAVIOUR** Open country, forest edge, lightly wooded areas; up to 2,565 m. Often seen in flocks during migration. **RANGE & STATUS** Breeds S Siberia, E Palearctic, NE China, N,S Korea, Japan; winters Sundas, Philippines, Sulawesi. **SE Asia** Uncommon to fairly common passage migrant (except SW,W,N,C,E Myanmar, N Annam). Winters locally S(south) Thailand, Peninsular Malaysia, Singapore.

195 BESRA *Accipiter virgatus* Plate **23**

IDENTIFICATION 26–32 cm. *A.v.affinis*: **Male** Distinguished by combination of very dark slate-greyish upperside (sometimes slightly purplish-tinged), prominent dark mesial streak, broad blackish and rufous-chestnut streaks on centre of upper breast, and broad rufous-chestnut bars on lower breast, upper belly and thighs. Eyes orange to deep red; closed wing-tips fall less than one-third along tail, uppertail usually with four visible dark bands broader or equal to intervening pale bands. Resembles miniature Crested Goshawk but lacks crest, darker above. In flight, wings appear relatively short, rounded and blunter than other small accipiters; undertail shows three complete dark bands but 5–6 narrower ones on outermost feathers (distal one broader). See Shikra and Japanese Sparrowhawk. **Female** Larger, with yellow eyes, and browner-tinged upperside with contrasting blackish crown and nape. **Juvenile** Difficult to separate from Shikra and Japanese Sparrowhawk but has more prominent dark mesial streak, blacker breast-streaks and even-width dark and pale tail-bands. In flight also shows blunter wings. Also differs from Japanese by lack of any contrast between hindcrown and mantle and, in flight, by broader, more pronounced dark bands across upperside of flight-feathers and tail. See Chinese Sparrowhawk. **VOICE** A loud squealing ***ki-weeer*** and rapid ***tchew-tchew-tchew...*** **HABITAT & BEHAVIOUR** Broadleaved evergreen and mixed broadleaved forest; up to 2,000 m, commoner at higher levels. Generally hunts birds inside wooded habitats and less likely to be seen soaring overhead than other accipiters. **RANGE & STATUS** Resident (subject to local movements) SW,NW,N,NE India, Nepal, Bhutan, Sri Lanka, C and southern China, Taiwan, Greater Sundas, Flores, Philippines. **SE Asia** Scarce to uncommon resident (except Tenasserim, S Thailand, Peninsular Malaysia, Singapore, E Tonkin, N Annam, Cochinchina). Vagrant Singapore. **BREEDING** April–May. **Nest** Bulky, depressed platform or old nest of other bird, in tree; 15–25 m above ground. **Eggs** 2–5; white to bluish-white, typically freckled, blotched and smudged reddish-brown (towards broader end); 38.2 x 30.5 mm (av.).

196 EURASIAN SPARROWHAWK
Accipiter nisus Plate **23**

IDENTIFICATION 28–38 cm. *A.n.nisosimilis*: **Male** Distinguished by combination of size, slaty-grey upperside, orange-rufous wash on cheeks, narrow orange-rufous streaks on throat, faint orange-rufous bars on breast, belly and flanks, and faint darker bands on uppertail (subterminal one noticeably broader). Often shows faint narrow pale supercilium, eyes orange-red to orange-yellow, has no isolated mesial streak. In flight appears relatively long-winged and -tailed. *A.n.melaschistos* (Myanmar except SW) has darker grey upperside with almost blackish crown and mantle and stronger rufescent barring on underparts. See Japanese Sparrowhawk and Besra. Note habitat and range. **Female** Larger, with more prominent whitish supercilium, somewhat browner-tinged upperside, more obvious dark uppertail-bands and darker, more pronounced markings below. Combination of size, prominent supercilium, unstreaked breast and tail pattern rule out Japanese and Besra. See Northern Goshawk. **Juvenile** Best separated from other smaller accipiters by combination of size, heavy rufous-chestnut to blackish barring on underparts, and tail pattern (like female). Some can show more streak-like markings or arrowhead shapes on breast. **VOICE** May give a loud, shrill ***kyi-kyi-kyi...*** when alarmed. **HABITAT** Forested and open areas; up to 3,000 m (mainly in mountains), breeding above 1,400 m in Himalayas. **RANGE & STATUS** Breeds NW Africa, Palearctic, northern Pakistan, NW,N,NE Indian subcontinent, S,E Tibet, NW,W,NE China; some northern populations winter south to N Africa, Middle East, Indian subcontinent (except Sri Lanka), southern China. **SE Asia** Scarce to uncommon winter visitor Myanmar, NW Thailand, N,C Laos, W,E Tonkin, N,C Annam. Likely to breed N Myanmar. Scarce to uncommon passage migrant W,E Tonkin. Vagrant NE,C,S Thailand, Peninsular Malaysia, Singapore. **BREEDING** India: April–June. **Nest** Untidy shallow cup or platform, or old nest of other bird, in tree. **Eggs** 4–6; buffy reddish-white to bluish-white, spotted and blotched reddish-brown to blackish-brown over pale reddish to lavender-grey undermarkings; 39.1 x 32.6 (av.; *melaschistos*).

197 NORTHERN GOSHAWK
Accipiter gentilis Plate **24**

IDENTIFICATION 48–62 cm. *A.g.schvedowi*: **Male** Most likely to be confused with female Eurasian Sparrowhawk. Usually considerably larger, white supercilium more pronounced, crown and sides of head darker than mantle, lower sides of head and throat rather uniform whitish, rest of underparts barred brownish-grey. In flight appears proportionately longer-winged, shorter-tailed and heavier-chested, underwing has much finer, less pronounced dark barring on coverts and much less contrasting darker bands across flight-feathers. **Female** Considerably larger than male, grey of upperside browner-tinged, eyes orange-yellow (vs orange-red), banding on underwing more obvious. **Juvenile** Distinctive. Upperside darker and browner than adult, with buffish to whitish fringes, supercilium less distinct and buffish, neck and underparts buff with strong dark brown streaks, uppertail with distinct irregular broad dark bands. In flight also shows more bulging secondaries and heavier markings on underwing, with dark streaks/spots on pale buffish coverts. Eyes yellow. See Crested Goshawk and Grey-faced Buzzard. **VOICE** May give a loud, guttural ***kyee-kyee-kyee...*** when alarmed. **HABITAT** Wooded habitats, sometimes more open areas; up to 2,800 m. **RANGE & STATUS** Breeds Holarctic, NW Africa, NW,N India, Nepal, Bhutan, E Tibet, NW,W,NE China, N Korea; some northern populations winter south to S Europe, Pakistan, northern India, southern China, N,S Korea, S Japan. **SE Asia** Rare winter visitor W,N Myanmar, NW Thailand, W,E Tonkin. Scarce passage migrant W Tonkin. Vagrant Cambodia, S Annam, Cochinchina.

198 WHITE-EYED BUZZARD
Butastur teesa Plate **22**

IDENTIFICATION 41–43 cm. Monotypic. **Adult** Similar to Grey-faced Buzzard but eyes whitish, head browner, has more extensively pale lores, broad pale area across wing-coverts, and rufescent rump to uppertail, latter with a single dark subterminal band and sometimes numerous indistinct, narrower dark bars. In flight, differs primarily by pale band across upperwing-coverts, darker axillaries and lesser underwing-coverts, rump to uppertail colour, and narrower bands across undertail. Note habitat and range. See Rufous-winged Buzzard. **Juvenile** Head, neck and underparts creamy to deep rich buff with narrow dark streaks on crown, hindneck, breast and belly; has very faint mesial streak, prominent pale band across wing-coverts, rufous-chestnut tinge to uppertail (greyer than adult) and rich buff base colour to underwing-coverts. **VOICE** A plaintive mewing ***pit-weeer pit-weeer pit-weere...*** **HABITAT & BEHAVIOUR** Semi-desert, dry open country; lowlands. Perches in small trees and bushes for long periods. **RANGE & STATUS** Resident Indian subcontinent (except Sri Lanka). **SE Asia** Uncommon to common resident SW,W,N,C,S Myanmar, north Tenasserim. **BREEDING** March–April. **Nest** Fairly bulky shallow cup, in tree; 6–12 m above ground. **Eggs** 2–4; plain greyish- to bluish-white, rarely spotted reddish-brown; 43–49.9 x 35–39.1 mm (av.).

199 **RUFOUS-WINGED BUZZARD**
Butastur liventer Plate **22**

IDENTIFICATION 38–43 cm. Monotypic. **Adult** Size and shape recalls Grey-faced Buzzard but has mostly greyish head and underparts, contrasting with mostly rufous-brown to rufous-chestnut upperside. Has indistinct dark streaks on crown, neck and breast, lacks dark mesial stripe, uppertail strongly rufescent. In flight, mostly uniform rufous-chestnut upperside of flight-feathers and rump to tail, coupled with relatively plain whitish underwing-coverts and indistinctly patterned undertail, diagnostic. **Juvenile** Head browner with narrow white supercilium; upperparts, lesser wing-coverts, breast and belly duller and more brownish. **VOICE** A shrill ***pit-piu*** with higher first note. **HABITAT** Dry deciduous forest, secondary growth; up to 1,525 m (800 m in Thailand). **RANGE & STATUS** Resident SW China (SW Yunnan), Java, Sulawesi. **SE Asia** Scarce to uncommon resident Myanmar (except southern Tenasserim), W,NW,NE,C Thailand, Cambodia, C,S Laos, S Annam, Cochinchina. **BREEDING** February–May. **Nest** Bulky shallow cup, in tree. **Eggs** 2–4; bluish-white (rarely spotted reddish-brown and grey); 42.8–48.7 x 35.3–40 mm (av.).

200 **GREY-FACED BUZZARD**
Butastur indicus Plate **22**

IDENTIFICATION 41–49 cm. Monotypic. **Male** Size, rather slim build and relatively large-headed appearance recalls both accipiters and *Buteo* buzzards. Crown, upperside and breast mostly rather plain greyish-brown with greyer sides of head, yellow eyes, white throat, blackish submoustachial and mesial stripes, greyish-brown and white bars on belly and three dark bands across tail. At rest, closed wing-tips fall not far short of tail-tip. In flight, wings straighter and longer than in accipiters and much narrower than *Buteo* buzzards; upperwing shows rufescent bases to primaries, underwing rather pale with darker trailing edge (primary tips blackish), some dark markings on coverts, and indistinct narrow dark bands across flight-feathers; pale undertail with three dark bands and plainer outer feathers. See Crested Goshawk and Jerdon's Baza. **Female** Tends to have more prominent supercilium, browner-tinged ear-coverts and more whitish barring on breast. **Adult dark morph** Rare. Head, body and wing-coverts (above and below) dark brown, rest of wings and tail typical. **Juvenile** Crown and neck brown, narrowly streaked white, broad white supercilium contrasts with greyish-brown sides of head, underparts dull whitish with pronounced dark streaks, mesial streak narrower, upperparts and wing-coverts browner with pale feather-tips (particularly latter). Underwing and tail similar to adults. See Northern Goshawk. **Juvenile dark morph** Differs primarily by dark brown eyes. **VOICE** A plaintive high-pitched ***tik HWEEER*** or ***tik H'WEEER*** (introductory note audible at fairly close range). **HABITAT & BEHAVIOUR** Open coniferous, broadleaved and mixed forest, secondary growth, open areas; up to 1,800 m. Flight rather direct with fast wingbeats and interspersed glides. Soars with wings held level. **RANGE & STATUS** Breeds SE Siberia, Ussuriland, NE China, N,S Korea, Japan; winters Greater Sundas, Philippines, Sulawesi, N Moluccas, rarely New Guinea (W Papuan Is). **SE Asia** Uncommon to common winter visitor Tenasserim, W,NW,NE,S Thailand, Peninsular Malaysia, Singapore (rare), Cambodia, S Laos, C,S Annam, Cochinchina. Fairly common to common passage migrant Tenasserim, Thailand, Peninsular Malaysia, Singapore, Cambodia, N,C Laos, W,E Tonkin, N Annam. Vagrant W Myanmar (one sight record).

201 **HIMALAYAN BUZZARD**
Buteo burmanicus Plate **24**

IDENTIFICATION 51–57 cm. Monotypic. **Adult pale morph** Very variable. Combination of size, robust build, rather large head, mostly dark brown upperside and mostly whitish underparts with variable, large dark brown patch across belly distinctive. Typically shows dark brown and whitish streaks on crown and neck, more whitish sides of head with dark eyestripe, heavy brown throat-streaks (particularly at sides), sparse brown breast-streaks and greyish-brown uppertail with numerous faint narrow darker bars. Often has pale thighs, unlike Common and Long-legged Buzzards, and also shows distinctive, nearly unbarred uppertail. Tarsi more than half feathered (mostly unfeathered in Common and Long-legged). In flight shows broad rounded wings and shortish rounded tail; upperwing usually has paler area across primaries, underwing with contrasting blackish outer primary tips and blackish carpal patches, rest of underwing rather pale with whiter primaries and rather narrow dark trailing edge; has numerous indistinct narrow dark bars across underside of flight-feathers and tail. See Long-legged Buzzard and Osprey. **Adult dark morph** Head, body and wing-coverts (above and below) rather uniform blackish-brown, tail with broader terminal band; rest of underwing similar to pale morph. **Juvenile pale morph** Similar to adult but has pale eyes, initially a strong buffy wash below, narrower, paler and more diffuse trailing edge to underwing and evenly barred tail, without broader subterminal band. **Juvenile dark morph** Eyes and tail differ similarly to pale morph. **VOICE** Relatively clear, thin, high-pitched ***FWEEEOU***, rising higher though less quickly than call of Common, then falling more steeply. **HABITAT & BEHAVIOUR** Open country, open forest and forest edge, cultivation; up to 3,660 m. Soars with wings in shallow V and tail spread; glides on level wings; often hovers. **RANGE & STATUS** Breeds SE Siberia (east of c.98°E), N Mongolia, Ussuriland, Sakhalin, N Pakistan, NW India, Nepal, SE,E Tibet, W,N,NE China; ? Bhutan, NE India, S Tibet, SW China; some populations winter to NW,S Indian subcontinent, southern China, rarely Philippines, ? Java and Bali. **SE Asia** Scarce to uncommon winter visitor (except SE Thailand, S Laos, W Tonkin). Also recorded on passage W,NW,S Thailand, Peninsular Malaysia, Singapore, W,E Tonkin.

202 **COMMON BUZZARD** *Buteo buteo* Plate **24**

IDENTIFICATION 39–47 cm. *B.b.vulpinus*: **Adult rufous morph** The commonest morph. May appear almost identical in plumage to Long-legged, but smaller and weaker-billed, lacking long-necked appearance at rest; upperside more solidly brown, with narrow rufous edgings, and tail usually crisply and narrowly barred (at least lightly). **Adult pale morph** Head to underparts more heavily dark-marked than Long-legged; less rufescent overall. A *Buteo* with heavily barred underparts will be this species. See larger Himalayan Buzzard. **Adult dark morph** On current knowledge, inseparable on plumage from Himalayan, and only told from Long-legged by size and proportions. **Juvenile pale morph** Differs from adult in similar way to Himalayan. Apparently told from Himalayan by streakier head to breast, barring on belly and thighs, less distinct carpal patches on underwing. **Juvenile dark morph** Not safely separable from Himalayan. **VOICE** Squealing, emphatic, strongly downslurred, whistled ***NYAARhh*** (non-tremolo to strongly tremolo). **HABITAT** Open country, open forest, cultivation. **RANGE & STATUS** Breeds W Palearctic (east to Yenisei R), NW Iran; some northern populations winter to S Africa, Middle East, SW Indian subcontinent, ? rarely Java and Bali. **SE Asia** Vagrant NW,S Thailand.

203 **LONG-LEGGED BUZZARD**

Buteo rufinus Plate **24**

IDENTIFICATION 57–65 cm. *B.r.rufinus*: **Adult pale morph** Very variable. Similar to Common Buzzard but larger and larger-headed, with long-necked appearance at rest, and rufous on upperparts, underparts and tail. Typically has pale cream-coloured head, neck and breast, with indistinct darker streaks on crown, hindneck and breast, slight darker eyestripe and malar line, rufous thighs and belly-patch, and rather plain pale rufous to cinnamon-coloured tail (above and below). Rufous on underparts almost always heavier on belly than breast. In flight shows longer, more eagle-like wings, and rather long square-tipped tail; underwing-coverts often tinged rufescent. See Himalayan Buzzard. **Adult dark morph** Extremely similar to Himalayan and best separated by size and shape, but typically shows broader dark tail-barring and heavier dark bands on undersides of secondaries. **Adult rufous morph** Head, body and wing-coverts (above and below) deep rufous to rufous-chestnut; underwing pattern (including carpal patches) and tail similar to pale morph. **Juvenile pale morph** Upperparts browner, underparts less rufous, more brownish, tail has faint narrow dark bars towards tip. In flight shows narrower, paler and more diffuse trailing edge to underwing. **Juvenile dark morph** Broader dark bars on flight-feathers and tail than on Himalayan. **VOICE** May give a plaintive, slightly descending ***eeeeaaah***, mellower and lower-pitched than Common. See Himalayan Buzzard. **HABITAT & BEHAVIOUR** Open country; recorded at 450 m. Soars with wings in deeper V than Common, recalling Western Marsh Harrier; glides with inner wing slightly raised; often hovers. **RANGE & STATUS** Resident N Africa, Turkey, Middle East, N Pakistan, NW,N India. Breeds south-east Europe, N Caspian region, C Asia, NW China; some populations winter south to NE Africa, Indian subcontinent. **SE Asia** Vagrant N Myanmar, Peninsular Malaysia.

204 **INDIAN SPOTTED EAGLE**

Aquila hastata Plate **25**

IDENTIFICATION 59–65.5 cm. Monotypic. **Adult** Like Greater Spotted but somewhat smaller; has proportionately smaller bill with larger gape (shows pronounced wide flange or 'thick lips'; extending further under eye than on Greater Spotted). Head, upperparts and upperwing-coverts very dark brown with black shaft-streaks. In flight, has more prominent whitish flash at base of primaries on upperwing than Greater Spotted, double whitish crescent in vicinity of primary underwing-coverts. Underwing-coverts are uniform apart from paler brown lesser coverts. **Juvenile** Head, body and wing-coverts noticeably paler than Greater; has less pronounced white spotting on tertials, median and lesser upperwing-coverts, buffish-brown below with darker streaks. Uppertail-coverts are very pale brown with white barring. In flight, differs from Greater Spotted in same way as adult. **Subadult** Exhibits a mixture of adult and juvenile characters. **VOICE** Repeated high-pitched yapping ***kyek*** in breeding season. **HABITAT & BEHAVIOUR** Wooded areas and open country; lowlands. Glides and soars on level to slightly raised wings with primaries angled downwards. **RANGE & STATUS** Resident north and peninsular India; formerly Bangladesh. **SE Asia** Recorded (formerly resident?) SW,S Myanmar. **BREEDING** India: March–July. **Nest** Large cupped platform, in tree; 9–12 m above ground. **Eggs** 1–2 (usually 1); white, with rather faint reddish-brown freckles or blotches over grey undermarkings; 58.5–66.6 x 47.3–54.4 mm. **NOTE** Best treated as distinct from the Lesser Spotted Eagle (Parry *et al.* 2002).

205 **GREATER SPOTTED EAGLE**

Aquila clanga Plate **25**

IDENTIFICATION 65–72 cm. Monotypic. **Adult** Differs from similar large dark eagles by relatively shorter wings, smallish bill and shortish tail. In flight, wings appear relatively short and broad, upperside rather uniformly dark with paler area on bases of primaries and narrow pale band across uppertail-coverts; underwing rather uniformly dark with distinctive whitish patch or crescent at base of outer primaries. **Adult pale ('*fulvescens*') morph** Rare. Has mostly buffish to rufous body, median and lesser upper- and underwing-coverts. See Indian Spotted, Steppe and Tawny Eagles. **Juvenile** Blackish, with pronounced whitish spots on upperwing-coverts, and whitish tips to scapulars, tertials, most other wing-feathers and tail; has broad pale buffish streaks on belly and thighs (sometimes also or only on breast or lacking altogether). In flight, whitish bands across upperwing-coverts and whitish band across uppertail-coverts contrast sharply with dark upperparts, rest of wing-coverts and secondaries; underwing shows darker coverts than flight-feathers, paler bases of primaries than adult and faint dark bars across secondaries and inner primaries. See Indian Spotted. **Juvenile pale morph** Differs from adult by paler rufous colour that fades to buffy or creamy on head, body and wing-coverts, and by whitish tips to upperwing-coverts, secondaries, inner primaries and tail. See Tawny Eagle. **VOICE** Series of short quick high-pitched notes: ***hi-hi-hi-hi-hi...***; ***hihi hihi-hi...*** etc. **HABITAT & BEHAVIOUR** Marshes, lakes, rivers, open country; up to 2,135 m (winters below 800 m). Soars on flattish wings, glides with primaries distinctly angled downwards. **RANGE & STATUS** Local resident S Pakistan, northern India. Breeds north-eastern W Palearctic, Siberia, Ussuriland, NW,NE China; winters NE Africa, S Europe, Middle East, northern Indian subcontinent, S China. **SE Asia** Rare to scarce winter visitor N,S Myanmar, Tenasserim, Thailand, Peninsular Malaysia, Cambodia (has remained until July), N,C Laos, E Tonkin, Cochinchina. Former winter visitor (current status unknown) SW,C,E Myanmar. Scarce passage migrant S Thailand, Cambodia, W,E Tonkin. Vagrant Singapore.

206 **TAWNY EAGLE** *Aquila rapax* Plate **25**

IDENTIFICATION 64–71 cm. *A.r.vindhiana*: **Adult pale morph** Difficult to separate from pale morph Greater Spotted Eagle but has somewhat larger bill, longer neck, fuller and longer 'trousers' and yellowish eyes. In flight, underwing shows paler inner primaries and bases of outer primaries, without well-defined whitish patch or crescent. **Adult dark morph** Rare. Difficult to separate from Steppe Eagle at rest but somewhat more powerfully built, lacks rufescent nape-patch and has dark throat. At very close range, gape typically ends below centre of eye (extends behind eye in Steppe). In flight, shows distinctive underwing pattern, with greyish inner primaries and bases of outer primaries contrasting with uniformly dark secondaries. Best separated from spotted eagles by combination of size, structure, underwing pattern and yellowish eyes. **Juvenile** Similar to respective adults but with creamy tips to upperwing-coverts, secondaries, inner primaries and tail. When fresh, juvenile pale morph has strongly rufescent head, body and wing-coverts. See Eastern Imperial and Steppe Eagles. **VOICE** May give a repeated barking ***kowk***. **HABITAT & BEHAVIOUR** Open country; recorded in lowlands. Often adopts a more upright posture than Steppe Eagle. Glides with inner wing slightly raised and primaries slightly angled downwards. **RANGE & STATUS** Resident Africa, Pakistan, India (except NE), S Nepal. **SE Asia** Vagrant N Myanmar (two recent sight records).

207 **STEPPE EAGLE** *Aquila nipalensis* Plate **25**

IDENTIFICATION 76–80 cm. *A.n.nipalensis*: **Adult** Larger, and larger-billed, than other large, uniformly dark eagles. Best separated from spotted eagles and dark morph Tawny Eagle by combination of size, structure, full 'trousers', rufous-buff nape-patch, somewhat paler throat and distinct dark bars on secondaries. At very close range, gape typically ends behind eye. In flight, wings and tail distinctly longer than spotted eagles. Underwing pattern distinctive, with mostly blackish primary coverts, and contrasting darker trailing edge and dark bars across flight-feathers. See Eastern Imperial and Black Eagles. **Juvenile** Head, body and wing-coverts paler and more grey-brown, with whitish tips to greater and primary upperwing-coverts, secondaries and tail-feathers, and broader, whiter band across uppertail-coverts. In flight, easily separated from other large dark eagles by broad whitish trailing edge to underwing-coverts (wide band across underwing). **Subadult** Lacks obvious whitish band on underwing but is already developing distinctive adult-like underwing. **VOICE** Slightly hoarse ***akh akh akh akh...*** May give a repeated, deep barking ***ow***. **HABITAT & BEHAVIOUR** Open country; up to 2,100 m. Soars on flatter wings than spotted eagles. **RANGE & STATUS** Breeds east W Palearctic, C Asia, S Siberia, Mongolia, NW,N China; mostly winters E,S Africa, Middle East, C and northern Indian subcontinent, S,SE Tibet, S China. **SE Asia** Rare winter visitor/vagrant W,C,S,E Myanmar, Tenasserim, W(coastal),NW,C,S Thailand, Peninsular Malaysia, Singapore, E Tonkin. Scarce to uncommon passage migrant N Myanmar.

208 **EASTERN IMPERIAL EAGLE**
Aquila heliaca Plate **26**
IDENTIFICATION 72–83 cm. Monotypic. **Adult** Unmistakable large blackish-brown eagle with large head and bill, golden-buff supercilium, ear-coverts, nape and hindneck, prominent white markings on upper scapulars, pale undertail-coverts and broadly black-tipped greyish tail. In flight shows almost uniformly blackish underwing and pale undertail-coverts. See Steppe and Black Eagles. **Juvenile** Recalls pale morph Greater Spotted and Tawny Eagles but larger, with diagnostic distinct dark streaks on nape, neck, breast and wing-coverts (above and below). In flight also shows pale greyish wedge on inner primaries. **VOICE** Repeated deep, barking ***owk***. **HABITAT & BEHAVIOUR** Open country, cultivation; up to 400 m. In flight, wings held more level than other eagles. **RANGE & STATUS** Breeds central and eastern W Palearctic, Iran, C Asia, SC Siberia, Mongolia, NW China; formerly Pakistan; more northern populations winter to south, north-eastern Africa, Middle East, Pakistan, W and northern Indian subcontinent, southern China. **SE Asia** Rare winter visitor Thailand (except S), Peninsular Malaysia, Cambodia. Vagrant E,S Myanmar, north Tenasserim, Singapore, N,S Laos, E Tonkin.

209 **BONELLI'S EAGLE** *Aquila fasciata* Plate **26**
IDENTIFICATION 65–72 cm. *A.f.fasciata*: **Adult** Dark brown upperparts, whitish underparts with dark streaks on foreneck to breast and dark-barred thighs recall some hawk-eagles. Differs primarily by combination of size, whitish patch on mantle, relatively uniform greyish tail with broad blackish subterminal band, and lack of crest. In flight shows distinctive, faintly barred greyish underwing with whitish leading edge to mostly black coverts, and grey tail with pronounced black subterminal band. See buzzards. **Juvenile** Head, upperparts and upperwing-coverts paler and browner, uppertail browner with even, narrow darker bars; throat and underparts warm buffish with dark streaks on lower throat and breast. In fresh plumage, head, underparts and underwing-coverts are rufous; this fades to dull rufous in time, later to buffy and, in some cases, to creamy. In flight, from below, shows warm buffish wing-coverts usually with distinctive dark tips to greater and primary coverts forming a narrow dark line across underwing; undertail evenly and finely dark-barred. See hawk-eagles. **VOICE** A repeated, shrill melodious ***iuh*** and longer whistled ***eeeoo*** (lower-pitched at end) during display-flight. **HABITAT & BEHAVIOUR** Forested areas, often near cliffs, karst limestone country; 500–1,900 m. Glides on flat wings. **RANGE & STATUS** Resident (subject to some movements) Africa, S Europe, Turkey, Middle East, Turkmenistan, Afghanistan, Pakistan, India (except NE), Nepal, Bhutan, C,S,SE China, Lesser Sundas. **SE Asia** Scarce resident W,C,E Myanmar, NW Thailand, N Laos, E Tonkin. Vagrant Cochinchina. **BREEDING** November–February. **Nest** Large bulky structure with fairly shallow cup, on rock ledge or in tall tree. **Eggs** 1–3; white, sometimes faintly flecked pale reddish (rarely well marked with light reddish-brown); 62–76.5 x 48–57.3 mm.

210 **BOOTED EAGLE** *Aquila pennata* Plate **26**
IDENTIFICATION 50–57 cm. Monotypic. **Adult pale morph** Distinctive. Relatively small with pale crown, dark sides of head and mostly whitish underparts. Pale scapulars, uppertail-coverts and band across wing-coverts contrast with mostly dark remainder of upperside. In flight, whitish underwing-coverts and pale wedge on inner primaries contrast sharply with otherwise blackish underwing, undertail greyish with indistinct dark terminal band and central feathers. Combination of size, narrow wings, longish tail and wing pattern rules out larger *Aquila* eagles. See Changeable Hawk-eagle. **Adult dark morph** Head and body rather uniform dark brown. In flight, can resemble Black Kite but shows distinctive tail (as in pale morph), contrasting paler scapulars and uppertail-coverts, and sharply contrasting pale band across upperwing-coverts. **Adult rufous morph** Shows strongly rufescent head, body and underwing-coverts. **Juvenile** Like adult but shows prominent white trailing edge to wings and tail in fresh plumage. **VOICE** May give a clear shrill chattering ***ki-ki-ki...*** or longer ***kee-kee-kee...*** **HABITAT & BEHAVIOUR** Open and wooded areas, cultivation; lowlands. When gliding and soaring, holds wings slightly forwards and level, or with primaries angled slightly downwards; sometimes twists tail like Black Kite. **RANGE & STATUS** Breeds NW Africa, C and southern W Palearctic, C Asia, SC Siberia, northern Pakistan, NW India, Nepal, NW China; mostly winters to south, sub-Saharan Africa, Indian subcontinent. **SE Asia** Rare to scarce passage migrant north Tenasserim, S Thailand, Peninsular Malaysia, Singapore. Vagrant N,C,S,E Myanmar, elsewhere Thailand, Cambodia, S Laos (probably regular passage migrant in at least some of these areas). Recorded in summer coastal W Thailand.

211 **BLACK EAGLE** *Ictinaetus malayensis* Plate **26**
IDENTIFICATION 69–81 cm. *I.m.malayensis*: **Adult** Blackish overall with yellow cere and feet. Has rather long tail with indistinct narrow pale bands; closed wing-tips fall close to or beyond tail-tip. In flight shows unmistakable long broad wings with pinched-in bases and well-spread 'fingers', and longish tail. See *Aquila* eagles and dark morph Changeable Hawk-eagle. Note habitat and range. **Juvenile** Head, neck and underparts pale buffish with heavy blackish streaks; some pale tips on upperparts and wing-coverts. In flight also shows pale buffish underwing-coverts with blackish streaks and more obviously pale-barred underside of flight-feathers and tail. **Other subspecies in SE Asia** *I.m.perniger* (Myanmar). **VOICE** A shrill yelping ***wee-a-kwek*** etc., particularly during display-flight. **HABITAT & BEHAVIOUR** Broadleaved evergreen forest and nearby open areas; up to 3,170 m. Soars with wings in V, often spiralling gently downwards into forest clearings. Display-flight involves steep dives through U-loop, up to near-vertical stall. **RANGE & STATUS** Resident Indian subcontinent (except Pakistan), SW,SE China, Taiwan, Greater Sundas, Sulawesi, Moluccas. **SE Asia** Uncommon to fairly common resident (except SW,C,S,E Myanmar, C Thailand, Singapore). Also recorded on passage E Tonkin. **BREEDING** October–April. **Nest** Large but rather compact with deepish cup, high in tree, c.10–18 m above ground. **Eggs** 1 (rarely 2); whitish to pinkish, finely and densely stippled or blotched pale brick-red to rich brown; 55–65.2 x 48–53.6 mm.

212 **RUFOUS-BELLIED EAGLE**
Lophotriorchis kienerii Plate **26**
IDENTIFICATION 53–61 cm. *L.k.formosae*: **Adult** Blackish sides of head and upperside, white throat and upper breast and otherwise rufous-chestnut underparts diagnostic. Has slight crest and narrow dark streaks on breast to undertail-coverts. In flight, from below, shows mostly rufous-chestnut wing-coverts, dark trailing edge to wings, and narrow dark bars across rather pale greyish flight-feathers and tail (latter with blackish subterminal band). **Juvenile** Upperside brown, sides of head whitish with blackish eyeline, underparts whitish with black area on flank. In flight, underside

shows whitish wing-coverts with broken dark trailing edge and narrower, less pronounced trailing edge to flight-feathers and tail. Separated from hawk-eagles and smaller *Aquila* eagles by combination of dark upperside with brown primaries, blackish eyeline and flank-patch, indistinctly barred tail, underwing-covert pattern and lack of obvious crest. See Short-toed Snake-eagle and Crested Serpent-eagle. **VOICE** A fairly clear but low-pitched series of notes terminated by a very thin breathless note: ***WHI-WHI-WHI-WHI yii***. See Changeable Hawk-eagle. **HABITAT & BEHAVIOUR** Broadleaved evergreen forest; up to 2,000 m. Usually glides and soars with wings held level. **RANGE & STATUS** Resident SW India, N,NE Indian subcontinent, Sri Lanka, S China (Hainan), Greater Sundas, W Lesser Sundas, Philippines, Sulawesi, Moluccas. **SE Asia** Scarce to uncommon resident N,S Myanmar, Thailand (except C), Peninsular Malaysia, Cambodia, C,S Laos, Vietnam (except W Tonkin). Also recorded on passage Peninsular Malaysia. Scarce non-breeding visitor Singapore. **BREEDING** February–March. **Nest** Large structure with fairly deep cup, in tree; 24–30 m above ground. **Eggs** 1; white, almost unmarked or variably blotched pale reddish-brown over lavender-grey undermarkings; 53.8–67 x 44.9–53.3 mm.

213 BLYTH'S HAWK-EAGLE

Nisaetus alboniger Plate **27**

IDENTIFICATION 51–58 cm. Monotypic. **Adult** Distinctive. Sides of head and crown to mantle distinctly blacker than other hawk-eagles, underparts whitish with prominent black mesial streak, bold blackish breast-streaks and bold blackish bars on belly, thighs and vent, tail blackish with pale greyish broad central band and narrow tip. Crest long and erectile, blackish or with fine white tip. In flight shows whitish underwing with heavy blackish bars on coverts, blackish bands across flight-feathers and distinctive tail pattern. See Crested Serpent-eagle. **Juvenile** Crown, hindneck and ear-coverts sandy-rufous, upperside browner than adult with whitish fringes, underparts plain pale buff to whitish with buffish breast and flanks, tail whitish with two or three medium-width dark bands and slightly broader terminal dark band, underwing-coverts plain creamy-whitish; crest similar to adult. From Mountain Hawk-eagle by smaller size, more rufescent unstreaked head and hindneck, and different tail pattern, with four dark bands on upperside. In flight, also differs by shorter, more parallel-edged wings. Moults directly into adult plumage. See very similar Wallace's Hawk-eagle and Changeable Hawk-eagle. **VOICE** A very high-pitched, fast, slightly metallic ***wiiii-hi***, ***eeee'ha***, ***wiii'a*** or ***wee'ah***, and shrill ***pik-wuee*** with slightly rising second note. **HABITAT & BEHAVIOUR** Broadleaved evergreen forest; up to 1,980 m. Soars on level wings. **RANGE & STATUS** Resident Sumatra, Borneo. **SE Asia** Uncommon to fairly common resident S Thailand, Peninsular Malaysia. Vagrant Singapore. **BREEDING** September–June. **Nest** Large bulky cupped structure, high in large tree. **Eggs** 1; otherwise undocumented.

214 MOUNTAIN HAWK-EAGLE

Nisaetus nipalensis Plate **27**

IDENTIFICATION 66–75 cm. *N.n.nipalensis*: **Adult** Similar to pale morph Changeable Hawk-eagle but has long erectile white-tipped blackish crest, whitish bars on rump, broader dark mesial streak, broad dark bars on belly, and equal-width dark and pale tail-bands. In flight from below, wings somewhat broader and more rounded (more pinched-in at base of secondaries), tail relatively shorter, wing-coverts with heavy dark markings, dark tail-bands broader. See Blyth's and Wallace's Hawk-eagles. **Juvenile** From Changeable by distinctive crest, pale to warm buff head and underparts, darker-streaked crown, hindneck and sides of head, and buff-barred rump. In flight from below also differs by shape, and equal-width pale and dark tail-bands. Underwing-coverts plain creamy-white. See Blyth's and Wallace's. **VOICE** A shrill ***tlueet-weet-weet***. **HABITAT & BEHAVIOUR** Broadleaved evergreen, deciduous and mixed forests; up to 2,565 m. Glides with wings held level; soars with wings in shallow V. **RANGE & STATUS** Resident (subject to local movements) N Pakistan, NW,N,NE,C,SW India, Nepal, Bhutan, Sri Lanka, SW,S,SE China, Taiwan, Japan. Breeds NE China; some winter to south. **SE Asia** Scarce to uncommon resident (subject to some movements) W,N Myanmar, Tenasserim, W,NW,S Thailand, north-west Peninsular Malaysia (Langkawi I). Scarce to uncommon (status uncertain but probably widespread resident) SW,C,S Myanmar, NE,SE Thailand, Cambodia, Laos, Vietnam. **BREEDING** February–March. **Nest** Large structure with fairly deep cup, in tree; 12–25 m above ground. **Eggs** 1–2; white, reddish-white or pale clay-coloured, blotched and spotted red to reddish-brown; 65–72.7 x 51.2–57.4 mm.

215 CHANGEABLE HAWK-EAGLE

Nisaetus limnaeetus Plate **27**

IDENTIFICATION 61–75 cm. *N.l.limnaeetus*: **Adult pale morph** Rather nondescript brown above, whitish below with dark mesial streak, dark streaks on breast and belly, and faint narrow rufous barring on thighs and undertail-coverts; has four dark bands on tail (terminal one broader), crest distinctly short. In flight shows broad, rather parallel-edged wings. From other hawk-eagles by combination of lack of belly-barring, relatively plain underwing-coverts, and undertail pattern with three narrowish dark bands and broader dark terminal band. *N.l.andamanensis* (Coco Is, off S Myanmar) is smaller and darker. See Bonelli's Eagle and Oriental Honey-buzzard. **Adult dark morph** Blackish with greyer, broadly dark-tipped tail. In flight, underwing shows greyer bases to flight-feathers. Recalls Black Eagle but wings much more parallel-edged, no bars on underwing and undertail (apart from terminal band). See buzzards. **Juvenile pale morph** Head, neck and underparts almost unmarked whitish (may show dark nuchal half-collar and eye-patch); has prominent whitish fringes to upperpart feathers (particularly wing-coverts) and narrower, more numerous dark tail-bands (lacks wide dark terminal band). In flight, underside similar to adult apart from paler body and wing-coverts and tail pattern. **Juvenile dark morph** Like adult but may show some dark barring on underside of flight-feathers and tail. See Black Eagle. **VOICE** A somewhat ascending series of loud, shrill, high-pitched whistles, terminated by a thin, stressed, high-pitched note: ***wi-wiwiwiwi-hii***; ***wi-wi-wi-wi-wi-wi-wi-hiii***; ***kwi-kwi-kwi-kwiii*** etc. Also ***k'wi-wi*** or ***kerWI-WI*** recalling Crested Serpent-eagle. Juveniles give shrill ***klit-klit*** and ***klit-kli*** with stressed second note. **HABITAT & BEHAVIOUR** Broadleaved evergreen and deciduous forest; up to 2,440 m. Soars and glides with wings held level. During display-flight, stretches neck forwards, lifts tail and holds wings in shallow V. **RANGE & STATUS** Resident except NW,N,NE India, Nepal, Bangladesh, Andaman Is, Greater Sundas, Philippines. **SE Asia** Uncommon to fairly common resident (except C Thailand, W Tonkin). **BREEDING** November–May. **Nest** Large bulky structure with deep cup, in tree; 6–45 m above ground. **Eggs** 1; white to greyish-white, sometimes sparsely and faintly freckled pale reddish (usually around broader end); 69.8 x 51.6 mm (av.; *limnaeetus*).

216 WALLACE'S HAWK-EAGLE

Nisaetus nanus Plate **27**

IDENTIFICATION 46 cm. *N.n.nanus*: **Adult** Similarly patterned to Blyth's Hawk-eagle but smaller, browner overall, sides of head and hindneck rufescent-brown with blackish streaks, crest broadly white-tipped, tail greyish with three dark bands (terminal one slightly broader). In flight, underside differs by buffish-white base colour to flight-feathers, warm buffish coverts with narrow dark barring and tail pattern. See Jerdon's Baza. Best separated from Mountain Hawk-eagle by much smaller size and tail pattern. Note habitat and range. **Juvenile** Very similar to Blyth's but has broad white tip to crest. See Mountain and Changeable Hawk-eagles. Note range. **VOICE** A shrill high-pitched ***yik-yee*** or ***kliit-kleeik***, with upward-inflected second note. Fledged juveniles give up to 8 high-pitched breathless whistles: ***yii-***

yii-yii-yii... and ***ee-ee-ee-ee-eeee.*** **HABITAT** Broadleaved evergreen forest; up to 580 m. **RANGE & STATUS** Resident Sumatra, Borneo. **SE Asia** Rare to uncommon resident south Tenasserim, S Thailand, Peninsular Malaysia. **BREEDING** November–May. **Nest** Large bulky cupped structure, in tree; 20–40 m above ground. **Eggs** 1; otherwise undocumented.

OTIDIDAE: Bustards

Worldwide 26 species. SE Asia 2 species. Large with stout bodies (held rather horizontal), long erect necks, strong legs, smallish bills and largely cryptically patterned plumage. Walk on ground, flight strong, on broad wings with neck outstretched. Feed on grasshoppers and other large insects, young birds, shoots, leaves, seeds and fruit.

217 GREAT BUSTARD *Otis tarda* Plate 14

IDENTIFICATION 75–105 cm *O.t.dybowskii*: **Male non-breeding** Distinctive. Very large, body heavy-looking, bill rather short and stout, legs medium length and robust, head and neck bluish-grey, rest of upperparts warm buffish with prominent blackish vermiculations, rest of underparts mostly white. Shows some rufous-chestnut on lower hindneck. In flight, extensively white to pale grey upperwing contrasts with broad blackish band across tips of flight-feathers (lesser wing-coverts as upperparts), underwing all whitish with broad blackish trailing edge. **Male breeding** Has long white moustachial whiskers, neck thicker with more rufous-chestnut at base extending to upper breast. **Female** Similar to male non-breeding but up to 50% smaller, with proportionately narrower bill, thinner neck and less white on upperwing (coverts more extensively patterned as upperparts). See Bengal Florican. **First winter** Similar to female (both sexes). Males reach adulthood between 3rd and 5th summers. **VOICE** Short nasal *ock* when agitated. **HABITAT & BEHAVIOUR** Open country, grassland, cultivation; recorded at 455 m. Strong direct flight, rarely glides. Wary; conceals itself in dead ground, long grass, crops. **RANGE & STATUS** Breeds Morocco, southern W,C Palearctic, S Siberia, Iran, Mongolia, NW,N,NE China; some (mainly eastern) populations winter to south, south-east W Palearctic, C,E China. **SE Asia** Vagrant N Myanmar.

218 BENGAL FLORICAN
Houbaropsis bengalensis Plate 14

IDENTIFICATION 66–68.5 cm. *H.b.blandini*: **Male** Unmistakable. Largely black with mostly white wings. Has some black on outer webs and tips of outer primaries and inner webs of inner primaries and secondaries, and fine buff markings on mantle, scapulars, tertials and uppertail. **Female** Rather nondescript buffish-brown with blackish markings and vermiculations on upperside, blackish sides of crown, warm buff supercilium and narrow median crown-stripe, pale buff neck and upper breast with fine dark brown vermiculations, underparts otherwise rather plain buff (whiter in centre) with sparse dark markings on flanks; more plain buff on wing-coverts than male, outer primaries mostly blackish. Thin neck, short bill, rather rounded head and longish legs distinctive. See Great Bustard and Indian Thick-knee. Note range. **Juvenile** Like female (both sexes) but buff of plumage noticeably richer and warmer. **First-summer male** Resembles adult male but reverts to female-like plumage in second winter; attains full adult plumage by second summer. **VOICE** Croaks and strange deep humming during courtship display. May utter shrill metallic ***chik-chik-chik*** when disturbed. **HABITAT & BEHAVIOUR** Grassland, nearby cultivation; lowlands. In spring display, male leaps 8–10 m above tall grass, hovers on extended quivering wings, then floats down. Approaches female with raised, spread tail and trailing wings. **RANGE & STATUS** Very local resident (subject to minor movements) S Nepal, N,NE India; formerly Bangladesh. **SE Asia** Rare resident (subject to local movements) north-west, central and south-east Cambodia, western and north-west Cochinchina. **BREEDING** March–September. Otherwise not described in region. India: **Nest** Slight depression in ground, amongst tall grass. **Eggs** 2; glossy olive-green, blotched purple and purple-brown, with some pale purplish-grey undermarkings; 64.3 x 45.8 mm (av.).

RALLIDAE: Rails, crakes, gallinules & coots

Worldwide c.145 species. SE Asia 18 species. Mostly robust with stout-based short to longish bills, long legs and toes, short tails (often cocked) and rounded wings. Flight appears weak (often with legs dangling) but many species highly migratory. Feed on insects, crustaceans, amphibians, fish and vegetable matter. **Crakes** (*Rallina, Crex, Amaurornis, Porzana, Gallicrex*), **rails** (*Gallirallus, Rallus*), and **swamphens** (*Porphyrio*) skulk in swampy vegetation and undergrowth. **Common Moorhen** and **Common Coot** spend much time swimming in open water.

219 RED-LEGGED CRAKE
Rallina fasciata Plate 28

IDENTIFICATION 22–25 cm. Monotypic. **Adult** Chestnut-tinged brown upperside, mostly dull chestnut head and breast, black-and-whitish bars on wing-coverts, primaries and secondaries, and bold blackish-and-whitish bars on lower breast to vent distinctive. Head-sides paler, grading to more whitish on centre of throat; eyes, broad eyering, legs and feet red; bill blackish. See Slaty-legged and Band-bellied Crakes. **Juvenile** Upperside and breast rather dark brown with little chestnut, pale bars and spots on wing-coverts buffier and less bold, underpart bars duller and much less distinct, centre of abdomen more uniformly whitish, throat whiter, legs and feet brownish-yellow. **VOICE** Male territorial call is a loud rapid hard (6–9 note) ***UH-UH-UH-UH-UH-UH...***, repeated every 1.5–3 s. Females sometimes join in with sudden quacking nasal ***brrr***, ***brr'ay*** or ***grr'erh*** notes. Often calls during night. **HABITAT** Streams and wet areas in broadleaved evergreen forest, forest edge, clearings and secondary growth, sometimes wet areas in cultivation; up to 200 m but recorded up to 1,220 m on migration. **RANGE & STATUS** Resident Greater Sundas, Flores, Philippines. Winter visitor Greater Sundas, rarely Palau Is, NW Australia. Recorded (status uncertain) NE India, Lesser Sundas, Moluccas. **SE Asia** Uncommon resident W(south),S Thailand, Singapore. Uncommon breeding visitor W(north),NW Thailand, N Laos. Uncommon resident or breeding visitor SE,S Thailand, S Laos, Cochinchina. Uncommon winter visitor Peninsular Malaysia (probably also breeds), Singapore. Uncommon passage migrant C Myanmar, S Thailand, Peninsular Malaysia. Recorded (status uncertain) W,E(south),S Myanmar, Tenasserim. **BREEDING** December–September. **Nest** Like other crakes; one being situated low in *Pandanus* clump. **Eggs** 3–6; fairly glossy, chalky-white, sometimes with obscure sepia or darker markings (possibly staining); 29.8–34.1 x 23.2–25.4 mm.

220 SLATY-LEGGED CRAKE
Rallina eurizonoides Plate 28

IDENTIFICATION 26–28 cm. *R.e.telmatophila*: **Adult** Similar to Red-legged Crake but larger and considerably larger-billed, upperside (apart from head and neck) distinctly duller and colder brown, chestnut parts somewhat paler (more rufous-chestnut), white throat contrasts sharply with sides of head, lacks obvious bars or spots on wings (rarely some vague pale barring on wing-coverts), flight-feathers appear uniform at rest (barred white on inner webs only), legs greenish-grey or deep slate-coloured to black. Eyes usually reddish, eyering dull greyish-pink to reddish. **Juvenile/first winter** Similar to juvenile Red-legged; best separated by plain wings and larger size and bill, paler face. Eyes brown. See Band-bellied Crake. **VOICE** Territorial call is said to be a persistent ***kek-kek kek-kek kek-kek kek-kek...*** Subdued ***krrrr*** when alarmed and long drumming ***krrrrrrr-ar-kraa-kraa-kraa-kraa***. Often calls at night. **HABITAT** Streams and wet areas in broadleaved evergreen forest and secondary growth, grassy vegetation in plantations, also marshes and gardens on passage; up to 1,830 m. **RANGE & STATUS** Resident Philippines, Sulawesi, Sula Is, Palau Is. Breeds N,NE and peninsular India, S China, Taiwan, Japan (Nansei Is); some populations winter to south Sri Lanka, Sumatra, rarely W Java. **SE Asia** Scarce breeding visitor E Tonkin. Scarce winter visitor C,S Thailand, Peninsular Malaysia, S Annam. Scarce passage migrant W,NE,C,S Thailand, Peninsular Malaysia, E Tonkin, C Annam. Vagrant Singapore. Recorded (status uncertain) E,S Myanmar, NW Thailand, S Laos, Cambodia, Cochinchina. **BREEDING** July–September. **Nest** Untidy pad with slight depression, on ground amongst grass etc., or in bamboo clump or tangled vegetation up to 1 m. **Eggs** 4–8; creamy-white; 30.9–36.5 x 25–28.1 mm.

221 SLATY-BREASTED RAIL
Gallirallus striatus Plate **28**

IDENTIFICATION 26–31 cm. *G.s.albiventer*. **Adult** Most widespread rail in region. Dull olive-brown upperside with black mottling/barring and white bars and spots, contrasting chestnut crown and nape and mostly plain grey sides of head and breast diagnostic. Rest of underparts whitish with greyish-brown to black bars (centre of abdomen plainer), crown often streaked blackish, bill relatively long, reddish-pink with darker tip, legs and feet dark brownish to slaty or slaty-green. **Juvenile** Upperparts duller and paler with fewer blackish and white markings but distinctly streaked blackish; almost lacks any chestnut on crown and nape; sides of head to breast paler and browner, throat and centre of underparts more extensively whitish to buffish-white with somewhat duller, less distinct bars, bill duller. **Other subspecies in SE Asia** *G.s.gularis* (Peninsular Malaysia, Singapore, Indochina). **VOICE** Territorial call is a series of sharp, metallic stony ***kerrek*** or ***trrrik*** notes, which may run together to form a kind of song (lasting up to 30 s). Males also utter low ***kuk*** and ***ka-ka-kaa-kaa*** during courtship rituals. **HABITAT** Marshes, reedbeds, mangroves, wet rice paddies, waterside vegetation, sometimes more open, drier areas; up to 1,300 m, mostly lowlands. **RANGE & STATUS** Resident Indian subcontinent (except NW and Pakistan), C,E and southern China, Taiwan, Greater Sundas, Philippines, Sulawesi. Recorded (status uncertain) Lesser Sundas (Lombok, Sawu). **SE Asia** Common resident, subject to some movements (except W Myanmar, NE,SE Thailand, S Annam). Also recorded on passage E Tonkin, N Annam. **BREEDING** All year. **Nest** Thick pad or saucer, amongst long grass etc.; on ground or up to 30 cm. **Eggs** 2–9; fairly glossy, whitish to pinkish-buff, sparsely blotched and spotted deep red to reddish- or purplish-brown over greyish undermarkings, or profusely speckled and spotted pale reddish (markings often denser around broader end) 33.7 x 25.8 mm (av.; *albiventer*).

222 EASTERN WATER RAIL
Rallus indicus Plate **28**

IDENTIFICATION 27–30 cm. Monotypic. **Adult** Similar to Slaty-breasted Rail but bill somewhat longer and narrower, crown and upperparts a buffy-tinged olive-brown with broad blackish streaks (crown may appear mostly blackish), sides of head grey, broad blackish-brown line through eye, no white bars/spots on upperparts, wings (except coverts) or tail, grey of underparts washed brown (particularly upper breast), bars on underparts much broader. Legs and feet fleshy-brown. **First winter** Browner with some pale scales on breast. **VOICE** Series of well-spaced, drawn-out, squealing, grunting ***grui*** or ***krueeh***, rising in crescendo to series of longer high-pitched whistles (like squealing piglets). Also sharp ***tic*** notes, followed by wheezy screaming ***wheee-ooo*** and sharp ***krribk***. **HABITAT** Freshwater marshes, reedbeds; up to 700 m. **RANGE & STATUS** Breeds E Siberia, N Mongolia, Ussuriland, Sakhalin I, NE China, N,S Korea, Japan; mostly winters south to N,NE India, Nepal, Bangladesh, southern China, Taiwan, southern Japan. **SE Asia** Rare winter visitor SW,N,E Myanmar, W(coastal),NW,C,NE Thailand, N Laos, E Tonkin.

223 CORNCRAKE *Crex crex* Plate **28**

IDENTIFICATION 24.5–28.5 cm. Monotypic. **Male non-breeding** Resembles female Watercock but considerably smaller and smaller-billed, crown distinctly streaked; shows some grey on supercilium, cheeks and often lower foreneck and breast; lacks dark bars/vermiculations on breast and centre of underparts; flanks and undertail-coverts broadly barred dark rufous and whitish (some narrow blackish bars); wing-coverts dark rufous with some variable pale buff to whitish bars. In flight, largely rufous upperwing diagnostic. Legs and feet pale greyish-flesh. **Male breeding** Has more defined blue-grey supercilium, cheeks and sides and front of neck; upperparts and breast tinged blue-grey. **Female non-breeding** Has less grey on supercilium than male non-breeding; often lacks grey on cheeks, neck and breast. **Female breeding** Upperparts buffier and less greyish than male breeding, supercilium narrower and greyer, often has less grey on cheeks, neck and breast. **First winter** Like female non-breeding but usually with fewer pale bars on wing-coverts. **VOICE** Usually silent but may give a short, loud ***tsuck*** when alarmed. Male territorial call is a loud, monotonous, dry, rasping ***krek-krek krek-krek krek-krek krek-krek krek-krek...***; likened to sound of fingernail drawn rapidly across teeth of comb. **HABITAT** Grassland, rice paddies and other cultivation; lowlands. **RANGE & STATUS** Breeds Palearctic (east to L Baikal region), N Iran, NW China; mostly winters southern Africa; vagrant N Pakistan, NW India, Sri Lanka, SE Australia, etc. **SE Asia** Vagrant Cochinchina.

224 WHITE-BREASTED WATERHEN
Amaurornis phoenicurus Plate **28**

IDENTIFICATION 28.5–36 cm. *A.p.phoenicurus*. **Adult** Unmistakable. Upperside dark slaty olive-brown, face and throat to upper belly white, bordered by blackish line from side of crown to side of breast, lower flanks and vent deep rufous-chestnut. Bill yellowish-green, upper mandible with blackish central area and red base, eyes red, legs and feet yellowish-green to yellowish. *A.p.insularis* (Coco Is, off S Myanmar) tends to have more extensively white forecrown, flanks broadly dark slate-coloured, white of abdomen restricted to rather narrow band and bordered black, vent deeper and richer chestnut. Smaller (less than 30.5 cm) in Peninsular Malaysia, Singapore. **Juvenile** Somewhat browner above, with dark lores and forehead, dark spots on ear-coverts, less white over eye, broken dark bars on breast and belly, brown eyes and greyish-brown bill. **VOICE** Weird loud, bubbling, roaring, quacking, grunting and croaking: ***kru-ak kru-ak kru-ak-a-wak-wak, krr-kwaak-kwaak krr-kwaak-kwaak*** etc., and long series of ***kwaak, kuk*** or ***kook*** notes. Contact call is a series of quick ***pwik*** notes. **HABITAT** Various kinds of well-vegetated smaller wetlands, pools and streams in open forest (including islands), mangroves and adjacent open areas, tracks, roadsides etc.; up to 1,525 m. **RANGE & STATUS** Resident Indian subcontinent, C,E and southern China, Taiwan, Japan, Sundas, Christmas I, Philippines, Wallacea (except N Moluccas); some Chinese populations winter to south. **SE Asia** Common resident throughout. Fairly common winter visitor S Thailand, Peninsular Malaysia, Singapore, N Laos (presumably elsewhere). Also recorded on passage S Thailand, Peninsular Malaysia. **BREEDING** All year. **Nest** Shallow saucer or pad, on ground amongst vegetation, in bush, bamboo clump or tree; up to 3 m above ground (often above water). **Eggs** 3–8; cream-coloured or pinkish-white to pale buff, spotted, blotched and sometimes streaked with shades of reddish-brown and pale purple (often capped at broader end); 35.2–45 x 27.1–32.2 mm (combined range).

225 **BROWN CRAKE** *Porzana akool* Plate **28**
IDENTIFICATION 23.5–29 cm. *P.k.akool*: Male Darkish olive-brown above, mostly grey on sides of head and underparts with white centre to throat and upper foreneck and brownish flanks and vent. Faint dull chestnut wash from side of neck to side of breast, bill greenish with bluish tip, eyes blood-red, legs and feet dark fleshy-brown to dark red or dark purplish-red (redder when breeding). *P.k.coccineipes* (E Tonkin) is larger and said to have redder legs and feet. See Black-tailed Crake. Note habitat and range. **Female** Averages smaller, bill has darker upper mandible and darker or lavender-coloured tip. **Juvenile** Like female but upperparts slightly darker, underparts slightly browner-tinged, eyes brown, legs and feet duller. **VOICE** Territorial call is a long high rippling trill, gradually falling slightly in pitch; similar to that of Black-tailed. Calls with spaced, stressed, rather high-pitched ***tuc*** or ***puc*** notes. **HABITAT & BEHAVIOUR** Swampy areas, overgrown watercourses; up to 800 m. Typically skulking but regularly feeds in open. **RANGE & STATUS** Resident (subject to relatively local movements) NE Pakistan, northern and C India, S Nepal, Bangladesh, S,E China. **SE Asia** Scarce to uncommon resident SW Myanmar, E Tonkin. **BREEDING** India: March–October. **Nest** Pad with slight depression, on ground amongst grass etc., rarely flotsam in tree up to 1.5 m. **Eggs** 5–6; creamy-buff to pale salmon-pink, flecked and blotched pale reddish-brown, purple-brown or pale brick-red (sometimes more densely around broader end); 33–39.5 x 25.4–29 mm (*akool*). **NOTE** Better placed in the genus *Porzana*, though formerly often included in *Amaurornis*.

226 **BLACK-TAILED CRAKE**
Porzana bicolor Plate **28**
IDENTIFICATION 21–24 cm. Monotypic. **Adult** Similar to Brown Crake but smaller, upperside distinctly rufescent with contrasting blackish uppertail-coverts and tail, grey of head, neck and underparts darker and more slate-coloured with white restricted to centre of throat or almost lacking. Bill green, often with darker culmen ridge and tip and some red at base (brighter with more obvious red when breeding), eyes red, eyering pinkish, legs and feet brick-red. Female may have duller bill. **Juvenile** Once independent, similar to adult but upperparts mixed with blackish-brown, underparts brownish-tinged, bill paler, eyes brown. Distinctly darker, above and below, than Brown, presence of rufous on upperparts distinctive. **VOICE** Male territorial call is a long trill, up to 13 s, like Ruddy-breasted Crake but more obviously descending and usually preceded by subdued harsh rasping ***waak-waak*** (not audible at distance); also recalls Little Grebe. Contact call is a high-pitched ***kek*** or ***kik***. **HABITAT** Small swampy areas and overgrown streams in or near broadleaved evergreen forest and secondary growth, adjacent wet cultivation; 1,000–1,830 m, locally down to c.200 m in N Myanmar and 300 m in NW Thailand. **RANGE & STATUS** Resident NE India, Bhutan, SW China; possibly E Nepal. **SE Asia** Rare to scarce resident W,N,S(east),E Myanmar, NW Thailand, N Laos, W Tonkin. **BREEDING** April–August. **Nest** Rough pad with depression, on ground amongst grass etc., sometimes in tree or bush up to 2 m, rarely 6 m. **Eggs** 5–8; pale cream-coloured or buff to salmon-pink, boldly spotted and blotched deep red-brown, purplish-brown or brick-red over grey and lavender undermarkings; 31.3–36.3 x 24.5–27 mm.

227 **BAILLON'S CRAKE** *Porzana pusilla* Plate **28**
IDENTIFICATION 19–20.5 cm. *P.p.pusilla*: **Adult** Smallest crake in region. Shows distinctive combination of short greenish bill, warm-tinged olive-brown upperside with broad blackish streaks and narrow white streaks and speckles, blue-grey supercilium, lower sides of head and breast, and grey-brown lower flanks and vent with black-and-white bars. Centre of throat whiter, eyes red, eyering greenish-yellow, legs and feet greenish, yellowish, brownish or pinkish. In flight, shows narrow white leading edge to primaries and marginal coverts. See Spotted Crake. **Juvenile** Like adult but sides of head mostly pale rufous, throat to centre of belly mostly whitish, washed rufous-buff to pale rufous-brown on upper breast and upper flanks, brown bars on breast-sides/upper flanks. Moults directly into adult plumage during late autumn and first winter. **VOICE** Usually silent but may give a short ***tac*** or ***tyiuk*** when alarmed. Territorial call is a creaky rasping ***trrrrr-trrrrr*** (like sound of nail along teeth of comb), repeated every 1–2 s and sometimes preceded by dry ***t*** sounds. **HABITAT** Freshwater marshes, reedbeds, vegetated wetlands; up to 1,400 m. **RANGE & STATUS** Resident N,E,S Africa, Madagascar, S Iran, N Sulawesi, New Guinea, Australia, New Zealand; possibly Sumatra, SE Borneo, Philippines. Breeds Palearctic, NW India, NW,NC,NE China, N Korea, N Japan; mostly winters to south, N Africa, Middle East, Indian subcontinent, S Japan, Greater Sundas, Philippines; possibly Sulawesi. Recorded (status unknown) Lesser Sundas (Flores), S Moluccas (Seram). **SE Asia** Scarce to common winter visitor SW,C,E,S Myanmar, Tenasserim, W,NW,C,NE(south-west),S Thailand, Peninsular Malaysia, Singapore, Cambodia, N Laos, C,S Annam, Cochinchina. Uncommon to fairly common passage migrant NW Thailand, Peninsular Malaysia, N Laos, E Tonkin (winters?), C,S Annam, Cochinchina.

228 **SPOTTED CRAKE** *Porzana porzana* Plate **28**
IDENTIFICATION 21–24 cm. Monotypic. **Male non-breeding** Similar to Baillon's Crake but larger, bill thicker, base colour of upperparts darker olive-brown, with diagnostic white speckles and spots on head, neck and breast, and plain buff undertail-coverts. Shows broader, slatier-grey supercilium, whitish upper lores, blackish lower lores/cheeks, grey throat with white speckles, browner base colour of breast and more white markings on upperside generally. Bill greenish-yellow with orange-yellow to orange-red base, eyes yellowish- to reddish-brown, legs and feet dull green. In flight, white leading edge of wing slightly more prominent than Baillon's. **Male breeding** Supercilium, cheeks, throat and much of breast distinctly bluish-grey, with fewer white spots, base of bill brighter red. **Female** Like male non-breeding but tends to show less grey on head and more white speckles. **First winter** Similar to female but throat largely whitish, head and sides of neck more densely speckled whitish, bill duller, eyes brownish. **VOICE** Usually silent but may utter sharp ***tck*** notes when alarmed. Territorial call is a short, loud, rather high-pitched ascending ***kwit*** or ***whitt***, repeated about once every second, often for several minutes; sometimes given in winter. **HABITAT** Marshes, reedbeds; lowlands. **RANGE & STATUS** Breeds W,C Palearctic, Iran, NW China, W Mongolia; winters sub-Saharan Africa, Egypt, Middle East, Indian subcontinent (except E and islands). **SE Asia** Vagrant SW Myanmar, W(coastal),C Thailand.

229 **RUDDY-BREASTED CRAKE**
Porzana fusca Plate **28**
IDENTIFICATION 21–26.5 cm. *P.f.fusca*: **Adult** Most widespread crake in region. Combination of uniform, rather cold dark olive-brown hindcrown and upperside, deep reddish-chestnut forecrown and sides of head and mostly deep reddish-chestnut underparts diagnostic. Throat whitish, lower flanks and vent dark greyish-brown to blackish with indistinct narrow whitish bars (sometimes only barred on undertail-coverts); bill blackish with greenish base; eyes, narrow eyering, legs and feet dull salmon-red to red. *P.f.erythrothorax* (visitor) is larger, a shade paler and less deep-toned overall, underparts less reddish. **Juvenile** Crown and centre of upperparts initially all blackish, sides of head, foreneck and underparts whitish with dense dull brownish-grey vermiculations/mottling, bill dull brown with paler lower mandible, eyes brown, legs and feet dull brown. Attains adult plumage after c.6 months. See Black-tailed Crake. **VOICE** Territorial call consists of hard ***tewk*** or ***kyot***, repeated every 2–3 s, with sequence often speeding up and typically followed by long, high-pitched, slightly descending trill, lasting 3–4 s. Trill recalls Black-tailed Crake, Little Grebe. Contact calls

include low ***chuck***. **HABITAT** Freshwater marshes, vegetated wetlands, reedbeds, sometimes more open wetlands, wet rice paddies, scrub, drier cultivation and mangroves (mainly on passage); up to 1,450 m. **RANGE & STATUS** Resident (subject to some movements) N Pakistan, NW,N,NE Indian subcontinent, SW India, Sri Lanka, southern China, Taiwan, S Japan, Greater Sundas (except Borneo?), W Lesser Sundas, Sulawesi, Philippines. Breeds C,NE(southern) China, N,S Korea, Ussuriland, N Japan; mostly winters to south, possibly Borneo, Philippines, rarely Christmas I. **SE Asia** Common resident, subject to some local movements (except W Myanmar, Tenasserim, inland W Thailand, C Laos, E Tonkin, N,S Annam); probably breeding visitor only in some northern areas. Common winter visitor E Myanmar, Thailand (except S), E Tonkin, Cochinchina. Also common passage migrant C Thailand, E Tonkin, C,S Annam. **BREEDING** February–June. **Nest** Pad or shallow saucer on ground amongst grass etc. or in sedge clump or bush up to 1 m. **Eggs** 3–9; whitish- to pinkish-cream or milky coffee-colour, liberally speckled and blotched rufous- to reddish-brown over purplish-grey undermarkings (often more heavily marked at broader end); 29–34.2 x 21.8–25.2 mm (*fusca*).

230 BAND-BELLIED CRAKE
Porzana paykullii Plate **28**

IDENTIFICATION 27 cm. Monotypic. **Adult** Similar to Slaty-legged Crake but crown and upperside uniform and darker, colder brown, chestnut parts paler, extending slightly further down underparts (onto upper belly); has some narrow whitish and dark bars on wing-coverts, completely unbarred primaries and secondaries, and salmon-red legs and feet. Bill greenish- to bluish-slate with pea-green base (possibly less obvious on female). See Red-legged Crake. **First winter** Like adult but sides of head and breast to upper belly paler and washed out with little pale chestnut, breast and upper belly indistinctly barred darker (centre of breast and abdomen plainer and whiter), legs and feet purplish-brown. See Slaty-legged and Red-legged. **VOICE** Usually silent. Territorial call is a loud metallic clangour running into brief trills (likened to sound made by wooden rattle). **HABITAT & BEHAVIOUR** Freshwater swamps, vegetated wetlands, wet areas in or near broadleaved evergreen forest, up to 1,220 m; winters mostly in lowlands. Very skulking. **RANGE & STATUS** Breeds SE Siberia, Ussuriland, NE China; winters N Sumatra, Java, N Borneo, rarely S Sulawesi. **SE Asia** Rare passage migrant/vagrant C Thailand, Peninsular Malaysia (has wintered), E Tonkin, Cochinchina.

231 WHITE-BROWED CRAKE
Porzana cinerea Plate **28**

IDENTIFICATION 19–21.5 cm. Monotypic. **Adult** Easily identified by combination of small size, grey sides of neck, breast and upper flanks, and blackish cheeks and orbital area bordered above and below by broad white streak. Crown slaty-grey with blackish feather-centres, rest of upperparts greyish-brown with broad blackish and narrow buffish streaks; throat and centre of underparts whitish, lower flanks and vent pinkish-cinnamon, bill greenish-yellow with some red at base, legs and feet greenish. **Juvenile** Grey parts somewhat browner. Crown to mantle paler and browner, sides of head to flanks paler grey, face pattern less distinct, lacks red at base of bill, legs and feet greyer. **VOICE** Both sexes utter a loud nasal chattering ***chika***, rapidly repeated 10–12 times. Also a loud ***kek-kro*** when foraging, repeated nasal ***hee*** notes and quiet, repeated ***charr-r*** when alarmed. **HABITAT** Well-vegetated freshwater lakes, floating mats of vegetation, marshes, sometimes overgrown ditches and rice paddies; lowlands. **RANGE & STATUS** Resident Sundas, Philippines, Wallacea, New Guinea, N Australia, Melanesia to Micronesia and Samoa. **SE Asia** Uncommon to locally common resident Thailand (except SE), Peninsular Malaysia, Singapore, Cambodia, N(south) Laos, E Tonkin, Cochinchina. **BREEDING** December–March, June–October. **Nest** Pad or shallow saucer, sometimes canopied by bent-over grasses etc., amongst waterside vegetation; up to 1 m. **Eggs** 3–7; glossy creamy-buff to greyish-white or olive-buff, heavily speckled and blotched reddish-brown and dull purplish-brown (particularly around broader end); 27.5–31.2 x 21–23.7 mm.

232 WATERCOCK *Gallicrex cinerea* Plate **28**

IDENTIFICATION Male 41–43 cm; female 31–36 cm. Monotypic. **Male breeding** Relatively large size, blackish plumage with whitish bars on vent, yellow bill with red base, long red 'frontal shield' and rather long red legs diagnostic. Has grey to buff feather-fringing on mantle, scapulars and wing-coverts (appears scaled) and warm brown fringing on rump and uppertail-coverts. In flight, shows white leading edge to wing. See Common Moorhen. **Male non-breeding** Like female non-breeding, but possibly shows broader, more defined underpart-bars. **Female breeding** Smaller and smaller-billed than male, upperside dark brown with buff feather-fringes and greyer fringing on hindneck, upper mantle and wing-coverts; crown fairly uniform, sides of head pale buff with brownish cheeks; underparts generally buffy-white, but deeper buff on lower foreneck/upper breast and undertail-coverts, and whiter on throat and centre of abdomen, marked overall with narrow wavy greyish-brown bars (denser on lower foreneck, breast, flanks and undertail-coverts); bill, legs and feet greenish. In flight, shows white leading edge to wing like male. See Corncrake. **Female non-breeding** Underparts somewhat more heavily barred. **Juvenile** Like female non-breeding (both sexes) but upperparts more broadly streaked warmer buff, crown narrowly streaked buff, sides of head and underparts rich buff with narrow dark bars (may be almost lacking), white of throat restricted to centre; respective sexes smaller than adults. **VOICE** Territorial call is 10–12 ***ogh*** or ***kok*** notes, then 10–12 quicker, deeper, hollower metallic booming ***ootoomb*** or ***utumb*** notes, ending with 5–6 ***kluck*** notes. May give a harsh nasal ***krey*** in alarm. **HABITAT & BEHAVIOUR** Freshwater marshes and swamps, wet grassland, rice paddies, rarely mangroves on passage; up to 1,220 m (mostly lowlands). Usually rather skulking. When giving territorial call, neck and often whole body is puffed out, head lowered when giving booming notes. **RANGE & STATUS** Breeds NE,SE Pakistan, Indian subcontinent, E,NE,C and southern China, Taiwan, N,S Korea, S Ussuriland, Japan (S Nansei Is), Sumatra, Philippines; many northern populations winter to south, Greater Sundas, W Lesser Sundas, Sulawesi. **SE Asia** Uncommon to fairly common resident C Myanmar, W,C,NE(west),SE,S Thailand, Peninsular Malaysia, Cambodia, S Laos, C,S Annam, Cochinchina. Uncommon to locally common breeding visitor N(some winter),E Myanmar, NW,C(north),NE Thailand, N Laos (some winter), E Tonkin. Fairly common winter visitor S Thailand, Peninsular Malaysia, Singapore, Cambodia, Cochinchina. Uncommon to fairly common passage migrant Peninsular Malaysia, N Laos, E Tonkin. Recorded (status uncertain) SW,S Myanmar, Tenasserim, C Laos. **BREEDING** May–September. **Nest** Large cupped pad, sometimes loosely canopied, on or near ground amongst marshy vegetation etc. **Eggs** 3–10; glossy whitish or buff to deep brick-pink, marked with reddish-brown blotches and spots over pale purple undermarkings (often denser around broader end); 39–46.6 x 28.1–33.1 mm.

233 GREY-HEADED SWAMPHEN
Porphyrio poliocephalus Plate **29**

IDENTIFICATION 42 cm. *P.p.poliocephalus*: **Adult** Large and robust with huge red bill, broad red frontal shield and medium-length red legs. Dark purple-blue, with largely dark turquoise throat, foreneck, upper breast and wings (tertials greener), dull blackish centre of abdomen and white undertail-coverts. Head and upper neck often has strong silvery wash. **Juvenile** Duller than adult, hindneck to rump mixed with brown, foreneck and breast duller, lower belly whiter, bill and 'frontal shield' smaller and duller (latter almost lacking), legs and feet much duller. **VOICE** Has varied vocabulary, and groups may give wailing, squawking, and cackling chorus.

Calls include series of strongly downslurred, very nasal sighing, mooing wails: ***HNYAarh NYaaar NYaaar***, that may start and end with shorter moos (longest ones before end). Loud, harsh, scraping, croaking slightly rising ***krAARRK***, recalling Common Moorhen but harsher and less trilled. Contact calls are hard clucks. Sudden outbursts of cackling, grating and variable staccato notes in alarm/aggression. **HABITAT & BEHAVIOUR** Freshwater marshes, swamps, reedbeds, well-vegetated lakes; up to 1,000 m. Usually quite conspicuous. **RANGE & STATUS** Resident Middle East, Caspian Sea region, Indian subcontinent, ? N Sumatra. **SE Asia** Scarce to locally common resident, subject to some minor movements Myanmar (except W), NW,C Thailand (south to Nakhon Sawan). **BREEDING** February–August. **Nest** Large built-up bowl, amongst/on marshy or floating vegetation; up to 1 m. **Eggs** 3–7; pale buff to reddish-buff, scattered with rich brown to dark brown blotches, often mostly around broader end; 45–54.6 x 32–37.8 mm. **NOTE** Split from extralimital Western (or Purple) Swamphen *Porphyrio porphyrio* following Sangster *et al.* (1998).

234 BLACK-BACKED SWAMPHEN
Porphyrio indicus Plate **29**

IDENTIFICATION 28–29 cm. *P.i.viridis*: **Adult** Smaller than similar Grey-headed Swamphen, with mostly blackish-brown upperside (apart from head and neck) and turquoise shoulder-patch. *P.i.indicus* (Singapore?) has hindcrown, sides of head and upper throat blacker, never shows silvery wash on head and upper neck. **Juvenile** Differs from adult in similar way to Grey-headed. **VOICE** Calls include a whining cackle, a loud raucous ***gaark gaark gaark...***, a creaking note (like a rusty hinge), and loud ***kwank*** in alarm. **HABITAT & BEHAVIOUR** Freshwater marshes, swamps, reedbeds, well-vegetated lakes; up to 1,000 m. Usually quite conspicuous. **RANGE & STATUS** Resident Greater Sundas, Sulawesi. **SE Asia** Scarce to locally common resident (subject to some minor movements) W(coastal),C(north to Nakhon Sawan),NE,SE,S Thailand, Peninsular Malaysia, Singapore, Indochina (except W Tonkin); ? south Tenasserim. **BREEDING** November–August. **Nest** Large built-up bowl, amongst/on marshy or floating vegetation. **Eggs** 4–7; pale olive-grey, with cinnamon-brown blotches, mostly around broader end; 49–54 x 35.9–36.1 mm (*viridis*). **NOTE** Split from extralimital Western (or Purple) Swamphen *Porphyrio porphyrio* following Sangster *et al.* (1998).

235 COMMON MOORHEN
Gallinula chloropus Plate **29**

IDENTIFICATION 30–35 cm. *G.c.chloropus*: **Adult** Overall dark plumage, contrasting white line along flanks and lateral undertail-coverts, yellow-tipped red bill and red 'frontal shield' distinctive. Upperside dark slaty-brown, head and neck slaty-blackish, underparts otherwise greyer with whitish belly-centre, legs and feet yellow-green, turning orange-red above joint. When worn (post-breeding), appears duller and may lose flank-line; bill, frontal shield, legs and feet duller. *G.c.orientalis* (southern S Thailand southwards) is smaller and darker with relatively large shield. **Juvenile** Upperside paler and browner, sides of head and underside much paler and browner with whitish throat and centre of abdomen (undertail-coverts as adult), flank-line buffy-white and less distinct (may be lost when moulting), bill and frontal shield greenish-brown, legs duller. **First winter/first summer** Head, neck and underparts browner than adult. **VOICE** Territorial call is a sudden loud ***krrrruk, kurr-ik*** or ***kark***. Other calls include a soft muttering ***kook***, loud ***kekuk*** or ***kittick*** and loud hard ***keh-keh*** when alarmed. **HABITAT & BEHAVIOUR** Freshwater lakes and pools, marshes, flooded rice paddies, irrigation ditches; up to 1,000 m. Habitually cocks tail, revealing white on undertail-coverts. Mostly seen swimming; walks on land. **RANGE & STATUS** Mostly resident Africa, Madagascar, Palearctic, Middle East, Indian subcontinent, southern Tibet, China, Taiwan, N,S Korea, Japan, Sundas, Philippines, Sulawesi, USA, C,S America; some northerly populations winter south to N Africa, Middle East, Indian subcontinent, southern China, Philippines, possibly N Borneo, C,S America. **SE Asia** Uncommon to locally common resident, subject to some movements (except W,N Myanmar, N Laos, W Tonkin, N Annam). Locally common winter visitor N,C,E Myanmar, S Thailand, Cambodia, N Laos (breeds?), N Annam, Cochinchina. **BREEDING** All year. **Nest** Simple to well-built bowl or saucer in shallow water, often amongst vegetation or in low tree or bush branches. **Eggs** 4–14; glossy whitish-grey to buffish, spotted and blotched deep reddish-brown, purple and blackish, with grey to pale purple undermarkings; 38–44 x 28–32 mm.

236 COMMON COOT *Fulica atra* Plate **29**

IDENTIFICATION 40.5–42.5 cm. *F.a.atra*: **Adult** Slaty-black plumage with paler brownish-slate breast and belly, white bill and white frontal shield diagnostic. In flight, shows fairly broad white trailing edge to secondaries. Legs and feet greenish. **Juvenile** Plumage browner-tinged, throat and centre of foreneck whitish, belly paler, frontal shield smaller. **VOICE** Male combat call is an explosive ***pssi*** or ***pyee***, becoming sharper in alarm; female equivalent is a short croaking ***ai*** or ***u***, becoming a rapid ***ai-oeu-ai-ai-oeu...*** etc. in alarm. Typical contact call is a short ***kow, kowk, kut*** or sharper ***kick*** (sometimes two notes combined). Male also utters a series of hard, smacking ***ta*** or ***p*** notes. **HABITAT & BEHAVIOUR** Freshwater lakes, marshes; up to 800 m. Usually seen swimming; patters along surface when disturbed. Sometimes walks on land. **RANGE & STATUS** Breeds N Africa, Palearctic, Middle East, Pakistan (occasional), India, Bangladesh, Sri Lanka, S Tibet, northern China, N Korea, Japan; some northern populations winter to south, N Africa, Indian subcontinent, Philippines, rarely N Borneo. Mostly resident New Guinea, Australia, New Zealand, formerly Java. **SE Asia** Rare to uncommon winter visitor Myanmar (locally common in N), NW,NE,C Thailand, Cambodia, N,S Laos, E Tonkin, N,C Annam, Cochinchina. Vagrant S Thailand, Peninsular Malaysia, Singapore.

HELIORNITHIDAE: Finfoots

Worldwide 3 species. SE Asia 1 species. Rather slender with long necks and stiff tails, bills stout and dagger-like with nostril near centre, legs strong, the toes laterally fringed with lobed webs. Feed on small fish and aquatic invertebrates, including small crabs. Mainly swim in water but can also walk, in ungainly manner, on land.

237 MASKED FINFOOT
Heliopais personata Plate **29**

IDENTIFICATION 52–54.5 cm. Monotypic. **Male** Unmistakable. Hindcrown and hindneck grey, throat and upper foreneck black with white border, forecrown and line along side of crown black, bill thick and yellow with small horn at base (spring only?). Rest of upperside mostly brown with greyer mantle, underparts mostly whitish, flanks and undertail-coverts brown with some whitish bars, eyes dark brown, legs and feet bright green. **Female** Throat, upper foreneck and much of lores whitish, less black on forecrown, no horn on bill and eyes yellow. **Juvenile** Similar to female but somewhat browner above, with no black on forecrown, less distinct and more mottled black on neck-side, creamy-yellow bill. **Male first winter** Like adult but has white centre to throat and upper foreneck and small amount of white on lores. **VOICE** Poorly documented. Series of rather high-pitched bubbling sounds (like air being blown through tube into water), possibly followed by series of clucks which increase in tempo. **HABITAT & BEHAVIOUR** Rivers in broadleaved evergreen forest, mangroves, swamp forest, sometimes pools and lakes away from forest; up to 700 m but recorded up to 1,220 m on migration. Usually secretive. Head jerks back and forth when swimming. **RANGE & STATUS** Breeds (resident?) NE India, S Bangladesh. Recorded (status unknown) Sumatra, W Java. **SE Asia** Scarce to locally fairly common resident N,S Myanmar, Cambodia. Scarce winter visitor and passage migrant (probably also local resident) S Thailand, Peninsular Malaysia. Vagrant Singapore. Recorded (status unknown) SW,C,E Myanmar, Tenasserim, W,NW,NE,SE Thailand, S Laos, C,S Annam. **BREEDING** February–October. **Nest** Large structure, resembling crow *Corvus* nest, in waterside tree, bush or large fern; 0.5-5 m above water. **Eggs** 5–7; pale cream, sparsely blotched reddish-brown to medium brown over lavender-grey undermarkings; 50.3 x 43.2 mm (av.).

GRUIDAE: GRUINAE: Typical cranes

Worldwide 13 species. SE Asia 4 species. Large to very large with long necks and legs, long, drooping inner secondaries and mostly grey plumage. Flight strong, with neck and legs extended; flocks often fly in V-formation. Diet varied, including seeds, roots, shoots, crustaceans, frogs, lizards, small fish and large insects.

238 DEMOISELLE CRANE *Grus virgo* Plate 13

IDENTIFICATION 90–100 cm. Monotypic. **Adult** Resembles Common Crane but smaller with smaller bill, shorter neck, longer, more pointed tertials; also mostly slaty-black sides of head and neck, long black feathers on lower foreneck/upper breast, grey crown and centre of hindneck, and long white tuft of feathers extending from behind eye. May be difficult to separate from Common in flight at distance. Legs blackish. **First winter** Black of head and neck duller, crown, hindneck, body, wing-coverts and tertials brownish-grey, feathers of lower foreneck and inner secondaries shorter, post-ocular tufts shorter and tinged grey. **VOICE** Loud ***garroo*** in flight, somewhat higher-pitched than similar call of Common. Courtship call is rasping ***krruorr***, more guttural than in Common. **HABITAT** Marshes, open areas, cultivation; up to 800 m. **RANGE & STATUS** Breeds Morocco (almost extinct), E Turkey, W(east),C Palearctic, Mongolia, NW,N,NE China; winters NC Africa, W,NW,N,C India; formerly Pakistan. **SE Asia** Vagrant SW,W,E Myanmar, NW Thailand.

239 SARUS CRANE *Grus antigone* Plate 13

IDENTIFICATION 152–156 cm. *G.a.sharpii*: **Adult** Huge size, rather uniform grey plumage and mostly naked red head and upper neck diagnostic. Primaries and primary coverts blackish, secondaries mostly grey, bill longish and pale greenish-brown, legs pale reddish. Red of head and neck brighter when breeding. *G.a.antigone* (reported SW Myanmar) is somewhat paler, particularly below, and has whiter midneck and tertials. **Juvenile** Head and upper neck buffish and feathered, overall plumage duller with brownish-grey feather-fringes, those of upperparts more cinnamon-brown. **VOICE** Very loud trumpeting, usually by duetting pairs. **HABITAT & BEHAVIOUR** Marshy grassland, open country, rice paddies; up to 1,000 m. Pairs perform dancing courtship display, which includes wing-spreading, bowing, head-lowering and leaps into air. **RANGE & STATUS** Resident northern and C India, S Nepal, NW Bangladesh, NE Australia; formerly Pakistan, elsewhere Bangladesh, Philippines (Luzon). **SE Asia** Rare resident (subject to local movements) SW,N,C,S,E Myanmar, Cambodia, S Laos, S Annam, Cochinchina (still breeds?); formerly NW,C,NE,S Thailand, north-west Peninsular Malaysia, C Laos. Formerly recorded (probably resident, but current status unknown) W Myanmar, Tenasserim. **BREEDING** March–October. **Nest** Large pile of vegetation, often in shallow water. **Eggs** 1–3; glossy, pinkish-cream to greenish-white, sometimes spotted or clouded pale yellowish-brown, purple or purplish-pink; 88.9–114.3 x 61–71.1 mm.

240 COMMON CRANE *Grus grus* Plate 13

IDENTIFICATION 110–120 cm. *G.g.lilfordi*: **Adult** Huge size, greyish plumage, mostly blackish head and upper neck, with red patch on crown and broad white band from ear-coverts down side of upper neck diagnostic. Long drooping tertials mixed with black. In flight, blackish primaries and secondaries contrast with grey wing-coverts. See Demoiselle and Black-necked Cranes. **First winter** Head and upper neck warm buffish to grey, rest of plumage (particularly upperparts) often mixed with brown; may show whitish to buffy-whitish patch behind eye. From Sarus Crane by smaller size, much shorter bill, dark markings on inner wing-coverts, tertials and secondaries, and dark legs and feet. More like adult by first spring, fully adult by third winter/spring. **VOICE** Flight calls include loud, flute-like bugling ***krooh***, ***krrooah*** and ***kurr*** etc., higher-pitched ***klay***, croaking ***rrrrrrrer*** and various honking sounds. Similar calls also given on ground, including duets by bonded pairs: ***kroo-krii-kroo-krii...*** **HABITAT** Marshes, open areas, less disturbed cultivation; up to 3,100 m. **RANGE & STATUS** Breeds Palearctic, NW China; mostly winters N Africa, southern W Palearctic, Iran, extreme SE Pakistan, W,C and northern India, Nepal, C and southern China; formerly Bangladesh. **SE Asia** Uncommon to locally common winter visitor W,N,C Myanmar, E Tonkin; formerly N,C Annam. Recorded (status uncertain) SW,E Myanmar. Vagrant NW Thailand.

241 BLACK-NECKED CRANE *Grus nigricollis* Plate 13

IDENTIFICATION 139 cm. Monotypic. **Adult** Like Common Crane but even larger, head and upper neck all black, apart from small whitish patch behind eye and mostly dull red naked crown, overall plumage paler and whiter with contrasting all-blackish tertials. In flight, shows stronger contrast between wing-coverts and rest of wing. **Juvenile/immature** Poorly documented. **Juvenile** has buff-washed and buff-scaled plumage. At three months crown-feathered and buffish, black of neck mixed with whitish. At eight months, similar to adult but shows some buff scales on upperparts, neck greyish-black. Like adult after one year. **VOICE** Similar to Common but somewhat higher-pitched. **HABITAT** Marshy areas, cultivation; lowlands. **RANGE & STATUS** Breeds NW India (Ladakh), Tibet, W,NW(south-east) China; winters Bhutan, NE India (Arunachal Pradesh), S Tibet, SW China. **SE Asia** Formerly a rare to scarce winter visitor E Tonkin.

TURNICIDAE: Buttonquails

Worldwide 17 species. SE Asia 3 species. Small, plump, very short-tailed and short-legged. Similar to quail but bills longer, narrower and straighter; lack hind toes. Sexual roles are reversed. Found in relatively open habitats; feed on insects, seeds and other vegetable matter.

242 SMALL BUTTONQUAIL *Turnix sylvaticus* Plate 1

IDENTIFICATION 13–14 cm. *T.s.mikado*: **Female** Similar to Yellow-legged Buttonquail but smaller, wing-coverts pale chestnut with buff and black streaks, defined buff lines along outer edge of mantle and inner edges of tertials, lighter buff breast-patch, slaty-blue to blackish bill and greyish to flesh-coloured legs and feet. *T.s.dussumier* (Myanmar) is a shade paler above with more prominent rufous-chestnut and buff markings on hindneck/upper mantle. From similar female quails by long bill and unmarked centre of breast. See Barred Buttonquail. **Male** Mantle less rufescent. **Juvenile** Like female but buff breast-patch less distinct, has blackish spots across breast. **Other subspecies in SE Asia** *T.s.davidi* (Cochinchina). **VOICE** Female territorial call is a half-minute-long, far-carrying, ventriloquial series of droning, booming ***hoooooo*** notes (each 1 s long with 1–3 s intervals). **HABITAT** Dry grassland, scrub and grass bordering cultivation; up to 1,150 m. **RANGE & STATUS** Resident (subject to some movements) northern Africa, SW Europe, Pakistan, India, S Nepal, Sri Lanka, S China, Taiwan, Java, Bali, Philippines. **SE Asia** Scarce to local resident C,S,E Myanmar, NW,C,SE Thailand, north-west Peninsular Malaysia, Cambodia, N,C Laos, E Tonkin, C Annam, Cochinchina. **BREEDING** June–September. **Nest** Sparsely lined scrape on ground. **Eggs** 4; greyish-white, finely speckled (sometimes blotched) yellowish-brown, reddish-brown or black; 21.3 x 16.8 mm (av.).

243 **YELLOW-LEGGED BUTTONQUAIL**

Turnix tanki Plate **1**

IDENTIFICATION 16.5–18 cm. *T.t.blanfordii*: **Female** Yellow legs and distinctly sandy-buff upperwing-coverts with pronounced large black spots diagnostic. Has rufous nuchal collar, distinctly deep buff breast-band and large round black spots on upper flanks, lacks bars on underparts, bill extensively yellowish. In flight shows more contrast between upperwing-coverts and rest of wing than other buttonquails. See Small and Barred Buttonquails and quails. **Male** Lacks female's rufous nuchal collar. **Juvenile** Duller than male with less distinct breast-patch, faint narrow brownish-grey bars on lower throat and breast and somewhat less pronounced wing-covert spots. **VOICE** Female territorial call is described as series of low-pitched hooting notes, gradually increasing in strength and turning into a human-like moan. **HABITAT** Grassy areas, slightly marshy grassland, scrub, cultivation, secondary growth; up to 2,135 m. **RANGE & STATUS** Breeds N,SE Pakistan, India, Nepal, C,E,NE and southern China, N Korea, SE Siberia, Ussuriland; some northern populations winter to south, S China. **SE Asia** Uncommon to common resident (except N Myanmar, southern S Thailand, Peninsular Malaysia, Singapore). Uncommon passage migrant E Tonkin, N Annam. Recorded in winter (status unknown) N Myanmar. **BREEDING** May–August (probably all year). **Nest** Sparsely lined scrape on ground. **Eggs** 4; like those of Barred; 25.5 x 20.8 mm (av.; *blanfordii*).

244 **BARRED BUTTONQUAIL**

Turnix suscitator Plate **1**

IDENTIFICATION 15–17.5 cm. *T.s.thai*: **Female** The most widespread buttonquail in the region, told by: black throat and centre of upper breast, heavy blackish bars across lower throat, breast and upper flanks, heavy black and pale buff bars on wing-coverts, and deep rufous-buff vent. Upperparts mostly greyish-brown with blackish markings and vermiculations; crown and sides of head densely speckled whitish-buff. In flight shows more contrast between upperwing-coverts and rest of wing than quails and Small Buttonquail, but less than Yellow-legged Buttonquail. *T.s.blakistoni* (recorded NW Thailand, northern Indochina) has strikingly rufous-chestnut base colour of upperparts and buffier underparts (both sexes); *atrogularis* (S Thailand southwards) has distinctly richer buff underparts. See Blue-breasted and other quails. **Male** Similar to female but lacks black patch on throat and breast, vent slightly duller. **Juvenile** Similar to male. Females gradually attain black patches on throat. **Other subspecies in SE Asia** *T.s.plumbipes* (SW,W Myanmar), *pallescens* (C,S Myanmar). **VOICE** Female territorial call is a series of soft, ventriloquial, reverberating booming notes, gradually increasing in volume before ending abruptly. **HABITAT** Dry grassy areas, scrub and cultivation, secondary growth; up to 1,650 m. **RANGE & STATUS** Resident Indian subcontinent (except NW and Pakistan), Taiwan, Japan (Nansei Is), Sumatra, Java, Bali, W Lesser Sundas, Philippines, Sulawesi. Breeds southern China; some winter to south. **SE Asia** Common resident throughout. **BREEDING** December–September. **Nest** Scrape (often with light roof) on ground, amongst grass. **Eggs** 3–5; greyish-white, profusely marked with tiny brown specks of various shades, often more densely towards broader end; 24.9 x 20.2 mm (av.; *plumbipes*).

BURHINIDAE: Thick-knees

Worldwide 10 species. SE Asia 3 species. Medium-sized to largish with thick to very thick bills, long legs and striking yellow eyes. Mostly crepuscular and nocturnal, spending much of day hidden. Feed on crabs, molluscs, insects, worms, small reptiles, amphibians, fish and small mammals etc.

245 **INDIAN THICK-KNEE**

Burhinus indicus Plate **29**

IDENTIFICATION 40–44 cm. Monotypic **Adult** From other thick-knees by smaller size, much smaller bill, streaked upperparts, neck and breast, and lack of prominent black head markings. Upperparts and scapulars pale sandy-brown, with blackish and white bands along smaller wing-coverts and mostly grey greater coverts. In flight shows blackish secondaries and primaries, latter with relatively small white patches. **Juvenile** Scapulars, inner wing-coverts and tertials fringed rufous-buff, with less obvious dark bars along wing-coverts and whiter tips to greater coverts. **VOICE** Normally calls at dusk and during night. A long series of short, upslurred, piping whistles, with series rising at first, then slowing, falling in pitch and dropping in volume: ***skipSKIPSKIP-SKIP-SKIP...SKIPSKEEPskeep, s'skelpit, s'skelpit...*** **HABITAT & BEHAVIOUR** Dry barren areas, semi-desert, sand-dunes, scrub, riverine sandbanks; lowlands. Mostly crepuscular and nocturnal. Rests in shade during daytime. **RANGE & STATUS** Resident (subject to minor movements) Pakistan (east of Indus R), India, Nepal, SC Bhutan, Sri Lanka. **SE Asia** Scarce to uncommon resident Myanmar (except W,E), Thailand (except S), Cambodia, S Laos, C,S Annam. Vagrant NW,S Thailand. **BREEDING** India: February–September. **Nest** Shallow unlined or poorly lined scrape on ground. **Eggs** 2–3; pale stone-coloured to buff (sometimes tinged purplish or reddish), blotched blackish-brown over grey undermarkings; 47.6 x 34.7 mm (av.). **NOTE** Split from Eurasian Thick-knee *B. oedicnemus*, following Rasmussen & Anderton (2005).

246 **GREAT THICK-KNEE**

Esacus recurvirostris Plate **29**

IDENTIFICATION 49–54 cm. Monotypic. **Adult** Very distinctive. Quite large with long legs and longish, thick, slightly upturned black bill with yellow base. Forehead white, crown and upperparts pale sandy-greyish, head-sides and throat white with sharply contrasting black lateral crown-stripe, ear-coverts and submoustachial patch, rest of underparts whitish with brownish wash on foreneck and upper breast, and buff-washed undertail-coverts. Has narrow blackish and whitish bands along lesser coverts. In flight, upperwing shows contrasting, mostly grey greater coverts, mostly black secondaries, black primaries with large white patches, underwing mostly white with black-tipped flight-feathers and primary coverts. See Beach and Indian Thick-knees. Note range. **Juvenile** Initially has buffish fringes and spots on upperparts. **VOICE** Territorial call is a series of wailing whistles with rising inflection: ***kree-kree-kree kre-kre-kre-kre-kre...*** etc. Gives a loud harsh ***see-eek*** when alarmed. **HABITAT & BEHAVIOUR** Shingle and sandbanks along large rivers, sand-dunes, dry lake shores, sometimes coastal mud- and sand-flats, saltpans; lowlands. Mainly crepuscular and nocturnal. **RANGE & STATUS** Resident Indian subcontinent, SW,S China. **SE Asia** Scarce to uncommon resident Myanmar. Rare to scarce resident NW,NE Thailand (now along Mekong R only), Cambodia (north-east only?), Laos, C,S Annam (mainly coastal); formerly C Thailand. Recorded (status uncertain) E Tonkin. **BREEDING** India: February–August. **Nest** Shallow scrape on ground, (eggs sometimes on bare rock). **Eggs** 2; like Eurasian; 54.4 x 41 mm (av.).

247 BEACH THICK-KNEE
Esacus neglectus Plate **29**

IDENTIFICATION 53–57 cm. Monotypic. **Adult** Similar to Great Thick-knee but slightly larger, bill bulkier with less upcurved upper mandible, crown and nape blackish, forehead and lores black, upper lesser coverts blacker. In flight, upperwing shows mostly grey secondaries and mostly white inner primaries, underwing all white, apart from black-tipped secondaries and outer primaries. See Indian Thick-knee. Note habitat and range. **Juvenile** Upperparts slightly paler with buffy fringes, duller bands along upper lesser coverts, median and greater coverts grey-brown with narrow buffish-white fringes and faint dark subterminal markings. **VOICE** Territorial call is a repeated harsh wailing ***wee-loo***. Other calls include an occasional rising ***quip-ip-ip*** and weak ***quip*** or ***peep*** when alarmed. **HABITAT & BEHAVIOUR** Undisturbed sandy beaches and sandflats, often near mangroves. Mostly crepuscular and nocturnal. **RANGE & STATUS** Resident coasts of Andaman and Nicobar Is, Sundas, Philippines, Wallacea, New Guinea region to N Melanesia, northern Australia and New Caledonia. **SE Asia** Rare to scarce resident S Myanmar (Coco Is), southern Tenasserim, S Thailand (mainly west coast islands), north-west Peninsular Malaysia (Langkawi I), Singapore; formerly elsewhere west Peninsular Malaysia. **BREEDING** March–April. **Nest** Shallow scrape on ground. **Eggs** 1–2; creamy-white, unevenly streaked, spotted and blotched black to blackish-brown; 63.7 x 45 mm (av.).

PLUVIALIDAE: *Pluvialis* plovers

Worldwide 4 species. SE Asia 2 species. Medium-sized with rounded heads and relatively short bills. Typically forage by running for short distance, pausing, then stooping to secure prey. Feed on worms, crustaceans, molluscs, insects etc.

248 PACIFIC GOLDEN PLOVER
Pluvialis fulva Plate **32**

IDENTIFICATION 23–26 cm. Monotypic. **Adult non-breeding** Medium-sized, relatively short-billed and rather nondescript, with distinctive golden spangling on upperside, and pale buffish-grey head-sides, neck and breast with dusky-grey streaks and mottling. Has a prominent dark patch on rear ear-coverts. In flight shows rather uniform upperside (lacking the whitish rump to uppertail-coverts shown by Grey Plover), but upperwing has an indistinct narrow whitish bar (finer than on Grey) and the underwing is dull, with dusky-greyish coverts and axillaries. See Grey Plover. **Male breeding** Face, ear-coverts, foreneck, centre of breast and belly black, with a broad white band that extends from the lores over and behind the ear-coverts and down the breast-sides; flanks and vent white, with black markings. **Female breeding** Tends to have less black on underparts than male. See Grey. **Juvenile** Similar to adult non-breeding but crown, nape and upperparts much more boldly patterned with gold or yellowish-buff, head-sides, neck and breast strongly washed golden, the latter finely spotted and streaked darker. **VOICE** In flight utters a clear rapid ***chu-it***, recalling Spotted Redshank. Also a more drawn-out klu-ee and more extended ***chu-EE***. **HABITAT** Cultivated lowlands, dry areas, coastal habitats. **RANGE & STATUS** Breeds NC,NE Palearctic, NW USA (W Alaska); winters E Africa, inner Arabian Gulf, Iran, S Pakistan, Indian subcontinent, S China, Taiwan, Sundas, Philippines, Wallacea, New Guinea, Australia, New Zealand, Oceania, SW USA. **SE Asia** Uncommon to fairly common winter visitor throughout. Also recorded on passage Cambodia, Laos, W,E Tonkin. Recorded in summer Peninsular Malaysia, Singapore.

249 GREY PLOVER *Pluvialis squatarola* Plate **32**

IDENTIFICATION 27–30 cm. Monotypic. **Adult non-breeding** Only likely to be confused with Pacific Golden Plover but larger, stockier and bigger-headed with stouter bill and relatively shorter legs, upperside more uniform and distinctly greyish with whitish speckling and spangling, has whitish (rather than buffish) supercilium, mostly greyish sides of head and whiter base colour to neck and breast. In flight shows prominent white bar across upperwing, white uppertail-coverts, boldly barred tail and largely whitish underwing with diagnostic black axillaries. **Male breeding** Face, ear-coverts, foreneck, centre of breast and belly black, with very broad white band extending from lores over and behind ear-coverts and down breast-sides; has black mid-flanks and white vent. Upperside boldly spangled silvery-white. See Pacific Golden. **Female breeding** Typically somewhat browner above, with less black on sides of head, and variable amounts of white admixed with black of underparts. **Juvenile** Similar to adult non-breeding but upperside blacker with more defined yellowish-white to yellowish-buff speckles and spangling, neck and breast washed yellowish-buff and more distinctly dark-streaked. **VOICE** Flight call is a loud plaintive melancholy ***tlee-oo-ee***. Sometimes gives a shorter ***tloo-ee***. **HABITAT & BEHAVIOUR** Mud- and sandflats, beaches, coastal pools. Gregarious. **RANGE & STATUS** Breeds north Palearctic, north N America; winters along coasts of Africa, Madagascar, W,S Europe, Middle East, Indian subcontinent, S,E China, Taiwan, southern Japan, Sundas, Philippines, Wallacea, New Guinea region, W,S Australia, USA, C,S America. **SE Asia** Uncommon to fairly common coastal winter visitor and passage migrant throughout. Rare passage migrant/vagrant (inland) N Myanmar, C Thailand, Cambodia, N Laos. Also recorded in summer S Thailand, Peninsular Malaysia, Singapore.

RECURVIROSTRIDAE: Stilts & avocets

Worldwide c.11 species. SE Asia 3 species. Black and white with very long legs and long thin bills (straight in *Himantopus*, upturned in *Recurvirostra*). Feed on insects, small crustaceans, molluscs and other aquatic invertebrates and fish.

250 BLACK-WINGED STILT
Himantopus himantopus Plate **29**

IDENTIFICATION 37.5 cm. *H.h.himantopus*. **Male non-breeding** Black and white plumage, slim build, medium-long needle-like blackish bill and very long pinkish-red legs diagnostic. Mostly white with grey cap and hindneck, black ear-coverts and black upper mantle, scapulars and wings. See White-headed Stilt. **Female non-breeding** Paler grey cap than male, often not much darker than hindneck; grey-brown mantle with pale edgings. **Male breeding** Head and neck typically all white. Can have variable amounts of grey and black on head and hindneck. Rarely, shows blackish hindneck similar to White-headed, but apparently not solidly jet-black, not as broad on the 'half-collar', and not combined with a pure white crown and head-sides. **Female breeding** Like male but mantle and scapulars browner, dusky hindcrown and ear-patch. **Juvenile** Crown and hindneck brownish-grey, upperparts and wing-coverts greyish-brown with buffish fringes. **VOICE** Typically utters sharp nasal ***kek*** and yelping ***ke-yak***. Monotonous high-pitched ***kik-kik-kik-kik...*** when alarmed. **HABITAT & BEHAVIOUR** Borders of open wetlands, saltpans, coastal pools, large rivers; lowlands. Gregarious. **RANGE & STATUS** Breeds Africa, Madagascar, southern W Palearctic, C Asia, Middle East, Indian subcontinent, NW China; some northern populations winter to south, N Africa,

Arabian Peninsula, S China, rarely Philippines, N Borneo. **SE Asia** Scarce to local resident C,S(some non-breeding movements/winter immigration) Myanmar, C,S(east) Thailand (coastal), north-west Peninsular Malaysia, Cambodia, N,C Laos, Cochinchina. Scarce to fairly common winter visitor Thailand, Peninsular Malaysia, Singapore, E Tonkin. Recorded (status uncertain) SW(winter at least),W,N(winter at least),E Myanmar, N,C,S Annam. **BREEDING** April–August. Often semi-colonial. **Nest** Lined depression on ground. **Eggs** 3–5; light drab to chestnut-brown, densely blotched black; 44 x 31 mm (av.).

251 **WHITE-HEADED STILT**
Himantopus leucocephalus Plate **29**
IDENTIFICATION 35 cm. Monotypic. **Adult** Like male breeding Black-winged Stilt, but shows long black 'mane' on back of neck and an otherwise white head; slightly smaller, but wing and bill longer relative to body size. **First immature** On current knowledge, possibly not distinguishable from adult non-breeding Black-winged. **VOICE** Feeble puppy-like ***yap-yap-yap...*** (incessant when agitated). Clear mournful piping. **HABITAT** Similar to Black-winged. **RANGE & STATUS** Breeds (largely resident) Greater Sundas, Wallacea, New Guinea, Australia, New Britain, New Zealand; some populations migrate to north/north-west during non-breeding season. **SE Asia** Vagrant Cochinchina (sight record).

252 **PIED AVOCET** *Recurvirostra avosetta* Plate **29**
IDENTIFICATION 42–45 cm. Monotypic. **Male** Black and white plumage, long, narrow, strongly upturned blackish bill and long bluish-grey legs diagnostic. Predominantly white with black face, crown and upper hindneck and mostly black scapulars, median and lesser coverts and primaries. **Female** Bill shorter and more strongly upturned. **Juvenile** Dark parts of plumage obscured with dull brown, white of mantle and scapulars mottled pale greyish-brown. **VOICE** Clear melodious liquid ***kluit***, often repeated. Harsher, more emphatic ***kloo-eet*** and shrill ***krrree-yu*** when alarmed. **HABITAT & BEHAVIOUR** Coastal pools, mud- and sandflats, lakes, large rivers; lowlands. Usually feeds in shallow water. **RANGE & STATUS** Breeds Africa, southern W,C Palearctic, S Siberia, NW,N China, rarely Pakistan; mostly winters Africa, W,S Europe, Middle East, Indian subcontinent, S China. **SE Asia** Local winter visitor C Myanmar. Vagrant N,S Myanmar, Thailand (except SE), E Tonkin, Cochinchina. Also recorded in summer C Thailand.

HAEMATOPODIDAE: Oystercatchers

Worldwide 11 species. SE Asia 1 species. Medium-sized with black or black and white plumage and long, thick, usually red bills, adapted for opening shells of bivalve molluscs.

253 **EURASIAN OYSTERCATCHER**
Haematopus ostralegus Plate **30**
IDENTIFICATION 40–46 cm. *H.o.osculans*. **Adult non-breeding** Told by robust shape, shortish pink legs, long thick orange-red bill with duller tip, and black plumage with white back to uppertail-coverts and lower breast to vent. May show a little white on lower throat. In flight shows broad white band across upperside of secondaries, greater coverts and inner primaries. **Adult breeding** Throat all black, bill uniform orange-red. **First winter** Like adult non-breeding but black of plumage tinged brown, bill much duller and somewhat narrower. **VOICE** Usual call is a loud shrill piping ***kleep*** or ***ke-beep*** (often repeated), particularly in flight. Also utters a soft ***weep*** and repeated sharp ***pik*** when alarmed. **HABITAT** Mud- and sandflats, rocky coasts. **RANGE & STATUS** Breeds Europe, W(east),C Palearctic, N Iran, SE Siberia, Ussuriland, NE China, N,S Korea, Kamchatka, Kuril Is; mostly winters along coasts of northern Africa, southern Europe, Middle East, S Pakistan, W,S India, Sri Lanka, S,E China. **SE Asia** Vagrant SW,S Myanmar, Peninsular Malaysia, E Tonkin.

DROMADIDAE: Crab-plover

Monotypic family. Mostly feeds on crabs but also mudskippers and crustaceans.

254 **CRAB-PLOVER** *Dromas ardeola* Plate **30**
IDENTIFICATION 38–41 cm. Monotypic. **Adult** Very distinctive. Medium-sized to largish, bill very thick, pointed and blackish, legs long and bluish-grey, plumage predominantly white with contrasting black mantle and scapulars and mostly black upperside of primaries, primary coverts, secondaries and outer greater coverts. Sometimes has blackish speckles on hindcrown and nape. **Juvenile** Hindcrown and nape more heavily speckled blackish, upperparts mostly greyish with paler brownish-grey scapulars and wing-coverts, dark of upperwing greyer. **VOICE** Flight call is a nasal yappy ***kirruc***. Also gives a repeated barking ***ka*** or ***ka-how*** and ***kwerk-kwerk-kwerk-kwerk...*** A sharp whistled ***kew-ki-ki*** and ***ki-tewk*** also recorded from breeding grounds. **HABITAT & BEHAVIOUR** Undisturbed sandy beaches, dunes, mud- and sandflats. Mainly a crepuscular and nocturnal feeder. Likely to be found in family parties or small flocks. **RANGE & STATUS** Breeds Eritrea, Somalia, Oman, Kuwait, Iran, N Sri Lanka, formerly Iraq. Non-breeding visitor along coasts of E Africa, Madagascar, Indian Ocean Is, Arabian Peninsula, Iran, Pakistan, W,S India, Andaman and Nicobar Is. **SE Asia** Rare non-breeding visitor S Thailand. Vagrant S Myanmar (sight record Let Kok Kon Beach), west Peninsular Malaysia.

IBIDORHYNCHIDAE: Ibisbill

Monotypic family. Feeds on aquatic invertebrates, including large waterbeetles.

255 **IBISBILL** *Ibidorhyncha struthersii* Plate **30**
IDENTIFICATION 38–41 cm. Monotypic. **Adult non-breeding** Unmistakable. Fairly large, with mainly grey head, neck and upperparts, blackish face, black crown, narrow white and broad black breast-bands and strongly down-curved reddish bill. Black of face partly obscured by white-tipped feathers. **Adult breeding** Face all black. **Juvenile** Upperparts browner with extensive warm buff fringes, face whitish to dark brown with numerous white feather-tips, lacks white breast-band, lower breast-band dark brown. **VOICE** Repeated ringing ***klew-klew*** and loud rapid ***tee-tee-tee-tee***. **HABITAT & BEHAVIOUR** Rocky rivers, stony riverbeds; 800–3,355 m. Usually found in pairs or family groups. Often well hidden amongst stones and boulders. **RANGE & STATUS** Resident (subject to altitudinal movements) C Asia, NW,N,NE India, Nepal, Bhutan, Tibet, W,NW,N,NC China. **SE Asia** Scarce to uncommon winter visitor N Myanmar.

VANELLIDAE: Lapwings & allies

Worldwide 24 species. SE Asia 5 species. Small to medium-sized with rounded heads and relatively short bills. Typically forage by running for short distance, pausing, then stooping to secure prey. Feed on beetles, worms, grasshoppers and other insects.

256 NORTHERN LAPWING
Vanellus vanellus Plate **29**

IDENTIFICATION 28–31 cm. Monotypic. **Adult non-breeding** Unmistakable. Appears black and white at a distance with distinctive long, thin, swept-back crest. Crown and crest black, sides of head mixed buffish and whitish with black facial patch and line through ear-coverts, upperparts and wing-coverts mostly dark glossy green with some buffish fringes, underparts white with broad blackish breast-band and orange-rufous undertail-coverts. In flight shows distinctive, very broad wings with black flight-feathers, pale band across tips of outer primaries and sharply contrasting white underwing-coverts. **Male breeding** Buff of sides of head replaced by white, lores and throat black (connecting with breast-band). **Female breeding** Like breeding male but lores and throat marked with white. **Juvenile** Similar to adult non-breeding but upperparts and wing-coverts more prominently fringed warm buff, crest shorter. **VOICE** Typical calls are a loud shrill ***cheew*** and more plaintive ***cheew-ip*** or ***wee-ip***. **HABITAT** Open country, cultivation, marshes; up to 500 m. **RANGE & STATUS** Breeds W,C Palearctic, S Siberia, Mongolia, northern China, S Ussuriland, N Korea; more northerly populations winter to south, N Africa, south W Palearctic, Middle East, Pakistan, NW India, Nepal, southern China, Taiwan, N,S Korea, Japan. **SE Asia** Rare to locally fairly common winter visitor W(north),N,C Myanmar, NW,NE,C,SE Thailand, N Laos, E Tonkin, N Annam.

257 RIVER LAPWING
Vanellus duvaucelii Plate **31**

IDENTIFICATION 29–32 cm. Monotypic. **Adult** Combination of black crown and long pointed crest, black face, black strip down centre of throat to uppermost breast and whitish remainder of head and neck diagnostic. Bill, legs and feet black, underparts white with sandy greyish-brown breast-band and small black belly-patch; black spur on bend of wing (often hidden). In flight, upperwing shows broad white band and narrow black patch on outer median and lesser coverts, tail-tip black. Note habitat. **Juvenile** Black of head partly obscured by brownish feather-tips, upperparts and wing-coverts sandy-brown with buff fringes and slightly darker subterminal markings. **VOICE** Typical call is a sharp high-pitched ***tip-tip*** or ***did did did...***, sometimes ending with ***to-weet*** or ***do-weet***. **HABITAT & BEHAVIOUR** Large rivers and surrounds; up to 600 m. Breeding display (on ground) includes stooping, spinning and upstretching. **RANGE & STATUS** Resident (subject to some movements) NW,N,NE India, Nepal, Bhutan, Bangladesh, SW,S China. **SE Asia** Scarce to locally common resident Myanmar, Thailand (except SE), Indochina (except N,S Annam). Former resident (current status unknown) Cochinchina. **BREEDING** India: March–June. **Nest** Shallow scrape on ground. **Eggs** 3–4; olive- to yellowish-stone colour, blotched and spotted brown and black; 41.1 x 29.4 mm (av.).

258 YELLOW-WATTLED LAPWING
Vanellus malabaricus Plate **31**

IDENTIFICATION 26–28 cm. Monotypic. **Adult non-breeding** Blackish crown, long yellow wattle in front of eye, long white post-ocular band and uniform sandy greyish-brown remainder of head, neck and upper breast diagnostic. Crown mixed with brown, has narrow blackish breast-band and yellow legs and feet. In flight recalls Red-wattled Lapwing but has more prominent white bar on upperwing and more extensively white underwing. **Adult breeding** Crown all black. **Juvenile** Crown brown with buffish speckling, upperparts and wing-coverts with buff fringes and dark subterminal bars, chin and throat mostly whitish, wattles smaller and duller, no blackish breast-band. **VOICE** Strident ***chee-eet*** or ***tchee-it*** and hard sharp ***tit-tit-tit...*** or ***whit-whit-whit...*** when alarmed. **HABITAT** Dry grassland and open areas, agriculture, wetland margins; lowlands. **RANGE & STATUS** Breeding visitor S Pakistan. Resident (subject to some movements) India, Sri Lanka. Non-breeding visitor Nepal, Bangladesh. **SE Asia** Vagrant S Myanmar, Peninsular Malaysia.

259 GREY-HEADED LAPWING
Vanellus cinereus Plate **31**

IDENTIFICATION 34–37 cm. Monotypic. **Adult non-breeding** Relatively large size, mostly plain brownish-grey head, neck and upper breast, rather long yellowish bill with black tip, and yellowish legs and feet diagnostic. Chin and centre of throat whitish, has broad, partly obscured, blackish breast-band. In flight shows diagnostic white greater coverts and secondaries (above and below). See Pheasant-tailed Jacana (in flight). **Adult breeding** Head, neck and upper breast grey, with neat broad blackish breast-band. **Juvenile** Head, neck and breast brownish, breast-band vague or absent, upperparts and wing-coverts neatly fringed buffish. **VOICE** Plaintive ***chee-it***, often repeated and, when alarmed, a rasping ***cha-ha-eet*** and sharp ***pink***. **HABITAT & BEHAVIOUR** Marshes, wet rice paddies, cultivation; up to 1,250 m. Gregarious in main wintering areas. **RANGE & STATUS** Breeds NE China, Japan; mostly winters Nepal, NE India, Bangladesh, S China. **SE Asia** Scarce to locally common winter visitor (except SE Thailand, Singapore, W Tonkin, S Annam). Also recorded on passage Cambodia, E Tonkin. Vagrant Singapore.

260 RED-WATTLED LAPWING
Vanellus indicus Plate **31**

IDENTIFICATION 31.5–35 cm. *V.i.atronuchalis*: **Adult** Black hood and upper breast, white patch on ear-coverts, red facial skin and black-tipped red bill diagnostic. Has cold sandy greyish-brown upperside with narrow white band across upper mantle, whitish remainder of underparts and long yellow legs. In flight shows prominent white band along greater upperwing-coverts, blackish flight-feathers, white band across rump and uppertail-coverts and mostly black tail with white corners. See Yellow-wattled and River Lapwings. **Juvenile** Dark of head, neck and breast duller, throat whitish, facial skin duller and reduced, bill, legs and feet duller. **VOICE** Typical call is a loud rapid ***did-ee-doo-it*** (first note often repeated). During display-flight utters frenzied series of typical calls, interspersed with ***did-did-did...*** or ***kab-kab-kab...*** Sharp incessant ***trint*** and high-pitched ***pit*** notes when alarmed. **HABITAT & BEHAVIOUR** Margins of lakes and large rivers, marshes, agriculture, wasteland; up to 1,525 m. Usually found in pairs or family parties. Display-flight involves short, dipping wingbeats, downward swoops and acrobatic tumbling dives. **RANGE & STATUS** Resident (subject to some movements) SE Turkey, Iraq, E Arabian Peninsula, Iran, Afghanistan, SE Turkmenistan, Indian subcontinent, SW China. **SE Asia** Uncommon to common resident throughout (may no longer breed Singapore). **BREEDING** January–June. **Nest** Natural depression or scrape, sometimes lined with bits of mud, stones, dung; on dry earth or stony ground. **Eggs** 4; grey-brown to drab, blotched blackish; 41.5 x 29.8 mm (av.).

CHARADRIIDAE: *Charadrius* plovers & allies

Worldwide c.38 species. SE Asia 8 species. Small to medium-sized with rounded heads and relatively short bills. Typically forage by running for short distance, pausing, then stooping to secure prey. Feed on beetles, worms, grasshoppers and other insects.

261 COMMON RINGED PLOVER
Charadrius hiaticula Plate **32**

IDENTIFICATION 18–20 cm. *C.h.tundrae*: **Adult non-breeding** Similar to Little Ringed Plover but slightly larger and more robust, has broader and less even dark breast-band, dull orange base to lower mandible and orange legs and feet. In flight shows prominent white bar across upperwing. Note voice. See Long-billed Plover. **Male breeding** From Little Ringed by orange bill with black tip, orange legs and feet and broader, less even black breast-band, no obvious yellow eyering or white band across midcrown. See Long-billed. **Female breeding** Like male but ear-coverts and breast-band brownish-black. **Juvenile** Like adult non-breeding but breast-band narrower, upperparts and wing-coverts with buffish fringes and narrow dark subterminal markings, bill all dark, legs and feet duller and more yellowish. Best separated from Little Ringed by size, proportions, colder-toned upperside, whiter forehead and supercilium, lack of obvious eyering, often more orange-tinged legs and feet, and white wing-bar (in flight). **VOICE** Mellow rising ***too-lee***, usually given in flight. Also a short ***wip*** and, when alarmed, soft low ***too-weep***. **HABITAT & BEHAVIOUR** Mud- and sandflats, saltpans, coastal pools, large rivers; up to 450 m. Feeding action and wingbeats slightly slower than Little Ringed. **RANGE & STATUS** Breeds N Palearctic, northeast N America, Greenland; mostly winters along coasts of Africa, Madagascar, W,S Europe, Middle East, S Pakistan, rarely India, Sri Lanka, S,E China, Japan. **SE Asia** Vagrant SW,N,C Myanmar, NW,C Thailand, Peninsular Malaysia, Singapore, E Tonkin.

262 LONG-BILLED PLOVER

Charadrius placidus Plate **32**

IDENTIFICATION 19–21 cm. Monotypic. **Adult non-breeding** Recalls Little Ringed Plover but larger and even more slender and attenuated in shape, has longer bill, broader dark band across forecrown, broader buff supercilium and narrower, more even breast-band. Bill blackish with some dull yellow at base of lower mandible, legs and feet pinkish-yellow. In flight, upperwing shows narrow white central bar, narrow white trailing edge to secondaries, greyish area on outer secondaries and inner primaries (forming faint panel) and more contrasting blackish primary coverts. Note voice. See Common Ringed Plover. **Adult breeding** Has white forehead and supercilium, broad black band across forecrown and narrow even black breast-band. Lacks black lores and ear-coverts of Little Ringed and has less prominent eyering. See Common Ringed. **Juvenile** Initially differs from adult by neat warm buff fringes to upperparts and wing-coverts, no dark band across midcrown, more buffish supercilium and greyish-brown breast-band. Likely to resemble adult non-breeding once it reaches the region. **VOICE** Clear rising ***piwee*** or ***piwii-piwii-piwii...*** and musical ***tudulu***. **HABITAT** Larger rivers, dry fields, sometimes mud- and sandflats, beaches; up to 1,830 m. **RANGE & STATUS** Breeds W,NC,NE China, Ussuriland, N Korea, Japan; mostly winters Nepal, NE India, southern China, Taiwan, N,S Korea. **SE Asia** Scarce to uncommon winter visitor W(north),N Myanmar. Rare winter visitor C Myanmar, NW,C Thailand, N,C Laos, E Tonkin, C Annam. Vagrant Peninsular Malaysia.

263 LITTLE RINGED PLOVER

Charadrius dubius Plate **32**

IDENTIFICATION 14–17 cm. *C.d.jerdoni*: **Adult non-breeding** From other small plovers by combination of rather dainty proportions, small-headed appearance, rather attenuated rear end, slender dark bill, pinkish to yellowish legs and feet, white collar, complete greyish-brown breast-band (may be slightly broken in centre) and uniform upperwing (in flight). Rest of head and upperside greyish-brown, apart from buffish-white forehead and indistinct supercilium; has narrow pale yellowish eyering. Note voice. See Common Ringed, Long-billed and Kentish Plovers. **Male breeding** Lores, ear-coverts and breast-band black, forehead and supercilium white; prominent black band across forecrown, backed by distinctive narrow white band and pronounced broad yellow eyering. See Common Ringed and Long-billed. **Female breeding** Eyering slightly narrower, ear-coverts and breast-band tinged brownish. **Juvenile** Like adult non-breeding but upperside somewhat sandier-brown with warm buff fringes and indistinct darker subterminal markings, breast-band browner and restricted more to lateral patches, supercilium more buffish. **Other subspecies in SE Asia** *C.d.curonicus* (wintering subspecies). **VOICE** Territorial call is a repeated harsh ***cree-ah***, often given during display-flight. Typical call is a plaintive ***pee-oo*** (first syllable stressed), particularly in flight. Also utters shorter ***peeu*** and rapid insistent ***pip-pip-pip-pip...*** when alarmed. **HABITAT & BEHAVIOUR** Large rivers, lakes, marshes, rice paddies, coastal pools; up to 1,450 m. Has gliding display-flight with wings raised in shallow V. Courtship display on ground involves puffing-out of breast, tail-fanning and prancing. **RANGE & STATUS** Breeds N Africa, Palearctic, Middle East, Indian subcontinent, SW,W,NW,N,NE China, Taiwan, N,S Korea, Japan, Philippines, New Guinea, Bismarck Archipelago; most northern populations winter N and equatorial Africa, Middle East, Indian subcontinent, southern China, Taiwan, S Japan, Sundas, Philippines, Wallacea, rarely New Guinea, N Australia. **SE Asia** Scarce to locally fairly common resident Myanmar, NW,NE Thailand, Cambodia, Laos, N Annam, Cochinchina. Common winter visitor (except SW,E Myanmar). Also recorded on passage E Tonkin. **BREEDING** March–April. **Nest** Shallow depression on ground. **Eggs** 4; buffish to olive-grey, spotted, scrawled and squiggled dark brown, over purplish undermarkings; 27.5 x 20.7 mm (av.).

264 KENTISH PLOVER

Charadrius alexandrinus Plate **32**

IDENTIFICATION 15–17.5 cm. *C.a.alexandrinus*: **Adult non-breeding** From other small plovers by combination of plain upperside, white nuchal collar, well-defined and rather narrow dark lateral breast-patches, blackish bill, and rather long bluish-grey to greyish (sometimes distinctly olive or pinkish) legs. In flight shows obvious white bar across upperwing and white outertail-feathers. Worn, faded individuals can be very pale and greyish above ('bleached'). See other small plovers, particularly Malaysian and Lesser Sand. **Male breeding** Distinctive. Has white forehead and short supercilium, black patch on midcrown, strongly rufous-washed remainder of crown and nape, broad blackish eyestripe and well-defined narrow black lateral breast-patches. See Malaysian Plover. **Female breeding** Usually has little or no rufous on crown and nape, may lack black on head and breast markings shown by male (more like adult non-breeding). **Juvenile** Like adult non-breeding but somewhat paler-headed, forehead and supercilium washed buff, upperparts and wing-coverts neatly fringed buff, lateral breast-patches somewhat paler and more diffuse. **VOICE** In flight may utter soft ***pit*** or ***pi*** notes and hard trilled ***prrr*** or ***prrrtut***, harsher than similar call of Lesser Sand. Plaintive ***too-eet*** or ***pweep*** when alarmed. During display-flight, utters repeated sharp rattling ***tjekke-tjekke-tjekke...*** **HABITAT & BEHAVIOUR** Beaches, sand- and mudflats, coastal pools, large rivers, dry lake margins; up to 450 m. In breeding season male performs stiff-winged display-flights with body tilting from side to side. Pair displays include wing- and tail-spreading. **RANGE & STATUS** Breeds N Africa, southern W,C Palearctic, Middle East, S Pakistan, N and coastal India, Sri Lanka, E,NW,N China, Taiwan, Ussuriland, N,S Korea, Japan; more northerly populations winter (mainly along coasts) N and equatorial Africa, Arabian Peninsula, Indian subcontinent, southern China, Taiwan, S Korea, Greater Sundas, Philippines. **SE Asia** Local coastal resident S Myanmar, C,S Annam; probably N Annam. Uncommon to common (mainly coastal) winter visitor (except E Myanmar, W Tonkin). Also recorded on passage S Myanmar, S Thailand, Peninsular Malaysia, Singapore, Cambodia, E Tonkin. **BREEDING** February–June. **Nest** Scrape on ground or saltpan bund (may be lined with fragments). **Eggs** 2–4; sandy-brown (sometimes olive-tinged), spotted, blotched and/or streaked and scrawled blackish-brown; 32.5 x 23.5 mm (av.; *alexandrinus*).

264A WHITE-FACED PLOVER

Charadrius (a.) dealbatus Not illustrated

IDENTIFICATION c.16–18 cm. **Adult non-breeding** Similar to Kentish Plover, but somewhat larger, larger-headed, heavier-breasted and flatter-backed, bill proportionately longer and heavier with pale-based lower mandible; distinctly paler sandy-brown above, lateral breast-patches shorter and typically somewhat narrower, legs longer and distinctly pinkish-toned. Generally shows more white on the forehead, and more white on face overall (loral area being largely white on some [males?], with faint dusky feathering below eye). In flight, has three white bars on upperwing: narrow trailing edge to secondaries, conspicuous central bar (broader and more prominent than on Kentish), and third formed by white tips to outer median coverts. Resembles Malaysian Plover in its pale upperparts, pink-toned legs and conspicuous wing-bar, but considerably larger, crown plainer, mantle and scapulars relatively uniform and unmarked (or feathers with just a dark shaft-streak) rather than variegated in appearance; lacks black collar below white collar shown by male Malaysian all year, and lacks rufous tones characteristic of female Malaysian. Lateral breast-patches shorter and narrower than on Kentish and Malaysian. **Male breeding** Acquires unique head pattern, with entirely white forehead, lores and orbital area (hint of brownish post-ocular line), solid black frontal bar above white forehead (wider than in Kentish), crown as bright or brighter orange than Kentish, extending to upper nape and hooked around to partially encircle rear ear-coverts. Black lateral breast-patches invariably smaller, narrower and less extensive than on Kentish (may be almost absent, with a few blackish feathers immediately in front of carpal bend). Legs may darken to pale lead-grey, but still slightly paler than Kentish. **Female breeding** More similar to non-breeding than male, though crown, lores, ear-coverts strongly rufous- to orange-toned. Lateral breast-patches generally slightly more extensive than on male, and typically paler than on Kentish (similar in colour to mantle), almost always washed rufous to orange, and therefore brighter than on Kentish. Head pattern most closely resembles Malaysian, with most showing dull rufous-brown line across lores and behind the eye, (conspicuous but rather blurred eyestripe). Rufous-brown tone extends over entire crown, becoming duller towards centre and brighter around rear. Note conspicuous pinkish legs and plain-looking upperparts. **HABITAT** Found in similar coastal situations to Kentish. **RANGE & STATUS** Breeds S,SE China; at least some winter to south, Sumatra (Jambi Province). **SE Asia** Scarce winter visitor W(coastal),C,S Thailand, Peninsular Malaysia, Singapore, C,S Annam. **NOTE** This highly distinctive plover has only recently been properly described, and may prove to be a distinct species (Bakewell & Kennerley 2008, Kennerley *et al.* 2008). The taxon *dealbatus* was previously considered to be a race of Kentish Plover.

265 **MALAYSIAN PLOVER**

Charadrius peronii Plate **32**

IDENTIFICATION 14–16 cm. Monotypic. **Male** Similar to breeding male Kentish Plover but slightly smaller and shorter-billed, upperparts and wing-coverts with prominent pale fringes (appear scaly or mottled), narrower black lateral breast-patches which extend in complete band below white nuchal collar; legs and feet often tinged yellowish or pinkish. **Female** Lacks black markings. From Kentish by rufous-washed ear-coverts and lateral breast-patches and scaly upperparts; crown always washed rufous. **Juvenile** Slightly duller than female. **VOICE** Soft ***whit*** or ***twik***, recalling Kentish. **HABITAT & BEHAVIOUR** Undisturbed sandy, coralline and shelly beaches, sometimes nearby mudflats. Usually encountered in pairs. **RANGE & STATUS** Resident Sundas (except Java), Philippines, Sulawesi, Sula Is. **SE Asia** Rare to scarce coastal resident Thailand, Peninsular Malaysia, Singapore, Cambodia, Cochinchina. **BREEDING** January–August. **Nest** Scrape on ground. **Eggs** 2–3; creamy to pale stone-coloured, profusely speckled and squiggled black and brown over lavender-grey undermarkings; 30.6 x 22.6 & 31.5 x 23.6 mm (means).

266 **LESSER SAND-PLOVER**

Charadrius mongolus Plate **31**

IDENTIFICATION 19–21 cm. *C.m.schaeferi* (Thailand, Peninsular Malaysia; ? parts of Myanmar and Indochina) **Adult non-breeding** From other small plovers (except Greater Sand) by broad lateral breast-patches and lack of white nuchal collar. Upperside sandy greyish-brown with whitish forehead and supercilium. Best separated from Greater by combination of following subtle features: smaller size, neater proportions, rounder head, shorter and blunter-tipped bill (length roughly equal to distance from base of bill to rear of eye), tibia obviously shorter than tarsus, dark grey to greenish-grey legs and feet, toes only project slightly beyond tail-tip in flight. Bill length varies from longest in this subspecies to shortest in *C.m.mongolus*. **Male breeding** Distinctive, with black forehead, lores and ear-coverts and deep orange-rufous sides of neck and broad breast-band. May show very small whitish markings on sides of forehead. *C.m.atrifrons* (recorded Myanmar, Peninsular Malaysia; ? Thailand) often has small white patch on side of forehead; *mongolus* (recorded Indochina; ? Thailand) has white forehead, bisected by vertical black line and narrow blackish upper border to breast-band. See Greater. **Female breeding** Much duller, with less orange-rufous on neck and breast and browner forehead and mask. **Juvenile** Like adult non-breeding but upperparts and wing-coverts fringed buffish, supercilium washed buff and often less pronounced, lateral breast-patches mixed with buffish. **VOICE** In flight, rather short sharp hard ***kruit*** or ***drrit*** notes; also a hard ***chitik*** and ***chi-chi-chi***. **HABITAT** Mud- and sandflats, coastal pools, rarely large rivers and other inland wetlands; up to 1,065 m. **RANGE & STATUS** Breeds C Asia, S,E Siberia, NE Palearctic, NW,N India, Tibet, NW,N China; winters along coasts of S,E Africa, Madagascar, Arabian Peninsula, Iran, Indian subcontinent, S China, Taiwan, S Japan, Sundas, Philippines, Wallacea, New Guinea region, Australia. **SE Asia** Common coastal winter visitor and passage migrant throughout. Scarce inland winter visitor and passage migrant W,N,C Myanmar. Scarce inland passage migrant NE,C Thailand, N,C Laos. Also recorded in summer C,S Thailand, Peninsular Malaysia, Singapore.

267 **GREATER SAND-PLOVER**

Charadrius leschenaultii Plate **31**

IDENTIFICATION 22–25 cm. *C.l.leschenaultii*: **Adult non-breeding** Very similar to Lesser Sand-plover but differs by combination of following subtle features: larger size, longer appearance, squarer head, longer bill with more tapered tip (length greater than distance from base of bill to rear of eye), longer tibia (may appear almost as long as tarsus), typically somewhat paler legs and feet (usually tinged greenish or yellowish), toes project distinctly beyond tail-tip in flight. **Male breeding** Plumage like Lesser (*C.m.mongolus*) but has narrower orange-rufous breast-band, with no black upper border. **Female breeding** Like adult non-breeding but with narrow orange-brown breast-band. **Juvenile** Differs in same way as Lesser. **VOICE** A trilled ***prrrirt***, ***kyrrrr trr*** and ***trrri*** etc., softer and longer than similar calls of Lesser, recalling Ruddy Turnstone. **HABITAT & BEHAVIOUR** Mud- and sandflats, beaches, saltpans, coastal pools, rarely large rivers on passage. Often associates with Lesser, but generally less common. **RANGE & STATUS** Breeds Turkey, Jordan, C Asia, S Siberia, Mongolia, NW,N China; winters along coasts of S,E,NE Africa, Madagascar, Middle East, Indian subcontinent, S China, Taiwan, S Japan (Nansei Is), Sundas, Philippines, Wallacea, New Guinea, N Melanesia, Australia, New Zealand. **SE Asia** Uncommon to locally common coastal winter visitor and passage migrant throughout. Rare to scarce inland passage migrant C Myanmar, Cambodia, N,C Laos. Also recorded in summer C Thailand, Peninsular Malaysia, Singapore.

268 **ORIENTAL PLOVER**

Charadrius veredus Plate **31**

IDENTIFICATION 22–25 cm. Monotypic. **Adult non-breeding** Recalls sand-plovers but larger and slimmer-looking with longer neck, legs and wings and slenderer bill, typically also longer and more pronounced supercilium and buffish-brown upper breast. When fresh, upperparts and wing-coverts have narrow rufous to warm buff fringes. Legs and feet yellow to orange (tinged pinkish to greenish). In flight, shows distinctive, very long, all-dark wings with paler upperwing-coverts. See Pacific Golden Plover. **Male breeding** Unmistakable. Head and neck largely whitish with greyish-brown cap, rufous-chestnut breast-band with broad black lower border. **Female breeding** Like adult non-breeding but upper breast with rufescent wash. **Juvenile**

Like fresh adult non-breeding but has more pronounced and paler buff fringing on upperparts and wing-coverts. **VOICE** In flight may give sharp whistled ***chip-chip-chip***, short piping ***klink*** and various trilled notes. **HABITAT & BEHAVIOUR** Dry mud near fresh or brackish water, short grassy areas, sometimes saltpans, mud- and sandflats; lowlands. Runs very fast, flight fast, high and rather erratic. **RANGE & STATUS** Breeds S Siberia, Mongolia, east N China; winters N Australia. Occurs widely in intervening areas. **SE Asia** Rare passage migrant W(coastal),C,SE Thailand, Cambodia, E Tonkin. Vagrant S Thailand, Peninsular Malaysia, Singapore, C,S Annam.

JACANIDAE: Jacanas

Worldwide 8 species. SE Asia 2 species. Resemble gallinules but have remarkably long toes and claws, enabling them to walk on floating vegetation. Polyandrous, females often mating with more than one male, which tend eggs and young. Feed on seeds, roots and aquatic invertebrates.

269 PHEASANT-TAILED JACANA
Hydrophasianus chirurgus Plate **30**

IDENTIFICATION 29–31.5 cm (breeding adult tail up to 25 cm more). Monotypic. **Adult breeding** Unmistakable. Mostly glossy blackish-brown with white head and foreneck, shiny yellow-buff hindneck (bordered black), very long, pointed blackish tail and mostly white wings. Outer primaries and primary tips largely blackish, the latter with strange pendant-like extensions, upperparts often slightly paler and faintly glossed purple. **Adult non-breeding** Crown, centre of hindneck and upperparts drab brown with whitish bars on crown and hindneck, white to pale buff supercilium extending in broad buff band down neck-side, underparts all white with black eyestripe extending down neck-side and broadly across breast, median and lesser coverts largely greyish-brown with some blackish bars, tail shorter. In flight mostly white wings distinctive. See juvenile Bronze-winged Jacana and Grey-headed Lapwing (in flight). **Juvenile** Resembles adult non-breeding but crown mostly rather pale rufous-chestnut with blackish-brown feather-centres, no obvious supercilium, rufous to rufous-buff fringes to upperparts, neck washed pale rufous-chestnut (less distinctly buff on sides), median and lesser coverts more rufescent, black on breast ill-defined. **VOICE** During breeding season gives a series of deep rhythmic ***t'you***, ***me-e-ou*** or ***me-onp*** notes, and nasal mewing ***jaew*** or ***tewn*** notes. Other calls include a sharp, high-pitched ***tic-tic-tic...***, nasal ***brrr-brrp*** and high-pitched, whining ***eeeaaar*** when alarmed. **HABITAT & BEHAVIOUR** Well-vegetated freshwater marshes, swamps, lakes and pools, sometimes apparently less suitable habitats on passage, including mangroves and large rivers; up to 1,000 m. Gregarious outside breeding season. **RANGE & STATUS** Resident (subject to some seasonal movements) Indian subcontinent, Philippines; ?Borneo. Breeds C,E and southern China; more northerly populations winter to south, Sumatra, Java, ?Borneo, rarely Bali, NW Australia. **SE Asia** Uncommon to locally common resident (subject to some movements) Myanmar (except W), C,NE Thailand, N Laos, Cambodia. Scarce to locally fairly common winter visitor Thailand, Peninsular Malaysia, Cambodia, E Tonkin (formerly bred), N,S Annam, Cochinchina. Scarce to uncommon passage migrant C Thailand, Peninsular Malaysia, N Laos, E Tonkin, C Annam. Vagrant Singapore. Recorded (status uncertain) C,S Laos. **BREEDING** May–September. **Nest** Flimsy pad or heap, either floating or on islet (sometimes none, with eggs on floating vegetation). **Eggs** 4; very glossy rich deep bronze with rufous to olive tinge (becoming bleached); 37.4 x 27.6 mm (av.).

270 BRONZE-WINGED JACANA
Metopidius indicus Plate **30**

IDENTIFICATION 26.5–30.5 cm. Monotypic. **Adult** Distinguished by black head, neck and underparts (glossed green), broad white supercilium, bronze-olive lower mantle, scapulars and wing-coverts and chestnut-maroon back to uppertail and vent; vivid purple to green gloss on lower hindneck. **Juvenile** Crown, nape and loral eyestripe dull chestnut, neck-sides, lower foreneck and upper breast deep rufous-buff, rest of underparts white with dark-barred thighs, uppertail-coverts barred blackish and pale chestnut, tail black and white with buff and/or bronze-green tinge. See Pheasant-tailed Jacana. **VOICE** Short low harsh guttural grunts and a wheezy piping ***seek-seek-seek...*** **HABITAT & BEHAVIOUR** Well-vegetated freshwater marshes, swamps, lakes, pools; up to 1,000 m. **RANGE & STATUS** Resident India, Nepal, Bangladesh, SW China, S Sumatra, Java. **SE Asia** Uncommon to locally common resident Myanmar (except W), Thailand (except SE), Indochina (except W,E Tonkin, N,C Annam); formerly north-west Peninsular Malaysia. **BREEDING** All year. **Nest** Floating pad with slight depression (sometimes none, with eggs on floating vegetation). **Eggs** 4; very glossy buff to deep red-brown with numerous long black lines and intricate scrawls; 36.4 x 25.1 mm (av.).

ROSTRATULIDAE: Painted-snipes

Worldwide 2 species. SE Asia 1 species. Superficially similar to snipe but bills slightly down-turned at tip, wings much broader and rounder. Sexually dimorphic and polyandrous. Relatively secretive and crepuscular.

271 GREATER PAINTED-SNIPE
Rostratula benghalensis Plate **30**

IDENTIFICATION 23–26 cm. *R.b.benghalensis*: **Female** Recalls snipes but much darker and more uniform with sharply contrasting white spectacles, white belly and white band around shoulder; bill droops slightly at tip. Crown blackish with narrow buffish median stripe, rest of head, neck and upper breast dark maroon-chestnut, with narrow rich buff lines bordering mantle, blackish breast-band and broad buff tail-bands. In flight appears relatively uniform with pale-barred flight-feathers and sharply contrasting whitish belly and mostly white underwing-coverts. **Male** Plumage somewhat paler and much more variegated, spectacles buffish, head-sides, neck and upper breast mostly greyish-brown, with large rich buffish markings on wing-coverts. **Juvenile** Resembles male but wing-coverts greyer with smaller buff markings. **VOICE** Female territorial call is a series of 20–80 or so short ***kook***, ***oook*** or ***koh*** notes (like sound made by blowing into empty bottle), usually given at dusk and night-time. Also gives a single ***kook*** during roding display. Usually silent when flushed but may give a loud explosive ***kek***. **HABITAT & BEHAVIOUR** Marshy and swampy areas, wet rice paddies (particularly when overgrown); up to 1,525 m. Quite secretive and mainly crepuscular. Feeds by probing in soft mud and sweeping bill from side to side in shallow water. Spends long periods standing motionless. When flushed, flies short distance, with rather slow and erratic wingbeats and legs trailing. Female has 'roding' display-flight, low over ground; both sexes perform spread-wing displays. **RANGE & STATUS** Resident (subject to some movements) Africa, Madagascar, Indian subcontinent, C,E and southern China, Taiwan, southern Japan, Greater Sundas, W Lesser Sundas, Philippines, Australia. **SE Asia** Uncommon resident (except N Annam). **BREEDING** January–September. **Nest** Slightly depressed pad amongst marsh vegetation or on bank, sometimes raised mound amongst vegetation in shallow water. **Eggs** 3–4; pale buff with heavy blackish-brown spots and blotches; 35.9 x 25.5 mm (av.).

SCOLOPACIDAE: SCOLOPACINAE: Woodcocks

Worldwide 8 species. SE Asia 1 species. Medium-sized and thickset with relatively shorter and thicker bill than snipes, and typically in more wooded habitats. Feed on earthworms, insect larvae and other invertebrates.

272 EURASIAN WOODCOCK
Scolopax rusticola Plate **33**

IDENTIFICATION 33–35 cm. Monotypic. **Adult** Resembles some snipes but larger, hindcrown broadly barred black, upperparts more softly patterned with rufescent-brown and buff (no lengthwise pale stripes), underparts entirely barred brown-and-buff. In flight appears heavy and robust with broad, rounded wings. See Wood Snipe. **Juvenile** Similar to adult. **VOICE** Usually silent when flushed but occasionally gives harsh snipe-like ***scaap*** notes. During roding display, utters weak, high-pitched ***chissick*** or ***pissipp***, interspersed with low, guttural ***aurk-aurk-aurk***. **HABITAT & BEHAVIOUR** Forest, secondary growth, dense cover along streams; up to 2,565 m (breeds above 1,830 m elsewhere in Himalayas). Prefers damp areas. Usually solitary. May be flushed from concealed position in daytime; flies with heavier wingbeats than snipes. During breeding season, males perform roding display at dawn and dusk, flying in regular circuits low over tree-tops, with slow deliberate wingbeats. **RANGE & STATUS** Breeds Palearctic, N Pakistan, NW,N,NE Indian subcontinent, NW,W China, northern Japan; some populations winter south to N Africa, south W Palearctic, Middle East, S India, S China, S Korea, S Japan, rarely Philippines, N Borneo. **SE Asia** Uncommon resident N Myanmar. Uncommon to fairly common winter visitor Myanmar, Thailand (except S), south-west Cambodia, Laos, Vietnam (rare Cochinchina). Also recorded on passage migrant E Tonkin. Vagrant S Thailand, Peninsular Malaysia, Singapore. **BREEDING** March–May. **Nest** Lined depression in ground. **Eggs** 4; pale clay-coloured to deep buff, boldly blotched pale reddish-brown and grey; 44.5 x 33.3 mm (av.).

SCOLOPACIDAE: GALLINAGININAE: Snipes

Worldwide 20 species. SE Asia 7 species. Smallish to medium-sized and thickset with long straight bills and rather cryptically patterned plumage. Relatively secretive and difficult to flush. Feed by probing and picking from surface, consuming worms, small molluscs, larvae and aquatic invertebrates.

273 JACK SNIPE *Lymnocryptes minimus* Plate **33**

IDENTIFICATION 17–19 cm. Monotypic. **Adult** Resembles other snipe but much smaller and shorter-billed, has all-dark crown-centre, distinctive 'split supercilium', very dark mantle and scapulars with purple and green gloss and prominent buff lengthwise lines and unbarred whitish underparts with dark streaks on foreneck, upper breast and flanks. In flight shows all-brown, slightly wedge-shaped tail and unbarred white panel on underwing-coverts. **Juvenile** Similar to adult. **VOICE** Usually silent when flushed but sometimes utters low, weak, barely audible ***gah***. **HABITAT & BEHAVIOUR** Marshy areas, ditches; up to 1,500 m. Often solitary. Feeds with nervous rocking action. Freezes when approached and difficult to flush. Flight often appears rather weak and fluttery, typically only covering short distance. **RANGE & STATUS** Breeds northern Palearctic; winters N and equatorial Africa, W Palearctic, Middle East, Indian subcontinent, NW,S China. **SE Asia** Scarce winter visitor SW,N,C,E Myanmar, NW,C Thailand, W Tonkin. Vagrant S Myanmar, N Tenasserim, coastal W Thailand, Cochinchina.

274 SOLITARY SNIPE *Gallinago solitaria* Plate **33**

IDENTIFICATION 29–31 cm. *G.s.solitaria*: **Adult** Shows distinctive combination of relatively large size, whitish face and mantle-lines, extensive rufous vermiculations on upperparts, gingery patch on side of breast (contrasts with whitish face), white edges to median coverts (forming pale panel) and bright rufous-chestnut on tail. Legs and feet pale yellowish-green to yellowish-brown. In flight, toes do not project beyond tail-tip. **Juvenile** Similar to adult. **VOICE** May give harsh ***kensh*** when flushed, somewhat deeper than call of Common. **HABITAT & BEHAVIOUR** Streams, swampy pools, sometimes marshes and rice paddies; 1,000–1,675 m. Usually solitary. Flight style similar to Common Snipe but somewhat slower and heavier. **RANGE & STATUS** Breeds C Asia, S,SE Siberia, Ussuriland, Kamchatka, NW,N,NE India, Nepal, Bhutan, Tibet, NW,NE China; some populations winter to south, NE Iran, northern Pakistan, S Tibet, southern China, N,S Korea, Japan. **SE Asia** Rare winter visitor W,N,E Myanmar.

275 WOOD SNIPE *Gallinago nemoricola* Plate **33**

IDENTIFICATION 28–32 cm. Monotypic. **Adult** Similar to Solitary Snipe but somewhat larger with broader-based bill, much darker mantle and scapulars, bolder, buffier stripes and scales on upperparts, uniform dark brown barring on lower breast to vent and no gingery coloration on breast. Legs and feet greyish-green. In flight, wings appear broader and more rounded, upperwing rather uniform apart from paler panel on coverts, no whitish trailing edge to secondaries and coverts, virtually no white on tail, dense dark brown bars over all underwing-coverts. See Eurasian Woodcock and Great Snipe. **Juvenile** Mainly differs from adult by finer, whiter fringes to mantle and scapulars and paler buff fringes to median coverts. **VOICE** Often silent when flushed but may give a deep, guttural croak or ***che-dep che-dep...*** **HABITAT & BEHAVIOUR** Streams, rivers and other wet areas in or near broadleaved evergreen forest and secondary growth, marshes and swamps with thick cover; 520–1,830 m. Usually solitary; rarely flies far when flushed. Flight slower and more laboured than Solitary. **RANGE & STATUS** Breeds N,NE India, Nepal, Bhutan, S,E Tibet, W China; some populations winter locally south to S India, SW China; rarely Sri Lanka. **SE Asia** Rare winter visitor W,N,C,E,S Myanmar, north Tenasserim, W,NW Thailand, N,C Laos, W Tonkin.

276 PINTAIL SNIPE *Gallinago stenura* Plate **33**

IDENTIFICATION 25–27 cm. Monotypic. **Adult** Very similar to Common Snipe but somewhat shorter-billed, buffish supercilium always broader than dark eyestripe at base of bill, has narrower pale brown to whitish edges to scapulars, tail only projects slightly beyond closed wing-tips, primaries only project slightly beyond tertials. Legs and feet greyish-green to brownish-green. In flight, tail appears somewhat shorter, almost entire length of toes extend beyond tail-tip, has paler, more contrasting sandy-buff lesser and median upperwing-coverts, much narrower and less distinct whitish trailing edge to secondaries, darker and more heavily barred underwing-coverts and almost no white on tail-corners. In hand, tail usually has 24–28 feathers, with 6–9 outer pairs pin-like. Note voice. See Swinhoe's Snipe. **Juvenile** Has narrower whitish edges to mantle and scapulars, and narrower, more even whitish-buff fringes to wing-coverts. **VOICE** When flushed, may utter short, rasping ***squok***, ***squack*** or ***squick***, rather weaker and lower-pitched than call of Common. **HABITAT & BEHAVIOUR** Open marshy areas, rice paddies; up to 2,135 m. Often in drier areas than Common; tends to feed more by picking and typically undertakes shorter, more direct flights. **RANGE & STATUS** Breeds C,E Palearctic; winters Indian subcontinent, S China, Sumatra, Borneo, W Lesser Sundas, Philippines, Sulawesi, S Moluccas (Buru), Cocos-Keeling Is, rarely Java, Bali, NW

Australia. **SE Asia** Fairly common to common winter visitor (except N Annam). Also recorded on passage S Myanmar, Peninsular Malaysia, Cambodia.

277 **SWINHOE'S SNIPE**

Gallinago megala Plate **33**

IDENTIFICATION 27–29 cm. Monotypic. **Adult** Extremely difficult to separate from Pintail Snipe at rest but bill longer (relatively like Common Snipe), tail projects more beyond wings, may show obvious projection of primaries beyond tertials. In flight, wings appear slightly longer and more pointed, toes project less beyond tail-tip, shows more white on tail-corners (less than Common). Differs from Pintail and Common by slightly larger size, slightly larger and squarer head with crown-peak more obviously behind eye (eye also set further back) and heavier, more barrel-chested appearance. Legs thicker than Common, usually thicker than Pintail and often rather yellow. In spring, face, sides of neck and flanks (sometimes breast) may appear rather dusky and quite heavily barred, creating dark coloration distinct from both Pintail and Common. In hand, usually has 20 (18–26) tail-feathers, with outer ones broader than Pintail. Legs and feet greenish to greenish-yellow. Note voice and behaviour. **Juvenile** Mainly differs from adult by narrower whitish edges to mantle and scapulars, and clear whitish-buff fringes to wing-coverts and tertials. **VOICE** Quieter than Pintail and Common. Usually silent when flushed but may give similar call to Pintail, on same pitch but possibly slightly less hoarse, rather thinner and quite nasal with a slight rattling quality. **HABITAT & BEHAVIOUR** Marshy areas, rice paddies and their margins. Shows preference for drier areas and possibly occurs in less open, more wooded areas than Pintail and Common. When flushed, take-off is rather slow and laboured compared to Pintail and Common. Flight rather direct and usually for short distance only. **RANGE & STATUS** Breeds SC Siberia, Ussuriland; winters S,NE India, S,SE China, Java, Bali, Borneo, Philippines, Wallacea, New Guinea region, N Melanesia, N Australia. **SE Asia** Vagrant N,C,S Myanmar, C Thailand, Peninsular Malaysia, Singapore, S Laos.

278 **GREAT SNIPE** *Gallinago media* Plate **33**

IDENTIFICATION 27–29 cm. Monotypic. **Adult** Similar to Common Snipe but larger and bulkier with relatively shorter bill, has bold white tips to wing-coverts, relatively indistinct pale lines on upperparts and more extensively barred lower breast to vent. In flight shows distinctly dark greater upperwing-coverts, narrowly bordered white at front and rear, narrower white trailing edge to secondaries, more uniformly dark-barred underwing-coverts and extensive unmarked white tail-corners. In hand, tail usually has 16 feathers, with three outer pairs mostly white. Note voice. See Solitary and Swinhoe's Snipe. **Juvenile** Similar to adult but often shows narrower whitish lines on mantle and scapulars, narrower white tips to wing-coverts and a little brown barring on outertail-feathers. **VOICE** Usually silent but may give weak croaking ***etch*** or ***aitch*** notes when flushed. **HABITAT & BEHAVIOUR** Open marshy areas, also drier ground, short grassy areas; recorded at sea level. Flight relatively short, direct and level. **RANGE & STATUS** Breeds north and north-eastern W and C Palearctic; winters sub-Saharan Africa; vagrant S Indian subcontinent. **SE Asia** Vagrant north Tenasserim.

279 **COMMON SNIPE**

Gallinago gallinago Plate **33**

IDENTIFICATION 25–27 cm. *G.g.gallinago*: **Adult** Robust with relatively long bill and shortish legs, plumage cryptically patterned with boldly striped head and upperparts and whitish underparts with dark streaks on neck and breast and dark bars on flanks. Resembles several other snipes (particularly at rest) and best distinguished by combination of following features: long bill, broad dark loral stripe (broader than pale supercilium at base of bill), pronounced pale buff to buffish-white lengthwise stripes on upperparts, white edging on outer scapular fringes only, largely white belly and vent, no obvious white tips to wing-coverts and obvious projection of tail beyond closed wing-tips. Legs and feet yellowish-green to greenish-grey. In flight shows prominent white trailing edge to secondaries, panel of unbarred white on underwing-coverts and mostly rufous-chestnut tail-tip with a little white on outer feathers; toes project only slightly beyond tail-tip. In hand, tail usually has 12–18 feathers, outermost broad and not pin-like. Note voice and behaviour. See Pintail, Swinhoe's and Great Snipes. **Juvenile** Has finer whitish lines on mantle and scapulars and narrower buff fringes to wing-coverts. **VOICE** When flushed, regularly gives rasping, often slightly rising ***scaaap***; more drawn out than Pintail and Swinhoe's. **HABITAT & BEHAVIOUR** Open marshy areas, rice paddies; up to 1,220 m. Flight-style fast and erratic, often zig-zagging. **RANGE & STATUS** Breeds Palearctic, NE Afghanistan, NW India, NW,NE China; some populations winter south to equatorial Africa, Middle East, Indian subcontinent, southern China, Taiwan, S Korea, southern Japan, Java, Borneo, Philippines. **SE Asia** Uncommon to common winter visitor (except W Tonkin, N Annam). Also recorded on passage Cambodia.

SCOLOPACIDAE: PHALAROPODINAE: Phalaropes

Worldwide 3 species. SE Asia 2 species. Small to largish, mostly with long, slender bills (downcurved in some species), long legs and often intricately mottled plumage. Mostly gregarious. Feed on small molluscs, crustaceans, insects and worms, mostly in freshwater and coastal wetlands.

280 **RED-NECKED PHALAROPE**

Phalaropus lobatus Plate **36**

IDENTIFICATION 17–19 cm. Monotypic. **Adult non-breeding** Small size, needle-like blackish bill, isolated blackish mask, mostly grey upperparts and mostly white underparts diagnostic. Has indistinct long whitish mantle-lines, whitish fringes to scapulars and wing-coverts, grey-washed breast-sides and grey streaks on flanks. In flight, upperwing appears rather blackish with contrasting narrow white band along tips of greater coverts and primary coverts. Note habitat and behaviour. See Red Phalarope. **Female breeding** Striking. Head, neck and upper breast slate-grey with broad rufous-chestnut band running from behind ear-coverts to upper foreneck. Face blackish, small mark over eye and throat white, upperparts blackish-slate with broad warm buff lines bordering mantle and scapulars. **Male breeding** Similarly patterned to female breeding but duller and more washed out. **Juvenile** Similar to adult non-breeding but crown and upperparts blackish, with broad rufous-buff to ochre-buff edgings and fringes, shows lengthwise rufous-buff to ochre-buff mantle- and scapular-lines ('mantle-V' and 'scapular-V'); neck-sides and upper breast washed vinous-pinkish, dark streaks on neck and breast-sides. See Red Phalarope. **First winter** Like adult non-breeding but has blacker crown, upperparts and wing-coverts, with only few grey feathers. See Red Phalarope. **VOICE** In flight may give a short sharp ***kip*** or ***twick***, harsher ***cherp*** or squeaky ***kirrik*** or ***kerrek***. **HABITAT & BEHAVIOUR** Open sea, coastal pools, sometimes mudflats, rarely inland rivers and pools; lowlands. Habitually swims, often spinning around, also wades when feeding. Normally gregarious. **RANGE & STATUS** Breeds N Holarctic; mostly winters at sea, particularly Arabian Sea, area from S Philippines, N Borneo and northern Wallacea to north coast of New Guinea and Bismarck Archipelago and offshore Peru. **SE Asia** Rare to scarce coastal passage migrant Thailand, Peninsular Malaysia, Singapore, Cambodia, E Tonkin, C,S Annam. Inland vagrant NW Thailand, Cambodia, N,S Laos, C,S Annam, Cochinchina.

281 **RED PHALAROPE**
Phalaropus fulicarius Plate **36**
IDENTIFICATION 20–22 cm. Monotypic. **Adult non-breeding** Similar to Red-necked Phalarope, but somewhat larger and heavier-bodied, shorter and thicker bill, whiter crown (thinly edged black), has paler grey and plainer upperside. In flight, wings proportionately larger than Red-necked, wingbeats slower, white wing-bar more pronounced. Combination of rather pale plumage, strong wing-bar, and call can recall Sanderling. Bill remarkably broad when viewed from above, dark with yellowish to yellowish-brown tinge. Legs and feet proportionately very short, greyish to brownish, with large, rounded yellowish lobes along toes. **Female breeding** Unmistakable, with blackish crown and face, white head-sides, and chestnut-red neck and underparts. Rest of upperside blackish-brown with mainly buff to rufous edgings. Bill rich yellow with black tip. Legs and feet yellowish-brown. **Male breeding** Duller than female, with pale-streaked crown, buff-washed head-sides (may show whitish supercilium), and often much white on belly. **Juvenile** Apart from size and proportions, told from Red-necked by less pronounced ochre-buff 'mantle-V', lack of obvious 'scapular-V', and weaker pinkish-buff wash on face and neck. Acquires the first pale grey scapulars of first winter plumage rather quickly (moults somewhat earlier than Red-necked). **First winter** Not well documented, but similar to adult non-breeding. **VOICE** Shrill ***wit*** or ***pit*** contact call, recalling Sanderling and Little Stint. Disyllabic twittering calls when breeding, a buzzing far-carrying ***brrreep*** and excited ***bip bip bip....*** **HABITAT & BEHAVIOUR** As for Red-necked. Mostly found singly or in small parties. **RANGE & STATUS** Breeds N Holarctic; winters at sea, off west coasts of Africa and S America. Rare to scarce passage migrant Japan, Taiwan. **SE Asia** Vagrant C Thailand.

SCOLOPACIDAE: TRINGINAE: Godwits, dowitchers, curlews, sandpipers & allies

Worldwide 34 species. SE Asia 18 species. Small to largish, mostly with long, slender bills (downcurved in some species), long legs and often intricately mottled plumage. Mostly gregarious. Feed on small molluscs, crustaceans, insects and worms, mostly in freshwater and coastal wetlands.

282 **BLACK-TAILED GODWIT**
Limosa limosa Plate **34**
IDENTIFICATION 36–40 cm. *L.l.melanuroides*: **Adult non-breeding** Told by combination of fairly large size, long blackish legs, rather long neck, long straight bicoloured bill with pinkish basal half, and rather plain brownish-grey plumage with somewhat paler head-sides, neck and breast, short whitish supercilium and whitish belly and vent. In flight shows pronounced white upperwing-bar, broad white band across rump and uppertail-coverts contrasting with mostly black tail, and white underwing with dark border. Legs and feet dark grey. See Bar-tailed Godwit and Long-billed Dowitcher. **Male breeding** Sides of head, neck and upper breast mostly reddish-rufous, mantle and scapulars boldly marked blackish with some chestnut, rest of underparts whitish with variable blackish bars. See Bar-tailed, and Asian and Long-billed Dowitchers. **Female breeding** Larger, longer-billed and duller than male breeding, retaining greater proportion of greyish non-breeding plumage. **Juvenile** Like female breeding but crown streaked brown and cinnamon, mantle and scapulars rather boldly marked dark grey-brown and chestnut, wing-coverts dark grey-brown with cinnamon-buff fringing, neck and breast initially washed rufous-buff. See Bar-tailed. **VOICE** Flocks utter a constant, excited, twittering babble. Individual calls described as yapping ***kip*** and chattering ***kett*** and ***chuk*** notes. Display-type call is a relatively shrill and slow-paced ***kitititititiw***. **HABITAT & BEHAVIOUR** Mud- and sandflats, coastal pools, marshes, sometimes wet rice paddies; lowlands. Usually found in flocks. Feeds by picking and forward-probing, often in deeper water than Bar-tailed. **RANGE & STATUS** Breeds Palearctic, NW,NE China; winters W,S Europe, N and equatorial Africa, Middle East, S Pakistan, Indian subcontinent, S China, Taiwan, Greater Sundas, Philippines, New Guinea, Australia, New Zealand. **SE Asia** Uncommon to fairly common coastal winter visitor and passage migrant (except N Annam); also locally inland N Myanmar, C Thailand. Scarce to uncommon inland passage migrant C,S Myanmar, C,NE Thailand, Peninsular Malaysia. Also recorded in summer C Thailand, Peninsular Malaysia.

283 **BAR-TAILED GODWIT**
Limosa lapponica Plate **34**
IDENTIFICATION 37–41 cm. *L.l.lapponica*: **Adult non-breeding** Similar to Black-tailed Godwit but upperparts buffier and distinctly streaked, bill slightly upturned, legs somewhat shorter. In flight lacks white wing-bar, back to uppertail-coverts white, tail white with blackish bars, underwing duller and less distinctly dark-bordered. *L.l.baueri* (recorded C Thailand southward) has dark back and rump, dark-barred uppertail-coverts and dark-barred underwing-coverts. See dowitchers. **Male breeding** Head, neck and underparts almost entirely reddish-chestnut, upperparts boldly marked with chestnut. See dowitchers and Red Knot. **Female breeding** Larger and longer-billed than male, but modest change from non-breeding plumage, with darker-marked upperparts and deep apricot wash on head-sides, neck and breast. **Juvenile** Like adult non-breeding but crown and upperparts have brown centres and broad bright buff edges, has buffish-brown wash and fine dark streaks on neck and breast. See Black-tailed Godwit. **VOICE** Flight calls include abrupt high-pitched ***kik*** or ***kiv-ik*** (often repeated), barking ***kak-kak*** and nasal ***ke-wuh*** or ***kirruc***. **HABITAT & BEHAVIOUR** Mud- and sandflats, beaches, coastal pools. Often feeds on open mud and sand and in shallower water than Eastern Black-tailed. **RANGE & STATUS** Breeds N Palearctic, NW N America; winters along coasts of Africa, Madagascar, W,S Europe, Middle East, Indian subcontinent, S China, Sundas, Philippines, Wallacea, New Guinea, Australia, New Zealand. **SE Asia** Scarce to fairly common coastal winter visitor and passage migrant (except Tenasserim, N Annam). Also recorded in summer S Thailand, Peninsular Malaysia.

284 **LONG-BILLED DOWITCHER**
Limnodromus scolopaceus Plate **35**
IDENTIFICATION 27–30 cm. Monotypic. **Adult non-breeding** Similar to Asian Dowitcher but smaller, basal half of bill greenish, upperside somewhat plainer, neck and upper breast rather plain brownish-grey, legs shorter, paler, greyish- to brownish-green. In flight shows unmarked white back, heavily dark-barred rump to uppertail, somewhat darker inner primaries and secondaries with narrow, defined white trailing edge and finely dark-barred underwing-coverts. See godwits and Grey-tailed Tattler. **Adult breeding** Similar to Asian but upperparts boldly mottled rather than streaked, supercilium and sides of head noticeably paler, underparts paler with dark speckles and bars on neck, breast, flanks and vent. Bill and leg colour similar to adult non-breeding. See godwits and Red Knot. **Juvenile** Resembles adult non-breeding but mantle and scapulars dark brown with fine chestnut fringes, sides of head and breast washed buff. **VOICE** Typical flight and contact call is a single or repeated sharp ***kik*** or ***keek***. When alarmed, may give a longer, shriller ***keeek*** (singly or in short series). **HABITAT & BEHAVIOUR** Coastal marshes, freshwater and brackish pools, sometimes mudflats. Feeds by continuous vertical probing, usually in knee-deep water. **RANGE & STATUS** Breeds NE Palearctic, north-west N America; winters west and southern USA, C America, rarely S China, Japan. **SE Asia** Coastal vagrant W,C Thailand, Singapore, E Tonkin.

285 **ASIAN DOWITCHER**

Limnodromus semipalmatus Plate **35**

IDENTIFICATION 34–36 cm. Monotypic. **Adult non-breeding** Most likely to be confused with godwits. Similar to Bar-tailed Godwit (race *baueri*) but somewhat smaller, shorter-necked and shorter-legged, with rather flattish forehead and straight all-black bill slightly swollen at tip (may be paler at extreme base). In flight, upperside of inner primaries, secondaries and greater coverts appear somewhat paler than rest of wing (may appear rather translucent), underwing pale with unmarked white coverts (similar to Bar-tailed race *lapponica*), back to uppertail-coverts white with dark markings, uppertail largely unbarred (somewhat paler than Bar-tailed race *baueri* and quite unlike race *lapponica*). See Long-billed Dowitcher. **Male breeding** From similar Bar-tailed Godwit by size, shape and distinctive bill; often shows largely white vent. See Long-billed and Red Knot. **Female breeding** Head, neck and underparts slightly duller than male. **Juvenile** Similar to adult non-breeding but mantle, scapulars and tertials blacker with narrow, neat pale buff fringes, neck and breast strongly washed buff and indistinctly dark-streaked. **VOICE** In flight utters occasional airy ***chaow*** or ***chowp***, yelping ***chep-chep*** and soft ***kiaow***. **HABITAT & BEHAVIOUR** Mudflats, coastal marshes and pools. Feeds by continuous vertical probing, usually in knee-deep water. Gregarious, often associates with godwit flocks. **RANGE & STATUS** Breeds southern Siberia, N Mongolia, NE China, Ussuriland; winters along coasts of SE,E India, S Bangladesh, locally Greater Sundas, Philippines, S New Guinea, NW Australia. **SE Asia** Rare to scarce coastal passage migrant (except Myanmar). Small numbers winter C Thailand, west Peninsular Malaysia. Also recorded in summer Peninsular Malaysia. Recorded (status unknown) S Myanmar.

286 **LITTLE CURLEW**

Numenius minutus Plate **34**

IDENTIFICATION 29–32 cm. Monotypic. **Adult** Similar to Whimbrel but much smaller (not much larger than Pacific Golden Plover) and finer-billed, pale supercilium contrasts more sharply with blackish eyestripe, underparts more buffish. In flight, back to uppertail-coverts concolorous with mantle, underwing-coverts mostly buffish-brown. Legs and feet yellowish to bluish-grey. **Juvenile** Similar to adult. **VOICE** In flight, excited whistled 3–4 note ***weep-weep-weep...*** or ***qwee-qwee-qwee...***, recalling Whimbrel but sharper and higher-pitched. Also rougher ***tchew-tchew-tchew*** and harsh ***kweek-ek*** when alarmed. **HABITAT & BEHAVIOUR** Short grassland, barren cultivation, margins of freshwater and sometimes coastal wetlands; lowlands. May be tame. Feeds mostly by picking. **RANGE & STATUS** Breeds NC,NE Siberia; winters N Australia; erratic passage migrant Java, Bali, Borneo, Philippines, Wallacea, New Guinea (common), New Britain, Solomon Is (Bougainville). **SE Asia** Vagrant C Thailand, Singapore.

287 **WHIMBREL** *Numenius phaeopus* Plate **34**

IDENTIFICATION 40–46 cm (female slightly larger and longer-billed than male). *N.p.phaeopus*: **Adult** Distinctive. Fairly large, bill longish and markedly kinked downward towards tip, upperside rather cold greyish-brown with whitish to pale buffish mottling, prominent blackish lateral crown-stripes and eyestripe, broad whitish supercilium and buffish-white neck and breast with heavy dark streaks. In flight, upperside appears all dark with contrasting clean white back and rump, underwing-coverts mostly plain whitish. Legs and feet dull bluish-grey. Note voice. *N.p.variegatus* (widespread but scarcer) has lower back and rump concolorous with mantle and shows heavy dark bars on underwing-coverts. Intergrades occur. See other curlews. **Juvenile** Has clear buff markings on scapulars and tertials, breast buffier with slightly finer streaks. **VOICE** In flight utters diagnostic clear whinny or titter: ***dididididididi...*** or ***puhuhuhuhuhuhu...***, varying in intensity. Also single plaintive ***curlee*** notes. **HABITAT & BEHAVIOUR** Coastal wetlands, mangroves, marshes, large rivers; lowlands. Gregarious. Feeds mostly by picking rather than probing. **RANGE & STATUS** Breeds Holarctic; winters along coasts of Africa, Madagascar, SW Europe, Middle East, Indian subcontinent, S China, Taiwan, Sundas, Philippines, Wallacea, New Guinea region, Melanesia, Micronesia, Australia, New Zealand, N(south),C,S America. **SE Asia** Fairly common to common winter visitor and passage migrant to coastal areas throughout. Scarce to uncommon inland passage migrant N,C,E Myanmar, N Thailand, Cambodia, N,S Laos. Also recorded in summer C,S Thailand, Peninsular Malaysia, Singapore.

288 **EURASIAN CURLEW**

Numenius arquata Plate **34**

IDENTIFICATION 50–60 cm (female larger and longer-billed than male). *N.a.orientalis*: **Adult non-breeding** Like Whimbrel but larger, bill longer and more strongly and gently downcurved, head more uniform with slightly darker crown and indistinct supercilium, upperparts rather more coarsely marked with whitish to pale buff, has more pronounced blackish streaks on neck, breast and upper belly. In flight shows stronger contrast between outer and inner upperwing, contrasting white back and rump (latter with some dark mottling) and largely white underwing-coverts. Legs and feet bluish-grey. Note voice. **Adult breeding** Has more buffish fringing on upperparts and somewhat more buffish base colour to sides of head, neck and breast. See Far Eastern Curlew. **Juvenile** Like adult breeding but even more buffish, breast and flanks more narrowly dark-streaked, more contrasting tertial pattern, and somewhat shorter and less downcurved bill. **VOICE** Loud, rising, ringing ***cour-lee*** or ***cour-loo***, uttered with varying emphasis, and low ***whaup*** or ***were-up***. Loud, rapid stammering ***tyuyuyuyu...*** or ***tutututu...*** when agitated. **HABITAT & BEHAVIOUR** Mud- and sandflats, coastal wetlands, large rivers; lowlands. Gregarious. Feeds mostly by deep probing into soft mud. **RANGE & STATUS** Breeds Palearctic (east to SC Siberia), Mongolia, NE China; some populations winter south (mostly on coasts) to S Africa, Madagascar, Middle East, Indian subcontinent, S,E China, Taiwan, S Korea, S Japan, Greater Sundas, Philippines. **SE Asia** Uncommon to fairly common coastal winter visitor and passage migrant (except N,C Annam); also locally inland N Myanmar. Rare to scarce inland passage migrant N,C,E Myanmar, C,NE,SE Thailand, Cambodia, N,C Laos. Also recorded in summer C Thailand.

289 **FAR EASTERN CURLEW**

Numenius madagascariensis Plate **34**

IDENTIFICATION 60–66 cm (female larger and longer-billed than male). Monotypic. **Adult non-breeding** Like Eurasian Curlew but larger and longer-billed, appears more uniform, with browner/buffier underparts. In flight shows rather uniform rufescent-tinged greyish-brown back and rump and dense dark bars on underwing-coverts. Legs and feet dull blue-grey. See Whimbrel. **Adult breeding** Upperpart fringes and back to uppertail distinctly washed rufous, sides of head and neck washed rufous, rest of underparts warmer-tinged. **Juvenile** Similar to adult non-breeding but has extensive neat buffish-white markings on upperparts and wing-coverts, finer dark streaks on underparts, and shorter bill. **VOICE** Call is ***coor-ee***, similar to Eurasian but flatter-sounding and less fluty. Also gives strident ***ker-ee ker-ee...*** or ***carr-eeir carr-eeir...*** when alarmed, and occasional bubbling trills. **HABITAT & BEHAVIOUR** Mud- and sandflats. Feeds mostly by deep probing. **RANGE & STATUS** Breeds NE China, SE Siberia, Ussuriland, Kamchatka; winters Sundas, Philippines, Wallacea, New Guinea, Australia, New Zealand, rarely S China, Taiwan. **SE Asia** Rare coastal passage migrant W,C,S Thailand, Peninsular Malaysia, Singapore, E Tonkin, S Annam, Cochinchina. Occasionally winters west Peninsular Malaysia, E Tonkin; ? Cochinchina.

290 **TEREK SANDPIPER** *Xenus cinereus* Plate **35**

IDENTIFICATION 22–25 cm. Monotypic. **Adult non-breeding** Smallish size, long upturned blackish bill with yellowish base, and shortish orange-yellow legs diagnostic. Upperside rather plain brownish-grey with darker median and lesser coverts, with suggestion of dark lines on scapulars; underparts white with some faint streaks on neck and breast. In flight, upperside appears rather uniform, with distinctive broad white trailing edge to secondaries (recalls Common Redshank). Legs and feet bright orange to orange-yellow, sometimes greenish-yellow. **Adult breeding** Upperparts much clearer grey with prominent black lines on scapulars, bill all dark. **Juvenile** Similar to adult non-breeding but upperside darker and browner, scapulars narrowly fringed buffish, wing-coverts finely fringed light buffish, has indistinct short blackish lines on scapulars and yellower base of bill. **VOICE** In flight utters low rippling trilled ***du-du-du-du-du...*** and shorter mellow ***chu-du-du***. Sharp ***tu-li*** when alarmed. **HABITAT & BEHAVIOUR** Mud- and sandflats, saltpans and other coastal wetlands, rarely wet rice paddies on passage; up to 1,150 m (mostly lowlands). Runs rather fast. **RANGE & STATUS** Breeds northern Palearctic; winters along coasts of Africa, Madagascar, Middle East, Indian subcontinent, S China, Taiwan, S Japan, Sundas, Philippines, Wallacea, New Guinea, Australia. **SE Asia** Uncommon to fairly common coastal winter visitor and passage migrant throughout. Vagrant N Laos. Also recorded in summer Peninsular Malaysia, Singapore.

291 COMMON SANDPIPER
Actitis hypoleucos Plate **35**

IDENTIFICATION 19–21 cm. Monotypic. **Adult non-breeding** Told by smallish size, medium-short straight bill, shortish legs, plain brownish upperside and white underparts with greyish-brown lateral breast-patches, separated from shoulder by prominent white 'spur'. Tail longish, extending well beyond closed wing-tips, legs and feet greyish-olive to dull yellowish-brown. In flight appears rather uniformly dark above with contrasting long white wing-bar across greater covert-tips and middle of inner primaries. See Green Sandpiper and much smaller Temminck's Stint. **Adult breeding** Upperparts slightly glossy greenish-brown with faint dark streaks and dark bars on larger feathers, lateral breast-patches browner with dark streaks, almost meeting across upper breast. **Juvenile** Like adult non-breeding but upperparts narrowly fringed buff with some darker subterminal markings, wing-coverts with prominent buff tips and subdued dark barring. **VOICE** Call is a high-pitched plaintive ringing ***tsee-wee-wee...*** or ***swee-swee-swee...***, mainly given in flight. Sometimes gives single ***sweet*** or longer ***sweeee-eet*** when alarmed. Song is an excited, rising, repeated ***kittie-needie*** (often given on wintering grounds). **HABITAT & BEHAVIOUR** Various wetlands, mudflats, tidal creeks, coastal rocks, lakes, rivers; up to 2,100 m. Often 'bobs' rear end of body. Flies with short flicking wingbeats. **RANGE & STATUS** Breeds Palearctic, N Iran, Afghanistan, N Pakistan, NW,N India, Tibet, Mongolia, northern China, Taiwan, N Korea, Japan; mostly winters Africa, Madagascar, SW Europe, Middle East, Indian subcontinent, southern China, Taiwan, southern Japan, Sundas, Philippines, Wallacea, New Guinea, N Melanesia, Australia. **SE Asia** Common winter visitor and passage migrant throughout. Also recorded in summer Peninsular Malaysia, Singapore.

292 GREEN SANDPIPER
Tringa ochropus Plate **35**

IDENTIFICATION 21–24 cm. Monotypic. **Adult non-breeding** Smallish and quite short-legged; crown, hindneck, sides of breast and upperparts blackish-brown with olive tinge, dull buff speckling on mantle, scapulars and wing-coverts, prominent white lores and eyering. Similar to Wood and Common Sandpipers. Differs from former by lack of prominent supercilium behind eye, plainer crown and hindneck, darker upperparts, more demarcated breast-band and greener legs and feet. From latter by more blackish upperparts and lack of white spur between breast-band and shoulder. In flight, white rump and uppertail-coverts contrast very sharply with almost uniformly dark upperside, tail white with two or three broad blackish bands. Legs and feet dull greyish-green. Note voice. **Adult breeding** Has bold streaks on crown, neck and breast and quite distinct white speckles on upperparts and wing-coverts. **Juvenile** Like adult non-breeding but upperparts and breast browner-tinged, with less distinct small deep buff spots on scapulars and tertials. **VOICE** Loud, sharp ***klU-Uweet-wit-wit*** and ***tluee-tueet***, usually given in flight. Sharp ***wit-wit-wit*** when alarmed. **HABITAT & BEHAVIOUR** Various lowland wetlands, rarely mudflats; up to 800 m. Occasionally bobs rear end of body. **RANGE & STATUS** Breeds northern Palearctic, NW,NE China; winters south to Africa, southern W Palearctic, Middle East, Indian subcontinent, S Tibet, southern and E China, Taiwan, S Korea, southern Japan, Sumatra, Borneo, Philippines, rarely Java. **SE Asia** Scarce to fairly common winter visitor throughout; rare S Thailand southwards. Also recorded on passage Cambodia, E Tonkin.

293 GREY-TAILED TATTLER
Tringa brevipes Plate **35**

IDENTIFICATION 24–27 cm. Monotypic. **Adult non-breeding** Resembles some larger *Tringa* sandpipers but upperside uniform grey, sides of head and upper breast rather plain pale grey, with prominent white supercilium (extending behind eye), rather stout straight bill with yellowish basal half, shortish yellow legs. Throat and rest of underparts unmarked white. In flight, completely grey upperside with somewhat darker outer wing distinctive. Note voice. See Red Knot and Long-billed Dowitcher. **Adult breeding** Ear-coverts, sides of throat and neck finely streaked grey, breast and upper flanks scaled grey, changing to more V-shaped markings on lower flanks, base of bill duller. **Juvenile** Like adult non-breeding but upperparts and wing-coverts have neat whitish spots and fringes; outer fringes of tail-feathers notched white. **VOICE** Plaintive ***tu-weet*** or ***tu-whip*** (sometimes repeated), usually given in flight. Alarm calls are more hurried ***tu-wiwi***, ***twi-wi*** and ***twiwiwi***. **HABITAT & BEHAVIOUR** Reefs, rocky shores, mud- and sandflats, sometimes saltpans and prawn-ponds. Gait recalls Common Sandpiper. Occasionally gregarious. **RANGE & STATUS** Breeds NC,NE Palearctic; winters Java, Bali, Borneo, Philippines, Wallacea, Micronesia, New Guinea, N Melanesia, Australia, New Zealand. **SE Asia** Rare to uncommon coastal passage migrant C,SE,S Thailand, Peninsular Malaysia, Singapore, E Tonkin, C,S Annam, Cochinchina. Occasionally winters S Thailand, Peninsular Malaysia, Singapore. Inland vagrant NE Thailand, C Laos (same record).

294 SPOTTED REDSHANK
Tringa erythropus Plate **36**

IDENTIFICATION 29–32 cm. Monotypic. **Adult non-breeding** Resembles Common Redshank but distinctly paler overall, bill longer, more slender and finer-tipped; more distinct white supercilium, no streaks on underparts, longer legs. In flight, upperside appears rather uniform apart from unbarred white back. See Marsh Sandpiper and Common Greenshank. **Adult breeding** Very distinctive. Upperside blacker, head, neck and underparts almost uniform blackish. Some (mainly females) have faint pale scales on head, neck and underparts. **Juvenile** Distinctive. Recalls adult non-breeding but brownish-grey overall with finely white-speckled upperparts and wing-coverts, dark-streaked neck and closely dark-barred breast to undertail-coverts. **VOICE** Loud, rising ***chu-it***, given in flight. Also a conversational ***uck*** when feeding, and short ***chip*** when alarmed. **HABITAT & BEHAVIOUR** Freshwater marshes, flooded rice paddies, coastal pools, large rivers; up to 450 m. Gregarious. Often feeds in quite deep water. **RANGE & STATUS** Breeds N Palearctic; winters N and equatorial Africa, southern W Palearctic, Middle East, Indian subcontinent, rarely N Borneo. **SE Asia** Scarce to locally common winter visitor (except SE,S Thailand, W Tonkin, S Annam). Vagrant S Thailand.

295 COMMON GREENSHANK
Tringa nebularia Plate **36**

IDENTIFICATION 30–34 cm. Monotypic. **Adult non-breeding** Distinctive. Medium-sized to largish with rather long neck, stout-based and slightly upturned greenish-grey bill with darker tip, and long greenish legs. Crown, ear-coverts, hindneck, mantle and sides of breast prominently streaked, rest of upperside greyish with somewhat darker lesser coverts, underparts white, has distinctly dark loral stripe and no white supercilium behind eye. In flight appears rather uniform above, back to uppertail contrastingly white with dark bars on longer uppertail-coverts and tail, toes extending beyond tail-tip. Legs and feet greyish-green, sometimes greenish-yellow or dull yellowish, basal half of bill often tinged bluish-grey. See Nordmann's Greenshank and Marsh Sandpiper. **Adult breeding** Some scapulars have prominent blackish centres, crown, neck and breast heavily streaked and spotted blackish. See Nordmann's and Marsh Sandpiper. **Juvenile** Like adult non-breeding but upperside slightly browner-tinged with clear pale buff fringes, median and lesser coverts darker, neck- and breast-streaks somewhat bolder. **VOICE** Distinctive flight call is a loud clear ringing ***teu-teu-teu*** or ***chew-chew-chew***. When alarmed, utters a throaty ***kiu kiu kiu*** or ***kyoup-kyoup-kyoup***, recalling Common Redshank, and a sharp ***tchuk*** or ***chip***. **HABITAT** Various wetlands, mudflats, large rivers; up to 450 m. **RANGE & STATUS** Breeds northern Palearctic; winters Africa, Madagascar, southern W Palearctic, Middle East, Indian subcontinent, Tibet, southern China, Taiwan, S Japan, Sundas, Philippines, Wallacea, New Guinea, Australia. **SE Asia** Fairly common to common winter visitor and passage migrant throughout. Also recorded in summer C Thailand, Peninsular Malaysia, Singapore.

296 NORDMANN'S GREENSHANK
Tringa guttifer Plate **36**

IDENTIFICATION 29–32 cm. Monotypic. **Adult non-breeding** Very similar to Common Greenshank but legs shorter (particularly above joint) and yellower, neck shorter, bill distinctly bicoloured, crown, nape and sides of breast more uniform and only faintly streaked, upperside much plainer, without obvious dark markings, has more white above eye and paler lores. In flight shows all-white uppertail-coverts and rather uniform greyish tail, toes do not extend beyond tail-tip. Legs and feet yellow to greenish-yellow or brownish-yellow. See Marsh Sandpiper. **Adult breeding** Distinctive. Much more boldly marked, feathers of upperside largely blackish with whitish spots and spangling, head and upper neck heavily dark-streaked, with distinctive broad blackish crescentic spots on lower neck and breast and darker lores. **Juvenile** Like adult non-breeding but crown and upperparts tinged pale brown, fringes of scapulars and tertials have whitish notching, wing-covert fringes pale buff, breast has slight brown wash and faint dark streaks at sides. **VOICE** Flight call is a distinctive ***kwork*** or ***gwaak***. **HABITAT** Mud- and sandflats, sometimes other coastal wetlands. **RANGE & STATUS** Breeds E Palearctic (Sakhalin I); winters coastal Bangladesh, Sumatra, possibly Borneo, Philippines. **SE Asia** Rare coastal winter visitor west C,S Thailand, west Peninsular Malaysia, E Tonkin. Rare coastal passage migrant C,W,S Thailand, Cambodia, E Tonkin, Cochinchina (may winter). Vagrant Singapore (formerly wintered). Recorded (status unknown) S Myanmar, Tenasserim.

297 MARSH SANDPIPER
Tringa stagnatilis Plate **36**

IDENTIFICATION 22–25 cm. Monotypic. **Adult non-breeding** Resembles Common Greenshank but smaller and slimmer with distinctly thin straight blackish bill and proportionately longer legs, no bold streaks on head, neck or mantle, no obvious dark loral stripe, but shows broad white supercilium. Upperside fairly uniform greyish, blacker on lesser and median coverts, underside white, legs and feet greenish. In flight, upperside appears fairly uniform with contrasting white back to uppertail-coverts. See Nordmann's Greenshank. **Adult breeding** Upperside boldly patterned with black, with distinct dark speckles and streaks on crown, neck and breast, and dark arrow-shapes extending along flanks; legs and feet often yellowish-tinged. **Juvenile** Like adult non-breeding but mantle to wing-coverts browner-tinged with narrow dark subterminal bars and pale buff fringes. **VOICE** Repeated plaintive mellow ***keeuw*** or ***plew***, higher-pitched, thinner and less ringing than Common Greenshank. Loud ***yip***, ***yup*** or ***chip*** (often repeated rapidly) when alarmed. **HABITAT** Various wetlands, mudflats, marshes, large rivers; lowlands. **RANGE & STATUS** Breeds W(east),SC Palearctic, S Siberia, Mongolia, NE China; winters Africa, SW Europe, Middle East, Indian subcontinent, S China, Taiwan, Sundas, Philippines, Wallacea, Australia. **SE Asia** Uncommon to common (mostly coastal) winter visitor and passage migrant (except NW Thailand, W Tonkin, C Annam). Also recorded in summer Peninsular Malaysia, Cochinchina.

298 WOOD SANDPIPER *Tringa glareola* Plate 35

IDENTIFICATION 18.5–21 cm. Monotypic. **Adult non-breeding** Told by combination of smallish size, longish straight bill, long broad whitish supercilium, dark brown upperparts with faint buffish speckles, lightly streaked foreneck and upper breast, and pale greenish-yellowish legs and feet. In flight appears quite uniform above, with contrasting darker flight-feathers and white rump. See Green and Marsh Sandpipers. **Adult breeding** Upperparts more blackish with much bolder whitish speckles and fringes, crown, neck and upper breast clearly streaked. **Juvenile** Similar to adult non-breeding but upperparts and wing-coverts browner with finer and denser buff speckles, has more defined streaking on foreneck and upper breast. **VOICE** Nervous ***chiff-if*** or ***chiff-iff-iff***, particularly in flight. Sharp ***chip*** (often rapidly repeated) when alarmed. **HABITAT & BEHAVIOUR** Marshes, flooded rice paddies, lake margins, large rivers, rarely mudflats; up to 800 m. Gregarious. **RANGE & STATUS** Breeds northern Palearctic, NE China; winters Africa, Madagascar, Middle East, Indian subcontinent, S China, Taiwan, southern Japan, Sundas, Philippines, Wallacea, uncommonly New Guinea, Australia. **SE Asia** Uncommon to common winter visitor and passage migrant (except W Tonkin).

299 COMMON REDSHANK
Tringa totanus Plate **36**

IDENTIFICATION 27–29 cm. *T.t.eurhinus*. **Adult non-breeding** Told by combination of size, plain-looking brownish-grey upperside, whitish underside with fine dark breast-streaks, stoutish straight red bill with dark distal half, and bright red legs and feet. Sides of head greyish with neat whitish eyering, but lacks pronounced supercilium. In flight shows diagnostic white secondaries and white-tipped inner primaries. See Spotted Redshank and Terek Sandpiper. **Adult breeding** Upperparts browner with small blackish markings, head, neck and underparts more heavily dark-streaked. **Juvenile** Similar to adult breeding but has neat pale buffish spotting and spangling on upperparts and wing-coverts, more narrowly streaked breast and often more yellowish-orange legs and feet. **Other subspecies in SE Asia** *T.t.craggi*, *terrignotae*, *ussuriensis* (ranges in region unclear but all recorded Peninsular Malaysia); only separable by minor detail in breeding plumage). **VOICE** Typical call is a plaintive ***teu-hu-hu***, particularly in flight. When alarmed, gives a long mournful ***tyuuuu*** and rapid repetition of call. **HABITAT** Coastal wetlands, lowland marshes, large rivers. **RANGE & STATUS** Breeds W and southern Palearctic, NW,N India, Tibet, Mongolia, NW,N,NE China, N Japan; some populations winter south to Africa, Middle East, Indian subcontinent, southern China, Taiwan, Sundas, Philippines, Wallacea, rarely New Guinea, N Australia. **SE Asia** Common winter visitor and passage migrant in coastal areas throughout. Scarce to uncommon inland passage migrant N,E,C,S Myanmar, C,NE Thailand, N,C Laos. Recorded in winter north W Myanmar. Vagrant NW Thailand. Also recorded in summer C,S Thailand, Peninsular Malaysia, Singapore.

SCOLOPACIDAE: CALIDRIDINAE: *Calidris* sandpipers & allies

Worldwide 24 species. SE Asia 15 species. Small to medium-sized, mostly with shortish, slender bills (downcurved in some species), variable leg length and generally marked non-breeding and breeding plumages. Mostly gregarious, on coastal wetlands; feeding mainly on insects, worms, molluscs and crustaceans etc.

300 **GREAT KNOT** *Calidris tenuirostris* Plate **38**
IDENTIFICATION 28.5–29.5 cm. Monotypic. **Adult non-breeding** Shows distinctive combination of medium size, rather attenuated shape with longish, broad-based, slightly downward-tapering blackish bill and shortish dark legs, grey upperside with dark shaft-streaks, rather nondescript pale greyish head, neck and upper breast with darker streaks and grey to black spots on sides of breast and upper flanks. Has indistinct white supercilium and white throat, belly and vent. In flight, upperside appears rather uniform but with blacker primary coverts, white uppertail-coverts contrasting with dark tail, underwing-coverts mostly white. See Red Knot and Curlew Sandpiper. **Adult breeding** Very distinctive. Upperside boldly marked with black, scapulars bright chestnut with black markings, head and neck have bold black streaks, and breast and flanks show dense black spots (centre of breast may be solidly black). **Juvenile** Similar to adult non-breeding but crown to mantle more prominently streaked blackish, scapulars blackish-brown with whitish to buffish-white fringes, wing-coverts with blackish arrow-shapes and broad whitish to whitish-buff fringes, breast washed buffish and more distinctly dark-spotted. **VOICE** In flight may give muffled ***knut*** or ***nyut*** notes and a harsher ***chak-chuka-chak*** and ***chaka-ruk-chak*** when flushed or alarmed. **HABITAT & BEHAVIOUR** Mud- and sandflats, sometimes coastal pools and saltpans. Feeds mainly by probing. Gregarious. **RANGE & STATUS** Breeds NE Palearctic; winters along coasts of Indian subcontinent, S China, Taiwan, Sundas, Philippines, Wallacea, S New Guinea, Australia. **SE Asia** Local winter visitor C,S(west) Thailand, west Peninsular Malaysia, E Tonkin. Rare to uncommon coastal passage migrant (except Cambodia, N,C,S Annam, Cochinchina). Recorded (status uncertain, but probably winters) Cochinchina. Also recorded in summer Peninsular Malaysia.

301 **RED KNOT** *Calidris canutus* Plate **38**
IDENTIFICATION 23–25 cm. *C.c.canutus*: **Adult non-breeding** Similar to Great Knot but smaller and more compact with relatively larger head and shorter neck, shorter, straighter bill, usually more distinct dark loral stripe, better defined whitish supercilium, more uniformly grey upperside, no black spots on underparts and usually smaller, more V-shaped flank markings. In flight shows uniform dark scales/bars on rump and uppertail-coverts, greyer tail, and less white on underwing-coverts. Legs and feet dull olive-green. See Grey-tailed Tattler and Curlew Sandpiper. **Adult breeding** Very distinctive. Crown, nape and mantle boldly streaked blackish, scapulars boldly patterned blackish and chestnut, face and underparts deep reddish-chestnut, vent whiter with dark markings. Note size and shortish bill and legs. See dowitchers. Post-breeding adult has largely blackish scapulars and moults to non-breeding plumage head first. **Juvenile** Similar to adult non-breeding but mantle, scapulars and wing-coverts browner with prominent buffish-white fringes and dark subterminal markings, breast and flanks washed buff and finely dark-streaked. **Other subspecies in SE Asia** *C.c.rogersi* (recorded Peninsular Malaysia, Vietnam). **VOICE** When foraging and in flight, may give occasional soft nasal ***knut*** or ***wutt*** notes. Sudden ***kikkik*** when flushed or alarmed. **HABITAT & BEHAVIOUR** Mostly mud- and sandflats. Feeds by probing but also picks from surface. Gregarious. **RANGE & STATUS** Breeds NC,NE Palearctic, north N America, Greenland; winters along coasts of N,W,S Africa, W,S Europe, Sri Lanka, Australia, New Zealand, south USA, south S America, rarely Bangladesh, S China, Philippines, Wallacea. **SE Asia** Scarce to uncommon coastal passage migrant S Myanmar, C,W,S Thailand, Peninsular Malaysia, Singapore, E Tonkin, C Annam. Occasionally winters C,W Thailand, E Tonkin, Cochinchina (may winter).

302 **SANDERLING** *Calidris alba* Plate **37**
IDENTIFICATION 18–21 cm. Monotypic. **Adult non-breeding** Combination of smallish size, shortish straight black bill, pale grey ear-coverts and upperside and snowy-white underside distinctive. Often shows contrasting dark area at bend of wings, legs and feet blackish, lacks hind toe. In flight shows very prominent broad white bar across upperwing. See Spoon-billed Sandpiper, Red-necked Stint and Dunlin. **Adult breeding** Brighter individuals are very similar to Red-necked Stint but larger, with dark streaks on somewhat duller, more chestnut sides of head, throat and breast (including lower breast-sides), mostly chestnut centres of scapulars and no hind toe. In flight shows much broader white wing-bar. Duller (fresh) individuals have only faint rufous or chestnut wash on head and breast and more contrasting dark breast markings. See Little Stint. **Juvenile** Similar to adult non-breeding but has darker streaks on crown and hindneck, blackish mantle and scapulars, boldly patterned with buffish-white and buff-washed breast with dark streaks at sides. **VOICE** Typical flight call is a quiet liquid ***klit*** or ***twik*** (often repeated), sometimes extending to short trill. **HABITAT & BEHAVIOUR** Sand- and mudflats, beaches, sometimes saltpans, large rivers. Feeds by rapid probing and pecking, runs very fast. **RANGE & STATUS** Breeds Svalbard, NC,NE Palearctic, N Canada, Greenland; winters along coasts of Africa, Madagascar, west and southern W Palearctic, Middle East, Indian subcontinent, S China, Taiwan, southern Japan, Sundas, Philippines, Wallacea, New Guinea, Australia, New Zealand, Oceania, N(southern),C,S America. **SE Asia** Scarce to uncommon coastal winter visitor and passage migrant (except SW Myanmar, Tenasserim, SE Thailand, Cambodia, N,C Annam). Rare inland passage migrant C Myanmar, S Laos.

303 **SPOON-BILLED SANDPIPER**
Calidris pygmeus Plate **37**
IDENTIFICATION 14–16 cm. Monotypic. **Adult non-breeding** Resembles Red-necked and Little Stints but has diagnostic spatulate-shaped bill (less obvious in profile). Also has somewhat bigger head, whiter forehead and breast and broader white supercilium. Note behaviour. **Adult breeding** Apart from bill, very similar to Red-necked Stint but scapulars more uniformly fringed rufous-buff. See Sanderling. **Juvenile** Very similar to Red-necked and Little Stints (apart from bill) but has whiter forehead and face, somewhat darker and more contrasting lores and ear-coverts and more uniform buff to buffish-white fringes to mantle and scapulars. **VOICE** Quiet rolling ***preep*** and shrill ***wheet***, usually in flight. **HABITAT & BEHAVIOUR** Mud- and sandflats, sometimes sandy beaches, saltpans and prawn-ponds. Often stands more upright than Red-necked Stint. Feeds in shallow water and on soft, wet mud, either sweeping bill from side to side or patting surface of wet mud with bill. **RANGE & STATUS** Breeds extreme NE Palearctic; winters along coasts of Bangladesh, rarely SE,E India, S China. **SE Asia** Scarce and local winter visitor SW,S Myanmar, C Thailand, E Tonkin. Coastal vagrant Tenasserim, C,W,S Thailand, Peninsular Malaysia, Singapore, S Annam. Recorded (status uncertain, but may winter) Cochinchina.

304 **LITTLE STINT** *Calidris minuta* Plate **37**
IDENTIFICATION 14–15.5 cm. Monotypic. **Adult non-breeding** Very similar to Red-necked Stint. At close range, appears slightly slimmer and longer-legged with somewhat finer, often slightly drooping bill, has broader dark centres to feathers of upperparts, grey areas of head and breast usually more streaked, sometimes a more diffuse, greyish breast-band. Legs and feet blackish. **Adult breeding** Like Red-necked but chin and throat whitish, more orange-rufous (less red) breast (which is usually completely dark-streaked or speckled), base

colour of lower breast-sides usually rufous, wing-coverts and tertials mostly blackish with distinct bright rufous fringes, often has narrow pale buffish lateral crown-stripes and mantle-stripes. See Sanderling and Spoon-billed Sandpiper. **Juvenile** From Red-necked, on average, by more contrastingly patterned head, darker-centred and more rufous-edged upperpart feathers (especially lower scapulars, tertials and wing-coverts) and more coarsely streaked breast-sides. Recalls Long-toed but has less contrasting head pattern, much less heavily streaked neck and breast, shorter neck and dark legs and feet. See Spoon-billed Sandpiper and Long-toed Stint. **VOICE** In flight may give sharp staccato ***kip*** or ***tit*** notes, sometimes extending to short trill. **HABITAT & BEHAVIOUR** Saltpans, prawn-ponds, mudflats, large rivers; lowlands. Feeding action similar to Red-necked. **RANGE & STATUS** Breeds N Palearctic; winters Africa, Madagascar, south W Palearctic, Middle East, Indian subcontinent. **SE Asia** Scarce to uncommon coastal winter visitor and passage migrant SW,S Myanmar, north Tenasserim. Scarce to uncommon passage migrant N,C Myanmar. Coastal vagrant W,C Thailand, Peninsular Malaysia, E Tonkin.

305 RED-NECKED STINT

Calidris ruficollis Plate **37**

IDENTIFICATION 14–16 cm. Monotypic. **Adult non-breeding** Distinguished by small size, shortish straight black bill, rather uniform greyish upperside with dark shaft-streaks, white underparts and fairly distinct greyish lateral breast-patches with slight dark streaking. Supercilium whitish, legs and feet blackish. See very similar Little Stint and Spoon-billed Sandpiper; also Sanderling, Long-toed Stint and Dunlin. **Adult breeding** Variable. Cheeks, ear-coverts, lower throat and centre of upper breast unstreaked rufous to brick-red, supercilium whitish to brick-red, face whiter around base of bill, mantle and scapulars with blackish central markings and rufous to brick-red fringes, tertials and wing-coverts mostly fringed greyish-white, breast-sides (below rufescent area) usually whitish with dark spots/streaks. Fresh-plumaged birds (pre-breeding) have little rufous or chestnut on head and breast and greyer fringes to mantle and scapulars. See Little Stint, Sanderling and Spoon-billed Sandpiper. **Juvenile** Crown and upperparts darker than adult non-breeding, mantle and upper scapulars blackish with pale warm fringes, lower scapulars grey with dark subterminal markings and whitish fringes, tertials grey with whitish to buffish fringes, wing-coverts edged whitish, breast-sides washed pinkish-grey and faintly streaked; sometimes shows faint whitish mantle-lines, rarely shows obvious 'split supercilium'. See very similar Little; also Long-toed and Spoon-billed Sandpiper. **VOICE** Typical flight call is a thin ***kreep*** or ***creek***. Sometimes gives shorter ***krep***, ***kiep*** or ***klyt***. May give short trill when flushed or alarmed. **HABITAT & BEHAVIOUR** Saltpans, coastal pools, mud- and sandflats, sometimes large rivers and other inland wetlands; up to 450 m. Feeds with rapid pecking action, sometimes probes. **RANGE & STATUS** Breeds NE Palearctic, NW USA (Alaska); winters (mostly on coasts) Bangladesh, S China, Taiwan, Greater Sundas, Philippines, Wallacea, New Guinea, N Melanesia, Australia, New Zealand, sparingly SE,E India, southern Japan. **SE Asia** Uncommon to common coastal winter visitor and passage migrant throughout. Rare to scarce inland passage migrant N,C Myanmar, NW,C Thailand, Laos, Cambodia. Also recorded in summer S Thailand, Peninsular Malaysia, Singapore.

306 TEMMINCK'S STINT

Calidris temminckii Plate **37**

IDENTIFICATION 13.5–15 cm. Monotypic. **Adult non-breeding** From other stints by combination of attenuated shape with rather long tail (often projecting slightly beyond closed wing-tips), relatively uniform cold greyish-brown upperside, plain-looking, drab brownish-grey ear-coverts and breast (sometimes slightly paler in centre) and typically greenish-yellow legs and feet. Resembles miniature Common Sandpiper. In flight shows distinctive white sides of tail. Legs and feet greenish-yellow to yellow or yellowish-brown. See Long-toed Stint. **Adult breeding** Upperparts more dull olive-brown, with irregular large blackish feather-centres and rufous fringes on mantle and scapulars, sides of head and breast washed brown, indistinct darker streaks on sides of head and neck and breast. **Juvenile** Similar to adult non-breeding but mantle, scapulars and wing-coverts browner with narrow buff fringes and blackish subterminal markings, sides of head and breast washed buffish. **VOICE** In flight utters distinctive rapid stuttering ***tirrr*** (often repeated) or longer ***tirrr'r'r*** or ***trrrrrit***. **HABITAT & BEHAVIOUR** Muddy freshwater wetlands and rice paddies, large rivers, saltpans, prawn-ponds, rarely mudflats; up to 450 m. **RANGE & STATUS** Breeds N Palearctic; winters N and equatorial Africa, Middle East, Indian subcontinent, S China, southern Japan, N Borneo, Philippines. **SE Asia** Scarce to locally common winter visitor (except W Tonkin, N,S Annam). Also recorded on passage C Myanmar, E Tonkin.

307 LONG-TOED STINT

Calidris subminuta Plate **37**

IDENTIFICATION 14–16 cm. Monotypic. **Adult non-breeding** Resembles Red-necked and Little Stints but somewhat longer-necked, has finer bill with pale-based lower mandible, upperparts browner with much larger dark feather-centres, neck-sides and breast washed brown and distinctly streaked darker, legs and feet usually yellowish-brown to greenish (sometimes pale orange-yellow) with distinctly longer toes. See Temminck's Stint. **Adult breeding** Crown rufous with dark streaks, upperparts and tertials broadly fringed rufous, neck-sides and breast washed creamy-buff and very distinctly dark-streaked (breast may be slightly paler in centre). See juvenile Little Stint and much larger Sharp-tailed Sandpiper. **Juvenile** Similar to breeding adult but usually shows slight 'split supercilium' and prominent white mantle-lines, lower scapulars somewhat greyer. See Little. **VOICE** In flight gives a soft liquid ***kurrrip*** or ***chirrup*** and shorter ***prit***. **HABITAT & BEHAVIOUR** Freshwater marshes, wet rice paddies, saltpans, coastal pools, rarely mudflats; lowlands. Often feeds amongst vegetation. When alarmed may stand erect with neck stretched up. **RANGE & STATUS** Breeds C,E Palearctic; winters SE,E,NE India, Sri Lanka, Bangladesh, S China, Taiwan, S Japan, Sundas, Philippines, Wallacea, New Guinea, Australia. **SE Asia** Uncommon to common winter visitor (except W,N Myanmar, C Laos, W Tonkin, N Annam). Also recorded on passage Peninsular Malaysia, Singapore, Cambodia, Cochinchina. Recorded in summer Singapore.

308 PECTORAL SANDPIPER

Calidris melanotos Plate **38**

IDENTIFICATION 19–23 cm. Monotypic. **Adult non-breeding** Similar to Sharp-tailed Sandpiper but head pattern less striking with less distinct supercilium and no obvious eyering, dark streaks on foreneck and breast more distinct and sharply demarcated from white belly. Legs and feet dull greenish, brownish or yellowish. **Male breeding** Like female breeding but foreneck and breast may be mostly blackish-brown with whitish mottling (particularly when worn). **Female breeding** Mantle, scapulars and tertials blackish-brown with dull chestnut to brownish-buff fringes, foreneck and breast washed buffish, slight flank-streaking. See Sharp-tailed. **Juvenile** Similar to female breeding but usually has clear white lengthwise lines on mantle and scapulars, wing-coverts neatly fringed buffish to whitish-buff. From Sharp-tailed primarily by less distinct supercilium, duller sides of head and neck and duller breast with much more extensive and bolder dark streaks. **VOICE** Flight call is a reedy ***kirrp*** or ***chyrrk***, recalling that of Curlew Sandpiper but somewhat harsher. **HABITAT & BEHAVIOUR** Marshes, saltpans, coastal wetlands. Often feeds on drier margins of wet habitats. Feeds by pecking and shallow probing. **RANGE & STATUS** Breeds N,NE Palearctic, north N America; winters SE Australia, New Zealand, S America, rarely S Africa. **SE Asia** Vagrant Peninsular Malaysia, Singapore.

309 **SHARP-TAILED SANDPIPER**
Calidris acuminata Plate **38**

IDENTIFICATION 17–21 cm. Monotypic. **Adult non-breeding** Distinguished by size, medium-length, slightly down-tapering dark bill, rich dark brown crown, prominent whitish supercilium, dull foreneck and breast with diffuse dark streaks, and dull greenish to yellowish legs and feet. Has fairly prominent pale eyering, upperparts rather nondescript, mostly dull greyish-brown with pale buffish to whitish fringes. Recalls much smaller Long-toed Stint. See Pectoral Sandpiper. **Adult breeding** Crown distinctly rufous with dark streaks, mantle and scapulars blackish-brown with mostly bright rufous fringes, streaks on neck and upper breast more distinct, bold dark brown arrowhead markings on lower breast, upper belly and flanks. **Juvenile** Striking. Resembles breeding adult but supercilium and throat plainer and whiter, neck-sides and breast rich buff with little dark streaking, rest of underparts unmarked white. **VOICE** Flight call is a soft ***ueep*** or ***wheep*** (often repeated). Also gives a twittering ***teet-teet-trrt-trrt*** or ***prtt-wheet-wheet***. **HABITAT & BEHAVIOUR** Marshes, fishponds, mudflats. Often feeds on drier margins of wet habitats. **RANGE & STATUS** Breeds NE Palearctic; winters New Guinea, N Melanesia, Australia, New Zealand. **SE Asia** Vagrant S Myanmar, C,S Thailand, Peninsular Malaysia, Singapore, E Tonkin, S Annam.

310 **DUNLIN** *Calidris alpina* Plate **38**

IDENTIFICATION 17–21 cm. *C.a.sakhalina*: **Adult non-breeding** Like Curlew Sandpiper but somewhat shorter-necked and shorter-legged, bill slightly shorter and less strongly downcurved, supercilium less distinct, foreneck and upper breast duller with fine streaking. In flight shows dark centre to rump and uppertail-coverts. Note voice. Also resembles Broad-billed Sandpiper but slightly larger and less elongated, legs relatively longer and always blackish, bill slightly narrower and more gently downcurved (rather than distinctly kinked at tip), never shows obvious 'split supercilium', wing-coverts more uniform. See Sanderling, Red Knot and stints. **Adult breeding** Has diagnostic large black belly-patch, neck and breast whitish with distinct dark streaks, supercilium whitish, mantle and scapulars mostly bright rufous-chestnut with some black subterminal markings and whitish fringes. **Juvenile** Quite different to adult non-breeding. Mantle and scapulars blackish with rufous, buff and whitish fringing, wing-coverts neatly fringed buff, sides of head, neck and upper breast washed buffish-brown, with extensive blackish streaks on hindneck, breast and belly. See much larger Sharp-tailed Sandpiper. **VOICE** Flight call is a characteristic harsh ***treeep*** or ***kreee***. Utters soft twittering notes when foraging. **HABITAT & BEHAVIOUR** Mudflats, saltpans, prawn-ponds, large rivers. Feeding action similar to Curlew Sandpiper. **RANGE & STATUS** Breeds north Holarctic; winters south to northern Africa, Middle East, Pakistan, Indian subcontinent (sparingly), southern China, Taiwan, S Korea, S Japan, Mexico. **SE Asia** Rare to scarce winter visitor NW,C,NE Thailand, N Laos, coastal E Tonkin (locally common). Vagrant N Myanmar, S Thailand, Peninsular Malaysia, Singapore. Recorded on coast (status uncertain) C Annam, Cochinchina.

311 **CURLEW SANDPIPER**
Calidris ferruginea Plate **38**

IDENTIFICATION 19–21.5 cm. Monotypic. **Adult non-breeding** Distinguished by size, relatively long downcurved blackish bill, fairly long blackish legs, rather plain greyish upperside, prominent white supercilium and white underparts with indistinct streaky greyish wash on breast. In flight, upperside appears rather uniform with narrow white bar along tips of primary and greater coverts and distinctive white band on lower rump and uppertail-coverts. Note voice. See Dunlin and Broad-billed and Stilt Sandpipers. **Male breeding** Head and underparts deep reddish-chestnut with dark-streaked crown, whitish facial markings and chin and largely white vent; mantle and scapulars boldly patterned with chestnut, black and whitish. See Red Knot and Stilt Sandpiper. **Female breeding** Like male breeding but underparts slightly paler, with more whitish fringes. **Juvenile** Similar to adult non-breeding but upperparts and wing-coverts browner with neat pale buff fringes and dark subterminal markings, sides of head and breast washed pale peachy-buff and faintly dark-streaked. See Stilt Sandpiper. **VOICE** Flight call is a distinctive rippling ***kirrip*** or ***prrriit***. **HABITAT & BEHAVIOUR** Mud- and sandflats, coastal pools, large rivers; lowlands. Feeds by pecking and vigorous probing, often in deeper water than Dunlin. **RANGE & STATUS** Breeds N,NE Palearctic; winters (mostly on coasts) Africa, Madagascar, Middle East, Indian subcontinent, S China, Taiwan, Greater Sundas, Australia, New Zealand, sparingly Philippines, Wallacea, New Guinea. **SE Asia** Fairly common to common coastal winter visitor and passage migrant (except N Annam). Rare to scarce inland passage migrant N,C,S Myanmar, C,NE Thailand, S Laos. Also recorded in summer C,S Thailand, Peninsular Malaysia.

312 **STILT SANDPIPER**
Micropalama himantopus Plate **38**

IDENTIFICATION 18–23 cm. Monotypic. **Adult non-breeding** Similar to Curlew Sandpiper but bill longer and straighter, less pointed and only slightly down-turned at tip, legs distinctly longer and dull yellowish to dull greenish, foreneck, breast and flanks streaked greyish. In flight shows contrasting white lower rump and uppertail-coverts but has almost no wing-bar, feet project prominently beyond tail-tip. See Long-billed Dowitcher. **Adult breeding** Very distinctive. Crown and sides of head flushed bright chestnut with contrasting white supercilium, neck and upper breast heavily streaked/spotted blackish, lower breast to vent boldly barred blackish, neck and underparts variably washed buffish-pink, mantle and scapulars blackish with bold rufous, pinkish and white fringing; white of rump and uppertail-coverts may be partly obscured by dark markings. **Juvenile** Apart from structure, differs from Curlew Sandpiper by darker-centred upperparts and wing-coverts, rufescent-fringed upper scapulars, lightly streaked flanks and more contrasting head pattern. **VOICE** Usual flight calls are a soft rattled ***kirr*** or ***drrr*** and hoarser whistled ***djew***. **HABITAT & BEHAVIOUR** Coastal wetlands. Feeds in similar fashion to dowitchers, often in belly-deep water. **RANGE & STATUS** Breeds north N America; winters south USA, S America; recorded widely as vagrant, including Taiwan, Japan, N Australia. **SE Asia** Vagrant Singapore.

313 **BROAD-BILLED SANDPIPER**
Limicola falcinellus Plate **37**

IDENTIFICATION 16–18 cm. *L.f.sibirica*: **Adult non-breeding** Resembles some stints but larger, bill longer and noticeably kinked downwards at tip. Upperside rather uniform grey with dark streaks and somewhat darker lesser coverts, has prominent white supercilium, typically shows narrow white lateral crown-streak ('split supercilium'), underside crisp white with narrow dark streaks on breast (mostly at sides). In flight shows dark leading edge to upperwing-coverts. Smaller and much shorter-legged than Dunlin and Curlew Sandpiper. **Adult breeding** Crown blackish, offsetting white 'split supercilium', ear-coverts tinged pinkish-brown, upperparts and wing-coverts blackish with rufous to pale chestnut and whitish fringes, prominent lengthwise white lines along edge of mantle and scapulars, and boldly dark-streaked neck and breast. When worn, upperparts may appear largely blackish, neck and breast washed pinkish-brown and even more densely streaked. **Juvenile** Similar to adult breeding (fresh) but white mantle-lines more prominent, wing-coverts broadly fringed buff. **VOICE** In flight gives a dry trilled ***trrreet*** or ***chrrreeit*** and shorter ***trett***. **HABITAT & BEHAVIOUR** Mud- and sandflats, saltpans, coastal pools. Feeding action similar to Dunlin and Curlew Sandpiper; slower than that of stints. **RANGE & STATUS** Breeds N Palearctic; winters along coasts of S,E Africa, Arabian Peninsula, Iran, Indian subcontinent, S China, Taiwan, Greater Sundas, Philippines, New Guinea, Australia.

SE Asia Uncommon to fairly common coastal winter visitor and passage migrant (except SE Thailand, N Annam). Also recorded in summer C Thailand.

314 **RUFF** *Philomachus pugnax* Plate **34**
IDENTIFICATION Male 29–32 cm, female 22–26 cm. Monotypic. **Adult non-breeding** Told by combination of size, rather hunch-backed and pot-bellied appearance, longish neck, relatively small head, shortish, slightly drooping bill, and longish, usually orange to yellowish legs. Plumage rather variable. Crown and upperside greyish-brown with pale buff to whitish fringes, underside mainly whitish with greyish wash and mottling on foreneck and upper breast, no obvious supercilium but whiter face. In flight shows broad-based wings and shortish tail, toes project beyond tail-tip, upperside rather nondescript with broad white sides to lower rump and uppertail-coverts and narrowish white wing-bar. Legs and feet pinkish-red or orange-red to dull yellowish or dull greenish. See Common Redshank. **Male breeding** Unmistakable. In full plumage, attains remarkable long head-, neck- and breast-plumes (forming broad loose 'ruff'), varying in colour from black or white to rufous-chestnut, with or without bars and streaks. Has greenish or yellowish to reddish naked, warty facial skin, usually mostly pinkish bill and typically orange to reddish-orange legs and feet. **Female breeding** Lacks plumes and ruff. Rather nondescript greyish-brown with whiter face, blackish markings on centres of upperparts and wing-coverts, and variable bold blackish markings on neck and breast; belly and vent whitish. **Juvenile** Upperside similar to female breeding but with neater warm buff to whitish fringing; head-sides, foreneck and breast rather uniform buff, legs and feet dull yellowish-brown to dull greenish. Male tends to be lighter overall with paler buff head-sides, foreneck and breast. **VOICE** Often silent but may give occasional low ***kuk*** or ***wek*** notes in flight. **HABITAT & BEHAVIOUR** Marshes, grassy areas, rice paddies, coastal pools, rarely mudflats; lowlands. Feeds mostly by surface pecking but often wades in fairly deep water. **RANGE & STATUS** Breeds northern Palearctic; winters Africa, west and south W Palearctic, Middle East, S Pakistan, India, Sri Lanka, Bangladesh, S China, rarely Japan, Greater Sundas, Philippines, Sulawesi, Australia. **SE Asia** Rare to scarce winter visitor SW,N,S Myanmar, Tenasserim, Thailand (except SE), Peninsular Malaysia, Singapore, Cambodia, Cochinchina, E Tonkin. Vagrant S Laos.

SCOLOPACIDAE: ARENARIINAE: Turnstones & allies

Worldwide 2 species. SE Asia 1 species. Fairly small, stocky and relatively short-billed and -legged, with striking blackish plumage patches, and marked non-breeding and breeding plumages. Quite strictly coastal, feeding mainly on insects, crustaceans, molluscs and worms, by turning over stones, seaweed and debris.

315 **RUDDY TURNSTONE**
Arenaria interpres Plate **35**
IDENTIFICATION 21–24 cm. *A.i.interpres*: **Adult non-breeding** Unmistakable. Smallish and robust with short stout blackish bill, short orange-red legs and feet and complex blackish pattern on sides of head and breast. Head mostly dark brown with paler lores/cheeks and supercilium, upperside mixed dark brown, blackish and dull rufous-chestnut, remainder of underparts white. **Male breeding** Head crisply patterned black and white, upperparts boldly marked blackish and orange-chestnut. **Female breeding** Head pattern duller than male, crown more streaked and washed brown, upperpart pattern less clear-cut. **Juvenile** Resembles adult non-breeding but has reduced dark pattern on sides of head and breast, whitish submoustachial stripe, broad buffish to buffish-white fringing on upperparts and wing-coverts and somewhat duller legs and feet. **VOICE** Usual call is a rapid rattling ***tuk tuk-i-tuk-tuk***, ***trik-tuk-tuk-tuk*** or ***tuk-e-tuk***. Utters low ***tuk*** notes when foraging and sharp ***chik-ik*** and ***kuu*** or ***teu*** when flushed or alarmed. **HABITAT & BEHAVIOUR** Mud- and sandflats, beaches, saltpans and other coastal wetlands. Turns over seaweed and small stones and digs holes in search of food; also scavenges. Walks with distinctive rolling gait. **RANGE & STATUS** Breeds N Holarctic; winters along coasts of Africa, Madagascar, W,S Europe, Middle East, Indian subcontinent, S China, Taiwan, southern Japan, Greater Sundas, Philippines, New Guinea, Australia, New Zealand, Oceania, N(south),C,S America, probably Wallacea. **SE Asia** Scarce to fairly common coastal winter visitor and passage migrant (except N Annam). Also recorded in summer Peninsular Malaysia, Singapore.

GLAREOLIDAE: GLAREOLINAE: Pratincoles

Worldwide 17 species. SE Asia 2 species. Small to smallish with short, arched and pointed bills, wide gapes, very short legs, long pointed wings and shortish tails. Flight action recalls terns. Feed by hawking, on larger insects, including moths, beetles and flying ants.

316 **ORIENTAL PRATINCOLE**
Glareola maldivarum Plate **30**
IDENTIFICATION 23–24 cm. Monotypic. **Adult breeding** Distinctive. Graceful and rather tern-like with short bill, long pointed wings and short forked tail. Crown and upperparts uniform warmish grey-brown, neck and breast paler and more buffish, lores blackish, throat and upper foreneck buff with narrow black border, prominent red base to lower mandible. In flight appears uniform above with sharply contrasting white rump and uppertail-coverts, underwing blackish-brown with mostly chestnut coverts. **Adult non-breeding** Neck, breast and throat duller, lores paler, throat-border broken into indistinct dark streaks, less red on base of bill. See Small Pratincole. **Juvenile** Similar to non-breeding adult but crown and upperside greyish-brown with prominent whitish to buff fringes and blackish subterminal markings, throat paler, neck-sides and breast streaked/mottled greyish-brown, bill all dark. **VOICE** A sharp tern-like ***kyik*** or ***kyeck***, ***chik-chik*** and ***chet*** etc., usually in flight. Also a loud ***cherr*** and rising ***trooeet***. **HABITAT** Marshes, large rivers, lakes, dry rice paddies and open country, coastal pools; up to 1,200 m (2,600 m on migration). **RANGE & STATUS** Breeds India, Bangladesh, Sri Lanka, S,E,N,NE China, Taiwan, SE Siberia, Japan, sporadically S Pakistan, N Borneo, N Philippines; more northern populations winter to south, India, Greater Sundas, New Guinea, N Australia. Passage migrant and/or winter visitor Philippines, Wallacea. **SE Asia** Scarce to locally common breeding visitor (with some irregular overwintering) Myanmar, Thailand, Peninsular Malaysia, Cambodia, N Laos, C,S Annam, Cochinchina. Possibly mostly resident in parts of Myanmar, Cochinchina. Scarce to fairly common passage migrant S Myanmar, NW,S Thailand, Peninsular Malaysia, Singapore, Cambodia, Laos, E Tonkin, C Annam. **BREEDING** April–June. Often colonial. **Nest** None or shallow scrape on ground or amongst stubble or short grass. **Eggs** 2–3; usually pale yellowish stone-colour to yellowish-buff, densely blotched black over grey undermarkings; 30.8 x 23.9 mm (av.).

317 **SMALL PRATINCOLE**
Glareola lactea Plate **30**
IDENTIFICATION 16–19 cm. Monotypic. **Adult breeding** Easily distinguished from Oriental Pratincole by smaller size, much paler and greyer upperside, paler throat with no dark border and pale buffish-grey breast. Lores black. In

flight (above and below), shows broad white band across secondaries and inner primaries and black trailing edge to secondaries, tail almost square-cut with broad black terminal band. **Adult non-breeding** Lores paler, throat faintly streaked. **Juvenile** Similar to adult non-breeding but chin white, lower throat bordered by brownish spots/streaks, upperparts and wing-coverts with indistinct buffish fringes and brownish subterminal bars, tail tipped buffish-brown. **VOICE** In flight gives a high-pitched ***prrip*** or ***tiririt***. Also a short ***tuck-tuck-tuck...*** **HABITAT & BEHAVIOUR** Large rivers, dry margins of lakes and marshes, rarely coastal pools and sandy areas; up to 450 m. Highly gregarious. **RANGE & STATUS** Resident (subject to local movements) Indian subcontinent, SW China. **SE Asia** Uncommon to locally common resident (subject to some movements) Myanmar, NW,NE,C(northern) Thailand, north-east Cambodia, Laos. Vagrant C(south),W(coastal) Thailand, Singapore, Cochinchina. **BREEDING** May. Usually colonial. India: **Nest** None, or shallow scrape on ground. **Eggs** 2–4; sandy-buff to sandy-grey with small light grey-brown to reddish-brown blotches over lavender to greyish undermarkings; 25.9 x 20.5 mm (av.).

STERCORARIIDAE: Jaegers & allies

Worldwide 7 species. SE Asia 3 species. Largish to medium-sized. Resemble dark-plumaged gulls but flight swifter and more agile. Parasitic, feeding mainly by chasing gulls, terns and other seabirds until they drop their food; also scavenge.

318 POMARINE JAEGER

Stercorarius pomarinus Plate **42**

IDENTIFICATION 47–61.5 cm (including tail-streamers up to 11.5 cm). Monotypic. **Adult pale morph breeding** Shows distinctive combination of relatively large size and heavy build, mostly blackish-brown plumage with broad whitish collar, large whitish patch on lower breast and upper belly, and long broad twisted central tail-feathers (often broken or absent after breeding). Bill bicoloured, forehead blackish-brown (as crown), sides of neck washed yellowish, flanks sometimes barred whitish. Male often has breast-band reduced to lateral patches. In flight shows dark wings with whitish shaft-streaks at base of primaries and narrow white crescent at base of primaries on underwing. See Parasitic and Long-tailed Jaegers. **Adult pale morph non-breeding** Variable, may be similar to breeding or immature plumage. Shows less contrast between cap and collar, neck-sides and throat mottled dark brown, tail-coverts barred whitish, central tail-feathers shorter and usually not twisted. **Adult dark morph** Rare. Blackish-brown overall with similar primary pattern to pale morph. Intermediates are very rare. **Juvenile pale morph** Usually mostly cold medium brown with rather plain head and contrasting dark face; mantle and back narrowly barred buff, scapulars and wing-coverts darker with buff tips, vent broadly barred whitish. In flight shows contrasting pale-barred uppertail-coverts, underwing-coverts and axillaries, distinctive double pale patch on underwing (pale-based primary-coverts and primaries) and only slightly protruding, thumb-shaped central tail-feathers (tail looks triangular). Combination of dark head, prominently pale-barred uppertail-coverts and underwing pattern diagnostic. Some individuals have mostly pale head, paler lower breast to belly and pale-barred flanks, and are best distinguished (on plumage) by contrasting dark face and double pale patch on underwing. See Parasitic and Long-tailed. **Juvenile dark morph** Head and body rather uniform blackish-brown with pale-barred tail-coverts, otherwise similar to pale morph but scapulars and upperwing-coverts darker with less distinct pale tips, underwing with narrower pale markings on coverts but similar double pale patch on primary-coverts and primaries. **First-winter pale morph** Like juvenile but with pale hindcollar and no pale tips on upperside. **Second-winter pale morph** Similar to adult non-breeding but underwing-coverts like juvenile. **Third-winter pale morph** Lower breast and belly whiter than adult, may lack bars on greater underwing-coverts. **HABITAT & BEHAVIOUR** Open seas, sometimes close inshore, particularly when hunting. Usually seen in flight but rests on water. Normal flight rather slow and gull-like, piratical flight relatively laboured. **RANGE & STATUS** Breeds N Holarctic; winters (and occurs as non-breeder) widely in southern oceans (north to c.40°N), mainly Caribbean Sea, W,E Atlantic Ocean, Arabian Sea, SW,NC and south-eastern Pacific Ocean. **SE Asia** Rare to scarce offshore non-breeding visitor Tenasserim, C,SE,S Thailand, Peninsular Malaysia, Singapore, Cambodia (passage migrant).

319 PARASITIC JAEGER

Stercorarius parasiticus Plate **42**

IDENTIFICATION 42–54.5 cm. (including tail-streamers up to 10.5 cm). Monotypic. **Adult pale morph breeding** Similar to Pomarine Jaeger but smaller and slimmer, with narrower, darker bill, small white patch on extreme forehead, somewhat paler greyish-brown scapulars and wing-coverts, paler yellowish wash on sides of neck, mostly white underparts with variable clean grey breast-band (sometimes only lateral patches), dusky vent and pointed tail-streamers. In flight, wing pattern like Pomarine but wings somewhat slimmer and narrower-based. See Long-tailed Jaeger. **Adult pale morph non-breeding** Cap less distinct, has more dark markings on underparts, including breast-band of dark mottling, pale-edged mantle and pale-barred tail-coverts; tail-streamers may be lacking. **Adult dark morph** Best separated from Pomarine by size, shape, tail and behaviour. A range of intermediates with pale morph occur. **Juvenile pale morph** Highly variable. Head and underbody vary from rather plain rufescent- or cinnamon-tinged brown to predominantly pale. Some show contrasting paler head. On plumage, in general, best separated from Pomarine by presence of small pale forehead-patch, pale rusty to bright rufous nuchal band, streaks on head and neck and/or rufescent or warm tinge to head and underbody, lacks contrasting dark face, rarely shows double pale patch on underside of primary coverts and primary bases. Dark-headed birds lack pronounced whitish tail-covert barring shown by Pomarine. Some have diagnostic broad whitish crescent (rather than shaft-streaks) on upperside of primary bases. See Long-tailed. **Juvenile dark morph** On plumage, best separated from Pomarine by all-dark underwing-coverts, including primary coverts. See Long-tailed. **First-winter pale morph** As juvenile but develops pale areas on head and body; warmer-tinged individuals lose such tones. **Second-/third-winter pale morph** Gradually becomes more like adult non-breeding. **HABITAT & BEHAVIOUR** Open seas, sometimes close inshore. Usually seen in flight. Compared to Pomarine, normal flight typically faster and more falcon-like with occasional shearwater-like glides; piratical flight faster and more acrobatic, often chases terns and gulls for up to several minutes but rarely attacks birds themselves. **RANGE & STATUS** Breeds N Holarctic; mainly winters (and occurs as non-breeder) SW,E,SE Atlantic Ocean, Arabian Sea, offshore S,E Australia and New Zealand, SW,SE Pacific Ocean. **SE Asia** Scarce non-breeding visitor/passage migrant Cambodia. Offshore vagrant Thailand, Peninsular Malaysia, Singapore. Inland vagrant C Thailand.

320 LONG-TAILED JAEGER

Stercorarius longicaudus Plate **42**

IDENTIFICATION 47–67.5 cm (including tail-streamers up to 26.5 cm). *S.l.pallescens*: **Adult breeding** Similar to Parasitic Jaeger but somewhat slighter and more tern-like in build, with shorter but relatively thicker bill, noticeably paler, greyer upperparts and wing-coverts and longer central tail-feathers. In flight appears smaller with more elongated rear end, upperwing-coverts contrast with blackish primaries, only the outermost two of which have white shafts (forming narrow line on forewing), underwing all dark (apart from white shaft on leading primary), tail-streamers usually longer than width of inner wing, breast and upper belly whitish, darkening to grey on lower belly and vent. **Adult non-breeding** Tail-streamers shorter or lacking. Best separated from Parasitic and Pomarine Jaegers by size, shape and virtual lack of any whitish flash on primaries (particularly below). **Juvenile pale morph** Very variable. Similar to Parasitic but generally greyer and more contrastingly patterned, never showing rufescent or cinnamon plumage tones, upperparts and wing-coverts have whitish to buffish barring/feather-tips. In flight, appears slightly longer-bodied (behind wings) and longer-tailed, producing noticeably attenuated rear end, upperside of primaries usually have only two white shaft-streaks, underwing-coverts and axillaries normally strongly pale-barred, has more protruding blunt-tipped central tail-feathers. Individuals showing strikingly greyish plumage with whitish head and/or belly, or greyish head and breast and whitish belly, can only be this species. **Juvenile dark morph** On plumage, best separated from Parasitic by two (normally) white shaft-streaks on upperside of primaries and bolder black-and-white barring on tail-coverts and often axillaries. **First winter** Similar to juvenile but upperparts and wing-coverts plain greyish-brown, belly cleaner and whiter, central tail-feathers more pointed and usually longer. **Second winter** Similar to adult non-breeding but underwing-coverts somewhat barred. **Third winter** As adult non-breeding but may show a few pale bars on underwing-coverts. **HABITAT & BEHAVIOUR** As Parasitic but normal flight lighter and less purposeful, often recalling smallish gull or tern; piratical attacks less confident and briefer. **RANGE & STATUS** Breeds N Holarctic; winters (and occurs as non-breeder) southern oceans (mainly between 20 and 70°S), particularly S Atlantic, SE Pacific. **SE Asia** Scarce to uncommon offshore passage migrant Cambodia. Offshore vagrant S Thailand (also once inland), Peninsular Malaysia (also once inland).

RYNCHOPIDAE: Skimmers

Worldwide 3 species. SE Asia 1 species. Tern-like, with long pointed wings, short forked tails and unusual long thickish bills with longer lower mandibles. Feed mainly on fish.

321 **INDIAN SKIMMER**
Rynchops albicollis Plate **41**

IDENTIFICATION 40–43 cm. Monotypic. **Adult breeding** Unmistakable. Black crown, nape and upperside contrast with white forehead, collar and underparts, with long, thick, yellow-tipped, deep orange bill with noticeably longer lower mandible. In flight shows white trailing edge to secondaries and inner primaries, and short forked tail with mostly blackish central feathers. **Adult non-breeding** Upperside somewhat duller and browner-tinged. **Juvenile** Bill dusky-orange with blackish tip, crown and nape paler, brownish-grey with darker mottling, mantle, scapulars and wing-coverts paler and more greyish-brown with whitish to pale buffish fringes. **VOICE** Rather high, nasal ***kap*** or ***kip*** notes, particularly in flight. **HABITAT & BEHAVIOUR** Large rivers, lakes, rarely coastal wetlands; lowlands. Feeds in flight, skimming lower mandible through water. Flies with slow graceful wingbeats. Usually encountered singly, in pairs or small parties. **RANGE & STATUS** Resident (subject to local movements) Pakistan, C and northern India, Nepal, Bangladesh. **SE Asia** Rare to scarce resident W,N,E,C,S Myanmar; formerly SW Myanmar, C,S Laos. Vagrant C Thailand. Formerly recorded (status uncertain) Cambodia. **BREEDING** India: February–April. **Nest** Scrape on ground. **Eggs** 3–4; pale buff or pinkish-buff to greyish- or greenish-white, boldly blotched and streaked dark brown or reddish-brown, sometimes with purplish undermarkings; 41 x 30 mm (av.).

STERNIDAE: Noddies & terns

Worldwide 45 species. SE Asia 21 species. Small to medium-sized, gull-like but typically more delicately built with more slender, pointed wings and more buoyant, graceful flight. **Noddies & White Tern** (*Anous, Gygis*) Pelagic. Former have mostly dark plumage and wedge-shaped tails, latter all white with slightly forked tail. Normally fly not far above water. Feed on small fish, squid, crustaceans and plankton, usually by surface-picking. **Typical terns** (*Onychoprion, Sternula, Gelochelidon, Hydroprogne, Sterna, Thalasseus*) Generally larger than other terns, with deeply forked tails. Inhabit various wetland habitats, many feeding primarily by plunge-diving. Feed on fish, crustaceans, tadpoles, crabs etc. **Marsh terns** (*Chlidonias*) Generally smaller and more compact, tails shorter, without obvious fork. Flight more erratic, with stiffer wingbeats. Inhabit inland and coastal wetlands, rarely found far offshore. Normally feed on insects by hawking and surface-picking.

322 **BROWN NODDY** *Anous stolidus* Plate **39**

IDENTIFICATION 40–45 cm. *A.s.pileatus*: **Adult** Rather uniform dark chocolate-brown plumage with whitish forehead, grey crown, broken white eyering, and long wedge-shaped tail (slightly cleft when spread) distinctive. In flight shows paler brownish band across upperwing-coverts and paler brownish underwing-coverts. See very similar Black Noddy. **Juvenile** Feathers of crown, mantle, scapulars and wing-coverts have indistinct pale buffish fringes, forecrown somewhat browner, rarely whitish. See Black Noddy and juvenile Sooty Tern. **VOICE** Occasionally utters a harsh ***kaark*** and ***kwok kuok...*** **HABITAT** Open seas, islets. **RANGE & STATUS** Breeds tropical and subtropical Indian, Pacific and Atlantic Oceans, including offshore islands S Red Sea, SW India (Lakshadweep and Maldive Is), Taiwan, Philippines, E Wallacea, northern Australia; wanders widely outside breeding season, including Sri Lanka, Andaman and Nicobar Is, Wallacea. **SE Asia** Former local offshore breeder SE Thailand, Peninsular Malaysia. Scarce coastal/offshore non-breeding visitor Peninsular Malaysia. Scarce to uncommon off Cambodia (may breed). Coastal/offshore vagrant Tenasserim, west S Thailand, Cochinchina. **BREEDING** May–August. **Nest** Depression on ground or rock. **Eggs** 1; white to pale sand-colour, variably flecked brown and blotched and smeared lavender; 48.7–53.8 x 34.6–38.2 mm.

323 **BLACK NODDY** *Anous minutus* Plate **39**

IDENTIFICATION 34–39 cm. *A.m.worcesteri*: **Adult** Like Brown Noddy but smaller and slimmer with narrower and proportionately longer bill (noticeably longer than head), with blacker plumage and sharply contrasting white forehead and midcrown, and shorter tail. In flight shows uniform dark upperwing and underwing. **Juvenile** Best separated from Brown by size, structure, bill and all-dark underwing. Additionally, shows narrower brownish-buff feather-tips on upperside and more sharply defined whitish forehead and crown (rarely similar on Brown). See Sooty Tern. **VOICE** Querulous ***krrrk*** and rapid laughing ***k'k'k'k'...*** **HABITAT** Sea coasts, open ocean, islets. **RANGE & STATUS** Breeds tropical Pacific and Atlantic Oceans, including offshore islands, W Africa, E Sumatra, Java Sea, N Borneo, SW Philippines, E New Guinea, NE Australia; wanders to some extent outside breeding season. **SE Asia** Offshore vagrant south-west Peninsular Malaysia ('Malacca').

324 **WHITE TERN** *Gygis alba* Plate **39**

IDENTIFICATION 28–33 cm. *G.a.monte*: **Adult** Small size, all-white plumage with dark primary shafts and blackish bill, legs, feet and area encircling eye diagnostic. **Juvenile** Like adult but has dark flecks on crown and nape, brown and buff bars/scales on upperparts and wing-coverts, and brownish-tipped tail-feathers. See Little Tern. **VOICE** May give a guttural ***heech heech...*** **HABITAT & BEHAVIOUR** Open seas, islets. Flight light and buoyant. **RANGE & STATUS** Breeds tropical and subtropical Indian, Pacific and Atlantic Oceans, including islands off SW India (Seenu Atoll), E Australia; wanders to some extent outside breeding season. **SE Asia** Coastal vagrant Cochinchina.

325 **SOOTY TERN** *Onychoprion fuscatus* Plate **39**
IDENTIFICATION 42–45 cm. *O.f.nubilosa*: **Adult non-breeding** Similar to Bridled Tern but larger, upperparts and upperwing blacker and more uniform, no whitish eyebrow or nuchal band. On underwing in flight, whitish coverts contrast more sharply with uniformly blackish flight-feathers. **Adult breeding** From Bridled by concolorous blackish crown, hindneck and upperside, broader square-cut white forehead-patch (without white eyebrow), wing pattern (as adult non-breeding) and blackish tail with more contrasting white edge. **Juvenile/first winter** Distinctive. Mostly sooty-blackish, with contrasting whitish vent, whitish tips on mantle to uppertail-coverts, scapulars, tertials and wing-coverts, and mostly whitish underwing-coverts. **First summer** Intermediate between juvenile and adult non-breeding, underside variably dark-marked (particularly breast and flanks). **VOICE** Typical calls include a high-pitched ***ker-wacki-wah*** and shorter ***kraark***. **HABITAT** Open seas, islets. **RANGE & STATUS** Breeds tropical and subtropical Indian, Pacific and Atlantic Oceans, including coasts and offshore islands, E Africa, Madagascar, Indian Ocean, S,E Arabian Peninsula, W,S India (off Maharashtra, Lakshadweep Is, probably Maldive Is), Taiwan, SE Borneo, Philippines, E Lesser Sundas, New Guinea region, Australia; many populations disperse widely outside breeding season. **SE Asia** Offshore/coastal vagrant S Myanmar, Tenasserim, C,S Thailand, Peninsular Malaysia.

326 **BRIDLED TERN**
Onychoprion anaethetus Plate **39**
IDENTIFICATION 37–42 cm. *O.a.anaethetus*: **Adult non-breeding** Combination of size, dark brownish-grey upperside, thin whitish nuchal band, whitish forehead-patch (speckled darker) and blackish crown, nape and mask distinctive. Has uneven paler tips to upperparts and wing-coverts, deeply forked tail and blackish bill, legs and feet. In flight shows dark upperwing and mostly whitish underwing with darker secondaries, inner primaries and outer primary tips. See Sooty Tern. **Adult breeding** Clean white forehead and short eyebrow contrast with black eyestripe, crown and nape, lacks paler feather-tips on brownish-grey of upperparts and upperwing. In flight, leading edge of upper lesser coverts may appear slightly darker. See Sooty and Aleutian Terns. **Juvenile** Has whitish tips and dark subterminal bars on upperparts and wing-coverts, breast-sides sullied brownish-grey, head pattern similar to adult non-breeding. See Sooty. **VOICE** Usual calls include a staccato yapping ***wep-wep...*** **HABITAT & BEHAVIOUR** Open seas, islets. Flight very buoyant with elastic wingbeats. **RANGE & STATUS** Mostly resident (subject to dispersive movements) along coasts and offshore islands W,NE Africa, Madagascar, Indian Ocean, Arabian Peninsula, W,S India (off Maharashtra, Lakshadweep Is, probably Maldive Is), S,SE China, Taiwan, Sundas, Philippines, Palau Is, New Guinea region, N Melanesia, N Australia, Caribbean, C America, north S America. **SE Asia** Scarce to locally common offshore resident (subject to some movements) SE,S Thailand, Peninsular Malaysia, Cambodia, Cochinchina. Rare to uncommon non-breeding offshore/coastal visitor Thailand, Peninsular Malaysia, Singapore, Cambodia, C,S Annam. Vagrant inland N Annam. **BREEDING** May–July. **Nest** Scrape on ground or depression in rock. **Eggs** 1 (rarely 2); whitish to stone-grey or pinkish-buff, flecked pale grey, lilac-grey and light and dark purple (rarely with some black); 43–49 x 31–34.5.

327 **ALEUTIAN TERN**
Onychoprion aleutica Plate **40**
IDENTIFICATION 32–34 cm. Monotypic. **Adult non-breeding** Similar to Roseate and Common Terns but combination of whitish crown and defined narrow dark band along underside of secondary tips diagnostic. Rump and uppertail-coverts white, tail with grey centre (paler than mantle) and white outer two pairs of feathers. Some show whitish median upperwing-coverts, which appears as a midwing-bar in flight. See Black-naped Tern. **Adult breeding** Very distinctive. Bill, legs and feet blackish, white forehead contrasts with black eyestripe, hindcrown and nape, whitish cheeks and lower ear-coverts contrast with grey underparts. In flight, shows white leading edge of upperwing-coverts, white trailing edge of secondaries and pale inner primaries contrast with rest of upperwing. Underwing shows distinctive narrow blackish band along secondary tips. See Bridled Tern. **Juvenile** Similar to Common Tern but forehead more uniform warm brown, feathers of mantle, back, scapulars, tertials, lesser and median coverts mostly dark brown with buff tips, contrasting sharply with grey on remainder of wing. Rump, uppertail-coverts and tail centre grey. **First winter** Similar to non-breeding adult but has gingery wash on lower nape (as in juvenile); grey central uppertail contrasts with white outer feathers and rump/uppertail-coverts. In early part of winter may show retained juvenile feathers on tertials, secondaries and lesser coverts. **Second winter** As adult non-breeding but upperwing shows darker bar across lesser coverts and darker tips to secondaries. **VOICE** Flight call is a distinctive soft whistled ***twee-ee-ee***. **HABITAT & BEHAVIOUR** Open seas. Often perches on flotsam, floating objects. **RANGE & STATUS** Breeds along coasts of Sakhalin, Kamchatka, Kuril Is, north-west N America; wintering grounds not fully known but include pelagic areas of Greater Sundas, Wallacea. Occurs on passage off S China (Hong Kong) and Philippines (Pamilacan I). **SE Asia** Scarce offshore/coastal winter visitor/passage migrant Peninsular Malaysia, Singapore, Cambodia.

328 **LITTLE TERN** *Sternula albifrons* Plate **39**
IDENTIFICATION 22–25 cm. *S.a.sinensis*: **Adult non-breeding** Small size, long-billed and short-tailed appearance distinctive. Bill, legs and feet dark, lores and forecrown white, hindcrown blackish with white streaks, band through eye and nape black, rump to tail-centre white (sometimes grey). In flight shows rather uniform grey upperside and upperwing with narrow dark band on leading edge of lesser coverts and blackish outermost primaries with white shafts. *S.a.albifrons* (recorded in winter west S Thailand, west Peninsular Malaysia) has slightly darker grey on upperside, longer tail-streamers and dark shafts on outer primaries. **Adult breeding** Bill yellow with black tip, legs and feet orange to yellow, has black crown, central nape and eyestripe and well-defined, contrasting white forehead-patch. **Juvenile** Similar to non-breeding adult but with dark subterminal markings on mantle, scapulars, tertials and wing-coverts. In flight shows dark leading edge to upperwing and often has largely whitish secondaries and inner primaries; outer primaries less sharply contrasting. **First winter/first summer** Like adult non-breeding. **VOICE** Sharp high-pitched ***kik*** or ***ket*** notes, particularly in flight, and a harsh rasping ***kyik*** when alarmed. Also a rapidly repeated ***kir-rikikki kirrikikki...*** **HABITAT & BEHAVIOUR** Coasts, beaches, saltpans, mud- and sandflats, large rivers; lowlands. Flies with rapid wingbeats, often hovers. **RANGE & STATUS** Breeds W,N Africa, W,C Palearctic, E Arabian Peninsula, Iran, Pakistan, W,C and northern India, Nepal, Bangladesh, Sri Lanka, China, Taiwan, N,S Korea, Japan, Java, Bali, Philippines, Wallacea, N Melanesia, Australia; northern populations winter widely at sea and along coasts south to S Africa, Indian Ocean, Indian subcontinent, Sundas, Philippines, Wallacea, New Guinea, N Australia. **SE Asia** Scarce to uncommon coastal resident Myanmar, SE,C,W,S(east) Thailand, east Peninsular Malaysia, Singapore, Cambodia, C,S Annam; also inland C,S,E Myanmar, Cambodia (now rare), S Laos (now

rare). Uncommon to fairly common coastal winter visitor (except Myanmar?, N Annam?). **BREEDING** April–September. Colonial. **Nest** Depression on ground or saltpan bund. **Eggs** 1–3; olive- to stone-buff, blotched and spotted dark chestnut and black over pale greyish to lavender undermarkings; 31–34 x 22.5–24.6 mm.

329 GULL-BILLED TERN
Gelochelidon nilotica Plate **41**

IDENTIFICATION 34.5–37.5 cm. *G.n.affinis*: **Adult non-breeding** Combination of size, white head with blackish ear-coverts, pale silvery-grey rump and uppertail and shallow tail-fork distinctive. Recalls Whiskered Tern but much larger and larger-billed, with longer, slenderer wings. See Roseate, Black-naped and Common Terns. **Adult breeding** Forehead to nape black. **First winter** Similar to adult non-breeding but secondaries and upper primary coverts somewhat darker, tail dark-tipped; may show odd darker centres and whiter tips to upperwing-coverts. **Other subspecies in SE Asia** *G.n.nilotica* (SW,C,E,S Myanmar, ? parts of Thailand). **VOICE** In flight may give a low nasal ***ger-erk*** and ***kay-vek*** and loud metallic ***kak-kak***. Nasal ***kvay-kvay-kvay...*** when alarmed. **HABITAT** Coasts, mud- and sandflats, coastal pools, lakes, large rivers; lowlands. **RANGE & STATUS** Breeds southern W,C Palearctic, S Siberia, Iran, Pakistan, E India, NW,S,SE,E China, S Australia, USA, N Mexico, Caribbean, north and east S America; most northern breeders winter south to Africa, Middle East, Indian subcontinent, Sundas, Philippines, Wallacea, New Guinea, N Australia, northern S America. **SE Asia** Uncommon to fairly common winter visitor (mostly coastal) SW,C,E,S Myanmar, Tenasserim, W,C,SE,S Thailand, Peninsular Malaysia, Singapore, Cambodia, E Tonkin, Cochinchina. Also recorded in summer C Thailand, Peninsular Malaysia, Singapore.

330 CASPIAN TERN *Hydroprogne caspia* Plate **41**

IDENTIFICATION 48–55 cm. Monotypic. **Adult non-breeding** Unmistakable. Larger than other terns with diagnostic thick red bill. Crown, nape and mask mostly black, with whiter, dark-mottled forecrown, black subterminal marking on bill-tip. In flight, upperwing has grey-edged blackish outer primaries, underwing with contrasting, mostly blackish outer primaries. See Great Crested Tern. **Adult breeding** Forehead to nape and mask all black. **First winter** Similar to adult non-breeding but secondaries, primary upperwing-coverts and tail somewhat darker; may show slightly darker centres to upperwing-coverts. **VOICE** A loud deep croaking ***kraah*** and ***kra-krah*** and hoarse ***kretch***. **HABITAT** Coastal pools, mud- and sandflats, sometimes lakes, large rivers; lowlands. **RANGE & STATUS** Breeds W,S,E,NE Africa, Madagascar, north and south-east W Palearctic, southern C Palearctic, Middle East, S Pakistan, W India (Gujarat), Sri Lanka, S,E China, Australia, New Zealand, N America; most northern populations winter along coasts of northern and E Africa, Madagascar, S Europe, Middle East, Indian subcontinent, S China. Uncommon non-breeding visitor (probably from Australia) Timor, S New Guinea. **SE Asia** Rare to scarce coastal winter visitor S Myanmar, C,W,S Thailand, Peninsular Malaysia, Singapore, Cambodia (also inland along Mekong R), E Tonkin, Cochinchina. Scarce passage migrant Cambodia. Also recorded in summer Peninsular Malaysia, Cochinchina (inland).

331 BLACK TERN *Chlidonias niger* Plate **39**

IDENTIFICATION 22–26 cm. *C.n.niger*: **Adult non-breeding** Like White-winged Tern but with more solidly black hindcrown, larger ear-patch, distinctive dark smudge on side of breast and darker grey on upperside, including rump and uppertail-coverts. See Whiskered Tern. **Adult breeding** Rather uniform greyish plumage with darker grey body, blackish hood and white vent diagnostic. Legs and feet blackish. In flight shows pale grey underwing-coverts. See White-winged Tern. **Juvenile** Differs from White-winged by greyer, more clearly barred/scaled 'saddle' (contrasts less with darker grey upperwing), grey rump and uppertail-coverts, and prominent dark smudge on side of breast. **First winter** Similar to adult non-breeding. **VOICE** In flight may give weak sharp ***kik*** notes (sometimes doubled), and a short shrill nasal ***kyeh*** or ***kja***. **HABITAT** Marshes, coastal pools. **RANGE & STATUS** N Africa, W,C Palearctic, NW China, N America; winters NE and sub-Saharan Africa, Caribbean, C America, north S America. Vagrant India, Sri Lanka, Japan, E Australia. **SE Asia** Vagrant Singapore.

332 WHITE-WINGED TERN
Chlidonias leucopterus Plate **39**

IDENTIFICATION 20–24 cm. Monotypic. **Adult non-breeding** Similar to Whiskered Tern but has finer bill and distinctive head pattern, with blackish area restricted to hindcrown, thin line on centre of nape and isolated roundish ear-patch ('headphones'). In flight shows white rump and uppertail-coverts, upperwing with darker outer primaries and dark bands across lesser coverts and secondaries. See Black Tern. **Adult breeding** Black head, body and underwing-coverts and whitish upperwing-coverts, rump and tail-coverts diagnostic. **Juvenile** Differs from Whiskered in same way as adult non-breeding; additionally has distinctly darker and more uniform 'saddle'. See Black. **First winter** Similar to adult non-breeding. **VOICE** May give a harsh high-pitched ***kreek*** and harsh creaking ***kesch***, particularly in flight. **HABITAT** Coasts, coastal pools, marshes, wet rice paddies, lakes, large rivers; up to 450 m. **RANGE & STATUS** Breeds central, east and south-east W Palearctic, C Palearctic, S Siberia, S Ussuriland, N,NE China; mainly winters Africa, India, Sri Lanka, N Australia. **SE Asia** Scarce to locally common passage migrant S,E Myanmar, W(coastal),NW,C,NE,S Thailand, Peninsular Malaysia, Singapore, Cambodia, Laos, E Tonkin, C,S Annam, Cochinchina. Winters locally S Myanmar, C,S Thailand, Peninsular Malaysia, Singapore. Also recorded in summer S Thailand, Peninsular Malaysia, Singapore. Recorded in early winter (status uncertain) Tenasserim.

333 WHISKERED TERN
Chlidonias hybrida Plate **39**

IDENTIFICATION 24–28 cm. *C.h.javanicus*: **Adult non-breeding** Smallish and rather compact with relatively short, blackish bill, blackish mask and hindcrown/nape, white crown with dark streaks at rear, dark reddish legs and feet, and shortish tail. In flight shows relatively short, broad wings, rather uniform grey upperside, and upperwing with darker secondaries and outer primaries and somewhat darker inner primaries and shallow tail-fork. Superficially resembles smaller *Sterna* terns but note size, shape, bill and habitat. See White-winged and Black Terns and much larger Gull-billed Tern. **Adult breeding** Distinctive, with dark red bill, legs and feet, black forehead to nape, white throat and lower sides of head and dark grey remainder of underparts, contrasting with white vent and underwing-coverts. Grey of underside often appears blackish at distance. See Black-bellied Tern. **Juvenile** Mantle, scapulars, tertials and innermost wing-coverts fawn-brown with blackish subterminal markings and narrow buff fringes, forming distinct 'saddle' effect, otherwise similar to non-breeding adult but forecrown and face washed brownish-buff. See White-winged and Black. **First winter** Similar to adult non-breeding. **VOICE** Short, hoarse rasping ***kersch*** and repeated short ***kek*** notes, mainly given in flight. **HABITAT** Coastal pools, mud- and sandflats, marshes, lakes, large rivers, wet rice paddies; up to 800 m. **RANGE & STATUS** Breeds N,S,E Africa, southern W,C Palearctic, Iran, NW,NE India, N,NE,E and southern China, S Ussuriland, southern Australia. Northern populations mostly winter Africa, Indian subcontinent, Greater Sundas, Philippines, Sulawesi; Australian populations spend southern winter as far north as Greater Sundas, Philippines, New Guinea. **SE Asia** Uncommon to common winter visitor (except W,N Myanmar, SE Thailand, Laos, W Tonkin);

scarce Singapore. Rare to uncommon passage migrant S Myanmar, Peninsular Malaysia, Laos; ? Cochinchina. Also recorded in summer C,S Thailand.

334 **RIVER TERN** *Sterna aurantia* Plate **40**
IDENTIFICATION 38–46 cm (including outertail-feathers up to 23 cm). Monotypic. **Adult non-breeding** Distinctive. Medium-sized with rather thick dark-tipped yellow bill, reddish legs and feet and only very little dark coloration at tips of outer primaries (more when worn). Has mostly black mask and nape and greyish crown with dark streaks; underparts greyish-white. Note habitat. See Black-bellied Tern. **Adult breeding** Bill uniform orange-yellow, forehead to nape and mask black, upperwing all grey with paler primaries and primary coverts (forms striking flash), with noticeably long streamer-like outertail-feathers. Some individuals occur in this plumage during non-breeding season. **Juvenile** Has dark-tipped yellow bill, blackish mask, blackish streaks on crown, nape, ear-coverts and throat-sides, whitish supercilium, blackish-brown fringes to mantle, scapulars, tertials and wing-coverts, and blackish-tipped primaries. **First winter** Similar to non-breeding adult but retains juvenile tertials, primaries and tail-feathers. **VOICE** In flight gives a rather high nasal ***kiaah*** or ***hiaah***. When displaying gives more extended, rapidly repeated disyllabic calls: ***kierr-wick kierrwick-kierr-wick...***, often accelerating to crescendo. **HABITAT & BEHAVIOUR** Large rivers, sometimes lakes; up to 1,200 m (mostly below 450 m where breeding). Flight strong and purposeful. **RANGE & STATUS** Resident Pakistan, India, S Nepal, Bangladesh, SW China. **SE Asia** Scarce to locally fairly common resident (subject to some movements) Myanmar, east Cambodia, S Laos; rare (may no longer breed) along Mekong R in NW,NE Thailand, N,C Laos. Vagrant C Thailand. **BREEDING** April–May. Sometimes loosely colonial. **Nest** Depression in ground. **Eggs** 2–3; pale greenish-grey to buffy stone-colour, blotches and streaks dark brown over pale inky-purple undermarkings; 42 x 31.4 mm (av.).

335 **ROSEATE TERN** *Sterna dougallii* Plate **40**
IDENTIFICATION 33–39 cm (including tail-streamers up to 11 cm). *S.d.bangsi*: **Adult non-breeding** Very similar to Common but more slimly built, grey of upperside paler, has longer tail (unless broken) with all-white outer feathers which project well beyond closed wing-tips, bill somewhat longer and slenderer. May show pinkish-flushed underparts. When fresh, inner edges of primaries white, forming defined white line along inner edge of closed wing (lacking on Common). In flight shows paler, more uniform upperparts and upperwing-coverts (back contrasts less with rump and uppertail-coverts), with whiter secondaries and less dark on primaries; underwing only has faint dark markings on outer primary tips. See Black-naped Tern. **Adult breeding** Bill black with red on base to basal half only (sometimes all red), underparts often flushed pink, tail-streamers very long. In flight differs from Common by paler and plainer upperwing, whiter secondaries and inner primaries, darker greyish outermost primaries, no dark wedge on mid-primaries. *S.d.korustes* (Myanmar, ? to west Peninsular Malaysia) tends to be larger and paler grey above with less black on bill. **Juvenile** From Common by all-blackish bill, legs and feet, darker forecrown (initially all dark), usually bolder blackish subterminal markings on mantle feathers, scapulars and tertials (latter may be almost completely blackish) and broad white upper edge to closed primary tips. See Black-naped. **First winter/first summer** Similar to non-breeding adult, but in flight upperwing shows dark bands across lesser coverts and secondaries and somewhat darker leading edge of outer wing. **VOICE** A distinctive, incisive, clicky ***dju-dik***, particularly in flight, and low rasping ***kraak***, ***zraaach*** or ***aaahrk*** when agitated. **HABITAT & BEHAVIOUR** Open seas, coasts, mud- and sandflats, beaches, islets. Typically flies with shallower, stiffer and faster wingbeats than Common. **RANGE & STATUS** Breeds along coasts and offshore islands S,E Africa, Madagascar, Azores, W Europe, Indian Ocean, SE Arabian Peninsula, W,SE India, Maldive Is, Sri Lanka, S,SE China, Taiwan, S Japan, Java, N Borneo, Philippines, Moluccas, New Guinea region, Australia, New Caledonia, E USA, Caribbean; northern populations winter at sea and along coasts to south, including western Africa, Greater Sundas. **SE Asia** Very local coastal and offshore resident (subject to some movements) SE,S Thailand, Peninsular Malaysia, Cochinchina (Con Dao Is); probably Tenasserim. Scarce to uncommon non-breeding offshore visitor (except SW Myanmar, Cambodia, E Tonkin, N,S Annam). Vagrant Singapore. **BREEDING** June–August. Colonial. **Nest** Unlined/lined depression on ground or rock. **Eggs** 1–3; pinkish- or greyish-buff to biscuit-brown, boldly spotted and blotched reddish-brown, dark brown or blackish over grey or lavender undermarkings; 40.2 x 29.3 mm (av.; *korustes*).

336 **BLACK-NAPED TERN**
Sterna sumatrana Plate **40**
IDENTIFICATION 30–35 cm. *S.s.sumatrana*: **Adult non-breeding** Combination of white head with narrow black band from eyes to hindnape, blackish bill, legs and feet and very pale grey of upperside distinctive. Has some dark streaks on hindcrown. In flight shows uniform very pale grey upperwing with blackish outer edge of outermost primary. Wings look very whitish at distance. See adult non-breeding Roseate, Common and Gull-billed Terns. **Adult breeding** Crown all white and nape-band sharply defined; can show pinkish flush on underparts. **Juvenile** Bill initially dusky-yellow but soon turns blackish, has white forehead and lores, variable dark streaks on crown, well-defined black band from eye to nape, blackish subterminal markings on mantle feathers, scapulars, tertials and wing-coverts, dark-centred tail-feathers, darker grey secondaries and darker grey primaries with white inner edges, which form defined line along inner edge of closed wing. Generally similar to Roseate but with different head pattern, dark markings on upperside narrower, more crescent-shaped and typically paler. **First winter** Similar to adult non-breeding but with darker band across lesser upperwing-coverts, retained juvenile primaries and darker greyish markings on tertials and tail. **VOICE** May give a sharp ***kick***, ***tsii-chee-ch-chip*** and hurried ***chit-chit-chit-er***, particularly in flight. **HABITAT** Open seas, coasts, beaches, islets. **RANGE & STATUS** Coastal and offshore resident (subject to fairly localised movements) Indian Ocean, Andaman and Nicobar Is, S,SE China, South China Sea, Taiwan, S Japan, Sundas, Philippines, Wallacea, New Guinea region, NE Australia, SW Pacific Ocean. **SE Asia** Local coastal and offshore resident (subject to some movements) SE,S Thailand, Peninsular Malaysia, Singapore, Cambodia, C Annam, Cochinchina. Scarce to locally fairly common non-breeding offshore visitor elsewhere (except SW Myanmar, C Thailand, E Tonkin). **BREEDING** April–October. Loosely colonial. **Nest** Scrape on ground (sometimes lined with fragments) or eggs laid on bare rock. **Eggs** 1–3; greyish stone-colour to buffish or pale olive with numerous dark or light brown spots, sometimes warmer with reddish-brown blotches; 39.6 x 28.6 mm (av.).

337 **COMMON TERN** *Sterna hirundo* Plate **40**
IDENTIFICATION 33–37 cm (including tail-streamers up to 8 cm). *S.h.tibetana*: **Adult non-breeding** Medium-sized with fairly slender, pointed blackish bill, dark red legs and feet, white forehead and lores, blackish mask and nape and medium grey mantle, back, scapulars and wing-coverts. Mostly likely to be confused with Roseate Tern but somewhat thickerset, lacks white upper edge to closed primary tips, has shorter tail with noticeably grey outer fringes to feathers and somewhat shorter bill. In flight, darker grey of upperparts and upperwing contrasts with white rump and uppertail-coverts (more so in breeding plumage), upperwing has mostly grey secondaries with only narrow white trailing edge and more dark on primaries, lesser and primary coverts; underwing shows more extensive dark tips to outer primaries. See Whiskered Tern. **Adult breeding** Bill orange-red with black tip, legs and feet dark red, forehead to nape black; develops

long tail-streamers (outer feathers). Best distinguishing features from Roseate are slight grey wash to underparts (lower ear-coverts and cheeks slightly whiter), more uniform closed primary tips and shorter tail-streamers (not projecting beyond closed wing-tips). In flight, upperwing differs by darker secondaries, somewhat paler outermost primaries (apart from tips) and distinctive dark wedge on mid-primaries; underwing shows dark trailing edge to outer primaries. *S.h.longipennis* (recorded C Thailand, Peninsular Malaysia, Cambodia, C Annam, Cochinchina at least) has mostly black bill, greyer upperparts and underparts, more contrasting white cheek-stripe and dark reddish-brown legs and feet. **Juvenile** Base of bill extensively orange (blackens with age), legs and feet orange, head pattern similar to adult non-breeding; mantle, scapulars, tertials and wing-coverts have dark brown subterminal markings and, initially, buffish fringes. In flight, upperwing shows pronounced blackish band at leading edge of lesser coverts; dark secondaries, outer primaries and primary coverts contrast with rest of upperwing. *S.h.longipennis* has mostly blackish bill and dark legs and feet. See Roseate and Black-naped Terns. **First winter/first summer** Similar to adult non-breeding but upperwing has somewhat bolder dark bands across lesser coverts and secondaries. **VOICE** Typical calls include a harsh ***kreeeah*** or ***kreeeerh***, short ***kik*** notes, a rapid ***kye-kye-kye-kye...*** and ***kirri-kirri-kirri***. **HABITAT & BEHAVIOUR** Coastal habitats, open ocean, occasionally large rivers and inland lakes. **RANGE & STATUS** Breeds W,N Africa, Palearctic, NW India (Ladakh), Tibet, W and northern China, N Korea, N America, Caribbean; winters widely at sea and offshore south to S Africa, Indian Ocean, Indian subcontinent, New Guinea region, N Melanesia, N,E Australia, S America. **SE Asia** Scarce to fairly common non-breeding coastal visitor (except SW,S Myanmar, N Annam). Inland vagrant north W Myanmar (Chindwin R). **NOTE** Sight records of breeding-plumaged adults in Melaka Straits, with whitish underside and all-blackish bills, have tentatively been identified as *S.h.hirundo* x *longipennis* intergrades ('*minussensis*').

338 ARCTIC TERN *Sterna paradisaea* Plate 40

IDENTIFICATION 33–39 cm. Monotypic. **Adult breeding** Shortish dark red bill, shorter neck than Common Tern, very short legs. In flight, shows whiter flight-feathers (apart from black-tipped outer primaries) which appear translucent from below; tail-streamers longer than Common (extend a little beyond wing-tips at rest, rather than being roughly equal). **Adult non-breeding** As adult breeding but has white forehead, blackish bill, shorter tail. Unlikely to occur in region in this plumage. **Juvenile/first winter** From Common Tern by less patterned upperwing, whitish secondaries and inner primaries; bill all black from August/September. **First summer** Darker leading-edge to upperwing than non-breeding adult. **VOICE** Piping ***pi*** and ***pyu*** notes, ringing ***prree-eh*** or ***keeeyurrr***, hard rattled ***kt-kt-kt-krrr-kt...*** and ***krri-errrrr*** in alarm. **HABITAT & BEHAVIOUR** Coastal habitats, open ocean. **RANGE & STATUS** Breeds N Holarctic; winters subequatorial seas, including pelagic S Australasia. Rare passage migrant Japan. **SE Asia** Vagrant Cochinchina (sight record).

339 BLACK-BELLIED TERN
Sterna acuticauda Plate 40

IDENTIFICATION 30–33 cm. Monotypic. **Adult non-breeding** Due to habitat, most likely to be confused with River Tern but much smaller in size, bill slenderer and orange with darker tip, often shows some blackish mottling on belly and vent; forehead and crown sometimes all black. See Whiskered Tern. **Adult breeding** Forehead to nape and mask black, bill uniform orange, with grey breast and blackish belly and vent, contrasting with whitish head-sides and throat. Combination of bill, upperwing (similar to that of River), deeply forked tail and dark underparts diagnostic. See Whiskered. **Juvenile** Best separated from River by size, bill (as adult non-breeding), lack of whitish supercilium and whiter sides of neck and lower sides of head. **VOICE** Flight call is a clear piping ***peuo***. **HABITAT** Large rivers, sometimes large lakes, marshes; up to 450 m. **RANGE & STATUS** Resident (subject to some movements) Pakistan, India, Nepal, Bangladesh, SW China. **SE Asia** Scarce to uncommon resident Myanmar, north-east Cambodia (very rare). Formerly resident (now very rare and may not breed) NW Thailand, Laos. Vagrant C,S Thailand, C Annam. Formerly occurred (status unknown) Cochinchina. **BREEDING** February–April. **Nest** Depression on ground. **Eggs** 2–4; similar to River Tern; 32.4 x 24.9 mm (av.).

340 LESSER CRESTED TERN
Thalasseus bengalensis Plate **41**

IDENTIFICATION 35–40 cm. *T.b.bengalensis*: **Adult non-breeding** Like Great Crested Tern but smaller, bill narrower and yellowish-orange, has more solid black hindcrown and 'mane' on nape, and paler grey on upperside. See Chinese Crested Tern. **Adult breeding** Crown to nape black, including extreme forehead (white on Great Crested), bill more orange. **Juvenile** From Great Crested, apart from size and bill, by darker upperside of inner primaries. **First winter** Greater and median upperwing-coverts plainer grey than Great, head pattern as adult non-breeding. **VOICE** Typical flight call is a harsh ***krrrik-krrrik***. **HABITAT** Open seas, coasts, mud- and sandflats. **RANGE & STATUS** Breeds along coasts of north-eastern Africa, Arabian Peninsula, Iran, S Pakistan, New Guinea, northern Australia, possibly Maldive and Lakshadweep Is (India); non-breeding offshore and coastal visitor N,E Africa, Madagascar, Arabian Peninsula, Indian subcontinent, Sundas, Wallacea, New Guinea region. **SE Asia** Scarce to uncommon non-breeding coastal visitor (mostly winter) Myanmar, Thailand (except SE), west Peninsular Malaysia, Singapore, Cambodia, Cochinchina.

341 GREAT CRESTED TERN
Thalasseus bergii Plate **41**

IDENTIFICATION 45–49 cm. *T.b.velox* (east in region to west S Thailand; ? Peninsular Malaysia): **Adult non-breeding** Combination of largish size, rather stocky build, long, thickish, cold yellow to greenish-yellow bill and darkish grey upperside distinctive. Has white face and forecrown, blackish mask, blackish hindcrown and nape with white streaks and well-defined whitish tertial fringes. In flight shows mostly uniform wings with darker-tipped outer primaries. Worn feathers on upperwing are darker and their presence creates patchy appearance. *S.b.cristata* (west in region to Peninsular Malaysia; ? Gulf of Thailand) is paler grey on upperside. See Lesser and Chinese Crested Terns. **Adult breeding** Bill brighter yellow, extreme forehead white, crown to nape black, the latter with shaggy crest. **Juvenile** Bill slightly duller than non-breeding adult, feathers of mantle, back, scapulars and tertials brownish-grey with whitish fringes, dark brownish-grey to blackish centres to wing-coverts, secondaries and outer primaries, and darker tail. In flight shows four dark bands across upperwing-coverts and secondaries and rather uniformly dark outer primaries and primary coverts. See Lesser. **First winter** Similar to juvenile but with uniform grey mantle and scapulars. Juvenile wing-coverts are gradually lost, often resulting in patchy appearance. See Lesser. **Second winter** As non-breeding adult but retains juvenile outer primaries, primary coverts and secondaries, tail typically darker with variable dark subterminal band. **VOICE** Usual calls include a harsh grating ***krrrik***, ***kerrer*** and ***kerrak***, particularly in flight. **HABITAT** Open seas, coasts, mud- and sandflats, islets. **RANGE & STATUS** Breeds along coasts and offshore islands S,E Africa, Madagascar, Indian Ocean, Arabian Peninsula, Iran, S Pakistan, W,E India, Sri Lanka, S,SE China, Taiwan, S Japan (S Nansei and Ogasawara Is), South China Sea, Sundas, Philippines, Wallacea, New Guinea region, Australia, W,C Pacific Ocean; non-breeding visitor in intervening areas. **SE Asia** Rare to local coastal (mostly offshore) resident (subject to some movements) SW Myanmar, Tenasserim, Cambodia, Cochinchina, formerly SE Thailand. Uncommon to locally common non-breeding coastal visitor (except N,C Annam). **BREEDING** May–June. Colonial. **Nest** None or shallow scrape on ground. **Eggs** 1–2; white to salmon-pink or purplish, blotched, spotted and streaked blackish-brown, dark chestnut or pale reddish-brown; c.60 x 40 mm (av.; *velox*).

342 CHINESE CRESTED TERN
Thalasseus bernsteini Plate **41**

IDENTIFICATION 43 cm. Monotypic. **Adult non-breeding** Similar to Great Crested Tern but smaller with prominently blackish-tipped yellow bill and much paler grey on upperside. Size intermediate between Lesser and Great Crested Terns. In flight differs from both by sharp contrast between pale grey upperwing and blackish outer primaries. **Adult breeding** Forehead to nape black, with similar shaggy nuchal crest to Great. Black-tipped yellow bill diagnostic. **Juvenile/immature** Undocumented. **HABITAT** Open seas, coasts. **RANGE & STATUS** Very rare breeder on Jiushan Is off Zhejiang, E China, and Matsu Is between Fujian, SE China and north-west Taiwan; current winter range unknown. Very rare passage migrant Taiwan. Formerly bred more widely E,SE China; wintering south as far as Borneo, Philippines, N Moluccas. **SE Asia** Recorded once in winter off east S Thailand (Nakhon Si Thammarat): three birds in 1923.

LARIDAE: Gulls & allies

Worldwide c.53 species. SE Asia 14 species. Medium-sized to largish with relatively long narrow wings, usually stout legs and webbed feet. Spend much time airborne but also walk and swim. Mostly opportunistic feeders with varied diet, generally feed on aquatic invertebrates and fish but also scavenge around habitation, harbours, fishing boats etc.

343 BLACK-TAILED GULL
Larus crassirostris Plate **42**

IDENTIFICATION 45–48 cm. Monotypic. **Adult non-breeding** Combination of size, dark grey upperside and white uppertail with broad black subterminal band diagnostic. Head, neck and underparts white with greyish streaks on hindcrown and nape, bill longish and yellow with red tip and black subterminal band, legs and feet greenish-yellow, eyes pale yellowish, eyering red. In flight (from above) shows contrasting, mostly blackish outer primaries and a broad white trailing edge to the secondaries and inner primaries. **Adult breeding** Head and neck completely white. **First winter** From other medium-sized gulls by combination of dark-tipped pinkish bill, rather uniform greyish-brown body with contrasting whitish forehead, throat, rump, uppertail-coverts and vent, and diagnostic blackish tail with narrow white terminal band. **Second winter** More similar to adult non-breeding but head and neck duller, mantle and scapulars are mixed with some brown, upperwing somewhat paler and browner, with blackish flight-feathers, tail mostly black. **VOICE** A deep mewing ***kaoo kaoo***, ***kau-kau***, or ***yark-yark-yark***. Also, a relatively high, clear, rasping mewing. **HABITAT & BEHAVIOUR** Mud- and sandflats, coasts, coastal pools. Often associates with other gulls. **RANGE & STATUS** Breeds SE,E China, N Korea, Ussuriland, Sakhalin, Kuril Is, Japan; some populations winter south to coasts of S,E,NE China, Taiwan, S Korea. **SE Asia** Rare coastal winter visitor C Thailand, E Tonkin.

344 MEW GULL *Larus canus* Plate **43**

IDENTIFICATION 43–46 cm. *L.c.kamtschatschensis*: **Adult non-breeding** Recalls Heuglin's Gull but considerably smaller and slimmer with more rounded head, bill shorter, slenderer and pale yellowish with dark subterminal mark, upperparts paler. In flight, more slender-winged, wing-tip with larger white 'mirrors'. Shares extensive dark streaking on head and hindneck. **Adult breeding** Bill uniform yellow, head and neck all white. **First winter** From similar medium-sized gulls by combination of black-tipped pinkish bill, pinkish legs and feet, heavy dark markings on head, neck, breast-sides and flanks, mostly plain grey mantle, back and scapulars, and mostly plain brownish-grey greater coverts, which contrast with dark brown pale-fringed wing-coverts. In flight, uppertail shows clear-cut broad blackish subterminal band. See Relict Gull and second-winter Heuglin's. **Second winter** Similar to adult non-breeding but upper primary coverts marked with black, primaries having smaller white 'mirrors'. Differs from third-winter Heuglin's in same way as adult non-breeding and additionally by more extensively plain grey upperwing and all-white tail. **VOICE** May give a shrill nasal high-pitched ***glieeoo***, much higher than similar call of Vega Gull. Also ***gleeu-gleeu-gleeu...*** when alarmed and a nasal ***keow***. **HABITAT & BEHAVIOUR** Coastal pools, sand- and mudflats. Often associates with other gulls. **RANGE & STATUS** Breeds northern Palearctic, north-west and north N America; mostly winters west and south-eastern W Palearctic, Middle East, southern and eastern China, Taiwan, N,S Korea, Japan, western N America, rarely Pakistan. **SE Asia** Coastal vagrant C Thailand, E Tonkin.

Large 'white-headed' gulls: *Larus mongolicus*, *L.fuscus* & *L.heuglini*

The systematics and identification of all taxa formerly considered as races of either Herring Gull *L.argentatus* and Lesser Black-backed Gull *L.fuscus* occurring worldwide have received much attention in recent years. There is growing evidence, based on the study of qualitative differences in morphology, moult, behaviour, ecology, voice and mitochondrial DNA sequences, that several forms, previously regarded as subspecies of either one of the above-mentioned species may, in fact, be better considered full species. Among the most perplexing taxa of all are the those that occur or are likely to occur in South-East Asia.

Recent study of these gulls on their breeding grounds, combined with examination of museum specimens, is providing a more scientifically reliable foundation for the identification of these birds in various parts of their wintering ranges. Unfortunately, it is beyond the scope of this book to cover the identification of these most difficult birds at the level of detail that would be required to treat the subject comprehensively. Some of the basic points to bear in mind, while considering the identity of large gulls that fall into this category, are:

First, a gathering of several or more similar-looking birds constitutes a much more important sample than do single individuals, however distinctive the latter may appear. Consequently it is often possible to assign homogeneous groups of birds to a particular form with more confidence. When it is suspected that two or more forms are present side by side, it may help to begin by looking for the common features among them, rather than devoting too much time to an individual bird that looks 'different' from all the others; it may be that the latter is an extreme or atypical variant which will prove to be more difficult to identify than the majority type.

Second, an assessment of moult, in particular the state of primary moult (with respect to time of year), will often provide a more reliable clue to identification than subjective assessments of size and mantle shade, etc.

345 MONGOLIAN GULL
Larus mongolicus Plate **43**

IDENTIFICATION 60–67 cm. Monotypic. **Adult non-breeding** Averages larger than Heuglin's, distinctly paler grey above (similar to north European Herring Gull *L.argentatus*, but perhaps slightly bluer-tinged), and typically whiter-headed. Eyes usually look yellowish in field, legs yellowish-flesh to flesh or pink (rarely yellow). In February/March, already has white head of breeding plumage (while Heuglin's still shows head-streaking). **Adult breeding** Similar to adult non-breeding, but head and neck all white. **First winter** Best separated from Heuglin's by much paler overall coloration, and distinctly 'frosty' upperside; paler greater coverts. In flight, note paler inner primaries. **Second winter** From Heuglin's by same features as first winter; additionally shows paler grey on upperparts. **HABITAT** Coasts, sand- and mudflats, coastal pools. **RANGE & STATUS** Breeds S Siberia, N Mongolia, NE China; winters E,S China, N,S Korea, southern Japan. **SE Asia** Rare winter visitor C Thailand, E Tonkin. **NOTE** On current knowledge, best treated as a distinct species; otherwise as a race of extralimital Vega Gull *L. vegae* (Yésou 2001,2002).

346 **LESSER BLACK-BACKED GULL** *Larus fuscus* Plate **43**

IDENTIFICATION 49–57 cm. *L.f.fuscus*: **Adult non-breeding** Similar to Heuglin's Gull, but smaller, slimmer and slightly smaller-billed, with proportionately longer wings, upperparts almost jet-black, head and neck only weakly streaked; legs yellow. **Adult breeding** White head and neck, bright yellow legs. **Juvenile** Apart from size and build, possibly inseparable on current knowledge from Heuglin's. **First winter** Difficult to separate from Heuglin's. Apart from size and build, somewhat darker overall, mantle and scapulars with typically less contrasting dark 'anchor-marks' on centres; perhaps shows less clean white on head and underparts. **Second winter** Upperpart coloration close to that of adult, but head and neck more heavily streaked than adult non-breeding. **HABITAT** Coasts, sand- and mudflats, coastal pools. **RANGE & STATUS** Breeds coastal W Palearctic, locally Greenland; winters mainly on coasts W Europe, N,W,E Africa, east N America (small numbers summer), rarely Caribbean. Passage migrant Middle East. **SE Asia** Vagrant C Thailand.

347 **HEUGLIN'S GULL** *Larus heuglini* Plate **43**

IDENTIFICATION 58–65 cm. *L.h.taimyrensis*: **Adult non-breeding** From Mongolian Gull by distinctly darker grey upperparts, and typically heavily streaked head and hindneck. Legs usually yellowish (sometimes pinkish-tinged), eyes usually pale. Moult starts later than Mongolian; large gulls with white head and retained, faded and worn primaries in September and October should be this species. By February/March, still has head-streaks of non-breeding plumage and may have outermost primaries growing, while most, if not all Mongolian will have white heads. See Lesser Black-backed Gull. **Adult breeding** Head and neck all white. **Juvenile** Heavy dull brownish streaks and mottling on white of head, underside, rump and uppertail-coverts, dark brownish upperparts with whitish to brownish-white fringing and notching, mostly blackish tail with dark-speckled white base and narrow white terminal band, blackish bill, dull pinkish legs. In flight, shows quite blackish upperwing with contrasting paler median coverts and slightly paler, dark-tipped inner primaries (paler when worn). See Lesser Black-backed Gull. **First winter** Moults in new scapulars and tertials, which are greyer with more diffuse darker central markings and less defined pale borders; head, neck and underside somewhat whiter. Shows contrasting dark 'anchor-marks' on feather-centres of mantle and scapulars. See Lesser Black-backed Gull. **Second winter** Like first winter, but mantle, back and scapulars mostly grey. **VOICE** Nasal ***gagaga***, deeper and stronger than voice of Lesser Black-backed Gull. **HABITAT** Coasts, sand- and mudflats, coastal pools, lakes, large rivers; lowlands. **RANGE & STATUS** Breeds north-east W and north C Palearctic; winters coasts of E Africa, Arabian Peninsula, S Iran, S Pakistan, W,S India, Sri Lanka, S China. **SE Asia** Rare to locally fairly common (mostly coastal) winter visitor N,S Myanmar, W,C Thailand, E Tonkin. Vagrant Singapore? **NOTE** It has recently been proposed that *taimyrensis* is an invalid taxon, comprising hybrids between extralimital *L.h.heuglini* and Vega Gull *L. vegae* (Yésou 2002). It seems highly unlikely, however, that homogeneous flocks wintering in the region are of hybrid origin, as some variation would be expected if this were the case. *L.h.heuglini* is currently unknown from the region.

348 **PALLAS'S GULL** *Larus ichthyaetus* Plate **43**

IDENTIFICATION 58–67 cm. Monotypic. **Adult non-breeding** Combination of large size, yellowish bill with broad blackish subterminal band, mask of dark streaks, comparatively pale grey mantle, back, scapulars and wing-coverts and small amount of black on outer primaries diagnostic. Legs and feet dull yellowish, hindcrown faintly dark-streaked; has longer sloping forehead than other large gulls and bill is thicker. In flight, upperwing shows white outer primaries with black subterminal markings and white outer primary coverts. See much smaller Relict Gull. **Adult breeding** Head black with broken white eyering, bill with yellower base and more red near tip, legs and feet bright yellow. **First winter** Similar to second-winter Heuglin's and Vega Gulls but has dark mask and hindcrown streaking (as adult non-breeding), densely dark-marked lower hindneck and breast-sides, paler grey on mantle, back and scapulars, mostly grey, unbarred greater coverts, and white rump, uppertail-coverts and tail-base, contrasting sharply with broad blackish subterminal tail-band. **Second winter** More similar to adult non-breeding but shows remnants of dark markings on median, lesser and primary coverts, mostly black outer primaries and narrow blackish subterminal tail-band. **VOICE** Occasionally gives a deep low ***kyow-kyow*** and nasal crow-like ***kraagh*** or ***kra-ah***. **HABITAT & BEHAVIOUR** Coasts, sand- and mudflats, coastal pools, large rivers. Regularly scavenges and parasitises other birds. **RANGE & STATUS** Breeds N Black and Caspian Sea regions, SW Russia, Kazakhstan, Kirgizstan, N China, N Mongolia; mostly winters (particularly coasts) NE Africa, south-east W Palearctic, Middle East, Indian subcontinent, locally S China. **SE Asia** Uncommon to locally common winter visitor N,C,S Myanmar. Rare to scarce coastal winter visitor W,C Thailand, E Tonkin. Vagrant NW Thailand, N Laos. Recorded (status uncertain) SW,E Myanmar, north Tenasserim.

349 **LAUGHING GULL** *Larus atricilla* Plate **42**

IDENTIFICATION 36–41 cm. *L.a.megalopterus*?: **Adult non-breeding** Told by medium size, relatively dark grey upperside, longish dark bill (tip often reddish), reddish-black legs and feet, and dark greyish-smudged ear-coverts and hindcrown (no defined ear-spot). In flight, uniform grey upperwing with black tips and bold white trailing edge to secondaries and inner primaries distinctive. **Adult breeding** Black hood (including nape) with broken white eyering, dark red bill and legs (former usually with black subterminal band). **First winter** Head, mantle, scapulars and underparts similar to adult non-breeding, but shows dusky nape to breast and flanks; blackish bill, legs and feet. Centres of wing-coverts and tertials mostly dark greyish-brown. In flight, shows mostly blackish flight-feathers, with similar trailing edge to adult, dark markings across mid-part of underwing; tail with broad black subterminal band and dark greyish sides. **Second winter** Like adult non-breeding but has greyer nape to breast-sides and flanks, some dark markings on primary coverts, may show faint suggestion of tail-band. **VOICE** Yelping ***kee-agh*** or ***kiiwa*** recalling Mew Gull. Also, cackling, rather goose-like, nasal, laughing ***ha-ha-ha***, deep ***greek***, and whining ***kerook***. **HABITAT** Coasts, mud- and sandflats. **RANGE & STATUS** Breeds coastal E,S USA, E Mexico, Caribbean, Venezuela, Surinam; some (more northerly) populations winter south along coasts to northern S America; also regular casual/vagrant Hawaii, Greenland, and Atlantic coast of W Palearctic, rarely reaching Japan, Australia. **SE Asia** Coastal vagrant west Peninsular Malaysia (sight record).

350 **RELICT GULL** *Chroicocephalus relictus* Plate **44**

IDENTIFICATION 44–45 cm. Monotypic. **Adult non-breeding** Recalls Brown-headed and Black-headed Gulls but larger, stockier, thicker-billed and longer-legged, with more uniformly dark-smudged ear-coverts and hindcrown (no obvious dark ear-spot), white tips to primaries. Bill dark red. In flight, upperwing shows white-tipped primaries, prominent, separated black subterminal markings on outer primaries and no white leading edge to outer wing. See much smaller Saunders's Gull. **Adult breeding** Has blackish hood (including nape) and broad broken white eyering. See Saunders's and much larger Pallas's Gull. **First winter** Combination of size, blackish bill, legs and feet, mostly rather pale grey upperside, and sharply contrasting dark markings on wing-coverts and tertials distinctive. Has dark speckles and markings on nape, neck and breast-sides and paler, greyish basal half of lower mandible. In flight, easily separated from similar-sized gulls by distinctive upperwing, with solid black outer primaries, only small dark markings on inner primary tips and secondaries and no obvious white on outer wing; shows narrow black subterminal tail-band. See Mew Gull. **Second winter** Poorly documented. Similar to adult non-breeding but bill black with dark red base, tertials blackish-brown with white edges, primaries with slightly broader white tips. **VOICE** A nasal downward-inflected ***kyeu*** and low-pitched, drawn-out, slightly disyllabic ***ke'arr***. **HABITAT & BEHAVIOUR** Estuaries, mud- and sandflats. Likely to be found associating with other gulls. Often walks with a very upright gait, neck upstretched. **RANGE & STATUS** Breeds E Kazakhstan, S Siberia, Mongolia, N China; winters coastal China, S Korea. **SE Asia** Coastal vagrant E Tonkin.

351 BROWN-HEADED GULL

Chroicocephalus brunnicephalus Plate **44**

IDENTIFICATION 42–46 cm. Monotypic. **Adult non-breeding** Like Black-headed Gull but slightly larger, bulkier, thicker-billed and thicker-necked, with pale eyes. In flight shows broader, more rounded wings with distinctive broadly black-tipped outer primaries, enclosing up to three white 'mirrors' and more extensively blackish underside of primaries with white 'mirrors' near wing-tip. **Adult breeding** Has dark brown hood with broken white eyering and paler face, bill uniform dark red. See Relict Gull. **First winter** From Black-headed, on plumage, by black outer primaries, broader black tips to inner primaries (above and below) and whitish patch extending over primary coverts and inner primaries on upperwing. **VOICE** Like Black-headed but deeper and gruffer. **HABITAT** Large rivers and lakes, coasts, coastal pools; up to 1,830 m. **RANGE & STATUS** Breeds SE Tajikistan, NW India (Ladakh), Tibet, NW,N China; winters Indian subcontinent, sparingly SW,S China. **SE Asia** Uncommon to locally common winter visitor SW,N,C,E,S Myanmar, Tenasserim, Thailand, Peninsular Malaysia, Indochina (except W Tonkin, N,C,S Annam). Also recorded on passage S Myanmar, Cambodia, N Laos. Vagrant W Myanmar, Singapore. Recorded in summer C Thailand.

352 BLACK-HEADED GULL

Chroicocephalus ridibundus Plate **44**

IDENTIFICATION 35–39 cm. Monotypic. **Adult non-breeding** Relatively small and slim with rather narrow black-tipped red bill, dark red legs and feet and pale grey upperside. Head mainly white, with prominent dark ear-spot and dark smudges on side of crown. Eyes dark. In flight, upperwing appears very pale with prominent white leading edge to outer wing and smallish black tips to outer primaries; underwing pattern mirrors this but coverts all greyish, outer primaries having more black and less white. See very similar Slender-billed Gull and Brown-headed Gull. **Adult breeding** Has dark brown hood with broken white eyering and uniform darker red bill. See Brown-headed, Saunders's and Relict Gulls. **First winter** At rest resembles non-breeding adult but bill paler and duller with more contrasting dark tip, legs and feet duller and more pinkish, has greyish-brown centres to median and lesser coverts, inner greater coverts and tertials. In flight, upperwing shows broad greyish-brown band across coverts and broad blackish band along secondaries and tips of primaries; blackish subterminal tail-band. See very similar Slender-billed Gull and Brown-headed. **VOICE** Typical calls are a high-pitched screaming ***kyaaar*** and ***karrr***. Contact calls include short ***kek*** and deeper ***kuk*** notes. **HABITAT** Large rivers, lakes, coasts, coastal pools; up to 800 m. **RANGE & STATUS** Breeds Palearctic, NW,NE China, E Canada; mostly winters northern Africa, west and south W Palearctic, Middle East, Indian subcontinent, southern and E China, Taiwan, S Korea, Japan, Philippines, eastern N America. **SE Asia** Scarce to locally common winter visitor Myanmar (except W), W(coastal),NW,C Thailand, Peninsular Malaysia, Singapore, Cambodia, N Laos, Vietnam (except W Tonkin). Also recorded on passage Cambodia. Vagrant NE,S Thailand.

353 SLENDER-BILLED GULL

Chroicocephalus genei Plate **44**

IDENTIFICATION 37–42 cm. Monotypic. **Adult non-breeding** Like Black-headed Gull but is longer-necked and longer-billed, with longer, more sloping forehead, all-white head (sometimes with faint ear-spot) and pale eyes. **Adult breeding** Head white, bill dark, sometimes appearing almost blackish (particularly at distance); underparts may be washed pink. **First winter** Apart from shape, differs from Black-headed by paler, more orange bill with less obvious dark tip, much fainter head markings, pale eyes and longer, paler legs. In flight shows less contrasting band across upperwing-coverts (feather-centres paler) and a little more white on outer primaries. **VOICE** Like that of Black-headed but slightly deeper, lower-pitched and more nasal. **HABITAT** Coastal pools. **RANGE & STATUS** Breeds NW Africa, Egypt, S Europe, south-eastern W Palearctic, W,C Kazakhstan, W Turkmenistan, Middle East, S Pakistan; mostly winters (particularly along coasts) south to northern Africa, Arabian Peninsula, Middle East, W,S India, rarely Nepal. **SE Asia** Vagrant N Myanmar, coastal C Thailand.

354 SAUNDERS'S GULL

Chroicocephalus saundersi Plate **44**

IDENTIFICATION 33 cm. Monotypic. **Adult non-breeding** At rest, similar to Black-headed Gull but smaller and more compact with shorter, thicker, blackish bill, white tips to primaries. In flight has similar wing pattern but upperside of outer primaries have white tips and small black subterminal markings. **Adult breeding** Has black hood (including nape) and broad, broken white eyering. **First winter** Apart from size and bill, best separated from Black-headed in flight, by different upperwing pattern, with smaller isolated black tips to inner primaries, narrower and more broken dark band along secondaries, and lack of white leading edge to outer wing; has narrower black tail-band. See much larger Relict Gull. **VOICE** A harsh, tern-like ***kip***. Also a harsh ***chao***, recalling *Chlidonias* terns. **HABITAT & BEHAVIOUR** Estuaries, mud- and sandflats. Gregarious. **RANGE & STATUS** Breeds coastal E,NE China, erratically S Korea; winters along coasts of S China, Taiwan, S Korea, S Japan. **SE Asia** Local winter visitor E Tonkin.

355 LITTLE GULL *Hydrocoloeus minutus* Plate **44**

IDENTIFICATION 24–28 cm. Monotypic. **Adult non-breeding** The smallest gull in the region. Told by size and neat proportions, weak blackish bill, pinkish legs and feet and, in flight, by rounded white wing-tip and extensively blackish underwing. Dark head markings recall several other small gulls but has a more capped appearance. **Adult breeding** Black hood, reddish-brown bill (looks black), scarlet legs and feet, pink-flushed underparts. **First winter** Broad blackish diagonal band across upperwing-coverts, and contrasting whitish secondaries and greater coverts rules out other small gulls, but recalls Black-legged Kittiwake. Smaller than latter, with no black nuchal band, and faint darker subterminal band along secondaries and inner primaries (also visible on underwing, along with dark tertial mark). Note grey hindneck and slender blackish bill; legs and feet dull flesh. **Second winter** Similar to adult non-breeding but with variable thin black subterminal markings on primaries and somewhat paler underwing. Most similar to Saunders's Gull, but note dark cap, grey hindneck, thin bill, darker underside of flight-feathers, and whiter and rounder extreme wing-tip etc.

VOICE Short, hard, nasal ***keck*** or ***kek*** (sometimes quickly repeated). Noisy, high-pitched, rhythmic ***KAY-ke-KAY-ke-KAY-ke...*** or ***kjaae-ki...kjaae-ki...kjaae-ki...*** during breeding season. **HABITAT & BEHAVIOUR** Coastal pools and marshes. Indirect, buoyant flight-style recalls terns. **RANGE & STATUS** Breeds W to EC Palearctic, N America (Great Lakes region since 1919); winters mainly along coasts of W Palearctic, N Africa, Caspian Sea, E China, E USA; rarely Japan. **SE Asia** Vagrant C Thailand.

356 BLACK-LEGGED KITTIWAKE

Rissa tridactyla Plate **42**

IDENTIFICATION 37–42 cm. *R.t.pollicaris*: **Adult non-breeding** Distinctive, with relatively dark grey upperside, grey nape, vertical blackish bar behind eye, yellowish bill, shortish dark brown to blackish legs (rarely tinged pinkish to reddish), slightly notched tail, and rather narrow outer wing which turns whitish before neat black tip. **Adult breeding** Head all white. **First winter** Differs from adult non-breeding by upperwing pattern, with broadly black outer primaries and black diagonal band across coverts, contrasting sharply with largely whitish secondaries and inner primaries. Black-tipped tail, black bill (may be slightly paler at base); head as adult non-breeding but may show black band across hindneck. **VOICE** May give short nasal ***kya*** or ***kja*** in flight, or short knocking ***kt kt kt...*** in alarm. **HABITAT** Open seas, rarely inshore. **RANGE & STATUS** Breeds N Holarctic; winters at sea and along coasts south to N Africa, NE China, N,S Korea, Japan, Pacific Mexico, Atlantic USA. **SE Asia** Vagrant C Thailand.

ALCIDAE: Auks & allies

Worldwide 24 species. SE Asia 1 species. Superficially grebe- or duck-like, small to medium-sized seabirds. Short tails and necks, and mostly pointed bills. Sexes alike, but distinct non-breeding and breeding plumages. Highly pelagic, spending most of time at sea. Wings used for underwater propulsion, when hunting fish and aquatic invertebrates.

357 ANCIENT MURRELET

Synthliboramphus antiquus Plate **9**

IDENTIFICATION 24–27 cm. *S.a.antiquus*: **Adult non-breeding** Highly distinctive, Little Grebe-sized, swimming seabird. Combination of dark-based pale pinkish to yellowish bill, blackish face, crown and nuchal collar, slaty-grey upperside, and white breast extending in band to behind ear-coverts diagnostic. Upper throat mostly dusky, flanks dark greyish with whitish mottling. **Adult breeding** Streaky white supercilium, white streaks on nuchal collar, black throat and ear-coverts. **Juvenile** Similar to adult non-breeding, but whiter throat, smaller pale flank markings. **VOICE** Short whistled ***teep***. Various chirps and trills at breeding sites. **HABITAT & BEHAVIOUR** Open seas, coastal waters, bays. Frequently dives for food. Flight direct, with rapid, whirring wingbeats. **RANGE & STATUS** Breeds coastal NE Palearctic (Ussuriland, Sakhalin I, Kamchatka, Kuril Is etc.), northwest N America, locally NE China, N Korea, N Japan; winters south to S China, N,S Korea, Japan, California. **SE Asia** Vagrant E Tonkin.

COLUMBIDAE: COLUMBINAE: Typical pigeons & doves

Worldwide c.182 species. SE Asia 14 species. Generally rather plump and round-bodied with small heads and bills and short legs. Flight strong and direct. Generally gregarious and frugivorous, some species travelling long distances in search of food. **Woodpigeons** (*Columba*) Medium-sized to large with relatively broad wings and tail. **Turtle-doves & collared-doves** (*Streptopelia*) and **Zebra Dove** Smaller and slenderer with more pointed wings, mostly with relatively long tails. **Cuckoo-doves** (*Macropygia*) Similar but predominantly brown with conspicuously long, graduated tails. **Emerald Dove & Nicobar Pigeon** Compact, broad-winged, short-tailed and essentially terrestrial.

358 ROCK PIGEON *Columba livia* Plate **45**

IDENTIFICATION 33 cm. *C.l.intermedia*: **Adult** Pure stock are predominantly grey with noticeably darker hood and breast, blackish tail-tip and paler wing-coverts with two broad blackish bars. Neck glossed green and purple. Feral stock may be highly variable with patches of white and brown in plumage; some are entirely blackish. In flight shows silvery-whitish underwing-coverts. **Juvenile** Pure stock are duller than adult and browner overall, head, neck and breast greyish-brown with gloss reduced or lacking, wing-coverts mostly pale greyish-brown. **VOICE** Song is a soft, guttural ***oo-roo-coo***. **HABITAT** Cliffs, ruins, groves in open and cultivated places, urban areas; up to 1,450 m. **RANGE & STATUS** Resident N Africa, Palearctic, Middle East, Indian subcontinent, northern China. Feral populations in most parts of world. **SE Asia** Common resident throughout. Possibly pure stock in SW,C and perhaps elsewhere in Myanmar. **BREEDING** All year. Multi-brooded. Often colonial. **Nest** Pad; on ledge or in cavity/recess in rock-face, building or tree. **Eggs** 2; white; 36.9 x 27.8 mm (av.).

359 SNOW PIGEON *Columba leuconota* Plate **45**

IDENTIFICATION 34.5 cm. *C.l.gradaria*: **Adult** Slaty-grey head sharply contrasting with white collar and underparts diagnostic. Has distinctive blackish tail with broad whitish central band (conspicuous in flight), white patch on lower back, blackish rump and uppertail-coverts, and greyish upperwing-coverts with three dark bands. **Juvenile** Head somewhat duller, collar and breast dull pinkish-grey, has paler tips to scapulars and wing-coverts. **VOICE** Double hiccup-like note followed by ***kuck-kuck*** and ending with another hiccup. **HABITAT** Open alpine areas, cultivation, cliffs, rocky screes; 3,355–4,570 m. **RANGE & STATUS** Resident (subject to local movements) S Kazakhstan, Kirghizstan, SW Tajikistan, NE Afghanistan, N Pakistan, NW,N India, Nepal, Bhutan, S,SE Tibet, SW,W China. **SE Asia** Recorded (possibly resident) N Myanmar. **BREEDING** India: May–July. Often colonial. **Nest** Pad; on ledge or in rock fissure. **Eggs** 2; white; 40.3 x 29.1 mm (av.).

360 SPECKLED WOODPIGEON

Columba hodgsonii Plate **45**

IDENTIFICATION 38 cm. Monotypic. **Male** Told by combination of pale grey head, neck and upper breast, dark maroon mantle, dark maroon scapulars and lesser coverts with bold whitish speckles, and dark belly and vent. Has maroon and pale grey streaks and scales on lower hindneck, upper mantle, lower breast and belly, and dark greenish to greyish legs and feet. **Female** Head and breast darker grey than male; mantle cold and rather slaty dark brown without maroon, lacks maroon on scapulars and lesser coverts, base colour of underparts dark brownish-grey without maroon. **Juvenile** Like female but dark parts of body and wings browner with indistinct speckles on scapulars and wing-coverts. **VOICE** Deep, throaty ***whock-whroooo whrrrooo***. **HABITAT** Broadleaved evergreen forest; 1,675–2,565 m, down to 1,350 m in winter. **RANGE & STATUS** Resident (subject to relatively local movements) NW,N,NE Indian subcontinent, S,SE Tibet, SW,W China. **SE Asia** Uncommon resident W,N,C,E Myanmar. Scarce to uncommon winter visitor (perhaps local resident) NW Thailand, west N Laos. **BREEDING** February–June. **Nest** Flimsy platform; in tree; 3–8 m above ground. **Eggs** 1; white; 39.4 x 30.2 mm (av.).

361 **ASHY WOODPIGEON**

Columba pulchricollis Plate **45**

IDENTIFICATION c.36 cm. Monotypic. **Adult** Dark slaty upperside and breast and contrasting grey head with broad buffish neck-collar and whitish throat diagnostic. Upper mantle and upper breast glossed green, legs and feet red. In flight, told from Speckled Woodpigeon and Mountain Imperial-pigeons by relatively small size, dark breast, pale belly and vent and all-dark tail. **Juvenile** Buffish neck-collar little developed and largely pale grey, no green gloss on upper mantle and upper breast, crown darker grey, wings and breast browner (breast obscurely barred dull rufous), lower breast and central abdomen tinged rufous. **VOICE** Song is a deep resonant ***whoo***, given singly or repeated up to 5 times. **HABITAT** Broadleaved evergreen forest; 1,400–2,745 m (down to 575 m C Laos, winter). **RANGE & STATUS** Resident Nepal, NE Indian subcontinent, S Tibet, SW China, Taiwan. **SE Asia** Uncommon to locally fairly common resident SW,W,N,E Myanmar, NW Thailand, N Laos, W Tonkin. Recorded (status uncertain) C Laos. **BREEDING** April–June. **Nest** Flimsy platform in small tree. **Eggs** 1; white; 37.6 x 27.4 mm (av.).

362 **PALE-CAPPED PIGEON**

Columba punicea Plate **45**

IDENTIFICATION 36–40.5 cm. Monotypic. **Male** Overall dark plumage with contrasting whitish-grey crown diagnostic. Upperparts purplish-maroon with faint green gloss on sides and back of neck, more strongly iridescent mantle and back and dark slate-coloured rump and uppertail-coverts. Ear-coverts, throat and underparts vinous-brown, undertail-coverts slaty-grey, tail and flight-feathers blackish, orbital skin and base of pale bill red. **Female** Like male but crown generally greyish. **Juvenile** Wing-coverts and scapulars duller than adult with rufous fringes, crown initially concolorous with mantle, gloss on upperparts much reduced, underparts greyer. **VOICE** Not adequately documented. **HABITAT** Broadleaved evergreen forest, secondary growth, locally mangroves, island forest, and more open areas (mainly during migratory movements); up to 1,400 m. **RANGE & STATUS** Resident (subject to local movements) E,NE India, E Bangladesh, S Tibet, S China (Hainan). **SE Asia** Rare to locally common resident (subject to local, nomadic movements) Myanmar (except W,N), east Cambodia, C,S Laos, E Tonkin, C,S Annam, Cochinchina. Recorded (status uncertain) W,NE,SE,S Thailand (probably breeds locally, perhaps on offshore islands in S). **BREEDING** June–July. **Nest** Flimsy platform in small tree, tall bush or bamboo; up to 6 m above ground. **Eggs** 1; white; 37.6 x 29.2 mm (av.).

363 **ORIENTAL TURTLE-DOVE**

Streptopelia orientalis Plate **46**

IDENTIFICATION 31–33 cm. *S.o.agricola*: **Adult** Resembles Spotted Dove but larger, bulkier, shorter-tailed and darker overall, with rufous fringes to lower mantle and wing-coverts, broader rufous fringes to scapulars, barred rather than spotted sides of neck, and bluish-slate rump and uppertail-coverts. Crown bluish-grey (forehead paler and more buffish), sides of head, neck, upper mantle and throat to belly rather uniform pale vinous-brownish, undertail-coverts grey. In flight, shows greyish tail-tips and lacks prominent pale bar across upperwing-coverts. *S.o.orientalis* (wintering race) is larger and greyer, less vinous on head to upper mantle and underparts, and has rather more distinct breast-band/collar, contrasting with the paler creamy throat and paler buffish belly. **Juvenile** Somewhat paler with narrower, paler rufous fringes to lower mantle, scapulars and wing-coverts, paler fringes to breast, much smaller or absent neck-patch. **VOICE** Song is a husky ***wu,whrroo-whru ru*** (sometimes without last note) or faster ***er-her-herher***. **HABITAT** Open forest, secondary growth, scrub, cultivation; up to 2,135 m. **RANGE & STATUS** Breeds Siberia (Urals east to Sakhalin), Afghanistan, N Pakistan, NW,N,NE Indian subcontinent, C,E India, southern Tibet, China, Taiwan, N,S Korea, Japan; northern populations winter to south, Indian subcontinent. **SE Asia** Locally common resident Myanmar, W,NW,NE Thailand. Uncommon winter visitor N Myanmar, NW,NE Thailand, Indochina (may also be resident in north). **BREEDING** February–April. **Nest** Flimsy platform in tree, bush or bamboo; 1.5–3.6 m above ground. **Eggs** 2; white; 31 x 23.6 mm (av.).

364 **EURASIAN COLLARED-DOVE**

Streptopelia decaocto Plate **46**

IDENTIFICATION 33 cm. *S.d.xanthocyclus*: **Adult** Resembles female Red Collared-dove but larger and longer-tailed, with paler mantle, no slaty-grey on rump and uppertail-coverts, paler and more pinkish-grey breast, grey vent and extensive pale grey on wing-coverts. See Spotted Dove. **Juvenile** Crown, mantle and underparts duller and browner-tinged than adult, no hindneck-bar, narrow buff fringes to upperparts, wings and breast. From Red by same features as adult plus less distinct buff fringing, no dull rufous tips to primaries etc. **VOICE** Song is a soft repeated ***coo-cooo cu*** or ***wu-hooo hu***. Call is a husky ***vvrrrrr*** or ***vvrrrroo***. **HABITAT** Dry open country, scrub, cultivation; lowlands. **RANGE & STATUS** Resident N Africa, W Palearctic, Middle East, C Asia, Indian subcontinent, northern China, N,S Korea. Introduced Japan, USA. **SE Asia** Common resident SW,N(south-east),C,S(north-west) Myanmar. **BREEDING** All year. **Nest** Flimsy platform in tree or bush, sometimes roof of building. **Eggs** 2–3; white; 29.5 x 22.9 mm (av.).

365 **RED COLLARED-DOVE**

Streptopelia tranquebarica Plate **46**

IDENTIFICATION 23–24.5 cm. *S.t.humilis*: **Male** Relatively small and compact with distinctive brownish vinous-red plumage, pale bluish-grey head (except throat), black hindneck-bar, grey rump and uppertail-coverts, rather short square-cut dark tail with broad white tips on outer feathers, and blackish flight-feathers. Undertail-coverts whitish. **Female** Similar pattern to male but body and wing-coverts mostly brownish, less grey on head, whitish vent. See Eurasian Collared-dove. **Juvenile** Like female but hindneck-bar absent, upperparts, wing-coverts and breast fringed buffish, primaries, primary coverts and alula tipped dull rufous, and crown rufescent-tinged. **VOICE** Song is a soft, throaty, rhythmically repeated ***croodle-oo-croo***. **HABITAT** Drier open country, scrub, cultivation; up to 1,200 m. **RANGE & STATUS** Resident Indian subcontinent (except S; mostly breeding visitor Pakistan), China (except NW,NE), Taiwan, Philippines. **SE Asia** Common resident, subject to relatively local movements (except S Thailand, Peninsular Malaysia, Singapore). Feral resident (population expanding) Peninsular Malaysia, Singapore. Vagrant S Thailand. **BREEDING** All year. **Nest** Flimsy platform or saucer in tree or bush; 3–8 m above ground. **Eggs** 2–3; white; 25.9 x 20.3 mm (av.).

366 **SPOTTED DOVE**

Streptopelia chinensis Plate **46**

IDENTIFICATION 30–31 cm. *S.c.tigrina*: **Adult** Told by broad black collar from sides to back of neck with conspicuous white spots, long graduated tail with extensive white tips to outer feathers, and broad pale greyish bar across outer greater coverts to carpal (prominent in flight). Rest of upperside greyish-brown with indistinct dark streaks and narrow light edging, crown and ear-coverts pale grey, neck and underparts pale vinous-brownish, throat and vent whitish, primaries and secondaries blackish. *S.c.chinensis* (north E Tonkin) has bluer-grey crown, unstreaked brown upperparts, deeper pinkish neck and breast, greyer undertail-coverts and darker slaty-grey on wing-coverts. See Red and Eurasian Collared-doves. **Juvenile** Much browner with warmer, less vinous-pink underparts, distinct buff fringes to upperparts, wing-coverts and breast, almost no grey on crown and wing-coverts, and much less distinct dark brown neck-collar with buffish-brown bars. **VOICE** Song is a soft repeated ***wu hu'crrroo***; ***wu-crrroo*** or ***wu huuu-croo***, or more hurried ***wu-hwrrroo..wu-hwrrroo..wu-hwrrroo*** etc. **HABITAT** Open areas, open woodland, scrub, cultivation, parks and gardens; up to 2,040 m. **RANGE & STATUS** Resident Indian subcontinent (N Pakistan only), China (except NW,N), Taiwan, Sundas, Philippines. Introduced Mauritius, Sulawesi, Moluccas, Australia, New Zealand, New Caledonia, Fiji, Hawaii, N America. **SE Asia** Common resident throughout. **BREEDING** All year. Multi-brooded. **Nest** Flimsy platform in tree, bush or bamboo. **Eggs** 2–3; white; 26.9 x 20.8 mm (av.).

367 BARRED CUCKOO-DOVE
Macropygia unchall Plate **46**

IDENTIFICATION 38–41 cm. *M.u.tusalia*: **Male** Slender proportions, long graduated tail with no white or grey markings and dark rufescent upperside (including tail) with broad blackish bars distinctive. Head paler brown and unbarred, underparts buffish-brown, breast vinous-tinged and finely barred blackish, has a violet and green gloss on nape, upper mantle and (less intensely) breast; primaries and secondaries all dark. General plumage tones quite variable. **Female** Like male but underparts paler buffish with dense blackish bars (throat and vent plainer). **Juvenile** Similar to female but darker, with all-barred head and neck. **Other subspecies in SE Asia** *M.u.minor* (N Indochina), *unchall* (Peninsular Malaysia). **VOICE** Song is a deep ***who-OO*** or ***wu-OO***, repeated every 1–2 s up to 12 times or more; sometimes a quicker ***wuOO*** or longer ***wuOOO***. **HABITAT** Broadleaved evergreen and semi-evergreen forest, forest edge, clearings; 140–1,800 m. **RANGE & STATUS** Resident Nepal, Bhutan, NE India, southern China, Sumatra, Java, Bali, Lombok, Flores. **SE Asia** Locally common resident, subject to some movements (except SW Myanmar, C,S Thailand, Singapore). **BREEDING** December–September. **Nest** Flimsy saucer in tree; 2–8 m above ground. **Eggs** 1–2; pale buffish; 35.3 x 25.4 mm (av. *tusalia*) and 32.8 x 24.9 mm (av. *unchall*).

368 LITTLE CUCKOO-DOVE
Macropygia ruficeps Plate **46**

IDENTIFICATION 28–33 cm. *M.r.assimilis*: **Male** Resembles Barred Cuckoo-dove in shape but smaller, crown distinctly rufous-chestnut, upperparts and tail without bars, underparts more uniform, breast rufous-buff with heavy whitish scales, underwing-coverts rufous-buff, upperwing-coverts dark brown with chestnut fringes. *M.r.malayana* (Peninsular Malaysia; presumably extreme S Thailand) has blackish mottling on breast and is somewhat darker overall, with broader chestnut fringes to wing-coverts. **Female** Resembles male but breast heavily mottled blackish, wing-coverts more distinctly fringed chestnut. *M.r.malayana* has heavier black mottling on lower throat and upper breast; otherwise differs as male. **Juvenile** Similar to female but mantle and belly a little more barred, has bolder markings on lower throat and breast. **Other subspecies in SE Asia** *M.r.engelbachi* (N Indochina). **VOICE** Song is a soft monotonous ***wup-wup-wup-wup-wup...***, with c.2 notes per second. Each bout consists of up to 40 notes. **HABITAT** Broadleaved evergreen forest, sometimes adjacent deciduous forest, forest edge; 500–1,830 m. **RANGE & STATUS** Resident SW China, Sundas. **SE Asia** Scarce to locally common resident S(east),E Myanmar, Tenasserim, W,NW,NE, extreme S Thailand, Peninsular Malaysia, N Laos, W Tonkin, N Annam. **BREEDING** January–November. **Nest** Flimsy platform, usually in small tree or bamboo; up to 8 m above ground. **Eggs** 1–2; white; 32 x 21.3 mm.

369 EMERALD DOVE
Chalcophaps indica Plate **46**

IDENTIFICATION 23–27 cm. *C.i.indica*: **Male** Unmistakable: metallic green mantle and wings, blue-grey crown and nape, white forehead and eyebrow and dark vinous-pinkish lower head-sides and underparts; whitish to pale grey double band on back (conspicuous in flight), white patch on lesser coverts, red bill. **Female** Like male but head, mantle and breast much browner, belly paler and more buffish, grey restricted to forehead and eyebrow, no white on wing. **Other subspecies in SE Asia** *C.i.maxima* (Cocos Is, off S Myanmar). **Juvenile** Resembles female but crown, mantle and breast darker brown, most of plumage barred rufous-buff (less on centre of crown and mantle, boldest on breast), almost lacks green on upperside. **VOICE** Song is a deep soft ***tit-whoooo*** or ***tik-whooOO*** (short clicking introductory note barely audible), repeated at c.1 s intervals up to 25 times. See Barred Cuckoo-dove. **HABITAT & BEHAVIOUR** Broadleaved evergreen, semi-evergreen and mixed deciduous forest, mangroves, coastal woodland/scrub; up to 1,500 m. Often flushed from forest tracks, trails and streambeds; darts through forest. **RANGE & STATUS** Resident Indian subcontinent (except W,NW and Pakistan), SW,S China, Taiwan, Greater Sundas and satellite islands, Lesser Sundas (east to Alor) Philippines, Sulawesi, Moluccas, West Papuan Islands, Christmas I. **SE Asia** Uncommon to common resident, subject to some movements (except C Thailand). **BREEDING** April–November. **Nest** Fairly compact platform in tree, bush or bamboo; 2–8 m above ground. **Eggs** 2; buff; 26.9 x 21.1 mm (av.).

370 ZEBRA DOVE *Geopelia striata* Plate **46**

IDENTIFICATION 21–21.5 cm. Monotypic. **Male** Resembles a miniature Spotted Dove but upperparts greyer with dark bars rather than streaks, hindneck, sides of neck and flanks barred black and white, centre of breast unbarred vinous-pink, forehead and face distinctly pale bluish-grey, orbital skin pale grey-blue. **Female** Like male but bars extend further onto breast, possibly with less distinctly blue-grey forecrown. **Juvenile** Duller, less contrasting bars on hindneck, rather uniform dark brownish and buffish-brown bars on crown, upperparts and wing-coverts, less distinct bars on underparts but extending further across breast (which almost lacks vinous-pink), and warm buffish fringes to tail and flight-feathers. **VOICE** Song is a high-pitched soft trilling, leading to a series of rapidly delivered short ***coo*** notes. **HABITAT** Scrub in open country and along coasts, parks, gardens, cultivation; up to 2,030 m (usually lowlands). **RANGE & STATUS** Resident Sumatra, Java, Bali, Lombok. Introduced Madagascar, W Indian Ocean islands, Borneo, Philippines, Sulawesi, S Moluccas (Ambon), Tahiti, Hawaii, St. Helena. **SE Asia** Common resident south Tenasserim, S Thailand, Peninsular Malaysia, Singapore. Uncommon to common feral resident rest of Thailand, north-west Cambodia, N Laos (Vientiane). **BREEDING** All year. **Nest** Flimsy platform in bush. **Eggs** 1–2; white; 22.1 x 16.5 mm (av.).

371 NICOBAR PIGEON
Caloenas nicobarica Plate **45**

IDENTIFICATION 40.5–41 cm. *C.n.nicobarica*: **Adult** All-dark plumage with white uppertail-coverts and short white tail diagnostic. Head, neck (including long hackles) and breast blackish-slate with golden-green and blue gloss, rest of plumage mostly blue and green with copper highlights on upperparts, bill blackish with short 'horn' near base of upper mandible. **Juvenile** Duller and browner with rather uniform dark greenish-brown head, mantle and underparts, no neck-hackles and very dark brownish-green tail with blue tinge (retained for several years). **VOICE** Harsh guttural croaking or barking ***ku-RRAU*** and deep low reverberating ***rrr-rrr-rrr-rrr..***. Usually silent. **HABITAT & BEHAVIOUR** Small wooded islands, dispersing to but rarely seen in mainland coastal forest. Mostly terrestrial, runs from danger or flies up to hide in trees. **RANGE & STATUS** Resident (subject to local movements) Andaman and Nicobar Is, Greater Sundas, W Lesser Sundas, Philippines, Sulawesi, Moluccas, Palau Is, N New Guinea region, N Melanesia. **SE Asia** Scarce to local resident (subject to local movements) on islands off S Myanmar (Coco Is), Tenasserim, west S Thailand, Peninsular Malaysia, Cambodia, Cochinchina. Visits mainland coasts. **BREEDING** April–May. Often colonial. **Nest** Rough platform in tree; 3–8 m above ground. **Eggs** 1; white; 48 x 33.9 mm (av.).

Worldwide c.126 species. SE Asia 17 species. Similar to those species previously mentioned under Columbinae. **Green-pigeons** (*Treron*) and **Jambu Fruit-dove** Small to medium-sized with rather broad wings, short tails and predominantly green plumage. **Imperial-pigeons** (*Ducula*) Medium-sized to large with relatively broad wings and tail.

372 CINNAMON-HEADED GREEN-PIGEON

Treron fulvicollis Plate **47**

IDENTIFICATION 25.5–26 cm. *T.f.fulvicollis*: **Male** Rufous-chestnut head and neck diagnostic. **Female** Similar to Thick-billed Green-pigeon but red-based bill distinctly narrower, eyering much narrower, crown greener, has yellowish thighs and streaked (not barred) undertail-coverts. **Juvenile** Both sexes initially similar to female; male soon shows patches of adult plumage. **VOICE** Song is similar to Little Green-pigeon but less whining, more syllabic. **HABITAT** Freshwater swamp forest, mangroves, coastal forest and secondary growth; lowlands (rarely up to 1,250 m in Peninsular Malaysia). **RANGE & STATUS** Resident Sumatra, Borneo. **SE Asia** Scarce to uncommon resident (subject to some movements) south Tenasserim, S Thailand (rare), Peninsular Malaysia. Rare non-breeding visitor (formerly resident) Singapore. **BREEDING** January–June. **Nest** Flimsy platform in tree; 4.5 m above ground. **Eggs** 2; white.

373 LITTLE GREEN-PIGEON

Treron olax Plate **47**

IDENTIFICATION 20–20.5 cm. Monotypic. **Male** Combination of bluish-grey hood (throat whiter), maroon mantle, scapulars and lesser coverts, and broad orange patch on upper breast diagnostic. Tail blackish-slate with paler grey terminal band, bill slate-coloured with yellowish-green tip. **Female** From other green-pigeons by combination of dark grey crown, dark green upperparts, pale greyish throat, dull green underparts, pale buffish undertail-coverts with dark green streaks, and bill (as male). **Juvenile** Both sexes initially like female but a little darker above, crown less distinctly grey, scapulars, tertials and lesser coverts tipped chestnut (obscurely on mantle); male soon shows patches of adult plumage. **VOICE** Song is high-pitched, rather nasal and well structured: roughly ***wiiiiii-iiu-iiu iiu-iiui iiui-iiuwu***. Repeated after shortish intervals. **HABITAT** Broadleaved evergreen forest, freshwater swamp forest, secondary growth; up to 1,220 m. **RANGE & STATUS** Resident Greater Sundas. **SE Asia** Scarce to fairly common resident (subject to local movements) S Thailand, Peninsular Malaysia. Scarce non-breeding visitor (formerly resident) Singapore. **BREEDING** April–July. **Nest** Flimsy platform in tree; 4 m above ground. **Eggs** 2; white.

374 PINK-NECKED GREEN-PIGEON

Treron vernans Plate **47**

IDENTIFICATION 26.5–32 cm. *T.v.griseicapilla*: **Male** Like Orange-breasted Green-pigeon but has grey head grading to vinous-pink nape, neck and sides of breast (central breast orange), greyer-green upperparts, and grey uppertail with complete blackish subterminal band and much narrower pale grey tips to outer feathers. Undertail blackish with narrow pale grey tips, undertail-coverts dark chestnut. **Female** From Orange-breasted by rather uniform greyer-green upperparts (including nape) and tail pattern (as male). **Juvenile** Both sexes initially like female but tertials browner, scapulars and tertials fringed whitish to buffish, and primaries tipped browner; male soon shows patches of adult plumage. **VOICE** A series of bubbling and gargling notes, leading to a series of harsh grating sounds. Foraging flocks utter a hoarse rasping ***krrak, krrak...*** **HABITAT** Scrub, cultivated areas, mangroves, peatswamp and freshwater swamp forest, island forest; lowlands. **RANGE & STATUS** Resident Greater Sundas, W Lesser Sundas, Philippines, Sulawesi, N Moluccas. **SE Asia** Common, mostly coastal resident (subject to local movements) south Tenasserim, Thailand (rare NE), Peninsular Malaysia, Singapore, Cambodia, S Annam, Cochinchina. **BREEDING** November–August. **Nest** Flimsy platform in tree, 1+ m above ground. **Eggs** 2; white; 27.4 x 21.6 mm (av.).

375 ORANGE-BREASTED GREEN-PIGEON

Treron bicincta Plate **47**

IDENTIFICATION 29 cm. *T.b.bicincta*: **Male** Combination of green head with grey nape, green upperparts (tinged brownish), vinous-pink and orange breast-patches and grey central tail-feathers diagnostic. Undertail-coverts dull rufous-brown, undertail blackish with broad grey terminal band. See Pink-necked and Large Green-pigeons. **Female** Lacks vinous-pink and orange breast-patches. From other green-pigeons by contrasting grey nape, strong brownish cast to green of upperparts, lack of obvious eyering, dark-based bill (with no red) and tail pattern (as male). **Juvenile** Both sexes initially like female but grey on nape less distinct, scapulars and tertials fringed dull buffish (vaguely on mantle feathers), some rufescent fringes on lesser coverts and pale-tipped primaries; male soon shows patches of adult plumage. **VOICE** Song is a mellow wandering whistle and subdued gurgling. Calls include ***ko-WRRROOOK, ko-WRRROOOK, ko-WRRROOOK*** and ***kreeeew-kreeeew-kreeeew***. **HABITAT** More open deciduous and semi-evergreen forest, secondary growth, sometimes broadleaved evergreen forest and mangroves; up to 1,400 m. **RANGE & STATUS** Resident N,NE Indian subcontinent, S,E India, Sri Lanka, S China (Hainan), Java, Bali. **SE Asia** Uncommon to locally fairly common resident (except C Thailand, Singapore, N,C Laos, W,E Tonkin). **BREEDING** February–June. **Nest** Shallow platform in tree; 2–8 m above ground. **Eggs** 2; white; 29.5 x 22.9 mm (av.).

376 ASHY-HEADED GREEN-PIGEON

Treron phayrei Plate **47**

IDENTIFICATION 25.5–26 cm. Monotypic. **Male** Like Thick-billed Green-pigeon but without broad eyering, bill slenderer and greyish, has orange wash on breast and yellower-tinged throat. Undertail-coverts brick-red. See Andaman Green-pigeon, but note range. **Female** From Thick-billed by same features (except breast-wash) as male plus short streaks (not scales) on undertail-coverts. **Juvenile** Both sexes initially like female but darker above with less grey on crown, pale tips to primaries and indistinct paler tips to mantle, scapulars and tertials; male rapidly shows patches of adult plumage. **VOICE** Song is a series of wandering pleasant, mellow, fluty whistles, ascending and descending scale, first note low and level, second reaching highest, and most notes low and relatively level. **HABITAT** Broadleaved evergreen and semi-evergreen forest; up to 800 m. **RANGE & STATUS** Resident NE Indian subcontinent, south-western India, Sri Lanka, SW China, Philippines, S Moluccas (Buru). **SE Asia** Scarce to locally common resident (subject to some movements) Myanmar, Thailand (except C,S), Cambodia, Laos, Cochinchina. Vagrant (status uncertain) C Thailand. **BREEDING** April–May. **Nest** Shallow platform in tree or bamboo; up to 13 m above ground. **Eggs** 2; white; 27.4 x 21.8 mm (av.).

377 ANDAMAN GREEN-PIGEON

Treron chloropterus Plate **47**

IDENTIFICATION 27–28 cm. Monotypic. **Male** Similar to Ashy-headed Green-pigeon but considerably larger and heavier-billed, has green lesser coverts, bright lime-green rump and grey-green undertail-coverts with pale yellow tips. Note range. **Female** Differs from Ashy-headed by size, larger bill, and bright lime-green rump and undertail-coverts (as male). **Juvenile** Duller and paler than adult female, with rounded yellow tips to wing-coverts, broad yellow fringe to greener tertials, and distinct white outer fringes to primaries. **VOICE** Song is similar to Ashy-headed but more nasal, and following a somewhat different pattern; first two notes short and steeply up- then downslurred, third note highest (c.3 downslurred nasal notes per strophe). **HABITAT**

Broadleaved evergreen and semi-evergreen; up to 800 m. **RANGE & STATUS** Resident Andaman and Nicobar Is. **SE Asia** Resident (status unknown) Coco Is, off S Myanmar. **BREEDING** February–June. Otherwise undocumented.

378 THICK-BILLED GREEN-PIGEON

Treron curvirostra Plate **47**

IDENTIFICATION 25.5–27.5 cm. *T.c.nipalensis*: **Male** Identified by combination of thick pale greenish bill with red base, broad greenish-blue eyering, grey crown, maroon mantle, scapulars and lesser coverts, and all-green throat and underparts. Undertail-coverts dull chestnut. See Little, Pompadour and Wedge-tailed Green-pigeons. **Female** Like male but lacks maroon on upperparts and wings, undertail-coverts creamy-buff with dark green scales, thighs dark green with whitish scales. See Cinnamon-headed and Pompadour Green-pigeons. **Juvenile** Both sexes initially like female but has some rusty fringes on tips of scapulars, tertials and primaries; male soon shows patches of adult plumage. **Other subspecies in SE Asia** *T.c.curvirostra* (S Thailand southwards). **VOICE** Song is similar to Little Green-pigeon but fuller, lower-pitched and broken more into separate phrases. Calls include guttural hissing or growling notes when foraging. **HABITAT** Broadleaved evergreen, semi-evergreen and mixed deciduous forest, secondary growth, sometimes mangroves; up to 1,280 m. **RANGE & STATUS** Resident Nepal, NE Indian subcontinent, SW,S China, Sumatra, Borneo, Philippines. **SE Asia** Common resident, subject to local movements (except C Thailand, Singapore, W Tonkin). Rare to uncommon non-breeding visitor C Thailand, Singapore (formerly resident). **BREEDING** January–September. **Nest** Flimsy platform in tree or bamboo 5–12 m above ground. **Eggs** 2; white; 28.7 x 22.6 mm (av.).

379 LARGE GREEN-PIGEON

Treron capellei Plate **47**

IDENTIFICATION 35.5–36 cm. Monotypic. **Male** Large size, very stout bill, yellowish eyering, legs and feet, green upperparts, all-green head (greyish around face) and yellow-orange breast-patch distinctive. Undertail-coverts dark chestnut-brown. **Female** Like male but breast-patch yellowish, undertail-coverts creamy-buff with dark green mottling. **Juvenile male** Like female but breast-patch more orange-tinged, undertail-coverts pale rufous. **Juvenile female** Like female. **VOICE** Song is a series of variable deep nasal creaking notes: ***oo-oo-aah oo-oo-aah aa-aa-aah*** and ***oooOOah oo-aah*** etc. Calls include deep conversational grumblings and growlings. **HABITAT** Broadleaved evergreen forest, freshwater swamp forest, forest edge and clearings, specialising on figs; up to 200 m, rarely to 1,220 m. **RANGE & STATUS** Resident Greater Sundas. **SE Asia** Rare to uncommon resident (subject to some movements) south Tenasserim, S Thailand, Peninsular Malaysia. **BREEDING** February–July. **Nest** Flimsy platform on tree-branch; 3–4 m above ground. **Eggs** 1; white.

380 YELLOW-FOOTED GREEN-PIGEON

Treron phoenicopterus Plate **47**

IDENTIFICATION 33 cm. *T.p.annamensis*: **Male** Unmistakable with grey crown and nape, pale green throat, yellowish-green neck and upper breast, pale grey-green upperparts, grey lower breast and belly and yellow legs and feet. Has small pinkish-maroon shoulder-patch, bright yellowish-olive tail with grey terminal half and dark maroon undertail-coverts with creamy-buff bars. *T.p.viridifrons* (Myanmar) has paler, yellower-green throat and breast (washed golden on latter), greenish-golden hindneck (collar), green forecrown, paler and greener (less greyish) upperparts and slightly paler grey on underparts. **Female** Tends to show less distinct shoulder-patch. **Juvenile** Both sexes paler and duller than female, with little or no shoulder-patch. **VOICE** Series of c.10 beautiful modulated mellow musical whistles, recalling Orange-breasted Green-pigeon but louder and lower-pitched. **HABITAT** Mixed deciduous forest, secondary growth; lowlands, sometimes up to 1,220 m. **RANGE & STATUS** Resident (subject to local movements) NE Pakistan, Indian subcontinent, SW China. **SE Asia** Scarce to locally common resident Myanmar, W,NW,NE Thailand, Cambodia, C,S Laos, C,S Annam, north Cochinchina. **BREEDING** March–June. **Nest** Flimsy platform in tree. **Eggs** 2; glossy, white; 31.7 x 24.1 mm (av.).

381 PIN-TAILED GREEN-PIGEON

Treron apicauda Plate **47**

IDENTIFICATION 30.5 cm (tail-prongs up to 10 cm more). *T.a.apicauda*: **Male** Wedge-shaped grey tail (outer feathers blackish near base) with greatly elongated and pointed central feathers (prongs) diagnostic. Has bright blue naked lores and base of rather slender bill, mostly rather bright green body, apricot flush on breast, green belly (any white markings restricted to vent) and chestnut undertail-coverts with outer webs fringed buffish-white. In hand has distinctive lobe on inner fringe of third and fourth primaries. *T.a.lowei* (southern Indochina) has duller green head with feathered lores; mantle and wing-coverts washed light brownish-grey; contrasting greenish-yellow rump and uppertail-coverts; and uniformly green underparts. In hand has distinctive indentation on inner web of third primary (also on female). **Female** Like male but breast all green, may show less chestnut on undertail-coverts (and more whitish), central tail-feathers shorter (still distinctly elongated and pointed). *T.a.lowei* differs by same features as male and also lacks chestnut on undertail-coverts. **Juvenile male** Tail-prongs shorter and blunter, wing-coverts rounder, creating different pattern of yellow fringes on wing; primary tips faintly tinged pale grey-green. **Other subspecies in SE Asia** *T.a.laotinus* (northern Indochina). **VOICE** Song is a series of musical, wandering whistles: ***ko-kla-oi-oi-oi-oilli-illio-kla***, possibly produced by duetting pair. Said to be more tuneful and less meandering than that of Wedge-tailed Green-pigeon. **HABITAT** Broadleaved evergreen forest; 600–1,830 m, sometimes down to 300 m or lower. **RANGE & STATUS** Resident (subject to local movements) N,NE Indian subcontinent, SW China. **SE Asia** Uncommon resident (subject to local movements) Myanmar, north Tenasserim, W,NW,NE Thailand, east Cambodia, Laos, Vietnam (except Cochinchina). Recorded (status uncertain) north Cochinchina. **BREEDING** March–August. **Nest** Flimsy platform in small tree, tall bush or bamboo. **Eggs** 2; white; 31.7 x 23.9 mm (av.; *apicauda*).

382 YELLOW-VENTED GREEN-PIGEON

Treron seimundi Plate **47**

IDENTIFICATION 26–28 cm (tail-prongs up to 5 cm more). *T.s.seimundi*: **Male** Like Pin-tailed Green-pigeon but has much shorter tail-prongs, generally darker green plumage, maroon shoulder-patch, whitish central belly and mostly yellow undertail-coverts with narrow green centres. Has golden-tinged forecrown, pinkish-orange wash across upper breast, blue bill with horn-grey tip, and blue naked lores and eyering. *T.s.modestus* (Indochina) has all-green crown and breast. See Wedge-tailed and White-bellied Green-pigeons. **Female** Like male but undertail-coverts have broader green centres, no maroon shoulder-patch, breast greener. **VOICE** Song is a high-pitched ***pooaah po-yo-yo-pooaah***. **HABITAT** Broadleaved evergreen forest, forest edge, exceptionally mangroves; up to 1,525 m (usually above at least 250 m). **RANGE & STATUS** Endemic. Rare to locally fairly common resident (subject to local movements) W,NW,SE Thailand, Peninsular Malaysia, Cambodia, C,S Laos, Vietnam (except W Tonkin). Recorded (status uncertain) C Thailand. **BREEDING** February–August. **Nest** Flimsy platform in tree; 6–25 m above ground. **Eggs** Undocumented?

383 WEDGE-TAILED GREEN-PIGEON

Treron sphenura Plate **47**

IDENTIFICATION 33 cm. *T.s.sphenura*: **Male** From other green-pigeons by combination of broad wedge-shaped tail with uniformly grey underside, maroon upper mantle, upper scapulars and lesser coverts and green head and underparts with strong apricot wash on crown and breast; very long undertail-coverts pale cinnamon, base of bill and narrow eyering blue. In hand has distinctive lobe on inner fringe of third and fourth primaries. *T.s.robinsoni* (Peninsular Malaysia) and *delacouri* (S Laos, C,S Annam) are smaller (26.5–30.5 cm) and darker with little or no apricot on crown and breast, maroon of upperparts restricted to shoulder-patch; former has undertail-coverts like female. In hand, both *robinsoni* and *delacouri* show an indentation on inner web of third primary (also on females). **Female** Crown, upperparts and breast all green (forehead slightly yellower), undertail-coverts creamy-buff with dark green centres. *T.s.robinsoni* and *delacouri* are smaller and darker with darker, less contrasting forehead. **Juvenile** Both sexes initially similar to female. **Other subspecies in SE Asia** *T.s.yunnanensis* (W Tonkin). **VOICE** Song is long and rather high-pitched, with a rolling introduction: ***phruuuuah-po phuu phuuuu phuu-phu phuo-oh po-oh-oh-po-po-ohpopopo puuuuuuuah puuooaha wo-pi-ohaauah*** or similar ***phuu phuuoh puuuUH pu-wu-pupupupupupu puuuuuo-aow pwaaaaAH pwaaaoah-aow***, without introduction. **HABITAT** Broadleaved evergreen forest, forest edge; 600–2,565 m (locally down to 350 m), rarely in plains outside breeding season (N Laos, C Annam). **RANGE & STATUS** Resident (subject to local movements) NE Pakistan, NW,N,NE Indian subcontinent, S,SE Tibet, south-western China, Sumatra, Java, Bali, Lombok. **SE Asia** Locally common resident, subject to local movements (except C,S Thailand, Singapore, E Tonkin, Cochinchina). **BREEDING** April–June. **Nest** Flimsy platform in tree, 6–12 m above ground. **Eggs** 2; white; 31.5 x 23.1 (av.; *sphenura*).

384 WHITE-BELLIED GREEN-PIGEON
Treron sieboldii Plate **47**

IDENTIFICATION 33 cm. *T.s.murielae*: **Male** Very similar to Wedge-tailed Green-pigeon but belly mostly greyish-white, undertail-coverts creamy-whitish with dark green centres, undertail blackish with very narrow grey terminal band, maroon of upperparts confined to upper scapulars and shoulder-patch, bill brighter blue. See Yellow-vented Green-pigeon. **Female** From Wedge-tailed by underpart colour and undertail pattern (as male). **VOICE** A mournful, protracted ***o-aooh*** or ***oo-whooo***, first note higher-pitched; repeated several times. Also a short ***pyu*** in alarm. **HABITAT** Broadleaved evergreen forest, forest edge, clearings; 200–900 m (to 2,000 m Thailand). **RANGE & STATUS** Breeds C,S,E China, Taiwan, Japan; some disperse to west and south-west in winter. **SE Asia** Local resident W,NW,NE(north-west) Thailand, C Laos, E Tonkin, N,C Annam. **BREEDING** Season not documented in region. **Nest** Flimsy platform in small tree or tall shrub. **Eggs** 2; white.

385 JAMBU FRUIT-DOVE
Ptilinopus jambu Plate **45**

IDENTIFICATION 26.5–27 cm. Monotypic. **Male** Greenish upperparts, crimson face, white eyering and white underparts with pink flush on foreneck and upper breast diagnostic. Bill orange-yellow, undertail-coverts chestnut. **Female** Mostly green with greyish-purple face, white eyering, maroon central stripe on throat, paler vent and buffish undertail-coverts. See green-pigeons. **Juvenile** Like female but face brownish and central throat whitish (washed dull rufous). Initially has warm brown fringing on upperparts, wing-coverts and tertials. **VOICE** Soft ***hooo***, repeated after short intervals. Usually silent. **HABITAT** Broadleaved evergreen forest, rarely mangroves; up to 1,280 m. **RANGE & STATUS** Resident Sumatra (possibly also erratic non-breeding visitor), Borneo. Recorded (status uncertain) W Java. **SE Asia** Scarce to uncommon resident (subject to local movements) southern S Thailand, Peninsular Malaysia. Scarce and irregular non-breeding visitor Singapore. **BREEDING** May–September. **Nest** Flimsy platform in tree; 2.7–5 m above ground. **Eggs** 1; white.

386 GREEN IMPERIAL-PIGEON
Ducula aenea Plate **45**

IDENTIFICATION 42–47 cm. *D.a.sylvatica*: **Adult** Similar to Mountain Imperial-pigeon but upperparts mostly dark metallic green (may be hard to discern) with variable rufous-chestnut gloss; head, neck and underparts rather uniform vinous-tinged pale grey, undertail-coverts dark chestnut, tail all dark. **Juvenile** Duller above than adult; head, neck and underparts paler, virtually without vinous tones. **Other subspecies in SE Asia** *D.a.polia* (southern S Thailand southwards). **VOICE** Song is a very deep, repeated ***wah-whhoo***; ***wah-whhrrooo*** or ***wah-wahrroo*** etc. Also gives a deep ***hoooo*** or ***huuooo*** and rhythmic purring ***crrhhoo***. **HABITAT** Broadleaved evergreen, semi-evergreen, mixed deciduous and island forest, mangroves, sometimes secondary habitats; up to 915 m. **RANGE & STATUS** Resident S,E,NE India, east Bangladesh, Sri Lanka, SW,S China, Greater Sundas, W Lesser Sundas, Philippines, Sulawesi. **SE Asia** Scarce to locally fairly common resident (except C Thailand); mostly coastal in Peninsular Malaysia. Scarce non-breeding visitor (formerly resident and perhaps just still) Singapore. **BREEDING** January–May and September. **Nest** Flimsy platform in small tree, sometimes bamboo; up to 10 m above ground. **Eggs** 1–2; white; 45.5 x 33.5 mm (av.).

387 MOUNTAIN IMPERIAL-PIGEON
Ducula badia Plate **45**

IDENTIFICATION 43–51 cm. *D.b.griseicapilla*: **Adult** The largest pigeon in the region. Distinguished by mostly purplish-maroon mantle and wing-coverts, bluish-grey crown and face, white throat, vinous-tinged pale grey neck (more vinous at rear) and underparts, whitish-buff undertail-coverts and dark tail with contrasting broad greyish terminal band. Red eyering and red bill with pale tip. ***D.b.badia*** (south Tenasserim and S Thailand southward) has upperparts more extensively and intensely purplish-maroon (including rump), crown and face duller and more vinous-grey (contrasts less with hindneck) and darker, stronger vinous-pink tinge to underparts, contrasting more with (buffish) undertail-coverts. **Juvenile** Like adult but less pink on hindneck, rusty-brown fringes to mantle, wing-coverts and flight-feathers. **VOICE** Song is a loud, very deep ***uh, WROO-WROO*** or ***uhOOH-WROO-WROO*** (introductory note only audible at close range) or just ***uOOH-WROO***. Repeated after rather long intervals. **HABITAT** Broadleaved evergreen forest; up to 2,565 m (mostly mountains). **RANGE & STATUS** Resident Nepal, Bhutan, NE,SW India, SW,S China, Greater Sundas. **SE Asia** Fairly common to common resident SW,W,N,C,E,S(east) Myanmar, Tenasserim, W,NW,NE,SE,extreme S Thailand, Peninsular Malaysia, Indochina (except Cochinchina). Recorded (status uncertain) north Cochinchina. **BREEDING** All year (March–August in north). **Nest** Flimsy platform in tree; 5–8 m above ground. **Eggs** 1–2; quite glossy, white; 46.2 x 33.5 mm (av.).

388 PIED IMPERIAL-PIGEON
Ducula bicolor Plate **45**

IDENTIFICATION 38–41 cm. Monotypic. **Adult** Unmistakable. White with contrasting black primaries and secondaries and black tail with much white on outer feathers. **Juvenile** White feathers have buffish tips, particularly on upperside. **VOICE** A deep but rather quiet ***cru-croo*** or ***croo croo-oo***. Deep, resonant purring ***rruuu*** or ***wrrooom***, repeated after 1–3 s intervals. Also ***whoo whoo whoo hoo hoo***, with notes descending in pitch and becoming progressively shorter. **HABITAT & BEHAVIOUR** Island forest, sometimes mangroves and coastal mainland forest; lowlands. Gregarious, generally seen in small flocks; larger aggregations on offshore islands for roosting and breeding. **RANGE & STATUS** Resident (subject to local movements) Nicobar and S Andaman Is, Greater Sundas, Lesser Sundas (very local), Philippines, Sulawesi, Moluccas, Western Papuan and Aru Is. **SE Asia** Locally common coastal resident (subject to local movements) SW Myanmar, Tenasserim, SE,S Thailand, Peninsular Malaysia, Cambodia, Cochinchina (Con Dao I only?). Mostly on offshore islands. Scarce non-breeding visitor southern W,S(Chumphon) Thailand (mostly coastal), Singapore. Recorded (status uncertain) Coco Is, S Myanmar. **BREEDING** February–September. **Nest** Flimsy platform on tree-branch; 7–10 m above ground. **Eggs** 1; white; 45.7 x 30.5 mm (av.).

PSITTACIDAE: LORICULINAE: Hanging-parrots

Worldwide 11 species. SE Asia 2 species. Small to smallish, mostly green with pointed wings and fast, direct flight.Thick, rounded, hooked bills and short legs with zygodactyl feet. Mostly arboreal, climbing branches and feeding mainly on fruit, seeds, buds, nectar and pollen. Sexually dimorphic.

389 VERNAL HANGING-PARROT
Loriculus vernalis Plate **48**

IDENTIFICATION 13–15 cm. Monotypic. **Male** Distinguished by tiny size, short tail, bright green plumage (mantle duller), contrasting red back to uppertail-coverts and light blue flush on lower throat/upper breast. Bill red, legs and feet dull yellow to orange, eyes usually whitish to pale yellow. In flight shows turquoise underwing with green coverts. **Female** Throat/breast has little or no blue, head and underparts somewhat duller, red of lower upperparts duller and mixed with some green on back and rump. **Juvenile** Both sexes like female but back to uppertail-coverts mixed with green, eyes, legs and feet duller. See Blue-crowned Hanging-parrot. **VOICE** High-pitched squeaky ***tsee-sip*** or ***pi-zeez-eet***, usually given in flight. **HABITAT & BEHAVIOUR** Broadleaved evergreen, semi-evergreen and deciduous forest, clearings; up to 1,525 m. Has strong direct flight; often hangs upside-down; fond of fruiting trees. **RANGE & STATUS** Resident Nepal, NE Indian subcontinent, S,E India, Andaman and Nicobar Is, south-western China. **SE Asia** Common resident Myanmar (except W,N), Thailand (except C), north Peninsular Malaysia, Indochina (except W,E Tonkin, N Annam). **BREEDING** January–February. **Nest** Natural hole (often adapted) in tree or stump; up to 10 m above ground. **Eggs** 3–4; white (often stained brownish); 19.1 x 15.8 mm (av.).

390 BLUE-CROWNED HANGING-PARROT
Loriculus galgulus Plate **48**

IDENTIFICATION 12–14.5 cm. Monotypic. **Male** Like Vernal Hanging-parrot but more vivid green, with dark blue crown-patch, golden 'saddle' on mantle, bright golden-yellow band across lower back, red patch on lower throat/upper breast, black bill, usually brown to grey eyes and greyish-brown to yellowish legs and feet. **Female** Duller, lacks red on lower throat/upper breast, crown and mantle-patches less distinct (particularly former), patch on lower back golden (lacking yellow tone), red of lower upperparts duller. **Juvenile** Both sexes like female but mantle all green, only faint tinge of blue on crown, lower back less golden, bill dusky-yellowish to blackish with yellowish tip, legs and feet dull yellowish. From Vernal by duller bill, obvious blue tinge on crown (if present), golden patch on lower back and slightly darker underparts. **VOICE** Shrill, high-pitched ***tsi*** or ***tsrri***, sometimes ***tsi-tsi-tsi...*** etc. **HABITAT** Broadleaved evergreen forest, clearings, wooded gardens and plantations, mangroves; up to 1,280 m. **RANGE & STATUS** Resident Sumatra, Borneo, W Java (at least some introduced). **SE Asia** Fairly common to common resident (subject to some movements) extreme south Tenasserim (scarce), southern S Thailand, Peninsular Malaysia, Singapore (scarce). **BREEDING** February–June. **Nest** Natural hole (often adapted) in tree or stump; up to 7.6 m above ground. **Eggs** 3–4; dull white (staining brownish); c.18 x 15 mm.

PSITTACIDAE: CACATUINAE: Cockatoos

Worldwide 21 species. SE Asia 2 species. Relatively large with rounded wings and mostly whitish plumage. Sexes similar. Large-headed, with thick, rounded, hooked bills and short legs with zygodactyl feet. Mostly arboreal, climbing branches and feeding mainly on fruit, seeds, buds, nectar and pollen.

390A YELLOW-CRESTED COCKATOO
Cacatua sulphurea Plate **58**

IDENTIFICATION 33–35 cm. *C.s.sulphurea*: **Male** Predominantly white plumage, yellow flush on ear-coverts and long, erectile yellow crest distinctive. Eyering pale bluish-white, eyes dark brown, legs and feet grey. In flight, shows lemon-yellow suffusion on underwing and undertail. See Tanimbar Corella. **Female** Bill slightly smaller, eyes reddish. **Juvenile** Eyes brownish-grey (both sexes), bill, legs and feet paler. Sulphur-crested Cockatoo *C. galerita*, which has also escaped in Singapore (not established), is larger (50 cm), with blue eyering. **VOICE** Very loud, harsh, raucous screeching, a variety of less harsh whistles and squeaks, and a series of 2–6 fairly high-pitched quavering, nasal screeches. Calls are typically longer and louder than those of Tanimbar Corella. **HABITAT & BEHAVIOUR** Open forest, plantations, cultivation, parks, gardens; lowlands. Found in groups of up to 10. **RANGE & STATUS** Resident Java Sea (Salembu Basar), Bali (Penida I), Lesser Sundas, Sulawesi. Introduced S China (Hong Kong). **SE Asia** Scarce to uncommon feral resident Singapore. **BREEDING** Season not documented in region. **Nest** Tree-hollow. **Eggs** 2–3; white; 38.1–44 x 25.7–28.4 mm.

390B TANIMBAR CORELLA
Cacatua goffini Plate **58**

IDENTIFICATION 32 cm. Monotypic. **Male** Mostly white plumage, salmon-pinkish lores and greyish-white bill distinctive. Short crest, eyering bluish-white, eyes dark brown, legs and feet greyish. In flight, shows yellow wash on underwing and undertail. See Yellow-crested Cockatoo. **Female** Eyes reddish-brown. **Juvenile** Eyes dark grey. **VOICE** Various loud harsh screeches. When excited, utters harsh, nasal screeches, each note varying in volume and length. In flight often gives a single longer, more quavering screech and slightly sweeter squabbling/screeching notes. **HABITAT** Open forest, cultivation; lowlands. **RANGE & STATUS** Resident E Lesser Sundas (Tanimbar Is). Also (status uncertain) SE Moluccas (Kai Kecil). **SE Asia** Fairly common to common feral resident Singapore. **BREEDING** September–December. **Nest** Excavated or natural tree-hollow. **Eggs** 2–3; white; 37.6–39.6 x 27.8–29.7 mm.

PSITTACIDAE: PSITTACINAE: Parrots & parakeets

Worldwide c.264 species. SE Asia 7 species. Large-headed, with thick, rounded, hooked bills and short legs with zygodactyl feet. Sexually dimorphic. Mostly arboreal, climbing branches and feeding mainly on fruit, seeds, buds, nectar and pollen. **Blue-rumped Parrot** Smallish, mostly green with pointed wings and fast, direct flight. Sexually dimorphic. **Parakeets** (*Psittacula*) Medium-sized to largish, generally green with long extended central tail-feathers.

391 BLUE-RUMPED PARROT
Psittinus cyanurus Plate **48**

IDENTIFICATION 18.5–19.5 cm. *P.c.cyanurus*: **Male** Only likely to be confused with hanging-parrots. Differs by larger size and stockier build, mostly greyish-blue head, blackish mantle, deep purplish-blue back to uppertail-coverts, yellowish-green fringing on wing-coverts, pale greyish- to brownish-olive breast and flanks (tinged blue, particularly on latter). Has narrow dark red patch on inner lesser wing-coverts and red bill with dark brownish lower mandible. In flight shows blackish underwing with largely red coverts and axillaries. **Female** Head mostly brown with paler sides and yellower throat, narrow dark brown streaks on lower sides of head and throat, blue of upperparts restricted to back, mostly green breast, belly and flanks, less dark red on inner wing-coverts, dark brown bill. **Juvenile** Both sexes like female but crown and sides of head green, little or no dark red on inner wing-coverts. Male may show blue tinge on forehead and sides of head. **VOICE** Sharp, high-pitched ***chi chi chi...*** and ***chew-ee***, mainly in flight. Melodious trilling. **HABITAT** More open broadleaved evergreen forest, clearings, sometimes plantations, rarely mangroves; up to 700 m, rarely 1,300 m. **RANGE & STATUS** Resident Sumatra, Borneo. **SE Asia** Scarce to locally fairly common resident (subject to some movements) south Tenasserim, W(south),S Thailand, Peninsular Malaysia, Singapore. **BREEDING** January–June. **Nest** Natural tree-hole; 30 m above ground, or higher. **Eggs** 1–3; white; 23.1–26.5 x 20.3–21.3 mm.

392 ALEXANDRINE PARAKEET

Psittacula eupatria Plate **48**

IDENTIFICATION 50–58 cm. *P.e.siamensis*: **Male** Relatively large with massive red bill, green to yellowish-green head with pale blue wash on hindcrown/nape to upper ear-coverts and distinctive broad maroon-red shoulder-patch. Has narrow collar, black at front and deep pink at rear, rest of plumage mostly green with duller mantle and blue wash on upperside of tail (underside yellowish), legs and feet yellow. In flight shows green underwing-coverts. *P.e.avensis* (Myanmar) has blue on head restricted to band bordering collar, rest of head more uniformly green. See Rose-ringed Parakeet. **Female** Lacks collar and obvious blue wash on head, shoulder-patch smaller and a little paler, tail-streamers average shorter. See Rose-ringed. **Juvenile** Both sexes like female but duller with smaller shoulder-patch, shorter tail, duller bill. **Other subspecies in SE Asia** *P.e.magnirostris* (Coco Is, off S Myanmar). **VOICE** Loud, ringing ***trrrieuw***, loud ***kee-ah*** and ***keeak*** and resonant ***g'raaak g'raaak...*** **HABITAT** Mixed deciduous forest, temple groves; up to 915 m. **RANGE & STATUS** Resident Indian subcontinent, SW China (W Yunnan). **SE Asia** Scarce to locally common resident Myanmar (except southern Tenasserim), C Thailand (local and possibly feral), Cambodia, S Laos, S Annam, north-west Cochinchina. Formerly resident (current status unknown) W,NW,NE Thailand, C Laos, C Annam. Formerly recorded south N Laos (once). **BREEDING** February–April. Loosely colonial. **Nest** Excavated tree-hole, sometimes old woodpecker or barbet nest. **Eggs** 2–5; white; c.34 x 28 mm (av.; *avensis*).

393 ROSE-RINGED PARAKEET

Psittacula krameri Plate **48**

IDENTIFICATION 40–42 cm. *P.k.borealis*: **Male** Like Alexandrine Parakeet but smaller, bill much smaller with largely blackish lower mandible, no maroon-red shoulder-patch, narrow but distinct black loral line (crosses extreme forehead), lighter green scapulars and wing-coverts, and yellowish- to greenish-grey legs and feet. Throat all black, eyes yellowish-white. **Female** Has indistinct dark green collar, no black loral line, tail-streamers average shorter; otherwise differs from Alexandrine as male. **Juvenile** Both sexes like female but bill slightly paler, eyes greyish, tail shorter. Male attains collar in third year. See Grey-headed Parakeet. **VOICE** Variable loud, shrill, rather harsh ***kee-ak kee-ak kee-ak...***, higher-pitched and less guttural than Alexandrine. Also rasping ***kreh kreh kreh kreh...*** given by flocks, and chattering ***chee chee...*** in flight. **HABITAT** Open mixed deciduous forest, edges of cultivation, groves, sometimes plantations, parks and gardens; up to 915 m. **RANGE & STATUS** Resident equatorial Africa, Indian subcontinent, SW China (W Yunnan). Introduced or escaped N Egypt, Kenya, S Africa, W,C Europe, Mauritius, Arabian Peninsula, S China (Hong Kong, Macao), USA. **SE Asia** Locally common resident Myanmar (except Tenasserim). Uncommon feral resident Peninsular Malaysia (Pulau Pinang), Singapore. Recorded (presumed local feral resident) C Thailand. **BREEDING** February–May. Loosely colonial. **Nest** Excavated or existing hole in tree, rock-face or building, sometimes old woodpecker or barbet nest; 3–10 m above ground. **Eggs** 4–6; white; 29.3 x 24 mm (av.).

394 GREY-HEADED PARAKEET

Psittacula finschii Plate **48**

IDENTIFICATION 36–40 cm. Monotypic. **Male** Slaty-grey head and mostly red bill with yellow lower mandible diagnostic. Body green, throat black extending in band behind ear-coverts and very narrowly across border of hindcrown, distinct light blue nuchal collar, small maroon shoulder-patch, tail-streamers very long, purplish-blue basally and pale yellowish distally (undertail all yellow), eyes creamy-white. In flight shows turquoise-green underwing-coverts. See female Blossom-headed Parakeet. **Female** Lacks shoulder-patch and black on centre of throat, body slightly darker green, tail-streamers average shorter. **Juvenile** Both sexes similar to female but have deep green (sometimes faintly blue-tinged) crown, bluish-green head-sides, pale green throat and hindneck; bill yellowish usually with ruddy upper mandible, eyes dark, tail initially shorter. See Rose-ringed Parakeet but note range. **First-summer** Has mostly slaty head but paler than adult with no black markings. See Blossom-headed Parakeet. **VOICE** Loud, short, shrill, high-pitched whistles, upwardly inflected: ***dreet dreet...***, ***sweet sweet...***, and ***swit*** etc. **HABITAT** Mixed deciduous, pine and open broadleaved evergreen forest; visits cultivations; up to 1,910 m. **RANGE & STATUS** Resident NE India, NE Bangladesh, SW China. **SE Asia** Uncommon to locally common resident Myanmar, W,NW,NE Thailand, Cambodia, Laos, Vietnam (south to north Cochinchina). **BREEDING** March–April. Loosely colonial. **Nest** Existing or excavated tree-hole, or old nest of woodpecker or barbet; 6–18 m above ground. **Eggs** 3–5; white; 27.1 x 21.5 mm (av.).

395 BLOSSOM-HEADED PARAKEET

Psittacula roseata Plate **48**

IDENTIFICATION 30–36 cm. *P.r.juneae*: **Male** Deep rosy-pink forehead and sides of head, rather pale violet-grey hindcrown, black throat and narrow collar, and orange-yellow to orange bill with black lower mandible diagnostic. Has small maroon shoulder-patch and mostly deep turquoise tail-streamers with pale yellow tips (undertail all yellow). In flight, shows green underwing-coverts. *P.r.roseata* (N Myanmar?) is generally greener with smaller maroon shoulder-patch. **Female** Head uniformly darker violet-grey, duller and paler on forehead and sides of head, lacks black collar but has blackish malar patch, central throat green, tail-streamers average shorter. Differs from similar Grey-headed Parakeet by bill colour (as male), maroon shoulder-patch, less slaty head with only small amount of black on malar area, more turquoise uppertail and underwing-coverts (as male). **Juvenile** Both sexes like female but hindcrown green, forecrown and sides of head duller and paler, no shoulder-patch, bill all yellowish, tail initially shorter. From Grey-headed by smaller size and bill, pale vinous-greyish forecrown and ear-coverts, dark malar patch, more turquoise uppertail and underwing-coverts (as adult). **VOICE** A rather soft ***pwi*** and watery ***driii***. **HABITAT** Mixed deciduous and open broadleaved evergreen forest; visits cultivation, temple groves; up to 915 m. **RANGE & STATUS** Resident NE India, Bangladesh, SW,S China; formerly E Nepal. **SE Asia** Uncommon to fairly common resident Myanmar (except N), W,NW,NE,C,S(extreme north) Thailand, Indochina (except N Laos, W,E Tonkin). **BREEDING** January–May. Loosely colonial. **Nest** None, in excavated or existing tree-hole, sometimes old nest of woodpecker or barbet. **Eggs** 4–6; white; 25.4 x 20.3 mm (av.).

396 RED-BREASTED PARAKEET

Psittacula alexandri Plate **48**

IDENTIFICATION 33–37 cm. *P.a.fasciata*: **Male** Unmistakable. Stocky and relatively short-tailed, crown and sides of head mostly pale lilac-grey to lilac-blue, black loral line (crossing forehead), very broad black malar band/throat-sides, and violet-tinged deep pink breast. Bill thick, with yellow-tipped red upper mandible and blackish lower mandible, belly turquoise-green, tail largely turquoise (yellowish below), strong yellowish wash on wing-coverts. **Female** Bill all black, breast richer pink (without violet tinge), strong light blue wash on crown and sides of head, tail-streamers average shorter. **Juvenile** Both sexes like female but forehead and sides of head dull vinous-grey, black head markings duller, rest of crown and breast to upper belly green. **VOICE** A shrill ***ek ek...*** and short sharp nasal ***kaink***, repeated rapidly in alarm. Rather nasal honking ***cheent cheent...*** interspersed with more grating notes. Also a raucous ***kak-kak-kak-kak-kak...*** **HABITAT & BEHAVIOUR** Open broadleaved evergreen, semi-evergreen and deciduous forest, temple groves; visits cultivation; up to 1,220 m. Large flocks raid crops. **RANGE & STATUS** Resident N,NE Indian subcontinent, Andaman Is, SW,S China, W Sumatran Is, Java, S Borneo. **SE Asia** Common resident Myanmar, Thailand (except southern S), Indochina. Formerly recorded (status unknown) north Peninsular Malaysia. Common feral resident Singapore. **BREEDING** December–April. Loosely colonial. **Nest** Existing tree-hole (often adapted) or old woodpecker or barbet nest; 3–10 m above ground. **Eggs** 3–4; white; 30.9 x 25.6 mm (av.).

397 LONG-TAILED PARAKEET
Psittacula longicauda Plate **48**

IDENTIFICATION 40–42 cm. *P.l.longicauda*: **Male** Reddish-pink sides of head and nuchal collar, deep green crown and long broad black malar band diagnostic. Has dark green loral stripe, very pale blue-green mantle, pale turquoise back, grading to vivid green uppertail-coverts, yellowish-green underparts, dark blue wash on flight-feathers, rather golden-green wing-coverts and very long dark purplish-blue tail-streamers (undertail dull olive). Bill has yellow-tipped red upper mandible and dark lower mandible. In flight shows distinctive blackish underwing with yellow coverts. *P.l.tytleri* (Coco Is, off S Myanmar) is larger (up to 49 cm), with lighter, brighter green crown, no reddish-pink nuchal collar, bright flame-red sides of head, black loral stripe, less turquoise on back and rump, deeper and darker green underparts, turquoise-blue and green uppertail, more turquoise-blue on wings and turquoise-tinged green underwing-coverts. **Female** Nape and rear sides of head green, malar band dark green, crown and upperparts fairly uniformly darker green with lighter, more vivid green rump and uppertail-coverts (almost lacks turquoise), bill dull brown, tail-streamers average shorter; shows reddish-pink on supercilium. *P.l.tytleri* has lighter, brighter crown, paler sides of head, darker loral stripe and malar band and no reddish-pink on supercilium; uppertail and wings differ as male. **Juvenile** Both sexes like female but sides of head greener with less pink, malar band duller. See Alexandrine and Rose-ringed Parakeets but note range. **VOICE** A high-pitched, rather melodious ***pee-yo pee-yo pee-yo...***, nasal quavering ***graak graak graak*** and bursts of scolding ***cheet*** notes. **HABITAT** More open broadleaved evergreen forest, freshwater swamp forest, clearings, plantations, mangroves; lowlands. **RANGE & STATUS** Resident Andaman and Nicobar Is, Sumatra, Borneo. **SE Asia** Fairly common to common resident (subject to local movements), Coco Is (off S Myanmar), Peninsular Malaysia, Singapore. **BREEDING** December–August. Loosely colonial. **Nest** Excavated tree-hole; 4–45 m above ground. **Eggs** 2–3; white; 30.6 x 24.7 mm (av.; *tytleri*).

CUCULIDAE: CUCULINAE: Old World cuckoos

Worldwide c.69 species. SE Asia 25 species. **Parasitic cuckoos** (*Clamator, Hierococcyx, Cuculus, Cacomantis, Chrysococcyx, Surniculus, Eudynamys*) Rather small to fairly large, generally with shortish, slightly downcurved bills; feet zygodactyl. Mostly shy and arboreal. Feed on insects, other invertebrates and berries, some larger species also taking small vertebrates. Lay eggs in nests of other birds (brood-parasitic). **Coral-billed Ground-cuckoo** Large, rather long-necked, long-tailed and long-legged, with zygodactyl feet. Terrestrial, skulking, walks and runs along forest floor. Feeds on small mammals and other vertebrates, insects, snails, fruit and seeds.

398 PIED CUCKOO *Clamator jacobinus* Plate **50**

IDENTIFICATION 31.5–33 cm. *C.j.jacobinus*: **Adult** Recalls Chestnut-winged Cuckoo but somewhat smaller, with glossy black hindneck and wings and all-white underparts. Has distinctive short white bar across base of primaries and broadly white-tipped outertail-feathers. **Juvenile** Upperparts and wings drab brown (crown and uppertail darker), throat and breast initially dull greyish, rest of underparts tinged buff, crest shorter, bill browner with yellow-based lower mandible, has narrow whitish tips to wing-feathers and scapulars. **VOICE** Frequently repeated loud metallic, ringing ***kleeuw*** or ***keeu*** notes: ***kleeuw kleeuw kleeuw kleeuw...*** (both sexes); sometimes preceded by shrill ***kiu-kewkew...kiu-kewkewkew...kiu-kewkew...*** Male often adds fast series of short, rising notes: ***kwik-kwik-kweek.*** Also, an abrupt ***kweek.*** **HABITAT** Open deciduous woodland, scrub, cultivation; lowlands. **RANGE & STATUS** Breeds sub-Saharan Africa; undertakes intra-African movements. Breeds Indian subcontinent; mostly winters E,C Africa. **SE Asia** Fairly common breeding visitor C,S Myanmar. Rare passage migrant SW Myanmar. Vagrant NW,C,S Thailand, Cambodia. **BREEDING** Brood-parasitic. **Eggs** 1–2; in nests of White-throated and Striated Babblers, and Lesser Necklaced Laughingthrush; like those of host but usually larger, rounder, paler blue and glossless; 23.9 x 18.6 mm (av.).

399 CHESTNUT-WINGED CUCKOO
Clamator coromandus Plate **50**

IDENTIFICATION 38–41.5 cm. Monotypic. **Adult** Distinctive. Slender and long-tailed, with glossy blackish upperparts and crest, white hindcollar, largely rufous-chestnut wings, whitish underparts with buffy-rufous throat, blackish vent and narrowly white-tipped outertail-feathers. Could be confused with Lesser Coucal in flight. **Juvenile** Upperparts dark greenish-brown with pale chestnut to buff feather-tips, wings similar to adult but feathers tipped pale chestnut to buff, hindcollar buffish, underparts all whitish, crest much shorter, bill paler. Gradually attains patches of adult plumage. **VOICE** Territorial call is a series of metallic whistled paired-notes: ***thu-thu...thu-thu...thu-thu...***, very similar to that of Moustached Hawk-cuckoo but each couplet has less well-spaced notes. Also a rapid grating woodpecker-like ***critititit.*** **HABITAT** Secondary growth, scrub, bamboo thickets; broadleaved evergreen forest and mangroves in winter and on migration. Up to 1,525 m. **RANGE & STATUS** Breeds N,NE Indian subcontinent, C,E and southern China; winters S India, Sri Lanka, Greater Sundas, Philippines, Sulawesi. **SE Asia** Uncommon breeding visitor Myanmar (except S and Tenasserim), W,NW,NE,C(rarely winters) Thailand, Laos, E Tonkin, N,C Annam. Uncommon breeding visitor and/or resident S Myanmar, Cambodia, Cochinchina. Scarce to uncommon winter visitor and passage migrant S Thailand, Peninsular Malaysia, Singapore. Also recorded on passage C Thailand, E Tonkin. Recorded (status uncertain) Tenasserim, SE Thailand, S Annam. **BREEDING** Brood-parasitic. **Eggs** Often more than one; primarily in nests of laughingthrushes, including Lesser and Greater Necklaced; like those of host but rounder, thicker-shelled and glossless; 26.9 x 22.8 mm (av.).

400 LARGE HAWK-CUCKOO
Hierococcyx sparverioides Plate **49**

IDENTIFICATION 38–41.5 cm. Monotypic. **Adult** Superficially resembles some *Accipiter* hawks (rounded wings, barred tail etc.) but flight weaker, tail more rounded, bill longer and slenderer. From other hawk-cuckoos by combination of size, slaty-grey crown, nape and sides of head, contrasting brownish-grey mantle and wings, extensively dark chin, rather dark rufous breast-patch (extent variable), prominent dark streaks on lower throat and breast, dark bars on lower breast, belly and flanks, and broad dark tail-bands (above and below). Underwing-coverts white with dark brown bars. See Common and Dark Hawk-cuckoos. Note voice and range. **Juvenile** Upperparts dark brown with rufescent barring, nape pale rufous with dark brown streaks, underparts white with variable buffish wash and bold dark brown drop-like streaks on breast, becoming bars on belly and flanks. See other hawk-cuckoos. Upperside more uniform and chin and breast browner than in Large. Note breeding range of latter. **Immature/subadult** Crown greyish-brown, dull pale rufous bars on upperparts and wings, underparts more like juvenile but has pale rufous breast-patch. See Common. Resembles adult Malaysian Hawk-cuckoo but larger, with narrower penultimate dark tail-band and narrow greyish band between that and last dark tail-band, buffier tail-tip (narrowly whitish at very tip), more extensively dark chin and heavier throat-streaks; flanks tend to be barred. **VOICE** Territorial call is a very loud, shrill, spaced, stressed ***pwi pwee-wru*** or ***PEEE FWIuu***, steadily repeated on rising pitch to screaming crescendo. Sequence may be preceded by a slow, pulsing trill. Secondary call (apparently by female) is a series of short, even ***dRUu-dRUu*** whistles, that rise in volume, but little in pitch. **HABITAT & BEHAVIOUR** Broadleaved evergreen and deciduous forest, more open habitats, gardens, mangroves etc. on migration; up to 2,565 m (mainly breeds above 650 m). Secretive and often difficult to observe. **RANGE & STATUS** Resident Sumatra, Borneo. Breeds N,NE Indian subcontinent, S Tibet, C and southern China; northern populations mostly winter south to S India, Greater Sundas, Philippines, Sulawesi. **SE Asia** Fairly common resident (subject to some movements) Myanmar (except S and southern Tenasserim), W,NW,NE Thailand, Cambodia, Laos, W,E Tonkin. Scarce to uncommon winter visitor W,C,NE(south-west),SE,S Thailand, Peninsular Malaysia, Cochinchina. Uncommon passage migrant S Thailand, Cambodia (also winters?). Recorded (status uncertain) S Myanmar, N,C,S Annam. Vagrant Singapore. **BREEDING** Brood-parasitic. **Eggs** Laid singly in nests of Streaked Spiderhunter, also Little Spiderhunter and Lesser Shortwing; usually like those of host but larger and rounder; 26.6 x 18.6 mm (av.).

401 DARK HAWK-CUCKOO

Hierococcyx bocki Plate **49**

IDENTIFICATION 33 cm. Monotypic. **Adult** Smaller, darker and richer-coloured than Large Hawk-cuckoo, with dark streaks below restricted to breast and broader richer orange-rufous breast-band; less dark on chin. **Juvenile** Crown and nape slaty-brown with a few white feathers on latter, rest of upperparts brownish-black, chin slaty-brown, upper breast slaty-brown with white feather-bases, lower breast to undertail-coverts white with dark bars; wings dark brown with indistinct rufous notches on outer fringes of primaries, and buff tips, secondaries uniform grey-brown. **Immature/subadult** Apart from size, like Large, but shows richer rufescent barring on upperside, whiter throat with less dark on chin. **VOICE** Territorial call is a very loud, shrill, spaced, stressed ***pi-phu***, ***pi-pi*** or ***pi-ha***, steadily repeated on rising pitch to shrill climax. Secondary call (by female?) is a rapid sequence of more hurried paired-notes: ***pipi-pipi-pipi-pipi...*** or ***phuphu-phuphu-phuphu...***, rising to fever-pitch, then falling away. **HABITAT** Broadleaved evergreen forest; 900–1,800 m (locally down to 450 m). **RANGE & STATUS** Resident Sumatra, Borneo. **SE Asia** Fairly common resident (subject to some movements) Peninsular Malaysia. **BREEDING** Brood-parasitic. **Eggs** Laid singly in nests of Chestnut-capped Laughingthrush. **NOTE** Recently split from *sparverioides* (see Payne 2005).

402 COMMON HAWK-CUCKOO

Hierococcyx varius Plate **49**

IDENTIFICATION 33–37 cm. Monotypic. **Adult** Like Large Hawk-cuckoo but smaller, usually with paler, more ashy-grey upperparts (crown more concolorous with mantle); has less black on chin, less distinct streaks on lower throat and breast (sometimes lacking), rufous of underparts often paler and more extensive, reaching belly and flanks, underpart bars less distinct, dark tail-bands narrower. Underwing-coverts washed rufous and only faintly dark-barred. See Hodgson's Hawk-cuckoo. Note voice and range. **Juvenile/immature/subadult** Difficult to separate from Large at all stages but smaller, usually with narrower dark tail-bands (particularly penultimate), tail more strongly tinged slaty-grey, base colour of underparts less clean white with much browner streaks, flank bars usually less distinct. See Hodgson's. **VOICE** Territorial call is shriller and thinner than that of Large; a series of 4–6 loud, high, shrill, shrieking ***wee-PIWhit*** (***PIW*** stressed) phrases, progressing to frantic shrillness, and then ending abruptly. Series also transcribed as consisting of ***bur FEEWEr*** strophes (the first note becoming longer and louder in successive repetitions) which gradually rise in pitch and lengthen, and are repeated after increasingly shorter intervals, sounding more and more frantic, before abruptly ending. Secondary call (apparently by female) described as a strident trilling scream, or series of burry notes starting with a few subdued level notes, gradually rising in pitch and volume, then crescendoing before falling away rapidly. **HABITAT** Open deciduous forest, secondary growth; up to 915 m. **RANGE & STATUS** Resident (subject to some movements) NE Pakistan, Indian subcontinent. **SE Asia** Fairly common resident SW Myanmar. Vagrant W(south),S(northern) Thailand. **BREEDING** Brood-parasitic on Jungle Babbler *Turdoides striata* in India. Otherwise unclear.

403 MOUSTACHED HAWK-CUCKOO

Hierococcyx vagans Plate **49**

IDENTIFICATION 28–30 cm. Monotypic. **Adult** From other hawk-cuckoos by combination of relatively small size and long-tailed appearance and conspicuous dark moustachial/cheek-stripe, contrasting with whitish upper throat and centre of ear-coverts. Has dark slate-coloured crown and nape, creamy-whitish underparts with blackish-brown streaks on lower throat to belly and flanks, and white-tipped tail. See Malaysian and Hodgson's Hawk-cuckoos and juvenile/immature plumages of other hawk-cuckoos. Note voice, habitat and range. **Juvenile** Poorly documented. Probably like adult but with brownish crown and nape. **VOICE** Territorial call is a loud, well-spaced ***chu-chu***, repeated monotonously (c.1 call every 2 s). Secondary call is an ascending sequence of mellow notes, first singly after short intervals, then paired and accelerating to fever-pitch, ending abruptly. **HABITAT** Broadleaved evergreen forest, secondary growth; up to 915 m. **RANGE & STATUS** Resident Greater Sundas. **SE Asia** Uncommon resident Tenasserim, W(south),SE,S Thailand, Peninsular Malaysia, S Laos. **BREEDING** Brood-parasitic. Hosts include Abbott's Babbler. Otherwise undocumented.

404 MALAYSIAN HAWK-CUCKOO

Hierococcyx fugax Plate **49**

IDENTIFICATION 29 cm. Monotypic. **Adult** Similar to Moustached Hawk-cuckoo but somewhat stockier and proportionately shorter-tailed, with all-dark sides of head, dark grey upper chin and mostly pale chestnut tail-tip (extreme tip narrowly whitish). From Large Hawk-cuckoo by smaller size, variable white markings on sides of nape (sometimes forming broken hindcollar), creamy-white underparts with blackish streaks from lower throat to belly and flanks, mostly pale chestnut tail-tip, and even tail-banding. See juvenile/immature of that species. Note voice, habitat and range. **Juvenile** Upperparts rather uniform dull dark brown with faint pale feather-fringes, underpart-streaks broken more into spots. Best separated from Large and Common by size, darker and plainer upperside and different uppertail pattern. Note habitat and range. **VOICE** Territorial call is a series of loud, shrill, very high-pitched ***pi-pwik*** or ***pi-pwit*** phrases (c.8 every 10 s), often followed by rapid sequence of ***ti-tu-tu*** phrases, accelerating and ascending to shrill crescendo climax (or climaxes).

Climax is followed by a slower, more even ***tu-tu-tu-tu...*** before tailing off. **HABITAT & BEHAVIOUR** Broadleaved evergreen forest; up to 250 m. Typically secretive and difficult to observe. **RANGE & STATUS** Resident Greater Sundas. **SE Asia** Uncommon resident Tenasserim, S Thailand, Peninsular Malaysia, Singapore. **BREEDING** Brood-parasitic. Only recorded host is White-rumped Shama.

405 HODGSON'S HAWK-CUCKOO

Hierococcyx nisicolor Plate **49**

IDENTIFICATION 27–31 cm. Monotypic. **Adult** Much greyer above than Malaysian Hawk-cuckoo, lacks warm brown bars on primaries and secondaries, shows little white on nape, usually has one inner tertial strikingly whiter than rest (plain or barred/notched darker), penultimate dark tail-band obviously narrower than rest, shows variable amount of pinkish-rufous on breast and upper belly (appears rather uniform if streaks narrow); dark underpart streaks variable, often split with whitish. Resembles Common Hawk-cuckoo but smaller with deeper-based bill, distinctive tertial pattern, no bars on underparts and relatively shorter tail with mostly pale chestnut tip. Note voice, habitat and range. **Juvenile** Warmer than Malaysian, buffier fringing on upperparts, usually shows paler innermost tertial. Best separated from Large and Common by size, darker and plainer upperside and different uppertail pattern/colour. Note habitat and range. **VOICE** Territorial call is very similar to that of Malaysian, but is followed by a rapid ***trrrrr-tititititititrrrtrrr...*** **HABITAT** Broadleaved evergreen and mixed deciduous forest; up to 1,300 m, sometimes to 1,550 m (mainly breeds above 500 m). **RANGE & STATUS** Breeds Bhutan, NE India, southern China; winters Greater Sundas. **SE Asia** Uncommon resident (subject to some movements), S,E Myanmar, Tenasserim, Thailand (except C,S), N,C Laos, W Tonkin, C Annam. Uncommon winter visitor and passage migrant S Thailand, Peninsular Malaysia, Singapore. Scarce to uncommon passage migrant W,C Thailand, Cochinchina. Recorded (status uncertain) Cambodia, S Laos, E Tonkin, N,S Annam. **BREEDING** Brood-parasitic. **Eggs** Probably this species (India) found in nests of shortwings and flycatchers; similar to those of host; 23.8 x 15.8 mm (av.). **NOTE** Recently split from *fugax* (see King 2002).

406 INDIAN CUCKOO

Cuculus micropterus Plate **49**

IDENTIFICATION 31–33 cm. *C.m.micropterus*: **Male** From other *Cuculus* cuckoos by distinctive brownish tinge to mantle, wings and tail (contrasts with grey head), prominent broad dark subterminal tail-band and dull yellowish to greyish-green eyering. *C.m.concretus* (resident extreme S Thailand southwards) is somewhat smaller and more darker-toned. **Female** Has rufescent wash across breast. **Juvenile** Crown and sides of head browner than adult with very broad buffish-white feather-tips (different from juvenile and hepatic morphs of other *Cuculus* cuckoos), upperparts and wings with prominent rufous or buffish to whitish tips, underparts buffish with broken dark bars, particularly on throat, breast and flanks; may have rufous wash on throat and breast. **VOICE** Male territorial call is a loud ***whi-whi-whi-wu*** or ***wa-wa-wa-wu***, either with a lower last note or alternating high and low (may omit last note). Also a loud hurried bubbling (probably female only). **HABITAT** Broadleaved evergreen and deciduous forest, secondary growth; up to 1,830 m (below 760 m in Peninsular Malaysia). **RANGE & STATUS** Resident Greater Sundas. Breeds Indian subcontinent (except W and Pakistan), China (except NW), SE Siberia, Ussuriland, N,S Korea; northern populations winter south to Greater Sundas, Philippines. **SE Asia** Fairly common to common resident (except C,SE Thailand, Singapore). Fairly common winter visitor and passage migrant W,C,S Thailand, Peninsular Malaysia, Singapore. Uncommon to fairly common passage migrant Cambodia. Recorded early spring (status uncertain) SE Thailand. **BREEDING** Brood-parasitic. **Eggs** Laid in nests of Black-and-yellow Broadbill, and Black, Ashy and Greater Racket-tailed Drongos; very like those of host; 26.2 x 17.9 mm (one egg).

407 EURASIAN CUCKOO

Cuculus canorus Plate **49**

IDENTIFICATION 32.5–34.5 cm. *C.c.bakeri*: **Male** Difficult to separate from Oriental Cuckoo but usually slightly larger, with cleaner white underparts, often with somewhat fainter (less blackish) and sometimes narrower bars, has grey bars on white leading edge of wing (difficult to see in field); typically shows whitish undertail-coverts with prominent blackish bars. See Lesser and Indian Cuckoos. Note voice. **Female** Like Oriental but rufous-buff wash across grey of upper breast (when present) is usually less extensive. **Female hepatic morph** Like Oriental but has narrower and less prominent blackish bars, particularly on back to uppertail and breast. **Juvenile** Possibly indistinguishable from Oriental but usually has narrower dark bars on underparts. See Lesser. **Juvenile female hepatic morph** Like Oriental but has less prominent dark bars on upperside and less warm buffish underparts with less prominent dark bars. See Lesser. **VOICE** Male territorial call is a loud, mellow ***cuc-coo***, with lower-pitched second note. Both sexes also give a loud bubbling trill. **HABITAT** Open broadleaved evergreen forest, secondary growth, more open habitats on migration; up to 2,195 m (probably only breeds above 600 m). **RANGE & STATUS** Breeds N Africa, Palearctic, Pakistan, northern and C Indian subcontinent, S,E Tibet, China, N,S Korea, Japan; winters mainly sub-Saharan Africa; rarely S Indian subcontinent. **SE Asia** Uncommon breeding visitor W,N,C,E Myanmar, N Laos, W,E Tonkin, C Annam. Scarce to uncommon passage migrant SW Myanmar. Recorded (status uncertain) S Myanmar, Tenasserim, NW Thailand (breeds?). **BREEDING** Brood-parasitic. **Eggs** Laid singly in nests of Paddyfield Pipit, Crested Bunting, Eastern Stonechat, Pied and Grey Bushchats, Striated Grassbird, cisticolas, and prinias (probably also some laughingthrushes); generally like those of host but larger and rounder; 24.2 x 17.9 mm (av.).

408 ORIENTAL CUCKOO

Cuculus horsfieldi Not illustrated

IDENTIFICATION 30–33 cm. Monotypic. Wing: Male 194–212, female 186–212 mm. **Male** On current knowledge, indistinguishable in the field from Himalayan Cuckoo, differing only in its larger size, relatively longer wing, and voice. **Female/female hepatic morph** On current knowledge, indistinguishable in the field from Himalayan Cuckoo. **Juvenile/juvenile female hepatic morph** On current knowledge, indistinguishable in the field from Himalayan Cuckoo. **VOICE** Male territorial call is a deep, booming, usually two-noted (but often preceded by several quick notes), on one pitch: ***poop-oop poop-oop...***, or ***hoop-hoop..hoop-hoop...*** Also hoarse croaks, chuckles and a harsh ***gaak-gaak-gak-ak-ak***. Females give a rapid series of sharp, upturned ***kwik-wik*** couplets, rising in pitch and volume, not decelerating but ending with an offset lower note. Sharper and less bubbling than corresponding call of Eurasian Cuckoo. **HABITAT** Open wooded country, secondary growth, recorded in lowlands. **RANGE & STATUS** Breeds C,E Palearctic, N,NE China, N,S Korea, Japan; mostly winters Sundas, Philippines, Wallacea, New Guinea, N,E Australia, N Melanesia. **SE Asia** Scarce passage migrant/vagrant Peninsular Malaysia, Singapore. **NOTE** Recently split from *Cuculus saturatus* (King 2002), but retaining the English name.

409 HIMALAYAN CUCKOO

Cuculus saturatus Plate **49**

IDENTIFICATION 29 cm. Monotypic. Wing: male 174–192, female 172–185 mm. **Male** Medium-sized cuckoo with grey head, breast and upperside, and otherwise whitish underparts with prominent blackish bars. Very similar to Eurasian Cuckoo but usually slightly smaller, underparts typically buff-tinged often with somewhat bolder (more blackish) and sometimes slightly broader bars; has plain buffish-white to whitish leading edge of wing (difficult to see in field), typically less obvious or no blackish bars on undertail-coverts. See Oriental, Lesser and Indian Cuckoos. Note voice. **Female** Similar to male but usually has rufous-buff wash across grey of upper breast, typically more extensive than on female Eurasian. **Female hepatic morph** Head and upperside rufescent-brown, throat and breast buffy-rufous, rest of underparts white; strongly barred blackish-brown overall (including leading edge of wing). Like Eurasian but has broader and more prominent blackish barring, particularly on back to uppertail and breast. **Juvenile** Resembles adult but head and upperside darker with whitish fringes to mantle, scapulars and wing-coverts and white nuchal patch (when fresh). Very similar to Eurasian but usually has broader dark bars on underparts. See Lesser. **Juvenile female hepatic morph** Differs in same way as typical juvenile but additionally has whiter throat and upper breast. From Eurasian by warmer buffish underparts with more prominent dark bars above and below. See Lesser. **VOICE** Male territorial call is a loud series of 2–4 mellow notes preceded by a softer, shorter note ***kuk PUP-PUP-PUP*** or ***kuk HU-HU-HU***. Also transcribed as ***hoop-HOOP'OOP'OOP'(OOP)***. Also a nervous rapid uneven ***wuk-wuk-wuk-wuk-wuk-wuk-uk...*** (possibly by female only). **HABITAT** Broadleaved evergreen forest, open wooded country, secondary growth; breeds 800–2,030 m, down to sea level on passage and in winter. **RANGE & STATUS** Breeds NE Pakistan, NW,N,NE Indian subcontinent, southern China, Taiwan; winters Andaman and Nicobar Is, Sundas, Philippines, Wallacea, New Guinea, N,E Australia. **SE Asia** Uncommon to fairly common breeding visitor W,N Myanmar, N,C Laos, W Tonkin. Scarce to uncommon passage migrant C,S Myanmar, Tenasserim, Thailand, Peninsular Malaysia (? winters), E Tonkin (may breed locally); ? (Oriental not ruled-out) Singapore, Cambodia, S Annam. **BREEDING** Brood-parasitic. **Eggs** Laid in nests of warblers, including Blyth's Leaf-warbler; white or whitish-buff with small reddish-brown speckles and tiny lines (usually ringed at broader end), sometimes mimicking those of host; 21.1 x 15.6 mm (av.).

410 **SUNDA CUCKOO** *Cuculus lepidus* Plate **49**

IDENTIFICATION 26 cm. Monotypic. Wing: Male 147–169, female 143–152 mm. **Male** Smaller and somewhat darker overall than Oriental and Himalayan Cuckoos, with broader dark bands below, and more rusty-buff undertail-coverts; eyes brown to dark reddish. See Lesser and Indian Cuckoos. Note voice and range. **Female** Similar to male but usually has rufous-buff wash across grey of upper breast. **Female hepatic morph** Apart from size, differs from Oriental and Himalayan by darker rufous upperparts with broader black barring, and broader black barring below. **Juvenile/juvenile female hepatic morph** Apart from size, perhaps not distinguishable in the field from Oriental and Himalayan. **VOICE** Male territorial call is similar to Himalayan, but higher-pitched, and with 2–3 ***PUP***, ***HU*** or ***HOOP*** notes. **HABITAT** Broadleaved evergreen forest, forest edge, secondary growth; 900–1,700 m. **RANGE & STATUS** Resident Greater and Lesser Sundas. **SE Asia** Fairly common resident Peninsular Malaysia. **BREEDING** Brood-parasitic. **Eggs** Laid in nests of Chestnut-crowned Warbler and Mountain Leaf-warbler; creamy-white, finely speckled and peppered dull brown, with a denser-marked ring around broader end; 18.2–19 x 11.3–11.8 mm.

411 **LESSER CUCKOO**
Cuculus poliocephalus Plate **49**

IDENTIFICATION 26–26.5 cm. Monotypic. **Male** Very similar to Oriental Cuckoo but smaller, with finer bill, usually darker rump and uppertail-coverts, contrasting less with blackish tail, usually more buffish underparts with wider-spaced and bolder black bars. See Eurasian and Indian Cuckoos. Note voice. **Female hepatic morph** Females typically occur in this plumage but sometimes similar to male. Like Oriental but usually more rufous, sometimes with almost no bars on crown, nape, rump and uppertail-coverts. See Eurasian and Grey-bellied Cuckoos. **Juvenile** Dark grey-brown above with variable narrow whitish to rufous bars and a few whitish spots on nape. From Oriental by somewhat darker, more uniform crown, sides of head and mantle and whiter underparts with bolder dark bars. See Eurasian. **Juvenile female hepatic morph** Has more prominent dark bars on crown and mantle and whitish nuchal patch. See Oriental and Eurasian. **VOICE** Male territorial call is a very loud, quite shrill series of 5–6 whistled notes (third prolonged and stressed): rather even ***wit-wit-witi-wit wit-wit-witi-wit wit-wit-witi-wit...*** or rising and falling ***wit,it-iti-witu wit,it-iti-witu wit,it-iti-witu...*** Repeated up to 7 times or more, higher-pitched at first but gradually becoming slower and deeper or lower-pitched. **HABITAT** Broadleaved evergreen forest, secondary growth, also deciduous forest on migration; 915–2,285 m; sometimes lower on migration. **RANGE & STATUS** Breeds N Afghanistan, N Pakistan, NW,N,NE Indian subcontinent, SE Tibet, China (except NW,SE), Ussuriland, N,S Korea, Japan; winters E Africa, perhaps occasionally S India, Sri Lanka. **SE Asia** Scarce to uncommon breeding visitor W,N Myanmar, NW Thailand, N Laos, W,E(north-west) Tonkin. Scarce to uncommon passage migrant C Myanmar. Recorded (status uncertain) S(east) Myanmar, C Annam, Cochinchina. **BREEDING** Brood-parasitic. **Eggs** Mainly or entirely laid in nests of *Cettia* bush- and leaf-warblers; closely mimic those of host but larger; 21.2 x 15.6 mm (av.).

412 **BANDED BAY CUCKOO**
Cacomantis sonneratii Plate **50**

IDENTIFICATION 23–24 cm. *C.s.sonneratii*: **Adult** Similar to female hepatic morph Plaintive Cuckoo but has broad whitish supercilium and area behind ear-coverts (finely scaled darker) isolating broad dark mask, outer edges of tail-feathers unbarred (appear more rufous in profile) and darker on either side of shafts, underparts usually whiter with narrower dark bars. *C.s.malayanus* (Tenasserim and S Thailand southwards) is somewhat smaller and more rufescent. **Juvenile** Has whitish to pale buff bars on sides of head, crown and mantle, less contrasting mask and coarser, more spaced dark bars on head and body. **VOICE** Male territorial call is a loud hurried ***pi,hi-hi-hi***, ***hee hew-hew-hew*** or ***pihu-hihu***, repeated on gently descending scale. Secondary call is an excited rapid ***pi pi pihihi-pi pihihi-pi*** or ***pi pi pi pi pi pi-hu-hu-hi pi-hu-hu-hi*** or ***pi pi pi pi-hi-hi-hi pi-hi-hi-hi pi-hi-hi-hi*** etc. on rising scale. **HABITAT** Broadleaved evergreen and deciduous forest, secondary growth; up to 1,500 m. **RANGE & STATUS** Resident N,NE Indian subcontinent, C and southern India, Sri Lanka, SW China, Greater Sundas; recorded Philippines (Palawan). **SE Asia** Uncommon to fairly common resident (except C Thailand, W Tonkin). **BREEDING** Brood-parasitic. **Eggs** Laid in nests of White-bellied Erpornis, minivets, Bar-winged Flycatcher-shrike, and Common Iora; white or pinkish, speckled and blotched with reddish- or purplish-brown over somewhat greyer undermarkings; 18.8 x 14.4 mm (av.; Sumatra).

413 **GREY-BELLIED CUCKOO**
Cacomantis passerinus Plate **50**

IDENTIFICATION 22–23 cm. Monotypic. **Male** Grey overall plumage with white vent distinctive. See cuckooshrikes. **Female hepatic morph** Typically occurs in this plumage. Resembles Plaintive Cuckoo but crown, mantle, rump to uppertail, throat and upper breast mostly plain deep bright rufous to rufous-chestnut, tail unbarred. Note range. **Female grey morph** Like male but slightly paler. **Juvenile male** Upperside, throat and breast more sooty-brown with fine pale to buffish fringing; belly and vent whiter. **Juvenile female hepatic morph** Similar to brown plumages of Plaintive but lacks distinct bars on tail, has much brighter, more rufous-chestnut upperside, plainer rump and uppertail-coverts and whiter base colour on belly and vent. **VOICE** Male territorial call is a loud clear ***phi phi phi wi-wihi*** or ***phi phi phi wi-hi***, well spaced and strongly stressed at start; sometimes shorter ***wi-pihui***. Secondary calls include clear ***peee peee peee-tcho-cho peee-tcho-cho...*** and ***pi-pipee pi-pipee pi-pipee***. Also a plaintive ***pi-wiuu pi-wiuu pi-wiuu pi-wiuu pi-wiuu...*** **HABITAT** Secondary growth, open woodland, scrub and grass, cultivated areas; up to 1,800 m. **RANGE & STATUS** Breeds NE Pakistan, India (except NE), Nepal, Bhutan, Bangladesh; some northern populations winter to south, Sri Lanka. **SE Asia** Recorded (status uncertain) N Myanmar. **BREEDING** Brood-parasitic. **Eggs** Mainly laid in nests of tailorbirds, cisticolas, and prinias; variable but usually like those of host; 19.9 x 14 mm (av.).

414 PLAINTIVE CUCKOO

Cacomantis merulinus Plate **50**

IDENTIFICATION 21.5–23.5 cm. *C.m.querulus*: **Male** Distinctive small cuckoo with grey head, throat and upper breast, with remaining underparts peachy-rufous (intensity varies). Rest of upperside greyish-brown; tail graduated and tipped whitish. *C.m.threnodes* (Peninsular Malaysia, Singapore) averages somewhat smaller and paler, head contrasts more with upperparts. See Rusty-breasted Cuckoo. **Female hepatic morph** Usually occurs in this plumage but sometimes similar to male. Variable. Combination of rufescent upperside with blackish bars, paler underside with less pronounced dark bars, somewhat paler supercilium and completely barred tail distinctive. See Rusty-breasted, Banded Bay and Grey-bellied Cuckoos and larger female hepatic morph *Cuculus* cuckoos. Note habitat and range. **Juvenile** Resembles female hepatic morph but paler rufous to more buffish (whiter below), with prominent blackish streaks on crown to upper mantle, throat and upper breast, indistinct barring on rest of underparts. **VOICE** Male territorial call is a plaintive series of high whistles: ***phi phi phi phi phi-pipipi*** or ***pi pi pi pi pi-hihihi***, hurried and fading at end. Secondary call is an ascending ***pii-pi-pui pii-pi-pui pii-pi-pui...***, gradually accelerating and sounding more agitated. **HABITAT** Secondary growth, open woodlands, scrub, grassland, cultivated areas, parks and gardens; up to 1,830 m. **RANGE & STATUS** Resident NE India, Bangladesh, SE Tibet, southern China, Greater Sundas, Philippines, Sulawesi. **SE Asia** Common resident throughout (scarce Singapore). **BREEDING** Brood-parasitic. **Eggs** Laid in nests of tailorbirds, cisticolas, and prinias; generally matching those of host but larger; 19.8 x 13.8 mm (av.; *querulus*).

415 RUSTY-BREASTED CUCKOO

Cacomantis sepulcralis Plate **50**

IDENTIFICATION 21.5–24 cm. *C.s.sepulcralis*: **Adult** Similar to male Plaintive Cuckoo but grey of upperparts a shade slatier, peachy-rufous of underparts extends onto central throat, eyering yellow (greyish on Plaintive). **Adult hepatic morph** Rare and probably only occurs among females. Difficult to separate from female hepatic morph Plaintive but somewhat larger and proportionately longer-tailed, with much broader blackish bars on upperside, throat and breast, and distinctive uppertail pattern with rufous bars restricted to notches along outer fringes. See Banded Bay Cuckoo. **Juvenile** Similar to adult hepatic morph but more uniformly dark above with only narrow, rather broken buffish to rufous-buff barring. **VOICE** Male territorial call is a series of 6–15 even melancholy whistled notes, gradually descending scale: ***whi whi whi whi whi...*** Secondary call is a gradually accelerating series of ***whi-wihu*** or ***whi-w'hu*** phrases. **HABITAT** Broadleaved evergreen forest, forest edge, secondary growth, mangroves, sometimes gardens; up to 1,115 m. **RANGE & STATUS** Resident Greater Sundas, W Lesser Sundas, Philippines, Sulawesi, S Moluccas. **SE Asia** Uncommon resident south Tenasserim, W(south),S Thailand, Peninsular Malaysia, Singapore. **BREEDING** Brood-parasitic. **Eggs** Sundas: laid in nests of Pied Fantail, Long-tailed Shrike, Olive-backed Sunbird, Pied Stonechat, Chestnut-naped Forktail, flycatchers, and tailorbirds; usually mimicking those of host; 19.5 x 14.8 mm (av.).

416 HORSFIELD'S BRONZE CUCKOO

Chrysococcyx basalis Plate **50**

IDENTIFICATION 16 cm. Monotypic. **Adult** Similar to Little Bronze Cuckoo but somewhat browner upperside, no pale forehead, rather plainer whitish supercilium, slightly more prominent dark on ear-coverts, pale brownish-white fringing on scapulars and wings, strong dark brownish streaks on throat, unbarred centre of abdomen and more extensive plain rufous-chestnut on outertail-feathers. Eyering greyish. See female Violet Cuckoo. **Juvenile** Unlikely to occur in region. Like Little Bronze but crown and mantle usually greyer brown, no pale forehead, more distinct brownish-white to buffish feather-fringing on scapulars, back, rump and wings, and plain rufous-chestnut on basal half of outer-tail-feathers. **VOICE** Male territorial call is a descending ***tseeeuw***, incessantly repeated (just under once per second). **HABITAT** Secondary growth, open woodland, coastal scrub, mangroves; lowlands. **RANGE & STATUS** Breeds Australia; some spend southern winter Sundas, Christmas I, Aru Is, S New Guinea; rarely Sulawesi. **SE Asia** Vagrant (July–August) Peninsular Malaysia, Singapore.

417 LITTLE BRONZE CUCKOO

Chrysococcyx minutillus Plate **50**

IDENTIFICATION 16 cm. *C.m.peninsularis*: **Male** Resembles female Violet Cuckoo but has black bill, pale forehead, isolated blackish patch on upper ear-coverts, more uniform upperparts with glossy bottle-green crown and bronzy-green mantle, and little rufescent coloration on outertail-feathers. Eyering red. See Horsfield's Bronze Cuckoo. **Female** Crown pale and less glossy, mantle less bronzy; less conspicuous pale forehead, eyering duller. **Juvenile** Similar to female but duller and browner above, pale forehead and supercilium less striking, underparts almost uniform greyish- to brownish-white with few or no dark bars on flanks and vent only. See Horsfield's. **VOICE** Territorial call is a descending series of 3–5 thin tremulous notes: ***rhew rhew rhew rhew...*** or ***eug eug eug eug...***; sometimes interrupted by rising, screeching ***wireeg-reeg-reeg***. Secondary call is a high-pitched, drawn-out trill, on a descending scale. **HABITAT & BEHAVIOUR** Mangroves, coastal scrub, secondary growth, forest edge; locally parks and gardens; up to 250 m. Very unobtrusive. **RANGE & STATUS** Resident Greater Sundas, E Lesser Sundas, Sulawesi (Banggai Is), Moluccas, New Guinea, N,E Australia. **SE Asia** Scarce to fairly common resident W(south),S Thailand, Peninsular Malaysia, Singapore, Cochinchina; ?Cambodia. **BREEDING** Brood-parasitic. **Eggs** Laid singly in nests of Golden-bellied Gerygone; dark olive-green, darker at narrower end and over broader half; 21 x 14 mm (av.; Java).

418 ASIAN EMERALD CUCKOO

Chrysococcyx maculatus Plate **50**

IDENTIFICATION 17 cm. Monotypic. **Male** Glossy gold-tinged emerald-green plumage, with white bars on lower breast to vent and dark-tipped orange-yellow bill diagnostic. Eyering red. **Female** Plain rufous crown and nape (fading onto mantle and ear-coverts) and contrasting coppery-green remainder of upperside distinctive. Underparts entirely barred dark greenish and white, bill rich yellowish with dark tip. See Violet Cuckoo. **Juvenile** Like female but less green above with some dark and buff bars on crown to mantle, rufous-tipped upperpart feathers and wing-coverts, darker tail with more rufous on outer feathers (lacking white bars) and less white at tip, throat and upper breast with rufous wash. Bill mostly dark with paler base to lower mandible. See Violet and Little Bronze Cuckoos. **VOICE** Male territorial call is a loud, clear descending 3–4 note ***kee-kee-kee...*** Flight call is a sharp ***chweek*** or ***chut-week***. **HABITAT** Broadleaved evergreen forest and secondary growth, also freshwater swamp forest, plantations and gardens on passage and in winter; up to 2,440 m (may only breed above 600 m). **RANGE & STATUS** Breeds N,NE Indian subcontinent, SE Tibet, SW,C,S China; some populations winter to south, Andaman and Nicobar Is, Sumatra. **SE Asia** Uncommon resident Myanmar, W,NW Thailand, N,C Annam. Uncommon winter visitor Thailand, Cambodia, C,S Laos, S Annam, Cochinchina; probably parts of Myanmar. Recorded (status uncertain) N Laos, W,E Tonkin. Vagrant Peninsular Malaysia, Singapore. **BREEDING** Brood-parasitic. **Eggs** Laid in nests of Crimson and Mrs Gould's Sunbirds and Little Spiderhunter; pale to warm buff with light olive-brown spots and specks (often distinctly ringed at broader end); 17.6 x 12.3 mm (av.).

419 **VIOLET CUCKOO**

Chrysococcyx xanthorhynchus Plate **50**

IDENTIFICATION 16.5–17 cm. *C.x.xanthorhynchus*: **Male** Glossy violet-purple plumage with white bars on lower breast to vent and orange bill with red base diagnostic. **Female** Resembles Asian Emerald Cuckoo but upperside browner, crown to upper mantle mostly dark bronzy-brown, bill yellowish with some red at base. See Little and Horsfield's Bronze Cuckoos. **Juvenile** Similar to Asian Emerald but has darker, rufous-chestnut base colour of crown to upper mantle, bold rufous-chestnut and dull dark greenish bars on upperparts and wing-coverts, rufous-chestnut fringes to primaries and secondaries and no rufous on throat and upper breast. **VOICE** A loud, sharp, spaced ***TEE-WIT***, given during undulating flight. Secondary call is a shrill accelerating, descending trill preceded by a triple note: ***seer-se-seer, seeseeseesee***. **HABITAT** Broadleaved evergreen, semi-evergreen and deciduous forest, sometimes parks and gardens during non-breeding movements; up to 1,300 m (mainly breeds below 600 m). Often in canopy of tall trees. **RANGE & STATUS** Resident (subject to some movements) NE Indian subcontinent, Andaman Is, SW China, Greater Sundas, Philippines; winter visitor Sumatra. **SE Asia** Scarce to fairly common resident (subject to some movements) W,S,E Myanmar, Tenasserim, W,NW,NE(south-west),SE,S Thailand, Peninsular Malaysia, Singapore, Cambodia, S Laos, Cochinchina. Scarce winter visitor C Thailand. Recorded (status uncertain) south N Laos. **BREEDING** Brood-parasitic. **Eggs** Laid in nests of Brown-throated, Crimson and Ruby-cheeked Sunbirds and Little Spiderhunter; fairly glossy, whitish-buff to pink, profusely speckled and blotched vinaceous-red or violet over more olive-brown undermarkings (sometimes loosely ringed); 16.4 x 12.3 mm (av.).

420 **DRONGO CUCKOO**

Surniculus lugubris Plate **51**

IDENTIFICATION 24.5 cm. *S.l.dicruroides*: **Adult** Shape and glossy greenish-black plumage recalls some drongos but has slender, slightly downcurved bill, white bars on undertail-coverts and underside of shorter outertail-feathers, and square-cut to slightly forked tail. May show some white feathers on nape and thighs (often concealed). In flight, underwing shows a white band across base of secondaries and inner primaries. **Juvenile** Somewhat browner (less glossy) with distinct white spots on body, wing-coverts and tail-tips. **Other subspecies in SE Asia** *S.l.brachyurus* (southern S Thailand southwards). **VOICE** Territorial call is a fairly quick 5–7 note ***pi pi pi pi pi...***, on steadily rising scale. Secondary call is a shrill ***phew phew phewphewphewphewphew phew phew...***, speeding up and rising, then falling away. **HABITAT** Broadleaved evergreen and deciduous forest, secondary growth, occasionally parks, gardens, mangroves etc. (mainly on migration); up to 1,300 m. **RANGE & STATUS** Breeds N,NE Indian subcontinent, S India, Sri Lanka, SE Tibet, W and southern China, Greater Sundas, Philippines, Sulawesi, N Moluccas; northern populations winter to south, Sumatra. **SE Asia** Fairly common resident throughout. Possibly a breeding visitor to some northern areas. Uncommon winter visitor C(south),S Thailand, Peninsular Malaysia, Singapore; probably Cochinchina. Uncommon passage migrant S Thailand, Peninsular Malaysia, Singapore, N Laos, E Tonkin. **BREEDING** Brood-parasitic. **Eggs** Laid in nests of Nepal Fulvetta, Chestnut-winged Babbler, Pin-striped Tit-babbler, Sooty-capped and Horsfield's Babblers, and Common Tailorbird; generally similar to those of host; 19.8 x 15.5 mm (av.; India). **NOTE** Taxonomy of *Surniculus* species in region confusing, and number of species involved remains unclear.

421 **ASIAN KOEL** *Eudynamys scolopaceus* Plate **52**

IDENTIFICATION 40–44 cm. *E.s.chinensis*: **Male** Relatively large size, long tail, glossy blue-black plumage, stout greenish bill and red eyes diagnostic. See crows. **Female** Variable but distinctive. Overall blackish to blackish-brown with heavy whitish streaks, spots and bars, or dark brown with predominantly rufous to buffish markings (including rufous malar stripe). Bill and eye colour like male. *E.s.malayana* (Resident Myanmar [except N,E] and W[south],C,SE Thailand southwards) typically shows rufous to buff markings. **Juvenile** Dull brownish-black with faint gloss (both sexes), whitish to buff tips on mantle to uppertail-coverts and wing-coverts, dull rufous bars on tail and broad whitish to pale buff bars on breast to vent (variable); bill greyish-buff, eyes brown. Female attains white spots on throat and breast during transition to adult plumage. **VOICE** Territorial call of male is a very loud ***ko-el*** (second note stressed), repeated with increasing emphasis. Secondary call is a loud, descending, bubbling ***wreep-wreep-wreep-wreep-wreepwreepwreep...*** or ***breep-breep-breep-breepbreepbreepbreep...*** **HABITAT & BEHAVIOUR** Open woodland, secondary growth, scrub, cultivated areas, parks and gardens; up to 1,220 m. Noisy but surprisingly skulking. **RANGE & STATUS** Resident Greater Sundas, W Lesser Sundas, Philippines, Sulawesi, Moluccas, New Guinea region, N Melanesia. Breeds Indian subcontinent, C and southern China; some northern populations winter south to Greater Sundas. **SE Asia** Fairly common to common resident, subject to some movements (except W,E Tonkin). Possibly a breeding visitor to some northern areas. Uncommon to fairly common winter visitor and passage migrant S Thailand, Peninsular Malaysia, Singapore. Fairly common passage migrant E Tonkin. Recorded (status uncertain) W Tonkin. **BREEDING** Brood-parasitic. **Eggs** Laid primarily in nests of Large-billed and House Crows, Common Myna and Black-collared Starling. Like those of host but (in case of crows at least) smaller; 32.5 x 24.2 mm (av.; *chinensis*).

422 **CORAL-BILLED GROUND-CUCKOO**

Carpococcyx renauldi Plate **51**

IDENTIFICATION 68.5–69 cm (including 36 cm tail). Monotypic. **Adult** Large size, overall greyish appearance with contrasting glossy blackish head, neck, upper breast, primaries and tail distinctive. Upperparts and tail glossed greenish and purplish, has fine blackish vermiculations on pale parts of underside, stout red bill, legs and feet, and violet and red facial skin. **Juvenile** Crown and nape blackish-brown, rest of upperparts dark brown with slight greenish tinge, forehead, throat and breast dull rufous-chestnut, fading onto lower breast and flanks; few dark bars on lower flanks only, browner wings and tail with slight purple tinge, and dull rufous-chestnut tips to scapulars and wing-feathers. Facial skin greyish, legs and feet dark brownish. **VOICE** Male territorial call is a loud, mellow, moaning ***woaaaah***, ***wooaa*** or ***wohaaau***, repeated every 5–10 s. Also a shorter ***pohh-poaaah*** and loud, vibrant, rolling ***wh ohh-whaaaao-hu***. Other calls include a deep, low, grumbling ***grrrro grrrro...*** or ***whrrro whrrro...*** and ***grrroah grrroah...*** **HABITAT & BEHAVIOUR** Broadleaved evergreen forest, secondary forest; up to 1,000 m, exceptionally 1,500 m. Spends most of time on ground, runs from danger. **RANGE & STATUS** Endemic. Scarce to locally fairly common resident NW,NE,SE Thailand, Cambodia, N(south),C,S Laos, E(south) Tonkin, N,C Annam. **BREEDING** May–August. **Nest** In captivity: simple cupped structure on ground or in tree, 3–4 m above ground. **Eggs** 2–4; white; 44 x 34 mm (av.).

CUCULIDAE: PHAENICOPHAEINAE: Malkohas & allies

Worldwide 13 species. SE Asia 6 species. Generally fairly large, with long to very long tails, relatively short, rounded wings and zygodactyl feet. Arboreal, usually in forest, clambering and creeping amongst dense foliage in mid-storey. Feed on small vertebrates and fruit.

423 BLACK-BELLIED MALKOHA
Rhopodytes diardi Plate **51**

IDENTIFICATION 35.5–38 cm (including 23 cm tail). *R.d.diardi*: **Adult** Similar to Green-billed Malkoha but smaller and proportionately shorter-tailed, throat and breast darker with fainter streaks, no whitish surround to facial skin, and narrower white tips to tail-feathers. Eyes pale blue or dark brown (any sex difference undocumented). See Chestnut-bellied Malkoha. **Juvenile** Crown and mantle browner, throat a little paler and whiter with no shaft-streaks, bill smaller and darker, eyes brown. **VOICE** A gruff ***gwaup***, more hurried ***gwagaup*** and louder, more emphatic ***pauk***. **HABITAT** Broadleaved evergreen forest, forest edge, secondary growth, sometimes plantations; up to 1,220 m (mainly below 200 m). **RANGE & STATUS** Resident Sumatra, Borneo. **SE Asia** Fairly common to common resident south Tenasserim, W(south),S Thailand, Peninsular Malaysia; formerly Singapore. **BREEDING** January–July. **Nest** Flimsy shallow saucer, in bush. **Eggs** 2; chalky-white; c.31.5 x 25.1 mm.

424 CHESTNUT-BELLIED MALKOHA
Rhopodytes sumatranus Plate **51**

IDENTIFICATION 40–40.5 cm (including 23 cm tail). *R.s.sumatranus*: **Adult** Like Black-bellied Malkoha but has diagnostic chestnut belly and undertail-coverts (may appear blackish in field), somewhat thicker bill and more orange facial skin. Eyes whitish, pale blue, brown or red (any sex differences not yet documented). Note habitat. **Juvenile** Like adult but tail-feathers narrower with less white at tip. **VOICE** Low ***tok tok...*** Also thin high-pitched mewing. **HABITAT** Mangroves, broadleaved evergreen forest, sometimes secondary growth and mature plantations; up to 1,005 m (mainly lowlands). **RANGE & STATUS** Resident Sumatra, Borneo. **SE Asia** Uncommon resident south Tenasserim, W(south),S Thailand, Peninsular Malaysia, Singapore. **BREEDING** February–September and November–December. **Nest** Flimsy saucer, in low shrub or tree; 1.5–6 m above ground. **Eggs** 2; white; 28.2–30.3 x 23–23.4 mm.

425 GREEN-BILLED MALKOHA
Rhopodytes tristis Plate **51**

IDENTIFICATION 52–59.5 cm (including 38 cm tail). *R.t.longicaudatus*: **Adult** Combination of relatively large size, very long tail, greyish head and underparts (vent darker) with blackish shaft-streaks, and red facial skin with narrow whitish surround distinctive. Upperside greyish-green, tail-feathers broadly tipped white. See Black-bellied and Chestnut-bellied Malkohas. Note range. **Juvenile** Crown, mantle and wings browner-tinged, vent browner, tail shorter, bill initially smaller and blackish, facial skin reduced and duller with less obvious whitish surround. **Other subspecies in SE Asia** *R.t.tristis* (W Myanmar); *saliens* (N,E[north] Myanmar, NW[east] Thailand, north Indochina). **VOICE** Territorial call is a mellow, slightly nasal, well-spaced ***oh oh oh oh...*** Typical call is a clucking, croaking ***ko ko ko...***, sometimes with an added gruff flurry: ***co-co-co-co...*** Also utters harsh chuckles when agitated. **HABITAT & BEHAVIOUR** Broadleaved evergreen, deciduous, freshwater swamp and peatswamp forest, secondary growth, coastal scrub, bamboo, sometimes plantations; up to 1,600 m, locally to 2,450 m. Forages amongst dense foliage, sometimes staying concealed for lengthy periods. **RANGE & STATUS** Resident N and north-eastern Indian subcontinent, SW,S China, Sumatra. **SE Asia** Common resident (except south Peninsular Malaysia, Singapore). **BREEDING** December–June. **Nest** Flimsy shallow saucer, in bush, bamboo clump or amongst creepers; 2.5–7 m above ground. **Eggs** 2–3; chalky-white (becoming stained); 33.8 x 25.8 mm (av.; *longicaudatus*).

426 RAFFLES'S MALKOHA
Rhinortha chlorophaeus Plate **51**

IDENTIFICATION 35 cm (including 19 cm tail). *R.c.chlorophaeus*: **Male** Relatively small size, rufescent head and breast, rufous-chestnut mantle and wings and blackish tail with white tips diagnostic. Bill pale green, facial skin turquoise, back, rump and tail-coverts blackish, belly greyish, tail finely barred dull bronze. **Female** Head, neck and breast pale grey, tail rufous-chestnut with black subterminal band and white tips, belly buffish. **Juvenile** Resembles adult. **VOICE** A slow descending 3–6 note series of strained, mournful mewing notes: ***kiau kiau kiau...*** or ***hiaa hiaa hiaa***. Also a hoarse, strained ***heeah*** or ***haaeew*** (singly or doubled) and harsh, strained croaking sounds. **HABITAT & BEHAVIOUR** Broadleaved evergreen forest, forest edge, sometimes plantations; up to 975 m. Forages slowly, amongst dense mid-storey foliage. **RANGE & STATUS** Resident Sumatra, Borneo. **SE Asia** Fairly common to common resident south Tenasserim, W,S Thailand, Peninsular Malaysia; formerly Singapore. **BREEDING** January–June. **Nest** Flimsy saucer, in dense foliage; 3 m above ground. **Eggs** 2–3; white; 33 x 25 mm (av.).

427 RED-BILLED MALKOHA
Zanclostomus javanicus Plate **51**

IDENTIFICATION 45–45.5 cm (including 26 cm tail). *Z.j.pallidus*: **Adult** Red bill, pale blue orbital skin, mostly greyish upperparts and mostly rufescent underparts diagnostic. Throat and upper breast rusty-rufous, belly and undertail-coverts chestnut, rest of underparts pale greyish, wings and tail glossed green, tail broadly tipped white, eyes whitish, brown or red (any sex differences not yet documented). **Juvenile** Tail-feathers narrower with less white at tip, primary coverts washed rufous, with pale fringes. **VOICE** An even, hard, frog-like ***uc uc uc uc uc uc...***, sometimes ending in a flurry: ***uc-uc-uc...*** **HABITAT** Broadleaved evergreen forest, forest edge, secondary growth; up to 1,200 m. **RANGE & STATUS** Resident Sumatra, Java, Borneo. **SE Asia** Fairly common to common resident south Tenasserim, south W,S Thailand, Peninsular Malaysia. **BREEDING** May–July. **Nest** Flimsy shallow saucer, in bush. **Eggs** 2; chalky-white; c.30 x 23.1 mm (av.).

428 CHESTNUT-BREASTED MALKOHA
Zanclostomus curvirostris Plate **51**

IDENTIFICATION 45.5–46 cm (including 26 cm tail). *Z.c.singularis*: **Male** Unmistakable. Upperside dark oily-green, most of underparts and broad tail-tip (upperside) dark chestnut. Crown, nape and cheeks mid-grey, vent blackish, no white on tail, bill rather thick and pale yellowish to pale greenish with mostly red lower mandible and nasal area, facial skin red, eyes bright pale blue. **Female** Eyes golden-yellow to whitish. **Juvenile** Similar to adult but facial skin reduced, bill mostly blackish on lower mandible and nasal area (initially all dark), eyes brown to grey, tail-feathers narrower and shorter with little chestnut. **VOICE** Low, clucking ***kuk kuk kuk...***, faster ***kok-kok-kok...*** when disturbed, and harsh, cat-like ***miaou*** when foraging. **HABITAT & BEHAVIOUR** Broadleaved evergreen forest, secondary growth, sometimes mangroves, mature plantations and gardens; up to 975 m. Forages in dense mid-storey foliage. **RANGE & STATUS** Resident Greater Sundas, W Philippines. **SE Asia** Fairly common to common resident south Tenasserim, W,S Thailand, Peninsular Malaysia. **BREEDING** January–October. **Nest** Broad, shallow saucer or cup, in sapling tree; 2.5–10 m above ground. **Eggs** 2–3; white (often becoming stained); 39 x 28 mm (av.).

CUCULIDAE: CENTROPODINAE: Coucals

Worldwide 28 species. SE Asia 4 species. Rather large and long-tailed with thickish bills and strong legs with zygodactyl feet. Mainly terrestrial, flight rather weak. Feed on small mammals, reptiles, amphibians, insects, crustaceans, molluscs, nestlings and eggs.

429 SHORT-TOED COUCAL

Centropus rectunguis Plate **51**

IDENTIFICATION 37 cm. Monotypic. **Adult** Resembles Greater Coucal but considerably smaller and relatively shorter-tailed. Underwing-coverts black. See Lesser Coucal. Note voice, habitat and range. **Juvenile** Crown and mantle dark chestnut-brown with blackish bars (more defined on mantle), wings similar to adult but with narrow blackish bars on tertials and coverts, dark brown lores and underparts (blacker ventrally) with whitish to buff bars and shaft-streaks overall, bill browner, underwing-coverts barred black and whitish; may show some pale bars on blackish tail. See Greater. **VOICE** Territorial call is a series of 4–5 slow, deep, melancholy, resonant notes: ***whu huup-huup-huup-huup***, descending somewhat towards end and repeated every 6–7 s. Occasionally a more rapid series on rising scale. **HABITAT & BEHAVIOUR** Broadleaved evergreen forest; up to 600 m (mainly lowlands). Shy and skulking, usually on ground. **RANGE & STATUS** Resident Sumatra, Borneo. **SE Asia** Scarce to uncommon resident extreme S Thailand, Peninsular Malaysia. **BREEDING** April–September. **Nest** Untidy ball, in palm; 2 m above ground. **Eggs** white; 37 x 30 mm (av.).

430 GREATER COUCAL

Centropus sinensis Plate **51**

IDENTIFICATION 48–52 cm. *C.s.intermedius*: **Adult** Large size and glossy purplish blue-black plumage with contrasting chestnut back and wings distinctive. Underwing-coverts black. See Short-toed and Lesser Coucals. Note habitat and range. **Juvenile** Dark of parts of upperside almost glossless with dull rufous-chestnut streaks and short bars, chestnut of upperside and wings duller with heavy blackish bars, tail with well-spaced narrow brownish- to greyish-white bars, underparts all blackish with whitish bars, whitish shaft-streaks on sides of head, lower throat and breast, browner bill and blackish and whitish bars on underwing-coverts. First-summer birds may show bars on secondaries. **Other subspecies in SE Asia** *C.s.bubutus* (south Peninsular Malaysia, Singapore). **VOICE** Territorial call is a series of loud, deep, far-carrying, mournful notes: ***puup puup puup puup...*** or ***wuup-uup-uup-uup-uupuupuupuupuup***, speeding up and ascending scale, before becoming lower and more even; sometimes ascends scale again during longer sequence. Also a more spaced, even 3–4 note series, introduced by a higher note: ***hi huup—huup—huup...*** Alarm call is a scolding, hissing ***shaeoooo*** or ***scheeeoh***. **HABITAT** Open forest, forest edge, secondary growth, scrub, grassland, mangroves; up to 1,525 m. **RANGE & STATUS** Resident Indian subcontinent, southern China, Greater Sundas, Philippines. **SE Asia** Common resident throughout (uncommon Singapore). **BREEDING** January–August. **Nest** Large dome with side-entrance, sometimes bowl, in bush, grass, bamboo or tree; up to 2.5 m above ground. **Eggs** 2–5; white (becoming stained); 35.7 x 28.6 mm (av.; *intermedius*).

431 ANDAMAN COUCAL

Centropus andamanensis Plate **51**

IDENTIFICATION 45.5–48.5 cm. Monotypic. **Adult** Unmistakable. Head, upper mantle and underparts buffy greyish-brown (duller on vent); back, scapulars and wings dark reddish-chestnut. Body-plumage quite variable in tone; uppertail may be uniform dark brownish, or darker towards tip and paler towards base. **Juvenile** Indistinctly barred paler and darker on supercilium, sides of head and underparts. May differ more substantially during early stages. **VOICE** Territorial call is a series of very deep, resonant ***hoop*** notes, running down and up scale, often followed by more ***hoop*** notes. Other calls include single ***tok*** and scolding cat-like ***skaaah*** when alarmed. habitat Forest edge, mangroves, plantations, gardens and cultivation; lowlands. **RANGE & STATUS** Resident Andaman Is. **SE Asia** Resident Great and Little Coco Is and Table I, off S Myanmar. **BREEDING** May–July. **Nest** Like that of Greater Coucal. **Eggs** 2–4; whitish; 34.7 x 28 mm (av.).

432 LESSER COUCAL

Centropus bengalensis Plate **51**

IDENTIFICATION 38 cm. *C.b.bengalensis*: **Adult breeding** Like Greater Coucal but much smaller with duller back, darker wing-tips and chestnut underwing-coverts. See Short-toed Coucal. Note voice, habitat and range. **Adult non-breeding** Distinctive. Crown, mantle, scapulars and sides of head dark brown, mixed with rufous and streaked whitish-buff; back to uppertail-coverts blackish-brown with dull rufous to rufous-buff bars; underparts dull buff with blackish-brown bars (broken on breast, lacking on centre of throat and centre of belly) and whitish-buff shaft-streaks; bill horn with dark ridge on upper mandible. Develops unusual elongated uppertail-coverts (may almost reach tail-tip). **Juvenile** Resembles adult non-breeding but upperside strongly rufescent, with broad blackish-brown streaks on crown and broad blackish-brown bars on rest of upperparts and wings; blackish tail with rufescent bars, more rufescent underparts with broader dark bars (centre of abdomen paler). Moults directly into non-breeding plumage in first winter, but may retain some barring, particularly on wings and tail. **Other subspecies in SE Asia** *C.b.javanensis* (S Thailand southwards). **VOICE** Territorial call is a series of 3–5 jolly, hollow notes (shorter than those of Greater), followed by 2–5 staccato phrases: ***huup huup huup-uup tokalok-tokalok*** or ***huup huup huup-uup-uup tokaruk-tokaruk-tokaruk...*** etc. Also gives a series of metallic cluckings, repeated quickly at first, then slowing and falling slightly before speeding up again and tailing off: ***thicthicthicthicthicthicthic-thuc-thuc-thuc-thuc-thucthucthucucucucuc...*** **HABITAT & BEHAVIOUR** Grassland, including marshy areas, scrub; up to 1,830 m. Secretive but often ascends grass stems or perches in bushes to sun itself. **RANGE & STATUS** Resident SW India, N and north-eastern Indian subcontinent, S,E China, Taiwan, Sundas, Philippines, Wallacea; some Chinese populations move south in winter. **SE Asia** Fairly common to common resident throughout. Uncommon winter visitor northern Thailand? Uncommon passage migrant E Tonkin, N Annam. **BREEDING** December–October. **Nest** Like that of Greater Coucal but smaller, in dense bush or tall grass; up to 1.5 m above ground. **Eggs** 2–4; white; 28.2 x 23.8 mm (av.; *bengalensis*).

TYTONIDAE: TYTONINAE: Barn- and grass-owls

Worldwide c.15 species. SE Asia 2 species. Similar to typical owls but characterised by heart-shaped faces, rather slender build and longer legs. Relatively large-headed and small-eyed. Feed on small vertebrates, particularly rodents.

433 COMMON BARN-OWL *Tyto alba* Plate **53**

IDENTIFICATION 34–36 cm. *T.a.stertens*: **Adult** Distinctive. Medium-sized to fairly large, upperside pale buffy-grey with golden-buff markings and blackish and whitish speckles (rather uniform in flight), with pale heart-shaped facial discs and white to buffy-white underparts, variably speckled blackish. Uppertail golden-buff with dark bars and greyish vermiculations. See Grass and Bay Owls. **Juvenile** Similar to adult.

Other subspecies in SE Asia *T.a.javanica* (S Thailand southwards). **VOICE** Song is a slightly rising, prolonged, dry, hissing ***ssssbHREEEIT***, building in volume and ending abruptly. Flight calls include a short, abrupt, down- then upslurred ***SHCREEIt***. Various shrill, metallic hisses. **HABITAT** Cultivation, open country, saltpans, marsh and swamp borders, plantations, urban areas; up to 1,220 m. **RANGE & STATUS** Resident Africa, Madagascar, W Palearctic, Middle East, Iran, Indian subcontinent, SW China, Sumatra, Java, Bali, Lesser Sundas, Flores Sea Is, E New Guinea, Australia, N Melanesia to Polynesia, Americas. **SE Asia** Uncommon to locally common resident (except SW,W,N Myanmar, W Tonkin, N Annam). **BREEDING** All year. **Nest** On floor in roof space of building or limestone cave, sometimes hole in bank or hollow tree. **Eggs** 4–7; white; 40.7 **x** 32.5 mm (av.; *stertens*).

434 EASTERN GRASS-OWL

Tyto longimembris Plate **53**

IDENTIFICATION 35.5 cm. *T.l.longimembris*: **Male** Like Common Barn-owl but upperside mostly dark brown to blackish-brown, with deep golden-buff markings and whitish speckles, neck-sides and breast washed pale rufous. Facial discs white, has narrow dark bars on predominantly buffish-white tail. Note range. In flight shows much more contrast between upper- and underside than Common Barn-owl, deep golden-buff patch at base of outer primaries contrasts with broad dark tips to primary coverts, underwing has much darker tips to primaries, with dark-tipped outer primary coverts (forming short bar). Short-eared Owl has similar underwing pattern but wings longer, dark bar on primary coverts longer and broader, no dark spots on underwing-coverts. **Female** Like male but facial discs and underparts usually suffused buff to brownish, tail more broadly banded. **Juvenile** At four weeks already very similar to adult with buffier body down, vinous-brown facial discs (particularly female). **VOICE** Song (given in flight) is a series of 3–6 rapid, musical high-pitched chirruping sounds (c. 12 notes per sec.), with most bursts rising then falling; the series repeated after short pauses. Calls include a harsh, downslurred, screeching ***SCHREEEoo***. **HABITAT** Grassland; up to 1,450 m. **RANGE & STATUS** Resident N,NE,E,S India, Nepal, S,SE China, Taiwan, Philippines, W Lesser Sundas, S(and E?) Sulawesi, New Guinea, Australia, New Caledonia, Fiji; formerly Bangladesh. **SE Asia** Rare to scarce resident C,S,E Myanmar, NW Thailand, E Tonkin, C,S Annam, Cochinchina. **BREEDING** December–February. **Nest** Simple pad, on ground amongst grass. **Eggs** 4–6; white; 39.9 **x** 32.8 mm (av.).

TYTONIDAE: PHODILINAE: Bay owls

Worldwide 3 species. SE Asia 1 species. Similar to barn- and grass-owls, but smaller and larger-eyed, relatively short-tailed and -legged; facial disc broadly divided on forehead and forming vestigial ear-tufts. Mainly forest-dwelling, feeding on small rodents, bats, birds, reptiles, amphibians and large insects.

435 ORIENTAL BAY OWL

Phodilus badius Plate **52**

IDENTIFICATION 29 cm. *P.b.badius*: **Adult** Recalls Common Barn-owl but smaller; has strange, almost triangular, pink-tinged buffy-whitish facial discs with rudimentary ear-tufts and dark smudge through eyes, dull rufous-chestnut upperside with black speckles and sparse white markings (crown mostly unmarked), largely golden-buff nuchal band and scapulars, and pinkish-washed underparts with buffier breast and round blackish spots. *P.b.saturatus* (N Myanmar?) has darker brown parts of plumage. **Juvenile** Little information. **VOICE** Eerie musical upward-inflected whistles, rising then fading away: ***oo hlii hoo hu-i-li hu-i-li hu-i-li hu-i-li*** (in full or in part). **HABITAT & BEHAVIOUR** Broadleaved evergreen forest, plantations, landward edge of mangroves; up to 1,220 m, rarely to 2,200 m. Often perches quite low down on vertical plant stems. **RANGE & STATUS** Resident NE India, SW,S China, Greater Sundas, Philippines (Samar); formerly Nepal. **SE Asia** Uncommon resident (except W Myanmar, C Thailand, Singapore, W Tonkin, S Annam). Formerly recorded Singapore. **BREEDING** India: March–May. **Nest** Tree-hollow; 2–5 m above ground. **Eggs** 3–5; white; 34.5 **x** 30 mm (av.; *saturatus*).

STRIGIDAE: STRIGINAE: Typical owls

Worldwide c.187 species. SE Asia 24 species. Compact with dense soft plumage, rounded heads, large forward-facing eyes surrounded by broad feathered facial discs and short tails. Most are brownish with cryptically patterned plumage. Generally nocturnal, roosting by day, so hard to observe. Feed on birds, rodents and insects, some species also taking other invertebrates, amphibians, reptiles and fish.

436 WHITE-FRONTED SCOPS-OWL

Otus sagittatus Plate **52**

IDENTIFICATION 27–29 cm. Monotypic. **Adult** Much larger and longer-tailed than other scops-owls. Recalls Reddish but upperside and sometimes upper breast brighter, dark rufous to rufous-chestnut, has distinctive broad, sharply demarcated whitish area on forehead and eyebrows (extending to ear-tufts) with fine brown vermiculations; underparts have dark vermiculations and white markings as well as dark spots, throat whiter; lacks dark bars on primaries. Bill bluish-white, eyes brown, no obvious nuchal collar. See Mountain and Collared Scops-owls and Large Frogmouth. Note habitat and range. **Juvenile** Undocumented. **VOICE** Vibrating hollow tremolo, lasting 13–14 s: ***wuwuwuwuwuwu...*** or ***w'w'w'w'w'w'w...***, rising somewhat in volume and ending quite abruptly. **HABITAT** Broadleaved evergreen forest; up to 610 m. **RANGE & STATUS** Resident north Sumatra. **SE Asia** Rare resident Tenasserim, W,S Thailand, Peninsular Malaysia. **BREEDING** Not reliably documented.

437 REDDISH SCOPS-OWL

Otus rufescens Plate **52**

IDENTIFICATION 19 cm. *O.r.malayensis*: **Adult** Small size and dark, dull rufescent-tinged plumage recalls Mountain Scops-owl but has pale buffish forehead and eyebrows with some blackish spots, plainer upperparts, warm buff and less distinct scapular markings, plain warm buffish to dull rufescent-buff underparts (breast sometimes darker) with distinct blackish spots, slightly highlighted above with pale buff, brown eyes (rarely dull amber) and distinctive broad dark and pale buffish bars on primaries. Bill pale flesh-coloured. Note voice, habitat and range. See White-fronted Scops-owl. **Juvenile** Undocumented. **VOICE** Territorial call is a hollow whistled ***hoooo***, fading at the end, repeated every 7–11 s. **HABITAT** Broadleaved evergreen forest; up to 200 m. **RANGE & STATUS** Resident Sumatra, Java, Borneo. **SE Asia** Scarce to uncommon resident southern S Thailand, Peninsular Malaysia. **BREEDING** April–May. **Eggs** 3+. Otherwise undocumented.

438 MOUNTAIN SCOPS-OWL

Otus spilocephalus Plate **52**

IDENTIFICATION 20 cm. *O.s.siamensis*: **Adult** Variable. Told by rather dull rufescent to rufous-chestnut upperparts, contrasting large white scapular markings, buffish to rufous underparts with white markings and dark vermiculations but no streaks, and yellow eyes. Eyebrows usually whitish-buff to warm buff, bill yellowish-brown. *O.s.spilocephalus* (SW,W,N Myanmar) ranges from less rufescent to strongly rufescent; *vulpes* (Peninsular Malaysia) is more deeply rich rufescent, particularly below, with coarser blackish upperpart markings and broad blackish crown-streaks; in northern Vietnam (*O.s.latouchei*?) often rather plain above and very rufescent with dark crown-streaks. See Reddish Scops-owl. Note voice, habitat and range. **Juvenile** Head and body paler and buffier with dark bars, no scapular markings. **VOICE** Territorial call is a clear, distinctly spaced ***phu-phu*** or ***toot-too***, repeated every 5–7 s. **HABITAT** Broadleaved evergreen forest; up to 2,200 m, commoner in mountains (above 800 m in Peninsular Malaysia). **RANGE & STATUS** Resident N Pakistan, NW,N,NE Indian subcontinent, S,SE China, Taiwan, Sumatra, N Borneo. **SE Asia** Fairly common to common resident (except C Thailand, Singapore). **BREEDING** February–June. **Nest** Natural tree-hole or old woodpecker or barbet nest; 2–7 m above ground. **Eggs** 2–5; white; 32.5 x 28.2 mm (av.; *spilocephalus*).

439 COLLARED SCOPS-OWL

Otus lettia Plate **52**

IDENTIFICATION 23 cm. **IDENTIFICATION** 23 cm. *O.l.lettia*: **Adult** Most widespread scops-owl in the region. Very variable, possibly exhibits greyish and buffish morphs but many intermediates occur. Tends to be smaller and possibly more buffish in south of region. Combination of densely vermiculated and streaked plumage, prominent broad whitish to buffish eyebrows and nuchal collar, pronounced ear-tufts and dark eyes distinctive. Upperparts rather greyish-brown, variably marked with buffish; underparts paler, whitish to deep buff, finely vermiculated and sparingly streaked darker. Bill dusky-yellowish to whitish-horn. *O.l.erythrocampe* (E Tonkin) is larger, tends to have darker mantle and scapulars, has much bolder dark underpart streaks. See White-fronted and Oriental Scops-owls. **Juvenile** Head and body paler (often warmer) with darker bars. Wings and tail less rufescent than Mountain Scops-owl. **Other subspecies in SE Asia** Uncertain; situation complex and unresolved. **VOICE** A rather soft but clear, slightly falling ***bouu***, repeated every c.12 s; pitch varies (possibly also between sexes). Occasionally a strident ***kuuk-kuuk-kuuk*** and rather quiet, moaning ***weyoo***. **HABITAT** Broadleaved evergreen, semi-evergreen and deciduous forest, clearings, wooded cultivation, gardens, plantations, island forest; up to 2,200 m. **RANGE & STATUS** Resident NE Pakistan, NW,N,NE Indian subcontinent, E Bangladesh, China (except NW,N), Taiwan, N,S Korea, Ussuriland, Sakhalin, Japan; some north-eastern populations winter to south. **SE Asia** Fairly common to common resident throughout. Unconfirmed winter visitor Peninsular Malaysia. **BREEDING** December–August. **Nest** Natural tree-hole or old woodpecker or barbet nest; 1.3–5 m above ground. **Eggs** 2–5; fairly glossy white; 32.3 x 28.1 mm (av.; *lettia*).

440 ORIENTAL SCOPS-OWL *Otus sunia* Plate 52

IDENTIFICATION 19 cm. *O.s.distans*: **Adult greyish morph** Variable. Recalls Collared Scops-owl but smaller and slimmer, eyes yellow, usually distinctly greyer with bolder dark streaks on underparts, less pale on forehead, eyebrows narrower and whiter, no pale nuchal collar, prominent white (not buff) scapular markings. Note voice. **Adult rufous morph** Variable. Resembles Mountain Scops-owl but has contrasting white belly and vent with blackish to rufous streaks and vermiculations, crown (sometimes mantle and scapulars) narrowly and distinctly streaked blackish, ear-tufts more pronounced. *O.s.stictonotus* (visitor Thailand, Cambodia, Cochinchina) is paler and more rufous, less rufous-chestnut. Note voice. **Juvenile** Head and body paler with more dark bars and fewer streaks, no scapular markings. **Other subspecies in SE Asia** Uncertain. Situation complex and unresolved; certainly at least *O.s.malayanus* (visitor Tenasserim, Peninsular Malaysia, Singapore). **VOICE** Territorial call is a loud, clear, measured ***toik toik'to-toik*** or shortened ***toik'to-toik*** (*O.s.stictonotus*). **HABITAT** Broadleaved evergreen and mixed deciduous forest and clearings; migrants also occur in island forest, mangroves, plantations etc.; up to 2,000 m (breeds below 1,000 m?). **RANGE & STATUS** Resident (subject to some movements) NE Pakistan, Indian subcontinent. Breeds SE Siberia, Ussuriland, Sakhalin, China (except NW,N), Taiwan, N,S Korea, Japan; northern populations winter to south, rarely Sumatra. **SE Asia** Uncommon to fairly common resident Myanmar (except S), W,NW,NE Thailand, Cambodia, S Laos, N,S Annam. Uncommon winter visitor S Myanmar, Thailand, Peninsular Malaysia, Cambodia. Scarce to uncommon passage migrant Peninsular Malaysia, E Tonkin. Vagrant Singapore. Recorded (status uncertain) N,C Laos, W Tonkin, C Annam. **BREEDING** February–June. **Nest** Hole in tree or wall. **Eggs** 3–4; white; 32.8 x 27 mm (av.; India).

441 INDIAN EAGLE-OWL

Bubo bengalensis Plate **53**

IDENTIFICATION 54.5–56 cm. Monotypic. **Adult** From other eagle-owls by prominent black border to facial discs, warm buff underparts, whitish gorget bisected by band of dark streaks, heavy dark breast-streaks, narrow dark streaks and cross-bars on belly, and yellow to orange-yellow eyes. Crown streaked blackish-brown and rich buff with some blackish and white bars, broadly rich buff nape/upper mantle with blackish-brown streaks, rest of upperparts dark to blackish-brown with buff to rich buff markings. See fish-owls; note range. **Juvenile** Head and body uniform creamy-buffish, evenly striated and vermiculated slightly darker. **VOICE** Loud, deep, full, resonant, almost two-noted ***WOo'HAOOo***. Voice of female reported to be higher. Also growls, hisses and coughs. **HABITAT** Bush-covered rocky country, ravines, wooded semi-desert and cultivation; lowlands. **RANGE & STATUS** Resident Indian subcontinent (except islands). **SE Asia** Recorded (status uncertain) SW Myanmar (record perhaps doubtful). **BREEDING** India: October–May. **Nest** Scrape on rock ledge, rarely on ground. **Eggs** 2–4; white, faintly tinged cream; 53.6 x 43.8 mm (av.).

442 SPOT-BELLIED EAGLE-OWL

Bubo nipalensis Plate **53**

IDENTIFICATION 61 cm. *B.n.nipalensis*: **Adult** Distinctive. Very large, upperside predominantly dark brown with pale buff markings (mainly on scapulars and wing-coverts), with long part-barred ear-tufts, whitish underparts with distinctive blackish-brown heart-shaped spots (upper breast and sides of neck appear more barred), dark eyes and pale yellow bill. See Barred Eagle-owl. Note range. **Juvenile** Head and body whitish to buffy-white with prominent blackish-brown bars (more prominent on upperside). **VOICE** Deep ***HOO HOO*** or ***HU HUU***, usually with about 2 s between notes, repeated every 1–2 minutes. Also a loud, eerie, rather nasal moaning scream: ***waayaoaah*** etc., repeated every 3–16 s and shorter, quieter ***aayao*** from roosting birds. **HABITAT** Broadleaved evergreen, semi-evergreen and deciduous forest, clearings; up to 1,200 m. **RANGE & STATUS** Resident N,NE Indian subcontinent, S India, Sri Lanka, SW China. **SE Asia** Scarce to uncommon resident S,E Myanmar, north Tenasserim, Thailand (except C,S), Cambodia, N,S Laos, W,E Tonkin, C,S Annam, Cochinchina. **BREEDING** February–June. **Nest** Tree-hollow, cave or old raptor nest. **Eggs** 1; white; 61.2 x 49.8 mm (av.).

443 BARRED EAGLE-OWL

Bubo sumatranus Plate **53**

IDENTIFICATION 45.5–46.5 cm. *B.s.sumatranus*: **Adult** Resembles Spot-bellied Eagle-owl but much smaller, with narrow brown-and-buffy (pale chestnut-tinged) bars on upperparts, no buff markings on scapulars or wing-coverts, narrow blackish-and-whitish barring (not heart-shaped spots) on underparts and legs, denser and washed browner on breast. Note range. **Juvenile** Like Spot-bellied but dark bars narrower and denser. **VOICE** Loud, deep ***uk OOO OO*** (introductory note barely audible), usually with about 2 s between notes and repeated after lengthy intervals. Also a loud quacking ***gagaga-gogogo***. **HABITAT** Broadleaved evergreen forest, mature plantations, coastal woodland, clearings; up to 610 m, locally up to 1,400 m. **RANGE & STATUS** Resident Greater Sundas. **SE Asia** Scarce to uncommon resident south Tenasserim, W(south),S Thailand, Peninsular Malaysia; formerly Singapore (now vagrant?). **BREEDING** November–April. **Nest** Tree-hollow; 10 m above ground. **Eggs** 1–2; white.

444 **DUSKY EAGLE-OWL**

Bubo coromandus Plate **53**

IDENTIFICATION 54–58 cm. *B.c.klossii*: **Adult** Resembles Brown Fish-owl but upperparts rather uniform drab dark greyish-brown, with no obvious markings, underparts much darker and greyer with less distinct streaks and denser dark vermiculations/cross-bars, no whitish gorget, ear-tufts more erect, primaries unbarred, legs feathered. Eyes yellow. **Juvenile** Head and body creamy-whitish with very faint darker striations and vermiculations, has distinctive, rather plain greyish wings and tail (like adult). **VOICE** Loud, hollow series of notes given at diminishing intervals: ***gwuk GWUK GWUK-WUK-WUK-WUK'UG'ug'g'ggg*** or ***WO WO WO WO-O-o-o-o***, recalling ping-pong ball bouncing to a halt. **HABITAT** Open woodland, usually near water; lowlands. **RANGE & STATUS** Resident Indian subcontinent (except islands), SE China. **SE Asia** Rare to scarce resident SW,C Myanmar, Tenasserim, Peninsular Malaysia (Perak and Selangor); formerly W(coastal),S(north) Thailand. **BREEDING** December–March. **Nest** Large rough cup, in tree fork, sometimes tree-hollow or old large raptor nest; 9–12 m above ground. **Eggs** 1–4; white; 59.2 x 48.3 mm (av.; India).

445 **BROWN FISH-OWL**

Ketupa zeylonensis Plate **53**

IDENTIFICATION 49–54 cm. *K.z.leschenault*: **Adult** Distinguished by combination of rather warm buffish-brown base colour, bold blackish-brown streaks on crown and upperparts, floppy streaked ear-tufts, strong blackish-brown and whitish/buff wing markings, whitish scapular markings and gorget, and long blackish-brown streaks and narrow brown cross-bars (sometimes faint) on underparts. Eyes golden-yellow, underparts paler than upperparts, no white on forehead and eyebrows, bill greenish-grey with darker tip. See Tawny and Buffy Fish-owls and Dusky Eagle-owl. **Juvenile** Head and body pale creamy-buffish with long dark brown streaks, wings paler than adult. **Other subspecies in SE Asia** *K.z.orientalis* (Indochina). **VOICE** Territorial call is a deep, rapid, hollow moaning ***HU WHO-hu*** or similar ***HUP-HUP-hu*** with last note barely audible. Also described as deep, booming, muffled, twangy humming ***hOOo BOOOOo-hOOo***. Also a series of deep mutterings, rising to maniacal laughter: ***hu-hu-hu-hu-hu hu ha*** or ***oof uh-oof uh-oof uh-oof uh-oof uh-oof u-uh-h-HA-oo-oo-oof***, with laughter before end. A mournful scream, hoarser than Spot-bellied Eagle-owl, is also reported. **HABITAT** More open broadleaved evergreen, semi-evergreen and deciduous forest near water; up to 915 m. **RANGE & STATUS** Resident Middle East, Indian subcontinent, S China. **SE Asia** Uncommon to fairly common resident Myanmar, Thailand (except C,SE), north-west Peninsular Malaysia, Indochina (except W Tonkin). **BREEDING** December–March. **Nest** Rock ledge, rock cleft, tree-hollow, platform-like epiphytic fern in tree or old large raptor nest; up to 30 m above ground. **Eggs** 1–2; white, faintly tinged cream; 58.4 x 48.9 mm (av.; *leschenault*).

446 **TAWNY FISH-OWL** *Ketupa flavipes* Plate **53**

IDENTIFICATION 58.5–61 cm. Monotypic. **Adult** Very similar to Buffy Fish-owl but much larger, upper- and particularly underparts more orange-buff (less so compared to *K.k.ketupu*), tends to show more unmarked buff on scapulars and greater coverts (making upperside less uniform), dark underpart-streaks broader, particularly on breast (where typically distinctly broader than belly-streaks), pale wing- and tail-bars rich buff (rather than pale buff), ear-tufts broader with thicker dark streaks. Bill blackish. Note range. **Juvenile** Possibly not separable from Buffy except on size. **VOICE** Song of male (Taiwan) is quick, very deep, booming, muffled, twangy humming ***huh-huhWOOo*** (***WOOo*** rising quickly, then falling). Females accompany (in Taiwan) with mewing ***hew***. Sibilant, slightly hissing ***fshhht*** or ***hshhht***, and high, weak, falling whistled ***pheeeeooo***, ***piii'iooo*** or ***pheee'eeooo*** from roosting birds. **HABITAT** Broadleaved evergreen and semi-evergreen forest along rivers or near water, freshwater swamp forest; up to 600 m. **RANGE & STATUS** Resident N,NE Indian subcontinent, C and southern China, Taiwan. **SE Asia** Rare to local resident W(north),N Myanmar, N,C Laos, W Tonkin, Cochinchina. **BREEDING** India: December–February. **Nest** Old large raptor nest or hole in bank. **Eggs** 1–2; white; 57.1 x 46.9 mm (av.).

447 **BUFFY FISH-OWL** *Ketupa ketupu* Plate **53**

IDENTIFICATION 45.5–47 cm. *K.k.aagaardi*: **Adult** Resembles Brown Fish-owl but overall plumage richer buff with broad blackish upperpart streaks and wing markings (contrasting more with underparts), distinctive (but variable) white area on forehead and eyebrows, less obvious white gorget and no cross-bars on underparts. Eyes yellow to golden-yellow. *K.k.ketupu* (Peninsular Malaysia, Singapore) is warmer, richer buff above and below. See Tawny Fish-owl. Note range. **Juvenile** Like Brown but body and wings richer buff. **VOICE** Long, monotonous ***bup-bup-bup-bup-bup-bup...*** or ***hup-hup-hup-hup-hup-hup...*** like a generator, and high-pitched screeching yelps: ***yiark, yark, yark, yeek*** etc. Subdued, hoarse rather hissing ***hyiiii*** or ***hyiiii-ih*** uttered by roosting birds. **HABITAT** Broadleaved evergreen forest near water, mangroves, plantations, wooded gardens, cultivation; up to 800 m. **RANGE & STATUS** Greater Sundas; formerly NE India. **SE Asia** Scarce to fairly common resident SW,N,S Myanmar, Tenasserim, W(south),NE(south),SE,S Thailand, Peninsular Malaysia, Singapore, Cambodia, C Annam, Cochinchina. **BREEDING** All year. **Nest** Cavity in rock or tree, sometimes old nest of other bird or clump of epiphytes in tree; 3–18 m above ground. **Eggs** 1; white; 47.3–49 x 42.5–43.3 mm. **NOTE** Birds in S Thailand (including type of *aagaardi*) apparently not different from *K.k.ketupu* (Wells 1999); new name for *aagaardi* may be required and review of ranges of different forms.

448 **SPOTTED WOOD-OWL**

Strix seloputo Plate **53**

IDENTIFICATION 44.5–48 cm. *S.s.seloputo*: **Adult** Resembles Brown Wood-owl but upperparts speckled and spotted white (barred on sides of neck), underparts white to buffy-white, mixed rich buff and with well-spaced bold blackish-brown bars (denser on breast), facial discs plain rufous-buff. Eyes dark. See Mottled Wood-owl. **Juvenile** Initially similar to Brown but lacks dark around eyes and contrasting even bars on wings and tail. **VOICE** Loud, abrupt booming ***WHO*** or ***UUH*** (possibly differing slightly between sexes), repeated every 8–11 s. Also a loud, deep quavering ***WRRRROOH WRRRROOH WRRRROOH...*** **HABITAT** Edge of broadleaved evergreen forest, logged forest, plantations, wooded parks, cultivation, sometimes mangroves; up to 305 m. **RANGE & STATUS** Resident Java, Philippines (Palawan, Busuanga); probably Sumatra. **SE Asia** Uncommon to locally fairly common resident S,E Myanmar, Tenasserim, NE,C,S Thailand, Peninsular Malaysia, Singapore, Cambodia, S Laos, Cochinchina. **BREEDING** January–August. **Nest** Tree-hollow or platform-like epiphytic fern in tree; 2–18 m above ground. **Eggs** 2–3; white; 50.3 x 42.8 mm (mean).

449 **MOTTLED WOOD-OWL**

Strix ocellata Plate **53**

IDENTIFICATION 44–48.5 cm. *S.o.grisescens*?: **Adult** Similar to Spotted Wood-owl but facial discs mostly dusky-white with blackish vermiculations and some rufous at sides; upperparts, wings and tail extensively vermiculated greyish; white spots restricted to crown, where more densely speckled, crown and breast mixed with rufous, underpart-bars narrower and broken on breast, eyelids reddish. Eyes dark. Note range. **VOICE** Loud, eerie, quavering ***UUWAHRRRR*** and shorter ***W'RROH W'RROH W'RROH W'ROHH...*** by respective sexes of pair. **HABITAT** Open woodland, wooded gardens and cultivation; lowlands. **RANGE & STATUS** Resident India (except NW,NE). **SE Asia** Scarce to uncommon resident SW Myanmar. **BREEDING** India: February–March. **Nest** Natural tree-hollow, rarely a bulky platform or shallow cup in tree. **Eggs** 2–3; creamy-white; 51.1 x 42.6 mm (av.).

450 **BROWN WOOD-OWL**

Strix leptogrammica Plate **53**

IDENTIFICATION 47–53 cm. *S.l.laotiana*: **Adult** Distinguished by largish size, rounded head with buffy-brown facial discs, dark eyes surrounded by dark brown to black, mostly dark brown upperparts, and pale buff to deep buff underparts with dense dark brown bars. Has whitish to buff bars on scapulars and, narrowly, across upper mantle (broken in centre); breast often dark brown, facial discs sometimes barred darker. Apparently tends to be smaller, darker above and buffier below, to the south and east of region (regardless of subspecies). **Juvenile** Head, body and wing-coverts whitish-buff to pale buffish with narrow dark bars; distinctive, contrasting, even dark brown and dull rufescent bars on remainder of wings and tail, and dark around eyes (as adult). **Other subspecies in SE Asia** *S.l.newarensis* (N,C Myanmar); *rileyi* (Tenasserim); *ticehursti* (S,E Myanmar); *maingayi* (S Thailand southwards). **VOICE** Fairly loud deep vibrating ***HU-HU-HU'HUHRRROO***, repeated every 1–5 s, (***oo***) ***WU'U'U'WOOo*** or ***HOO HOO-HOO-HOO-(HOO)***, with pause after first note. Sometimes 2–3 well-spaced hoots: (***U***) ***HUUU***, *UWUTUH*, or ***WUT'UU***. Also a loud eerie scream, ***eeeeooow***, or a rather more subdued, slightly vibrating ***ayaarrrh***. **HABITAT** Broadleaved evergreen, semi-evergreen and mixed deciduous forest; up to 2,590 m. **RANGE & STATUS** Resident NW,N,NE and peninsular India, Nepal, Bhutan, Bangladesh, Sri Lanka, SW,S,SE China, Taiwan, Sumatra, Java, Borneo. **SE Asia** Uncommon to fairly common resident (except SW Myanmar, C Thailand, Singapore, W Tonkin, Cochinchina). **BREEDING** February–May. **Nest** On rock ledge, in cave or tree-hollow etc. **Eggs** 1–2; white; 56.1 x 46 mm (av.; India).

451 **HIMALAYAN WOOD-OWL**

Strix nivicola Plate **53**

IDENTIFICATION 43 cm. *S.n.nivicola*: **Adult pale morph** Shape and proportions recall Brown Wood-owl but appears rather uniformly pale overall; upperparts mottled, streaked and vermiculated blackish-brown, pale buff and pale greyish, facial discs pale greyish with fine darker markings, underparts white to buffy-white with heavy blackish-brown streaks and vermiculations, has prominent buffy-white markings on scapulars and wing-coverts. **Adult dark morph** Patterned as pale morph but base colour of upperparts blacker brown with richer rufous-buff markings, base colour of underparts rich buff, mixed with rufous and white, facial discs rufous-buff, scapular and wing-covert markings pale warm buff. Note range. **Juvenile** Head, body and wing-coverts paler and more uniform with even dark bars overall; much more heavily barred than Brown but wings and tail less heavily barred. **VOICE** Territorial call is a loud, resonant ***HU-HU*** or ***COO-COO***. **HABITAT** Broadleaved evergreen and coniferous forest; 2,450–3,080 m. **RANGE & STATUS** Resident N,NE India, Nepal, Bhutan, S,SE Tibet, China (except NW,N,NE), Taiwan, N,S Korea. **SE Asia** Uncommon resident W,N,E Myanmar, W Tonkin. **BREEDING** India: January–April. **Nest** Natural tree-hollow, sometimes rock crevice. **Eggs** 2–3; white; 48.2 x 41.6 mm (av.).

452 **COLLARED OWLET**

Glaucidium brodiei Plate **52**

IDENTIFICATION 16–16.5 cm. *G.b.brodiei*: **Adult** Resembles Barred Owlet but much smaller, with diagnostic buff-and-blackish imitation face pattern on nape/upper mantle, crown greyer and often more speckled than barred, usually narrower dark band across throat, more teardrop-shaped streaks on belly and lower flanks, no prominent white markings on wing-coverts. See Jungle Owlet. Note voice. **Adult rufous morph** (northern Vietnam at least): Very rufous, except for white of eyebrows and underparts. **Juvenile** Crown to mantle unmarked brown, apart from whitish shaft-streaks on forecrown. **VOICE** Territorial call is a loud, rhythmic, hollow piping ***pho pho-pho pho*** (about two seconds duration), repeated every 1–2 s; often heard in daytime. **HABITAT** Broadleaved evergreen forest, occasionally deciduous forest; up to 3,100 m, mostly above 600 m in Thailand, only above 395 m (exceptionally lower) in Peninsular Malaysia. **RANGE & STATUS** Resident N Pakistan, NW,N,NE Indian subcontinent, C and southern China, Taiwan, Sumatra, Borneo. **SE Asia** Fairly common to common resident (except SW Myanmar, C Thailand, Singapore, Cochinchina). **BREEDING** March–July. **Nest** Natural tree-hole or old woodpecker or barbet nest; 2–10 m above ground. **Eggs** 2–5; white; c.29 x 24 mm.

453 **ASIAN BARRED OWLET**

Glaucidium cuculoides Plate **52**

IDENTIFICATION 20.5–23 cm. *G.c.bruegeli*: **Adult** Distinctive. Small and robust with broadly rounded head (no ear-tufts) and neckless appearance; dull brown with pale buffish to whitish bars, white ventral line (breast may be all barred), and whitish belly and lower flanks with broad brown streaks (some bars at sides). Has narrow whitish eyebrows and yellow eyes. Tends to be larger, darker and more rufescent to west and north (*G.c.rufescens*: Myanmar except northern N and Tenasserim, and *austerum*: west N Myanmar), and darker and more rufescent, with rustier belly-streaks and contrasting greyer head with whiter bars, to the south-east (*deignani*: south-eastern Thailand, southern Indochina; particularly latter). See Jungle, Collared and Spotted Owlets. **Juvenile** Pale barring on upperparts more diffuse and spot-like, underpart-bars broken and more diffuse, streaks more diffuse, crown more speckled. **Other subspecies in SE Asia** *G.c.delacouri* (N Laos, W Tonkin), *whitelyi* (east N Myanmar, E Tonkin). **VOICE** Long, descending, eerie quavering trill: ***wu'u'u'u'u'u'u'u'u'u...*** (lasting about 10 s) and gradually increasing in volume. Long series of raucous double notes, uttered at increasing pitch and volume, preceded by mellow ***hoop*** notes, recalling a barbet. **HABITAT & BEHAVIOUR** Relatively open broadleaved evergreen, semi-evergreen and deciduous forest, open areas with clumps of trees; up to 1,980 m. Often seen perched in open during daytime, wags tail from side to side when agitated. **RANGE & STATUS** Resident NE Pakistan, NW,N,NE Indian subcontinent, SE Tibet, C and southern China. **SE Asia** Common resident Myanmar, W,NW,NE,SE,C,north S Thailand, Indochina. **BREEDING** March–June. **Nest** Natural tree-hole or old woodpecker or barbet nest. **Eggs** 3–5; white; 36.5 x 30.5 mm (av.; *rufescens*). **NOTE** Subspecific distinctions often poorly marked, unresolved and clouded by individual variation.

454 **JUNGLE OWLET**

Glaucidium radiatum Plate **52**

IDENTIFICATION 20.5 cm. *G.r.radiatum*: **Adult** Like Asian Barred Owlet but smaller, somewhat greyer and more densely barred overall, with prominent contrasting dull rufous bars on flight-feathers, smaller white scapular markings and entirely barred underparts. See Collared and Spotted Owlets. **Juvenile** Less distinct barring below, browner barring on tail. **VOICE** Unusual, slightly raucous ***PRAA-PRAA-PRAA-praa-pruu*** or ***prr-prr-prr-praa-praa-praa-praa-praa-praa-praa'praa'praa*** etc., repeated after shortish intervals. **HABITAT** Mixed deciduous forest, secondary growth; up to 1,220 m. **RANGE & STATUS** Resident Indian subcontinent (except Pakistan). **SE Asia** Uncommon to fairly common resident SW,W Myanmar (old records, and perhaps doubtful). **BREEDING** India: March–May. **Nest** Natural tree-hole or old woodpecker or barbet nest; 2–8 m above ground. **Eggs** 3–4; white; 31.5 x 26.8 mm (av.).

455 **SPOTTED OWLET** *Athene brama* Plate **52**

IDENTIFICATION 20–20.5 cm. *A.b.mayri*: **Adult** Within range, the most likely small owl to be seen in non-forest habitats. Upperparts brownish-grey with white spots, broad white eyebrows, broken white nuchal collar, broken dark foreneck collar and whitish underparts with broken dark bars and no streaks. Eyes yellow. **Juvenile** Apparently more washed out and less spotted above, underpart bars more diffuse; belly may be lightly streaked. **Other subspecies in SE Asia** *A.b.pulchra*

(Myanmar). **VOICE** Harsh, screeching ***chirurr-chirurr-chirurr...*** etc., followed by or alternating with ***cheevak cheevak cheevak...*** Also complex high-pitched screeching and chuckling. **HABITAT** Open woodland, semi-desert, cultivation, gardens, buildings, urban areas; up to 1,220 m (mostly lowlands). **RANGE & STATUS** Resident S Iran, Indian subcontinent (except islands). **SE Asia** Common resident SW,W,C,S,E Myanmar, Thailand (except S), Cambodia, Laos, C Annam, Cochinchina. **BREEDING** February–May. **Nest** Hole in tree, building or rock cleft. **Eggs** 3–6; white; 31.6 x 27.4 mm (av.; S India).

456 **BROWN BOOBOOK**

Ninox scutulata Plate **52**

IDENTIFICATION 30–31 cm. *N.s.burmanica*: **Adult** Distinctive. Medium-sized, slim, small-headed and relatively long-tailed, upperside uniform-looking dark slaty-brown (greyer on sides of head, crown and nape), with whitish patch between eyes and whitish to buffish-white underparts, streaked with very broad drab chestnut-brown heart-shaped spots (often denser on breast). Lacks obvious facial discs, tail pale brownish-grey with broad blackish bars and dusky-white tip, eyes golden-yellow. *N.s.scutulata* (Peninsular Malaysia, Singapore) has darker, more uniform sides of head and upperparts (vaguely warm-tinged), markings on underparts darker and denser on belly and vent, where slightly more chestnut-tinged. In flight, shape recalls some *Accipiter* species. See Northern Boobook (note range and status) and Barred Owlet. **Juvenile** Upperparts paler and somewhat warmer, underpart markings more diffuse. **VOICE** Territorial call is a loud, haunting, rather deep but rising ***whu-UP*** or ***whoo-WUP***, repeated every 0.6–0.9 sec. **HABITAT & BEHAVIOUR** Open forest, mangroves, tall secondary growth, occasionally parkland, wooded gardens; up to 1,200 m. May be seen hawking insects at dusk. **RANGE & STATUS** Resident Indian subcontinent (except NW, Pakistan & islands), SW,S China, Sumatra, W Java, Borneo, Philippines, N Sulawesi. **SE Asia** Fairly common to common resident (except C Thailand, W Tonkin, S Annam). **BREEDING** March–June. **Nest** Natural tree-hole; up to 6 m above ground. Eggs 2–5; white; 35.1 x 29.5 mm (av.).

457 **NORTHERN BOOBOOK**

Ninox japonica Not illustrated

IDENTIFICATION 30–31 cm. *N.j.japonica*: **Adult** On current knowledge, very difficult to separate from Brown Boobook, but slightly darker above and has colder, darker underpart markings, with somewhat more intervening white (breast less solidly brown); wings more pointed and tail proportionately shorter. Note range and status. **VOICE** Territorial call (unlikely to be heard in winter) is mellow, hollow couplet (occasionally triplet), consisting of 2(–3) ***whoop*** notes (separated by 0.25–0.5 sec gap), repeated every 0.4–0.9 sec. **HABITAT** Open forest, mangroves, also wooded parks and gardens; up to 1,200 m. **RANGE & STATUS** Breeds SE Siberia, Ussuriland, W,C,NE,SE China, N,S Korea, Japan (except Nansei Is); winters Greater Sundas, Philippines, Wallacea. Resident Taiwan, Nansei Is (S Japan). **SE Asia** Uncommon winter visitor and passage migrant W(coastal),C,S Thailand, Peninsular Malaysia, Singapore, Cambodia.

458 **LONG-EARED OWL** *Asio otus* Plate **53**

IDENTIFICATION 35–37 cm. *A.o.otus*: **Male** Similar to Short-eared Owl but ear-tufts distinctly longer (obvious when held erect), upperparts greyer, underparts completely and more evenly dark-streaked and with indistinct dark vermiculations, facial discs rufescent with less dark around eyes, eyes orange, underwing-coverts white. In flight, upperside of primaries, secondaries and tail less boldly barred, base of primaries distinctly rufous-buff, underwing similar to Short-eared with less black on primary-tips. See much larger eagle and fish-owls. **Female** Facial discs and underparts usually more richly coloured, upperparts more boldly dark-streaked, underwing-coverts rich buff. **First winter** Shows 4–6 dark bars on outer primaries (3–5 on adult) and 5–6 dark bars on tail, beyond uppertail-coverts (4 on adult). **VOICE** Territorial males give a series of soft, muffled, quite far-carrying ***ooh*** notes, recalling sound of air being blown into bottle. Female utters weak, nasal ***paah***. Sharp barking ***kvik kvik kvik...*** and yelping or squealing sounds when alarmed. **HABITAT & BEHAVIOUR** Woodland, wooded cultivation, plantations; recorded at 1,065–1,385 m. Normally roosts in trees but forages in open areas. Generally flies lower and with more flickering wingbeats than Short-eared and banks less often. **RANGE & STATUS** Breeds Holarctic, N Africa, Middle East, W,N Pakistan (erratic), NW India (erratic), northern China, N Korea, Japan. Some winter irregularly to south, Pakistan, NW India, S,SE China, S Korea, S Japan, C Mexico. **SE Asia** Vagrant N Myanmar, N Laos.

459 **SHORT-EARED OWL** *Asio flammeus* Plate **53**

IDENTIFICATION 37–39 cm. *A.f.flammeus*: **Male** At rest, only likely to be confused with fish-owls, but much smaller and rounder-headed, lacking obvious ear-tufts (may show short points), facial discs form rough circle or heart shape with darker centre and pale cross-like pattern between eyes, breast-streaks broader and denser. Eyes yellow. In flight recalls Common Barn- and Eastern Grass-owls, but longer-winged, with broadly dark-barred upperside of flight-feathers and tail and plain pale buff bases of primaries, which contrast sharply with mostly blackish primary coverts; underwing appears mostly whitish with contrasting black bar along tips of primary coverts and black and white markings on primary tips. See Long-eared Owl. Note habitat. **Female** Base colour of plumage distinctly deeper buff, dark markings bolder, facial discs and underwing-coverts deeper buff, base colour of underparts more uniformly buff. **VOICE** Male territorial call (mainly given during display-flight) is a repeated, low-pitched, hollow, rather muffled ***boo-boo-boo-boo-boo-boo...*** Also a hoarse rasping ***cheeee-op*** (mostly by females). Harsh barking ***chef-chef-chef*** when alarmed. **HABITAT & BEHAVIOUR** Grassland, marshes, open areas; up to 1,830 m. Only flies in daylight; easily flushed from rather open habitat, flying off with rather stiff rowing wingbeats. **RANGE & STATUS** Breeds Holarctic, NE China, S America, locally Caribbean and Pacific Is; some northern populations winter south to N Africa, C,S Europe, Middle East, Indian subcontinent, China, Taiwan, N,S Korea, Japan, Philippines, C America. **SE Asia** Rare to scarce winter visitor W,N,C,S Myanmar, NW Thailand, N Laos, E Tonkin, C Annam. Vagrant C Thailand, Peninsular Malaysia, Singapore.

PODARGIDAE: BATRACHOSTOMINAE: Asian frogmouths

Worldwide 12 species. SE Asia 4 species. Smallish to medium-sized. Have wide gape and soft, cryptically patterned plumage like nightjars but bills thick and wide, wings shortish and rounded; habits more arboreal, adopting upright posture when perched. Catch prey at night by foliage-gleaning or pouncing from tree-branches. Feed on cicadas, grasshoppers, moths, butterflies, beetles, insects, ants, termites and other insects, caterpillars etc.

460 **LARGE FROGMOUTH**

Batrachostomus auritus Plate **54**

IDENTIFICATION 39–42 cm. Monotypic. **Adult** From other frogmouths by much larger size and large white tips to upperwing-coverts and lower scapulars. Has buffy-white and blackish-barred nuchal collar and rather uniform warm brown throat and breast with a few small white markings. Eyes brown. See much smaller Gould's. Females are apparently usually rather duller and plainer. **Juvenile** Paler and plainer, with no nuchal collar or spotting on upperparts, scapulars and wing-coverts. **VOICE** Territorial call is an unmistakable series of 4–8 very loud bubbling trills: ***prrrrrooh prrrrrooh prrrrrooh prrrrrooh...***, either rising or even-pitched, each separated by 3–6 s pauses. **HABITAT** Broadleaved evergreen forest; up to 200 m. **RANGE**

& STATUS Resident Sumatra, Borneo. **SE Asia** Rare to scarce resident S Thailand, Peninsular Malaysia. **BREEDING** February–June. **Nest** Thick downy pad attached to upperside of horizontal branch, 1.2–1.3 m above ground. **Eggs** 1.

461 GOULD'S FROGMOUTH

Batrachostomus stellatus Plate **54**

IDENTIFICATION 23–26.5 cm. Monotypic. **Adult** Similar to females of other small frogmouths but has white spots on tips of wing-coverts, dark rufous-brown scales and no blackish markings on underparts (belly and undertail-coverts whiter), and more protruding bill. Eyes bronze to pale yellow. Some (possibly mainly females) are much darker, colder and browner on upperparts, underpart-scaling colder and browner. **Juvenile** When newly fledged, rather uniform with dark bars on upperparts and paler streaks on central underparts. Gradually attains adult body features. **VOICE** Male territorial call is an eerie, rather weak whistled ***woah-weeo***, with falling second note; occasionally only gives ***weeo*** notes. Female utters growling notes and rapid series of high-pitched yapping ***wow*** notes, 3–5 higher-pitched ***wek*** notes and descending whistled ***weeeoh***. **HABITAT** Broadleaved evergreen forest; up to 185 m. **RANGE & STATUS** Resident Sumatra, Borneo. **SE Asia** Scarce to fairly common resident S Thailand, Peninsular Malaysia. **BREEDING** February–August. **Nest** Small downy pad, attached to upperside of horizontal branch, 1.2–1.5 m above ground. **Eggs** 1; white; 28–30.5 x 20–21.7 mm.

462 HODGSON'S FROGMOUTH

Batrachostomus hodgsoni Plate **54**

IDENTIFICATION 24.5–27.5 cm. *B.h.indochinae*: **Male** Similar to Blyth's (particularly *B.a.affinis*) but more heavily vermiculated and marked with black on upperparts and breast; breast lacks rufous, bill smaller, less protruding. Eyes light brown to yellowish-brown. Note voice and range. **Female** Very like Blyth's but paler, with much more prominent white markings on underparts, whiter lores and smaller bill. **Juvenile** Upperparts barred blackish and pale brown with warm tinge, no nuchal collar, underparts similar to upperparts, grading to plainer and whiter from lower breast to vent. **Other subspecies in SE Asia** *B.h.hodgsoni* (W,N Myanmar). **VOICE** Series of up to 10 variable, soft, slightly trilled rising whistles: ***whaaeee***, ***whaaow***, ***wheeow*** or ***wheeow-a***, each one separated by 1–7 s intervals. Also a series of soft, chuckling ***whoo*** notes. **HABITAT** Broadleaved evergreen and mixed coniferous and evergreen forest, secondary growth; 900–1,900 m, rarely down to 305 m. **RANGE & STATUS** Resident NE India, E Bangladesh, SW China. **SE Asia** Uncommon resident W,N,E Myanmar, NW Thailand, Laos, C,S Annam. **BREEDING** March–September. **Nest** Small downy pad, attached to upperside of horizontal branch, 1.5–10 m above ground. **Eggs** 1–2; white; 23.6–31.1 x 16.3–22 mm.

463 BLYTH'S FROGMOUTH

Batrachostomus affinis Plate **54**

IDENTIFICATION 23–24 cm. *B.a.continentalis*: **Male** Warmish brown above, speckled, vermiculated and spotted buff, white and blackish, scapulars marked with whitish spots; has narrow white and black nuchal collar, underside buffy-white with more rufous throat and breast and with dark vermiculations/scales and large white markings on breast and belly. Eyes yellow. *B.a.affinis* (extreme S Thailand southward) is generally less rufous-tinged (particularly throat and breast), usually has more black markings on upperparts and underparts, more whitish vermiculations on crown, and typically a more contrastingly barred tail. See Hodgson's Frogmouth. Note voice, habitat and range. **Female** Rather dark rufous-chestnut with white and black marked nuchal collar, large white markings on scapulars and black-fringed white markings on lower throat and breast, tail faintly banded. See Hodgson's and Gould's Frogmouths. **Juvenile** Not well documented. **VOICE** Male utters a series of mournful, wavering whistles, ***tee-loo-eee*** (descending in middle and rising at end), or shorter ***loo-eee***. Also a series of ***KWAH-a*** or ***e-ah*** notes or more drawn-out ***kwaaha***, and loud falling ***whah*** or ***gwaa*** notes. Female utters a series of unusual descending laughing notes: ***grra-ga-ga-ga*** or ***kerrr-ker-ker***. **HABITAT** Broadleaved evergreen and mixed deciduous forest, forest edge, secondary growth; up to 800 m. **RANGE & STATUS** Resident Sumatra, Borneo. **SE Asia** Uncommon resident Tenasserim, Thailand (except C), Peninsular Malaysia, south-west Cambodia, N,S Laos, Cochinchina. **BREEDING** February–June. **Nest** Small downy pad, attached to upperside of horizontal branch, 1–7 m above ground. **Eggs** 1–2; white; 25–31.9 x 16–21.2 mm.

CAPRIMULGIDAE: EUROSTOPODINAE: Eared-nightjars

Worldwide 7 species. SE Asia 2 species. Like typical nightjars but characterised by lack of pronounced rictal bristles around gape, thickly feathered legs, strong feet, lack of whitish wing and tail markings, and feathers of rear sides of crown elongated into 'ear-tufts'.

464 MALAYSIAN EARED-NIGHTJAR

Eurostopodus temminckii Plate **54**

IDENTIFICATION 25–28 cm. Monotypic. **Male** Darkest nightjar in region. Similar to Great Eared-nightjar but much smaller, crown darker, ear-tufts less pronounced, tail darker and less contrastingly barred. Note voice, habitat and range. **Female** Apparently tends to be more rufescent. **Juvenile** Upperparts somewhat paler, warmer and less heavily vermiculated, pale bars on underparts duller. **VOICE** Similar to Great but introductory note louder and always audible, second note shorter: ***tut wee-ow***, repeated 5–7 times after shortish intervals. **HABITAT** Open areas and clearings in or near broadleaved evergreen forest; up to 1,000 m. **RANGE & STATUS** Resident Sumatra, Borneo. **SE Asia** Scarce to uncommon resident extreme S Thailand, Peninsular Malaysia, Singapore. **BREEDING** January–August. **Nest** None; eggs laid on ground. **Eggs** 1–2; white with grey and brown spots, speckles and scrawls; 34.4–34.5 x 25.5–27.8 mm (av.).

465 GREAT EARED-NIGHTJAR

Eurostopodus macrotis Plate **54**

IDENTIFICATION 40.5–41 cm. *E.m.cerviniceps*: **Adult** Much larger, longer-winged and longer-tailed than other nightjars, lacking whitish or pale markings on wings and tail. Buffish-grey crown with a few dark central markings, contrasting with dark sides of head; has narrow collar (white on throat, pale buff on nape), rest of upperside relatively dark and uniform, with slightly contrasting paler scapulars with small black markings and distinctly chestnut-tinged shoulders; uppertail distinctly and broadly barred blackish and buffish-brown, throat and upper breast blackish-brown, rest of underparts barred blackish and pale buff; has pronounced ear-tufts (sometimes visible when perched). See Malaysian Eared-nightjar. **Juvenile** Paler, plainer and buffier above than adult; has fewer, more contrasting markings, paler and plainer and more pale chestnut-tinged wing-coverts, diffuse barring below. **VOICE** Long double whistle, introduced by a short well-separated note (audible at close range): ***put PEE-OUW***. **HABITAT & BEHAVIOUR** Open areas and clearings in or near broadleaved evergreen and deciduous forest, freshwater swamp forest; up to 1,220 m. Flies with slow leisurely wingbeats, resembling smaller harriers *Circus*; often feeds high in air. **RANGE & STATUS** Resident NE,S India, Bangladesh, SW China, W Sumatra (Simeulue I), Philippines, Sulawesi. **SE Asia** Fairly common resident Myanmar, Thailand (except C), north Peninsular Malaysia, Indochina (except N,S Annam). **BREEDING** January–June. **Nest** None; eggs laid on ground. **Eggs** 1; creamy-white to deep salmon-pink, sparsely marbled and blotched pale reddish-brown over pale lavender to grey undermarkings; 42.1 x 30.5 mm (av.).

CAPRIMULGIDAE: CAPRIMULGINAE: Typical nightjars

Worldwide c.74 species. SE Asia 4 species. Medium-sized with long, pointed wings, large gapes surrounded by extensive rictal bristles and soft, cryptically patterned plumage. Middle toe pectinated. Crepuscular and nocturnal. Perch on ground or lengthwise on branch in daytime but difficult to detect due to well-camouflaged plumage. Food is caught in flight and consists mainly of moths, beetles, crickets and other flying insects.

466 GREY NIGHTJAR

Caprimulgus jotaka Plate **54**

IDENTIFICATION 28–32 cm. *C.j.hazarae*: **Male** A relatively variable nightjar. Somewhat similar to Large-tailed but somewhat smaller and darker crowned; lacks any rufescent tinge on nape, has somewhat heavier black vermiculations on upperparts, duller and darker ear-coverts and throat, usually less whitish coloration on lower throat (may take the appearance of two distinct whitish patches or 'headlights'), breast darker, scapulars less contrasting, with black, buff and whitish bars and vermiculations overall; shows less obvious whitish to buff bars across wing-coverts. In flight shows smaller white wing- and tail-patches. See Malaysian Eared, Indian and Savanna Nightjars. **Female** Wing-patches smaller than on male and buff (may be very indistinct), lacks any obvious pale tail-patches (outer feathers are narrowly tipped with brownish-white to brownish-buff). See Savanna. **Juvenile** Like female but somewhat paler; primaries and secondaries narrowly tipped pale warm buff. **Other subspecies in SE Asia** *C.j.jotaka* (wintering form). **VOICE** Male territorial call is a rapid ***tuctuctuctuctuctuctuc...***, in bursts of up to 16 notes (3–4 per second), repeated monotonously after short pauses. Also a fast, ***deep quor-quor-quor*** (possibly by females only). **HABITAT** Open broadleaved evergreen and coniferous forest, secondary growth; also open areas and gardens (non-breeders); up to 2,565 m (breeds above 600 m). **RANGE & STATUS** Breeds S,SE Siberia, Ussuriland, NE Pakistan, NW,N,NE Indian subcontinent, S,SE Tibet, China (except NW,N), N,S Korea, Japan, Palau Is; more northerly populations winter to south, Greater Sundas, Philippines. **SE Asia** Uncommon to fairly common resident (? breeding visitor to some northern areas) W,N,S(east),E Myanmar, NW,W Thailand, east S Laos, W Tonkin; ? Peninsular Malaysia, E Tonkin. Uncommon winter visitor (except SW,C,S Myanmar, SE Thailand). Scarce to uncommon passage migrant Singapore (may winter), E Tonkin. **BREEDING** March–July. **Nest** None; eggs laid on ground. **Eggs** 1–2; creamy-white, spotted and marbled greyish-brown, dark grey and umber; 30.7 x 22.7 mm (av.).

467 LARGE-TAILED NIGHTJAR

Caprimulgus macrurus Plate **54**

IDENTIFICATION 31.5–33 cm. *C.m.bimaculatus*: **Male** Quite variable. Told by combination of size, relatively pale crown with dark median stripe, prominent row of black scapulars with pronounced broad buff to whitish-buff outer fringes, and prominent large white patches on primaries and distal part of outertail-feathers (obvious in flight). Has rather prominent whitish to buff bars across wing-coverts, large area of white to buffish-white across lower throat, brownish-grey tail with rather uneven dark bars, and pale buffish-brown remainder of underparts with blackish bars; often shows strong rufescent tinge to nape. See Grey, Indian and Savanna Nightjars. **Female** Tends to be paler and greyer on upperparts and breast, wing-patches smaller and buff, tail-patches much duller, buffish to buffish-white. See Savanna. **Juvenile** Paler and buffier than female, with duller tail-patches. **VOICE** Territorial males utter a monotonous series of loud, resonant ***chaunk*** notes, roughly one per second (variable); may be preceded by low grunting or croaking notes. May give deep harsh ***chuck*** when flushed. **HABITAT & BEHAVIOUR** Open forest, secondary growth, cultivation; up to 2,135 m. During display, male runs around female, while wagging tail from side-to-side, and puffs-out throat. **RANGE & STATUS** Resident NE Pakistan, N and north-eastern Indian subcontinent, SW China, Sundas, Flores Sea Is, Philippines (Palawan), Moluccas, New Guinea region, Bismarck Archipelago, N Australia. **SE Asia** Common resident throughout. **BREEDING** January–November. Sometimes multi-brooded. **Nest** None; eggs laid on ground. **Eggs** 2; deep salmon-pink to pale buff, indistinctly blotched pale reddish-brown to dark pink over pale lavender undermarkings; 31.3 x 22.6 mm (av.).

468 INDIAN NIGHTJAR

Caprimulgus asiaticus Plate **54**

IDENTIFICATION 23–24 cm. *C.a.asiaticus* (incl. *siamensis*): **Male** Smallest and palest nightjar in region. Resembles Large-tailed Nightjar but much smaller and shorter-tailed, crown a shade paler with more contrasting dark median stripe; has distinct buff nuchal collar with dark markings, dark-centred scapulars even more extensively fringed on both sides with whitish-buff, typically has large round white patch on each side of throat ('headlights'), rest of throat and breast generally paler with fewer blackish markings. In flight, white to buffy-white wing- and tail-patches slightly smaller. See Grey and Savanna Nightjars. **Female** Tends to have slightly smaller wing- and tail-patches than male. **Juvenile** Somewhat paler and plainer; upperpart streaking restricted to hindcrown and nape, scapular fringes more rufous. **VOICE** Territorial call of male is a distinctive, knocking ***chuk-chuk-chuk-chuk-k'k'k'roo*** (2–4 ***chuk*** notes), like ping-pong ball bouncing to rest on hard surface. Short sharp ***quit-quit*** or ***chuk-chuk*** may be given in flight. **HABITAT** Open dry forest, semi-desert, dry scrub and cultivation; up to 915 m. **RANGE & STATUS** Resident Indian subcontinent. **SE Asia** Uncommon to common resident SW,W,C,E,S Myanmar, north Tenasserim, Thailand (except S), Cambodia, N,S Laos, C,S Annam. **BREEDING** February–April. **Nest** None; eggs laid on ground. **Eggs** 2; cream-coloured or pale pink to salmon-pink, spotted and blotched reddish-brown and inky-purple; 26.5 x 19.9 mm (av.).

469 SAVANNA NIGHTJAR

Caprimulgus affinis Plate **54**

IDENTIFICATION 25–25.5 cm. *C.a.monticolus*: **Male** Quite variable. From other nightjars by rather uniform, heavily vermiculated brownish-grey upperparts, lack of defined dark median crown-stripe, ill-defined scapular pattern (though often with some contrasting warm buff feather-fringes) and white outertail-feathers (often dusky-tipped). Has indistinct, broken buffish nuchal band, little or no pale moustachial line, normally distinct roundish white to buffish-white patch on each side of throat, and large white wing-patches. **Female** Wing-patches slightly smaller and buff, no obvious pale or whitish tail markings. See Grey and Large-tailed Nightjars. **Juvenile** Somewhat paler. **Other subspecies in SE Asia** *C.a.affinis* (south Peninsular Malaysia, Singapore) is smaller and more heavily marked with smaller wing-patches. **VOICE** Male territorial call is a constantly repeated, loud rasping ***chaweez*** or ***chweep***. **HABITAT** Open dry dipterocarp, pine and broadleaved evergreen forest, grassland, scrub; up to 915 m. **RANGE & STATUS** Resident N Pakistan, Indian subcontinent (except islands), S China, Taiwan, Sundas, Philippines, Sulawesi. **SE Asia** Scarce to uncommon resident Myanmar (except SW,E), Thailand (except S), Peninsular Malaysia (north to Perak, Pulau Pinang), Singapore, Cambodia, Laos, C,S Annam, Cochinchina. **BREEDING** March–August. **Nest** None; eggs laid on ground. **Eggs** 2; pale salmon-pink to deep salmon-red, spotted and blotched deep red and reddish-brown over lavender-pink undermarkings; 30.2 x 22.1 mm (av.; *monticolus*).

APODIDAE: APODINAE: Typical swifts

Worldwide c.82 species. SE Asia 18 species. Aerial habits distinctive. Compact and thick-necked with streamlined bodies, short legs, long wings, small broad bills and large gapes adapted for catching airborne insects. **Swiftlets** (*Hydrochous, Collocalia, Aerodramus*) Mostly small with rather fluttering flight on bowed, paddle-shaped wings; have only slightly forked to almost square-ended tails. Most utter echo-locating clicks when flying inside dark caves etc. **Needletails** (*Rhaphidura, Hirundapus*) Broad-winged and heavy-bodied with short squarish tails that have needle-like projections (not visible in field). Typical swifts (*Cypsiurus, Tachymarptis, Apus*) Medium-sized with long, tapered wings and deeply forked to square-cut tails.

470 **WATERFALL SWIFT**
Hydrochous gigas Plate **55**

IDENTIFICATION 16–16.5 cm. Monotypic. **Adult** Wing 156–168 mm, tail 58–66 mm (outer), 51–52 mm (inner). Distinctly larger than other swiftlets (larger and longer-winged than House Swift) and darker overall, rump concolorous with rest of upperparts, tail more deeply forked. **Juvenile** In hand, has less obvious greyish-white undertail-covert fringes and pointed central tail-feathers. **VOICE** Sharp wicker notes and loud twittering. **HABITAT & BEHAVIOUR** Waterfalls and nearby areas of broadleaved evergreen forest; 800–1,500 m. Gregarious in breeding areas. **RANGE & STATUS** Resident (subject to some movements) W Sumatra, W Java, N Borneo. **SE Asia** Found (status uncertain) Peninsular Malaysia; ? Singapore. **BREEDING** April–May. **Nest** Truncated cone made of liverworts, moss, fern material and some feathers (fixed with saliva), attached to rock ledge, within permanent spray zone of waterfall. **Eggs** 1; white; 26–31.5 x 17.7–19.1 mm.

471 **GLOSSY SWIFTLET**
Collocalia esculenta Plate **55**

IDENTIFICATION 10 cm. *C.e.cyanoptila*: **Adult** Tiny size, blackish upperside with variable dark blue to dark greenish gloss and whitish vent diagnostic. Throat and upper breast dark greyish, sometimes with some paler feather-fringing. *C.e.elachyptera* (Mergui Archipelago, Tenasserim) is supposed to have greener gloss on upperside, paler throat, narrow pale fringes to rump feathers and extensive white rami on upperparts (particularly nape and rump). **Juvenile** Said to have stronger greenish gloss on upperside, and pale grey/buff fringing on wing-feathers. **VOICE** Short, grating, twittering sounds at the nest. **HABITAT & BEHAVIOUR** Forested and open areas; up to 1,900 m. Flight rapid and rather bat-like. **RANGE & STATUS** Resident Andaman and Nicobar Is, Sumatra, Borneo, Philippines, Wallacea (except Lombok), New Guinea region, New Caledonia, Vanuatu; occasionally NE Australasia. **SE Asia** Scarce to common resident south Tenasserim, S Thailand, Peninsular Malaysia. Scarce non breeding visitor (formerly resident) Singapore. **BREEDING** All year. **Nest** Cup of vegetable matter, moss etc., bound with saliva, attached to rock-face, building wall, underside of bridge or culvert. **Eggs** 2; white; 17.5 x 11.2 mm (av.; Andaman and Nicobar Is).

472 **HIMALAYAN SWIFTLET**
Aerodramus brevirostris Plate **55**

IDENTIFICATION 13–14 cm. *A.b.brevirostris* (N Myanmar): **Adult** Wing 128–137 mm, tail 56–62 mm (outer), 46–52 mm (inner). Blackish-brown above with faint blue-green gloss and paler greyish rump-band, rather uniform throat and breast with darker chin (slightly paler than ear-coverts); belly and vent mid-brownish-grey. In hand, has darker shaft-streaks on breast, belly and undertail-coverts, variable leg-feathering and white rami (sometimes lacking?). *A.b.inopina* (winter visitor N Myanmar?, W Tonkin) has longer wings and dark rump-band (almost same as rest of upperparts); so-called *innominata* (visitor) is intermediate between *brevirostris* and *inopina*; *rogersi* (breeds E Myanmar, NW,W Thailand; visitor S Thailand; recorded C Laos) is like *innominata* but smaller and shorter-winged; differs (in hand) by lack of white rami and unfeathered or lightly feathered legs. Very similar to several other swiftlets. Larger and longer-winged than Germain's and Edible-nest with darker underparts and deeper tail-notch. From Black-nest by obvious tail-notch. Note range. **Juvenile** Rump-band somewhat less defined, legs more sparsely feathered. **VOICE** Low rattling twitter. **HABITAT** Forested and open areas; up to 3,100 m (mostly mountains). **RANGE & STATUS** Breeds (mostly resident) N,NE Indian subcontinent, S Tibet, SW,W,C China; recorded in winter Maldive Is, Andaman Is, Java, possibly Sumatra. **SE Asia** Uncommon resident W,E Myanmar, W,NW Thailand. Uncommon winter visitor Thailand (except SE), Peninsular Malaysia, Singapore, E Tonkin, N,C,S Annam, Cochinchina. Also recorded on passage Peninsular Malaysia, Singapore. Recorded (status uncertain) N,C,S(east) Myanmar, N Laos, W Tonkin. **BREEDING** April–May. Colonial. **Nest** Small cup of moss and saliva, attached to vertical rock-face. **Eggs** 2; white; 21.8 x 14.6 mm (av.; *brevirostris*). **NOTE** *C.b.rogersi* is sometimes treated as a full species.

473 **BLACK-NEST SWIFTLET**
Aerodramus maximus Plate **55**

IDENTIFICATION 12–13.5 cm. *A.m.maximus*: **Adult** Wing 126–133 mm, tail 47–52.5 mm (outer), 43.5–48 mm (inner). Throat and breast rather uniform, or grading to darker chin (all a little paler than ear-coverts). In hand, shows faint darker shaft-streaks on throat and densely feathered legs. Like Germain's Swiftlet but somewhat bulkier, bigger-headed and longer-winged; little or no tail-notch, tends to show narrower, duller rump-band (can be almost same). From paler-rumped subspecies of Himalayan Swiftlet by typically more clearly defined pale rump-band and lack of obvious tail-notch. See Edible-nest Swiftlet. *A.m.lowi* (recorded Gunung Benom, Peninsular Malaysia) has rump concolorous with upperparts. See Himalayan (*A.b.inopina*). **VOICE** Similar to Himalayan. **HABITAT** Open areas, sometimes over forest, offshore islets, urban areas; up to 1,830 m. **RANGE & STATUS** Resident (subject to minor movements) Sumatra, W Java, Borneo. Recorded Philippines (Palawan). **SE Asia** Uncommon to locally common resident Tenasserim, S Thailand, Peninsular Malaysia, Singapore. **BREEDING** February–September. Colonial, mainly on offshore islets. **Nest** Bracket of parents' feathers and saliva, attached to vertical rock-face. **Eggs** 1; white; 21.9–23.5 x 15.3–16 mm.

474 **EDIBLE-NEST SWIFTLET**
Aerodramus fuciphaga Plate **55**

IDENTIFICATION 11–12 cm. *A.f.inexpectata*: **Adult** Wing 113–121 mm, tail 47–53 mm (outer), 41–46 mm (inner). In hand, legs always unfeathered. Difficult to separate from Germain's Swiftlet but rump-band narrower and darker, less contrasting (but see Germain's *A.g.amechana*), lower breast, belly and undertail-coverts somewhat darker. Smaller and shorter-winged than Himalayan and Black-nest (nominate) Swiftlets, tail-notch shallower than former and deeper than latter, underparts slightly paler than former and slightly darker and browner than latter. **HABITAT** Open areas, sometimes over forest and mangroves, offshore islets; lowlands. **RANGE & STATUS** Resident Andaman and Nicobar Is, Sundas, Tanahjampea Is (Flores Sea). **SE Asia** Recorded Tenasserim (vagrant or offshore breeder). **BREEDING** Andaman and Nicobar Is: March–April. Colonial, mostly on offshore islets. **Nest** Bracket of solidified saliva, attached to vertical rock-face. **Eggs** 2; white; 20.2 x 13.6 mm (av.).

475 **GERMAIN'S SWIFTLET**

Aerodramus germani Plate **55**

IDENTIFICATION 11.5–12.5 cm. *A.g.germani*: **Adult** Wing 113–123.5 mm, tail 50–53 mm (outer), 43–46 mm (inner). Has palest underparts (particularly lower throat and upper breast) and rump-band (whitish-grey with blackish shaft-streaks) of any swiftlet in region. Lower throat and upper breast paler than chin and obviously paler than ear-coverts. In hand, legs always unfeathered, has similar (but fainter) dark shaft-streaks on throat to Black-nest Swiftlet. *A.g.amechana* (extreme S Thailand, Peninsular Malaysia [except NW], Singapore) is said to have slightly duller rump-band. See Black-nest and Edible-nest Swiftlets. **HABITAT** Open areas, sometimes over forest, offshore islets; up to 1,300 m, mostly lowlands. **RANGE & STATUS** Resident N Borneo, S Philippines. **SE Asia** Locally common to common resident (mostly near coasts) S Myanmar, Tenasserim, W(south),C,S Thailand, Peninsular Malaysia, Singapore, Cambodia, Vietnam (except W Tonkin). Recorded (status uncertain) N,C Laos. **BREEDING** February–December. Colonial; mostly on coast and offshore islets. **Nest** Like Edible-nest. **Eggs** 2; white; 21.1 x 13.2 mm (av.; *germani*).

476 **SILVER-RUMPED NEEDLETAIL**

Rhaphidura leucopygialis Plate **55**

IDENTIFICATION 11 cm. Monotypic. **Adult** Unmistakable: small but robust, blackish with prominent silvery-white lower back, rump and uppertail-coverts, short square-cut tail and very broad paddle-shaped wings, deeply pinched-in at base and pointed at tips. Has dark bluish gloss on upperparts and bare shafts (spines) extending from tail-tip (not usually visible in field). **Juvenile** Less glossy. **VOICE** High-pitched ***tirrr-tirrr*** and rapid chattering, recalling House Swift. **HABITAT & BEHAVIOUR** Broadleaved evergreen forest, clearings; up to 1,250 m. Flight fluttery and erratic. **RANGE & STATUS** Resident Sumatra, Java, Borneo. **SE Asia** Uncommon to common resident south Tenasserim, S Thailand, Peninsular Malaysia; formerly Singapore. **BREEDING** February–April. Sometimes (perhaps always) colonial. **Nest** Tree-hollow. Otherwise undocumented.

477 **WHITE-THROATED NEEDLETAIL**

Hirundapus caudacutus Plate **55**

IDENTIFICATION 21–22 cm. *H.c.caudacutus*: **Adult** Like Silver-backed Needletail but with clearly defined white throat and short white band from extreme forehead to upper lores; pale saddle tends to be more extensive and more whitish; has distinctive white tertial markings (not usually visible in field). *H.c.nudipes* (recorded NW,SE Thailand, Cambodia at least) has all-blackish forehead and lores, while whitish on saddle tends to be restricted to lower mantle/upper back. See Brown-backed Needletail. **Juvenile** Forehead and lores greyish-brown, upperparts less glossy, white of lower flanks and undertail-coverts marked with blackish. **VOICE** Rapid insect-like chattering: ***trp-trp-trp-trp-trp-trp...*** **HABITAT** Forested and open areas; up to 2,300 m. **RANGE & STATUS** Breeds S,SE Siberia, Ussuriland, Sakhalin, Kuril Is, NE China, N Korea, Japan; winters New Guinea, Australia, occasionally New Zealand; widespread passage migrant in intervening areas. Mostly resident (subject to relatively local movements), NE Pakistan (breeding visitor only), NW,N,NE Indian subcontinent, SW,W China. **SE Asia** Scarce passage migrant Tenasserim, W,NW,NE,SE Thailand, Peninsular Malaysia, Cambodia, N,C Laos, W,E Tonkin, S Annam. Vagrant Singapore. Recorded (status uncertain) W Myanmar. **BREEDING** April–May (India). **Nest** Tree-hollow. **Eggs** 2–7; white; 31.2 x 22.4 mm (*nudipes*).

478 **SILVER-BACKED NEEDLETAIL**

Hirundapus cochinchinensis Plate **55**

IDENTIFICATION 20.5–22 cm. *H.c.cochinchinensis*: **Adult** Similar to Brown-backed Needletail but centre of saddle distinctly brownish-white (whiter when bleached/worn), throat paler, brownish-grey (may appear whitish), lacks white spot on lores. Can show whitish inner webs to tertials when bleached/worn. See White-throated Needletail. **Juvenile** Has some dark brown markings on white of lower flanks and undertail-coverts. **VOICE** Soft, rippling trill. **HABITAT** Forested and open areas, large rivers in or near forest; up to 3,355 m. **RANGE & STATUS** Breeds Nepal, NE India, S China, Taiwan; some populations migratory. Passage migrant S China (Hong Kong at least). Recorded (status uncertain) Sumatra, W Java. **SE Asia** Uncommon resident W Tonkin, S Annam, Cochinchina. Uncommon winter visitor and passage migrant Peninsular Malaysia. Scarce passage migrant Singapore, W,E Tonkin. Scarce to uncommon (status uncertain) N Myanmar, Thailand (except W,S), Cambodia, Laos, N,C Annam. **BREEDING Nest** Undocumented. **Eggs** White; 28.1 x 21 mm (in oviduct).

479 **BROWN-BACKED NEEDLETAIL**

Hirundapus giganteus Plate **55**

IDENTIFICATION 21–24.5 cm. *H.g.indicus*: **Adult** Very large and bulky swift, blackish above and dark brown below with distinctive brown saddle on lower mantle to back and pronounced white V on lower flanks and vent. Has white spot on lores (visible at close range), chin and centre of upper throat often somewhat paler. See White-throated and Silver-backed Needletails. *H.g.giganteus* (resident south Tenasserim, S Thailand, Peninsular Malaysia) is larger (24.5–26.5 cm) and lacks white spot on lores. **Juvenile** White loral spot less pronounced, white of lower flanks and vent faintly marked darker. **VOICE** Rippling trill, similar to White-throated but slower. Also, squeaky ***cirrwiet***, repeated 2–3 times, and thin squeaky ***chiek***. **HABITAT & BEHAVIOUR** Forested and open areas; up to 2,000 m. Has fast gliding flight, wings make loud whooshing sound when zooming overhead. **RANGE & STATUS** Resident SW,NE India, SE Bangladesh, Sri Lanka, Andaman Is, Greater Sundas, Philippines (Palawan). **SE Asia** Uncommon to common resident throughout. Possibly only breeding visitor to some (more northerly) areas. Also uncommon winter visitor and passage migrant Peninsular Malaysia, Singapore. breeding February–March. **Nest** Natural tree-hollow or old woodpecker nest; 15 m above ground, or higher. **Eggs** S India: 3–5; white, becoming stained; 29.6 x 22.2 mm (av.; *indicus*).

480 **ASIAN PALM-SWIFT**

Cypsiurus balasiensis Plate **55**

IDENTIFICATION 11–12 cm. *C.b.infumatus*: **Adult** Small size, rather uniform greyish-brown plumage, long slender wings and long, deeply forked tail distinctive. Tail appears long, narrow and pointed when closed; rump, sides of head, breast and belly somewhat paler, throat paler still. Resembles some swiftlets when viewed distantly and tail closed, but wings and tail much more slender, never glides with wings held stiffly below horizontal. See treeswifts. **Juvenile** Tail somewhat less sharply and deeply forked. **VOICE** Frequently uttered, high-pitched trilled ***sisisi-soo-soo*** or ***deedle-ee-dee***. **HABITAT & BEHAVIOUR** Open country, urban areas, often near palm trees; up to 1,525 m. Often found in small, highly active groups. **RANGE & STATUS** Resident India, Nepal, Bhutan, Bangladesh, Sri Lanka, SW,S China, Greater Sundas, Philippines. **SE Asia** Uncommon to common resident throughout. **BREEDING** December–August. Semi-colonial. **Nest** Bracket of seed-down and saliva, attached to palm leaf, sometimes thatched roof. **Eggs** 2–3; white; 17.1 x 11.7 mm (av.; *infumatus*).

481 **ALPINE SWIFT** *Tachymarptis melba* Plate **56**

IDENTIFICATION 20–23 cm. *T.m.nubifuga*: **Adult** Highly distinctive, large, long-winged swift; dark brown, with white throat (often hard to see at distance) and breast-/belly-patch. **Juvenile** Slightly darker above; extensive white fringing, particularly on wing-coverts (also shown by fresh adults in winter/spring). **VOICE** Twittering ***ti ti titititititititititititi-ti-ti-ti-ti ti tu tu***, accelerating, then decelerating/dropping slightly in pitch. Single ***zri*** and ***ziiu***. **HABITAT & BEHAVIOUR** Forested and open areas; found in plains. Rather slow, deep wingbeats compared to *Apus* swifts. **RANGE & STATUS** Breeds Africa, Madagascar, southern W Palearctic, C Asia, Middle East, Iran, Afghanistan, NW,N,NE Indian subcontinent; some northern populations winter W,E Africa, south to peninsular India, and some southern populations in E Africa. **SE Asia** Vagrant N Myanmar.

482 FORK-TAILED SWIFT
Apus pacificus Plate **55**

IDENTIFICATION 18–19.5 cm. *A.p.cooki*: **Adult** Relatively large size, long sickle-shaped wings, sharply forked tail (harder to see when closed) and blackish plumage with clear-cut narrow white rump-band (often hard to see at distance) diagnostic. Has slightly paler throat, indistinct whitish scales on rest of underparts, and darker shaft-streaks on rump-band and throat (all hard to see in field). *A.p.pacificus* (widespread passage migrant) has upperparts browner and almost glossless, head and nape slightly paler than mantle with narrow greyish-white feather margins, white rump-band broader, dark shaft-streaks on rump-band and throat narrower and less obvious, throat whiter; *kanoi* (visitor Indochina, Peninsular Malaysia) is said to be intermediate. See Dark-rumped and House Swifts. **Juvenile** Secondaries and inner primaries narrowly tipped whitish. **VOICE** Shrill ***sreee***. **HABITAT** Forested and open areas; up to 2,750 m. **RANGE & STATUS** Breeds C,E Palearctic, NE Pakistan, NW,N,NE Indian subcontinent, S,E Tibet, China (except NW), Taiwan, N,S Korea, Japan, possibly N Philippines; more northerly populations winter to south, Lesser Sundas, Australia, possibly Sumatra, Java, S New Guinea; widespread on passage in intervening areas. **SE Asia** Uncommon resident (subject to some movements) E Myanmar, W,NW Thailand, N,C Laos, W Tonkin, C Annam. Uncommon to locally common winter visitor Thailand (except C), Peninsular Malaysia, Cambodia. Uncommon to locally common passage migrant C,S Thailand, Peninsular Malaysia, Singapore, Cambodia, W,E Tonkin, N,S Annam, Cochinchina. Recorded (status uncertain) Myanmar (except E), S Laos. **BREEDING** April–July. Colonial. **Nest** Half-cup of grass and vegetable matter, attached to sloping rock-face, sometimes old house-martin *Delichon* nest. **Eggs** 2–3; white; 22.7 x 15 mm (av.; India).

483 DARK-RUMPED SWIFT
Apus acuticauda Plate **55**

IDENTIFICATION 17–18 cm. Monotypic. **Adult** Very similar to Fork-tailed Swift but lacks white rump-band, tends to have darker, more heavily marked throat, sharper tail-fork with narrower and more pointed outer feathers. **Juvenile** Probably differs as Fork-tailed. **VOICE** Very high-pitched, rapid, sibilant, quavering ***tsrr'i'i'i*** and ***tsrr'i'i'i'is'it*** etc. at nest sites. **HABITAT** Forested areas, cliffs; 1,000–2,300 m. **RANGE & STATUS** Breeds Bhutan, NE India; winter movements little known. **SE Asia** Scarce (status unknown, but could be resident) N Myanmar. Recorded in winter (vagrant?) NW Thailand. **BREEDING** India: March–April. Colonial. **Nest** Shallow cup of grass and saliva, on ledge in cliff fissure. **Eggs** 2–4; white; 26 x 16.3 mm (av.).

484 HOUSE SWIFT *Apus affinis* Plate **55**

IDENTIFICATION 14–15 cm. *A.a.subfurcatus*: **Adult** Overall blackish plumage with whitish throat and broad clear white rump-band and only slightly notched, squarish tail-tip diagnostic. Has narrow dark shaft-streaks on rump-band (not visible in field). See Fork-tailed Swift and swiftlets. **Juvenile** Tends to have paler-fringed wing-feathers. **VOICE** Harsh rippling trilled ***der-der-der-dit-derdiddidoo***, rapid shrill ***siksik-siksik-sik-sik-siksiksiksik...*** etc., and staccato screaming. **HABITAT** Urban and open areas, sometimes over forest; up to 2,300 m. **RANGE & STATUS** Resident (subject to some relatively local movements) Africa, Middle East, Iran, Afghanistan, SE Uzbekistan, Tajikistan, Indian subcontinent, southern China, Taiwan, S Korea, S Japan, Sundas, Philippines, Sulawesi. **SE Asia** Common resident, subject to some local movements (except SW,W Myanmar). Recorded (status uncertain) SW,W Myanmar. **BREEDING** All year. Colonial. **Nest** Untidy globular mass (several nests often joined), sometimes old swallow nest, on various man-made structures or natural rock-face. **Eggs** 2–4; white; 22.7 x 14.9 mm (av.).

APODIDAE: HEMIPROCNINAE: Treeswifts

Worldwide 4 species. SE Asia 3 species. Similar to other swifts but very slender and long-tailed, more colourful and sexually dimorphic. Less exclusively aerial in habits, often perching on tree-branches and undertaking sallies for insect food.

485 CRESTED TREESWIFT
Hemiprocne coronata Plate **56**

IDENTIFICATION 21–23 cm. Monotypic. **Male** Distinctive. Slightly bluish-tinged grey with darker forehead crest (often held erect) and wings, paler grey lower throat and breast, whitish belly and vent, and pale rufous sides of head, upper throat and sides of throat. Primaries and secondaries browner than blue-black wing-coverts, tertials paler, brownish-grey. In flight shows very long, slender wings and long, slender, very deeply forked tail; nape to rump appears uniformly grey, underwing-coverts (apart from leading edge) concolorous with rest of underwing. See Grey-rumped Treeswift. Note range. **Female** Lacks rufous on head, has blackish lores, blackish-slate ear-coverts, very narrow dusky-whitish line along edge of crown (over eye), thin dusky-whitish moustachial line (to below rear ear-coverts) and uniform grey throat. **Juvenile** Has extensive white feather-fringing on upperparts (less obvious on mantle), paler lower back and rump, dusky-whitish underpart feathers with grey-brown subterminal bands and white tips, and broadly white-tipped tertials and flight-feathers. **VOICE** Harsh, rather explosive ***kee-kyew***, second note lower. When perched, ***kip-KEE-kep***. **HABITAT & BEHAVIOUR** Open deciduous forest, forested and open areas; up to 1,400 m. Regularly perches upright on exposed branches. **RANGE & STATUS** Resident Indian subcontinent (except W,NW and Pakistan), SW China. **SE Asia** Uncommon to common resident Myanmar, NW,W,NE Thailand, Cambodia, Laos, C,S Annam, Cochinchina. **BREEDING** March–June. **Nest** Very small, shallow cup/platform, attached to upperside or side of horizontal branch; 4–18 m above ground. **Eggs** 1; greyish-white; 23.7 x 17.1 mm (av.).

486 GREY-RUMPED TREESWIFT
Hemiprocne longipennis Plate **56**

IDENTIFICATION 18–21.5 cm. *H.l.harterti*: **Male** Similar to Crested Treeswift but crown, nape and mantle dark glossy green, lores blackish, ear-coverts dull dark chestnut, throat all grey, tertials mostly whitish-grey contrasting sharply with rest of wing, tail-tip falls short of primary tips at rest. In flight, grey back and rump contrast with dark mantle; has contrasting blackish underwing-coverts. **Female** Ear-coverts blackish. **Juvenile** Upperpart feathers extensively fringed rusty-brown (less so on rump), underpart feathers off-white with irregular brown subterminal bands and white tips; scapulars, flight-feathers and tail broadly tipped whitish. **VOICE** Harsh, piercing ***ki, ki-ki-ki-kew*** and staccato ***chi-chi-chi-chew***, sometimes a disyllabic ***too-eit***, with more metallic second note. **HABITAT & BEHAVIOUR** Forested and open areas; up to 1,220 m. Regularly perches upright on exposed branches. **RANGE & STATUS** Resident Greater Sundas, Lombok, Sulawesi, Sula Is, SW Philippines (Sibutu I). **SE Asia** Fairly common to common resident south Tenasserim, W(south),S Thailand, Peninsular Malaysia, Singapore. **BREEDING** February–September. **Nest** Like Crested; 4.5–12 m above ground. **Eggs** 1; greyish-white; 25.9 x 17.3 (av.) mm.

487 **WHISKERED TREESWIFT**
Hemiprocne comata Plate **56**
IDENTIFICATION 15–16.5 cm. *H.c.comata*: **Male** Recalls other treeswifts but smaller, body mostly olive-bronze with dark glossy blue crown, nape, upper throat, sides of throat and wings, long white supercilium (to nape) and malar/moustachial streak (chin to nape) and white vent. Ear-coverts dull rufous-chestnut, only slight non-erectile crest. In flight, shows dark underwing with contrasting, broadly white-tipped secondaries and inner primaries, and demarcated white vent. **Female** Ear-coverts blackish. **Juvenile** Similar to respective adults, but supercilium and malar/moustachial less bold; male with smaller, duller ear-patch. **VOICE** High-pitched shrill chattering ***she-she-she-she-shoo-shee***, with higher penultimate note. Plaintive ***chew*** when perched. **HABITAT & BEHAVIOUR** Clearings in broadleaved evergreen forest, forest edge; up to 1,200 m. Spends much more time perched than other treeswifts, normally flying only short distances to feed. **RANGE & STATUS** Resident Sumatra, Borneo, Philippines. **SE Asia** Uncommon to common resident south Tenasserim, W(south),S Thailand, Peninsular Malaysia. Rare non-breeding visitor (formerly resident) Singapore. **BREEDING** February–September. **Nest** Like Crested; 9–40 m above ground. **Eggs** 1; white.

TROGONIDAE: Trogons

Worldwide 40 species. SE Asia 7 species. Medium-sized with stout, broad bills, short legs, short rounded wings and long, straight, square-ended tails. Sexually dimorphic with males brightly coloured. Generally shy and retiring, typically adopting upright posture on branch, spending long periods sitting motionless. Feed on caterpillars, beetles, grasshoppers, cicadas and other large insects; also some leaves and berries. Make short sorties to catch flying insects.

488 **RED-NAPED TROGON**
Harpactes kasumba Plate **56**
IDENTIFICATION 31.5–34.5 cm. *H.k.kasumba*: **Male** Like Diard's Trogon but crown all black, has broad red nuchal patch meeting broad cobalt-blue facial skin, well-demarcated narrow white breast-band, dull golden-buffish upperparts, usually redder underparts and unmarked white on undertail. **Female** Resembles Diard's but lower breast to vent brownish-buff, white on underside of tail unmarked. See Cinnamon-rumped Trogon. **Juvenile** Both sexes similar to female. Male soon attains patches of adult plumage. **VOICE** Male territorial call is a subdued but rather harsh, evenly pitched, 3–6 note ***kau kau kau kau...***, lower-pitched and more spaced than that of Diard's. Female gives a quiet whirring rattle. **HABITAT** Broadleaved evergreen and freshwater swamp forest, bamboo; up to 550 m. **RANGE & STATUS** Resident Sumatra, Borneo. **SE Asia** Scarce to locally fairly common resident S Thailand, Peninsular Malaysia; formerly Singapore. **BREEDING** July–August. **Nest** Rounded cavity in dead tree (stump); 1.2 m above ground. **Eggs** 2; white.

489 **DIARD'S TROGON**
Harpactes diardii Plate **56**
IDENTIFICATION 32.5–35 cm. *H.d.sumatranus*: **Male** Distinctive. Forehead, throat and upper breast black, faint narrow breast-band pale pink, rest of underparts reddish-pink, hindcrown blackish with maroon wash, narrow nuchal band pale pink, rest of upperparts dark warm brown. Has violet to violet-blue orbital skin, undertail white with dark vermiculations/speckles and black border. See Red-naped Trogon. **Female** Combination of rather uniform dull brown head, upper breast and mantle, contrasting reddish-pink to pink belly and undertail pattern (like male) distinctive. Back to uppertail-coverts more rufescent, lower breast more buffish-brown, undertail-coverts buffish-brown mixed with pink. See Red-naped and Red-headed Trogons. **Juvenile** Both sexes similar to female. Male soon attains patches of adult plumage. **VOICE** Male territorial call is a series of 10–12 ***kau*** notes, either with the second somewhat higher than the first and the rest descending, with the last few slower, or else all evenly spaced. **HABITAT & BEHAVIOUR** Middle storey of broadleaved evergreen forest; up to 600 m, rarely to 915 m in Peninsular Malaysia. Very unobtrusive. **RANGE & STATUS** Resident Sumatra, Borneo. **SE Asia** Uncommon to locally common resident S Thailand, Peninsular Malaysia; formerly Singapore. **BREEDING** February–June. **Nest** Cavity in dead tree (stump); 1.2–3 m above ground. **Eggs** Glossy, cream-white. Otherwise undocumented.

490 **CINNAMON-RUMPED TROGON**
Harpactes orrhophaeus Plate **56**
IDENTIFICATION 25.5–28 cm. *H.o.orrhophaeus*: **Male** Like Scarlet-rumped Trogon but larger and thicker-billed, with no pinkish-red on rump and uppertail-coverts but much pinker underparts. **Female** Recalls Scarlet-rumped but lacks pink in plumage, brown of head darker and richer with contrasting deep rufous lores and orbital area, throat mostly blackish-brown with dull rufous centre, upperparts slightly duller, less rufescent. See Red-naped Trogon. **Juvenile** Both sexes similar to female but pale wing vermiculations much broader, upperparts duller, greyer-tinged and more uniform, with more extensive deep rufous on lores, sides of head and throat. From Scarlet-rumped by extensive and contrasting deep rufous on sides of head, more uniform upperparts (including rump and uppertail-coverts) and darker, less buffish breast. **VOICE** Territorial call of male is a weak, descending, 3–4 note ***taup taup taup...*** or ***ta'up ta'up ta'up ta'up***, with each note inflected downwards. **HABITAT & BEHAVIOUR** Lower to middle storey of broadleaved evergreen forest; up to 200 m. Very shy. **RANGE & STATUS** Resident Sumatra, Borneo. **SE Asia** Scarce to uncommon resident S Thailand, Peninsular Malaysia. **BREEDING** March–July. **Nest** Cavity in dead tree; 1–1.5 m above ground. **Eggs** 2; white. Otherwise undocumented.

491 **SCARLET-RUMPED TROGON**
Harpactes duvaucelii Plate **56**
IDENTIFICATION 23.5–26.5 cm. Monotypic. **Male** Combination of black head and bright, deep pinkish-red rump, uppertail-coverts and underparts diagnostic. Has prominent pale blue skin above/in front of eye, and white (black-bordered) undertail. See Cinnamon-rumped Trogon. **Female** Much duller with rather drab dark brown head, paler and slightly warmer lores, orbital area and throat, mostly buffy-brown rump and uppertail-coverts (mixed with pink), deep buffy-brown breast, and pinkish to reddish-pink belly and vent. **Juvenile** Both sexes like female but lack obvious pink in plumage; rump and uppertail-coverts paler and more rufescent than rest of upperparts, underparts rufescent-buff with pale buff belly-centre. Males show some pink below and soon attain patches of adult plumage. **VOICE** Male territorial call is a distinctive, accelerating, descending ***teuk teuk teuk-euk-euk-euk-euk-euk-euk-euk-euk-euk...*** Otherwise, vibrating ***chrrrrrrr***, ***charr***, and ***chowrrr*** etc. **HABITAT** Middle to lower storey of broadleaved evergreen forest; up to 400 m, sometimes to 1,065 m in Peninsular Malaysia. **RANGE & STATUS** Resident Sumatra, Borneo. **SE Asia** Uncommon to locally common resident southern Tenasserim, W(south),S Thailand, Peninsular Malaysia. **BREEDING** February–June. **Nest** Cavity in dead tree. **Eggs** 2; whitish; 24.1 x 19.8 mm (av.).

492 **ORANGE-BREASTED TROGON**
Harpactes oreskios Plate **56**
IDENTIFICATION 26.5–31.5 cm. *H.o.stellae*: **Male** Unmistakable. Head and upper breast greenish-olive (throat and central breast more yellowish-tinged), upperparts chest-

nut-maroon, underparts yellowish-orange with paler, yellower vent. **Female** Head, mantle and back uniform drab olive-brownish, rump and uppertail-coverts duller and paler, throat and upper breast distinctly grey-washed, rest of underparts paler and yellower, has broader pale bars on wings. **Juvenile male** Similar to juvenile female but mantle, scapulars and back chestnut. **Juvenile female** Head, upperparts and breast tinged rufous-chestnut, pale wing-bars much broader (broader than dark bars), belly and vent paler (may appear whitish). **Other subspecies in SE Asia** *H.o.uniformis* (south Tenasserim and S Thailand southward). **VOICE** Territorial call of male is a subdued, even, rather rapidly delivered 3–5 note ***teu-teu-teu...*** or ***tu-tau-tau-tau...*** Female may utter a slower, lower-pitched version. **HABITAT** Middle to upper storey of broadleaved evergreen, semi-evergreen and mixed deciduous forest, bamboo; up to 1,220 m. **RANGE & STATUS** Resident SW China, Greater Sundas. **SE Asia** Common resident (except W,N,C Myanmar, C Thailand, Singapore, W,E Tonkin, N Annam). **BREEDING** January–May. **Nest** Excavated hollow/cavity in dead tree (usually stump) or large bamboo; 0.6–4.5 m above ground. **Eggs** 2–3; cream-coloured to pale brownish-buff; 26.2–26.4 x 20.3–20.8 mm (*uniformis*).

493 RED-HEADED TROGON
Harpactes erythrocephalus Plate **56**

IDENTIFICATION 31–35.5 cm. *H.e.erythrocephalus*: **Male** Dark red head and upper breast, pinkish-red remainder of underparts and narrow white breast-band diagnostic. Wing vermiculations whitish, undertail white with black border. Largest in N Myanmar (*H.e.helenae*), smallest in S Thailand and Peninsular Malaysia (*H.e.chaseni*). **Female** Head and upper breast brown, concolorous with mantle, vermiculations on wings pale warmish brown. See Diard's Trogon. Note habitat and range. **Juvenile** Both sexes similar to female but head, breast and upperparts rufescent; has less red on underparts. Male has broad buff bars on wings and soon attains red on head and upper breast. **Other subspecies in SE Asia** *H.e.klossi* (SE Thailand, Cambodia), *annamensis* (NE Thailand, N[south-east],C,S Laos, N,C,S Annam, Cochinchina), *intermedius* (north-east N Laos, W,E Tonkin). **VOICE** Male territorial call is a deep, well-spaced, descending 4–5 note ***taup taup taup taup taup...*** Call is a coarse, rattling ***tewirr***. **HABITAT & BEHAVIOUR** Middle to upper storey of broadleaved evergreen forest, 305–2,590 m; 700–1,680 m in Peninsular Malaysia, locally down to 50 m in Indochina. Unobtrusive, spending long periods sitting motionless. **RANGE & STATUS** Resident N,NE Indian subcontinent, southern China, Sumatra. **SE Asia** Common resident (except C Thailand, Singapore); local S Thailand. **BREEDING** March–July. **Nest** Cavity in dead tree (often stump) or large bamboo, sometimes old woodpecker hole; 1.5–5 m above ground. **Eggs** 2–4; glossy whitish to pale buffish (possibly staining); 28.6 x 24 mm (av.; *erythrocephalus*).

494 WARD'S TROGON *Harpactes wardi* Plate 56

IDENTIFICATION 38 cm. Monotypic. **Male** Unmistakable. Head, upperparts and breast slaty with pinkish-maroon wash, forehead and sides of forecrown dark pinkish-red, lores and upper throat blackish, rest of underparts reddish-pink. Tail blackish with largely pink outer feathers (appears mostly pink from below), bill red (purplish at gape), orbital skin pale blue. **Female** Equally distinctive. Head, upperparts and breast mostly olive-brownish; has bright yellow forehead and sides of forecrown, pale yellow belly and vent, mostly pale yellow outertail-feathers, and yellowish-brown bill with mostly dark upper mandible (purplish at gape). **Juvenile female** Has warmer mantle and back, much broader warmish brown wing vermiculations, and less yellow on forecrown. Details not available for male. **VOICE** Male territorial call is a rapid series of loud, mellow ***klew*** notes, often accelerating and dropping in pitch towards end. Call is a harsh ***whirrur***. **HABITAT** Broadleaved evergreen forest; 1,830–2,620 m, sometimes down to 1,220 m in winter. **RANGE & STATUS** Resident Bhutan, NE India, SW China (W Yunnan). **SE Asia** Scarce to uncommon resident N Myanmar, W Tonkin. **BREEDING** April–May. Otherwise unrecorded.

CORACIIDAE: Rollers

Worldwide 13 species. SE Asia 2 species. Medium-sized and large-headed with stout bills. Usually solitary or in pairs. Strong fliers, with acrobatic rolling and tumbling display-flight. Feed on insects, reptiles and other small animals. Flying insects are taken during sorties from exposed perch.

495 INDIAN ROLLER
Coracias benghalensis Plate **58**

IDENTIFICATION 31.5–34.5 cm. *C.b.affinis*: **Adult** Brilliant dark purplish-blue and light turquoise wings and outertail-feathers diagnostic. When perched at distance, often appears rather drab brownish but has mostly turquoise crown (duller and darker in centre), greenish-olive (brownish-tinged) mantle, upper back, scapulars and tertials, dark purplish-blue rump, light turquoise uppertail-coverts and vent, and vinous-brown sides of head and underparts with more distinctly light purple throat (narrowly streaked paler) and lower belly; bill blackish. **Juvenile** Browner above than adult, including forehead and central crown (turquoise restricted to supercilium), sides of head and throat paler and browner (almost completely lacking purple) with whitish streaks, breast and belly paler and browner, the former buffier than in adult. **VOICE** Harsh, retching ***kyak***. **HABITAT & BEHAVIOUR** Open country, semi-desert, cultivation, coastal scrub, urban areas; up to 1,525 m. Often on telegraph wires and other exposed perches; drops to ground in search of food. **RANGE & STATUS** Resident Oman, S Iraq, Iran, Indian subcontinent, SW China. **SE Asia** Common resident (except Singapore, E Tonkin); coastal north-west and east Peninsular Malaysia only. **BREEDING** March–May. **Nest** None or pad in hole in dead tree, top of palm, building etc., sometimes old nest of other bird. **Eggs** 3–5; glossy, white; 34.7 x 27.9 mm (av.).

496 DOLLARBIRD *Eurystomus orientalis* Plate 58

IDENTIFICATION 27.5–31.5 cm. *E.o.orientalis*: **Adult** Distinctive. Head and mantle dark brown, rest of upperparts dark greenish-brown, throat mostly dark bluish-purple with lighter, bluer streaks, rest of underparts dark turquoise with darker greenish-brown breast, has thick red bill. Wing-coverts mostly dull turquoise, rest of wings and uppertail mostly black. Often appears all dark at distance. In flight, shows distinctive pale silvery-turquoise patch on primaries (may appear white). *E.o.abundus* (visitor Indochina, Peninsular Malaysia; breeding visitor E Tonkin at least) tends to have blacker-brown crown, nape and sides of head and mostly green mantle, producing more hooded appearance, and is supposed to show distinctive dark blue wash on flight-feathers. **Juvenile** Bill largely blackish, upperparts browner, head (including throat) initially all brown with little or no dark turquoise on throat-centre. **Other subspecies in SE Asia** *E.o.deignani* (NW,NE Thailand). **VOICE** A hoarse sharp rasping ***kreck, kreck...***; ***kak***, ***kiak*** etc., sometimes in rapid series. **HABITAT & BEHAVIOUR** Open broadleaved evergreen, semi-evergreen and deciduous forest, forest edge, clearings, plantations, mangroves, island forest; up to 1,500 m (mostly below 1,220 m). Often perches on dead tree-tops; mostly an aerial feeder. **RANGE & STATUS** Mostly resident N,NE Indian subcontinent, SW India, Sri Lanka, Andaman Is, Sundas, Philippines, Wallacea, New Guinea region, N Melanesia. Breeds China (except NW), SE Siberia, Ussuriland, N,S Korea, Japan; winters Greater Sundas. Breeds N,E Australia; non-breeding visitor Wallacea, New Guinea region, Bismarck Archipelago. **SE Asia** Uncommon to common resident Myanmar, Thailand (except C), Peninsular Malaysia, Singapore (scarce), Cambodia, S Laos, C,S Annam, Cochinchina. Fairly common breeding visitor N,C Laos, W,E Tonkin, N Annam. Fairly common winter visitor S Thailand, Peninsular Malaysia, Singapore. Also recorded on passage Peninsular Malaysia. **BREEDING** February–June. **Nest** Tree-hole, sometimes old woodpecker or barbet nest; 8–20 m above ground. **Eggs** 2–4; glossy, white; 36.3 x 28.2 mm (av.; *orientalis*).

ALCEDINIDAE: HALCYONINAE: Larger kingfishers

Worldwide 61 species. SE Asia 9 species. Smallish to medium-sized with long, heavy bills, large heads, fairly short tails and short legs. Plumages variable. Flight powerful and direct. Typically sit upright on exposed, mainly waterside perches. Most feed on fish and other aquatic animals but some spend much of their time away from water.

497 RUFOUS-COLLARED KINGFISHER

Actenoides concretus Plate **57**

IDENTIFICATION 24–25 cm. *A.c.concretus*: **Male** Combination of green crown, dark blue submoustachial stripe and upperside, and rufous nuchal collar and breast distinctive. Long black eyestripe (reaches nape), short buffish supercilium, paler blue back and rump and yellowish bill with dark ridge to upper mandible. **Female** Lower mantle, scapulars and wings mostly dull green with pale buffish speckles and teardrops on scapulars, wing-coverts and tertials. **Juvenile** Duller than respective adults, bill greyish-brown with dull yellowish tip and base of lower mandible. Males have heavily spotted mantle. **Other subspecies in SE Asia** *A.c.peristephes* (Tenasserim, S Thailand south to Trang). **VOICE** Territorial call is a long series of rising whistles: ***kwi-i kwi-i kwi-i kwi-i kwi-i...***, at rate of c.1 note per second. Also gives softer, more tremulous ***kwi-irr kwi-irr kwi-irr...*** etc. **HABITAT & BEHAVIOUR** Broadleaved evergreen forest, usually fairly close to water; up to 1,200 m. Unobtrusive, often sitting motionless for long periods. **RANGE & STATUS** Resident Sumatra, Borneo. **SE Asia** Scarce to uncommon resident south Tenasserim, W(south),S Thailand, Peninsular Malaysia; formerly recorded Singapore. **BREEDING** April–August. **Nest** Excavated burrow in sloping bank, sometimes hole in rotten tree. **Eggs** 2; white; 37 x 31 mm (Borneo).

498 BANDED KINGFISHER

Lacedo pulchella Plate **57**

IDENTIFICATION 21.5–24.5 cm. *L.p.amabilis*: **Male** Unmistakable. Tail relatively long, hindcrown and nape blue with blackish and whitish markings, rest of upperside barred blue to whitish and blackish, underparts whitish with warm buffish breast-band. Forecrown and sides of head chestnut, bill thick, red. **Female** Equally distinctive. Sides of head and upperparts barred rufous and blackish, underparts white with blackish scales on breast and flanks. **Juvenile male** Duller with fine dusky scales/bars on ear-coverts and breast, bill with pale tip, mainly brown upper mandible and orange lower mandible. **Juvenile female** Somewhat duller with heavy blackish scales/bars on underparts; bill differs as juvenile male. **Other subspecies in SE Asia** *L.p.deignani* (most of S Thailand), *pulchella* (extreme S Thailand, Peninsular Malaysia). **VOICE** Male territorial call is a long whistle, followed by series of up to 15 slow whistled disyllables: ***wheeeoo chi-wiu chi-wiu chi-wiu chi-wiu chi-wiu...***, gradually dying away. Other calls include a sharp ***wiak wiak...*** **HABITAT & BEHAVIOUR** Broadleaved evergreen and mixed deciduous forest, bamboo, often away from water; up to 1,100 m. Typically sits still for long periods. Slowly raises and lowers crown and nape feathers when alarmed. **RANGE & STATUS** Resident Sumatra, Java, Borneo. **SE Asia** Fairly common resident S,E(south) Myanmar, Tenasserim, Thailand (except C), Peninsular Malaysia, Cambodia, Laos, C,S Annam, Cochinchina. **BREEDING** March–June. **Nest** Excavated tunnel in bank, rotten tree, or arboreal termitarium; up to 3 m above ground. **Eggs** 2; white; 24.6 x 20.8 mm (av.; Borneo).

499 STORK-BILLED KINGFISHER

Pelargopsis capensis Plate **57**

IDENTIFICATION 37.5–41 cm. *P.c.burmanica*: **Male** Relatively large size, huge red bill, pale dull brownish crown and sides of head, greenish-blue mantle, wings and tail, and mostly warm buffish nuchal collar and underparts distinctive. Back to uppertail-coverts pale turquoise. In flight lacks whitish wing-patches shown by White-throated and Black-capped Kingfishers. *P.c.malaccensis* (S Thailand southward) has somewhat darker and more contrasting crown and sides of head and bluer (less turquoise) upperparts, wings and tail. See Brown-winged Kingfisher. **Female** Tends to be duller above. **Juvenile** Nape, neck, breast and flanks vermiculated dusky brownish (broader and denser on breast). **VOICE** Territorial call is a melancholy whistled ***iuu-iuu iuu-iuu iuu-iuu iuu-iuu iuu-iuu...***, with higher ***i*** syllables; often in duet with rasping calls. Loud whistled ***tree-trew***, with lower second note and explosive, dry cackling ***ke-ke-ke-ke-ke...*** or ***kek-ek-ek-ek...*** **HABITAT** Rivers and large waterbodies in or near broadleaved evergreen and mixed deciduous forest or open woodland, mangroves in some areas (particularly outside range of Brown-winged); up to 800 m. **RANGE & STATUS** Resident Indian subcontinent (except W and Pakistan), SW China, Greater Sundas, W Lesser Sundas, Philippines. **SE Asia** Uncommon to locally common resident (except W,E Tonkin, N Annam). **BREEDING** March–May. **Nest** Excavated tunnel in bank or ant-hill, sometimes natural cavity in tree; up to 6 m above ground. **Eggs** 2–5; glossy white; 36.3 x 31 mm (av.; *burmanica*).

500 BROWN-WINGED KINGFISHER

Pelargopsis amauroptera Plate **57**

IDENTIFICATION 36–37 cm. Monotypic. **Adult** Like Stork-billed Kingfisher but head orange-buff like underparts; mantle, wings and tail dark brown. **Juvenile** Nape, neck, breast and flanks vermiculated dusky brown, wings-coverts with narrow pale fringes. **VOICE** A loud, tremulous, descending ***tree treew-treew*** etc., and a loud deep ***cha-cha-cha-cha...*** Dry cackling sounds, similar to Stork-billed. **HABITAT** Mangroves, particularly old growth. **RANGE & STATUS** Resident E India, Bangladesh. **SE Asia** Fairly common to locally common coastal resident SW,S Myanmar, Tenasserim, west S Thailand, north-west Peninsular Malaysia. **BREEDING** India: March–April. **Nest** Excavated tunnel in bank; up to 4 m above ground/water. **Eggs** 3–4; white; 33.1–36 x 29.1–30.2 mm.

501 RUDDY KINGFISHER

Halcyon coromanda Plate **57**

IDENTIFICATION 26.5–27 cm. *H.c.coromanda*: **Adult** Overall bright rufescent plumage, violet-tinged upperside, bluish-white patch on lower back and rump, and thick, bright red bill diagnostic. *H.c.minor* (resident S Thailand southwards) has darker upperparts and much darker underparts, breast also violet-tinged, rump-patch larger. **Juvenile** Much browner above, lower back and rump brilliant blue (less whitish); has narrow dark brown bars on underparts and brownish-orange bill. **VOICE** Territorial call is a soft, rather hoarse tremulous ***tyuurrrrr*** or ***quirrr-r-r-r-r***, repeated after short intervals. **HABITAT** Mangroves, forest on islands, broadleaved evergreen forest near water; up to 900 m. **RANGE & STATUS** Resident Andaman Is, Sumatra, Java, Borneo, SW Philippines, Sulawesi, Sula Is. Breeds SC Nepal, NE Indian subcontinent, Andaman Is, SW,NE China, Taiwan, N,S Korea, Japan; more northerly populations winter to south, Sumatra, Borneo, Philippines, N Sulawesi. **SE Asia** Uncommon resident W,SE,S Thailand,

Peninsular Malaysia, Singapore. Uncommon breeding visitor N,C Annam. Uncommon winter visitor and passage migrant S Thailand, Peninsular Malaysia, Singapore (rare). Uncommon passage migrant NW,NE,C Thailand, N,C Laos, E Tonkin (may breed). Recorded (status uncertain, but probably resident in some coastal area) C,S,E Myanmar, Tenasserim, Cambodia. **BREEDING** March–July. **Nest** Excavated tunnel in bank. **Eggs** 4–6; glossy white; 27.3 x 23.2 mm (av.; *coromanda*).

502 WHITE-THROATED KINGFISHER

Halcyon smyrnensis Plate **57**

IDENTIFICATION 27.5–29.5 cm. *H.s.perpulchra*: **Adult** Dark chestnut head and belly and contrasting white throat and centre of breast diagnostic. Mantle to tail and most of wings turquoise-blue, wing-coverts dark chestnut and black, bill thick, dark red. In flight, shows large whitish patch on base of primaries. **Juvenile** Upperparts, wings and tail duller, has dark vermiculations on throat and breast, browner bill. **VOICE** Territorial call is a loud whinnying ***klilililili***. Also utters shrill, staccato, descending laughter: ***chake ake ake-ake-ake-ake...*** **HABITAT** Open habitats, secondary growth, cultivation; up to 1,525 m. **RANGE & STATUS** Resident NE Egypt, S Turkey, Middle East, Indian subcontinent, southern China, Taiwan, Philippines, Sumatra, W Java. **SE Asia** Common resident (subject to some dispersive movements) throughout. **BREEDING** October–June (possibly all year). **Nest** Excavated tunnel in bank. **Eggs** 3–7; glossy white; 29.2 x 26.7 mm (av.).

503 BLACK-CAPPED KINGFISHER

Halcyon pileata Plate **57**

IDENTIFICATION 29–31.5 cm. Monotypic. **Adult** Combination of black crown and sides of head, white collar, throat and breast and thick red bill diagnostic. Mantle to tail and wings deep blue with mostly black wing-coverts. In flight, shows very large white patch on primaries. **Juvenile** Blue parts duller, with small rufous-buff loral spot, collar buff-tinged, has dark vermiculations on sides of throat and breast (throat-sides sometimes streaked) and brownish-orange bill. **VOICE** A ringing, cackling ***kikikikikiki...***, higher-pitched than similar call of White-throated Kingfisher. **HABITAT** Mangroves, sea coasts, various inland and coastal wetlands, gardens; up to 1,525 m (mostly lowlands). **RANGE & STATUS** Breeds C,E and southern China, Taiwan, N,S Korea; more northerly populations winter to south, Indian subcontinent (except W,NW), Sumatra, Borneo, Philippines, rarely Java, N Sulawesi. **SE Asia** Fairly common to common winter visitor throughout. Also recorded on passage N Laos, E Tonkin, Peninsular Malaysia. Recorded oversummering W Thailand, Peninsular Malaysia, Singapore. **BREEDING** April–May. **Nest** Excavated tunnel in bank. **Eggs** 4–5; white; 29.7 x 26.4 mm (av.).

504 COLLARED KINGFISHER

Todiramphus chloris Plate **57**

IDENTIFICATION 24–26 cm. *T.c.armstrongi*: **Adult** Blue crown, sides of head, mantle to tail and wings and contrasting white collar and underparts distinctive. Crown, sides of head, mantle and scapulars strongly but variably washed turquoise; has whitish supercilium from eye forwards, faint creamy-buffish wash on lower flanks and no white on primaries. Bill black with extensively pale yellowish lower mandible. *T.c.humii* (Myanmar and S Thailand southwards) is bluer above; *davisoni* (Coco Is, off S Myanmar) is smaller with blackish-green sides of head, dusky-olive upper mantle and buff-tinged underparts. See Sacred Kingfisher. **Juvenile** Upperparts duller and greener-tinged, wing-coverts narrowly fringed buff, has dark vermiculations on collar and breast; often slight buffish-brown tinge to collar, sides of breast and flanks. **VOICE** Deliberate, loud, nasal shrieking ***kick kyew, kick kyew...*** (***kick*** notes rising, ***kyew*** notes falling). Loud, shrill ***krerk krerk krerk krerk...*** or ***kek-kek-kek-kek...***, descending slightly and often ending with characteristic ***jee-jaw*** notes. **HABITAT** Mangroves, various coastal wetland habitats, cultivation, gardens and parks, sometimes large rivers and marshes; lowlands but up to 1,300 m on migration. **RANGE & STATUS** Resident southern Red Sea coast, UAE, Oman, SW,E India, Bangladesh, Andaman and Nicobar Is, Greater Sundas, Philippines, Wallacea, Micronesia, New Guinea region and N Australia to Samoa and Tonga. **SE Asia** Common coastal resident (except E Tonkin, N Annam). Also locally inland NE Thailand, Peninsular Malaysia and along Mekong R in Cambodia, Cochinchina; formerly C,S Laos. Recorded on passage Peninsular Malaysia. Recorded (status uncertain) E Tonkin. **BREEDING** December–August. **Nest** Excavated hole in tree-ant nest, termitaria, dead tree or bank; up to 6–10 m above ground. **Eggs** 3–7; glossy white; 29.3 x 24.5 mm (av.; *davisoni*).

505 SACRED KINGFISHER

Todiramphus sanctus Plate **57**

IDENTIFICATION 18–23 cm. *T.s.sanctus*: **Adult** Like Collared Kingfisher but considerably smaller, sides of head blackish-green (but see *T.c.davisoni*), loral patch distinctly buff, nape and flanks obviously washed buff. **Juvenile** Differs from Collared as adult. **VOICE** Rapid short high-pitched ***ki-ki-ki-ki*** (usually 4–5 notes) or high-pitched, squealing, reeling ***schssk-schssk-schssk*** (usually 1–4 notes), and drawn-out low squealing ***kreee-el kreee-el kreee-el***. **HABITAT** Mangroves, more open coastal habitats, cultivation, gardens; lowlands. **RANGE & STATUS** Breeds Australia, New Zealand, Lord Howe Is, Norfolk Is, New Caledonia, Loyalty Is, E Solomon Is (occasional). Breeds Australia; southern birds spend southern winter to north, N Melanesia, New Guinea region, Wallacea, Greater Sundas. **SE Asia** Vagrant Singapore (sight record).

ALCEDINIDAE: ALCEDININAE: Smaller kingfishers

Worldwide 24 species. SE Asia 6 species. **Dwarf kingfishers** (*Ceyx*) Similar to 'blue kingfishers' but with predominantly rufous to reddish plumage. Often feed on insects and other small animals some distance from water. **Blue kingfishers** (*Alcedo*) Small to smallish with long dagger-like bills, rather large heads, compact bodies and short tail and legs. All have brilliant blue, rufous and white in plumage. Flight rapid and direct. Typically sit upright on waterside perches. Catch fish and other aquatic animals by plunge-diving.

506 BLACK-BACKED KINGFISHER

Ceyx erithaca Plate **57**

IDENTIFICATION 12.5–14 cm. *C.e.erithaca*: **Adult** Very small size, bright red bill and rufous, lilac and yellowish plumage with blackish-blue mantle and scapulars and dark wings distinctive. Has blue patches on forehead and to rear of ear-coverts. See Rufous-backed Kingfisher. **Juvenile** Underparts duller, more whitish, often with brownish wash across breast, bill duller. **VOICE** Sharp metallic piping, weaker and higher-pitched than Blue-eared Kingfisher; usually given in flight. Contact calls include a weak, shrill ***tit-sreet*** and ***tit-tit***. **HABITAT** Vicinity of small streams and pools in broadleaved evergreen forest, sometimes gardens and mangroves on migration; up to 915 m. **RANGE & STATUS** Breeds SW,NE India, Bhutan, Bangladesh, Sri Lanka, Andaman and Nicobar Is, SW,S China, Sumatra, Java, Borneo, Philippines; some northern populations winter to SE and S. **SE Asia** Uncommon resident Myanmar (except W,E), Thailand (except C and southern S), Laos, E Tonkin, N,C Annam, Cochinchina (likely to be breeding visitor only to some northern areas). Uncommon to fairly common winter visitor S Thailand, Peninsular Malaysia, Singapore. Uncommon passage migrant S Myanmar, C Thailand, Peninsular Malaysia, Singapore. Recorded (status uncertain) Cambodia. **BREEDING** April–August. **Nest** Excavated tunnel in bank, termite mound or ant-hill. **Eggs** 2–7; glossy white; 18.9 x 15.6 mm (av.).

507 **RUFOUS-BACKED KINGFISHER**
Ceyx rufidorsa Plate **57**
IDENTIFICATION 12.5–14.5 cm. *C.r.rufidorsa*: **Adult** Like Black-backed Kingfisher but mantle and scapulars all rufous, lacks blackish-blue on forehead and ear-coverts. **Juvenile** Underparts duller, more whitish, often with brownish wash across breast; bill duller. **VOICE** Similar to Black-backed. A soft high insect-like ***tjie-tjie-tjie***, usually in flight, and a shrill ***tsriet-siet***. **HABITAT** Vicinity of small streams and pools in broadleaved evergreen forest, sometimes mangroves; up to 455 m. **RANGE & STATUS** Resident Greater Sundas, W Lesser Sundas, Philippines. **SE Asia** Uncommon resident south Tenasserim, W,S Thailand, Peninsular Malaysia. Uncommon non-breeding visitor Singapore. Recorded (status uncertain) NE Thailand (Khao Yai NP). **BREEDING** April–August. **Nest** Excavated tunnel in bank, termite mound or ant-hill. **Eggs** 2–3; glossy white; 20–22.1 x 16.7–17.9 mm.

508 **BLUE-BANDED KINGFISHER**
Alcedo euryzona Plate **57**
IDENTIFICATION 20–20.5 cm. *A.e.peninsulae*: **Male** Relatively small size, mostly dull dark brownish wings and blue breast-band diagnostic. Bill largely blackish. **Female** Similar to Common Kingfisher but larger and bulkier, crown, scapulars and wings much duller and browner, blue stripe down centre of upperparts whiter, lacks rufous-chestnut on ear-coverts behind eye. Bill has mostly dull reddish lower mandible. See Blyth's and Blue-eared Kingfishers. Note range. **Juvenile male** Shows more rufous on belly than adult. **VOICE** Similar to Common but less shrill. **HABITAT** Medium-sized and larger streams in broadleaved evergreen forest, sometimes smaller streams; up to 825 m. **RANGE & STATUS** Resident Sumatra, Java, Borneo. **SE Asia** Scarce to uncommon resident Tenasserim, W,S Thailand, Peninsular Malaysia. **BREEDING** January–June. **Nest** Excavated tunnel in bank. **Eggs** 3–5; white.

509 **BLUE-EARED KINGFISHER**
Alcedo meninting Plate **57**
IDENTIFICATION 15.5–16.5 cm. *A.m.verreauxii*: **Male** Similar to Common Kingfisher but slightly smaller, ear-coverts blue, upperside generally much deeper blue without turquoise tinge, underparts much deeper orange-rufous. Bill mostly blackish. See female Blue-banded and much larger Blyth's Kingfisher. *A.m.coltarti* (W,N,C,S Myanmar, NW,NE[northwest] Thailand, Indochina) has strongly turquoise bars on crown and nape, spots on scapulars and wing-coverts and stripe down centre of upperparts; *scintillans* (S[east] Myanmar, Tenasserim, W,SE,NE[south-west] Thailand) is roughly intermediate. **Female** Bill has mostly scarlet-reddish lower mandible. **Juvenile** Cheeks and ear-coverts rufous, breast distinctly dark-scaled, bill initially mostly reddish. **VOICE** Typical call is higher-pitched and shorter than in Common, and often given singly. **HABITAT** Streams, smaller rivers and pools in broadleaved evergreen and mixed deciduous forest, mangroves; up to 915 m. **RANGE & STATUS** Resident N,NE Indian subcontinent, SW,E India, Sri Lanka, Andaman Is, SW China, Greater Sundas, Philippines. **SE Asia** Fairly common resident (except E Myanmar, C Thailand, W,E Tonkin); scarce Singapore. **BREEDING** February–August. **Nest** Excavated tunnel in bank or termite mound. **Eggs** 3–8; glossy white; 20.3 x 17.6 mm (av.; *coltarti*).

510 **COMMON KINGFISHER**
Alcedo atthis Plate **57**
IDENTIFICATION 16–18 cm. *A.a.bengalensis*: **Male** Combination of smallish size, rufous ear-coverts, strong turquoise tinge to blue of upperparts and relatively pale rufous underparts distinctive. In flight, turquoise strip down upperparts very conspicuous. Bill mostly blackish. See Blue-eared, Blyth's and female Blue-banded Kingfishers. **Female** Base or most of lower mandible orange-reddish. **Juvenile** Both sexes have underparts paler with dusky wash across breast, base of lower mandible orange-reddish. **VOICE** Usually 2–3 shrill, high-pitched piping notes, particularly in flight. **HABITAT** Streams in open and wooded areas, various inland and coastal wetlands (tends to avoid denser forest); up to 1,830 m. **RANGE & STATUS** Breeds N Africa, Palearctic, Indian subcontinent, SE Tibet, China, Taiwan, N,S Korea, Japan, Wallacea, New Guinea region, N Melanesia; some northern populations winter to south, N Africa, Middle East, Greater Sundas, Philippines, N Sulawesi, Sula Is, N Moluccas. **SE Asia** Uncommon resident Myanmar, W,NW,NE Thailand, Peninsular Malaysia, N Laos, E Tonkin, N Annam. Common winter visitor throughout. Also recorded on passage NW Thailand, Peninsular Malaysia. **BREEDING** January–July. **Nest** Excavated tunnel in bank. **Eggs** 2–7; glossy white; 20.9 x 17.6 mm (av.).

511 **BLYTH'S KINGFISHER**
Alcedo hercules Plate **57**
IDENTIFICATION 22–23 cm. Monotypic. **Male** Resembles Blue-eared Kingfisher (*A.m.verreauxii*) but larger with much longer, heavier, black bill, darker and duller crown and wings but with light blue speckles on crown, nape and wing-coverts. See Common Kingfisher. **Female** Base of lower mandible reddish. **Juvenile** Unrecorded. **VOICE** Call is hoarser than that of Common, closer to Blue-eared but much louder. **HABITAT** Larger streams or smaller rivers in broadleaved evergreen forest, secondary growth; 50–1,220 m. **RANGE & STATUS** Resident E Nepal, NE Indian subcontinent, SW,S China. **SE Asia** Scarce to locally common resident W,N,S Myanmar, Laos, W Tonkin, N,C,S Annam. Recorded (status uncertain) NW Thailand. **BREEDING** February–May. **Nest** Excavated tunnel in riverbank; 1.2–2.5 m above water. **Eggs** 4–6; glossy white; 26.1 x 21.7 mm (av.).

ALCEDINIDAE: CERYLINAE: Pied kingfishers

Worldwide 9 species. SE Asia 2 species. Characterised by predominantly black-and-white plumage, and prominent crests.

512 **CRESTED KINGFISHER**
Ceryle lugubris Plate **57**
IDENTIFICATION 38–41.5 cm. *C.l.guttulata*: **Male** Distinctive. Relatively very large; upperside densely speckled and barred blackish and white, underparts white with dark grey and blackish streaks and speckles along malar line and across breast, and blackish bars on flanks. Has uneven tufted crest, often some pale, washed-out chestnut markings on malar area and breast. In flight, wings appear rather uniform with white underwing-coverts. See Pied Kingfisher. **Female** Underwing-coverts pale rufous; tends to lack pale chestnut markings on underparts. **Juvenile** Similar to female but sides of neck, breast, flanks and undertail-coverts washed pale rufous. **VOICE** A loud squeaky ***aick*** or indignant ***kek***, particularly in flight; rapidly repeated raucous grating notes. **HABITAT** Large streams, medium-sized rivers, lakes, in or near forested areas; up to 1,830 m. **RANGE & STATUS** Resident NE Pakistan, NW,N,NE Indian subcontinent, C,NE and southern China, N,S Korea, Japan. **SE Asia** Scarce to uncommon resident Myanmar, W,NW Thailand, Laos, Vietnam (except Cochinchina). **BREEDING** March–May. **Nest** Excavated hole in bank. **Eggs** 4–5; white; 38.5 x 32.5 mm (av.).

513 **PIED KINGFISHER** *Ceryle rudis* Plate **57**
IDENTIFICATION 27–30.5 cm. *C.r.leucomelanura*: **Male** Complex black-and-white plumage, pronounced flattened crest, long white supercilium and white underparts with two black bands on side of breast distinctive. In flight, upperwing shows large white patches on flight-feathers and coverts, underwing mostly whitish, tail black with white tips and bases of outer feathers. See much larger Crested Kingfisher.

Female Has only one black breast-band. **Juvenile** Like female but feathers of lores, throat and breast fringed brownish, breast-band greyish-black, bill shorter. **VOICE** A high-pitched, chattering, rather squeaky ***kwik*** or ***kik***, repeated at irregular intervals, loud shrill ***chirruk chirruk...*** and high ***TREEtiti TREEtiti...***; particularly in flight. **HABITAT & BEHAVIOUR** Rivers, canals and lakes in open country, flooded fields; up to 915 m. Often hovers above water. **RANGE & STATUS** Resident sub-Saharan Africa, Egypt, S Turkey, Middle East, Indian subcontinent, southern China. **SE Asia** Fairly common resident (except southern S Thailand, Peninsular Malaysia, Singapore). **BREEDING** October–May. **Nest** Excavated hole in bank. **Eggs** 4–6; glossy white; 29.9 x 21.4 mm (av.).

MEROPIDAE: Bee-eaters

Worldwide 26 species. SE Asia 6 species. Smallish to medium-sized with long, narrow downcurved bills, rather long pointed wings and short legs. Sexes similar, all brightly coloured. Feed on bees and other insects caught mainly in flight. **Bearded bee-eaters** (*Nyctyornis*): Occur in wooded areas, sluggish and not gregarious, lack tail-streamers. **Typical bee-eaters** (*Merops*): Mostly in open country, graceful in flight and gregarious, all but one species with elongated central tail-feathers.

514 RED-BEARDED BEE-EATER
Nyctyornis amictus Plate **58**

IDENTIFICATION 32–34.5 cm. Monotypic. **Male** Similar to Blue-bearded but shaggy feathers of throat and central breast red, forecrown purplish-pink (sometimes with red on forehead), belly and vent pale green, undertail pale olive-yellow with broad black terminal band, green of plumage deeper-toned. **Female** Like male but forehead usually red. **Juvenile** Rather uniform deep green head, throat and breast, pale bluish line from lores across extreme forehead (indistinct), whiter belly and vent with dull yellowish tinge, duller undertail with narrower and duller terminal band. **VOICE** A loud hoarse gruff ***chachachacha...***, ***quo-qua-qua-qua*** and slightly descending, chattering ***kak kak-ka-ka-ka-ka...*** Also a deep guttural croaking ***aark*** and ***kwow*** or ***kwok*** and rattling ***kwak-wakoogoogoo***. **HABITAT & BEHAVIOUR** Broadleaved evergreen forest, rarely wooded gardens; up to 1,525 m (below 1,220 m in Peninsular Malaysia). Often in lower canopy. **RANGE & STATUS** Resident Sumatra, Borneo. **SE Asia** Uncommon to locally common resident Tenasserim, W,S Thailand, Peninsular Malaysia. **BREEDING** January–August. **Nest** Excavated tunnel in bank or termite mound. **Eggs** 3; white; 33–34 x 27.9–28.4 mm.

515 BLUE-BEARDED BEE-EATER
Nyctyornis athertoni Plate **58**

IDENTIFICATION 33–37 cm. *N.a.athertoni*: **Adult** Large size, thickset appearance and largely green plumage with blue forecrown (extent variable) and shaggy blue centre of throat and breast diagnostic. Upperparts variably washed light blue, belly and vent pale buffish-yellow with broad green streaks, undertail golden-yellowish with indistinct narrow dark terminal band. **Juvenile** Apparently shows some brown markings on crown and centre of throat, undertail more golden-brown. **VOICE** A loud deep guttural croaking and harsh cackling, including a purring ***grrew-grrew-grrew...***, harsh ***kow kow-kow kowkowkow...*** and repeated ***gikhu*** and ***gikh*** notes. **HABITAT & BEHAVIOUR** Broadleaved evergreen, semi-evergreen and mixed deciduous forest, freshwater swamp forest, rarely wooded gardens; up to 2,200 m. Usually in lower canopy. **RANGE & STATUS** Resident N,NE Indian subcontinent, E and peninsular India, SW,S China. **SE Asia** Fairly common to common resident (except south Tenasserim, C,S Thailand, Peninsular Malaysia, Singapore). **BREEDING** December–October. **Nest** Excavated tunnel in bank; up to 8 m above ground. **Eggs** 4–6; white; 28.2 x 25.4 mm (av.).

516 LITTLE GREEN BEE-EATER
Merops orientalis Plate **58**

IDENTIFICATION 19–20 cm (tail-prongs extend up to 6 cm more). *M.o.ferrugeiceps*: **Adult** Told by relatively small size, coppery-rufous crown to upper mantle (washed yellowish-green), green throat with light blue chin and cheeks (sometimes most of throat) and isolated black patch on upper breast. Uppertail green, variably edged pale blue. **Juvenile** Upperparts duller and paler, crown to upper mantle mostly green, throat and cheeks mostly pale creamy yellowish-buff, breast duller and paler with no black patch (may show vague dark blue line), belly and vent mixed with cream, no elongated central tail-feathers. See Chestnut-headed Bee-eater. **VOICE** A pleasant, rather monotonous trilling ***tree-tree-tree-tree...*** and staccato ***ti-ic*** or ***ti-ti-ti*** when alarmed. **HABITAT** Drier open country and cultivation, semi-desert, beach slacks and dunes; up to 1,600 m. **RANGE & STATUS** Resident W,NE Africa, Middle East, Iran, Indian subcontinent, SW China. **SE Asia** Common resident, subject to some minor movements (except south Tenasserim, S Thailand, Peninsular Malaysia, Singapore, W,E Tonkin). **BREEDING** February–May. Often loosely colonial. **Nest** Excavated tunnel in bank or bare ground. **Eggs** 4–7; white; 19.3 x 17.3 mm (av.).

517 BLUE-TAILED BEE-EATER
Merops philippinus Plate **58**

IDENTIFICATION 23–24 cm (tail-prongs up to 7.5 cm more). Monotypic. **Adult** Combination of bronze-green crown to back, pale yellowish upper throat and dull chestnut wash on lower throat and upper breast distinctive. Has pale blue line below broad black mask, pale blue line from above eye to extreme forehead (where faint), mid-blue rump and uppertail-coverts and blue (green-tinged) uppertail. See Blue-throated. **Juvenile** Crown and mantle darker and more bluish-green, yellowish-white upper throat with more restricted, washed-out chestnut on lower throat and upper breast, paler and bluer breast, no elongated central tail-feathers. **VOICE** Typical contact calls (particularly in flight) are a loud ***rillip rillip rillip...***, shorter ***trrrit trrrit trrrit...*** and rapid ***tri-tri-trip***, sometimes interspersed with stressed ***chip*** notes; sharp ***pit*** notes when perched. **HABITAT & BEHAVIOUR** Open country, cultivation, beach slacks, dunes, borders of large rivers, mangroves; up to 2,850 m (breeds in lowlands). Gregarious, often in large flocks at traditional roosting sites and during migration (when may occur over forest). **RANGE & STATUS** Resident Philippines, Sulawesi, Flores, New Guinea, New Britain. Breeds NE Pakistan, NW,N,NE,E India, Nepal, Bangladesh, E Sri Lanka, SW,S China; most populations winter to south, India, Sri Lanka, Sundas. **SE Asia** Uncommon to locally common resident (subject to some movements) SW,S Myanmar, Tenasserim, W,C,SE,S Thailand, Peninsular Malaysia (Penang I only), Cambodia, S Laos, S Annam, Cochinchina. Fairly common breeding visitor W,C,N(some winter),E Myanmar, NW Thailand. Uncommon to locally common resident or breeding visitor N,C Annam (mostly coastal). Fairly common to common winter visitor and passage migrant Tenasserim, NE(south),S Thailand, Peninsular Malaysia, Singapore, Cambodia. Fairly common to common passage migrant SW,W,S Myanmar, C Thailand, N Laos, E Tonkin. Recorded (status unclear) C Laos. **BREEDING** February–May. Colonial. **Nest** Excavated tunnel in bank or bare ground. **Eggs** 5–7; white; 23.2 x 20.1 mm (av.).

518 BLUE-THROATED BEE-EATER
Merops viridis Plate **58**

IDENTIFICATION 22.5–23.5 cm (tail-prongs extend up to 9 cm more). *M.v.viridis*: **Adult** Combination of dark chestnut crown to mantle and blue throat diagnostic. Breast plain bright green, rump and uppertail-coverts distinctly pale blue, uppertail blue. See Blue-tailed Bee-eater. **Juvenile** Crown to mantle deep green (crown and nape may be washed blue), throat paler than adult with whitish to pale buff chin, breast duller, lacks elongated central tail-feathers. Starts to attain adult plumage during first winter. **VOICE** Typical contact calls (richer than similar calls of Blue-tailed) include a liquid ***terrip-terrip-terrip...***, faster ***terrip-rrip-rrip*** and deeper ***trrurrip***; sharp ***chip*** when alarmed. **HABITAT & BEHAVIOUR** Open country, borders of large rivers, cultivation, sometimes parks and gardens; also forest clearings and edge, and mangroves (mainly non-breeders); up 800 m. Usually in flocks; often roosts in large numbers. **RANGE & STATUS** Resident (subject to local movements) Greater Sundas, Philippines. Breeds S,EC,SE China; winters to south. **SE Asia** Uncommon resident SE Thailand, Cambodia, S Annam (mostly coastal), Cochinchina (mostly coastal). Uncommon to locally common resident (with some local movements) and breeding visitor S Thailand, Peninsular Malaysia, Singapore. Uncommon to locally common resident or breeding visitor N,C Annam. Fairly common to common winter visitor and passage migrant SE,S Thailand, Peninsular Malaysia, Singapore, Cambodia. Uncommon to fairly common passage migrant W(south),C(southern),NE(south-west) Thailand, C Laos, E Tonkin, C Annam, Cochinchina. **BREEDING** March–August. Usually colonial. **Nest** Excavated tunnel in bare ground or bank. **Eggs** 3–6; white; 23.3 x 20.5 mm (mean).

519 **CHESTNUT-HEADED BEE-EATER**
Merops leschenaulti Plate **58**
IDENTIFICATION 21–22.5 cm. *M.l.leschenaulti*: **Adult** Combination of dull chestnut crown to mantle, pale yellow upper throat and lack of elongated central tail-feathers diagnostic. Centre of lower throat dull chestnut; has narrow black gorget along border of lower throat, greenish to pale bluish remainder of underparts, narrow black mask, pale blue rump and uppertail-coverts (particularly former) and green uppertail, edged blue. *M.l.andamanensis* (Coco Is, off S Myanmar) has dull chestnut on rear part of mask and some dull chestnut on upper sides of breast. **Juvenile** Crown and mantle green, with slightly paler band across upper mantle, variable amount of dull chestnut on hindcrown/nape (may be restricted to nape-sides), very washed-out chestnut on lower throat, less distinct gorget (mixed with bluish-green), paler underparts. See Blue-tailed Bee-eater. **VOICE** In flight a repeated soft bubbling ***pruuip***, ***pruik***, ***churit*** or ***djewy***; a soft airy ***chewy-chewy-chewy*** when perched. **HABITAT** Open broadleaved evergreen, semi-evergreen and mixed deciduous forest and bamboo (often along rivers), forest edge and clearings, coastal scrub, mangroves, island forest, sometimes plantations; up to 1,830 m. **RANGE & STATUS** Resident (subject to some movements) N,NE Indian subcontinent, E and peninsular India, Sri Lanka, Andaman Is, SW China, Sumatra, Java, Bali. **SE Asia** Fairly common to common resident (subject to some local movements) south to northern Peninsular Malaysia (except C[southern] Thailand, E Tonkin, N Annam). Uncommon non-breeding visitor C(southern) Thailand, southern Peninsular Malaysia. **BREEDING** January–May. Loosely colonial. **Nest** Excavated tunnel in sloping bank or level bare ground. **Eggs** 4–6; white; 21.7 x 19 mm (av.).

UPUPIDAE: Hoopoes

Worldwide 2 species. SE Asia 1 species. Medium-sized with erectile fan-like crests, broad rounded wings and slender, slightly downcurved bills. Flight lazy and undulating. Probes ground and animal dung for insect larvae etc.

520 **COMMON HOOPOE** *Upupa epops* Plate **58**
IDENTIFICATION 27–32.5 cm. *U.e.longirostris*: **Adult** Unmistakable. Bill long, narrow and downcurved; has dull pale rufous crown and long black-tipped crest (often held erect, fan-like), black-and-white to buff bars on back and rump, black wings and tail with broad white bars and dull dark pinkish throat to upper belly. Mantle pale warm brown (grey-tinged), uppertail-coverts white, flanks streaked black. **Juvenile** Somewhat duller and paler above and browner below. **Other subspecies in SE Asia** *U.e.saturata* (visitor). **VOICE** Soft ***hoop-hoop-hoop*** (sometimes two notes); recalls Himalayan Cuckoo. **HABITAT** Open country, semi-desert, scrub, open woodland, cultivation, gardens; up to 1,525 m, rarely 2,450 m. **RANGE & STATUS** Breeds Palearctic, Indian subcontinent, Tibet, China, N,S Korea; some northern populations winter to south, rarely Taiwan, Sumatra, N Borneo, Philippines (Palawan). **SE Asia** Scarce to common resident (except Peninsular Malaysia, Singapore). Scarce to uncommon winter visitor N,S Myanmar, NW,C,S Thailand. Also recorded on passage E Tonkin. Vagrant Peninsular Malaysia (formerly resident in north-west). **BREEDING** January–May. **Nest** Slight, in hole in tree or building. **Eggs** 5–7; pale greenish-blue to pale olive-brown (staining to dirty brown); 24.9 x 17.9 mm (av.).

BUCEROTIDAE: Hornbills

Worldwide 54 species. SE Asia 13 species. Relatively large to very large with broad wings, fairly long to long tails, short legs and large bills (mostly downcurved with distinctive casque). Plumage generally black and white, the white parts often stained yellowish to brownish by preen-oil. Flight slow and graceful with wings of larger species producing swan-like whistling sound. Mostly frugivorous, often gathering to feed in fruiting trees. Some species also take small animals, including snakes, other reptiles, snails and insects, as well as eggs and nestlings of other birds.

521 **NORTHERN BROWN HORNBILL**
Ptilolaemus austeni Plate **59**
IDENTIFICATION 73–74 cm. Monotypic. **Male** Combination of relatively small size, mostly brownish plumage with whitish throat, foreneck and upper breast and brownish-rufous lower breast and belly diagnostic. Bill pale yellowish with small casque, orbital skin pale blue, outertail-feathers tipped white. In flight, shows narrow white tips to secondaries and outer primaries. See Southern Brown and Bushy-crested Hornbills, but note range. **Female** Casque smaller, throat and underparts dark brownish (foreneck paler). **Juvenile** Similar to female (both sexes) but underparts dull greyish-brown, lacks white tips to primaries, bill smaller, orbital skin pinkish. **VOICE** Loud, piercing, airy yelps and squeals. Loud, yelping, upward-inflected ***klee-ah***. **HABITAT & BEHAVIOUR** Broadleaved evergreen forest, sometimes adjacent mixed deciduous forest, secondary forest; locally pine forest in Laos. Up to 1,500 m, locally 1,800 m. Usually found in flocks, sometimes quite large. **RANGE & STATUS** Resident NE India, SW China. **SE Asia** Uncommon to locally common resident W Myanmar, NW,NE Thailand, Cambodia, Laos, W,E Tonkin, N,C,S(north) Annam. **BREEDING** February–June. Co-operative. **Nest** Natural tree-cavity or old woodpecker hole, partially sealed with food debris, wood pulp etc.; 3.5–18 m above ground. **Eggs** 2–5; white, becoming stained; 46–57 x 33–35.4 mm.

522 **SOUTHERN BROWN HORNBILL**
Ptilolaemus tickelli Plate **59**
IDENTIFICATION 73–74 cm. Monotypic. **Male** Like Northern Brown Hornbill, but shows overall bright brownish-rufous throat and underparts. **Female** From Northern Brown by horn-coloured to blackish bill. **Juvenile** Differs from adults in similar way to Northern Brown. **VOICE** Similar to Northern Brown. **HABITAT** Broadleaved evergreen forest, sometimes adjacent mixed deciduous forest, secondary forest;

up to 1,500 m. **RANGE & STATUS** Endemic. Uncommon to locally common resident Tenasserim, W Thailand. **BREEDING** February–June. Co-operative. **Nest** Natural tree-cavity or old woodpecker hole, partially sealed with food debris, wood pulp etc.; 3.5–18 m above ground. **Eggs** 2–5; white, becoming stained; 42.3–51.2 x 32.2–35.5 mm.

523 **BUSHY-CRESTED HORNBILL**
Anorrhinus galeritus Plate **59**

IDENTIFICATION 89 cm. Monotypic. **Male** Completely dark plumage with thick drooping crest, somewhat paler and greyer vent and paler dirty brownish-grey basal two-thirds of tail distinctive. Bill and small casque blackish, bare orbital and gular skin pale bluish. See Black and Southern Brown Hornbills. Note range. **Female** Casque smaller, bill usually pale yellowish with mostly blackish basal half. **Juvenile** Browner with more whitish belly, bill pale olive, skin of head pale yellowish with pink eyering. **VOICE** Loud, excited, rising and falling yelps, ***klia-klia-klia kliu-kliu...***; ***wah wah wohawaha*** etc.; often given by all members of a group and building into crescendo. High-pitched ***aak aak aak*** when alarmed. **HABITAT & BEHAVIOUR** Broadleaved evergreen forest; up to 1,220 m. Usually in flocks of 5-15. **RANGE & STATUS** Resident Sumatra, Borneo. **SE Asia** Fairly common to common resident south Tenasserim, S Thailand, Peninsular Malaysia. **BREEDING** January-August. Co-operative. **Nest** Natural tree cavity, partially sealed with droppings and fruit pulp etc.; 10-25 m above ground. **Eggs** 2. Otherwise unrecorded.

524 **ORIENTAL PIED HORNBILL**
Anthracoceros albirostris Plate **59**

IDENTIFICATION 68–70 cm. *A.a.albirostris*: **Male** Unmistakable. Relatively small and mostly black, with white facial markings, belly and vent, and pale yellowish bill and casque with dark markings. In flight, black wings with broad white trailing edge and black tail with broadly white-tipped outer feathers diagnostic. *A.a.convexus* (extreme S Thailand southwards) has mostly white outertail-feathers (tail may appear all white from below). See White-crowned Hornbill. **Female** Bill and casque smaller and more extensively dark distally. **Juvenile** Duller, bill and casque uniformly pale yellowish, casque poorly developed; often has less white on tips of outertail. **VOICE** Loud, high-pitched yelping laughter: ***kleng-keng kek-kek-kek-kek-kek*** and ***ayip-yip-yip-yip...*** etc. **HABITAT & BEHAVIOUR** Broadleaved evergreen and mixed deciduous forest, island forest, secondary growth, sometimes coastal scrub, plantations and gardens; up to 1,400 m (below 150 m S Thailand southwards). Usually in flocks, sometimes quite large. **RANGE & STATUS** Resident N,NE Indian subcontinent, E India, SW,S China, Greater Sundas. **SE Asia** Uncommon to locally common resident (except C Thailand); rare Singapore. **BREEDING** January–June. **Nest** Natural tree-cavity, sometimes old woodpecker hole, partially walled up with droppings, fruit pulp, plant fibre and mud etc.; 1.2–50 m above ground. **Eggs** 1–3; white, becoming stained; 43.6–54 x 30–38 mm (*albirostris*).

525 **BLACK HORNBILL**
Anthracoceros malayanus Plate **59**

IDENTIFICATION 76 cm. Monotypic. **Male** Relatively small size, black plumage with broadly white-tipped outertail and unmarked yellowish-white bill and casque diagnostic. Facial skin blackish. Often has broad white to greyish supercilium. **Female** Bill and casque smaller and blackish, orbital skin and submoustachial patch pinkish. See Bushy-crested Hornbill. **Juvenile** Bill pale greenish-yellow (darker on very young birds) with undeveloped casque, facial skin dull yellowish with orange around eye, white tail-tips flecked black. **VOICE** Unusual and distinctive loud harsh retching sounds and grating growls. **HABITAT & BEHAVIOUR** Broadleaved evergreen forest; up to 215 m. Usually found in pairs or small flocks, occasionally larger flocks of up to 30 or more. **RANGE & STATUS** Resident Sumatra, Borneo. **SE Asia** Rare to locally fairly common resident S Thailand, Peninsular Malaysia. **BREEDING** January–April. **Nest** Natural tree-cavity, partially walled up with droppings and fruit pulp etc.; 4.5–25 m above ground. **Eggs** 2–3; white, becoming stained; 46–49.5 x 32–33 mm (Borneo).

526 **GREAT HORNBILL**
Buceros bicornis Plate **59**

IDENTIFICATION 119–122 cm. Birds in S Thailand and Peninsular Malaysia average smaller. *B.b.homrai*: **Male** Unmistakable. Very large with huge, mostly yellowish bill and casque, mostly blackish plumage with white nape and neck, white vent and white tail with broad black central band. In flight (above and below) shows diagnostic broad white band across greater coverts and broadly white-tipped secondaries and primaries. Neck and greater covert bar variably stained yellowish, casque marked with black, eyes reddish, eyering blackish. **Female** Smaller, lacks black casque markings, eyes whitish, eyering red. **Juvenile** Bill much smaller, casque barely developed, eyes pale blue-grey, eyering pinkish. **VOICE** Series of very loud, deep ***gok*** or ***kok*** notes (given by duetting pairs), leading up to a loud harsh roaring and barking. Also a deep coarse ***who*** given by male and ***whaa*** by female; double ***who-whaa*** at take-off or in flight is a duet of these calls. **HABITAT & BEHAVIOUR** Broadleaved evergreen and mixed deciduous forest, forest on some larger islands; up to 1,525 m. Usually found in pairs or small groups. **RANGE & STATUS** Resident SW India, N,NE Indian subcontinent, SW China, Sumatra. **SE Asia** Scarce to locally common resident (except C Thailand, Singapore). **BREEDING** January–August. **Nest** Natural tree-cavity, partially walled up with droppings, fruit and wood pulp, plant debris and some mud etc.; 8–25 m above ground. **Eggs** 1–3; glossy white to creamy-white, becoming stained; 59.8–72.2 x 42–50 mm.

527 **RHINOCEROS HORNBILL**
Buceros rhinoceros Plate **59**

IDENTIFICATION 91–122 cm. *B.r.rhinoceros*: **Male** Very large with diagnostic bright red and yellow upward-curved casque with black base. Mostly black with white lower belly and undertail-coverts, and white tail with broad black band across centre. Bill mostly pale yellowish to whitish with black line at bill/casque junction and along edge of mandibles, eyes reddish, eyering blackish. **Female** Smaller, lacks black on casque and at bill-casque junction, eyes whitish, eyering reddish. **Juvenile** Bill yellow with orange base, casque barely developed, eyes pale blue-grey, eyering blue-grey. **VOICE** Male utters deep, forceful ***hok*** notes, female gives a higher ***hak***, often in duet: ***hok-hak hok-hak hok-hak...*** Also a loud throaty ***GER-RONK*** by both sexes when flying (often simultaneously or antiphonally). **HABITAT & BEHAVIOUR** Broadleaved evergreen forest; up to 1,220 m. Usually found in pairs or small groups, occasionally larger flocks of up to 25. **RANGE & STATUS** Resident Greater Sundas. **SE Asia** Scarce to locally common resident southern S Thailand, Peninsular Malaysia; formerly recorded Singapore. **BREEDING** January–June. **Nest** Natural tree-cavity, partially walled up with droppings and fruit pulp etc.; 9–15 m above ground. **Eggs** 1–2; white, becoming stained; 63–64.8 x 44–44.6 mm (Borneo).

528 **HELMETED HORNBILL**
Rhinoplax vigil Plate **59**

IDENTIFICATION 127 cm (central tail-feathers extend up to 50 cm more). Monotypic. **Male** Very large with diagnostic elongated central tail-feathers and bare dark red skin on sides of head, throat and neck. Has short, straight, yellowish bill with reddish base, short and rounded reddish casque with yellow tip and mostly blackish plumage with white rump, vent and tail-coverts. Tail whitish with black central and subterminal bands. In flight, additionally shows broadly white-tipped secondaries and primaries. **Female** Smaller, bill speckled black at tip, skin of face and neck tinged pale lilac. **Juvenile** Bill yellowish-olive, casque poorly developed, skin of head and neck pale greenish-blue, central tail-feathers shorter. **VOICE** A loud, resonant, protracted series of notes, beginning with spaced ***hoop*** notes, slowly quickening to ***ke-hoop*** and ending with manic, comical laughter. Also loud clanking ***ka-hank***

ka-hank..., usually uttered in flight. **HABITAT & BEHAVIOUR** Broadleaved evergreen forest; up to 1,400 m. Usually found singly or in pairs. **RANGE & STATUS** Resident Sumatra, Borneo. **SE Asia** Scarce to uncommon resident Tenasserim, W(south),S Thailand, Peninsular Malaysia. **BREEDING** Sumatra: January–May. **Nest** Natural tree-cavity, sealed with fruit pulp and mud. **Eggs** 1–2. Otherwise unrecorded.

529 WHITE-CROWNED HORNBILL
Berenicornis comatus Plate **59**

IDENTIFICATION 90–101 cm. Monotypic. **Male** All-whitish head, neck, breast and tail and long shaggy whitish crest diagnostic. Upperparts, lower belly, vent and wings black, flight-feathers broadly tipped white, bill blackish with paler base and small casque, bare facial skin pale blue. **Female** Neck and underparts black. See Black Hornbill. **Juvenile** Similar to female but somewhat browner with blackish bases and shafts to crest feathers, base of tail black, bill yellowish-brown with dark patches; may show whitish tips to greater coverts. Male gradually attains white patches on underparts. **VOICE** Deep, resonant hooting with lower first note: ***hoo hu-hu-hu hu-hu-hu...***; often dying away. **HABITAT & BEHAVIOUR** Broadleaved evergreen forest; up to 1,000 m, rarely to 1,675 m. Usually found in small flocks, often low down in forest. Less frugivorous than other hornbills. **RANGE & STATUS** Resident Sumatra, Borneo. **SE Asia** Uncommon to locally common resident south Tenasserim, S Thailand, Peninsular Malaysia. **BREEDING** March–September. **Nest** Natural tree-cavity, sealed with droppings, fruit pulp and mud etc. **Eggs** 1–2. Otherwise unrecorded.

530 RUFOUS-NECKED HORNBILL
Aceros nipalensis Plate **59**

IDENTIFICATION 117 cm. Monotypic. **Male** Very large with diagnostic bright rufous head, neck and underparts. Rest of upperside black with white-tipped outer primaries, tail white with black basal half/third, bill pale yellowish with row of vertical dark ridges on upper mandible and almost no casque, orbital skin blue, gular skin red. **Female** Smaller; head, neck and underparts black, orbital skin a little duller. From Wreathed and Plain-pouched Hornbills by bill structure and pattern, opposite colour of bare head-skin, less inflated gular pouch, lack of obvious crest, white-tipped outer primaries and black base of tail. **Juvenile** Both sexes like male but bill smaller with no dark ridges, tail-feathers may be narrowly dark-tipped. **VOICE** Barking ***kup*** notes; less deep than similar calls of Great Hornbill. **HABITAT & BEHAVIOUR** Broadleaved evergreen forest; 600–2,900 m. Usually in pairs, sometimes small groups. **RANGE & STATUS** Resident NE India, Bhutan, SE Tibet, SW China; formerly Nepal, Bangladesh. **SE Asia** Rare to locally uncommon resident W,N,C,S(east),E Myanmar, north Tenasserim, W,NW Thailand, N,C Laos, W Tonkin, N Annam. **BREEDING** February–March. **Nest** Natural tree-cavity, partially sealed with droppings, fruit pulp and mud etc.; 10–30 m above ground. **Eggs** 1–2; white, becoming stained; 53.3–68 x 39.9–46.5 mm.

531 WRINKLED HORNBILL
Aceros corrugatus Plate **59**

IDENTIFICATION 81–82 cm. Monotypic. **Male** Resembles Wreathed Hornbill but smaller, has smaller yellow bill with reddish base, somewhat squarer-looking reddish casque, blue orbital skin, unmarked and less bulging gular pouch, blacker centre of crown and nape and black base of tail (often difficult to see). White of tail usually strongly stained buffish to yellowish. See Plain-pouched Hornbill. **Female** From Wreathed by smaller size, smaller plain yellowish bill with shorter and squarer casque, blue orbital skin, plain, less bulging gular pouch, less pronounced crest and black base of tail. See Plain-pouched and Rufous-necked Hornbills. Note habitat and range. **Juvenile** Like male (both sexes) but bill unridged and pale yellow with orange wash at base, casque undeveloped, orbital skin pale yellow; may have blackish base to upper mandible. **VOICE** Sharp, barking ***kak kak-kak*** etc. **HABITAT & BEHAVIOUR** Broadleaved evergreen forest, freshwater swamp forest; up to 800 m. Usually found in pairs or small flocks. **RANGE & STATUS** Resident Sumatra, Borneo. **SE Asia** Rare to uncommon resident S Thailand, Peninsular Malaysia. Formerly recorded Singapore. **BREEDING** Unrecorded in wild. **Eggs** 3; 52.5–54.5 x 39–41 mm (in captivity).

532 PLAIN-POUCHED HORNBILL
Aceros subruficollis Plate **59**

IDENTIFICATION Male 86.5–89.5 cm; female 76–84 cm. Monotypic. **Male** Very similar to Wreathed Hornbill but somewhat smaller, bill shorter with warm brownish base and no corrugations, casque slightly more peaked with more dark ridges, lacks blackish streak on gular pouch. **Female** Like Wreathed but differs as male. **Juvenile** Both sexes like male but casque undeveloped, bill uniform pale yellowish. Possibly indistinguishable from Wreathed, apart from plain gular pouch. **VOICE** Loud ***keh-kek-kehk***, higher-pitched and more quacking than similar call of Wreathed; or ***ehk-ehk-EHK***, with accentuated end-note. **HABITAT & BEHAVIOUR** Broadleaved evergreen and mixed deciduous forest; up to 915 m. Usually found in pairs or small groups, but large flocks in autumn recorded north Peninsular Malaysia. **RANGE & STATUS** Endemic. Rare to uncommon resident (subject to local movements) S Myanmar, Tenasserim, W and southern S Thailand, north Peninsular Malaysia. **BREEDING** February–May. **Nest** Natural tree-cavity, partially sealed with droppings, fruit pulp and mud etc.; 18–21 m above ground. **Eggs** 1–3; white, becoming stained; 52.9–60.3 x 38.5–47 mm.

533 WREATHED HORNBILL
Aceros undulatus Plate **59**

IDENTIFICATION Male 100.5–115 cm; female 84–98 cm. Birds in S Thailand and Peninsular Malaysia are smaller. Monotypic. **Male** Mostly blackish plumage with brownish-white sides of head, neck and breast, all-white tail and bulging yellow gular pouch with long blackish lateral streak diagnostic. Centre of crown to hindneck shaggy warmish dark brown, bill pale dull yellowish with darker corrugated base (not always obvious), casque small and short with dark ridges, orbital skin reddish. Tail often lightly stained yellowish to brownish. See Plain-pouched and Wrinkled Hornbills. **Female** Head, neck and breast black, gular pouch blue. See Plain-pouched, Wrinkled and Rufous-necked Hornbills. **Juvenile** Both sexes like male but casque undeveloped, bill uncorrugated, dark streak on gular pouch fainter. **VOICE** A loud, rather breathless ***kuk-kwehk***, with emphasis on higher second note. **HABITAT & BEHAVIOUR** Broadleaved evergreen and mixed deciduous forest, forest on islands; up to 1,830 m (rarely to 2,500 m N Myanmar). Usually found in pairs or small flocks. May be seen in very large flocks flying to and from roosts; even over deforested areas. **RANGE & STATUS** Resident NE India, Bhutan, SW China, Greater Sundas; formerly Bangladesh. **SE Asia** Uncommon to locally common resident (except C Myanmar, C Thailand, Singapore, W,E Tonkin). **BREEDING** February–August. **Nest** Natural tree-cavity, partially sealed with droppings, fruit pulp and mud etc.; 5–30 m above ground. **Eggs** 1–3; white, becoming stained; 49.5–72.1 x 38–47.1 mm.

RAMPHASTIDAE: MEGALAIMINAE: Asian barbets

Worldwide 29 species. SE Asia 17 species. Thickset, big-headed and rather short-tailed with large bills and short legs with zygodactyl feet. Most species largely green with brightly coloured head markings and similar sexes. Slow-moving, often in canopy where difficult to observe. Frugivorous, often gathering in numbers to feed in fruiting trees.

534 **FIRE-TUFTED BARBET**

Psilopogon pyrolophus Plate **60**

IDENTIFICATION 28 cm. Monotypic. **Adult** Largish size, pale yellowish-green bill with vertical blackish central band, grey ear-coverts and yellow and blackish bands on upper breast diagnostic. Forehead and lores black with prominent tuft of reddish spines extending over base of bill; hindcrown, nape and post-ocular stripe brownish-maroon; narrow white band across crown runs into short pale green supercilium. **Juvenile** Brownish-maroon of head replaced by dull olive-brown, supercilium duller and reduced to eyebrow. **VOICE** Male territorial call is an unusual cicada-like buzzing, starting with spaced notes, then speeding up and rising in pitch toward end. Recalls some broadbills. **HABITAT** Broadleaved evergreen forest; 1,070–2,010 m. **RANGE & STATUS** Resident Sumatra, Java (Gunung Gede-Pangrango NP; status uncertain). **SE Asia** Uncommon to common resident extreme S Thailand, Peninsular Malaysia. **BREEDING** January–April. **Nest** Tree-hole. Otherwise undocumented.

535 **GREAT BARBET** *Megalaima virens* Plate **60**

IDENTIFICATION 32–33 cm. *M.v.virens*: **Adult** Relatively large size, large, mostly pale yellowish bill and dark bluish head diagnostic. Mantle, scapulars and breast very dark brownish, belly yellow with broad dark greenish-blue streaks, undertail-coverts red. **Juvenile** Similar to adult. **Other subspecies in SE Asia** *M.v.magnifica* (SW,W,west S Myanmar), *clamator* (east N Myanmar). **VOICE** Male territorial call is a very loud strident ***pIYAAo*** or ***KAY-oh***, repeated about once a second. Also (by female?) a more rapid, continuous ***piou-piou-piou-piou...***, often given with former in duet by pair. Alarm call is a harsh grating ***keeah***. **HABITAT** Broadleaved evergreen and occasionally deciduous forest; 600–2,800 m (locally down to 440 m N Myanmar). **RANGE & STATUS** Resident NE Pakistan, NW,N,NE Indian subcontinent, S,SE Tibet, southern China. **SE Asia** Common resident Myanmar (except southern Tenasserim), W,NW,NE (north-west) Thailand, N,C Laos, W,E Tonkin, N Annam. **BREEDING** February–July. **Nest** Excavated or disused woodpecker hole in tree; 3–5 m above ground. **Eggs** 2–4, white; 34.5 x 26.9 mm (av.; *virens*).

536 **RED-VENTED BARBET**

Megalaima lagrandieri Plate **60**

IDENTIFICATION 29.5–34 cm. *M.m.lagrandieri*: **Adult** Relatively large size, large bill and rather plain, mostly greenish plumage with contrasting red undertail-coverts distinctive. Head brownish with paler buffy-greyish sides and throat, orange-red forehead-tuft, narrow pale blue eyebrow and some pale blue and orange-red streaks on sides of nape, bill mostly dark greyish with pale yellowish-brown cutting edge and tip. **Juvenile** Head more uniformly greyish-brown, apart from orange-red forehead-tuft. **Other subspecies in SE Asia** *M.l.rothschildi* (W,E Tonkin, N Annam). **VOICE** Male territorial call is a very loud, strident, throaty ***CHOA*** or ***CHORWA***, repeated at c.1–2 s intervals. Also a descending series often given antiphonally by mate: ***uk uk-ukukukukukuk..*** Alarm call is a harsh, grating, rather high-pitched ***grrric..grrric..*** or ***brrret..brrret...*** **HABITAT** Broadleaved evergreen and semi-evergreen forest; up to 1,900 m. **RANGE & STATUS** Endemic. Uncommon to common resident north and east Cambodia, C,S Laos, Vietnam. **BREEDING** April–July. Otherwise unknown.

537 **LINEATED BARBET**

Megalaima lineata Plate **60**

IDENTIFICATION 27–28 cm. *M.l.hodgsoni*: **Adult** Distinctive. Medium-sized with thick yellowish bill and mostly dark brown head and breast with broad whitish streaks. Has broad yellow orbital skin, whitish cheeks to throat, whitish streaks on upper mantle, and whitish and dark streaks extending onto belly. **Juvenile** Similar to adult. **VOICE** Male territorial call is a very loud, mellow ***POO-POH***, with higher second note, repeated about once a second. Also a rapid, dry, bubbling ***koh-koh-koh-koh-koh...*** **HABITAT** Deciduous forest, scattered trees in open areas, coastal scrub, plantations; up to 1,220 m (below 800 m in Thailand, lowlands only in Peninsular Malaysia). **RANGE & STATUS** Resident N,NE Indian subcontinent, SW China (S Yunnan), Java, Bali. **SE Asia** Common resident (except south Peninsular Malaysia, Singapore, W,E Tonkin, N Annam). **BREEDING** September–May. **Nest** Excavated tree-hole; 3–12 m above ground. **Eggs** 2–4, white; 30.7 x 23 mm (av.).

538 **GREEN-EARED BARBET**

Megalaima faiostricta Plate **60**

IDENTIFICATION 24.5–27 cm. *M.f.faiostricta*: **Adult** Similar to Lineated but smaller, bill smaller and mostly dark; has restricted dark greyish orbital skin, green cheeks and ear-coverts and all-green mantle. **Juvenile** Like adult. **Other subspecies in SE Asia** *M.f.praetermissa* (E Tonkin). **VOICE** Male territorial call is a loud throaty ***took-a-prruk***, rapidly repeated more than once per second. Also a mellow, fluty, rising ***pooouk***. **HABITAT** Broadleaved evergreen and semi-evergreen forest, mixed deciduous forest, scattered trees in more open areas; up to 1,015 m. **RANGE & STATUS** Resident S China. **SE Asia** Common resident Thailand (except C and S), Indochina. **BREEDING** February–June. **Nest** Excavated hole in tree. Otherwise unknown.

539 **GOLD-WHISKERED BARBET**

Megalaima chrysopogon Plate **60**

IDENTIFICATION 30 cm. *M.c.laeta*: **Adult** Combination of largish size, strong blackish bill, broad blackish-brown eyestripe, large yellow patch on lower sides of head, and pale greyish-buff throat with narrow blue lower border diagnostic. See Golden-throated, Black-browed and Red-crowned Barbets. Note habitat and range. **Juvenile** Has duller yellow patch on lower sides of head. **VOICE** Male territorial call is a very loud, rather deep, rapid ***tehoop-tehoop-tehoop-tehoop-tehoop...*** Also a repeated, long, low-pitched trill on one note, gradually slowing and eventually breaking up into 3–4 note phrases. **HABITAT** Broadleaved evergreen forest; up to 1,065 m, occasionally to 1,525 m. **RANGE & STATUS** Resident Sumatra, Borneo. **SE Asia** Uncommon to common resident south Tenasserim, W(south),S Thailand, Peninsular Malaysia. **BREEDING** October–June. **Nest** Excavated tree-hole; 9–13 m above ground. Otherwise undocumented.

540 **RED-CROWNED BARBET**

Megalaima rafflesii Plate **60**

IDENTIFICATION 25–27 cm. *M.r.malayensis*: **Adult** All-red crown and blue throat diagnostic. Has blue supercilium, smallish yellow neck-patch, and small red patch on side of upper breast. See Gold-whiskered, Red-throated and Black-browed Barbets. Note habitat and range. **Juvenile** Much duller with less defined head pattern. **VOICE** Male territorial call is a loud series of 1–2 ***took*** notes followed, after a pause, by up to 20 rapidly repeated, shorter ***tuk*** notes. **HABITAT** Broadleaved evergreen forest; up to 200 m. **RANGE & STATUS** Resident Sumatra, Borneo. **SE Asia** Scarce to locally fairly common resident south Tenasserim, S Thailand, Peninsular Malaysia, Singapore. **BREEDING** March–May. Excavated tree-hole; 4.5 m above ground. **Eggs** 2+. Otherwise unknown.

541 **RED-THROATED BARBET**

Megalaima mystacophanos Plate **60**

IDENTIFICATION 23 cm. *M.m.mystacophanos*: **Male** Red throat diagnostic. Forehead yellow, crown red; has broad black line through eye, blue cheeks and band across uppermost breast, and prominent red patch on side of upper breast. See Red-crowned Barbet. **Female** Unusual. Head mostly greenish with faint red patches on lores, hindcrown and upper breast-sides, faint bluish tinge to face, and yellowish-tinged upper throat. See juvenile Moustached, Black-browed and Blue-throated Barbets. **Juvenile** Head all green with yellower-tinged forehead and throat. See Blue-eared Barbet. **VOICE** Male territorial call is a slow series of 1–2 (sometimes 3) deep notes given at uneven intervals: ***chok..chok-chok.chok...chok-chok..chok...*** etc. Also a repeated high-pitched trill, which gradually shortens. **HABITAT** Broadleaved evergreen forest; up to 760 m. **RANGE & STATUS** Resident Sumatra, Borneo. **SE Asia** Common resident south Tenasserim, S Thailand, Peninsular Malaysia. **BREEDING** January–August. **Nest** Excavated hole in tree, arboreal ant-nest or termitarium; 3–6 m above ground. **Eggs** 2–4; rather glossy, white; 27.7 x 20.5–21.2 mm.

542 GOLDEN-THROATED BARBET

Megalaima franklinii Plate **60**

IDENTIFICATION 20.5–23.5 cm. *M.f.ramsayi*: **Adult** Combination of medium size, red forehead and hindcrown, yellow midcrown, grey-streaked blackish lower ear-coverts, yellow upper throat and pale greyish lower throat distinctive. Broad black band through eye streaked with grey, outer fringes of secondaries and outer wing-coverts distinctly blue-washed. *M.f.franklinii* (N Myanmar, NW[east],NE Thailand, N Indochina) and *minor* (Peninsular Malaysia; presumably extreme S Thailand) have all-black band through eye and deeper yellow on throat, latter also has some blue behind ear-coverts; *auricularis* (S Laos, S Annam) has all-black band through eye, violet-washed lower ear-coverts, mostly yellow throat with dark border and narrow blue lower border to throat continuing in narrow line to rear of black eyestripe. See Black-browed and Gold-whiskered Barbets. Note habitat and range. **Juvenile** Duller with less distinct head pattern. **Other subspecies in SE Asia** *M.f.trangensis* (S Thailand). **VOICE** Male territorial call is a very loud, ringing ***puKWOWK***, repeated about once a second. **HABITAT** Broadleaved evergreen forest; 800–2,565 m (above 1,280 m in Peninsular Malaysia, rarely down to 500 m in Laos and 225 m in C Annam). **RANGE & STATUS** Resident Nepal, NE Indian subcontinent, SE Tibet, SW China. **SE Asia** Common resident (except C,SE Thailand, Singapore, Cambodia, Cochinchina). **BREEDING** February–August. **Nest** Excavated tree-hole; 2–18 m above ground. **Eggs** 3–4; white; 27.4 x 21.1 mm (av.; *franklinii*).

543 BLACK-BROWED BARBET

Megalaima oorti Plate **60**

IDENTIFICATION 21.5–23.5 cm. Monotypic. **Adult** Similar to Golden-throated but sides of head blue, throat yellow with broad blue lower border, has red spot on side of upper breast, no blue tinge to wing-feathers; medium-width black eyestripe/supercilium. **Juvenile** Duller with less distinct head pattern. **VOICE** Male territorial call is a relatively slow and deliberate, rolling ***pt'A'PROR***, ***put'A'PRRR*** or ***pu'TA'TRRR*** (introduction can be hard to discern) repeated about once a second. **HABITAT** Broadleaved evergreen forest; 600–1,450 m, rarely down to 250 m. **RANGE & STATUS** Resident Sumatra. **SE Asia** Fairly common to common resident extreme S Thailand, Peninsular Malaysia. **BREEDING** February–May and October. **Nest** Excavated tree-hole; 9–10.5 m above ground. Otherwise undocumented. **NOTE** For taxonomy, see Annam Barbet.

544 ANNAM BARBET

Megalaima annamensis Plate **60**

IDENTIFICATION 21.5–23.5 cm. Monotypic. **Adult** As Black-browed Barbet, but has broader black supercilium and more blue on lower throat. Note range. **Juvenile** Duller with less distinct head pattern. **VOICE** Male territorial call is longer and more rolling than Black-browed: ***ut'TRR'RR'UP***; ***tup'ARA'RR'UP***; ***ut'TUPA'RR'UT***; or ***ut'TUP'ARA'RUP*** etc. **HABITAT** Broadleaved evergreen forest; 600–1,450 m, rarely down to 250 m. **RANGE & STATUS** Endemic. Fairly common to common resident east Cambodia, S Laos, C,S Annam. **BREEDING** Undocumented. **NOTE** Split from Black-browed Barbet, along with extralimital taxa, following Collar (2006) and Feinstein *et al.* (2008).

545 BLUE-THROATED BARBET

Megalaima asiatica Plate **60**

IDENTIFICATION 23 cm. *M.a.davisoni*: **Adult** Smallish size and all-blue sides of head and throat diagnostic. Crown red with blue band across centre, has narrow black supercilium and small red patch at side of upper breast. *M.a.asiatica* (Myanmar; except Tenasserim) has black and yellow bands across midcrown; *chersonesus* (S Thailand) has more extensively blue crown. See Moustached and Black-browed Barbets. **Juvenile** Duller with ill-defined head pattern. See Blue-eared Barbet. **VOICE** Male territorial call is a very loud, quickly repeated ***took-arook***. **HABITAT** Broadleaved evergreen forest, secondary growth; 400–2,400 m (mainly 600–1,830 m). **RANGE & STATUS** Resident NE Pakistan, NW,N,NE Indian subcontinent, SW,S China. **SE Asia** Common resident Myanmar, Thailand (except C,SE), N Laos, W,E Tonkin, N Annam. **BREEDING** April–July. **Nest** Excavated tree-hole; 2–8 m above ground. **Eggs** 3–4, white; 27.8 x 20.5 mm (av.; *asiatica*).

546 MOUSTACHED BARBET

Megalaima incognita Plate **60**

IDENTIFICATION 23 cm. *M.i.elbeli*: **Adult** Similar to Blue-throated Barbet but crown greenish with small red patch at rear; has diagnostic long black submoustachial stripe. See Black-browed Barbet. **Juvenile** Duller with greener sides of head and throat, and narrower black submoustachial stripe. **Other subspecies in SE Asia** *M.i.incognita* (Tenasserim, W Thailand); *euroa* (SE Thailand, Indochina). **VOICE** Male territorial call similar to Blue-throated but notes more spaced and deliberate: ***u'ik-a-ruk u'ik-a-ruk u'ik-a-ruk...*** **HABITAT** Broadleaved evergreen forest; 600–1,700 m. **RANGE & STATUS** Endemic. Fairly common to common resident northern Tenasserim, W,NW,NE,SE Thailand, Cambodia, Laos, E Tonkin (local), C Annam (local). **BREEDING** May–June. Otherwise undocumented.

547 YELLOW-CROWNED BARBET

Megalaima henricii Plate **60**

IDENTIFICATION 22–23 cm. *M.h.henricii*: **Adult** Smallish size, yellow forecrown and sides of crown, green sides of head and blue throat diagnostic. Lores and short eyestripe black, centre of hindcrown blue, has small red patches on foreneck and neck-sides. **Juvenile** Duller with more washed-out head pattern. See Blue-eared Barbet. **VOICE** Male territorial call consists of 4–6 loud ***tok*** notes introduced by short trill: ***trrok....tok-tok-tok-tok...***, with one phrase about every 2 s. **HABITAT** Broadleaved evergreen forest; up to 975 m. **RANGE & STATUS** Resident Sumatra, Borneo. **SE Asia** Uncommon to fairly common resident S Thailand, Peninsular Malaysia. **BREEDING** March–September. **Nest** Excavated hole in tree; 9 m above ground. Otherwise unknown.

548 BLUE-EARED BARBET

Megalaima australis Plate **60**

IDENTIFICATION 17–18 cm. *M.a.cyanotis*: **Male** Combination of small size, black forehead, orange-red cheek-patch, red patches above and below blue ear-coverts and blue throat with narrow black lower border distinctive. Has blue midcrown and green hindcrown. *M.a.duvaucelii* (Peninsular Malaysia) has black ear-coverts with larger red patches above and below, red cheek-patch and much broader black lower border to throat. **Female** Head pattern somewhat duller than male. **Juvenile** Uniform dull green with blue-tinged ear-coverts and throat. See Red-throated, Blue-throated, Moustached and Yellow-crowned Barbets. **Other subspecies in SE Asia** *M.a.stuarti* (Tenasserim, W,S Thailand). **VOICE** Male territorial call is a loud, monotonous, rapidly repeated ***KO-TEK*** or ***DAJIK***. Also a series of shrill whistled ***pleow*** notes, about one per second. **HABITAT** Open broadleaved evergreen and semi-evergreen forest, mixed deciduous forest, secondary growth; up to 1,525 m (below 975 m in Peninsular Malaysia). **RANGE & STATUS** Resident E Nepal, NE Indian subcontinent, SW China, Greater Sundas. **SE Asia** Common resident (except C Thailand); formerly Singapore. **BREEDING** January–August. **Nest** Excavated hole in tree; 3–12 m above ground. **Eggs** 2–4, white; 24.5 x 18.3 mm (av.; *cyanotis*).

549 COPPERSMITH BARBET

Megalaima haemacephala Plate **60**

IDENTIFICATION 17 cm. *M.h.indica*: **Adult** Unmistakable. Small with yellow head-sides and throat, black eyestripe and submoustachial stripe and pale greenish underparts with broad dark green streaks. Crown red at front and black towards rear, has red band across upper breast. **Juvenile** Dark parts of head duller, yellow of sides of head and throat paler, no red on crown or breast. **VOICE** Male territorial call is a series of up to 100 or more loud, resonant, quickly repeated ***TONK*** or ***TUUK*** notes. **HABITAT** Deciduous forest, forest edge, mangroves, scattered trees in open areas, parks and gardens, plantations; up to 915 m. **RANGE & STATUS** Resident Indian subcontinent, SW China, Sumatra, Java, Bali, Philippines. **SE Asia** Common resident (except W,E Tonkin). **BREEDING** All year. **Nest** Excavated hole in tree. **Eggs** 2–4; white; 25.2 x 17.5 mm (av.).

550 **BROWN BARBET**
Calorhamphus fuliginosus Plate **60**
IDENTIFICATION 20 cm. *C.f.hayii*: **Adult** Brown plumage with paler, more whitish breast to vent diagnostic. Legs and feet pinkish-orange. **Juvenile** Similar to adult. **VOICE** Thin, forced ***pseeoo*** notes. **HABITAT & BEHAVIOUR** Broadleaved evergreen forest, secondary growth, scattered trees in more open areas; up to 1,065 m. Often forages in small parties. **RANGE & STATUS** Resident Sumatra, Borneo. **SE Asia** Fairly common to common resident south Tenasserim, S Thailand, Peninsular Malaysia; formerly Singapore. **BREEDING** Partly communal. February–September. **Nest** Excavated hole in tree or tree termitarium; 2.5–20 m above ground. **Eggs** 3; white; 26.2 x 20.1 mm (av.).

INDICATORIDAE: Honeyguides

Worldwide 17 species. SE Asia 2 species. Largely nondescript with stout bills, long pointed wings and zygodactyl feet. Very unobtrusive. Feed on beeswax and insects but, unlike certain African species, Asian species do not lead humans and other animals to stores of honey. Asian species are thought to have parasitic nesting habits.

551 **MALAYSIAN HONEYGUIDE**
Indicator archipelagicus Plate **61**
IDENTIFICATION 18 cm. Monotypic. **Male** Nondescript, thick-billed and passerine-like, resembling some bulbuls. Upperparts cold dark olive-brown with narrow olive-green streaks, mainly on wings, scapulars and uppertail-coverts; underparts whitish with greyish wash across breast and broad dark streaks on belly/lower flanks. Has long lemon-yellow shoulder-patch (often obscured) and reddish eyes. **Female** Lacks shoulder-patch. **Juvenile** Like female but underparts indistinctly streaked, eyes brown. **VOICE** Song is a mewing note followed by a nasal ascending rattle: ***miaw-krrrruuu*** or ***miaw-miaw-krrwuu***. **HABITAT & BEHAVIOUR** Broadleaved evergreen forest; up to 915 m. Sits motionless on exposed perch for long periods; sometimes seen around bee nests. **RANGE & STATUS** Resident Sumatra, Borneo. **SE Asia** Scarce to uncommon resident W,S Thailand, Peninsular Malaysia. **BREEDING** January–February. Otherwise unknown but possibly parasitic on Brown Barbet.

552 **YELLOW-RUMPED HONEYGUIDE**
Indicator xanthonotus Plate **61**
IDENTIFICATION 15 cm. Monotypic. **Male** Similar to Malaysian Honeyguide but somewhat smaller, darker above, has diagnostic bright orange-yellow forecrown, cheeks and band down centre of lower back and rump and white inner fringes of tertials. Throat broadly but indistinctly streaked greyish and whitish, rest of underparts largely dusky-whitish with broad dark streaks. Note range. See female Scarlet Finch. **Female** Duller with less yellow on forehead and cheeks. **Juvenile** Undocumented. **VOICE** Single ***weet*** is said to be uttered in flight. **HABITAT & BEHAVIOUR** Cliffs and adjacent evergreen forest; recorded at 2,285 m. Sits motionless for long periods; usually near bee nests (thought to be exclusively those of Giant Rock Bee *Apis dorsata* in Indian subcontinent). **RANGE & STATUS** Resident NW,N,NE Indian subcontinent, SE Tibet; formerly NE Pakistan. **SE Asia** Rare resident N Myanmar. **BREEDING** April–May (Nepal). Otherwise unknown but possibly parasitic on barbets.

PICIDAE: JYGNINAE: Wrynecks

Worldwide 2 species. SE Asia 1 species. Highly distinctive relative of piculets and woodpeckers. Has passerine-like appearance with relatively long square-cut tail and unusual cryptic plumage; commonly found on ground, feeding on ants and other insects.

553 **EURASIAN WRYNECK**
Jynx torquilla Plate **61**
IDENTIFICATION 16–18 cm. *J.t.chinensis*: **Adult** Cryptically patterned plumage distinctive. Typically has greyish-brown upperside with dark vermiculations and broad dark central stripe on nape and mantle, whitish underparts (often washed buff, particularly on throat and breast) with dark vermiculations, and heavily vermiculated wings and tail. **Juvenile** Upperparts duller, darker and more barred, underpart-barring less distinct. **VOICE** Male territorial call is a repeated series of clear, ringing notes, each falling in pitch at end: ***quee-quee-quee-quee-quee...*** Otherwise gives repeated ***tak*** or ***kek*** notes. **HABITAT & BEHAVIOUR** Open dry country and secondary growth, scrub and grass, cultivation; up to 2,285 m. Commonly feeds on ground, often sitting motionless for long periods, particularly when disturbed. **RANGE & STATUS** Breeds Palearctic, W,N,NE China, N Korea, northern Japan; mostly winters equatorial Africa, northern Indian subcontinent, southern China, southern Japan. **SE Asia** Uncommon to fairly common winter visitor Myanmar, Thailand (except southern S), N,C Laos, W,E Tonkin; scarce S Laos, N Annam, Cochinchina. Also recorded on passage N Myanmar, N Laos, W Tonkin.

PICIDAE: PICUMNINAE: Piculets

Worldwide 30 species. SE Asia 3 species. Tiny, short-tailed relatives of woodpeckers, *Sasia* strongly favouring bamboo. Feed on grubs and pupae of wood-boring beetles etc.

554 **SPECKLED PICULET**
Picumnus innominatus Plate **61**
IDENTIFICATION 9–10.5 cm. *P.i.malayorum*: **Male** Unmistakable. Very small and short-tailed with largely olive-slate crown, ear-coverts and submoustachial stripe, broad white supercilium and moustachial stripe and whitish underparts with bold blackish spots and bars. Forehead rufous-buff with blackish scales, rest of upperparts olive-green, tail white centrally. *P.i.chinensis* (Tonkin) is a little larger with dark cinnamon-coloured forehead, crown, ear-coverts and submoustachial stripe. **Female** Forehead and crown concolorous. **Juvenile** Resembles respective adults but has pale bill. **VOICE** Territorial call is a high ***ti-ti-ti-ti-ti***. Also produces loud tinny drumming. Calls include sharp ***tsit*** and squeaky ***sik-sik-sik***. **HABITAT & BEHAVIOUR** Broadleaved evergreen and mixed deciduous forest, secondary growth, bamboo; up to

1,935 m (915–1,370 m in Peninsular Malaysia). Often joins bird waves. **RANGE & STATUS** Resident N Pakistan, NW,N,NE Indian subcontinent, SW,E India, SE Tibet, C and southern China, Sumatra, Borneo. **SE Asia** Common resident Myanmar (except SW), W,NW,NE,extreme S Thailand, Peninsular Malaysia, Cambodia, Laos, Vietnam. **BREEDING** November–April. **Nest** Excavated hole in bamboo or small tree; 1–8 m above ground. **Eggs** 2–4, quite glossy, white; 14.8 x 12 mm (av.; *malayorum*).

555 **RUFOUS PICULET** *Sasia abnormis* Plate **61**
IDENTIFICATION 8–9.5 cm. *S.a.abnormis*: **Male** Like White-browed Piculet but lacks white supercilium, has darker olive upperparts and darker rufous underparts. Eyering dull pinkish-red, forehead yellowish. **Female** Forehead dark rufous. **Juvenile** Distinctive. Upperparts duller olive (washed slaty on mantle), head and underparts dull brownish-slate, may show a little dull rufous on chin, belly and vent. Bill all dark. **VOICE** Male territorial call is a high-pitched ***kik-ik-ik-ik-ik-ik***. Also drums like White-browed. Call is a sharp ***tic*** or ***tsit***. **HABITAT & BEHAVIOUR** Bamboo, broadleaved evergreen forest, secondary growth; up to 1,370 m. Often joins bird waves. **RANGE & STATUS** Resident Greater Sundas. **SE Asia** Fairly common to common resident south Tenasserim, S Thailand, Peninsular Malaysia. **BREEDING** April–August. **Nest** Excavated hole in bamboo or dead branch. **Eggs** 2; white.

556 **WHITE-BROWED PICULET**
Sasia ochracea Plate **61**
IDENTIFICATION 8–9.5 cm. *S.o.reichenowi*: **Male** Very small size, very short tail, prominent white supercilium behind eye and buffish-rufous underparts (darker on head/neck-sides) diagnostic. Forehead yellow, crown and wing-coverts dull olive, mantle and scapulars olive-rufous, eyering dull crimson to brownish-red. *S.o.ochracea* (syn. *querulivox*, *kinneari*: N,E Myanmar, NW,NE Thailand, N Indochina) has darker olive crown, dark olive wash on mantle (creating rufous-collared appearance) and darker, deeper rufous sides of head and underparts; *hasbroucki* (Tenasserim, S Thailand) has blackish eyering. **Female** Forehead rufous. **Juvenile** Similar to Rufous Piculet. **VOICE** Male territorial call is a rapid high-pitched trill starting with call-note: ***chi rrrrrrrra***. Also produces a loud tinny drumming on bamboo: ***tit..trrrrrrrrrit***. Call is a short sharp ***chi***. **HABITAT & BEHAVIOUR** Bamboo, broadleaved evergreen and mixed deciduous forest, secondary growth; up to 1,910 m. Often joins bird waves. **RANGE & STATUS** Resident N,NE Indian subcontinent, SE Tibet, SW,S China. **SE Asia** Fairly common to common resident (except C,SE,southern S Thailand, Peninsular Malaysia, Singapore). **BREEDING** February–July. **Nest** Excavated hole in bamboo or small tree; 50 cm above ground. **Eggs** 2–4, quite glossy, white; 15.9 x 12.7 mm (av.; *ochracea*).

PICIDAE: PICINAE: Typical woodpeckers

Worldwide c.183 species. SE Asia 39 species. Typically small to fairly large with chisel-shaped bills, stiff pointed tails (used for support), strong legs and zygodactyl feet. Flight strong and undulating. Most species hammer very rapidly against tree-trunks or bamboo, creating rattling sound known as 'drumming', which is used for territorial advertisement. Mostly arboreal, clinging to trunks and branches of trees and bamboo; some also forage on ground. Feed on grubs and pupae of wood-boring beetles, ants, termites and other insects, sometimes fruit.

557 **RUFOUS-BELLIED WOODPECKER**
Hypopicus hyperythrus Plate **61**
IDENTIFICATION 19–23 cm. *H.h.hyperythrus*: **Male** Unmistakable, with white-barred black upperparts and dull dark rufous sides of head and underparts. Crown, nape and vent red, orbital area, cheeks and chin whitish. *H.h.subrufinus* (visitor Tonkin) is larger (25 cm) with paler and browner sides of head and underparts and pinker vent; *annamensis* (NE Thailand, S Indochina) has paler rufous sides of head and underparts; both have less red on nape. **Female** Crown and nape black with white speckles, vent paler red. **Juvenile** Has dark streaks on sides of head and duller/paler underparts with heavy blackish bars. Both sexes have orange-red tips to crown-feathers (fewer on females); subadults have blackish bars and whitish mottling on throat and breast. **VOICE** Male territorial call is a rattling, trilled ***ki-i-i-i-i-i-i*** or ***chit-chit-chit-r-r-r-h***. Both sexes drum. Calls with a fast ***ptikitititit...*** and ***tik-tik-tik-tik...*** **HABITAT** Open oak forest, pine forest, mixed broadleaved evergreen and coniferous forest, locally deciduous forest; 600–3,100 m (below 1,460 m Thailand). **RANGE & STATUS** Resident NE Pakistan, NW,N,NE Indian subcontinent, S,SE Tibet, SW,W China. Breeds NE China, Ussuriland; winters S China. **SE Asia** Scarce to common resident W,N,C,E Myanmar, W(local),NW,NE(southern) Thailand, Cambodia, S Laos, S Annam. Scarce winter visitor W,E Tonkin. **BREEDING** March–May. **Nest** Excavated tree-hole; 5–6 m above ground. **Eggs** 4–5; white; 22.2 x 16.5 mm (av.; *hyperythrus*).

558 **SUNDA PYGMY WOODPECKER**
Dendrocopos moluccensis Plate **61**
IDENTIFICATION 12.5–13 cm. *D.m.moluccensis*: **Male** Very similar to Grey-capped Pygmy Woodpecker but somewhat smaller, with distinctly brownish crown, contrasting less with blackish crown-sides and nape, dark greyish-brown ear-coverts and more defined submoustachial stripe, base colour above generally browner, streaking below generally broader and more diffuse. Has short red streak on side of hindcrown. **Female** Lacks red streak on side of hindcrown. **Juvenile** Breast and belly duller and browner with less distinct streaks, lower mandible pale with dark tip. Male has more orange-red streak on side of hindcrown. **VOICE** Male territorial call is a sharp, wheezy, trilled ***kikikikikiki*** or whirring ***trrrrr-i-i***. **HABITAT** Mangroves, coastal scrub, sometimes parks and gardens; lowlands. **RANGE & STATUS** Resident Greater Sundas, W Lesser Sundas. **SE Asia** Uncommon to locally common resident (mostly coastal) Peninsular Malaysia, Singapore (also occurs inland). **BREEDING** March–August. **Nest** Excavated hole in tree; 4.5 m above ground. **Eggs** 2; white; 17–17.5 x 14 mm.

559 **GREY-CAPPED PYGMY WOODPECKER**
Dendrocopos canicapillus Plate **61**
IDENTIFICATION 13–15.5 cm. *D.c.canicapillus*: **Male** Combination of small size, pale greyish crown, contrasting black crown-sides, nape and faint submoustachial stripe, white-barred blackish upperside and dark streaked underparts diagnostic. Crown often vaguely brown-tinged, has short red streak on sides of hindcrown, sides of head and neck whitish with distinctive dark stripe through ear-coverts to side of neck. *D.c.kaleensis* (syn. *obscurus*, *tonkinensis*: N Myanmar, N Indochina) is somewhat larger with more extensive black mantle and darker buffish-brown wash on underparts. See Sunda Pygmy Woodpecker. Note range. **Female** Lacks red on crown. **Juvenile** Somewhat darker overall, underpart streaks heavier. Male often has more extensive and orange-red (rather than red) on nape and rear crown-sides. **Other subspecies in SE Asia** *D.c.delacouri* (south-eastern Thailand, Cambodia, Cochinchina), *auritus* (S Thailand southwards). **VOICE** Male territorial call is a rattling ***tit-tit-erh-r-r-r-r-h***, usually introduced by call. Drumming fairly subdued. Call is a short ***kik*** or ***pit*** and squeaky ***kweek-kweek-kweek***. **HABITAT & BEHAVIOUR** Broadleaved evergreen and deciduous forest, secondary growth, coastal scrub; up to 1,830 m (lowlands only Peninsular Malaysia). Joins bird waves. **RANGE & STATUS** Resident NE Pakistan, NW,N,NE Indian subcontinent, China (except NW,N), Ussuriland, N,S Korea, Taiwan, Sumatra, Borneo. **SE Asia** Common resident (except Singapore). **BREEDING** December–April. **Nest** Excavated tree-hole; 5–10 m above ground. **Eggs** 3–5; glossy white; 18.6 x 14.4 mm (av.; *canicapillus*).

560 **FULVOUS-BREASTED WOODPECKER**
Dendrocopos macei Plate **61**

IDENTIFICATION 18.5–21 cm. *D.m.macei*: **Male** Combination of red crown to upper nape, black upperside with close white bars, and black central tail diagnostic. Buffy-whitish below with short dark streaks on breast, distinctly dark-barred lower flanks, and reddish vent. See Spot-breasted, Stripe-breasted and Yellow-crowned Woodpeckers. Note habitat and range. **Female** Crown and nape all black. See Grey-capped Pygmy Woodpecker. **Juvenile** Duller, red of undertail-coverts paler and more restricted, both sexes with some red on crown-centre (particularly male). **VOICE** Drumming short, subdued and moderately accelerating, becoming weaker near end. Call is an explosive, very short, sharp, slightly nasal ***SKIK***, repeated every few seconds. Occasionally a grouchy, growling ***kik-i-derr***. **HABITAT** Deciduous woodland, scattered trees in open country, bamboo, secondary growth; up 1,220 m. **RANGE & STATUS** Resident NE Pakistan, NW,N,NE Indian subcontinent, E India. **SE Asia** Uncommon to fairly common resident W,N Myanmar. **BREEDING** December–June. **Nest** Excavated hole in tree or bamboo; 1–3 m above ground. **Eggs** 3–5, white; 22.2 x 16.4 mm.

561 **SPOT-BREASTED WOODPECKER**
Dendrocopos analis Plate **61**

IDENTIFICATION 17–18 cm. *D.a.longipennis*: **Male** Smaller than Fulvous-breasted Woodpecker, lacks red on hindcrown/upper nape, whiter below, with more spot-like breast markings, only vaguely barred lower flanks, and pinker undertail-coverts; central tail-feathers barred white. Note habitat and range. **Female** Differs from male by all black crown and nape. **Juvenile** Duller, pink of undertail-coverts paler and more restricted, both sexes with some red on crown-centre (particularly male). **VOICE** Drumming short and weak. Calls with an explosive ***TCHICK***. Also, a resonant ***chu-ik***, a less disyllabic ***kui***, and a loud, rising, chattering ***kut ku kut kutt-rrr-it***, and softer chattering ***chik-a-chik-a-chit***. **HABITAT** Deciduous woodland, scattered trees in open country, gardens, plantations; up to 600 m, locally 1,220 m. **RANGE & STATUS** Resident Andaman Is, S Sumatra, Java. **SE Asia** Uncommon to fairly common resident SW,C,S,E Myanmar, Tenasserim, NW,W,C Thailand, Cambodia, C,S Laos, S Annam, Cochinchina. **BREEDING** December–June. **Nest** Excavated hole in tree or bamboo; 1–3 m above ground. **Eggs** 3–5, white; 21.1 x 15.7 mm (av.).

562 **STRIPE-BREASTED WOODPECKER**
Dendrocopos atratus Plate **61**

IDENTIFICATION 20.5–22 cm. Monotypic. **Male** Like Fulvous-breasted Woodpecker, but a little larger, upper mantle unbarred, underparts more uniformly and distinctly streaked, base colour of belly darker, dusky buffish golden-brown. Note habitat and range. **Female** Crown and nape all black. See Grey-capped Pygmy Woodpecker. **Juvenile** Much paler and greyer below with less distinct streaks on belly and paler, more flame-red undertail-coverts. Male has paler red on crown, female some red on centre of crown. **VOICE** Male territorial call is a whinnying rattle. Call is a loud ***tchik***, similar to Great Spotted Woodpecker. **HABITAT** Broadleaved evergreen forest; 800–2,200 m (locally down to 230 m). **RANGE & STATUS** Resident NE India, SW China, possibly Bangladesh. **SE Asia** Fairly common to common resident SW,W,S,E Myanmar, Tenasserim, W,NW,NE Thailand, Laos, southern C Annam. **BREEDING** February–May. **Nest** Excavated tree-hole; up to 20 m above ground. **Eggs** 4–5; white; 21.3 x 16.9 mm (av.).

563 **YELLOW-CROWNED WOODPECKER**
Dendrocopos mahrattensis Plate **61**

IDENTIFICATION 17–19 cm. *D.m.aurocristatus*: **Male** Dull brownish-yellow forecrown, grading to bright red hindcrown and dense white bars and spots on upperside diagnostic. Uppertail-coverts largely white with dark central streaks, has faint dull brownish head markings, indistinct streaks on underparts and red centre of belly but no red on undertail-coverts. See Fulvous-breasted and Spot-breasted Woodpeckers. **Female** Crown dull brownish-yellow, grading to blackish-brown nape. **Juvenile** Browner above, underpart-streaks more diffuse, belly-patch pinker. Male has some orange-red on hindcrown, female a few orange-red feathers on centre of crown. **VOICE** Male territorial call is a rapid ***kik-kik-kik-r-r-r-h***. Also drums. Calls with a sharp ***click-click*** and feeble ***peek*** notes. **HABITAT** Deciduous woodland, scattered trees in open country, scrub; up to 915 m. **RANGE & STATUS** Resident Indian subcontinent. **SE Asia** Rare to uncommon resident SW,W,C,S Myanmar, C Thailand, east Cambodia, S Laos, S Annam. **BREEDING** February–April. **Nest** Excavated tree-hole; 1–10 m above ground. **Eggs** 3; glossy, white; 22.2 x 16.4 mm (av.).

564 **CRIMSON-BREASTED WOODPECKER**
Dendrocopos cathpharius Plate **61**

IDENTIFICATION 17–17.5 cm. *D.c.tenebrosus*: **Male** Relatively small with distinctive red breast-patch. Has black crown, red patch on hindcrown and nape, all-black mantle and scapulars, large white patch on wing-coverts, heavy dark streaks on underparts, pinkish-red centre of lower belly and red undertail-coverts. Black submoustachial band continues down side of breast but does not join shoulder. *D.c.pyrrhothorax* (W Myanmar) has entirely scarlet-red nape, more scarlet-red breast-patch and more black on side of neck. See Darjeeling and Great Spotted Woodpeckers. **Female** Lacks red on hindcrown and nape, red breast-patch duller and smaller. *D.c.pyrrhothorax* has more black on neck-side. **Juvenile** Upperparts duller, underparts whiter with more diffuse streaks and no red breast-patch, red of undertail-coverts paler (may be lacking), both sexes have orange-red on hindcrown and nape (less on female). **VOICE** Male territorial call is a fast, descending rattle. Also drums. Calls include a loud ***chip*** or ***tchik*** (higher-pitched than Darjeeling) and shrill ***kee-kee-kee***. **HABITAT** Broadleaved evergreen forest; 1,200–2,800 m. **RANGE & STATUS** Resident Nepal, Bhutan, NE India, SE Tibet, SW,W,C China. **SE Asia** Scarce to uncommon resident W,C,N,E Myanmar, NW Thailand, N Laos, W Tonkin. **BREEDING** India: April–May. **Nest** Excavated tree-hole. **Eggs** 2–4; white; 23.3 x 16.8 mm (av.).

565 **DARJEELING WOODPECKER**
Dendrocopos darjellensis Plate **61**

IDENTIFICATION 23.5–25.5 cm. Monotypic. **Male** Distinctive. Relatively large and long-billed, with small scarlet-red patch on hindcrown/upper nape, all-black mantle and scapulars, large white patch on wing-coverts, bright golden-buff sides of neck, rich buff breast and belly with bold blackish streaks and red undertail-coverts. See Stripe-breasted and Great Spotted Woodpeckers. **Female** Lacks red on hindcrown/upper nape. **Juvenile** Lacks golden-buff on sides of neck, underparts duller with faint dark throat-streaks, undertail-coverts paler and duller red. Male has pale flame-red markings on most of crown, female some on centre of crown or none. **VOICE** Male territorial call is a fast rattling ***di-di-di-d-dddddt***. Both sexes drum (similar to Great Spotted). Calls with a loud ***tsik***, like Great Spotted. **HABITAT** Broadleaved evergreen forest; 1,525–2,800 m. **RANGE & STATUS** Resident Nepal, Bhutan, NE India, S,SE Tibet, SW,W China. **SE Asia** Uncommon resident W,N Myanmar, north W Tonkin. **BREEDING** India: April–May. **Nest** Excavated tree-hole; 1–2 m above ground. **Eggs** 2–4; white; 27.8 x 19.8 mm (av.).

566 **GREAT SPOTTED WOODPECKER**
Dendrocopos major Plate **61**

IDENTIFICATION 25.5–28 cm. *D.m.cabanisi*: **Male** Combination of relatively large size, all-black mantle and scapulars, large white patch on wing-coverts and unstreaked brownish-white underparts diagnostic. Patch on hindcrown/upper nape and vent red. *D.m.stresemanni* (W,N Myanmar) is darker and browner on sides of head and underparts, may show some red on breast. See Darjeeling and Crimson-breasted Woodpeckers. **Female** Crown and nape all black. **Juvenile** Crown-feathers red with narrow black edges (both sexes), upperparts duller, submoustachial band less defined, post-auricular band sometimes broken and may not extend to nape, underparts duller, often with dusky streaks on flanks, lower flanks sometimes barred, vent pinker, occasionally buffy or whitish. **VOICE** Male territorial call is a ***kix-krrarraarr***. Both sexes drum. Call is a sharp ***kix***. **HABITAT** Broadleaved evergreen forest; 1,000–2,745 m, sometimes down to 450 m in winter. **RANGE & STATUS** Resident (subject to some minor movements) Palearctic, NE India, SE Tibet, China, N,S Korea, Japan. **SE Asia** Scarce to uncommon resident W,N,E Myanmar, N Laos, W,E Tonkin. Winter visitor only in parts of E Tonkin. **BREEDING** March–April. **Nest** Excavated hole in tree; 2–15 m above ground. **Eggs** 3; white; 27.6 x 19.6 mm (av.; *cabanisi*).

567 RUFOUS WOODPECKER

Micropternus brachyurus Plate **63**

IDENTIFICATION 25 cm. *M.b.phaioceps*: **Male** Distinctive smallish rufescent-brown woodpecker with short stout blackish bill and blackish bars on upperside and flanks. Head dull brown with red patch on cheeks, and dark-speckled pale throat. *M.b.squamigularis* (S Thailand southwards) has more heavily marked throat and barred belly; *fokiensis* (Tonkin, N Annam) has cream-coloured head with heavy streaks on crown, nape and throat, dark dusky underparts and broader dark barring on upperside; *annamensis* (southern Indochina, N Laos) is similar to *fokiensis* but has darker head. See Bay Woodpecker. **Female** Head paler, lacks red on ear-coverts. *M.b.fokiensis* has creamy base colour on head and darker base colour on underparts; *annamensis* is similar to *fokiensis* but has darker head. **Juvenile** Like adult. **VOICE** Male territorial call is a nasal laughing ***kweep-kweep-kweep***. Also utters a long, slightly descending and accelerating series of notes. Drumming (both sexes) diagnostic, decelerating to a halt: ***bdddddd d d d dt***. **HABITAT** Broadleaved evergreen and deciduous forest, forest edge, secondary growth; up to 1,450 m (below 1,050 m in Thailand and Peninsular Malaysia). **RANGE & STATUS** Resident N,NE Indian subcontinent, SW,C,E India, Sri Lanka, SE Tibet, southern China, Greater Sundas. **SE Asia** Fairly common to common resident throughout; uncommon Singapore. **BREEDING** January–June. **Nest** Excavated hole in occupied nest of tree ants, occasionally in tree; 4–5 m above ground. **Eggs** 2–3; white (staining to pale brown); 27.9 x 19.6 mm (av.; *phaioceps*).

568 WHITE-BELLIED WOODPECKER

Dryocopus javensis Plate **62**

IDENTIFICATION 37.5–43 cm. *D.j.feddeni*: **Male** Very large size and black plumage with white rump, lower breast and belly diagnostic. Crown and submoustachial patch red. *D.j.forresti* (N Myanmar, north W Tonkin) is larger (43.5–47 cm); *javensis* (S Thailand southwards) has black rump (juvenile may show some white). **Female** Red on head restricted to hindcrown. **Juvenile** Duller than adult with paler throat. Male has black mottling on forecrown and much smaller red submoustachial patch. **VOICE** Staccato ***kek-ek-ek-ek-ek*** and ***kiau-kiau-kiau***. Loud accelerating drumming (both sexes). Typical call is a loud explosive ***keer*** or ***kyah***. **HABITAT** Deciduous and broadleaved evergreen forest, sometimes coniferous forest and mangroves. Up to 915 m (to 1,450 m S Annam); above 1,525 m N Myanmar and north W Tonkin. **RANGE & STATUS** Resident peninsular India, SW China, S Korea, Greater Sundas, Philippines; formerly Japan (Tsushima). **SE Asia** Scarce to locally fairly common resident (except C Thailand, N Laos, E Tonkin, N Annam); rare Singapore. **BREEDING** December–June. **Nest** Excavated tree-hole; 8–16 m above ground. **Eggs** 2–4; white; 33.5 x 24.9 mm (av.; *feddeni*).

569 BANDED WOODPECKER

Chrysophlegma miniaceus Plate **62**

IDENTIFICATION 25.5–27 cm. *C.m.malaccensis*: **Male** Resembles Crimson-winged Woodpecker but ear-coverts reddish-rufous with faint pale streaks, red of crown extends to sides of nape, mantle and scapulars more distinctly scaled pale yellowish, underparts brownish-white with dull rufescent-brown wash on side of neck and breast, broad but diffuse dark bars on breast to vent, and pale bars on primaries. *C.m.perlutus* (W Thailand) has narrower dark bars on underparts. **Female** Lacks reddish-rufous on ear-coverts and has whitish speckles on forehead and face. **Juvenile** Duller than respective adults, forehead and crown initially dull brown with red restricted to rear, mantle plainer, underparts less distinctly barred. **VOICE** Male territorial call is a series of 1–7 mournful, falling ***peew*** or ***kwee*** notes. Drumming unrecorded. Call is a short ***keek***. **HABITAT** Broadleaved evergreen forest, secondary growth, plantations, sometimes mangroves; up to 1,225 m. **RANGE & STATUS** Resident Greater Sundas. **SE Asia** Uncommon to fairly common resident south Tenasserim, W(south),S Thailand, Peninsular Malaysia, Singapore. **BREEDING** November–September. **Nest** Excavated tree-hole; 8–20 m above ground. **Eggs** 2–3; glossy white; 26.9–27.7 x 19.5–20.5 mm (av.; *malaccensis*).

570 GREATER YELLOWNAPE

Chrysophlegma flavinucha Plate **62**

IDENTIFICATION 31.5–35 cm. *C.f.lylei*: **Male** Distinctive. Relatively large with pronounced yellow nuchal crest, crown dull rufous, upperparts unmarked olive-green, throat yellow, lowermost throat and uppermost breast white with broad blackish streaks; head sides and breast plain dark olive, rest of underparts unmarked olive-whitish, primaries reddish-rufous with black bars. *C.f.flavinucha* (Myanmar, W Thailand) tends to only have dull rufous-chestnut on forecrown and darker sides of neck and breast; *archon* (NW[east],NE[north] Thailand, Laos, W Tonkin, N,C,S Annam) and *pierrei* (NE[south],SE Thailand, Cambodia, Cochinchina) have black streaks extending further up throat; *wrayi* (Peninsular Malaysia) is smaller (29.5–31.5 cm), darker overall, throat more extensively dark (leaving defined yellow chin and submoustachial band), rearmost crown duller and more olive, crown indistinctly washed dull rufous-chestnut, nuchal crest shorter with less yellow. **Female** Hindcrown olive-tinged, chin and submoustachial stripe dull rufous-chestnut, upper throat streaked/mottled blackish and whitish. *C.f.wrayi* is darker overall with more uniform underparts, crest differs as male. **Juvenile** Underparts duller, may appear faintly barred on belly. Male initially has olive scales on crown, yellow of throat more buffy-whitish. **Other subspecies in SE Asia** *C.f.styani* (E Tonkin). **VOICE** Male territorial call is an accelerating ***kwee-kwee-kwee-kwee-kwee-kwee-kwee-kwee-kwi-kwi-kwi-kwi-wi-wi-wi-wi-wik***. Weak, even drumming is infrequently heard. Call is a loud, disyllabic ***kyaa*** or ***kiyaep***. **HABITAT** Broadleaved evergreen and deciduous forest, native pine forest; up to 2,745 m (above 915 m in Peninsular Malaysia). **RANGE & STATUS** Resident N,NE,E Indian subcontinent, SE Tibet, SW,S,SE China, Sumatra. **SE Asia** Common resident (except C,S Thailand, Singapore). **BREEDING** February–May. **Nest** Excavated hole in tree; 2–6 m above ground. **Eggs** 2–4; white; 28.8 x 22.2 mm (av.; *flavinucha*).

571 CHECKER-THROATED WOODPECKER

Chrysophlegma mentalis Plate **62**

IDENTIFICATION 26.5–29.5 cm. *C.m.humii*: **Male** Similar to Crimson-winged Woodpecker but crown olive-coloured, sides of neck and upper breast chestnut, submoustachial stripe speckled blackish-brown and rufous-whitish, rest of throat whitish with broad dark streaks, belly and flanks unbarred, primaries blackish with rufous bars. See Banded Woodpecker. **Female** Submoustachial stripe, sides of neck and upper breast dull chestnut. **Juvenile** Crown and underparts browner, wings duller. **VOICE** Territorial male utters a long series of ***wi*** notes, similar to Greater Yellownape.

Drums in short bursts. Calls include a single ***kyick*** and ***kiyee..kiyee..kiyee...***, with stressed first syllable. **HABITAT** Broadleaved evergreen forest, sometimes mangroves; up to 1,220 m. **RANGE & STATUS** Resident Greater Sundas. **SE Asia** Uncommon to common resident south Tenasserim, W(south),S Thailand, Peninsular Malaysia; formerly Singapore. **BREEDING** March–August. **Nest** Excavated hole in tree, 5–6 m above ground. **Eggs** 2–3; white.

572 LESSER YELLOWNAPE
Picus chlorolophus Plate **62**

IDENTIFICATION 25–28 cm. *C.c.chlorolophus* (syn. *chlorolophoides*): **Male** Similar to Greater Yellownape but smaller, with narrowly red crown-sides, red submoustachial stripe, narrow white moustachial stripe, no yellow on throat, olive and whitish bars on lower breast to undertail-coverts, and unbarred primaries. *C.c.rodgeri* (Peninsular Malaysia) has red crown with blackish feather-tips and blackish centre, and much darker side of head, neck and underparts, with only narrow pale bars on lower breast to vent; *laotianus* (NW[east],NE[north] Thailand, N Laos) has more red on crown; *citrinocristatus* (Tonkin, N Annam) has more red on crown, almost lacks red submoustachial stripe, and has duller, darker underparts with pale barring restricted to flanks; *annamensis* (NE[south-west],SE Thailand, southern Laos, C,S Annam) and *krempfi* (Cochinchina) have more red on crown, darker upperparts and whiter lower underparts. **Female** Red on head restricted to rear crown-sides. **Juvenile** Similar to respective adults but crown/nape duller, breast more distinctly barred. Male lacks red submoustachial stripe. **VOICE** Male territorial call is a far-carrying, plaintive ***peee-uu*** or ***pee-a***. Also utters a slightly descending series of up to 10 ***kwee*** or ***kee*** notes. Occasionally drums. Call is a short ***chak***. **HABITAT** Broadleaved evergreen and deciduous forest; up to 1,830 m (above 1,065 m in Peninsular Malaysia). **RANGE & STATUS** Resident N,NE Indian subcontinent, E and peninsular India, Sri Lanka, S China. **SE Asia** Common resident (except C,S Thailand, Singapore). **BREEDING** January–June. **Nest** Excavated hole in tree; 1.5–20 m above ground. **Eggs** 2–5; white; 24.3 x 19 mm (av.; *chlorolophus*).

573 CRIMSON-WINGED WOODPECKER
Picus puniceus Plate **62**

IDENTIFICATION 24–28 cm. *P.p.observandus*: **Male** Olive-greenish body, yellow nuchal crest and deep red crown, broad submoustachial stripe and wings diagnostic. Flanks and belly mottled/barred whitish, primaries unbarred, eyering pale greyish-blue. See Banded and Checker-throated Woodpeckers. **Female** Lacks red submoustachial stripe, underparts paler and plainer. **Juvenile** Body duller and greyer; red of head restricted to hindcrown; sides of head and neck and underparts speckled whitish. Male has much smaller red submoustachial patch (may be lacking). **VOICE** Male territorial call is a ***pee-bee***, with emphasis on first syllable, sometimes extended to ***pee-hee-hee-hee***. Occasionally ***peep*** or falling ***pi-eew***. Drums weakly. **HABITAT** Broadleaved evergreen forest, secondary growth, plantations; up to 860 m (below 600 m in Thailand). **RANGE & STATUS** Resident Greater Sundas (except Bali). **SE Asia** Fairly common to common resident south Tenasserim, W(south),S Thailand, Peninsular Malaysia. Rare (formerly resident, but current status unclear) Singapore. **BREEDING** February–June. **Nest** Excavated tree-hole; up to 18 m above ground. **Eggs** 2–3; white.

574 GREY-HEADED WOODPECKER
Picus canus Plate **62**

IDENTIFICATION 30.5–34.5 cm. *P.c.hessei* (syn. *gyldenstolpei*): **Male** Distinctive. Forecrown red, hindcrown and centre of nape black, sides of head plain grey with black loral stripe and solid black submoustachial stripe, throat dusky whitish with brown tinge, rest of underparts unmarked greyish-olive. *P.c.robinsoni* (Peninsular Malaysia) has much darker body, darker grey sides of head, and longer, thinner bill; *sobrinus* (E Tonkin) has golden-tinged upperparts and greener underparts. **Female** Crown all black with grey streaks. **Juvenile** Upperparts, breast and belly duller, mantle and scapulars appear slightly mottled, submoustachial stripe less well defined, sometimes has bars/mottling on belly. Male has less red on forecrown. See Streak-breasted, Laced and Streak-throated Woodpeckers. **Other subspecies in SE Asia** *P.c.sordidior* (north E Myanmar). **VOICE** Male territorial call is a far-carrying, descending ***kieu... kieu...kieu...kieu*** (3–4 or more notes). Drums in long rolls. Calls with a short ***kik*** and ***keek..kak-kak-kak***. **HABITAT** Open forest of various types, including deciduous and native pine forest; up to 2,135 m (915–1,830 m in Peninsular Malaysia). **RANGE & STATUS** Resident Europe, Turkey, C Palearctic, Siberia, NE Pakistan, NW,N and north-eastern Indian subcontinent, S,SE Tibet, China, Taiwan, N,S Korea, N Japan, Sumatra. **SE Asia** Scarce to fairly common resident (except C,S Thailand, Singapore, N Annam). **BREEDING** April–June. **Nest** Excavated or natural tree-hole; 0.15–8 m above ground. **Eggs** 4–5; white; 29.5 x 22.8 mm (av.; *hessei*).

575 BLACK-HEADED WOODPECKER
Picus erythropygius Plate **62**

IDENTIFICATION 31–35 cm. *P.e.nigrigenis*: **Male** Black head and band down centre of nape to upper mantle, red patch on centre of crown, yellow throat and sides of nape, and red rump diagnostic. Eyes pale yellow, bill blackish, lower breast and belly whitish with fairly distinct dark scales. *P.e.erythropygius* (NE Thailand, Indochina) has smaller red crown-patch. **Female** Crown all black. **Juvenile** Duller above, throat paler, upper breast more buffish, underpart scales more diffuse. Male has red crown-patch duller and very washed-out. **VOICE** Territorial male utters an undulating, yelping laugh ***ka-tek-a-tek-a-tek-a-tek...*** or rapid ***cha-cha-cha...cha-cha-cha*** with stressed first notes. Call is a loud double note. **HABITAT** Dry dipterocarp, deciduous and pine forest; up to 1,000 m. **RANGE & STATUS** Scarce to uncommon resident C,S,E Myanmar, north Tenasserim, W,NW,NE Thailand, Cambodia, Laos (north to Vientiane area), C(south-west),S Annam, north Cochinchina. **BREEDING** February–June. **Nest** Excavated hole in tree. **Eggs** 3–4; white; 27.9 x 20.3 mm (av.; *nigrigenis*).

576 STREAK-BREASTED WOODPECKER
Picus viridanus Plate **62**

IDENTIFICATION 30.5–32.5 cm. *P.v.viridanus*: **Male** Very similar to Laced Woodpecker but sides of neck and throat duller and more olive-tinged, throat with faint whitish streaks, upper breast paler with dark olive streaks. Crown and nape red. *P.v.weberi* (southern Tenasserim and S Thailand southward) is smaller (28–31 cm) with somewhat darker body. See Streak-throated Woodpecker. Note habitat and range. **Female** Forehead to nape black, throat and breast duller. **Juvenile** Duller than respective adult with more indistinct and diffuse underpart markings (particularly on throat and breast); flanks and belly appear more scaled. Male has more orange-red crown. **VOICE** Drumming unrecorded. Call is an explosive ***kirrr***. Also gives a series of four or more ***tcheu*** notes. **HABITAT** Broadleaved evergreen forest, coastal scrub, mangroves; lowlands. **RANGE & STATUS** Resident SW Bangladesh (Sundarbans). **SE Asia** Uncommon to locally fairly common resident SW,S,C,south E Myanmar, Tenasserim, W,S Thailand, extreme north-west Peninsular Malaysia. **BREEDING** February–May. **Nest** Excavated hole in tree. **Eggs** 4; white; 28.2 x 21.3 mm (av.; *viridanus*).

577 LACED WOODPECKER
Picus vittatus Plate **62**

IDENTIFICATION 27–33 cm. Monotypic. **Male** Combination of unmarked, mostly olive-green upperside, unmarked buffy-olive throat, sides of neck and upper breast, and red crown and nape distinctive. Rest of underparts olive-whitish with dark olive streaks/loops; has grey ear-coverts, narrow black line from extreme forehead to above eye, narrow whitish post-ocular and moustachial streaks and broad black submoustachial stripe with whitish mottling. Subject to some individual variation; southern birds are smaller. See Streak-breasted and Streak-throated Woodpeckers. **Female** Forehead to nape black, sides of neck, throat and upper breast more olive-tinged. **Juvenile** Duller than respective adults with more scaled appearance on belly; may show some faint streaks on lower throat and breast. Red on crown of male paler and less extensive. **VOICE** Male territorial call is a long series of notes similar to Grey-headed Woodpecker but delivered faster, with each note shorter and lower-pitched. Drums in steady rolls. Call is a loud, short, abrupt ***ik*** or ***yik***. **HABITAT** Broadleaved evergreen and deciduous forest, secondary growth, coastal scrub, mangroves, gardens, plantations, bamboo; up to 1,525 m (lowlands only Peninsular Malaysia, Singapore). **RANGE & STATUS** Resident Sumatra, Java, Bali, Kangean Is. **SE Asia** Common resident E Myanmar, Tenasserim, Thailand (except S), southern Peninsular Malaysia (also Langkawi Is), Singapore (uncommon), Indochina (except W,E Tonkin). **BREEDING** December–July. **Nest** Excavated hole in tree, 0.5–9 m above ground. **Eggs** 3–4; white; 27.7 x 21.3 (mean).

578 STREAK-THROATED WOODPECKER
Picus xanthopygaeus Plate **62**

IDENTIFICATION 27.5–30 cm. Monotypic. **Male** Similar to Streak-breasted and Laced Woodpeckers but has pale eyes, clearly defined white supercilium (from eye back), less black on submoustachial stripe and prominently streaked ear-coverts, throat, sides of neck and breast. Crown and nape red. **Female** Forehead to nape black with distinctive grey streaks on forehead and crown. **Juvenile** Duller than respective adult with less distinct underpart markings; appears somewhat more scaled/barred on belly. Male has less red on crown and nape, female less distinct crown-streaks. **VOICE** Drums. Call is a sharp ***queemp***. **HABITAT** Deciduous forest, scattered trees in open areas; up to 500 m, locally higher in Myanmar. **RANGE & STATUS** Resident India, Nepal, Bhutan, Bangladesh, Sri Lanka, SW China. **SE Asia** Scarce to common resident W,C,S,E Myanmar, W,NW(west),NE(south-west) Thailand, Cambodia, S Laos, Cochinchina. **BREEDING** March–May. **Nest** Excavated tree-hole; 0.6–8 m above ground. **Eggs** 3–5; white; 26.2 x 20.1 mm (av.).

579 RED-COLLARED WOODPECKER
Picus rabieri Plate **62**

IDENTIFICATION 30–32 cm. Monotypic. **Male** All-red head and upper breast, apart from largely olive-greyish sides of head and buffish olive-grey throat diagnostic. Otherwise resembles Laced Woodpecker but underparts unstreaked greenish-olive with vague broken whitish bars on belly. **Female** Forehead and crown blackish, less red on submoustachial stripe. **Juvenile** Similar to female but duller overall. **Immature male** has more orange-red on head and upper breast, crown extensively mixed with black. **VOICE** Drums in fast rolls. **HABITAT & BEHAVIOUR** Broadleaved evergreen forest, secondary growth; up to 700 m, exceptionally 1,050 m. Often feeds on or near ground; joins bird waves, which typically include laughingthrushes and other woodpeckers. **RANGE & STATUS** Resident SW China (extreme S Yunnan). **SE Asia** Scarce to uncommon resident north-east Cambodia, Laos, W,E Tonkin, N,C Annam. **BREEDING** May–June. Otherwise unknown.

580 OLIVE-BACKED WOODPECKER
Dinopium rafflesii Plate **63**

IDENTIFICATION 28 cm. *D.r.rafflesii* (syn. *peninsulare*): **Male** Similar to flamebacks but easily separated by olive-green mantle, lack of red rump and unscaled dull olive-coloured underparts. Forehead to nape and crest red. Bold black and white head pattern eliminates other green woodpeckers. **Female** Forehead to nape and crest black. **Juvenile** Duller. Male has red of head restricted to crest, forehead sometimes spotted red. **VOICE** Male territorial call is a slow, variable ***chak chak chak chak chak-chak*** (6–30 or more notes) or faster, more regular series of 10–50 notes. Drumming unrecorded. Calls include a single ***chak***, soft trilling ***ti-i-i-i-i*** and squeaky ***tiririt***. **HABITAT** Broadleaved evergreen forest; up to 1,200 m. **RANGE & STATUS** Resident Sumatra, Borneo. **SE Asia** Scarce to uncommon resident south Tenasserim, W(south),S Thailand, Peninsular Malaysia; formerly Singapore. **BREEDING** March–May. **Nest** Excavated hole in tree, 4.5–9 m above ground; otherwise unknown.

581 HIMALAYAN FLAMEBACK
Dinopium shorii Plate **63**

IDENTIFICATION 30 cm. *D.s.anguste*: **Male** Like Common and Greater Flamebacks but has pale brownish submoustachial stripe and redder mantle. From Common also by less heavily marked breast, and from Greater by all-black nape and upper mantle and lack of black on sides of forecrown and lores. Crown and crest red. See Black-rumped Flameback. **Female** Crown and crest black with white streaks. From Common by black submoustachial loop. From Greater by paler lores, streaked (rather than spotted) crown and black hindneck. **Juvenile** Browner with more obscurely marked underparts. Male has red of head restricted to crest, forehead and crown brownish-buff with paler streaks; female has brown crown and crest with broad pale streaks. **VOICE** A rapid, tinny ***klak-klak-klak-klak-klak***, slower and quieter than Greater. **HABITAT** Deciduous and semi-evergreen forest; up to 1,220 m. **RANGE & STATUS** Resident N,NE Indian subcontinent, SE India. **SE Asia** Uncommon resident SW,W,N,C,S Myanmar. **BREEDING** India: April–May. **Nest** Excavated hole in tree. **Eggs** 2–3, white; 29.9 x 20.8 mm (av.).

582 COMMON FLAMEBACK
Dinopium javanense Plate **63**

IDENTIFICATION 28–30 cm. *D.j.intermedium*: **Male** Like Greater and Himalayan Flamebacks but has prominent black submoustachial line (not looped). From Greater by all-black nape and upper mantle and lack of black on sides of forecrown and lores. From Himalayan by less reddish mantle and more heavily marked breast. Crown and crest red. See Black-rumped Flameback. **Female** Crown and crest black with white streaks. From Greater by paler lores and streaked (not spotted) crown. **Juvenile** Breast more blackish-brown with white spots, lower underparts more obscurely marked. Male has mostly black forehead and crown and red crest; female has crown more spotted than streaked. **Other subspecies in SE Asia** *D.j.javanense* (S Thailand southwards). **VOICE** A long, trilled ***ka-di-di-di-di-di-di...***, recalling some *Porzana* crakes; faster and less metallic than Greater. Drumming softer than Greater. Calls include a single or double ***kow*** note and ***kowp-owp-owp-owp***, uttered in flight. **HABITAT** Open deciduous forest, scrub, gardens, plantations, sometimes mangroves; up to 800 m, locally higher in Myanmar. **RANGE & STATUS** SW,NE India, Bangladesh, SW China, Greater Sundas, Philippines. **SE Asia** Uncommon to common resident (except N Myanmar). **BREEDING** January–October. **Nest** Excavated tree-hole; 1.4–10 m above ground. **Eggs** 2–3, white; 29 x 19.1 mm (av.; *intermedium*).

583 BLACK-RUMPED FLAMEBACK
Dinopium benghalense Plate **63**

IDENTIFICATION 26–29 cm. *D.b.benghalense*: **Male** From other flamebacks by white-streaked black throat, black lower back to uppertail-coverts and white spots on black outer lesser and median coverts. Also shows whitish bars on primaries, white streaks on black post-ocular stripe, and black sides of forehead. Crown and crest red. **Female** Forecrown black with white speckles; has broader white streaks on throat. **Juvenile** Duller with black parts browner, underparts duller and more obscurely marked. Male has narrowly red-tipped crown-feathers and sometimes some white spots; female has little or no spotting on forecrown. **VOICE** A whinnying ***kyi-kyi-kyi-kyi...*** Contact call is a single strident ***kierk***. **HABITAT** Light and more open woodland (mainly deciduous), plantations, groves of trees bordering villages and cultivation; lowlands. **RANGE & STATUS** Resident Indian subcontinent. **SE Asia** Common resident SW Myanmar. **BREEDING** India: February–July. **Nest** Excavated tree-hole; 1.5–15.2 m above ground. **Eggs** 2–3, glossy white; 28.1 x 20.9 mm (av.).

584 GREATER FLAMEBACK

Chrysocolaptes lucidus Plate **63**

IDENTIFICATION 29–32 cm. *C.l.guttacristatus*: **Male** Distinctive largish, long-billed woodpecker with red crown and pointed crest, bold black and white head pattern and olive-golden upperparts with red lower back and rump. Narrow black moustachial and malar lines form distinctive broad loop; has black lores, narrow black crown-sides, broad black band from eye to side of neck, and blackish-scaled white lower nape and upper mantle, breast and belly. *C.l.chersonesus* (Peninsular Malaysia) is smaller, slightly darker and more olive above with more red on back, has a broader white supercilium and broader black markings below. See other flamebacks. Note habitat and range. **Female** Crown and crest black with white spots/streaks. See other flamebacks. **Juvenile** Upperparts more olive-coloured, underparts duller and more obscurely marked. Male has less red on crown and variable pale spots on forehead and crown. **VOICE** A sharp, metallic, monotone ***tibitititititit...*** Drums loudly. Other calls include single ***kik*** notes. **HABITAT** Deciduous and broadleaved evergreen forest, forest edge, mangroves, old plantations; up to 1,200 m. **RANGE & STATUS** N,NE Indian subcontinent, SW,E India, SW China, Greater Sundas, Philippines. **SE Asia** Common resident (except C Thailand); local and coastal in Peninsular Malaysia. **BREEDING** March–May. **Nest** Excavated tree-hole; 2–20 m above ground. **Eggs** 4–5, white; 30 x 22.1 mm (av.; *guttacristatus*).

585 PALE-HEADED WOODPECKER

Gecinulus grantia Plate **63**

IDENTIFICATION 25 cm. *G.g.indochinensis*: **Male** Similar to Bamboo Woodpecker but mantle, scapulars, wings and tail maroon-chestnut, wings and tail have obvious pale barring, crown pale greenish with pinkish-red patch in centre. *G.g.grantia* (Myanmar) has more crimson crown-patch and redder upperside, yellower sides of head, more olive-coloured underparts and less extensive wing-barring. See Maroon Woodpecker. **Female** Lacks pinkish-red crown-patch. **Juvenile** Like female (both sexes) but upperparts mostly dark brown, underparts very dark brown to grey-brown. **VOICE** Male territorial call is a loud, strident laughing ***yi wee-wee-wee***, with shorter, stressed first note, recalling Bay Woodpecker. Drums in loud, fairly short, even-pitched bursts. Call is a harsh, high, quickly repeated ***grrrit-grrrit-grrrit*** and ***grridit grrit-grrit...*** etc. **HABITAT** Bamboo, broadleaved evergreen and semi-deciduous forest; up to 1,900 m. **RANGE & STATUS** Resident E Nepal, Bhutan, NE India, E Bangladesh, S China. **SE Asia** Rare to uncommon resident SW,W,N,C Myanmar, north NW Thailand (rare), east Cambodia, Laos, Vietnam. **BREEDING** India: March–May. **Nest** Excavated hole in tree or bamboo; 1–6 m above ground. **Eggs** 3; white; 25.7 x 19.2 mm (av.; *grantia*).

586 BAMBOO WOODPECKER

Gecinulus viridis Plate **63**

IDENTIFICATION 25–26 cm. *G.v.viridis*: **Male** Relatively small, bill pale, upperside greenish-olive, underside olive-brown, crown and nape red; head-sides and neck more buffish-olive, has red tips to rump feathers and uppertail-coverts. See Pale-headed Woodpecker. **Female** Head yellowish-olive overall. See Pale-headed. **Juvenile** Similar to female but darker and browner above, very dark, often grey-tinged below. **Other subspecies in SE Asia** *G.v.robinsoni* (S Thailand southwards). **VOICE** Male territorial call is a loud, shrill, monotone ***kyeek-kyeek-kyeek-kyeek***; ***keep-kee-kee-kee-kee-kee-kee*** or ***kwi-kwi-week-kweek-kweek***. Drums in short loud bursts. Calls with a dry undulating cackle, recalling Bay Woodpecker but slower; occasionally single ***bik*** notes. **HABITAT** Bamboo, broadleaved evergreen and deciduous forest; up to 1,400 m. **RANGE & STATUS** Uncommon resident S,E Myanmar, Tenasserim, Thailand (except C), Peninsular Malaysia, south-west N Laos. **BREEDING** April–May. **Nest** Excavated hole in bamboo, up to 5 m above ground. Otherwise unknown.

587 MAROON WOODPECKER

Blythipicus rubiginosus Plate **63**

IDENTIFICATION 23–24 cm. Monotypic. **Male** Similar to Bay Woodpecker but smaller, upperparts and wings unbarred maroon-chestnut (except flight-feathers and tertials), tail blackish with faint pale bars. Has red neck-patch and may show some red on submoustachial area. **Female** Lacks red on head. **Juvenile** Upperparts more rufescent; may show some red on crown. See Pale-headed and Rufous Woodpeckers. Note habitat and range. **VOICE** Male territorial call is a shrill, descending ***keek-eek-eek-eek-eek-eek***, higher-pitched than Bay. Drumming unrecorded. Calls include a wavering high-pitched ***kik-kik-kik-kik-kik-kik-kik-kik...***, slowing somewhat towards end (slower than similar call of Bay), and high-pitched, nervously repeated ***kik*** notes, sometimes ***kik-ik...kik-ik*** with second note upwardly inflected. **HABITAT** Broadleaved evergreen forest, secondary growth, bamboo; up to 1,525 m (below 900 m in Thailand). **RANGE & STATUS** Resident Sumatra, Borneo. **SE Asia** Common resident south Tenasserim, W(south),S Thailand, Peninsular Malaysia; formerly Singapore (one recent unconfirmed record). **BREEDING** December–May. Otherwise unknown.

588 BAY WOODPECKER

Blythipicus pyrrhotis Plate **63**

IDENTIFICATION 26.5–29 cm. *B.p.pyrrhotis*: **Male** Distinctive. Medium-sized with long pale bill and mostly dark rufescent plumage, has large red neck-patch, mantle and scapulars barred, wings heavily barred rufous and black, tail rufous with some black bars on outer feathers. Head somewhat paler with narrow rufous streaks on crown. *B.p.cameroni* (Peninsular Malaysia) and *annamensis* (S Annam) are darker and have smaller red neck-patch. See Rufous and Maroon Woodpeckers. Note habitat and range. **Female** Lacks red on head. **Juvenile** Head darker with paler crown-streaks, has more prominent barring above, underparts darker with faint rufous bars. Male has duller, more restricted red on neck. **VOICE** Loud harsh descending laughter: ***keek keek-keek-keek-keek-keek***. Drumming unrecorded. Calls include an undulating dry cackling ***dit-d-d-di-di-di-di-dit-d-d-di-di-di...***, harsh squirrel-like ***kecker-rak-kecker-rak...*** and loud chattering ***kerere-kerere-kerere...*** when agitated. **HABITAT** Broadleaved evergreen forest, sometimes semi-evergreen and mixed deciduous forest. Up to 2,770 m (mostly above 1,000 m); only above 1,065 m in Peninsular Malaysia. **RANGE & STATUS** Resident Nepal, NE Indian subcontinent, SE Tibet, S China. **SE Asia** Common resident (except C Myanmar, C,SE,S Thailand, Singapore). **BREEDING** March–June. **Nest** Excavated tree-hole; 1–4 m above ground. **Eggs** 2–4, white; 29.7 x 21.2 mm (av.; *pyrrhotis*).

589 ORANGE-BACKED WOODPECKER

Reinwardtipicus validus Plate **63**

IDENTIFICATION 30 cm. *R.v.xanthopygius*: **Male** Unmistakable. Medium-sized to large with red crown, nape and peaked crest, red underparts with brownish vent, and blackish-brown upperside with broad whitish to orange-buff stripe down centre of upperparts. Has broad rufous bands on flight-feathers. **Female** Crown and crest blackish-brown, back and rump whitish, underparts dull brownish. **Juvenile** Both sexes like female. Male may show some red on crown and orange-buff on rump. **VOICE** Rapid trilled ***ki-i-i-i-i-i-ik***. Drumming quite weak, in short bursts. Also squeaky anxious ***kit kit kit kit kit-it*** (with sharply rising last note) in alarm. **HABITAT** Broadleaved evergreen forest; up to 730 m. **RANGE & STATUS** Resident Sumatra, Java, Borneo. **SE Asia** Uncommon resident S Thailand, Peninsular Malaysia; formerly Singapore. **BREEDING** January–July. **Nest** Excavated hole in tree, 5 m above ground. **Eggs** 1–2; white; 25.9 x 20.5 mm.

590 BUFF-RUMPED WOODPECKER
Meiglyptes tristis Plate **63**

IDENTIFICATION 17 cm. *M.t.grammithorax*: **Male** Small and short-tailed with distinctive dense pale barring on blackish plumage, contrasting whitish-buff lower back and rump, buff loral line and eyering, and short red submoustachial stripe. See Buff-necked Woodpecker. **Female** Lacks red submoustachial stripe. **Juvenile** Somewhat darker with narrower pale body barring and more obscurely marked underparts. **VOICE** Male territorial call is a rapid trilled ***ki-i-i-i-i-i-i***. Drums in weak bursts. Calls with a single sharp ***pit*** (sometimes repeated) and longer ***pee*** notes. **HABITAT** Broadleaved evergreen forest, forest edge, secondary growth, sometimes plantations; up to 760 m. **RANGE & STATUS** Resident Sumatra, Java, Borneo. **SE Asia** Common resident south Tenasserim, S Thailand, Peninsular Malaysia; formerly Singapore. **BREEDING** March–July. **Nest** Excavated tree-hole; 1.5–15 m above ground. **Eggs** 2; white.

591 BLACK-AND-BUFF WOODPECKER
Meiglyptes jugularis Plate **63**

IDENTIFICATION 17–19.5 cm. Monotypic. **Male** Similar to Heart-spotted Woodpecker but nape and hindneck white, malar area and throat predominantly blackish, tertials white with black bars, short submoustachial stripe reddish. Forecrown and sides of head blackish with fine buffish-white bars, throat black with white speckles. **Female** Lacks reddish submoustachial stripe. Otherwise differs from Heart-spotted as male and also by dark forecrown. **Juvenile** Duller, head more clearly barred. **VOICE** Male territorial call is a high rattling ***titititititit'weerk'weerk'weerk...***, sometimes interspersed with nasal ***ki'yew*** notes. **HABITAT** Relatively open broadleaved evergreen and semi-evergreen forest, bamboo; up to 915 m. **RANGE & STATUS** Uncommon to locally fairly common resident SW,S,E Myanmar, Tenasserim, W,NW,NE,SE Thailand, Cambodia, Laos, Vietnam (except W Tonkin). **BREEDING** March–June. Otherwise unknown.

592 BUFF-NECKED WOODPECKER
Meiglyptes tukki Plate **63**

IDENTIFICATION 21 cm. *M.t.tukki*: **Male** Similar to Buff-rumped Woodpecker but somewhat larger and longer-tailed, darker overall with narrower pale bars on body and wings, unbarred crown and head-sides, sharply contrasting pale buff neck-patch and longer red submoustachial stripe. Upper breast blackish and more uniform (offsets neck-patch). **Female** Lacks red submoustachial stripe. **Juvenile** Pale barring broader, upper breast less contrastingly dark. Male may show some red on forehead and crown. **VOICE** Male territorial call is a high-pitched, monotone, trilled ***kirr-r-r***, recalling Buff-rumped. Both sexes drum. Other calls include a high-pitched ***ti ti ti ti...***, ***ki-ti ti ti ti...*** and single ***pee*** notes, like Buff-rumped. **HABITAT** Broadleaved evergreen forest; up to 1,250 m. **RANGE & STATUS** Resident Sumatra, Borneo. **SE Asia** Uncommon to fairly common resident south Tenasserim, S Thailand, Peninsular Malaysia; formerly Singapore. **BREEDING** February–July. **Nest** Excavated tree-hole; 1.5–5 m above ground. **Eggs** 2; white.

593 GREY-AND-BUFF WOODPECKER
Hemicircus concretus Plate **63**

IDENTIFICATION 14 cm. *H.c.sordidus*: **Male** Unmistakable. Small and very short-tailed with prominent triangular-shaped crest. Dark sooty-greyish with red crown, bold whitish-buff scales on mantle, scapulars, vent and wing-coverts, and whitish rump. Tertials boldly patterned whitish-buff and black. **Female** Lacks red crown. **Juvenile** Scaling buffier and more prominent, crown-feathers cinnamon-rufous with narrow black tips; both sexes show some red on crown. **VOICE** Drums weakly. Calls include a high-pitched drawn-out ***ki-yow*** or ***kee-yew***, sharp ***pit*** notes and vibrating ***chitterr***. **HABITAT** Broadleaved evergreen forest; up to 1,130 m. **RANGE & STATUS** Resident Greater Sundas. **SE Asia** Uncommon resident south Tenasserim, W(south),S Thailand, Peninsular Malaysia; formerly Singapore. **BREEDING** December–July. **Nest** Excavated tree-hole; 9–30 m above ground. Otherwise unknown.

594 HEART-SPOTTED WOODPECKER
Hemicircus canente Plate **63**

IDENTIFICATION 15.5–17 cm. Monotypic. **Male** Distinctive. Relatively small and short-tailed with prominent triangular crest. Mostly blackish with white throat, malar area and sides and front of neck and white lower scapulars and tertials with prominent black heart-shaped markings. **Female** Forecrown white. **Juvenile** Like female but whitish parts buffier, often with some black bars on forehead. **VOICE** Drumming weak. Calls include a nasal ***ki-yew***, with stressed second note, high-pitched ***kee-kee-kee-kee***, drawn-out grating ***chur-r*** and squeaky ***chirrick*** (often given in flight). **HABITAT** Deciduous, broadleaved evergreen and semi-evergreen forest, forest edge, bamboo; up to 915 m. **RANGE & STATUS** Resident India. **SE Asia** Fairly common resident Myanmar (except W,N,C), Thailand (except C and southern S), Cambodia, Laos, S Annam, Cochinchina. **BREEDING** September–April. **Nest** Excavated tree-hole; 1–12 m above ground. **Eggs** 2–3, white; 24.9 x 18 mm (av.).

595 GREAT SLATY WOODPECKER
Mulleripicus pulverulentus Plate **62**

IDENTIFICATION 45–51 cm. *M.p.harterti*: **Male** Unmistakable. Relatively very large and long-necked, mostly slaty-grey with golden-buff throat and foreneck and broad red submoustachial stripe. Head and neck speckled white. *M.p.pulverulentus* (southern S Thailand southwards) is much more blackish-slate overall. **Female** Lacks red submoustachial stripe. **Juvenile** Duller, head less distinctly speckled, throat and foreneck whitish. Males have larger red submoustachial stripe and may show red on crown. **VOICE** A loud, rapid wavering 2–5 note whinny: ***woi-kwoi-kwoi-kwoik...woi-kwoi-kwoi-kwoik...***, often given in flight. Drumming unrecorded. Calls with a single ***dwot*** and soft ***whu-ick***. **HABITAT** Deciduous, broadleaved evergreen and semi-evergreen forest, forest edge, mangroves; up to 1,065 m (below 215 m in Peninsular Malaysia). **RANGE & STATUS** Resident N,NE Indian subcontinent, SW China, Greater Sundas, Philippines (Palawan). **SE Asia** Uncommon to fairly common resident (except C Thailand, W,E Tonkin); formerly Singapore. **BREEDING** March–September. At least partly co-operative. **Nest** Excavated tree-hole; 9–45 m above ground. **Eggs** 2–4; white; 39.1 x 29.4 mm (av.; *harterti*).

EURYLAIMIDAE: CALYPTOMENINAE: Asian green broadbills

Worldwide 3 species. SE Asia 1 species. Rather small, robust and short-tailed with mostly green plumage and tuft of feathers covering nostrils. Sexes differ. Feed mainly on fruit but also insects.

596 **GREEN BROADBILL**

Calyptomena viridis Plate **64**

IDENTIFICATION 15–17 cm. *C.v.continentis*: **Male** Very distinctive. Chunky, rounded and short-tailed, bright deep green with black patch on rear ear-coverts and broad black bars on wing-coverts. Has broad tuft of feathers on top of bill and small black spot in front of eye. **Female** Duller, with no black markings, tuft of feathers on top of bill much smaller. **Juvenile** Like female but breast paler, vent greenish-white. **VOICE** Soft bubbling trill, starting quietly and increasing in tempo: ***toi toi-oi-oi-oi-oick***. Other calls include ***goik-goik*** and ***goik-goik-doyik*** (last note faster and higher), a loud ***oik***, a frog-like bubbling rattled ***oo-turrr***, and mournful whistles. **HABITAT & BEHAVIOUR** Broadleaved evergreen forest; up to 1,300 m (mostly below 800 m). Unobtrusive, in middle storey; visits fruiting figs. Male's display includes wing-flashing, head-bobbing, gaping and ricocheting from perch to perch. **RANGE & STATUS** Resident Sumatra, Borneo. **SE Asia** Uncommon to locally common resident (subject to minor movements) Tenasserim, W,S Thailand, Peninsular Malaysia; formerly Singapore. **BREEDING** February–September. **Nest** Laterally compressed oval with side-entrance, suspended from twig; 1–2 m above ground. Eggs 1–3; fairly glossy white, sometimes tinged cream or yellowish; 28.4–31.5 x 19.7–22 mm.

EURYLAIMIDAE: EURYLAIMINAE: Typical broadbills

Worldwide 8 species. SE Asia 6 species. Rather thickset with wide bills, large heads and strikingly patterned plumage. Mostly arboreal and slow-moving, feeding on a wide variety of insects, as well as grubs, larvae and snails; *Cymbirhynchus* also takes some aquatic prey, such as small crabs and fish.

597 **LONG-TAILED BROADBILL**

Psarisomus dalhousiae Plate **64**

IDENTIFICATION 24–27 cm. *P.d.dalhousiae*: **Adult** Distinctive, with mostly green body, long blue tail (black below), black crown, nape and ear-coverts and yellow lores, throat, narrow collar and patch on rear crown-side. Has blue patch on centre of crown, blue markings on rear of collar, blue outer fringes of primaries, blue-tinged breast to vent (sometimes mostly blue) and mostly pale greenish bill. In flight shows white patch at base of primaries. **Juvenile** Crown, nape and ear-coverts dark green, underparts more uniformly green, bill with darker upper mandible. **Other subspecies in SE Asia** *P.d.psittacinus* (extreme S Thailand, Peninsular Malaysia); *cyanicauda* (SE Thailand, Cambodia, S Annam). **VOICE** Series of loud, high-pitched, piercing whistles, ***tseeay*** or ***pseew***, repeated 5–8 times. Recalls Dusky Broadbill but less frantic with downward- rather than upward-inflected notes. Sometimes gives single sharp ***tseeay*** and short rasping ***psweep***. **HABITAT & BEHAVIOUR** Broadleaved evergreen forest; 500–2,000 m, locally down to 50 m. Usually in flocks, rather shy. **RANGE & STATUS** Resident N,NE Indian subcontinent, SW China, Sumatra, Borneo. **SE Asia** Uncommon to locally common resident (except C Thailand, Singapore, Cochinchina). **BREEDING** February–September. **Nest** Bulky, tailed oval with side-entrance, suspended by threads from branch or creepers; up to 30 m above ground. **Eggs** 3–8; white to deep pink, blotched pinkish, red and reddish-brown over smaller lilac-grey undermarkings (sometimes unmarked); 25–29.6 x 17–20.5 mm.

598 **DUSKY BROADBILL**

Corydon sumatranus Plate **64**

IDENTIFICATION 25–28.5 cm. *C.s.laoensis*: **Adult** Distinguished by blackish-brown plumage with contrasting buffish throat and upper breast, massive dark reddish bill with greyish tip. Has dull purplish orbital skin, orange to flame-coloured streaks on back (difficult to see), white subterminal markings on outertail-feathers and white patch at base of primaries (prominent in flight). **Juvenile** Browner, with darker, less contrasting throat and upper breast, less white on wings and tail, more pinkish bill and orbital skin, no orange streaks on back. **Other subspecies in SE Asia** *C.s.sumatranus* (extreme S Thailand, Peninsular Malaysia). **VOICE** Series of 6–8 shrill, screaming, upward-inflected thin whistles: ***hi-ky-ui ky-ui ky-ui...*** or ***ky-ee ky-ee ky-ee ky-ee...*** Also a shrill thin falling ***pseeoo*** and piercing high-pitched ***tsiu***; a repeated quavering ***ch wit*** may be given in flight. **HABITAT & BEHAVIOUR** Broadleaved evergreen and semi-evergreen forest, wetter areas in mixed deciduous forest; up to 1,220 m. Usually travels in small flocks through middle and upper storey; sits rather still on branches for long periods. **RANGE & STATUS** Resident Sumatra, Borneo. **SE Asia** Uncommon resident S Myanmar, Tenasserim, Thailand (except C), Peninsular Malaysia, Cambodia, Laos, Vietnam (except W,E Tonkin). **BREEDING** November–September. Co-operative breeder. **Nest** Untidy extended oval with side-entrance, suspended from branch, rattan or bamboo; 4–15 m above ground (sometimes over water). **Eggs** 2–4; pale dull cream to pale reddish, densely freckled reddish-brown (sometimes more sparsely marked); 27.2–34.9 x 20–24 mm.

599 **SILVER-BREASTED BROADBILL**

Serilophus lunatus Plate **64**

IDENTIFICATION 16–17 cm. *S.l.lunatus*: **Male** Pale brownish upperparts, greyish-white forehead and underparts and prominent broad black supercilium distinctive. Rump and uppertail-coverts dark rufous, wings black with mostly warm, light brown tertials and broad greyish-blue bases to flight-feathers, tail black with white tips and fringes to outer feathers. In flight, from below, shows whitish band along base of flight-feathers. *S.l.rubropygius* (SW,W,N Myanmar) has all-grey crown to upper mantle, greyer back, dark grey supercilium, mid-grey head-sides and underparts with darker ear-coverts and whitish vent, more rufescent tertials and less blue at base of flight-feathers; *rothschildi* (extreme S Thailand, Peninsular Malaysia) has greyer crown, head-sides, throat and breast and more rufous-chestnut back to uppertail-coverts and tertials. **Female** Like male but shows thin whitish necklace across upper breast; broken in centre on *S.l.rubropygius*. **Juvenile** Similar to adult with duller bill. **Other subspecies in SE Asia** *S.l.elisabethae* (E Myanmar, NE,SE Thailand, northern Indochina); *impavidus* (S Laos); *stolidus* (southern Tenasserim, northern S Thailand). **VOICE** Curious melancholy ***KI-uu*** or ***PEE-uu***, with lower second syllable. Also a high-pitched, staccato, insect-like trill: ***kitikitikit...***, particularly in flight. **HABITAT & BEHAVIOUR** Broadleaved evergreen forest, sometimes mixed deciduous forest, bamboo; 50–2,230 m. Usually in small, slow-moving parties in lower to middle storey. **RANGE & STATUS** Resident NE Indian subcontinent, SW,S China, Sumatra; formerly Nepal. **SE Asia** Uncommon to locally common resident (except C Thailand, Singapore, Cochinchina). **BREEDING** February–August. **Nest** Extended oval with side-entrance, suspended from branch, bamboo or palm frond; 1–7 m above water or ground. **Eggs** 4–7; white to cream or warm pink, sparsely to profusely speckled or spotted reddish-purple to purplish-brown (often more towards broader end), or unmarked white; 22.2–26.7 x 16.2–18.1 mm.

600 **BLACK-AND-RED BROADBILL**

Cymbirhynchus macrorhynchos Plate **64**

IDENTIFICATION 21–24 cm. *C.m.malaccensis*: **Adult** Unmistakable. Upperparts, chin and cheeks black, rump, uppertail-coverts and underparts dark maroon-red with narrow black breast-band, has long white streak along scapulars to tertials and brightly coloured bill with turquoise-blue upper mandible and yellowish lower mandible. In flight, also shows whitish band at base of flight-feathers. *C.m.affinis* (SW,S Myanmar) is smaller with red spots on tertials, larger white patch at base of primaries (on closed wing) and dark fringes to rump feathers; red plumage-parts are redder and less maroon, paler on belly and vent. See Banded Broadbill. **Juvenile** Browner overall, with some dark red on rump to uppertail-coverts, lower throat and vent, white streak restricted to scapulars, whitish tips to upperwing-coverts. **VOICE** Accelerating series of short grating cicada-like notes. Rasping ***wiark*** and rapid series of ***pip*** notes when alarmed. **HABITAT & BEHAVIOUR** Broadleaved evergreen and semi-evergreen forest and forest edge near water, freshwater swamp forest, mangroves; up to 300 m. Unobtrusive, often sitting still for long periods. **RANGE & STATUS** Resident Sumatra, Borneo. **SE Asia** Uncommon to locally common resident SW,S Myanmar, Tenasserim, W,SE,S Thailand, Peninsular Malaysia, Cambodia, S Laos, Cochinchina; formerly NE(south) Thailand, Singapore. **BREEDING** January–September. **Nest** Untidy, tailed oval with side-entrance, suspended by threads from branch or creepers; 1.5–8 m above water or ground. **Eggs** 2–4; pale pinkish, cream or fawn-coloured, freckled dull pale reddish-brown, sometimes whitish with claret to purplish-red (or fewer, darker) markings; 25–29.3 x 18.2–20.7 mm.

601 BANDED BROADBILL

Eurylaimus javanicus Plate **64**

IDENTIFICATION 21.5–23.5 cm. *E.j.harterti*: **Male** Told by rather uniform dull vinous-reddish head and underparts, dark upperparts with prominent yellow markings, and blackish-brown wings with broad yellow band on flight-feathers. Has greyer-tinged head and breast, narrow blackish band across upper breast and turquoise-blue bill. See Black-and-yellow Broadbill. **Female** Lacks breast-band. **Juvenile** Underparts much paler, mostly buffish to yellowish, with indistinct dark streaking on lower throat, breast and flanks, browner crown and sides of head, narrow yellow eyebrow, pale yellow spots/streaks on mantle, yellow tips to wing-coverts and mostly horn-coloured bill. **VOICE** Brief sharp whistled ***wheeoo***, followed by loud, rising, frantic series of notes (see Black-and-yellow). Birds often duet, with one bird starting shortly after the other. Also gives a brief, rather nasal ***whEE-u*** (***EE*** stressed), falling ***kyeeow***, rolling ***keowrr*** and yelping ***keek-eek-eek***. **HABITAT & BEHAVIOUR** Broadleaved evergreen and semi-evergreen forest, wetter areas in mixed deciduous forest; up to 1,100 m. Usually in small, slow-moving parties in middle storey. **RANGE & STATUS** Resident Greater Sundas. **SE Asia** Uncommon to locally common resident S Myanmar, Tenasserim, Thailand (except C), Peninsular Malaysia, Cambodia, Laos, S Annam, Cochinchina; formerly Singapore. **BREEDING** February–August. **Nest** Like Black-and-red Broadbill but often away from water; up to 21 m above ground. **Eggs** 2–3; white to creamy-white, speckled deep purple, dark reddish-brown or pale reddish-lavender; 26.1–28 x 17.1–20 mm.

602 BLACK-AND-YELLOW BROADBILL

Eurylaimus ochromalus Plate **64**

IDENTIFICATION 13.5–15 cm. *E.o.ochromalus*: **Male** Similar to Banded Broadbill but much smaller with black head and diagnostic white collar. Underparts pinkish-white with neat narrow black breast-band and yellow vent, eyes yellow. **Female** Breast-band broken in centre. **Juvenile** Similar to female but black of plumage duller, throat whitish with dark chin, eyebrow pale yellowish and breast with indistinct dark streaks and no band. **VOICE** Loud, frantic, rapid series of notes, beginning slowly and downslurred, before gradually gaining speed and momentum. Similar to Banded but lacks introductory note and ends very abruptly. Has similar ***kyeeow*** and ***keowrr*** notes to Banded. **HABITAT & BEHAVIOUR** Broadleaved evergreen forest; up to 925 m. Usually in small slow-moving groups in middle to upper storey. Male display involves wing-stretching and tail-wagging. **RANGE & STATUS** Resident Sumatra, Borneo. **SE Asia** Fairly common to common resident Tenasserim, W,S Thailand, Peninsular Malaysia. **BREEDING** February–September. **Nest** Untidy oval, suspended from tree-branch or other vegetation; 5–18 m above ground (sometimes above water). **Eggs** 2–3; mushroom-pink, flecked brown and purple-brown, mostly in darker ring around broader end; 23.5–23.7 x 16.5–17.4 mm. Brood-parasitised by Indian Cuckoo.

PITTIDAE: Pittas

Worldwide: c.32 species. SE Asia 14 species. Plump with stout bills, short tails and longish legs. Mainly terrestrial, bounding along, often with relatively erect posture, secretive and difficult to observe but often habituated to tracks and trails. Feed mainly on ants and other insects, worms and grubs, often smashing snails on stones.

603 HOODED PITTA *Pitta sordida* Plate **64**

IDENTIFICATION 16.5–19 cm. *P.s.cucullata*: **Adult** Black head, dark brown crown to nape-centre, green body and greater coverts diagnostic. Has turquoise-blue lower rump, uppertail-coverts and median/lesser covert patch and red vent. In flight shows large white patch on primaries. *P.s.muelleri* (extreme S Thailand, north Peninsular Malaysia) has black crown and more white on primaries. See larger Blue-winged Pitta. **Juvenile** Head-sides and upperparts duller and browner, throat mostly white, rest of underparts dull brownish with pinkish-red vent; reddish base and tip of bill. See Blue-winged and Mangrove Pittas. **VOICE** Song is a loud, perky ***WHEP-WHEP*** or ***WHEW-WHEW*** (sometimes three notes), repeated after shortish intervals. Short, squeaky ***skyew*** when alarmed. **HABITAT & BEHAVIOUR** Broadleaved evergreen forest, secondary forest, sometimes moist mixed deciduous forest, old rubber plantations near forest, bamboo; up to 915 m. Often perches on tree-branches and vines to sing. **RANGE & STATUS** Breeds N,NE Indian subcontinent, Nicobar Is, SW China, Greater Sundas, Philippines, Sulawesi, New Guinea region; northern populations winter to south, Sumatra, Java. **SE Asia** Fairly common resident extreme S Thailand, north Peninsular Malaysia. Uncommon to locally common breeding visitor Myanmar, W,NE,SE,S Thailand, Cambodia, N,S Laos, W Tonkin. Uncommon to fairly common passage migrant NW,C Thailand, Peninsular Malaysia (also winters), Singapore. Recorded (status uncertain) Cochinchina. **BREEDING** May–October. **Nest** Flattened dome, on ground amongst leaf-litter. **Eggs** 2–5; glossy white, with sparse dark brown to dark purple spots and clouds (often a few lines) over pale dull lilac undermarkings; 23–28 x 19.6–22.5 mm.

604 FAIRY PITTA *Pitta nympha* Plate **64**

IDENTIFICATION 16–19.5 cm. Monotypic. **Adult** Similar to Blue-winged Pitta but smaller, crown-sides rufous, contrasting with narrow whitish-buff supercilium; has paler blue on rump and uppertail-coverts, much paler buff underparts and paler, much more restricted blue on wing-coverts. In flight shows much smaller, rounder white patch on primaries. **VOICE** Song is a repeated clear whistled ***kwah-he kwa-wu***, longer and slower than Blue-winged. **HABITAT** Broadleaved evergreen forest; up to 1,000 m. **RANGE & STATUS** Breeds S,E China, Taiwan, South Korea, southern Japan; winters Borneo. **SE Asia** Rare passage migrant (mainly coastal) E Tonkin, N,C Annam, Cochinchina.

605 BLUE-WINGED PITTA

Pitta moluccensis Plate **64**

IDENTIFICATION 18–20.5 cm. Monotypic. **Adult** Distinguished by blackish head with buff crown-sides/supercilium and white throat, green mantle and scapulars and plain buff underparts with red vent. Upper chin blackish, rump, uppertail-coverts and tail-tip deep violet-blue, tail blackish, upperwing-coverts mostly deep violet-blue. In flight shows very large white patch on primaries. See

Mangrove, Fairy and Hooded Pittas. Note habitat and range. **Juvenile** Head and upperparts duller, crown-sides/supercilium dark-scaled, blue of plumage duller, chin whiter, vent washed-out pinkish, tail-tip green; orangey-red base and tip of bill. In flight shows smaller white wing-patch. See Mangrove and Hooded. **VOICE** Song is a loud, clear ***TAEW-LAEW TAEW-LAEW*** (with stressed ***LAEW*** notes), repeated every 3–5 s. Harsh ***skyeew*** when alarmed. **HABITAT & BEHAVIOUR** Relatively open broadleaved evergreen and mixed deciduous forest, secondary growth, bamboo; also parks, gardens and mangroves on migration; up to 800 m. Often sings from quite high in tree. **RANGE & STATUS** Breeds SW China; winters Sumatra, Borneo, rarely Java, Bali. **SE Asia** Fairly common to common breeding visitor (rarely overwinters) SW,C,S,E Myanmar, Tenasserim, Thailand (except C), Peninsular Malaysia (south to Pahang), Cambodia, Laos, C,S Annam, Cochinchina. Uncommon to fairly common winter visitor and passage migrant Peninsular Malaysia, Singapore. Scarce to fairly common passage migrant W,C,S Thailand, Cambodia, E Tonkin. **BREEDING** April–September. **Nest** Large dome with side-entrance, on ground; sometimes in tree or palm up to 7.6 m. **Eggs** 3–7; glossy whitish with purple and purplish-brown specks, spots and scrawls (sometimes more at broader end); 24–28.9 x 20–22.9 mm.

606 MANGROVE PITTA

Pitta megarhyncha Plate **64**

IDENTIFICATION 18–21 cm. Monotypic. **Adult** Like Blue-winged Pitta but bill much longer, crown drabber and more uniform brown (almost lacks black in centre), underparts slightly duller, chin whitish. See Fairy and Hooded Pittas. Note habitat and range. **Juvenile** Duller. Separated from similar Blue-winged by same characters as adult. **VOICE** Song is similar to Blue-winged but more slurred and hurried: ***WIEUW-WIEUW.*** **HABITAT** Mangroves. **RANGE & STATUS** Resident S Bangladesh, Sumatra. **SE Asia** Scarce to locally common coastal resident Myanmar, west S Thailand, Peninsular Malaysia, Singapore. **BREEDING** March–August. **Nest** Ball or dome on ground, sometimes in vegetation up to 1 m. **Eggs** 2–4; whitish with dark brown to dull maroon spots, blotches and streaks over pale inky-purple to greyish undermarkings (sometimes mainly towards broader end); 29–31.7 x 23.5–24.7 mm.

607 GIANT PITTA *Pitta caerulea* Plate **65**

IDENTIFICATION 28–29 cm. *P.c.caerulea*: **Male** Large size, pale greyish head, black centre of crown and nape, black eyestripe and blue upperparts diagnostic. Pale throat and breast grade to pale buffish underparts; shows narrow, broken blackish necklace across upper breast. **Female** Similarly patterned to male but base colour of head warm buffish-brown, crown-centre darker-scaled, has narrow black nuchal collar, rufescent-brown upperside, with blue lower rump to tail and deeper buff underparts. See Rusty-naped Pitta. Note habitat and range. **Juvenile** Resembles female but upperparts cold, dull dark brown, tail dull dark blue, underparts dirty whitish to pale buff with broad, smudgy dark breast-band, head and upper breast indistinctly scaled, bill reddish at base and tip. See Rusty-naped. **VOICE** Song is a loud, airy ***hwoo-er*** or ***whee-er***, repeated every 5–10 s. Also gives short, tuneless ***phreew***. **HABITAT** Broadleaved evergreen forest, bamboo; up to 245 m; once at 885 m in S Thailand. **RANGE & STATUS** Resident Sumatra, Borneo. **SE Asia** Rare to scarce resident south Tenasserim, W(south),S Thailand, Peninsular Malaysia. **BREEDING** May–November. **Nest** Large ball with platformed side-entrance, in palm or sapling; 0.8–3 m above ground. **Eggs** 2; off-white with brown speckles (sometimes ringed towards broader end); c.32–35 x 26–30 mm.

608 BLUE PITTA *Pitta cyanea* Plate **65**

IDENTIFICATION 19.5–24 cm. *P.c.cyanea*: **Male** Rather variable but unmistakable, with blue upperside, broadly orange-scarlet rear crown-sides and nape, and pale bluish-whitish underparts, spotted and barred black. In flight, shows small white patch at base of primaries. *P.c.aurantiaca* (SE Thailand, south-west Cambodia) has more yellowish-orange colour on head; *willoughbyi* (C Laos, S Annam) is brighter and often shows some red on breast. **Female** Duller overall and browner above, orange-scarlet restricted more to nape and often duller, underparts more buffish to whitish with heavier dark markings. **Juvenile** Crown and nape buffish-brown with dark scales and darker median stripe, has dark brown eyestripe, almost lacks darker malar stripe, mantle, back and scapulars dark brown with warm buff streaks, uppertail-coverts and tail duller blue, breast and belly dark brown with warm buff streaks and some blackish and whitish barring on flanks and centre of abdomen; bill initially reddish. See Bar-bellied and Gurney's Pittas. Note habitat and range. **VOICE** Song is a repeated, loud ***peroo-WHIT***, with long-drawn-out first note and louder, sharper, shriller second note. Sometimes gives shorter ***priaw-PIT***. Rasping ***skyeew*** when alarmed. **HABITAT** Broadleaved evergreen forest, moister areas in mixed deciduous forest; up to 1,890 m. **RANGE & STATUS** Resident NE India, Bangladesh. **SE Asia** Uncommon to locally common resident SW,S,E Myanmar, Tenasserim, W,NW,NE,SE, north S Thailand, Cambodia, Laos, E Tonkin, N,C,S Annam. **BREEDING** February–September. **Nest** Large dome with side-entrance (sometimes platformed), on ground, or in sapling or tree-cavity up to 4 m. **Eggs** 3–7; glossy white, with fairly numerous deep purple-black, reddish-brown or pale reddish spots and blotches (sometimes short tangled lines); 24–28.2 x 20.1–22 mm.

609 RUSTY-NAPED PITTA *Pitta oatesi* Plate **65**

IDENTIFICATION 21–25 cm. *P.o.oatsei*: **Male** Mostly deep rufous head and underparts (sharply demarcated from dull green upperparts) and well-defined narrow blackish post-ocular stripe diagnostic. Throat paler, underparts may have pinkish tinge, rump sometimes blue-tinged. *P.o.bolovenensis* (S Laos, probably S Annam) has bright blue on back and rump (sometimes also wash on mantle) and pinker-tinged underparts; *deborah* (Peninsular Malaysia; ? extreme S Thailand) has blue rump and strongly pinkish throat and underparts. See Blue-naped and Blue-rumped Pittas. Note range. **Female** Somewhat duller, with brown-tinged upperparts and indistinctly dark-scaled lower throat and upper breast. **Juvenile** Crown dark brown with whitish streaks, upperparts, breast and wing-coverts dark brown with whitish-buff spots; has whitish-buff supercilium, whitish sides of head with indistinct dark streaks, and whitish throat. See Blue-naped and Blue-rumped. **Other subspecies in SE Asia** *P.o.castaneiceps* (C Laos, W,E Tonkin, probably N Annam). **VOICE** Song is a sharp ***CHOW-WIT***, recalling Blue Pitta. Sometimes gives liquid, falling ***poouw***, similar to Blue-rumped. **HABITAT** Broadleaved evergreen forest, bamboo; 760–2,565 m (locally down to 300 m Thailand, 380 m E Tonkin). **RANGE & STATUS** Resident SW China. **SE Asia** Uncommon to fairly common resident N,S,E Myanmar, Tenasserim, W,NW,NE,extreme S Thailand, Peninsular Malaysia, Laos, W,E(north-west) Tonkin, N,S Annam. **BREEDING** December–September. **Nest** Rough dome, situated in palm, up to 1.7 m above ground. **Eggs** 2–6; slightly glossy, white, variably speckled purple-brown and chestnut-red, mainly towards broader end; dimensions 25.9–31.3 x 23–25.3 mm.

610 BLUE-RUMPED PITTA *Pitta soror* Plate **65**

IDENTIFICATION 20–22 cm. *P.s.petersi*: **Male** Similar to Rusty-naped Pitta but shows distinctly blue lower back and rump and pale blue (slightly green-tinged) crown and nape, has mostly lilac-pinkish sides of head and paler, more buffish breast and belly. *P.s.tonkinensis* (northern Tonkin) has green crown and nape with faint blue tinge. See Blue-naped Pitta. Note range. **Female** Somewhat duller with mostly greenish crown and nape, upperparts browner-tinged with less blue on rump. See Blue-naped. **Juvenile** Similar to Rusty-naped but head-sides and throat more buffish, markings on upperside richer buff, breast less heavily marked. See Blue-naped. **Other subspecies in SE Asia** *P.s.flynnstonei* (SE Thailand, Cambodia); *annamensis* (S Laos, C Annam); *soror* (S Annam, Cochinchina). **VOICE** Song is a full-sounding ***weaoe*** or ***weeya*** (slightly inflected), repeated after longish intervals. Call is a short, rather quiet ***PPEU, EAU*** or ***CHO***. Sometimes gives a sharp explosive ***hwip*** or ***hwit*** when agitated. **HABITAT** Broadleaved evergreen and semi-evergreen forest, sometimes mixed deciduous forest, secondary forest; up to 1,700 m (above 900 m in SE Thailand; below 1,000 m in Indochina, except W Tonkin). **RANGE & STATUS** Resident S China. **SE Asia** Fairly common to common resident SE Thailand (local), Indochina (except N Laos). **BREEDING** March–July. **Nest** Round dome with large, platformed side-entrance, in palm or sapling; 1.3–2.4 m above ground. **Eggs** 3; pinkish-white with large chocolate-brown speckles and blotches (mainly towards broader end); c.29 x 22 mm.

611 **BLUE-NAPED PITTA** *Pitta nipalensis* Plate **65**

IDENTIFICATION 22–26 cm. *P.n.nipalensis*: **Male** Similar to Blue-rumped Pitta but blue of hindcrown and nape frequently extends onto upper mantle, forehead and sides of head strongly rufous, lacks blue on lower back to uppertail-coverts (but sometimes tinged blue). *P.n.hendeei* (north Indochina) has blue confined to neat patch on nape. See Rusty-naped Pitta. Note range. **Female** Hindcrown to upper mantle green (often tinged blue on nape). From Blue-rumped by lack of obvious blue patch on lower back and rump and more rufescent forehead and sides of head. *P.n.hendeei* has browner hindcrown and green of head confined to neat patch on nape. See Rusty-naped. **Juvenile** From very similar Blue-rumped by paler, buff (less rufescent) spots on wing-coverts and less boldly marked crown, which is dark brown with buff streaks. From Rusty-naped by buff ear-coverts and more buffish crown and wing-covert markings. **VOICE** Song is clear ***uk-WUIP*** or ***ip-WUI'IP***. **HABITAT** Broadleaved evergreen forest, secondary growth; up to 1,400 m. **RANGE & STATUS** Resident (subject to local movements) Nepal and NE Indian subcontinent, SW China. **SE Asia** Scarce to uncommon resident SW,W,N,C(western),S Myanmar, N,C Laos, W,E Tonkin. **BREEDING** May–June. **Nest** Large oval with entrance at end, amongst leaves on ground or in bamboo, sapling or bush up to 2 m. **Eggs** 3–7; glossy white, usually sparingly spotted and blotched (may show lines) purple-black or blackish-brown to reddish-brown over lilac to pale lavender undermarkings; 26.1–32.6 x 21.8–25.6 mm.

612 **BAR-BELLIED PITTA** *Pitta elliotii* Plate **65**

IDENTIFICATION 19.5–21 cm. Monotypic. **Male** Very distinctive, with vivid green crown and upperparts, blackish lores to nape-sides, pale green lower throat, greenish-yellow breast and yellow remainder of underparts with narrow dark bars and dark bluish centre of abdomen. Tail and undertail-coverts blue. **Female** Crown and breast warm buffy-brown, sides of head blackish-brown with buffish streaks on cheeks, no dark blue on underparts. **Juvenile** Initially rather uniform dark brown with darker sides of head, paler throat, pale buff spots on crown, mantle, wing-coverts and breast, and orangey-red bill with darker base. Similar to Gurney's Pitta but note range. **VOICE** Song is a loud whistled ***CHAWEE-WU*** (only ***WEE-WU*** audible at distance) repeated every 9–12 s. Occasionally a mellow ***HHWEE-HWHA***. Harsh, shrill ***jeeow*** or ***jow*** when alarmed. **HABITAT** Broadleaved evergreen, semi-evergreen and mixed deciduous forest, bamboo; up to 800 m. **RANGE & STATUS** Endemic. Resident NE(south-east),SE Thailand (scarce), Indochina (fairly common to common). **BREEDING** April–June. **Nest** Dome with large platformed side-entrance, in tree, sapling or palm; 1.3–7 m above ground. **Eggs** 3; creamy-white with brown to chestnut speckles (sometimes only at broader end) or unmarked; 28–29 x 22–25 mm.

613 **BANDED PITTA** *Pitta guajana* Plate **65**

IDENTIFICATION 21–24 cm. *P.g.irena*: **Male** Distinctive. Crown-centre and sides of head black, crown-sides and supercilium bright yellow, turning vivid orange-red on nape, breast to vent blue-black with orange-rufous bars on breast-sides, has broad white band along inner wing-coverts and secondaries. Upperparts warm brown, uppertail-coverts and tail blue, throat white with yellower sides. **Female** Breast to undertail-coverts whitish to buffish with narrow dark barring; less orange-red on nape. **Juvenile** Duller than female, with all-buffish crown-sides, supercilium and lower nape (scaled darker), browner sides of head, dark brown breast with buff spots/streaks, duller whitish wing-band and (initially) orange-red base and tip of bill. **VOICE** Song is a short ***POUW*** or ***POWW***, repeated after shortish intervals. Whirring, slightly explosive ***kirrr*** or ***pprrr*** when alarmed. Sometimes gives hollow ***whup*** or soft, moaning ***who-oo*** when agitated. **HABITAT** Broadleaved evergreen forest, secondary forest; up to 610 m (possibly higher locally). **RANGE & STATUS** Resident Greater Sundas. **SE Asia** Uncommon to locally common resident S Thailand, Peninsular Malaysia. **BREEDING** February–November. **Nest** Ball with side-entrance, in palm or sapling; up to 3 m above ground. **Eggs** 2–5; fairly glossy white, thickly spotted and streaked dark purplish-brown (mainly towards broader end); 24.2–25.2 x 20.7–21 mm.

614 **GURNEY'S PITTA** *Pitta gurneyi* Plate **65**

IDENTIFICATION 18.5–20.5 cm. Monotypic. **Male** Blue crown and nape, black forecrown and sides of head, yellow underparts with black centre of breast and belly to undertail-coverts, and black-barred flanks diagnostic. Upperside warm dark brown, tail turquoise-tinged deep blue. **Female** Crown and nape buffy-rufous, sides of head blackish-brown with paler streaks on lores and cheeks, throat whitish, rest of underparts pale buffy-whitish with dark bars, tail blue. See Banded Pitta. Note range. **Juvenile** Forehead to nape, breast and upper belly dark brown with buff streaks (broader and more spot-like on nape), rest of upperside, head-sides and tail similar to female; initially has fleshy-orange base and tip of bill. See Bar-bellied Pitta but note range. **VOICE** Song is a short, explosive ***LILIP***, repeated every 2–6 s. Female may give more subdued ***llup***. Harsh falling ***skyeew*** when alarmed. **HABITAT** Broadleaved evergreen forest, secondary forest, old rubber plantations near forest; up to 160 m. **RANGE & STATUS** Endemic. Rare and local resident south Tenasserim, S Thailand. **BREEDING** May–October. **Nest** Ball or slightly flattened dome, in palm or sapling; 0.95–2.4 m above ground. **Eggs** 3–5; glossy white, sparsely speckled, spotted and squiggled brown over larger grey undermarkings (mainly towards broader end); 25.3–27 x 20–22.4 mm.

615 **GARNET PITTA** *Pitta granatina* Plate **65**

IDENTIFICATION 14–16.5 cm. *P.g.coccinea*: **Adult** Black head-sides and throat, scarlet-red crown and nape and crimson-red belly and vent diagnostic. Has narrow pale blue line along hindcrown/nape-side, deep purplish upperparts, purplish-black breast, and iridescent azure to light blue patch on wing-coverts. **Juvenile** Mostly dark brown with paler throat and some red on nape and vent, wing-coverts and tail duller blue than adult, bill initially reddish at base and tip. **VOICE** Song is a drawn-out monotone whistle (lasting about 1.5 s) which swells in volume. Very similar to Rail-babbler but has slight upward inflection and ends very abruptly. Occasionally gives purring ***prrr prrr prrr...*** when agitated. **HABITAT** Broadleaved evergreen forest; up to 300 m. **RANGE & STATUS** Resident Sumatra, Borneo. **SE Asia** Uncommon to locally common resident south Tenasserim, extreme S Thailand, Peninsular Malaysia; formerly Singapore. **BREEDING** March–August. **Nest** Flattened dome, on ground. **Eggs** 2; glossy white, dotted and blotched pinkish-brown and rich brown (mainly towards broader end); 26–27 x 20–21 mm.

616 **EARED PITTA** *Anthocincla phayrei* Plate **65**
IDENTIFICATION 20–24 cm. Monotypic. **Male** Very distinctive, with rather long, slender bill, blackish crown-centre, head-sides and nape, broad buffy-whitish (blackish-scaled) elongated feathers on crown-sides and supercilium which often protrude beyond nape (like ears), blackish submoustachial stripe, and deep buffish underparts (throat paler) with dark spots/scales on flanks; upperside rather plain warm brown with pale buff and blackish wing markings. **Female** Crown-centre, nape and sides of head browner than male, underparts more heavily dark-marked. **Juvenile** Duller brown than female with all-buffish crown-sides and supercilium, shorter 'ears', no dark submoustachial stripe and dark brown breast, with some lighter, rufous shaft-streaks. **VOICE** Song is an airy whistled ***wheeow-whit***, repeated after rather lengthy intervals. Short dog-like whine when alarmed. **HABITAT & BEHAVIOUR** Broadleaved evergreen and mixed deciduous forest, bamboo; up to 900 m, sometimes 1,500 m. Territorial males make rapid wing-flicking sound. **RANGE & STATUS** Resident SW China. **SE Asia** Uncommon resident C,S,E Myanmar, north Tenasserim, W,NW,NE,SE Thailand, Cambodia, Laos, W,E Tonkin, C,S Annam. **BREEDING** April–September. **Nest** Dome with platformed side-entrance, on ground. **Eggs** 4; glossy white, with a few minute purple-black specks, mostly at broader end; 27.4–27.9 x 21.6–23.3 mm.

ACANTHIZIDAE: Gerygones & allies

Worldwide c.60 species. SE Asia 1 species. Small and warbler-like; feed mainly on insects.

617 **GOLDEN-BELLIED GERYGONE**
Gerygone sulphurea Plate **66**
IDENTIFICATION 10–10.5 cm. *G.s.sulphurea*: **Adult** Most likely to be confused with certain warblers and female sunbirds. Can be distinguished by combination of rather short straight blackish bill, plain greyish-brown upperparts, pale yellow underparts and whitish lores and subterminal spots on tail-feathers. Darker cheeks/sides of head contrast sharply with yellow throat. See Yellow-bellied Warbler. Note habitat and range. **Juvenile** Has complete narrow whitish eyering, plain greyish sides of head, slightly paler yellow underparts, and flesh-coloured base of bill. **VOICE** Song consists of up to ten high-pitched musical, wheezy, glissading, rising or descending whistles: ***zweee***, ***zrriii*** and ***zrii-i'i'i'i'uu*** etc. Call is a musical, rising ***chu-whee***. **HABITAT** Mangroves, coastal scrub, also inland in freshwater swamp and peatswamp forest, plantations (particularly rubber) and sometimes other types of forest, secondary growth, parks and gardens; up to 915 m (mostly lowlands). **RANGE & STATUS** Resident Greater Sundas, W Lesser Sundas, Sulawesi, Philippines. **SE Asia** Common resident (scarcer inland) coastal south Tenasserim, coastal and S Thailand, Peninsular Malaysia, Singapore, Cambodia, Cochinchina. **BREEDING** December–October. **Nest** Purse-shaped with lateral entrance, suspended from tree-branch. **Eggs** 2–3; whitish with tiny reddish-brown specks mainly around broader end; 15.2 x 10.9 mm (av). Brood-parasitised by Little Bronze Cuckoo.

EUPETIDAE: Rail-babbler & allies

Worldwide 10 species. SE Asia 1 species. Fairly small to medium-sized, with longish necks and legs. Terrestrial, with rather rail-like gait. Feed mainly on insects, grubs, worms etc.

618 **RAIL-BABBLER** *Eupetes macrocerus* Plate **66**
IDENTIFICATION 29 cm. *E.m.macrocerus*: **Adult** Unmistakable. Rather slender, with long neck, bill and tail. Overall warm brown, with buffy-rufous forehead, chestnut-red crown, hindneck, throat and foreneck, long black band from base of bill through ear-coverts to neck-side, bordered above by long white supercilium/neck-stripe and reddish-rufous breast. Has strip of blue skin on neck-side (more obvious when calling). **Juvenile** Resembles adult but shows dull chestnut crown and hindneck, warmer upperparts, somewhat duller head/neck-stripes, orange-rufous foreneck and breast (without red), whitish throat and greyer belly. **VOICE** Long, thin, drawn-out monotone whistle, 1.5–2 s in duration. Very similar to Garnet Pitta but purer and higher-pitched and not rising at end. Also, popping frog-like notes when agitated. **HABITAT & BEHAVIOUR** Broadleaved evergreen forest; up to 1,270 m. Walks on forest floor, jerking head like chicken; very shy. **RANGE & STATUS** Resident Sumatra, Borneo. **SE Asia** Scarce to locally fairly common resident S Thailand, Peninsular Malaysia. **BREEDING** January–July. **Eggs** 2.

VIREONIDAE: Shrike-babblers, White-bellied Erpornis & allies

Worldwide c.58 species. SE Asia 6 species. Smallish to small, with relatively thick bills, thickset bodies, and generally boldly patterned plumage. Sexually dimorphic in the case of shrike-babblers; White-bellied Erpornis with somewhat more slender bill and slight crest. All strongly arboreal, often joining bird-waves.

619 **BLACK-HEADED SHRIKE-BABBLER**
Pteruthius rufiventer Plate **66**
IDENTIFICATION 21 cm. *P.r.rufiventer*: **Male** Unmistakable, with black head, wings and tail, rufous-chestnut upperparts, pale grey throat with whitish line at side, grey breast and deep pinkish belly and vent. Has yellowish patch at side of breast and chestnut-tipped secondaries and tail. See Rufous-backed Sibia. **Female** Crown scaled grey, head-sides grey, upperparts, wing-feather fringes and uppertail mostly bright olive-green; shows dark markings on mantle, chestnut-tipped tail, and darker underparts than male. **Juvenile** Similar to female (both sexes). **Immature male** Crown and nape duller and browner than adult, mantle feathers mixed with yellowish-green and tipped with blackish bars and spots, uppertail-coverts paler; has less chestnut and some whitish on tips of secondaries. **Other subspecies in SE Asia** *P.r.delacouri* (W Tonkin). **VOICE** Song similar to White-browed Shrike-babbler. Full, rather mellow ***wip-wiyu*** (last syllable slightly stressed), repeated every 2–5 s; or similar but with longer pause after first note: ***wip wu-yu***, repeated every 1–3 s. Also a more evenly spaced, slightly descending ***yu-wu-uu***. Calls include a curious quick, nervous, tremulous ***ukuk-wrrrrii-yiwu*** (first note low and throaty, second long and high-pitched), repeated after short intervals, and harsh scolding gruff 5–7 note ***rrrrt-rrrrt-rrrr-rrrrt-rrrrt...*** when agitated. **HABITAT** Broadleaved evergreen forest; 1,220–2,600 m. **RANGE & STATUS** Resident Nepal, Bhutan, NE India, SW China. **SE Asia** Scarce to uncommon resident W,N Myanmar, W Tonkin. **BREEDING** Undocumented.

620 **WHITE-BROWED SHRIKE-BABBLER**
Pteruthius flaviscapis Plate **66**
IDENTIFICATION 16–17.5 cm. *P.f.aeralatus* (C[east], S[east],E Myanmar, north Tenasserim, W,NW[west] Thailand): **Male** Told by black crown, nape and ear-coverts, broad white supercilium, grey rest of upperparts (mantle may be faintly mottled black), black tail and wings, pale grey throat and underparts with vinous wash on lower flanks and

thighs. Has whiter line along throat-side, golden-rufous and chestnut tertials (tipped black), white-tipped flight-feathers. *P.f.validirostris* (W,N[west] Myanmar) and *annamensis* (S Annam) have whiter underparts, former also all-chestnut tertials, latter no black tertial tips; *ricketti* (N[east] Myanmar, NW[east] Thailand, N Indochina) is darker grey above, black mottling extending to back, rear ear-coverts greyer, throat to vent darker grey. See Cutia. **Female** Crown and head-sides greyish, supercilium duller, rest of upperparts greyish-brown, wings and tail mostly golden-olive, tertials with little chestnut, flight-feathers with smaller white tips, underparts creamy-buffish, throat paler. *P.f.validirostris* and *annamensis* have chestnut tertials, darker crown, nape and head-sides, and whiter throat and upper breast; *schauenseei* (S Thailand) and *cameranoi* (Peninsular Malaysia) have browner crown, nape and head-sides and almost no chestnut on tertials; *ricketti* has grey throat and upper breast. See Black-headed Shrike-babbler. **Juvenile male** Like juvenile female but nape to uppertail-coverts more rufescent-tinged, lores and ear-coverts blacker, wings similar to adult male but with golden-olive fringing on median and greater coverts. **Juvenile female** Like adult but crown and upperparts rather uniform greyish-brown, initially with buffish-white shaft-streaks, underparts much whiter. **VOICE** Sings with a repeated loud strident rhythmic series of notes with short weak introduction. Subject to considerable, probably regional/subspecific variation. In W Myanmar and NW Thailand ***ip ch-chu ch-chu*** or ***itu chi-chu chi-chu***, in Peninsular Malaysia ***ip chip chip ch-chip***, and in S Annam ip ***chu ch-chu***. Calls include a short ***pink***, and grating churring sounds when alarmed. **HABITAT** Broadleaved evergreen forest, mixed evergreen and coniferous forest; 700–2,500 m. **RANGE & STATUS** Resident NE Pakistan, NW,N,NE Indian subcontinent, S,SE Tibet, W and southern China, Greater Sundas. **SE Asia** Common resident (except SW Myanmar, C Thailand, Singapore, Cochinchina). **BREEDING** November–June. **Nest** Cradle or cup, suspended from tree-branch; 4–13 m above ground. **Eggs** 2–5; whitish to pale pink, finely spotted dark purplish with more profusely spotted wide ring around broader end; 21.8 x 16.3 mm (av.; *aeralatus*).

621 GREEN SHRIKE-BABBLER

Pteruthius xanthochlorus Plate **66**

IDENTIFICATION 12–12.5 cm. *P.x.hybrida*: **Male** Relatively nondescript but has distinctive short stout bill, greyish crown and head-sides, darker lores and cheeks, white eyering, greyish-white throat and breast and yellowish-buff flanks and vent. Rest of upperparts and wing-feather fringes greenish-olive, shows narrow pale wing-bar, blackish primary coverts and narrow whitish tips to tail-feathers (hard to see in field). Could be confused with Yellow-browed and Fire-capped Tits or some warblers but note robust shape, stubby bill and eyering. *P.x.pallidus* (N Myanmar) has smooth slaty-grey crown to upper mantle and ear-coverts, blacker lores and cheeks, very conspicuous broad white eyering, much brighter, rich yellowish-buff lower flanks and vent, largely grey fringes to wing-feathers (whiter on primaries and tips of outer greater coverts) and prominent whitish tips to flight-feathers. **Female** Has paler, less obvious grey on head, and duller, more brownish-olive upperparts. *P.x.pallidus* is slightly paler above than male, with vague brownish tinge on crown and mantle and mostly greenish-olive fringes to wing-feathers. **VOICE** Song is a variable, monotonous, fairly high-pitched series of notes, repeated every 2–8 sec: fairly well-spaced ***whitu-whitu-wheet***, ***wheet-wheet-wheet*** and ***chuwi-chuwi***; well-spaced 2–4 note ***chiew-chiew...***; faster, slightly tremulous 4–6 note ***whitwhitwhitwhit...***; fairly fast, shorter, 4–5 note ***whiwhiwhiwhi...***; and rather fast 4–5 note ***chiwichiwichiwichiwi...*** Calls include a tit-like ***jerr***, ***jerr-jerr***, ***jer-ri*** and higher ***jerri*** etc., as well as soft ***ik***, ***uk*** and ***jep*** contact notes. **HABITAT & BEHAVIOUR** Broadleaved evergreen forest; 1,700–2,800 m. Usually found singly or in pairs, often associating with mixed-species feeding flocks; movements rather slow and heavy. **RANGE & STATUS** Resident N Pakistan, NW,N,NE Indian subcontinent, SE Tibet, SW,W,SE China. **SE Asia** Common resident W,N Myanmar. **BREEDING** April–July. **Nest** Deep purse or cradle, suspended hammock-like from tree-branch; 1.5–8 m above ground. **Eggs** 2–4; cream-coloured, typically with reddish-brown blotches (mainly around broader end); 18.8 x 14.7 mm (av.; India).

622 BLACK-EARED SHRIKE-BABBLER

Pteruthius melanotis Plate **66**

IDENTIFICATION 11.5–12 cm. *P.m.melanotis*: **Male** Like Chestnut-fronted Shrike-babbler but forehead yellowish (without chestnut), chestnut of throat paler, nape slate-grey, ear-coverts with broad black rear border, and flight-feathers with uniform slaty-grey fringes. Note habitat and range. *P.m.tahanensis* (extreme S Thailand, Peninsular Malaysia) has darker upperparts, less yellow on forehead, paler yellow underparts with chestnut restricted to throat, and greener fringes to flight-feathers. **Female** Duller above than male with rufous-buff (not white) wing-bars, all-yellowish underparts with pale chestnut chin and malar area, and duller fringes to flight-feathers. *P.m.tahanensis* is more whitish-yellow below with paler chestnut on chin and malar area. Told from Chestnut-fronted by mostly greenish-olive forehead, grey nape, blackish rear border to ear-coverts, distinct pale chestnut chin and malar area, yellower rest of underparts and duller, more greyish-olive flight-feather fringes. **Juvenile** Like female (both sexes) but has less distinct pattern on head-sides, no pale grey on nape or chestnut on chin and malar area, whiter supercilium, much paler yellow underparts, greener wing-fringing. *P.m.tahanensis* is whiter below with warm buff malar area. **VOICE** Song is a repeated, variable, very monotonous series of notes, apparently subject to some geographical/subspecific variation. In NW Thailand and W Tonkin, a fairly slowly delivered 4–13 note ***twi-twi-twi-twi-twi...***, even slower ***dwit-dwit-dwit-dwit-dwit-dwit...*** or rapid rattling ***dr'r'r'r'r'r'r'r'r'r'r***. In Peninsular Malaysia, a rapid 12–19 note ***whiwhiwhiwhiwhiwhiwhiwhiwhiwhiwhiwhi...*** or ***jujujujujujujujujujuju...*** Calls include a short ***chid-it***, first note stressed. **HABITAT & BEHAVIOUR** Broadleaved evergreen forest; 1,220–2,200 m (down to 1,050 m in Laos, locally 700 m N Myanmar). Often joins mixed-species feeding flocks. **RANGE & STATUS** Resident Nepal, NE Indian subcontinent, SE Tibet, SW China. **SE Asia** Uncommon resident W,N,S(east),E Myanmar, W,NW,extreme S Thailand, Peninsular Malaysia, Laos, W,E(north-west) Tonkin, N,C Annam. **BREEDING** February–June. **Nest** Small cradle, suspended hammock-like from branch of tree or bush; 2–9 m above ground. **Eggs** 2–6; pinkish-white to pale cream, stippled purplish to pale rufous or reddish-brown, particularly towards broader end (may have pale lilac to grey undermarkings); 17.9 x 13.5 mm (av.; *melanotis*).

623 CHESTNUT-FRONTED SHRIKE-BABBLER

Pteruthius aenobarbus Plate **66**

IDENTIFICATION 11.5–12 cm. *P.a.intermedius*: **Male** Small, robust and stout-billed with distinctive chestnut then yellow forehead, whitish-grey supercilium, very broad white eyering, deep yellow underparts with dark chestnut centre to throat and variable wash on breast, and two broad white wing-bars. Rest of upperparts greenish-olive, primaries fringed greyish-white. *P.a.indochinensis* (S Annam) has narrower chestnut forehead and throat-patches and lacks chestnut wash on breast. See Black-eared Shrike-babbler. **Female** Duller green above with pale rufous-chestnut forehead, no yellow band on forecrown, creamy-whitish underparts with pale rufous-chestnut wash on throat, lemon-yellow tinge to undertail-coverts, and variable yellowish wash on breast and belly; wing-bars dull pale rufous-buff, olive (in place of black) on wings and pale rufous-buff fringes to primaries. See Black-eared and Fire-capped Tit. **VOICE** Sings with various notes, repeated monotonously: ***chip-chip-chip-chip...***, ***wheet-wheet-wheet wheet-wheet-wheet-wheet...***, ***whit-whit-whit-whit whit-whit-whit-whit-whit-whit-whit...*** and ***wchip-wchip-wchip-wchip...*** etc., or more hurried ***wchipwchipwchipwchip-wchip-wchip***. Calls include a short, buzzy, slightly nasal ***jer-jer-jer...*** and ***jwi-jwi-jwi...***, chattering ***chr'r'r'r'uk*** and sharp ***pwit***. **HABITAT & BEHAVIOUR**

Broadleaved evergreen forest; 700–2,500 m. Often joins mixed-species flocks. **RANGE & STATUS** Resident NE India (Garo Hills), SW,S China, Java. **SE Asia** Fairly common resident Myanmar (except SW), W,NW,NE Thailand, Laos, W Tonkin, N,C,S Annam. **BREEDING** January–April. **Nest** Cradle, suspended from high tree-branch. **Eggs** 2; pale greyish, speckled purplish and grey; 18.5 x 13.3 mm (av.).

624 **WHITE-BELLIED ERPORNIS**

Erpornis zantholeuca Plate **66**

IDENTIFICATION 12–13.5 cm. *E.z.zantholeuca*: **Adult** Unmistakable, with rather light yellowish olive-green upperside, pale greyish lores and ear-coverts, pale greyish underside with whiter throat and centre of abdomen, and bright yellow vent. Crown and shortish erect crest have scaly appearance, bill pinkish with browner culmen. *E.z.tyrannulus* (NW[east],NE Thailand, N Indochina) has deeper green upperparts, darker and greyer lores and ear-coverts, and greyer underparts (particularly breast-sides and flanks); *canescens* (SE Thailand, west Cambodia) has distinct pale grey wash on crown and crest (green limited to rear or almost absent), rest of upperparts faintly grey-washed. **Juvenile** Most of upperside strongly washed light warm golden-brown, lacks scales on crown/crest, crest shorter. **VOICE** Song is a short high-pitched descending trill: ***si'i'i'i'i'i***. Typically calls with nasal ***nher-nher*** and short ***nhi*** notes. **HABITAT & BEHAVIOUR** Broadleaved evergreen and mixed deciduous forest; up to 2,000 m, rarely to 2,650 m in N Myanmar (below 1,220 m in Peninsular Malaysia). Often in small groups; regularly joins mixed-species feeding flocks. **RANGE & STATUS** Resident Nepal, NE Indian subcontinent, southern China, Taiwan, Sumatra (one record?), Borneo. **SE Asia** Common resident (except C Thailand). **BREEDING** January–August. **Nest** Small deep cradle, suspended from tree, bush or bamboo twigs; 0.5–2 m above ground. **Eggs** 2–3; white to creamy-white (may be pinkish-tinged), speckled pale pinkish-red to pale red-brown (often with ill-defined ring around broader end); 16.7 x 12.7 mm (av.; *zantholeuca*).

CAMPEPHAGIDAE: Cuckooshrikes, trillers, minivets & allies

Worldwide c.82 species. SE Asia 18 species. Mostly small to medium-sized, well-patterned and sexually dimorphic. Frequent higher levels of forest. **Cuckooshrikes** (*Coracina*), **Pied Triller** and **flycatcher-shrikes** (*Hemipus*) Mostly grey or black and white with longish rounded tails. Feed mainly on insects in upper storey or canopy of forest; or lower down in open areas (Pied Triller). **Minivets** (*Pericrocotus*) Slender and long-tailed, with upright posture. Most species have bright red or yellow in plumage. Mostly gregarious, feeding on insects in upper storey and canopy of forest, or lower down in more open areas (Jerdon's).

625 **LARGE CUCKOOSHRIKE**

Coracina macei Plate **67**

IDENTIFICATION 27–30.5 cm. *C.m.siamensis*: **Male** Relatively large size, thick bill and mostly rather pale grey plumage, with blackish lores, dark grey ear-coverts and whitish vent distinctive. Primaries blackish with pale grey to whitish outer fringes, tail grey to blackish with white tip, eyes reddish-brown. In flight shows broad, rounded wings and whitish underwing-coverts with a little dark barring. See very similar Javan Cuckooshrike and Bar-bellied Cuckooshrike. Note range. *C.m.rexpineti* (recorded E Myanmar, N Indochina) has darker forehead, head-sides and throat. **Female** Paler, with paler lores and ear-coverts, whiter breast and belly with variable grey barring, variable pale barring on rump and uppertail-coverts and more dark bars on underwing-coverts. **Juvenile** Head, upperparts, and breast heavily barred/scaled buffy-whitish and dark dusky-brownish; has whitish rump, broad whitish fringes to wing-coverts, tertials and secondaries, blackish and whitish subterminal markings on tertials, browner eyes. **Immature** Like female but retains some juvenile wing-coverts and tertials; possibly shows more distinct darker barring on breast to undertail-coverts. **VOICE** Loud, shrill ***kle-eep***. Variable chuckling notes. **HABITAT & BEHAVIOUR** Open broadleaved evergreen and mixed deciduous forest, pine forest, open areas with scattered trees; up to 2,710 m. Usually high in trees or flying over forest. Has habit of flicking up wings alternately when perched. **RANGE & STATUS** Resident (subject to some movements) Indian subcontinent (except W and Pakistan), southern China, Taiwan. **SE Asia** Common resident (except S Thailand, Peninsular Malaysia, Singapore). Also recorded on passage E Tonkin. **BREEDING** March–May. **Nest** Rather small, shallow cup, on tree-branch; 13–18 m above ground. **Eggs** 2–3; greenish to buffish, boldly blotched pale inky-purple; 32 x 22.6 mm (av. *siamensis*).

626 **JAVAN CUCKOOSHRIKE**

Coracina javensis Plate **67**

IDENTIFICATION 27.5–29 cm. *C.j.larutensis*: **Male** Very similar to Large Cuckooshrike but darker grey, primaries and primary coverts blacker and more uniform, has less white on tail-tip, grey underwing-coverts. Note range. See Bar-bellied Cuckooshrike. **Female** Differs from Large in same way as male. **Juvenile** Probably similar to Large Cuckooshrike. **VOICE** Loud, thin nasal ***yiee***, ***yi'ik*** or ***yi'ee*** and lower, scratchier nasal ***yerrk yerrk...*** and ***yererr'erk*** etc. **HABITAT & BEHAVIOUR** Broadleaved evergreen forest; above 1,000 m. Usually in tree-tops. **RANGE & STATUS** Resident Java, Bali. **SE Asia** Common resident extreme S Thailand, Peninsular Malaysia. **BREEDING** March–August. **Nest** Shallow cup, situated on horizontal tree-branch. **Eggs** 2; buffish, spotted with brown and grey.

627 **BAR-BELLIED CUCKOOSHRIKE**

Coracina striata Plate **67**

IDENTIFICATION 27.5–30 cm. *C.s.sumatrensis*: **Male** Similar to Javan Cuckooshrike but has whitish to yellowish-white eyes, rather uniform head with only slightly darker lores, pale bars and some dark bars on rump and uppertail-coverts, and faint grey bars on vent. In flight shows blacker bars on underwing-coverts. Note habitat. **Female** Rump, uppertail-coverts and lower breast to undertail-coverts broadly barred blackish and whitish. **Juvenile** Head and body heavily scaled whitish, blackish and dusky-brownish, has whiter fringes to wing-feathers, blackish and whitish subterminal markings on tertials, and brownish eyes. **VOICE** Clear whinnying ***kliu-kliu-kliu...*** or shrill ***kriiu-kriiu***. **HABITAT** Broadleaved evergreen forest, freshwater swamp forest, forest edge, sometimes mangroves and old plantations; lowlands. **RANGE & STATUS** Resident Sumatra, Borneo, Philippines. **SE Asia** Scarce to locally fairly common resident southern S Thailand, Peninsular Malaysia; formerly Singapore. **BREEDING** April–May. **Nest** Shallow cup, on tree-branch; up to 21 m above ground. **Eggs** 2.

628 **INDOCHINESE CUCKOOSHRIKE**

Coracina polioptera Plate **67**

IDENTIFICATION 21.5–22 cm. *C.p.indochinensis*: **Male** Much smaller, slimmer, longer-tailed and thinner-billed than Large Cuckooshrike, has prominent white tips to underside of tail-feathers. Difficult to separate from Black-winged Cuckooshrike but shows more pronounced pale fringing to wing-feathers, tail less graduated (central feathers less than 25 mm longer than outer ones), greyer above and with somewhat broader white feather-tips below (typically appearing barely separated). In flight, often shows whitish area on underside of primaries. *C.p.polioptera* (W Thailand, Cambodia, Cochinchina) is a shade paler (paler, with paler wing-feather fringing, than any subspecies of Black-winged); *jabouillei* (N,C Annam) is much darker with more uniform blacker upperwing, no whitish patch on underside of primaries, and only little grey at base of uppertail. Note voice and range. See Lesser Cuckooshrike. **Female** Head-sides and throat to belly paler with darker and paler bars/scales, undertail-coverts whitish with blackish scales/bars; has indistinct whitish, dark-scaled supercilium and broken whitish eyering. Underparts typically more contrastingly barred than Black-winged; in flight shows larger area of whitish on underside of primaries. Also differs by tail shape and pattern (as male). See Lesser and Black-headed Cuckooshrikes. **Juvenile** Best separated from Black-winged by tail shape and pattern and larger whitish patch on underside of primaries. **VOICE** Song is a distinctive series of 5–7, generally descending, loud, high-pitched whistles: ***wi-wi-wi-wi-wu*** and ***wi-wi-wi-wi-wiu-wu*** etc.; more quickly delivered than those of Black-winged. Also utters nasal chuntering ***uh'uh'uh'uh-ik*** and ***uh'uh'uh'uh...*** **HABITAT & BEHAVIOUR** Deciduous, semi-deciduous and pine forest, locally peatswamp forest; up to 1,400 m. Relatively slow and deliberate feeder in middle and upper storey of trees. May join mixed-species feeding flocks. **RANGE & STATUS** Endemic. Fairly common resident W,C,E,S Myanmar, Tenasserim, W,NW,NE Thailand, Indochina (except W,E Tonkin). **BREEDING** April–May. **Nest** Shallow cup, on tree-branch. **Eggs** 2–3. Otherwise poorly documented.

629 BLACK-WINGED CUCKOOSHRIKE
Coracina melaschistos Plate **67**

IDENTIFICATION 21.5–25.5 cm. *C.m.avensis*: **Male** Very similar to Indochinese Cuckooshrike but wings uniformly blackish (also shown by Indochinese subspecies *C.p.jabouillei*), tail more graduated (central feathers always more than 25 mm longer than outer ones), blackish above and with somewhat narrower white feather-tips below (usually appearing well separated). In flight lacks or shows only small whitish area on underside of mid-primaries. *C.m.melaschistos* (W,N Myanmar, recorded in winter NW Thailand) is darker (similar to Indochinese *C.p.jabouillei*); *saturata* (E Tonkin, N,C Annam, recorded in winter NW Thailand, Cambodia, S Laos) is much darker, showing little contrast between mantle and wings. Note voice and range. See Lesser Cuckooshrike. **Female** Slightly paler overall, with vaguely paler supercilium (in front of eye) and ear-coverts, faint darker-and-paler bars/scales on underparts, blackish bars/scales on whitish undertail-coverts and some pale feather-fringing on wings. Throat to belly typically more uniform than Indochinese; usually shows less obvious pale fringing on wings and less grey on upperside of central tail-feathers; in flight shows much smaller (or no) whitish area on underside of mid-primaries. Also differs by tail shape and pattern (as male). See Lesser and Black-headed Cuckooshrikes. **Juvenile** Head and body paler, heavily barred/scaled buffish to whitish and dark sooty-brownish, has whitish to buffish-white tips to wing-feathers. **Immature** Similar to female but shows scaling/barring on rump and uppertail-coverts and more prominently barred underparts. Best separated from Indochinese by tail shape and pattern; whitish patch on underside of primaries smaller or absent. **Other subspecies in SE Asia** *C.m.intermedia* (visitor Tenasserim, W,C,NE,SE Thailand). **VOICE** Series of 3–4 clear, well-spaced, high-pitched whistles: ***wii-wii-jeeu-jeeu***, ***wi'i-wii-wii-juu*** and ***witi-jeeu-jeeu-jeeu*** etc.; slower and more measured than Indochinese. **HABITAT & BEHAVIOUR** Broadleaved evergreen forest; 300–1,920 m. Wintering birds also occur in lowlands in gardens and more open areas, sometimes deciduous forest. Usually breeds at higher levels than Indochinese Cuckooshrike. **RANGE & STATUS** Breeds NE Pakistan, NW,N,NE India, Nepal, Bhutan, SE Tibet, C,E and southern China; some populations winter to south, India, Bangladesh. **SE Asia** Fairly common resident (subject to some movements) W,N,E Myanmar, W,NW,NE Thailand, N Laos, W,E Tonkin, N,C Annam. Fairly common to common winter visitor C,S Myanmar, Tenasserim, Thailand (except southern S), Indochina. Uncommon passage migrant C,NE Thailand, E Tonkin. **BREEDING** April–May. **Nest** Shallow cup, on tree-branch; 3–8 m above ground. **Eggs** 2–4; olive-grey, longitudinally blotched pale brown overall; 24.3 x 17.4 mm (av.; *melaschistos*).

630 LESSER CUCKOOSHRIKE
Coracina fimbriata Plate **67**

IDENTIFICATION 19–20.5 cm. *C.f.neglecta*: **Male** Very similar to Black-winged Cuckooshrike but smaller with smaller white tips to undertail, tail less graduated, with central feathers less than 25 mm longer than outer ones. Note voice and range. See Indochinese Cuckooshrike. **Female** Like Indochinese Cuckooshrike but underparts more uniformly barred, tends to show more pronounced pale supercilium, dark eyestripe and pale ear-covert streaking, has smaller white tips to undertail and usually lacks whitish area on underside of primaries. See Black-headed Cuckooshrike. Note range. **Juvenile** Best separated from Indochinese by range, smaller size and tail shape and pattern. **Other subspecies in SE Asia** *C.f.culminata* (extreme S Thailand southwards). **VOICE** Song is a repeated, loud, clear ***whit-it-it-chui-choi*** etc. Also gives more rapid ***whit-whit-whit-whit-whit-whit***. When alarmed, utters squeaky nasal, bulbul-like ***wherrrh-wherrrh-wherrrh...*** and high ***whit-weei***. **HABITAT** Broadleaved evergreen forest, secondary growth, plantations; up to 975 m. **RANGE & STATUS** Resident Greater Sundas. **SE Asia** Common resident south Tenasserim, W(south),S Thailand, Peninsular Malaysia. Rare (status uncertain) Singapore. **BREEDING** April-August. **Nest** Small, neat cup, situated on horizontal tree-branch. **Eggs** 2; pale bluish with brown speckles.

631 BLACK-HEADED CUCKOOSHRIKE
Coracina melanoptera Plate **67**

IDENTIFICATION 19–20.5 cm. *C.m.sykesi*: **Male** Easily identified from other cuckooshrikes by black hood. **Female** Similar to Indochinese Cuckooshrike but less pure grey; has more prominent pale fringing on wing-coverts, more pronounced whitish supercilium and streaking on ear-coverts, and white underparts with more distinct, blacker barring, and unbarred white belly-centre and undertail-coverts. **Juvenile** Like female but upperparts scaled blackish and buffish-white, wing-coverts and tertials have dark subterminal markings, bill dark brown. See Black-winged Cuckooshrike. Note range. **VOICE** Song is a series of 6–9 quickly delivered loud whistles: ***twit-wit-wee-TWY-TWY-TWY-TWY*** or ***twit-wit-wit-wee-TWEE-TWEE-TWY-TWY-TWY***; hurried and rising, then falling and decelerating slightly, sometimes followed by quick ***pit-pit-pit***. Also an even series of loud, upturned whistles: ***KWEe-KWEe-KWEe...*** **HABITAT** Open deciduous and semi-deciduous forest, secondary growth, sometimes orchards and gardens; up to 1,525 m. **RANGE & STATUS** Resident (subject to seasonal movements in north) India, S Nepal, Bangladesh, Sri Lanka. **SE Asia** Scarce winter visitor SW,W Myanmar.

632 PIED TRILLER *Lalage nigra* Plate **67**

IDENTIFICATION 17–18 cm. *L.n.nigra*: **Male** Recalls Bar-winged Flycatcher-shrike but has diagnostic whitish supercilium, black eyestripe and grey back to uppertail-coverts; underparts white with grey wash on breast. Note habitat and range. **Female** Dark parts of plumage greyish-brown, underparts tinged buffish (browner on breast and flanks) with extensive but indistinct dark scales/bars, patch on wing-coverts restricted more to broad bars. **Juvenile** Crown and upperparts paler and browner than female with pale buff bars/scales, underparts paler with dark brown streaks on lower throat and breast (sometimes all underparts), wing-feather fringes and tips tinged buff. **VOICE** Disyllabic whistle with lower second note and descending series of nasal ***chack*** notes. **HABITAT** Coastal scrub, plantations, gardens; lowlands. **RANGE & STATUS** Resident Andaman and Nicobar Is, Greater Sundas, Philippines. **SE Asia** Uncommon to locally common resident S Thailand, Peninsular Malaysia, Singapore. **BREEDING** February–September. Multi-brooded. **Nest** Shallow cup, on tree-branch; 2.5–9 m above ground. **Eggs** 2–3; brownish-white to brownish-green, blotched and streaked brown to reddish-brown (mostly towards broader end), sometimes with grey undermarkings; 21.8–22.1 x 15–16 mm (Nicobar Is).

633 **ROSY MINIVET** *Pericrocotus roseus* Plate **68**
IDENTIFICATION 18–19.5 cm. *P.r.roseus*: **Male** Combination of brownish-grey upperparts with darker, greyer crown, reddish rump and red uppertail-coverts and rosy-pink underparts with whitish throat diagnostic. Has broad red wing-patch and red lines extending along fringes of tertials and some primaries. Some (possibly older birds) have all-red rump and uppertail-coverts and more uniformly grey upperparts; others have only pinkish-tinged rump and uppertail-coverts. Individuals of the form '*stanfordi*' (recorded in winter NE Thailand, Cambodia, Cochinchina) belong to a little known, highly variable population breeding in S China. They resemble Swinhoe's Minivet but have pale pink to salmon-pink forecrown and can show pinkish to flame-red on outertail-feathers and wing-patch (latter sometimes larger), pinkish to reddish tinge on rump and uppertail-coverts and variable pink tinge to underparts. **Female** Pattern similar but red replaced by pale yellow. From other minivets by combination of isolated yellow wing markings, pale yellowish-white throat and washed-out pale yellow remainder of underparts (often duller-tinged on breast and flanks). Rump and uppertail-coverts have variable, faint yellowish-olive wash (particularly latter). See Grey-chinned Minivet. Individuals of the '*stanfordi*' form resemble Swinhoe's but differ by prominent yellow wing markings, yellow on tail and yellowish tinge to uppertail-coverts. **Juvenile** Like female (both sexes) but feathers of crown and upperparts densely scaled/barred yellowish-white to whitish with dark subterminal markings, wing-coverts and tertials tipped same colour, some obscure scaling/mottling on breast. **Immature male** Like female but has some orange to flame-red on uppertail-coverts and tail-fringes, some orange on wing markings and variable pink suffusion on underparts. **VOICE** Whirring trill, similar to Ashy Minivet. **HABITAT & BEHAVIOUR** Deciduous and semi-deciduous forest, sometimes open broadleaved evergreen forest, forest edge; up to 1,525 m. May occur in large flocks in winter. **RANGE & STATUS** Breeds N Pakistan, NW,N,NE India, Nepal, Bhutan, SW,S China; some populations winter south to S India, Bangladesh. **SE Asia** Fairly common to common resident Myanmar. Breeds E Tonkin. Scarce to fairly common winter visitor Thailand, Cambodia, N,S Laos, Cochinchina. Also recorded on passage E Tonkin. Vagrant Peninsular Malaysia. Recorded (status uncertain) W Tonkin. **BREEDING** India: April–June. **Nest** Small, neat cup, attached to upperside of tree-branch; 3–10 m above ground. **Eggs** 2–4; whitish to buffish-white, blotched dark brown and lavender (mainly towards broader end); 19.4 x 15.2 mm (av.).

634 **SWINHOE'S MINIVET**
Pericrocotus cantonensis Plate **68**
IDENTIFICATION 19 cm. Monotypic. **Male** Similar to Ashy Minivet but hindcrown and nape dark grey, white forehead-patch extends in marked short eyebrow to just behind eye, upperparts and wings (mainly greater coverts and tertials) tinged brown, breast and belly washed pale vinous-brownish, rump and particularly uppertail-coverts pale drab brownish, clearly paler than rest of upperparts; wing-patch (if present) tinged very pale yellowish-buff (rather than whitish). **Female** Upperparts paler, rump and uppertail-coverts less sharply contrasting, wing-patch (if present) may be yellower-tinged. From Ashy by paler rump and uppertail-coverts, slightly browner-tinged rest of upperparts, no obvious dark band across extreme forehead and, usually, more extensively pale forehead extending in line to just behind eye; breast and belly typically less clean whitish. See Rosy Minivet. **First winter** Shows similar characters to Ashy. Males have breast and belly recalling adult and a hint of adult head pattern. **HABITAT** Broadleaved evergreen, semi-evergreen and deciduous forest, forest edge; up to 1,200 m. **RANGE & STATUS** Breeds C,SC,SE China; mostly winters to south. **SE Asia** Uncommon winter visitor C,S Myanmar, Tenasserim, Thailand, Cambodia, Laos, C Annam, Cochinchina. Uncommon passage migrant Cambodia, E Tonkin.

635 **ASHY MINIVET**
Pericrocotus divaricatus Plate **68**
IDENTIFICATION 18.5–20 cm. *P.d.divaricatus*: **Male** Black lores, crown and nape, all-grey mantle to uppertail-coverts and white to creamy-white forecrown and underparts diagnostic. Wing markings indistinct and whitish or almost lacking, but whitish band across base of flight-feathers is prominent in flight. See Swinhoe's Minivet. **Female** Crown and upperparts pale grey (a little darker on crown); has whitish band across forehead to eye, and blackish loral stripe extending narrowly across extreme forehead. See Swinhoe's. **First winter** Like female but shows white fringing and blackish subterminal markings on tertials, and white tips to greater coverts. **VOICE** Metallic jingling trill. Flight call rather unmelodious, a slightly hesitant ascending ***tchu-de tchu-dee-dee tchu-dee-dee***. **HABITAT & BEHAVIOUR** Open forest, areas with scattered trees, mangroves, coastal vegetation, casuarinas, plantations; up to 1,200 m; usually lowlands. Usually in small to quite large, restless flocks. **RANGE & STATUS** Breeds SE Siberia, Ussuriland, NE China, N,S Korea, Japan; winters Sumatra, Borneo, Philippines. **SE Asia** Uncommon to locally common winter visitor (except SW,W,N,E Myanmar, W Tonkin, N Annam). Also recorded on passage Cambodia, E Tonkin.

636 **SMALL MINIVET**
Pericrocotus cinnamomeus Plate **68**
IDENTIFICATION 14.5–16 cm. *P.c.vividus*: **Male** Easily identified by combination of small size, grey head and upperparts with darker ear-coverts and cheeks (sometimes throat), reddish-orange rump, uppertail-coverts, breast and flanks, orange-yellow vent and orange and yellow wing-patch. *P.c.sacerdos* (Cambodia, Cochinchina) has red-orange of plumage redder and more vivid (both sexes). See Grey-chinned Minivet. Note habitat. **Female** Combination of pale dull grey crown and upperparts, bright reddish-orange rump and uppertail-coverts (mixed yellowish on former), greyish-white throat and upper breast and pale yellow belly and vent diagnostic. Wing-patch orange-yellow. See Fiery Minivet. Note range. *P.c.separatus* (west S Thailand) has more washed-out, less yellow underparts. **Juvenile** Like female but crown to back browner with pale yellowish-white to whitish bars/scales, wing-coverts tipped and tertials fringed same colour, breast very faintly mottled darker. **First-winter male** Like female but often shows some orange on flanks. **Other subspecies in SE Asia** *P.c.thai* (C,E Myanmar, NW,NE Thailand, Laos). **VOICE** Continuously repeated, very thin, sibilant, drawn-out, high-pitched ***tswee-eet*** and ***swee swee...*** etc. **HABITAT & BEHAVIOUR** Deciduous forest, forest edge, more open areas with trees, parks and gardens, locally peatswamp forest and coastal scrub; up to 1,525 m. Usually in small flocks. **RANGE & STATUS** Resident Indian subcontinent, Java, Bali. **SE Asia** Common resident (except N Myanmar, southern S Thailand, Peninsular Malaysia, Singapore, W,E Tonkin, N Annam). **BREEDING** January–September. **Nest** Small neat cup, attached to upperside of tree-branch; up to 12 m above ground. **Eggs** 2–4; pale greenish-white or creamy-buff, speckled and blotched reddish-brown (mostly at broader end); 16.8 x 13.8 mm (av.; *vividus*).

637 **FIERY MINIVET** *Pericrocotus igneus* Plate **68**
IDENTIFICATION 15–15.5 cm. *P.i.igneus*: **Male** Like a miniature Scarlet Minivet but wing-patch orange to red-orange, lacks isolated markings on tertials and inner secondaries, underwing has orange (not red) coverts and band across base of flight-feathers, has much more orange-coloured tail markings, and all-black central tail-feathers; remaining red of plumage more orange-tinged. See Grey-chinned Minivet race *montanus*. Note habitat and range. **Female** From other minivets by combination of size, orange-yellow band across forehead to eye, bright red-orange rump and uppertail-coverts, and all-yellow throat and underparts (tinged orange). Crown and upperparts slaty-grey, wing-patch pale orange to orange-yellow. See Small Minivet. Note range. **Juvenile** Like

female but browner above, with narrow whitish fringing/barring (mainly crown to upper mantle) and dark subterminal feather markings, wing-coverts and tertials tipped/fringed paler, indistinct dark mottling on breast. Male attains patches of adult plumage during first winter. **VOICE** Calls with very thin ***tit tit swiiii*** or simple ***swiiiii*** notes. **HABITAT** Broadleaved evergreen forest, forest edge; up to 1,220 m, mostly below 610 m. **RANGE & STATUS** Resident Sumatra, Borneo, Philippines (Palawan). **SE Asia** Uncommon resident south Tenasserim, S Thailand, Peninsular Malaysia; formerly Singapore. **BREEDING** April–May. **Nest** Small neat cup, attached to upperside of tree-branch. **Eggs** 2; pale yellowish, profusely marked grey and brown; 20.6 x 15.2 mm.

638 JERDON'S MINIVET

Pericrocotus albifrons Plate **68**

IDENTIFICATION 14.5–16 cm. Monotypic: **Male** Unmistakable. Upperparts glossy black with white supercilium and band across forehead, white wing markings, and white lower head-sides, underparts with orange (sometimes orange-red) breast-patch. Rump white with orange (sometimes orange-red) centre. **Female** Crown to back mousy grey-brown, underparts white with grey wash on breast. **Juvenile** Like female but crown (and to lesser extent mantle and back) scaled/barred white, with blackish subterminal feather markings, wing-coverts tipped white, some dark speckles/mottling on breast (mainly sides). Male attains patches of adult plumage during first winter. **VOICE** Various spaced, sweet, high-pitched notes: ***thi***, ***tuee***, ***chi***, ***tschi*** and ***tchu-it*** etc. Contact call is a soft ***tchip***. **HABITAT & BEHAVIOUR** Semi-desert, dry cultivation with scrub and scattered trees, thorn scrub; lowlands. Occurs in pairs or small parties. **RANGE & STATUS** Endemic. Uncommon resident C,S(north) Myanmar. **BREEDING** Season undocumented. **Nest** Small neat cup, on branch of bush or small tree; 1–2 m above ground. **Eggs** 3; greyish-brown with lengthwise dark brown streaks overlaying a few pale blue-grey streaks; 17.3 x 13.7 mm (av.). **NOTE** Split from White-bellied Minivet *P.erythropygius* of Indian subcontinent, following Rasmussen & Anderton (2005).

639 GREY-CHINNED MINIVET

Pericrocotus solaris Plate **68**

IDENTIFICATION 17–19 cm. *P.s.rubrolimbatus*: **Male** Recalls Long-tailed Minivet but head and mantle dark grey, with paler grey lower sides of head, pale grey to whitish chin and orange-yellow wash on lower throat; has more orange underparts, rump, uppertail-coverts and wing-patch (latter with single line along primary fringes). *P.s.griseogularis* (N[east] Laos, northern Vietnam), *nassovicus* (south-west Cambodia) and *deignani* (C,S Annam) have more uniform lower head-sides and throat, variable yellowish-olive tint on throat (darker, less contrasting on latter two) and redder-orange parts of plumage (particularly latter two), crown and mantle also slightly darker on latter; *montanus* (extreme S Thailand, Peninsular Malaysia) is much redder, with black (slightly glossy) upperparts and ear-coverts and very dark grey throat (hardly contrasts). See Fiery and Long-tailed Minivets. Note habitat and range. **Female** Uniform grey forehead, crown, head-sides and mantle and greyish-white chin diagnostic. Rest of underparts and wing-patch yellow, rump and uppertail-coverts olive-yellow. *P.s.griseogularis* has uniform pale greyish-white throat, slight olive cast to yellow of underparts, and slightly more olive rump and uppertail-coverts; *montanus* has distinctive blackish-slate (slightly glossy) crown and mantle and sharply contrasting orange-yellow on underparts, wing-patch, rump and uppertail-coverts. See Rosy Minivets. **Juvenile** Similar to female (both sexes) but upperpart feathers slightly darker with pale yellow scales/bars (most obvious on crown, nape and scapulars) and blackish subterminal markings; has pale yellow wing-covert tips and tertial fringes, and indistinct dark mottling/barring on sides of breast. **Other subspecies in SE Asia** *P.s.solaris* (W,N,C,S Myanmar). **VOICE** Call is a thin, repeated ***tswee-seet*** and more slurred ***swirrririt***. Soft ***trip*** notes and more sibilant ***trii-ii*** when foraging. **HABITAT & BEHAVIOUR** Broadleaved evergreen forest, forest edge, sometimes pine forest; 300–2,350 m. Usually found in small noisy parties, often associating with mixed-species feeding flocks. **RANGE & STATUS** Resident Nepal and NE Indian subcontinent, southern China, Taiwan, Sumatra, Borneo. **SE Asia** Common resident Myanmar (except SW), W,NW,NE,extreme S Thailand, Peninsular Malaysia, Cambodia, Laos, Vietnam (except Cochinchina). **BREEDING** February–June. **Nest** Small neat cup, attached to upperside of tree-branch; 13.7 m above ground, or higher. **Eggs** 3; very pale green, finely marked pale rufous; 19.3 x 14.2 mm (av.; *solaris*).

640 LONG-TAILED MINIVET

Pericrocotus ethologus Plate **68**

IDENTIFICATION 17.5–20.5 cm. *P.e.ethologus* (visiting race): **Male** Glossy black with bright red lower back to uppertail-coverts, breast to vent and wing-patch and extensively red outertail-feathers. Wing-patch distinctive, with two spaced lines along tertial and primary fringes. From Scarlet Minivet by slimmer build, lack of isolated red spots on tertials and secondaries, and less extensively red tail. *P.e.annamensis* (S Annam) may show similar wing and tail markings to Short-billed Minivet; see that species and Grey-chinned Minivet. **Female** Upperparts greyish-olive, with greyer crown, paler yellow-tinged extreme forehead, pale grey lower sides of head (faintly washed olive-yellowish) and yellowish olive-green lower back to uppertail-coverts; underparts fairly bright yellow (faintly washed olive) with paler lower throat and whitish upper throat. Yellow wing-patch lacks line along tertial fringes. *P.e.yvettae* (N,E[north] Myanmar) has more uniform grey crown to mantle and often yellower forehead; *mariae* (W Myanmar) and *annamensis* have more uniform grey crown to mantle, yellow parts of plumage deeper and brighter, slightly orange-tinged (distinctly so in *annamensis*), much yellower extreme forehead, cheeks and upper throat. *P.e.annamensis* can have forecrown and tail resembling Short-billed, or no yellow on forehead, just a yellow eyebrow. *P.e.ripponi* (E Myanmar, NW,?NE Thailand) is roughly intermediate between *mariae* and *ethologus*. See Short-billed, Grey-chinned and Scarlet. **Juvenile** Similar to female but upperpart feathers have prominent scaly yellowish-white tips and blackish subterminal markings, wing-coverts tipped yellowish-white, throat-sides, breast and flanks dark spotted and/or barred and washed drab olive-greyish, yellow of underparts typically restricted to centre of throat/upper breast and vent. **Immature male** Variable. First-summer birds typically have orange extreme forehead, darker crown to mantle (than female), deep orange to red-orange rump to uppertail-coverts and wing-patch and extensive bright/deep orange on underparts. **VOICE** Sings with loud, rich, musical, slow ascending or descending chattery trill, usually terminated by a short, upslurred ***SWEEET***. Loud, sweet, clear, see-sawing ***SFWEEEfweet..SFWEEEfweet..*** and ***SWEEET'SWEEET..*** Also, sweet, rolling ***prrr'wi***, ***prrr'i-wi*** and ***prrr'i-prrr'i*** and thin, sibilant, rather hurried ***swii-swii swii-swii-swii..*** and shrill ***seeseesee...seeseeSEET'SEET***. **HABITAT & BEHAVIOUR** Broadleaved evergreen and pine forest, forest edge; 900–3,100 m, down to 450 m in winter. Gregarious; may form quite large flocks in winter. **RANGE & STATUS** Resident (subject to some movements) NE Afghanistan, NW,N,NE Indian subcontinent, S Tibet, SW,W,NC,NE China. **SE Asia** Common to locally common resident (subject to some winter movements) W,N,E Myanmar, NW,NE Thailand, C,S Annam. Uncommon to locally common winter visitor C,E,S Myanmar, NW,NE Thailand, E Tonkin. Recorded (status uncertain) N,C Laos, W Tonkin. **BREEDING** December–June. **Nest** Neat cup, attached to upperside of tree-branch; 15–24 m above ground. **Eggs** 3–4; white to greenish-white, spotted and blotched light brown to blackish-brown over pale lavender to inky-grey undermarkings; 19.8 x 15.1 mm (av.).

641 **SHORT-BILLED MINIVET**
Pericrocotus brevirostris Plate **68**
IDENTIFICATION 17.5–19.5 cm. *P.b.neglectus*: **Male** Like Long-tailed Minivet but wing-patch shows single red line extending along primary fringes, has much more red on outertail-feathers (all red from below, or almost so), black of throat extends in semi-circle onto upper breast (not sharply cut off). Note voice and range. See Scarlet Minivet. **Female** Like Long-tailed race *ethologus* but extreme forehead tinged golden-yellow, forecrown washed yellow to golden-yellow, mantle greyer (but similar to Long-tailed race *mariae* etc.), throat always yellow, shows much more yellow on outertail-feathers (in similar pattern to male). See Scarlet and Grey-chinned. **Juvenile** Differs from adult in same way as Long-tailed. **Other subspecies in SE Asia** *P.b.affinis* (W,N Myanmar); *anthoides* (W,E Tonkin). **VOICE** Sings with irregularly repeated, loud, high, gently falling, sad whistled ***TSEEEr***, and slow, alternating series of sad, thin, sweet piping whistles: ***tSUTSEEET tSUTSEEET..*** or ***tsuuuit tsuuuit tsuuuit...*** Also utters dry ***tup*** contact notes, and a dry, upturned ***chweet***. **HABITAT & BEHAVIOUR** Broadleaved evergreen forest, forest edge, sometimes pine forest; 915–2,135 m (locally down to 750 m). Usually in pairs. **RANGE & STATUS** Resident Nepal and NE Indian subcontinent, SE Tibet, SW,S China. **SE Asia** Fairly common to locally common resident W,N,E Myanmar, N Tenasserim, W,NW Thailand, Laos, W,E Tonkin. **BREEDING** March–April. **Nest** Neat cup on outer tree-branch; 10 m above ground. Otherwise undocumented.

642 **SCARLET MINIVET**
Pericrocotus speciosus Plate **68**
IDENTIFICATION 17–21.5 cm. *P.s.semiruber*: **Male** Larger, larger-billed, more robust and shorter-tailed than Long-tailed and Short-billed Minivets, with more vivid flame-red plumage. Shows diagnostic isolated red markings near tips of tertials and inner secondaries and mostly flame-red tail (pronounced in flight). Outer webs of central tail-feathers usually red. Rarely has red of plumage replaced by orange-yellow to deep yellowish-orange. *P.s.elegans* (northern Myanmar, N Indochina and wintering range) is larger, with mostly black central tail-feathers; *flammifer* (W,S Thailand, north-east/east Peninsular Malaysia) and *xanthogaster* (west/south Peninsular Malaysia) are smaller (within size range given). See Fiery and Grey-chinned Minivets. **Female** From other minivets by combination of size, build, rich yellow forehead, prominently yellow-washed forecrown, slaty-grey hind-crown to mantle, uniform, rather deep bright yellow throat and underparts (sometimes washed olive on breast and flanks) and distinctive wing pattern (like male but yellow). Central tail-feathers all dark, second pair broadly yellow-tipped. See Short-billed and Long-tailed Minivets. Note habitat and range. **Juvenile** Similar to Long-tailed and Short-billed but has yellower underparts and suggestion of adult tertial/inner secondary pattern. **Immature male** Differs from female in similar way to Long-tailed. **VOICE** Loud, piercing whistles: ***swEEEP-swEEEP-swEEEP...*** and ***wEEEP-wEEEP-wEEEP-WIT-WIP*** etc. Faster ***kapitit-kapitit-kapitit-kapitit-kapitit...*** when agitated. **HABITAT & BEHAVIOUR** Broadleaved evergreen, semi-evergreen, deciduous and peatswamp forest; up to 1,700 m (below 915 m in Peninsular Malaysia). Gregarious; regularly joins mixed-species feeding flocks. Often seen in noisy swooping flight, with tail fanned. **RANGE & STATUS** Resident (subject to some movements) N,E India, Nepal, Bhutan, Bangladesh, southern China, Greater Sundas, Philippines. **SE Asia** Common resident (except C Thailand); scarce Singapore. Uncommon winter visitor E Myanmar, NW Thailand. **BREEDING** March–July. **Nest** Small neat cup, attached to upperside of tree-branch; 6–18.5 m above ground. **Eggs** 2–3; pale green or blue-green, spotted and blotched pale blue and dark brown; 22.9 x 17 mm (av.; *semiruber*). **NOTE** Split from *P. flammeus* (Orange Minivet) of S India and Sri Lanka, following Rasmussen & Anderton (2005).

PACHYCEPHALIDAE: Whistlers & allies

Worldwide c.41 species. SE Asia 1 species. Rather robust, with thickset rounded heads and strong, thick bills. Usually solitary or in pairs. Glean insects from foliage.

643 **MANGROVE WHISTLER**
Pachycephala cinerea Plate **70**
IDENTIFICATION 15.5–17 cm. *P.g.cinerea*: **Adult** Nondescript, with drab brown upperparts, rather slaty-grey crown (contrasts somewhat with mantle) and white underparts with duller throat and greyish-washed breast. Upright posture and dull plumage may cause confusion with some flycatchers, but has thick black bill, no obvious head/wing markings or rufous tones in plumage. *P.g.vandepolli (*east S Thailand, Peninsular Malaysia [except Langkawi I] and Singapore eastwards) tends to have slightly richer brown upperside (not crown). See White-chested Babbler and Brown Fulvetta. Note habitat, behaviour and range. **Juvenile** Tertials and secondaries prominently rufescent-fringed, bill flesh-coloured to brown. **VOICE** Song is loud, deliberate and high-pitched, consisting of variable phrases introduced by 2–4 short notes: ***tit tit PHEW-WHIU-WHIT; chi chi chi WIT-PHEW-CHEW; tit tit tit TOO-WHIT*** etc. Last note often louder, shriller and more explosive. Also, similar phrases with undulating latter notes: ***tit tit tit CHEWY-CHI-CHEWY*** etc. **HABITAT & BEHAVIOUR** Mangroves and adjacent coastal vegetation, locally plantations and wooded gardens, island forest, locally freshwater swamp forest and similar vegetation inland. Unobtrusive and sluggish; sits still for relatively long periods amongst foliage. **RANGE & STATUS** Resident coastal E India, S Bangladesh, Andaman Is, Greater Sundas, Lombok, Philippines (Palawan). **SE Asia** Uncommon to locally common coastal resident (except E Tonkin, N,C Annam); scarce Singapore. Also inland Cambodia (Tonle Sap area), south-west S Laos, Cochinchina (Tay Ninh). **BREEDING** March–June. **Nest** Flimsy open cup, in tree; 1–15 m above ground. **Eggs** 2; lightly glossed creamy-white to pale buff, spotted sepia to blackish-brown over pale grey to light purple undermarkings (usually ringed at broader end); 21.7 x 15.7 mm (av. *cinerea*).

ORIOLIDAE: Orioles & allies

Worldwide c.29 species. SE Asia 8 species. Smallish to medium-sized, well-patterned and sexually dimorphic. Robust and stout-billed with colourful plumage. Frequent higher levels of forest, feeding mainly on insects and fruit; flight strong and undulating.

644 **DARK-THROATED ORIOLE**
Oriolus xanthonotus Plate **69**
IDENTIFICATION 20–20.5 cm. *O.x.xanthonotus*: **Male** Small size, black hood and wings and white lower breast to belly with prominent black streaks diagnostic. Upperparts and tail-coverts yellow, tail mostly black, has some indistinct narrow yellow and whitish fringing on flight-feathers. See Black-hooded Oriole. **Female** Upperside olive-green with brighter lower rump and uppertail-coverts and darker, greyer crown and sides of head, underparts whitish with bold blackish streaks and yellow undertail-coverts. Could be confused with juvenile Black-naped Oriole but much smaller, bill plain fleshy-orange, crown

and sides of head darker, underparts more boldly patterned. **Juvenile** Similar to female but with duller bill; apparently has (on skins) rufous fringes/tips to wing-coverts and sometimes also narrow tips to outer wing-feathers. **Immature male** Similar to female but upperparts yellower-green, with sooty crown and nape and greyish wash on throat. See juvenile Black-hooded. **VOICE** Song is a repeated fluty liquid ***PHU PHI-UU***; ***PHU-PHU-PHU WO***; ***PHU'PHU-WIU-UU*** and ***PHU-PUI*** etc. Call is a high-pitched piping ***kyew***, ***pheeu*** or ***ti-u***, less harsh than other orioles. **HABITAT** Canopy of broadleaved evergreen forest, forest edge; up to 1,220 m (mostly below 300 m). **RANGE & STATUS** Resident Greater Sundas, Philippines (Palawan). **SE Asia** Uncommon to common resident south Tenasserim, S Thailand, Peninsular Malaysia; formerly Singapore. **BREEDING** February–March. **Nest**. Cradle, suspended from forked tree-branch. **Eggs** 2; pinkish, with chestnut and lavender markings; 25.6 x 19.6 mm (av.).

645 INDIAN GOLDEN ORIOLE

Oriolus kundoo Plate **69**

IDENTIFICATION 22–25 cm. Monotypic. **Male** Somewhat smaller and more slender-billed than Black-naped Oriole. Combination of yellow nape, restricted black eye-patch, and black shoulders diagnostic. Wings black with bold yellow markings. **Female** From similar Slender-billed and Black-naped Orioles by lack of blackish head-band and blacker wing-coverts. **Juvenile** More similar to Slender-billed and Black-naped (both of which often show shadowy head-band). Apart from size, streaking of underparts is thinner, and tends to be more lacking on throat and vent; wing-coverts with darker centres? **VOICE** Calls include loud, nasal, upslurred, miaowing ***nyEEARSCh***, with scratchy, harsh ending. Song (unlikely to be heard in region) is said to recall Maroon Oriole. **HABITAT** Recorded in island forest. **RANGE & STATUS** Breeds west C Asia, N,E Afghanistan, NW,N Indian subcontinent; northern populations winter south to C,S India, Sri Lanka, rarely Andaman Is. **SE Asia** Vagrant Peninsular Malaysia (Langkawi I). **NOTE** Split from extralimital *Oriolus oriolus* (European Golden Oriole) following Rasmussen & Anderton (2005).

646 SLENDER-BILLED ORIOLE

Oriolus tenuirostris Plate **69**

IDENTIFICATION 23–26 cm. *O.t.tenuirostris*: **Male** Very similar to female Black-naped Oriole but bill slightly longer and considerably thinner, black nape-band narrower, roughly equal to width of black surrounding eye (obviously broader on Black-naped). Note voice, habitat, range and status. **Female** Like male but yellow of plumage tinged greener, shows some indistinct, narrow darker streaks on underparts. **Juvenile/immature** Difficult to separate from Black-naped, except by longer, thinner bill. **Other subspecies in SE Asia** *O.t.invisus* (S Annam). **VOICE** Song is a repeated loud fluty ***WIP-WI'U'WOW'WOW*** or ***WI WI'U-WU-WU*** and variants (more hurried than Black-naped). Also a single fluty ***TCHEW*** or ***TCHI'U***. Alarm call is a harsh, slightly nasal, grating ***kyERRRRRh*** or ***ey'ERRRRRh***. **HABITAT** Open pine forest, sometimes mixed pine/oak forest, also broadleaved evergreen forest, forest edge and clearings with scattered trees on wintering grounds; 1,000–1,900 m; down to at least 600 m in winter. **RANGE & STATUS** Resident (subject to relatively local movements) NE Indian subcontinent, SW China. **SE Asia** Fairly common to common resident Myanmar (except N), N,S Laos, S Annam. Fairly common winter visitor W,NW,NE Thailand (probably breeds locally). Recorded (status uncertain) W Tonkin, N Annam. **BREEDING** April–June. **Nest** Cradle, suspended from forked pine branch. **Eggs** 2–4; pale pink, speckled and spotted black, the markings surrounded by reddish halos; 27.9 x 20.7 mm (av. *tenuirostris*).

647 BLACK-NAPED ORIOLE

Oriolus chinensis Plate **69**

IDENTIFICATION 24.5–27.5 cm. *O.c.diffusus*: **Male** Relatively large size, golden-yellow body and wing-coverts and contrasting broad black band from lores to hindcrown and nape distinctive. Rest of wings and tail patterned black and yellow, bill thick and fleshy-orange. **Female** Upperparts and wing-coverts mostly olive-yellow. See very similar Slender-billed Oriole. **Juvenile** Much duller than female, lacking black head-band (often shows shadow of it). Crown and upperparts yellowish-olive, sides of head yellow with faint dark eyestripe, underparts creamy to yellowish-white with narrow blackish streaks, yellow-washed flanks and plain yellow vent, wings mostly yellowish-green, tail greener, bill mostly blackish. Possibly not separable in field from Slender-billed, except by thicker and slightly shorter bill. Note voice, habitat, range and status. See female Dark-throated Oriole. **Immature** Gradually attains dark head-band. **Other subspecies in SE Asia** *O.c.maculatus* (west Peninsular Malaysia, Singapore). **VOICE** Song consists of repeated, loud, liquid, fluty phrases: ***KWIA-LU***; ***U-DLI-U***; ***U-LI-U***; and ***U-LIU*** etc. Call is a long harsh, rasping, nasal ***kyEHHR*** or ***grWEEESH***. **HABITAT** Open broadleaved evergreen and deciduous forest, open areas with scattered trees, parks, gardens, plantations, mangroves; up to 1,525 m. **RANGE & STATUS** Resident Andaman and Nicobar Is, Taiwan, Greater Sundas, W Lesser Sundas, Philippines, Sulawesi, Sula Is, N Moluccas (Mayu). Breeds China (except NW), SE Siberia, Ussuriland, N,S Korea; northern populations winter to south, C,S,NE India, Bangladesh, S China. **SE Asia** Uncommon to common resident SW Myanmar, C,S Thailand (local and scarce), Peninsular Malaysia, Singapore. Uncommon breeding visitor N Laos. Fairly common to common winter visitor (except SW,W,N Myanmar, Singapore, W,E Tonkin, N Annam). Fairly common passage migrant east Cambodia, W,E Tonkin. **BREEDING** December–July. **Nest** Cradle, suspended from forked tree-branch. **Eggs** 2; fairly glossy, white to pinkish-white, sparingly blotched and spotted grey and black or purple-brown (mainly at broader end); 27.3–34.2 x 20.1–23.2 mm.

648 BLACK-HOODED ORIOLE

Oriolus xanthornus Plate **69**

IDENTIFICATION 22–25 cm. *O.x.xanthornus*: **Male** Uniform golden-yellow body and wing-coverts and contrasting black hood diagnostic. Rest of wings black with extensive yellow, bill fleshy-orange. See Dark-throated Oriole. Note range and habitat. **Female** Lower mantle to rump washed olive, underparts and wing markings slightly paler and less rich yellow, wing markings slightly smaller. **Juvenile** Similar to female but black of plumage duller, crown often streaked olive, has yellowish forehead with blackish streaks, yellowish-white eyering, whitish throat with blackish streaks (extending and fading onto breast), less obvious wing markings (mostly fringing only) and blackish bill; shows some darker streaks on mantle. **VOICE** Song is a repeated, measured, clear fluty ***h HWI'UU*** and ***h wu'CHI-WU*** etc. Also loud mellow ***tcheo*** or ***tchew*** notes. Harsh ***chEEEEah*** or ***kwAAAAh*** when alarmed. **HABITAT** More open dry dipterocarp, mixed deciduous and broadleaved semi-evergreen forest, forest edge, secondary growth, mangroves, freshwater swamp forest; up to 915 m. **RANGE & STATUS** Resident Indian subcontinent (except W and Pakistan), SW China, Sumatra, Borneo. **SE Asia** Common resident Myanmar, Thailand, north-west Peninsular Malaysia (Langkawi I), Indochina (except W Tonkin, N Annam). **BREEDING** February–August. **Nest** Cradle suspended from forked tree-branch; 4–10 m above ground. **Eggs** 2–4; warm salmon-pink, spotted chestnut-brown to deep red-brown over dark inky-purple undermarkings; 28 x 19.4 mm (av.).

649 BLACK-AND-CRIMSON ORIOLE

Oriolus cruentus Plate **69**

IDENTIFICATION 23–24.5 cm. *O.c.malayanus*: **Male** Unmistakable, with glossy bluish-black plumage, large dark crimson patch on lower breast and upper belly, dark crimson primary coverts and pale bluish bill, legs and feet. **Female** Lacks crimson in plumage, has pale greyish lower breast and belly. **Juvenile** Similar to female but initially shows pale warmish-brown streaks on lower breast and upper belly, and very narrow pale warmish-brown fringes on at least some wing-coverts. **First-winter male** Similar to female. Gradually attains odd crimson feathers. **VOICE** Unusual thin strained ***hhsssu*** or ***hsiiiu***. **HABITAT** Broadleaved evergreen forest; 915–1,280 m, occasionally down to 610 m. **RANGE & STATUS** Resident Greater Sundas. **SE Asia** Uncommon to fairly common resident Peninsular Malaysia. **BREEDING** March–July. **Nest** Deep, compactly built hammock cup, suspended from forked tree-branch in canopy foliage; 8–10 m above ground. **Eggs** 2+.

650 **MAROON ORIOLE** *Oriolus traillii* Plate **69**
IDENTIFICATION 24–28 cm. *O.t.traillii*: **Male** Unmistakable, with dark maroon body, black hood and wings, pale dull maroon tail, bluish-grey bill and pale yellowish eyes. *O.t.nigellicauda* (visitor SE Thailand, N Indochina at least) has much redder body and tail. See Black-and-crimson Oriole. Note range. **Female** Crown, nape and sides of head blackish-brown, upperparts dark brown with variable maroon tinge and dark reddish-chestnut rump and uppertail-coverts, tail brownish-maroon (pale reddish-maroon below), underparts whitish with heavy dark streaks and pale reddish-maroon undertail-coverts. Throat and upper breast may be darker and less distinctly streaked, sometimes mostly blackish. *O.t.nigellicauda* has reddish wash on mantle and back, distinctly deep reddish rump, tail-coverts and tail. See Silver Oriole. **Juvenile** Similar to paler-throated females but has pale streaks on forehead, rufescent tips to wing-coverts and scales on mantle, back and scapulars, dark streaks on rump, narrower dark streaks on underparts, washed-out undertail-coverts with dark streaks and browner eyes; often has pale rufous wash on lower throat and upper breast. **Immature male** Attains black hood and variable maroon wash on body while underparts still streaked. **Other subspecies in SE Asia** *O.t.robinsoni* (S Laos, C,S Annam). **VOICE** Song is a rich, fluty ***PI-LOI-LO*** and ***PI-OHO-UU*** etc. Call is a nasal, drawn-out ***hwyERRRRh*** or ***nyaAOOOOOW***. **HABITAT & BEHAVIOUR** Broadleaved evergreen forest, forest edge, sometimes deciduous forest; 450–2,710 m, locally down to 150 m. Usually in upper storey of forest; often joins mixed-species feeding flocks. **RANGE & STATUS** Resident N,NE Indian subcontinent, SW,S (Hainan) China, Taiwan. **SE Asia** Fairly common to common resident Myanmar, W,NW,NE Thailand, Cambodia, Laos, Vietnam (except Cochinchina). Uncommon winter visitor south NE,SE Thailand, N Laos, W,E Tonkin, N Annam. **BREEDING** March–June. **Nest** Deep cup, suspended from forked tree-branch; 4–10 m above ground. **Eggs** 2–3; pinkish-white, spotted black or reddish-brown; 29.4 x 20.6 mm (av.; *traillii*).

651 **SILVER ORIOLE** *Oriolus mellianus* Plate **69**
IDENTIFICATION 28 cm. Monotypic. **Male** Silvery-whitish body-plumage, contrasting black hood and wings and dull maroon tail diagnostic. Has dull maroon centres to body feathers (hard to see in field), dull maroon undertail-coverts with narrow silvery-whitish fringes and silvery-whitish outer edges of tail-feathers. Bill, legs and feet bluish-grey. **Female** Similar to Maroon Oriole but has mostly greyish mantle to rump and paler pinkish undertail-coverts with whitish fringes; tends to have narrower dark streaks on underparts. **Juvenile** Little information. Presumably differs in similar way to Maroon. **HABITAT & BEHAVIOUR** Broadleaved evergreen and semi-evergreen forest; up to 800 m. Sometimes joins mixed-species feeding flocks. **RANGE & STATUS** Breeds W,S China. **SE Asia** Rare to scarce winter visitor W,NW,NE,SE Thailand, south-west Cambodia.

ARTAMIDAE: Woodswallows

Worldwide 10 species. SE Asia 2 species. Woodswallows are stocky and large-headed, with long wings, short tails and wide gapes. Feed on insects, mostly caught in flight.

652 **ASHY WOODSWALLOW**
Artamus fuscus Plate **69**
IDENTIFICATION 16–18 cm. Monotypic. **Adult** Reminiscent of some starlings at rest but combination of bulky head, pale bluish bill, brownish-grey plumage with paler, browner lower breast and belly and whitish undertail-coverts and band across uppertail-coverts distinctive. In flight shows broad-based pointed wings, rather short tail and mostly pale underwing. See White-breasted Woodswallow. **Juvenile** Browner above with pale brownish-white to whitish feather-tips, uppertail-coverts duller with dark bars, lower throat to belly paler with vague darker vermiculations, bill duller. See White-breasted. **VOICE** Song is a drawn-out twittering, interspersed with harsh ***chack*** notes. Calls include a sharp nasal ***ma-a-a ma-a-a...*** and repeated shrill, nasal ***chreenk*** and ***chek***. **HABITAT & BEHAVIOUR** Open areas with scattered trees, cultivation, sometimes over forest; up to 2,135 m. Gregarious, often found perched in huddled groups. Spends much time gliding and circling in search of food. **RANGE & STATUS** Resident (subject to local movements) Indian subcontinent (except W and Pakistan), SW,S China. **SE Asia** Fairly common to common resident, subject to some movements (except south Tenasserim, S Thailand, Peninsular Malaysia, Singapore). Also recorded on passage E Tonkin. **BREEDING** March–June. **Nest** Loose shallow cup, in existing tree-hole, on branch, telegraph pole, or in crown of palm; 9–12 m above ground. **Eggs** 2–3; white to greenish-white, spotted light brown (mostly at broader end); 23.4 x 17.1 mm (av.).

653 **WHITE-BREASTED WOODSWALLOW**
Artamus leucorynchus Plate **69**
IDENTIFICATION 17.5–19.5 cm. *A.l.leucorynchus*: **Adult** Very similar to Ashy Woodswallow but has much broader white band across rump and uppertail-coverts and clean white breast to undertail-coverts. In flight shows cleaner white underwing-coverts. **Juvenile** Similar to Ashy but underparts whiter, with dark chin and throat-sides; broader white band across rump and uppertail-coverts, whiter feather-tips on upperside. **Other subspecies in SE Asia** *A.l.humei* (Coco Is, S Myanmar). **VOICE** Has chattering song, incorporating avian mimicry. Calls include rasping ***wek-wek-wek...*** and sharp, metallic ***pirt pirt...*** **HABITAT** Open areas with scattered trees, cultivation. **RANGE & STATUS** Resident Andaman Is, Sundas, Philippines, Wallacea, Palau Is, New Guinea region, Australia, New Caledonia, Vanuatu. **SE Asia** Uncommon to fairly common resident S Myanmar (Coco Is), coastal west Peninsular Malaysia (local). **BREEDING** March–May. **Nest** Loose, shallow cup, in hollow-topped pylon pole or on tree-branch; at least 3.5 m above ground. **Eggs** 3; white, spotted light brown or fawn-colour over grey undermarkings (mostly at broader end); 23.8 x 17.2 mm (av. *humei*).

GENERA INCERTAE SEDIS: Woodshrikes, flycatcher-shrikes & philentomas

Worldwide 8 species. SE Asia 6 species. Smallish to medium-sized, stocky and stout-billed. Rather sluggish and slow-moving. Feed mainly on insects. **Woodshrikes** (*Tephrodornis*) Sexes similar, mostly brownish and whitish, often in small flocks in relatively open situations. **Philentomas** (*Philentoma*) Sexes markedly different with predominantly dark bluish males; inhabit middle to lower storey of forest, usually singly or in pairs.

654 LARGE WOODSHRIKE

Tephrodornis gularis Plate **70**

IDENTIFICATION 18.5–22.5 cm. *T.g.jugans*: **Male** Combination of greyish crown and nape, pale greyish-brown rest of upperside and contrasting blackish mask distinctive. Has whitish rump-band and whitish underparts with pale brownish-grey wash on throat and breast. See Common Woodshrike. *T.g.annectens* (south Tenasserim, S Thailand [except south-east], north-west Peninsular Malaysia) has more greyish-washed, less contrasting upperparts; *fretensis* (south-east S Thailand, elsewhere Peninsular Malaysia) is smaller (within size range given) and has fairly uniform pale bluish-slate crown, mantle and scapulars. **Female** Crown duller and brown-streaked, mask and bill browner, wash on throat and breast more buffish. *T.g.fretensis* is somewhat darker on crown, throat and breast. **Juvenile** Browner above than female with whitish shafts and tips to feathers of crown, mantle and scapulars, and buffish and whitish bars on wing-coverts and tertials; has dark buffish borders to tail-feathers. **Other subspecies in SE Asia** *T.g.pelvicus* (SW,W,N,E[north],C,S Myanmar), *vernayi* (south E Myanmar, N Tenasserim, W Thailand), *mekongensis* (NE,SE Thailand, southern Indochina), *hainanus* (east N Laos, northern Vietnam). **VOICE** Song is a loud airy ringing ***pi-pi-pi-pi-pi-pi...*** Typical call is a harsh, scolding ***chreek chreek chreek...*** **HABITAT & BEHAVIOUR** Broadleaved evergreen and mixed deciduous forest, forest edge, secondary growth, sometimes mangroves and overgrown plantations; up to 1,500 m. Usually in groups, often associating with mixed-species flocks. Moves rather sluggishly, often high in trees. **RANGE & STATUS** Resident E India, N,NE Indian subcontinent, S China, Greater Sundas. **SE Asia** Common resident (except C Thailand); formerly Singapore. **BREEDING** March–June. **Nest** Shallow cup, on tree-branch; 3–9 m above ground. **Eggs** 2–4; greenish- to buffish-white, with heavy brown spots and lines over grey undermarkings; 22 x 17.6 mm (av.; *pelvicus*).

655 COMMON WOODSHRIKE

Tephrodornis pondicerianus Plate **70**

IDENTIFICATION 14.5–17.5 cm. *T.p.pondicerianus*: **Adult** Similar to Large Woodshrike but smaller, with prominent broad dull whitish supercilium, whitish outertail-feathers and smaller, less distinct whitish rump-band. See female Grey Bushchat. **Juvenile** Has similar head pattern to adult but upperparts browner and spotted whitish-buff (particularly crown and nape), wing-coverts broadly tipped whitish-buff; has pale and dark markings on tertials and dusky-brown speckles/mottling on sides of throat and breast. **Other subspecies in SE Asia** *T.p.orientis* (Cambodia, Vietnam). **VOICE** Song is an accelerating trill: ***pi-pi-i-i-i-i-i***. Contact calls include weak ***tue*** and ***tee*** notes and a harsher, slightly ascending ***wih-wih-whee-whee*** and variants. **HABITAT & BEHAVIOUR** Dry dipterocarp and mixed deciduous forest, dry open country with scattered trees; up to 1,100 m. Often in small groups. **RANGE & STATUS** Resident Indian subcontinent (except Sri Lanka). **SE Asia** Common resident Myanmar, W,NW,NE,C(north) Thailand, Cambodia, Laos (north to Vientiane area), C(south-west),S Annam, Cochinchina. **BREEDING** February–June. **Nest** Small neat cup, in tree; 2–9 m above ground. **Eggs** 2–4; cream-coloured to greenish- or buffish-white, spotted and blotched yellowish-brown to reddish-brown over pale purple undermarkings (often in zone around broader end); 19 x 15.1 mm (av.; *pondicerianus*).

656 BAR-WINGED FLYCATCHER-SHRIKE

Hemipus picatus Plate **67**

IDENTIFICATION 12.5–14.5 cm. *H.p.picatus*: **Male** Black upperside, whitish underside and prominent long white wing-patch distinctive. Lower throat to belly dusky-washed (more vinous on breast), has white rump-band and prominent white on outertail-feathers. *H.p.capitalis* (N,C,E Myanmar, NW Thailand, N Laos, W Tonkin) has dark brown mantle, back and scapulars. See Black-winged Flycatcher-shrike, Pied Triller and Little Pied Flycatcher. **Female** Similar to male but black of upperside replaced by brown, underparts somewhat paler. *H.p.intermedius* (S Thailand southwards) has more blackish-brown upperparts and darker underparts (more similar to male). **Juvenile** Like female but crown and upperparts barred/scaled buff, wing-patch buffish with dark bars/scales, underparts whiter with brown wash across breast. **VOICE** Rapid, thin musical ***swit'i'wit-swit'i'wit...***, ***sitti-wittit*** and ***sit-tititit*** etc. **HABITAT & BEHAVIOUR** Broadleaved evergreen, mixed deciduous and peatswamp forest, forest edge, secondary growth, bamboo; up to 1,980 m. Usually in small groups, often associating with mixed-species feeding flocks. **RANGE & STATUS** Resident Indian subcontinent east to SW China, Sumatra, Borneo. **SE Asia** Common resident (except C Thailand, Singapore). **BREEDING** February–May. **Nest** Small neat cup, on tree-branch; 3–12 m above ground. **Eggs** 2–3; pale greenish-white to pinkish-white, boldly blotched inky-black and sometimes brick-red, over grey to lavender undermarkings; 16 x 12.8 mm (av. *capitalis*).

657 BLACK-WINGED FLYCATCHER-SHRIKE

Hemipus hirundinaceus Plate **67**

IDENTIFICATION 13.5–14.5 cm. Monotypic. **Male** Similar to Bar-winged Flycatcher-shrike but easily separated by lack of white wing-patch. Upperparts always black, tail almost all black, breast lightly washed greyish. **Female** Similar to male but upperside browner. **Juvenile** Crown and upperparts paler and browner than female with buff bars/scales, upperwing-coverts mostly buffish, with dark brown markings, underparts whiter with browner wash across upper breast. **VOICE** Song is rapid, sustained, rather thin, high, undulating ***tiwu'itiwu'itiwu'itiwu'itiwu'itiwu'itiwu...*** or ***tiwi'iti-wi'itiwi'itiwi'itiwi'itiwi'itiwi...*** Calls include a quick, sharp, quite even ***tee-tee*** or ***tee-tu***, or more extended ***tee-swi'swi'ti***. **HABITAT** Broadleaved evergreen forest, freshwater swamp forest, forest edge, sometimes mangroves, old plantations; up to 275 m. **RANGE & STATUS** Resident Greater Sundas. **SE Asia** Scarce to locally common resident south Tenasserim, extreme S Thailand, Peninsular Malaysia. **BREEDING** April–June. **Nest** Open cup, on horizontal tree-branch; up to 40 m above ground. **Eggs** 2; greenish with brown spots.

658 RUFOUS-WINGED PHILENTOMA

Philentoma pyrhoptera Plate **70**

IDENTIFICATION 16.5–17 cm. *P.p.pyrhoptera*: **Male typical morph** Distinguished by rather stocky, stout-billed appearance, mostly dull blue head and body, buffy-whitish belly and vent, and sharply contrasting reddish-chestnut greater coverts, tertials, secondaries and tail. Eyes red to dark red. **Male blue morph** Scarce. All dull blue, with greyish vent (mixed whitish). See Pale Blue Flycatcher and female Maroon-breasted Philentoma. **Female** Lacks blue in plumage, crown and sides of head cold dark greyish-brown (sometimes tinged blue), rest of upperparts mid-brown, underparts buffy-whitish (throat and breast more buffish), breast-sides (sometimes across breast) and flanks mid-brown. Shape, bill structure, reddish eyes and prominent reddish-chestnut of wings and tail eliminate flycatchers. See Asian Paradise-flycatcher. **Juvenile** Poorly documented. Fledglings have rather uniform pale reddish-chestnut head and body. Sexes are said to be separable soon after fledging. **VOICE** Song is a clear mellow piping whistled ***tu-tuuu*** (second note slightly lower). Also harsh scolding notes. **HABITAT & BEHAVIOUR** Middle to lower storey of broadleaved evergreen forest; up to 915 m. Rather sluggish. **RANGE & STATUS** Resident Sumatra, Borneo. **SE Asia** Fairly common to common resident south Tenasserim, W(south),S Thailand, Peninsular Malaysia; formerly Singapore. **BREEDING** March–September. **Nest** Small open cup, in sapling or small tree; 1.5–5 m above ground. **Eggs** 2; pink with reddish markings.

659 MAROON-BREASTED PHILENTOMA

Philentoma velata Plate **70**

IDENTIFICATION 19–21 cm. *P.v.caesia*: **Male** Told by dull blue plumage with black forehead, upper throat and sides of head and dark maroon lower throat and breast (often appears blackish). Vent greyer (mixed whitish), eyes crimson. **Female** Duller above; has dull blackish lores, cheeks, throat and upper breast and duller and darker lower breast and upper belly. See Asian Fairy-bluebird and blue morph male Rufous-winged Philentoma. **Juvenile** Poorly documented. According to specimens, fledglings have dull chestnut head and body with dark streaks/mottling on crown and upperparts. Soon after fledging, males show some chestnut-maroon patches on breast. However, dependent juveniles observed in the field were reported to be overall dull slaty-blue with black eyes. **VOICE** Long series of clear bell-like whistles: ***phu phu phu phu phu phu...*** Also rather powerful clear ***chut-ut chut-ut chut-ut chut-ut...*** **HABITAT & BEHAVIOUR** Middle to upper storey of broadleaved evergreen forest; up to 1,060 m. Very sluggish, often sitting motionless. **RANGE & STATUS** Resident Greater Sundas. **SE Asia** Scarce to fairly common resident Tenasserim, W,S Thailand, Peninsular Malaysia. **BREEDING** April–July. **Nest** Shallow open cup, in sapling or small tree. **Eggs** 2. Otherwise undocumented.

AEGITHINIDAE: Ioras

Worldwide 4 species. SE Asia 3 species. Small and robust, stout-billed and fairly short-tailed, with mostly green and yellow plumage. Feed on insects and caterpillars gleaned from foliage.

660 COMMON IORA *Aegithina tiphia* Plate 73

IDENTIFICATION 12–14.5 cm. *A.t.philipi*: **Male breeding** Bright yellow head-sides and underparts, rather deep olive-green upperparts and mostly black wings and tail with two pronounced white to yellowish-white wing-bars diagnostic. May show some black on mantle to rump, has yellowish-white fringes to tertials, secondaries and primaries. See Great Iora. *A.t.tiphia* (SW Myanmar) may have black mottling on crown; *deignani* (W,N,C,S[north] Myanmar) frequently has black crown to back with pale yellow admixed on mantle; *horizoptera* (S Myanmar, Tenasserim, W,S Thailand southwards) frequently has black hindcrown to mantle (black less often present on rest of upperparts); *cambodiana* (SE Thailand, Cambodia, S Annam, Cochinchina) sometimes has black on crown and nape. **Male non-breeding** Lacks black on upperparts, yellow of head-sides and underparts less vivid, flanks washed olive. See Green Iora. **Female breeding** Similar to male non-breeding but upperparts somewhat paler green, uppertail olive-green, wings greyer with somewhat less distinct bars, yellow of underparts paler, flanks broadly washed olive. **Female non-breeding** Yellow of underparts paler. See Green Iora. **Juvenile** Like female non-breeding (both sexes) but body a shade paler, upper wing-bar less distinct. **VOICE** Song is a thin drawn-out note (or short series of notes) terminating in an abrupt downward slide to a note about an octave lower (briefly held): ***whiiiiii piu***. Other calls include a whistled ***di-di-dwiu***; ***dwi-o dwi-o dwi-o***; ***du-i du-i*** and ***di-du di-du***, and subdued harsh chattering. **HABITAT** Open forest, mangroves, freshwater swamp and peatswamp forest, secondary growth, parks, gardens, plantations; up to 1,500 m. **RANGE & STATUS** Resident Indian subcontinent (except Pakistan), SW China, Greater Sundas, Philippines (Palawan). **SE Asia** Common resident throughout. **BREEDING** January–July. **Nest** Neat cup, in branch fork of bush or small tree; 3–9.1 m above ground. **Eggs** 2–4; pink with reddish markings, creamy-white to greyish-white with a few lengthwise grey markings over neutral-coloured undermarkings or greenish-white with brown spots; 17.5 x 13.5 mm (av.; *deignani*).

661 GREEN IORA *Aegithina viridissima* Plate 73

IDENTIFICATION 12.5–14.5 cm. *A.v.viridissima*: **Male** Resembles Common Iora but body dark olive-green with paler belly and yellow vent; has dark lores and distinctive broad broken yellow eyering. **Female** Similar to Common but upperparts deeper green, breast and flanks darker and greener, rest of underparts more greenish-yellow, has greenish lores and indistinct broken eyering; wing-bars always yellow. **Juvenile** Like female but somewhat paler and duller, more washed out. **VOICE** Song is a very thin, high-pitched *tsiiiu tsii-tu* (with stressed third note) or ***itsu tsi-tu tsi-tu*** (with stressed first note) etc. Also chattering ***tit-teeer*** phrases and subdued ***chitititit***. **HABITAT** Broadleaved evergreen forest, forest edge, locally mangroves; up to 825 m. **RANGE & STATUS** Resident Sumatra, Borneo. **SE Asia** Uncommon to common resident south Tenasserim, W(south),S Thailand, Peninsular Malaysia; formerly Singapore. **BREEDING** April–June. **Nest** Neat cup, amongst branches or in tree-branch fork; 8–12 m above ground. **Eggs** 2.

662 GREAT IORA *Aegithina lafresnayei* Plate 73

IDENTIFICATION 15.5–17 cm. *A.l.innotata*: **Male** From other ioras by larger size, thicker, longer bill and lack of wing-bars. Upperparts and wing-coverts dark olive-green, rest of wings blackish with green feather-fringes, underparts bright rich yellow. *A.l.lafresnayei* (S Thailand, Peninsular Malaysia) has variable glossy bluish-black scales on head-sides and upperparts (except forehead), mostly glossy bluish-black wings and tail, some narrow pale green fringes on flight-feathers, darker green admixed on mantle, and yellow-white streaks on rump and uppertail-coverts; upperparts sometimes entirely black apart from yellow forehead. *A.l.xanthotis* (Cambodia, southern Vietnam) has paler, brighter green upperparts (both sexes). **Female** Like male but upperparts slightly paler green, underparts less vivid yellow. **Juvenile** Like female but underparts duller and washed olive, particularly on flanks. **VOICE** Song is a clear ***chew chew chew chew...*** or ***tieu tieu tieu tieu***. **HABITAT** Broadleaved evergreen, semi-evergreen and mixed deciduous forest, forest edge; up to 900 m. **RANGE & STATUS** Resident SW China. **SE Asia** Common resident SW Myanmar, Tenasserim, Thailand, Peninsular Malaysia, Indochina. **BREEDING** January–September. **Nest** Rather deep cup, situated on tree or bush branch; 4.6 m above ground. **Eggs** 3; greyish-white, with lengthwise grey streaks; 20.1 x 15 mm (av.; *lafresnayei*).

RHIPIDURIDAE: Fantails

Worldwide 42 species. SE Asia 5 species. Slender, with short rounded wings and long graduated tails, usually held cocked and fanned. Very active, often associating with mixed-species feeding flocks. Feed mostly on insects caught in mid-air.

663 YELLOW-BELLIED FANTAIL *Chelidorhynx hypoxantha* Plate 70

IDENTIFICATION 11.5–12.5 cm. Monotypic. **Male** Dull greenish upperparts, long dark white-tipped graduated tail, blackish mask and bright deep yellow forehead, supercilium and underparts diagnostic. Has whitish tips to greater coverts, bill short and triangular when viewed from below. See Black-faced Warbler. **Female** Mask same colour as crown. **Juvenile** Like adult but upperparts duller, yellow parts paler, less yellow on forehead and in front of eye. **VOICE** Song is a series of thin, sweet ***sewit***, ***sweeit*** and ***tit*** or ***tsit*** notes, followed by a high-pitched trill. **HABITAT & BEHAVIOUR** Broadleaved evergreen forest; 1,500–3,655 m, locally down 180 m in winter. Fans tail and nervously twitches from side to side. **RANGE & STATUS** Resident NW,N,NE Indian subcontinent, SE Tibet, SW China. **SE Asia** Uncommon to common resident W,N,S(east),E Myanmar, W,NW Thailand, N Laos, W Tonkin, N Annam. **BREEDING** April–July. **Nest** Compact deep cup, attached to upperside of branch; 3–6 m above ground. **Eggs** 3;

cream to pinkish-cream with tiny dark reddish speckles (usually restricted to ring around broader end); 14.4 x 11.3 mm (av.).

664 WHITE-THROATED FANTAIL
Rhipidura albicollis Plate **70**

IDENTIFICATION 17.5–20.5 cm. *R.a.celsa*: **Adult** Mostly dark greyish to blackish-slate plumage (crown blacker) with contrasting white supercilium and throat and fan-shaped white-tipped tail diagnostic. Chin blackish. See White-browed, Spotted and Pied Fantails. Note range and habitat. **Juvenile** Resembles adult but upperparts browner-tinged, scapulars and back to uppertail-coverts scaled/barred paler brown to warm brown, wing-coverts indistinctly tipped same colour; shows indistinct paler barring on underparts and almost no white on throat, supercilium often tinged buffish. **Other subspecies in SE Asia** *R.a.cinerascens* (southern Indochina), *stanleyi* (W and northern Myanmar), *atrata* (S Thailand southwards). **VOICE** Song is a series of 4–8 widely and unevenly spaced, clear, high-pitched whistled notes, in mostly descending sequence: ***tsu sit tsu sit sit sit sit-tsu***. Call is a squeaky, harsh ***jick*** or ***wick***. **HABITAT** Broadleaved evergreen forest, locally groves and bamboo in cultivated areas, parks, wooded gardens, some coastal habitats (Indochina); 460–3,050 m, locally down to sea-level in Indochina. **RANGE & STATUS** Resident NE Pakistan, NW,N,NE Indian subcontinent, SE Tibet, SW,S,C China, Sumatra, Borneo. **SE Asia** Common resident (possibly subject to some movements) Myanmar, W,NW,NE(north-west),SE,S(central) Thailand, Peninsular Malaysia, Indochina (except Cochinchina). **BREEDING** February–June. Multi-brooded. **Nest** Neat deep cup (sometimes with short 'tail'), in fork of tree or on bush branch; 1.2–3 m above ground. **Eggs** 2–4; white to rich or dingy cream-coloured, variously spotted and speckled yellowish-brown, grey-brown, pale grey or black (often mainly in ring around broader end); 17.3 x 13 mm (av.: *stanleyi*).

665 WHITE-BROWED FANTAIL
Rhipidura aureola Plate **70**

IDENTIFICATION 16–18.5 cm. *R.a.burmanica*: **Adult** From other fantails by combination of long broad white supercilium (crossing forehead) and mostly whitish underparts. Upperside and breast-sides brownish-grey, crown and uppertail blacker, the latter broadly tipped and edged white, throat-centre greyish (feathers black, tipped/scaled white); has pale creamy-buffish wash on belly, flanks (fainter on breast), and small white spots on wing-covert tips. **Juvenile** Similar to adult but throat initially darker, upperparts browner with pale warm brown scales/bars, mainly on scapulars, back and rump, broad whitish to dull pale rufous tips to wing-coverts and fringes to tertials. **VOICE** Series of 6–7 well-spaced melodious whistles, usually with first few ascending, rest descending: ***chee-chee-cheweechee-vi*** etc. Call is a harsh ***chuck***. **HABITAT** Dry dipterocarp and mixed deciduous forest; up to 1,065 m. **RANGE & STATUS** Resident Pakistan, India (except NE), S Nepal, Sri Lanka, SW China (W Yunnan). **SE Asia** Uncommon to fairly common resident SW,W,N,C,E(south),S Myanmar, north Tenasserim, W,NW,NE(south-west) Thailand, Cambodia, C,S Laos, C,S Annam. **BREEDING** February–August. Multi-brooded. **Nest** Like White-throated but usually lacks 'tail'; 2–12 m above ground. **Eggs** 2–4; white or dingy cream-coloured to buffish with zone of greyish-brown specks and spots around broader end and pale grey to inky-purple undermarkings; 17.2 x 12.8 mm (av.).

666 PIED FANTAIL *Rhipidura javanica* Plate **70**

IDENTIFICATION 17.5–19.5 cm. *R.j.longicauda*: **Adult** From other fantails by pale underparts with contrasting blackish breast-band. Throat and lower breast white, rest of underparts pale creamy-buffish, upperparts dark brown to blackish-brown with blacker crown, only small amount of white above eye (may be concealed). Note habitat and range. **Juvenile** Browner, more uniform upperparts, duller breast-band, dull rufescent scales/bars on upperparts (mainly scapulars and back to uppertail-coverts) and dull rufescent tips to wing-coverts and tips and narrow fringes to tertials. **VOICE** Squeaky, measured ***chew-weet chew-weet chew-weet-chew***, last note falling. Various squeaky chattering and squawking calls: ***chit*** and ***cheet*** etc. **HABITAT** Mangroves, freshwater and peatswamp forest, parks, gardens, plantations, secondary growth, scrub; usually near water; up to 455 m, locally 825 m. **RANGE & STATUS** Resident Greater Sundas, Lombok, Philippines. **SE Asia** Common resident (mostly coastal) Tenasserim, Thailand (except NW), Peninsular Malaysia, Singapore, Cambodia, N Laos (Mekong R), Cochinchina. **BREEDING** January–August. Multi-brooded. **Nest** Neat deep cup (often 'tailed'), in small tree, bush, creeper or bamboo; up to 3 m above ground. **Eggs** 2; glossy buffish with fine brown speckles, often forming ring around broader end; 17.8 x 12.7 mm (av.).

667 SPOTTED FANTAIL
Rhipidura perlata Plate **70**

IDENTIFICATION 17–18 cm. Monotypic. **Adult** Easily identified from other fantails by uniform blackish-slate plumage, white vent and prominent white spots/streaks on throat and breast. Outertail-feathers broadly tipped white, only small white eyebrow; usually shows small white spots on wing-covert tips. **Juvenile** Browner above with warm brownish tips to wing-coverts. **VOICE** Song variously described as rather metallic, upswinging ('mini' drongo-like) ***chu chirri CHEE***; ***chi chu chirri CHEE***; ***chilip peCHILIP-chi*** (second phrase rising sharply); and jingly ***chap go HEE***. Calls with sparrow-like chirps when foraging. **HABITAT & BEHAVIOUR** Broadleaved evergreen forest; up to 1,130 m. Does not fan tail. **RANGE & STATUS** Resident Sumatra, Borneo, W Java. **SE Asia** Scarce to locally fairly common resident southern S Thailand, Peninsular Malaysia. **BREEDING** April–August. **Nest** Neat cup, attached to branch fork, creeper or vine etc.; 3–10 m above ground. **Eggs** 2.

DICRURIDAE: Drongos

Worldwide c.23 species. SE Asia 8 species. Medium-sized with heavy bills, long forked or racketed tails and mostly glossy blackish plumage. Often seen sallying from exposed perch. Feed mainly on large insects but also nectar and occasionally small birds, reptiles and mammals.

668 BLACK DRONGO
Dicrurus macrocercus Plate **71**

IDENTIFICATION 27–28.5 cm. *D.m.thai*: **Adult** All-blackish plumage (slightly glossed bluish) and long, deeply forked tail distinctive. Often shows small white spot below/in front of eye at base of bill. *D.m.cathoecus* (Myanmar [except W,N], northern Thailand, northern Indochina; also only visiting form) is larger; *albirictus* (W,N Myanmar) is larger still (29–32 cm). See Ashy and Crow-billed Drongos. Note habitat. **Juvenile** Duller with sooty-brown tinge, no gloss, vague paler scales on breast and belly. **First winter** Similar to adult but throat duller (sometimes streaked greyish/whitish), upper breast and wings duller; has prominent whitish scales on uppertail-coverts, breast, belly and undertail-coverts, whitish markings on underwing-coverts and wing-bend. See Drongo Cuckoo. **VOICE** Harsh ***ti-tiu***, rasping ***jeez*** or ***cheece*** and ***cheece-cheece-chichuk*** etc. **HABITAT & BEHAVIOUR** Open country, cultivation, roadsides, scrub; up to 1,220 m (breeds mostly in lowlands). May roost in large numbers in winter. **RANGE & STATUS** Resident SE Iran, Afghanistan, Indian subcontinent, Java, Bali. Breeds China (except NW,W), Taiwan; some populations winter to south, S China. **SE Asia** Common resident, subject to some movements (except S Thailand, Peninsular Malaysia, Singapore).

Uncommon to common winter visitor SW,S Myanmar, Tenasserim, W(south),C,SE,S Thailand, Peninsular Malaysia, Singapore, Cambodia, Cochinchina. Also recorded on passage C,S Thailand, N Laos, E Tonkin. **BREEDING** April–August. **Nest** Broad shallow cup, suspended from outer tree-branches; 4–12 m above ground. **Eggs** 2–5; white to pinkish-cream, spotted and blotched black to reddish-brown or purplish-brown (sometimes plain); 25.4 x 18.5 mm (av. *cathoecus*), 27.1 x 19.8 mm (av. *albirictus*). Brood-parasitised by Asian Koel and Indian Cuckoo.

669 ASHY DRONGO
Dicrurus leucophaeus Plate **71**

IDENTIFICATION 25.5–29 cm. *D.l.mouhoti* (breeds SW,W,C,S,E[south-western] Myanmar, NW,NE Thailand, N[southern],C Laos, C Annam; winter visitor/passage migrant south only to S[north] Thailand, Cambodia; ? Cochinchina): **Adult** Variable. Resembles Black Drongo but dark steely-grey with slight bluish gloss (mainly upperparts and wings), has somewhat paler underparts, darker lores, longer bill and shallower tail fork. *D.l.nigrescens* (resident south Tenasserim, S Thailand, Peninsular Malaysia; visitor Singapore) is distinctly blacker above and darker below, particularly on throat and breast (intergrades occur in north); *leucogenis* (visitor S Myanmar, Tenasserim, Thailand, Cambodia, N Laos and S Annam southwards) is very pale ashy-grey with black forehead, whitish lores to ear-coverts and vent and broad blackish tips to flight-feathers; *salangensis* (visitor Tenasserim, C,SE,S Thailand, Peninsular Malaysia, Cambodia, E Tonkin, Cochinchina) is like *leucogenis* but a shade darker with mostly grey ear-coverts. **Juvenile** Like adult but glossless and slightly paler overall with paler throat and vent and shorter tail. *D.l.nigrescens* is more sooty-tinged. **Other subspecies in SE Asia** *D.l.bondi* (resident W,NE[south],SE Thailand, S Indochina); *hopwoodi* (northern Myanmar, N Indochina; recorded in winter southern Myanmar, NW,NE Thailand, S Laos). **VOICE** Thin, slow, wheezy whistled ***phuuuu*** and ***hieeeeeer*** and loud sprightly ***tchik wu-wit tchik wu-wit***. Also more varied and complex vocalisations, including rapid chattering interspersed with shrill whistles, harsh notes and sometimes mimicry. **HABITAT** Open areas and clearings in various kinds of forest, forest edge, secondary growth; mangroves and coastal scrub in S Thailand and Peninsular Malaysia. Up to 2,750 m. **RANGE & STATUS** Resident Greater Sundas, Philippines (Palawan). Breeds NE Afghanistan, NE Pakistan, NW,N,NE India, Nepal, Bhutan, N,C and southern China; some northern populations winter to south, India, Sri Lanka, Bangladesh, S China. **SE Asia** Common resident (except Singapore). Fairly common to common winter visitor S Myanmar, Tenasserim, Thailand, Peninsular Malaysia, Singapore (scarce), Laos, Cambodia, Cochinchina. Also recorded on passage C Thailand, E Tonkin, C Annam. **BREEDING** February–June. **Nest** Broad shallow cup, suspended from outer tree-branches; 2–20 m above ground. **Eggs** 3–4; white to pinkish, boldly blotched and spotted (sometimes also streaked) reddish-pink to reddish-brown over faint inky-purple undermarkings (latter often mainly around broader end); 24.1 x 18.8 mm (av.; *mouhoti*?), 23 x 18.5 (mean of 3; *nigrescens*). Brood-parasitised by Indian Cuckoo.

670 CROW-BILLED DRONGO
Dicrurus annectans Plate **71**

IDENTIFICATION 27–32 cm. Monotypic. **Adult** Resembles Black Drongo but bulkier, tail relatively shorter and broader with shallower fork and more strongly upcurled outer feather-tips, bill much thicker-based and longer, plumage (except head) distinctly glossed greenish-blue, no white loral spot. Note habitat. Like juvenile Hair-crested Drongo but smaller, has shorter, relatively thick straight bill, obvious tail-fork, more uniformly glossed head and body, no head plumes and no greenish iridescence on wings or tail. **Juvenile** Browner than adult. **First winter** Like adult but has variable white spots/scales on underparts (except throat) and brown eyes (red-brown on adult). Distinct white spots on breast help rule out Black. **VOICE** Varied loud musical whistles, churrs and chattering, including a thin rising ***hee'weeiit*** and ***heeer-wu-wit-it***, interspersed with harsher throaty chatters. **HABITAT** Broadleaved evergreen forest, sometimes mixed deciduous forest; also mangroves, coastal scrub, wooded areas, plantations and secondary growth in winter and on passage. Up to 1,400 m, rarely 1,700 m. **RANGE & STATUS** Breeds N,NE Indian subcontinent, SW,S China; winters Greater Sundas. **SE Asia** Local breeding visitor W,NW(some winter),northern C Thailand, C,S Laos, E Tonkin, N,C Annam. Uncommon winter visitor south Tenasserim, SE,S Thailand, Peninsular Malaysia, Singapore. Uncommon passage migrant Tenasserim, Thailand, Peninsular Malaysia, Singapore, Cambodia, E Tonkin, Cochinchina. Recorded (status uncertain) N,E,S Myanmar, S Annam. **BREEDING** March–August. **Nest** Shallow cup, suspended from tree-branch; 5–12 m above ground. **Eggs** 3–4; pale cream to salmon-buff, streaked reddish-brown or purple-brown over pale grey to lavender understreaks (markings denser towards broader end); 26.3 x 19.4 mm (av.).

671 BRONZED DRONGO
Dicrurus aeneus Plate **71**

IDENTIFICATION 22–23.5 cm. *D.a.aeneus*: **Adult** Small size, relatively small bill and black plumage with very strong dark blue to greenish-blue gloss on upperparts, throat and breast distinctive. **Juvenile** Sooty-brown with some gloss on crown, mantle, wings and upper breast. **First winter** Less brilliantly glossed above than adult, gloss below mostly on upper breast. **Other subspecies in SE Asia** *D.a.malayensis* (south Peninsular Malaysia, Singapore). **VOICE** Song varied, with much mimicry. Calls include mixture of dry gravelly notes and quick, sharp, nasal notes: ***GRZZCHIKZEet*** etc. **HABITAT** Broadleaved evergreen, semi-evergreen and deciduous forest, forest edge, secondary growth; up to 2,135 m. **RANGE & STATUS** Resident N,NE,E,S India, Nepal, Bhutan, Bangladesh, SW,S China, Taiwan, Sumatra, Borneo. **SE Asia** Common resident (except C Thailand, Singapore). Vagrant Singapore (formerly resident). **BREEDING** February–July. **Nest** Small shallow cup, suspended from tree-branch; 10–15 m above ground. **Eggs** 2–4; deep pink to pale salmon-pink with irregular ring or cap of reddish to purplish cloudy spots at broader end (sometimes blotched and spotted reddish-brown); 21.1 x 16 mm (av.).

672 LESSER RACKET-TAILED DRONGO
Dicrurus remifer Plate **71**

IDENTIFICATION 25–27.5 cm (pendant-tipped tail-shafts extend up to 40 cm more). *D.r.tectirostris*: **Adult** Relatively small size and square tail-tip with very long, extended bare shafts of outer feathers terminating in longish pendants (often broken or absent) diagnostic. Has short tufted feathers on forehead which overlap bill and create distinctive flat-headed appearance; plumage strongly glossed dark blue to greenish-blue on upperside, throat and breast. *D.r.peracensis* (south of 16°N, except south-west Cambodia) has narrower, much longer ribbon-like tail-pendants (intergrades in north of range); *lefoli* (south-west Cambodia) similar to *peracensis*, but with longer ribbon-like tail-pendants. See Greater Racket-tailed Drongo. **Juvenile** Duller, with no tail-feather extensions. **VOICE** Loud, musical and very varied, including much mimicry. **HABITAT** Broadleaved evergreen forest, sometimes semi-evergreen forest; 500–2,590 m (locally down to 140 m). **RANGE & STATUS** Resident N,NE India, Nepal, Bhutan, NE Bangladesh, SW China, Sumatra, Java. **SE Asia** Locally common resident (except C Thailand, Singapore). **BREEDING** March–June. **Nest** Broad shallow cup, suspended from forked tree-branch; 3–7 m above ground. **Eggs** 2–4; white or pale pink to terracotta, thickly blotched, streaked and spotted dark brown and purple (often more at broader end); 25.5 x 18.4 mm (av. *tectirostris*).

673 GREATER RACKET-TAILED DRONGO

Dicrurus paradiseus Plate **71**

IDENTIFICATION 33–35.5 cm (outertail-feather-tips and their pendant-tipped shafts extend up to 30 cm more). *D.p.rangoonensis*: **Adult** Told by relatively large size, broad, tall tuft of feathers on forehead, and shallowly forked tail with very long bare shafts of outer feathers terminating in (usually) twisted pendants (often broken or absent). Upperside glossed dark blue (no obvious green tinge), gloss below restricted to spots/streaks on lower throat and upper breast. *D.p.platurus* (Peninsular Malaysia except Tioman I) smaller (30–31 cm) with much shorter crest; *hypoballus* (S Tenasserim, S Thailand, N Peninsular Malaysia) intermediate; *grandis* (N,E[north] Myanmar, N Indochina) larger (up to 37 cm) with longer crest. **Juvenile** Has much shorter crest than adult, lacks tail extensions, upperparts a little browner and less glossy, underparts brownish-black and almost glossless. **First winter** Like adult but has prominent white bars on undertail-coverts, indistinct white bars on belly, and some white specks on lower breast. **Other subspecies in SE Asia** *D.p.microlophus* (Tioman I, Peninsular Malaysia), *paradiseus* (Tenasserim, W[south],C,SE Thailand, S Indochina; intergrades with other subspecies in north and south of range). **VOICE** Highly varied. Loud, incorporating musical whistles, harsh screeching and churring and much avian mimicry. **HABITAT** Broadleaved evergreen, semi-evergreen, deciduous and peatswamp forest, secondary growth, plantations; up to 1,700 m (mainly lowlands). **RANGE & STATUS** Resident Indian subcontinent (except W,NW, Pakistan and Sri Lanka), SW,S China, Greater Sundas. **SE Asia** Common resident throughout. **BREEDING** December–July. **Nest** Cradle, suspended from or wedged in tree-branch fork; 4.5–15.2 m above ground. **Eggs** 3–4; white to rich cream or pinkish, blotched and speckled reddish-brown or purple over pale blue to pinkish-grey undermarkings; 27.8 x 20.2 mm (av.; *paradiseus*). Brood-parasitised by Indian Cuckoo.

674 HAIR-CRESTED DRONGO

Dicrurus hottentottus Plate **71**

IDENTIFICATION 29–33 cm. *D.h.hottentottus*: **Adult** Told by large size, rather long, slender, slightly downcurved pointed bill, and rather triangular, squarish-ended tail with strongly up- and inward-curled tips of outer feathers. Has very strong waxy dark greenish gloss on wings, tail and uppertail-coverts, pointed glossy dark blue feathers on crown, lower throat, neck-sides and breast, all-black mantle to rump, head-sides, belly and vent, and unusual hair-like plumes extending from forehead (usually draped over nape and difficult to see). See Crow-billed and juvenile Greater Racket-tailed Drongos. **Juvenile** Body blackish-brown, wings and tail less shiny, tips of outertail-feathers less strongly curled, forehead plumes much shorter. **Other subspecies in SE Asia** *D.h.brevirostris* (N Indochina, perhaps N Myanmar; also only visiting form). **VOICE** Loud ***chit-wiii***, with stressed first note and rising second note, and ***wiii*** notes given singly. **HABITAT & BEHAVIOUR** Broadleaved evergreen, semi-evergreen and deciduous forest, secondary growth; sometimes parks and gardens on migration; up to 2,440 m (mostly lowlands). Often in flocks, feeding in flowering trees. **RANGE & STATUS** Resident (subject to local movements) N,NE,E,S India, Nepal, Bhutan, Bangladesh, C and southern China, Sundas, Philippines, Sulawesi, Sula Is, C Moluccas, Kai Is. **SE Asia** Uncommon to locally common resident Myanmar, W,NW,NE,SE Thailand, Indochina. Scarce to fairly common winter visitor Thailand (except S). Also recorded on passage C Thailand, E Tonkin. **BREEDING** December–June. **Nest** Broad shallow cup, suspended from outer tree-branch; 5–10 m above ground. **Eggs** 2–5; pale cream or white to deep salmon-pink, profusely and minutely freckled pale red to purplish-red; 29.2 x 21.2 mm (av. *hottentottus*).

675 ANDAMAN DRONGO

Dicrurus andamanensis Plate **71**

IDENTIFICATION 31.5–33.5 cm. *D.a.dicruriformis*: **Adult** Resembles Black Drongo but larger, bill much longer and thicker (ridge of upper mandible strongly arched), tail much longer with slightly shallower fork and upward- and inward-curling outer feather-tips; has short wire-like shafts extending from forehead (difficult to see in field). See Crow-billed, Hair-crested and Greater Racket-tailed Drongos. Note range. **Juvenile** Less glossy and slightly browner than adult, tail initially squarer-ended. **VOICE** Jangling song trips evenly up scale (first notes wooden/clipped, last metallic, bell-like and dropping: ***TIR'RIP'CHEEIPit-TIR'RIP'CHEEIPit...***. High-pitched, sharp, metallic, downslurred ***tschew*** or ***sIEEEUW*** notes, and quick ***SHLEEUp*** (hissing buzz then jangle). **HABITAT & BEHAVIOUR** Broadleaved evergreen forest, forest edge; lowlands. Often in small flocks. **RANGE & STATUS** Resident Andaman Is (India). **SE Asia** Resident Coco Is, off S Myanmar. **BREEDING** April–May. **Nest** Cradle, suspended from or wedged in tree-branch fork. **Eggs** 2–3; pale salmon-pink, spotted pale brownish overall with some pale grey undermarkings; 24.8 x 18.3 mm (av.).

MONARCHIDAE: Monarchs, paradise-flycatchers & allies

Worldwide c.85 species. SE Asia 3 species. Small to smallish and flycatcher-like, with stout bills and longish to very long tails. Feed mainly on insects gleaned from foliage.

676 BLACK-NAPED MONARCH

Hypothymis azurea Plate **70**

IDENTIFICATION 16–17.5 cm. *H.a.styani*: **Male** Predominantly blue plumage with black nuchal tuft/bump, narrow black band across uppermost breast, and whitish belly and vent distinctive. Extreme forehead and uppermost chin black. *H.a.tytleri* (Coco Is, S Myanmar) and *prophata* (Peninsular Malaysia, Singapore) tend to be deeper, more purplish blue, with less white on vent, particularly former, which has greyish vent (tinged blue). **Female** Like male but blue duller and restricted mainly to head, no black nuchal patch or breast-band, rest of upperside warm-tinged greyish-brown, breast greyish (tinged blue). *H.a.tytleri* and *prophata* have darker, deeper brown upperside and deeper blue parts, particularly former which also shows greyish belly and vent (tinged blue). **Juvenile** Very similar to female (both sexes). **Other subspecies in SE Asia** *H.a.forrestia* (Mergui Archipelago, Tenasserim), *montana* (W,NW,NE Thailand), *galerita* (W[south],SE Thailand). **VOICE** Song is a clear, monotonous, ringing ***wii'wii'wii'wii'wii'wii'wii...***, at a rate of about 3 notes per s. Usual calls are a harsh ***shweh-shweh*** or ***chwe-wi*** and ***chit-whit-whit...*** and a high-pitched metallic rasping ***tswit*** and ***tswit-wit***. **HABITAT & BEHAVIOUR** Middle to lower storey of broadleaved evergreen, semi-evergreen, deciduous and peatswamp forest, secondary growth, overgrown plantations, island forest; up to 1,520 m (below 1,065 m in Peninsular Malaysia). Holds tail stiffly, often slightly spread, recalling fantails. **RANGE & STATUS** Resident Indian subcontinent (except NW and Pakistan), SW,S China, Taiwan, Greater Sundas, W Lesser Sundas, Sulawesi, Sula Is, Philippines. **SE Asia** Common resident, subject to relatively local movements (except C Thailand, E Tonkin); rare Singapore. Uncommon to fairly common winter visitor S Myanmar, C,NE,SE Thailand. Uncommon passage migrant C Thailand, E Tonkin. **BREEDING** March–August. **Nest** Neat deep cup (sometimes 'tailed'), in fork of tree-branches, bush or bamboo; 0.6–9 m above ground. **Eggs** 2–4; cream-coloured to pinkish-white, spotted light red to reddish-brown over pale grey undermarkings (mainly at broader end); 17.4 x 13.3 mm (av.; *styani*).

677 **ASIAN PARADISE-FLYCATCHER**
Terpsiphone paradisi Plate **70**
IDENTIFICATION 19.5–23.5 cm (male tail extends up to 27 cm more). *T.p.indochinensis*: **Male** Bright rufous-chestnut upperside and extremely long rufous-chestnut tail diagnostic. Head and breast slaty-grey, crown black and crested with dark bluish-green to greenish gloss, rest of underparts whitish; has stout blue bill and broad blue eyering. May lack long central tail-feathers. Birds with all-black heads are probably immatures of the white 'morph'. *T.p.saturatior* (breeds N Myanmar; winters Tenasserim and S Thailand southwards) and *nicobarica* (Coco Is, S Myanmar) have duller, paler, more rufous (less chestnut) upperside and buffish-tinged belly and vent (particularly undertail-coverts); *affinis* (resident Peninsular Malaysia) tends to have richer, more chestnut upperside and darker throat and breast; *incei* (visitor E Myanmar, Tenasserim, Thailand, Cambodia and N Laos southward) has all-glossy black head contrasting sharply with grey breast, and much darker, deeper chestnut (often violet-tinged) upperside. **Male white 'morph'** Stunning, with glossy black head and all-white rest of plumage (apart from black feather shafts and white-fringed black flight-feathers). It is not clear whether this plumage is age-related. **Female** Like rufous male but lacks tail-streamers, crest shorter, eyering usually somewhat duller. *T.p.incei* has similarly coloured upperparts but throat obviously darker than breast. **Juvenile** Similar to female but upperparts paler and more rufous (including crown and sides of head), underparts whitish, initially scaled/mottled dull rufous on breast, bill dark brownish with flesh-coloured base, no pronounced eyering. Male begins breeding when still in female-like plumage (before developing longer tail) and may only be separable in field by brighter blue eyering. **Other subspecies in SE Asia** *T.p.burmae* (resident SW,S,C Myanmar). **VOICE** Song is a clear rolling ***chu'wu'wu'wu'wu'wu...***, similar to Black-naped Monarch. Also a repeated harsh ***whiwhi-chu-whiwhi-chu*** and variants. Typical call is a repeated harsh shrill rasping ***whii***, ***whi-whu*** and ***whi-whu'whu*** etc. **HABITAT** Broadleaved evergreen forest, secondary growth, sometimes mangroves, island forest; also parks and wooded gardens on migration. Up to 1,500 m (below 1,220 m in Peninsular Malaysia). **RANGE & STATUS** Breeds Turkestan, E Afghanistan, Indian subcontinent, China (except NW), Ussuriland, Greater Sundas, W Lesser Sundas; most northern populations winter to south, S India, Sri Lanka, Sumatra. **SE Asia** Fairly common to common. Resident (except N Myanmar, C Thailand, N Laos, W,E Tonkin, N Annam); formerly Singapore. Breeding visitor (small numbers may overwinter) N Myanmar, N Laos, W,E Tonkin, N Annam. Winter visitor C,S,E Myanmar, W,NE(south-west),SE Thailand. Passage migrant S Myanmar, N Laos, E Tonkin. Winter visitor and passage migrant Tenasserim, C,S Thailand, Peninsular Malaysia, Singapore. **BREEDING** March–August. **Nest** Neat deep cup, in fork of tree-branches or bush; 1–3 m above ground (sometimes up to 15 m). **Eggs** 2–4; whitish to warm pink, spotted brownish-red over some pale inky-purple undermarkings (often ringed or capped at broader end); 20.2 x 15.1 mm (av.; *saturatior*).

678 **JAPANESE PARADISE-FLYCATCHER**
Terpsiphone atrocaudata Plate **70**
IDENTIFICATION 17.5–20.5 cm (male tail extends up to 23 cm more). *T.a.atrocaudata*: **Male** Shape recalls Asian Paradise-flycatcher but plumage all black, apart from glossy dark purple mantle to rump and whitish belly and vent. **Female** Very similar to Asian but crown duller (lacking obvious gloss), contrasting less with head-sides and throat, which are typically duller and darker (more brownish-tinted), breast usually appears more sharply cut off from whitish belly, upperparts duller, tail darker and browner, never distinctly bright rufous/chestnut. Some (possibly spring only) show more purplish-chestnut upperparts. **First winter** Similar to female, possibly with browner tint to throat and breast. Male possibly shows blacker crown, slightly darker, warmer upperparts and darker throat and breast. **HABITAT** Broadleaved evergreen forest, also mangroves, parks and wooded gardens on migration; up to 1,200 m. **RANGE & STATUS** Breeds Japan, N,S Korea, Taiwan, N Philippines; winters Sumatra, N Borneo, Philippines (Mindoro at least). **SE Asia** Rare to scarce winter visitor S Thailand, Peninsular Malaysia. Rare to scarce passage migrant Thailand, Peninsular Malaysia, Singapore, Cambodia, C,S Laos, W,E Tonkin, C Annam.

CORVIDAE: Crows, nutcrackers, magpies, jays, treepies & allies

Worldwide c.121 species. SE Asia 23 species. Very varied but mostly medium-sized to fairly large and stocky with strong bills, legs and feet and tuft of rictal bristles extending over base of upper mandible. Sexes alike. Typically gregarious and noisy. Diet varied, including large invertebrates, reptiles, amphibians, nestlings, eggs, fruit and carrion.

679 **HOUSE CROW** *Corvus splendens* Plate **71**
IDENTIFICATION 40–43 cm. *C.s.insolens*: **Adult** All-blackish plumage with broad dull greyish collar encompassing nape, upper mantle, rear ear-coverts, neck and breast diagnostic. Can look all black at distance. *C.s.protegatus* (feral range) has paler, more contrasting brownish-grey collar; *splendens* (SW Myanmar) has much paler, sharply contrasting grey collar. Smaller, slimmer and thinner-necked than Eastern and Southern Jungle Crows, with much shorter, slenderer bill. See Slender-billed and other crows. Note habitat and range. **Juvenile** Somewhat duller and browner, particularly head and body. **VOICE** Flat toneless dry ***kaaa-kaaa*** (weaker than Eastern and Southern Jungle), rasping ***ka*** or down-turned ***kow*** and low-pitched ***kowk***. **HABITAT & BEHAVIOUR** Open and urban areas, cultivation; up to 1,525 m. Roosts communally. **RANGE & STATUS** Resident Indian subcontinent, SW China. Introduced NE,E,S Africa, Middle East. **SE Asia** Common resident Myanmar; formerly W(south) Thailand. Fairly common to common feral resident Peninsular Malaysia, Singapore. Vagrant (of feral origin) southern S Thailand. **BREEDING** All year. **Nest** Fairly large deep cup, in tree; 3–6 m above ground. **Eggs** 3–7; fairly glossy, blue-green, blotched, speckled and streaked dull reddish and brown over some grey undermarkings; 34.8 x 25.6 mm (av.; *protegatus*). Brood-parasitised by Asian Koel.

680 **SLENDER-BILLED CROW**
Corvus enca Plate **71**
IDENTIFICATION 43–47 cm. *C.e.compilator*: **Adult** Very similar to Southern Jungle Crow but slimmer with somewhat shorter, squarer-ended tail, relatively long slender bill with less strongly arched upper mandible (beware female Southern Jungle with less arched bill), smoother, less steep forehead, less baggy throat (lacking specialised hackle-feathers), and non-glossy underparts. In flight, proportionately shorter, more rounded wing-tip and faster wingbeats. In hand, lacks rictal bristles at base of ridge of upper mandible. See House and Carrion Crows. Note voice, habitat and range. **Juvenile** Somewhat duller and browner, particularly body. **VOICE** Much more high-pitched and nasal than Southern Jungle: ***ka ka ka-a-a***, dry ***ahk-ahk-ahk***, explosive, throaty, croaking ***krok kok-kok*** etc. When excited, prolonged series of ***caaaw*** or ***aaaaw*** notes, interspersed with unusual (rather comical) resonant twanging nasal ***pe-yong*** and ***ne-awh***. **HABITAT & BEHAVIOUR** Broadleaved evergreen forest, sometimes mangroves; up to 520 m. Often in small flocks in forest canopy. **RANGE & STATUS** Resident Greater Sundas, Philippines, Sulawesi, Sula Is, S Moluccas. **SE Asia** Uncommon resident Peninsular Malaysia. **BREEDING** February–March. **Nest** Large, untidy shallow cup in tree; up to 26 m above ground. **Eggs** 2–4; blue, pale bluish-green or whitish, spotted and speckled olive-brown to greyish and blackish.

681 **CARRION CROW** *Corvus corone* Plate **71**
IDENTIFICATION 52–56 cm. *C.c.orientalis*: **Adult** Similar to Large-billed Crow but bill obviously shorter, slenderer and more pointed; has flatter forehead profile and somewhat shorter, squarer-ended tail. See House and Slender-billed Crows. Note voice, habitat and range. **Juvenile** Initially duller and glossless. Unlikely to occur. **VOICE** Calls are usually longer, harsher and lower-pitched than Large-billed: vibrant dry ***kraaa*** (often repeated) and hollower ***konk-konk***. **HABITAT** Open country, cultivation; recorded at c.700 m. **RANGE & STATUS** Mostly resident Palearctic, N Pakistan, NW India, W and northern China, N,S Korea, Japan; some northern populations winter to south, N Pakistan, NW India, southern China. **SE Asia** Vagrant E Tonkin.

682 **LARGE-BILLED CROW**
Corvus japonensis Plate **71**
IDENTIFICATION 48–59 cm. *C.j.tibetosinensis*: **Adult** Large and all black with strong purplish to dark bluish gloss; has rather long, high-arched bill, steep forehead, peaked crown, distinctly wedge-shaped tail, and heavy feet. See Eastern and Southern Jungle and Carrion Crows. Note range. *C.j.colonorum* (N Indochina) is somewhat smaller, and almost unglossed on neck and underparts. **Juvenile** Duller and less glossy. **VOICE** Calls of *tibetosinensis* include full-sounding, 'soft-edged' ***KAA*** or ***kYARRh kYARRh...***, and more forceful ***KOWH KOWH KOWH...*** In central Nepal, ***KWAA***; ***KWA'AA***; ***KWAA'WA*** and ***KWAOW*** etc. **HABITAT & BEHAVIOUR** Open forest and woodland, open areas and cultivation; up to 3,660 m. Usually seen in pairs but tends to roost communally. **RANGE & STATUS** Resident NE Afghanistan, NE Pakistan, N,NE India, Nepal, Bhutan, S,E Tibet, China (except NW), Taiwan, N,S Korea, SE Siberia, Ussuriland, Sakhalin I, Japan. **SE Asia** Uncommon to common resident N Myanmar (mountain areas), north N Laos, W(north),E Tonkin. **BREEDING** Himalayan *intermedius*: February–June. **Nest** Large, fairly deep cup, in tree; 7–18 m above ground. **Eggs** 3–6; pale blue-green, streaked and blotched umber- and blackish-brown, sometimes with neutral-coloured undermarkings; 44.8 x 31.3 mm (av.).

683 **EASTERN JUNGLE CROW**
Corvus levaillantii Plate **71**
IDENTIFICATION 45–48 cm. *C.l.levaillantii*: **Adult** Smaller than Large-billed Crow, with squarer-ended and relatively longer tail, less steep forehead, less high-arched bill, and less heavy feet. See Southern Jungle and House Crows. Note range. **Juvenile** Duller and less glossy, with paler-based bill. **VOICE** Higher, more prolonged, and more nasal than Large-billed: extended ***KAAAA KAAA***; ***KAAA KAAA KAAA KAAA...***; ***KRRRR KRRRR-KRRRR-KRRRR-KRRRR-KRRRR...*** etc. Sometimes, rather deeper ***OW'RRR OW'RRR OW'RRR***. **HABITAT** Open forest and woodland, open areas and cultivation, urban areas, mangroves; up to 2,565 m. **RANGE & STATUS** Resident SE Nepal, Bangladesh, lowland NE India, Andaman Is. **SE Asia** Uncommon to common resident Myanmar (lowlands in N), W,NW,NE,C Thailand. **BREEDING** December–May. **Nest** Large, fairly deep cup, typically in tree. **Eggs** 4–6; greenish blue to pale blue or bright green to pale yellowish-green, variably blotched, spotted and streaked blackish-brown (often in lengthwise direction), sepia or olive-brown, sometimes with pale purple undermarkings; typically averaging 42 x 29 mm (largest clutch 46.5 x 31.5 mm, smallest 37 x 28.5) in C Thailand. Brood-parasitised by Asian Koel. **NOTE** Split from Large-billed following Dickinson *et al.* (2004), Rasmussen & Anderton (2005), etc.

684 **SOUTHERN JUNGLE CROW**
Corvus macrorhynchos Plate **71**
IDENTIFICATION c.47–50 cm. *C.m.macrorhynchos* (? monotypic): **Adult** All black with purplish to dark bluish gloss, rather long bill with strongly arched ridge of upper mandible, peaked crown and noticeably rounded tail-tip. Hard to separate from very similar Eastern Jungle Crow, though apparently somewhat larger, with relatively shorter tail that has more rounded tip, plumage somewhat glossier below. See Slender-billed and House Crows. Note range and habitat. **Juvenile** Duller and less glossy, underparts matt sooty-black. **VOICE** Fairly deep, throaty, quite harsh ***KHERR KHERR*** or ***KWER KWER*** and ***KHARR KHARR KHARR*** etc.; and rather more urgent ***KUWOI-KUWOI-KUWOI***. **HABITAT** Open forest and woodland, open areas and cultivation, urban areas, mangroves; up to 1,525 m. **RANGE & STATUS** Resident Sundas; ? Philippines (taxonomy-dependent). **SE Asia** Uncommon to common resident S Thailand, Peninsular Malaysia, Singapore, Indochina (northern limit unknown); ? south Tenasserim. **BREEDING** December–June. **Nest** Large, fairly deep cup, in tree, palm etc.; 12–20 m above ground. **Eggs** 2–5; blue-green to grey-green, variably flecked brown, dark brown and purple, with small black dots mostly at broader end; 41.5–46.2 x 29.5–31 mm. Brood-parasitised by Asian Koel. **NOTE** Split from Large-billed following Dickinson *et al.* (2004), Rasmussen & Anderton (2005), etc. Identity of Indochina population debated; may be Eastern Jungle Crow.

685 **COLLARED CROW** *Corvus torquatus* Plate **71**
IDENTIFICATION 52–55 cm. Monotypic. **Adult** Resembles Large-billed and Carrion Crows but has diagnostic white collar, encompassing nape, hindneck and lower breast. See House Crow. Note range. **Juvenile** Body browner and glossless, collar somewhat duller with dark feather-tips. **VOICE** Calls include loud ***kaaarr*** (often repeated), ***kaar-kaar*** and cawing, creaking and clicking sounds. **HABITAT** Open areas and cultivation with scattered trees (particularly near water), coasts, sometimes urban areas; lowlands. **RANGE & STATUS** Resident W,C,E,S China. **SE Asia** Rare resident E Tonkin. Former resident (current status unknown) N,C Annam. **BREEDING** January–February. **Nest** Bulky cup, in tree. **Eggs** 2–6; pale bluish-green to blue, speckled and blotched olive-brown.

686 **SPOTTED NUTCRACKER**
Nucifraga caryocatactes Plate **72**
IDENTIFICATION 32–35 cm. *N.c.macella*: **Adult** Unmistakable. Dark brown with large white spots on mantle, breast and upper belly, broad white streaks on upper throat and sides of head, white vent and mostly white outertail-feathers. Crown, wings and rest of tail blackish-brown, bill rather pointed. In flight, underwing shows small whitish tips to coverts. **Juvenile** Paler, crown and sides of head concolorous with mantle; has whitish tips to wing-coverts, much narrower and generally buffier spots and streaks on upperparts (extending to crown) and paler, mottled underparts. **VOICE** Song consists of various quiet musical piping, squeaking, clicking and whistling notes, interspersed with mimicry. Call is a dry, harsh ***kraaaak***, sometimes quickly repeated in a discordant rattle. Also a short weak ***zhree***. **HABITAT** Open coniferous and mixed broadleaved/coniferous forest; 2,285–3,660. **RANGE & STATUS** Resident (subject to some movements) Palearctic, N,NE Indian subcontinent, S,SE Tibet, W,NC and northern China, Taiwan, N,S Korea, Japan. **SE Asia** Uncommon resident N Myanmar. **BREEDING** April–May. **Nest** Shallow cup or platform, in tree; 6–18 m above ground. **Eggs** 3–4; bluish-white, speckled and blotched dull brown to inky-brown (often mainly at broader end); c.35 x 26 mm.

687 **BLACK-BILLED MAGPIE** *Pica pica* Plate **72**
IDENTIFICATION 43–48 cm (including tail up to 26 cm). *P.p.sericea*: **Adult** Slim build, long tail and mostly black plumage with white scapulars and belly diagnostic. Has purplish-blue and green gloss on wings, green, purple and blue gloss on tail, and narrow white rump-band (sometimes indistinct). In flight shows largely white primaries. **Juvenile** Dark body-plumage duller and browner. **VOICE** Call is a harsh 4–12 note ***chak-chak-chak-chak-chak...*** Also an enquiring ***ch'chack*** and more squealing ***keee-uck***.

HABITAT & BEHAVIOUR Forest edge, vegetated waterways, cultivation, plantations; up to 2,255 m. Often in small parties. **RANGE & STATUS** Resident N Africa, Palearctic, Middle East, Pakistan, NW India, Bhutan, Tibet, China, Taiwan, N,S Korea, S Japan. **SE Asia** Scarce to locally common resident N,C,E Myanmar, N Laos, W,E Tonkin, C,S Annam (rare), Cochinchina. Former resident (current status unknown) N Annam. Vagrant NW,C Thailand (possibility of escapees). **BREEDING** February–May. **Nest** Large dome with side-entrance, in tree or bush; 2–3 m above ground. **Eggs** 3–8; pale blue-green, profusely blotched and spotted dull reddish-brown (more towards broader end); 35.6 x 24.4 mm (av.).

688 **EURASIAN JAY** *Garrulus glandarius* Plate **72**
IDENTIFICATION 31–34 cm. *G.g.leucotis*: **Adult** Combination of black cap and broad submoustachial band, white head-sides and throat and buffish-grey upperparts diagnostic. Forehead white with black streaks, lower rump and uppertail-coverts white (prominent in flight), underparts light buffish with greyer upper breast; has blue, black and whitish barring on wing-coverts and secondaries. *G.g.oatesi* (northern W, north C Myanmar) has pinkish-brown crown with indistinct blackish streaks and slightly duller white head-sides and throat; *sinensis* (east N Myanmar) has warm pinkish-buff crown, nape and sides of head, more pinkish-buff mantle, smaller submoustachial band and pinker underparts with pinkish-buff breast; *haringtoni* (S Chin Hills, W Myanmar) resembles *sinensis* but is more buffish-grey above. **Juvenile** Darker, with more rufescent body-plumage. **VOICE** Song consists of a variety of subdued musical notes, including mimicry and clearer mewing notes. Call is a harsh, rasping, screeching ***skaaaak skaaaak***... **HABITAT & BEHAVIOUR** Open broadleaved evergreen, pine, mixed evergreen/pine and deciduous forest, forest edge; up to 2,700 m. Usually in small, rather slow-moving parties. **RANGE & STATUS** Resident N Africa, Palearctic, Middle East, NE Pakistan, NW,N,NE Indian subcontinent, China, Taiwan, N,S Korea, Japan. **SE Asia** Fairly common to common resident Myanmar, W,NW,NE Thailand, Cambodia, C,S Laos, E Tonkin (local), S Annam, Cochinchina. **BREEDING** February–May. **Nest** Untidy wide cup, in sapling or shrub; up to 8 m above ground. **Eggs** 3–6; pale yellowish to pale greenish, finely freckled and blotched brown to olive-brown (sometimes mostly at broader end); 32.8 x 22.9 mm (av.; *leucotis*).

689 **COMMON GREEN MAGPIE**
Cissa chinensis Plate **72**
IDENTIFICATION 37–40.5 cm. *C.c.chinensis*: **Adult** Distinctive, with mostly bright green plumage, largely reddish-chestnut wings, black band from lores to nape-sides and bright red bill, eyering, legs and feet. Has prominent black and whitish markings on tips of tertials, inner secondaries and outertail-feathers and whitish-tipped central tail-feathers. Worn birds may have strongly bluish plumage and browner wings. *C.c.margaritae* (S Annam) has bright golden-yellow crown; *klossi* (C Laos, C Annam) has yellow forehead and yellowish-green crown. See Indochinese Green Magpie. **Juvenile** Somewhat duller with paler lower breast to vent, smaller dark subterminal markings on tertials and inner secondaries, browner bill and duller eyering, legs and feet. **Other subspecies in SE Asia** *C.c.robinsoni* (extreme S Thailand, Peninsular Malaysia). **VOICE** Loud and highly variable. Penetrating series of high-pitched notes (sometimes ending with harsh note): ***wi-chi-chi jao***; ***wi-chi-chi jao wichitchit wi-chi-chi jao...*** etc. When agitated, gives hoarse manic scolding chatter: ***chakakakakakak*** or ***chakakak-wi*** with higher-pitched last note. Also, softer chattering ***churrk chak-chak-chak*** and high-pitched ***weeer-wit*** and rising ***wieeee*** etc. Complex high shrill whistles, combined with avian mimicry, probably constitutes the song. **HABITAT & BEHAVIOUR** Broadleaved evergreen and mixed deciduous forest; up to 2,075 m. Often in small parties, regularly associating with mixed-species feeding flocks. Shy. **RANGE & STATUS** Resident N,NE Indian subcontinent, SW China, Sumatra, Borneo. **SE Asia** Common resident Myanmar, W,NW,NE,extreme S Thailand, Peninsular Malaysia, Indochina (except Cochinchina). **BREEDING** January–August. **Nest** Broad, rather shallow cup, in bush, sapling or bamboo. **Eggs** 3–6; greyish or pale greenish, blotched and freckled reddish-brown (often more so at broader end); 30.2 x 22.9 mm (av.; *chinensis*).

690 **INDOCHINESE GREEN MAGPIE**
Cissa hypoleuca Plate **72**
IDENTIFICATION 31–35 cm. *C.h.hypoleuca*: **Adult** Very similar to Common Green Magpie but tertials and inner secondaries appear mostly pale greenish (with no black or white), has strong lemon-yellow wash on underparts; upperside slightly darker than underside, tail shorter. Can also show strongly bluish plumage. Note range. *C.h.concolor* (E Tonkin, N Annam) shows greener underparts, darker upperparts and buffish tinge to tail-tips; *chauleti* (C Annam) has yellower-green on head, deeper yellow underparts, and brownish-buff washed tail (particularly outer feather). **Juvenile** Somewhat duller with paler vent, browner bill and duller eyering, legs and feet. **VOICE** Similar to Common Green and equally variable. Loud shrill whistled ***peeeoo-peeeoo peeeoo-peeeoo...***, more clipped, shrill ***peu-peu-peu*** and clear whistles terminated by harsh note: ***po-puueeee chuk*** and rising ***eeeoooeeep graak*** etc. Also long piercing falling ***peeeeooo*** and abrupt ***weep*** notes. When agitated, utters very noisy, harsh, high-pitched, scolding chatters. **HABITAT** Broadleaved evergreen and semi-evergreen forest, bamboo; up to 1,870 m. **RANGE & STATUS** Resident SW,S China. **SE Asia** Fairly common to common resident SE Thailand, Indochina (except N Laos, W Tonkin, Cambodia). **BREEDING** March–July. **Nest** Bulky, untidy, shallow cup, in small tree; 2.5–3 m above ground. **Eggs** 3.

691 **YELLOW-BILLED BLUE MAGPIE**
Urocissa flavirostris Plate **72**
IDENTIFICATION 61–66 cm (including tail up to 48 cm). *U.f.flavirostris*: **Adult** Resembles Red-billed Blue Magpie but has diagnostic yellow bill and all-blackish hood with small white nuchal patch; blue of plumage is paler and greyer; has orange to orange-yellow legs and feet. In fresh plumage, has faint olive wash on upperparts and faint pale yellowish wash on underparts. *U.f.robini* (W Tonkin) has brighter olive wash on upperparts and yellowish wash on underparts (when fresh). **Juvenile** Similar to adult but initially browner on head, mantle and breast; has somewhat duller nuchal patch, dull yellowish-olive bill, and duller legs and feet. **Other subspecies in SE Asia** *U.f.schaferi* (W Myanmar). **VOICE** Complex subdued squeaky chattering and whistles, and more measured raucous ***tsii-trrao tsii-trrao...*** and ***shitu-charrh shitu-charrh...*** When alarmed, gives scolding ***tcheh-he-he-he-he-he...*** and abrupt scratchy ***tcherr***. **HABITAT & BEHAVIOUR** Broadleaved evergreen forest, open pine forest; 1,220–3,100 m. Often shy and elusive. **RANGE & STATUS** Resident N Pakistan, NW,N,NE India, Nepal, Bhutan, SE Tibet, SW China. **SE Asia** Locally uncommon to fairly common resident W,N Myanmar, W Tonkin. **BREEDING** India: May–June. **Nest** Rather shallow cup, in tree; 5–6 m above ground. **Eggs** 3–4; pale cream-coloured, blotched bright reddish-brown over sparse neutral-coloured undermarkings; 34.8 x 23.4 mm (av.; *flavirostris*).

692 **RED-BILLED BLUE MAGPIE**
Urocissa erythrorhyncha Plate **72**
IDENTIFICATION 65–68 cm (including tail up to 47 cm). *U.e.magnirostris*: **Adult** Red bill, black hood, blue upperparts and very long white-tipped blue tail diagnostic. Has broad white band from hindcrown to hindneck, whitish remainder of underparts and red legs and feet. *U.e.alticola* (N Myanmar) and *erythrorhyncha* (N Indochina) have rather more extensive, very pale bluish-grey hindcrown/hindneck-band, somewhat

greyer upperparts and smaller bill. See Yellow-billed Blue Magpie. **Juvenile** Dark areas of head and upper breast much duller and browner, upperparts and wing-coverts browner (latter pale-tipped); has more white on crown, greyer bill, and duller legs and feet. **VOICE** Sharp raucous ***chweh-chweh-chweh-chweh...*** or ***chwit-wit-wit...*** and shrill ***shrii*** and subdued ***kluk*** notes. **HABITAT & BEHAVIOUR** Deciduous forest, secondary growth, bamboo, sometimes open broadleaved evergreen forest, conifer plantations; up to 1,940 m. Usually found in flocks, hunting at relatively low levels. Flies with a few flaps and a glide. **RANGE & STATUS** Resident N,NE India, Nepal, E Bangladesh, NE,C and southern China. **SE Asia** Uncommon to common resident Myanmar (except Tenasserim), Thailand (except S), Cambodia, Laos, Vietnam. **BREEDING** February–May. **Nest** Rough, flimsy cup, placed in tree; at least 6–8 m above ground. **Eggs** 3–6; clay- or stone-coloured, boldly blotched dark brown or reddish-brown (often ringed or capped at broader end); 33.9 x 23.9 mm (av.; *magnirostris*).

693 **WHITE-WINGED MAGPIE**

Urocissa whiteheadi Plate **72**

IDENTIFICATION 45–46 cm (including tail up to 25 cm). *U.w.xanthomelana*: **Adult** Unmistakable, with dark brown head, upper breast and upperside, orange bill and large white markings on wing-coverts, tertials and tail-tips. Has pale yellowish eyes, dusky whitish lower breast to vent and mostly white uppertail-coverts. White patches on wings and tail very striking in flight. **Juvenile** Paler greyer head, nape and breast, greyish to brownish bill and eyes, strong yellowish wash on pale of tail (except tip), belly and vent. **VOICE** Hoarse rising ***shureek***, low hoarse ***churrree***, soft liquid rippling ***brrriii brrriii...*** and harsher rising ***errreep errreep...*** **HABITAT** Broadleaved evergreen forest, forest edge, secondary growth; up to 1,300 m. **RANGE & STATUS** Resident SW,S China. **SE Asia** Locally fairly common to common resident Laos, W,E Tonkin, N,C Annam. **BREEDING** April–August. **Nest** Large shallow cup, in tree; 12 m above ground. **Eggs** 6; pale greenish-blue, flecked and spotted brown.

694 **RUFOUS TREEPIE**

Dendrocitta vagabunda Plate **72**

IDENTIFICATION 46–50 cm (including tail up to 30 cm). *D.v.kinneari*: **Adult** Dark rufescent-brown upperparts, deep buff underparts and contrasting blackish-grey hood and upper breast diagnostic. Has large pale grey area across wing-coverts, tertials and secondaries and black-tipped pale grey uppertail. *D.v.saturatior* (Tenasserim, W Thailand) and *sakeratensis* (NE,SE Thailand, Indochina) have darker hood. See Grey Treepie. **Juvenile** Hood paler and browner, underparts paler buff, greater coverts, tertials and tail-feathers tipped buffish. **Other subspecies in SE Asia** *D.v.sclateri* (SW,W,S[west] Myanmar). **VOICE** Loud metallic flute-like ***koku-lii*** or ***koku-wli***. Pairs utter loud intermingled ***kuki-uii***; ***akuak*** and ***ekhekhekh*** calls. When alarmed, utters harsh ***herh-herh-herh-herh hah-hah-hah herh-herh-herh...*** etc. **HABITAT** Dry dipterocarp and mixed deciduous forest, secondary growth, sometimes cultivation; up to 2,135 m (mostly below 1,000 m). **RANGE & STATUS** Resident Pakistan, Indian subcontinent (except islands). **SE Asia** Locally common resident Myanmar, W,NW,NE,SE Thailand, Cambodia, Laos (north to Vientiane area), S Annam, Cochinchina. **BREEDING** March–April. **Nest** Shallow cup, in bush or tree; 6–8 m above ground. **Eggs** 2–5; pale greenish, blotched and spotted grey-brown or pinkish with dark brown and reddish blotches over lilac and greyish undermarkings; 29.7 x 22.1 mm (av.; Myanmar).

695 **GREY TREEPIE**

Dendrocitta formosae Plate **72**

IDENTIFICATION 36–40 cm (including tail up to 23 cm). *D.f.assimilis*: **Adult** Recalls Rufous Treepie but has paler grey hindcrown and nape, contrasting with blackish forecrown and face, all-blackish wings with white patch at base of primaries, and dull greyish underparts with deep rufous undertail-coverts. Mantle and scapulars brown, rump and uppertail-coverts pale grey, uppertail grey becoming blacker towards tip. *D.f.sinica* (E Tonkin) has darker brown mantle and scapulars, whiter rump and uppertail-coverts and all-blackish uppertail. See Collared Treepie. **Juvenile** Has less black on forecrown, paler hindcrown, nape, head-sides, lower throat and breast with warm buffish infusion and whiter belly; may show some rufous tips to wing-coverts and tertials. **Other subspecies in SE Asia** *D.f.himalayensis* (W,N,C,E[north] Myanmar). **VOICE** Rather varied. Loud, ringing, metallic, comical repeated ***koh-kli-ka***; ***koh-kli-koh-koh***; ***kokli-kaka***; ***ko-kiki*** and ***kuh'kuh'kuh'ki-kuh*** etc. Harsh scolding chatters when agitated. **HABITAT & BEHAVIOUR** Broadleaved evergreen forest, secondary growth; 700–2,285 m, locally down to 450 m. Usually in small noisy flocks. **RANGE & STATUS** Resident NE Pakistan, E India, NW,N,NE Indian subcontinent, W,C and southern China, Taiwan. **SE Asia** Fairly common to common resident Myanmar, W,NW,NE Thailand, N,C Laos, W,E Tonkin, N Annam. **BREEDING** April–June. **Nest** Flimsy, shallow cup, in bush or sapling; 3–7 m above ground. **Eggs** 3–5; pale cream to pale reddish or pale bluish, boldly blotched dark brown to reddish-brown, sometimes with inky undermarkings (often ringed or capped at broader end); 28.8 x 20.1 mm (av. *himalayensis*).

696 **COLLARED TREEPIE**

Dendrocitta frontalis Plate **72**

IDENTIFICATION 38 cm (including tail up to 26 cm). Monotypic. **Adult** Shows unmistakable combination of very pale grey hindcrown, collar, breast and upper belly, rufous remainder of body (paler below) and sharply demarcated black forecrown, ear-coverts and throat. Has all-black tail, tertials, secondaries and primaries, and grey median and greater coverts. Distinctly smaller-bodied, slimmer and deeper-billed than Grey Treepie. See Rufous Treepie. **Juvenile** Black of head duller, breast colour mixed with warm brown, upperwing-coverts browner grey. **VOICE** Throaty, clicking ***u-WIP***, and sudden shrill, high, metallic ***driii*** or ***dreet***. Piping ***phewt*** in flight. **HABITAT** Broadleaved evergreen forest, bamboo; up to 1,220 m. **RANGE & STATUS** Resident Bhutan, NE India, SW China (W Yunnan). **SE Asia** Resident N Myanmar (locally common), W,E Tonkin (rare). **BREEDING** India: April–July. **Nest** Well-made cup, in tall bush or bamboo; above ground. **Eggs** 3–4; like Grey Treepie but rather more profusely marked; 27 x 19.9 mm (av.).

697 **RACKET-TAILED TREEPIE**

Crypsirina temia Plate **72**

IDENTIFICATION 30.5–32.5 cm (including tail up to 20 cm). Monotypic. **Adult** Relatively small and slim with distinctive all-blackish plumage and long, straight, spatulate-tipped tail. Has dark bronze-green gloss on body-plumage, blacker face and light bluish eyes. See Ratchet-tailed Treepie. Could be confused with drongos when tail worn but note short, thick bill and eye colour. **Juvenile** Head and body duller and browner, has less contrasting dark face, brown eyes and narrower tail-tip. **VOICE** Short, ringing ***chu***, deep rasping ***churg-churg***, harsh ***chraak-chraak*** or ***chrrrk-chrrrk***, more rising, questioning ***churrrk*** and higher ***grasp-grasp***. **HABITAT & BEHAVIOUR** Mixed deciduous woodland and open broadleaved evergreen and semi-evergreen forest (particularly near water), secondary growth, bamboo, mangroves, coastal scrub; up to 915 m. Often in pairs or small parties, working through vegetation. **RANGE & STATUS** Resident SW China, Java, Bali. **SE Asia** Fairly common to common resident E(south),S Myanmar, Tenasserim, Thailand, north-west Peninsular Malaysia, Indochina. **BREEDING** April–August. **Nest** Shallow cup, in bamboo, bush or small tree; 2–6 m above ground. **Eggs** 2–4; whitish to greenish-buff, marked with various shades of brown; 24.9 x 18.3 mm (av.).

698 **HOODED TREEPIE**

Crypsirina cucullata Plate **72**

IDENTIFICATION 30–31 cm (including tail up to 20 cm). Monotypic. **Adult** Smallest treepie in region with distinctive pale greyish body and wing-coverts and contrasting black head, primaries and central tail-feathers. Has narrow whitish collar, black and white secondaries and pronounced spatulate tip to tail. Note range. **Juvenile** Head much paler and browner, body browner-tinged, no whitish collar. **VOICE** Contact calls include a quiet, purring ***drrrriiii'k***. **HABITAT & BEHAVIOUR** Open deciduous woodland, thorn-scrub jungle, bamboo, cultivation borders; up to 915 m. Found in pairs or small flocks of up to 12. **RANGE & STATUS** Endemic. Uncommon to locally common resident W,N,C,S Myanmar. **BREEDING** April–July. **Nest** Shallow cup, in bush or tree. **Eggs** 2–4; creamy- or greenish-white to greyish stone-colour, spotted brown to olive (often ringed at broader end); 23.1 x 18 mm (av.).

699 BLACK MAGPIE

Platysmurus leucopterus Plate **71**

IDENTIFICATION 39–41 cm. *P.l.leucopterus*: **Adult** All-black plumage with prominent white band along greater coverts and tertials and fairly long, broad tail diagnostic. Has short, tufted crest on forehead, reddish eyes. **Juvenile** Body somewhat browner, crest shorter, has dark (instead of reddish) eyes. **VOICE** A wide variety of amazing, often comical sounds. Loud, discordant, metallic ***keh'eh'eh'eh'eh***, resonant, bell-like ***tel-ope*** and ***kontingka-longk*** and xylophone-like ***tok-tok tek-lingk-klingk-klingk*** etc. **HABITAT & BEHAVIOUR** Broadleaved evergreen forest, forest edge, sometimes mangroves; up to 215 m. Quite shy, but betrayed by highly distinctive voice. Wings make soft throbbing sound in flight. **RANGE & STATUS** Resident Sumatra, Borneo. **SE Asia** Uncommon resident south Tenasserim, S Thailand, Peninsular Malaysia. Rare (status unclear) Singapore. **BREEDING** December–May. **Nest** Rough shallow cup, in bush, sapling or palm; 0.9–3 m above ground. **Eggs** 2–4; whitish to fawn-grey, peppered with dirty brown or light brown freckles; 30.4–33 x 23.3–24.4 mm.

700 RATCHET-TAILED TREEPIE

Temnurus temnurus Plate **72**

IDENTIFICATION 32–35.5 cm (including tail up to 18 cm). Monotypic. **Adult** Resembles Racket-tailed Treepie but has diagnostic broad tail with long spikes projecting from tips of outer webs of feathers. Overall greyish-black with black face and dark red to brown eyes. **First year** Has narrower tail-feathers with blunter spikes, almost absent on outer feathers. **VOICE** Usual calls include a loud ringing ***clee-clee-clee...***, harsh rhythmic grating ***graak-graak-graak...*** and short squeaky rising ***eeup-eeup-eeup...*** Also gives ringing ***CLIPeeee*** (first syllable stressed) and hollow-er-starting ***pupueeee*** (sometimes during pair duets) and short, rather high-pitched rasping, rippling ***rrrrrrr***. **HABITAT** Broadleaved evergreen forest, forest edge, bamboo, secondary growth; 50–1,500 m. **RANGE & STATUS** Resident S China (Hainan). **SE Asia** Uncommon to common resident southern Tenasserim (local), W(south) Thailand (local), north-east Cambodia, C Laos, W,E Tonkin, N,C,S Annam, north Cochinchina. **BREEDING** February–June. **Nest** Shallow cup, in bamboo. **Eggs** 2.

701 CRESTED JAY

Platylophus galericulatus Plate **71**

IDENTIFICATION 31–33 cm. *P.g.ardesiacus*: **Adult** Unmistakable, with short tail, overall blackish plumage, large white patch on neck-side and unusual long, slightly forward-pointing, narrow, erect crest. **Juvenile** Upperparts tinged warm brown, underparts duller and barred whitish; has warm buff spots on wing-coverts and short crest with buffish tips. **Immature** Like adult but paler bars on underparts and whitish shaft-streaks on throat and breast. **VOICE** Song is a strange fluty phrase, preceded by an abrupt shrill high-pitched whistle: ***psSSSsiu HI-WU*** (repeated every few seconds). Usual call is a diagnostic, very rapid, grating metallic rattle: ***tit'it'it'it'it'it'it'it...*** **HABITAT & BEHAVIOUR** Broadleaved evergreen forest; up to 1,220 m. Usually occurs in small parties, frequenting middle to lower storey of forest. **RANGE & STATUS** Resident Greater Sundas. **SE Asia** Uncommon to locally common resident south Tenasserim, S Thailand, W(south),S Thailand, Peninsular Malaysia. **BREEDING** February–June. **Nest** Sturdy shallow cup, on tree-branch; 2–3 m above ground. **Eggs** 1–2; whitish to pale blue-green, coarsely spotted purple and brown or reddish-brown (mainly towards broader end); 30.2 x 22.8 mm (av. Java).

LANIIDAE: Shrikes

Worldwide c.31 species. SE Asia 6 species. Medium-sized and rather large-headed, with long tails and stout hooked bills. Solitary, mostly frequenting open country. Predatory, feeding on large insects, nestlings and small mammals, birds and reptiles. Prey may be impaled on thorns or barbed wire, to be consumed later.

702 TIGER SHRIKE *Lanius tigrinus* Plate **73**

IDENTIFICATION 17–18.5 cm. Monotypic. **Male** Grey crown and nape, black forehead and mask and all-whitish underparts distinctive. Resembles Burmese Shrike but has diagnostic blackish bars/scales on deep rufous-brown upperparts, warm brown uppertail and no white wing-patch. In autumn, mostly shows cocoa-brown crown and nape, partial white ring at rear of eye, and lacks obvious mask. Some show grey feathers on crown/nape and shadowy mask, and a few resemble dull spring/summer birds. By mid-winter, when most worn, shows pale-streaked crown to mantle, black-barred sides of underparts and wing-coverts (sometimes across breast). **Female** Somewhat duller with more prominently barred/scaled upperparts and buff-tinged flanks with blackish scales; has whitish patch on lores and narrow white supercilium. Believed to show similar variation as male in autumn/winter. **First winter** Duller than female, crown and sides of head uniform warmish brown with blackish bars/scales, pinkish base to lower mandible. Similar to Brown Shrike but head and body more contrastingly dark-scaled, no obvious dark mask. **VOICE** Alarm call is a harsh, scolding chatter. Also gives a subdued sharp ***tchick***. **HABITAT & BEHAVIOUR** Forest edge, overgrown clearings, secondary growth; up to 1,220 m. Skulking. **RANGE & STATUS** Breeds C,E,NE China, Ussuriland, N,S Korea, Japan; winters Greater Sundas, rarely Philippines. **SE Asia** Scarce to uncommon passage migrant E Myanmar, south Tenasserim, Thailand (has wintered south W), Peninsular Malaysia (also winters), Singapore (also winters), Cambodia, N Laos, W,E Tonkin, C,S Annam, Cochinchina.

703 BULL-HEADED SHRIKE

Lanius bucephalus Plate **73**

IDENTIFICATION 19.5–20.5 cm. *L.b.bucephalus*: **Male** Resembles Brown Shrike but shows distinctive combination of rufescent-brown crown and nape, greyish mantle to uppertail-coverts with a variable warm brown wash (particularly on mantle), and a white patch at the base of the primaries. Also has blacker base colour to wings, blacker tail, variable faint dark brown scaling on the breast and flanks, and a pinkish base to the lower mandible. **Female** Head browner and more uniform with a much narrower supercilium and duller mask, upperparts duller brownish-grey, underparts indistinctly scaled darker, lacks white wing-patch. **First winter** Like female but upperparts and sides of body are more rufous-tinged, tail mid-brown, almost lacks dark mask, has more extensive pinkish base to lower mandible. **VOICE** Noisy chattering ***ju-ju-ju*** or ***gi-gi-gi***. **HABITAT** Forest edge, clearings; up to 750 m. **RANGE & STATUS** Breeds NC,NE China, N,S Korea, Ussuriland, Sakhalin I, Kuril Is, Japan; some populations winter south to S China. **SE Asia** Vagrant C Annam (one record, perhaps doubtful).

704 **BROWN SHRIKE** *Lanius cristatus* Plate **73**

IDENTIFICATION 19–20 cm. *L.c.confusus*: **Male** Uniform warmish brown upperside, black mask, prominent whitish supercilium and whitish underparts with rich buffish wash along sides diagnostic. Lacks white wing-patch, has greyish-white forehead and brighter rufous-chestnut tinge to rump/uppertail-coverts. *L.c.superciliosus* (east and south of region) has rich chestnut upperparts (brightest on crown), sharply demarcated white forehead and supercilium, and rich rufous-chestnut rump; *lucionensis* (widespread in region) has pale grey crown, less distinct supercilium and pale grey wash on mantle and scapulars. See Grey-backed Shrike. **Female** Often slightly duller, with cream-tinged supercilium and fine dusky vermiculations on breast and flanks. **Juvenile** Duller, with narrow blackish and some buffish scales on upperparts, whiter underparts with dark scales/bars along sides, shorter supercilium and pinkish base to lower mandible. Combination of blackish mask, whitish supercilium and all-brown base colour to upperside distinctive. Separation of subspecies difficult but *L.c.superciliosus* typically shows brighter rufous-chestnut crown. **First winter** Like female but has dark subterminal borders to scapulars, wing-coverts and tertials and more prominent dark scales on sides of underbody. **Other subspecies in SE Asia** *L.c.cristatus* (throughout). **VOICE** Harsh staccato ***chak-ak-ak-ak-ak*** when alarmed. High-pitched squawking. Song consists of rich and varied chattering. **HABITAT** Open country, cultivation, gardens, secondary growth, forest edge; up to 2,000 m (mainly lowlands). **RANGE & STATUS** Breeds C,E Palearctic, C,E,NE China, N,S Korea, Japan; winters Indian subcontinent (except Pakistan), S China, Taiwan, Sundas, Philippines, N Sulawesi. **SE Asia** Common winter visitor throughout. Also recorded on passage SW,S Myanmar, C Thailand, Peninsular Malaysia, Cambodia, N Laos, E Tonkin.

705 **BURMESE SHRIKE**
Lanius collurioides Plate **73**

IDENTIFICATION 19–21 cm. *L.c.collurioides*: **Male** Resembles Tiger Shrike, with black forehead and mask, rufous-chestnut upperparts and whitish underparts, but crown and nape slaty-grey, upperparts unbarred, has white patch at base of primaries and blackish, strongly white-edged uppertail. *L.c.nigricapillus* (C,S Annam, Cochinchina) has much darker, blackish-slate crown and nape and darker chestnut upperparts. **Female** Duller above, whitish lores and sometimes a very narrow whitish supercilium (broken over eye). *L.c.nigricapillus* appears to differ less obviously from male, with indistinct pale lores, only slightly paler grey hindcrown and nape and often somewhat duller chestnut upperparts. **Juvenile** Like female (both sexes) but crown and nape warm brownish to greyish-brown with blackish-and-buff bars, mask duller with narrow buffish streaks, rest of upperpart feathers, lesser and median coverts paler with blackish subterminal scales/bars, rest of wing-feathers fringed paler, white wing-patch smaller or almost absent; has wavy blackish-brown bars on breast, flanks and thighs. **VOICE** Song is subdued, quiet, rapid and scratchy, including much repetition and varied mimicry. Alarm call consists of a loud, rapid harsh chattering: ***chikachikachitchit***; ***chekoochekoochititititititit***; and ***chetetetetet*** etc. and a harsh single ***JAO***. **HABITAT** Clearings and open areas in various kinds of forest (particularly pine), cultivation; breeds 600–1,995 m (locally down to sea-level); widespread down to sea-level in winter. **RANGE & STATUS** Breeds NE India, SW,S China; some move south in winter. **SE Asia** Locally fairly common to common resident S(northwest),W,N,C,E Myanmar, W,NW Thailand, Cambodia, Laos, Vietnam (except W,E Tonkin). Fairly common winter visitor SW,E(south) Myanmar, Tenasserim, Thailand (except S), E Tonkin (also passage migrant). Recorded spring (status uncertain) W Tonkin. **BREEDING** March–June. **Nest** Neat cup, in tree or sapling; 1.2–10 m above ground. **Eggs** 3–6; white, cream-coloured, pinkish or greenish, speckled and blotched reddish-brown over neutral to pale blue undermarkings; 21.1 x 16.5 mm (av. *collurioides*).

706 **LONG-TAILED SHRIKE**
Lanius schach Plate **73**

IDENTIFICATION 25–28 cm. *L.s.longicaudatus*: **Adult** Black head with white throat diagnostic. Relatively large, with long blackish tail, rufous-chestnut rest of upperparts, rufescent flanks and vent, and prominent white patch at base of primaries. *L.s.schach* (Vietnam; vagrant NW Thailand) is bulkier but shorter-tailed, has grey midcrown to upper mantle, paler and more rufous rest of upperparts and much smaller wing-patch; '*fuscatus*' morph of *L.s.schach* (E Tonkin) typically has brownish-black throat; brownish-grey crown, nape and body (paler below); mostly blackish scapulars and back; warm brown tinge to rump, uppertail-coverts, flanks and belly; dull chestnut undertail-coverts; and blackish wings with browner flight-feathers and no white patch; *bentet* (Peninsular Malaysia, Singapore) resembles *schach* but is much smaller (24–24.5 cm), has variable black on forecrown (may extend behind eye), paler grey hindcrown to back, pale warm buff scapulars with whitish outer edges, rufous-buff rump and uppertail-coverts and whitish-fringed tertials. See Grey-backed Shrike. **Juvenile** Crown and nape whitish-buff to pale greyish, extreme forehead and supercilium whiter, lores and ear-coverts browner than adult, mostly buffish mantle, black bars/scales on crown, mantle and scapulars (vaguely on back to uppertail-coverts); wavy blackish-brown bars on breast, flanks and vent. Also, browner tail and browner wings with mostly buffish, black-scaled lesser and median coverts, rufous-buff-fringed greater coverts, and smaller white patch. *L.s.schach* has browner crown to upper mantle with indistinct dark bars/scales; '*fuscatus*' morph almost lacks markings on crown to upper mantle and is darker overall. **Other subspecies in SE Asia** *L.s.tricolor* (Myanmar, NW Thailand, N Laos; recorded once W Tonkin). **VOICE** Song consists of subdued, scratchy warbling notes, incorporating mimicry. A loud, scolding drawn-out ***chaak-chaak*** when alarmed. **HABITAT** Open country, cultivation, gardens, roadsides, secondary growth, sometimes forest edge; up to 2,135 m. **RANGE & STATUS** Resident (mostly sedentary) C Asia, Afghanistan, Indian subcontinent, C and southern China, Taiwan, Sundas, Philippines, E New Guinea. **SE Asia** Uncommon to common resident, locally subject to minor movements (except S Myanmar, Tenasserim, S Thailand, Cambodia, S Annam, Cochinchina). Vagrant Cambodia. Recorded (status uncertain) SW,S Myanmar, Tenasserim. **BREEDING** April–February (probably all year). **Nest** Large compact cup, in tree, bush or sapling; 0.9–6 m above ground. **Eggs** 3–6; pink to buff or pale green to pale blue, blotched and spotted greyish-brown, reddish-brown or purplish-brown (often ringed or capped at broader end); 23.6 x 17.9 mm (av. *tricolor*).

707 **GREY-BACKED SHRIKE**
Lanius tephronotus Plate **73**

IDENTIFICATION 22.5–25.5 cm. *L.t.tephronotus*: **Adult** Resembles Long-tailed Shrike (*L.s.schach*) but mantle and scapulars uniform grey (concolorous with crown), has less black on forehead, no white wing-patch and shorter, browner tail. See Brown Shrike (*L.c.lucionensis*). **Juvenile** Crown to back cold brownish-grey, scaled/barred blackish and dull pale buff to dull rufous, mask duller, back to uppertail-coverts more rufescent with black bars, underparts more buffish with blackish scales (plainer on upper throat and vent), tertials and wing-coverts fringed rufescent. Best separated from Long-tailed by greyish mantle and scapulars and lack of white wing-patch. **First winter** More similar to adult but mask duller, wings and underparts similar to juvenile, may have slightly paler mantle with brown tinge. **VOICE** Call is harsh and grating. Song is subdued and scratchy, incorporating much mimicry. **HABITAT** Open country, cultivation, secondary growth; up to 2,135 m. **RANGE & STATUS** Breeds NW,N,NE India, Nepal, Bhutan, S,E Tibet, SW,W,N,NC China; some populations winter to south, north India, Bangladesh. **SE Asia** Uncommon winter visitor Myanmar (except southern Tenasserim), W,NW,NE,C Thailand, Cambodia, Laos, W,E Tonkin, N Annam.

REGULIDAE: Crests & allies

Worldwide c.6 species. SE Asia 1 species. Tiny and warbler-like with fine bills and strongly patterned heads and wings. Very active, feeding mainly on small insects.

708 **GOLDCREST** *Regulus regulus* Plate **107**
IDENTIFICATION 9 cm. *R.r.yunnanensis*: **Male** Resembles small *Phylloscopus* warblers but has diagnostic bright yellowish-orange median crown-stripe with blackish border and broad pale area around eye; lacks pale supercilium, shows distinctive blackish and whitish markings at base of flight-feathers. **Female** Median crown-stripe yellow. **Juvenile** Like adult but crown unmarked, upperparts somewhat browner. **VOICE** Song is a very high-pitched, undulating ***seeh sissisyu-see sissisyu-see siss-seeitueet***. Usual call is a thin, high-pitched ***see-see-see***. **HABITAT** Coniferous and broadleaved evergreen forest; 2,200–2,800 m. **RANGE & STATUS** Breeds W,C Palearctic, S,SE Siberia, N Pakistan, NW,N India, Nepal, Bhutan, S,E Tibet, SW,W,C,NW,NE China, N Korea, Japan; some populations winter to south, southern Europe, E,NE China, N,S Korea, S Japan. **SE Asia** Recorded in winter (status uncertain) N,E Myanmar. **BREEDING** India: May–July. **Nest** Pouch, hanging from tree-branch; 2–12 m above ground. **Eggs** 4–7; white, with large reddish spots, or pale creamy-buff, speckled pinkish-buff around broader end; 14.3 x 10.7 mm (av.; India).

NECTARINIIDAE: Sunbirds & spiderhunters

Worldwide c.129 species. SE Asia 24 species. Fairly small to very small, with slender, mostly downcurved bills. Serrations near bill-tip and long tongue are adapted for nectar-feeding; also feed on insects and spiders etc. **Sunbirds** (*Anthreptes, Leptocoma, Cinnyris, Aethopyga, Chalcoparia, Hypogramma*) Relatively small and short-billed, males with strongly iridescent plumage. **Spiderhunters** (*Arachnothera*) Relatively large and long-billed, sexes alike.

709 **PLAIN SUNBIRD** *Anthreptes simplex* Plate **75**
IDENTIFICATION 13 cm. Monotypic. **Adult** Told from other sunbirds by nondescript female-like plumage, relatively short and straight bill, lack of obvious yellow in plumage and dark iridescent blue-green forehead-patch. Upperparts olive-green, underparts pale greyish-olive (greyer on throat and breast), tail browner than rest of plumage. **Female** Lacks forehead-patch. **Juvenile** Similar to female, but with orange-pink bill, legs and feet. **VOICE** Inadequately documented. **HABITAT & BEHAVIOUR** Broadleaved evergreen forest, forest edge, sometimes coastal scrub, rarely mangroves; up to 915 m. Often feeds in middle storey. Moves in slow and methodical manner, reminiscent of Arctic Warbler. **RANGE & STATUS** Resident Sumatra, Borneo. **SE Asia** Uncommon to fairly common resident south Tenasserim, W(south),S Thailand, Peninsular Malaysia, Singapore (scarce). **BREEDING** March–August. **Nest** Oval to globular, suspended from tree-branch; 3.5 m above ground. **Eggs** 2; very pale lilac, with purplish clouds over broader end, and sparse purplish-black splashes and squiggles elsewhere; 18.5–19 x 12.8 mm.

710 **BROWN-THROATED SUNBIRD**
Anthreptes malacensis Plate **75**
IDENTIFICATION 14 cm. *A.m.malacensis*: **Male** Relatively robust, with glossy dark green and purple upperparts, dull brownish head-sides and throat, and yellow lower breast to vent distinctive. Very like Red-throated Sunbird but head-sides and throat browner, has extensive iridescent purple shoulder-patch bordered by small amount of maroon-chestnut on wing-coverts. Note habitat and range. **Female** From other sunbirds (apart from Red-throated) by robust shape, relatively straight bill, all-yellow underparts, broken broad pale yellowish eyering, and lack of white on tail. From Red-throated by brighter yellow underparts and more pronounced eyering. **VOICE** Song is an irregular, tailorbird-like ***wrick-wrick-wrick wrah wrick-wrick-wrick wrick-wrick wrah...*** Calls with a sharp ***too-wit*** (recalling Green Sandpiper). **HABITAT** Forest edge, mangroves, freshwater swamp forest, secondary growth, coastal scrub, plantations, gardens; lowlands. **RANGE & STATUS** Resident Greater Sundas, W Lesser Sundas, Philippines, Sulawesi, Sula Is. **SE Asia** Common resident (mostly coastal) SW,S Myanmar, Tenasserim, Thailand (except NW,NE), Peninsular Malaysia, Singapore, Cambodia, N(south),C,S Laos, S Annam, Cochinchina. **BREEDING** December–August. **Nest** Pendant- or pear-shaped with side-entrance towards top; 1.2–12 m above ground. **Eggs** 2; white to purplish-pink, squiggled purple-brown to purple-black over lavender to lilac-grey underblotches (mostly at broader end); c.17.3–17.8 x 12.7 mm. Brood-parasitised by Violet Cuckoo.

711 **RED-THROATED SUNBIRD**
Anthreptes rhodolaema Plate **75**
IDENTIFICATION 12.5–13 cm. Monotypic. **Male** Told from very similar Brown-throated Sunbird by pale brick-red throat, maroon-red head-sides and mostly chestnut-maroon wing-coverts, with only small metallic purple shoulder-patch. **Female** Difficult to separate from Brown-throated but plainer-faced, with greyish-tinged and less prominent eyering, breast-sides tinged greenish, underparts generally less bright yellow; may show orange tinge on throat, breast and occasionally wing-coverts. **VOICE** Rising high-pitched ***uu'is*** or ***tsuu'i***. **HABITAT** Broadleaved evergreen forest, forest edge; up to 900 m. **RANGE & STATUS** Resident Sumatra, Borneo. **SE Asia** Scarce to uncommon resident south Tenasserim, S Thailand, Peninsular Malaysia. **BREEDING** April–August. **Nest** Pendant-shaped or oval, with porched entrance towards top; 9–21 m above ground. **Eggs** Pale lavender-purple, marked with lines and a few large blotches of dark brown over some paler undermarkings; 16.4 x 12.4 mm.

712 **PURPLE-RUMPED SUNBIRD**
Leptocoma zeylonica Plate **75**
IDENTIFICATION 11.5 cm. *L.z.flaviventris*: **Male** Unmistakable, with mostly maroon head, upper breast and upperparts, iridescent green crown and shoulder-patch, purple throat and rump, and yellow lower breast and upper belly. **Female** Similar to Olive-backed and Purple Sunbirds but has pale greyish-white throat, upper breast and flanks. From Copper-throated by browner upperparts and shorter all-dark tail. **Juvenile** Like female but underparts completely pale lemon-yellow. **VOICE** Song is a sharp twittering ***tityou tityou tityou trr-r-r-tit tityou...*** or ***tu-witou-titou-switou...*** Calls include high-pitched ***ptsee*** or ***sisiswee*** notes and metallic ***chit***. **HABITAT** Forest edge, gardens, cultivation; lowlands. **RANGE & STATUS** Resident India (except W,NW), Bangladesh, Sri Lanka. **SE Asia** Fairly common resident SW Myanmar. **BREEDING** India: all year. **Nest** Pendant- or pear-shaped, usually with porched entrance; 2–15 m above ground. **Eggs** 2–3; greenish-grey to buffish-grey, speckled greyish-brown (mostly around broader end); 16.4 x 11.8 mm (av.; India).

713 **VAN HASSELT'S SUNBIRD**
Leptocoma brasiliana Plate **75**
IDENTIFICATION 10 cm. *L.b.brasiliana*: **Male** Small size, dark plumage, iridescent green crown, purple throat and dark red lower breast and upper belly diagnostic. **Female** From other sunbirds by combination of small size, dull olive-coloured upperside, plain head and dull yellow underparts with olive-washed throat and upper breast. **Other subspecies**

in **SE Asia** *N.s.emmae* (Indochina). **VOICE** Song is a high-pitched, rather discordant series of ***swit*** or ***psweet, psit-it*** and ***trr'rr*** notes. Also, high-pitched ***ti-swit titwitwitwitwitit...***, repeated, very thin ***tisisisit*** and ***si-si-si-si-si-si...*** **HABITAT** Open broadleaved evergreen forest, peatswamp forest, forest edge, secondary growth, coastal scrub, gardens, cultivation, sometimes mangroves; up to 1,200 m. **RANGE & STATUS** Resident NE India, Bangladesh, Sumatra, Java, Borneo. **SE Asia** Uncommon to fairly common resident SW,S Myanmar, Tenasserim, C,NE,SE,S Thailand, Peninsular Malaysia, Singapore, Cambodia, S Laos, S Annam, Cochinchina. Some local movements noted Singapore. **BREEDING** February–July. **Nest** Pendant- or pear-shaped; 1.5–6 m above ground. **Eggs** 2; glossy brown, finely spotted dark brown (often ringed or capped at broader end); 13–14 x 11 mm.

714 COPPER-THROATED SUNBIRD

Leptocoma calcostetha Plate **75**

IDENTIFICATION 14 cm. Monotypic. **Male** Relatively long-tailed; appears all dark with distinctive iridescent green crown, shoulder-patch and uppertail-coverts, and iridescent copper-red throat and upper breast. **Female** From other sunbirds by combination of greyish crown and head-sides, white throat, mostly yellow underparts, and longish tail with prominently white-tipped outer feathers. **VOICE** Inadequately documented. **HABITAT** Mangroves, coastal scrub, rarely away from coast, in secondary growth and cultivation. Up to 915 m. **RANGE & STATUS** Resident Sumatra, Java, Borneo, Philippines (Palawan). **SE Asia** Uncommon to locally common (mostly coastal) resident S Tenasserim, SE,S Thailand, Peninsular Malaysia, Singapore, Cambodia, Cochinchina. **BREEDING** January–September. **Nest** Pendant- or pear-shaped with oval, slightly porched entrance towards top; 0.6–3.5 m above ground. **Eggs** 2; sandy-grey to pale brown, densely freckled chocolate-brown (sometimes ringed towards broader end); 17.4 x 14.9 mm (av.).

715 PURPLE SUNBIRD

Cinnyris asiaticus Plate **75**

IDENTIFICATION 10.5–11.5 cm. *C.a.intermedia*: **Male** Uniformly dark; mostly iridescent dark bluish to purplish plumage diagnostic. **Male eclipse** Similar to Olive-backed Sunbird but shows mostly iridescent dark blue wing-coverts and darker wings. **Female** Similar to Olive-backed but belly and undertail-coverts paler yellow, often whitish, shows much less white on tail-tips. **Juvenile** Entire underparts yellow. **VOICE** Song is a pleasant descending ***swee-swee-swee swit zizi-zizi***. Calls with a buzzing ***zit*** and high-pitched, upward-inflected, slightly wheezy ***swee*** or ***che-wee***. **HABITAT** Deciduous woodland, bushy semi-desert, coastal scrub, gardens, cultivation; up to 1,525 m. **RANGE & STATUS** Resident Iran, Indian subcontinent, SW China (S Yunnan). **SE Asia** Common resident Myanmar, Thailand (except SE,S), Cambodia, N(south),C,S Laos, C,S Annam, north-west Cochinchina. **BREEDING** February–June. **Nest** Pendant- or pear-shaped, usually suspended from twig; 1–6 m above ground. **Eggs** 2–3; greyish to greenish-white, grey or brown, minutely freckled grey, brown and dull purple (sometimes ringed or capped at broader end); 16.3 x 11.6 (av.; India).

716 OLIVE-BACKED SUNBIRD

Cinnyris jugularis Plate **75**

IDENTIFICATION 11.5 cm. *C.j.flamaxillaris*: **Male** Told by plain olive-green upperparts, iridescent blue-black forehead, throat and breast, and yellow belly and vent. Has dark reddish-chestnut to rufous band across upper breast and extensive white on tail. *C.j.ornatus* (Peninsular Malaysia south of 5°30 in west, 3°50 in east) lacks rufescent breast-band, upper- and underparts yellower (also female); *rhizophorae* (northern Vietnam) has more whitish belly and duller upperparts. **Male eclipse** Iridescent blue-black restricted to stripe on centre of throat and breast. **Female** From other sunbirds by combination of obviously downcurved bill, all-yellow underside and extensive white on tail. May show faint yellow supercilium. **Other subspecies in SE Asia** *C.j.andamanicus* (Coco Is, off S Myanmar). **VOICE** Song consists of alternating, varied, thin, high-pitched phrases, rapidly delivered after intervals. Phrases include: ***tswi-tswit-titititit***; ***tswi-switswitswitswitswit***; ***tuittuittuittuit***; ***tuchi-tuchi-tuchi-tuchi***; ***tswit-tswit-tswit-tswit***; ***tswi-tswi-tswi-tswi-tswi-tswi***; ***tswitswitswi-trr-trr***; ***tswi-tisisisisisisisis*** and ***tswittswittswittswittswittswit-tit-titchu-chi*** etc. Calls with a loud rising ***sweet***. **HABITAT** Deciduous woodland, open forest, swamp forest, mangroves, coastal scrub, gardens, cultivation; up to 915 m. **RANGE & STATUS** Resident Andaman and Nicobar Is, SW,S China, Greater Sundas, W Lesser Sundas, Philippines, Sulawesi, Moluccas, New Guinea region, N Melanesia, NE Australia. **SE Asia** Common resident (except N Myanmar). **BREEDING** All year. Multi-brooded. **Nest** Untidy, pendant- or pear-shaped, with porched entrance; 1–12 m above ground. **Eggs** 2–3; pale grey, heavily marked with yellowish-brown or pale brown overall; 16.6 x 11.5 mm (av.; *andamanica*).

717 CRIMSON SUNBIRD

Aethopyga siparaja Plate **74**

IDENTIFICATION 11–13.5 cm. *A.s.seheriae*: **Male** Red head, mantle and upper breast, iridescent dark green crown and tail and greyish belly diagnostic. Has purple malar streak and yellow rump-band. See Temminck's Sunbird. **Male eclipse** Like female but with red throat and breast. **Female** Dullest sunbird in region; overall dull olive-coloured with slightly yellower-tinged underside; lacks paler rump-band and white on tail. **Juvenile** Like female. Males have red wash on throat and breast. **Other subspecies in SE Asia** *A.s.siparaja* (extreme S Thailand southwards), *tonkinensis* (E Tonkin), *mangini* (south-east NE Thailand, southern Indochina), *cara* (C,S,E[south] Myanmar, Tenasserim, NW[east],C,NE,SE Thailand), *trangensis* (W[south],S Thailand), *insularis* (Cochinchina [Phu Quoc Is]). **VOICE** Has rapid, tripping, sharp, high-pitched 3–6 note song: ***tsip-it-sip-it-sit*** etc. Calls with sharp ***whit***, ***tit*** and ***wit-it***. **HABITAT** Broadleaved evergreen and deciduous forest, forest edge, secondary growth, gardens; up to 1,370 m (lowlands only in south). **RANGE & STATUS** Resident E India, N,NE Indian subcontinent, Nicobar Is, SW,S China, Sumatra, Java, Borneo, Philippines, Sulawesi. **SE Asia** Common resident (except C Thailand). Mostly coastal and on offshore islands Peninsular Malaysia, Singapore. **BREEDING** February–August. **Nest** Oval to globular or pouch-shaped, attached to overhanging roots or suspended from branch or ginger-plant frond; 0.15–1.2 m above ground. **Eggs** 2–3; white or cream-coloured, speckled brown or reddish-brown (mainly at broader end); 15.1 x 11.4 mm (av.; India). Brood-parasitised by Asian Emerald Cuckoo.

718 TEMMINCK'S SUNBIRD

Aethopyga temminckii Plate **74**

IDENTIFICATION 10–12.5 cm. *A.t.temminckii*: **Male** Similar to Crimson Sunbird but head-sides, nape and mantle, wing-coverts, throat, breast and tail all scarlet. **Female** Similar to female Crimson but has yellowish-olive underparts and reddish-rufous fringes to wing- and tail-feathers. **VOICE** Song is a monotonous rhythmic ***tit-it tit-it tit-it tit-it tit-it tit-it...*** **HABITAT** Broadleaved evergreen forest, forest edge, secondary growth; up to 1,525 m. **RANGE & STATUS** Resident Sumatra, Borneo. **SE Asia** Scarce to fairly common resident S Thailand, Peninsular Malaysia. **BREEDING** February–August. **Nest** Typical of genus. **Eggs** 3.

719 MRS GOULD'S SUNBIRD

Aethopyga gouldiae Plate **74**

IDENTIFICATION 11–16.5 cm. *A.g.dabryii*: **Male** Combination of red upperparts, uniform red breast, and yellow rump-band and belly diagnostic. Crown, ear-coverts, throat and tail iridescent purple-blue. *A.g.isolata* (W

Myanmar) has yellow breast; *annamensis* (S Laos, S Annam) has yellow breast and iridescent blue rump (no yellow). See Black-throated (*A.s.johnsi*) and Green-tailed Sunbirds. **Male eclipse** Like female but retains red on breast (*dabryii*) and yellow belly. **Female** From other sunbirds (except Black-throated) by prominent yellow rump-band. From Black-throated by broader, brighter rump-band, brighter yellow belly, and broader pale tips on undertail. *A.g.annamensis* has distinctly greyish hood and upper breast, rump yellowish but not forming prominent band. **Juvenile** Like female. Male has yellower underparts. **VOICE** Song is monotonous series of high-pitched ***twit*** or ***tzip*** notes, at rate of c.2 per sec. Calls include a quickly repeated ***tzip***, lisping ***squeEEEee*** (rising in middle) and ***tshi-stshi-ti-ti-ti...*** when alarmed. **HABITAT** Broadleaved evergreen forest, forest edge, secondary growth; 1,000–2,565 m. **RANGE & STATUS** Resident N,NE Indian subcontinent, S,SE Tibet, SW,W,C,S China. **SE Asia** Common resident W,N,C,E,S(east) Myanmar, north Tenasserim, Laos, W,E(north-west) Tonkin, C(southern),S Annam. Fairly common winter visitor W,NW,NE(north-west) Thailand. Recorded (status uncertain) N Annam. **BREEDING** March–June. **Nest** Pear-shaped, with entrance towards top, suspended from tree-branch; 10 m above ground. **Eggs** 2–3; white, scantily freckled pale reddish-brown (mainly at broader end); 14.6 x 11.2 mm (av.; *isolata*).

720 GREEN-TAILED SUNBIRD

Aethopyga nipalensis Plate **74**

IDENTIFICATION 11–13.5 cm. *A.n.koelzi* (N,E Myanmar, W Tonkin): **Male** Similar to Mrs Gould's Sunbird but has iridescent dark green crown, throat and tail, dark red mantle, and yellow breast with orange-scarlet streaks. *A.n.karensis* (south E Myanmar) has yellow breast and red of upperparts restricted to upper mantle-sides; *angkanensis* (NW Thailand) has more defined scarlet patch on lower breast. **Female** Differs from Mrs Gould's and Black-throated Sunbirds by lack of sharply defined pale rump-band, tail graduated with more distinct whitish tips to blackish underside. From Crimson by yellower belly and tail pattern, rump brighter green than rest of upperparts. *A.n.karensis* is very similar to Mrs Gould's (*A.g.annamensis*) but bill longer, upperparts greener, rump less yellow, lower breast more solidly olive-green, tertial and secondary fringes less golden, tail less graduated. Note range. **Juvenile** Apparently as female but tail squarer with less white on tips. Male may have orange wash on breast. **Other subspecies in SE Asia** *A.n.victoriae* (W Myanmar), *australis* (S Thailand), *blanci* (south-east N Laos), *ezrai* (C,S Annam; presumably east S Laos). Race in W Thailand undocumented (new or *koelzi*?). **VOICE** Sings with a monotonous high-pitched, metallic ***wit-iritz wit-iri wit-iritz wit-iritz wit-iri wit-iritz...*** and ***tu-tsri tu-tsri tu-tsri tu-trsi*** etc. Calls with loud, high-pitched ***chit*** notes. **HABITAT** Broadleaved evergreen forest, forest edge, secondary growth; 1,400–2,745 m, down to 1,140 m in winter (above 915 m in S Thailand). **RANGE & STATUS** Resident N,NE Indian subcontinent, S,SE Tibet, SW China. **SE Asia** Locally common resident W,N,S(east),E Myanmar, W(Doi Mokoju),NW(Doi Inthanon),S(Khao Nong, Khao Luang) Thailand, N,C,S(east) Laos, W Tonkin, N,C(southern),S Annam. **BREEDING** February–June. **Nest** Similar to Mrs Gould's; up to 2 m above ground. **Eggs** 2–3; white, ringed with reddish-brown spots near broader end; 16.5 x 12.7 mm (Myanmar).

721 BLACK-THROATED SUNBIRD

Aethopyga saturata Plate **74**

IDENTIFICATION 11–15 cm. *A.s.petersi*: **Male** Only likely to be confused with Mrs Gould's and Green-tailed Sunbirds; differs by black throat and upper breast (former mixed with iridescent blue), pale yellow lower breast with light orange-scarlet streaks, and dull whitish-yellow belly and rump-band. Crown and tail iridescent dark purple-blue, mantle dark red. *A.s.johnsi* (S Annam) has brighter red mantle, orange-scarlet breast with narrow pale yellow streaks and entirely iridescent purple-blue throat; *assamensis* (W,N Myanmar) has all-black breast and darker, greyish belly; *wrayi* (Peninsular Malaysia) and *anomala* (S Thailand) are intermediate, latter lacking yellow rump-band. **Female** Greyer overall than Mrs Gould's, with paler, narrower rump-band, and distinctly duller underparts. From other sunbirds by sharply demarcated pale yellow rump-band. Undertail shows indistinct pale tips. **Other subspecies in SE Asia** *A.s.galenae* (southern NW Thailand), *sanguinipectus* (E[south],S[east] Myanmar, Tenasserim), *ochra* (S Laos, C Annam), *cambodiana* (south-west Cambodia). **VOICE** Song is an uneven series of sharp high-pitched ***swi***, ***tis*** and ***tsi*** notes, interspersed with rapid metallic trills: ***swi'it'it'it'it'it*** and ***swi'i'i'i'i'i'i*** etc. Calls with repeated, quick, high-pitched, thin ***tit***; ***tit-it*** and ***tiss-it*** etc. **HABITAT** Broadleaved evergreen forest, forest edge; 200–1,700 m. **RANGE & STATUS** Resident N,NE Indian subcontinent, SE Tibet, SW China. **SE Asia** Uncommon to common resident W,N,C(east),E,S(east) Myanmar, Tenasserim, Thailand (except C), Peninsular Malaysia, Indochina (except Cochinchina). **BREEDING** January–October. **Nest** Pear-shaped with porched entrance, suspended from twig; 1–3.6 m above ground. **Eggs** 1–2; white, speckled and spotted (occasionally blotched) inky-black over inky-grey undermarkings; 14.6 x 11.3 mm (av.; *assamensis*).

722 FIRE-TAILED SUNBIRD

Aethopyga ignicauda Plate **74**

IDENTIFICATION 11.5–19 cm. *A.i.ignicauda*: **Male** Similar to Mrs Gould's Sunbird but uppertail-coverts and tail red, tail-streamers much longer. No eclipse plumage. **Female** Similar to Green-tailed Sunbird but rump yellower, tail squarer and almost without pale tips below; often shows trace of brownish-orange on tail-sides. **First-winter male** Similar to female, but larger, has yellow back-patch and orange rump and tail. Can have orange tail-streamers when body still dull. **Other subspecies in SE Asia** *A.i.flavescens* (W Myanmar). **VOICE** Sings with repeated short bland high-pitched ***it'i'tit-tit'tut'tutututut*** etc. **HABITAT** Oak/rhododendron forest, rhododendron and juniper scrub, open broadleaved evergreen forest, forest edge, secondary growth; 1,220–3,960 m (breeds above 2,745 m), down to 700 m in winter N Myanmar. **RANGE & STATUS** Resident (subject to altitudinal movements) N,NE Indian subcontinent, S Tibet, SW China. **SE Asia** Common resident W,N Myanmar. Vagrant C Myanmar, NW Thailand. **BREEDING** India: April–June. **Nest** Similar to Green-tailed. **Eggs** 2–3; white with tiny brown blotches, coalescing in ring near broader end; 15.6 x 11.8 mm (av.; *ignicauda*).

723 FORK-TAILED SUNBIRD

Aethopyga christinae Plate **74**

IDENTIFICATION 10–12 cm. *A.c.latouchii*: **Male** Easily identified by combination of small size, iridescent green crown and tail, dark olive-coloured mantle, yellow rump-band, and dark crimson throat and upper breast. **Female** Like female Crimson Sunbird but crown greyish, belly and undertail-coverts yellowish (paler than breast), has prominent white tips to undertail. From Green-tailed by smaller size, shorter bill, yellow-olive throat and upper breast and less graduated tail. **VOICE** Calls with loud, sharp high-pitched, metallic ***tswit*** or ***twis*** notes and nervous ***wi'wi'wi'wi'wi'i'i*** and ***ts-wi'i'i'i'i'i'i***. **HABITAT** Broadleaved evergreen and semi-evergreen forest, forest edge, secondary growth; up to 1,400 m. **RANGE & STATUS** Resident W and southern China. **SE Asia** Fairly common resident C,S(east) Laos, W,E Tonkin, N,C,S Annam, north Cochinchina. **BREEDING** China: April–June. **Nest** Ball- or oval-shaped, in tree; 3 m above ground. **Eggs** 2-4; greenish-greyish, marked purplish, with reddish-brown tinge, and with dusky dots.

724 RUBY-CHEEKED SUNBIRD

Chalcoparia singalensis Plate **75**

IDENTIFICATION 10.5–11 cm. *C.s.assamensis*: **Male** Combination of mostly iridescent dark green upperside, iridescent copper-red ear-coverts, pale orange-rufous throat and upper breast, and yellow belly and vent diagnostic. *C.s.koratensis* (NE,SE Thailand, Indochina) has orange-rufous restricted and sharply demarcated on upper breast. **Female** Upperside mostly greenish-olive, underside similar to male. **Juvenile** Like female but underparts mostly yellow. **Other subspecies in SE Asia** *C.s.singalensis* (Peninsular Malaysia, Singapore), *internotus* (W[south],C[south] Thailand; Tenasserim?), *interposita* (S Thailand). **VOICE** Song is a rapid, high-pitched ***switi-ti-chi-chu tusi-tit swit-swit switi-ti-chi-chu switi-ti-chi-chu...*** Other calls include a thin rising ***swiiii***, sometimes extended to ***swit-si-swiiii***, thin tinny metallic ***tit-swit-swi*** and ***twee-twee***, and disyllabic rising ***wee-eest***, reminiscent of Yellow-browed Warbler. **HABITAT** Deciduous and broadleaved evergreen forests, peatswamp forest, secondary growth, mangroves, occasionally gardens. Up to 1,370 m. **RANGE & STATUS** Resident Nepal, NE Indian subcontinent, SW China, Sumatra, Java, Borneo. **SE Asia** Common resident (except E Myanmar, C Thailand, Singapore). Recorded (status uncertain) Singapore. **BREEDING** February–September. **Nest** Pendant- or pear-shaped with overhanging porch, suspended from twigs; 1.2–8 m above ground. **Eggs** 2; white, mottled and clouded with pale purplish-grey and a few purplish-black spots; 16.9 x 12 mm (av.; India). Brood-parasitised by Violet Cuckoo.

725 **PURPLE-NAPED SUNBIRD**
Hypogramma hypogrammicum Plate **74**
IDENTIFICATION 14–15 cm. *H.h.lisettae*: **Male** Easily told by relatively large size, plain greenish-olive upperside with iridescent purple-blue nuchal band, rump and uppertail-coverts, and boldly streaked underparts. *H.h.nuchale* (south Tenasserim and S Thailand southwards) has bluer nuchal band, rump and uppertail-coverts. See Streaked Spiderhunter. **Female** Lacks purple-blue markings on upperparts. **Other subspecies in SE Asia** *H.h.mariae* (Cochinchina). **VOICE** Song is a subdued short rapid forced high-pitched tinkling trill, repeated after longish intervals and interspersed with a short series of typical call-notes. Calls with repeated high ***CHIP*** or ***TCHU*** notes. **HABITAT & BEHAVIOUR** Broadleaved evergreen forest; up to 1,160 m. Frequents lower storey; movements relatively slow for a sunbird. **RANGE & STATUS** Resident SW China, Sumatra, Borneo. **SE Asia** Uncommon to fairly common resident (except SW,W,S,E Myanmar, C,NE,SE Thailand, Singapore). Vagrant Singapore. **BREEDING** December–October. **Nest** Pendant-shaped, attached to leaf; up to 6 m above ground. **Eggs** 2–3; pale lilac-grey, with squiggly lines and a few blotches of purple-black (mostly at larger end); 18 x 13.2 mm (Borneo).

726 **GREY-BREASTED SPIDERHUNTER**
Arachnothera affinis Plate **74**
IDENTIFICATION 18 cm. *A.a.modesta*: **Adult** Easily told by distinctive combination of bright olive-green upperparts and paler olive-greyish underparts, narrowly dark-streaked on throat and breast. **Juvenile** Underparts unstreaked. **VOICE** Calls with (often quickly) repeated harsh ***CHITTICK***, sometimes extending to more chattering ***TCHITITITEW*** and ***TCHEW-TEW-TEW*** when agitated. **HABITAT & BEHAVIOUR** Broadleaved evergreen forest, secondary growth, gardens, sometimes cultivation; up to 1,130 m. Often feeds in lower storey, particularly around banana plants. **RANGE & STATUS** Resident Sumatra, Java, Borneo. **SE Asia** Fairly common to common resident Tenasserim, W(south),S Thailand, Peninsular Malaysia; formerly Singapore. **BREEDING** March–July. **Nest** Long cup, sewn to underside of large leaf. **Eggs** 2; deep olive-brown, mottled and clouded grey, sometimes with ring of black spots and blotches around broader end; 21 x 15 mm.

727 **STREAKED SPIDERHUNTER**
Arachnothera magna Plate **74**
IDENTIFICATION 17–20.5 cm. *A.m.musarum*: **Adult** Heavily streaked head and body and orange legs and feet diagnostic. Base colour of upperside yellowish-olive, that of underside whitish. **Juvenile** Streaking somewhat less distinct. **Other subspecies in SE Asia** *A.m.magna* (SW,W,N Myanmar), *aurata* (central C,S Myanmar), *pagodarum* (central Tenasserim, south W Thailand), *remota* (S Annam). **VOICE** Has strident chattering song. Calls with loud, strident ***CHIT-IK***, particularly in flight; longer ***CHITITITITITIK*** when agitated. **HABITAT & BEHAVIOUR** Broadleaved evergreen and mixed deciduous forest, secondary growth. Up to 1,830 m (scarcer in lowlands); above 820 m in Peninsular Malaysia. Flight strongly undulating. **RANGE & STATUS** Resident Nepal, NE Indian subcontinent, SE Tibet, SW China. **SE Asia** Common resident Myanmar, W,NW,NE,extreme S Thailand, Peninsular Malaysia, east Cambodia, Laos, Vietnam. **BREEDING** March–October. **Nest** Inverted dome, fixed to underside of large leaf; c.2 m above ground. **Eggs** 2–3; brown or olive-brown with darker ring around broader end; 22.7 x 15.9 mm (av.; *magna*). Brood-parasitised by Large Hawk-cuckoo.

728 **LITTLE SPIDERHUNTER**
Arachnothera longirostra Plate **74**
IDENTIFICATION 16–16.5 cm. *A.l.longirostra*: **Adult** Combination of long downcurved bill, slaty-grey head with whitish lores/cheeks and broken eyering, whitish throat and yellow rest of underparts distinctive. Has dark moustachial stripe and white tips to outertail-feathers. **Juvenile** Throat tinged yellowish-olive. **Other subspecies in SE Asia** *A.l.sordida* (NW[east],NE[north-west] Thailand, N Indochina), *pallida* (SE Thailand, Cambodia), *cinereicollis* (S Thailand southwards). **VOICE** Song is a monotonous, rapidly repeated ***wit-wit-wit-wit-wit-wit...*** Calls with a loud, sharp, abrasive ***itch*** or ***chit***. **HABITAT & BEHAVIOUR** Broadleaved evergreen and semi-evergreen forest, forest edge, secondary growth, gardens, cultivation; up to 1,670 m. Usually hunts in lower storey, often around banana plants. **RANGE & STATUS** Resident SW,E India, Nepal, NE Indian subcontinent, SW China, Greater Sundas, Philippines. **SE Asia** Common resident (except C Thailand). **BREEDING** December–October. **Nest** Elongated structure with side-entrance, fixed to underside of leaf. **Eggs** 2–3; white to pink, finely spotted with pinkish-chestnut or reddish-brown (usually ringed at broader end); 18.4 x 13.1 mm (av.; *longirostra*) (Myanmar). Brood-parasitised by Large Hawk-cuckoo.

729 **THICK-BILLED SPIDERHUNTER**
Arachnothera crassirostris Plate **74**
IDENTIFICATION 16.5–17 cm. Monotypic. **Adult** Similar to Little Spiderhunter but bill thicker, throat and upper breast greyish-olive, has broad broken yellowish eyering and lacks white tips to tail-feathers. See Long-billed Spiderhunter. **Juvenile** Underparts greyer. **VOICE** Song is a monotonous stressed ***whit whit whit whit whit...*** or faster ***whit-whit-whit-whit-whit...*** Calls include chattering ***chit*** notes (softer than Grey-breasted Spiderhunter) and repeated worrisome, rather high ***ut-u-it-it-it-it-it-it***. **HABITAT & BEHAVIOUR** Broadleaved evergreen forest, forest edge, clearings, banana groves; up to 1,220 m (lowlands only in Thailand). Usually in canopy but descends to feed. **RANGE & STATUS** Resident Sumatra, Borneo. **SE Asia** Uncommon resident S Thailand, Peninsular Malaysia. Recorded (status uncertain) Singapore. **BREEDING** April–June. **Nest** Trough-shaped, fixed to underside of leaf. **Eggs** 2; white with black lines and spots (ringed at broader end); 17.9–18.4 x 12.5–13.1 mm.

730 **LONG-BILLED SPIDERHUNTER**
Arachnothera robusta Plate **74**
IDENTIFICATION 21.5–22 cm *A.r.robusta*: **Adult** From other spiderhunters by very long, strongly downcurved bill, uniform head-sides, yellowish-olive throat and breast with faint dark streaks and yellow belly; underside of tail blackish, tipped white. **Juvenile** Unstreaked. **VOICE** Series of up to 7 loud, quite deep ***CHUT***, ***CHIT*** or ***CHIP*** notes: ***CHUT CHUT CHUT...*** or ***CHUT-CHUT-CHUT-CHUT-CHUT...*** etc. **HABITAT** Broadleaved evergreen forest (usually canopy), forest edge; up to 1,280 m. **RANGE & STATUS** Resident

Sumatra, Java, Borneo. **SE Asia** Scarce to uncommon resident extreme south Tenasserim, south S Thailand, Peninsular Malaysia; unconfirmed south W Thailand. **BREEDING** February–July. **Nest** Trough-shaped, fixed to underside of leaf; 2–5 m above ground. **Eggs** 2; white with black lines and spots (ringed at broader end); 21.3–21.5 x 15.2 mm.

731 SPECTACLED SPIDERHUNTER
Arachnothera flavigaster Plate **74**

IDENTIFICATION 21.5–22 cm. Monotypic. **Adult** Relatively large size, robust shape, proportionately short thick broad-based bill, broad yellow eyering, and prominent isolated yellow patch on ear-coverts distinctive. See similar Yellow-eared Spiderhunter. **Juvenile** Bill-base, legs and feet flushed orange-red. **VOICE** Calls, particularly in flight, with a loud, rather deep ***CHIT'IT***, ***CHUT'UT***, or ***CHA-TAK***; sometimes extended ***CHA-TA-TAK***. **HABITAT** Broadleaved evergreen forest, forest edge, secondary growth; lowlands, occasionally up to 610 m. **RANGE & STATUS** Resident Sumatra, Borneo. **SE Asia** Fairly common to common resident W(south),S Thailand, Peninsular Malaysia; formerly Singapore. **BREEDING** February–September. **Nest** Round basket attached to underside of leaf; 5.5–10 m above ground. **Eggs** 2; pale clay-grey with slight greenish tinge, heavily flecked dark grey, or sepia and grey-brown (larger marks mostly at broader end); 23–25.6 x 17–17.7 mm.

732 YELLOW-EARED SPIDERHUNTER
Arachnothera chrysogenys Plate **74**

IDENTIFICATION 18 cm. *A.v.chrysogenys*: **Adult** Similar to Spectacled but smaller with proportionately longer, thinner bill, narrower yellow eyering and slightly larger yellow patch on cheeks/ear-coverts. Note voice. **Juvenile** Duller, cheek/ear-covert patch smaller or almost absent. **VOICE** Call is a harsh ***TCHICK*** or rough ***CHIT***. **HABITAT** Broadleaved evergreen forest (mostly canopy), forest edge, secondary growth, gardens; up to 1,830 m (mainly lowlands). **RANGE & STATUS** Resident Sumatra, Java, Borneo. **SE Asia** Fairly common to common resident S Tenasserim, W(south),S Thailand, Peninsular Malaysia. Formerly resident (current status unclear, but rare) Singapore. **BREEDING** January–May. Otherwise, not definitely documented.

DICAEIDAE: Flowerpeckers

Worldwide c.45 species. SE Asia 12 species. Very small to tiny, with shortish bills and tails and tongue adapted for nectar-feeding. Strictly arboreal and hyperactive, typically in tree-tops. Feed on small fruits, berries and nectar, also some small insects and spiders etc.

733 THICK-BILLED FLOWERPECKER
Dicaeum agile Plate **76**

IDENTIFICATION 10 cm. *D.a.modestum*: **Adult** Only likely to be confused with Brown-backed Flowerpecker. Differs by reddish to orange eyes, somewhat narrower bill, olive-green upperparts and paler, more clearly darker-streaked underparts; white-tipped underside of tail-feathers (sometimes showing on upperside) difficult to see in field. See Yellow-breasted Flowerpecker. **Juvenile** Bill more pinkish, underpart streaking less distinct. **Other subspecies in SE Asia** *D.a.remotum* (Peninsular Malaysia, Singapore). **VOICE** Thin ***pseeou*** (often in flight), unlike other flowerpeckers in region. **HABITAT & BEHAVIOUR** Broadleaved evergreen, semi-evergreen and mixed deciduous forest, secondary growth; up to 1,500 m. Usually in tree-tops. Has habit of twisting (wagging) tail from side to side. **RANGE & STATUS** Resident NE Pakistan, India, S Nepal, E Bangladesh, Sri Lanka, Sumatra, Java, Borneo, Lesser Sundas. **SE Asia** Uncommon to common resident (except SW,W,N Myanmar, Singapore, W,E Tonkin). Scarce non-breeding visitor Singapore. **BREEDING** January–June. **Nest** Bag-shaped, with entrance towards top, in bush or tree; up to 3 m above ground. **Eggs** 2–4; pale to deep pink, freckled bright reddish-brown over lavender to pale purplish-brown undermarkings; 15.9 x 11.5 mm (av.; India).

734 BROWN-BACKED FLOWERPECKER
Dicaeum everetti Plate **76**

IDENTIFICATION 10 cm. *D.e.sordidum*: **Adult** Similar to Thick-billed but eyes paler, bill somewhat thicker, head-sides browner, without obvious dark moustachial line, upperparts dark earth-brown, underpart-streaking fainter, belly greyer/browner, wing-feather fringes less contrastingly pale; almost completely lacks whitish tips on underside of tail-feathers. **Juvenile** Unknown. **VOICE** Inadequately documented. **HABITAT** Edge of broadleaved evergreen forest, swamp forest; lowlands. **RANGE & STATUS** Sumatra (Riau Archipelago), Borneo. **SE Asia** Rare resident Peninsular Malaysia. **BREEDING** Not known in region. Borneo: **Nest** Small pouch, suspended from end of tree-branch; 5-6 m above ground. Otherwise undocumented.

735 YELLOW-BELLIED FLOWERPECKER
Dicaeum melanoxanthum Plate **75**

IDENTIFICATION 13 cm. Monotypic. **Male** Unmistakable. Relatively large, crown, upperparts, sides of head and breast black, throat and centre of breast white, rest of underparts yellow. **Female** Patterned like male but head and breast-sides dull olive-greyish, upperparts dull greyish-brown, yellow of underparts duller and paler. **Juvenile male** Similar to female but has brighter yellow on underparts and blue-black cast to mantle and back. **VOICE** Calls with agitated ***zit*** notes. **HABITAT & BEHAVIOUR** Broadleaved evergreen forest, forest edge; 1,200–2,500 m. Often sits upright on exposed perch for prolonged periods. Sallies for insects. **RANGE & STATUS** Resident N,NE Indian subcontinent, SW China. **SE Asia** Scarce to uncommon resident W,E Myanmar, NW,NE(north-west) Thailand, N Laos, W Tonkin. **BREEDING** Unknown.

736 YELLOW-BREASTED FLOWERPECKER
Dicaeum maculatus Plate **76**

IDENTIFICATION 9.5–10 cm. *D.m.septentrionalis*: **Adult** Easily identified by yellowish underparts with broad olive-green streaks. Upperparts distinctly greenish, with orange crown-patch, bill thick and all dark. See Yellow-vented and Thick-billed Flowerpeckers. **Juvenile** Lacks crown-patch, bill pinkish with darker culmen, underparts paler yellow and less boldly streaked. **Other subspecies in SE Asia** *D.m.oblitus* (Peninsular Malaysia, Singapore). **VOICE** Repeated sibilant, high-pitched, silvery ***tisisisit***. **HABITAT** Lower to middle storey of broadleaved evergreen forest, secondary growth; up to 1,600 m. **RANGE & STATUS** Resident Sumatra, Borneo. **SE Asia** Common resident south Tenasserim, W(south),S Thailand, Peninsular Malaysia; formerly Singapore. **BREEDING** April–October. **Nest** Oval or purse-shaped, suspended from bush or sapling branch; 2.5 m above ground. **Eggs** 2; glossy white, thickly blotched and spotted brown, with zone around broader end.

737 CRIMSON-BREASTED FLOWERPECKER
Dicaeum percussus Plate **75**

IDENTIFICATION 10 cm. *D.p.ignicapilla*: **Male** Unmistakable, with slaty-blue upperparts, bright yellow underparts and red crown- and breast-patches. Shows prominent white submoustachial stripe. **Female** Rather nondescript. Told from other flowerpeckers by combination of thick bill, dull orange crown-patch, unstreaked greyish-olive underparts with bright yellow central stripe and indistinct whitish submoustachial stripe. **Juvenile** Duller and more uniformly olive-coloured than female, bill mostly pinkish, centre of belly whitish with slight yellow tinge. **VOICE** Repeated sharp high-pitched ***teez tit-tit***, with stressed first note, and buzzy ***whit-whit*** or ***vit-vit***. **HABITAT** Broadleaved evergreen forest (usually canopy), forest edge; up to 1,200 m. **RANGE & STATUS** Resident Sumatra, Java, Borneo. **SE Asia** Uncommon to fairly common resident south Tenasserim, S Thailand, Peninsular Malaysia. **BREEDING** March–June. **Nest** Oval or purse-shaped. **Eggs** White.

738 SCARLET-BREASTED FLOWERPECKER
Dicaeum thoracicus Plate **75**

IDENTIFICATION 10 cm. Monotypic. **Male** Easily told by black head, red crown-patch, yellowish rump and uppertail-coverts and bright yellow underparts with large red breast-patch encircled by black. **Female** From other flowerpeckers by combination of thick bill, mostly greyish head, yellowish-tinged rump and uppertail-coverts, whitish throat and submoustachial stripe, and yellow underparts with variable reddish-orange wash and greyish-olive mottling on centre of breast. **Juvenile** Much duller and more greyish-olive below than female, with narrow yellowish strip down centre of abdomen. Very like Crimson-breasted but pale strip on centre of abdomen tends to be more clearly defined and yellower; wing-feather fringes less green. **VOICE** Undocumented. **HABITAT & BEHAVIOUR** Broadleaved evergreen forest (usually canopy), swamp forest; up to 1,280 m. **RANGE & STATUS** Resident Sumatra, Borneo. **SE Asia** Rare to uncommon resident southern S Thailand, Peninsular Malaysia. **BREEDING** April–September and December. **Nest** Oval or purse-shaped, suspended from bush branch. Otherwise undocumented.

739 YELLOW-VENTED FLOWERPECKER
Dicaeum chrysorrheum Plate **76**

IDENTIFICATION 10–10.5 cm. *D.c.chrysochlore*: **Adult** From other streaked flowerpeckers by narrower, slightly downcurved bill, whitish loral stripe, clean-looking whitish underparts with bold, well-defined blackish streaks and yellow to orange-yellow undertail-coverts; upperparts brighter yellowish-olive. **Juvenile** Underparts greyer with less contrasting greyish-brown streaks, undertail-coverts pale yellow. **Other subspecies in SE Asia** *D.c.chrysorrheum* (S Thailand southwards). **VOICE** Briskly repeated ***zet zet zet zet zet zet...***, stretching to ***zeet zeet zeet...*** in flight. Short harsh ***dzeep***. **HABITAT** Broadleaved evergreen, semi-evergreen and mixed deciduous forest, forest edge, secondary growth, gardens; up to 1,300 m. **RANGE & STATUS** Resident Nepal, NE Indian subcontinent, SW China, Greater Sundas. **SE Asia** Fairly common resident throughout; rare Singapore. **BREEDING** December–August. **Nest** Oval-shaped, suspended from bush or tree-branch; up to 12 m. **Eggs** 2–3; white; 15.3 x 11 m (av.; India).

740 ORANGE-BELLIED FLOWERPECKER
Dicaeum trigonostigma Plate **75**

IDENTIFICATION 9 cm. *D.t.rubropygium*: **Male** Distinguished by slaty-blue crown, nape, wings and tail, grey throat and upper breast, and orange lower breast, lower mantle and back, grading to orange-yellow on uppertail-coverts and vent. **Female** Much duller overall. Told by combination of rather slender, slightly downcurved bill, plain head, orange-yellow to yellowish rump, greyish throat and upper breast, and dull yellow to orange-yellow belly-centre and vent. **Juvenile** Similar to female but throat and breast more olive-tinged. **Other subspecies in SE Asia** *D.t.trigonostigma* (southern S Thailand southwards). **VOICE** Song is a high-pitched, slightly descending ***tsi-si-si-si-sew***. Calls with harsh ***dzip*** notes. **HABITAT** Edge of broadleaved evergreen forest, secondary growth, gardens, cultivation; up to 915 m, locally 1,525 m. Usually in tree-tops. **RANGE & STATUS** Resident S Bangladesh, Greater Sundas, Philippines. **SE Asia** Fairly common to common resident SW,S,E(south) Myanmar, Tenasserim, W(south),S Thailand, Peninsular Malaysia, Singapore. **BREEDING** All year. **Nest** Pendant with entrance towards top, suspended from leaf or branch; 1.5–12 m above ground. **Eggs** 2–3; white, sometimes with few to many dark speckles at broader end; 15.5 x 11.2 mm (av.; *rubropygium*).

741 PALE-BILLED FLOWERPECKER
Dicaeum erythrorynchos Plate **76**

IDENTIFICATION 8.5–9 cm. *D.e.erythrorynchos*: **Adult** Similar to Plain Flowerpecker but has pinkish bill, paler, greyer upperparts and slightly paler underparts. Could be confused with juvenile Scarlet-backed Flowerpecker but has completely pinkish bill, uniform paler, greyer upperparts and paler head-sides. **Juvenile** Bill slightly yellowish-tinged, somewhat greyer above and below. **VOICE** Short, sharp, ticking ***chik*** or ***SHTIk*** and high-pitched ***pseep***. **HABITAT & BEHAVIOUR** Open broadleaved forest, plantations, gardens; lowlands. Usually in tree-tops. **RANGE & STATUS** Resident Indian subcontinent (except W and Pakistan). **SE Asia** Fairly common resident SW,W,S Myanmar. Records need substantiation. **BREEDING** February–June. **Nest** Oval purse, suspended from twig; 3–12 m above ground. **Eggs** 2–3; white; 14.4 x 10.5 mm (av.).

742 PLAIN FLOWERPECKER
Dicaeum minullum Plate **76**

IDENTIFICATION 8–8.5 cm. *D.m.olivaceum*: **Adult** Nondescript, with greenish-olive upperside, pale olive-greyish underside, cream-coloured belly-centre and sometimes throat. From Pale-billed and female Fire-breasted and Scarlet-backed Flowerpeckers by combination of dark bill, relatively pale head-sides, greenish-olive rump and uppertail-coverts, lack of buff on underparts and yellowish-white pectoral tufts (if visible). **Juvenile** Bill pinkish to yellowish-pink with darker culmen. **Other subspecies in SE Asia** *D.m.borneanum* (Peninsular Malaysia). **VOICE** Sings with repeated, high-pitched ***tsit tsi-si-si-si-si*** or ***tsit'tsit'tsit-si-si-si-si***. Also utters monotonous ***tu-wit tu-wit tu-wit tu-wit...*** Calls include a rapid, very dry, thin ticking, at rate of c.3 per sec. **HABITAT** Open broadleaved evergreen, semi-evergreen and deciduous forest, secondary growth; up to 1,700 m. Usually in tree-tops. **RANGE & STATUS** Resident N,NE Indian subcontinent, SE Tibet, southern China, Taiwan, Greater Sundas. **SE Asia** Fairly common resident (except S Thailand, Cochinchina); formerly Singapore. **BREEDING** February–August. **Nest** Oval purse, suspended from twig; 6–12 m above ground. **Eggs** 2–3; white; 14.5 x 10.6 mm (av.). **NOTE** Split from *D.concolor* (Nilgiri Flowerpecker) following Rasmussen & Anderton (2005).

743 FIRE-BREASTED FLOWERPECKER
Dicaeum ignipectus Plate **76**

IDENTIFICATION 8–9 cm. *D.i.ignipectus*: **Male** Easily told by glossy dark greenish-blue upperparts and buffish underparts with red breast-patch and black line on belly-centre. Sides of head and breast blackish. *D.i.cambodianum* (SE Thailand, western Cambodia) lacks red breast-patch. **Female** May be confused with several other flowerpeckers. Best identified by combination of dark bill, rather dark greenish-olive upperparts and head-sides, and uniform buffish underside with olive-tinged flanks. **Juvenile** Like female but basal half of lower mandible paler, underparts lack buff and appear greyish across breast, belly-centre more yellowish. From Plain by darker bill and head-sides. See Scarlet-backed Flowerpecker. Note habitat. **Other subspecies in SE Asia** *D.i.dolichorhynchum* (S Thailand, Peninsular Malaysia). **VOICE** Song is a high-pitched shrill ***tissit tissit tissit tissit...*** or ***titty-titty-titty-titty...***, similar to Scarlet-backed. Also, a high, strident ***see-bit see-bit see-bit....*** Calls with sharp ***dik*** or ***tsit*** notes. **HABITAT & BEHAVIOUR** Broadleaved evergreen forest, secondary growth; 450–2,565 m. Usually in tree-tops. **RANGE & STATUS** Resident N,NE Indian subcontinent, SE Tibet, C and southern China, Taiwan, Sumatra, Philippines. **SE Asia** Common resident (except SW Myanmar, C Thailand, Cochinchina). **BREEDING**

March–August. **Nest** Pendant-shaped, suspended from branch; 3–9 m above ground. **Eggs** 2–3; white; 14.8 x 10.3 mm (av.; *ignipectus*).

744 SCARLET-BACKED FLOWERPECKER
Dicaeum cruentatum Plate **76**

IDENTIFICATION 8.5–9 cm. *D.c.cruentatum*: **Male** Easily identified from other flowerpeckers by bright red crown and upperparts, which contrast sharply with blackish sides of head and breast and glossy blue-blackish wings. **Female** Bright red rump and uppertail-coverts diagnostic. Otherwise similar to Plain Flowerpecker but shows greyer head and mantle and white throat and centre of underparts. **Juvenile** Similar to female but basal half of bill mostly reddish-pink, upperparts uniform with orange-tinged uppertail-coverts (lacks red on rump and uppertail-coverts), throat and breast more uniform greyish. See Plain and Fire-breasted Flowerpeckers. Note habitat. **VOICE** Song is a thin ***tissit tissit tissit...*** Calls with a hard metallic ***tip*** (often quickly repeated) and thin metallic ***tizz*** and ***tsi*** notes (latter may be given in long series). **HABITAT** Open forests, forest edge, secondary growth, parks, gardens, cultivation; up to 1,220 m. Usually in tree-tops. **RANGE & STATUS** E Nepal, NE Indian subcontinent, southern China, Sumatra, Borneo. **SE Asia** Common resident throughout. **BREEDING** All year. **Nest** Oval purse, suspended from leaf or branch; 2–20 m above ground. **Eggs** 2–3; white; 14 x 10.3 mm (av.).

CHLOROPSEIDAE: Leafbirds

Worldwide 10 species. SE Asia 5 species. Small to medium-sized, with fairly slender bills and shortish tails. Sexually dimorphic. Largely bright green, males with mostly black faces and throats. Vocalisations complex, incorporating much mimicry. Arboreal, frequenting thick foliage and canopy. Feed on insects, berries and nectar.

745 GREATER GREEN LEAFBIRD
Chloropsis sonnerati Plate **76**

IDENTIFICATION 20.5–22.5 cm. *C.s.zosterops*: **Male** Stout bill, all-green plumage apart from black face and bib and purple-blue malar band distinctive. Has blue shoulder-patch (usually concealed in field). Very similar to Lesser Green Leafbird but larger, with much longer, heavier bill, no yellowish forehead or border to black bib. **Female** From other leafbirds by size, stout bill and all-green plumage with sharply demarcated yellow throat and eyering. Blue malar band rather faint. See juvenile Lesser Green. **Juvenile** Similar to female but blue malar band faint or absent; has yellowish submoustachial band. **VOICE** Song is a series of liquid musical whistles, interspersed with brief chattering notes, ***wi-i chaka-wiu chi-wiu...*** etc. **HABITAT & BEHAVIOUR** Broadleaved evergreen forest, sometimes mangroves; up to 915 m. Normally frequents middle to upper storey of forest, like other leafbirds. **RANGE & STATUS** Resident Greater Sundas. **SE Asia** Fairly common to common resident south Tenasserim, W(south),S Thailand, Peninsular Malaysia, Singapore (scarce). **BREEDING** March–July. **Nest** Cup-shaped, near end of branch; 6 m above ground. Otherwise undocumented.

746 LESSER GREEN LEAFBIRD
Chloropsis cyanopogon Plate **76**

IDENTIFICATION 16–19 cm. *C.c.septentrionalis*: **Male** Very similar to Greater Green Leafbird but smaller and much smaller-billed; has yellowish forehead and border to black bib and lacks blue shoulder-patch. *C.c.cyanopogon* (extreme S Thailand southwards) has much less yellow on forehead and bordering bib; more similar to Greater Green. **Female** All-green with golden-green forehead and blue to purplish-blue malar band. Somewhat deeper green than Blue-winged, with no blue on shoulder, wings or tail. From juvenile Golden-fronted by greener forehead and shorter, less curved bill. See Greater Green and Orange-bellied Leafbirds. Note habitat and range. **Juvenile** Very similar to Greater Green but smaller and much smaller-billed, yellow of throat less well defined, no obvious eyering. Males gradually attain blackish patches on bib. **VOICE** Loud, rich and varied warbling song, intermixed with deep, mellow notes. **HABITAT** Broadleaved evergreen forest, open forest, forest edge; up to 700 m. **RANGE & STATUS** Resident Sumatra, Borneo. **SE Asia** Common resident south Tenasserim, W(south),S Thailand, Peninsular Malaysia, Singapore (scarce). **BREEDING** March–July. Otherwise undocumented.

747 BLUE-WINGED LEAFBIRD
Chloropsis cochinchinensis Plate **76**

IDENTIFICATION 16.5–18.5 cm. *C.c.chlorocephala*: **Male** From other green-bodied leafbirds by extensive glossy blue to turquoise-blue on outer fringes of secondaries and primaries, leading edge of wing- and tail-feathers, and bright, lighter blue shoulder-patch. Head pattern similar to other leafbirds, with yellowish border to black face and bib, but has more yellowish-bronze head and upper breast. *C.c.moluccensis* (extreme S Thailand southwards) has even yellower forecrown and border to bib and more defined yellowish-bronze or golden nape; *serithai* (S Thailand, possibly southern Tenasserim) is more or less intermediate. **Female** Head all greenish with diffuse blue malar band, throat and cheeks often tinged blue, has less blue on wings and tail (much reduced or lacking on wing-coverts, except shoulder). Similar to other leafbirds but blue on wings and tail diagnostic. Crown and nape more golden-tinged than Lesser Green, malar band less defined. *C.c.moluccensis* tends to show more defined yellowish-bronze or golden nape. **Juvenile** Similar to female but almost lacks any blue on face and malar area, crown and nape greener. See Orange-bellied. **Other subspecies in SE Asia** *C.c.cochinchinensis* (SE,NE[south] Thailand, southern Indochina); *kinneari* (NE(north) Thailand, northern Indochina). **VOICE** Song consists of various musical liquid notes, ***pli-pli-chu-chu*** and ***chi-chi-pli-i*** etc. Also utters a high ***chi-chi-chi*** and ***chi'ii*** and slightly rattling ***pridit***. **HABITAT & BEHAVIOUR** Broadleaved evergreen and mixed deciduous forest, forest edge, secondary growth; up to 1,500 m. Often seen in small parties, regularly associating with mixed-species feeding flocks. **RANGE & STATUS** Resident NE India, south-east Bangladesh, SW China, Greater Sundas. **SE Asia** Common resident throughout; scarce (status uncertain) Singapore. **BREEDING** April–August. **Nest** Fairly deep cup, suspended (hammock-like) from twigs of tree-branch; 6–9 m above ground. **Eggs** 2–3; pale cream to pinkish-white, sparingly speckled, blotched and streaked blackish, purplish and reddish-brown (mainly around broader end); 22.3 x 15.5 mm (av.; *chlorocephala*).

748 GOLDEN-FRONTED LEAFBIRD
Chloropsis aurifrons Plate **76**

IDENTIFICATION 18–19 cm. *C.a.pridii*: **Adult** Similar to male of other green-bodied leafbirds (both sexes) but shows diagnostic shining golden-orange forecrown, purple-blue upper throat and malar area, and rather slender, slightly downcurved bill. Has broad yellowish lower border to bib and turquoise-blue shoulder-patch; lacks blue elsewhere on wings and tail. *C.a.incompta* (W[south] Thailand, S Laos, C,S Annam) and *inornata* (W,C,NE,SE Thailand, Cambodia, Cochinchina) lack obvious yellow lower border to bib. **Juvenile** All green, with purple-blue and black malar band and blue on shoulder. From other female and juvenile leafbirds by contrasting golden-yellowish forehead (duller than adult), bill shape, and lack of blue on wings (apart from shoulder) and tail. Gradually attains patches of adult plumage. See juvenile Orange-bellied Leafbird. Note habitat and range. **Other subspecies in SE Asia** *C.a.aurifrons* (SW,W,N,C,S[west] Myanmar). **VOICE** Song is complex, squeaky and scratchy but quite melodious, incorporating mimicry. **HABITAT** Dry dipterocarp and mixed deciduous forest, sometimes broadleaved evergreen and semi-evergreen forest, secondary growth; up to 1,220 m, rarely 1,560 m. **RANGE & STATUS** Resident peninsular India, NW,N,NE Indian subcontinent, Sri Lanka, SW China, Sumatra. **SE Asia** Common resident (except S Thailand, Peninsular Malaysia, Singapore, W,E Tonkin). **BREEDING** February–September. **Nest** Rather small shallow cup, suspended from outer branches of tree; 9–12 m above ground. **Eggs** 2–3; pale cream to buffy-cream, freckled pale reddish (usually more around broader end); 23.4 x 15.5 mm (av.; *aurifrons*).

749 ORANGE-BELLIED LEAFBIRD

Chloropsis hardwickii Plate **76**

IDENTIFICATION 18.5–20.5 cm. *C.h.hardwickii*: **Male** Easily separated from other leafbirds by dull yellowish-orange lower breast to vent. Black of bib extends broadly onto upper breast (where glossed dark bluish-purple), has deep bluish-purple outer fringes to primaries, secondaries and outer wing-coverts, pale turquoise-blue shoulder-patch and mostly bluish-purple tail. *C.h.melliana* (C Laos, E Tonkin, N,C Annam) is slightly duller, with greyish-blue tinge to crown, darker blue shoulder-patch and more strongly bluish-purple upper breast. **Female** Similar to other leafbirds, with overall green plumage, but has diagnostic dull yellowish-orange centre to abdomen and undertail-coverts and broad purplish-blue malar band. Has much less bluish-purple on wings and tail than male. *C.h.melliana* has uniform green underparts and dull turquoise cast to forecrown and ear-coverts. From other leafbirds by bluish colour on crown and ear-coverts, broader and purpler malar band, darker green upperside and slightly downcurved bill. Note habitat and range. **Juvenile** Underparts uniform light green, very little blue on malar area, no blue shoulder-patch. Similar to Blue-winged Leafbird but upperside darker green, without obvious pale blue on wings and tail, bill longer and slightly downcurved. Shows some dull blue on outer fringes of outer primaries and outertail-feathers. See Lesser Green and Golden-fronted Leafbirds. Note habitat and range. **Other subspecies in SE Asia** *C.h.malayana* (Peninsular Malaysia). Birds in S Annam may be of an undescribed subspecies. **VOICE** Song is highly variable. Different types include: jumpy phrases of mixed *chip*, *tsi*, *chit* and *chi* notes; monotonous mixture of shrill ***shrittitit*** and ***shrit*** notes; monotonous ***chit-wiu chit-wiu chit-wiu...*** and rippling, melodious ***chip-chip-chip-chip-irr chirriwu-i pichu-pi*** etc. Also includes a wide variety of avian mimicry. Calls include a loud, stressed ***whip whip dzrii dzrii dzrii dzrii*** when alarmed and loud ***CHISSICK*** in flight, recalling Streaked Spiderhunter. **HABITAT** Broadleaved evergreen forest, forest edge; 600–2,135 m, locally down to 200 m. **RANGE & STATUS** Resident N,NE Indian subcontinent, southern China. **SE Asia** Uncommon to locally common resident Myanmar, W,NW,NE Thailand, Peninsular Malaysia, N,C Laos, Vietnam (except Cochinchina). **BREEDING** March–June. **Nest** Shallow cup, suspended from outer branches of tree; 3–9 m above ground. **Eggs** 2–3; similar to Golden-fronted; 22.8 x 15.9 mm (av.; *hardwickii*).

IRENIDAE: Fairy-bluebirds

Worldwide 2 species. SE Asia 1 species. Medium-sized, with extensively iridescent blue plumage. Sexually dimorphic. Arboreal, frequenting thick foliage and canopy. Simple vocalisations. Feeds mostly on fruit and nectar.

750 ASIAN FAIRY-BLUEBIRD

Irena puella Plate **76**

IDENTIFICATION 24.5–26.5 cm. *I.p.puella*: **Male** Unmistakable. Bulbul-like with shining deep blue upperparts and undertail-coverts and blackish sides of head, remainder of underparts, tail and wings (apart from median and lesser coverts). Eyes red. **Female** Overall dull turquoise-blue with blackish tail and flight-feathers. Eyes reddish. See Maroon-breasted Philentoma. **Juvenile** Similar to female but duller and browner-winged. Male gradually attains patches of adult plumage. **Other subspecies in SE Asia** *I.p.malayensis* (extreme S Thailand southwards). **VOICE** Loud, liquid ***tu-lip wae-waet-oo***. Various other loud liquid notes, ***wi-it***; ***wait***; ***pip*** etc. and quavering ***u-iu***. **HABITAT & BEHAVIOUR** Broadleaved evergreen forest, less often in mixed deciduous forest; up to 1,525 m. Often encountered near fruiting trees in small flocks. **RANGE & STATUS** Resident S India, NE Indian subcontinent, Sri Lanka, Andaman Is, SW China, Greater Sundas, Philippines (Palawan). **SE Asia** Fairly common to common resident (except C Thailand). **BREEDING** February–July. **Nest** Shallow cup, in tree or bush; 1–5 m above ground, or higher. **Eggs** 2–3; olive-grey to olive-white, irregularly streaked and blotched (often mainly at broader end); 28.2 x 20.2 mm (av.; *puella*).

PRUNELLIDAE: Accentors

Worldwide 13 species. SE Asia 3 species. Small and pipit-like with slender pointed bills but relatively short tails. Unobtrusive and largely terrestrial; hop or move with shuffling gait. Feed mostly on insects in summer and seeds in winter.

751 ALPINE ACCENTOR

Prunella collaris Plate **78**

IDENTIFICATION 18 cm. *P.c.nipalensis*: **Adult** Robust and rounded with brownish-grey head and breast and distinctive chestnut-streaked flanks and black-and-white barred throat. Rest of upperparts buffish-brown with broad blackish streaks, shows two white wing-bars and whitish tail-tips. **Juvenile** Similar to adult but head, breast and upperparts browner, flank-streaks dark brownish. **VOICE** Song is a varied, well-sustained, chattering warble. Calls with a clear, rippling ***truiririp*** and ***turrr***, shorter ***zuju*** and ***tju-tju-tju*** etc. and sparrow-like chirping sounds. **HABITAT & BEHAVIOUR** Rocky places, stony gullies, sparsely vegetated open areas; above 2,440 m. Slow-moving, usually creeping and hopping on ground, often very tame. **RANGE & STATUS** Resident (subject to local movements), NW Africa, southern Palearctic, N Pakistan, NW,N,NE Indian subcontinent, Tibet, W and northern China, Taiwan. **SE Asia** Recorded (status uncertain) N Myanmar. **BREEDING** India: June–August. **Nest** Compact cup, amongst rocks or stones, usually on or near ground. **Eggs** 3; blue; 21–23 x 16 mm.

752 RUFOUS-BREASTED ACCENTOR

Prunella strophiata Plate **78**

IDENTIFICATION 15 cm. *P.s.strophiata*: **Adult** Dark-streaked plumage recalls some pipits but shows diagnostic orange-rufous supercilium, contrasting dark lores and ear-coverts, orange-rufous breast and rufous-buff flanks. **Juvenile** Lacks orange-rufous on supercilium and breast. Supercilium buffish-white, ear-coverts paler than adult, upperparts less contrastingly streaked, breast washed buff and heavily dark-streaked. Small size, streaked throat, rufescent fringing on wings, and habitat and behaviour help eliminate small pipits. **VOICE** Has melodious song, recalling Winter Wren but less shrill and vehement, and with some pretty trills and warbling. Calls with penetrating rattling ***trrrrrt*** and ***trrr'rit***. **HABITAT & BEHAVIOUR** Forest edge, scrub and thickets, borders of cultivation; 2,135–2,590 m. Usually on ground or skulking in thick bushes. **RANGE & STATUS** Resident (subject to altitudinal movements) E Afghanistan, northern Pakistan, NW,N,NE Indian subcontinent, S,E Tibet, W China. **SE Asia** Recorded (status uncertain) N Myanmar. **BREEDING** India: May–August. **Nest** Deep cup, in shrub or lower branch of conifer; up to 3 m above ground. **Eggs** 3–5; turquoise-blue; 19.4 x 14.4 mm (av.).

753 MAROON-BACKED ACCENTOR
Prunella immaculata Plate **78**

IDENTIFICATION 16.5 cm. Monotypic. **Adult** Unmistakable, with dark grey head, blackish face, contrasting whitish to yellowish-white eyes and extensively dark chestnut scapulars and wings with contrasting grey greater coverts. Rest of upperparts brownish-olive with maroon-tinged lower back, underparts grey with rufescent flanks and undertail-coverts. **Juvenile** Head olive-greyish, upperparts warm brownish with dark streaks/mottling, throat whitish with blackish spots, breast pale buff with dark brown streaks; wings similar to adult. **VOICE** Calls with feeble, thin, metallic ***tzip*** and ***zieh-dzit***. **HABITAT & BEHAVIOUR** Forest edge, scrub; 1,675–2,135 m. Unobtrusive, usually on ground. **RANGE & STATUS** Resident Nepal, Bhutan, NE India, SE Tibet, W China. **SE Asia** Recorded (status uncertain) N Myanmar. **BREEDING** India: May–July: **Nest** Cup, on ground or low down in bush or brambles. **Eggs** 3–5; blue; 19.5 x 14.6 mm (av.).

PLOCEIDAE: Weavers & allies

Worldwide c.108 species. SE Asia 4 species. Small, plump and finch-like, with large conical bills. Sexually dimorphic, males with yellow in plumage when breeding. Highly gregarious. Feed mainly on seeds and grain, supplemented with small invertebrates.

754 BLACK-BREASTED WEAVER
Ploceus benghalensis Plate **82**

IDENTIFICATION 14 cm. Monotypic. **Male breeding** Combination of yellow crown and broad blackish breast-band diagnostic. Shows blackish to whitish head-sides and throat. See Streaked Weaver. **Male non-breeding** Similar to female but has variable broad blackish breast-band, scaled buffish. **Female** Similar to Streaked but crown and nape plainer and greyer, breast scaled blackish (mainly at sides) and unstreaked, rump plainer (contrasting with heavily streaked mantle), usually brighter yellow supercilium (in front of eye), submoustachial stripe and neck-patch. **Juvenile** Similar to female. **VOICE** Song is a very subdued, barely audible ***tsi tsi tsisik tsisik tsik tsik***. Calls with soft ***chit*** notes. **HABITAT** Marshes and grassland near water, cultivation; lowlands. **RANGE & STATUS** Resident Indian subcontinent (except islands). **SE Asia** Doubtful historical records from S Myanmar. **BREEDING** India: June–October. **Nest** Similar to Baya Weaver but with shorter entrance tube and not suspended by threads; in grass or reeds near water. **Eggs** 2–5; white.

755 STREAKED WEAVER
Ploceus manyar Plate **82**

IDENTIFICATION 13.5 cm. *P.m.williamsoni*: **Male breeding** Combination of yellow crown and prominent blackish streaks on breast diagnostic. Has blackish head-sides, narrowly dark-streaked flanks and bold buffish-brown and blackish streaks on upperparts. *P.m.peguensis* (Myanmar) has much broader blackish streaking above and below. See Black-breasted Weaver. **Male non-breeding** Similar to female but often shows more distinct breast-streaking. **Female** Well-defined (but often fine) blackish streaks on breast diagnostic. Has distinctive dark head-sides, with pronounced yellowish-white supercilium, submoustachial stripe and neck-patch; crown and upperparts boldly streaked buffish-brown and blackish. See Black-breasted Weaver. Resembles some buntings but much thicker-billed, tail shorter and more rounded, without white. **Juvenile** Similar to female. **VOICE** Sings with soft continuous trill: ***see-see-see-see-see...*** ending with ***o-chee***. Also ***tre tre cherrer cherrer***. Calls with loud ***chirt*** notes. **HABITAT & BEHAVIOUR** Grassland, reedbeds, marshes, cultivation, often near water; up to 915 m. Gregarious, often in mixed flocks with other weavers. **RANGE & STATUS** Resident Indian subcontinent, SW China, Java, Bali. **SE Asia** Locally common resident Myanmar (except Tenasserim), NW, C,W(coastal) Thailand, Cambodia, C,S Annam, Cochinchina. Uncommon feral resident Singapore (perhaps not yet established). **BREEDING** February–August. Usually colonial. **Nest** Like Baya Weaver but entrance tube much shorter, not suspended by threads, amongst grass or reeds near water. **Eggs** 2–5; white; 20.6 x 14.9 mm (av.; India).

756 BAYA WEAVER *Ploceus philippinus* Plate **82**

IDENTIFICATION 15 cm. *P.p.angelorum*: **Male breeding** Told from other weavers by combination of yellow crown and unstreaked warm buffish-brown breast. Rest of upperparts similar to Streaked Weaver; breast may appear faintly mottled or show faint dark streaks at sides. *P.p.infortunatus* (south Tenasserim and S Thailand southwards; C Annam? Cochinchina?) has more rufescent base colour to upperparts and darker, more rufescent breast and flanks. **Male non-breeding** Like female. **Female** From Black-breasted and Streaked by no pronounced blackish streaks or scales on breast and no defined blackish and yellowish markings on head-sides. Breast rather plain warm brown, sometimes faintly streaked darker on sides and flanks; has whitish throat and vent. See very similar Asian Golden Weaver for differences. **Juvenile** Like female but breast and flanks rather deeper buff, crown-streaking broader and more broken. **Other subspecies in SE Asia** *P.p.burmanicus* (Myanmar, except Tenasserim). **VOICE** Song is a series of chattering notes, ending with a wheezy rattling drawn-out ***cher-wiu***. Calls include harsh repeated ***chit*** notes. **HABITAT & BEHAVIOUR** Cultivation, grass and reeds, open areas, secondary growth; up to 1,220 m (mostly lowlands). Gregarious; roosts communally. Often associates with other weavers. **RANGE & STATUS** Resident Indian subcontinent, SW,S China, Sumatra, Java, Bali. **SE Asia** Uncommon to locally common resident (except C Laos, W,E Tonkin, N Annam). **BREEDING** All year; mainly December–June. Usually colonial. **Nest** Tightly woven ball with long down-hanging entrance tube, suspended by threads from tree-branch or bamboo etc., sometimes in grass; up to 30 m above ground (often near water). **Eggs** 2–5; white; 21.6 x 14.7 mm (av.; *burmanicus*).

757 ASIAN GOLDEN WEAVER
Ploceus hypoxanthus Plate **82**

IDENTIFICATION 15 cm. *P.h.hymenaicus*: **Male breeding** Unmistakable, with largely yellow head and body, black head-sides and throat, and blackish streaks on mantle, scapulars and back. **Male non-breeding** Like female but often tinged yellow on supercilium and underparts (sometimes upperparts). **Female** Difficult to separate from Baya Weaver but shows shorter, more conical bill (as deep as long), almost lacks obvious forehead, lacks mottling on breast, crown-streaking broader (making crown contrast more with supercilium). **Juvenile** Like female. **VOICE** Similar to Baya Weaver. **HABITAT** Marshes, grass and reeds, rice paddies, invariably close to water; lowlands. **RANGE & STATUS** Resident Sumatra, Java. **SE Asia** Scarce to locally common resident SW,N,C,S Myanmar, Tenasserim, NW,NE,C Thailand, Cambodia, S Laos, S Annam, Cochinchina. **BREEDING** April–October. Semi-colonial. **Nest** Untidy ball with side-entrance, in small tree or waterside grasses. **Eggs** 2–4; greyish-white, sometimes with some darker markings around broader end; 20.1 x 13.7 mm (av.).

ESTRILDIDAE: ESTRILDINAE: Avadavats & allies

Worldwide c.39 species. SE Asia 1 species. Small and round-bodied with conical bills and rather short rounded wings; tails rounded or pointed. Highly varied but mostly well patterned. Gregarious, sometimes in large flocks; flight fast and direct. Feed mainly on seeds in grassland, crops and bamboo. Many species are exploited for cagebird trade.

758 RED AVADAVAT
Amandava amandava Plate 77

IDENTIFICATION 10 cm. *A.a.punicea*: **Male breeding** Bright red plumage with white spotting on scapulars, uppertail-coverts and underparts diagnostic. Crown and mantle washed brown, wings blackish with white spots on tips of coverts and tertials, tail and vent blackish, bill red. *A.a.flavidiventris* (Myanmar) has whitish-orange to yellowish-buff belly and less deep red head and breast-sides. **Male non-breeding** Similar to female but shows white spots on uppertail-coverts and larger spots on wing. **Female** Mostly greyish-brown (paler below), with distinctive bright red rump, uppertail-coverts and bill; wings blackish with small white spots on tips of coverts and tertials. **Juvenile** Like female but upperparts browner, rump and uppertail-coverts brown, wing markings more extensive and buffish, bill dark brown to blackish. **VOICE** Song is a feeble, very high-pitched warble, intermingled with sweeter twittering notes. Calls with shrill thin ***pseep***, ***teei*** or ***tsi***, particularly in flight. Also a variety of high-pitched chirps and squeaks. **HABITAT & BEHAVIOUR** Grassland, marshes, secondary growth and scrub, open areas; up to 1,525 m (mainly lowlands). Typically in small flocks; may be secretive. **RANGE & STATUS** Resident Indian subcontinent, SW,S China, Java, Bali, Lesser Sundas. Introduced Sumatra, Philippines (Luzon), Fiji, Hawaii and elsewhere. **SE Asia** Scarce to locally common resident Myanmar (except Tenasserim), NW,NE,C,SE Thailand, Cambodia, W,E Tonkin, Cochinchina. Scarce feral resident Singapore. **BREEDING** January–September. **Nest** Neat ball, with side-entrance (sometimes slightly spouted), amongst grass or reeds or in bush; up to 90 cm above ground. **Eggs** 5–10; white; 14.4 x 11.2 mm (av.; *punicea*).

ESTRILDIDAE: LONCHURINAE: Java Sparrow, munias, parrotfinches & allies

Worldwide c.49 species. SE Asia 10 species. Small and round-bodied with conical bills and rather short rounded wings; tails rounded or pointed. Highly varied but mostly well patterned. Gregarious, sometimes in large flocks; flight fast and direct. Feed mainly on seeds in grassland, crops and bamboo. Many species are exploited for cagebird trade.

758A JAVA SPARROW *Padda oryzivora* Plate 77

IDENTIFICATION 16 cm. Monotypic. **Adult** Easily told by grey plumage, black head with whitish ear-coverts, large red bill and vinous-pinkish belly. Rump/uppertail-coverts and tail black, legs reddish. **Juvenile** Upperparts and wings mostly brown, crown and tail duller, ear-coverts duller and less contrasting, underparts pale dull buffish with darker breast-streaking, bill only reddish at base. **VOICE** Sings with series of soft bell-like notes, followed by trilling and clucking sounds and often ending with a whining, drawn-out, metallic ***ti-tui***. Calls include a soft liquid ***tup***, ***t-luk*** or ***ch-luk*** (particularly in flight) and sharp ***tak***. **HABITAT** Cultivation, rice paddies, margins of human habitation; lowlands. **RANGE & STATUS** Resident Java, Kangean, Bali. Introduced India (Calcutta, Madras), Sri Lanka, SE China, Sumatra, Borneo, Lombok, Sumbawa, Philippines, Sulawesi and elsewhere. **SE Asia** Introduced. Local resident C,S Thailand (northern Bangkok, Phuket I), west Peninsular Malaysia, Singapore. Formerly established (current status unknown) SW Myanmar, Tenasserim, S Annam, Cochinchina. **BREEDING** May–July, October and December. **Nest** Ball with side-entrance, in hole in building, cliff or tree; 3.6–17 m above ground. **Eggs** 3–5; white.

759 WHITE-RUMPED MUNIA
Lonchura striata Plate 77

IDENTIFICATION 11–11.5 cm. *L.s.subsquamicollis*: **Adult** Dark brownish plumage with contrasting whitish rump and belly diagnostic. See White-bellied and Javan Munias. **Juvenile** Dark parts of plumage paler and browner, rump and belly tinged buffish. **Other subspecies in SE Asia** *L.s.acuticauda* (Myanmar [except Tenasserim], north NW Thailand, N Laos, W Tonkin), *swinhoei* (E Tonkin). **VOICE** Song is a twittering ***pit pit pit spee boyee*** or ***prt prt prt spee boyee***, with distinctly down-turned end-note. Calls with a tinkling metallic ***prrrit***, ***pirit*** or ***tr-tr-tr*** etc., particularly when flying. **HABITAT & BEHAVIOUR** Clearings, secondary growth, scrub and grass, cultivation; up to 1,830 m. Usually found in flocks. **RANGE & STATUS** Resident Indian subcontinent (except W,NW and Pakistan), C and southern China, Taiwan, Sumatra. **SE Asia** Common resident throughout; local Singapore. **BREEDING** All year. Multi-brooded. **Nest** Untidy ball with slightly spouted side-entrance, in bush or tree, sometimes grass; 1.5–18.3 m above ground. **Eggs** 3–6; white; 15.3 x 10.9 mm (av.; *acuticauda*).

759A JAVAN MUNIA
Lonchura leucogastroides Plate 77

IDENTIFICATION 11.5 cm. Monotypic. **Adult** Recalls White-bellied Munia but upperparts paler brown (without whitish shaft-streaks), tail dark brown, face, throat and upper breast blackish, flanks white. **Juvenile** Paler, with brownish throat and upper breast, dark barring on undertail-coverts and pale buffish belly. From other munias by combination of uniform upperparts, two-tone underparts and barred undertail-coverts. **VOICE** Song is a pleasant purring or series of ***prreet*** notes. Usual calls are a shrill ***pi-i*** and ***pee-ee-eet*** or ***tyee-ee-ee***. **HABITAT & BEHAVIOUR** Secondary growth, scrub, gardens; lowlands. Often in small flocks. **RANGE & STATUS** Resident S Sumatra, Java, Bali, Lombok. **SE Asia** Introduced. Uncommon to scarce (declining) resident Singapore. Vagrant (or recent escapee) south Peninsular Malaysia (Johor Baru). **BREEDING** March–October. **Nest** Ball, with side-entrance, in tree, bush, creeper or fern etc.; up to 3 m above ground. **Eggs** 4–6; white.

760 SCALY-BREASTED MUNIA
Lonchura punctulata Plate 77

IDENTIFICATION 12–12.5 cm. *L.p.topela*: **Adult** Whitish breast and flanks, with darkish brown scaling diagnostic. Shows rather uniform drab brown upperside, yellowish-olive markings on rump and uppertail-coverts and fringes to tail-feathers, chestnut-tinged head-sides, somewhat darker cheeks and throat, whitish vent and black bill. *L.p.subundulata* (Myanmar [except southern S and Tenasserim]) and *fretensis* (southern S Thailand southwards) show bolder, blacker scaling below and slightly paler, warmer upperparts. **Juvenile** Upperside paler and more uniform, underside pale, slightly buffish drab brown, with slightly more whitish vent, lower mandible paler than upper. Similar to Black-headed and White-headed Munias but colder-tinged above, has more drab brown throat, breast and flanks, and paler vent; bill dark greyish-brown to blackish, with pinkish base to lower mandible. **Other subspecies in SE Asia** *L.p.yunnanensis* (north E Myanmar). **VOICE** Song is a very quiet soft series of high flute-like whistles and low-pitched slurred notes. Calls include a sibilant piping ***ki-dee ki-dee...*** or ***kitty-kitty-kitty*** (particularly when flying), rapidly repeated harsh ***chup*** or ***tret*** notes and ***kit-eeeeee*** or ***ki-ki-ki-ki-ki-teeee*** when alarmed. **HABITAT & BEHAVIOUR** Cultivation, scrub, secondary growth; up to 1,915 m. Usually in flocks. **RANGE & STA-**

TUS Resident Indian subcontinent, SE Tibet, southern China, Taiwan, Sundas, Philippines, Sulawesi. Introduced Mascarene Is, Seychelles, Australia, Caroline Is, Hawaii. **SE Asia** Common resident throughout. **BREEDING** All year. **Nest** Ball with slightly spouted side-entrance, in bush, tree or creepers etc.; 0.8–13 m above ground. **Eggs** 3–6; white; 14.5–15.5 x 10.5–12 mm (*fretensis*).

761 WHITE-BELLIED MUNIA

Lonchura leucogastra Plate 77

IDENTIFICATION 11–11.5 cm. *L.l.leucogastra*: **Adult** Similar to White-rumped Munia but lacks whitish rump, plumage very dark chocolate-brown with black uppertail-coverts, olive-yellow tail-feather fringes and sharply contrasting whitish belly. Shows paler shaft-streaks on upperparts, flanks dark brown. See Javan Munia. **Juvenile** Dark plumage-parts browner, lacks shaft-streaks on upperparts, tail duller, belly buffier. **VOICE** Song is a rapidly repeated ***di-di-ptcheee-pti-pti-pti-pteep***. Calls with a piping ***prrip prrip...*** and soft cheeping ***chee-ee-ee***. **HABITAT & BEHAVIOUR** Open and secondary broadleaved evergreen forest, forest edge, scrub and cultivation near forest; up to 455 m. Often in small flocks. **RANGE & STATUS** Resident Sumatra, W Java, Borneo, Philippines. **SE Asia** Locally common resident south Tenasserim, S Thailand, Peninsular Malaysia. **BREEDING** March–June. **Nest** Ball, with side-entrance (often slightly spouted), in bush or sapling; 1.5–3 m above ground. **Eggs** 4–5; white; 14.7 x 11.2 mm (av.).

762 CHESTNUT MUNIA

Lonchura atricapilla Plate 77

IDENTIFICATION 11–11.5 cm. *L.a.deignani*: **Adult** Unmistakable, with dark rufous-chestnut plumage and contrasting black hood. Belly-centre and vent darker, bill bright blue-grey. *L.a.atricapilla* (Myanmar [except Tenasserim]) has orange-yellowish fringes to uppertail-coverts and tail and can show blacker belly-centre and vent. **Juvenile** Rather uniform rich brown above and paler uniform buff below. From Scaly-breasted Munia by bluish bill, uniform rich buff underparts and warmer upperparts. From White-headed by darker, warmer crown and more strongly buff underparts. **Other subspecies in SE Asia** *L.a.sinensis* (Tenasserim and south-west Thailand southwards). **VOICE** Song is a very quiet series of bill-snapping notes followed by 'silent' singing (no sound audible), ending with faint drawn-out whistled notes. Call is a weak reedy ***pee pee...*** (particularly in flight). **HABITAT & BEHAVIOUR** Grassland, marshes, scrub, cultivation, rice paddies; up to 1,525 m (mostly lowlands). Usually in flocks, sometimes very large. **RANGE & STATUS** Resident N,NE India, Nepal, Bangladesh, SW,S China, Taiwan, Greater Sundas, Philippines, Sulawesi, Moluccas. Introduced Japan, Australia, Palau, Guam, Hawaii, Jamaica. **SE Asia** Locally common resident (except S Laos, W Tonkin, N Annam). **BREEDING** December–October. Sometimes semi-colonial. **Nest** Ball with slightly spouted side-entrance, in bush, tree or grass. **Eggs** 4–6; white; 16.3 x 11.5 mm (av.; *atricapilla*).

762A WHITE-CAPPED MUNIA

Lonchura ferruginosa Plate 77

IDENTIFICATION 10.5–11.5 cm. Monotypic. **Adult** Like Chestnut, but has whitish head with black throat and black from lower mid-breast to belly. See White-headed Munia. **Juvenile** Apparently indistinguishable from White-headed. **VOICE** Song is an almost inaudible series of clicks and wheezes, followed by a long drawn-out ***wheeee***. Calls variously described as ***psitt psitt***; ***pseet pseet***; and ***veet veet***. **HABITAT & BEHAVIOUR** Grassland, marshes, scrub, cultivation, rice paddies; lowlands. Usually in flocks, sometimes large. **RANGE & STATUS** Resident Java, Bali. **SE Asia** Introduced. Resident (status unknown) Singapore (? established). **BREEDING** Season unknown in region. **Nest** Typical of genus, in grasses or reeds; up to 1 m above ground. **Eggs** 3–7. Otherwise undocumented.

763 WHITE-HEADED MUNIA

Lonchura maja Plate 77

IDENTIFICATION 11.5 cm. *L.m.maja*: **Adult** Mostly white head and broad pale vinous-brownish collar diagnostic. Otherwise similar to Black-headed Munia. **Female** tends to have white of head more restricted to face and forecrown. *L.m.vietnamensis* (Vietnam) has mostly pale brownish head with white around eye only. **Juvenile** Like Black-headed Munia but crown duller and paler (though often slightly darker than upperparts), head-sides paler, underparts duller buff. **VOICE** Song involves bill-clicking followed by a high-pitched, tinkling ***weeeeee heeheeheeheehee***, constantly repeated. Calls with thin piping ***puip***, ***peekt*** and ***pee-eet*** (mainly in flight); higher-pitched and less reedy than Black-headed. **HABITAT & BEHAVIOUR** Grassland, cultivation, rice paddies, scrub; up to 500 m. Usually in flocks, sometimes quite large. **RANGE & STATUS** Resident Sumatra, Java. **SE Asia** Uncommon to locally fairly common resident S Thailand, Peninsular Malaysia, Singapore. Scarce resident S Annam, Cochinchina. **BREEDING** February–October. **Nest** Ball with slightly spouted side-entrance, in tall grass, bush, hedge or tree; 0.6–6 m above ground. **Eggs** 4–5; white.

764 TAWNY-BREASTED PARROTFINCH

Erythrura hyperythra Plate 77

IDENTIFICATION 10 cm. *E.h.malayana*: **Male** Undescribed. In extralimital *E.h.hyperythra*, male as female but rather brighter green above, with black patch on extreme forehead and more extensive, brighter/paler blue on forecrown. **Female** Resembles Pin-tailed Parrotfinch but has shorter, rounded and green tail, uniform warm buff lores, ear-coverts, lower rump, uppertail-coverts and underparts, small dark brown patch on extreme forehead, and dull/dark blue forecrown. Bill black. **Juvenile** Paler overall than adults, with all-green crown, pale yellowish to pinkish bill with dark tip. From Pin-tailed by buffish lower rump and uppertail-coverts, green tail and mostly pale upper mandible. **VOICE** Song is a series of very quiet soft notes, followed by four musical or bell-like notes. Calls with a high-pitched hissing ***tzit-tzit*** or ***tseet-tseet***. **HABITAT & BEHAVIOUR** Bamboo, edge of broadleaved evergreen forest, scrub; 790–2,000 m. Skulking. **RANGE & STATUS** Resident W Java, N Borneo, W Lesser Sundas, Philippines, Sulawesi. **SE Asia** Rare resident Peninsular Malaysia. **BREEDING** Java: February–March. **Nest** Ball, with side-entrance, in tree, often amongst ferns, orchids or moss etc.; 4–12 m above ground. **Eggs** Undescribed.

765 PIN-TAILED PARROTFINCH

Erythrura prasina Plate 77

IDENTIFICATION 12.5–13 cm (male tale up to 3 cm more). *E.p.prasina*: **Male** Unmistakable, with green upperparts, blue face and throat, warm buff underparts and bright red lower rump, uppertail-coverts and long pointed tail. Often shows pale red patch on centre of abdomen and pale bluish wash on breast. **Female** Much shorter-tailed, no blue face but often some pale powder-blue on head-sides, uniform washed-out buffish underside. **Adult yellow morph** In this rare morph, the red lower rump to tail (both sexes) and on underparts (male) is replaced by golden-yellow. **Juvenile** Like female but lower rump, uppertail-coverts and tail dull orange- to brownish-red, lower mandible largely yellowish to pinkish. See Tawny-breasted Parrotfinch. **VOICE** Song is described as less musical than Tawny-breasted, with more clinking or chirping notes. Calls include a high-pitched sharp ***zit***; ***tseet-tseet*** or ***tsit-tsit*** and sharp ***teger-teter-terge***. **HABITAT & BEHAVIOUR** Bamboo, open forest, forest edge, secondary growth; up to 1,500 m. Semi-nomadic, depending on availability of seeding bamboo. Often in flocks, sometimes quite large. **RANGE & STATUS** Resident Sumatra, Java, Borneo. **SE Asia** Scarce to locally fairly common resident (subject to unpredictable movements) Tenasserim, Thailand (except C), Peninsular Malaysia, Cambodia, N,C Laos, C,S Annam, north Cochinchina. **BREEDING** February–September. **Nest** Ball with side-entrance, in bamboo or sapling. **Eggs** 4–6; white.

PASSERIDAE: Sparrows & allies

Worldwide c.40 species. SE Asia 4 species. Small, with thick conical bills. Gregarious, often close to human habitation. Feed on seeds, shoots, buds and insects.

766 **HOUSE SPARROW** *Passer domesticus* Plate 77
IDENTIFICATION 15 cm. *P.d.indicus*: **Male breeding** Easily told by combination of grey crown, whitish head-sides and greyish-white underparts with broad black bib. Has broad chestnut band from eye to nape-sides, dull chestnut mantle with pale brownish and blackish streaks, pale brownish-grey rump and uppertail-coverts and black bill. **Male non-breeding** Shows browner-tinged crown and greyer mantle, bib and chestnut of crown and nape-sides obscured by pale feather-tips, bill mostly pale horn-coloured. **Female** Rather nondescript brownish, with paler supercilium and underparts, blackish-brown streaks on mantle and scapulars and pale horn-coloured bill. Superficially resembles some weavers and buntings but lacks streaks on head and underparts. See Russet and Plain-backed Sparrows and Plain Mountain-finch. Note habitat and range. **Juvenile** Like female but upperparts initially paler, with plainer mantle and scapulars, supercilium often better defined, bill paler. **VOICE** Sings with a monotonous series of call-notes: ***chirrup cheep chirp...*** etc. Calls include a familiar ***chirrup***, ***chissick*** or ***tissip***, soft ***swee swee*** or ***dwee***, shrill ***chree*** and rolling ***chur-r-r-it-it-it*** when alarmed. **HABITAT & BEHAVIOUR** Towns and villages, cultivation, scrub, particularly in dry areas. Roosts communally. **RANGE & STATUS** Resident N Africa, Palearctic, Middle East, Indian subcontinent, SW Tibet, NW,NE China. Introduced S Africa, Australia, New Zealand, Americas. **SE Asia** Locally common to common resident (still spreading east) Myanmar, W,NW,NE,C Thailand, Cambodia, Laos, southern Cochinchina. Uncommon feral resident Singapore (one site; declining). **BREEDING** All year. Multi-brooded. **Nest** Untidy ball with side-entrance, in hole in building, in tree or in old nest or burrow of other bird. **Eggs** 3–6; greyish- or greenish-white, speckled light and dark ash-grey and brown; 20.6 x 14.9 mm (av.).

767 **PLAIN-BACKED SPARROW**
Passer flaveolus Plate 77
IDENTIFICATION 13.5–15 cm. Monotypic. **Male** Shows diagnostic unstreaked upperparts, with rufous-chestnut band extending from eye to nape-side, rufous-chestnut scapulars and lower mantle, and yellowish lower head-sides and vent. Crown, upper mantle, back, rump and uppertail-coverts greenish-grey, has narrow black bib. **Female** Told from other sparrows by unstreaked upperparts and pale yellowish-tinged underparts (throat often buffish), with drab pale greyish wash across breast. **Juvenile** Like female but crown and lower mantle slightly darker, throat duller. **VOICE** Loud clear ***filip*** or ***chirrup***, less harsh but more metallic than Eurasian Tree-sparrow. Also ***chu-chu-weet***, when alarmed. **HABITAT** Open woodland, coastal scrub, dry open areas and cultivation, margins of human habitation; up to 1,525 m (below 800 m in Thailand). **RANGE & STATUS** Endemic. Locally common to common resident SW,W,C,E,S Myanmar, Thailand, Peninsular Malaysia, Cambodia, Laos, C,S Annam, Cochinchina. **BREEDING** All year. Semi-colonial. **Nest** Ball, in hole in tree or dead bamboo etc., or in lamp-post junction box, etc.; 4.5–20 m above ground. **Eggs** 2–4; very pale bluish-white, heavily speckled grey-brown; 18.3 x 14 mm (av.).

768 **EURASIAN TREE-SPARROW**
Passer montanus Plate 77
IDENTIFICATION 14–14.5 cm. *P.m.malaccensis*: **Adult** Whitish head-sides with isolated blackish patch diagnostic. Crown and nape dull chestnut, small bib blackish. **Juvenile** Duller; crown paler brown with dark markings on forecrown, ear-covert patch and bib less defined, pale base to lower mandible. **VOICE** Song is a repeated series of call-notes, interspersed with ***tsooit***, ***tsreet*** and ***tswee-ip*** notes. Calls with harsh ***chip*** and ***chissip***, sharp ***tet*** and metallic ***tsooit***. In flight, gives dry ***tet-tet-tet...*** **HABITAT** Urban areas, human habitation, cultivation; up to 1,830 m. **RANGE & STATUS** Resident (subject to some minor movements) Palearctic, Iran, Afghanistan, Pakistan, E India, N,NE Indian subcontinent, Tibet, China, Taiwan, N,S Korea, Japan, Sumatra, Java, Bali. Introduced Borneo, Philippines, Wallacea, Micronesia, SE Australia, USA, Bermuda. **SE Asia** Common resident throughout. **BREEDING** All year. Multi-brooded. **Nest** Pad, in hole in building or tree, sometimes in old weaver nest. **Eggs** 3–6; fairly glossy, pale bluish-white, spotted, flecked and streaked brown and grey-brown; 19.2 x 14.2 mm (av.).

769 **RUSSET SPARROW** *Passer rutilans* Plate 77
IDENTIFICATION 13.5–14 cm. *P.r.intensior*: **Male** Unmistakable, with rufous-chestnut crown and upperparts, dingy pale yellowish head-sides and centre of abdomen, narrow black bib and prominent white wing-bar. Mantle streaked blackish, breast and flanks greyish. **Female** Resembles House Sparrow but shows darker eyestripe, more contrasting pale supercilium, rufescent rump, whiter bar on median coverts, creamy-yellowish throat with dusky central stripe and creamy-yellowish belly-centre and vent. **Juvenile** Like female but crown, mantle and scapulars tinged warmer brown, supercilium buffier, bill paler. **First-winter male** Duller than adult, upperparts mixed with greyish-brown, bib dusky mixed with black. **VOICE** Song is similar to House: ***cheep-chirrup-cheweep*** or ***chwe-cha-cha*** etc., frequently repeated. Calls with ***cheeep*** or ***chilp*** notes, sweeter and more musical than House. Also ***swee swee***, when alarmed. **HABITAT** Open forest, cultivation; 1,100–2,680 m, locally down to 700 m. **RANGE & STATUS** Resident (subject to altitudinal movements) N Pakistan, NW,N,NE India, Nepal, Bhutan, S,SE Tibet, China (except NW,N,NE), Taiwan, N,S Korea, Japan. **SE Asia** Fairly common resident SW,W,N,S(east),E Myanmar, N Laos, W,E(north-west) Tonkin. Rare winter visitor NW,NE Thailand. **BREEDING** April–August. Multi-brooded. **Nest** Pad, in hole in tree or building; up to 9 m above ground. **Eggs** 4–5; similarly coloured to House but usually more heavily marked; 19.1 x 13.9 mm (av.; India).

MOTACILLIDAE: Wagtails & pipits

Worldwide c.66 species. SE Asia 16 species. Relatively small, slender-billed and long-legged, with white or pale outertail-feathers. Largely terrestrial; walk and run on ground. Flight quite strong and undulating. Mostly gregarious. Feed primarily on insects. **Pipits** (*Anthus*) Shorter-tailed, mostly dull brownish and streaked. **Wagtails** (*Dendronanthus, Motacilla*) Tail markedly long and often wagged, plumage strikingly patterned.

770 **BUFF-BELLIED PIPIT**
Anthus rubescens Plate 78
IDENTIFICATION 16.5 cm. *A.r.japonicus*: **Adult non-breeding** Resembles Rosy Pipit but cold greyish-brown upperparts have only faint streaking and appear very plain at distance (a little more prominently streaked when worn), lores paler, base colour to underparts whitish to pale buffish. See Tree and Red-throated Pipits. **Adult breeding** Upperparts greyer, underparts plainer and buffier (warmer on breast) with fine dark streaks restricted to breast. **VOICE** Flight call is a thin high-pitched ***tseep*** or ***zzeeep***, frequently repeated. **HABITAT** Marshy areas, wetland fringes, cultiva-

tion; up to 2,440 m. **RANGE & STATUS** Breeds E Palearctic, northern N America; winters south to northern Indian subcontinent, southern China, S Korea, Japan, N(southern),C America. **SE Asia** Vagrant N,C Myanmar, NE Thailand, W,E Tonkin.

771 **ROSY PIPIT** *Anthus roseatus* Plate **78**

IDENTIFICATION 16.5 cm. Monotypic. **Adult non-breeding** Similar to first-winter Red-throated Pipit but upperparts greyer-brown (more olive-tinged when fresh) and somewhat less boldly streaked, has darker head-sides and prominent whitish supercilium, less distinct (if any) whitish mantle 'braces', more olive fringing on tertials and secondaries (particularly when fresh), rather plain rump and uppertail-coverts, blacker bill. Note voice. From Buff-bellied Pipit by heavily dark-streaked upperparts. See Tree Pipit. **Adult breeding** Loses dark malar streak and breast-streaking and shows diagnostic vinous-pinkish flush on supercilium, throat and breast. **Juvenile** Similar to non-breeding adult but browner above and less heavily streaked below. **VOICE** Song (usually given in display-flight) involves twittering ***tit-tit-tit-tit-tit teedle teedle*** (during ascent), followed by long fading ***tsuli-tsuli-tsuli-tsuli...*** or ***sweet-sweet-sweet*** (during descent). Flight call is a thin ***tsip tsip tsip...*** or ***seep-seep...***, very similar to extralimital Meadow Pipit *A. pratensis.* **HABITAT & BEHAVIOUR** Marshy areas, rice paddies, open, short grassy areas and clearings. Up to 1,300 m (winter), breeds above 3,000 m in Himalayas. During song-flight, rises high in air, then parachutes down on outstretched quivering wings. **RANGE & STATUS** Breeds E Afghanistan, N Pakistan, NW,N,NE Indian subcontinent, Tibet, western and C China; some populations winter to south, northern Indian subcontinent, south-western China. **SE Asia** Probably breeds N Myanmar. Uncommon winter visitor SW,N,C,E Myanmar, NW Thailand, N Laos, W,E Tonkin. Vagrant north Tenasserim, C Thailand, Cochinchina. **BREEDING** India: June–July. **Nest** Cup, in depression on ground. **Eggs** 3–4; grey to brownish-buff, finely but very densely speckled various shades of brown; 22 x 15.6 mm (av.).

772 **RED-THROATED PIPIT** *Anthus cervinus* Plate **78**

IDENTIFICATION 15–16.5 cm. Monotypic. **Adult** Brick-red to pinkish-red head-sides, throat and upper breast diagnostic. Often lacks dark markings on head-sides and breast-centre; reddish coloration on head tends to be paler and less extensive on females and in autumn/winter. See Rosy Pipit. **First winter** Lacks reddish coloration on head. Similar to Tree Pipit but upperparts and flanks more boldly dark-streaked (including rump), usually shows pronounced whitish to pale buff mantle 'braces' and whiter wing-bars, underparts whiter with bolder blackish streaks on breast and flanks. Note voice. For differences from similar Rosy Pipit, see that species. **VOICE** Has rhythmic ringing song, consisting of sharp drawn-out notes and dry buzzing sounds. Flight call is similar to Tree Pipit but longer, more drawn-out and higher-pitched: ***pseeoo*** or ***pssiih***. Also gives short ***chupp*** when alarmed. **HABITAT & BEHAVIOUR** Open areas, drier cultivation, often near water; up to 1,500 m. Usually in flocks. **RANGE & STATUS** Breeds N Palearctic, N America (Alaska); winters Africa, Middle East, northern Indian subcontinent, Andaman and Nicobar Is, southern China, Taiwan, S Japan, N Borneo, Philippines, Sulawesi. **SE Asia** Uncommon to locally common winter visitor (except W Myanmar, Peninsular Malaysia, W Tonkin); rare Singapore. Uncommon passage migrant E Tonkin. Vagrant Peninsular Malaysia.

773 **OLIVE-BACKED PIPIT** *Anthus hodgsoni* Plate **78**

IDENTIFICATION 16–17 cm. *A.h.yunnanensis*: **Adult** The most widespread pipit in more wooded habitats. Told by combination of rather plain greenish-olive upperparts, broad whitish supercilium (typically more buff-tinged in front of eye), distinct whitish spot and blackish patch on rear ear-coverts (indistinct on some birds) and prominent blackish streaking/spotting on buff breast and flanks. More greyish-olive above and somewhat whiter below when worn. *A.h.hodgsoni* (widely recorded N Myanmar to Indochina) shows more heavily streaked upperparts and flanks, and more extensive streaking on lower underparts. See Tree, Red-throated, Rosy and Buff-bellied Pipits. **Juvenile** Initially differs from adult by browner, more boldly streaked upperparts. **VOICE** Song is faster, slightly higher-pitched and a little softer than Tree Pipit. Trills are slightly harder and drier, recalling Winter Wren. Call is a thin hoarse ***teez*** or ***spiz*** in flight, very similar to Tree. Alarm call is a very quiet short ***tsi tsi...*** **HABITAT & BEHAVIOUR** Open forest, forest tracks and edge, secondary growth, wooded cultivation; up to 3,050 m. Secretive, often found walking amongst undergrowth and leaf-litter; gently pumps tail up and down. Typically sings from tree- or bush-top; also in song-flight, ascending, then making angled gliding descent with wings and tail spread. **RANGE & STATUS** Breeds C,E Palearctic, N,NE Indian subcontinent, SE Tibet, SW,W,N,NE China, Taiwan, N Korea, Japan; winters Indian subcontinent (except W and Pakistan), southern China, Philippines. **SE Asia** Uncommon resident/breeding visitor N Myanmar; W Myanmar? Common winter visitor (except S Thailand, Singapore); rare Peninsular Malaysia. Also recorded on passage SW,S Myanmar, Cambodia, E Tonkin. Vagrant S Thailand. **BREEDING** India: May–July. **Nest** Cup, on ground. **Eggs** 3–5; dark brown, densely spotted darker, sometimes pale greyish with dark grey-brown blotches; 21.4 x 15.8 mm (av.).

774 **TREE PIPIT** *Anthus trivialis* Plate **78**

IDENTIFICATION 15–16.5 cm. *A.t.trivialis*?: **Adult** Like Olive-backed Pipit but crown and upperparts buffier-brown to greyish (lacking greenish-olive cast) and more heavily streaked, breast- and flank-streaking often narrower and lighter, supercilium more uniformly buff and less contrasting (sometimes more whitish above/behind eye when worn), usually lacking (or with much fainter) pale spot and dark marking on rear ear-coverts. Upperparts browner, underparts whiter when worn. See Red-throated, Rosy and Buff-bellied Pipits. **VOICE** Song is louder, slower and more far-carrying than Olive-backed: ***zit-zit-zit-zit cha-cha-cha-cha surrrrrrrrr siiiii-a tvet-tvet-tvet-tvet siva siiva siiiva siiihva cha-cha-cha*** etc. (given from perch or during undulating song-flight). Flight call is a short incisive ***zeep*** or ***spzeep***, sometimes a more scolding ***speez***; similar to Olive-backed but averages slightly louder. Alarm call is a soft ***syt***. **HABITAT & BEHAVIOUR** Open woodland, scrub and grass; up to 1,000 m (could occur higher). Prefers grassier habitats than Olive-backed; often gently pumps tail up and down in similar fashion. **RANGE & STATUS** Breeds Palearctic, N Iran, N Pakistan, NW India, NW China; winters Africa, Middle East, S Pakistan, India, Nepal. **SE Asia** Vagrant S Myanmar.

775 **BLYTH'S PIPIT** *Anthus godlewskii* Plate **78**

IDENTIFICATION 17 cm. Monotypic. **Adult** Very difficult to separate from Richard's Pipit but, at close range, shows slightly shorter and more pointed bill, slightly shorter legs and tail, rather more contrasting upperpart-streaking and different median-covert pattern, with more square-cut (less pointed) and well-defined centres; hindclaw shorter and more arched, penultimate outertail-feathers typically have white of inner web restricted to wedge on terminal third of feather (normally mostly white on Richard's). Note voice. See Paddyfield Pipit. **First winter** Differs from adult in same way as Richard's and cannot be separated by median-covert pattern (unless freshly moulted adult feathers present). **VOICE** Calls are quieter and less rasping than Richard's. Short ***tchu***; ***dju***; ***chep***; or ***tchupp*** (sometimes doubled), softer ***chewp*** or ***cheep***, slightly nasal ***tchii*** and shriller ***psheet*** or ***pshreu*** when alarmed. **HABITAT & BEHAVIOUR** Open country, dry cultivation; up to 3,050 m. Rarely hovers before landing;

often adopts distinctly horizontal (more wagtail-like) posture than Richard's. **RANGE & STATUS** Breeds S Siberia, Mongolia, northern China; winters Indian subcontinent (except Pakistan). **SE Asia** Uncommon winter visitor W,C,S Myanmar, north Tenasserim. Vagrant NE Thailand.

776 **RICHARD'S PIPIT** *Anthus richardi* Plate **78**
IDENTIFICATION 18–20.5 cm. Monotypic. **Adult** Larger than other pipits (apart from Long-billed). From very similar Blyth's and Paddyfield by combination of size, posture, subtle plumage details and voice. From Blyth's by slightly longer bill, legs and tail, slightly less contrastingly streaked upperparts, slightly more extensive white on outertail-feathers, and pattern of median coverts, which show more pointed and less clear-cut dark centres; hindclaw longer, straighter and less arched. Much larger than Paddyfield, with longer bill and tail, usually heavier and more extensive dark breast-streaking. Worn adults appear greyer above with more pronounced dark streaking, and paler below. **First winter** Like adult but retains whiter-fringed juvenile wing-coverts; median covert fringes more even and very similar to Blyth's. **VOICE** Calls with a loud, harsh ***schree-ep*** or ***shreep*** in flight, particularly when flushed. Song is a simple grinding ***tschivu-tschivu-tschivu-tschivu-tschivu...*** (given in undulating song-flight). **HABITAT & BEHAVIOUR** Open country, grassy areas, cultivation; up to 1,830 m. Frequently hovers before landing; typically adopts more upright posture than Blyth's. **RANGE & STATUS** Breeds southern Siberia, Mongolia, China (except SW); winters Africa, S Spain, Israel, Indian subcontinent, S China, N Borneo. **SE Asia** Fairly common winter visitor and passage migrant (except Peninsular Malaysia, Singapore); rare S Thailand. Vagrant Singapore.

777 **PADDYFIELD PIPIT** *Anthus rufulus* Plate **78**
IDENTIFICATION 15–16 cm. *A.r.rufulus*: Adult Very similar to Richard's Pipit but much smaller (this may be difficult to discern without direct comparison), with shorter bill and tail, usually less distinct and more restricted breast-streaking. Note voice. Worn birds differ as Richard's. See Blyth's Pipit. **Juvenile** Upperparts appear more scalloped, has heavy dark spotting on breast. **Other subspecies in SE Asia** *A.r.malayensis* (south Tenasserim and south Thailand southwards; S Indochina). **VOICE** Song is a simple, repetitive ***chew-ii chew-ii chew-ii chew-ii...***; also ***chik-a-chik*** sometimes in song-flight descent. Call is an explosive but relatively subdued ***chip***, ***chup*** or ***chwist***, usually given in flight. **HABITAT & BEHAVIOUR** Open areas, drier cultivation; up to 1,500 m. Typically has weaker, more fluttering flight pattern than Richard's; rarely hovers before landing. Has undulating song-flight; also rises high in air and parachutes down. **RANGE & STATUS** Resident Indian subcontinent, SW China, Sundas, Philippines, Sulawesi. **SE Asia** Common resident throughout. **BREEDING** January–August. **Nest** Cup (sometimes semi-dome), on ground, often in depression. **Eggs** 2–4; whitish to brownish or pale greyish, moderately to densely streaked, clouded and spotted dull brownish-red to purplish-red or various shades of brown and purplish-grey (often capped at broader end); 20.2 x 15.4 mm (av.; *rufulus*). Brood-parasitised by Eurasian Cuckoo.

778 **LONG-BILLED PIPIT** *Anthus similis* Plate **78**
IDENTIFICATION 20.5 cm. *A.s.yamethini*: **Adult** Told from other pipits by relatively large size, long bill and tail, rather plain, indistinctly streaked upperparts, and rather uniform dark buff underparts with whitish upper throat and indistinct dark breast-streaking; outertail-feathers fringed buffish (may appear whitish in field when worn). **Juvenile** Upperpart feathers have rounded dark centres and paler fringes, breast more prominently dark-streaked. **VOICE** Song is a simple ***ureee-tsur***, repeated after short intervals. Calls with flat-sounding ***chup***, ***chip*** or ***klup*** etc. and slightly lower-pitched, often repeated ***djup***. **HABITAT & BEHAVIOUR** Scrubby, often hilly semi-desert with ravines, dry open areas and cultivation; up to 2,745 m (mainly lowlands). Flight strong and bounding; has undulating song-flight. **RANGE & STATUS** Resident Africa, Middle East, Pakistan, India, W Nepal, NW Bangladesh. **SE Asia** Fairly common to common resident W,C,S(north),E Myanmar. **BREEDING** July–August. **Nest** Cup, in depression on ground. **Eggs** 3; whitish, heavily spotted and blotched brown (more at broader end) over grey and purplish-brown undermarkings; 20.6 x 16.3 mm.

779 **FOREST WAGTAIL**
Dendronanthus indicus Plate **78**
IDENTIFICATION 17–18 cm. Monotypic. **Adult** Unmistakable, with brownish-olive crown and upperparts, whitish underparts with double dark breast-band (lower one broken) and strongly contrasting blackish and whitish wing pattern. **VOICE** Calls with subdued, metallic ***pink*** or ***dzink-dzzt*** (particularly in flight). Song is an intense, repetitive see-sawing series of usually 3–6 notes: ***dzi-chu dzi-chu dzi-chu dzi-chu...*** **HABITAT & BEHAVIOUR** Open broadleaved evergreen and deciduous forest, forest tracks and trails, wooded cultivation, gardens, mangroves; up to 1,500 m. Often found walking on leafy ground, where well camouflaged. Roosts communally, locally in large numbers. **RANGE & STATUS** Breeds SE Siberia, Ussuriland, Sakhalin, C,E,NE China, N,S Korea, S Japan; winters India, Bangladesh, S China, Greater Sundas, Philippines. **SE Asia** Uncommon to locally common winter visitor (except N Myanmar, N,C Laos, W Tonkin, N Annam). Uncommon to fairly common passage migrant SW,S Myanmar, Thailand, Peninsular Malaysia, Cambodia, N,C Laos, E Tonkin. Recorded (status uncertain) N Annam.

780 **WHITE WAGTAIL** *Motacilla alba* Plate **79**
IDENTIFICATION 19 cm. *M.a.leucopsis* (visitor throughout): **Male non-breeding** Unmistakable, with white head and underparts, black hindcrown, nape and isolated breast-patch, black upperside and wings with broad white fringes to wing-coverts and tertials, and white outertail-feathers. *M.a.alboides* (only breeding race and visitor C,S Myanmar, Tenasserim, NW Thailand, Indochina) has black ear-coverts and neck-sides joining broader breast-patch, and less white on forehead (supercilium well defined); *personata* (visitor N Myanmar) has black hood with white forecrown, eyering and upper throat, and grey upperparts; *baicalensis* (visitor Myanmar [except SW,W], NW,NE Thailand) has grey upperparts and larger black breast-patch; *ocularis* (visitor N,C,S Myanmar, NW,NE,C Thailand, Indochina, Peninsular Malaysia) is like *baicalensis* but shows black eyestripe. *M.a.lugens* (vagrant SE Thailand, E Tonkin; sometimes treated as distinct species, Black-backed Wagtail) differs by combination of black on upperparts, white forehead and supercilium, black eyestripe, and white cheeks and ear-coverts. Upperparts apparently most often with large black patches, rather than solidly black. Shows smudgy black breast-patch, extensive white on wing. **Male breeding** Black breast-patch extends up to lower throat and joins black of mantle. *M.a.alboides* and *personata* have all-black throat; *baicalensis* and *ocularis* have black lower throat and upper breast. *M.a.lugens* has solid black lower throat to upper breast and upperparts, white head-side connected to white neck-patch. **Female** Like non-breeding/breeding male but has slaty-grey upperparts and narrower breast-patch. *M.a.alboides* is greyer above than male; *personata* has hindcrown and nape greyer than male; *baicalensis* is like *leucopsis* but upperparts much paler grey, crown also duller than male; *ocularis* has black of crown/nape greyer and less contrasting than male. *M.a.lugens* has slate-grey nape to upperparts (female breeding notably greyer above than male), though noticeably darker above than similar *ocularis*. **Juvenile** Like female but crown and nape grey, breast-patch more diffuse. *M.a.personata* and *alboides* appear more like very washed-out adult. *M.a.lugens* has distinct, complete dark eyestripe, remainder of ear-coverts somewhat paler than on *ocularis*, making rear eyestripe stand-out more, median coverts mostly white with only narrow dark central marking, greater coverts

much more extensively whitish, greyish breast-patch faint or absent. **VOICE** Has simple twittering and chattering song. Calls (*baicalensis, ocularis*) with clear disyllabic ***tche-rip***; ***tchree-lit***; ***tse-lit***; or ***tchle-wit***. Somewhat sharper and higher-pitched ***tchi-tchik*** from *personata, leucopsis*, and *alboides* is more reminiscent of Grey Wagtail. **HABITAT** Various open habitats, often near water; up to 2,000 m. **RANGE & STATUS** Breeds NW Africa, Palearctic, Iran, Afghanistan, N Pakistan, NW,N,NE Indian subcontinent, S,E Tibet, China, Taiwan, N,S Korea, Japan, N America (Alaska), SE Greenland; some northerly populations winter south to northern Africa, Middle East, Indian subcontinent, southern China, Taiwan, N,S Korea, C,S Japan, Borneo, Philippines. **SE Asia** Uncommon resident/breeding visitor N Myanmar, N Laos, W,E Tonkin. Common winter visitor and passage migrant (except S Thailand, Singapore); rare Peninsular Malaysia. Vagrant S Thailand, Singapore. **BREEDING** March–June. **Nest** Cup, in hole in rock or wall, amongst bush roots or on bank etc. **Eggs** 4–6; white to greyish-white, speckled brownish-grey; 21.3 x 15.5 mm (av.).

781 MEKONG WAGTAIL

Motacilla samveasnae Plate **79**

IDENTIFICATION 19 cm. Monotypic. **Male** Combination of black forehead and head-sides, broad white supercilium, and white throat and enclosed neck-patch diagnostic. **Female** Distinctly paler and greyer above, but still showing distinctive combination of features as male. **Juvenile** Like washed-out female, but with less distinct head pattern, dark malar stripe, and smudgy greyish breast. **VOICE** Short sharp ***dzeer*** (sometimes doubled) in flight; otherwise, thin, soft ***tsit***; ***tseeup*** and ***tsriu*** etc. **HABITAT** Sandy and rocky banks and islands with grass and bushland, in Mekong R and its tributaries. **RANGE & STATUS** Endemic. Local resident south-east NE Thailand, north-east Cambodia, S Laos, north-west S Annam. **BREEDING** Undocumented.

782 GREY WAGTAIL *Motacilla cinerea* Plate **79**

IDENTIFICATION 19 cm. *M.c.cinerea*: **Male non-breeding** Slaty-grey crown, ear-coverts and upperparts, narrow whitish supercilium and bright yellow vent distinctive. Resembles female and non-breeding male Western and Eastern Yellow Wagtails but has slatier-grey upperparts, more uniform blackish wing-coverts, and yellowish rump and uppertail-coverts; in flight shows diagnostic white bar along base of secondaries and inner primaries. See Citrine Wagtail. **Male breeding** Underparts all yellow with diagnostic black throat and upper breast. **Female** Like non-breeding male. May show blackish throat-mottling in spring. **Juvenile** Similar to female/non-breeding male but upperparts browner-tinged, supercilium buff-tinged, has dark mottling on breast-sides. **VOICE** Song is a short mechanical series of sharp notes: ***ziss-ziss-ziss-ziss...***, often alternated with higher ***si si si siu*** etc. Calls with a loud disyllabic ***tittick*** or ***tzit-tzit***, recalling White Wagtail but clearly sharper and higher-pitched. **HABITAT** Various open habitats and open forest, often near flowing streams; up to 2,565 m. **RANGE & STATUS** Breeds NW Africa, Palearctic, northern Middle East, northern Pakistan, NW,N India, Nepal, Bhutan, northern China, N,S Korea, Japan; some populations winter south to Africa, Middle East, Indian subcontinent, southern China, Sundas, Philippines, Wallacea, New Guinea, N Australia. **SE Asia** Fairly common to common winter visitor and passage migrant throughout. Arrives as early as 26 June.

783 WESTERN YELLOW WAGTAIL

Motacilla flava Plate **79**

IDENTIFICATION 18 cm. *M.f.thunbergi* [incl. '*plexa*']: **Male breeding** Combination of grey-tinged olive-green upperparts, yellow underparts (throat whiter), and mid-grey crown, nape and rear head-sides, with blackish-grey face to ear-coverts distinctive. Blackish-grey of lores typically extends up onto forehead, lacks supercilium or shows short whitish one behind eye, shows clay-cream edging/tipping on wing-coverts, sometimes has necklace of darkish flecks across upper breast. Told from female Grey Wagtail by obvious pale fringes and tips to wing-coverts, lack of bright yellowish rump, uppertail-coverts and vent, and less pure grey upperparts. See Eastern Yellow Wagtail (particularly subspecies *macronyx*). **Male non-breeding** Head much duller (more concolorous with upperside), underparts whiter. **Female breeding** Duller than male, upperparts dull greyish- to brownish-olive, yellow of underparts duller and less extensive. Male-like birds are rare. **Female non-breeding** Duller still. **First winter** Like female but upperparts tend to be greyer, underparts whiter. See Citrine Wagtail. **VOICE** Not well documented for population that winters in region. Calls of European ***thunbergi*** transcribed as rather loud ***pseeu***; ***pslie***; or ***psie*** etc. Song-types include simple series of very short, scratchy, explosive strophes ***kizZIK-kizZIK*** etc., repeated after short pauses, rather weak, prolonged twittering, and a fast, high ***zi-zi-zi-zi-zi-zi*** or ***zri-zri-zri-zri-zri-zri***. **HABITAT & BEHAVIOUR** As for Eastern Yellow. **RANGE & STATUS** Breeds N Africa, Palearctic (east to W,N Siberia), Middle East, W Mongolia, NW China; winters Africa, Middle East, Indian subcontinent; ? east to southern China. **SE Asia** Status in region unclear, due to confusion with Eastern Yellow, but thought to occur Myanmar, and perhaps as far east as Thailand and Peninsular Malaysia.

784 EASTERN YELLOW WAGTAIL

Motacilla tschutschensis Plate **79**

IDENTIFICATION 18 cm. *M.t.macronyx* (throughout region, west at least to Thailand, Peninsular Malaysia[uncommon], Singapore): **Male breeding** Very difficult to separate from Western Yellow Wagtail but, on average, has slightly paler, cleaner grey forehead to nape (blackish-grey of lores usually not extending up onto forehead as in most *thunbergi* Western), slightly cleaner grey (less extensively blackish-grey) ear-coverts; tends to be brighter and greener above, slightly cleaner yellow underparts appear to only rarely show necklace of darkish flecks; has slightly wider, more clear-cut, and yellower wing-bars. Typically lacks any supercilium. *M.t.tschutschensis* [incl. '*angarensis*', '*simillima*'] (Peninsular Malaysia[commonest form], Singapore, E Tonkin, Cochinchina; ? SW,N,C,S Myanmar) shows pronounced, full-length whitish supercilium, face to ear-coverts grey to blackish, crown and nape bluish-grey, upperparts purer olive-green; *taivana* (W,C,S Myanmar, C,S Thailand, Peninsular Malaysia[rarest form], Singapore, Cambodia, S Laos, E Tonkin, C Annam) has distinctive olive-green crown and head-sides and broad yellow supercilium (probably best treated as distinct species). **Male non-breeding** Head duller (more concolorous with upperside), underparts paler. **Female breeding** Usually like dull version of male, with some apparently inseparable by plumage; typically shows narrow whitish supercilium. **Female non-breeding** duller, but not well documented. *M.t.tschutschensis* typically has more prominent supercilium; *taivana* has broad, clear-cut, yellowish supercilium. **First winter** Like female but upperparts tend to be greyer, underparts whiter. **VOICE** Not well documented for forms/populations that winter in region. Calls somewhat drier and harsher than Western Yellow: ***tzreep***; ***tccreep***; or ***tseerp*** etc.; similar to Citrine Wagtail. Song-types include fast, high ***zi-zi-zi-zi-zi-zi*** or ***zri-zri-zri-zri-zri-zri***. **HABITAT** Various open habitats, often near water; up to at least 1,200 m. **RANGE & STATUS** Breeds C,E Siberia, E Mongolia, NE China, N Japan, N America (west Alaska); winters southern China, Taiwan, Greater Sundas, Philippines, Wallacea, W New Guinea, Bismarck Archipelago, Australia; unconfirmed in Indian subcontinent. **SE Asia** Fairly common to common winter visitor Thailand, Peninsular Malaysia, Singapore, Indochina. Status in Myanmar unclear, due to confusion with Western Yellow. **NOTE** Molecular studies indicating former Yellow Wagtail (*M. flava*) is paraphyletic (Ödeen & Alström 2001, Alström & Ödeen 2002, Voelker 2002), led to division into two species (Banks *et al.* 2004). Subspecies *macronyx* and *taivana* not considered, however, and may possibly constitute one or more additional species. Taxonomic status of eastern *thunbergi* (Western Yellow) populations perhaps not resolved, and sometimes considered allied to this species instead.

785 **CITRINE WAGTAIL**
Motacilla citreola Plate **79**
IDENTIFICATION 18–19 cm. *M.c.citreola*: **Male non-breeding** Resembles female Western and Eastern Yellow Wagtails but shows distinctive yellow lores, supercilium, throat and breast, grey upperparts and whitish undertail-coverts; ear-coverts have dark border and paler centre (often washed yellow) and are encircled by yellow; usually shows prominent white wing-bars. **Male breeding** All-yellow head and underparts diagnostic. Has black nuchal band (often extending to breast-side) and grey mantle, back and rump. *M.c.calcarata* (SW,S,N Myanmar; NW Thailand?) has black mantle to uppertail-coverts. **Female** Similar to non-breeding male. **First winter** Similar to Yellow but upperparts purer grey (Yellow sometimes as grey), prominent whitish supercilium extends down behind ear-coverts (or slightly interrupted) and often bordered above by distinctive blackish line; shows buff wash on forehead, pale lores, all-dark bill and usually more prominent white wing-bars. Gradually develops patches of adult non-breeding plumage. **VOICE** Song is a simple repetition of sounds resembling call-notes. Usually calls with harsh ***dzeep*** or ***brrzreep***, very similar to Eastern Yellow Wagtail. **HABITAT** Marshes, banks of lakes and larger rivers, wet cultivation; up to 450 m. **RANGE & STATUS** Breeds eastern W Palearctic east to C Siberia, Iran, northern Pakistan, NW India, Tibet, W and northern China; winters Indian subcontinent, S China. **SE Asia** Uncommon to locally common winter visitor Myanmar, NW,NE,C Thailand, N Laos. Uncommon passage migrant E Tonkin. Vagrant coastal W Thailand, Singapore, Cochinchina.

FRINGILLIDAE: CARDUELINAE: Siskins, serins, finches, grosbeaks & allies

Worldwide c.141 species. SE Asia 26 species. Variable, small to medium-sized, mostly compact in build with conical bills, rounded wings and notched tails. Most are sexually dimorphic, with colourful males and dull females. Many are at least partly migratory and gregarious, forming flocks outside breeding season. Flight strong, often undulating. Forage in trees and on ground, where typically progress by hopping or short shuffling motion. Feed mainly on various kinds of seeds, for which bills are variously adapted.

786 **EURASIAN SISKIN** *Spinus spinus* Plate **80**
IDENTIFICATION 12 cm. Monotypic. **Male** Small and boldly patterned, with diagnostic black cap and chin contrasting with yellow breast and ear-covert surround. Upperparts dark green with blackish streaks and yellow rump; has broad yellow wing-bars, yellow fringing to flight and tail-feathers, white belly and vent, and dark-streaked flanks and undertail-coverts. See Yellow-breasted Greenfinch. **Female** Lacks black head markings; crown, upperparts, rump and breast prominently dark-streaked, base colour to throat and breast whiter, wing-bars duller and whiter. From Tibetan Serin and juvenile greenfinches by whiter underside and broad, well-defined wing-bars. **Juvenile** Like female but head and upperparts rather browner, ear-coverts, throat and breast prominently dark-streaked. **VOICE** Song is a flowing series of twittering and trilling notes (incorporating avian mimicry), interspersed with occasional drawn-out choking or wheezing notes. Calls with a clear ringing ***tluih*** and ***tilu***, dry ***tet*** or ***tetete***, trilling ***tirrillilit*** or ***tittereee*** and sharp ***tsooeet*** when alarmed. **HABITAT & BEHAVIOUR** Open forest, forest edge, secondary growth; recorded at c.1,500 m. Usually in small flocks. **RANGE & STATUS** Breeds N Africa, Palearctic, NE China, N Japan; some populations winter south to N Africa, Middle East, southern China, N,S Korea, Japan. **SE Asia** Vagrant W Tonkin.

787 **TIBETAN SERIN** *Serinus thibetana* Plate **80**
IDENTIFICATION 12.5 cm. Monotypic. **Male** Small size and unstreaked greenish-yellow plumage distinctive. Shows yellowish supercilium and ear-covert surround, yellow to greenish-yellow rump, yellowish underparts (darker on breast, paler on vent) and prominent greenish-yellow fringing to wing- and tail-feathers. **Female** Upperparts darker greyish-green with blackish streaks, belly and vent whiter, lower breast, flanks and undertail-coverts dark-streaked, yellowish-green wing-bars better defined. See Eurasian Siskin. **Juvenile** Rather duller than female, rump duller, wing-coverts fringed and tipped buffish, underparts paler, breast more heavily streaked. **VOICE** Song is a nasal buzzing 2–3 note ***zeezle-eezle-eeze*** etc., mixed with trills. Flocks give a continuous tremulous twittering. **HABITAT & BEHAVIOUR** Open forest, forest edge; 610–2,135 m. Probably breeds above 1,500 m. Usually in flocks. **RANGE & STATUS** Resident (subject to local movements) Nepal east to Bhutan, SE Tibet, SW/W China. **SE Asia** Recorded (status uncertain) N Myanmar. **BREEDING** Undocumented.

788 **RED CROSSBILL** *Loxia curvirostra* Plate **80**
IDENTIFICATION 16.5–17.5 cm. *L.c.meridionalis*: **Male** Heavy bill with crossed mandibles, big head and largely red plumage, with whitish vent and rather uniform dark brownish wings and tail diagnostic. Crown, mantle, scapulars and back variably mottled/streaked brown and blackish. *L.c.himalayensis* (Myanmar) is much smaller (15 cm) and smaller-billed (mandibles more obviously crossed). See Scarlet Finch. **Female** Mostly dull greenish-grey, with paler, more yellowish rump, breast and belly, and darker streaking overall. **Juvenile** Head and body paler than female, with more prominent dark streaking; shows paler fringing on wing-coverts and tertials. From greenfinches by bigger head, shorter tail, distinctive bill shape and lack of pronounced markings on wings and tail. Males gradually attain red patches on plumage. **VOICE** Song of *meridionalis* is a repeated series of 3–4 well-structured, quite buzzy couplets, ***wi'DRR-wi'DRR-wi'DRR...***, interspersed with quick, uneven series of ***chih*** or ***whiH*** call-notes. In flight, typically calls with Rain Quail-like *whit-it*, or more extended ***whit-whit-whit*** or more spaced ***whit whit whit....*** Sometimes gives chippier ***chi-chu*** and ***chit***, and somewhat deeper ***whiH***, sometimes repeated in series from family groups etc. *L.c.himalayensis* calls with short, quite high and stressed ***chit***, singly or in even or uneven series. **HABITAT & BEHAVIOUR** Mature pine forest, sometimes alder woodland; 1,370–2,900 m. Usually in small flocks. **RANGE & STATUS** Resident (subject to local movements) NW Africa, Holarctic, N Pakistan, N India, Nepal, Bhutan, S Tibet, W and northern China, Japan, Philippines; winter visitor E China, N,S Korea. **SE Asia** Locally common resident (subject to some movements) N Myanmar, S Annam. **BREEDING** January–March. **Nest** Rather shallow cup, on pine branch; 9 m above ground, or higher. **Eggs** 5; pale grey-green, with a few black specks around broader end; dimensions 22.7–24 x 16.6–17 mm (*himalayensis*). **NOTE** The large size and bill of *meridionalis*, along with its distinct vocalisations, suggest that it should be regarded as a full species, 'Vietnamese Crossbill'.

789 **GREY-CAPPED GREENFINCH**
Chloris sinica Plate **80**
IDENTIFICATION 14 cm. *C.s.sinica*: **Male** Told from other greenfinches by greyish crown and nape, rich brown mantle and scapulars, and warm brownish breast and flanks; secondaries broadly fringed greyish-white to buffish-white, rump yellowish. **Female** Like male but head more uniform greyish-brown, upperparts duller, with dark shaft-streaks, rump duller, underparts more washed out. **Juvenile** From similar

Yellow-breasted and Black-headed Greenfinches by rather plain brown wing-coverts and tertials and uniform pale greyish fringing on secondaries; somewhat paler overall. Note range. See Eurasian Siskin. **VOICE** Song is a repeated series of phrases incorporating call-notes, usually beginning with a dry nasal trill and including ***kirr*** and ***korr*** notes. Calls include a nasal ***dzweee*** or ***djeeen*** and twittering ***dzi-dz-i-dzi-i...*** in flight. **HABITAT & BEHAVIOUR** Parks, gardens, coastal casuarinas, open woodland; lowlands. Gregarious. **RANGE & STATUS** Resident (subject to local movements) southern E Palearctic, China (except NW), N,S Korea, Japan. **SE Asia** Local resident E Tonkin, coastal C and S Annam. **BREEDING** March–April. **Nest** Deep cup, in bush, tree or bamboo. **Eggs** 2–5, pale green, sparsely speckled and streaked black and reddish to yellowish, occasionally over reddish-grey underspots; 18.6 x 13.5 mm (av.).

790 BLACK-HEADED GREENFINCH

Chloris ambigua Plate **80**

IDENTIFICATION 13–13.5 cm. *C.a.ambigua*: **Male** Dull olive-green upperparts, mottled dull olive-green throat, breast and flanks and blackish crown and ear-coverts distinctive. Shows slightly yellower rump, greyish-white greater covert bar and incomplete fringing on tertials and secondaries, and paler belly-centre and vent. See Vietnamese Greenfinch. Note range. **Female** Like male but crown and ear-coverts duller, throat, breast and flanks paler and more uniform, flanks tinged brown. From Yellow-breasted by lack of obvious supercilium, plainer upperparts and greyish wing-bars. **Juvenile** Like Yellow-breasted but body-streaking heavier and darker, base colour to upperparts darker and greener, wing-bars buffier and less defined, yellow slash on flight-feathers broader. See Grey-capped Greenfinch. **VOICE** Song is a long wheeze: ***wheeeeeu***, ***wheeeeee*** or ***jiiiiii***, usually repeated at intervals and punctuated by call-notes. Typical call is a jumbled jingling ***titutitu*** and ***titu-titu titu-tittrititititit*** etc., mixed with harder ***chututut*** or ***jututut***, quiet rising buzzy ***jieuu*** and soft ***chu-chu*** etc. **HABITAT & BEHAVIOUR** Open forest (including pines), secondary growth and scrub, cultivation; 1,200–2,565 m, locally down to 1,010 m in winter. Often in flocks, sometimes quite large. **RANGE & STATUS** Resident (subject to local movements) extreme NE India, SE Tibet, SW China. **SE Asia** Locally common resident N,S(east),E Myanmar, N Laos (scarce), W Tonkin. Scarce winter visitor NW Thailand. **BREEDING** July–October. **Nest** Neat cup, on tree-branch (often pine). **Eggs** 3–4; pale greenish-blue, sparingly spotted black at broader end; 17.8 x 13.2 mm (av.).

791 VIETNAMESE GREENFINCH

Chloris monguilloti Plate **80**

IDENTIFICATION 13.5 cm. **Male** Told from other greenfinches by black head, blackish-green nape to back and yellow throat and underparts, with dark mottling on breast. Lacks pale fringing on tertials and secondaries. See Black-headed Greenfinch. Note range. **Female** Similar to male but duller and paler, with darker-streaked nape to back and more extensive dark mottling on breast and flanks. **Juvenile** Similar to female but head duller and greener, underparts paler and duller. **VOICE** Song is a slowly rising ***seeuuu-seeuuu-seeuuu*** or ***teoo-teoo-teoo***, followed by a dry nasal ***weeeee*** or ***chweee***. Calls include a twittering ***chi-chi-chi...*** and dry nasal ***zweee***. **HABITAT & BEHAVIOUR** Open pine forest; 1,050–1,900 m. Typically in small flocks. **RANGE & STATUS** Endemic. Locally common resident S Annam. **BREEDING** December–May. **Nest** Neat cup, on pine branch. Otherwise undocumented.

792 YELLOW-BREASTED GREENFINCH

Chloris spinoides Plate **80**

IDENTIFICATION 13.5–14 cm. *C.s.heinrichi*: **Male** Blackish crown and head-sides and yellow supercilium, throat and underparts diagnostic. Nape to back dark greenish, rump yellow; has two broad yellow wing-bars and diagonal yellow band across flight-feathers. **Female** Similarly patterned to male but much duller; nape to back vaguely streaked darker, yellow of body-plumage paler. **Juvenile** Most likely to be confused with Black-headed Greenfinch but body-streaking paler and browner, base colour to upperparts somewhat paler and browner (less greenish), wing-bars yellower and rather more sharply defined, yellow slash on flight-feathers narrower. See Grey-capped Greenfinch and Eurasian Siskin. Note range. **VOICE** Song is said to be similar to extralimital European Greenfinch *C. chloris*, but higher-pitched. Calls with light twittering, interspersed with harsh ***dzwee***, ***beez*** or ***zeez*** notes and drawn-out ***sweee-tu-tu*** (dropping slightly at end). **HABITAT & BEHAVIOUR** Forest edge, secondary growth, alder groves, cultivation; 1,220–2,550 m. Often in flocks. Has stiff-winged, bat-like display flight. **RANGE & STATUS** Resident N Pakistan, NW,N,NE Indian subcontinent, S,SE Tibet. **SE Asia** Locally common resident W Myanmar. **BREEDING** June. **Nest** Neat cup, on tree-branch; 2–20 m above ground. **Eggs** 3–5; whitish (slightly green-tinged), usually with irregular ring of tiny dark brown spots around broader end and a few scattered elsewhere; 18.7 x 13.7 mm (av.; India).

793 PINK-RUMPED ROSEFINCH

Carpodacus eos Plate **81**

IDENTIFICATION 15 cm. Monotypic. **Male** Shows distinctive combination of pinkish-tinged greyish crown and upperparts with bold blackish streaking and contrasting, rather uniform, reddish-pink rump, supercilium, lower head-sides, throat and underparts. Face darker and redder. See Sharpe's Rosefinch. **Female** Resembles juvenile Common but has defined paler supercilium, greyer upperparts and whiter underparts, with denser, regular, darker streaking overall; wing-bars and tertial fringes indistinct, bill narrower. From Dark-rumped and Spot-winged by overall paler coloration, whiter underparts with broader dark streaks, no tertial markings. **VOICE** Calls with an assertive ***pink*** or ***tink*** and bunting-like ***tsip*** or ***tsick***. Sometimes gives a harsh ***piprit*** or tinny rattling ***tvitt-itt-itt-itt***. **HABITAT** Forest edge, secondary growth and scrub, cultivation borders; recorded in lowlands (could occur in mountains). **RANGE & STATUS** Resident (subject to local movements) E Tibet, W China. **SE Asia** One specimen record from NW Thailand in October 1968 (may have been misidentified).

794 VINACEOUS ROSEFINCH

Carpodacus vinaceus Plate **81**

IDENTIFICATION 15 cm. *C.v.vinaceus*: **Male** Resembles Dark-breasted Rosefinch but smaller and stubbier-billed; body almost uniform dark red, with contrasting pink supercilium, paler red rump and undertail-coverts and distinctive whitish to pale pinkish tertial spots. **Female** Crown and upperparts rather dull dark brown with darker streaks, rather plain rump and uppertail-coverts; underparts dull, deep, dark buffy-brown with dark streaking, mainly on throat and breast. From Dark-rumped and Sharpe's Rosefinches by plainer head-sides (without dark ear-coverts and contrasting supercilium), less distinctly dark-streaked upper- and underparts. See Dark-breasted Rosefinch. **Juvenile** Like female but more boldly streaked above and below; closer to female Dark-rumped but still separable by smaller size and plainer head. **VOICE** Song is a simple repeated ***pee-dee*** or ***do-do*** (c.2 s duration). Calls include a hard assertive ***pwit*** or ***zieh***, with whiplash-like quality (often repeated and given in introduction to song), thin high-pitched tip, faint ***tink***, ***pink*** and bunting-like ***zick***. **HABITAT** Edge of broadleaved evergreen forest, secondary growth and scrub, bamboo; 1,830–2,745 m. **RANGE & STATUS** Resident N India, Nepal, SW,W,C China, Taiwan. **SE Asia** Scarce to uncommon resident N,E(north) Myanmar. **BREEDING** India: June–July. Otherwise undocumented?

795 DARK-RUMPED ROSEFINCH

Carpodacus edwardsii Plate **81**

IDENTIFICATION 17 cm. *C.e.rubicunda*: **Male** Recalls Dark-breasted Rosefinch but shows more defined upperpart-streaking, more extensively dark ear-coverts and contrasting, complete pale pink supercilium, pink lower ear-coverts and throat, pinker (less red) belly, paler, pinkish markings on outer webs of tertials and shorter, thicker bill. See Vinaceous Rosefinch. **Female** Very similar to Sharpe's Rosefinch but larger and bulkier, mantle lacks odd whitish-buff streaks, supercilium somewhat duller and less pronounced, underparts darker, deeper buffish-brown with less contrasting streaks (narrower on lower throat and breast). See Pink-rumped and juvenile Vinaceous Rosefinches. **Juvenile** Similar to female. **First-summer male** Similar to female but shows dark red tinge on crown to uppertail-coverts and paler, pinker supercilium and face; breast heavily washed deep pink. **VOICE** Call is an abrupt, rather shrill, high-pitched metallic ***zwiiih*** or ***tswii***. **HABITAT** Undergrowth in more open broadleaved evergreen forest, forest edge, secondary growth and scrub; 1,980–3,050 m. **RANGE & STATUS** Resident Nepal, Bhutan, NE India, S Tibet, SW,W China. **SE Asia** Local resident N Myanmar. **BREEDING** Not definitely documented.

796 SHARPE'S ROSEFINCH
Carpodacus verreauxii Plate **81**

IDENTIFICATION 16.5 cm. Monotypic. **Male** Told from other rosefinches by combination of rather plain dark reddish crown, odd narrow pinkish-white streaks on mantle, pink rump, broad pink supercilium, deep pink underparts with darker mottling/streaking on throat and breast, and prominent pinkish-white tertial markings. See Pink-rumped Rosefinch. **Female** Very similar to Dark-rumped Rosefinch but smaller and slimmer, shows some thin whitish-buff streaks on mantle, paler and more pronounced supercilium and paler underparts with more contrasting dark streaks (which are also broader on lower throat and breast). See Vinaceous and Pink-rumped Rosefinches. **Juvenile** Like female. **VOICE** Calls include short, piercing, metallic, emphasised ***spink..spink....*** **HABITAT** Edge of evergreen forest, secondary growth, scrub, bamboo; 2,135–2,895 m. **RANGE & STATUS** Resident SW,W China. **SE Asia** Uncommon resident N,E(north) Myanmar. **BREEDING** Undocumented. **NOTE** Split from extralimital Spot-winged Rosefinch *C. rodopeplus* following Rasmussen & Anderton (2005).

797 COMMON ROSEFINCH
Carpodacus erythrinus Plate **81**

IDENTIFICATION 16–16.5 cm. *C.e.roseatus*: **Male breeding** Easily told from other rosefinches by red head and body, with darker, vaguely streaked/mottled mantle, scapulars and uppertail-coverts and whitish vent. Shows darker line through eye, dark brown wings with two red bars on coverts, and pale fringing on tertials and flight-feathers. *C.e.erythrinus* (visitor N Myanmar, NW Thailand; Indochina?) is less vivid red and shows whiter (more contrasting) belly, lower flanks and vent, paler lower breast, greyer-brown upperparts and duller wing-bars. **Male non-breeding** Like female. **Female** Rather nondescript, with greyish-brown upperparts and dull whitish throat and underparts (duller on lower throat and breast); has delicate darker streaking on crown, mantle, lower throat, breast, upper belly and flanks, and two narrow buffy-whitish wing-bars. From other streaked female rosefinches by rather plain head-sides (producing beady-eyed appearance), lack of pale supercilium, more lightly streaked upperparts and plain rump. Resembles some buntings but has plainer head-sides, upperparts and wings, no white on tail, thicker bill. **Juvenile** Browner above and below than female, with stronger dark streaking above, broader, buffier wing-bars, and pale buffish outer fringes of tertials. See Pink-rumped, Dark-rumped and Sharpe's Rosefinches. **VOICE** Song is a monotonously repeated, slowly rising whistle: ***weeeja-wu-weeeja*** or ***te-te-wee-chew***. Typically calls with a clear whistled ***ooeet*** or ***too-ee*** and sharp nasal ***chay-eeee*** when alarmed. **HABITAT** Forest edge, secondary growth and scrub, cultivation; up to 2,565 m. **RANGE & STATUS** Breeds Palearctic, Iran, Afghanistan, northern Pakistan, N,NE India, Nepal, Bhutan, southern and E Tibet, W,C and northern China; winters Indian subcontinent (except islands), southern China. **SE Asia** Scarce to locally common winter visitor Myanmar, W,NW,NE Thailand, N Laos, W,E Tonkin. May breed N Myanmar. **BREEDING** India: June–August. **Nest** Cup, in bush or small tree; up to 2 m above ground. **Eggs** 3–5; deep blue, sparsely blotched dark brown and black, particularly at broader end (sometimes very lightly marked or spotless); 20.8 x 14.5 mm (av.; *roseatus*).

798 DARK-BREASTED ROSEFINCH
Carpodacus nipalensis Plate **81**

IDENTIFICATION 16–16.5 cm. *C.n.intensicolor*: **Male** Dark brownish-red upperparts and breast and contrasting red forecrown, pinkish-red rear supercilium, lower ear-coverts, throat and belly diagnostic. Lacks distinct pale wing markings. From Dark-rumped Rosefinch by red forehead and belly, redder supercilium and lower ear-coverts, dark-centred undertail-coverts and narrower bill. See Vinaceous Rosefinch. **Female** From other rosefinches by unstreaked drab brown plumage. Shows very vague darker mantle-streaking, warm buffish-brown wing-bars and slightly buffier-brown outer fringes to tertials; underparts paler and greyer than upperparts. **Juvenile** Similar to female. **VOICE** Song is a series of monotonous chipping notes. Calls with a fairly quick, harsh, slightly nasal ***eeehh*** or ***yeeh***, repeated after shortish intervals. **HABITAT** Underbrush in open broadleaved evergreen forest, forest edge, secondary growth and scrub, cultivation borders; 1,525–2,900 m, rarely down to 500 m NW Thailand. **RANGE & STATUS** Resident (subject to altitudinal movements) N,NE India, Nepal, Bhutan, S Tibet, W China. **SE Asia** Recorded (status uncertain) N Myanmar (probably resident), W Tonkin. Rare winter visitor E Myanmar, NW Thailand. **BREEDING** Undocumented.

799 SCARLET FINCH
Haematospiza sipahi Plate **80**

IDENTIFICATION 19 cm. Monotypic. **Male** Unmistakable, with uniform bright red head and body and pale yellowish bill. Wings and tail black with scarlet feather-fringes. See Red Crossbill. **Female** Rather uniform scaly brownish-olive, with paler underparts, distinctive pale bill and sharply defined bright yellow rump. See Yellow-rumped Honeyguide. **Juvenile** Similar to female. First-year male Differs from female by orange-rufous tinge to head, upperparts and wing-feather fringes, more rufous-tinged throat and underparts, and bright orange rump. **VOICE** Song is a clear liquid ***par-ree-reeeeeee***. Calls with a loud clear ***too-eee*** or ***pleeau*** and ***kwee'i'iu*** or ***chew'we'auh***. **HABITAT** More open broadleaved evergreen forest, forest edge, secondary growth; 1,160–2,100 m, locally down to 200 m in winter. **RANGE & STATUS** Resident (subject to altitudinal movements) N,NE Indian subcontinent, S,SE Tibet, SW China. **SE Asia** Uncommon resident W,N Myanmar. Recorded (status uncertain) NW Thailand, N Laos, W Tonkin. **BREEDING** India: May–June. **Nest** Bulky cup, in tree; 7–12 m above ground. **Eggs** 4; pale blue, with small red-brown or light brown and purplish-brown blotches (often ringed or capped at broader end); 22.1–25.4 x 17–18 mm.

800 GREY-HEADED BULLFINCH
Pyrrhula erythaca Plate **81**

IDENTIFICATION 15 cm. *P.e.erythaca*: **Male** Unmistakable, with mid-grey head and upperparts, black face with narrow whitish border, and deep orange breast and belly. Wings and tail glossy black, greater coverts with broad greyish bar, lower back with black band, rump white and vent whitish. **Female** Resembles male but mantle, scapulars and lower throat to belly dull brown tinged pinkish-buff, band on lower back grey rather than black, wing-bar browner. **Juvenile** Like female but head and body more uniform buffish-brown, face black but no black on forehead. **First-winter male** Red of adult replaced by dull yellow-orange (also splashed on throat). **VOICE** Calls with soft ***soo-ee*** or ***poo-ee***, often repeated. **HABITAT** Open broadleaved evergreen forest, forest edge, secondary growth, scrub; 2,135–3,655 m. **RANGE & STATUS** Resident Nepal, Bhutan, NE India, SE Tibet, SW,W,NC China, Taiwan. **SE Asia** Uncommon resident N Myanmar. Vagrant E Tonkin. **BREEDING** Himalayas: June–July. **Nest** Frail cup, on tree-branch; up to 3.6 m above ground, or higher.

801 BROWN BULLFINCH

Pyrrhula nipalensis Plate **81**

IDENTIFICATION 16.5 cm. *P.n.waterstradti*: **Adult** Rather uniform greyish-brown head and body, with blackish face and chin, whitish ear-coverts and blackish wings and tail diagnostic. Shows faint dark scales on crown, black band on back and white rump-band; wings and tail glossed bluish to purplish (greater coverts mostly similar to upperparts), vent whitish. *P.n.ricketti* (N Myanmar,W Tonkin) and *victoriae* (W Myanmar) show darker crown with pale scaling and white of ear-coverts reduced to crescent or strip below/behind eye; unnamed subspecies (S Annam) appears (from field observation) to have mostly whitish head (apart from throat). **Juvenile** Lacks contrasting dark head markings of adult, body-plumage tinged buffish (particularly above), shows buffish bar on median coverts. **VOICE** Song is a quickly repeated, mellow ***u'iih pi-huu*** with a rather clicking nasal first note; in W Myanmar a similar ***ip-pr'ipi-piru*** (mellow throughout), in S Annam a quick mellow ***ip'ipi-you.*** In S Annam, calls with a repeated ***pirr-pirru*** (lower, rolling first note), and repeated hurried ***per-you*** (more forced first note). **HABITAT & BEHAVIOUR** More open broadleaved evergreen and mixed evergreen and coniferous forest, forest edge, secondary growth; 1,050–2,520 m. Often in small flocks. **RANGE & STATUS** Resident N,NE Indian subcontinent, SE Tibet, SW,SE China, Taiwan. **SE Asia** Scarce to uncommon resident W,N Myanmar, Peninsular Malaysia, W Tonkin, S Annam. **BREEDING** January–July. **Nest** Cup, on tree-branch (often conifer); 1.8–9 m above ground. **Eggs** 2; pale blue, with small blackish spots at broader end.

802 CRIMSON-BROWED FINCH

Propyrrhula subhimachala Plate **80**

IDENTIFICATION 20.5 cm. Monotypic. **Male** Larger and deeper-billed than rosefinches and virtually unstreaked; distinctive warm to reddish-brown head and upperparts, greyish underparts with contrasting red rump, forehead, throat and breast. Shows some darker streaking on crown, mantle and scapulars, pinkish speckling on throat and breast, and greenish fringing on flight-feathers and tertials. See breeding male Common Rosefinch. **Female** Upperside dull green with more yellowish rump, grey head-sides and nape and yellowish forehead and upper breast. Throat whitish, streaked grey. **Juvenile** Similar to female. **First-year male** Red of head and breast more orange than adult, nape and rump to uppertail-coverts warm rufous. **VOICE** Song is a bright, varied warble. Calls with a simple ***ter-ter-tee.*** **HABITAT** Broadleaved evergreen and coniferous forest, forest edge, secondary growth and scrub; 1,830–3,050 m. **RANGE & STATUS** Resident Nepal, Bhutan, NE India, S,SE Tibet, W China. **SE Asia** Uncommon resident W(sight record),N Myanmar. **BREEDING** Undocumented.

803 PLAIN MOUNTAIN-FINCH

Leucosticte nemoricola Plate **82**

IDENTIFICATION 17 cm. *L.n.nemoricola*: **Adult** Upperparts drab brownish with broad blackish-brown streaks, rump plainer and greyer, uppertail-coverts broadly scaled whitish; underparts rather plain greyish-brown with whiter vent and dark streaking on upper flanks and undertail-coverts. Shows broad pale supercilium, whitish outer fringes to tail-feathers, and black-marked brownish wing-coverts and tertials, with two narrow but prominent whitish bars on former. Could possibly be confused with some female sparrows or buntings. Note habitat and range. **Juvenile** Warmer brown overall, crown, nape and ear-coverts much more uniform, warmer-coloured and unstreaked (darker on crown). **VOICE** Sings with a sharp twittering ***rick-pi-vitt*** or ***dui-dip-dip-dip.*** Call is a soft sparrow-like twittering ***chi-chi-chi-chi..*** **HABITAT & BEHAVIOUR** Open, often rocky areas, cultivation; 1,830–2,285 m (winter). Breeds above 3,000 m in Himalayas. Usually in flocks (sometimes large), foraging on open ground. **RANGE & STATUS** Resident (subject to altitudinal movements) N Pakistan, N,NE India, Nepal, Bhutan, Tibet, W,NW China. **SE Asia** Recorded (status uncertain) N Myanmar. **BREEDING** India: July–August. **Nest** Shallow cup, amongst rocks or in animal burrow. **Eggs** 3–4; white; 20.5 x 15.1 mm (av.).

804 JAPANESE GROSBEAK

Eophona personata Plate **81**

IDENTIFICATION 21–23 cm. *E.p.magnirostris*?: **Male** Like Yellow-billed Grosbeak, but larger, with heavier, yellow-tipped bill, black of head restricted more to crown and face, body much greyer, wings with white patch on mid-primaries only. Tertials as brownish-grey mantle, or washed browner. **Female** Like male but perhaps with smaller white wing-patch and browner tertials. **Juvenile** Browner overall than adults (but not as brown as Yellow-billed), wing-coverts tipped pale buffish, black of head reduced to mask; flight-feathers similar. **VOICE** Short, hard ***tak..tak..*** in flight. Sings with short series of 4–5 fluty whistled notes. **HABITAT** Parks and gardens, edges of cultivation, open woodland; recorded in lowlands. **RANGE & STATUS** Breeds Ussuriland, NE China, N Korea, Japan (resident in C,S); winters southern China, rarely Taiwan. **SE Asia** Vagrant N Laos, E Tonkin.

805 YELLOW-BILLED GROSBEAK

Eophona migratoria Plate **81**

IDENTIFICATION 19–20 cm. *E.m.migratoria*: **Male** Easily told by combination of greyish-brown body (darker on mantle and back), glossy bluish-black head and tail, and glossy bluish-black wings with prominent white markings on tertials, flight-feather tips and primary coverts. Bill yellow with dark tip, base and cutting edge, rump and uppertail-coverts paler than rest of upperparts, flanks buffy-rufous, vent white. **Female** Head brownish-grey with darker face, has less white on wings, underparts more uniformly greyish, with duller flanks. **Juvenile** Similar to female but head and underparts initially buffier-brown with whitish throat, upperparts browner; shows two narrow, buffish wing-bars. **VOICE** Song consists of various whistles and trills. Call is a loud ***tek-tek.*** **HABITAT** Open woodland, secondary growth, scrub, cultivation, parks and gardens; up to 800 m. **RANGE & STATUS** Breeds SE Siberia, Ussuriland, C,E,NE China, N,S Korea; some winter south to S China, Japan. **SE Asia** Vagrant N Myanmar, NE Thailand, N Laos, E Tonkin.

806 WHITE-WINGED GROSBEAK

Mycerobas carnipes Plate **81**

IDENTIFICATION 24 cm. *M.c.carnipes*: **Male** Sooty-black plumage, with greenish-yellow back, rump, belly and vent, and white patch on base of primaries diagnostic. Shows greenish-yellow tips to greater coverts, outer webs of tertials and lower scapulars. **Female** Paler, greyer version of male, with grey of breast merging into yellowish-green of vent; has whitish streaks on ear-coverts, throat and upper breast, reduced white and yellowish wing markings, and greener rump to uppertail-coverts. **Juvenile** Like female but browner, with pale fringes to feathers of head and mantle. Male gradually attains dark patches on crown, upperparts and breast. **VOICE** Song usually consists of 4–5 notes introduced by call: ***add-a-dit un-di-di-di-dit*** or ***add-a-dit dja-dji-dji-dju*** etc. Calls include a soft nasal ***shwenk*** or ***chwenk***, squawking ***wit*** or ***wet*** (often repeated) and ***wet-et-et*** and ***add-a-dit*** etc. **HABITAT** Stunted vegetation at high altitudes, scrub, cultivation; recorded at 3,655 m. **RANGE & STATUS** Resident (subject to local movements) C Asia, Iran, Afghanistan, Pakistan, NW,N,NE India, Nepal, Bhutan, Tibet, W,N,NW China. **SE Asia** Uncommon resident N Myanmar. **BREED-**

ING June–August. **Nest** Deep cup, in bush or tree; up to 20 m above ground. **Eggs** 2–3; pale greenish-grey (tinged pink), streaked and spotted deep purple-black over scant grey to pale purple undermarkings; 27 x 19.1 mm (av.).

807 COLLARED GROSBEAK
Mycerobas affinis Plate **81**

IDENTIFICATION 24 cm. Monotypic. **Male** Unmistakable; large with black head, mantle-sides, scapulars, wings and tail, and bright yellow nape, mantle-centre, back, rump and underparts. Nape and rump flushed orange-rufous. **Female** Head grey, upperparts pale greyish-green (greener on upper mantle, rump and uppertail-coverts), underparts plain yellowish-olive, wings similar to upperparts, primaries and tail blackish. **Juvenile male** Similar to adult but yellow plumage-parts tinged olive, head duller, with greyish-brown mottling on throat. **Juvenile female** Similar to adult but chin and throat paler, underparts duller, rump shows variable amount of yellow. **VOICE** Song is a loud, clear, rising, piping, 5–7 note whistle: ***ti-di-li-ti-di-li-um*** etc. Has a second song consisting of loud creaky sounds, interspersed with musical notes (constantly repeated). Calls include a rapid mellow ***pip-pip-pip-pip-pip-pip-ugh*** and sharp ***kurr*** (often rapidly repeated) when alarmed. **HABITAT** Broadleaved evergreen and mixed broadleaved and coniferous forest, forest edge, secondary growth; 2,500–3,655 m. **RANGE & STATUS** Resident (subject to some local movements) N,NE India, Nepal, Bhutan, SE Tibet, W China. **SE Asia** Uncommon resident N Myanmar. Vagrant NW Thailand. **BREEDING** Undocumented.

808 SPOT-WINGED GROSBEAK
Mycerobas melanozanthos Plate **81**

IDENTIFICATION 23 cm. Monotypic. **Male** Black plumage, with yellow breast to undertail-coverts and whitish tips to greater coverts, secondaries and outer webs of tertials diagnostic. Black of upperparts glossed bluish-grey. See White-winged Grosbeak. **Female** Paler, with yellow streaks on crown to mantle, streaky yellow supercilium and lower head-sides, and all-yellow underparts with blackish malar line and blackish streaking on breast and belly. Shows buffy-white wing-bars, tips to secondaries and outer webs of tertials. **Juvenile** Similar to female but dark plumage-parts a shade paler, yellow parts paler and less bright, head-sides, throat, breast and flanks tinged buffish. **VOICE** Song is a loud, melodious ***tew-tew-teeeu***. Also gives some mellow oriole-like whistles: ***tyop-tiu*** or ***tyu-tio*** and rising human-sounding ***ah***. Calls include a rattling ***krrr*** or ***charrarauk***. Feeding flocks produce a cackling chorus. **HABITAT & BEHAVIOUR** Broadleaved evergreen forest, mixed broadleaved and coniferous forest, forest edge, secondary growth; 1,400–2,440 m, locally down to 300 m in winter. Usually found in flocks, often fairly large. **RANGE & STATUS** Resident (subject to some movements) NE Pakistan, N,NE Indian subcontinent, SW,W China. **SE Asia** Uncommon resident W,N,E Myanmar, NW Thailand, N Laos, W Tonkin. **BREEDING** March–April. **Nest** Cup, in tree; 4 m above ground or higher. **Eggs** 2–3; pale green, blotched and streaked dark reddish-brown over pale grey to brown undermarkings; 26.6–30 x 20.5–20.6 mm.

809 GOLD-NAPED FINCH
Pyrrhoplectes epauletta Plate **80**

IDENTIFICATION 15 cm. Monotypic. **Male** Easily identified by small size and blackish plumage, with contrasting golden-orange crown/nape-patch and shoulder-spot and white inner fringes to tertials. **Female** Differs markedly but shows distinctive olive-green hindcrown and nape, grey forehead, face and upper mantle and dull rufescent-brown to drab rufous-chestnut remaining plumage, with dark flight-feathers and tail, and white inner fringes to tertials. **Juvenile** Similar to female but deeper, richer brown, with more uniformly brownish-grey head. Male gradually attains patches of adult plumage. **VOICE** Song is a rapid high-pitched ***pi-pi-pi-pi***. Usual calls are a thin, high-pitched ***teeu***, ***tseu*** or ***peeuu***; ***purl-ee***; and squeaky ***plee-e-e***. **HABITAT** Undergrowth in open broadleaved evergreen forest, forest edge, secondary growth, scrub; 1,705–2,135 m. **RANGE & STATUS** Resident N,NE India, Nepal, Bhutan, S,SE Tibet, SW China. **SE Asia** Uncommon resident N Myanmar. **BREEDING** Undocumented.

FRINGILLIDAE: FRINGILLINAE: Brambling & chaffinches

Worldwide 3 species. SE Asia 2 species. Smallish with strong conical bills, used for shelling seeds, which form bulk of diet. Plumage strongly patterned. Forage on ground, with quick walk interspersed with hops.

810 BRAMBLING *Fringilla montifringilla* Plate 79

IDENTIFICATION 15.5–16 cm. Monotypic. **Male non-breeding** Combination of blackish head and mantle with heavy grey to brown scaling, pale orange throat, breast, flanks and scapulars, and white rump diagnostic. Nape and neck mostly greyish, bisected by blackish line, bill yellowish with dark tip; has rufous-orange bar on greater coverts, dark streaks/spots on lower flanks, and white vent. **Male breeding** Uniform black head and mantle, brighter orange throat, breast, flanks and scapulars, whiter greater covert bar, blackish bill. **Female** Like non-breeding male but crown and ear-coverts plainer greyish-brown, nape and neck smoother grey with more contrasting dividing line, throat, breast, flanks and scapulars slightly duller and patchier. **Juvenile** Resembles female but pale parts of head buffier. **VOICE** Song is a simple buzzing ***rrrrrhuh***. Calls include a hard nasal ***te-ehp***, slightly nasal ***yeck*** in flight, sharp rasping ***zweee*** or ***tsweek*** and repeated silvery ***slitt*** notes when agitated. **HABITAT** Open forest, cultivation; recorded at c.1,400–1,500 m (could occur lower). **RANGE & STATUS** Breeds northern Palearctic; mostly winters south to N Africa, W and southern Europe, Middle East, China, N,S Korea, Japan. **SE Asia** Vagrant NW Thailand, W Tonkin.

811 CHAFFINCH *Fringilla coelebs* Plate 79

IDENTIFICATION 16 cm. *F.c.coelebs*. **Male non-breeding** From similar Brambling by blue-grey crown and nape with dark brownish lateral bands, deep vinous-pinkish face, ear-coverts and underparts, dark vinous-brown mantle with vague dark streaks, grey and brown scapulars and greenish-grey rump. **Male breeding** Crown and nape smooth blue-grey, forehead black, face, ear-coverts and underparts brighter and more rufous-tinged, scapulars mostly blue-grey. **Female** Duller than Brambling with greyish-brown mantle and scapulars, dull greyish-buff (unstreaked) underparts (vent whiter), dull buffish-white bar on greater coverts, and greenish rump; darker band along side of crown and nape much less contrasting. **Juvenile** Like female but nape and rump browner. **VOICE** Song is a bright, loud, almost rattling phrase, introduced by rapidly repeated sharp notes: ***zitt-zitt-zitt-zitt-sett-sett-chatt-chiteri-idia***. Call is a loud sharp ***fink***, unobtrusive ***yupp*** in flight (softer than similar call of Brambling) and sharp fine ***ziih*** notes when agitated. **HABITAT** Open forest, secondary growth, cultivation; lowlands. **RANGE & STATUS** Breeds N Africa, W,C Palearctic, N Iran; more northerly populations winter to south, N Africa, Middle East, northern Pakistan, sporadically N India, Nepal, Bhutan. **SE Asia** Vagrant NW Thailand.

EMBERIZIDAE: Buntings & allies

Worldwide c.321 species. SE Asia 15 species. Small to smallish, with strong conical bills designed for shelling grass seeds, which form bulk of diet (also includes some insects). Gregarious, with many species highly migratory. Plumages generally streaky and cryptic (particularly females).

812 CRESTED BUNTING

Emberiza lathami Plate **83**

IDENTIFICATION 16.5–17 cm. Monotypic. **Male breeding** Unmistakable, with blackish head and body, chestnut wings and tail and long erect crest. **Male non-breeding** Blackish body feathers edged buffish-grey. **Female breeding** Much paler and browner, with less chestnut wings and tail and shorter crest. Upperparts olive-brown, streaked darker; underparts paler with faint dark breast-streaking. Largely chestnut wings and tail, short crest and lack of white on tail rules out other buntings. **Female non-breeding** A little darker than breeding, mantle sandy-brown and more diffusely streaked. **Juvenile** Like female breeding but slightly darker, with shorter crest; underparts buffier with more extensive but fainter breast-streaking. Male develops black blotching on body-plumage. **VOICE** Song is a brief, monotonous, falling ***tzit dzit dzit see-see-suee*** or ***tzit dzit tzit-tzitswe-e-ee-tiyuh*** etc. (introductory notes hesitant and slightly grating). Call is a soft ***tip*** or ***tup***, particularly when flying. **HABITAT & BEHAVIOUR** Cultivation, scrub, tall grass; up to 2,565 m. Often roosts in sizeable numbers. **RANGE & STATUS** Resident (subject to local movements) N Pakistan, northern Indian subcontinent, SE Tibet, W and southern China. **SE Asia** Uncommon to locally common resident (subject to local movements) Myanmar, N,C Laos, W,E Tonkin. Locally common winter visitor W,NW Thailand. **BREEDING** April–August. **Nest** Cup, on ground or in hole in bank or wall. **Eggs** 3–4; whitish (tinged greenish to buffish), spotted shades of red, brown or purple (usually more densely around broader end); 20.1 x 16 mm (av.). Brood-parasitised by Eurasian Cuckoo.

813 RED-HEADED BUNTING

Emberiza bruniceps Plate **82**

IDENTIFICATION 16 cm. Monotypic. **Male non-breeding** Bright colours of male breeding obscured by pale buffish to greyish feather-tips. **Male breeding** Unmistakable, with chestnut-red crown, head-sides, throat and upper breast, and yellow rump and lower underparts. Mantle and scapulars yellowish-olive, with blackish streaks. See Chestnut Bunting. **Female non-breeding** Can be difficult to separate from Black-headed Bunting, but buffier head-sides contrast less with throat, shows more greenish-yellow lower back and uppertail-coverts, warm buffish breast, less yellow remainder of underparts; throat typically whiter than belly. Note primary projection, as for first winter. **Female breeding** May have rufous tinge to forehead and throat, and/or yellower rump and underparts. **First winter** Extremely similar to Black-headed, which see for differences. **VOICE** Calls with sparrow-like ***chupp*** and ***chuh***, and short, dry, buzzy ***zrit*** or ***zrip***, as well as fast series of clicks, ***ptr'r'r***. Song is very similar to Black-headed, but with shorter and higher introductory notes, most notes shorter, and strophe more strongly descending. **HABITAT** Open country, scrub, cultivation; lowlands. **RANGE & STATUS** Breeds C Asia, NE Iran, NW,E Afghanistan, WC Pakistan (N Baluchistan), W Mongolia, NW China; winters India (south of Gangetic Plain). Stragglers reach Europe, Israel, Nepal, NE India, Sri Lanka, SE Tibet, S,E China, S Korea, Japan. **SE Asia** Vagrant NW Thailand.

814 BLACK-HEADED BUNTING

Emberiza melanocephala Plate **82**

IDENTIFICATION 16–18 cm. Monotypic. **Male non-breeding** Duller version of male breeding, with pale fringes to head and upperpart feathers and duller underparts. **Male breeding** Black crown and head-sides, rufous-chestnut nape and upperparts and all-yellow underparts diagnostic. See Yellow-breasted Bunting. **Female non-breeding** From other buntings by relatively large size and long bill, plain washed-out appearance, and no white on outertail-feathers. Upperparts pale sandy-brown with indistinct fine darker streaks, underparts plain pale buffish (often tinged yellow), with pale yellow undertail-coverts; wings darker with whitish feather-fringes. See Red-headed Bunting. **Female breeding** May show darker head and some rufous on upperparts, resembling very washed-out version of male non-breeding. **First winter** Although sometimes almost impossible to separate from Red-headed in the field, the following features can confirm identification collectively: five (rarely six) primary tips clearly showing beyond longest tertial (usually only four [rarely five] in Red-headed), darker and more numerous crown-streaks, weaker dark streaking on darker, warmer scapulars and/or mantle, strongly chestnut-tinged rump to uppertail-coverts, presence of extensive yellow on underparts, bill proportions (typically somewhat longer/narrower than Red-headed). **VOICE** Calls with sparrow-like ***chleep*** or ***chlip***, metallic ***tzik*** or ***plutt*** and hard deep ***tchup*** in flight. Song is melodious, quite harsh, and introduced by several short notes: ***zrt zrt preepree chu-chiwu-chiwu ze-treeurr***. **HABITAT** Open country, scrub, cultivation; lowlands. **RANGE & STATUS** Breeds SE Europe, south-eastern W Palearctic, Iran; winters S Pakistan, W,C India. Stragglers reach E Nepal, E India, S China, Japan, N Borneo. **SE Asia** Vagrant NW,C,S(market specimen) Thailand, Singapore, N Laos, E Tonkin.

815 CHESTNUT-EARED BUNTING

Emberiza fucata Plate **83**

IDENTIFICATION 15–16 cm. *E.f.fucata*: **Male breeding** Shows unmistakable combination of grey, dark-streaked crown and nape, chestnut ear-coverts, and necklace of heavy black spots/streaks. Throat, submoustachial stripe and rest of underparts mostly white, with strong rufous-chestnut band across lowermost breast. Shows very little white on outertail-feathers. *E.f.arcuata* (recorded W,N Myanmar) shows much more solid black necklace and broader rufous-chestnut breast-band (also shows on females), extending onto flanks. **Male non-breeding** Similar to female but supercilium tinged buff; shows more contrast between whitish base colour of throat and breast and buffy flanks and upper belly. **Female** Similar to male breeding but crown and nape tinged brown, supercilium whiter, necklace and breast-band less distinct, underparts less white. **Juvenile** Initially (before September) duller than female, ear-coverts dull greyish-brown, with very pale centre and distinctive broad dark border; shows darker lateral crown-stripes, narrow pale median crown-stripe, whiter supercilium, sandy grey-brown scapulars and pale buff breast with fine dark spots/streaks. From Black-faced Bunting by better-marked head pattern, ochre-brown rump and uppertail-coverts and much less heavily streaked underparts. **First-winter male** Like non-breeding male. **First-winter female** Resembles adult but crown, nape, breast and flanks washed buffish. Dull example could be confused with juvenile Godlewski's Bunting but shows darker breast-streaking, darker malar line, whitish submoustachial stripe and much less white on outertail-feathers. **VOICE** Song consists of variable rapid twittering, beginning with staccato notes and often ending in a characteristic double note: ***zwee zwizwezwizizi trup-trup*** or ***zip zizewuziwiziriri chupee churupp*** etc. Calls with an explosive ***pzick***. **HABITAT** Open country, cultivation, rice-paddy stubble; scrubby hillsides and waterside thickets when breeding; up to 1,890 m. Breeds above 1,800 m in Himalayas. **RANGE & STATUS** Breeds S,SE Siberia, Ussuriland, N Pakistan, NW,N India, Nepal, SE Tibet, SW,C,SE,NE China, N,S Korea, Japan; winters south to NE Indian subcontinent, southern China. **SE Asia** Uncommon to locally common winter visitor Myanmar, NW,NE Thailand, Laos, W,E Tonkin, C,S Annam. Vagrant W Thailand, Peninsular Malaysia. May breed N Myanmar, N Indochina. **BREEDING** India: May–August. **Nest** Cup, in concealed position on ground. **Eggs** 3–5; whitish to yellowish-white or greenish-white, speckled (sometimes also streaked) dull reddish to purplish-brown (sometimes with grey undermarkings); 19.9 x 15.6 mm (av.).

816 GODLEWSKI'S BUNTING

Emberiza godlewskii Plate **83**

IDENTIFICATION 16.5–17 cm. *E.g.yunnanensis*: **Male** Grey hood, with dull chestnut lateral crown-stripes, eyestripe and rear border of ear-coverts diagnostic. Pre-ocular eyestripe and moustachial line blackish. Lower breast, flanks, scapulars and rump to uppertail-coverts rufous-chestnut; shows white on outertail-feathers. **Female** Like male but shows more prominent dark streaking on grey crown-centre, scapulars less extensively rufous-chestnut; underparts often paler with a few streaks on flanks. **Juvenile** Lacks striking head pattern of adult but otherwise similar. Head and breast buffish-brown, with darker streaking, crown-centre only slightly paler, throat and supercilium paler, eyestripe and rear border of ear-coverts slightly darker. From similar heavily streaked buntings by lack of prominent head markings. See Chestnut-eared, Yellow-throated and Crested Bunting. **VOICE** Song is a monotonous series of fairly high-pitched notes: ***chit-chit-chu-chitu-tsi-chitu-chu-chitrru*** and ***chit-situ-chit-tsi-situ-chi*** etc. (***tsi*** notes much thinner and higher-pitched). Calls include a thin drawn-out ***tzii*** and hard ***pett pett***. **HABITAT** Open country with bushes, cultivation; 745–2,285 m. **RANGE & STATUS** Resident (subject to local movements) C Asia, S Siberia, N Mongolia, NE India, S,E Tibet, SW,W,C,N China. **SE Asia** Uncommon (status uncertain) N,E(north) Myanmar. **BREEDING** SE Tibet: June–July. **Nest** Cup, on ground or amongst stones or rocks. **Eggs** 3–5; whitish, with purple-black to dark reddish-brown lines and squiggles in broad ring around broader end, and a few underlying light greyish, greenish or umber-yellow smears; 22.5 x 15.7 mm (av.).

817 GREY-NECKED BUNTING

Emberiza buchanani Plate **83**

IDENTIFICATION 15 cm. *E.b.neobscura*?: **Male non-breeding** Distinctive, with plain brownish-grey head and breast-sides, whitish to buffish-white eyering and submoustachial stripe, buffish- to pinkish-white throat, and pinkish-rufous underparts with pale feather-tips. Note pinkish bill (all plumages), and some rufous on scapulars. May show faint streaking on crown and nape. **Male breeding** Purer and cleaner blue-grey on head and breast-sides, more solidly pinkish-rufous below; largely rufous-chestnut scapulars. **Female non-breeding** Similar to male non-breeding, but more prominently streaked on crown and mantle, reddish-rufous of underparts mostly restricted to breast and admixed with some dark markings, scapulars browner and less conspicuous. **Female breeding** Usually duller than male, with head and nape buffier (showing little contrast with mantle); crown and nape often with some streaking, scapulars duller, breast slightly paler. **First-winter male** Similar to female non-breeding, but more blackish spots/streaks on breast to flanks. **First-winter female** Browner on crown and nape than first-winter male. **VOICE** Calls include a soft ***tsip*** in flight, and a high, falling ***tcheup*** or ***TCHREIp***. Song (*neobscura*) described as ***ti-ti-ti tiu-tiu-tiuu u*** with higher-pitched introductory and final notes. **HABITAT** Usually in dry open areas, particularly rocky ones, also weedy fields and crop-stubble; lowlands. **RANGE & STATUS** Breeds SW,C Asia, west Mongolia, NW China; mainly winters India (except E,NE); vagrants reaching Oman, Bhutan, south China (Hong Kong), Japan. **SE Asia** Vagrant Cochinchina.

818 YELLOW-THROATED BUNTING

Emberiza elegans Plate **83**

IDENTIFICATION 15 cm. *E.e.elegantula*: **Male breeding** Easily told from other buntings by black crown, ear-coverts, chin and breast-band and contrasting broad yellow supercilium and throat. Shows distinct erectile crest, plain greyish rump, and largely white remainder of underparts with lightly dark-streaked flanks. **Male non-breeding** Duller than breeding, with pale fringing to black of head and breast (retains distinctive pattern); rump tinged browner. **Female breeding** Similar to male non-breeding but crown, ear-coverts and breast-band dark warmish brown, supercilium and throat light yellowish-buff; lacks black on chin, breast-band less solid. **Female non-breeding** Has breast-band replaced by dark brown streaking. **Juvenile** Initially duller than female non-breeding, with very short crest; breast pale buffish with indistinct darker streaking. Combination of broad pale supercilium, unstreaked pale buffish-brown back to uppertail-coverts, pale underparts and rufescent fringing on tertials and secondaries distinctive. **VOICE** Song is a monotonous, twittering ***tswit-tsu-ri-tu tswee witt tsuri weee-dee tswit-tsuri-tu***, repeated after short intervals. Call is a sharp, rather liquid ***tzik***. **HABITAT** Forest edge, scrub and cultivation near forest; 1,010–2,135 m. **RANGE & STATUS** Breeds SE Siberia, Ussuriland, W,C,NE China, N,S Korea; some populations winter south to southern China, Japan. **SE Asia** Recorded (status uncertain) N Myanmar. **BREEDING** Extralimitally: May–June. **Nest** Cup, in concealed position on ground. **Eggs** 4–6; whitish, tinged lilac-grey, sparingly marked with dark spots and irregular lines.

819 PALLAS'S BUNTING

Emberiza pallasi Plate **83**

IDENTIFICATION 14 cm. *E.p.pallasi*?: **Male breeding** Black hood, contrasting white submoustachial stripe and mostly white collar, underparts, rump and uppertail-coverts distinctive. Upperparts whitish, broadly streaked blackish. See Reed Bunting. **Male non-breeding** Like non-breeding female but often shows some black mottling on throat, upper breast and ear-coverts (particularly in late winter). **Female breeding** Lacks black head markings of male and resembles other small brown-streaked buntings. Combination of largely pale buffish-brown crown and ear-coverts, white submoustachial stripe and throat, prominent dark malar line, indistinct underpart-streaking, greyish lesser coverts and whitish rump with vague dark streaks distinctive. **Female non-breeding** Similar to breeding but body washed warmer buff, ear-coverts rufescent-tinged. **Juvenile** Like non-breeding female but shows rufescent ear-coverts and heavier, darker body-streaking; rump pale sandy-brown and dark-streaked but still clearly paler than rest of upperparts. For differences from similar Reed Bunting, see that species. **First winter** Very similar to respective non-breeding adults. **VOICE** Calls include ***twi*** notes, recalling sparrow or distant Rose-ringed Parakeet. Low, husky chat-like notes and rapid ***thieu-thieu-thieu-thieu-thieu-thieu-thieu***. Also said to utter a fine ***chleep*** or ***tsilip***, recalling Eurasian Tree-sparrow but weaker. Song is a simple, rather shrill ***srrie srrie srrie srrie srrie srrie***. **HABITAT** Grass and scrub near water, reedbeds, cultivation; recorded at c.900 m. **RANGE & STATUS** Breeds C,E Palearctic, NE China; mainly winters northern and eastern China, S Korea, S Japan. **SE Asia** Vagrant C Myanmar.

820 REED BUNTING

Emberiza schoeniclus Plate **83**

IDENTIFICATION 15 cm. *E.s.pyrrhulina*: **Male breeding** Similar to Pallas's Bunting but bill distinctly thicker (culmen convex), upperparts washed pale greyish-brown and less boldly streaked, rump and uppertail-coverts faintly washed greyish-buff, lesser coverts rufescent-brown. **Male non-breeding/female/juvenile/first winter** Best separated from Pallas's by much thicker, all-dark bill (including lower mandible), rufescent-brown to rufous lesser coverts, and warmer-tipped median coverts. **VOICE** Song is a short, rather hesitant and uninspired ***zritt zreet zreet zritt zriuu*** or ***zreet zreet zreet zruhuhu***. Calls with a characteristic ***seeoo*** (falling in pitch) and hoarse ***brzee***. **HABITAT** Grass and scrub, cultivation; lowlands. **RANGE & STATUS** Breeds Palearctic, NW,NE China, Japan; some populations winter to south, N Africa, Middle East, Pakistan, NW India, northern and eastern China, S Korea, Japan. **SE Asia** Vagrant E Tonkin.

821 RUSTIC BUNTING *Emberiza rustica* Plate **83**

IDENTIFICATION 14–15 cm. *E.r.rustica*: **Male non-breeding** Combination of dark crown-sides, pale median crown-stripe, blackish-bordered ear-coverts, with prominent whitish spot at rear, and largely reddish-chestnut nuchal band, scapulars, rump, uppertail-coverts and bold underpart streaking diagnostic. **Male breeding** Bold black and white head pattern, more solidly dark reddish-chestnut on nuchal-band, scapulars, rump, uppertail-coverts, and breast. **Female non-breeding** Somewhat duller than male; less black on crown and ear-coverts, with former more uniformly streaked. **Female breeding** Similar to male non-breeding. **First-winter male** Like male non-breeding. **First-winter female** Buffier than female non-breeding, browner nape, neck, rump, and uppertail-coverts, shows some dark streaks on lower throat and upper breast. Note pale hindcrown and ear-covert spot. **VOICE** Calls with short, piercing ***zit*** or ***tzik***. Song is mellow ***dudeleu-dewee-deweea-weeu*** etc. **HABITAT** Secondary growth, scrub and grass, cultivation; found at c.900 m. **RANGE & STATUS** Breeds northern Palearctic; winters C Asia, China, Taiwan, N,S Korea, Japan. **SE Asia** Vagrant E Tonkin (sight record).

822 LITTLE BUNTING *Emberiza pusilla* Plate 83

IDENTIFICATION 12–14 cm. Monotypic. **Adult non-breeding** Small with finely dark-streaked breast and flanks and rather uniform base colour of upperside. Always separable from dull plumages of Chestnut-eared Bunting by distinctive bold broad blackish (streaky) lateral crown-stripes, eye-stripe and border to rufous-chestnut ear-coverts. Median crown-stripe, supercilium, submoustachial stripe and spot on rear ear-coverts contrastingly buffish; shows rufescent lores and forehead and whitish eyering. **Adult breeding** Attains diagnostic chestnut flush over most of head; lateral crown-stripes solid black. **Juvenile** Similar to adult non-breeding. Initially shows much duller lateral crown-stripes, less neat breast- and flank-streaking and browner-tinged underparts. **VOICE** Variable. Rather metallic ***zree zree zree tsutsutsutsu tzriiitu*** and ***tzru tzru tzru zee-zee-zee-zee zriiiiiru*** etc. Call is a hard ***tzik***. **HABITAT** Secondary growth, scrub and grass, cultivation, orchards; up to 2,610 m. **RANGE & STATUS** Breeds northern Palearctic; winters N,NE Indian subcontinent, southern and E China, Taiwan. **SE Asia** Fairly common to common winter visitor Myanmar, NW,NE Thailand, N,C Laos, W,E Tonkin. Also recorded on passage E Tonkin. Vagrant S Thailand.

823 BLACK-FACED BUNTING *Emberiza spodocephala* Plate 82

IDENTIFICATION 14–15 cm. *E.s.sordida*: **Male** Plain greenish-olive head and breast, blackish face and yellow belly and vent diagnostic. Non-breeders may show some paler, brownish fringing on head and breast. **Female breeding** Variable. Often resembles male but lacks blackish face and shows yellow throat and breast, darker malar line and darker streaking on breast and flanks. Duller individuals resemble non-breeding. **Female non-breeding** Similar to first winter but with buffier wash to head and pale buffish-yellow wash to underparts. **First winter** Lacks yellow on underparts and rather featureless. Rather plain greyish ear-coverts, neck-sides and lesser coverts, combined with dull greyish-brown rump, fairly prominent underpart-streaking and lack of strong rufous tones in plumage, diagnostic. Also lacks pronounced supercilium and shows contrasting whitish submoustachial stripe. Some males may resemble adult on neck-sides and lores. See Pallas's and Reed Buntings. **VOICE** Song is a variable, lively series of ringing chirps and trills: ***chi-chi-chu chirri-chu chi-zeee-chu chi-chi*** etc. Call is a sibilant sharp thin ***tzii***. **HABITAT** Scrub and grass, cultivation, often near water; lowlands. **RANGE & STATUS** Breeds E Palearctic, W,C,NE China, N,S Korea, Japan; winters Nepal, NE Indian subcontinent, southern China, Taiwan. **SE Asia** Scarce to uncommon winter visitor W,N Myanmar, NW Thailand, N Laos, W,E Tonkin, N Annam.

824 CHESTNUT BUNTING *Emberiza rutila* Plate 82

IDENTIFICATION 14–14.5 cm. Monotypic. **Male breeding** Bright chestnut plumage with yellow breast and belly diagnostic. **Male non-breeding** Duller than breeding, with pale fringes to chestnut feathers. **Female** Similar to juvenile Yellow-breasted Bunting but smaller, with contrasting plain rufous-chestnut rump and uppertail-coverts, rather plain ear-coverts without distinct dark border, no paler mantle 'braces' and almost no white on outertail-feathers. See juvenile Black-faced Bunting. **Juvenile** Resembles female but initially more contrastingly streaked with more marked head pattern. Rufous-chestnut rump and uppertail-coverts distinctive. **VOICE** Song is a rather high-pitched ***wiie-wiie-wiie tzrree-tzrree-tzrree zizizitt*** etc. Call is a ***zick*** similar to Little Bunting. **HABITAT** Underbrush in open forest, forest edge, scrub and grass, bamboo, cultivation; up to 2,590 m. **RANGE & STATUS** Breeds E Palearctic, NE China; winters NE India, S China. **SE Asia** Scarce to locally common winter visitor SW,W,N,C,S,E Myanmar, north Tenasserim, W,NW,NE Thailand, Cambodia, N,C Laos, Vietnam. Also recorded on passage E Tonkin.

825 YELLOW-BREASTED BUNTING *Emberiza aureola* Plate 82

IDENTIFICATION 15 cm. *E.a.ornata*: **Male breeding** Chestnut upperparts and breast-band (blacker at sides), blackish forecrown, head-sides and upper throat, and yellow rest of underparts diagnostic. Median and lesser coverts strikingly white, flanks dark-streaked. *E.a.aureola* (Myanmar) has only little black on forehead and all-chestnut breast-band. **Male non-breeding** Much less strikingly patterned, with obviously streaked mantle and scapulars. Yellow throat and underparts with warm brown breast-band and white median and lesser coverts distinctive. Supercilium and ear-coverts pale buffish, latter with broad dark border. **Female** Duller than non-breeding male. Lacks breast-band, underparts much paler yellow, streaked darker on breast-sides and flanks, upperparts paler and less rufescent, crown dark-streaked with paler median stripe. Often shows contrasting paler mantle 'braces' and usually lacks darker streaking on breast-centre. See Black-faced Bunting. **Juvenile** Like female but underparts paler still, breast finely dark-streaked, shows ill-defined dark malar line. From Chestnut Bunting by pale brownish rump with darker streaks, dark-bordered ear-coverts and white on outertail-feathers. **VOICE** Song is a fairly slow, high-pitched ***djuu-djuu weee-weee ziii-zii*** etc. Call is a short, metallic ***tic***. **HABITAT & BEHAVIOUR** Grass and scrub, cultivation, open areas, often near water; up to 1,370 m. Often roosts in large numbers, particularly in reedbeds. **RANGE & STATUS** Breeds northern Palearctic, NW,NE China, N Japan; winters N,NE Indian subcontinent, southern China. **SE Asia** Scarce to locally common winter visitor (except N,S Annam). Also recorded on passage Cambodia, N Laos, E Tonkin.

826 TRISTRAM'S BUNTING *Emberiza tristrami* Plate 83

IDENTIFICATION 15 cm. Monotypic. **Male breeding** Black head with striking white supercilium, median crown- and submoustachial stripes and ear-covert spot diagnostic. **Male non-breeding** Head pattern duller, with black feathers tipped paler and white feathers tinged buffish. Shows larger buffish-white spot on rear ear-coverts. **Female** Similar to non-breeding male but throat buffish-white and centre of ear-coverts, cheeks and lores pale brownish; breast and flanks streaked brown. From Little Bunting by larger size, whitish supercilium and median crown-stripe, duller, less rufous ear-coverts, with bolder blackish border, unstreaked rufous-chestnut rump and uppertail-coverts, and browner breast and flanks. **Juvenile** Similar to female. **VOICE** Has simple song with one or two introductory notes: ***hsiee swee-swee swee-tsirririri*** and ***hsiee swiii chew-chew-chew*** etc. Calls with explosive ***tzick***, sometimes repeated. **HABITAT** Dense underbrush in open forest, forest edge, secondary growth; 900–2,565 m. **RANGE & STATUS** Breeds SE Siberia, Ussuriland, NE China, N Korea; winters southern China. **SE Asia** Rare winter visitor N Myanmar, NW Thailand, N Laos, W,E Tonkin.

CERTHIIDAE: CERTHIINAE: Treecreepers

Worldwide 9 species. SE Asia 4 species. Small and slender with cryptic plumage, thin downcurved bills, and longish tails with stiff pointed feathers (used as prop when climbing). Creep up tree-trunks and along branches, often starting at bottom of tree and working upwards, before moving to another tree. Diet consists mostly of insects and spiders, gleaned from cracks and crevices in bark.

827 HODGSON'S TREECREEPER

Certhia hodgsoni Plate **84**

IDENTIFICATION 12.5–14 cm. *C.h.khamensis*: **Adult** Blackish-brown upperparts, streaked brown and buffish white, predominantly whitish supercilium and underparts, narrow, slightly downcurved bill, and long tail with sharply pointed feather-tips distinctive. From other treecreepers by combination of unbarred tail, relatively short bill, broad white supercilium and mostly whitish underparts. Note range. **Juvenile** Underparts a shade duller with some dark feather-fringing, mantle appears somewhat more spotted. **VOICE** Song (extralimital races) is a thin, high-pitched, silvery ***tsi-tih ti-tit-teeh*** or ***tsee-tsee tsissi-ssissi-seeeh***, with lower, slower introductory notes, or ***tseip-tseep-tsip tsink-tink-tink'tsip-TSINK TSCHEVIT***, starting weak and high, then gaining momentum as it descend in steps. Calls are not well documented. **HABITAT** Broadleaved evergreen and coniferous forest, mixed oak and rhododendron forest; recorded at 3,960 m. **RANGE & STATUS** Resident (subject to minor movements) Palearctic, Iran, NE Pakistan, NW,N,NE Indian subcontinent, S,SE Tibet, W,NW,N,NE China, N Korea, Japan. **SE Asia** Local resident N Myanmar. **BREEDING** India: April–June. **Nest** Loose mass, usually wedged behind flap of bark or in tree crevice; 3–12 m above ground. **Eggs** 5–6; white, speckled reddish (usually more at broader end); 15.8 x 11.9 mm (av.). **NOTE** Split from extralimital Eurasian Treecreeper *C. familiaris* following Martens & Tietze (2006).

828 BAR-TAILED TREECREEPER

Certhia himalayana Plate **84**

IDENTIFICATION 15–16 cm. *C.h.ripponi*: **Adult** Easily told from other treecreepers by rather pale greyish-brown uppertail with contrasting narrow dark cross-bars and relatively long and strongly downcurved bill. Throat and narrow supercilium whitish, rest of underparts whitish with pale, drab buffish wash. **Juvenile** White throat contrasts a little more with underparts, which show some faint dark feather-fringing, mantle more uniform, bill shorter. **Other subspecies in SE Asia** *C.h.yunnanensis* (N Myanmar). **VOICE** Song is a rather high-pitched, trilled series of 7–16 slightly disyllabic notes, with a short introduction: ***tsee tsui-tsui-tsui-tsui-tsui-tsui-tsui-tsui-tsui-tsui-tsuip***; sometimes a more slurred ***si-liu-liu-liu-liu-liu-liu-liu...*** or shorter more undulating ***tsee't-su-tsu'tsut'tut'tut'ti'tee*** etc. Calls include a thin, descending ***tsiu*** (sometimes in slow descending series), thin ***tsee*** or slightly rising ***tseeet***, rather full ***chit*** and sharp thin ***psit*** or ***tsit***. **HABITAT** Coniferous, mixed broadleaved and coniferous and broadleaved evergreen forest; 2,135–3,000 m. **RANGE & STATUS** Resident (subject to minor movements) C Asia, Afghanistan, northern Pakistan, NW,NE India, Nepal, SE Tibet, SW,W China. **SE Asia** Locally common resident W,N Myanmar. **BREEDING** March–May. **Nest** Loose mass, usually wedged behind flap of bark or in tree crevice; 1.2–15 m above ground. **Eggs** 4–6; dull white to pinkish-white, profusely freckled reddish-brown (often ringed at broader end); 15.8 x 12.2 mm (av.; India).

829 RUSTY-FLANKED TREECREEPER

Certhia nipalensis Plate **84**

IDENTIFICATION 15–16 cm. Monotypic. **Adult** Whitish to buffish-white throat and contrasting cinnamon-coloured breast-sides, flanks, belly and vent diagnostic. Otherwise resembles Hodgson's Treecreeper but broad whitish supercilium encircles dark ear-coverts. **Juvenile** Underparts washed dull buffish, with some faint dark feather-fringing, cinnamon colour of underparts more restricted to flanks. **VOICE** Song is a short, high-pitched, accelerating trill, introduced by clear silvery notes: ***si-si-sit-st't't't***. Usual call is a thin ***sit***. **HABITAT** Broadleaved evergreen and mixed broadleaved and coniferous forest; 2,285–3,050 m. **RANGE & STATUS** Resident N,NE Indian subcontinent, S,SE Tibet, SW China. **SE Asia** Local resident N Myanmar. **BREEDING** Nepal: April–May. **Nest** Situated behind loose flap of tree bark, c.12 m above ground. Otherwise undocumented.

830 HUME'S TREECREEPER

Certhia manipurensis Plate **84**

IDENTIFICATION 15–16 cm. *C.m.shanensis*: **Adult** Combination of drab greyish throat and underparts, buffy vent and indistinct supercilium diagnostic. Note range. *C.m.manipurensis* (W Myanmar) has distinctive deep buffish throat to upper belly; *meridionalis* (S Annam) and *laotiana* (N Laos) have darker, greyer throat to belly and warmer, darker brown upperparts (particularly *meridionalis*). **Juvenile** Shows faint darker scaling on throat and breast. **VOICE** Song is distinctive: a monotonous hesitant trotting rattle, ***tchi-tchi tchi-tchi tchi-tchi tchi-tchi tchichip*** etc. Calls with a loud, explosive ***chit*** or ***tchip***, sometimes extended to a short, rattling ***chi'r'r'it*** and higher, thinner, softer ***tsit*** or ***seep***. **HABITAT & BEHAVIOUR** Broadleaved evergreen and mixed broadleaved and pine forest; 1,370–3,000 m. Sometimes associates with mixed-species feeding flocks. **RANGE & STATUS** Resident N,NE Indian subcontinent, SE Tibet, SW China. **SE Asia** Uncommon to locally common resident W,N,S,E Myanmar, W,NW Thailand, N Laos, W Tonkin, S Annam. **BREEDING** January–July. **Nest** Loose mass, usually wedged behind flap of bark or in tree crevice; 2–9 m above ground. **Eggs** 3–5; whitish, well freckled and blotched pale reddish to reddish-brown; 15.9 x 12.3 mm (av.; *manipurensis*). **NOTE** Split from extralimital Brown-throated Treecreeper *C. discolor* following Martens & Tietze (2006)

SITTIDAE: Nuthatches

Worldwide c.26 species. SE Asia 10 species. Small, with compact bodies, short tails and relatively large, strong feet. Climb up and down tree-trunks and branches, extracting insects and spiders from bark. Some also feed on seeds and nuts in winter.

831 VELVET-FRONTED NUTHATCH

Sitta frontalis Plate **84**

IDENTIFICATION 12–13.5 cm. *S.f.frontalis*: **Male** Distinctive combination of red bill, violet-blue upperparts, black forehead and narrow post-ocular stripe, whitish throat and pale dull beige underparts (washed lavender on flanks, belly and vent). *S.f.saturatior* (S Thailand, Peninsular Malaysia) shows deeper cinnamon-buff to pinkish-buff underparts, washed lilac. See Yellow-billed Nuthatch. **Female** Lacks black post-ocular stripe, underparts slightly more cinnamon-tinged and less lilac, particularly breast and belly. **Juvenile** Similar to adult but bill blackish with fleshy to yellowish gape, upperparts slightly duller and greyer, underparts washed cinnamon-orange to warm buff without lilac tones, undertail-coverts pale pinkish-buff with fine dark cinnamon-brown barring. Males are slightly more orange-buff below. **VOICE** Song is a series (1.5–2 s in duration) of ***sit*** notes, repeated at intervals, sometimes becoming a fast hard rattle. Calls include a hard stony ***chit*** and thinner sibilant ***sit*** notes; former often doubled or given in short rattling series: ***chit-it'it'it...*** etc., or mixed with latter: ***chip chip***

sit-sit-sit-sit-sit-sit-sit... etc. **HABITAT & BEHAVIOUR** Broadleaved evergreen, semi-evergreen and mixed deciduous forest, mixed oak and pine forest; up to 1,800 m (below 1,450 m in S Annam). Often associates with mixed-species feeding flocks. **RANGE & STATUS** Resident Indian subcontinent (except W,NW and Pakistan), southern China, Greater Sundas, Philippines (Palawan). **SE Asia** Common resident (except Singapore). Recorded (status unclear) Singapore. **BREEDING** January–May. **Nest** Pad, in tree-cavity (entrance often enlarged, occasionally 'plastered'); 1–12 m above ground. **Eggs** 3–6; usually rather heavily spotted and blotched brick-red and sometimes purple; 17.2 x 13.2 mm (av.; *frontalis*).

832 YELLOW-BILLED NUTHATCH

Sitta solangiae Plate **84**

IDENTIFICATION 12.5–13.5 cm. *S.s.fortior*: **Male** Similar to Velvet-fronted Nuthatch but has diagnostic yellow bill, paler underparts (contrasting more with upperparts) and vague pale nuchal collar. *S.s.solangiae* (north W Tonkin) shows paler, more violet, less blue crown (contrasts less with nape), slightly paler, greyer-blue upperparts and drabber greyish underparts, without violet wash. **Female** Lacks obvious black post-ocular stripe. **Juvenile** Undescribed. **VOICE** Song is a fast ***sit'ti'ti'ti'ti'ti...*** (1–2.5 s in duration), slower and slightly lower-pitched towards end; a little faster and more whinnying than Velvet-fronted. Other calls also similar to Velvet-fronted: ***chit*** (often doubled or given in series: ***chit-it chit-it-it-it-it-it...*** etc.) and single ***sit***. **HABITAT & BEHAVIOUR** Broadleaved evergreen forest; 900–2,500 m. Often associates with mixed-species feeding flocks. **RANGE & STATUS** Resident S China (Hainan). **SE Asia** Scarce to uncommon resident east S Laos, W Tonkin, C,S Annam. **BREEDING** Undocumented.

833 BLUE NUTHATCH *Sitta azurea* Plate **84**

IDENTIFICATION 13.5 cm. *S.a.expectata*: **Adult** Unmistakable, with blackish plumage, sharply contrasting whitish throat and breast, broad bluish-white eyering and silver-blue and black wing markings. Has violet-blue tinge to upperparts, black crown, nape and ear-coverts and pale blue-grey undertail-coverts; bill pale with dark tip. **Juvenile** Like adult but bill blackish with some fleshy-pink at base, crown and ear-coverts tinged brown, undertail-coverts show some whitish tips and fringes. **VOICE** Calls include a thin squeaky ***zhe*** (sometimes doubled), nasal ***snieu*** or ***kneu***, mellow ***tup***, abrupt ***whit***, thin sibilant ***sit*** and fuller, harder ***chit***. Utters short, rapid repetitions when agitated: ***chi-chit chit-chit-chit*** or ***chir-ri-rit***; sometimes faster series, accelerating into staccato trilled ***tititititititik*** or rattling ***tr'r'r'r'r't***. **HABITAT & BEHAVIOUR** Broadleaved evergreen forest; 820–2,180 m. Sometimes associates with mixed-species feeding flocks. **RANGE & STATUS** Resident Sumatra, Java. **SE Asia** Locally fairly common resident extreme S Thailand, Peninsular Malaysia. **BREEDING** April–June. **Nest** Pad, in small tree-cavity. **Eggs** 3–4; dirty white to pale pinkish, heavily speckled rufous-brown over lavender-grey undermarkings; c.19.3 x 13.4 mm (Java).

834 BEAUTIFUL NUTHATCH

Sitta formosa Plate **84**

IDENTIFICATION 16.5 cm. Monotypic. **Adult** Unmistakable. Large; upperparts black, streaked brilliant blue to white on crown, nape and mantle, has broad blue band along scapulars to back and rump, black wings with two narrow white wing-bars and dull rufous-buff underparts, with paler throat and head-sides. In flight, from below, shows prominent white patch at base of primaries, contrasting with blackish underwing-coverts. **Juvenile** Very similar to adult but white streaks on upperparts may be bluer, underparts possibly paler and whiter, particularly breast. **VOICE** Rapid high, shrill tremulous ***chit'it'it'it'it'it'it'it...*** (1–5 s in duration). Also a shorter, hesitant ***chit-it chit-it chit-it...*** and ***chit'it-it chirririt-it*** etc. **HABITAT & BEHAVIOUR** Broadleaved evergreen and semi-evergreen forest; 700–2,290 m. Joins mixed-species feeding flocks. **RANGE & STATUS** Resident Bhutan, NE India, SW China. Recorded (status uncertain) Bangladesh. **SE Asia** Scarce to uncommon resident S(west),N,E Myanmar, NW Thailand, N,C Laos, W,E(north-west) Tonkin. **BREEDING** March–April. India: **Nest** Pad, in tree-cavity (entrance may be 'plastered'); 2–8 m above ground. **Eggs** 4–6; white, with small dark red specks and spots; 20.8 x 15.3 mm (av.).

835 CHESTNUT-VENTED NUTHATCH

Sitta nagaensis Plate **84**

IDENTIFICATION 12.5–13.5 cm. *S.n.montium*: **Male** Told from Chestnut-bellied and White-tailed Nuthatches by pale greyish-buff underparts (greyer when worn, after breeding), sharply contrasting with reddish-chestnut of lower flanks and vent. Undertail-coverts marked with white. *S.n.nagaensis* (W,N Myanmar) and *grisiventris* (W Myanmar, S Annam; S Laos?) have pale grey underparts with faint buff wash. See much larger Giant Nuthatch. **Female** Lower flanks more rufous-chestnut, underparts may be slightly duller buff and less grey. **Juvenile** Very similar to adult. **VOICE** Song is a fast tremolo or rattle: ***chichichichichi...*** or ***trr'r'r'r'r'r'r'ri...***; sometimes a shriller, more spaced ***chi-chi-chi-chi-chi...*** or much slower ***diu-diu-diu-diu-diu...*** Calls include slightly squeaky ***sit*** notes and a lower, drier ***chit***, often extended in a trilled ***chit'it'it'it'it...*** When alarmed, gives whining nasal ***quir*** or ***kner*** and hard metallic ***tsit*** notes. **HABITAT & BEHAVIOUR** Broadleaved evergreen, mixed broadleaved and coniferous forest, pine forest; 1,000–2,800 m. Often joins mixed-species feeding flocks. **RANGE & STATUS** Resident NE India, SE Tibet, SW,W,SE China. **SE Asia** Common resident W,N,C(east),E Myanmar, NW Thailand, N,S Laos, W Tonkin, S Annam, north Cochinchina. **BREEDING** January–June. **Nest** Pad, in tree-cavity (entrance often 'plastered'); up to 10 m above ground. **Eggs** 2–5; white, spotted red over reddish-violet undermarkings (mainly towards broader end); 18.6 x 13.8 mm (av.; *nagaensis*).

836 CHESTNUT-BELLIED NUTHATCH

Sitta cinnamoventris Plate **84**

IDENTIFICATION 13 cm. *S.c.cinnamoventris* (N Myanmar): **Male** Darker above than Neglected Nuthatch, underparts deep dark reddish-chestnut, contrasting sharply with white cheeks, undertail-coverts chestnut and grey, marked with white. *S.c.tonkinensis* (rest of range) has white cheeks finely barred with blackish, undertail-coverts blackish marked with white. **Female** From male Neglected by deeper pale chestnut underparts with more contrasting white cheeks; somewhat darker overall, notably on bases/centres of undertail-coverts. See Chestnut-vented Nuthatch. **Juvenile** Similar to adult. **VOICE** Songs include a short, musical, quivering, whinnying trilled ***wHR'R'R'R'Rr*** or ***wh'H'H'H'H'r***. Also, a wiry, musical, slightly nasal ***fu'ZIT-fu'zit***. Calls with mellow conversational ***tsup*** notes, and thin ***sit*** or ***sit-sit***. Also, a clipped, screechy, sparrow-like ***chreet chreet chreet...***, a paired ***chewi-chewi*** or triple ***tuwi-tuwi-tuwi tuwi-tuwi-tuwi...***, a fast, staccato ***si-lit*** or ***si-l-it***, and similar but longer, explosive rattles, e.g. ***si'li'li'li'lit***, which may be repeated many times in excitement. **HABITAT** Broadleaved evergreen forest; 1,000–2,200 m, rarely down to 365 m. **RANGE & STATUS** Resident NE Pakistan, NW,N,NE India, Nepal, Bhutan, SW China **SE Asia** Locally common resident N Myanmar, NW Thailand (Doi Hua Mot only), N,C(north-east) Laos, W,E Tonkin. **BREEDING** Himalayan foothills: April–June. **Nest** Pad, in tree-cavity or hole in wall or bank (entrance usually 'plastered'); up to 10 m above ground. **Eggs** 4–7; white to pale pink, finely spotted brick-red and reddish-violet (often mostly towards broader end); 19.8 x 14.1 mm (av.; *cinnamoventris*). **NOTE** Split from extralimital *S. castanea* (Indian Nuthatch) following Rasmussen & Anderton (2005).

837 NEGLECTED NUTHATCH

Sitta neglecta Plate **84**

IDENTIFICATION 13 cm. Monotypic. **Male** Told by medium bluish-grey upperside and pale buffish-chestnut underparts (darker on flanks), rather contrasting white cheeks and mostly dark grey undertail-coverts with large white markings. **Female** Similar to male but underparts pale, drab orange-buff, with whiter cheeks. See Chestnut-bellied and Chestnut-vented Nuthatches. **Juvenile** Similar to adult. **VOICE** Has similar whinnying song to Chestnut-bellied, but shorter and much faster, less wavering, ***whiHHHR***. Also utters loud, full, clear, mellow whistled ***ihU-ihU-ihU-ihU*** and ***UHi-UHi-UHi-UHi***. Other calls include a loud, screechy, sparrow-like ***chreet-chreet*** and hard, explosive rattled ***sri'i'i'i'i'i'i'***. **HABITAT & BEHAVIOUR** Dry dipterocarp and pine forest up to 1,525 m. Often accompanies mixed-species feeding flocks. **RANGE & STATUS** SW China. **SE Asia** Locally common resident Myanmar (but south-east of N only), W,NW,NE Thailand, Cambodia, Laos (north to Vientiane area), C(south-west),S(west) Annam, north Cochinchina. **BREEDING** March–June. **Nest** Simple pad, in tree-cavity (entrance usually 'plastered'); 10 m above ground. **Eggs** c.4–6; similar to Chestnut-bellied; 19.4 **x** 14.1 mm (av.; *neglecta*). **NOTE** Split from extralimital *S. castanea* (Indian Nuthatch) following Harrap (2008).

838 **GIANT NUTHATCH** *Sitta magna* Plate **84**
IDENTIFICATION 19.5 cm. *S.m.magna*: **Male** Resembles Chestnut-vented Nuthatch but much larger and bigger-billed, with much broader black head-bands, centre of crown paler grey than rest of upperparts; no chestnut on lower flanks, no buff wash on underparts. **Female** Underparts washed buff, head-bands duller, crown-centre less contrasting. **Juvenile** Like female but crown-feathers mid-grey with drab pale grey centres and narrow dark fringes, upperparts greyer, less bluish, head-bands duller, dark grey, tertials and greater upperwing-coverts fringed warm brown. **VOICE** Calls include a distinctive chuntering, rapidly repeated ***gd-da-da*** or ***dig-er-up***; sometimes a more melodic ***kid-der-ku*** or ***ge-de-ku***, with louder last note, or harsher ***gu-drr gu-drr gu-drr***. Also gives a very nasal trumpet-like ***naa***, and clear piping tree-frog-like ***kip*** or ***keep*** notes repeated at irregular intervals (may be song). **HABITAT** Open mature pine and mixed oak and pine forest; 1,200–1,830 m. **RANGE & STATUS** Resident SW China. **SE Asia** Scarce to locally common resident C,S(east),E Myanmar, NW Thailand. **BREEDING** February–April. **Nest** Pad/cup, in tree-cavity (entrance may be shaped but not 'plastered'); 1.5–3 m above ground. **Eggs** 3–6; cream-coloured and lightly spotted with brown.

839 **WHITE-TAILED NUTHATCH**
Sitta himalayensis Plate **84**
IDENTIFICATION 12 cm. Monotypic. **Male** Very similar to female Chestnut-bellied Nuthatch but slightly smaller with diagnostic plain cinnamon-orange undertail-coverts and white patch at base of central uppertail-feathers (difficult to see in field). See Chestnut-vented and White-browed Nuthatches. Note range and habitat. **Female** Ear-coverts and underparts slightly paler and duller. **Juvenile** Very similar to adult. **VOICE** Song is a fast crescendo of 3–10 whistled notes (5–7 per s): ***tiu-tiu-tiu...***, ***dwi-dwi-dwi...*** or ***pli-pli-pli...***; or slower 2–4 note (4 per s) ***tui-tui-tui...*** or ***pui-pui...*** Calls include a squeaky ***nit***, soft full ***tschak*** and harder sharper ***chak-kak*** or rattling ***chik-kak-ka-ka-ka-ka...*** When agitated, gives louder shrill ***tsik*** notes; also a thin ***sisisit*** and long, shrill quavering ***kreeeeeeeeee...*** or ***preeeeuh*** (often rising or falling in pitch). **HABITAT** Broadleaved evergreen and mixed broadleaved and coniferous forest; 1,800–2,900 m, locally down to 980 m in winter. **RANGE & STATUS** Resident N,NE Indian subcontinent, S,SE Tibet, SW China. **SE Asia** Resident (subject to local movements) W,N,E Myanmar, N Laos, W Tonkin. **BREEDING** April–May. **Nest** Pad, in tree-cavity (entrance usually 'plastered'); 1–15 m above ground. **Eggs** 4–7; white, densely speckled and spotted dark red (sometimes ringed at broader end); 18.6 **x** 13.4 mm (av.).

840 **WHITE-BROWED NUTHATCH**
Sitta victoriae Plate **84**
IDENTIFICATION 11.5 cm. Monotypic. **Adult** Similar to White-tailed Nuthatch but has whitish lores, long narrow white supercilium, white lower head-sides and underparts with contrasting (variable) orange-rufous patch on ear-coverts, and orange-rufous band along breast-side and flanks. Also has more white on central uppertail. **Juvenile** Similar to adult but orange-rufous of head-sides to flanks fainter. **VOICE** Song is a 9–12 note crescendo (9 per s): ***whi-whi-whi-whi-whi-whi-whi-whi-whi...*** Calls include subdued liquid ***pit*** or ***plit*** notes and an insistent ***pee pee pee pee pee...*** (2.5–3.5 notes per s), repeated for long periods. **HABITAT** Oak and oak/rhododendron forest; 2,285–2,800 m (mostly above 2,500 m in summer). **RANGE & STATUS** Endemic. Locally fairly common resident W Myanmar (south Chin Hills). **BREEDING** March–April. **Nest** In natural tree-cavity 4–10 m above ground. Otherwise undocumented.

TICHODROMADIDAE: Wallcreeper

Worldwide 1 species. SE Asia 1 species. Striking relative of nuthatches. Does not use tail for support or cache food, as in nuthatches. Exhibits breeding and non-breeding plumages, as well as sexual dimorphism. Forages for invertebrates amongst rocks and boulders, and notably on cliff-faces.

841 **WALLCREEPER** *Tichodroma muraria* Plate **84**
IDENTIFICATION 16.5 cm. *T.m.nepalensis*: **Adult non-breeding** Unmistakable, being largely grey, with whiter throat and upper breast, thin curved dark bill, pale-tipped undertail-coverts and tail, and crimson and black on wings. In flight, also shows white spots on flight-feathers. **Male breeding** Black face to upper breast. **Female breeding** Whitish chin, variably blackish on throat and upper breast. **Juvenile** Paler, more uniform grey below, straighter bill. **VOICE** Thin piping ***twee*** or ***tuee***, and ***tuweehtt*** etc. Rising or falling, whistled ***chuit dweeoo*** when agitated. Male song is ascending, clear, piping 4–5 note ***zee-zee-zee-zee-zwee*** (gradually slowing and with notes sometimes rising slightly), or shorter ***zizizitui***; ***ti tiu treeh*** or ***tiu-tueh-tee-ü*** etc. Female song similar but somewhat shorter, less emphatic, and sometimes faster. **HABITAT** Cliffs, rocks, stony riverbeds in winter; found in plains. **RANGE & STATUS** Breeds Europe, Middle East, C Asia, Mongolia, SE Siberia, N Pakistan, NW India, Nepal, Tibet, W and northern China; mostly winters to south, including N Indian subcontinent, S China. **SE Asia** Scarce winter visitor/vagrant N Myanmar.

TROGLODYTIDAE: Wrens

Worldwide c.75 species. SE Asia 1 species. Very small and plump, with rather short wings, strong legs and shortish tails, characteristically held cocked. Unobtrusive in undergrowth, feeding primarily on insects.

842 **WINTER WREN**
Troglodytes troglodytes Plate **84**
IDENTIFICATION 9.5–10 cm. *T.t.talifuensis*: **Adult** Small size, short tail and warm dark brown plumage with paler underparts, barred blackish on lower body, wings and tail, distinctive. Shows indistinct buffish supercilium and rather long slender bill. See wren-babblers. **Juvenile** Similar to adult but throat darker, underparts more heavily dark-barred. **VOICE** Song is a loud prolonged series of rapid, vibrant notes. Calls include a dry ***chek***, more rolling

cherrr, often drawn out into a hard rattle, and a mechanically repeated ***chet chet...*** **HABITAT** More open broadleaved evergreen and coniferous forest, forest edge and clearings, rocky areas, cultivation borders; 1,830–2,800 m. **RANGE & STATUS** Resident (subject to relatively local movements) Palearctic, N Africa, Himalayas, China (except S,E), N Korea. **SE Asia** Local resident N Myanmar. **BREEDING** India: May–July. **Nest** Ball with side-entrance, in hole in bank or amongst tree-roots or boulders etc.; up to 9 m above ground. **Eggs** 3–6; white with pale rusty-red speckles (sometimes unspotted); 17.4 x 12.7 mm (av.; India).

CINCLIDAE: Dippers

Worldwide 5 species. SE Asia 2 species. Robust with short tails (often held cocked). Flight low, fast and direct. Typically inhabit fast-flowing rivers and streams in hilly areas; often use wings to propel themselves under water. Feed on aquatic invertebrates.

843 WHITE-THROATED DIPPER

Cinclus cinclus Plate **84**

IDENTIFICATION 19–20.5 cm. *C.c.przewalskii*: **Adult** Resembles Brown Dipper but has greyer upperparts, and white throat and breast. **Juvenile** Upperparts greyish with dark feather-tips, underparts all whitish with dark scales; has white tips to wing- and tail-feathers. Unlikely to occur in region. **First winter** Similar to adult but has pale scales on belly and undertail-coverts. **VOICE** Song is a sustained lively high-pitched piercing warble, weaker and simpler than that of Brown. Call is a sharp, penetrating, rather rasping ***zink*** or ***zrets***, harsher than that of Brown. **HABITAT** Rivers and streams; recorded at 2,835 m. **RANGE & STATUS** Resident (subject to relatively local movements) NW Africa, Palearctic, N Pakistan, NW,N,NE Indian subcontinent, S,E Tibet, NW,W,SW China. **SE Asia** Scarce winter visitor N Myanmar.

844 BROWN DIPPER *Cinclus pallasii* Plate 84

IDENTIFICATION 21.5 cm. *C.p.dorjei*: **Adult** Robust shape, stout bill, short tail (which is often cocked) and uniform dark brown plumage distinctive. Note distinctive habitat, but see similar White-throated Dipper. **Juvenile** Paler than adult and greyer-brown, with fine blackish scaling on the body-plumage, pale greyish markings on the throat and belly and whitish fringes to the wing-feathers. **Other subspecies in SE Asia** *C.p.pallasii* (Indochina). **VOICE** Song is a strong, rich and full warbling. Call is an abrupt shrill high-pitched rasping ***dzzit*** or ***dzit-dzit***. **HABITAT** Rivers and streams; 200–2,835 m (below 1,000 m in Thailand). **RANGE & STATUS** Resident, subject to some movements, C Asia, E Palearctic, NE Afghanistan, N Pakistan, NW,N,NE Indian subcontinent, S,E Tibet, China, Taiwan, N,S Korea, Japan. **SE Asia** Uncommon to common resident W,N,E Myanmar, N,C Laos, W,E(north-west) Tonkin, N Annam. Recorded (status uncertain) NW Thailand. **BREEDING** December–May. **Nest** Globular with large entrance or cup, in hole in rocks, bridge or fallen tree; close to water. **Eggs** 4–5; white; 26.7 x 18.9 mm (av.; *dorjei*). Brood-parasitised by Eurasian Cuckoo.

STURNIDAE: STURNINAE: Mynas, starlings & allies

Worldwide c.114 species. SE Asia 23 species. Generally medium-sized, stocky and relatively short-tailed, with strong legs and strong, pointed bills. Gregarious, mostly in more open habitats. Feed on insects, fruit, nectar etc.; some species omnivorous. Smaller starlings have rather pointed wings and fast direct flight; larger starlings and mynas have more rounded wings and slower flight, with more deliberate wingbeats.

845 SPOT-WINGED STARLING

Saroglossa spiloptera Plate **86**

IDENTIFICATION 19–20 cm. Monotypic. **Male** Greyish, dark-scaled upperparts, rufescent uppertail-coverts, breast and flanks, blackish ear-coverts and dark chestnut throat diagnostic. Has pale yellow to whitish eyes and small white patch at base of primaries. **Female** From other starlings by slightly scaly brown upperparts, pale underparts with darker throat-streaking and breast-scaling, pale eyes and slender bill. In flight, upperparts and upperwing rather uniform, apart from small white wing-patch. **Juvenile** Similar to female but underparts more streaked. **VOICE** Song is a continuous, harsh, unmusical jumble of dry discordant notes and some melodious warbling. Calls include a scolding ***kwerrh***, nasal ***schaik*** or ***chek*** notes and noisy chattering from flocks. **HABITAT & BEHAVIOUR** Open areas with scattered trees, open deciduous woodland, cultivation; lowlands. Gregarious; habitually feeds on nectar in flowering trees. **RANGE & STATUS** Breeds N,NE Indian subcontinent; more westerly populations winter to east, NE India, formerly Bangladesh. **SE Asia** Rare to uncommon winter visitor N,E,C,S Myanmar, W,NW Thailand.

846 CRESTED MYNA

Acridotheres cristatellus Plate **85**

IDENTIFICATION 25.5–27.5 cm. *A.c.brevipennis*: **Adult** Like White-vented but bill ivory-coloured with rosy-red flush at base of lower mandible, eyes pale orange, crest shorter and fuller, undertail-coverts black with narrow white fringes. In flight shows very large white wing-patches and narrow white tips to outertail-feathers. **Juvenile** Very similar to White-vented but bill, legs and feet paler and duller, undertail-coverts darker, white wing-patch larger. **Other subspecies in SE Asia** *A.c.cristatellus* (recorded E Myanmar). **VOICE** Said to be similar to Common Myna. **HABITAT** Open areas, scrub, cultivation, rice paddies, urban areas; lowlands. **RANGE & STATUS** Resident C and southern China, Taiwan. Introduced Japan, Philippines (Manila), Canada (Vancouver), Argentina. **SE Asia** Uncommon to common resident C,S Laos, Vietnam (except Cochinchina). Recorded (status unknown) N(south),E Myanmar. Vagrant ? (origin unclear) C Thailand. Scarce to local feral resident Peninsular Malaysia (Penang I, Kuala Lumpur), Singapore. **BREEDING** April–August. **Nest** Untidy structure, in hole in building roof or old nest of other bird. **Eggs** 4–7; glossy pale blue to blue-green; 30 x 22 mm (av.).

847 WHITE-VENTED MYNA

Acridotheres grandis Plate **85**

IDENTIFICATION 24.5–27.5 cm. Monotypic. **Adult** Yellow bill, long, floppy tufted crest and uniform slaty-black plumage with sharply contrasting white undertail-coverts diagnostic. Eyes reddish-brown. In flight shows large white wing-patch (above and below), mostly blackish underwing-coverts and broadly white-tipped outertail-feathers. See Javan, Crested and Jungle Mynas. **Juvenile** Somewhat browner overall, with no obvious crest, undertail-coverts dark brown with pale scaling, little or no white on tail-tip, bill duller. Very similar to Crested, differing primarily by yellower bill, legs and feet, somewhat paler undertail-coverts and smaller white wing-patch. **VOICE** Song is a disjointed jumble of repeated tuneless phrases, very similar to Common Myna but perhaps coarser and harsher. Typical calls include a high-pitched ***chuur-chuur...***, harsh ***kaar*** when alarmed and soft ***piu*** when flushed. **HABITAT** Open country, cultivation, rice paddies, urban areas; up to 1,525 m (mostly lowlands). **RANGE & STATUS** Resident NE India, SE Bangladesh, SW China. **SE Asia** Common resident (except Peninsular Malaysia, Singapore); generally scarce in Vietnam (possibly decreasing), but increasing and spreading in south of region. Uncommon to local feral resident Peninsular Malaysia (Kuala Lumpur). **BREEDING** April–August. **Nest** Untidy structure in tree-hole, crown of palm, building roof etc. **Eggs** 4–6; deep turquoise-blue; 25.4–32 x 19–23 mm.

848 **JUNGLE MYNA** *Acridotheres fuscus* Plate **85**
IDENTIFICATION 24.5–25 cm. *A.f.fuscus*: **Adult** Similar to White-vented Myna but bill orange with deep bluish base, eyes yellow, very short crest; blackish head contrasts with greyer body-plumage, which grades to dull whitish on undertail-coverts. In flight shows smaller white wing-patch (above and below) and mostly greyish underwing-coverts. See Javan Myna. **Juvenile** Browner overall, head less contrasting, centre of throat or whole throat slightly paler, no obvious crest, bill yellowish. From Common by greyer-brown upperparts, lower breast and belly, lack of yellow facial skin and, in flight, smaller white wing-patch and mostly dark underwing-coverts. See Javan and Collared Mynas. **VOICE** Song is similar to Common. Typical calls include a repeated ***tiuck-tiuck-tiuck*** and high-pitched ***tchieu-tchieu***. **HABITAT** Open dry and grassy areas, often bordering wetlands and rivers, cultivation, roadsides; occasionally forest clearings, mangroves; up to 1,525 m (mostly lowlands). **RANGE & STATUS** Resident N Pakistan, India, Nepal, Bhutan, Bangladesh. **SE Asia** Uncommon to locally common resident Myanmar, C(coastal),W(coastal),S Thailand, Peninsular Malaysia. **BREEDING** February–July. Often colonial. Multi-brooded. **Nest** Simple structure, in hole in tree, crown of palm, rock, concrete bridge, embankment or building. **Eggs** 3–6; glossy turquoise-blue; 28.9 x 20.9 mm (av.).

848A **JAVAN MYNA** *Acridotheres javanicus* Plate **85**
IDENTIFICATION 24–25 cm. Monotypic. **Adult** Similar to Jungle Myna but body-plumage (apart from blackish crown and ear-coverts) dark slaty-grey with contrasting white undertail-coverts, bill all yellow. In flight shows blacker underwing-coverts and broader white tail-tip. See White-vented and Crested Mynas. **Juvenile** Very similar to Jungle but has greyer body, including belly. **VOICE** Very similar to Common Myna. **HABITAT** Urban areas, open country, cultivation; lowlands. **RANGE & STATUS** Resident Java, Bali. Introduced Sumatra. **SE Asia** Introduced. Common resident W and southern Peninsular Malaysia, Singapore. **BREEDING** March–June. **Nest** Untidy structure, in hole in tree, wall, bridge, or electricity box on utility pole, crown of palm etc. **Eggs** 2–5; pale bluish; 27.5–33.1 x 19.5–22.5 mm.

849 **COLLARED MYNA**
Acridotheres albocinctus Plate **85**
IDENTIFICATION 25.5–26.5 cm. Monotypic. **Adult** Similar to White-vented Myna but shows diagnostic broad white collar on neck-side (buff-tinged in winter), blackish-grey undertail-coverts with broad white tips, short crest and pale blue eyes. In flight shows similar wing pattern to Jungle Myna. See Crested Myna. **Juvenile** Browner than adult with less obvious neck-patch. **VOICE** Bouts of continuous, discordant, shrill chattering, incorporating some mimicry. **HABITAT** Open country, grassy areas, cultivation; up to 1,525 m. **RANGE & STATUS** Resident NE India (Manipur), SW China (W Yunnan). **SE Asia** Uncommon to locally common resident W,N,C,E Myanmar. **BREEDING** April–June. **Nest** Rough structure, in hole in tree, bank or wall. **Eggs** 4; pale blue; 27.9 x 20.3 mm.

850 **COMMON MYNA** *Acridotheres tristis* Plate **85**
IDENTIFICATION 24.5–27 cm. *A.t.tristis*: **Adult** Combination of brown plumage with greyish-black hood, whitish vent and yellow bill and facial skin diagnostic. In flight, shows large white patch on primary coverts and bases of primaries and distinctive white underwing-coverts. **Juvenile** Hood paler and more brownish-grey. Very similar to Jungle Myna but upperparts, lower breast and belly warmer brown, shows some yellow facial skin, much larger white wing-patch and distinctive white underwing-coverts. See Javan Myna. **VOICE** Song consists of repetitive tuneless, whistled, chattering and gurgling notes: ***hee hee chirk-a chirk-a chirk-a*** and ***krr krr krr ci ri ci ri krrup krrup krrup chirri chirri chirri weeu weeu...*** etc.; often combined with skilled avian mimicry. Typical calls include harsh, scolding ***chake-chake...*** when alarmed and weak ***kwerrh*** when flushed. **HABITAT** Open areas, scrub, cultivation, urban areas; up to 1,525 m. **RANGE & STATUS** Resident C Asia, SE Iran, Afghanistan, Indian subcontinent, SW China. Introduced S Africa, Middle East, S China (Hong Kong), Sumatra, Borneo (Brunei), Australia, New Zealand and many islands in tropical and subtropical oceans worldwide. **SE Asia** Common resident throughout. **BREEDING** All year. Multi-brooded. **Nest** Rather large untidy structure, in existing tree-hole, crown of palm, building etc.; occasionally old nest of other bird or squirrel. **Eggs** 4–6; glossy turquoise-blue; 30.8 x 21.9 mm (av.). Rarely brood-parasitised by Asian Koel?

850A **BLACK-WINGED MYNA**
Acridotheres melanopterus Plate **87**
IDENTIFICATION 23 cm. *A.m.melanopterus*: **Adult** Unmistakable all-white starling with black wings and tail. White parts are often stained/tinged pale cream to buffish; shows white primary coverts and tail-tip and yellowish bill, facial skin, legs and feet. *A.m.tricolor* (also recorded) has contrastingly dark mantle, white of wing-coverts restricted to lesser coverts. **Juvenile** Crown, mantle and scapulars mostly greyish, streaked with brownish-white, wings and tail duller. **VOICE** Repeated, downward-inflected ***cha*** notes, throaty ***tok*** or ***chok***, harsh ***kaar*** and drawn-out ***keeer***; high-pitched whistled ***tsoowit*** or ***tsoowee*** in flight. **HABITAT & BEHAVIOUR** Open areas, scrub; lowlands. Usually in pairs or small parties. **RANGE & STATUS** Resident Java, Bali, Lombok. **SE Asia** Introduced. Uncommon resident Singapore (may have died out or been trapped-out). **BREEDING** March–July. **Nest** Existing tree-hole. **Eggs** 3–4; clear blue; 26.5–27.9 x 19–20 mm.

851 **VINOUS-BREASTED MYNA**
Acridotheres burmannicus Plate **87**
IDENTIFICATION 22–25.5 cm. *A.b.leucocephalus*: **Adult** Combination of pale grey to whitish head, narrow naked black mask, yellow to orange-yellow bill (some red at base) and pale vinous-brownish underparts distinctive. Mantle, scapulars and back dark slate-grey, tail-coverts pale buffish, wings dark brownish with white patch on primary coverts and bases of primaries, tail dark with pale buffish-tipped outer feathers. *A.b.burmannicus* (Myanmar, except Tenasserim) is smaller (19.5–24 cm), has red bill with blackish base, mid-grey upperparts with only slightly paler rump and uppertail-coverts, more vinous-grey underparts, whiter undertail-coverts and tail-tip and paler upperwing-coverts, tertials and secondaries. **Juvenile** Browner overall with dull mask and bill and buffish-fringed wing-feathers. Resembles Red-billed and Black-collared Starlings but has warmer-tinged plumage, slightly dark narrow mask and browner wings and tail. **VOICE** Similar to Black-collared Starling. Loud, harsh ***tchew-ii tchew-tchieuw*** and ***tchew'iri-tchew'iri-tchieuw*** etc. **HABITAT** Semi-desert, dry open country, scrub, cultivation, large forest clearings; up to 1,500 m. **RANGE & STATUS** Endemic. Fairly common to common resident Myanmar, W,C,NE,SE,S(north) Thailand (spreading south), Cambodia, N(Vientiane),C,S Laos, C,S Annam, Cochinchina. Rare (perhaps not yet established as a breeder; ? of captive origin) Peninsular Malaysia, Singapore. **BREEDING** April–November. Multi-brooded. **Nest** Large, untidy structure, in existing tree-hole or building roof. **Eggs** Blue; 27.4 x 21.1 mm (av.; *burmannicus*).

852 BLACK-COLLARED STARLING

Gracupica nigricollis Plate **87**

IDENTIFICATION 27–30.5 cm. Monotypic. **Adult** Relatively large size and whitish head and underparts with broad blackish collar diagnostic. Bill, legs and feet blackish, prominent facial skin yellowish; upperside blackish-brown with white band across rump and uppertail-coverts and prominent white wing markings and tail border. **Juvenile** Lacks black collar, head and breast dull brownish, white plumage-parts duller. See Vinous-breasted Myna. **VOICE** Loud, shrill, harsh ***tcheeuw tcheeuw-tchew*** and ***tcheeuw-tchew-trieuw*** etc. **HABITAT** Open country, scrub, cultivation, urban areas; up to 1,525 m. **RANGE & STATUS** Resident S China. **SE Asia** Common resident (except SW,S Myanmar, southern S Thailand, Peninsular Malaysia, Singapore). Presumed rare feral resident (perhaps not yet established) west Peninsular Malaysia, Singapore. **BREEDING** February–August. Multi-brooded. **Nest** Large untidy dome, in tree. **Eggs** 3–5; blue to blue-green; 32.5 x 22.6 mm (av.). Brood-parasitised by Asian Koel.

853 ASIAN PIED STARLING

Gracupica contra Plate **87**

IDENTIFICATION 22–25 cm. *G.c.floweri*: **Adult** Distinctive. Black and white with rather long, pointed, red-based yellowish bill. Head, upper breast, upperparts, wings and tail mostly blackish, forecrown heavily streaked white; has white ear-coverts, narrow scapular band, uppertail-coverts and lower breast to vent. *G.c.superciliaris* (Myanmar) has pale grey underparts and less white streaking on forecrown; *contra* (recorded SW Myanmar) has duller, greyer-tinged underparts and almost no pale streaking on forehead. See White-cheeked and Black-collared Starlings. **Juvenile** Black of plumage replaced by dark brown, including entire crown; centre or whole of throat paler to whitish, pale plumage-parts duller, bill uniformly brownish. See White-cheeked. **VOICE** Song recalls Common Myna but more melodious. Calls include a myna-like ***cheek-cheurk***, descending ***treek-treek-treek*** and variety of high-pitched musical liquid notes from flocks. **HABITAT** Open areas, particularly near water, cultivation, towns; lowlands. **RANGE & STATUS** Resident NE Pakistan, N,NE Indian subcontinent, C,E India, SW China, Sumatra, Java. Introduced United Arab Emirates. **SE Asia** Fairly common to common resident Myanmar, W,NW,NE,SE,C,S(north) Thailand, Cambodia, west N Laos. Presumed rare feral resident (perhaps not yet established) west Peninsular Malaysia, Singapore. **BREEDING** March–November. Sometimes loosely colonial. **Nest** Untidy structure (usually domed) in tree; 5–10 m above ground. **Eggs** 4–6; glossy blue; 26.8 x 19.8 mm (av.; *superciliaris*).

854 WHITE-CHEEKED STARLING

Sturnus cineraceus Plate **86**

IDENTIFICATION 24 cm. Monotypic. **Male** Blackish head and breast with mostly white forehead and ear-coverts distinctive. Rest of plumage mostly dark with white band across uppertail-coverts, paler centre of abdomen and vent, whitish-fringed secondaries and whitish tail-border; bill orange with dark tip. See Asian Pied Starling. **Female** Upperparts somewhat paler, throat paler and mixed with whitish, base colour of breast and flanks paler and browner. **Juvenile** Paler and browner than female, greyish-brown overall with whitish ear-coverts and throat and darker crown and submoustachial stripe; bill duller, lacking obvious dark tip. From other starlings by combination of bare-part colours, rather uniform plumage and contrasting whitish ear-coverts, uppertail-covert band and tail border. **VOICE** Monotonous creaking ***chir-chir-chay-cheet-cheet...*** **HABITAT** Open country; lowlands. **RANGE & STATUS** Breeds NC,NE China, N,S Korea, Japan, SE Siberia, Ussuriland, Sakhalin, Kuril Is; winters south to S China, Philippines. **SE Asia** Uncommon winter visitor E Tonkin. Vagrant N Myanmar, NW,C Thailand, N Laos.

855 RED-BILLED STARLING

Sturnus sericeus Plate **86**

IDENTIFICATION 24 cm. Monotypic. **Male** Light slaty-greyish body, contrasting whitish hood (sometimes yellowish-tinged) and dark-tipped red bill diagnostic. Rump and uppertail-coverts paler than rest of upperparts, wings and tail black with white on primary coverts and bases of primaries, legs and feet orange. Easily separated from White-shouldered and Chestnut-tailed Starlings by bill and leg colour and all-dark wings and tail (apart from white wing-patch). **Female** Similar to male but body paler and brown-tinged. **Juvenile** Browner overall than female, wings and tail browner with slightly less distinct whitish patch, bill yellower. See White-shouldered Starling. **HABITAT & BEHAVIOUR** Scrub, cultivation, open areas; lowlands. Usually in flocks, sometimes associating with other starlings. Gathers in large numbers at communal roosts. **RANGE & STATUS** Breeds C and southern China; some populations winter to south, rarely S Japan (Nansei Is), Philippines. **SE Asia** Scarce to locally common winter visitor E Tonkin. Vagrant S Thailand, N Laos, C Annam. Recorded (escapees or vagrants) Singapore.

856 BRAHMINY STARLING

Sturnus pagodarum Plate **86**

IDENTIFICATION 19–21 cm. Monotypic. **Adult** Blackish crown and nape, greyish upperparts and salmon-pinkish head-sides, breast and belly diagnostic. Has lighter breast-streaking and blue-based yellow bill. **Juvenile** Duller above, crown-feathers duller, browner and shorter, underparts duller and paler with plainer breast, bill duller. **VOICE** Has short song, consisting of drawn-out gurgling sound followed by louder bubbling yodel: roughly ***gu-u-weerh-kwurti-kwee-ah***. **HABITAT & BEHAVIOUR** Dry open country; lowlands. May associate with other starlings. **RANGE & STATUS** Resident (subject to some movements) Indian subcontinent. **SE Asia** Vagrant W,S Thailand.

857 CHESTNUT-TAILED STARLING

Sturnus malabaricus Plate **86**

IDENTIFICATION 18.5–20.5 cm. *S.m.nemoricola*: **Adult** Combination of blue-based yellowish bill, greyish-white hood and rufous-chestnut outertail-feathers diagnostic. Upperside greyish, with blacker primaries and primary coverts and small white area on wing-bend, rump and uppertail-coverts often tinged rufous-chestnut; underparts pale with variable amount of salmon-buff (usually restricted to belly and flanks). *S.m.malabaricus* (recorded north W Myanmar; vagrant northern S Thailand) usually has very pale chestnut underparts (sometimes including throat) with deeper chestnut vent and usually less white at wing-bend. **Juvenile** Similar to adult but has browner fringes to upperwing-coverts, tertials and secondaries and less rufous-chestnut on tail. See White-shouldered Starling. **VOICE** Sharp disyllabic metallic notes and mild tremulous single whistles. **HABITAT & BEHAVIOUR** Open forest of various types, open country with scattered trees; up to 1,450 m. Gregarious; often feeds on nectar in flowering trees. **RANGE & STATUS** Breeds N,NE,E India, Nepal, Bhutan, Bangladesh, SW China; winter visitor W,C,S India. **SE Asia** Fairly common resident (subject to some movements) Myanmar, W,NW Thailand, Laos, Vietnam (except W Tonkin). Uncommon to common winter visitor Thailand (except S), Cambodia. Vagrant southern S Thailand. **BREEDING** March–June. **Nest** Small pad, in tree-hole, 6–15 m above ground. **Eggs** 3–5; glossy pale blue-green; 23.8 x 18.2 mm (av.; *malabaricus*).

858 WHITE-SHOULDERED STARLING

Sturnus sinensis Plate **86**

IDENTIFICATION 18.5–20.5 cm. Monotypic. **Male** Mostly grey plumage, contrasting black wings (glossed dark green), blackish tail with white border and wholly white upperwing-coverts and scapulars diagnostic. Has whiter rump and uppertail-coverts and mostly bluish-grey bill;

whiter plumage-parts occasionally washed salmon-buff. **Female** Wings almost glossless with smaller white patch, rump and uppertail-coverts duller. **Juvenile** Similar to female but initially lacks white wing-patch; upperparts, rump and uppertail-coverts more uniform, grey of plumage tinged brown, pale tail-feathers tips duller, bill duller. Combination of bare-part colour, all-dark upperwing and pale-bordered dark tail rules out similar starlings. **VOICE** Soft ***preep*** when flushed and harsh ***kaar*** when agitated. **HABITAT** Open areas with scattered trees, scrub, cultivation, coastal habitats; up to 400 m. **RANGE & STATUS** Breeds S China; mostly winters to south, Taiwan, S Japan. **SE Asia** Locally common resident E Tonkin, N,C Annam. Scarce to fairly common winter visitor (except SW,W,N,C,E Myanmar, Tenasserim, southern S Thailand). Also recorded on passage Cambodia, E Tonkin. Vagrant southern S Thailand. **BREEDING** March–June. **Nest** Pad or rough cup, in hole in tree or building etc. **Eggs** 3–5; very pale blue; 24–25.7 x 17.3–18 mm.

859 PURPLE-BACKED STARLING
Sturnus sturninus Plate **86**

IDENTIFICATION 17–19 cm. Monotypic. **Male** Distinctive, with pale greyish head and underparts, glossy dark purplish nape-patch and upperparts and glossy dark green upperwing, with whitish to pale buff scapular band and tips of median and greater coverts and tertials. Has blackish bill, legs and feet, pale buff uppertail-coverts and vent, glossy dark green tail, and buff fringing on flight-feathers. See Chestnut-cheeked Starling. **Female** Similarly patterned to male but glossy purple and green of plumage replaced with brown, crown duller and browner. See Chestnut-cheeked and juvenile Red-billed and White-shouldered Starlings. **Juvenile** Similar to female. **VOICE** Soft drawn-out ***chirrup*** or ***prrrp*** when flushed. **HABITAT & BEHAVIOUR** Secondary growth, forest edge, open areas, cultivation; lowlands. Highly gregarious, often associates with Asian Glossy Starling. **RANGE & STATUS** Breeds SE Siberia, Ussuriland, Sakhalin, Mongolia, N,NE China, N Korea; winters Sumatra, Java. **SE Asia** Scarce to uncommon passage migrant S Myanmar, Tenasserim, Thailand (except NW), Peninsular Malaysia, Singapore, Cambodia, N Laos, E Tonkin, N Annam, Cochinchina. Local winter visitor C,S Thailand, Peninsular Malaysia, Singapore. Vagrant NW Thailand.

860 CHESTNUT-CHEEKED STARLING
Sturnus philippensis Plate **86**

IDENTIFICATION 16.5–17 cm. Monotypic. **Male** Resembles Purple-backed Starling but head pale (tinged yellowish-buff to brownish) with diagnostic chestnut patch on ear-coverts and neck-side; has darker grey breast and flanks, whitish fringing on secondaries and lacks whitish tips to scapulars, greater coverts and tertials. **Female** Similar to Purple-backed but has whitish fringing on secondaries and lacks whitish tips to scapulars, greater coverts and tertials. See juvenile Red-billed and White-shouldered Starlings. **Juvenile** Similar to female. **VOICE** Song is described as a simple babbling sequence. Calls include ***airr*** or ***tshairr*** notes, a penetrating ***tshick*** when alarmed and a soft, melodious ***chrueruchu*** in flight. **HABITAT & BEHAVIOUR** Secondary growth, open areas, cultivation; lowlands. May be found in association with other starlings. **RANGE & STATUS** Breeds Sakhalin, Japan; winters S Japan (Nansei Is), Philippines, Borneo, rarely Sulawesi, Moluccas. **SE Asia** Vagrant NW,S Thailand, Peninsular Malaysia, Singapore.

861 ROSY STARLING *Sturnus roseus* Plate **87**

IDENTIFICATION 21–24 cm. Monotypic. **Adult non-breeding** Dull buffish-pink plumage with contrasting blackish hood, glossy greenish-black wings and blackish tail and vent diagnostic. Bill brownish-pink, vent scaled paler, has shaggy crest. **Adult breeding** Body-plumage cleaner and pinker, hood glossy purplish-black, bill pink with black base. **Juvenile** Overall pale sandy greyish-brown with darker wings and tail and paler rump and underparts; bill yellowish. Similar to White-shouldered Starling but has paler bill, legs and feet, paler wings, paler tail with no whitish tip and slight streaking on crown and breast. See Red-billed Starling. **VOICE** Song is a long series of bubbling, warbling and whistled phrases. Gives a loud, clear ***ki-ki-ki...*** in flight and harsh ***shrr*** and rattling ***chik-ik-ik-ik...*** when foraging. **HABITAT** Open areas, scrub; lowlands. **RANGE & STATUS** Breeds W(south-east),SC Palearctic, Iran, Afghanistan, NW China; winters Oman, W,C,S India, Sri Lanka. **SE Asia** Vagrant W,NW,C,S Thailand, north-west Peninsular Malaysia, Singapore, E Tonkin.

862 COMMON STARLING
Sturnus vulgaris Plate **87**

IDENTIFICATION 20.5–23 cm. *S.v.poltaratskyi*: **Adult non-breeding** Blackish plumage with heavy white to buff speckling and spotting diagnostic. Bill blackish. **Adult breeding** Plumage more uniform glossy purplish- and greenish-black with only sparse pale buffish speckling on mantle to uppertail-coverts, flanks, belly and vent (sometimes restricted to vent only); bill yellow. **Juvenile** Rather uniform dusky-brown with paler throat and vent, indistinct dark streaking on underparts and buffish-fringed wing-feathers; bill and legs dark. **VOICE** Song is a complex mixture of chirps, twittering, clicks, drawn-out whistles and skilled avian mimicry. Calls include a soft ***prurrp*** in flight, short metallic ***chip*** when alarmed and ***scree*** notes when foraging. **HABITAT & BEHAVIOUR** Open country, cultivation; lowlands. Gregarious, often in company with other starlings or mynas. **RANGE & STATUS** Breeds Atlantic Is, W,C Palearctic, Iran, Pakistan, NW India (Kashmir), NW China; some populations winter south to N Africa, Middle East, Pakistan, northern India, Nepal, S China, S Japan. Introduced S Africa, SE Australia, New Zealand, Polynesia, N America, Bermuda, Jamaica, Puerto Rico. **SE Asia** Vagrant/rare winter visitor N Myanmar, NW,NE,C,S Thailand, E Tonkin, C Annam.

863 ASIAN GLOSSY STARLING
Aplonis panayensis Plate **85**

IDENTIFICATION 19–21.5 cm. *A.p.strigata*: **Adult** Overall glossy blackish-green plumage (sometimes slightly bluish-tinged) and red eyes diagnostic. **Juvenile** Greyish-brown above, whitish to dull buffish-white below, with bold dark streaks; eyes often paler. **Other subspecies in SE Asia** *A.p.affinis* (SW Myanmar), *tytleri* (Coco Is, off S Myanmar). **VOICE** Shrill sharp ringing whistles: ***tieuu***, ***tseu*** etc. **HABITAT & BEHAVIOUR** Coastal scrub, secondary growth, cultivation, plantations, urban areas; lowlands. Gregarious. **RANGE & STATUS** Resident (subject to local movements) NE India, Bangladesh, Andaman and Nicobar Is, Greater Sundas, Philippines, Sulawesi. **SE Asia** Common (mostly coastal) resident (subject to local movements) SW,S Myanmar, Tenasserim, S Thailand, Peninsular Malaysia, Singapore. **BREEDING** February–September. Sometimes loosely colonial. **Nest** Rough cup, in tree-hole, crown of palm, building roof, pylon etc. **Eggs** 3–4; pale blue (sometimes whitish) speckled dark brown to reddish-brown; 26.7 x 19 mm (av.; *strigata*).

864 GOLDEN-CRESTED MYNA
Ampeliceps coronatus Plate **85**

IDENTIFICATION 22–24 cm. Monotypic. **Male** Glossy blackish plumage, with contrasting yellow crown, lores, cheeks, throat and patch at base of primaries distinctive. Can be confused with Common Hill-myna in flight, but much smaller, wing-patch pale yellow (can appear white at distance). **Female** Yellow on head restricted and patchy. **Juvenile** Duller and browner overall than adult with no yellow on crown, yellowish-white lores, throat and wing-patch, and faint streaking on underparts. **VOICE** Somewhat higher-pitched and more metallic than Common Hill, including bell-like notes. **HABITAT & BEHAVIOUR** Broadleaved evergreen and mixed deciduous forest, forest edge and clearings; up to 800 m. Gregarious; often perches in exposed tops of tall trees. **RANGE & STATUS** Resident NE India, SW China (SW Yunnan). **SE Asia** Scarce to locally common resident W,C,S Myanmar, Tenasserim, Thailand (except C), Cambodia, Laos, Vietnam (except W Tonkin). Recorded (status unclear) north-west Peninsular Malaysia. **BREEDING** April–August. **Nest** Roughly lined tree-hole; 6–15 m above ground. **Eggs** 3–4; blue-green with brown blotches (mainly at broader end); 24.8–28.8 x 19.8–21 mm.

865 COMMON HILL-MYNA

Gracula religiosa Plate **85**

IDENTIFICATION 27–31 cm. *G.r.intermedia*: **Adult** Combination of size, glossy black plumage, heavy deep orange bill (often more yellowish at tip), connected yellow wattles on ear-coverts and nape and prominent white wing-patch diagnostic. *G.r.religiosa* (S Thailand southwards) is larger (29–34.5 cm) and thicker-billed, with separated head-wattles (those on nape also longer). See much smaller Golden-crested Myna. **Juvenile** Duller and less glossy, shows naked pale yellow areas on head where wattles develop, bill duller. **Other subspecies in SE Asia** *G.r.andamanensis* (Coco Is, off S Myanmar). **VOICE** Extremely varied, including loud piercing whistles, screeches, croaks and wheezes. **HABITAT & BEHAVIOUR** Broadleaved evergreen and deciduous forest, forest edge and clearings; up to 1,370 m (mostly below 600 m). Often seen in pairs; regularly perches in exposed tops of tall trees. **RANGE & STATUS** Resident N,NE Indian subcontinent, Andaman and Nicobar Is, SW,S China, Greater Sundas, W Lesser Sundas, Philippines (Palawan). Introduced Hawaii, Florida, Puerto Rico. **SE Asia** Uncommon to locally fairly common resident (except C Thailand). Rare (perhaps of captive origin) C Thailand. **BREEDING** August–June. **Nest** Natural tree-hole; 3–16 m above ground (used year after year). **Eggs** 2–3; glossy pale blue to deep blue, or greenish-blue, sparsely spotted and blotched chocolate-brown and reddish-brown over pale purple or bluish undermarkings; 36.2 x 25.6 mm (av.; *intermedia*).

TURDIDAE: Thrushes, cochoas, Grandala & allies

Worldwide c.147 species. SE Asia 27 species. **Typical thrushes** (*Zoothera, Turdus*) Fairly small to medium-sized with strong bills and legs. *Zoothera* have broad white to buff bands on underwing. Sexes alike in some species. Mostly in wooded habitats. Hop and bound along ground, also perch in trees. Feed on invertebrates, snails, fruit etc. **Cochoas** (*Cochoa*) Medium-sized to fairly large with broad bills. Shy and unobtrusive, feeding in tree-tops and on ground; largely frugivorous. **Grandala** Medium-sized with very long, pointed wings, gregarious.

866 CHESTNUT-CAPPED THRUSH

Zoothera interpres Plate **89**

IDENTIFICATION 17.5–18.5 cm. *Z.i.interpres*: **Adult** Unmistakable, with chestnut crown and nape, black throat and breast, bold white markings on blackish sides of head and wings, and white remainder of underparts with black spots on lower breast and flanks. See Chestnut-naped Forktail. **Juvenile** Crown and mantle mostly dull chestnut with pale rufous streaks; head-sides, throat and breast rufous-buff (upper breast more rufescent) with two blackish bars on ear-coverts, and blackish malar line and breast-blotching; wing-patches smaller and tinged rufous-buff. **VOICE** Song is a series of rising flute-like whistles, interspersed with chirrups: ***see-it-tu-tu-tyuu*** etc., sometimes with some grating and higher-pitched notes. Recalls White-rumped Shama. Calls include harsh, hard ***tac*** notes and a very thin, high-pitched, falling ***tsi-i-i-i***. **HABITAT & BEHAVIOUR** Broadleaved evergreen forest; up to 760 m. Very secretive, usually near ground but often sings from quite high up in trees. **RANGE & STATUS** Resident Greater Sundas, W Lesser Sundas, SW Philippines. **SE Asia** Scarce resident S Thailand, Peninsular Malaysia. **BREEDING** May–August. **Nest** Fairly deep cup, on horizontal tree-branch, or in sapling or bamboo; 2–4 m above ground. **Eggs** 2–3; whitish with reddish-brown markings (varying shades), particularly numerous at broader end.

867 ORANGE-HEADED THRUSH

Zoothera citrina Plate **89**

IDENTIFICATION 20.5–23.5 cm. *Z.c.innotata* (breeds NW,NE Thailand, N Laos, S Indochina; south to Peninsular Malaysia in winter): **Male** Bright orange-rufous head and underparts (vent whitish) and bluish-grey upperparts and wings diagnostic. *Z.c.citrina* (breeds western & northern Myanmar) and *gibsonhilli* (breeds southern & eastern Myanmar, W Thailand; south to Peninsular Malaysia in winter) have white-tipped median coverts; *aurimacula* (N Laos, E Tonkin, C Annam) has white-tipped median coverts, buffy-whitish lores and sides of head, two dark brown bars on ear-coverts, indistinct darker malar line (see juvenile *innotata*) and whiter throat and belly. **Female** Upperparts and wings mostly greyish-olive; grey on back to uppertail-coverts, head and underparts duller orange-rufous. *Z.c.citrina* and *gibsonhilli* have white-tipped median coverts; *aurimacula* differs as male but also has even darker ear-covert bars. **Juvenile** Similar to female but crown and mantle darker and browner; has two blackish bars on ear-coverts, dark malar line, paler underparts with dark mottling/scales on breast and flanks, and some thin rufous-whitish streaks on crown, mantle and breast; may show rufous-tipped uppertail-coverts. **VOICE** Song is a sweet rich series of variable musical phrases: ***wheeper-pree-preeteelee wheeeoo-peeerper-wheechee-leet-wheeechee-leet pir-whoo-peer-rte-rate...*** Often employs mimicry. Calls include a subdued low ***tjuck***, loud screeching ***teer-teer-teer*** or ***kreeee...*** and thin ***tsee*** or ***tzzeet*** in flight. **HABITAT & BEHAVIOUR** Broadleaved evergreen forest, forest edge, secondary growth and thickets (particularly on migration); up to 1,525 m. Usually on ground but also feeds on fruit and sings from quite high in trees. **RANGE & STATUS** Breeds NE Pakistan, NW,N,NE Indian subcontinent, peninsular India, Andamans and Nicobar Is, S China, Greater Sundas; some populations winter to south, Sri Lanka. **SE Asia** Uncommon resident SW,W,S Myanmar, Tenasserim, W Thailand, Cambodia, C,S Annam. Uncommon breeding visitor N Myanmar, NW,NE Thailand, E Tonkin. Uncommon winter visitor W,NE,S Thailand, Peninsular Malaysia, Cochinchina; S Myanmar? Uncommon to fairly common passage migrant S Myanmar, C,S Thailand, Peninsular Malaysia, E Tonkin. Recorded (status uncertain) C,E Myanmar, SE Thailand, Laos, N Annam. Vagrant Singapore. **BREEDING** May–October. **Nest** Quite shallow cup, in tree or bush; 1–4 m above ground. **Eggs** 2–5; glossy, pale olive or pale stone-colour, speckled and blotched reddish-brown; 25.6 x 19.3 mm (av.; *citrina*).

868 PLAIN-BACKED THRUSH

Zoothera mollissima Plate **88**

IDENTIFICATION 25.5–27 cm. *Z.m.mollissima*: **Adult** Like Long-tailed Thrush but lacks paler area behind eye, and generally has less distinct dark patch on rear ear-coverts and warmer tinge to upperparts, more densely marked belly and richer buff base colour to throat and breast (contrasts more); has plainer wings with only faint bars on median and greater coverts and no contrasting band across primaries and bases of secondaries. *Z.m.griseiceps* (W Tonkin) has darker, greyer crown and more rufescent upperside. **Juvenile** Differs as Long-tailed; breast markings may be very dense, forming blackish patch. **VOICE** Song recalls Scaly Thrush but is slightly faster and more varied with rich melodious whistled phrases, ***plee-too*** or ***plee-chuu*** followed by ***ti-ti-ti*** or ***ch-up-ple-ooop***, punctuated by pauses of up to 10 s. Calls include a thin ***chuck*** and sharp rattle when alarmed. Normally silent in winter. **HABITAT** Rhododendron and coniferous forest, low vegetation and rocks above tree-line; 1,600–2,895 m. Breeds above 2,700 m in Himalayas. **RANGE & STATUS** Resident NE Pakistan, NW,N,NE Indian subcontinent, S,SE Tibet, SW,W China. **SE Asia** Scarce resident (subject to altitudinal movements) W Tonkin. Recorded (probably resident) N Myanmar. **BREEDING** India: April–July. **Nest** Cup, on ground or in low vegetation. **Eggs** 4; whitish, profusely spotted deep red and reddish-brown, more densely at broader end; 35.3–35.8 x 23.3–24.3 mm.

869 LONG-TAILED THRUSH
Zoothera dixoni Plate **88**

IDENTIFICATION 25.5–27 cm. Monotypic. **Adult** Bold head pattern and buffy-white underparts with blackish scales/bars recall Scaly Thrush but has plain olive-brown upperparts and two distinct buffy-whitish bars on wing-coverts. Sides of head mostly whitish with dark patch on rear ear-coverts, has pale area behind eye (above ear-coverts) and contrasting darker band from middle of inner primaries to base of secondaries. In flight shows prominent pale buffish bands on underwing. See very similar Plain-backed Thrush, also Chinese and female Siberian Thrushes. **Juvenile** Similar to adult but shows buff streaks on crown, mantle and scapulars, and dark bars on mantle and scapulars to uppertail-coverts; underpart markings somewhat darker and more extensive. **VOICE** Song is fairly slow and slurred (particularly last note), typically with rather dry ***wu-ut-cheet-sher*** or ***wut-chet-shuur*** phrases, interspersed with musical twitters and ***too-ee*** or ***ee-ee*** phrases. May be introduced by an upward-inflected ***w'i-it***, which is sometimes included in main song. Normally silent in winter. **HABITAT & BEHAVIOUR** Broadleaved evergreen forest, rhododendron and coniferous forest; 2,135–3,655 m, down to 1,000 m in winter. Often feeds along leafy tracks and roadsides. **RANGE & STATUS** Resident (subject to some movements), N,NE Indian subcontinent, SE Tibet, SW,W China. **SE Asia** Uncommon resident N Myanmar. Scarce to uncommon winter visitor E Myanmar, NW Thailand, N Laos. Recorded (status uncertain) W Myanmar, W Tonkin. **BREEDING** India: May–July. **Nest** Cup, in low vegetation or small tree; up to 3 m above ground. **Eggs** 3; dull greenish, blotched or stippled reddish-brown (may be ringed or capped at broader end) 30.5 x 21.6 mm (av.).

870 WHITE'S THRUSH *Zoothera aurea* Plate **88**

IDENTIFICATION 29–30.5 cm. *Z.a.aurea*: **Adult** Larger and somewhat paler and greyer above than Scaly Thrush; typically with longer bill, more prominent whitish eyering, well-mottled ear-coverts with less contrasting blackish patch at rear, stronger spots on malar area, and often bolder, broader and more rounded scales on upperparts. **VOICE** Song consists of slow melancholy thin whistles (repeated after long intervals): thin, slightly undulating ***huuwiieee...*** or ***weeeoooooooo***, sometimes more disyllabic ***pee-yuuuuu...*** (starts higher). Calls with drawn-out high-pitched ***tsee***, ***seeh*** or ***zeeea***. **HABITAT & BEHAVIOUR** Broadleaved evergreen, and occasionally mixed deciduous; up to 2,590 m. Very shy, often flushed from ground. **RANGE & STATUS** Breeds Siberia, Ussuriland, Mongolia, NE China, N,S Korea, Japan; winters S China, Philippines, ? E,NE India. **SE Asia** Scarce to uncommon winter visitor N,E Myanmar, W,NW,NE Thailand, N,C Laos, W,E Tonkin, S Annam, ? Cambodia. Also recorded on passage E Tonkin. **NOTE** Split from Scaly following Rasmussen & Anderton (2005).

871 SCALY THRUSH *Zoothera dauma* Plate **88**

IDENTIFICATION 27–30 cm. *Z.d.dauma*. **Adult** Mostly warm olive-brown to buffish upperbody and whitish underbody with heavy blackish scales distinctive. Has two buffy-white bands on underwing and white-tipped outertail-feathers (often prominent when taking off). Very similar to White's Thrush, which see for differences. **Juvenile** Upperparts and wings have warmer buff markings, dark scales on body more diffuse; initially has dense blackish spots on breast. **VOICE** Song is a slow series of short ***pur-loo-tree-lay***, ***dur-lee-dur-lee*** or ***drr-drr-chew-you-we-eeee*** phrases, with one phrase repeated twice or more before using another after a long pause. Also a more languid ***chirrup cheweee chueu wiow we erp chirrol chup cheweee wiop***. Soft squeaks, twitters or chuckles may be included in song. Calls when perched include strange, quiet, breathy rasping notes and ***chick*** notes. **HABITAT & BEHAVIOUR** Broadleaved evergreen and rhododendron/coniferous forest; c.365–2,590 m. Very shy, often flushed from ground. Sings from trees. **RANGE & STATUS** Breeds N Pakistan, NW,N,NE Indian subcontinent, S Tibet, SW,W,S China, Taiwan; some winter to south. **SE Asia** Scarce to uncommon local resident W,N Myanmar, W,NW,NE(north-west) Thailand, N Laos, W,E Tonkin, S Annam, ? south-west Cambodia. Recorded (status uncertain; may be resident) S Thailand, Peninsular Malaysia. **BREEDING** May–July. **Nest** Wide shallow cup, in tree or bush, sometimes on bank; 1–6 m above ground. **Eggs** 3–4; pale clay-coloured to buffish-olive, densely freckled pale reddish; 30.5 x 22.3 mm (av.).

872 LONG-BILLED THRUSH
Zoothera monticola Plate **88**

IDENTIFICATION 26.5–28 cm. *Z.m.monticola*: **Adult** Similar to Dark-sided Thrush but larger with even bigger bill; has much greyer upperparts (with indistinct dark scales) and wings, darker breast contrasts more with well-demarcated buffy-whitish throat, has spotted, rather than scaled/scalloped lower breast and belly. **Juvenile** Similar to Dark-sided but base colour of upperparts and wings colder brown, underpart markings blacker and more spot-like. **Other subspecies in SE Asia** *Z.m.atrata* (W Tonkin). **VOICE** Song is a loud mournful series of slow clear whistles: ***te-e-uw*** or ***sew-a-tew-tew*** (middle note higher); also ***weech-a-wee-wuu***. May be introduced by (or include) harsher, more rasping notes: ***rrraee ti tuu***; ***trrray tya tyee*** etc. Generally silent in winter but has a loud ***zaaaaaaaa*** alarm note. **HABITAT** Broadleaved evergreen forest, forest edge, bamboo, mainly near rocky streams; 915–2,135 m. **RANGE & STATUS** Resident (subject to some movements) N,NE Indian subcontinent. **SE Asia** Scarce resident W,N Myanmar, W Tonkin. **BREEDING** May–July. **Nest** Large deep cup, in tree; 2–7 m above ground. **Eggs** 3–4; pale green to grey-green or pale cream-coloured to warm buff, speckled or freckled reddish-brown; 30 x 21.3 mm (av.; *monticola*).

873 DARK-SIDED THRUSH
Zoothera marginata Plate **88**

IDENTIFICATION 23.5–25.5 cm. Monotypic. **Adult** Noticeably long bill, robust shape and relatively short tail distinctive. Upperparts warm olive-brown, sides of head and underparts resemble Plain-backed and Long-tailed Thrushes but flanks more uniform and dark olive-brown. Tail all dark. **Juvenile** Similar to adult but has rufous-buff streaks on crown, mantle and scapulars, rufous-buff triangular spots on tips of wing-coverts and tertials, and more rufescent to buffish breast with darker scales/mottling. **VOICE** Song is a thin monotone whistle, softer and shorter (0.5 s) than that of White's Thrush, and downward-inflected. Call is a soft deep ***tchuck***. Usually silent. **HABITAT** Broadleaved evergreen forest, usually near streams or wet areas; 600–2,565 m. **RANGE & STATUS** Resident N,NE Indian subcontinent, SW China. **SE Asia** Uncommon resident Myanmar (except N,C), W,NW,NE,SE Thailand, south-west Cambodia, Laos, W,E Tonkin, N,C,S Annam. **BREEDING** April–July. **Nest** Cup, in tree; up to 5 m above ground. **Eggs** 3–4; off-white to pale grey or greyish-green, spotted or blotched reddish-brown; 26.9 x 20.1 mm (av.).

874 SIBERIAN THRUSH
Zoothera sibirica Plate **90**

IDENTIFICATION 21.5–23.5 cm. *Z.s.sibirica*: **Male** Dark, slaty plumage, broad white supercilium and white vent with dark-scaled undertail-coverts diagnostic. *Z.s.davisoni* (recorded E Myanmar, Tenasserim, NW Thailand, Peninsular Malaysia, W,E Tonkin) is more blackish and usually has all-dark belly. **Female** From other thrushes and rock-thrushes by combination of prominent buffy-whitish supercilium (curves down behind ear-coverts), prominent dark eyestripe, plain brown upperparts and buffy-whitish underparts with dark brown scales/mottling on breast and flanks. Has dark markings on undertail-coverts, all-dark bill, two narrow pale buffish wing-bars and whitish bands on underwing (may be obvious in flight). Outertail-feathers have pale tips (both sexes). **First-winter male** Shows mixed characters of male and female. **VOICE** Song is a very hesitant, languid, rich, clear series of 2–3 note phrases, punctuated by long pauses. Phrases include: ***tvee-tring***, ***tvee-tryu***; ***tvee-kvee***; ***tvee-kwi-tring*** and ***yui'i-tss***. Notes vary in pitch. *Z.s.davisoni* utters less musical phrases: ***feep-tss***, ***tweet-tss*** or ***kleep-tss*** etc. Calls include a quiet, thin ***tsit***, stronger ***seep*** or ***tseee***, soft whistled ***tsip*** and soft ***tsss*** or ***chrsss*** when alarmed. **HABITAT & BEHAVIOUR** Broadleaved evergreen forest; up to 2,565 m (winters mainly in mountains). Very shy, often flushed from ground, sometimes visits fruiting trees. **RANGE & STATUS** Breeds E Palearctic, NE China, Japan; winters Sumatra, W Java. **SE Asia** Uncommon to fairly common winter visitor and passage migrant Peninsular Malaysia, Singapore, W Tonkin, C,S Annam. Uncommon passage migrant (may winter locally) C,E,S(east) Myanmar, Tenasserim, W,NW(occasionally winters),NE,S Thailand, N,C Laos, E Tonkin. Vagrant C Thailand.

875 CHINESE THRUSH

Turdus mupinensis Plate **88**

IDENTIFICATION 23 cm. Monotypic. **Adult** Resembles Plain-backed and Long-tailed Thrushes but smaller, bill shorter and stouter, breast and belly distinctly spotted overall; has dark-centred wing-coverts with two prominent buff bars (rows of spots), relatively plain rest of wings, and no white on outertail-feathers. In flight shows peachy-buff underwing-coverts and lacks buff bands across underwing. **First winter** Shows some ill-defined darker feather-fringes and pale buff shaft-streaks on mantle, back and scapulars, and buffier spots at tips of wing-coverts. **VOICE** Song is a measured series of pleasant, usually 3–5 note phrases, punctuated by 3–11 s intervals: ***drrip-dee-du dudu-du-twi dju-wu-wi chu-wii-wr'wup chu-wi'i-wu-wrrh dju-dju-weee'u dju-dju-weee'u...*** Notes mostly quite even-pitched, sometimes rising and occasionally slurred. **HABITAT** Plantations and thickets on migration; recorded at sea-level but could occur higher. **RANGE & STATUS** Resident (subject to relatively local movements) SW,W,C,N China. **SE Asia** Vagrant E Tonkin (sight record).

876 WHITE-COLLARED BLACKBIRD

Turdus albocinctus Plate **88**

IDENTIFICATION 26.5–28.5 cm. Monotypic. **Male** Blackish plumage with broad white collar diagnostic. Bill, legs and feet yellow. **Female** Distinctively patterned as male but largely warm-tinged dark brown with greyish- to dusky-white collar. **Juvenile male** Like juvenile female but blacker above and with mostly dark greyish base colour to breast and belly. **Juvenile female** Crown, ear-coverts and upperparts blackish-brown with buffish streaks, wings blackish-brown with rufous-buff tips to median and greater coverts, underparts warm brownish-buff with broad blackish malar stripe and heavy blackish blotches/scales on breast and belly; has buffish patch behind ear-coverts. See Grey-winged Blackbird and female Chestnut-bellied Rock-thrush. **VOICE** Song is a sad, mellow, simple series of descending whistles: ***tew-i tew-u tew-o***, sometimes varied with ***tew-eeo***; repeated 5–7 times per minute and occasionally interspersed with hissing and squeaking notes. Calls with a loud throaty ***chuck-chuck*** or ***tuck-tuck-tuck-tuck...*** **HABITAT** Oak, rhododendron and coniferous forest, meadows, open broadleaved evergreen forest and edge in winter; recorded at c.250 m (winter). Breeds above 2,100 m in Himalayas. **RANGE & STATUS** Resident (subject to relatively local movements) N,NE Indian subcontinent, S,E Tibet, W China. **SE Asia** Recorded in winter N Myanmar (once). Could breed at higher levels. **BREEDING** India: April–July. **Nest** Bulky cup, in tree, 1–3 m above ground; rarely on ground. **Eggs** 3–4; pale blue with pale reddish-brown blotches; 30.5 x 21.7 mm (av.).

877 CHINESE BLACKBIRD

Turdus mandarinus Plate **88**

IDENTIFICATION 28–29 cm. *T.m.mandarinus*: **Male** Overall sooty-black plumage and yellow bill diagnostic. See Grey-winged Blackbird and Blue Whistling-thrush. **Female** Very similar but a shade browner above and particularly below, throat often paler with narrow dark streaks. Similar to Grey-winged but darker, with darker bill and uniformly dark wings. **First winter** Similar to respective adults but has mostly dark bill. **VOICE** Song is a beautiful, mellow, leisurely series of melodious warbling and flute-like notes, with little phrase repetition. Calls with explosive deep ***chup-chup...***, higher ***whiiiik*** and soft ***p'soook***. **HABITAT & BEHAVIOUR** Open forest, secondary growth, clearings, cultivation; up to 1,015 m. Often in flocks. **RANGE & STATUS** Breeds W,C,E and southern China; some winter to south. **SE Asia** Scarce to common winter visitor Laos, W,E Tonkin, N,C Annam. Also recorded on passage E Tonkin. Vagrant N Myanmar, NW,NE Thailand, north Cambodia. **NOTE** Split from extralimital Common Blackbird *T. merula* following Rasmussen & Anderton (2005).

878 TICKELL'S THRUSH

Turdus unicolor Plate **90**

IDENTIFICATION 22 cm. Monotypic. **Male** Unmistakable smallish thrush, with ashy-grey plumage, white belly and vent, and orange-yellow to yellow bill. Eyering yellow, legs and feet brownish-yellow. **Female** Upperside olive-brown, breast and flanks clay-brown with warm buff wash; blackish streaks/spots on malar to upper breast but mostly whitish throat. Combination of relatively dull flanks with no markings rules out Grey-backed, Black-breasted and Japanese Thrushes. First-winter Black-breasted has darker bill and some flank streaking. Note range. **First winter** Similar to female, but with pale-tipped wing-coverts, and weaker streaking on malar. **VOICE** Calls include loud, full ***juk-juk***, and more chattering, decelerating ***juh'juk-juk juk***. Typical *Turdus* song; weaker and less musical than Chestnut, rather variable, relaxed series of phrases comprised of short, complex, mostly slurred notes. **HABITAT** Forest edge, more open woodland, orchards, groves; recorded below 300 m. **RANGE & STATUS** Breeds NW Pakistan, NW,N India, Nepal, Bhutan; winters south to WC Pakistan, peninsular India, Bangladesh. **SE Asia** Vagrant west N Myanmar (sight record).

879 GREY-BACKED THRUSH

Turdus hortulorum Plate **90**

IDENTIFICATION 24 cm. Monotypic. **Male** Unmistakable, with bluish-grey head, breast and upperside, broadly orange-rufous flanks and whitish remainder of underparts. Has faint streaking on throat and breast. See Eyebrowed Thrush. **Female** Very similar to Black-breasted Thrush but has brownish bill, paler and more olive-tinged head-sides and upperparts, paler breast with several rows of blackish arrowhead spots. See Japanese Thrush. **First-winter male** Throat-streaking darker than adult male; shows dark breast markings, and may show dark mottling/markings on flanks. **VOICE** Song is a loud, pure, versatile series of fluty whistled phrases: ***tvet-tvet-tvet qwee-qwee-qwee tveeu-tveeu-tveeu tveeu-tve tevetee-tevetee-tevetee k'yuu-qwo tvee-tvee-tvee trryuuu tevtee-tevtee-tevtee...*** etc. Calls include a soft low chuckle, harsh ***chack-chack*** and shrill whistled ***tsee*** or ***cheee*** when alarmed. **HABITAT** Broadleaved evergreen forest, open forest, secondary growth; up to 1,100 m. **RANGE & STATUS** Breeds SE Siberia, Ussuriland, NE China, N,S Korea; winters southern China. **SE Asia** Uncommon to fairly common winter visitor E Tonkin, N Annam. Also recorded on passage E Tonkin. Vagrant N Laos.

880 **BLACK-BREASTED THRUSH**
Turdus dissimilis Plate **90**
IDENTIFICATION 23–23.5 cm. Monotypic. **Male** Easily identified by black hood and upper breast, dark slaty upperside and orange-rufous lower breast and flanks. **Female** Combination of yellowish bill, plain brown head-sides and upperside, rather dull breast (mixed grey and pale rufous), blackish spots/blotches on throat and breast, and orange-rufous flanks distinctive. See Grey-backed and Japanese Thrushes. **Juvenile** Resembles female but upperpart feathers tipped darker and streaked pale to warm buff, wing-coverts tipped warm buff, breast lacks greyish wash and has heavier blackish blotches/bars, flanks duller rufous with indistinct dark bars/blotches. **VOICE** Song is sweet, mellow and unhurried, phrases consisting of 3–8 notes, e.g.: ***tew-tew weet***; ***tew-tew-tiwi***; ***pieu-pieu-pieu twi***; ***wui-ui'ui-tri-tri*** and ***wirriwi-wu iih*** etc. Calls with resounding ***tup-tup tup-tup-tup-tup-tup-tup...*** and thin ***seee***. **HABITAT** Oak and coniferous forest, broadleaved evergreen forest, secondary growth; 1,220–2,500 m, down to 200 m in winter. **RANGE & STATUS** Breeds NE India, SW,S China; some winter to south, Bangladesh. **SE Asia** Uncommon to fairly common resident (subject to relatively local movements) W,N,E Myanmar, NW Thailand (local). Scarce to uncommon winter visitor NW Thailand, C Laos, E Tonkin. Recorded (status uncertain, though probably breeds locally) N Laos, W Tonkin. **BREEDING** April–July. **Nest** Sturdy cup, in tree or bush, 1–6 m above ground; rarely on ground. **Eggs** 3–4; cream or buff to pale green or greenish-blue, blotched and spotted brown and lilac or deep reddish-brown to purplish-brown; 26.9 x 19.8 mm (av.).

881 **JAPANESE THRUSH** *Turdus cardis* Plate **90**
IDENTIFICATION 22.5 cm. Monotypic. **Male** Blackish head and upper breast, dark slaty to blackish upperside and white remainder of underparts with dark spots on lower breast and flanks diagnostic. Bill yellow. **Female** Recalls Black-breasted Thrush but lacks extensive orange-rufous on flanks, has dark brown to blackish spots on flanks (sometimes also belly); may show a little orange-rufous on upper flanks and often shows some warm buff on throat and breast. See Grey-backed and Black-throated Thrushes. **First-winter male** Paler and more uniformly slaty-greyish above than adult male; has whitish centre of throat with dark markings and grey breast with heavy blackish markings. **VOICE** Thin ***tsweee*** or ***tsuuu***. **HABITAT** Broadleaved evergreen forest, secondary growth; up to 1,100 m. **RANGE & STATUS** Breeds C,E China, Japan; winters S China. **SE Asia** Uncommon to locally common winter visitor N,C Laos, W,E Tonkin, N,C Annam. Also recorded on passage E Tonkin. Vagrant NW,NE Thailand, S Laos.

882 **GREY-WINGED BLACKBIRD**
Turdus boulboul Plate **88**
IDENTIFICATION 27.5–29 cm. Monotypic. **Male** Blackish plumage with contrasting large pale greyish wing-patch diagnostic. Bill orange. **Female** Rather plain, warm olive-brown with yellowish bill. Has similarly shaped wing-patch to male but colour only slightly paler (and warmer) brown than rest of wing. See Chinese Blackbird. **Juvenile male** Dark parts blacker than juvenile female. **Juvenile female** Upperparts and wings darker brown than adult female, crown, upperparts and scapulars streaked buff, wing pattern similar to adult but browner with buff spots/streaks on lesser and median coverts; has buffish throat and underparts with slightly paler submoustachial stripe, darker moustachial and malar lines, and heavy dark brown mottling on breast and belly. See White-collared Blackbird. **VOICE** Song is rich and melodious with generally few repeated phrases: ***tweee-toooh tweee-toooh chuiyui-twit weear-twit weear-trtrtrtt-whih-whih-which wheeeyar-wheeeyar***, ***chir-bles-we-bullie-dee we-put-kur-we-put-kur who-bori-chal-let-cha-he*** etc. Calls include a low ***chuck-chuck***, more emphasised ***chook-chook*** when alarmed and ***churi*** contact note. **HABITAT** Oak and rhododendron forest, broadleaved evergreen forest, clearings; 980–2,565 m. Breeds above 1,200 m in Himalayas. **RANGE & STATUS** Resident (subject to relatively local movements) NE Pakistan, NW,N,NE Indian subcontinent, SW,S China. **SE Asia** Uncommon resident N Laos, W Tonkin. Scarce to uncommon winter visitor S Myanmar, NW Thailand. Vagrant NE Thailand. Recorded (status uncertain) N,E Myanmar, E Tonkin. **BREEDING** March–July. Multi-brooded. **Nest** Bulky cup, in tree, 2–5 m above ground; sometimes on ground. **Eggs** 2–4; pale greenish, blotched and thickly streaked (sometimes very heavily) dull brownish-red; 29 x 20.9 mm (av.).

883 **CHESTNUT THRUSH**
Turdus rubrocanus Plate **88**
IDENTIFICATION 24–26.5 cm. *T.r.gouldi*: **Male** Largely chestnut body, contrasting dark brownish-grey hood and blackish wings and tail diagnostic. Undertail-coverts blackish with whitish feather-tips. *T.r.rubrocanus* (W Myanmar) has very pale grey hood (somewhat darker on crown, nape and sides of head) and pale and dark markings on belly-centre. **Female** Similar to male but body somewhat paler and more rufous (browner on mantle and scapulars), wings and tail browner. *T.r.rubrocanus* differs more markedly from male, being duller and paler with more uniform crown, nape and upper mantle. **VOICE** Song is a series of short phrases, each repeated 3–8 times: ***yee-bre yee-bre yee-bre-diddyit diddiy-it-yip bru yipbru-yip-bru***. Calls with deep ***chuk-chuk...*** and faster ***kwik-kwik*** when alarmed. **HABITAT** Broadleaved evergreen forest; 900–2,900 m, rarely down to 200 m. **RANGE & STATUS** Breeds N Pakistan, NW India, SW,W,NC China; some eastern populations winter south to N,NE India. **SE Asia** Scarce to uncommon winter visitor W,N,E Myanmar, NW Thailand, N Laos, W Tonkin. Vagrant C Laos, E Tonkin.

884 **BLACK-THROATED THRUSH**
Turdus atrogularis Plate **90**
IDENTIFICATION 23.5–27.5 cm. Monotypic. **Male non-breeding** Black of supercilium, throat and upper breast with whitish feather-fringing. Similarly patterned to Red-throated Thrush, but has black base colour to supercilium, throat and upper breast and lacks rufous-red on tail; averages larger. Intergrades occur. **Male breeding** Solid black supercilium, throat and upper breast, without pale fringing. **Female** Similar to male non-breeding but has whiter throat and submoustachial area, more scaled/mottled upper breast. **First-winter male** Initially resembles adult female but has pale fringes and tips to greater coverts. **First-winter female** Differs from Red-throated by lack of rufous on tail, and has darker (less warm) mottling/streaks on breast and flanks (first-winter male differs in same way). See Dusky and Japanese Thrushes. **VOICE** Sings with hoarse, raucous, whistled notes: ***t'eeee t'yuyuu teeu-eet*** etc. (first two phrases drawn out and falling, latter sharp and rising), with bouts sometimes preceded by more ponderous ***hweet*** or ***hweet-a***. Typical calls (often repeated) include soft ***jak*** or ***tsak***, ***chuck***, abrupt ***chk*** and single thin ***seee*** or ***ziep*** in flight. Contact calls include ***qui-kwea***, rapid ***hetetetet***, hoarser ***retet riep riep*** and rapid ***wiwiwi*** on take-off. **HABITAT & BEHAVIOUR** Open broadleaved evergreen forest and edge, clearings, cultivation; 150–2,565 m. Often in small flocks; associates with other thrushes. **RANGE & STATUS** Breeds extreme E Europe, W,WC Siberia, NW Mongolia, NW China; winters C Asia, east Saudi Arabia, S Iraq, Iran, Afghanistan, Pakistan, northern Indian subcontinent, S,SE Tibet, SW China. **SE Asia** Scarce to uncommon winter visitor N,C,E Myanmar, NW Thailand. Vagrant E Tonkin. **NOTE** Split from Red-throated following Dickinson (2003).

885 **RED-THROATED THRUSH**
Turdus ruficollis Plate **90**
IDENTIFICATION 22.5–26 cm. Monotypic. **Male non-breeding** Brownish-grey upperparts and white underparts with contrasting rufous-red supercilium, throat and upper

breast diagnostic. Has narrow whitish feather-fringes on supercilium, throat and upper breast, outertail-feathers largely reddish-rufous. See Black-throated Thrush. **Male breeding** Lacks whitish feather-fringing on supercilium, throat and upper breast. **Female** Similar to male non-breeding but has black streaks on throat-sides and upper breast, less rufous-red on upper breast and less distinct supercilium. **First-winter male** Initially resembles adult female but has pale fringes and tips to greater coverts. **First-winter female** Lacks obvious rufous-red on head and breast, supercilium whitish, throat and upper breast whitish with heavy blackish streaks (mainly at sides), lower breast washed greyish, breast and flanks mottled/streaked warmish brown; shows some rufous fringing on outertail-feathers. See Black-throated, Dusky and Japanese Thrushes. **VOICE** Song is a simple, rambling, cackling ***chve-che-chve-che chvya-chya-chvya-chya...***; ***chooee chooee whee-oo-ee oo*** or ***hoo-eee whee-oo-ee oo***. Typical calls are similar to Black-throated. Also has softer chuckling ***which-which-which***. **HABITAT & BEHAVIOUR** Open broadleaved evergreen forest and edge, clearings, cultivation; 1,000–2,565 m, sometimes down to 150 m in N Myanmar. Often in small flocks; associates with other thrushes. **RANGE & STATUS** Breeds SC Siberia, N Mongolia; winters S Turkmenistan, E Afghanistan, N Pakistan, NW,N,NE India, Nepal, Bhutan, S,SE Tibet, SW,NE China. **SE Asia** Scarce to uncommon winter visitor N,C,E Myanmar. Vagrant NW Thailand.

886 NAUMANN'S THRUSH
Turdus naumanni Plate **90**

IDENTIFICATION 22.5–25 cm. Monotypic. **Male** Similar to Dusky Thrush, but much greyer and almost unmarked above, rump to tail strongly rufescent (recalling Red-throated Thrush), pale parts of head-side washed reddish-rufous, breast and body-sides heavily mottled reddish-rufous. Shows reddish-rufous edgings on scapulars to median coverts. Intergrades with Dusky Thrush occur. **Female** Duller than male; lacks reddish-rufous feather edgings above, tail less strongly rufescent, rufous of head-side duller and more pinkish to buffish-tinged, upper throat whitish, with stronger dark malar marking. **First winter** Generally duller, faintly dark-mottled above, but with some bright rufous edgings on scapulars, pale parts of head-side more buffy-white, throat whiter with more contrasting malar marking, shows some blackish spotting on breast. Males probably tend to more brightly plumaged, while females have duller tones and heavier mottling on breast and flanks. Dullest females still show rufescent tinges on tail, head-side and underparts that are lacking on Dusky. **VOICE** Song is quite fluty and melodious, frequently ending in faint trill or twitter: rather flat, unhurried ***tyee-tryu-uu-tee tvee-tryuuuu-tvee*** (***tryuuuu*** longest and accentuated), ending with rising ***tsee-tsee-tee***. Calls include loud, shrill ***cheeh-cheeh***, rather harsh repetitions of ***ket***, ***cha*** or ***kra*** notes, and a chuckling ***chak-chak***. **HABITAT** Cultivation, forest edge; recorded at 980–2,900 m (perhaps lower in Thailand). **RANGE & STATUS** Breeds SC Siberia; winters W,C and southern China, rarely Tibet. **SE Asia** Rare to scarce winter visitor N Myanmar. Vagrant south-east NW Thailand.

887 DUSKY THRUSH *Turdus eunomus* Plate **90**

IDENTIFICATION 22.5–25 cm. Monotypic. **Male** Blackish ear-coverts and malar line, sharply contrasting clear whitish supercilium, submoustachial area and throat, largely blackish breast and flanks with white scales and (usually) extensively bright rufous-chestnut wings diagnostic. Crown and upperparts mostly dull greyish-brown with blackish feather-centres; often appears to show double breast-band. See Naumann's Thrush. **Female** Usually duller with less blackish coloration on ear-coverts and upperparts and less black on breast; tends to show more pronounced malar line. **First winter** Generally similar to female or duller, with extremes approaching respective adults. Has more restricted rufescent fringes to wing-feathers, and pale tips to tertials and some or all greater coverts (retained juvenile feathers). Dull females can resemble Black-throated Thrush but have darker, browner ear-coverts and upperparts (latter with obviously darker feather-centres), rufescent fringing on wings, and bolder dark markings on lower breast and flanks. See Naumann's. **VOICE** Song is similar to Naumann's, but with emphasis at beginning: ***tryuuuu-tvee-tryu tyuu-trrryu-uute tryuute tryuute frrrr***. Calls like Naumann's, but also a chattering ***quaawag*** or ***kvaevaeg***. **HABITAT** Open broadleaved evergreen forest and edge, cultivation; up to 2,900 m. **RANGE & STATUS** Breeds C,E Palearctic; winters China (except NW), Taiwan, N,S Korea, Japan; irregularly west along Himalayas to N Pakistan. **SE Asia** Rare to locally fairly common winter visitor N,C,S Myanmar, W,NW Thailand, W,E Tonkin.

888 GREY-SIDED THRUSH *Turdus feae* Plate **90**

IDENTIFICATION 23.5 cm. Monotypic. **Male** Very similar to Eyebrowed Thrush but upperparts usually warmer brown, particularly crown; has grey underparts with white chin, belly-centre and vent. **Female** Breast and flanks less extensively pure grey, usually has less grey on throat; has slight dark spots/streaks on sides of throat and upper breast and warm brownish fringes to breast feathers. **First winter** Like female but has warmer breast and flanks, darker streaks on throat-sides and pale buffish tips to greater coverts. See Eyebrowed. **VOICE** Call is slightly but distinctly thinner than that of Eyebrowed: ***zeeee*** or ***sieee***. **HABITAT** Broadleaved evergreen forest; 520–2,565 m. **RANGE & STATUS** Breeds east N China; winters NE India. **SE Asia** Scarce to uncommon winter visitor W,E Myanmar, north Tenasserim, W,NW Thailand, C Laos. Vagrant NE Thailand.

889 EYEBROWED THRUSH
Turdus obscurus Plate **90**

IDENTIFICATION 22.5–24.5 cm. Monotypic. **Male** Combination of grey hood, white supercilium, cheek-bar and chin, olive-brown upperside and pale orange-rufous lower breast and flanks distinctive. See Grey-sided Thrush. **Female** Base colour of head much browner than male, centre of throat and submoustachial line whitish, flanks more washed out. **First-winter male** Like first-winter female but often has greyer ear-coverts to upper breast and brighter flanks. **First-winter female** Similar to adult female but has pale buffish tips to greater coverts. Individuals which lack obvious orange-rufous on flanks recall Grey-sided but lack clear grey on underparts and usually show defined white submoustachial stripe. **VOICE** Song is a series of 2–3 clear mournful phrases: ***teveteu trrryutetyute trrryutetyutyu...***, followed by lower-pitched twittering and discordant warbling or chattering, punctuated by short pauses. May include mimicry. Flight call is a thin, drawn-out ***zieeh*** or harsher ***seee*** or ***tseee***. Contact calls include thin ***sip-sip***, ***che-e*** and chuckling ***dack-dack*** or ***tuck-tuck*** and ***tchup*** or ***tchuck***. **HABITAT & BEHAVIOUR** Forest, secondary growth, plantations, mangroves and gardens on migration; up to 3,100 m, mostly above 1,000 m. Often in flocks; fond of fruiting trees. **RANGE & STATUS** Breeds C,E Palearctic, NE China; winters NE India, SW,S China, Greater Sundas, Philippines. **SE Asia** Fairly common winter visitor (except SW Myanmar, Singapore, N Annam). Uncommon to fairly common passage migrant Peninsular Malaysia, Singapore (winters?), Cambodia, E Tonkin.

890 PURPLE COCHOA *Cochoa purpurea* Plate **88**

IDENTIFICATION 26.5–28 cm. Monotypic. **Male** Brownish-purple plumage with pale lavender-blue crown and black sides of head distinctive. Has broad pale lavender-purple band across base of black flight-feathers, and lavender-purple tail with black tip (all blackish below). **Female** Similarly patterned to male but brownish-purple of upperparts replaced by dark rufescent-brown, entire underparts deep buffish-rufous. **Juvenile** Resembles respective adult but crown blackish with bold white markings (or mostly white), upperpart feathers dark-tipped and streaked/spotted with warm buff, underparts rich buff boldly barred blackish, throat plainer with blackish malar line, ear-coverts with some white markings. **VOICE**

Song recalls Green Cochoa but deeper and clearer: ***whiiiii***. Contact calls include very thin, rather thrush-like ***sit*** and ***tssri*** notes. **HABITAT & BEHAVIOUR** Broadleaved evergreen forest; 1,000–2,135 m, rarely down to 400 m. Forages on ground and visits fruiting trees. **RANGE & STATUS** Resident N,NE Indian subcontinent, south-western China. **SE Asia** Scarce resident W,N,C,S(east),E Myanmar, north Tenasserim, NW Thailand, N Laos, W,E Tonkin. **BREEDING** April–July. **Nest** Shallow cup, in small tree; 2–6 m above ground. **Eggs** 2–4; pale sea-green with bright reddish-brown blotches over lavender to grey undermarkings (often capped at broader end); 31.3 x 21.6 mm (av.).

891 **GREEN COCHOA** *Cochoa viridis* Plate **88**

IDENTIFICATION 27–29 cm. Monotypic. **Male** Striking, with green plumage (variably mixed blue on underparts), bright blue crown and nape, extensive silvery-blue markings on black wings and blue tail with broad black tip. **Female** Like male but has extensive brownish wash on secondaries, tertials and inner greater coverts, no blue on underparts. **Juvenile** Crown blackish with bold white markings, rest of upperparts broadly fringed blackish and spotted buff, cheeks and ear-coverts largely whitish, wings like respective adults but with rich buff to rufescent spots on wing-covert tips, underparts rich buff with bold blackish scales; throat plainer with blackish malar line. Tail as adult. **First summer** Like respective adult but has whitish strip from chin to lower ear-coverts and neck-side, underparts washed rich golden-buff (also slight golden tinge above). **VOICE** Song is a series of loud, pure, monotone whistles: ***hiiiiii***, lasting c.2 s. **HABITAT** Broadleaved evergreen forest; 700–2,565 m, occasionally down to 400 m. **RANGE & STATUS** Resident N,NE Indian subcontinent, SW,SE China. **SE Asia** Scarce resident W,S(east),E Myanmar, W,NW,NE(west),SE Thailand, south-west Cambodia, Laos, W,E Tonkin, N,C,S Annam. **BREEDING** March–June. **Nest** Shallow cup, on tree-branch; 10 m above ground. **Eggs** 2–4; like those of Purple Cochoa; 30.4 x 21.3 mm (av.).

892 **GRANDALA** *Grandala coelicolor* Plate **89**

IDENTIFICATION 20.5–23 cm. Monotypic. **Male** Brilliant purplish-blue plumage with black lores, wings and tail diagnostic. Has very pointed bill and wings. See Blue Rock-thrush. **Female** Distinctive. Brownish with whitish streaks on head, upper mantle and underparts (mainly throat and breast), greyish-blue rump and uppertail-coverts, white tips/fringes to tertials, white tips to greater coverts (lost with wear) and white patch near base of secondaries and primaries. **Juvenile** Like female but has heavier whitish streaks on crown and mantle, extending to scapulars and uppertail-coverts. **VOICE** Song is described as a subdued, clear, soft ***tju-u tju-u ti-tu tji-u...*** Foraging flocks utter conversational ***dju-i***; ***djew*** and ***djwi*** notes. **HABITAT** Barren slopes and rocky areas above tree-line, open forest; recorded at 1,830 m in winter. Breeds above 3,900 m in Himalayas. **RANGE & STATUS** Resident (subject to minor movements) NW,N,NE Indian subcontinent, S,E Tibet, SW,W,N China. **SE Asia** Recorded in winter (status uncertain) N Myanmar. **BREEDING** Himalayas: June–July. **Nest** Largish cup, on rock ledge. **Eggs** 2; greenish-white with reddish-brown blotches over purplish undermarkings; 27.3–29.7 x 19.4–21 mm.

MUSCICAPIDAE: SAXICOLINAE: Shortwings, robins, redstarts, rock-thrushes, chats, forktails, whistling-thrushes & allies

Worldwide c.176 species. SE Asia 47 species. **Shortwings** (*Brachypteryx*) Small with short rounded wings, rather long legs and relatively short to short tails. Mainly inhabit forests, skulking on or near ground. Feed mainly on insects. **Robins** (*Luscinia, Myiomela, Cinclidium, Tarsiger*) Small to medium-sized; most skulk in understorey or on ground. **Redstarts** (*Rhyacornis, Chaimarrornis, Phoenicurus, Hodgsonius*) Relatively small, most with some rufous or chestnut on tail. Generally quite conspicuous, in more open situations; *Rhyacornis* and *Chaimarrornis* frequent streams, *Hodgsonius* skulks in understorey. **Rock-thrushes** (*Monticola*) Fairly small to medium-sized, similar to typical thrushes but with strong sexual dimorphism (males extensively blue). Mostly in more open habitats. **Chats** (*Oenanthe, Saxicola*) Small; usually adopt upright posture. Found in more open situations, grassland etc. **Forktails** (*Enicurus*) Striking black and white plumage, deeply forked tails. Inhabit streams and rivers. **Whistling-thrushes** (*Myophonus*) Like typical thrushes and relatively large, with uniformly dark plumage. Sexes alike. Usually near forest streams, feeding on invertebrates, small amphibians, berries and nestlings.

893 **GOULD'S SHORTWING**
Brachypteryx stellata Plate **91**

IDENTIFICATION 13–13.5 cm. *B.s.stellata*: **Adult** Unmistakable, with chestnut crown and upperside, black lores, fine grey and black vermiculations on underside and small white arrow- to star-shaped markings on lower breast to vent. **Juvenile** Crown and ear-coverts to back blackish-brown with rufous to rufous-chestnut streaks, underparts greyish-brown with larger but duller arrow-shapes and streaks on belly, throat with pale streaks but no bars, and breast-band blackish-brown with rufous-buff streaks. **Other subspecies in SE Asia** *B.s.fusca* (W Tonkin). **VOICE** Song is very high-pitched, with introductory notes gradually becoming louder and closer, running into a slightly undulating series of 26–34 very piercing high-frequency notes: ***tssiu tssiu tssiu tssiu-tssiu-tsitsitssi-utssiutssiutssiutsitsitssiutssiu...*** **HABITAT** Boulder-strewn gullies in rhododendron and conifer forest, bamboo, broadleaved evergreen forest near streams in winter; 1,800–2,450 m (winter). Breeds above 3,300 m in Himalayas. **RANGE & STATUS** Resident (subject to altitudinal movements) N,NE India, Nepal, Bhutan, SE Tibet, SW,W China. **SE Asia** Scarce resident N Myanmar, W Tonkin. **BREEDING** May–July (Himalayas). Otherwise unknown.

894 **RUSTY-BELLIED SHORTWING**
Brachypteryx hyperythra Plate **91**

IDENTIFICATION 12 cm. Monotypic. **Adult** Unmistakable, with short tail, dark slaty-blue upperside, orange-rufous underside (including vent), and short white eyebrow (mainly in front of eye). See Indian Blue-robin and White-browed Bush-robin. **Female** Dark brown above, and duller below than male, lacks eyebrow. **Juvenile** Undocumented. **VOICE** Song very similar to Lesser Shortwing, but faster, longer and more musical. **HABITAT** Thickets and grass near broadleaved evergreen forest; 980–1,050 m. **RANGE & STATUS** Resident (subject to altitudinal movements) NE India, Bhutan, SW China; ? Nepal. **SE Asia** Recorded in winter (status unknown) N Myanmar. **BREEDING** Undocumented.

895 **LESSER SHORTWING**
Brachypteryx leucophrys Plate **91**

IDENTIFICATION 11.5–12.5 cm. *B.l.carolinae* (W,S,E Myanmar, Tenasserim, W,NW,SE Thailand, northern Indochina): **Male** Rather nondescript. Upperside warm dark brown, underside paler and buffier with whitish centre to throat and abdomen, whitish mottling on breast and whitish undertail-coverts. Short white eyebrow distinctive (when unconcealed). Can be confused with several species, including White-browed Shortwing (female-type plumages), Rufous-browed Flycatcher, female Snowy-browed Flycatcher and Buff-breasted Babbler. Note voice, habitat, small size, short tail, long pale legs, relatively strong blackish bill, restricted whitish areas on underparts and breast-mottling. *B.l.langbianensis* (Cambodia, S Laos, S Annam) lacks buff on underparts, has grey breast-band, cheeks and rear supercilium, grey mixed in on flanks, white eyebrow extending further forward; *nipalensis* (N Myanmar) and *wrayi* (extreme S

Thailand, Peninsular Malaysia) have slaty-blue head-sides and upperside, white eyebrow extending further forward, brown of underparts replaced by bluish-grey (paler than upperparts); latter has darker, more contrasting slaty-blue and grey plumage-parts. See White-browed. **Female** Like male *B.l.carolinae* (all subspecies) but *wrayi* is warmer above with more chestnut-tinged wings and tail. **Juvenile** Like female but ear-coverts and upperparts a little darker with rufous to rufous-chestnut streaks; initially lacks white eyebrow, underparts more uniform buff, deeper-coloured on breast and flanks where scaled/scalloped blackish-brown, wing-coverts tipped rufous-chestnut. Males of *B.l.nipalensis* and *wrayi* apparently start to acquire adult plumage during first summer. **Other subspecies in SE Asia** *B.l.leucophrys* (most of S Thailand?). **VOICE** Song is brief, high-pitched and melodious, with pause after first note and ending with a rapid jumble. Typical calls are a subdued hard ***tack*** or ***tuck*** and thin high-pitched whistle. **HABITAT & BEHAVIOUR** Broadleaved evergreen forest; 975–2,550 m, locally down to 380 m (winter only?). Very skulking, on or near ground. **RANGE & STATUS** Resident N,NE Indian subcontinent, southern China, Sumatra, Java, Bali, Lesser Sundas. **SE Asia** Fairly common resident W,N,S(east),E Myanmar, Tenasserim, W,NW,NE,SE,S Thailand, Peninsular Malaysia, Cambodia, Laos, W,E Tonkin, N,C,S Annam. **BREEDING** February–June. **Nest** Oval ball with side-entrance, on bank, amongst moss on tree or rock or wedged in orchid clump or rattan; up to 90 cm above ground. **Eggs** 2–4; olive-green to sea-green, profusely speckled and freckled light reddish-brown; 19.5 x 14.6 mm (av.; *nipalensis*). Brood-parasitised by Large Hawk-cuckoo.

896 WHITE-BROWED SHORTWING

Brachypteryx montana Plate **91**

IDENTIFICATION 12.5–13.5 cm. *B.m.cruralis*: **Male** Uniform dull dark blue plumage with clear white supercilium (almost meets on forehead when flared) diagnostic. Bulkier and somewhat longer-tailed than Lesser Shortwing. **Female** Rather uniform brown with distinctive rufous forehead, lores, orbital area and (short, slight) supercilium. See Rufous-browed and female Snowy-browed Flycatchers. **Juvenile** Very like Lesser. Best told by more uniformly dark throat and breast, darker legs and longer tail; breast is streaked buff and tends to appear less scalloped (may be similar). **First-winter male** Like female but has white supercilium (similar to adult male) and darker lores. **VOICE** Song is a complex, monotone, meandering warble, usually introduced by 1–3 ***wheez*** notes. Usual call is a hard ***tack***. **HABITAT & BEHAVIOUR** Broadleaved evergreen forest; 1,400–2,746 m, locally down to 305 m in winter (N Myanmar). Skulks on or near ground. **RANGE & STATUS** Resident N,NE Indian subcontinent, SE Tibet, C and southern China, Taiwan, Greater Sundas, Flores, Philippines. **SE Asia** Uncommon to fairly common resident W,N,S(east),E Myanmar, W,NW Thailand, N Laos, W Tonkin, N,C,S Annam. **BREEDING** February–August. **Nest** Domed with side-entrance, amongst moss on rock-face or tree-trunk. **Eggs** 3–4; fairly glossy, white; 22.7 x 16 mm (av.). Brood-parasitised by Lesser Cuckoo.

897 JAPANESE ROBIN *Luscinia akahige* Plate 92

IDENTIFICATION 14–15 cm. *L.a.akahige*: **Male** Easily identified by uniform rufous-orange forehead, head-sides, throat and upper breast. Rufescent-brown above with brighter rufous-chestnut tail; has variable narrow blackish breast-band and broadly slate-greyish flanks. **Female** Resembles male but lower breast and flanks browner, rufous-orange of head and breast duller and a little less extensive, no breast-band. **VOICE** Song is a series of well-spaced, simple, mostly quavering phrases, each with a brief introduction: ***hi CH'H'H'H'H hi-tu CH'I'I'I'I hi CH'H'H'H'H ts-ti CH'U'U'U'U tsi CHUK'CHUK'CHUK...*** etc. Calls include thin, metallic ***tsip*** notes. **HABITAT & BEHAVIOUR** Broadleaved evergreen forest, sometimes parks and gardens; up to 1,525 m. Skulks on or close to ground. **RANGE & STATUS** Breeds Sakhalin, Kuril Is, Japan; some winter S China. **SE Asia** Rare winter visitor/vagrant Thailand (except S), C Laos, E Tonkin, C Annam.

898 SIBERIAN RUBYTHROAT

Luscinia calliope Plate **92**

IDENTIFICATION 15–16.5 cm. Monotypic. **Male non-breeding** Brilliant red throat, short white supercilium and white submoustachial stripe distinctive. Upperside all brown, lores and malar line blackish, breast brownish-grey, bill black with paler base to lower mandible; blackish border to throat and extent of grey on breast variable. See White-tailed Rubythroat. **Male breeding** Breast more solidly grey, bill all black. **Female** Resembles male but throat white (may be pink) with no blackish border, supercilium and submoustachial stripes less distinct, lores paler, breast usually browner. See White-tailed and Bluethroat. **First-winter male** Like adult but shows some buff tips on tertials and greater coverts, breast usually browner. **First-winter female** From adult mainly by buff tips on tertials and greater coverts; rarely shows pink on throat. **VOICE** Song is a sweet scratchy varied warble, with much avian mimicry. Calls include a loud, clear ***ee-uh*** or ***se-ic*** and deep ***tschuck***. **HABITAT & BEHAVIOUR** Grass, scrub, thickets, sometimes gardens; up to 1,555 m, rarely 2,375 m on passage. Skulks in dense vegetation, runs along ground, often cocks tail. **RANGE & STATUS** Breeds C,E Palearctic, W,N,NE China, N Korea, N Japan; winters northern and eastern Indian subcontinent, southern China, Taiwan, Philippines. **SE Asia** Fairly common to common winter visitor (except S Thailand, Peninsular Malaysia, Singapore). Also recorded on passage E Tonkin. Vagrant Peninsular Malaysia, Singapore.

899 WHITE-TAILED RUBYTHROAT

Luscinia pectoralis Plate **92**

IDENTIFICATION 15–17 cm. *L.p.tschebaiewi*: **Male breeding** Resembles Siberian Rubythroat but shows mostly black breast, slatier crown, head-sides and upperparts, and blackish tail with white on base and tips of outer feathers. **Male non-breeding** Has browner crown and upperparts and grey to whitish scaling on breast. **Female** Similar to Siberian but has distinctive white spots on tips of outertail-feathers, colder and darker upperparts and distinctly grey throat-sides and breast-band. **Juvenile** Feathers of ear-coverts and upperside bordered blackish and streaked pale buffish, wing-coverts and tertials tipped pale buffish, underparts uniform buffish-white with dark brownish-grey to blackish scales/streaks on throat and breast (fading to lower flanks), tips of outertail-feathers buffish. **First-winter male** Like first-winter female but has some white on base of outertail-feathers, often has darker sides of head and darker breast with some black; usually acquires red throat and black breast by first summer. **First-winter female** Like adult but has retained juvenile greater coverts and tertials, possibly tends to be more buffish below. **VOICE** Song is a complex series of undulating, warbling trills and twitters, often given in short bursts (sometimes more prolonged). Calls with a deep ***tchuk*** (similar to Siberian) and sparrow-like ***tchink***. **HABITAT & BEHAVIOUR** Dwarf rhododendron, juniper and other scrub, typically above tree-line; grass and scrub (often near water) in winter. Up to 4,420 m; breeds above 2,700 m in NE India. Skulks on or near ground but often sings from exposed perch. **RANGE & STATUS** Breeds C Asia, N Pakistan, NW,N,NE Indian subcontinent, S,E Tibet, NW,N,W,SW China; some winter south to N,NE Indian subcontinent. **SE Asia** Scarce resident N Myanmar. Vagrant NW,C Thailand. **BREEDING** June–July. Multi-brooded. **Nest** Dome with side-entrance or deep cup, on or near ground, amongst vegetation or rocks; up to 60 cm. **Eggs** 3–4; blue-green with faint ring of reddish freckles around broader end; 21.6 x 15.4 mm (av.).

900 BLUETHROAT *Luscinia svecica* Plate **92**

IDENTIFICATION 13.5–15 cm. *L.s.svecica*: **Male non-breeding** Pale underparts with scaly blue, black and rufous-red breast-bands, broad whitish supercilium and extensively rusty-rufous basal half of outertail-feathers (conspicuous in flight) distinctive. **Male breeding** Unmistakable. Throat and upper breast blue with red central patch and black lower border, and separate, solid rufous-red breast-band. **Female** Could be confused with Siberian Rubythroat but shows diagnostic rufous on outertail-feathers, pronounced blackish malar stripe and pronounced band of blackish markings across upper breast. Older birds can show variable amounts of blue on throat and breast and indistinct rufous-red breast-band (particularly in breeding plumage). **First winter** Similar to respective non-breeding adults but with warm buff tips to greater coverts. **VOICE** Song is a continuous, varied, clear, rapid series of fine ringing notes, intermixed with call-notes and avian mimicry. Typical calls are a twanging ***dzyink*** and low sharp ***tuck*** or ***tchak*** and ***tsee-tchak-tchak*** etc. **HABITAT & BEHAVIOUR** Grass and scrub, thickets, usually near water; up to 760 m. Skulking but often frequents open areas bordering dense vegetation. Cocks tail. **RANGE & STATUS** Breeds Palearctic, N Pakistan, NW India, NW,NE China, NW USA (W Alaska); winters N Africa, Middle East, Indian subcontinent, southern China. **SE Asia** Uncommon to locally common winter visitor Myanmar, NW,NE,C Thailand, Indochina (except W Tonkin, N,S Annam). Also recorded on passage N Laos, E Tonkin. Vagrant S Thailand.

901 RUFOUS-HEADED ROBIN
Luscinia ruficeps Plate **92**

IDENTIFICATION 15 cm. Monotypic. **Male** Striking, with orange-rufous crown, ear-coverts and nape, black lores and cheeks and clean white throat broadly bordered with black. Upperparts slaty-grey, upper breast and flanks grey, rest of underparts white, tail blackish, the outer feathers fringed rufous and tipped blackish. **Female** Similar to Siberian Blue Robin but lacks blue on rump to uppertail, tail browner with warmer-edged outer feathers, breast and flanks more olive-tinged and less buff, has more dark scaling on throat and all-dark bill. Legs and feet fleshy-coloured. See Blackthroat. **VOICE** Song is a series of well-spaced powerful rich phrases, each introduced by a thinner short note: ***ti CHO CHUK'UK'UK ti TCH-WR'RR'RR ti CHI-WRU-W'R'R'R ti CHR'R'R'R ti CHR'RIU'IU'IU...*** etc. Calls with a rather deep ***tuc*** or ***toc***, similar to Siberian Blue, and a soft thin high-pitched ***si***, similar to Indian Blue Robin. **HABITAT & BEHAVIOUR** Recorded once in montane heather-like scrub, at 2,010 m. Skulks on or near ground, cocks tail jerkily, sometimes pushes it downwards. **RANGE & STATUS** Breeds W,NC China; winter range unknown. **SE Asia** Vagrant Peninsular Malaysia (Gunung Brinchang), March 1963.

902 BLACKTHROAT *Luscinia obscura* Plate **92**

IDENTIFICATION 12.5–14.5 cm. Monotypic. **Male** Resembles Siberian Blue Robin but has diagnostic all-black throat and upper breast and largely white basal two-thirds of outertail-feathers. Legs and feet dark. See Firethroat. **Female** Like Siberian Blue Robin but lacks obvious scales on underparts; has pale warm buff undertail-coverts, rufescent-tinged uppertail-coverts, warm-tinged brown tail, dark bill with only slightly paler lower mandible, and dark (brownish-plumbeous) legs and feet. Colder and more olive above than Firethroat and much more whitish below, with paler undertail-coverts. See Rufous-headed and Indian Blue Robins. **VOICE** Song is a rather shrill, laid-back, cheerful series of phrases, each repeated after shortish intervals: ***whr'ri-whr'ri, chu'ti-chu'ti*** (second note higher), alternated with purring trills, ***hdrriiii-ju'ju*** and ***uu ji'uu*** etc. Contact call is a series of soft, subdued ***tup*** notes. **HABITAT** Dense thickets, grass, scrub, bamboo; recorded at 300–395 m. **RANGE & STATUS** Breeds W,NC China; winter range unknown. **SE Asia** Vagrant NW Thailand; February 1965 (specimen), and March 2000 (sight record).

903 FIRETHROAT *Luscinia pectardens* Plate **92**

IDENTIFICATION 14 cm. Monotypic. **Male breeding** Recalls Blackthroat but has diagnostic bright orange-red throat and centre of upper breast, and white neck-patch. See Indian Blue Robin. **Male non-breeding** Head-sides and underparts similar to female; may show hint of neck-patch. **Female** Similar to Indian Blue Robin but underside more uniform warm buff (including undertail-coverts) and unscaled, legs and feet dark (grey-black or black-brown to pale plumbeous). See Blackthroat and Siberian Blue Robin. **First-winter male** Resembles adult female but scapulars, back and uppertail-coverts dark slaty-blue, tail similar to adult male. **VOICE** Song is a long series of simple, well-spaced, rather subdued phrases, interspersed with husky, buzzing notes: ***wiu-wihui'wi wi'chu-wi'chu whiiiiii wi-chudu'chudu t'sii-sii wi'chu-wi'chu-wi'chu chu-tsri'sri...*** etc. **HABITAT** Broadleaved forest, bamboo, thickets; recorded at 150 m in winter. Breeds above 2,800 m in SE Tibet. **RANGE & STATUS** Breeds SE Tibet, SW,W,NC China; recorded in winter NE India, Bangladesh. **SE Asia** Recorded (status uncertain but could breed in high mountain areas) N Myanmar. **BREEDING** Undocumented.

904 INDIAN BLUE ROBIN
Luscinia brunnea Plate **92**

IDENTIFICATION 13.5–14.5 cm. *L.b.wickhami*: **Male** Dark blue upperparts, broad white supercilium and bright orange-rufous throat, breast and flanks diagnostic. Chin, short malar line and vent white. See White-browed Bush-robin. **Female** Similar to Siberian Blue Robin but has distinctly rich buff breast (no obvious scaling) and flanks, whitish areas on underparts restricted to throat, centre of belly and undertail-coverts, brown rump to uppertail. Legs and feet brownish-flesh to pale fleshy-white. See Firethroat. Note range. **Juvenile** Crown, sides of head and upperparts darker than female and streaked buffish, breast and flanks heavily scaled/mottled dark brown. Bill paler than adult with yellow at gape. Male has blue on tail. **First-winter male** Duller than adult with duller sides of head, buff-tinged supercilium, duller lower throat and breast. **VOICE** Song phrases are sweet but rather short, hurried and jumbled, introduced by 2–4 high-pitched thin whistles. Calls with hard ***tek*** notes when alarmed. **HABITAT** Bamboo, secondary growth, thickets, broadleaved evergreen forest; 1,480–2,040 m. **RANGE & STATUS** Breeds N Pakistan, NW,N,NE Indian subcontinent, SE Tibet, SW,W,NC China; some populations winter SW India, Sri Lanka. **SE Asia** Locally common resident W Myanmar. **BREEDING** May–July. **Nest** Untidy cup, situated on bank, on ground or among stones. **Eggs** 3–4; pale blue; 19.3 x 14.5 mm (av.).

905 SIBERIAN BLUE ROBIN
Luscinia cyane Plate **92**

IDENTIFICATION 13.5–14.5 cm. *L.c.cyane*: **Male** Dull dark blue upperparts, white underparts and broad black line from lores through cheeks to breast-side diagnostic. **Female** Crown and upperparts greyish-brown, often with some dull blue on rump and uppertail-coverts (sometimes tail), throat buffish-white, breast and (usually) flanks slightly deeper buff, throat-side and breast variably scaled or mottled darker. Legs distinctly pinkish. See other *Luscinia* robins. **First-winter male** Resembles female but rump to uppertail dull blue, usually shows some blue on scapulars and wing-coverts (sometimes most of upperparts and all wing-coverts), has rufous-buff tips to outer greater coverts. **First-winter female** Similar to adult but often lacks blue on rump to uppertail, outer greater coverts tipped rufous-buff; possibly tends to have richer buff wash on underparts. See Rufous-headed Robin. **Other subspecies in SE Asia** *L.c.bochaiensis* possibly occurs in Indochina. **VOICE** Song is a loud, rapid, rather explosive ***tri-tri-tri-tri, tjuree-tiu-tiu-tiu-tiu*** etc., usually introduced by fine, spaced ***sit*** notes. Calls with a subdued hard ***tuk, tak*** or ***dak*** and louder ***se-ic***. **HABITAT & BEHAVIOUR** Broadleaved evergreen and mixed deciduous forest, secondary growth, bamboo; parks, gardens and mangroves on migration; up to 1,830 m (mainly below 900 m). Skulks on or near ground, quivers tail. **RANGE & STATUS** Breeds E Palearctic, NE China, N,S Korea, Japan; winters S China, Sumatra, Borneo, rarely N,NE Indian subcontinent, Andamans. **SE Asia** Fairly common winter visitor (except N Myanmar, Cambodia). Also recorded on passage C Thailand, Peninsular Malaysia, Singapore, E Tonkin.

906 **RUFOUS-TAILED ROBIN**
Luscinia sibilans Plate **92**
IDENTIFICATION 14 cm. Monotypic. **Adult** Similar to female Siberian Blue Robin but has strongly rufescent uppertail-coverts and tail, no buff on underparts, usually much more distinct brownish-grey scales/scalloping on throat, breast and upper flanks. Bill black. **VOICE** Song is a repeated, accelerating, silvery trill, falling slightly in pitch towards end: ***tiuuuuuuuuuwwwww***. **HABITAT & BEHAVIOUR** Broadleaved evergreen and semi-evergreen forest; up to 1,200 m. Skulks on or near ground. **RANGE & STATUS** Breeds E Palearctic; winters S China. **SE Asia** Scarce to uncommon winter visitor NW,NE Thailand, Laos, Vietnam. Vagrant W Thailand.

907 **WHITE-TAILED ROBIN**
Myiomela leucura Plate **91**
IDENTIFICATION 17.5–19.5 cm. *M.l.leucura*: **Male** Overall blackish plumage (tinged blue on upperparts and belly), paler shining blue forehead, supercilium and shoulder-patch, and long white line on outertail-feathers diagnostic. White on tail often hard to see unless tail fanned. Also has distinctive white marking on neck-side (usually concealed). *M.l.cambodiana* (SE Thailand, Cambodia) almost lacks shining blue supercilium/forehead and shoulder-patch. See White-tailed Flycatcher. **Female** Nondescript. Upperparts cold olive-brown, underparts paler and buffish-tinged with paler buffish patch on lower throat and buffy-whitish belly-centre; has dull rufescent fringes to wing-feathers and distinctive white tail-lines (like male). Small white neck-patch always concealed. *M.l.cambodiana* is much darker and colder coloured overall. See White-tailed Flycatcher. **Juvenile** Like female (with similar tail) but ear-coverts and upperparts darker and warmer with dark feather-tips and warm buff streaks/spots, throat and breast (initially also belly and vent) warm buffish with heavy blackish-brown scales/streaks, wing-coverts tipped warm buff. Male is somewhat darker brown; starts to attain adult plumage during first winter/spring. **VOICE** Song is a short, rather hurried, clear, sweet, thin, quavering warble. Calls with thin whistles and low ***tuc***. **HABITAT** Broadleaved evergreen forest, bamboo; 1,000–2,480 m, locally down to 150 m, rarely sea-level E Tonkin (Hanoi). **RANGE & STATUS** Resident N,NE Indian subcontinent, SW,S,C China, Taiwan. **SE Asia** Uncommon resident, locally subject to some movements (except SW,C Myanmar, C Thailand, Singapore, Cochinchina). Also recorded (rarely) on passage E Tonkin. **BREEDING** March–September. **Nest** Cup-shaped or dome, in hole in bank or amongst rocks. **Eggs** 2–5; fairly glossy, whitish or pale pink to pinkish-brown, unmarked or faintly freckled slightly darker pinkish (often in vague ring around broader end); 22.9 x 17 mm (av.).

908 **BLUE-FRONTED ROBIN**
Cinclidium frontale Plate **91**
IDENTIFICATION 18–20 cm. *C.f.orientale*: **Male** Very similar to White-tailed Robin but tail longer with no white, rather ashier-blue overall (never blackish on body). Shows distinct light shining blue shoulder-patch, lacks concealed white neck-patch (both sexes). See Large Niltava. **Female** Difficult to separate from White-tailed but tail longer with no white, upperparts somewhat deeper, richer brown, underparts deeper, more russet-brown. **Juvenile** Best separated from White-tailed by lack of white on tail. Male starts to attain adult plumage during first winter/spring. **VOICE** Sings with a series of short melodic phrases (clearer, less watery than White-tailed): ***tuuee-be-tue*** and ***tuu-buudy-doo*** etc. Gives a harsh buzzy ***zshwick*** in alarm. **HABITAT &BEHAVIOUR** Bamboo, broadleaved evergreen forest; recorded at 1,850–2,100 m. Extremely hard to detect, except when singing. **RANGE & STATUS** Resident NE India, Bhutan, SW China. **SE Asia** Rare to scarce resident N Myanmar, NW Thailand, N Laos, W Tonkin. **BREEDING** Undocumented.

909 **PLUMBEOUS WATER-REDSTART**
Rhyacornis fuliginosa Plate **91**
IDENTIFICATION 15 cm. *R.f.fuliginosa*: **Male** Slaty-blue plumage with contrasting chestnut tail-coverts and tail diagnostic. **Female** Upperparts dark blue-grey, underparts scaled grey and whitish, wings brown with two whitish bars on coverts; has distinctive white uppertail-coverts and blackish-brown tail with white basal half of outer feathers. **Juvenile** Like female but upperparts and wing-coverts brown with buffish-white speckles and streaks, tertials tipped with buffish-white spots, underparts more buffish and more broadly mottled, vent paler. **First-year male** Like female. **VOICE** Song is a very thin, rapidly repeated, insect-like *streee-treee-tree-treeeh*. Calls with short sharp strident whistled ***peet*** notes. **HABITAT & BEHAVIOUR** Rocky rivers and streams, waterfalls, nearby wet areas; 300–2,285 m, rarely down to sea-level E Tonkin (Hanoi). Restless, fans tail, often makes flycatching sallies. **RANGE & STATUS** Resident (subject to altitudinal movements) N Pakistan, NW,N,NE Indian subcontinent, S,SE Tibet, China (except NW,NE), Taiwan. **SE Asia** Scarce to locally common resident (subject to some movements) W,N Myanmar, NW Thailand, Laos, W,E Tonkin, N,C,S Annam. Recorded in winter (status uncertain) SW,C,S(east),E Myanmar, north Tenasserim. **BREEDING** March–July. Multi-brooded. **Nest** Neat cup, in hole or on side branch of tree-stump near water. **Eggs** 3–5; pale olive-white or pale stone-colour, heavily speckled and freckled dingy yellowish or reddish-brown (often capped at broader end); 19.8 x 14.6 mm (av.).

910 **WHITE-CAPPED WATER-REDSTART**
Chaimarrornis leucocephalus Plate **91**
IDENTIFICATION 19 cm. Monotypic. **Adult** Unmistakable, with black plumage, sharply contrasting white crown and nape and chestnut-red lower body and tail-base. **Juvenile** Head-sides and upperparts browner than adult, indistinctly fringed brown on mantle and back, crown duller with blackish feather-fringing, rump blackish-brown with dull dark rufous feather-fringing, tail duller and darker, underparts blackish-brown with drab warmish brown scales; has small drab warmish-brown spots on tips of wing-coverts and tertials. **VOICE** Song is a weak, drawn-out, undulating ***tieu-yieu-yieu-yieu***. Typical call is a loud sharp upward-inflected ***tseeit*** or ***peeeiii***. **HABITAT & BEHAVIOUR** Rocky rivers and streams, waterfalls; 915–4,265 m, down to 215 m in winter. Conspicuous, cocks tail. **RANGE & STATUS** Resident (subject to altitudinal movements) C Asia, NE Afghanistan, N Pakistan, NW,N,NE Indian subcontinent, S,E Tibet, SW,W,N,C,E China. **SE Asia** Uncommon to fairly common resident (subject to local movements) N Myanmar, N,C Laos, W,E Tonkin. Recorded in winter (status uncertain) SW,W,C,S,E Myanmar, NW,NE Thailand, N Annam. **BREEDING** May–August. **Nest** Deepish cup, in hole in bank or wall, sometimes under stone or amongst tree-roots. **Eggs** 3–5; pale blue or blue-green (sometimes pinkish-tinged), speckled and spotted reddish-brown over greyish undermarkings (may be ringed or capped at broader end); 24.6 x 16.8 mm (av.).

911 **HODGSON'S REDSTART**
Phoenicurus hodgsoni Plate **93**
IDENTIFICATION 16 cm. Monotypic. **Male** Resembles Daurian Redstart but mantle and wings grey, upper breast black, white wing-patch narrower. **Female** Similar to Black Redstart but upperparts paler and greyer (less brownish), has

more obvious pale eyering and creamy-whitish underparts with broad greyish wash across breast extending to throat-side (sometimes most of throat and also flanks). Shows a large whitish belly-patch. **First-year male** Similar to female. **VOICE** Calls include a rattling ***prit*** and ***trr*** or ***tschrrr*** when alarmed. **HABITAT** Open, often rocky areas, scrub; recorded at 150–1,830 m. **RANGE & STATUS** Breeds SE,E Tibet, SW,W,N China; winters N,NE Indian subcontinent, C,SW China. **SE Asia** Uncommon winter visitor N Myanmar.

912 WHITE-THROATED REDSTART
Phoenicurus schisticeps Plate **93**

IDENTIFICATION 16 cm. Monotypic. **Male breeding** Mostly light blue crown, black throat, white patch on centre of lower throat/upper breast, white patch on median and inner greater coverts and white tertial fringes diagnostic. Has black mantle, rufous-chestnut rump and uppertail-coverts, blackish tail with rufous-chestnut restricted to basal third of outer feathers, reddish rufous-chestnut breast, paler belly with white centre and rufous vent. **Male non-breeding** Crown, mantle and breast fringed pale brownish. **Female** Easily identified from other redstarts by white throat/breast-patch, greyish-brown rest of throat and extensive white wing markings. **Juvenile** Like female (with similar wings and tail) but has dark-scaling and buff spots on upper and underparts. **VOICE** Calls include a drawn-out ***zieh*** followed by a rattling note. **HABITAT** Open forest, low vegetation above tree-line, rocky areas; 2,285–3,230 m. Breeds above 2,700 m in Himalayas. **RANGE & STATUS** Resident (subject to relatively local movements) Nepal, NE India (Sikkim), Bhutan, S,E Tibet, SW,W,NC China. Also winter visitor Nepal, Bhutan. **SE Asia** Recorded (status uncertain) N Myanmar. **BREEDING** Tibet: May–August. **Nest** Cup, in hole in tree, bank or rock; within 2 m of ground. **Eggs** 3; reddish-cream to pale greenish-grey, unmarked, or with fine, often faint orange to clay-coloured freckles (sometimes capped at broader end); 19.3 x 14.6 mm (av.).

913 DAURIAN REDSTART
Phoenicurus auroreus Plate **93**

IDENTIFICATION 15 cm. *P.a.leucopterus*: **Male breeding** Told by combination of grey crown to upper mantle, black lower mantle and throat, broad white wing-patch, rufous rump and uppertail-coverts and rufous tail with blackish central feathers. Remainder of underparts orange-rufous. **Male non-breeding** Crown to upper mantle mostly brownish-grey, has broad brown fringing on lower mantle and pale greyish fringing on breast. **Female** Rather uniform brown with paler, warmer underparts and distinctive rump to tail pattern (as male), broad white wing-patch and prominent pale buffish eyering. **Juvenile** Similar to female (including wings and tail) but has dark scaling and buff spots on upper and underparts. **First-winter male** Similar to non-breeding adult. **VOICE** Song is a scratchy trill followed by a short wheezy jingle. Calls include a high-pitched ***tseep*** or ***tsip*** (repeated when agitated), scolding ***tak*** or ***tuc*** notes and a rapid rattling ***tititik***. **HABITAT** Open forest, forest edge, orchards, scrub, thickets; up to 2,565 m. Breeds above 2,800 m in Tibet. **RANGE & STATUS** Breeds southern E Palearctic, NE India, SE,E Tibet, SW,W,N,NE China, N,S Korea; some populations winter south to Bhutan, NE India, S China, Taiwan, Japan. **SE Asia** Scarce to fairly common winter visitor Myanmar (except SW and Tenasserim), W,NW Thailand, Laos, W,E Tonkin, N,C(rare),S(rare) Annam. May be resident N Myanmar. **BREEDING** Tibet: May–August. **Nest** Cup, in hole in ground, bank or wall; up to 1.5 m above ground. **Eggs** 3–4; very variable, whitish to pale green with light to warm brown spots and speckles (sometimes ringed at broader end); 18 x 14.5 mm (av.).

914 BLUE-FRONTED REDSTART
Phoenicurus frontalis Plate **93**

IDENTIFICATION 16 cm. Monotypic. **Male breeding** Dark blue head, upper breast, mantle and back, and largely orange-rufous tail with blackish central feathers and tip diagnostic. Shows brighter blue eyebrow and rufous rump, uppertail-coverts and rest of underparts. See Black Redstart. **Male non-breeding** Dark blue feathers broadly fringed pale brown. **Female** From other redstarts by distinctive tail pattern (as male) and lack of white wing-patch; has contrasting buffy tips to greater coverts and fringes to tertials. See Black. **Juvenile** Recalls female (tail similar) but upperparts and throat to upper belly blackish, speckled/streaked buff above and mottled buffy below, has buff vent and wing-bars and buff-fringed tertials and secondaries. **First-winter male** Like non-breeding male. **VOICE** Song consists of 1–2 rather harsh trilled warbles followed by short whistled phrases; repeated with some variation. Calls included a thin ***ee-tit, ee-tit-tit*** etc. and single clicking ***tik***. **HABITAT** Open forest, forest edge, clearings, cultivation, thickets; 800–2,750 m. **RANGE & STATUS** Resident (subject to altitudinal movements) N Pakistan, NW,N,NE Indian subcontinent, S,E Tibet, SW,W,N,C China. **SE Asia** Scarce winter visitor E Myanmar, NW Thailand. Recorded in winter (status uncertain) W,N Myanmar, N Laos, W Tonkin. Likely to breed at high levels in N Myanmar at least. **BREEDING** May–July. **Nest** Cup, in hole in bank, amongst rocks or under dense vegetation, sometimes in tree-hole; up to 7 m above ground. **Eggs** 3–4; pale greyish-pink to light buffish, profusely stippled pale reddish; 19.4 x 14.7 mm (av.).

915 BLACK REDSTART
Phoenicurus ochruros Plate **93**

IDENTIFICATION 16 cm. *P.o.rufiventris*: **Male non-breeding** From other redstarts by brownish-grey crown to back with some black on mantle, blackish sides of head, blackish throat and breast with brownish-grey scales, and blackish wings with buff feather-fringes; lacks white wing-patch. **Male breeding** Head, mantle, back, breast and wings more uniformly blackish. **Female** Like female Daurian Redstart but lacks white wing-patch, underparts somewhat duller, has less distinct pale eyering. **First-year male** Resembles female. **VOICE** Song is a scratchy trill followed by a short wheezy jingle. Calls include a high-pitched ***tseep*** or ***tsip*** (repeated when agitated), scolding ***tak*** or ***tuc*** and rapid rattling ***tititik***. **HABITAT** Open country, semi-desert; lowlands. **RANGE & STATUS** Breeds NW Africa, W,SC Palearctic, Middle East, W,N Pakistan, NW,N India, Nepal, S,E Tibet, W,NW,N China; some populations winter south to N Africa, Indian subcontinent. **SE Asia** Scarce to uncommon winter visitor W,N,C,E Myanmar. Vagrant W,NW Thailand, W Tonkin.

916 BLUE-CAPPED ROCK-THRUSH
Monticola cinclorhynchus Plate **89**

IDENTIFICATION 18–19.5 cm. Monotypic. **Male non-breeding/breeding** Similar to White-throated Rock-thrush but has blue base colour to throat, deep rufous rather than chestnut base colour to rump, uppertail-coverts and underparts, and blackish lores. **Female** Similar to White-throated but upperparts plain apart from bars on rump and uppertail-coverts; has less defined white centre to throat and upper breast. **First winter** Differs as White-throated. **VOICE** Song is a clear ***rit-prileee-prileer*** or more extended ***tew-li-di tew-li-di tew-li-di***, sometimes with harsher ***tra-trree-trreea-tra*** or ***seer-twik-twik*** phrases; repeated monotonously with varying emphasis. Call is a sharp, slightly rising ***peri-peri***. When alarmed, utters loud ***goink*** notes, often introduced by a high-pitched ***tri***. **HABITAT** Open forest and edge, secondary growth, cultivation borders. Winters 610–2,380 m in Indian subcontinent. **RANGE & STATUS** Breeds NE Afghanistan, N Pakistan, NW,N,NE Indian subcontinent; winters mainly SW,NE(Assam) India. **SE Asia** Recorded (status uncertain) SW Myanmar.

917 WHITE-THROATED ROCK-THRUSH
Monticola gularis Plate **89**

IDENTIFICATION 18–19.5 cm. Monotypic. **Male non-breeding** Distinctive with blue crown and nape, blue-black mantle, chestnut lores, rump, uppertail-coverts and underparts, white patch on centre of throat and upper breast (occasionally lacking), and white wing-patch. Has greyish-brown feather-fringes on head and upperparts and buffish to whitish tips to greater coverts and tertials. See Blue-capped and other rock-thrushes. **Male breeding** Lacks greyish-brown fringing on head and upperparts; white wing-patch and blue lesser coverts contrast sharply with blackish remainder of wing. **Female** Upperparts greyish-brown, with black bars on mantle and scapulars to uppertail-coverts and some buffish scales (mainly on rump and uppertail-coverts); wing-coverts broadly fringed buff to whitish and inset with blackish; sides of head, sides of neck and underparts buffish with heavy blackish scales (belly-centre and vent unmarked); has well-defined white centre to throat and upper breast. See Blue-capped and Chestnut-bellied Rock-thrushes. **First winter** Like non-breeding adults. **VOICE** Song is a series of rather melancholy, long, flute-like, slightly rising whistles, interspersed with 1–2 more complex ringing phrases and short repeated ***chat-at-at***. Calls include soft ***queck-quack*** phrases interspersed with a sharp ***tack-tack*** and occasional thin ***tsip*** or ***tseep*** in flight. **HABITAT** More open deciduous and broadleaved evergreen forest, plantations, secondary growth; up to 1,220 m. **RANGE & STATUS** Breeds SE Siberia, Ussuriland, north-eastern China, N Korea; winters S China. **SE Asia** Scarce to uncommon winter visitor (except SW,N,S Myanmar, Tenasserim, C Thailand, Peninsular Malaysia, Singapore, W,E Tonkin). Uncommon passage migrant E Tonkin. Vagrant Peninsular Malaysia, Singapore.

918 BLUE ROCK-THRUSH
Monticola solitarius Plate **89**

IDENTIFICATION 21–23 cm. *M.s.pandoo*: **Male non-breeding** Rather uniform dull grey-blue plumage distinctive. Has whitish and blackish bars/scales on body and browner crown. *M.s.philippensis* (winter visitor throughout) has chestnut on lower breast to vent; differs from Chestnut-bellied Rock-thrush by dense bars/scales on body, head may appear largely greyish, has less blue above and less chestnut below. **Male breeding** Lacks bars/scales on body, bluer overall. See Blue Whistling-thrush. Cleaner plumage renders *M.s.philippensis* more similar to Chestnut-bellied. **Female non-breeding** More uniform than other rock-thrushes, without strongly patterned head-side and underparts. Usually appears rather bluish above, particularly back to uppertail-coverts (but variable), has dark bars and some pale bars on mantle and scapulars to uppertail-coverts, wing-coverts and tertials. **Female breeding** Upperparts plainer. **Juvenile** Like female but crown and mantle speckled dull pale buffish, underparts somewhat paler and more diffusely marked. **First winter** Similar to non-breeding adults. **Other subspecies in SE Asia** *M.s.madoci* (resident S Thailand, Peninsular Malaysia). **VOICE** Song is a fluty, rather high-pitched ***chu sree chur tee tee***; ***wuchee-trr-trrt*** etc. (some notes rather scratchy). Alarm calls include a harsh ***tak-tak***; low ***tchuck*** (often doubled), sometimes interspersed with a high ***tsee***, ***peet*** or ***tzick***; rapid ***chakerackack*** or ***schrrrackerr*** and ***chack chack chack eritchouitchouitchouit tchoo tchoo*** and harsh rattling ***trrr*** and ***trr ti chak chak***. **HABITAT & BEHAVIOUR** Open rocky areas, roadsides, cultivation; residents frequent limestone cliffs, sometimes urban buildings; up to 1,830 m (southern breeders mainly coastal). Often sits on exposed perch. **RANGE & STATUS** Breeds NW Africa, southern W Palearctic, Middle East, C Asia, northern Pakistan, NW India, Nepal, southern Tibet, China, Taiwan, Ussuriland, Japan, N Philippines (Batanes); some populations winter south to N Africa, Arabia, Indian subcontinent, S China, Sumatra, Borneo, Philippines, Sulawesi, Moluccas. **SE Asia** Uncommon resident S(west) Thailand, Peninsular Malaysia, W,E Tonkin; possibly N Myanmar, N Laos. Common winter visitor (except Singapore). Also recorded on passage SW Myanmar, E Tonkin. Vagrant Singapore. **BREEDING** January–June. **Nest** Cup, placed in hole in rock, wall or building. **Eggs** 3–5; rich to pale blue, sometimes spotted or blotched brown or rust-colour; 26 x 19.1 mm (av.; *pandoo*).

919 CHESTNUT-BELLIED ROCK-THRUSH
Monticola rufiventris Plate **89**

IDENTIFICATION 22–24 cm. Monotypic. **Male non-breeding** Deep bluish upperside (including rump and uppertail-coverts) and dark chestnut lower breast to vent distinctive. Mantle feathers and scapulars fringed pale greyish-brown, throat dark blue with some pale greyish-brown fringing. Recalls Blue Rock-thrush subspecies *philippensis* but sides of head blackish, chestnut of underparts deeper and more extensive with no pale barring, upperparts considerably darker. **Male breeding** Upperparts more shining cobalt-blue; lacks pale greyish-brown feather-fringing on mantle, scapulars and throat. **Female** From other rock-thrushes by large buffish-white patch behind dark ear-coverts, prominent whitish to pale buff eyering, and pale buffish to whitish underparts with uniform blackish bars/scales on breast to vent. Has some darker-and-paler barring/scaling on mantle and scapulars to uppertail-coverts. See female Siberian Thrush and juvenile White-collared Blackbird. **Juvenile male** Like juvenile female but wing-feathers (apart from fringes) and tail mostly blue, rump and uppertail-coverts washed dull chestnut, base colour to underparts warmer. **Juvenile female** From adult by sooty-blackish body and wing-coverts with broad pale buff to whitish spots, buff fringes to greater coverts, tertials and secondaries, dark rufous tips to uppertail-coverts and mostly whitish-buff throat with dark feather-tips. **First winter** Like respective non-breeding adults but retains some juvenile greater coverts. **VOICE** Song is a quite rich undulating series of short warbling, whistled phrases: ***twew-twi-er tre-twi teedle-desh*** or ***jero-terry-three fir-tar-ree***, occasionally varied or interspersed with ***tewleedee-tweet-tew*** or ***til-tertew***. Calls include a sharp querulous ***quach*** and thin shrill ***sit***, ***tick*** or ***stick***, often combined with deep rasping ***churr***, ***chhrrr*** or ***chaaaaa*** notes when alarmed. **HABITAT** More open broadleaved evergreen and coniferous forest, rock outcrops; 1,200–2,700 m, locally down to 900 m (winter only?). **RANGE & STATUS** Resident, subject to some movements, NE Pakistan, NW,N,NE Indian subcontinent, S,SE Tibet, C and southern China. **SE Asia** Uncommon to common resident (subject to some movements) W,N,E Myanmar, NW Thailand (Doi Inthanon), N Laos, W,E Tonkin. Uncommon winter visitor NW Thailand (should breed). **BREEDING** May–June. **Nest** Cup-shaped, situated in hole in rock-face or on ledge, sometime hole in bank. **Eggs** 3–6; glossy, pale pinkish creamy-white, speckled reddish to red-brown overall, more so at broader end; 26.8 x 19.9 mm (av.).

920 ISABELLINE WHEATEAR
Oenanthe isabellina Plate **92**

IDENTIFICATION 16.5 cm. Monotypic. **Adult** Robust and rather short-tailed and long-legged, with pale sandy-brown upperside, darker primaries (edged pale sandy-brown), contrasting blackish loral stripe and alula, white uppertail-coverts and blackish tail with white base. Has strong black bill, pale buff supercilium (whiter in front of eye) and whitish underparts, washed buff on breast and flanks. **First winter** Very similar to adult. **VOICE** Song is a variable mixture of hard and harsh notes, clear whistles and mimicry. Calls include a clear ***cheep*** and clicking ***chick*** or ***tshk*** when alarmed. **HABITAT & BEHAVIOUR** Semi-desert, dry cultivation and open areas; lowlands. Mainly terrestrial, typically adopting very upright posture. **RANGE & STATUS** Breeds W(south-east),SC Palearctic, S Siberia, Middle East, Pakistan, NW,N,NE(west) China, Mongolia; winters south/south-west to north-eastern Africa, Arabian Peninsula, Middle East, Pakistan, north-western India. **SE Asia** Vagrant C Myanmar, south W Thailand (sight records).

921 GREY BUSHCHAT *Saxicola ferreus* Plate **93**

IDENTIFICATION 14–15.5 cm. Monotypic. **Male breeding** Slaty-grey upperparts with dark streaks, black sides of head and contrasting white supercilium and throat distinctive. Breast and flanks greyish, has narrow white patch on wing-coverts. **Male non-breeding** Upperpart feathers broadly tipped warmish brown, supercilium and ear-coverts duller, underparts washed brownish. **Female breeding** Ear-coverts dark brown, crown and mantle brown with distinct greyish wash and fairly distinct dark streaks, uppertail-coverts rufescent, outertail-feathers edged dull chestnut; has broad buffish-white supercilium, whitish throat contrasts with dull greyish-brown breast. Lacks white on wing. See Pied and Jerdon's Bushchats. **Female non-breeding** Ear-coverts, crown, mantle and breast-band browner, upperparts less distinctly streaked. **Juvenile** Like non-breeding female (both sexes) but ear-coverts and upperparts darker with broad buff to rufous streaks and mottling, underparts buffish (whiter on throat) with blackish-brown scalloping on breast. **VOICE** Song is a brief repeated ***tree-toooh tu-treeeh-t't't't-tuhr***, with more emphatic beginning and trilled ending. Calls include a soft ***churr***, often followed by clear ***hew*** and harsher ***bzech***. **HABITAT** Open pine and broadleaved evergreen forest, scrub and grass, cultivation, locally parks and gardens in winter; up to 3,054 m, breeds above 1,220 m. **RANGE & STATUS** Resident (subject to relatively local movements) N Pakistan, NW,N,NE Indian subcontinent, S,SE Tibet, W,C,E and southern China; some winter south to northern India, S China, Taiwan. **SE Asia** Common resident SW,W,N,S,E Myanmar, NW Thailand (local), W Tonkin, S Annam. Uncommon to fairly common winter visitor C Myanmar, Thailand (except SE,S), Laos, E Tonkin (may breed in north/west). Recorded (status uncertain) Cambodia, Cochinchina. **BREEDING** March–June. Multi-brooded. **Nest** Cup, on ground or in hole in bank or wall etc. **Eggs** 4–6; pale blue with reddish freckles, overall or in ring or cap at broader end; 17.9 x 14.2 mm (av.). Brood-parasitised by Eurasian Cuckoo.

922 EASTERN STONECHAT
Saxicola maurus Plate **93**

IDENTIFICATION 14 cm. *S.t.stejnegeri*: **Male breeding** Distinctive, with black head, blackish upperparts and wings, broad white patch on side of nape and neck, white wing-patch, whitish rump and uppertail-coverts and orange-rufous breast. *S.t.przewalskii* (resident form) is slightly larger with rufous-chestnut on underparts extending to belly. **Male non-breeding** Resembles female but black feather-bases visible on head-sides and throat, lores and chin often all black. **Female breeding** Rather nondescript, with sandy-brown, dark-streaked crown and upperparts, whitish to buffish rump and uppertail-coverts (contrasting with dark tail), unmarked buffish underparts with paler throat and vent, and more rufescent breast, white patch on inner wing-coverts (smaller than on male). See Pied Bushchat. **Female non-breeding** Generally warmer overall, rump and tail-tip warm buffish to rufous-buff. **Juvenile** Similar to non-breeding female but upperparts blackish-brown, boldly streaked and spotted buff, ear-coverts boldly marked with buff, underparts more uniform buffish with blackish-brown streaks/mottling on breast, tail broadly tipped warm brownish. **First-winter male** Like non-breeding male. **VOICE** Song is a variable, rather scratchy series of twittering, warbling notes. Usual call is a repeated hard ***chack*** or ***tsak*** and thin ***hweet***. **HABITAT** Grass and scrub, cultivation, open areas; up to 2,470 m (breeds above 1,000 m). **RANGE & STATUS** Breeds C,E Palearctic, northern Pakistan, NW,N,NE Indian subcontinent, S,E Tibet, China (except E,SE), N,S Korea, Japan; some northerly populations winter south to Middle East, Indian subcontinent, China, Taiwan, Borneo, Philippines. **SE Asia** Local to fairly common resident N Myanmar, NW Thailand, N,S Laos, W,E Tonkin. Fairly common to common winter visitor throughout; uncommon Singapore. Also recorded on passage NW Thailand, N Laos, E Tonkin. **BREEDING** India: March–July. Multi-brooded. **Nest** Cup, on ground, in hole in bank, amongst stones or in low vegetation. **Eggs** 4–5; pale greyish-blue with pale reddish spots, forming ring at broader end; 16.9 x 13.5 mm (av.). Brood-parasitised by Eurasian Cuckoo.

923 WHITE-TAILED STONECHAT
Saxicola leucura Plate **93**

IDENTIFICATION 14 cm. Monotypic. **Male breeding** Extremely similar to Eastern Stonechat but central tail-feathers edged whitish, outertail-feathers have mostly whitish inner web and base of outer web; tends to show whiter belly, demarcated from orange-rufous breast-patch. Note habitat and range. **Male non-breeding** Very like Common but has distinctive tail pattern, usually paler, more greyish buffy-brown crown and mantle. **Female breeding** Very like Common but has paler greyish to whitish tail-feather fringes (pattern similar to male) and appears somewhat paler, greyer and plainer above and whiter below. **Female non-breeding** More distinctly pale/plain above and below. **Juvenile** Very like Eastern Stonechat and best separated by tail pattern. Note habitat and range. **VOICE** Not adequately described. **HABITAT** Grassland, nearby scrub and cultivation, often along larger rivers; lowlands. **RANGE & STATUS** Resident plains of Pakistan and northern Indian subcontinent. **SE Asia** Locally fairly common resident N,C,S,E Myanmar. **BREEDING** February–May. **Nest** Cup, under clod of earth or amongst grass roots. **Eggs** 3–5; pale blue, variably freckled light reddish-brown (may be ringed or capped at broader end); 18 x 14 mm (av.).

924 PIED BUSHCHAT *Saxicola caprata* Plate **93**

IDENTIFICATION 14 cm. *S.c.burmanica*: **Male breeding** All blackish with white rump, vent and wing-streak. **Male non-breeding** Blackish plumage-parts prominently fringed brownish (less so on head), rump and uppertail-coverts tipped rufous. **Female breeding** Much darker than similar Eastern Stonechat, upperparts broadly dark-streaked but less contrasting, has rusty uppertail-coverts and tinge to belly, dark mottling/streaks on underparts (except vent), lacks white wing-streak. **Female non-breeding** Much plainer with less distinct body streaks. See Grey Bushchat and Black Redstart. **Juvenile male** Darker than juvenile female with blacker wings and white wing-streak. **Juvenile female** Resembles non-breeding female but body speckled pale buffish. **VOICE** Song is a series of brisk whistled phrases, with short-noted introduction: roughly ***hiu-hiu-hiu u'wee'wipee'chiu*** etc. Calls include a clear ***chep*** or ***chep-hee***, ***chek chek trweet*** and clear whistled ***hew***. **HABITAT** Open areas, cultivation, grass and scrub; up to 1,600 m. **RANGE & STATUS** Resident C Asia, Iran, Afghanistan, Indian subcontinent, SW China, Java, Bali, Lesser Sundas, Sulawesi, Philippines, New Guinea, Bismarck Archipelago. **SE Asia** Uncommon to common resident (except SE,S Thailand, Peninsular Malaysia, Singapore, W,E Tonkin, N Annam). **BREEDING** March–June. Multi-brooded. **Nest** Small cup, on ground or bank, sometimes hole in rock or building. **Eggs** 3–5; bluish-white or pinkish-white, freckled and speckled pale to deep reddish-brown (often more at broader end); 16.8 x 13.9 mm (av.). Brood-parasitised by Eurasian Cuckoo.

925 JERDON'S BUSHCHAT
Saxicola jerdoni Plate **93**

IDENTIFICATION 15 cm. Monotypic. **Male** Uniform glossy blackish upperside and all-white underside diagnostic. **Female** Similar to non-breeding female Grey Bushchat but lacks supercilium, tail longer, without dull chestnut fringing on outer feathers, centre of underparts and undertail-coverts usually whiter. Note habitat and range. **Juvenile** Poorly documented but apparently spotted. **VOICE** Calls with a short, clear whistle. **HABITAT & BEHAVIOUR** Seasonally exposed bushland within river channels. Tall grass, scrub, thickets, particularly bordering larger rivers and lakes; up to 1,650 m. Relatively skulking. **RANGE & STATUS** Resident (subject to relatively local movements) N,NE Indian subcontinent, SW China. **SE Asia** Rare to scarce resident (subject to some movements) N,C,E,S Myanmar, NW,NE(north-west) Thailand, N,C Laos, E Tonkin. Recorded (status uncertain) C Thailand, W Tonkin (autumn). **BREEDING** March–May. **Nest** Small stout cup, amongst grass roots or in hole in bank. **Eggs** 3–4; deep turquoise; 16.2 x 13.3 mm (av.).

926 **WHITE-BROWED BUSH-ROBIN**

Tarsiger indicus Plate **94**

IDENTIFICATION 13.5–15 cm. *T.i.yunnanensis*: **Male** Plumage pattern recalls Indian Blue Robin but longer-tailed and darker-legged; head- and throat-sides and upperparts slaty blue-grey, supercilium longer, undertail-coverts buff to rufous. **Female** From other robins by combination of narrow white supercilium (often tinged buff), plain brown remainder of sides of head, lack of blue on upperside, deep rich buffy-brownish underparts with slightly paler throat and paler belly-centre and vent. Bill blackish, legs and feet dark. See Lesser and White-browed Shortwings. **Juvenile** Resembles Golden and Rufous-breasted Bush-robins and best separated by uniform brown tail. **First-summer male** Like female and can breed in this plumage. **VOICE** Song is a wispy slurred ***whi-wi'wich'u-wi'rr*** etc., repeated after rather short intervals. Calls with repeated ***trrrrr*** notes. **HABITAT & BEHAVIOUR** Rhododendron and conifer forest, bamboo; 2,000–3,355 m (breeds above 3,000 m in NE India). Quite skulking, usually near ground but sings from small trees. Adopts more upright posture than Indian Blue. **RANGE & STATUS** Resident N,NE Indian subcontinent, SW,W China, Taiwan. **SE Asia** Scarce resident N Myanmar, W Tonkin. **BREEDING** India: April–July. **Nest** Cup, in hollow in bank. **Eggs** 3–4; white, usually freckled pale pink at broader end; 17.7 **x** 13.6 mm (av.).

927 **RUFOUS-BREASTED BUSH-ROBIN**

Tarsiger hyperythrus Plate **94**

IDENTIFICATION 13–14 cm. Monotypic. **Male** Similar to White-browed Bush-robin but lacks white supercilium. Upperparts deep blue with shining ultramarine blue forehead, eyebrow, shoulder-patch and uppertail-coverts. Underparts orange-rufous with white belly-centre and vent. Recalls some male blue flycatchers but note orange-rufous throat and flanks, shape, habitat, behaviour and range. **Female** Resembles bluetails, but much darker above, lacks white throat-patch, shows little rufous-orange on flanks. See Himalayan and Red-flanked Bluetails. **Juvenile** Similar to Himalayan but much darker above and buffier below. See White-browed. **VOICE** Song is described as a lisping warble: ***zeew zee zwee zwee...*** Calls with low ***duk*** notes. **HABITAT & BEHAVIOUR** Rhododendron and conifer forest, broadleaved evergreen forest in winter; recorded at 1,105–1,525 m in winter. Breeds above 3,200 m in Himalayas. Usually on or close to ground, adopts upright posture. **RANGE & STATUS** Resident Nepal, Bhutan, NE India, SW China (north-west Yunnan). **SE Asia** Recorded (probably breeds) N Myanmar. **BREEDING** Nepal: May-June. **Nest** Fairly neat cup in recess in bank, sheltered by rock. **Eggs** <3. Otherwise undocumented.

928 **HIMALAYAN BLUETAIL**

Tarsiger rufilatus Plate **94**

IDENTIFICATION 13.5–15 cm. *T.r.rufilatus*: **Male** Deep dark blue upperside, throat- and breast-side and white remainder of underparts with contrasting rufous-orange flanks distinctive. Supercilium shining pale blue, rump and uppertail-coverts shining deep blue. See Red-flanked Bluetail and Ultramarine Flycatcher. **Female** Grey-brown above, with contrasting blue rump to uppertail; has narrow white throat-patch enclosed by grey-brown breast-band, and distinctive rufous-orange flanks; belly whitish. See similar Red-flanked. **Juvenile** Head-sides and upperparts like female but feathers fringed blackish and marked with buff specks and streaks, underparts whitish, throat and breast washed buff, flanks richer buff (suggesting adult pattern), breast and flanks scaled blackish; uppertail-coverts and tail mostly blue. **First winter** Like adult female (both sexes). **First-summer male** Like female but may show some blue on lesser coverts and scapulars; often breeds in this plumage. **VOICE** Song is a rather short, blurry, mellow, vibrato warbled ***chi-CHI-chiri*** or ***chri-CHU-chi*** etc.; sometimes descending somewhat. Calls with a curious, deep, slightly nasal ***akh-akh-akh***, ***ugh-ugh*** etc. **HABITAT & BEHAVIOUR** Shrubbery and understorey in rhododendron, oak, conifer and birch forest; 600–3,655 m (breeds above 2,400 m in Himalayas). Also edge of broadleaved evergreen forest and clearings etc. in winter. Not very skulking, usually fairly low down; erratically flicks tail downwards. **RANGE & STATUS** Breeds N Pakistan, NW,N,NE Indian subcontinent, S,E Tibet, SW,W,C China; some winter to south ? (NE India south of Brahmaputra R). **SE Asia** Recorded (probably breeds) N Myanmar. Uncommon to locally fairly common winter visitor W,C,E Myanmar, NW,NE Thailand. **BREEDING** India: May–July. **Nest** Cup, in hole in bank or steep slope, amongst fallen tree-roots etc. **Eggs** 3–5; white, usually freckled pink at broader end; 17.8 **x** 13.5 mm (av.).

929 **RED-FLANKED BLUETAIL**

Tarsiger cyanurus Plate **94**

IDENTIFICATION 13.5–15 cm. Monotypic. **Male** Like Himalayan Bluetail but has duller, lighter, almost turquoise-tinged upperside, paler supercilium (whitish in front of eye) and less pure white underparts. **Female** Like Himalayan, but breast paler, contrasting less with white throat-patch, possibly has greyer belly. Other differences unclear at present. **First winter** Like adult female (both sexes). **VOICE** Song is a rather constant, fast, short, clear melancholy ***itru-CHURR-tre-tre-tru-trurr***. Calls with an often repeated, straight whistled ***viht***, and muffled, hard, slightly throaty ***track***. **HABITAT & BEHAVIOUR** Broadleaved evergreen forest, forest edge; up to c.1,000 m. Behaves like Himalayan. **RANGE & STATUS** Breeds W(north-east),C,E Palearctic, NE China, N Korea, N Japan; winters NE India, S China, Taiwan. **SE Asia** Uncommon winter visitor NW Thailand, Laos, W,E Tonkin, N,C Annam; ? E Myanmar. Also recorded on passage E Tonkin. Vagrant S Annam.

930 **GOLDEN BUSH-ROBIN**

Tarsiger chrysaeus Plate **94**

IDENTIFICATION 13–14 cm. *T.c.chrysaeus*: **Male** Striking, with rufous-yellow supercilium, rump, uppertail-coverts and underparts, blackish sides of head, and blackish tail with largely rufous-yellow basal half of outer feathers. Crown, lower mantle and back rufescent-olive (sometimes blackish on lower mantle and back). **Female** Upperside uniform greenish-olive, tail brown with distinctive outertail-feathers (like male); has distinctive olive-yellow eyebrow and underparts, broad yellowish-white eyering. **Juvenile** Resembles Himalayan Bluetail but base colour to upperparts generally darker, spots/streaks on upperparts and base colour of underparts distinctly rich yellowish-buff; tail similar to respective adults. **VOICE** Sings with hurried wispy ***tze'du'tee'tse*** etc., ending with lower, rolling ***tew'r'r'r***. Typically calls with purring ***trrr'rr*** and harder ***tcheck*** notes. **HABITAT & BEHAVIOUR** Breeds in rhododendron and conifer forest, forest edge, clearings, scrub; winters in thickets, broadleaved evergreen forest; 3,050–3,655 m, down to 1,050 m in winter. Usually very skulking. **RANGE & STATUS** Resident (subject to relatively local movements) NE Pakistan, NW,N,NE Indian subcontinent, S,SE Tibet, SW,W,NC China. **SE Asia** Uncommon resident W,N Myanmar. Scarce winter visitor E Myanmar, NW Thailand. Recorded (status uncertain) W Tonkin. **BREEDING** May–July. **Nest** Cup, in hole in bank or steeply sloping ground. **Eggs** 3–4; pale verditer-blue; 19.7 **x** 14.8 mm (av.).

931 **LITTLE FORKTAIL** *Enicurus scouleri* Plate **94**

IDENTIFICATION 12.5–14.5 cm. *E.s.scouleri*: **Adult** Recalls White-crowned Forktail but much smaller, tail much shorter with white restricted to sides, has less black on breast. **Juvenile** Upperparts browner than adult, lacks white on forehead, throat and breast white with sooty scales. **VOICE** Song has been described as a loud, thin ***ts-youeee***. **HABITAT** Rocky rivers and streams, waterfalls; 450–2,590 m. **RANGE & STATUS** Resident NE Afghanistan, N Pakistan, NW,N,NE Indian subcontinent, S Tibet, W,C,E,S China, Taiwan. **SE Asia** Scarce to locally fairly common resident W,N Myanmar, W,E Tonkin. **BREEDING** India:

April–June. **Nest** Compact cup, in hole in rock or bank or on ledge. **Eggs** 2–4; white, sparsely speckled pale reddish-brown to brown; 20.1 x 15 mm (av.).

932 CHESTNUT-NAPED FORKTAIL

Enicurus ruficapillus Plate **94**

IDENTIFICATION 19.5–21 cm. Monotypic. **Male** From other forktails by chestnut crown to upper mantle and dark scales on white breast. See Chestnut-capped Thrush. **Female** Has dull chestnut extending from crown to back. **Juvenile** Duller above than female, has white submoustachial stripe and throat, black malar line and ill-defined breast markings. **VOICE** Calls with a series of thin, shrill metallic whistles and high ***dir-tee***, recalling Large-billed Leaf-warbler. **HABITAT** Rivers and streams, waterfalls; up to 915 m. **RANGE & STATUS** Resident Sumatra, Borneo. **SE Asia** Uncommon resident Tenasserim, W,S Thailand, Peninsular Malaysia. **BREEDING** All year. **Nest** Cup, on rock-face or bank; 0.5–2 m above ground/water. **Eggs** 2; white, spotted and slightly blotched reddish-brown over purplish-red undermarkings (markings densest over or in ring at broader end); 23.1 x 17.4 mm (mean).

933 BLACK-BACKED FORKTAIL

Enicurus immaculatus Plate **94**

IDENTIFICATION 20.5–23 cm. Monotypic. **Adult** Like Slaty-backed Forktail but crown and mantle black. From White-crowned Forktail by white breast and black mid-crown. **Juvenile** Ear-coverts, crown and upperparts duller and browner; no white forehead-patch, throat white, breast with faint sooty scales. **VOICE** Short, high-pitched, whistled ***zeee*** (slightly higher-pitched than Slaty-backed); sometimes preceded by hollow ***huu***. **HABITAT** Rivers and streams, waterfalls; up to 1,135 m. **RANGE & STATUS** Resident N,NE Indian subcontinent. **SE Asia** Uncommon to fairly common resident Myanmar (except Tenasserim), NW Thailand. **BREEDING** March–May. **Nest** Cup, in hole in rock or bank or on ledge. **Eggs** 3; pinkish, speckled and irregularly blotched reddish-brown (may be capped at broader end); 20.8 x 15.8 mm (av.).

934 SLATY-BACKED FORKTAIL

Enicurus schistaceus Plate **94**

IDENTIFICATION 22.5–24.5 cm. Monotypic. **Adult** Black and white plumage, long forked tail and slaty-grey crown, nape and mantle diagnostic. Has narrow band of white on forehead. See Black-backed Forktail. **Juvenile** Grey of upperparts tinged brown, lacks white on forehead, has white throat with greyish flecks and extensive dull greyish scales/streaks on breast and upper belly. **VOICE** Usual call is a thin shrill sharp metallic whistle: ***teenk***. **HABITAT** Rivers, streams, waterfalls; 400–2,200 m, lower locally (above 610 m in Peninsular Malaysia). **RANGE & STATUS** Resident N,NE Indian subcontinent, W and southern China. **SE Asia** Fairly common resident (except SW Myanmar, C Thailand, Singapore, Cochinchina). **BREEDING** February–June. **Nest** Cup or semi-dome, attached to rock or boulder with mud, or in hole near running water. **Eggs** 2–4; white to bluish-white, sparsely but boldly spotted or speckled dark reddish-brown over lavender undermarkings; 21.4 x 16.3 mm (av.).

935 WHITE-CROWNED FORKTAIL

Enicurus leschenaulti Plate **94**

IDENTIFICATION 28–28.5 cm. *E.l.indicus*: **Adult** From other forktails by combination of large size (most of range), steep white forehead, black mantle and breast, all-white rump and white tips to tail-feathers. *E.l.frontalis* (S Thailand, Peninsular Malaysia) is smaller (20.5 cm) and shorter-tailed; white extends slightly further back on crown, white wing-patch smaller. **Juvenile** Black plumage-parts tinged brown; initially lacks white on forehead and crown, shows indistinct white shaft-streaks on throat and breast. **VOICE** Song is an elaborate series of high-pitched whistles: ***tsswi'i'i-lli'i'i*** etc. Usual call is a harsh, shrill, whistled ***tssee*** or ***tssee chit-chit-chit*** etc. **HABITAT** Rivers, streams and adjacent forest, swampy forest; up to 2,400 m (below 760 m in S Thailand and Peninsular Malaysia). **RANGE & STATUS** Resident NE India, Bhutan, W,C,E and southern China, Greater Sundas. **SE Asia** Fairly common resident (except SW,C Myanmar, C,SE Thailand, Singapore, Cambodia, Cochinchina). **BREEDING** March–October. **Nest** Cup, in hole in bank, rock-face or tree or amongst rocks. **Eggs** 2–4; pale cream-coloured to rich buff with bold reddish-brown speckles over lavender underspecks; 24.6 x 17.7 mm (av.; *indicus*).

936 SPOTTED FORKTAIL

Enicurus maculatus Plate **94**

IDENTIFICATION 25–28.5 cm. *E.m.guttatus*: **Adult** Like White-crowned Forktail but mantle heavily spotted white. **Juvenile** Extremely similar to White-crowned but shows diagnostic large white tips on outer webs of tertials and secondaries. **Other subspecies in SE Asia** *E.m.bacatus* (W Tonkin), *robinsoni* (S Annam). **VOICE** Usual call is a repeated thin, very high-pitched ***tsueee***. **HABITAT** Rivers and streams; 915–2,560 m. **RANGE & STATUS** Resident NE Afghanistan, N Pakistan, NW,N,NE Indian subcontinent, W,S China. **SE Asia** Scarce to locally fairly common resident SW,W,S(west),N,E Myanmar, W,E Tonkin, S Annam. **BREEDING** April–July. **Nest** Cup, in various rock cavities or hole in bank. **Eggs** 3–4; pale cream to pale greenish, freckled or spotted light reddish-brown; 24.9 x 17.3 mm (av.; *guttatus*).

937 MALAYAN WHISTLING-THRUSH

Myophonus robinsoni Plate **89**

IDENTIFICATION 25.5–26 cm. Monotypic. **Adult** Very similar to Blue Whistling-thrush race *dicrorhynchus* but much smaller, smaller-billed and relatively longer-tailed; has bright metallic blue lesser coverts and bases of median coverts to wing-bend but no dull whitish tips to median coverts or shiny bluish speckles/spots on upperparts, neck-sides and throat. **Juvenile** Generally sootier-black. **VOICE** Song is similar to Blue but much softer. Call is a loud, thin, high-pitched ***tseeee***. **HABITAT & BEHAVIOUR** Broadleaved evergreen forest, usually near streams; 760–1,770 m. Very shy; occasionally frequents mountain roadsides at dawn and dusk. **RANGE & STATUS** Endemic. Uncommon resident Peninsular Malaysia (Selangor, Pahang). **BREEDING** March and September. **Nest** Large cup, in epiphytic fern on tree-branch. **Eggs** 2; very pale bluish-grey, peppered with pale pinkish-brown spots; 34 x 24 mm.

938 BLUE WHISTLING-THRUSH

Myophonus caeruleus Plate **89**

IDENTIFICATION 30.5–35 cm. *M.c.eugenei*: **Adult** Relatively large size and dark purplish-blue plumage with lighter blue spangling distinctive. Bill yellow. *M.c.caeruleus* (winter visitor; may breed Tonkin?) has smaller all-blackish bill; *dicrorhynchus* (south-east S Thailand, Peninsular Malaysia [except north-west]) is much duller and browner with dull whitish median covert tips and vague shiny bluish speckles/spots on mantle, scapulars, neck-side and throat. See Malayan Whistling-thrush, Blue Rock-thrush and Chinese Blackbird. **Juvenile** Body mostly unmarked dark brown, bill duller. **Other subspecies in SE Asia** *M.c.temminckii* (SW,W,N,C,E[north] Myanmar; winters NW Thailand), *crassirostris* (SE,S Thailand, north-west Peninsular Malaysia). **VOICE** Song is a pleasant mixture of mellow fluty and harsh scratchy notes. Calls include a loud harsh ***scree***, recalling White-crowned Forktail, and occasional shrill whistles. **HABITAT** Broadleaved evergreen and mixed deciduous forest, usually near rocky rivers and streams, waterfalls, caves, sometimes mangroves; up to 3,050 m, below 1,550 m in Peninsular Malaysia. **RANGE & STATUS** Breeds C Asia, Afghanistan, N Pakistan, NW,N,NE Indian subcontinent, S,SE Tibet, China (except NW,NE), Sumatra, Java; some northern populations winter to south, S China. **SE Asia** Common resident (except C Thailand). Uncommon winter visitor E Myanmar, NW,NE Thailand, Laos, W,E Tonkin, N Annam. Also recorded on passage E Tonkin. **BREEDING** January–August. Multi-brooded. **Nest** Bulky cup, in rock recess, cave or on mossy bank, sometimes in tree-hollow, eaves of building or on tree-branch. Eggs 2–5; pale olive-grey to pale creamy-buff, variably freckled pink to pale brown; 35.8 x 24.8 mm (av.; *temminckii*). Brood-parasitised by Large Hawk-cuckoo.

939 WHITE-BELLIED REDSTART
Hodgsonius phaenicuroides Plate **91**

IDENTIFICATION 18–18.5 cm. *H.p.ichangensis*: **Male** Dark slaty-blue plumage, white belly-centre, and rather long graduated blackish tail with conspicuous orange-rufous on basal half of outer feathers diagnostic. Has two white markings near wing-bend. *H.p.phaenicuroides* (W Myanmar) has paler blue plumage-parts. **Female** Nondescript brown with paler, buffier underparts and distinctive tail (browner than male but with similar but duller rufescent panel on outer feathers). Note size and behaviour. **Juvenile** Similar to female (with similar tail) but upperparts cold olive-brown with blackish feather-fringing and buff spots and streaks, small warm buff spots on wing-covert tips, underparts buffy with blackish-brown scales on throat and breast (fading to flanks and belly). **First-winter male** Resembles female but has brighter tail-patches (similar to adult male); starts to turn blue during first summer. **VOICE** Song is a 3–4 note whistle ***teuuh-tiyou-tuh*** etc., with second note rising and falling and last note lower-pitched. Calls include subdued, deep ***tuk*** and grating ***chack*** notes. **HABITAT & BEHAVIOUR** Rhododendron and coniferous forest and dense low vegetation above tree-line; scrub and grass, secondary growth, open broadleaved evergreen forest and edge, bamboo etc. in winter; above 2,075 m, down to 1,200 m, locally 150 m (N Myanmar) in winter. Very skulking, in low vegetation; usually adopts a very horizontal posture and often cocks tail. **RANGE & STATUS** Mostly resident (subject to relatively local movements) NE Pakistan, NW,N,NE Indian subcontinent, SE Tibet, SW,W,C,N China. **SE Asia** Uncommon resident (subject to relatively local movements) W,N Myanmar, N Laos, W Tonkin. Uncommon winter visitor E Myanmar, NW Thailand. **BREEDING** June–July. **Nest** Bulky cup, on ground or in low bush up to 50 cm above ground. **Eggs** 2–4; dark blue to deep blue-green; 22.7 x 16.1 mm (av.; *phaenicuroides*).

MUSCICAPIDAE: MUSCICAPINAE: Old World flycatchers & allies

Worldwide c.123 species. SE Asia 47 species. Very varied. Small with relatively narrow, flattened bills and well-developed rictal bristles, which help catch flying insects. Some make sallying, flycatching flights, or drop to ground, from a regular perch; others flit through foliage or glean leaves. Mostly found singly or in pairs, only a few joining mixed-species feeding flocks.

940 HAINAN BLUE FLYCATCHER
Cyornis hainanus Plate **95**

IDENTIFICATION 13.5–14 cm. Monotypic. **Male** Similar to other small *Cyornis* flycatchers but has dark blue upperside, throat and breast and very pale bluish-grey belly and flanks, grading to whiter vent. Lacks white on tail; belly and vent often vaguely buffish-tinged. Often shows white triangle/patch on centre of lower throat and/or white scaling on breast. Resembles Small Niltava but bill longer, plumage paler and less vivid blue, lacks shining blue neck-patch, rump and uppertail-coverts, belly much whiter and better demarcated. See White-tailed, Pale Blue and Blue-and-white Flycatchers. **Female** Very similar to Blue-throated Flycatcher but tends to be duller and darker above (*C.r.klossi* is similar), throat- and breast-sides similarly coloured to mantle (*C.r.rubeculoides* is similar), throat and breast duller buffy-rufous, lores and eyering usually less rich buff. From Hill Blue Flycatcher by usually paler throat than breast, lores and eyering more whitish, no orange-buff on flanks (may be washed light buff). Some (old birds?) show bluish-grey wash above and some blue on uppertail-coverts and tail. See Tickell's Blue Flycatcher. **Juvenile male** Like juvenile female but shows blue on wings and tail. **Juvenile female** Upperpart feathers dark-tipped and spotted buff (crown-centre streaked), has buff spots on wing-coverts and dark scaling on breast. **VOICE** Song is weaker and typically less complex than Hill Blue, consisting of rather short, hurried, slurred phrases. Typical call is a series of light ***tic*** notes. **HABITAT** Broadleaved evergreen, semi-evergreen and mixed deciduous forest, bamboo; up to 1,020 m. **RANGE & STATUS** Resident (some movements?) S China. **SE Asia** Fairly common to common resident C,S,E Myanmar, Tenasserim, Thailand (except C,S), Indochina (except S Annam). Also recorded on passage E Tonkin. **BREEDING** April–May. Otherwise undocumented?

941 PALE BLUE FLYCATCHER
Cyornis unicolor Plate **95**

IDENTIFICATION 16–17.5 cm. *C.u.unicolor*: **Male** Recalls Verditer Flycatcher (particularly female) but bill longer and less triangular from below, lacks turquoise in plumage, wings much duller, upperparts mid-blue with contrasting shining blue forecrown and eyebrow, brighter blue tail-fringes, paler throat and breast grading to paler and greyer on belly and vent; lores blackish-blue. See Hainan Blue, Blue-and-white and White-tailed Flycatchers. **Female** Distinctive combination of grey-tinged crown and nape, rufescent uppertail-coverts and tail, brownish-grey underparts with whiter belly-centre. Upperparts brownish, chin and throat-centre often paler than breast, vent faintly buff-tinged; some (old birds?) have strongly grey-washed upperparts, small amounts of pale blue on crown-sides, wing-coverts and scapulars, and slight bluish cast to grey of underparts. *C.u.harterti* (S Thailand, Peninsular Malaysia) has rather more rufescent upperparts, wings and tail, greyer crown and somewhat warmer brownish wash on underparts. **Juvenile male** Like juvenile female but has blue on wings and tail. **Juvenile female** Upperparts broadly tipped blackish and spotted buff, has buff spots on wing-coverts, uniformly dark-scaled underparts. **VOICE** Has more melodious, warbling song than other *Cyornis*, phrases often beginning with shorter ***chi*** notes and ending with buzzy ***chizz*** or ***wheez***. **HABITAT** Middle to upper storey of broadleaved evergreen forest; up to 1,600 m. **RANGE & STATUS** Mostly resident N,NE Indian subcontinent, SW,S China, Greater Sundas. **SE Asia** Uncommon resident Myanmar (except SW,S), Thailand (except C,SE), Peninsular Malaysia, Cambodia, Laos, E Tonkin, N,C,S Annam. Vagrant coastal E Tonkin. **BREEDING** April–June. **Nest** Cup or bracket, attached to rock, or in hole in bank or tree; up to 1.8 m above ground. **Eggs** 2–4; deep yellowish-buff, densely freckled brown to bright reddish (may appear almost uniform); 23.1 x 17.5 mm (av.; *unicolor*).

942 HILL BLUE FLYCATCHER
Cyornis banyumas Plate **95**

IDENTIFICATION 14–15.5 cm. *C.b.whitei*: **Male** Very similar to Tickell's Blue Flycatcher but orange-rufous of breast grades into white belly (not so sharply demarcated), often shows orange-rufous extending down flanks. Uppermost chin at base of bill dark or as rest of throat. *C.b.lekahuni* (south-west NE Thailand) and *deignani* (SE Thailand) have bigger bill and slightly deeper blue upperparts; *coerulifrons* (resident south Tenasserim, S Thailand, Peninsular Malaysia) has slightly deeper blue upperparts and possibly deeper, more rufous breast (may contrast slightly with throat). See Blue-throated and Mangrove Blue Flycatchers. Note habitat and range. **Female** Similar to Tickell's but shows same gradation of orange-rufous to white on underparts as male, upperparts warm-tinged olive-brown, uppertail-coverts, tail- and wing-feather fringes markedly rufescent. May show some blue on uppertail-coverts (older birds?). Best separated from Blue-throated by almost uniform deep orange-rufous throat and breast, somewhat darker lores, and dark remainder of head-side, which contrasts more strongly with throat. See Hainan Blue Flycatcher. **Juvenile male** Like juvenile female but has blue on wings and tail. **Juvenile female** Darker above than adult with darker feather-tips and warm buff spotting (crown more streaked); has warm buff spots on wing-coverts and buffish underparts with deeper buff, heavily dark-scaled breast, upper belly and flanks. **VOICE** Song is sweet, high-pitched and melancholy, typically starting with thin ***si*** or ***tsi*** notes. Recalls Tickell's Blue but phrases longer, more complex and more rapidly delivered. Calls include a hard ***tac*** and scolding ***trrt-trrt-trrt...*** **HABITAT** Broadleaved evergreen forest, sometimes more open areas, parks and gardens on migration; 400–2,515 m (below 1,220 m in Peninsular Malaysia). **RANGE & STATUS** Breeds south-western China, Java, Borneo. Recorded in winter (status uncertain) NE India. Some northerly populations winter to south and south-west. **SE Asia** Uncommon to common resident N,E Myanmar, Tenasserim, Thailand (except C), Peninsular Malaysia, N,C Laos, W,E Tonkin, N,C Annam. Uncommon passage migrant/non-breeding visitor C Thailand. Recorded (status uncertain) C Myanmar. **BREEDING** March–July. **Nest** Cup, in hollow on ground, stump or against mossy tree-trunk; usually quite low down. **Eggs** 4–5; pale sea-green or buffish stone-colour with brown speckles, forming broad ring around broader end; 19.1 x 14.6 mm (av.).

943 LARGE BLUE FLYCATCHER

Cyornis magnirostris Plate **95**

IDENTIFICATION 15 cm. Monotypic. **Male** Very similar to Hill Blue Flycatcher race *whitei* but has much bigger, hook-tipped bill, longer wings (primary projection), slightly deeper blue upperparts, orange-rufous of throat slightly paler than breast, with very little visible darkness on uppermost chin at base of bill. Harder to separate from Hill Blue races *lekahuni* and *deignani*, but note range. See Blue-throated Flycatcher. **Female** Like Hill Blue but has much bigger bill, longer wings, more strongly rufescent tail, and possibly paler throat (throat slightly paler than breast). **VOICE** Not yet definitely documented. **HABITAT** Broadleaved evergreen forest, sometimes more open areas; up to 600 m, locally c.1,200 m on passage. **RANGE & STATUS** Breeds Bhutan, NE India, ? Nepal; winters to south-east. **SE Asia** Uncommon winter visitor and passage migrant Tenasserim, S Thailand, Peninsular Malaysia. Uncommon passage migrant south E Myanmar.

944 TICKELL'S BLUE FLYCATCHER

Cyornis tickelliae Plate **95**

IDENTIFICATION 13.5–15.5 cm. *C.t.indochina*: **Male** Widespread lowland resident. From similar Chinese Blue and Hill Blue Flycatchers by sharp demarcation between orange-rufous breast and white belly, rarely showing obvious orange-rufous on flanks. Upperparts rather uniform deep blue with paler blue forehead and sides of forecrown. *C.t.tickelliae* (N,C Myanmar ?) tends to have paler blue upperparts; *sumatrensis* (Tenasserim and S Thailand southwards) has slightly deeper blue upperparts, throat usually paler and buffier than breast. **Female** From similar *Cyornis* flycatchers by greyish to bluish-grey tinge to upperparts (may be faint) and well-demarcated underpart colours (like male); tail usually shows some blue, throat usually paler and buffier than breast and sharply demarcated from sides of head. Lores broadly pale buff to buffy-whitish. *C.t.tickelliae* has dull, rather pale grey-blue upperparts (i.e. much bluer); *sumatrensis* ranges from similar to much bluer on mantle and scapulars to uppertail. **Juvenile male** Like juvenile female but shows blue on wings and tail. **Juvenile female** Resembles Blue-throated and Hill Blue but base colour to upperparts greyish-tinged, lacks rufous on tail, has narrow pale buffish crown-streaking. **VOICE** Song consists of quite slowly delivered, sweet, high-pitched, slightly descending (or rising then descending) phrases: ***tissis-swii'i'i'i-ui, sisis'itu'ii'iiw, tis-swiu'iiu'iiu.*** Typical calls are a hard ***tac*** and ***trrt*** notes. **HABITAT** Broadleaved evergreen, semi-evergreen and deciduous forest, bamboo; up to 915 m (below 600 m in Thailand). **RANGE & STATUS** Resident lowland India, S Nepal, Sri Lanka; recorded N Sumatra. **SE Asia** Common resident (except SW Myanmar, C Thailand, Singapore, N Laos, W,E Tonkin). **BREEDING** April–August. **Nest** Cup, in hole in various situations or in sapling or bamboo; usually within 2 m of ground (sometimes much higher). **Eggs** 2–5; similar to Blue-throated Flycatcher; 18.4 x 14.2 mm (av.; *tickelliae*).

945 CHINESE BLUE FLYCATCHER

Cyornis glaucicomans Plate **95**

IDENTIFICATION 14–15 cm. Monotypic. **Male** Recalls Blue-throated Flycatcher race *dialilaema* but has blue of throat restricted to sides and chin (visible at close range), deeper and duller orange-rufous throat and breast, extensive brown wash on flanks, somewhat darker blue upperparts and different song; from Hill Blue Flycatcher by much deeper blue upperparts, with contrasting shining azure shoulder-patch and uppertail-coverts, more azure-blue forehead and eyebrow. Always has dark blue chin. See Malaysian Blue Flycatcher. **Female** Darker and warmer-tinged above than Blue-throated race *dialilaema*, darker and less rufescent tail, deeper orange-rufous breast, contrasting pale buff throat, and brown-washed flanks. See Fulvous-chested Flycatchers. **VOICE** Song is richer, more sustained, varied and warbling than that of Blue-throated. **HABITAT** Broadleaved evergreen, semi-evergreen and mixed deciduous forest, secondary growth, bamboo, also gardens and mangroves on migration; up to 1,100 m. **RANGE & STATUS** Breeds SW,W,C China; winters to south. **SE Asia** Uncommon winter visitor Tenasserim, W,S Thailand, Peninsular Malaysia. Uncommon passage migrant N Myanmar, NW,NE,C,S Thailand, Peninsular Malaysia, C Laos, E Tonkin. Vagrant Singapore. **NOTE** Split from Blue-throated following King (1997).

946 BLUE-THROATED FLYCATCHER

Cyornis rubeculoides Plate **95**

IDENTIFICATION 14–15 cm. *C.r.dialilaema*: **Male** From other *Cyornis* flycatchers by dark blue throat with orange-rufous triangle/wedge extending up to its centre. Upperside fairly deep blue. *C.r.rubeculoides* (W Myanmar; visitor SW,C,S,E Myanmar) has uniform dark blue throat and somewhat darker upperparts; *klossi* (east Cambodia, S Laos, C,S Annam, Cochinchina) has paler rufous-orange breast and whitish point of throat-triangle; sometimes has pale buffy to whitish throat-triangle and whitish breast with warm buff wash/smudges. **Female** Like Hill Blue but throat and breast markedly paler and more rufous-buff, shows less sharp demarcation between head-side and throat, lores somewhat paler, crown and mantle usually paler and greyer with less rufescent forehead. Has strongly rufescent tail. Some (old birds?) show blue on forecrown, uppertail-coverts and tail. *C.r.klossi* has slightly darker upperparts, somewhat less rufescent tail, slightly deeper rufous-orange throat and breast, and often slightly rufescent forecrown; *rogersi* (SW Myanmar) is warmer above and has deeper rufous-orange breast. See Pale-chinned Flycatcher. Note range. **Juvenile male** Like juvenile female but has blue on wings and tail. **Juvenile female** Like Hill Blue but paler above, has paler, more rufous tail and lighter breast-scaling. **VOICE** Song consists of sweet trilling and slurred tinkling notes, mixed with fairly well-structured phrases: ***trrr-sweei-iu-iu***; ***tch'tch'tch-hiu'hiu'hiu'hiu*** and ***trr-trr-swiwiwiwi*** etc., often interspersed with metallic ***tit-it-trrt-rrt*** and ***tit-it-it*** calls. Recalls Tickell's Blue Flycatcher but more rapidly delivered and higher-pitched, with more trilling notes. Calls include hard ***tac*** and ***trrt*** notes. **HABITAT** Broadleaved evergreen, semi-evergreen and mixed deciduous forest, secondary growth, bamboo, also gardens and mangroves on migration; up to 1,700 m, breeds below 1,350 m. **RANGE & STATUS** Breeds NE Pakistan, NW,N,NE Indian subcontinent; some winter to south, eastern and southern India, Bangladesh, Sri Lanka. **SE Asia** Uncommon to fairly common resident Myanmar, W,NW,NE Thailand, S Laos, C,S Annam, Cochinchina. Uncommon winter visitor SW,C,S,E Myanmar. **BREEDING** April–May. **Nest** Cup, in various hollows, sometimes amongst ferns or orchids in tree. **Eggs** 3–5; pale olive to yellowish-stone, densely stippled olive-brown to reddish-brown; 18.7 x 14.3 mm (av.; *rubeculoides*).

947 MALAYSIAN BLUE FLYCATCHER
Cyornis turcosus Plate **95**

IDENTIFICATION 14 cm. *C.t.rupatensis*: **Male** Like Blue-throated Flycatcher nominate *rubeculoides* but throat bright deep blue, breast paler rufous-orange, upperparts somewhat deeper, brighter blue, with paler shining blue rump and uppertail-coverts. Note habitat and range. **Female** Like male but throat pale warm buff (whiter on chin and sides), upperparts less deep blue. From male Hill and Tickell's Blue Flycatchers by paler blue upperparts, shining blue rump and uppertail-coverts, and paler throat and breast. **Juvenile** Poorly documented but similar to other *Cyornis*. **VOICE** Song consists of relatively weak 5–6 note phrases. Female may give thin, strained ***swiii swii-swew***. **HABITAT** Broadleaved evergreen forest, near rivers and streams; up to 760 m. **RANGE & STATUS** Resident Borneo, Sumatra. **SE Asia** Scarce to uncommon resident extreme S Thailand, Peninsular Malaysia. **BREEDING** April–June. **Nest** Cup, in cavity of dead tree or tree-fern etc., often jutting from streambank; 1.2 m or more above ground. Otherwise undocumented.

948 MANGROVE BLUE FLYCATCHER
Cyornis rufigastra Plate **95**

IDENTIFICATION 14.5 cm. *C.r.rufigastra*: **Male** Resembles Tickell's and Hill Blue Flycatchers but upperparts darker, somewhat duller blue, with less obvious lighter forehead and eyebrow, underparts dull deep orange-rufous, with paler, buffier vent and whiter belly-centre. Note habitat and range. **Female** Like male but shows distinctive whitish lores, cheek-spot and chin. **Juvenile** Upperparts sooty-coloured with darker feather-tips and dull buff speckling (crown more streaked with plainer centre), has dull buff spots on wing-coverts and pale buff underparts (breast deeper buff, vent whiter) with prominently dark-scaled throat-sides, breast and upper belly. **VOICE** Song is like Tickell's but slightly slower and deeper. **HABITAT** Mangroves. **RANGE & STATUS** Resident Greater Sundas, Philippines, Sulawesi. **SE Asia** Uncommon to fairly common resident S Thailand (local), Peninsular Malaysia, Singapore (rare and local). **BREEDING** March–July. **Nest** Cup, in hollow dead palm stump etc.; 1.5 m above ground. **Eggs** 2.

949 WHITE-TAILED FLYCATCHER
Cyornis concretus Plate **95**

IDENTIFICATION 19 cm. *C.c.cyanea*: **Male** Relatively large and robust, with rather uniform blue head and body, contrasting white belly and vent and distinctive white lines running down whole length of outertail-feathers (obvious when tail spread). See White-tailed Robin and Hainan Blue and Blue-and-white Flycatchers. **Female** Combination of white crescent/patch on upper breast/lower throat (sometimes hidden) and white lines on outertail-feathers diagnostic. Otherwise rather uniform warm brown with paler throat and whitish belly and vent. See Blue-and-white and Rufous-bellied Niltava. ***C.c.concretus*** (S Thailand, Peninsular Malaysia) has brighter rufous-brown plumage, warm buff throat and whiter, more demarcated belly. **Juvenile** Little known. **Immature male** Head mostly buffy-rufous with blackish streaks, has large rufous-buff spots on rest of upperparts and wing-coverts. **VOICE** Song is very variable, sometimes incorporating skilled avian mimicry. Typically a repeated series of 3–7 rather shrill, high, piercing but tuneful notes: ***pieu pieu pieu***; ***pieu pieu pieu jee-oee***; ***ti ti ti teu tear-tear***; ***jo di di dear-dear***; ***phi phi phi phi ju-rit*** and ***pheu pheu jing*** etc. Calls include harsh ***scree*** notes. **HABITAT & BEHAVIOUR** Broadleaved evergreen forest; 100–1,360 m. Sometimes associates with bird-waves. **RANGE & STATUS** Resident NE India, SW China, Sumatra, Borneo. **SE Asia** Scarce to locally fairly common resident N Myanmar, Tenasserim, W,S Thailand, Peninsular Malaysia, Laos, W,E Tonkin, N,C Annam, Cochinchina. **BREEDING** April–August. **Nest** Cup, in hole in bank. **Eggs** 2+; pale buffish stone-coloured, stippled dark reddish overall (sometimes vaguely capped at broader end); 23.9 x 18 mm (*cyanea*).

950 PALE-CHINNED FLYCATCHER
Cyornis poliogenys Plate **98**

IDENTIFICATION 15–17 cm. *C.p.poliogenys*: **Adult** Similar to females of other *Cyornis* flycatchers but crown and sides of head greyer, throat whitish contrasting sharply with rufescent breast, rest of underparts more buffish. Has prominent pale buff eyering and loral line and strongly rufescent tail. Note range. *C.p.cachariensis* (N Myanmar) has more chestnut tail and deeper, purer rufous-buff on underparts. **Juvenile** Upperpart feathers darker-tipped and faintly speckled buffish, wing-coverts and tertials tipped buff, breast dark-scaled (fading to flanks). **VOICE** Typical *Cyornis* song is a high-pitched, well-structured, slightly undulating series of usually 4–11 notes; sometimes interspersed with harsh ***tchut-tchut*** call-notes. **HABITAT** Broadleaved evergreen forest; up to 1,500 m. **RANGE & STATUS** Resident N,NE Indian subcontinent, E India, SW China. **SE Asia** Uncommon resident SW,W,N,S(west) Myanmar. **BREEDING** NE India: April–June. **Nest** Cup, in hole in bank or tree or amongst rocks; within 1 m of ground. **Eggs** 3–5; pale olive to olive-buff, densely speckled reddish to red-brown (often appear uniform reddish-brown); 18.5 x 14.6 mm (av.; *poliogenys*).

951 VERDITER FLYCATCHER
Eumyias thalassinus Plate **98**

IDENTIFICATION 15–17 cm. *E.t.thalassinus*: **Male breeding** Overall bright turquoise-tinged pale blue plumage and black lores diagnostic. Has dark undertail-coverts with whitish tips. See Pale Blue Flycatcher. **Male non-breeding** Duller and more turquoise. **Female breeding** Similar to male but duller and slightly greyish-tinged, with dusky lores. **Female non-breeding** Underparts duller and greyer. All plumages of *E.t.thallasoides* (S Tenasserim, S Thailand, Peninsular Malaysia) are generally somewhat bluer and less turquoise. **Juvenile** Resembles non-breeding female but head and body greyer with buff to whitish speckling, wing-coverts tipped buff to whitish, feathers of throat and underparts dark-tipped, giving scaled appearance. **VOICE** Song is a rather hurried series of high-pitched, undulating musical notes, gradually descending scale. **HABITAT** More open broadleaved evergreen forest, clearings, also wooded gardens and mangroves on migration; up to 2,740 m (below 1,220 m in Peninsular Malaysia). **RANGE & STATUS** Breeds N Pakistan, NW,N,NE Indian subcontinent, SW,W,C,S China, Sumatra, Borneo; some winter south to south India, Bangladesh, SE China. **SE Asia** Common resident (except C,S Myanmar, C Thailand). Uncommon winter visitor C,NE,SE Thailand. Scarce to uncommon passage migrant E Tonkin, S Annam. Recorded (status uncertain) C,S Myanmar. **BREEDING** April–July. **Nest** Cup, in hole in bank, wall or building, sometimes amongst moss against tree-trunk; up to 6 m above ground. **Eggs** 3–5; pale creamy-pink to white, unmarked or with reddish-pink zone or cap of cloud-like markings at broader end; 19.3 x 14.7 mm (av.).

952 FUJIAN NILTAVA *Niltava davidi* Plate **96**

IDENTIFICATION 18 cm. Monotypic. **Male** Like Rufous-bellied Niltava but only front and sides of crown (to above eye) pale shining blue, rest of crown only slightly brighter than mantle, no lighter shoulder-patch, breast somewhat duller, darker and more rufous, usually contrasting with paler, buffier belly and vent. **Female** Very like Rufous-bellied but upperparts, wings and tail a shade darker and colder, lacking bright rufous tones on wings and tail; lower throat, upper breast and flanks tend to be slightly darker, making breast-patch (which may be larger) contrast more. **Juvenile** Undocumented? **VOICE** Very thin, high-pitched ***ssssew*** or ***siiiii***, repeated after shortish intervals. Sharp metallic ***tit tit tit...*** and ***trrt trrt tit tit trrt trrt...*** etc. **HABITAT** Broadleaved evergreen forest, also parks and gardens on migration; up to 1,700 m (probably only breeds at higher levels). **RANGE & STATUS** Breeds southern China; some winter to south. **SE Asia** Uncommon resident or breeding visitor W Tonkin. Scarce to fairly common winter visitor south-west Cambodia, Laos, E Tonkin, N,C Annam. Vagrant (or under-recorded) SE Thailand. **BREEDING** Undocumented?

953 **RUFOUS-BELLIED NILTAVA**
Niltava sundara Plate **96**

IDENTIFICATION 18 cm. *N.s.sundara*: **Male** Robust and rather large, upperside very dark blue with paler shining blue crown (appears capped), neck, shoulder-patches, rump and uppertail-coverts, bluish-black head-sides and throat and dark orange-rufous remainder of underparts. See Fujian, Rufous-vented and Vivid Niltavas and *Cyornis* flycatchers. **Female** From most niltavas and other flycatchers by strongly rufescent wings and tail (latter without white), blue neck-patch and prominent whitish patch on centre of uppermost breast (sometimes absent). Rest of underparts greyish-olive, with buff chin and paler buffy-grey to whitish belly-centre and undertail-coverts. Upperparts warm olive-brown, greyer on crown and nape; loral line and eyering rufous-buff. See Fujian and Rufous-vented Niltavas and White-tailed Flycatcher. **Juvenile** Resembles Large but mantle paler, uppertail-coverts and tail much more rufous; shows suggestion of breast-patch and whiter belly-centre and vent. Male soon attains orange-rufous patches below. **Other subspecies in SE Asia** *N.s.denotata* (recorded N Myanmar, NW Thailand, N Laos). **VOICE** Calls include thin metallic ***tsi tsi tsi tsi...***, hard ***tic*** and ***trrt*** notes and husky scolding rattles. **HABITAT** Lower to middle storey of broadleaved evergreen forest; 900–2,720 m, locally down to 450 m in winter. **RANGE & STATUS** Resident, subject to relatively local movements, NE Pakistan, NW,N,NE Indian subcontinent, SE Tibet, SW,W,C China. Winter visitor E Bangladesh. **SE Asia** Fairly common to common resident W,N,E Myanmar. Uncommon to fairly common winter visitor C,S(east) Myanmar, north Tenasserim, NW Thailand. Recorded (status uncertain) N,C Laos. **BREEDING** India: April–August. **Nest** Cup, in hole in bank, tree-stump or amongst rocks. **Eggs** 4; pale cream to buffish with pale reddish-brown speckles (sometimes ringed at broader end); 20.7 x 15.9 mm (av.; *sundara*).

954 **RUFOUS-VENTED NILTAVA**
Niltava sumatrana Plate **96**

IDENTIFICATION 15–15.5 cm. Monotypic. **Male** Like Rufous-bellied Niltava but smaller, shining blue parts duller, no shoulder-patch and much less obvious neck-patch, underparts a shade darker. Note range. **Female** Recalls Rufous-bellied but crown and nape much slatier-grey, rest of upperparts, wings and tail darker, more deeply rufescent, no neck-patch, little or no white breast-patch, upper throat warm buff with dark barring, breast and belly greyer with whitish centre to latter, undertail-coverts pale rufous. **Juvenile** Upperparts and wing-coverts sooty-brown, the feathers with large orange-buff centres and blackish tips, underparts rich buff with sooty fringing; flight-feathers and tail as respective adults. **VOICE** Song is a monotonous series of rather undulating clear whistles, reminiscent of Verditer Flycatcher. Also utters a rapid series of complex scratchy slurred notes. **HABITAT** Broadleaved evergreen forest; above 1,525 m. **RANGE & STATUS** Resident Sumatra. **SE Asia** Locally common resident Peninsular Malaysia. **BREEDING** February–May. **Nest** Deep cup, amongst moss on tree. Otherwise undocumented.

955 **VIVID NILTAVA** *Niltava vivida* Plate **96**

IDENTIFICATION 18.5–19 cm. *N.v.oatesi*: **Male** Like Rufous-bellied Niltava but larger and longer-winged, with diagnostic orange-rufous wedge/triangle extending from breast to centre of lower throat, sometimes almost to chin (not obvious when viewed from side); throat-sides bluer, crown, shoulder, rump and uppertail-coverts duller, no defined neck-patch. **Female** Lacks blue neck-patch of other niltavas in its range. Otherwise similar to Large but underparts somewhat paler and greyer, belly-centre often paler still or buffy-tinged, undertail-coverts pale buff, wings and tail a shade paler and less rufescent. Lacks white breast-patch. **Juvenile** Very similar to Large but upperpart spotting and base colour of underparts tend to be paler and more buff, undertail-coverts more uniform, paler buff, eyering buff. **VOICE** Song is a series of slow mellow whistles, usually interspersed with scratchier notes: ***heu wii riu chrt-trrt heu wii tiu-wii-u...*** **HABITAT & BEHAVIOUR** Broadleaved evergreen forest; 750–2,565 m. Rather arboreal, often in middle storey to canopy. **RANGE & STATUS** Resident NE India, SE Tibet, south-western China, Taiwan. **SE Asia** Scarce to uncommon winter visitor NW,NE,SE Thailand. Recorded (status uncertain) W,C,S(east),E Myanmar, north Tenasserim, N Laos, W Tonkin. **BREEDING** Undocumented?

956 **LARGE NILTAVA**
Niltava grandis Plate **96**

IDENTIFICATION 20–21.5 cm. *N.g.grandis*: **Male** Very distinctive. Relatively large, upperparts very dark blue with lighter and brighter crown, sides of neck, shoulder-patch, rump and uppertail-coverts; head-sides, throat and upper breast blackish, rest of underparts blue-black with greyer vent. See Blue-fronted Robin and Small Niltava. **Female** From other niltavas by combination of size, light blue neck-patch and rather uniformly dark brown underparts with contrasting narrow buffish patch on centre of throat, greyer centre of abdomen and no white breast-patch. Crown and nape greyish, undertail-coverts fringed buffish. See Blue-and-white Flycatcher. *N.g.decorata* (S Annam) and *decipiens* (S Thailand, Peninsular Malaysia; N Laos—possibly unnamed subspecies) have darker, warmer brown plumage, darker blue neck-patch and distinctive shining deep blue (former) or bluish-slate (latter) crown and nape. **Juvenile male** Like juvenile female but has mostly blue wings and tail, soon acquiring blue patches on upperparts. **Juvenile female** Feathers of upperside and wing-coverts tipped black and spotted buff to rufous-buff, underparts deep rich buff, paler on throat-centre and vent and narrowly scaled blackish on throat-sides, breast and upper belly (fading to vent). **VOICE** Ascending series of usually 3–4 softly whistled notes: ***uu-uu-du-di*** or ***uu'uu'di*** etc. Calls with rasping, scolding rattles and a soft ***chu-ii***, with higher second note. **HABITAT & BEHAVIOUR** Broadleaved evergreen forest; 900–2,565 m, locally down to 450 m in winter (above 1,220 m in Peninsular Malaysia). Usually inhabits middle storey; sometimes drops to ground. **RANGE & STATUS** Resident Nepal, Bhutan, NE India, SW China, Sumatra. **SE Asia** Fairly common resident (except SW,C,S Myanmar, C Thailand, Singapore, C Annam, Cochinchina). **BREEDING** February–August. **Nest** Bulky cup or dome, amongst moss on tree or boulder, in hole in bank or wall or amongst rocks; up to 6 m above ground. **Eggs** 2–5; pale cream to pale buff, spotted (often very densely) fawn to pinkish-brown (sometimes ringed at broader end); 24.7 x 18 mm (av.; *grandis*).

957 **SMALL NILTAVA**
Niltava macgrigoriae Plate **96**

IDENTIFICATION 13.5–14 cm. *N.m.signata*: **Male** Like a miniature Large Niltava but throat and upper breast blue-black, grading to much paler grey belly and whitish vent; shows brighter, lighter blue forehead and neck-patch (rest of crown like mantle). See Hainan Blue Flycatcher. **Female** Also like miniature Large but has darker, less contrasting throat-centre, browner (less greyish) crown and nape, and generally greyer underparts with whitish centre to abdomen and buffy undertail-coverts. **Juvenile** Resembles Large but much smaller, throat concolorous with breast, centre of abdomen and vent whitish. **VOICE** Song is a very thin, high-pitched, rising and falling ***swii-swii-ii-swii***, level ***tsii-sii*** or descending ***tsii-sii-swi***. Typical calls include harsh metallic churring and scolding notes. **HABITAT** Broadleaved evergreen forest; 700–2,565 m, locally down to 275 m in winter. **RANGE & STATUS** Resident (subject to minor movements) N,NE Indian subcontinent, SE Tibet, SW,S China. **SE Asia** Uncommon to fairly common resident N,E,S(east) Myanmar, north Tenasserim, W,NW Thailand, Laos, W,E Tonkin, N,C Annam. Also recorded on passage coastal E Tonkin. **BREEDING** March–August. **Nest** Cup, in hole in bank or amongst rocks. **Eggs** 3–5; creamy-white to pale greyish-yellow, blotched dull reddish (often vaguely ringed around broader end); 18.1 x 13.6 mm (av.). Probably brood-parasitised by Hodgson's Hawk-cuckoo.

958 YELLOW-RUMPED FLYCATCHER

Ficedula zanthopygia Plate **97**

IDENTIFICATION 13–13.5 cm. Monotypic. **Male** Distinctive, with black upperside, vivid yellow lower back, rump and underparts and pronounced white supercilium, long wing-patch and undertail-coverts. Often shows strong orange flush on throat and breast in spring. See Narcissus Flycatcher. **Female** Much more nondescript, with dull greyish-olive upperparts, but shows distinctive yellow rump (obvious in flight) and similar white wing-patch to male (though often with less white on greater coverts). Has buffy- to yellowish-white loral stripe and underparts and faint brownish scales/mottling on throat and breast. See Narcissus Flycatcher. **First-winter male** Resembles adult female but has largely blackish uppertail-coverts, white of wing-patch restricted to bar on inner greater coverts. With a few exceptions, moults almost completely into adult plumage before spring migration north. **First-winter female** Shows narrower white outer fringes/tips to greater coverts. **VOICE** Dry, rattled ***tr'r'r't***. **HABITAT** Broadleaved forests, forest edge, also plantations, parks, gardens and mangroves on migration; up to 950 m. **RANGE & STATUS** Breeds SE Siberia, Ussuriland, C,E,NE China, N,S Korea; winters locally Greater Sundas. **SE Asia** Scarce to uncommon passage migrant south Tenasserim, Thailand, Peninsular Malaysia, Singapore, N Laos, W,E Tonkin, N,C Annam. Scarce to uncommon winter visitor W(south),S Thailand, Peninsular Malaysia, Singapore.

959 NARCISSUS FLYCATCHER

Ficedula narcissina Plate **97**

IDENTIFICATION 13–13.5 cm. *F.n.narcissina*: **Male** Similar to spring Yellow-rumped Flycatcher but has yellow supercilium, yellowish-white belly and lacks white line along edge of tertials. Throat more red-orange when breeding. **Female** Similar to Green-backed, but much browner above (olive-tinged on lower back and rump) and whiter below, with variable dark scales/mottling on throat-sides and breast, greyish/brownish wash on breast and flanks, and sometimes a faint yellowish tinge on belly. **First-winter male** Resembles adult female. **HABITAT** Broadleaved forest, forest edge, plantations, parks and gardens; up to 1,400 m on passage. **RANGE & STATUS** Breeds Sakhalin, Japan (including Nansei Is); winters Borneo, Philippines. **SE Asia** Rare passage migrant/vagrant south Tenasserim, W(coastal) Thailand, Cambodia, E Tonkin, C,S Annam.

960 GREEN-BACKED FLYCATCHER

Ficedula elisae Plate **97**

IDENTIFICATION 13–13.5 cm. Monotypic. **Male** Very distinctive. Greyish olive-green above, wings and tail dark, has contrasting bright yellow loral stripe, eyering, rump and underparts and broad white wing-patch. **Female** Somewhat duller above with no yellow on rump or white wing-patch, subdued facial pattern, two faint narrow pale bars on wing-coverts, and rufescent-tinged uppertail-coverts and tail; underparts duller but still quite bright yellow. **First-winter male** Presumably resembles adult female. **HABITAT** Broadleaved evergreen forest, forest edge, also plantations, parks and gardens on migration; lowlands but up to 1,295 m on passage. **RANGE & STATUS** Breeds E/NE China. **SE Asia** Scarce winter visitor S Thailand, Peninsular Malaysia. Rare to scarce passage migrant NE,C Thailand, Peninsular Malaysia, E Tonkin. Vagrant Singapore.

961 MUGIMAKI FLYCATCHER

Ficedula mugimaki Plate **97**

IDENTIFICATION 13–13.5 cm. Monotypic. **Male** Very distinctive, with blackish-slate upperside, short white supercilium, large white wing-patch and tertial edges, and bright rufous-orange throat to upper belly. Rest of underparts whitish; has white patch at base of outertail-feathers. See Rufous-chested Flycatcher. Note range and habitat. **Female** Upperparts greyish-brown (greyer in spring), throat and breast buffish-orange, supercilium faint or lacking; usually lacks white on tail, has one or two narrow buffish to whitish bars on wing-coverts and faint pale edges to tertials. See Rufous-chested, Taiga, Sapphire and *Cyornis* Flycatchers. **First-winter male** Resembles adult female but lores and sides of head often greyer, with slight darker moustachial line, supercilium more pronounced, throat and breast brighter orange-rufous, uppertail-coverts mostly blackish, some whitish on base of outertail-feathers. **VOICE** Typical call is a rattled ***trrr'rr*** or ***trrrik***. **HABITAT & BEHAVIOUR** Broadleaved evergreen and pine forest, also plantations, parks and gardens on migration; up to 2,010 m, mainly above 800 m in winter. Often forages in middle storey to high canopy. **RANGE & STATUS** Breeds E Palearctic, NE China; winters Greater Sundas, Philippines, rarely Sulawesi. **SE Asia** Scarce to locally fairly common winter visitor NE,SE,W(south),S Thailand, Peninsular Malaysia, Cambodia, Laos, C,S Annam. Scarce to uncommon passage migrant NW,NE,C,S Thailand, Singapore, Cambodia, Laos, W,E Tonkin.

962 SLATY-BACKED FLYCATCHER

Ficedula hodgsonii Plate **97**

IDENTIFICATION 13–13.5 cm. Monotypic. **Male** Combination of dark, dull bluish-slate upperside, blackish tail with white at base of outer feathers and orange-rufous underparts (fading on vent) diagnostic. Larger and longer-tailed than Snowy-browed Flycatcher with no white eyebrow. See Slaty-blue and Hill Blue Flycatchers. **Female** Very nondescript with dull olive-brown upperside, dull greyish to buffish-grey throat and breast, rufescent uppertail-coverts and whitish eyering and loral line; often has narrow buffish to buffish-white bar on greater coverts, lacks white on tail. Most likely to be confused with Slaty-blue Flycatcher but lacks any obvious buff on underparts or rufous on tail. See Snowy-browed, Little Pied and Narcissus Flycatchers. **Juvenile** Little information but spotted like other flycatchers. **First winter** Similar to female but somewhat warmer brown above, with more prominent wing-bar. **VOICE** Song is a rather short, meandering, generally descending ditty of slurred-together whistled notes. Typical call is a rather deep hard rattling ***terrht*** or ***tchrt***. **HABITAT & BEHAVIOUR** Broadleaved evergreen forest, forest edge, secondary growth; 600–2,750 m. Breeds above 1,200 m in NE India. Frequents middle- and lower-storey vegetation. Sometimes gathers in flocks to feed on berries. **RANGE & STATUS** Resident (subject to relatively local movements) Nepal, NE Indian subcontinent, SE,E Tibet, SW,W China. **SE Asia** Uncommon to fairly common resident (subject to some movements) W,N,E Myanmar. Scarce to fairly common winter visitor C,S Myanmar, Tenasserim, W,NW,NE Thailand, south-west Cambodia, N,C Laos. **BREEDING** India: April–July. **Nest** Cup, in hole in bank, amongst rocks or exposed tree-roots, or in hanging moss; usually low down. **Eggs** 4–5; pale green to warm buff, stippled light reddish overall; 17.8 x 13.4 mm (av.).

963 **WHITE-GORGETED FLYCATCHER**
Ficedula monileger Plate **97**
IDENTIFICATION 12–13 cm. *F.m.leucops*: **Adult** Very distinctive. Small, robust, short-tailed and mostly brown, with clean white throat bordered by black. Upperside olive-brown with greyer sides of head, broad whitish eyebrow and warm-tinged uppertail-coverts, tail and wings; breast and flanks buffish olive-brown, rest of underparts whitish. See Rufous-browed Flycatcher. **Juvenile** Upperparts dark brown with warm buff streaks, has buff-tipped greater coverts and buffish underparts with diffuse dark brown streaks. Lacks white throat and black border but shows distinctive pinkish legs and feet like adult. **Other subspecies in SE Asia** *F.m.gularis* (SW Myanmar). **VOICE** Song consists of very high-pitched, wispy, slurred, rather forced scratchy phrases, sometimes running into a more structured, descending ***sii-siii-suuu.*** Typical call consists of metallic ***tik*** or ***trik*** notes, sometimes interspersed with very thin, stressed piercing ***siii***, ***swiii*** or ***siiiu*** notes. **HABITAT & BEHAVIOUR** Broadleaved evergreen forest, bamboo; 600–1,900 m. Skulks in low vegetation. **RANGE & STATUS** Resident Nepal, NE Indian subcontinent, SW China. **SE Asia** Uncommon to locally fairly common resident SW,W,S,N,E Myanmar, NW,NE Thailand, Laos, W,E Tonkin, N,C Annam. **BREEDING** March–August. **Nest** Dome, on sloping ground or in bank recess, occasionally just above ground in bush. **Eggs** 4; white, speckled and freckled pinkish-red to reddish-brown, forming ring around broader end; 18.2 x 13.8 mm (av.).

964 **RUFOUS-BROWED FLYCATCHER**
Ficedula solitaris Plate **97**
IDENTIFICATION 12–13 cm. *F.s.submonileger*: **Adult** Like White-gorgeted Flycatcher but usually lacks complete black throat-border, has distinctive bright rufous lores, eyering and crown-sides, overall plumage tones considerably more rufescent. *F.s.malayana* (southern S Thailand, Peninsular Malaysia) has richer, more rufous-chestnut brown parts of plumage, including lores. **Juvenile** Presumably very similar to White-gorgeted. **VOICE** Very similar to White-gorgeted, though may not give the structured, descending song-notes. **HABITAT & BEHAVIOUR** Broadleaved evergreen forest, bamboo; 760–1,400 m, locally down to 395 m. Behaves like White-gorgeted. **RANGE & STATUS** Resident Sumatra. **SE Asia** Uncommon to locally common resident Tenasserim, W,S Thailand, Peninsular Malaysia, east S Laos, S Annam. **BREEDING** February–July. **Nest** Dome, on ground or in recess in bank. **Eggs** 2–3; white with light reddish-brown speckles, often forming dense ring around broader end; c.19 x 14 mm.

965 **SNOWY-BROWED FLYCATCHER**
Ficedula hyperythra Plate **97**
IDENTIFICATION 11–13 cm. *F.h.hyperythra*: **Male** Small and relatively short-tailed, with dark slaty-blue upperparts, orange-rufous throat and breast and distinctive white eyebrow and white patch at base of outertail-feathers. Male *Cyornis* flycatchers are larger and longer-tailed, with no white on head or tail; Slaty-blue Flycatcher is similar but larger and longer-tailed, with no white eyebrow. See Pygmy Blue Flycatcher and Indian Blue Robin. **Female** Resembles several other flycatchers and best identified by combination of small size, shortish tail, relatively uniform plumage tone, distinctly buffy loral stripe, eyering and underparts (paler on throat and vent) and rufescent-tinged wings, without obvious markings. Upperparts rather cold greyish olive-brown. See Lesser Shortwing. *F.h.annamensis* (Cambodia, S Annam) has crown washed bluish-grey and is much more orange-buff below. **Juvenile** Upperparts darker than female, with broad blackish feather-fringes and warm buff streaks/speckles; throat and breast heavily streaked/scalloped blackish, fading on throat-centre, belly and vent. Male has slaty-blue tail. **Other subspecies in SE Asia** *F.h.sumatrana* (Peninsular Malaysia). **VOICE** Song consists of fairly well-structured but subdued, thin, high-pitched, wheezy phrases: ***tsit-sit-si-sii***; ***tsi-sii-swrri*** and ***tsi sii'i*** etc. Calls include thin ***sip*** notes. **HABITAT** Broadleaved evergreen forest; 1,000–2,750 m, locally down to 400 m in winter. **RANGE & STATUS** Resident (subject to some movements) N,NE Indian subcontinent, SW,S China, Taiwan, Sundas, Philippines, Wallacea. **SE Asia** Fairly common resident (subject to some movements) Myanmar (except SW), NW,NE Thailand, Peninsular Malaysia, south-west Cambodia, Laos, W,E Tonkin, S Annam. Also recorded on passage E Tonkin. **BREEDING** March–June. **Nest** Small deep cup, in hole in bank or tree, amongst rocks, exposed tree-roots or hanging moss; up to 12 m (usually low down). **Eggs** 4–5; pale yellowish-grey to deep pinkish-red, freckled reddish-brown (sometimes ringed or capped at broader end); 17.5 x 13.8 mm (av.).

966 **RUFOUS-CHESTED FLYCATCHER**
Ficedula dumetoria Plate **97**
IDENTIFICATION 11.5–12 cm. *F.d.muelleri*: **Male** Resembles Mugimaki Flycatcher but smaller, shorter-winged and shorter-tailed, upperparts blacker, white supercilium extends in front of eye, has long white streak across wing-coverts, all-dark tertials, pale buffy-rufous throat and paler orange-rufous breast. Note behaviour and range. **Female** From Mugimaki by size and proportions, paler throat (contrasts with breast), warm buff loral line and eyering and rusty-buffish tips to wing-coverts and edges to tertials. **Juvenile** Poorly documented but certainly speckled with warm buff. Male probably shows some adult characters at early age. **VOICE** Song phrases are very thin and wispy: ***sii'wi-sii***; ***si-wi-si-ii***; and ***si-wi-oo*** etc. **HABITAT** Broadleaved evergreen forest; up to 825 m. Inhabits low vegetation, often near streams. **RANGE & STATUS** Resident Sundas. **SE Asia** Scarce to uncommon resident S Thailand, Peninsular Malaysia. **BREEDING** January–August. **Nest** Long and globular, with rough 'tail' and large lateral entrance, suspended among dead palm leaflets hanging below frond, 1 m above ground. **Eggs** (W Java) 2; pale blue.

967 **LITTLE PIED FLYCATCHER**
Ficedula westermanni Plate **96**
IDENTIFICATION 11–12.5 cm. *F.w.australorientis*: **Male** Unmistakable. The only black and white flycatcher in region. See Bar-winged Flycatcher-shrike. **Female** Distinctive. Upperside distinctly greyish with contrasting rufescent uppertail-coverts, underparts white with greyish wash on breast-sides and flanks. Has thin pale wing-bar on greater coverts. See Ultramarine, Brown and Slaty-backed Flycatchers. *F.w.westermanni* (S Thailand, Peninsular Malaysia) has distinctly slaty-grey upperparts. **Juvenile male** Like juvenile female but wings and tail blacker, with similar white markings to adult (though reduced). **Juvenile female** Upperparts darker than adult with broad blackish feather-fringes and buff spotting, wing-coverts tipped buff, underparts whitish with only light blackish scaling. **Other subspecies in SE Asia** *F.w.langbianis* (Cambodia, S Laos, S Annam). **VOICE** Song is thin, sweet and high-pitched, often followed by a rattled call-note. Typical call is a sharp ***swit*** followed by a rattling ***trrrrt.*** Also tinkling ***swee-swee-swee...*** **HABITAT** Upper storey to canopy of broadleaved evergreen and pine forest; 600–2,565 m, locally down to 200 m in winter, only above 1,065 m in Peninsular Malaysia. **RANGE & STATUS** Resident, subject to relatively local movements, Nepal, NE Indian subcontinent, SW,S China, Sundas, Philippines, Wallacea. Some Himalayan populations winter irregularly to south. **SE Asia** Common resident (except C,SE Thailand, Singapore, E Tonkin, Cochinchina). **BREEDING** February–July. **Nest** Cup, nestled in moss or epiphytes in tree, in hole in bank or amongst stones on ground; up to 15 m. **Eggs** 3–4; creamy to warm buff, liberally specked pale brown to dark reddish-brown; 15.1 x 11.9 mm (av.; *australorientis*).

968 **ULTRAMARINE FLYCATCHER**
Ficedula superciliaris Plate **96**
IDENTIFICATION 12 cm. *F.s.aestigma*: **Male** Deep blue upperparts, dark blue sides of head and broad dark blue lateral breast-patches, contrasting sharply with white remainder

of underparts, diagnostic. See Slaty-blue Flycatcher. **Female** Very similar to Little Pied Flycatcher but larger, with no rufous on uppertail-coverts, has distinctly brownish-grey to grey sides of throat and breast (mirroring pattern of male); may show a little blue on uppertail-coverts and tail. **Juvenile male** Like juvenile female but shows blue on wings and tail. **Juvenile female** Darker and browner above than adult, with blackish feather-tips and buff spotting, underparts white to buffy-white with clear blackish scaling (vent plainer). **First-winter/first-summer male** Resembles female but scapulars and back to uppertail-coverts mostly blue, has extensive blue on wings, tail and mantle and broad buff greater covert bar and tertial fringes. **VOICE** Song is quite feeble, high-pitched and rather disjointed: ***tseep-te-e-te-e-te-e te-tih tseep tse-e-ep...*** etc. Calls include a low rattled ***trrrrt*** (slower and deeper than Little Pied) and ***chi trrrrt*** with a squeaky introductory note. **HABITAT** Open broadleaved evergreen, deciduous and pine forest; 915–1,700 m, locally down to 150 m in winter. Breeds above 1,500 m in NE India. **RANGE & STATUS** Breeds N Pakistan, NW,N,NE Indian subcontinent, W/SW China; some winter south to southern India, Bangladesh, SW China. **SE Asia** Scarce to uncommon winter visitor NW Thailand. Recorded (status uncertain) C,E Myanmar. **BREEDING** NE India: April–July. **Nest** Cup, in hole in tree or bank; up to 7 m above ground. **Eggs** 3–5; olive to dull stone-buff, densely freckled reddish-brown (sometimes capped at broader end); 16 x 12.2 mm (av.).

969 SLATY-BLUE FLYCATCHER

Ficedula tricolor Plate **97**

IDENTIFICATION 12.5–13 cm. *F.t.diversa* (Thailand, Indochina): **Male** Distinguished by dark slaty-blue upperside, blackish tail with prominent white at base of outer feathers and whitish to buffy-white throat, contrasting with blue-black sides of head and buffy blue-grey breast-band. Rest of underparts buffish to buffy-whitish with greyish wash on flanks, paler blue forehead and eyebrow. *F.t.cerviniventris* (W Myanmar) has pale of underparts all warm buff. See Slaty-backed Flycatcher. Identity of other Myanmar populations unknown. **Female** Nondescript. Like Slaty-backed but shows somewhat darker, warmer upperparts, distinctly rufous-chestnut tail (browner at tip) and distinctly buffish narrow loral line, eyering and underparts (throat and sometimes belly-centre slightly paler). *F.t.cerviniventris* has uniform rich buff underparts. See Snowy-browed Flycatcher. Note range. **Juvenile** Upperparts blackish-brown with large warm buff spots, underparts buffy with heavy blackish scales/streaks (vent plainer), wings tending to be warmer than adult female. **First-winter male** Like adult female. **VOICE** Song consists of 3–4 high-pitched notes, the first drawn out, the second short and emphasised, the rest low and more trilled: ***chreet-chrr-whit-it*** etc. Typical calls are a series of sharp ***tic*** notes and rolling ***trrri trrri trrri...*** **HABITAT & BEHAVIOUR** Secondary growth, scrub and grass, bamboo; 1,500–2,750 m, down to 450 m in winter. Skulks in often dense, low vegetation. Adopts distinctly horizontal posture and often cocks tail. **RANGE & STATUS** Resident (subject to relatively local movements) N Pakistan, NW,N,NE Indian subcontinent, SE Tibet, SW,W China. **SE Asia** Uncommon to fairly common resident (subject to some movements) W,N,S(east),E Myanmar, W Tonkin. Scarce to local winter visitor NW Thailand, N Laos, E Tonkin. Also recorded on passage E Tonkin. **BREEDING** India: May–July. **Nest** Small cup, in hole in bank or tree or amongst boulders, sometimes against tree-trunk; up to 2 m (rarely 6 m) above ground. **Eggs** 3–4; pale pinkish-cream, densely but minutely speckled pinkish-red (sometimes ringed or capped at broader end); 15.8 x 12 mm (av.; *cerviniventris*).

970 TAIGA FLYCATCHER

Ficedula albicilla Plate **97**

IDENTIFICATION 13–13.5 cm. Monotypic. **Male non-breeding** Widespread and common in winter. Nondescript greyish-brown upperside and whitish underparts with pale buffish-grey wash on breast, but distinctive blackish uppertail-coverts and tail with prominent white on basal half of outertail-feathers (very obvious in flight). Sides of head rather uniform apart from pale eyering, bill all blackish. See Asian Brown and Rufous-gorgeted Flycatchers. **Male breeding** Shows distinctive rufous-orange throat, surrounded by grey. **Female** Like male non-breeding. **First winter** Similar to female but has buffish tips to greater coverts and tertials. **VOICE** Song is a series of rhythmic notes, starting sharp and high-pitched but ending clear and descending: ***zri zri zri chee chee dee-cha dee-cha dee-cha chu chu chu tu tu tu tu too taa.*** Calls include soft rattling ***trrrt*** and less frequent dry clicking ***tek*** notes and a harsh ***zree*** or ***zee-it.*** **HABITAT & BEHAVIOUR** Open woodland, forest edge, secondary growth, plantations, parks, gardens; up to 2,135 m. Regularly cocks tail, often drops from perch to feed on ground. **RANGE & STATUS** Breeds C,E Palearctic (west to E European Russia); winters C,N,NE Indian subcontinent, S China, rarely Borneo, Philippines (Palawan). **SE Asia** Common winter visitor (except Peninsular Malaysia, Singapore). Vagrant Peninsular Malaysia. Also recorded on passage Cambodia, E Tonkin.

971 SAPPHIRE FLYCATCHER

Ficedula sapphira Plate **96**

IDENTIFICATION 11–11.5 cm. *F.s.laotiana*: **Male breeding** Resembles Ultramarine Flycatcher but blue of plumage much brighter and more vivid, shows orange-rufous centre of throat and breast. See *Cyornis* flycatchers. Note behaviour and range. **Male non-breeding** Resembles female but scapulars, back to uppertail-coverts, wings and tail as breeding male. **Female** Shows distinctive combination of small size, shortish tail, warm brown upperparts, strongly rufescent uppertail-coverts and distinctly deep buffish-orange throat and upper breast. Eyering and loral stripe buff, belly and vent white, wings and tail unmarked. See Mugimaki and Pygmy Blue Flycatchers and female *Cyornis* flycatchers. Note voice, habitat, behaviour and range. **Juvenile** Upperparts blackish-brown prominently spotted rich buff, wing-coverts tipped with buff spots, throat and upper breast buffier than female, finely scaled blackish from throat to belly (mostly on breast). **First-winter male** Like non-breeding adult but shows buff greater covert bar and pale tertial tips. **Other subspecies in SE Asia** *F.s.sapphira* (Myanmar). **VOICE** Call consists of short hard rattles, sometimes introduced by high thin note/s: ***tssyi tchrrrt*** and ***tchrrrt tchrrrt tchrrrt...*** etc. **HABITAT & BEHAVIOUR** More open broadleaved evergreen forest; 1,200–2,565 m. Breeds above 1,400 m (mainly 2,100 m) in NE India. Usually found rather high up in trees. **RANGE & STATUS** Resident (subject to minor movements) Nepal, NE Indian subcontinent, SW,W,C China. **SE Asia** Recorded (status uncertain but probably resident) N,S(east),E Myanmar, N,C Laos, W Tonkin. Scarce winter visitor NW Thailand. **BREEDING** March–May. **Nest** Cup, in hole in tree or bank. **Eggs** 4; pale yellowish-grey to warm buff, faintly or densely stippled reddish-brown; average dimensions 15.4 x 11.8 mm (av.; *sapphira*).

972 PYGMY BLUE FLYCATCHER

Muscicapella hodgsoni Plate **95**

IDENTIFICATION 9–9.5 cm. *M.h.hodgsoni*: **Male** Tiny size, short tail and small bill create flowerpecker-like appearance. Upperparts rather deep dark blue, with lighter, brighter forecrown, underparts rather uniform buffy rufous-orange (vent may be slightly paler). See Slaty-backed and Snowy-browed Flycatchers. **Female** Resembles Sapphire Flycatcher but smaller and shorter-tailed, upperparts warmer, particularly back to uppertail, underparts rather uniform pale rufescent-buff (sometimes whiter on vent). See Snowy-browed Flycatcher. *M.h.sondaica* (Peninsular Malaysia) appears to have paler underparts with whiter centre to abdomen. **Juvenile** Undescribed but certainly has paler speckling at least on crown and sides of head. **VOICE** Song is a weak, thin, high-pitched ***sii-su'u-siiii***, with quavering last note. Calls with subdued, weak ***tup*** or ***tip*** notes. **HABITAT & BEHAV-**

IOUR Broadleaved evergreen forest; 610–2,565 m. Usually inhabits middle to upper storey. Often flicks wings and cocks tail. **RANGE & STATUS** Resident Nepal, Bhutan, NE India, SW China (Yunnan), Sumatra, Borneo. **SE Asia** Uncommon resident W,N Myanmar, extreme S Thailand, Peninsular Malaysia, Laos, W Tonkin, N Annam. Recorded in winter (status uncertain; may breed) E Myanmar, W,NW Thailand. **BREEDING** February–June. **Nest** Small cup, in moss suspended from liana, in decayed root-ball of dead fern suspended upside-down on thin vine, or in creepers overgrowing stump; 3.5–6 m above ground. Otherwise undocumented.

973 FERRUGINOUS FLYCATCHER

Muscicapa ferruginea Plate **98**

IDENTIFICATION 12.5–13 cm. Monotypic. **Adult (fresh)** Recalls Dark-sided Flycatcher but shows distinctive slaty-grey cast to head, strongly rufescent rump, uppertail-coverts and tail, rusty-rufous fringes to greater coverts and tertials and rusty-buff breast and flanks (former mixed with brown). **Adult (worn)** Duller and greyer on breast and flanks; plainer-winged, lores and eyering possibly whiter. **Juvenile** Crown blackish with bold broad buff streaks, mantle to upper rump and scapulars blackish, boldly spotted/mottled rich buff, throat-sides and breast scaled/streaked blackish. **VOICE** Probable song consists of very high-pitched silvery notes introduced by a short sharp harsher note: ***tsit-tittu-tittu*** and ***tsit tittu-tittu tsit tittu-tittu*** etc. Calls include a short, sharp, high-pitched ***tssit-tssit*** and ***tssit tssit tssit..*** **HABITAT** Broadleaved evergreen forest, forest edge; up to 2,135 m. Breeds above 1,200 m in NE India. **RANGE & STATUS** Breeds Nepal, NE Indian subcontinent, SW,W China, Taiwan; winters Greater Sundas, Philippines. **SE Asia** Uncommon to fairly common breeding visitor W Tonkin. Scarce to uncommon winter visitor S Thailand, Peninsular Malaysia, S Annam, Cochinchina. Scarce to uncommon passage migrant E,S Myanmar, Thailand, Peninsular Malaysia, N,C Laos, C Annam. Vagrant Singapore, E Tonkin. Recorded (status uncertain) W,N Myanmar. **BREEDING** India: June–July. **Nest** Neat cup, on tree-branch or projection; 3–15 m above ground. **Eggs** 2–3; very similar to Dark-sided Flycatcher; 17.9 x 13.6 mm (av.).

974 RUFOUS-GORGETED FLYCATCHER

Muscicapa strophiata Plate **98**

IDENTIFICATION 13–14.5 cm. *M.s.strophiata*: **Male** Shows diagnostic combination of blackish-grey face and throat, short whitish eyebrow, slaty-grey breast with orange-rufous patch on upper centre (sometimes lacking) and blackish tail with white at base of outer feathers. Upperparts dark warm-tinged brown. *M.s.fuscogularis* (S Annam) has warmer upperparts, much larger gorget and slaty-grey rest of throat, lores and sides of head. **Female** Usually duller than male, with paler, grey throat and breast (lacking black), less distinct white eyebrow, smaller gorget, buffish chin. *M.s.fuscogularis* has warmer upperparts, much larger gorget and grey chin. **Juvenile** Crown and upperparts warmer brown than female with broad blackish feather-tips, crown streaked rich buff, rest of upperparts and median coverts spotted/streaked rich buff, throat, breast and belly buff heavily scaled/streaked blackish (fading to whiter on vent). **VOICE** Song is a thin, well-spaced, rather jolly ***zwi chir rri*** etc., with sharp first note. Calls include a rather deep, chat-like ***tchuk-tchuk-tchuk***, harsh ***trrt*** and sharp metallic ***zwi***. **HABITAT** Broadleaved evergreen forest; 1,500–3,050 m, down to 700 m in winter. **RANGE & STATUS** Resident (subject to minor movements) NW,N,NE India, Nepal, Bhutan, S,SE Tibet, SW,W,C China. Winter visitor Bangladesh, S China. **SE Asia** Fairly common to common resident W,N Myanmar, east S Laos, C,S Annam; probably locally in other northern areas. Uncommon to fairly common winter visitor C,E,S(east) Myanmar, Tenasserim, NW,NE Thailand, N Laos, W,E Tonkin. **BREEDING** March–June. **Nest** Cup, on sloping ground or in hole in bank or tree; up to 6 m above ground. **Eggs** 3–4; whitish, sometimes with faint brown ring around broader end; 18.1 x 13.5 mm (av.).

975 GREY-STREAKED FLYCATCHER

Muscicapa griseisticta Plate **98**

IDENTIFICATION 15 cm. Monotypic. **Adult** Resembles Dark-sided and Brown-streaked Flycatchers but larger, underparts clean whitish with extensive, well-defined broad dark greyish streaks across entire breast and flanks (often extending to belly), undertail-coverts always whitish. **First winter** Similar to adult. **HABITAT** Open forest, plantations, gardens; recorded in lowlands. **RANGE & STATUS** Breeds E Palearctic, NE China, N Korea; winters N Borneo, Philippines, Sulawesi, E Lesser Sundas, Moluccas, Palau Is, W New Guinea. **SE Asia** Vagrant Singapore, C Annam, S Annam, Cochinchina.

976 DARK-SIDED FLYCATCHER

Muscicapa sibirica Plate **98**

IDENTIFICATION 11.5–13 cm. *M.s.rothschildi* (Breeding range, apart from W Myanmar?; visitor at least to W,NW,S Thailand, Peninsular Malaysia; ? C,S,E Myanmar, Tenasserim [or *cacabata*]): **Adult** Similar to Asian Brown Flycatcher but has smaller, mostly dark bill and smudgy greyish-brown breast and flanks with ill-defined darker streaking and variable whitish line down centre of abdomen; upperparts darker and browner, whitish submoustachial stripe mottled darker, undertail-coverts dark-centred, wing-tips extending further towards tail-tip. *M.s.sibirica* (widespread visitor, except Myanmar) has more extensively white centre of abdomen and reduced dark centres to undertail-coverts; breast appears more distinctly streaked; worn individuals or first winters are more likely to be confused with darker individuals of Asian Brown. See Grey-streaked and Brown-streaked Flycatchers. Note range. **Juvenile** Blacker above with distinct pale buff spots/streaks, underparts whitish, boldly scaled/streaked blackish on throat-sides, breast and flanks (fading on belly), median and lesser coverts spotted warm buff, greater coverts and tertials fringed warm buff, tail narrowly tipped warm buff. **Other subspecies in SE Asia** *M.s.cacabata* (W Myanmar?). **VOICE** Song is a weak, subdued series of thin, high-pitched phrases, including a series of sibilant ***tsee*** notes, followed by quite melodious trills and whistles. **HABITAT** More open broadleaved evergreen, rhododendron, coniferous and mixed deciduous forest, secondary growth; also plantations, gardens and mangroves on migration; up to 3,660 m (breeds above 2,135 m). **RANGE & STATUS** Breeds E Palearctic, N Pakistan, NW,N,NE Indian subcontinent, S,E Tibet, SW,W,NE China, N Korea, Japan; northerly populations winter Bangladesh, S China, locally Greater Sundas, Philippines (Palawan). **SE Asia** Locally fairly common resident (subject to some movements) W,N Myanmar. Recorded (probably breeds) E Myanmar, W,E Tonkin. Uncommon to fairly common winter visitor (except SW,W,N Myanmar, N,C Laos, W,E Tonkin, N Annam). Uncommon to fairly common passage migrant W,C,S Thailand, Peninsular Malaysia, Cambodia, Laos, E Tonkin, N Annam. Recorded (status uncertain) SW Myanmar. **BREEDING** May–July. **Nest** Compact cup, on tree-branch, sometimes in hole in tree; 2–18 m above ground. **Eggs** 3–4; pale green, densely freckled pale reddish (mainly at broader end); 17.1 x 12.2 mm (av.; *cacabata*).

977 ASIAN BROWN FLYCATCHER

Muscicapa dauurica Plate **98**

IDENTIFICATION 12.5–13.5 cm. *M.d.dauurica*: **Adult** Widespread and common in winter. Upperparts plain brownish-grey, underparts (including undertail-coverts) whitish, variably suffused brownish-grey on upper breast (mainly sides) and usually lacking defined streaks (occasionally shows very vague streaking); has pale basal half of lower mandible (viewed side-on), whitish eyering and broad loral stripe, short whitish submoustachial stripe and pale greyish fringes to wing-coverts and tertials. When worn, appears greyer above and paler below; when fresh, has indistinct dull rufous edges and tips to wing-coverts and tertials. See Dark-sided, Grey-streaked and Taiga Flycatchers. *M.d.siamensis* (resident form; subject to some movements, e.g. C Thailand) has browner upperparts and wings, duller, more uniform underparts, paler bill with dusky-tipped dull yellowish lower mandible and less distinct pale eyering and lores. See Brown-streaked Flycatcher. **Juvenile** Similar to Dark-sided but has bolder, paler spotting on scapulars, whiter underparts with fine dark scaling on breast, and whiter markings on wing-coverts and tertials. **First winter** Similar to adult but has paler tips to greater coverts and paler tertial fringes. Fresh *siamensis* has broad warm buff wing-fringing and tail-tip, breast vaguely mottled, not distinctly streaked as in Brown-streaked, and overall plumage tones less rufescent-tinged than that species. **VOICE** Song is similar to Dark-sided but louder, consisting of short trills interspersed with 2–3 note whistled phrases. Calls include a thin sharp ***tse-ti-ti-ti-ti*** or ***sit-it-it-it*** and short thin ***tzi***. **HABITAT** Open forest, secondary growth, plantations, parks, gardens, mangroves; breeds in open forest (including deciduous and dry dipterocarp woodland) inland. Up to 1,585 m, breeds 600–1,400 m. **RANGE & STATUS** Resident SW India, Sumatra, N Borneo, W Lesser Sundas (Sumba). Breeds E Palearctic, NE Pakistan, NW,N India, Nepal, Bhutan, N,NE China, N Korea, Japan; winters C,E,S India, Sri Lanka, S China, Greater Sundas, Philippines. **SE Asia** Scarce to uncommon resident north Tenasserim, W,NW Thailand, S Annam. Common winter visitor (except N,E Myanmar). Also recorded on passage C,S Thailand, Peninsular Malaysia, Singapore, Cambodia, Laos, E Tonkin. **BREEDING** March–June. **Nest** Compact cup, on tree-branch; 2–9 m above ground. **Eggs** (Himalayas): 2–4; pale olive-grey, heavily stippled sienna-brown; 17.4 x 13.1 mm (av.).

978 BROWN-STREAKED FLYCATCHER
Muscicapa williamsoni Plate **98**

IDENTIFICATION 13 cm. Monotypic. **Adult** Very similar to Asian Brown Flycatcher. Much browner above than *M.d.dauurica*; always shows warm tinge, at least on lower upperparts; bill slightly larger with pale yellowish (usually dark-tipped) lower mandible, loral stripe buffier and less distinct, fringes of greater coverts and tertials tend to be more buffish; has variable but usually distinct brownish streaks on breast, upper belly and flanks. Worn (breeding) birds have vaguer underpart-streaking and plainer head and wing pattern. Apart from underparts, could easily be confused with resident subspecies of Asian Brown (*M.d.siamensis*). **Juvenile** Very similar to Asian Brown. Note range. **First winter** Fresh birds (August–early November) are very distinct, with broad buffy/rusty-white to orange-buff wing-fringing, distinctly rusty-tinged lower rump and uppertail-coverts, and relatively heavy streaks below; warm-tinged lower underparts. **VOICE** Song is apparently similar to Asian Brown. Calls include a thin sharp ***tzi*** and harsh slurred ***cheititit***. **HABITAT** Open broadleaved evergreen and semi-evergreen forest, clearings, also parks and wooded gardens on migration; up to 1,295 m. **RANGE & STATUS** Recorded (status uncertain) Sumatra. **SE Asia** Scarce to uncommon resident (subject to some movements) S Myanmar, S Thailand, north-west Peninsular Malaysia, northern and eastern Cambodia, Cochinchina. Scarce to uncommon winter visitor Peninsular Malaysia, Singapore (rare). Scarce passage migrant C,W(coastal) Thailand. **BREEDING** March–June. **Nest** Neat cup, on tree-branch; 6-18 m above ground. **Eggs** <3. Otherwise undocumented.

979 BROWN-BREASTED FLYCATCHER
Muscicapa muttui Plate **98**

IDENTIFICATION 13–14 cm. Monotypic. **Adult (worn)** Similar to Asian Brown Flycatcher but larger, bill bigger with uniformly pale yellowish lower mandible, has colder, darker, greyish-brown crown contrasting sharply with broad whitish eyering and loral patch, rufescent-tinged remainder of upperparts, dark greyish-brown ear-coverts connecting with dark malar (distinctly enclosing white submoustachial patch) and warm greyish-brown breast and flanks. **Adult (fresh)** In autumn, has more distinctly rufescent upperparts, richer and darker breast and flanks and deep buff fringes to greater coverts, tertials and secondaries. See Brown-streaked and Ferruginous Flycatchers and Fulvous-chested Jungle-flycatcher. Note range. **Juvenile** Undocumented? **VOICE** Song is said to be pleasant but feeble. Typical call is a thin ***sit***. **HABITAT** Broadleaved evergreen forest; 560–1,645 m. **RANGE & STATUS** Breeds NE India, SW,W,S China; winters SW India, Sri Lanka. **SE Asia** Scarce breeding visitor NW Thailand, W,E Tonkin. Scarce to uncommon passage migrant W,C Myanmar. Recorded (status uncertain, probably breeds) N,E Myanmar, N Laos. **BREEDING** May–June. **Nest** Compact cup, on tree-branch; 6–7.5 m above ground. **Eggs** 4–5; pinkish-brown to grey-blue, closely stippled reddish-brown; 16.9 x 13.8 mm (av.).

980 BLUE-AND-WHITE FLYCATCHER
Cyanoptila cyanomelana Plate **96**

IDENTIFICATION 18 cm. *C.c.cyanomelana*: **Male** Told by rather large size, azure to cobalt-blue upperside, blackish head-sides, throat and breast and contrasting white remainder of underparts and patch on base of outertail-feathers. *C.c.cumatilis* (widespread and perhaps more frequent) tends to be more turquoise-blue above (particularly crown and tail) and has bluer head-sides, throat and breast (sometimes strongly turquoise); differences somewhat clouded by individual variation. See White-tailed Flycatcher. **Female** Resembles White-tailed, some niltavas and Pale Blue but shows distinctive demarcated white belly and vent, has no blue on crown and neck-side or white on tail, throat and breast rather uniform pale brownish, usually with buffish to whitish vertical or horizontal patch on lower throat. Often appears greyer in spring, particularly crown, mantle and breast. **First-winter male** Resembles female but wings, tail, scapulars and back to uppertail-coverts similar to adult male. **VOICE** Subdued ***tic*** and ***tac*** notes. **HABITAT** Open broadleaved evergreen forest, island forest, plantations, parks, wooded gardens; up to 1,830 m (mostly lowlands). **RANGE & STATUS** Breeds south-east E Palearctic, NE China, N,S Korea, Japan; winters Greater Sundas, Philippines. **SE Asia** Scarce to uncommon passage migrant south Tenasserim, Thailand, Peninsular Malaysia (some winter), Singapore, Indochina (except N Annam; may occasionally winter S Annam).

981 BROWN-CHESTED JUNGLE-FLYCATCHER *Rhinomyias brunneata* Plate 98

IDENTIFICATION 15 cm. *R.b.brunneata*: **Adult** Very similar to other *Rhinomyias* flycatchers but shows diagnostic rather long, stout bill with pale yellow lower mandible and faint dark mottling/flecking on whitish throat. Upper breast dull brownish. See Brown-breasted Flycatcher. **First winter** Has dark tip to lower mandible and rufous-buff tips to greater coverts and tertials. **HABITAT** Broadleaved evergreen and mixed deciduous forest; up to 395 m. **RANGE & STATUS** Breeds S,SE China; winters to south. Vagrant N Borneo (Brunei). **SE Asia** Rare to scarce winter visitor S Thailand, Peninsular Malaysia, Singapore. Scarce passage migrant W,NW,C,S Thailand, Peninsular Malaysia, E Tonkin.

982 FULVOUS-CHESTED JUNGLE-FLYCATCHER *Rhinomyias olivacea* Plate 98

IDENTIFICATION 15 cm. *R.o.olivacea*: **Adult** Resembles other *Rhinomyias* and some female *Cyornis* flycatchers but shows distinctive combination of all-dark bill, pinkish legs and feet, unmarked whitish throat and warm buffish-brown upper breast and wash on flanks. Upperside plain brownish with warm-tinged uppertail-coverts and outertail-feather fringes. See Chinese Blue and Brown-breasted Flycatchers. Note range. **Juvenile** Upperparts and wings warmer, upperpart feathers tipped blackish-brown and speckled buff, wing-coverts and tertials tipped rufous-buff, breast-band and flanks mottled buff and dark brown. **VOICE** Song recalls

Cyornis flycatchers but phrases rather short (1–2 s) and slurred, including scratchy notes. Calls include drawn-out ***churr*** or ***trrt*** and a harsh ***tac***. **HABITAT** Broadleaved evergreen forest; up to 885 m. **RANGE & STATUS** Resident Greater Sundas. **SE Asia** Scarce to locally fairly common resident south Tenasserim, S Thailand. **BREEDING** May–August. **Nest** Cup, in recess in tree; 2.1 m above ground. **Eggs** 2–3; glossy capucine-buff, spotted light purplish-grey and chestnut (mostly at broader end); c.23 x 16 mm.

983 GREY-CHESTED JUNGLE-FLYCATCHER

Rhinomyias umbratilis Plate **98**

IDENTIFICATION 15 cm. Monotypic. **Adult** From other *Rhinomyias* flycatchers by grey to olive-grey upper breast, contrasting with clean, gleaming white throat and dark malar area (darker than ear-coverts). See Moustached Babbler. **Juvenile** Similar to Fulvous-chested Jungle-flycatcher but darker above, with larger, more pronounced rich buff spots, breast-mottling washed greyish. **VOICE** Song is a thin, sweet, well-structured, somewhat descending ***si ti-tu-ti'-tooee'u*** (with sharp second note), ***sii tu'ee'oo***; ***sii tee'oo'ee*** etc. Richer and more varied than Fulvous-chested; recalls *Cyornis* flycatchers. Alarm calls are a scolding ***chrrr-chrrr-chrrr*** and ***trrrt'it'it'it*** etc. **HABITAT** Broadleaved evergreen forest; up to 1,160 m. **RANGE & STATUS** Resident Sumatra, Borneo. **SE Asia** Scarce to fairly common resident extreme S Thailand, Peninsular Malaysia. **BREEDING** January–July. **Nest** Cup-shaped, situated inside dead, curled-up leaf, suspended vertically from vine; 2 m above ground. **Eggs** (Borneo) 2–3; brownish-white (tinged bluish), thickly but faintly spotted reddish-brown; c.23 x 15 mm.

984 ORIENTAL MAGPIE-ROBIN

Copsychus saularis Plate **91**

IDENTIFICATION 19–21 cm. *C.s.erimelas*: **Male** Large and conspicuous with glossy blackish head, upperside and upper breast, white lower breast to vent, broad wing-stripe and outer-tail-feathers. **Female** Patterned like male but black body-plumage replaced by dark grey. **Juvenile male** Like juvenile female but upperparts somewhat darker; soon shows some glossy black feathers. **Juvenile female** Duller above than adult, flight-feathers fringed brown, white on wing washed warm buff and scaled blackish on wing-coverts; throat and breast paler and more buffish with dark greyish scales (breaking up on lower breast and flanks). **Other subspecies in SE Asia** *C.s.musicus* (south Tenasserim and S Thailand southwards). **VOICE** Has varied musical warbling song, alternated with churrs and sliding whistles: ***si-or***; ***sui-i*** (upward-inflected) and ***su-u*** (lower-pitched) etc. Calls include a clear rising whistle and harsh rasping ***che'e'e'eh*** when alarmed. **HABITAT & BEHAVIOUR** Gardens, cultivated and urban areas, open woodland, mangroves, secondary growth; up to 1,830 m (mostly below 1,000 m). Conspicuous and confiding. Cocks tail sharply. **RANGE & STATUS** Resident Indian subcontinent (local Pakistan), C,E and southern China, Greater Sundas, Philippines. **SE Asia** Common resident throughout. **BREEDING** January–September. Multi-brooded. **Nest** Variable cup, in hole in tree, bank or building or in palm, bamboo clump etc.; 1–18 m above ground. **Eggs** 3–5; fairly glossy, greenish, streaked and mottled brownish-red over purplish-grey undermarkings (usually more densely marked at broader end); 22.1 x 16.8 mm (av.).

985 WHITE-RUMPED SHAMA

Copsychus malabaricus Plate **91**

IDENTIFICATION 21.5–28 cm (male tail up to 7 cm longer than females). *C.m.interpositus*: **Male** Very striking, with conspicuous white rump and uppertail-coverts and long blackish tail with white outer feathers. Head, upper breast and upperparts glossy blue-black, rest of underparts deep orange-rufous. **Female** Similarly patterned to male but blue-black replaced by dark greyish, underparts duller and paler rufous, tail shorter. **Juvenile** Head-sides and upperparts paler and browner than female with buff streaks and speckles, some brownish feather-fringes on rump and uppertail-coverts (initially lacks white), rich buff tips to wing-coverts and broad buff fringes to primaries, secondaries and tertials; throat and breast initially buffish, mottled/scalloped dark greyish. **Other subspecies in SE Asia** *C.m.pellogynus* (Tenasserim, S Thailand), *mallopercnus* (Peninsular Malaysia, Singapore). **VOICE** Song is highly variable, loud and distinctly rich and melodious, incorporating skilled avian mimicry. Calls include a harsh ***tschack***. **HABITAT & BEHAVIOUR** Broadleaved evergreen and mixed deciduous forest, secondary growth, bamboo; up to 1,525 m. Rather skulking. **RANGE & STATUS** Resident Indian subcontinent (except Pakistan, W,NW, and Andaman Is), SW,S China, Greater Sundas. **SE Asia** Fairly common to common resident throughout (scarce Singapore). **BREEDING** March–September. **Nest** Slight cup or pad, in tree-hole or base of bamboo clump; within 2 m of ground. **Eggs** 4–5; greenish to pale bluish-green, densely streaked and spotted brownish-red to umber-brown; 21.6 x 16.5 mm (av.).

986 RUFOUS-TAILED SHAMA

Trichixos pyrropyga Plate **91**

IDENTIFICATION 21–22.5 cm. Monotypic: **Male** Resembles female White-rumped Shama but shows diagnostic bright rufous rump, uppertail-coverts and basal two-thirds of tail; tail shorter and squarer, has small white mark above/in front of eye and paler-centred underparts. **Female** Head-sides and upperparts grey-brown with no white marking above eye, throat and breast buffy-rufous and belly whitish. **Juvenile** Resembles female but crown and upperparts heavily streaked rich buff, uppertail-coverts and basal part of tail buffier, extreme tip rich buff; has rich buff tips to wing-coverts and spots on tertial tips, sides of head and throat/breast buff, broadly streaked sooty-blackish. **VOICE** Song is a series of loud well-spaced glissading whistles: ***whi-ii*** and ***whi-uuu*** etc. Calls with a scolding, drawn-out ***tcherr***. **HABITAT & BEHAVIOUR** Broadleaved evergreen forest, freshwater swamp forest; up to 915 m. Often sits motionless for long periods; cocks tail. **RANGE & STATUS** Resident Sumatra, Borneo. **SE Asia** Scarce to uncommon resident extreme S Thailand, Peninsular Malaysia. **BREEDING** January–April. **Nest** Reportedly in "hole" in Thailand. Otherwise undocumented.

PARIDAE: Typical tits

Worldwide c.56 species. SE Asia 10 species. Generally small and robust with stout bills and strikingly patterned plumage. Gregarious and very active, occurring in varying levels of forest, feed mainly on insects, also some seeds and small fruits.

987 BLACK-BIBBED TIT

Poecile hypermelaena Plate **99**

IDENTIFICATION 11.5 cm. Monotypic. **Adult** Small size, olive-tinged greyish-brown upperparts, pale underparts and black head with broad whitish patch from lores through ear-coverts to nape-sides distinctive. Resembles Coal Tit but upperparts browner, lacks crest and wing-bars. Note range and habitat. **Juvenile** Dark parts of head browner, upperparts warmer, less olive, whitish patch on head-side tinged yellowish-buff, bib smaller, underparts paler, washed yellowish-buff. **VOICE** Calls include a thin ***stip***; ***si-si*** and explosive ***psiup***, sometimes combined into ***si-si psiup*** etc. Also a chattering ***chrrrrrr*** and scolding ***chay***. **HABITAT** Open broadleaved evergreen and pine forest, forest edge, scrub; 2,200–3,000 m. **RANGE & STATUS** Resident SW,W,C China. **SE Asia** Local resident W Myanmar (Mt Victoria). **BREEDING** March–April. **Nest** Pad or cup, in tree-cavity; up to 8 m above ground. **Eggs** c.4–6; white, speckled purplish; c.17.9 x 13.1 mm. **NOTE** Sometimes lumped in extralimital Marsh Tit *P. palustris*.

988 GREY-CRESTED TIT

Lophophanes dichrous Plate **99**

IDENTIFICATION 12 cm. *L.d.wellsi*: **Adult** Easily told by greyish upright crest and pale buffish throat, half-collar and underparts. Rest of upperside mid-greyish. **Juvenile** Crest shorter, upperpart feathers faintly darker-tipped, underparts paler. **VOICE** Song is a simple ***whee-whee-tz-tz-tz***. Calls include a high-pitched ***zai***, rapid stuttering ***ti-di or ti-ti-ti-ti-ti***, quiet ***sip-pi-pi***, clear ***pee-di*** and lisping ***sip sip sip...*** Also a rapid ***cheea cheea*** when agitated. **HABITAT & BEHAVIOUR** Coniferous and broadleaved evergreen forest; 2,745–3,200 m. Often in small flocks, associating with mixed-species feeding flocks. **RANGE & STATUS** Resident NW,N,NE India, Nepal, Bhutan, SE,E Tibet, SW,W,C China. **SE Asia** Local resident N Myanmar. **BREEDING** India: April–June. **Nest** Pad, in hole in tree or tree-stump; 3–6 m above ground. **Eggs** 4–5; white, densely spotted reddish; 17.1 x 12.8 mm (av.; NW India).

989 COAL TIT *Periparus ater* Plate **99**

IDENTIFICATION 11 cm. *P.a.aemodius*: **Adult** Similar to Black-bibbed and Rufous-vented Tits but shows diagnostic combination of greyish upperparts, pinkish- to buffish-white underparts and two pronounced whitish bars on upperwing-coverts. Also shows whitish nape-patch and pointed crest like Rufous-vented but underparts much paler. **Juvenile** Dark plumage-parts duller and browner with fainter dull olive-grey bib, no obvious crest and pale nape-patch, whitish of head-sides washed yellowish, wing-bars duller. **VOICE** Song is a variably paced series of 2–8 ***di***- or trisyllabic notes: ***chip-pe chip-pe***; ***peechoo-peechoo-peechoo***; ***tu-wa-chi tu-wa-chi...***; ***chi-chi-chi***; ***pe-twi pe-twi...*** and slurred ***sit'tui-sit'tui-sit'-tui...*** etc. Calls include a clear ***pwi***, cheerful ***tsueet***, flat emphatic ***sui*** or ***chuu***, rather breathless piping ***sih***; all may be repeated in slow series or combined with slightly hoarse, buzzing ***pi*** notes, or followed by an explosive twitter. Thin ***sisisi*** and hoarse ***pih*** and ***szee*** notes when agitated. **HABITAT & BEHAVIOUR** Coniferous and mixed coniferous and broadleaved forest; 2,745–3,445 m. Occurs in flocks outside breeding season, often joining mixed-species feeding flocks. **RANGE & STATUS** Resident (subject to minor movements) N Africa, Palearctic, Iran, Nepal, Bhutan, NE India, S,SE Tibet, China, Taiwan, N,S Korea, Japan. **SE Asia** Locally common resident N Myanmar. **BREEDING** India: May–June. **Nest** Pad, in tree-hole or crevice. **Eggs** 4–10; white, speckled reddish-brown, occasionally unmarked; 17.9 x 12.9 mm (av.).

990 RUFOUS-VENTED TIT

Periparus rubidiventris Plate **99**

IDENTIFICATION 12 cm. *P.r.beavani*: **Adult** Greyish plumage, black head with pointed upright crest and whitish patches on side and nape, and rufous vent diagnostic. Underparts washed buff. *P.r.saramatii* (west N Myanmar [Mt Saramati]) has buffy-olive tinge or faint isabelline cast to upperparts and deep olive-grey underparts with drabber vent. Recalls Coal Tit but much darker below, lacks whitish wing-bars. **Juvenile** Black of head duller, crest shorter, upperparts duller and washed olive, underparts drabber greyish-buff (including vent), whitish head-patches washed yellowish to yellowish-buff. **VOICE** Song is a variable 13–34 note stony rattle, recalling Brown-throated Treecreeper: ***chi-chi-chi-chi...*** or ***chip-chip-chip-chip-chip-chip...*** etc.; sometimes a slower 3–6 note series or in less structured whistled song-types, combining pure and slurred notes at varying pitches. Calls include a thin ***seet*** and ***psset***, sharp ***psit***, clear ***pee***, sharp clicking ***chip***, full ***tip***, mellow ***pwit*** and sharp ***chit***, sometimes extended to a scolding ***chit'it'it'it***. Various calls may be combined. **HABITAT & BEHAVIOUR** Broadleaved evergreen and coniferous forest; 2,745–3,660 m. Occurs in small flocks outside breeding season, often joining mixed-species feeding flocks. **RANGE & STATUS** Resident N,NE Indian subcontinent, S,E Tibet, SW,W,C China. **SE Asia** Locally common resident N Myanmar. **BREEDING** India: April–May. **Nest** Pad, in existing tree-hole, sometimes amongst tree-roots on bank; up to 6 m above ground. **Eggs** 2–3; white, spotted reddish; 17.6 x 12.9 mm (av.).

991 GREY TIT *Parus cinereus* Plate **99**

IDENTIFICATION c.14 cm. *P.m.ambiguus* (S Thailand, Peninsular Malaysia): **Male** Relatively large with grey upperparts, black head and ventral stripe, large white patch on head-side, small patch on nape and single broad whitish wing-bar. Rest of underparts drab whitish, pale outer fringes to flight-feathers, and prominent white on outertail-feathers. See very similar Japanese Tit and smaller Coal Tit. Note range and habitat. **Female** Ventral stripe narrower, black parts of plumage may be duller, particularly bib and ventral stripe. **Juvenile** Similar to female but dark parts of head duller and browner, upperparts tinged olive, head-sides and underparts may be very faintly tinged yellowish, ventral stripe much reduced. **Other subspecies in SE Asia** *P.m.nipalensis* (SW,W,N,C,S Myanmar), *templorum* (NE Thailand, southern Laos, southern Vietnam). **VOICE** Has wide vocabulary. Sings with rapidly repeated combinations of 2–4 whistled notes: ***pi-chi-chew pi-chi-chew...***; ***wit wit wit chirr wit wit wit chirr...***; ***spih-tui spih-tui spih-tui...*** etc. Calls include sharper ***wit-wit-wit*** in flight. **HABITAT** Dry dipterocarp forest, pines amongst broadleaved evergreen and deciduous forest, mangroves, coastal scrub, casuarina groves, cultivation; up to 1,100 m. Strictly coastal in S Thailand and Peninsular Malaysia. **RANGE & STATUS** Resident (subject to some relatively minor movements) south-west Turkmenistan, NE Iran, Afghanistan, Indian subcontinent, south China (Hainan I), Greater Sundas, W Lesser Sundas. **SE Asia** Locally common resident SW,W,N,C,S Myanmar, NE,S Thailand, Peninsular Malaysia, Cambodia, C,S Laos, C(north to Quangtri at least),S Annam, Cochinchina. **BREEDING** December–July. **Nest** Pad, in hole in tree, bank or wall etc. **Eggs** 2–6 (2–3 in south); white, with small reddish-brown speckles; 17 x 13.3 mm (av.; *nipalensis*).

992 JAPANESE TIT *Parus minor* Plate **99**

IDENTIFICATION c.15 cm. *P.m.nubicolus* **Male** Like Grey Tit, but somewhat larger, with yellowish-green wash on upper mantle, blue-grey fringes to flight-feathers and more white on outertail-feathers. *P.m.commixtus* (northern Vietnam) has these features less well marked, and appears somewhat intermediate with Grey. See Green-backed Tit. Note range and habitat. **Female** Ventral stripe narrower, black parts of plumage may be duller, particularly bib and ventral stripe. **Juvenile** Similar to female but dark parts of head duller and browner, upperparts tinged olive, head-sides and underparts may be very faintly tinged yellowish, ventral stripe much reduced. **VOICE** In NW Thailand, repeated ***tit WI-TCHU***; ***ti TRIH-tu***; ***si WIT-chu*** etc.; mixed with fairly harsh ***tchrr'rr'rrt*** and ***wutchit'ti'trr*** etc. when highly agitated. Calls in E Myanmar include ringing ***(ter)TSCHINK (ter)TSCHINK***. Songs in China include 2-note ***si-pwi si-pwi si-pwi...*** **HABITAT** Open woodland, including oak and pine, pine stands in broadleaved evergreen forest, coastal scrub, casuarina groves, cultivation; up to 2,135 m. **RANGE & STATUS** Resident SE,E Tibet, China (except NW,N and Hainan I), N,S Korea, Amurland, Ussuriland, Sakhalin, Kuril Is, Japan. **SE Asia** Locally common resident N(south-east),E Myanmar, NW Thailand, N Laos, W,E Tonkin, N Annam (possibly extending down coast as far as Da Nang in C Annam); east N Myanmar? **BREEDING** March–July. **Nest** Pad, in hole in tree, bank or wall etc. **Eggs** 4–6; white, speckled and spotted reddish-brown and pale purplish (tending to form ring around broader end); 17.8 x 13.7 mm (av.; *nubicolis*).

993 GREEN-BACKED TIT

Parus monticolus Plate **99**

IDENTIFICATION 12.5–14.5 cm. *P.m.legendrei*: **Male** Superficially resembles Great Tit but upperparts dull greyish olive-green (rump grey), underparts pale yellow with very broad black ventral stripe, shows two white wing-bars.

P.m.yunnanensis (range except S Annam) has greener upperparts, yellower underparts with narrower black ventral stripe and slightly broader wing-bars. **Female** Bib and ventral stripe tend to be narrower and duller. **Juvenile** Similar to female but crown browner, white of head-sides duller, nuchal patch and wing-bars yellowish, bib and ventral stripe brownish, yellow of underparts a little duller. **VOICE** Song is a variable repeated series of 2–6 (often disyllabic) rising-and-falling notes: ***seta-seta-seta...***; ***tu-weeh tu-weeh...***; ***whit-ee whit-ee...***; ***ti-ti-tee-ti***; ***tsing-tsing pi-diu...*** and ***psit-psit tutu...*** etc.; sometimes a purer, even series of rising or falling notes: ***tee-tee-tee...*** and ***pli-pli-pli...*** etc. Calls include a rapid thin ***si-si-si-si-si-li***, harsh ***shick-shick-shick***, clear ***te-te-whee*** and ***pling pling pling tee-eurp***. **HABITAT** Broadleaved evergreen and pine forest, oak and alder forest; 915–2,650 m, locally down to 315 m in winter (N Myanmar); resident at 220–500 m in C Laos. **RANGE & STATUS** Resident N Pakistan, NW,N,NE Indian subcontinent, S,SE Tibet, SW,W,C China, Taiwan. **SE Asia** Common resident W,N Myanmar, C Laos, north E Tonkin, S Annam. **BREEDING** February–June. **Nest** Pad or cup, in hole or crevice in tree, bank or wall etc.; up to 7 m above ground. **Eggs** 4–8; white, heavily speckled and spotted reddish-brown and pale purplish (often ringed at broader end); 16.7 x 13.1 mm (av.; India).

994 **YELLOW-CHEEKED TIT**
Parus spilonotus Plate **99**

IDENTIFICATION 13.5–15.5 cm. *P.s.subviridis*: **Male** Combination of black crest (yellow at rear), yellow sides of head and underparts, with black post-ocular stripe, bib and ventral stripe, and broad whitish wing-bars diagnostic. Upperparts yellowish-olive with black scaling, flanks greyish-olive. *P.s.rex* (N Indochina) has blue-grey upperparts with black streaks, much broader black bib (joining post-ocular stripe) and ventral stripe and greyish remainder of underparts; *basileus* (S Indochina) is roughly intermediate. See Green-backed Tit. **Female** Less black on upperparts (more uniform), bib and ventral stripe olive-yellow (sometimes absent). *P.s.rex* and *basileus* have less yellow on underparts (greyer on former) and greyer throat and ventral band. **Juvenile** Crown and bib duller than respective adults, crest shorter, upperparts plainer, underparts paler yellow, wing-bars washed yellow. **Other subspecies in SE Asia** *P.s.spilonotus* (W,N Myanmar). **VOICE** Song is a ringing 3-note phrase, rapidly repeated 2–6 times: ***chee-chee-piu chee-chee-piu chee-chee-piu...*** and ***dzi-dzi-pu dzi-dzi-pu...*** etc. Calls with thin ***sit si-si-si***, lisping ***tsee-tsee-tsee*** and ***si-si-pudi-pudi*** and ***witch-a-witch-a-witch-a***, often combined with harsh ***churr'r'r'r'r***. **HABITAT & BEHAVIOUR** Broadleaved evergreen forest; 800–2,745 m, locally down to 600 m (E Tonkin). Often joins mixed-species feeding flocks. **RANGE & STATUS** Resident Nepal, Bhutan, NE India, SE Tibet, southern China. **SE Asia** Fairly common to common resident Myanmar (except SW,C), W,NW,NE Thailand, Laos, W,E Tonkin, N,C,S Annam. **BREEDING** January–June. **Nest** Pad, in hole in tree, sometimes bank or wall; up to 15 m above ground. **Eggs** 4–7; white with bold red to reddish-brown blotches and spots, sometimes with reddish-lilac undermarkings; 17.6 x 14.1 mm (av.; *subviridis*).

995 **SULTAN TIT** *Melanochlora sultanea* Plate **99**

IDENTIFICATION 20.5 cm. *M.s.sultanea*: **Male** Size, longish tail and black plumage with floppy-crested yellow crown and yellow lower breast to vent diagnostic. Upperparts glossed greenish-blue. *M.s.gayeti* (S Laos, C Annam) has glossy blue-black crest (both sexes). **Female** Upperparts browner and less glossy, washed oily-green, head-sides and wings browner, throat and upper breast duller, yellowish olive-green. **Juvenile** Similar to female but crest shorter, greater coverts finely tipped whitish, dark parts of plumage duller and glossless, throat and upper breast sooty olive-brown. **Other subspecies in SE Asia** *M.s.flavocristata* (S Myanmar, Tenasserim, W[south],NE[south-west],SE,S Thailand, Peninsular Malaysia). **VOICE** Song is a clear mellow ***piu-piu-piu-piu-piu...***. Typical calls are a stony, rattling ***chi-dip*** or ***tji-jup*** and fast shrill squeaky ***tria-tria-tria***; ***tcheery-tcheery-tcheery*** and ***squear-squear-squear*** etc. (both types often combined when agitated). Also a quiet, rather squeaky ***whit*** or ***quit*** and loud abrupt ***vheet***. **HABITAT & BEHAVIOUR** Broadleaved evergreen, semi-evergreen and mixed deciduous forest; up to 1,680 m (below 1,200 m in Thailand and Peninsular Malaysia). Usually in small flocks, high in trees. **RANGE & STATUS** Resident Nepal, Bhutan, NE India, SW,S,SE China; formerly Bangladesh. **SE Asia** Uncommon to locally common resident (except C Thailand, Singapore, Cambodia, southern S Annam, Cochinchina). **BREEDING** March–July. **Nest** Thick pad, in tree-cavity; up to 15 m above ground. **Eggs** 5–7; white, liberally speckled reddish-brown over pinkish-grey undermarkings; 21.7 x 16.5 (av.; *sultanea*).

996 **YELLOW-BROWED TIT**
Sylviparus modestus Plate **99**

IDENTIFICATION 9–10 cm. *S.m.modestus*: **Adult** Distinctive. Small and short-tailed with greyish-olive to olive-green upperparts and pale greyish-olive underparts, variably washed olive-yellow to pale buffish-yellow on vent. Has slight tufted crest, narrow pale yellowish-olive eyering, narrow indistinct paler bar on greater coverts and short pale yellowish eyebrow (often concealed). *S.m.klossi* (S Annam) has brighter, yellower-olive upperparts and fringes to flight-feathers, tertials and tail, and brighter yellower underparts. See Fire-capped Tit and Green Shrike Babbler. Could be confused with some *Phylloscopus* warblers but lacks obvious head or wing markings, bill short and stubby. **Juvenile** Very similar to adult. **VOICE** Possible song is a ringing 1–5 note ***pli-pli-pli-pli...*** or ***pili-pili-pili...*** or mellower 1–6 note ***piu-piu-piu-piu...*** or ***tiu-tiu-tiu-tiu...***; sometimes more rapidly repeated with up to 15 notes. Contact calls include a thin emphatic ***psit*** or ***tis*** and fuller ***chip*** or ***tchup***, often irregularly mixed in series: ***psit tchup psit psit tchup psit...*** etc. Also gives a rapid trilling ***tszizizizi, tszizizi...*** or ***sisisisisisi***. **HABITAT & BEHAVIOUR** Broadleaved evergreen forest; 1,450–3,350 m, locally down to 1,100 m (N Myanmar). Often associates with mixed-species feeding flocks. **RANGE & STATUS** Resident NW,N,NE India, Nepal, Bhutan, S,SE Tibet, SW,W,SE China. **SE Asia** Uncommon to fairly common resident W,N,E Myanmar, NW Thailand, N,S(east) Laos, W Tonkin, C,S Annam. **BREEDING** Nepal: April–May. **Nest** Pad, in tree-hole or crevice; 0.4–7 m above ground. **Eggs** 4–6; white.

REMIZIDAE: Penduline tits & allies

Worldwide 12 species. SE Asia 2 species. Small and plump with rather short conical pointed bills and shortish tails. Normally gregarious, active and restless; feed mainly on insects and seeds.

997 **CHINESE PENDULINE TIT**
Remiz consobrinus Plate **99**

IDENTIFICATION 10.5 cm. Monotypic. **Male non-breeding** Very small size, short pointed dark bill and greyish-brown head with white-bordered black mask diagnostic. Crown tinged brown, nape mostly grey, throat and body mostly buffish-brown with dull chestnut band across upper mantle, wings mostly buffish-fringed with contrasting pale-tipped dull chestnut greater coverts. **Male breeding** Crown more uniform drab mid-grey, has narrower white border to mask, buffier upperparts with narrower but more contrasting mantle-band, pale-fringed greater and median coverts and somewhat darker flight-feathers. **Female** Similar to male but crown and mantle all brownish, mask dark brown, with border reduced to narrow post-ocular supercilium and cheeks. **First-year male** Duller than non-breeding adult with browner crown and nape, pale mottling on mask and less distinct dull chestnut mantle-band. **First-year female** Similar to adult but mask paler, hardly contrasting with crown. **VOICE** Calls with high-pitched soft thin penetrating ***tseee*** or ***pseee***. Also fuller ***piu*** and fast ***siu-siu-siu-siu...*** **HABITAT** Reedbeds and scrub in fresh- and salt-water marshes, sometimes plantations and scrub (some distance from water) on migration; lowlands. **RANGE & STATUS** Breeds NC,NE China; winters southern China, S Korea, S Japan. **SE Asia** Vagrant E Tonkin (K. Ozaki pers. comm.).

998 **FIRE-CAPPED TIT**
Cephalopyrus flammiceps Plate **99**

IDENTIFICATION 10 cm. *C.f.olivaceus*: **Male breeding** Tiny warbler-like bird with yellowish-green upperparts (yellower on rump), yellowish underparts and diagnostic bright reddish-orange forehead-patch. Has faint reddish wash on chin and centre of throat. **Female breeding** Similar to male breeding but forehead-patch golden-olive and smaller, throat and breast overall dull olive-yellow. **Adult non-breeding** Similar to breeding female but throat whitish; female generally somewhat duller than male. Recalls Yellow-browed Tit but lacks crest, upperparts much greener, underparts yellower; shows indistinct pale yellowish fringes to upperwing-coverts and tertials, and more pointed bill. Could possibly be confused with some *Phylloscopus* warblers. **Juvenile** Underparts paler than non-breeding adult, rather uniform whitish, without yellow. Daintier and thinner-billed than similar Green Shrike-babbler. Note range. **VOICE** Song is a variable deliberate series of 1–7 high-pitched notes: ***pitsu-pitsu...***, hurried ***pis-su-psisu-pissu-pissu...***, very thin ***tink-tink-tink-tink***, ringing ***psing-psing-psing...*** and sweet, clear ***tsui tsui-tsui...*** Often alternates between different types, repeating song for long periods. Calls include a high-pitched, abrupt ***tsit*** (repeated irregularly for long periods) and soft, weak, rather tit-like ***whitoo-whitoo***. **HABITAT & BEHAVIOUR** Broadleaved evergreen and semi-evergreen forest; 1,400–2,135 m. Often in small flocks. **RANGE & STATUS** Breeds NE Pakistan, NW,N,NE Indian subcontinent, SW Tibet, SW,W,NC China; some winter south to C India. **SE Asia** Rare winter visitor E Myanmar, NW Thailand, N Laos.

STENOSTIRIDAE: Canary-flycatchers & allies

Worldwide at least 8 species. SE Asia 1 species. Small, conspicuous and flycatcher-like 'sentinel-species' in bird-waves. Feed mainly by sallying for insects.

999 **GREY-HEADED CANARY-FLYCATCHER**
Culicicapa ceylonensis Plate **97**

IDENTIFICATION 11.5–13 cm. *C.c.calochrysea*: **Adult** Unmistakable. Head and breast grey, crown darker and tufted, upperside olive-green with yellower uppertail-coverts, rest of underparts bright yellowish with olive-washed flanks. *C.c.antioxantha* (southern Tenasserim, W[south] and S Thailand southwards) is darker grey on head and breast, less bright green on upperside and less bright yellow on belly. Could be confused with some warblers but upright posture, flycatcher-like behaviour and rather uniform head and wings distinctive. **Juvenile** Similar to adult but head and breast duller and browner, upperparts duller, underparts paler yellow. **VOICE** Song is a brief sharp clear ***wittu-wittu-wit***; ***wichu-wichu-wit***; ***chuit-it-ui*** or ***witti-wuti*** etc. Calls with sharp metallic trills and twitters. **HABITAT & BEHAVIOUR** Broadleaved evergreen, semi-evergreen and deciduous forest, also parks, wooded gardens and mangroves on migration. Up to 3,050 m (mainly breeds above 650 m), only below 1,280 m in Peninsular Malaysia. Usually in middle to upper storey, often accompanying mixed-species feeding flocks; performs regular sallies from perch. **RANGE & STATUS** Breeds N Pakistan, NW,N,NE Indian subcontinent, SW India, Sri Lanka, SE Tibet, SW,W,C China, Sundas; some northern populations winter to south, northern and C India, S China. **SE Asia** Common resident, subject to local movements Myanmar, W,NW,NE(north-west),S Thailand, Peninsular Malaysia, Indochina (except Cochinchina). Uncommon winter visitor S Myanmar, C,NE,SE Thailand. Also recorded on passage E Tonkin. **BREEDING** February–August. **Nest** Cup, attached to mossy or vine-covered rock or tree-trunk; up to 20 m above ground. **Eggs** 2–4; white, pinkish-white or pale buff, flecked and blotched mid-brown to greyish (often ringed at broader end); 15.2 x 12.2 mm (av.; *calochrysea*).

ALAUDIDAE: Larks

Worldwide 91 species. SE Asia 8 species. Nondescript, smallish and mostly terrestrial, with relatively stout bills, short legs and elongated hindclaws. Run and walk on ground. Feed on insects, molluscs, seeds etc.

1000 **AUSTRALASIAN BUSHLARK**
Mirafra javanica Plate **87**

IDENTIFICATION 14–15 cm. *M.j.williamsoni*: **Adult** Told from other bushlarks by rather weak dark streaking on strongly warm brown-washed breast, and diagnostic all-whitish outermost tail-feathers (penultimate pair also have whitish outer web). See Oriental Skylark. **Juvenile** Crown and mantle less clearly streaked, crown darker than mantle, with pale scaling, breast paler with more diffuse streaking, bill more uniformly pale. **VOICE** Song consists of short varied strophes, frequently including mimicry, repeated after short intervals (slower and less continuous than Oriental Skylark). When alarmed, utters sharp, rapid, rather explosive ***pitsi pitsi pitsipipipipi*** or ***tsitsitsitsi***. **HABITAT & BEHAVIOUR** Short grassland with bushes, dry edges of marshland, rice-paddy stubble; up to 915 m. Sings from perch or in towering song-flight, with flickering wings. **RANGE & STATUS** Resident S China, Java, Bali, Borneo, Philippines, Lesser Sundas, New Guinea, Australia. **SE Asia** Uncommon to locally common resident C,E Myanmar, north Tenasserim, W,NW,NE,C Thailand, Cambodia, N,C Laos, C,S Annam, Cochinchina. **BREEDING** April–August. **Nest** Cup, in depression on ground. **Eggs** 3–4; yellowish- or greyish-white, streaked and spotted rich brown, with yellowish-brown and inky-purple specks (mostly marked towards broader end); 18.8 x 14.7 mm (av.).

1001 **BENGAL BUSHLARK**
Mirafra assamica Plate **87**

IDENTIFICATION 15–16 cm. Monotypic. **Adult** Similar to Burmese Bushlark but somewhat larger and larger-billed, crown and upperparts greyer and less boldly streaked, breast less boldly streaked, breast and belly richer buff. See Indochinese and Singing Bushlarks and Oriental Skylark. Note voice and range. **Juvenile** From Burmese by greyer and less boldly patterned upperparts (crown also plainer) and less boldly marked breast. **VOICE** Song is of two types. (1) Thin, high-pitched, slightly hoarse, squeaky ***u-eez*** or ***uu-eez*** notes, repeated monotonously after short intervals (for up to a few minutes); syllables either evenly stressed or first or second stressed. Occasionally has faint third syllable. Short bouts of mimicry may also be included. (2) Slow-paced jingle of thin high-pitched notes and mimicry. Calls with variable thin short high-pitched notes, generally delivered in a short, rather explosive series: ***tzrep-tzi'tzee'tzee'tzee*** or ***tzep-tzi'tzi'tzi'tzu'tzi'tzi'tzu*** etc.; distinctive but reminiscent of Burmese song-types 1 and

2. **HABITAT & BEHAVIOUR** Dry areas with scrub and scattered trees, edge of open forest, cultivation. Strictly terrestrial, rarely perching above ground, flight weak and low. Song-type 1 is typically delivered in high prolonged display-flight, after which bird drops to ground; type 2 is normally given from ground or low perch. **RANGE & STATUS** Resident northern India, S Nepal, Bangladesh. **SE Asia** Common resident SW Myanmar. **BREEDING** May–August. **Nest** Flimsy pad, in depression on ground, sometimes slightly domed. **Eggs** 3–4, greyish-white, greenish-white or yellowish-white, speckled and blotched brown; 20.3 x 15.3 mm (av.).

1002 BURMESE BUSHLARK

Mirafra microptera Plate **87**

IDENTIFICATION 14 cm. Monotypic. **Adult** Similar to Indochinese Bushlark but somewhat warmer and more boldly streaked on upperparts, breast-streaks smaller and more spot-shaped, bill slightly smaller. See Bengal and Australasian Bushlarks. Note voice and range. **Juvenile** Possibly indistinguishable from Indochinese. **VOICE** Song is of three types. (1) 3–10 short, high-pitched, squeaky, jingling, varied notes, delivered at quick, almost explosive pace (entire strophe on average less than 1 s); each strophe generally repeated 2–4 times after rather long intervals. (2) Similar but strophes average more than twice as long (with more notes) and are less often repeated, phrases are commoner and the pauses, on average, distinctly shorter; given exclusively in high prolonged song-flight and frequently ending (during descent) with type 3. (3) 8–20 rather high-pitched, mostly drawn-out notes: ***tsi'tsi'tsiu'tsui'tsuu't-si'ee'tsuu'tsi'eee'tsuu'tsi'eee'tsi'tsuu'tsiu*** etc. Calls with short high-pitched ***heep*** notes, and quick high-pitched ***tsi-tsi-tsi-tsi-tsi-tsi-tsi-tsi-tsi...***; sometimes a very faint soft ***tsupp-tsupp-tsupp***. **HABITAT & BEHAVIOUR** Semi-desert with scrub and scattered trees, cultivation; up to 1,310 m. Mainly terrestrial but often perches in trees (up to 10 m) when flushed from ground; also in bushes and on telephone wires etc. Song-type 1 is usually given from elevated perch, type 2 during prolonged display-flight and type 3 from short display-flight. **RANGE & STATUS** Endemic. Common resident C,S(north),E(west) Myanmar. **BREEDING** June–October. **Nest** Dome with side-entrance, on ground. **Eggs** 2–4; white, speckled yellowish-brown, dark brown, ashy-purple and black.

1003 INDOCHINESE BUSHLARK

Mirafra erythrocephala Plate **87**

IDENTIFICATION 14–15 cm. Monotypic. **Adult** Relatively robust, stout-billed and broad-winged, without obvious crest. In flight, shows extensively rufous-chestnut flight-feathers and no white on outertail-feathers. Best told from Australasian Bushlark by more heavily marked breast and lack of white on outertail-feathers. Very similar to Burmese but upperparts somewhat less boldly streaked, breast-streaking heavier, bill slightly larger. See Bengal Bushlark and Oriental Skylark. Note voice and range. **Juvenile** Generally buffier above than adult; crown, mantle and scapulars scaled buff, breast-streaking more diffuse, shows more obvious pale nuchal collar. **VOICE** Song consists of phrases made up of 1–3 variable, thin, high-pitched, mostly drawn-out notes, repeated to form a strophe lasting 2–8 s, repeated every few seconds: ***tzu'eeez'eezu-eeez'eezu-eeez'eezu-eeez'eezu-eeez'eezu-eeez'eezu-eeez'eezu-eeez'-eeez' eez'piz'piz-eeez'piz'piz-eeez'piz'piz-eeez'piz'piz-eeez'piz'piz-tzueeez'piz'piz-tzueeez'piz'piz-tzueeez'piz'piz-tzueeez pizeeeu-pizeeeu-pizeeeu-pizeeeu-pizeeeu-pizeeeu-pizeeeu-pizeeeu...*** Resembles song-type 3 of Burmese. Calls with thin, high-pitched, metallic, drawn-out, fast, rattling trilled ***tirrrrrrrrrrrrrrrrrrrrrr...***; occasionally a hard, hammering ***tzet-tzet-tzet-tzet-tzet-tzet-tzet-tzet-tzet...*** Also, a short series of thin, high-pitched, variable whistles, similar to song elements and often combined with the rattling trill (may be song-type). **HABITAT & BEHAVIOUR** Dry areas with scrub and scattered trees, edge of open forest, cultivation; up to 900 m. Mainly terrestrial but often perches in bushes, trees and on telephone wires etc. Usually only flies short distance when flushed. Sings from ground, fence post, telephone wire or small tree etc., sometimes in short, relatively low song-flight, ending in parachuting, with wings slightly raised, tail fanned and legs dangling. **RANGE & STATUS** Endemic. Locally common resident E Myanmar, Tenasserim, Thailand (except S), Cambodia, Laos, N,C,S Annam, Cochinchina. **BREEDING** Undocumented?

1004 ORIENTAL SKYLARK

Alauda gulgula Plate **87**

IDENTIFICATION 16.5–18 cm. *A.g.herberti*: **Adult** Most widespread lark in region. From others by combination of relatively slender bill, prominent crest, pale brownish upperside with distinct dark streaking, and pale buffish to whitish underparts with strong breast-streaking. In flight, wings less rounded than bushlarks, with less obvious rufous on upperside of flight-feathers, shows narrow paler sandy or rusty trailing edge to wing. *A.g.vernayi* (W,N Myanmar) has broader, bolder, blacker markings above; *gulgula* (rest of Myanmar) is less boldly marked above; *inopinata* (visitor; ? N Myanmar only) tends to show buffier and more contrastingly dark-streaked upperparts, buffier and less heavily dark-streaked rump, whiter throat, belly and vent, and more strongly rufescent wing-fringing (mainly secondaries). See Paddyfield Pipit. **Juvenile** Crown, mantle and scapulars paler with narrow whitish feather-fringes, wing-coverts tipped whitish, breast-streaking more diffuse. **Other subspecies in SE Asia** *A.g.coelivox* (Vietnam). **VOICE** Song is incessant, high-pitched and sweet. Calls with a dry twangy ***chizz***, ***baz baz*** and ***baz-terrr*** etc. **HABITAT & BEHAVIOUR** Various open habitats, cultivation, larger forest clearings; up to 3,095 m. Flight stronger than bushlarks; has high soaring song-flight. **RANGE & STATUS** Breeds C Asia, E Iran, Afghanistan, Indian subcontinent, S,E Tibet, China (except NW,NE), Taiwan; some populations move south in winter. **SE Asia** Uncommon to fairly common resident (except southern Tenasserim, S Thailand, Peninsular Malaysia, Singapore, S Laos, W Tonkin). Uncommon winter visitor W,N,E Myanmar. **BREEDING** March–July. **Nest** Shallow cup, in depression on ground. **Eggs** 3–5; greyish to yellowish-white, very thickly freckled pale yellowish-brown, purplish-brown or pale inky-purple (may be ringed at broader end); 19.5 x 16.2 mm (av. *herberti*).

1005 GREATER SHORT-TOED LARK

Calandrella brachydactyla Plate **87**

IDENTIFICATION 16 cm. *C.b.dukhunensis*: **Adult** Told from other larks by rather plain breast, lacking (or almost lacking) obvious dark streaks, and small dark patch on breast-side (sometimes faint). Similar to Asian Short-toed Lark but somewhat warmer, browner and more prominently streaked above; at close range shows longer tertials, which typically extend to cloak tips of primaries (beware birds with worn tertials). Smaller, stubbier-billed and paler than Oriental Skylark, with less obvious crest, lacks heavy dark breast-streaking and warm coloration on fringes of flight-feathers. See Sand Lark. **VOICE** Calls (usually in flight) with dry ***trrrip***, ***trriep*** or ***prrit***, often in combination with short soft ***dju*** or ***djyp*** notes: ***trriep-dju*** etc. **HABITAT & BEHAVIOUR** Dry open habitats; lowlands. Flight strong. **RANGE & STATUS** Breeds N Africa, southern W,C Palearctic, Middle East, NW China; some populations winter south to northern Africa, Middle East, Indian subcontinent. **SE Asia** Vagrant/rare winter visitor W,N,C,S Myanmar.

1006 SAND LARK *Calandrella raytal* Plate **87**

IDENTIFICATION 14 cm. *C.r.raytal*: **Adult** Relatively small size, short tail and very pale, greyish upperside distinctive. Only likely to be confused with short-toed larks but smaller, paler and purer grey above, whiter below, with faint breast-streaking and no lateral breast-patches, bill slenderer. **Juvenile** Differs from adult by whitish tips and indistinct dark subterminal markings on upperpart feathers. **VOICE** Sings with repeated short bursts of disjointed tinkling notes (often including mimicry), usually in flight, sometimes from perch near ground. Typical calls are a subdued dry ***chrrru***, ***chrrt chu*** and ***chirrru*** etc. **HABITAT & BEHAVIOUR** Dry banks and sand-bars along larger rivers, rarely dry open areas; lowlands. Has rather fluttering, jerky flight action. **RANGE & STATUS** Resident E Iran, Pakistan, northern Indian subcontinent. **SE Asia** Common resident C,S Myanmar. **BREEDING** February–April. **Nest** Flimsy cup, in depression on ground. **Eggs** 2–3; pale grey or yellowish-white, with pale greyish- to reddish-brown freckles and patches; 20.1 x 14.6 mm (av.).

1007 **ASIAN SHORT-TOED LARK**
Calandrella cheleensis Plate **87**
IDENTIFICATION 16 cm. *C.c.kukunoorensis*: **Adult** Similar to Greater Short-toed Lark but upperparts usually less distinctly streaked, paler and more ashy-grey, shows fine darker streaking across breast (sometimes sparser in centre), lacks dark patch on breast-side, head less contrastingly patterned. At close range, tips of primaries protrude well beyond tips of tertials (on closed wing). Larger, more buffish above, with more distinct streaking (may be hard to judge when fresh) and shorter-billed than Sand Lark. See Oriental Skylark. **VOICE** Calls with dry rolling ***trrrrt*** or ***prrrt-up***. Dry, quiet, slightly rolled ***chrrr'rt***; ***jrrr'rh*** or ***chrrrrh***, with no intonation. **HABITAT** Dry open areas; recorded in lowlands. **RANGE & STATUS** Breeds C Asia, S Siberia, Mongolia, E Tibet, northern China; some populations move south in winter. Vagrant Pakistan, NW India, Japan. SE Asia Vagrant N Myanmar, E Tonkin.

PYCNONOTIDAE: Bulbuls

Worldwide 137 species. SE Asia 39 species. Smallish to medium-sized, with soft plumage, rather short, rounded wings, medium-length tails, fairly short, weak legs and rather slender bills (except finchbills). Otherwise variable, though generally rather nondescript, occurring in wide variety of habitats. Feed mainly on berries and other fruits and insects, also nectar and buds.

1008 **CRESTED FINCHBILL**
Spizixos canifrons Plate **100**
IDENTIFICATION 21.5 cm. *S.c.ingrami*: **Adult** Relatively large size, greenish to yellowish-green plumage, thick pale yellowish bill and erect pointed crest distinctive. Head mostly greyish with blacker lores, crest and throat; tail-tip blackish. See Collared Finchbill. **Juvenile** Crown and crest paler and mixed with green; forehead and throat yellowish-green, ear-coverts much paler, dark tail-band less distinct. Gradually attains adult head pattern. **Other subspecies in SE Asia** *S.c.canifrons* (W,N Myanmar). **VOICE** Long bubbling trilled ***purr-purr-prruit-prruit-prruit...*** and gentle scolding bubbling ***pri pri-pri prrrrrr*** and ***pri-pri*** etc. when alarmed. **HABITAT** Secondary growth, scrub and grass; 1,065–2,720 m. **RANGE & STATUS** Resident NE India, SW China. **SE Asia** Locally common resident W,N,C,E Myanmar, NW Thailand, N Laos, W Tonkin. **BREEDING** March–August. **Nest** Cup, in bush or small tree; 1.5–3 m above ground. **Eggs** 2–4; dull pink, very densely freckled various shades of red; 25.7 x 17.6 mm (av.).

1109 **COLLARED FINCHBILL**
Spizixos semitorques Plate **100**
IDENTIFICATION 23 cm. *S.s.semitorques*: **Adult** Resembles Crested Finchbill but lacks crest, has white patch on side of forehead, white streaks on ear-coverts and whitish half-collar on upper breast. **Juvenile** Dark head markings, upperparts and wings much browner than adult, ear-covert streaks less distinct. **VOICE** Undocumented. **HABITAT** Secondary growth, scrub; recorded at 1,200–1,500 m. **RANGE & STATUS** Resident C,S,SE China, Taiwan. **SE Asia** Local resident W,E Tonkin. **BREEDING** Season not recorded for region. **Nest** Flimsy shallow cup, in tree. **Eggs** 3–4; similar to Light-vented Bulbul but rather finely speckled; 25.5 x 20 mm.

1010 **STRAW-HEADED BULBUL**
Pycnonotus zeylanicus Plate **102**
IDENTIFICATION 29 cm. Monotypic. **Adult** Largest bulbul in region, with diagnostic golden-yellowish crown and cheeks and blackish eye and submoustachial stripes. Throat white; nape, mantle, scapulars and breast finely streaked whitish. **Juvenile** Head duller and browner. **VOICE** Song consists of bursts of extended loud, rich and very melodious warbling. **HABITAT** Broadleaved evergreen forest, secondary growth, scrub, plantations and cultivation, sometimes mangroves; almost exclusively along banks of larger rivers and streams. Up to 245 m. **RANGE & STATUS** Resident Greater Sundas. **SE Asia** Rare to local resident south Tenasserim, S Thailand (possibly extinct), Peninsular Malaysia, Singapore. **BREEDING** November–September. **Nest** Largish, untidy shallow cup, in bush; up to 1.5–4.6 m above ground. **Eggs** 2; off-white, speckled and smeared shades of reddish-brown, purplish-brown and grey (often more towards broader end); 27.9 x 19.6 mm (av.).

1011 **STRIATED BULBUL**
Pycnonotus striatus Plate **100**
IDENTIFICATION 23 cm. *P.s.paulus*: **Adult** Combination of heavy yellowish-white streaking on head and body, prominent crest, mostly yellow throat and yellow undertail-coverts diagnostic. Also has prominent yellow spectacles. **Other subspecies in SE Asia** *P.s.striatus* (Myanmar, except N,C), *arctus* (N Myanmar). **VOICE** Loud repeated short jolly phrases: ***chu-wip***; ***chi'pi-wi*** and ***chit-wrrri*** (slurred second note) etc. Also a harsh slurred ***djrrri*** or ***rrri***. **HABITAT** Broadleaved evergreen forest, forest edge, secondary growth; 1,200–2,900 m. **RANGE & STATUS** Resident Nepal, Bhutan, NE India, SW China. **SE Asia** Uncommon to fairly common resident SW,W,N,E,S Myanmar, north Tenasserim, W,NW,NE Thailand, N,C Laos, W Tonkin. **BREEDING** April–July. **Nest** Cup, in bush or bamboo; usually close to ground. **Eggs** 3; whitish (sometimes tinged brown), spotted and freckled reddish-brown and deep purple (often ringed at broader end); 22.4 x 16.3 mm (av.; *striatus*).

1012 **BLACK-AND-WHITE BULBUL**
Pycnonotus melanoleucos Plate **102**
IDENTIFICATION 18 cm. Monotypic. **Male** Unmistakable, with uniform blackish-brown plumage and contrasting, mostly white upper- and underwing-coverts. **Female** Somewhat browner than male, with slightly less white on upperwing-coverts. **Juvenile** Much paler and browner (particularly below), lacks white on upperwing-coverts. Best identified by size, overall cold brownish plumage, darker-centred upperpart feathers (giving scaly appearance) and dark mottling/streaks on breast. Gradually attains patches of adult plumage. See Red-eyed and Spectacled Bulbuls. **VOICE** Usual call is a tuneless ***pet-it***. **HABITAT & BEHAVIOUR** Broadleaved evergreen forest, forest edge; up to 1,830 m. Nomadic; often frequents canopy. **RANGE & STATUS** Resident Sumatra, Borneo. **SE Asia** Rare to uncommon resident S Thailand, Peninsular Malaysia; formerly Singapore. **BREEDING** May–July. **Nest** Cup-shaped. **Eggs** Pale fawn, faintly freckled darker fawn (slightly ringed at broader end).

1013 **BLACK-HEADED BULBUL**
Pycnonotus atriceps Plate **101**
IDENTIFICATION 18 cm. *P.a.atriceps*: **Adult** Mostly yellowish-green plumage, glossy black (crestless) head and broad black subterminal tail-band diagnostic. Yellowness of plumage varies. Typically shows distinctly yellower fringes to feathers of lower body, tail and wing, latter contrasting with blackish primaries and inner edges of tertials. Sometimes has greener wash overall. See Black-crested Bulbul. **Adult grey morph** Rare. Neck, breast and belly grey. See Grey-bellied Bulbul. **Juvenile** Generally duller, with less pronounced wing pattern, head largely dull greenish (gradually attains black in patches). **VOICE** Song is a series of short, spaced, tuneless whistles. Calls with distinctive repeated loud chipping ***chew***

or ***chiw***. **HABITAT** Broadleaved evergreen forest, mixed deciduous and evergreen forest, forest edge, secondary growth; up to 1,600 m (rarely to 2,440 m); commoner in lowlands. **RANGE & STATUS** Resident NE India, Bangladesh, SW China (SW Yunnan), Greater Sundas, Philippines (Palawan). **SE Asia** Uncommon to common resident (except W,E Tonkin, N Annam). **BREEDING** January–September. **Nest** Cup, in low vegetation or bamboo; 1.2–6 m above ground. **Eggs** 2–3; pale fleshy-pink to lilac, spotted and freckled reddish over pale grey and pale lilac undermarkings (often ringed or capped at broader end); 21.1 x 15.9 mm (av.).

1014 **BLACK-CRESTED BULBUL**
Pycnonotus flaviventris Plate **101**
IDENTIFICATION 18.5–19.5 cm. *P.f.caecilli* (S Tenasserim and S Thailand southwards): **Adult** Unmistakable with bright yellow underparts and glossy black head with tall erect crest. Upperside rather uniform greenish-olive. *P.f.johnsoni* (C,NE,SE Thailand, S Indochina) tends to have deeper yellow underparts (often with orange wash on breast) and shows variable amount of red on throat. See Black-headed Bulbul. **Juvenile** Head duller and browner, throat mixed with olive-yellow. **Other subspecies in SE Asia** *P.f.flaviventris* (SW,W,N,C,S[west] Myanmar), *vantynei* (E Myanmar, NW[north],NE[north-east] Thailand, N Indochina), *xanthops* (S,E Myanmar, north Tenasserim, W[north],NW[south] Thailand), *auratus* (NE[north] Thailand, C Laos), *elbeli* (islands off SE Thailand), *negatus* (central Tenasserim, W[south] Thailand). **VOICE** Song is a cheerful quick ***whitu-whirru-wheet***; ***whit-whaet-ti-whaet*** and ***whi-wiu*** etc. **HABITAT** Broadleaved evergreen and mixed deciduous forest, forest edge, secondary growth; up to 2,565 m. **RANGE & STATUS** Resident C,E India, N,NE Indian subcontinent, SW China, Greater Sundas. **SE Asia** Common resident (except Singapore). Uncommon feral resident (established?) Singapore. **BREEDING** January–September. **Nest** Flimsy cup, in bushy undergrowth or sapling; 30 cm-2.5 m above ground. **Eggs** 2–4; pinkish-white, profusely speckled pink, red and purple; 24.2 x 16.4 mm (av.; *flaviventris*).

1015 **SCALY-BREASTED BULBUL**
Pycnonotus squamatus Plate **101**
IDENTIFICATION 14–16 cm. *P.s.weberi*: **Adult** Unmistakable. Relatively small, with black head, white throat and white-scaled black breast and flanks. Undertail-coverts yellow, tail black with white-tipped outer feathers. **Juvenile** Has smudgy black upper breast and clean white lower breast. **VOICE** Calls with a series of sharp, high-pitched, chinking ***wit*** or ***tit*** notes. **HABITAT & BEHAVIOUR** Broadleaved evergreen forest; up to 1,000 m. Often frequents canopy. **RANGE & STATUS** Resident Greater Sundas. **SE Asia** Uncommon to fairly common resident south Tenasserim, S Thailand, Peninsular Malaysia. **BREEDING** April–July. **Nest** Presumably typical. **Eggs** 2. Otherwise undocumented.

1016 **GREY-BELLIED BULBUL**
Pycnonotus cyaniventris Plate **101**
IDENTIFICATION 16.5 cm. *P.c.cyaniventris*: **Adult** Relatively small size, grey head and underparts and contrasting yellow undertail-coverts diagnostic. Tail uniform, without black or yellow. See grey morph Black-headed Bulbul. **Juvenile** Undescribed. **VOICE** Clear minivet-like ***pi-pi-pwi...***; ***pi-pi-pwi-pwi...*** etc. and subdued ***wit wit wit...*** contact calls. Also a slightly descending, bubbling, trilled whistle, ***pi-pi-pi-pi-pi-pi-pi***. **HABITAT** Broadleaved evergreen forest, forest edge; up to 1,000 m. **RANGE & STATUS** Resident Sumatra, Borneo. **SE Asia** Uncommon to fairly common resident south Tenasserim, S Thailand, Peninsular Malaysia; formerly Singapore. **BREEDING** January–July. **Nest** Cup, in low vegetation. **Eggs** 2–3; like Red-whiskered Bulbul but smaller; 20.3 x 15.5 mm (av.).

1017 **PUFF-BACKED BULBUL**
Pycnonotus eutilotus Plate **101**
IDENTIFICATION 23 cm. Monotypic. **Adult** Dark rich brown upperside, rather clean whitish throat and underparts and short crest (often held erect) distinctive. Breast washed greyish, eyes dark red, outertail-feathers tipped white on inner web. See Red-eyed, Spectacled and Ashy Bulbuls. **Juvenile** Similar to adult, but lacks white tail-tips. **VOICE** Song is a loud, rather high-pitched, cheerful, slurred, quavering warble: ***tchui'uui tch'i-iwi'iwi***; ***iwu'iwi i'wu-u*** and ***tch'uwi'i'iwi*** etc. **HABITAT & BEHAVIOUR** Middle storey of broadleaved evergreen forest, forest edge; up to 210 m. **RANGE & STATUS** Resident Sumatra, Borneo. **SE Asia** Uncommon to fairly common resident south Tenasserim, S Thailand, Peninsular Malaysia; formerly Singapore. **BREEDING** March–June. **Nest** Small neat cup, in spiny palm; 60 cm above ground. **Eggs** 2; densely marked dull reddish (more so towards broader end).

1018 **STRIPE-THROATED BULBUL**
Pycnonotus finlaysoni Plate **101**
IDENTIFICATION 19–20 cm. *P.f.eous*: **Adult** Rather nondescript, apart from prominent broad yellow streaks on forecrown, ear-coverts, throat and upper breast, and yellow vent. Lower breast greyish with narrow whitish streaks. *P.f.davisoni* (S Myanmar) lacks obvious yellow streaks on forecrown and ear-coverts, has smaller throat-streaks, greener crown and rump, rather darker rest of upperparts, breast and flanks, and pale eyes. **Juvenile** Crown, upperparts and breast paler and browner, no yellow streaks on forecrown and centre of throat/upper breast, almost no breast-streaking. See Streak-eared. **Other subspecies in SE Asia** *P.f.finlaysoni* (S Tenasserim and S Thailand southwards). **VOICE** Song is a throaty, measured ***whit-chu whic-ic*** and ***whit whit-tu-iwhit-whitu'tu*** etc. **HABITAT** Secondary growth, scrub, more open areas and clearings in broadleaved evergreen and mixed deciduous and evergreen forest, forest edge; up to 1,300 m. **RANGE & STATUS** Resident SW China (S Yunnan). **SE Asia** Common resident, subject to some minor movements (except W,N,C Myanmar, C Thailand, Singapore). **BREEDING** February–September. **Nest** Small neat cup, in bush or sapling; 0.6–6 m above ground. **Eggs** 2–4; like richly marked version of Red-whiskered Bulbul; 20.8 x 15.7 mm (av.).

1019 **FLAVESCENT BULBUL**
Pycnonotus flavescens Plate **101**
IDENTIFICATION 21.5–22 cm. *P.f.vividus*: **Adult** Quite nondescript with dull olive upperside, variably yellow-washed underside, bright yellow undertail-coverts, distinctive greyish head with black lores, pronounced whitish pre-ocular supercilium and eyering and darker crown with slight tufted crest. Crown and upperparts obscurely scaled, breast indistinctly streaked darker. See Yellow-vented. **Juvenile** Crown and upperparts browner and more uniform, no obvious supercilium and eyering, lores duller, throat and breast brownish without streaked appearance, undertail-coverts less vivid yellow, bill paler and browner (blackish on adult). **Other subspecies in SE Asia** *P.f.flavescens* (south-western Myanmar), *sordidus* (S Laos, S Annam). **VOICE** Song consists of jolly, rather quickly delivered 3–6 note phrases: ***joi whiti-whiti-wit***; ***ti-chi whiti-whiti-whit-tu***; ***chi whiti-whiti-whi-tu chitiwit*** and ***brr whiti-chu*** etc. Alarm call is a harsh rapid buzzing ***djo djo drrrrrt***; ***dreet dreet drrrr dreet-dreet...*** etc. **HABITAT** Clearings in broadleaved evergreen forest, forest edge, secondary growth, scrub and grass; 900–2,590 m. **RANGE & STATUS** Resident NE India, E Bangladesh, SW China, Borneo. SE Asia Common resident Myanmar, W,NW,NE Thailand, Laos, W,E Tonkin, C,S Annam. **BREEDING** January–June. **Nest** Neat shallow cup, in bush, sapling or grass; up to 3 m above ground (usually low down). **Eggs** 2–3; pale cream to pinkish, profusely and minutely speckled light or dark reddish to chestnut (larger markings in broad ring around broader end), sometimes with grey to mauve undermarkings; 22.1 x 16.4 mm (av. *flavescens*).

1020 **YELLOW-VENTED BULBUL**

Pycnonotus goiavier Plate **101**

IDENTIFICATION 20–20.5 cm. *P.g.personatus*: **Adult** Combination of complete broad white supercilium, dark crown and loral stripe, white throat and yellow vent diagnostic. Upperside brownish, breast and belly whitish with vague darker streaking. See Flavescent Bulbul. **Juvenile** Supercilium much less distinct, crown paler and browner, upperparts and wing-feather fringes warmer brown, no obvious streaking on underparts, bill paler and browner (blackish on adult). **Other subspecies in SE Asia** *P.g.jambu* (C Thailand eastwards). **VOICE** Calls with rapid bubbling ***chic-chic-chic...*** and sharp harsh ***chwich-chwich***. Also bubbling ***tiddloo-tiddloo-tiddloo...*** or ***tud-liu tud-liu tud-liu...*** **HABITAT** Coastal scrub, mangroves, secondary growth, plantations, cultivation; lowlands but up to 1,830 m in Peninsular Malaysia. **RANGE & STATUS** Resident Greater Sundas, Philippines. **SE Asia** Common resident (subject to some dispersive movements) south Tenasserim, Thailand (just spreading to NW,NE), Peninsular Malaysia, Singapore, Cambodia, Laos, Cochinchina. **BREEDING** December–October. Multi-brooded. **Nest** Deep cup, in bush or sapling, sometimes creeper or grass tussock; 0.5–3 m above ground. **Eggs** 2; white to pink, profusely spotted reddish to reddish-brown, lavender and grey; 20.3 x 16 mm (av.; *personatus*).

1021 **OLIVE-WINGED BULBUL**

Pycnonotus plumosus Plate **101**

IDENTIFICATION 20–20.5 cm. *P.p.plumosus*: **Adult** Similar to Streak-eared but whitish streaks on ear-coverts less pronounced; shows contrasting yellowish-green fringes to flight-feathers, slightly contrasting darker lores and orbital area, dark red eyes, all-dark bill and dark buffish undertail-coverts. See Stripe-throated and Red-eyed Bulbuls. **Juvenile** Browner overall with even less obvious whitish shaft-streaks on ear-coverts, flight-feather fringes duller, eyes brown. **Other subspecies in SE Asia** *P.p.chiroplethis* (Pulau Tinggi, off SE Peninsular Malaysia). **VOICE** Song is a simple repeated chirruping ***whip wi-wiu wu-wurri'i*** etc. Calls include a throaty ***whip-whip...*** and ***wrrh wrrh wrrh...*** **HABITAT** Secondary growth, coastal scrub, mangroves; up to 610 m. **RANGE & STATUS** Resident Greater Sundas, Philippines (Palawan). **SE Asia** Common resident south Tenasserim, S Thailand, Peninsular Malaysia, Singapore. **BREEDING** January–August. **Nest** Deep cup, in sapling; 1.5–3 m above ground. **Eggs** 2; like Yellow-vented Bulbul but darker, with larger redder spots and blotches mixed with bluish-grey blotches; 22.4 x 17.8 mm (av.; *plumosus*).

1022 **STREAK-EARED BULBUL**

Pycnonotus blanfordi Plate **101**

IDENTIFICATION 17.5–19.5 cm. *P.b.conradi*: **Adult** Very nondescript brownish with paler throat and belly and yellowish-washed vent. Best identified by prominent whitish streaks on ear-coverts and pale greyish eyes. Bill brown, pinker at base. *P.b.blanfordi* (W,C,E,S Myanmar) has much less yellow-tinged vent. See Olive-winged Bulbul. **Juvenile** Crown and upperparts paler, ear-covert streaks less distinct, eyes brown. **VOICE** Calls with harsh, rasping ***which-which-which...*** and piping ***brink-brink-brink...*** **HABITAT** Semi-desert, scrub, cultivation, gardens, urban areas, open mixed deciduous forest; up to 915 m. **RANGE & STATUS** Endemic. Common resident Myanmar (except N), Thailand, north Peninsular Malaysia, Cambodia, Laos, C,S Annam, Cochinchina. **BREEDING** December–September. **Nest** Cup, in bush, small tree or creeper. **Eggs** 2–3; pale pink to pinkish-white, speckled and blotched deep red to reddish-brown over pale purple undermarkings (usually ringed or capped at broader end); 21.5 x 15.5 mm (av.; *conradi*).

1023 **CREAM-VENTED BULBUL**

Pycnonotus simplex Plate **101**

IDENTIFICATION 18 cm. *P.s.simplex*: **Adult** Easily told by striking whitish eyes. Otherwise similar to Red-eyed and Spectacled Bulbuls but upperside rather cold brown, underside somewhat cleaner, creamy-whitish (washed greyish-brown on breast and flanks). **Juvenile** Eyes pale warm brown to brown, crown, upperparts and wing-feather fringes warmer brown. **VOICE** Subdued quavering ***whi-whi-whi-whi-whi...*** interspersed with low ***pru-pru***, ***prrr*** and ***prr-pru***. **HABITAT** Broadleaved evergreen forest, forest edge, secondary growth; up to 1,220 m. **RANGE & STATUS** Resident Greater Sundas. **SE Asia** Uncommon to common resident south Tenasserim, S Thailand, Peninsular Malaysia, Singapore. **BREEDING** January–May. **Nest** Cup, in thick vegetation; up to 3 m above ground. **Eggs** 2; finely and evenly speckled dark reddish and grey; c.18 x 14 mm (Borneo).

1024 **RED-EYED BULBUL**

Pycnonotus brunneus Plate **101**

IDENTIFICATION 19 cm. *P.b.brunneus*: **Adult** Nondescript, apart from prominent red eyes. Upperside uniform warmish brown, underside pale brownish. Very similar to Spectacled Bulbul but somewhat larger and plumper, lacks orange eyering, has dark brown to dark reddish-brown bill, more uniform upperparts, duller throat and belly and browner breast and upper belly. See Olive-winged Bulbul. **Juvenile** Eyes brownish, upperparts and wings slightly paler, warmer brown, bill paler brown. **Other subspecies in SE Asia** *P.b.zapolius* (Pulau Tioman, off SE Peninsular Malaysia). **VOICE** Series of high-pitched bubbling notes, the last ones rising sharply: ***pri-pri-pri-pri-pri-pit-pit***. **HABITAT** Broadleaved evergreen forest, forest edge, secondary growth; up to 1,000 m. **RANGE & STATUS** Resident Sumatra, Borneo. **SE Asia** Fairly common to common resident south Tenasserim, S Thailand, Peninsular Malaysia, Singapore (scarce). **BREEDING** February–August. **Nest** Open cup, in sapling, tree-/bush-crown, ginger plant or bank; up to 5 m above ground. **Eggs** 2. Otherwise undocumented.

1025 **SPECTACLED BULBUL**

Pycnonotus erythropthalmos Plate **101**

IDENTIFICATION 16–18 cm. *P.e.erythropthalmos*: **Adult** Very like Red-eyed Bulbul but has diagnostic narrow orange to yellow-orange eyering (only obvious at close range); smaller, with blacker bill, duller, greyer crown and mantle, more rufescent rump and uppertail-coverts (contrasting with rest of upperparts), whiter throat and belly, and greyer breast and upper belly. **Juvenile** Eyes brown, eyering duller, upperparts and wings slightly paler and warmer brown, bill browner. **VOICE** Distinctive, repeated, rather high-pitched mechanical ***wip-wip-wi'i'i'i'i***. **HABITAT** Broadleaved evergreen forest, forest edge, secondary growth; up to 900 m. **RANGE & STATUS** Resident Sumatra, Borneo. **SE Asia** Fairly common to common resident south Tenasserim, S Thailand, Peninsular Malaysia; formerly Singapore. **BREEDING** February–June. **Nest** Open cup, in crown of sapling; 3 m above ground. **Eggs** 2; pale clay-buff, flecked and blotched pale grey-brown (most densely over broader end), and dotted and irregularly flecked dull chestnut overall, with larger markings concentrated in zone around broader end; 22.8 x 15.9 mm.

1026 **RED-WHISKERED BULBUL**

Pycnonotus jocosus Plate **100**

IDENTIFICATION 18–20.5 cm. *P.j.pattani*: **Adult** Combination of tall erect black crest, black moustachial line, whitish ear-coverts and underparts, and red ear-patch and undertail-coverts diagnostic. Upperside mostly dull brownish, tail dark towards tip with white-tipped outer feathers, blackish patch on breast-side. *P.j.emeria* (southern Myanmar, W Thailand) has larger red ear-patch; *monticola* (N Myanmar) and *hainanensis* (northern Indochina) are larger and darker above with darker red ear-patch. **Juvenile** Browner-tinged overall, crest shorter, red ear-patch lacking, undertail-coverts more orangey. **VOICE** Song consists of varied lively musical phrases: ***wit-ti-waet***, ***queep kwil-ya***, ***queek-kay*** etc. Call is a rolling ***prroop***. **HABITAT & BEHAVIOUR** Secondary growth, scrub, open areas, cultivation (parks and gardens where introduced); up to 2,285 m (mostly montane in southern Vietnam?). Sometimes forms large flocks. **RANGE & STATUS** Resident India, Nepal, Bhutan, Bangladesh, SE Tibet, SW,S China. Introduced Mauritius, Australia, N America. **SE Asia** Common resident (except southern Peninsular Malaysia, Singapore). Uncommon feral resident southern Peninsular Malaysia, Singapore. **BREEDING** January–August. **Nest** Cup, in bush; up to 1.5–4.5 m above ground. **Eggs** 2–4; whitish, reddish-white or pink, densely freckled and streaked various shades of red over pale inky-purple to blackish undermarkings (often more at broader end); 22.2 x 16.2 mm (av.; *emeria*).

1027 BROWN-BREASTED BULBUL
Pycnonotus xanthorrhous Plate **100**

IDENTIFICATION 20 cm. *P.x.xanthorrhous*: **Adult** Similar to Sooty-headed but upperparts browner (including rump and uppertail-coverts), has brown ear-coverts and breast-band, deep yellow undertail-coverts and little or no white on tail-tip. **Juvenile** Breast-band less distinct, dark head markings browner, upperparts somewhat paler. **VOICE** Song is a repeated quick simple ***chirriwu'i whi'chu whirri'ui*** etc. Calls with harsh ***chee*** notes and thinner ***ti-whi***. **HABITAT** Secondary growth, scrub and grass, clearings; 1,020–2,290 m; sometimes down to 610 m. **RANGE & STATUS** Resident SE Tibet, C and southern China. **SE Asia** Locally common resident N,C(east),S(east),E Myanmar, NW Thailand, N Laos, W Tonkin. **BREEDING** April–August. **Nest** Loose cup, in grass or dense low vegetation; up to 90 cm above ground. **Eggs** 2–3; like Red-vented Bulbul; 20.3 x 15.7 mm (av.).

1028 LIGHT-VENTED BULBUL
Pycnonotus sinensis Plate **100**

IDENTIFICATION 19 cm. *P.s.hainanus* (resident and visitor Vietnam): **Adult** Similar to Sooty-headed and Brown-breasted Bulbuls but has whitish undertail-coverts, yellowish-grey breast-band, yellowish-green fringes to wing- and tail-feathers and small whitish patch on otherwise dark ear-coverts; lacks white on rump, uppertail-coverts and tail. *P.s.sinensis* (visitor throughout) has distinctive broad white patch from rear supercilium to nape. See Ashy Bulbul. **Juvenile** Upperparts paler, blackish head markings replaced by rather pale greyish-brown, breast-band less distinct, wings and tail paler and duller. **VOICE** Calls with short ***jhieu*** and ***jhoit*** notes, the latter often doubled. **HABITAT & BEHAVIOUR** Cultivation, scrub, open woodland; up to 1,250 m. Often in large flocks in winter. **RANGE & STATUS** Breeds C,S,E China, Taiwan; some populations winter to south. **SE Asia** Uncommon to local resident E Tonkin (and possibly further south). Common winter visitor E Tonkin, N,C Annam. Also recorded on passage E Tonkin. Vagrant NW,NE Thailand, C Laos. **BREEDING** April–June. Multi-brooded. **Nest** Firm cup, in bush, hedge, bamboo or creeper etc. **Eggs** 3–4; mauve to pink, thickly speckled or spotted dark red and lavender-grey over pale grey underspots (often ringed at broader end); 23 x 17 mm (av.; China).

1029 RED-VENTED BULBUL
Pycnonotus cafer Plate **100**

IDENTIFICATION 23 cm. *P.c.melanchimus*: **Adult** Combination of mostly dark plumage, white rump and uppertail-coverts and red undertail-coverts distinctive. Head blackish with browner ear-coverts; upperparts, breast and flanks variably scaled whitish to pale brown, has white-tipped tail-feathers and slight crest. **Juvenile** Similar to adult but upperparts and wings browner. **Other subspecies in SE Asia** *P.c.stanfordi* (range except S Myanmar). **VOICE** Song described as a cheerful ***be care-ful*** or ***be quick-quick*** etc. Calls include a ***peep-peep-peep...*** and, when agitated, ***peep-a peep-a-lo*** and slow ***peet-wit-wit-wit-wit*** and rapid ***pitititit***. **HABITAT** Semi-desert, secondary growth, scrub and grass, cultivation; up to 1,910 m. **RANGE & STATUS** Resident Indian subcontinent, SW China (W Yunnan). Introduced Fiji. **SE Asia** Common resident Myanmar (except Tenasserim). **BREEDING** February–August. **Nest** Rather untidy cup, in bush or tree; 1–9 m above ground. **Eggs** 2–4; pinkish-white with numerous reddish and purplish markings of various shapes (sometimes ringed or capped at broader end); 22.6 x 16.2 mm (av.; *stanfordi*).

1030 SOOTY-HEADED BULBUL
Pycnonotus aurigaster Plate **100**

IDENTIFICATION 19–21 cm. *P.a.klossi*: **Adult** Combination of black cap and cheeks, short crest, prominent whitish patch on rump and uppertail-coverts, rather plain pale greyish underparts and red undertail-coverts distinctive. Tail prominently tipped whitish. Most populations of *P.a.thais* (C[south],NE[south-west],SE Thailand, C Laos), and all populations of *germani* (NE Thailand, S Indochina) and *aurigaster* (Singapore) show yellow undertail-coverts. See Brown-breasted, Red-whiskered and Red-vented Bulbuls. **Juvenile** Crown browner, cheeks paler, red or yellow of vent paler. **Other subspecies in SE Asia** *P.a.latouchei* (E Myanmar, NW[north] Thailand, N Laos, W Tonkin), *resurrectus* (E Tonkin, N Annam), *dolichurus* (C Annam), *schauenseei* (central Tenasserim, W Thailand). **VOICE** Song is a chatty *whi wi-wiwi-wiwi* etc., with stressed introductory note. Also utters repeated clear, shrill whistled ***u'whi'hi'hu*** or ***wh'i-i-wi*** with stressed ***wh*** sounds. **HABITAT** Secondary growth, scrub, grass, forest clearings, cultivation; up to 1,830 m. **RANGE & STATUS** Resident southern China, Java, Bali. Introduced Sumatra, S Sulawesi, Lesser Sundas (west Timor at least). **SE Asia** Fairly common to common resident E,S Myanmar, Tenasserim, Thailand (except S), Indochina. Rare feral resident Singapore. **BREEDING** March–June. Multi-brooded. **Nest** Flimsy cup, in tree; 3–3.4 m above ground. **Eggs** 2–6; heavily peppered purplish and grey; 23 x 17 mm (av.; China).

1031 OLIVE BULBUL *Iole virescens* Plate **102**

IDENTIFICATION 19 cm. *I.v.virescens*: **Adult** (Very) difficult to separate from Grey-eyed Bulbul, depending on subspecies. Always separable by dark reddish to brown eyes (but birds with white to grey eyes have been reported from W Myanmar) and call. Note range. Typically differs from Grey-eyed by more strongly olive-tinged upperparts, much yellower underparts and more yellowish-olive (less greyish) head-sides. *I.v.myitkyinensis* (N,E Myanmar) is less yellow below and probably indistinguishable on plumage from some Grey-eyed. See that species and also Buff-vented Bulbul. **Juvenile** Upperparts and wings more rufescent, undertail-coverts paler. **VOICE** Calls with a musical ***whe-ic***, like Buff-vented but less nasal than Grey-eyed. **HABITAT** Broadleaved evergreen and semi-evergreen forest; up to 915 m. **RANGE & STATUS** Resident NE India, E Bangladesh. **SE Asia** Locally common resident Myanmar, W,NW Thailand. **BREEDING** March–June. **Nest** Neat shallow cup, in bush; c.1.5 m above ground. **Eggs** 2–3; like Red-vented Bulbul but smaller; 22.2–23.1 x 16.2–16.5 mm (India).

1032 GREY-EYED BULBUL
Iole propinqua Plate **102**

IDENTIFICATION 19–19.5 cm. *I.p.propinqua*: **Adult** Combination of relatively small size, slim build, rather small head, narrow paler supercilium and deep rufous-buffish undertail-coverts distinctive. Can be very difficult to separate from Olive Bulbul but always differs by whitish-grey to grey eyes and call. Typically has browner-tinged upperparts, less yellow underparts and greyer (less yellowish-olive) head-sides. *I.p.aquilonis* (E Tonkin, N Annam) is as yellow below as northern subspecies of Olive but ranges widely separated; *cinnamomeoventris* (south Tenasserim, S Thailand) is smaller (17–18.5 cm), browner above, and less yellow below (mainly restricted to centre of abdomen), much less yellow below than Olive, and told from more similar Buff-vented by undertail-covert colour, more olive (less brown) wings, and call; usually shows some yellowish below, supercilium greyer. Note range. **Juvenile** As adult, but eyes light brown. **Other subspecies in SE Asia** *I.p.lekhakuni* (northern Tenasserim, W Thailand), *simulator* (SE Thailand, Cambodia, S Laos, S Annam), *innectens*

(Cochinchina). **VOICE** Distinctive, loud, very nasal ***uu-WIT***, with stressed second note. In Cochinchina gives a slightly different ***whii-IT*** and flatter ***wowh*** and ***weeao***. Birds in S Thailand are said to give a much less nasal, flatter ***prrrit*** or ***berret***. **HABITAT** Broadleaved evergreen forest, forest edge, secondary growth; up to 1,525 m. **RANGE & STATUS** Resident SW China. **SE Asia** Common resident E,S(east) Myanmar, Tenasserim, Thailand (except C), Indochina. Apparently scarce in S Thailand. **BREEDING** Undocumented.

1033 BUFF-VENTED BULBUL
Iole olivacea Plate **102**

IDENTIFICATION 20–20.5 cm. *I.o.cryptus*: **Adult** Difficult to separate from Grey-eyed Bulbul but has browner (less olive) wings; duller, greyer (less yellow-tinged) underparts and creamy-buffish to pale brownish-buff undertail-coverts (without rufous tinge); bill slightly longer and heavier. Eyes silvery-white. See Olive Bulbul. Note range. **Juvenile** As adult, but eyes light brown. **VOICE** Call is a musical ***er-whit*** or ***wher-it*** with a sharper, more metallic second note, similar to Olive Bulbul but thinner, higher-pitched and less nasal than Grey-eyed. Also gives a flatter ***whirr***. **HABITAT** Broadleaved evergreen forest, secondary growth; up to 825 m. **RANGE & STATUS** Resident Sumatra, Borneo. **SE Asia** Common resident south Tenasserim, W,S Thailand, Peninsular Malaysia. Rare (status unclear) Singapore. **BREEDING** January–June. **Eggs** 2. Otherwise undocumented.

1034 HAIRY-BACKED BULBUL
Tricholestes criniger Plate **102**

IDENTIFICATION 16.5–17 cm. *T.c.criniger*: **Adult** Relatively small size and pale yellowish lores and broad area surrounding eye distinctive. Breast mottled dirty greyish-olive, belly and undertail-coverts yellow, wings and tail strongly rufescent, eyes grey-brown to grey, feet yellowish olive-brown. Unusual long hair-like plumes extend from nape but are rarely visible in field. **Juvenile** Similar to adult, but eyes brown, feet pale pink. **VOICE** Presumed song consists of scratchy, chattering warbled phrases, interspersed with a quavering ***whirrrh***. Also utters a long, fairly high-pitched, husky, rising whistle: ***whiiii*** (repeated at longish intervals). **HABITAT** Lower to mid-storey of broadleaved evergreen forest; up to 915 m. **RANGE & STATUS** Resident Sumatra, Borneo. **SE Asia** Fairly common to common resident south Tenasserim, W(south),S Thailand, Peninsular Malaysia. **BREEDING** February–September. Otherwise undocumented.

1035 FINSCH'S BULBUL
Alophoixus finschii Plate **102**

IDENTIFICATION 16.5–17 cm. Monotypic. **Adult** Relatively small size, rather short blackish bill and overall brownish-olive to dark olive plumage with contrasting bright yellow throat and vent diagnostic. **Juvenile** Very similar to adult. **VOICE** Call consists of harsh grating ***scree*** notes. **HABITAT** Broadleaved evergreen forest; up to 760 m. **RANGE & STATUS** Resident Sumatra, Borneo. **SE Asia** Scarce to uncommon resident extreme S Thailand, Peninsular Malaysia. **BREEDING** March–August. Otherwise undocumented.

1036 YELLOW-BELLIED BULBUL
Alophoixus phaeocephalus Plate **102**

IDENTIFICATION 20–20.5 cm. *A.p.phaeocephalus*: **Adult** Lacks crest and shows unmistakable combination of bluish-grey head with contrasting whitish-grey lores, white throat and bright yellow underparts. Eyes rich red-brown (often looking dark in field). **Juvenile** As adult, but eyes light brown. **VOICE** Calls with subdued, harsh, slightly buzzy ***whi'ee whi'ee whi'ee...*** **HABITAT** Lower storey of broadleaved evergreen forest; up to 915 m. **RANGE & STATUS** Resident Sumatra, Borneo. **SE Asia** Fairly common to common resident south Tenasserim, S Thailand, Peninsular Malaysia; formerly Singapore. **BREEDING** February–August. **Eggs** 2. Otherwise undocumented.

1037 GREY-CHEEKED BULBUL
Alophoixus bres Plate **102**

IDENTIFICATION 21.5–22 cm. *A.b.tephrogenys*: **Adult** Similar to White-throated, Puff-throated and Ochraceous Bulbuls but lacks obvious crest (reduced to tuft on hind-crown), has strong dark olive wash across breast. Head-sides darker grey than White-throated, belly yellower than Ochraceous and Puff-throated (apart from *A.p.griseiceps*). Range does not overlap with Puff-throated. See Yellow-bellied. **Juvenile** Wings more rufescent-tinged, head-sides tinged brown. **VOICE** Very variable. Song consists of well-structured mournful phrases: ***whi'u wiu iwi*** and ***iiu you yuwi*** etc., regularly followed by shriller, high-pitched, discordant ***ii-wi tchiu-tchiu***; ***whii wi witchi-witchi-witchi*** and ***iiu witchu witchu*** etc. Also utters a clear ***prii chiu chew chew*** and longer ***uu-ii-chewi-chew-chew-chew***, interspersed with rattled ***trrit*** notes. **HABITAT** Broadleaved evergreen forest; up to 915 m. **RANGE & STATUS** Greater Sundas, Philippines (Palawan). **SE Asia** Uncommon to common resident Tenasserim, S Thailand, Peninsular Malaysia. **BREEDING** January–August. **Nest** Shallow cup, in low vegetation. **Eggs** 2; pinkish, blotched purplish and reddish.

1038 WHITE-THROATED BULBUL
Alophoixus flaveolus Plate **102**

IDENTIFICATION 21.5–22 cm. *A.f.burmanicus*: **Adult** Similar to Puff-throated Bulbul (particularly *A.p.griseiceps*) but breast to undertail-coverts all yellow, lores and ear-coverts more whitish-grey, crest longer and wispier. See Grey-cheeked and Yellow-bellied Bulbuls. Note range. **Juvenile** Initially rather uniform brown with whiter throat. Crown and upperparts brown, uppertail-coverts and wings more rufescent than adult, underparts suffused brown. **Other subspecies in SE Asia** *A.f.flaveolus* (SW,W,N,C Myanmar). **VOICE** Calls similar to Puff-throated but sharper and higher-pitched: ***chi-chack chi-chack chi-chack*** and nasal ***cheer*** etc. **HABITAT & BEHAVIOUR** Broadleaved evergreen forest; up to 1,525 m. Often in small groups in middle storey of forest. **RANGE & STATUS** Resident N,NE Indian subcontinent, SE Tibet, SW China (W Yunnan). **SE Asia** Uncommon to locally common resident Myanmar, W,NW Thailand. **BREEDING** March–June. **Nest** Sturdy cup, in thick understorey; up to 3 m above ground. **Eggs** 2–4; glossy deep pink (rarely lilac-tinged), spotted, blotched and irregularly streaked blood-red, blackish-red and neutral tint (often ringed or capped at broader end); 26.9 x 18.6 mm (av.; *flaveolus*).

1039 PUFF-THROATED BULBUL
Alophoixus pallidus Plate **102**

IDENTIFICATION 22–25 cm. *A.p.henrici*: **Adult** From White-throated Bulbul by much less yellow underparts, buffier undertail-coverts and darker grey head-sides; from Ochraceous by strong greenish-olive tinge to upperparts and darker underparts, usually with stronger yellow infusion (but see that species). Note range. *A.p.griseiceps* (Pegu Yomas, S Myanmar) is greener above with greyer crown, more whitish-grey lores and ear-coverts, much yellower breast and belly and deeper yellowish-buff undertail-coverts (best told from White-throated by greyer crown and supercilium, less brown-tinged upperparts, greener, less rufescent uppertail-coverts and buffish-tinged—even on throat—and less vivid yellow underparts); *robinsoni* (Tenasserim) is somewhat intermediate; *annamensis* (N,C Annam) and *khmerensis* (Cambodia, S Laos, C,S Annam) have darker breast and flanks with yellowish centre to abdomen, slatier-grey lores and head-sides. **Juvenile** Undocumented. **Other subspecies in SE Asia** *A.p.isani* (north-west NE Thailand). **VOICE** Harsh raucous abrupt ***churt churt churt...***; ***chutt-chutt-chutt...*** and ***chutt chutt chick-it chick-it...*** etc. **HABITAT** Broadleaved evergreen forest; up to 1,450 m. **RANGE & STATUS** Resident SW,S China. **SE Asia** Common resident S,E Myanmar, Tenasserim, NW,NE Thailand, Cambodia (except south-west), Laos, Vietnam (except Cochinchina).

BREEDING March–September. **Nest** Fairly deep neat cup, in small tree; 1–5 m above ground. **Eggs** 2–3; whitish to pale cream, with dark rusty-red blotches at larger end (sometimes forming cap), breaking into small blotches and fine streaks around middle.

1040 OCHRACEOUS BULBUL

Alophoixus ochraceus Plate **102**

IDENTIFICATION 19–22 cm. *A.o.sacculatus*: **Adult** Very like Puff-throated Bulbul but slightly smaller and shorter-crested, with duller, browner (less olive) upperparts and no yellow tinge below. *A.o.hallae* (S Annam, Cochinchina) does have yellowish-tinged underparts and is hard to separate from Puff-throated, differing primarily by size, crest, less pure grey head-sides, paler and slightly less yellow (rather buffier-tinged) underparts and slightly paler buff undertail-coverts. See Grey-cheeked Bulbul. **Juvenile** Wings and tail more rufescent than adult. **Other subspecies in SE Asia** *A.o.cambodianus* (SE Thailand, south-west Cambodia), *ochraceus* (central Tenasserim, south W Thailand), *sordidus* (S Tenasserim and S Thailand southwards). **VOICE** Similar to Puff-throated: harsh, raucous ***chrrt chrrt chrrt chrrt...***; ***chik-chik-chik-chik*** and ***chit'it-chit'it-chit'it-it*** (with higher ***it*** notes), normally preceded by distinctive fluty nasal ***eeyi*** and ***iiwu*** notes. **HABITAT & BEHAVIOUR** Broadleaved evergreen forest; up to 1,525 m. Frequents middle storey of forest. **RANGE & STATUS** Resident Sumatra, Borneo. **SE Asia** Common resident southern Tenasserim, W,SE,S Thailand, Peninsular Malaysia, south-west and north-east Cambodia, S Annam, Cochinchina. **BREEDING** February–July. **Nest** Fairly deep cup, on tree-branch fork or in sapling; 2.4–9 m above ground. **Eggs** 2; glossy pinkish-white with obsolete dull red blotches (heavier towards broader end) and dark reddish-chestnut spots; 25 x 17.5 mm (*sordidus*).

1041 STREAKED BULBUL

Ixos malaccensis Plate **102**

IDENTIFICATION 23 cm. Monotypic. **Adult** Combination of dull dark olive upperparts, lack of crest, greyish throat and breast with narrow (but distinct) whitish streaks, and white vent diagnostic. Bill rather long and slender. See Mountain Bulbul. Note range. **Juvenile** Upperparts and wings mostly warm rufescent-brown, breast less distinctly streaked. **VOICE** Sings with simple, short, rather high-pitched, slightly descending phrases: ***chiri-chiri-chu*** and ***chiru-chiru*** etc. Call is a loud, harsh rattle (often given in flight). **HABITAT & BEHAVIOUR** Canopy and upper storey of broadleaved evergreen forest; up to 915 m. Has distinctive wing-waving habit. **RANGE & STATUS** Resident Sumatra, Borneo. **SE Asia** Fairly common to common resident south Tenasserim, S Thailand, Peninsular Malaysia. Vagrant Singapore. **BREEDING** April–July. **Eggs** 2. Otherwise not reliably described.

1042 ASHY BULBUL *Hemixos flavala* Plate **100**

IDENTIFICATION 20.5–21 cm. *H.f.hildebrandi*: **Adult** Black crown, face and cheeks, grey upperparts with prominent bright yellowish-green wing-panel, and white underparts with grey-washed breast distinctive. Ear-coverts pale brownish. *H.f.flavala* (SW,W,N,E[north],S[west] Myanmar) has dark grey crown; *davisoni* (Tenasserim and W Thailand) has browner upperparts; *bourdellei* (NE Thailand, N Laos) has blacker crown; *cinereus* (S Thailand southwards) is much browner and more uniform above (including wings), has brownish crown with contrasting grey feather-fringes, narrow greyish supercilium and distinctive triangular black patch on lores and cheeks; *H.f.remotus* (S Laos, S Annam) has paler and browner dark head markings, mid-brown upperparts and brownish-tinged breast. **Juvenile** Similar to adult but upperparts browner-tinged. **VOICE** Song is a simple, repeated, rather high-pitched ***ii-wit'ti-ui*** etc., with second note higher. Calls include a loud ringing nasal ***tree-tree-tree...*** **HABITAT & BEHAVIOUR** Broadleaved evergreen forest, forest edge; up to 2,100 m (mostly above 600 m). Usually frequents upper middle storey and forest canopy. Often puffs out throat. **RANGE & STATUS** Resident N,NE Indian subcontinent, SE Tibet, southern China, Sumatra, Borneo. **SE Asia** Fairly common to common resident (subject to some movements) Myanmar, Thailand (except C,SE), Peninsular Malaysia, E Cambodia, Laos, W Tonkin, E Tonkin (Mt Ba Vi only), N,C,S Annam. Rare non-breeding visitor Singapore. **BREEDING** January–July. **Nest** Open, deepish cup, in small tree or dense undergrowth; 0.6–3 m above ground. **Eggs** 2–4; pearl-white to pale pink, profusely speckled pinkish-red or reddish-brown overall; 24.3 x 17.3 mm (av.; *flavala*).

1043 CHESTNUT BULBUL

Hemixos castanonotus Plate **100**

IDENTIFICATION 21.5 cm. *H.c.canipennis* (recorded north E Tonkin) **Adult** Similar to Ashy Bulbul but shows diagnostic chestnut head-sides and upperparts (apart from crown). Upperwing-coverts, tertials and secondaries fringed whitish-grey. *H.c.castanonotus* (resident E Tonkin) shows yellowish-green wing-feather fringes. **Juvenile** Head-sides and upperparts much duller, crown much paler brown, wings paler and more uniform, breast-band browner. Gradually attains black on crown. **VOICE** Sings with clear simple jolly phrases: ***whi i-wu*** etc. **HABITAT** Broadleaved evergreen forest, forest edge, secondary growth; up to 1,000 m. **RANGE & STATUS** Resident (subject to local movements) S China. **SE Asia** Local resident (and winter visitor?) E Tonkin. **BREEDING** China: May–June. **Nest** Cup, in tree or bush. **Eggs** 3; dull pink, spotted, blotched and streaked pale and dark carmine over violet underblotches; 24.5–25.5 x 17–17.5 mm (av.; *canipennis*).

1044 MOUNTAIN BULBUL

Ixos mcclellandii Plate **102**

IDENTIFICATION 21–24 cm. *I.m.tickelli*: **Adult** Combination of shaggy crown-feathers, greenish-olive upperparts and wings, narrow whitish streaks on brownish crown, greyish throat and upper breast and yellow vent distinctive. Head-sides dull rufous, bill rather long and slender, eyes chestnut. *I.m.similis* (east N Myanmar, N Indochina) has dull greyish-brown upperparts, and dull pale pinkish-chestnut ear-coverts, neck and breast; *mcclellandii* (west N Myanmar) and *ventralis* (SW,W,S[west] Myanmar) are like *similis* but have greenish-olive upperparts; *canescens* (south-west Cambodia) is duller and plainer than *tickelli*. See Streaked, which is unlikely to occur in same area, and Striated. **Juvenile** Browner above than adult with duller, browner wings; crown less shaggy, eyes brownish-grey. **Other subspecies in SE Asia** *I.m.loquax* (NW[east],NE[north-west] Thailand, S Laos), *griseiventer* (S Annam), *peracensis* (S Thailand, Peninsular Malaysia). **VOICE** Calls with a repeated shrill, squawking ***cheu*** or ***tscheu*** and ***tchi-chitu***. **HABITAT & BEHAVIOUR** Broadleaved evergreen forest; 800–2,590 m. Often puffs out throat. **RANGE & STATUS** Resident N,NE Indian subcontinent, SE Tibet, southern China. **SE Asia** Common resident Myanmar, Thailand (except C,SE), Peninsular Malaysia, Indochina (except Cochinchina). **BREEDING** March–November. **Nest** Neat, rather shallow cup, in tree, bamboo or bush; 1.2–12 m above ground. **Eggs** 2–4; white or pale cream, speckled (sometimes blotched) light reddish-brown overall; 25.7 x 18.1 mm (av.; *mcclellandii*).

1045 HIMALAYAN BLACK BULBUL

Hypsipetes leucocephalus Plate **100**

IDENTIFICATION 23.5–26.5 cm. *H.l.concolor* (resident S[east],E Myanmar, Thailand, Indochina): **Adult** Relatively large size, all-blackish to blackish-grey plumage (greyer below) and red bill, legs and feet diagnostic. Shows shaggy crown and throat, tail broadens towards tip. *H.l.ambiens* (N Myanmar), *perniger* (E Tonkin?) and *sinensis* (visitor NE Thailand, E Tonkin, S Laos) show overall blacker plumage (but latter sometimes greyer below); *leucocephalus* (visitor N,E Myanmar, northern Indochina) and *stresemanni* (visitor NW,NE Thailand, S Laos, E Tonkin) have all-white head, latter also with paler, greyer rump and underparts; *leucothorax* (visitor W,N,E Myanmar, NW,NE Thailand, W,E Tonkin, S Annam) has all-white head and breast. Birds of white-headed subspecies can be greyer below or show varied black and white head patterns, possibly relating to age. See White-headed Bulbul. **Juvenile** Dark plumage-parts browner overall, wings and tail much browner. **Other subspecies in SE Asia** *H.l.nigrescens* (SW,W,C,S[west] Myanmar). **VOICE** Song is a discordant but measured series of 4–5 notes: ***trip wi tit-i-whi*** etc., with higher ***wi/whi*** notes. Calls include a long mewing whistled ***hwiiii*** or ***hwieeer***, and abrupt, rather nasal ***her her hic-her her...*** **HABITAT & BEHAVIOUR** Broadleaved evergreen and mixed deciduous forests; 500–3,000 m, down to 120 m in winter. Often in large flocks. **RANGE & STATUS** Breeds N Pakistan, NW,N,NE Indian subcontinent, S,SE Tibet, C and southern China, Taiwan; some populations winter to west, south-west and south. **SE Asia** Fairly common to common resident Myanmar, W,NW,NE Thailand, Indochina (except Cochinchina). Uncommon to locally common winter visitor W,N,C,E Myanmar, NW,NE Thailand, Cambodia, C,S Laos, W,E Tonkin, N,C Annam. Also recorded on passage E Tonkin. Vagrant S Thailand. **BREEDING** March–June. **Nest** Rather shallow cup, in tree; 9 m above ground. **Eggs** 2–4; pinkish-white, profusely spotted, blotched or clouded various shades of red, brownish-red and purple (often ringed or capped at broader end); 27.1 x 19.9 mm (av.; *nigrescens*).

1046 WHITE-HEADED BULBUL

Cerasophila thompsoni Plate **100**

IDENTIFICATION 20 cm. Monotypic. **Adult** Similar to white-headed subspecies of Himalayan Black Bulbul but easily separated by smaller size and bill, lack of shaggy crown and throat feathers, uniform paler grey body-plumage and rufous-chestnut undertail-coverts. Lores black. **Juvenile** Browner overall, head initially brownish-grey (gradually turning white). **VOICE** Harsher and more buzzing than Himalayan Black Bulbul. Song consists of variable short squeaky phrases, including a rhythmic ***chit-chiriu chit-chiriu...*** **HABITAT** Secondary forest, scrub and grass with scattered trees, edge of broadleaved evergreen forest; 900–2,135 m. **RANGE & STATUS** Endemic. Uncommon to locally common resident N,C,E,S(east) Myanmar, north Tenasserim, W,NW Thailand. **BREEDING** March–April. **Nest** Fairly shallow cup, in recess in bank; 1.4 m above ground. **Eggs** 2–3; pinkish-white, densely freckled and blotched reddish-brown to maroon overall (sometimes capped at broader end); 23.1 x 16.8 mm (av.).

HIRUNDINIDAE: HIRUNDININAE: Martins, swallows & allies

Worldwide c.86 species. SE Asia 13 species. Small to smallish with streamlined bodies, small bills, legs and feet, pointed wings and forked tails (some species with streamers extending from outer feathers). Mostly aerial, flight agile and graceful. Feed on insects.

1047 NORTHERN HOUSE-MARTIN

Delichon urbicum Plate **103**

IDENTIFICATION 13–14 cm. *D.u.lagopoda*: **Adult** Similar to Asian House-martin but rump-patch whiter and larger (extends to uppertail-coverts), underparts whiter overall, has greyish-white underwing-coverts (often hard to discern in field) and more strongly forked tail. **Juvenile** Upperparts browner with pale-tipped tertials, underparts variably washed pale brownish-grey, often with duller breast-sides; shows some dark feather-centres on vent and tail-coverts. **VOICE** Sings with unstructured, chirpy twittering. Calls include a sharp ***d-gitt***, scratchy, dry twittering ***prrit*** notes and emphatic, drawn-out ***chierr*** when agitated. **HABITAT** Over forests and open areas; up to 2,565 m. **RANGE & STATUS** Breeds N Africa, Palearctic, Middle East, N Pakistan, NW India, SW Tibet, NW,NE China; winters sub-Saharan Africa, possibly S India. **SE Asia** Scarce to uncommon winter visitor N,C,E,S Myanmar, Tenasserim, NW,NE Thailand, Laos, C,S Annam, Cochinchina.

1048 ASIAN HOUSE-MARTIN

Delichon dasypus Plate **103**

IDENTIFICATION 12–13 cm. *D.d.dasypus*: **Adult** Smallish size, shallowly forked tail, blue-glossed black upperparts and whitish rump and underside distinctive. Has faint dark streaking on rump (difficult to see in field) and typically shows darker centres to undertail-coverts. Very similar to Northern House-martin but uppertail-coverts and upper chin blackish, underparts sullied greyish-brown, undertail-coverts darker than belly, underwing-coverts blackish, tail less forked (often looks almost square when spread). See Nepal House-martin. **Juvenile** Upperparts browner, tertials tipped whitish, tail less deeply forked. **Other subspecies in SE Asia** *D.d.cashmiriensis* (recorded Myanmar, NW Thailand). **VOICE** Song is similar to Northern. Calls include a reedy ***screeeel***. **HABITAT** Over forests and open areas; up to 3,100 m. **RANGE & STATUS** Breeds S,SE Siberia, NE Afghanistan, N Pakistan, NW,N,NE Indian subcontinent, SW,S,SE Tibet, W,C and southern China, Taiwan, N,S Korea, Japan; more northern populations winter to south, Greater Sundas, Philippines. **SE Asia** Uncommon to fairly common winter visitor W,N,C,S Myanmar, Thailand, Peninsular Malaysia, Singapore, Cambodia, Laos, C,S Annam. Rare to scarce passage migrant E Tonkin.

1049 NEPAL HOUSE-MARTIN

Delichon nipalense Plate **103**

IDENTIFICATION 11.5–12.5 cm. *D.n.nipalense*: **Adult** Similar to Asian House-martin but somewhat smaller and more compact with almost square-cut tail, mostly dark throat (sometimes only chin), black undertail-coverts and narrower white rump-band. **Juvenile** Upperparts browner, throat and undertail-coverts mixed with whitish, breast greyish. **Other subspecies in SE Asia** *D.n.cuttingi* (recorded N Myanmar). **VOICE** Calls with high-pitched ***chi-i***. **HABITAT** Over forested and open areas, often near cliffs; up to 1,830 m. **RANGE & STATUS** Resident N,NE Indian subcontinent, S Tibet. **SE Asia** Uncommon resident SW,W,N Myanmar, NW(south-east),NE Thailand, Laos, W Tonkin. Recorded (status uncertain) E Tonkin. **BREEDING** India: April–July. Colonial. **Nest** Mud dome with narrow entrance hole, attached to underside of rock overhang. **Eggs** 3–5; white; 18.6 x 12.8 mm (av.).

1050 COMMON SAND-MARTIN

Riparia riparia Plate **103**

IDENTIFICATION 11.5–13 cm. *R.r.ijimae*: **Adult** Larger and stockier than similar Grey-throated Sand-martin, underparts whitish with sharply contrasting broad brown breast-band, upperparts somewhat darker brown, tail somewhat more deeply forked. See very similar Pale Sand-martin. **Juvenile** Feathers of upperparts and wings fringed pale buffish (often most obvious on uppertail-coverts and tertials), throat tinged buff, rest of underparts may be less white. **VOICE** Calls with a dry rasping trrrsh and a higher chiir when alarmed. The probable song is an extended, harsh twittering ***trrrsh trre-trre-trre-rrerrerre...*** etc. **HABITAT & BEHAVIOUR** Lakes, large rivers, marshes, sometimes various open areas away from water; up to 1,830 m (mostly lowlands). Typically gregarious, often associating with other swallows and martins. **RANGE & STATUS** Breeds Holarctic, Egypt, Middle East, NE India, NE China, N Korea, N Japan; winters sub-Saharan Africa, S China, Borneo, S America, rarely Philippines. Recorded (status unclear) W and northern India. **SE Asia** Uncommon winter visitor (except S Laos, W Tonkin, N,S Annam). Also recorded on passage N Laos.

1051 **PALE SAND-MARTIN**

Riparia diluta Plate **103**

IDENTIFICATION 11–12.5 cm. *R.d.fokienensis*: **Adult** Difficult to separate from Common Sand-martin but slightly smaller and smaller-billed with slightly shallower tail-fork, upperparts paler and greyer, underparts less clean white, with creamy wash on belly and undertail-coverts, breast-band paler but neater, and often faint in centre. Dark smudging on centre of breast/upper belly indistinct or lacking, pale tertials and inner secondaries can contrast strongly with rest of wing. Paler than Grey-throated Sand-martin, but uniform above, without such contrastingly paler rump/uppertail-coverts; throat whiter (can be hard to see in flight). **Juvenile** Differs from adult in similar way to Common Sand-martin. **VOICE** Undocumented. **HABITAT** Lakes, rivers, marshes; lowlands. **RANGE & STATUS** Breeds C Asia, S Siberia, E Iran, Afghanistan, N Pakistan, NW,N India, S,E Tibet, W,NW,N,C,E China, Mongolia; winters south to Pakistan, northern India and S China at least. **SE Asia** Recorded in winter (exact status unknown) E Myanmar, E Tonkin. Vagrant Peninsular Malaysia.

1052 **GREY-THROATED SAND-MARTIN**

Riparia chinensis Plate **103**

IDENTIFICATION 10.5–12 cm. *R.p.chinensis*: **Adult** Similar to Common and Pale Sand-martins but smaller and daintier, with shallower tail-fork, no breast-band, throat and breast sullied greyish-brown (throat occasionally slightly paler than breast), rump and uppertail-coverts noticeably paler than mantle. **Juvenile** Feathers of upperparts and wings fringed warm buffish (most obviously on rump/uppertail-coverts and tertials), throat and breast paler and buffier. **VOICE** Song is a weak, high-pitched twitter. Usual calls are a subdued spluttering ***chrr'r*** and short, spaced, slightly explosive ***chit*** or ***chut*** notes. **HABITAT & BEHAVIOUR** Large rivers, lakes; up to 1,220 m. Gregarious. **RANGE & STATUS** Resident SE Uzbekistan, SW Tajikistan, N Afghanistan, Pakistan, NW,N,NE Indian subcontinent, W,C India, SW China, Taiwan, Philippines. **SE Asia** Locally common resident Myanmar, NW,NE Thailand, Cambodia, Laos, W,E Tonkin. Decreasing outside Myanmar. **BREEDING** November–April. Colonial. **Nest** Tunnel in sandbank, with chamber. **Eggs** 3–5; white; 17.3 x 12.2 mm (av.).

1053 **DUSKY CRAG-MARTIN**

Ptyonoprogne concolor Plate **103**

IDENTIFICATION 13–14 cm. Monotypic. **Adult** Uniform dark brown plumage diagnostic. Throat and breast slightly paler with thin dark streaks on former, tail barely forked and with row of whitish spots (apparent when tail spread). **Juvenile** Feathers of upperparts and wings narrowly fringed rufous-grey, throat paler with no streaks. **VOICE** Song consists of soft twittering sounds. Usually calls with soft ***chit*** notes. **HABITAT & BEHAVIOUR** Various habitats, normally in vicinity of cliffs, caves or buildings; up to 2,000 m. Often found in association with other swallows and martins. **RANGE & STATUS** Resident SE Pakistan, India, SW China (S Yunnan). **SE Asia** Local resident C,S(south-east),E Myanmar, Tenasserim, Thailand (except SE), Peninsular Malaysia, Laos, W Tonkin, E Tonkin, N,S Annam, Cochinchina. **BREEDING** March–April. **Nest** Oval mud saucer, attached to vertical rock-face, building or other artificial site; up to 30 m above ground. **Eggs** 2–4; white, finely speckled reddish-brown; 17.6 x 12.8 mm (av.).

1054 **BARN SWALLOW** *Hirundo rustica* Plate **103**

IDENTIFICATION 15 cm (outertail-feathers up to 5 cm more). *H.r.gutturalis*: **Adult breeding** Combination of glossy blue-black upperside, chestnut-red forehead and throat, blue-black breast-band, all-whitish remainder of underparts and underwing-coverts and deeply forked tail distinctive. Has row of whitish spots/streaks across tail, outertail-feathers extend to long narrow streamers. *H.r.tytleri* (recorded SW,N,E,C,S Myanmar, C,SE Thailand, Peninsular Malaysia, C Annam) has pale rufous underparts and underwing-coverts. **Adult non-breeding** Lacks tail-streamers. Moulting birds may be similar to juvenile. **Juvenile** Similar to non-breeding adult but upperparts browner, forehead and throat dull orange-buff, breast-band browner. **Other subspecies in SE Asia** *H.r.mandschurica* (recorded NW Thailand). **VOICE** Sings with a rapid twittering, interspersed with a croaking sound that extends to a dry rattle. Typical calls are a high-pitched sharp ***vit*** (often repeated), and louder, anxious ***vheet vheet...*** or ***flitt-flitt*** when agitated. **HABITAT & BEHAVIOUR** Open areas, often near water and habitation; up to 2,000 m. Often roosts in very large numbers. **RANGE & STATUS** Breeds Holarctic, N Africa, Middle East, W,N Pakistan, NW,N,NE Indian subcontinent, S,E Tibet, China, Taiwan, N,S Korea, Japan; winters Africa, Indian subcontinent, SW China, Sundas, Philippines, Wallacea, New Guinea, N Australia, C,S America. **SE Asia** Local resident/breeding visitor NW,NE Thailand, N Laos, W,E Tonkin. Common winter visitor and passage migrant throughout. Also recorded in summer (regularly in places) Peninsular Malaysia, Singapore, Cambodia, S Laos, C,S Annam, Cochinchina. **BREEDING** May–June. **Nest** Mud saucer, attached to vertical wall. **Eggs** 4–6; white with reddish- to purplish-brown speckles (denser towards broader end); 19.6 x 13.7 mm (av.; India).

1055 **WIRE-TAILED SWALLOW**

Hirundo smithii Plate **103**

IDENTIFICATION 13–14 cm (tail-streamers up to 12.5 cm more). *H.s.filifera*: **Adult** Recalls Barn Swallow but shows diagnostic combination of chestnut crown and completely snowy-white underparts (including throat) and underwing-coverts; apart from distinctive dark thigh-patch. Upperparts also bluer and tail square-cut, with much longer wire-like streamers. **Juvenile** Upperparts browner, crown duller and paler, throat faintly washed warm buffish. **VOICE** Has twittering song. Typical calls include ***chit-chit*** and ***chirrik-weet chirrik-weet...***, and ***chichip chichip...*** when alarmed. **HABITAT** Large rivers, lakes and nearby areas; up to 1,980 m. **RANGE & STATUS** Resident (subject to relatively local movements) sub-Saharan Africa, SE Turkmenistan, SE Uzbekistan, Afghanistan, Pakistan, India (except NE), SW Nepal, SW China (W Yunnan). **SE Asia** Scarce to uncommon resident (subject to local movements) Myanmar, NW,NE Thailand, Cambodia, Laos, C,S(now rare) Annam. **BREEDING** January–November. **Nest** Mud saucer, attached to rock-face, building or other artificial site. **Eggs** 3–4; white, speckled and blotched reddish-brown (usually mostly towards broader end); 18.4 x 13.1 mm (av.).

1056 **HOUSE SWALLOW**

Hirundo tahitica Plate **103**

IDENTIFICATION 13–14 cm. *H.t.abbotti*: **Adult** Similar to Barn Swallow but chestnut-red of forehead more extensive, upper breast chestnut-red (without blue-black band), breast and flanks dirty greyish-brown, centres to undertail-coverts dark brown, underwing-coverts dusky brownish and tail less forked (without streamers). **Juvenile** Browner above than adult with less chestnut-red on forehead, throat and upper breast paler. **Other subspecies in SE Asia** *H.t.javanica* (Coco Is, off S Myanmar). **VOICE** Short, high-pitched, slightly explosive ***swi*** or ***tswi***, sometimes in series: ***tswi-tswi-tswi*** etc. Also a lower ***swoo***. **HABITAT** Coastal habitats, often near habitation; also over forest, clearings and open areas inland, up to 2,010 m (Peninsular Malaysia). **RANGE & STATUS** Resident Andaman Is, Taiwan, S Japan (Nansei Is), Sundas, Philippines, Wallacea, New Guinea region, Melanesia and Polynesia east to Tahiti. **SE Asia** Common coastal resident (except C Thailand, E Tonkin, N,C Annam). Also common inland (subject to minor movements) Peninsular Malaysia. **BREEDING** January–September. **Nest** Mud saucer, attached to rock-face, building or other artificial site. **Eggs** 3–4; white, speckled reddish-brown and deep purple-brown over lavender and grey undermarkings; 17.5 x 12.7 mm (av.; *javanica*).

1057 **RED-RUMPED SWALLOW**
Cecropis daurica Plate **103**
IDENTIFICATION 16–17 cm (tail-streamers up to 3.5 cm more). *C.d.japonica*: **Adult** Easily told from Barn Swallow by orange-rufous neck-sides and rump and dark-streaked whitish underparts (including throat). Very similar to Striated Swallow subspecies *mayri* but slightly smaller, with almost complete orange-rufous nuchal collar (narrowly interrupted on centre of nape) and somewhat narrower dark streaks on rump and underparts. Breast-streaks possibly tend to be slightly heavier than belly streaks. *C.d.nipalensis* (widespread records, or Myanmar only?) has thinner, more even underpart-streaking. **Juvenile** Duller; nuchal collar, rump and underparts paler, tertials browner and pale-tipped, tail-streamers shorter. **VOICE** Song is similar to Barn Swallow but twittering lower-pitched, harsher, slower and shorter. Calls include a short nasal ***djuit*** or ***tveyk*** and sharp ***kiir*** notes when alarmed. **HABITAT** Open areas, often near water; up to 2,440 m. **RANGE & STATUS** Breeds Africa, southern W Palearctic, C Asia, S,SE Siberia, Middle East, Indian subcontinent, S,E Tibet, China (except NW), N,S Korea, Japan; northern populations winter sub-Saharan Africa, N,NE Indian subcontinent, S China. **SE Asia** Uncommon to fairly common winter visitor throughout. Also recorded on passage W,E Tonkin.

1058 **STRIATED SWALLOW**
Cecropis striolata Plate **103**
IDENTIFICATION 18–18.5 cm (tail-streamers up to 3.5 cm more). *C.s.stanfordi* (breeding range apart from N[west] Myanmar, north Tenasserim, W Thailand): **Adult** Like Red-rumped Swallow but slightly larger, lacks (near complete) orange-rufous nuchal collar, has much broader streaks on rump and underparts. Only shows a little reddish-rufous behind ear-coverts. *C.s.mayri* (wintering range; breeding only in west N Myanmar, ? and north W Myanmar) has narrower streaks (only slightly broader than on Red-rumped), *vernayi* (resident north Tenasserim, W Thailand) is like *stanfordi* but base colour of lores, ear-coverts, upper breast, flanks and undertail-coverts reddish-rufous (rest of underparts with paler reddish-rufous wash), underpart-streaks shorter (more like spot-streaks on throat and breast). **Juvenile** Duller above, rump paler, tertials browner and pale-tipped, tail-streamers shorter. **VOICE** Sings with soft twittering notes, including a low, metallic, rippling purr introduced by quick high note: ***her tr'r'r'r*** and ***hii tr'r'r'r***. Calls include quite high short ***chwi*** or ***trri***; ***chree*** or ***chrr*** and ***weet*** notes (which are sometimes doubled), and less high, more burry couplets, ***hwii-hu*** or ***wrri-wru*** and ***hrii-her***. **HABITAT** Open areas (often near water), cliffs; up to 2,565 m. **RANGE & STATUS** Resident NE India, NE Bangladesh, SW China (S Yunnan), Taiwan, Sundas, Philippines. **SE Asia** Common resident N,C,S,E Myanmar, north Tenasserim, W,NW,NE Thailand, Cambodia, Laos, N,C,S Annam, Cochinchina. Uncommon winter visitor N,S[east],E Myanmar, NW Thailand, C Annam; ? E Tonkin. Recorded (status uncertain, but probably breeds) north W Myanmar, W Tonkin. **BREEDING** February–June. **Nest** Dome-shaped with long tubular entrance, attached to underside of rock overhang or on building, bridge or other artificial site. **Eggs** 2–5; white; 21.4 x 14.7 mm (av.; *mayri*).

1059 **RUFOUS-BELLIED SWALLOW**
Cecropis badia Plate **103**
IDENTIFICATION c.19 cm (tail-streamers up to 5 cm more). Monotypic. **Adult** Recalls Striated Swallow but larger with deep rufous-chestnut underparts and underwing-coverts. **Juvenile** Lores and head-side duller/paler than adult, with some sooty-brown mottling on ear-coverts, duller above, with gloss restricted to feather-tips (uppertail-coverts unglossed), rump a shade paler, underparts paler/duller with narrower shaft-streaks and some sooty mottling; pale chestnut tips to tertials and secondaries. **VOICE** Song is similar to Striated. Calls include a single ***tweep*** or sharp ***cheenk*** and rather excited, tremulous, quavering ***chir'r'r'rw***, ***chwirrrr***, ***chwi'i'i'i*** or ***schwirrrr***. **HABITAT** Open areas (often near water), cliffs; up to 1,250 m. **RANGE & STATUS** Endemic. Common resident S Thailand, Peninsular Malaysia, Singapore. **BREEDING** February–June. **Nest** Dome-shaped with tubular entrance of variable length, attached to underside of rock overhang or on building, bridge or other artificial site. **Eggs** 2–3; white; 23.1 x 16.7 mm. **NOTE** Split from Striated following Dickinson & Dekker (2001).

HIRUNDINIDAE: PSEUDOCHELIDONINAE: River-martins

Worldwide 2 species. SE Asia 1 species. Recall martins and swallows but more robust and bigger-headed, with stouter bills and relatively large feet. Flight rapid and energetic with brief periods of gliding. Feed primarily on small beetles and other winged insects.

1060 **WHITE-EYED RIVER-MARTIN**
Pseudochelidon sirintarae Plate **103**
IDENTIFICATION 15 cm (tail-streamers up to 9 cm more). Monotypic. **Adult** Unmistakable. Robust and big-headed with stout yellow bill, white eyes and broad eyering, all-dark underparts and white rump-band. Has long narrow streamers extending from central tail-feathers. **Juvenile** Head and underparts browner with paler throat; lacks tail-streamers. **VOICE** Unknown. **HABITAT** Found in lakeside reeds during night (where possibly roosting); lowlands. **RANGE & STATUS** Only known from C Thailand (Bung Boraphet) in winter. May be extinct (last definite record 1980). **BREEDING** Unknown. Likely to be colonial, nesting in burrows in riverine sandbars.

CETTIIDAE: *Abroscopus* warblers, Mountain Tailorbird, Broad-billed Warbler, *Cettia* bush-warblers, tesias & allies

Worldwide at least 34 species. SE Asia 17 species. Small and active with relatively fine, pointed bills. Plumage and habitats vary widely. Feed mainly on insects and spiders etc. ***Abroscopus* warblers**, **Mountain Tailorbird**, and **Broad-billed Warbler** Similar to *Seicercus* warblers; generally well marked with varied head-patterns that generally include some rufous or chestnut, and with yellow on underparts; all in wooded habitats. ***Cettia* bush-warblers** Small, unstreaked and brownish with paler underparts, some of latter with throat-/breast-spotting), tail relatively square-ended with 10 feathers, very skulking, close to ground, mostly in grass and scrub. **Tesias** (*Tesia*) and **Asian Stubtail** Very small, almost tail-less or very short-tailed; mainly terrestrial, occurring in wooded habitats.

1061 **YELLOW-BELLIED WARBLER**
Abroscopus superciliaris Plate **105**
IDENTIFICATION 9.5–11.5 cm. *A.s.superciliaris*: **Adult** Plain greyish crown and ear-coverts, whitish supercilium, olive-green upperparts, whitish throat and upper breast and yellow remainder of underparts distinctive. Has darker eye-stripe, lacks wing-bars and white on tail. *A.s.smythiesi* (C,S Myanmar) has grey of crown paler and more limited to forehead, yellower-tinged upperparts; *sakaiorum* (southern S Thailand southwards) shows white on centre of belly. See Grey-hooded Warbler. **Juvenile** Paler yellow below. **Other subspecies in SE Asia** *A.s.drasticus* (SW,W Myanmar),

bambusarum (northern S Thailand), *euthymus* (Vietnam). **VOICE** Song thin, high-pitched and tinkling, with 3–6 notes per phrase: ***uu-uu-ti*** and ***uu-uu-ti-i*** etc., with higher ***ti*** notes. Occasionally gives a continuous, subdued thin twittering. **HABITAT** Bamboo in or near broadleaved evergreen, semi-evergreen and deciduous forest, secondary growth; up to 1,525 m. **RANGE & STATUS** Resident Nepal, NE Indian subcontinent, SE Tibet, SW China, Greater Sundas. **SE Asia** Common resident (except C,SE Thailand, Singapore). **BREEDING** February–August. **Nest** Pad, in hole in dead bamboo stem, sometimes tree-branch. **Eggs** 2–5; glossy whitish, boldly spotted or minutely freckled reddish-brown overall; 15.2 x 11.6 mm (av.; India).

1062 RUFOUS-FACED WARBLER

Abroscopus albogularis Plate **105**

IDENTIFICATION 8.5–9.5 cm. *A.a.hugonis*: **Adult** Rufescent-olive crown, blackish lateral crown-stripes, rufous head-sides, black-and-white mottled throat and whitish underparts with yellow breast-band and undertail-coverts diagnostic. Rump whitish-yellow, lacks wing-bars and white on tail. *A.a.fulvifacies* (Indochina) has more olive, less rufescent crown and very faint yellow breast-band. **Juvenile** Rufous of head-sides pale and washed out, lacks blackish lateral crown-stripes, throat-streaking much less obvious and restricted to a few fine dark shaft-streaks, green of upperparts tinged warmer and browner. **Other subspecies in SE Asia** *A.a.albogularis* (Myanmar). **VOICE** Sibilant, high-pitched, repetitive whistles: ***titiriiiii titiriiiii titiriiiii titiriiiii...*** **HABITAT** Broadleaved evergreen forest, bamboo, secondary growth; 450–1,800 m. **RANGE & STATUS** Resident E Nepal, NE Indian subcontinent, W,C and southern China, Taiwan. **SE Asia** Rare to local resident W,N Myanmar, NW Thailand, N,C Laos, W,E Tonkin, N,C Annam. **BREEDING** March–May. **Nest** Cupped pad, in hollow bamboo stem, sometimes rock crevice; 30 cm or so above ground. **Eggs** 3–5; similar to Yellow-bellied Warbler; 14.4 x 11.5 mm (av.; *albogularis*).

1063 BLACK-FACED WARBLER

Abroscopus schisticeps Plate **105**

IDENTIFICATION 10 cm. *A.s.ripponi*: **Adult** Slaty-grey crown, nape and upper breast-sides, blackish mask, yellow supercilium and yellow throat with sooty centre diagnostic. Has only little white on tail, whitish lower breast and belly and yellow undertail-coverts; may show grey breast-band. *A.s.flavimentalis* (W Myanmar) has all-yellow throat, almost lacks grey on breast and has paler grey on head. See Yellow-bellied Fantail. **Juvenile** Crown and nape washed greenish, yellow of throat paler. **VOICE** Song is a very thin, high-pitched, tinkling ***tirririr-tsii tirririr-tsii tirririr-tsii...*** and ***tit sirriri-sirriri sirriri tit sirriri-sirriri...*** etc. Contact calls are very subdued ***tit*** notes. **HABITAT** Broadleaved evergreen forest; 1,525–2,350 m. **RANGE & STATUS** Resident N,NE India, Nepal, Bhutan, S,SE Tibet, SW China. **SE Asia** Fairly common resident W,N,E Myanmar, W,E(north-west) Tonkin. **BREEDING** April–May. **Nest** Pad, in hole in tree or dead bamboo stem. **Eggs** 4–5; whitish to dull pinkish-white, mottled and streaked or densely freckled red to reddish-brown; 15.1–15.4 x 11.1–11.3 mm (*flavimentalis*).

1064 MOUNTAIN TAILORBIRD

Phyllergates cucullatus Plate **105**

IDENTIFICATION 10.5–12 cm. *P.c.coronatus*: **Adult** Easily told from other tailorbirds by golden-rufous forecrown, long yellowish-white supercilium, dark eyestripe and bright yellow belly and undertail-coverts. Resembles Broad-billed Warbler but has much longer bill, paler and more restricted rufous on crown, greyish hindcrown and nape, greener upperparts, yellowish supercilium, paler greyish-white throat and upper breast, and no white on tail. *P.c.malayanus* (extreme S Thailand, Peninsular Malaysia) has darker rufous-chestnut on crown, more extensive darker grey on hindcrown and nape, and deeper green upperparts. **Juvenile** Resembles adult but crown and mantle uniform dull green, grey of hindcrown/nape duller, pattern on head-sides fainter with yellower supercilium, underparts more uniform (less bright) yellow. Like adult Yellow-bellied Warbler but lacks grey on crown, bill much longer. Gradually attains rufous on forehead and crown-sides. **Other subspecies in SE Asia** *P.c.thais* (S Thailand). **VOICE** Song consists of 4–6 very high-pitched notes, glissading up and down the scale (usually introduced by 1–2 short notes). Gives thin ***trrit*** notes when agitated. **HABITAT** Undergrowth in broadleaved evergreen forest, bamboo, forest edge, scrub; 1,000–2,200 m, locally down to 450 m (N Myanmar). **RANGE & STATUS** Resident NE India, Bhutan, E Bangladesh, SW,S China, Greater Sundas, W Lesser Sundas, Philippines, Sulawesi, Moluccas. **SE Asia** Common resident (except SW,C Myanmar, C,SE Thailand, Singapore, Cochinchina). **BREEDING** February–May. **Nest** Deep flimsy pouch, in rattan, sapling, bamboo, ferns overhanging bank etc.; 1.5–8 m above ground. **Eggs** 3–4; pale blue with fine reddish specks; 15.5 x 11.3 mm (av.; *coronatus*). **NOTE** Not related to *Orthotomus* tailorbirds (Alström *et al.* 2006). Wells (2007) suggested *Phyllergates* as an available alternative.

1065 BROAD-BILLED WARBLER

Tickellia hodgsoni Plate **105**

IDENTIFICATION 10 cm. *T.h.hodgsoni*: **Adult** Resembles Mountain Tailorbird but has darker rufous crown, greyer supercilium, duller eyestripe, duller green upperparts, darker and greyer throat and upper breast and much shorter bill. **Other subspecies in SE Asia** *T.h.tonkinensis* (W Tonkin). **VOICE** Song is a very thin high-pitched ***si seeee-ee-eee si seeee-ee-eee siiiiiiiiiii siiiiiiiiiii...*** etc., often increasing in pitch and becoming difficult to hear. Also excited, rapid, rather metallic ***witiwiwitiwi-chu-witiwiwitiwit*** and ***chitiwichitiwit-chit-chitiwichitiwit-chit-chitiwichitiwit...*** etc. **HABITAT** Bamboo, broadleaved evergreen forest; 1,830–2,650 m, locally down to 1,050 m (N Myanmar). **RANGE & STATUS** Resident E Nepal, Bhutan, NE India, SW China. **SE Asia** Uncommon resident W,N Myanmar, W Tonkin. **BREEDING** India: May–June. **Nest** Oval dome with entrance near top, in sapling or bamboo. **Eggs** 3; pale claret-red, speckled and streaked darker claret-red (mainly at broader end); 16 x 11.9 mm (av.; *hodgsoni*).

1066 HUME'S BUSH-WARBLER

Cettia brunnescens Plate **104**

IDENTIFICATION 11.5 cm. Monotypic. **Adult** Smaller, smaller-billed and shorter-tailed than Brownish-flanked and Aberrant Bush-warblers, with more rufescent upperside and wing-feather fringes; yellowish belly contrasts with dull-coloured throat and breast. Rufous on upperparts variably intense; possibly paler and less yellow below when worn. **Juvenile** Poorly documented; perhaps similar to adult but with all underparts washed yellow. **VOICE** Song is an ascending (stepped) series of c.4 increasingly longer, drawn-out, very high-pitched whistles (becoming almost inaudible), abruptly followed by a rather slow series of melodious, oscillating whistles. Calls with doubled ***tret*** or ***trit*** (sometimes singly), and quickly repeated longer ***trrrt*** when agitated. **HABITAT** Bamboo and other undergrowth in or near broadleaved evergreen and mixed evergreen and coniferous forest. Occurs at 2,000–3,660 m in Indian subcontinent (once down to 1,350 m in winter). **RANGE & STATUS** Resident N,NE India, Nepal, Bhutan, SE Tibet. **SE Asia** Claimed for E Myanmar but specimen may have been misidentified. Should be resident N Myanmar. **BREEDING** Unknown in region. Indian subcontinent: **Nest** Dome with entrance near top, in low vegetation; up to 50 cm above ground. **Eggs** 2–3; deep terracotta, with darker ring or cap at broader end; 17 x 12.8 mm (av.). **NOTE** Split from Yellowish-bellied Bush-warbler *Cettia acanthizoides* of W,SE China and Taiwan (Alström *et al.* 2007).

1067 GREY-SIDED BUSH-WARBLER

Cettia brunnifrons Plate **104**

IDENTIFICATION 10.5–11.5 cm. *C.b.umbraticus*: **Adult** Superficially resembles Chestnut-crowned Bush-warbler but distinctly smaller, with whiter supercilium (extending to lores/sides of forehead) and largely dark grey breast and upper flanks; lower flanks and vent dull buffish-brown. **Juvenile** Markedly different. Upperparts similar to adult but crown same colour as mantle, underparts drab brownish-olive with paler buffish to slightly yellowish supercilium, centre of throat and centre of abdomen, and somewhat browner, buffier lower flanks, belly and vent. Gradually attains grey on breast and rufous-chestnut on crown. **VOICE** Song is a repeated, rapid, thin, high-pitched ***ti si'si'si'swi*** or ***ti sisisi'swi***, followed by unusual doubled nasal buzzing wheeze. Alarm call is an abrupt high-pitched metallic ***tizz*** or ***tiss***, sometimes a rapidly repeated ***tiss-ui***. **HABITAT** Scrub and grass bordering broadleaved evergreen and mixed evergreen and coniferous forest; breeds above 2,100 m, winters down to 460 m. **RANGE & STATUS** Resident (subject to local movements) N Pakistan, NW,N,NE Indian subcontinent, SE Tibet, SW,W China. **SE Asia** Uncommon resident W,N Myanmar. Scarce to uncommon winter visitor SW Myanmar. **BREEDING** May–July. **Nest** Dome or semi-dome, in low vegetation; within 50 cm of ground. **Eggs** 3–5; bright reddish-brown, usually with darker markings forming cap at broader end; 17.9 x 13 mm (av.; India).

1068 CHESTNUT-CROWNED BUSH-WARBLER *Cettia major* Plate **104**

IDENTIFICATION 12.5–13 cm. Monotypic. **Adult** Resembles Pale-footed Bush-warbler but more robust, with distinctive rufescent chestnut-brown crown and nape, darker olive-brown upperparts and olive-brownish suffusion on breast and flanks. From smaller Grey-sided by rufescent lores, buffier supercilium and largely whitish underparts. **Juvenile** Very drab, with grey-brown crown, and grey mottling on whitish throat and breast. **VOICE** Sings with rather hurried, shrill, slightly slurred ***i i-wi-wi-wirri-wi***. Call is apparently very similar to Grey-sided. **HABITAT** Grass and scrub near wet areas, sometimes under open forest; 1,500–2,200 m; may occur lower. **RANGE & STATUS** Breeds N,NE Indian subcontinent, SE Tibet, SW,W China; some winter to south. **SE Asia** Vagrant NW Thailand. Recorded in winter (status uncertain) N Myanmar.

1069 BROWNISH-FLANKED BUSH-WARBLER *Cettia fortipes* Plate **104**

IDENTIFICATION 11.5–12.5 cm. *C.f.fortipes*: **Adult** Upperparts deep rufescent-tinged brown, supercilium and underparts paler, has dark eyestripe and deep buffy-brownish flanks, vent and wash across breast. *C.f.davidiana* (N Laos, W,E Tonkin) is whiter on throat to belly-centre, with less buff mixed with breast and flank coloration. Difficult to separate from Sunda, Aberrant and Hume's Bush-warblers, but relatively shorter-tailed and stouter than former, and larger, larger-billed and relatively longer-tailed than latter; upperside and wing-feather fringes darker and more rufescent-tinged than Aberrant but usually less rufescent than Hume's; has darker, warmer flanks and vent than both and lacks any obvious yellowish coloration below. Note voice. See Manchurian and Pale-footed Bush-warblers. **Juvenile** Like adult but warmer on wing-feather fringes, possibly slightly warmer on upperparts, sometimes tinged yellowish below. **VOICE** Song is a short explosive phrase, introduced by a prolonged clear rising whistle: ***wheeeeee chiwiyou*** etc. Calls with a mixture of short, rather metallic ***tit*** and ***trrt*** notes. **HABITAT & BEHAVIOUR** Scrub and grass bordering broadleaved evergreen forest, overgrown clearings, sometimes undergrowth and bamboo inside forest; 900–2,135 m, locally down to 215 m in winter. Skulking. Flicks wings when alarmed, flashing pale underwing-coverts. **RANGE & STATUS** Resident (subject to altitudinal movements) N Pakistan, NW,N,NE Indian subcontinent, W,C and southern China, Taiwan. **SE Asia** Fairly common resident W,N Myanmar, N Laos, W,E Tonkin. **BREEDING** May–June. **Nest** Dome or cup, in low vegetation; up to 50 cm above ground. **Eggs** 3–5; rather glossy rich chocolate-colour to deep chestnut (obscure darker ring often at broader end); 17.3 x 13.4 mm (av.; *fortipes*).

1070 MANCHURIAN BUSH-WARBLER

Cettia canturians Plate **104**

IDENTIFICATION Male 18 cm, female 15 cm. Monotypic. **Male** Similar to female but large size distinctive. From larger reed-warblers by pronounced head pattern and rufescent forecrown. **Female** Similar to Brownish-flanked Bush-warbler but bulkier and stouter-billed with diagnostic contrasting rufescent forecrown; tends to show darker eyestripe and often buffier breast (compared to *C.f.davidiana*). Tinge of brown upperparts varies from rufescent to greyish/olive, less rufescent on nape to rump; extent of buff on breast and vent also variable (presumably with age or wear). See Chestnut-crowned Bush-warbler. **Juvenile** Undocumented. **VOICE** Song is a loud short fluty warble, usually introduced by a slurred liquid crescendo (similar to Brownish-flanked): ***wrrrrrr-whuchiuchi*** etc. Calls with stony ***tchet tchet tchet...***, sometimes interspersed with short rattling ***trrrt*** notes. **HABITAT & BEHAVIOUR** Scrub, secondary growth, cultivation borders, bamboo, forest edge; up to 1,525 m. Fairly skulking. **RANGE & STATUS** Breeds NC,E China; winters S China, N Philippines. **SE Asia** Uncommon to locally fairly common winter visitor NW Thailand, N,C Laos, Vietnam (except Cochinchina).

1071 SUNDA BUSH-WARBLER

Cettia vulcania Plate **104**

IDENTIFICATION 12–13 cm. *C.v.intricata* (Myanmar, Thailand): **Adult** Somewhat slimmer and longer-tailed than Brownish-flanked Bush-warbler, with paler, more olive upperside and wing-feather fringes (paler, warm olive on rump and uppertail-coverts), faintly pale yellowish-washed underparts and paler, more olive flanks. Note voice. *C.v.oblita* (Indochina) is slightly brighter olive above and more yellow-washed below, with warmer-washed flanks. See Hume's Bush-, Tickell's Leaf- and Buff-throated Warblers but note range. **Juvenile** Underparts more uniform buffish-yellow. **VOICE** Song of *oblita* is a long, high, thin wispy ***utu-WEEEOO'WEE*** or ***tt-wiIIUU'WIT*** (introductory note barely audible, middle part falling then rising). Calls are similar to Aberrant. **HABITAT & BEHAVIOUR** Scrub and grass bordering broadleaved evergreen forest, overgrown clearings, sometimes undergrowth and bamboo inside forest; 1,200–2,590 m (mainly breeds at higher levels). Skulking. Flicks wings when alarmed, flashing pale underwing-coverts. **RANGE & STATUS** Resident Greater and Lesser Sundas. Philippines (Palawan). **SE Asia** Fairly common resident (subject to some seasonal movements?) N(east),S(east),E Myanmar, N Laos, W Tonkin. Uncommon to locally fairly common winter visitor NW Thailand. Vagrant NE Thailand. **BREEDING** Undocumented in region. W Java: **Nest** Ball-shaped with large side entrance, on/near ground. **Eggs** 2; deep purplish-pink. **NOTE** The forms dealt with here were previously lumped in Aberrant Bush-warbler, but see Olsson *et al.* (2006).

1072 ABERRANT BUSH-WARBLER

Cettia flavolivacea Plate **104**

IDENTIFICATION 12–13 cm. *C.f.weberi*: **Adult** Somewhat more rufescent above than Sunda Bush-warbler (never as dark rufescent as Brownish-flanked), slightly yellower below and buffier on flanks (also lightly across breast). Note range. See Hume's Bush-, Tickell's Leaf- and Buff-throated Warblers. **Juvenile** Not well documented. **VOICE** Song is similar to Sunda: ***utut'wi'YIUUUU*** or ***tt'twi'IUUUU*** (introductory note barely audible, next part quick, last part rising slightly), or somewhat slower ***T'wi TUUUUU*** (last part quite level). In Nepal, second part of song may be highest. Calls with short, grating ***trrik***, ***trrrk***, or dry ***drrt*** notes; similar to Brownish-flanked. **HABITAT & BEHAVIOUR** Scrub and grass bordering broadleaved evergreen forest, overgrown clearings, sometimes undergrowth and bamboo inside forest; 1,200–2,700 m (mainly breeds at higher levels). Skulking. Flicks wings like Sunda. **RANGE & STATUS** Breeds N,NE

India, Nepal, Bhutan, SE Tibet; some winter to south. **SE Asia** Fairly common resident W Myanmar. **BREEDING** May–June. **Nest** Domed with lateral entrance, on ground. **Eggs** 3–4; pale to deep reddish-brown, often with darker cap at broader end; 17.2 x 12.6 mm (av.; India).

1073 **PALE-FOOTED BUSH-WARBLER**
Cettia pallidipes Plate **104**
IDENTIFICATION 11–12 cm. *C.p.laurentei*: **Adult** Told from other bush-warblers by striking pale supercilium and blackish eyestripe, rather cold olive-tinged upperparts, relatively short, square-ended tail, rather clean whitish underparts and very pale pinkish legs and feet. Shares some characters with Asian Stubtail but larger and longer-tailed, with less rufescent upperparts; lacks crown-scaling. See Dusky Warbler. **Juvenile** Presumably like adult. **VOICE** Song is a sudden loud explosive jumbled series of chattering notes: roughly ***wi wi-chi'ti'ti'chi*** and ***yi wi'ti'chi'iwi*** etc. Calls with loud, slightly spluttering ***twip*** and ***tup*** notes, and two-note ***tid-ip***. **HABITAT & BEHAVIOUR** Grass and scrub, sometimes in open broadleaved evergreen and pine forest, bracken-covered slopes; 550–2,135 m, locally down to 200 m in E Tonkin. Very skulking, in thick low vegetation. **RANGE & STATUS** Resident N,NE,E India, Nepal, Bhutan, Andaman Is, SW China (SE Yunnan). **SE Asia** Uncommon resident C,E,S(east) Myanmar, W,NW Thailand, N,C Laos, W,E Tonkin, S Annam. **BREEDING** May–June. **Nest** Domed or semi-domed, low down in bush. **Eggs** 3–4; purplish brick-red with darker markings forming cap or ring at broader end (sometimes deep purplish-chocolate); 17.1 x 13.1 mm (India).

1074 **ASIAN STUBTAIL**
Urosphena squameiceps Plate **104**
IDENTIFICATION 9.5–10 cm. Monotypic. **Adult** Very small size, very short tail, prominent long buffy-whitish supercilium, black eyestripe and pale pinkish legs and feet distinctive. Upperparts dark rufescent-tinged brown, underparts whitish with buffy vent; shows indistinct dark scales on crown (difficult to see in field). See Pale-footed Bush-warbler. **VOICE** Song is a repeated, extremely high-pitched, insect-like ***si'i'i'i'i'i'i...*** Also a slower ***si-si-si-si-si-si-si...***, gradually rising in volume. Calls with sharp ***sit*** notes when alarmed. **HABITAT & BEHAVIOUR** Undergrowth in broadleaved evergreen and mixed deciduous forest, bamboo; up to 2,285 m. Always on or close to ground; shy. **RANGE & STATUS** Breeds Ussuriland, Sakhalin, N,S Korea, Japan; winters S China. **SE Asia** Uncommon to fairly common winter visitor Myanmar (except SW,N), Thailand (except southern S), Indochina (except Cochinchina). Also recorded on passage E Tonkin.

1075 **GREY-BELLIED TESIA**
Tesia cyaniventer Plate **105**
IDENTIFICATION 8.5–10 cm. Monotypic. **Adult** Told by small size, tail-less appearance, dark olive-green crown and upperparts with contrasting black eyestripe and olive-yellow supercilium, slaty-grey underparts and pale grey on centre of throat, breast and abdomen; bill has paler, dull yellowish lower mandible. See Slaty-bellied Tesia. **Juvenile** Upperparts and wings dark, faintly olive, warmish brown, supercilium and eyestripe duller than adult; underparts drab dark olive, paler on centre of throat and abdomen (where may show greyish cast). Gradually acquires grey, starting with breast and belly. **VOICE** Sings with short, loud, rich slurred phrases, introduced by a few short-spaced high-pitched notes: ***ji ji ju ju ju-chewit***; ***ji ju'wi-iwi***; ***ji ji ji'wi-jui*** and ***ji ji ji ji'wiu*** etc. Calls with a repeated rattling ***trrrrrrk***. **HABITAT** Undergrowth in broadleaved evergreen forest, mainly near streams; 1,000–2,565 m, down to 60 m in winter (N Myanmar). **RANGE & STATUS** Resident N,NE Indian subcontinent, SW,S China. **SE Asia** Common resident W,N,E,S Myanmar, NW Thailand (scarce), south-west Cambodia, N,S Laos, W,E(scarce) Tonkin, C,S Annam. **BREEDING** India: May–July. **Nest** Ball-shaped, with entrance near top, in low bush, creeper or moss etc. growing against tree-trunk or rock; up to 30 cm above ground. **Eggs** 3–5; bright pale pink, profusely marked with tiny bright brick-red specks; 17.4 x 12.9 mm (av.).

1076 **SLATY-BELLIED TESIA**
Tesia olivea Plate **105**
IDENTIFICATION 9 cm. Monotypic. **Adult** Like Grey-bellied Tesia but forecrown or most of crown washed golden-yellow, underparts uniform dark slaty-grey, orange to orange-yellow lower mandible. Sometimes has quite dull crown and hint of brighter supercilium. Note voice. **Juvenile** Very like Grey-bellied but said to be more uniform dull olive-green below. **VOICE** Song is similar to Grey-bellied but phrases longer, more tuneless and jumbled, preceded by 4–11, more spaced and irregular introductory notes. Calls with a spluttering rattle: ***trrrrt trrrrt trrrrt...*** **HABITAT** Undergrowth in broadleaved evergreen forest, particularly near streams; 700–2,565 m, locally down to 455 m in winter. **RANGE & STATUS** Resident (subject to altitudinal movements) E Nepal, Bhutan, NE India, SW China. **SE Asia** Common resident W,N,E,S Myanmar, W,NW Thailand, N,C Laos, W,E Tonkin, N Annam. **BREEDING** May–July. Apparently undocumented otherwise.

1077 **CHESTNUT-HEADED TESIA**
Tesia castaneocoronata Plate **105**
IDENTIFICATION 8.5–9.5 cm. *T.c.castaneocoronata*: **Adult** Unmistakable. Very small and almost tail-less, with bright chestnut crown and head-sides and contrasting bright yellow throat and upper breast. Rest of plumage dark olive, with yellow extending down centre of abdomen; shows small whitish patch behind eye. **Juvenile** Upperparts colder, darker and browner, underparts deep dark rufous with more golden-buffish centre of throat and belly. **Other subspecies in SE Asia** *T.c.abadiei* (Tonkin). **VOICE** Sings with short phrases recalling Slaty-bellied Tesia but less hurried and more structured, lacking introductory notes: fairly quick ***ti tisu-eei***, slower ***tis-tit-ti-wu*** and more simple ***si tchui*** and ***si-ti-chui*** etc. Calls include high-pitched, shrill explosive ***whit*** notes (often doubled) and sharp stony ***tit*** notes. **HABITAT** Undergrowth in broadleaved evergreen forest, secondary growth, particularly near streams; 950–2,810, down to 450 m in winter. **RANGE & STATUS** Resident N,NE Indian subcontinent, S,SE Tibet, SW China. **SE Asia** Uncommon to locally common resident W,N,S(east),E Myanmar, NW Thailand, N Laos, W,E Tonkin. **BREEDING** India: May–July. **Nest** Rather neat but flimsy ball, suspended from small branch, within 2 m of ground. **Eggs** 2; uniform dark terracotta or dull chestnut with faint cap of darker mottling at broader end; 17.4 x 12.9 mm.

AEGITHALIDAE: Long-tailed tits

Worldwide c.9 species. SE Asia 4 species. Very small; similar to typical tits but with short stout bills and long tails. Gregarious and hyperactive, often in undergrowth, searching for insects.

1078 **BLACK-THROATED TIT**
Aegithalos concinnus Plate **99**
IDENTIFICATION 11–11.5 cm. *A.c.pulchellus*: **Adult** Unmistakable. Very small with long tail, long black mask, white bib and neck-side with large isolated black central patch, and rufous-chestnut breast-band extending narrowly along flanks. Grey above with paler, drab greyish crown, whitish centre of underparts and vent, yellowish-white eyes. *A.c.manipurensis* (W Myanmar) and *talifuensis* (N Myanmar, N Indochina) have orange-rufous to rufous-cinnamon crown with slight white lateral line towards rear. **Juvenile** Lacks black throat-patch (may show faint dark mottling), has row of dark smudges along border of lower throat/upper breast, crown paler. See Black-browed and Burmese Tits. **VOICE** Song is a

repeated twittering ***tir-ir-ir-ir-ir***, interspersed with well-spaced single chirping notes, or a very high thin ***tur-r-r-tait-yeat-yeat-yeat***. Contact calls are a thin ***psip psip*** and sibilant ***si-si-si-si...*** or ***si-si-si-si-li-u***. Gives a fuller ***sup*** when agitated, extending to a short rattling ***churr trrrt trrrt***. **HABITAT & BEHAVIOUR** Broadleaved evergreen and mixed broadleaved and pine forest, forest edge, secondary growth; 900–2,600 m. Usually in small fast-moving parties, sometimes in company with mixed-species feeding flocks. **RANGE & STATUS** Resident NE Pakistan, NW,N,NE India, Nepal, Bhutan, S,SE Tibet, C and southern China, Taiwan. **SE Asia** Locally common resident W,N,S(east),E Myanmar, NW Thailand, N,C Laos, W,E Tonkin, N Annam. **BREEDING** February–June. **Nest** Neat oval/ball with side-entrance towards/near top, in tree, bush or low tangled vegetation; 8–12 m above ground. **Eggs** 3–8; white to cream or pinkish with ring of reddish and purple specks around broader end (sometimes sparsely elsewhere or unmarked); 13.1 x 10.4 mm (av.; *manipurensis*).

1079 GREY-CROWNED TIT
Aegithalos annamensis Plate **99**

IDENTIFICATION 11–11.5 cm. Monotypic. **Adult** Told by combination of dull grey crown, narrow drab greyish breast-band and dirty white underparts with light greyish-pink wash on flanks. See Black-throated Tit, but note range. **Juvenile** Lacks black throat-patch (may show faint dark mottling), has row of dark smudges along border of lower throat/upper breast. **VOICE** Similar to Black-throated. **HABITAT & BEHAVIOUR** Broadleaved evergreen and mixed broadleaved and pine forest, forest edge, secondary growth; 490–2,000 m (perhaps higher). Usually in small fast-moving parties, sometimes joins bird-waves. **RANGE & STATUS** Endemic. Locally common resident east Cambodia, S Laos, C,S Annam, north Cochinchina. **BREEDING** February–March. **Nest** Similar to Black-throated, with entrance near top, situated in outer leafy sprigs of pine branch; 9 m above ground. **Eggs** Undocumented. **NOTE** Split from Black-throated following King (1997).

1080 BLACK-BROWED TIT
Aegithalos bonvaloti Plate **99**

IDENTIFICATION 11–12 cm. *A.b.bonvaloti*: **Adult** Recalls Black-throated Tit but black head-bands broader, offsetting narrow buffish median crown-stripe (whitish on forehead), lower head-sides pale cinnamon-rufous, throat white with triangle of dense black speckles on centre (extending to chin), breast-band broader and cinnamon-rufous, as flanks. Shows cinnamon tinge to uppermost mantle. **Juvenile** Duller black head-bands and throat, no cinnamon on upper mantle, paler underparts with greyish mottling on upper breast and belly-centre. **VOICE** Undocumented. **HABITAT & BEHAVIOUR** Open broadleaved evergreen and mixed broadleaved and coniferous forest, pine forest, forest edge, secondary growth; 1,830–3,110 m. Usually in small fast-moving flocks. **RANGE & STATUS** Resident SE Tibet, SW,W China. **SE Asia** Locally fairly common resident N Myanmar. **BREEDING** April–May. **Nest** Neat oval with side-entrance towards top, in tree; 0.8–5 m above ground. **Eggs** 4–5; white; c.14.4 x 10.6 mm (*bonvaloti*).

1081 BURMESE TIT *Aegithalos sharpei* Plate **99**

IDENTIFICATION 11–12 cm. Monotypic. **Adult** Recalls Black-browed Tit but has slightly narrower black head-bands, whiter median crown-stripe, white lower head-sides and lower throat, dark brownish breast-band and uniformly buffish remainder of underparts (whiter on undertail-coverts). **Juvenile** Undescribed. **VOICE** Very thin ***see-see-see-see...***, mixed with subdued ***tup*** and ***trrup*** or ***t'dup***. Shrill ***zeet*** and ***trr-trr-trr*** when agitated. **HABITAT & BEHAVIOUR** Open broadleaved evergreen forest, secondary growth; 2,100–3,300 m. Usually in small fast-moving flocks. **RANGE & STATUS** Endemic. Locally fairly common resident W Myanmar. **BREEDING** April–May; otherwise undocumented. **NOTE** Split from Black-browed Tit following Harrap (2008).

PHYLLOSCOPIDAE: *Seicercus* & *Phylloscopus* warblers

Worldwide at least 74 species. SE Asia 39 species. Small and active with relatively fine, pointed bills. Plumage and habitats vary widely. Feed mainly on insects and spiders. ***Seicercus* warblers** Similar to leaf-warblers; generally well patterned with olive-green, bright yellow and grey, many with dark lateral crown-stripes; all occur in wooded habitats. **Leaf-warblers** (*Phylloscopus*) Small and either brownish above without wing-bars or greenish above with wing-bars (except Mountain Leaf-warbler), all with prominent supercilium; mostly in wooded habitats.

1082 BIANCHI'S WARBLER
Seicercus valentini Plate **105**

IDENTIFICATION 12 cm. *S.v.valentini*: **Adult** Like Plain-tailed Warbler but forehead typically with less greenish admixed, lateral crown-stripes generally blacker and more extensive on sides of forehead, outertail-feathers with much more white (similar to Grey-crowned but only on outer two tail-feathers). Wing-bar usually distinct (sometimes lacking). See Whistler's Warbler. **VOICE** Song is almost identical to Whistler's but introduced by soft ***chu***. Call is a short, soft, deflected ***tiu*** or ***heu***, occasionally doubled. **HABITAT** Broadleaved evergreen forest, secondary growth; 1,760–1,900 m (at least), down to 200 m in winter. **RANGE & STATUS** Breeds SW,W,NC,S,SE China; some populations winter to south. **SE Asia** Fairly common to common resident W Tonkin. Uncommon to fairly common winter visitor C,S(east),E Myanmar, north Tenasserim, NW Thailand, N Laos, E Tonkin. **BREEDING** Undocumented?

1083 WHISTLER'S WARBLER
Seicercus whistleri Plate **105**

IDENTIFICATION 12.5–12 cm. *S.w.nemoralis*: **Adult** Told from Grey-crowned, Plain-tailed and Bianchi's Warblers by green median crown-stripe with sparse pale greyish streaks (hard to see in field) and green rear lower border of lateral crown-stripes. Wing-bar typically distinct (very rarely lacking); lateral crown-stripes relatively indistinct anteriorly (often fading out above mid-part of eye) and greyish-black throughout, has extensive white on outer three pairs of tail-feathers. Best told from very similar Bianchi's by amount of white on tail; apparently also always has slightly more orangey-yellow underparts. Note voice and range. **Juvenile** Undocumented. **VOICE** Song is very similar to Bianchi's, a series of short phrases, each with fewer than 10 short soft whistled notes, normally preceded by a short introductory note; individual phrases are often repeated either a few times or after some time. The call of this taxon is described as a soft low-pitched ***tiu'du***, but may be more similar to Bianchi's than this. **HABITAT** Broadleaved evergreen forest, secondary growth, bamboo; 2,000–2,740 m. **RANGE & STATUS** Resident (subject to some movements) N Pakistan, NW,N,NE Indian subcontinent. **SE Asia** Common resident W Myanmar. **BREEDING** May–June. **Nest** Dome with side-entrance, on ground. **Eggs** 4; white; 14.7–15.5 x 11.7 mm.

1084 MARTENS'S WARBLER
Seicercus omeiensis Not illustrated

IDENTIFICATION 11–12 cm. Monotypic. **Adult** Roughly intermediate between Grey-crowned and Plain-tailed. White on two outermost tail-feathers also intermediate (Grey-crowned shows much, Plain-tailed very little); sometimes a little on tip of next pair inwards? (never on Plain-tailed, often on Grey-crowned). Note voice and range. **Juvenile** Undocumented. **VOICE** Call is diagnostic wholesome ***chup***

or ***chut*** and ***chu-du*** or ***chu-tu***. **HABITAT** As Grey-crowned; 600–2,075 m. **RANGE & STATUS** Breeds W,C China; at least some winter to south. **SE Asia** Uncommon to fairly common winter visitor E Myanmar, W,NW,NE,SE Thailand, south-west Cambodia, N Laos; ? C Laos. **NOTE** Only recently described (Martens *et al.* 1999).

1085 **PLAIN-TAILED WARBLER**

Seicercus soror Plate **105**

IDENTIFICATION 10.5–11 cm. Monotypic. **Adult** Like Grey-crowned and Bianchi's Warblers but has considerably less white on outertail-feathers (very little on penultimate feathers), median crown-stripe dull pale grey with a little pale greenish admixed, forehead usually greenish, lateral crown-stripes shorter and more diffuse anteriorly (often fading out above mid-part of eye) and greyish-black throughout. Shows some grey along rear lower border of lateral crown-stripes, wing-bar typically lacking (sometimes obvious); upperparts appear somewhat duller green than other similar *Seicercus*. Note voice and range. See Whistler's Warbler. **Juvenile** Undocumented. **VOICE** Song is higher-pitched than Bianchi's and introduced by a soft ***chip***; each series of notes covers a wider frequency range. Call is a short, rather high-pitched thin ***tsrit*** or ***tsi-dit***. **HABITAT** Broadleaved evergreen forest, secondary growth; up to 1,500 m. **RANGE & STATUS** Breeds W,C,SE China; winters to south/south-west. **SE Asia** Uncommon to fairly common winter visitor S Myanmar, Tenasserim, Thailand (except NW,S), Cambodia, C Laos, S Annam, north Cochinchina. Vagrant S Thailand, north-west Peninsular Malaysia.

1086 **GREY-CROWNED WARBLER**

Seicercus tephrocephalus Plate **105**

IDENTIFICATION 11–12 cm. Monotypic. **Adult** Combination of rather bright yellowish-green upperparts, extensively grey crown with black lateral stripes, prominent (complete) yellow eyering, deep yellow underparts and lack of wing-bars distinctive. Forehead mostly grey with small amount of green just above bill-base (sometimes extensively green), lateral crown-stripes distinct and relatively clear-cut on forehead, reaching close to bill-base (occasionally less distinct on forehead), median crown-stripe pure grey (sometimes slightly streaked greenish); has grey along rear lower border of lateral crown-stripes and obvious white on outertail-feathers; normally lacks wing-bar. Difficult to separate from Bianchi's, Whistler's, Martens's and Plain-tailed Warblers; for differences see those species. Note voice and range. See White-spectacled Warbler race *intermedius*. **Juvenile** Lateral crown-stripes less pronounced, upperparts slightly darker, underparts duller with drab olive wash on head-sides, throat and breast (somewhat yellower on throat), belly and vent duller yellow. **VOICE** Song is easily told from similar *Seicercus* by presence of tremolos and trills. Also higher-pitched than Whistler's and Bianchi's, covering greater frequency range. Call is a sibilant ***ch'rr***, ***chrr'k*** or ***trr'uk***, recalling Buff-throated. **HABITAT & BEHAVIOUR** Broadleaved evergreen forest, secondary growth, bamboo; 1,400–2,500 m, down to sea level in winter. Quite skulking, in low vegetation. **RANGE & STATUS** Resident (subject to some movements) NE India, SW,W,NC China. **SE Asia** Common resident (subject to some movements) W,N Myanmar, W Tonkin. Scarce to fairly common winter visitor C,S,E Myanmar, Tenasserim, NW,NE Thailand, N Laos, E Tonkin (may breed), N Annam, Cochinchina; ? C Laos. **BREEDING** May–June. **Nest** Globular structure, in bush; 60 cm above ground. **Eggs** White; 16.5 x 12.2 mm (av.). **NOTE** For recent taxonomic revision of Golden-spectacled Warbler complex see Alström and Olsson (1999). Due to this revision, complete ranges of this and the following three species are currently little known.

1087 **WHITE-SPECTACLED WARBLER**

Seicercus affinis Plate **105**

IDENTIFICATION 11–12 cm. *S.a.affinis*: **Adult** Like Grey-crowned and Bianchi's Warblers but has slaty-grey crown, upper ear-coverts and neck-side, white eyering (faintly broken above anterior part of eye) and prominent short, broad yellowish wing-bar. From Grey-cheeked Warbler by slightly larger size, paler grey crown with more contrasting lateral stripes, yellow lores, chin and throat, slightly narrower/neater eyering, less grey on ear-coverts, and orange-flesh lower mandible. Note voice. *S.a.intermedius* (visitor) has crown more similar to Grey-crowned Warbler and shows very broad all-yellow eyering (but broken above anterior part of eye); some show slight second wing-bar. **Juvenile** Crown somewhat greener with less distinct lateral stripes, shows only little grey on crown and supercilium, duller green upperparts, paler, less intense yellow underparts and drab olive tinge to head-sides, throat and breast (centre of throat yellower). **Other subspecies in SE Asia** *S.a.ocularis* (S Annam; presumably also C Annam). **VOICE** Song of *ocularis* is rather sweet and variable, usually with 5–8 rather rapidly delivered notes, with phrases like ***sweet-sweet-sweet-sweet-sweet-sweet-sweet***; ***tutitutitutituti***; ***sweet-sweet-swititit***; ***tit-twi-twi-twi-twi***; ***tu-swit-swititititit*** and ***tu-swit-swee-sisisisisi*** etc. *S.a.affinis* calls with thin, shrill, rising then falling ***U-EE'SI*** or ***U-EE'ss***; *ocularis* with totally different hurried, tremulous ***chri-chri-chri*** or ***chuu-chuu-chuu...*** **HABITAT & BEHAVIOUR** Broadleaved evergreen forest, secondary growth; 1,375–2,285 m, down to 450 m in winter. Typically fairly secretive in lower to middle storey of forest. **RANGE & STATUS** Resident (subject to altitudinal movements) Bhutan, NE India, SW China. Breeds SE China, said to winter to SW. **SE Asia** Locally common resident N Myanmar, C,S Annam. Scarce winter visitor W,E Tonkin; ? N Laos. **BREEDING** April–June. **Nest** Dome, in hole in bank or dead tree. **Eggs** 4–5; white; 15.4 x 12.4 mm (av.). **NOTE** It is thought that the three taxa mentioned here represent different sibling species (Olsson *et al.* 2004).

1088 **GREY-CHEEKED WARBLER**

Seicercus poliogenys Plate **105**

IDENTIFICATION 9–10 cm. Monotypic. **Adult** Similar to White-spectacled Warbler but grey of head darker, lateral crown-stripes less contrasting, has mostly dark lores with a little greyish-white, chin greyish-white, orbital ring broader and more tear-shaped behind eye, lower mandible dark; duller lower throat and upper breast. **Juvenile** Olive-tinged forehead, pale yellow throat, bright caramel-buff breast-sides, slightly paler legs than adult. Younger birds lack crown-stripes. **VOICE** Song is similar to White-spectacled but thinner, notes less melodic and more slurred together, lacking tremolos and trills, with phrases (in S Annam) such as ***chwee-chwee-chwee-chwee-chwee***; ***titwi-titwi-titwi-titwi***; ***ittu-ittu-ittu-ittu***; ***itsi-itsi-itsi-it*** etc. In SE Tibet, phrases include ***itwu-itwu-itwu-itwu***; ***itchu-itchu-itchu-itchu-it***; ***iitwit-twi-twi-twi-twi***; in Meghalaya (India), ***titwu-titwu-wiitu-wiitu***; ***wit-wit-wittu-wittu-wittu***; ***tiswu-tiswu-tiswu*** etc Calls (in S Annam) include a thin, piercing ***tssiiu tssiiu tssiiu....*** In SE Tibet and the Himalayas, a quite different quick thin, shrill ***TU'iss*** or ***TU'i***, and in Meghalaya, ***U'iss***. **HABITAT** Broadleaved evergreen forest, secondary growth; 900–2,135 m. **RANGE & STATUS** Resident (subject to altitudinal movements) N,NE Indian subcontinent, SE Tibet, SW China. **SE Asia** Locally fairly common resident N Myanmar, Laos, W,E Tonkin, N,C,S Annam. Recorded (status uncertain) NW Thailand. **BREEDING** May–June. **Nest** Dome, on ground. **Eggs** 4; glossy white; 15.8 x 12.5 mm (av.). **NOTE** It is thought that two or three sibling species occur from the NE Indian subcontinent to SW China and SE Asia (Olsson *et al.* 2004), and there certainly appear to be at least two in SE Asia based on vocal differences.

1089 **CHESTNUT-CROWNED WARBLER**

Seicercus castaniceps Plate **105**

IDENTIFICATION 9–10.5 cm. *S.c.collinsi*: **Adult** Unmistakable, with rufous crown, blackish lateral crown-stripes, yellow rump, flanks, vent and wing-bars, and pale grey throat and upper breast. Head-sides and mantle dull greyish, has small whitish patch on nape-side, white eyering, olive-green back and scapulars, prominent white on outertail-feathers and whitish centre to abdomen. *S.c.sinensis* (northern

Indochina) and *annamensis* (S Annam) have all-yellow belly and vent, sharply demarcated from grey upper breast, green of upperparts extending to lower mantle; *butleri* (Peninsular Malaysia) has dark chestnut crown with indistinct lateral stripes, darker grey plumage-parts, indistinct pale marking on nape-side and no yellow rump-patch; *youngi* (S Thailand) is like *butleri* but has more rufous crown, more extensively grey upperparts (paler on rump) and less yellow on flanks and vent. **Juvenile** Crown duller and browner, whitish patch on nape-side indistinct, throat and breast sullied yellowish-olive, yellow of underparts paler. *S.c.butleri* is said to have brownish-grey crown with ashy-grey lateral stripes, pale grey supercilium and faint ashy-grey eyestripe; *youngi* is said to have dull olive-orange median crown-stripe and supercilium and faint olive-yellow wash on mantle, throat and breast. **Other subspecies in SE Asia** *S.c.castaniceps* (W,N Myanmar), *strresemanni* (south-west Cambodia, S Laos). **VOICE** Song is an extremely thin, high-pitched, metallic ***wi si'si'si-si'si'si*** etc., with upward-inflected ***si*** notes. Calls include soft subdued ***chit*** notes. **HABITAT & BEHAVIOUR** Broadleaved evergreen forest; 1,000–2,500 m, down to 825 m in Peninsular Malaysia, and 450 m in winter in N Myanmar. Highly active, often in mixed-species feeding flocks. **RANGE & STATUS** Resident (subject to minor movements) N,NE Indian subcontinent, SE Tibet, SW,C,SE China, Sumatra. **SE Asia** Fairly common to common resident W,N,E Myanmar, W,NW,S Thailand, Peninsular Malaysia, south-west Cambodia, Laos, W,E Tonkin, N,C,S Annam. Also recorded on passage E Tonkin **BREEDING** January–September. **Nest** Dome, on ground or in hole in bank. **Eggs** 3–5; glossy white; 14.6 x 11.6 mm (av.; *castaniceps*). Brood-parasitised by Oriental and Sunda Cuckoos.

1090 YELLOW-BREASTED WARBLER

Seicercus montis Plate **105**

IDENTIFICATION 9.5–10 cm. *S.m.davisoni*: **Adult** Recalls Chestnut-crowned Warbler but has rufous head-sides (extending to sides of breast), greenish mantle and all-yellow underparts. Note range. **Juvenile** Has more restricted, duller rufous on head and breast-sides, duller lateral crown-stripes and less intense yellow underparts. **VOICE** Song is very similar to Chestnut-crowned: ***wi si'si-si'si*** and ***wi si-si-si'si'si*** etc. **HABITAT & BEHAVIOUR** Broadleaved evergreen forest; 1,160–2,185 m. Highly active, often in mixed-species flocks. **RANGE & STATUS** Resident Sumatra, Borneo, Lesser Sundas, Philippines (Palawan). **SE Asia** Uncommon to locally common resident Peninsular Malaysia. **BREEDING** January–July. **Nest** Dome with side-entrance, in hole in bank. **Eggs** 2. Otherwise undocumented?

1091 EASTERN CROWNED WARBLER

Phylloscopus coronatus Plate **106**

IDENTIFICATION 12.5–13 cm. Monotypic. **Adult** Overall proportions recall Arctic Warbler but shows darker crown, with well-defined pale median stripe (normally not reaching forehead), usually slightly yellower-green upperparts (notably fringes of wing-feathers) and whiter ear-coverts and underparts with contrasting pale yellow undertail-coverts; bill shows uniform pale lower mandible. White-tailed Leaf-, Davison's, Claudia's, Blyth's Leaf- and Hartert's Warblers are smaller and shorter-billed, with two distinct wing-bars and varying amounts of white on outertail-feathers. See Emei Leaf-, Greenish and Two-barred Warblers. **VOICE** Song is a repeated series of clear notes, terminating with a harsh squeaky drawn-out note: ***tuweeu tuweeu tuweeu tuweeu tswi-tswi zueee*** etc.; sometimes shorter ***psit-su zueee***. Call is a harsh ***zweet*** (quieter than Arctic). **HABITAT** Broadleaved evergreen and mixed deciduous forest, also mangroves and gardens on migration; up to 1,830 m (mainly winters in lowlands). Shows strong preference for canopy of larger trees. **RANGE & STATUS** Breeds SE Siberia, Ussuriland, NE China, N,S Korea, Japan; winters Sumatra, Java. **SE Asia** Uncommon to common winter visitor Tenasserim, W(south),S Thailand, Peninsular Malaysia, Singapore, S Laos; occasionally south-west NE Thailand. Uncommon to fairly common passage migrant S Myanmar, Thailand, Peninsular Malaysia, Singapore, Laos, W,E Tonkin, N Annam. Recorded (status uncertain) Cambodia, Cochinchina.

1092 YELLOW-VENTED WARBLER

Phylloscopus cantator Plate **106**

IDENTIFICATION 11 cm. *P.c.cantator*: **Adult** Shows unmistakable combination of bright yellow head, upper breast and undertail-coverts, contrasting dark eye- and lateral crown-stripes and whitish lower breast and belly. See Sulphur-breasted Warbler. **Other subspecies in SE Asia** *P.c.pernotus* (N,C Laos). **VOICE** Song is quickly repeated, consisting of fast high-pitched phrases, introduced by a very short note and ending with a distinctive ***si-chu*** or ***si-chu-chu***: roughly ***sit siri'sii'sii si-chu***; ***sit siwiwi'siwiwi si-chu*** and ***sit weet'weet-weet'weet si-chu-chu*** etc. **HABITAT & BEHAVIOUR** Broadleaved evergreen and semi-evergreen forest; 500–1,700 m. Usually in midstorey of forest, often in mixed bird flocks. Hangs down head-first from branches. **RANGE & STATUS** Breeds SE Nepal, Bhutan, NE India; winters to SE, SW China (Yunnan). **SE Asia** Scarce winter visitor Myanmar (except Tenasserim), W,NW Thailand, N,C Laos, north-west E Tonkin. May be resident/breeding visitor N Myanmar, N,C Laos, E Tonkin. **BREEDING** India: April–June. **Nest** Ball-shaped; on ground or in hole in bank. **Eggs** 3–4; white; 14.5 x 11.9 mm (av.).

1093 SULPHUR-BREASTED WARBLER

Phylloscopus ricketti Plate **106**

IDENTIFICATION 11 cm. **Adult** Completely bright yellow head and underparts, with contrasting blackish eye- and median crown-stripes diagnostic. Has olive-bordered ear-coverts and two narrow yellow wing-bars; lacks white on tail. See very similar Limestone Warbler, as well as White-tailed Leaf- and Yellow-vented Warblers. **VOICE** Song is a very high-pitched series of short notes, speeding up towards end: ***sit'ti si-si-si-si'si'chu*** and ***sit si-si'si'si-chu*** etc. Call is higher-pitched than that described for Limestone, covering a broader frequency range, and often consisting of multiple elements. **HABITAT** Broadleaved evergreen and mixed deciduous forest, rarely more open habitats on migration; up to 1,520 m. **RANGE & STATUS** Breeds W,SC,SE China; winters to south-west. **SE Asia** Uncommon to locally fairly common winter visitor Thailand (except C,S), Cambodia, S Laos. Scarce to uncommon passage migrant Laos, W,E Tonkin, S Annam; ? N Annam. Vagrant C Thailand.

1094 LIMESTONE WARBLER

Phylloscopus sp. Plate **106**

IDENTIFICATION 11 cm. **Adult** Smaller and proportionately longer-billed than Sulphur-breasted Warbler, but perhaps not separable in field. Possible differences (slight and to be confirmed) are greyer-tinged upperparts, and paler, less intense yellow of plumage. Best differentiated by voice. **Juvenile** Slightly less intense yellow below, marginally whiter and more clear-cut pale tips to greater coverts, probably slightly more green admixed on anterior part of lateral crown-stripes. **VOICE** Song consists of quicker phrases than Sulphur-breasted, which end less hurriedly and with distinctive ***wetu*** or ***we'chu*** (single or doubled), roughly: ***sit-ititu-ititu-wetu***; ***sit-it ti'wetu-wetu***; ***sit-it si-we'chu-we'chu*** and ***sit-it-it we'chu-we'chu*** etc. Call is a subdued ***pi-chu*** or ***pit-choo*** or ***pi-tsu*** with louder second note. **HABITAT** Broadleaved evergreen forest on limestone; up to 600 m. **RANGE & STATUS** Endemic. Local resident northern C Laos, W,E Tonkin, north C Annam. **BREEDING** March–April, otherwise undocumented. **NOTE** Split from Sulphur-breasted following Alström *et al.* (in prep.).

1095 MOUNTAIN LEAF-WARBLER

Phylloscopus trivirgatus Plate **106**

IDENTIFICATION 11.5 cm. *P.t.parvirostris*: **Adult** Greenish upperparts, yellowish underparts and lack of wing-bars distinctive. Has similar head pattern to Sulphur-breasted Warbler but supercilium and median crown-stripe greenish-yellow. Note range. **Juvenile** Head-stripes duller than adult, underparts washed olive and less yellow. **VOICE** Sings with very rapid, slurred, high-pitched phrases, recalling some white-eyes. **HABITAT** Broadleaved evergreen forest, secondary growth; above 1,220 m. **RANGE & STATUS** Resident Greater Sundas, W Lesser Sundas, Philippines. **SE Asia** Common resident extreme S Thailand, Peninsular Malaysia. **BREEDING** February–August. **Nest** Dome, on ground or in hole in bank. **Eggs** 2–3; glossy white; 15–16.6 x 12–12.6 mm. Brood-parasitised by Sunda Cuckoo.

1096 WHITE-TAILED LEAF-WARBLER

Phylloscopus ogilviegranti Plate **106**

IDENTIFICATION 11–11.5 cm. *P.o.klossi* (SE Thailand, C,S Annam): **Adult** Recalls smaller *Phylloscopus* warblers, with yellowish-white supercilium, median crown-stripe and double wing-bars, and prominent white on outertail-feathers, but lacks pale rump-patch and whitish tertial markings; larger and longer-billed with mostly pale lower mandible. Difficult to separate from Davison's Warbler, but has less white on tail (which is yellow-tinged), much brighter yellow underparts (including throat), and much yellower-green upperparts; note the resident status and breeding range of that species. Also very similar to Claudia's, Blyth's Leaf- and Hartert's Warblers, but shows much more white on two outermost tail-feathers and yellower supercilium and median crown-stripe (less marked when worn). Note range, habitat, voice and behaviour. *P.o.ogilviegranti* (visitor ? Vietnam) is a little less yellow above and below, with little white on tail (similar to *disturbans*); *disturbans* (visitor? E Myanmar, northern Thailand, N Indochina) shows least amount of white on tail and least amount of yellow on pale head-stripes and underparts (very similar plumage to Blyth's race *assamensis*). **Juvenile** Presumably differs in same way as Blyth's. **VOICE** Song is similar to Blyth's but higher-pitched, more quickly delivered and more slurred (without clearly repeated notes): ***itsi'itsi-CHEE-yi itsi'itsi-CHEE-yi itsi'itsi-CHEE-yi...*** etc. Call is a rapid high-pitched ***ti'CHI'wi*** or ***ti'SI'wi*** and ***WI'CHI'rr*** etc., with first or second note somewhat more stressed. **HABITAT** Broadleaved evergreen and pine forest; 900–2,000 m at least, possibly down to 50 m in winter. **RANGE & STATUS** Breeds W,S China; some apparently winter to south/south-west. **SE Asia** Common resident SE Thailand, south-west Cambodia, C,S Laos, C,S Annam. Historically reported to winter E Myanmar, NW,NE(north-west) Thailand, N Laos, W,E Tonkin, S Annam. **BREEDING** January–February, otherwise undocumented. **NOTE** Split from *Phylloscopus davisoni* (but retaining the English name) following Olsson *et al.* (2005), but pending data on vocal differences.

1097 DAVISON'S WARBLER

Phylloscopus davisoni Plate **106**

IDENTIFICATION 11–11.5 cm. Monotypic. **Adult** Difficult to separate from White-tailed Leaf-warbler, which see for differences. Note range. **Juvenile** Presumably differs in same way as Blyth's. **VOICE** Song appears to be even more rapid and slurred than White-tailed: ***it'pi'chu-WI'pi'chu-WI'pi'chu-it*** and ***tis WIT'ti'chi-WIT'ti'chu-it*** etc. Call is with loud ***WIt-chi***, and rapid high ***wit-ti'chrr***. **HABITAT** Broadleaved evergreen and pine forest; 900–2,565 m. **RANGE & STATUS** Resident SW China. **SE Asia** Common resident N,E,C(east),S(east) Myanmar, Tenasserim, W,NW,NE Thailand, N Laos, W Tonkin; ? E Tonkin, N Annam. **BREEDING** February–May. **Nest** Ball or flattened dome, on ground or in low vegetation up to 40 cm. Otherwise undocumented.

1098 GREY-HOODED WARBLER

Phylloscopus xanthoschistos Plate **106**

IDENTIFICATION 9.5–11 cm. *P.x.tephrodiras*: **Adult** Told by rather pale grey crown, nape, upper mantle and head-sides, dark grey eye- and lateral crown-stripes and whitish supercilium. Recalls Yellow-bellied Warbler but easily separated by dark lateral crown-stripes, all-yellow underparts and extensive white on outertail-feathers. **Juvenile** Crown and upper mantle duller grey, lower mantle duller green, underparts slightly pale yellow. **Other subspecies in SE Asia** *P.x.flavogularis* (N Myanmar). **VOICE** Song is an incessantly repeated, brief, high-pitched warble ***ti-tsi-ti-wee-ti*** or ***tsi-weetsi-weetsi-weetu-ti-tu*** etc. Calls include a high-pitched ***psit-psit*** and clear ***tyee-tyee...*** when alarmed. **HABITAT & BEHAVIOUR** Broadleaved evergreen forest, mixed broadleaved and pine forest, secondary growth; 1,065–2,105 m. Very active in middle storey of forest; often joins mixed-species feeding flocks. **RANGE & STATUS** Resident N Pakistan, NW,N,NE Indian subcontinent, S Tibet. **SE Asia** Common resident W,N,S(west) Myanmar. **BREEDING** March–August. **Nest** Dome, often with rather large entrance, on ground or in hole in bank, up to 60 cm. **Eggs** 3–5; fairly glossy white; 15.7 x 11.9 mm (av. *tephrodiras*). **NOTE** Olsson *et al.* (2005) demonstrated that this taxon (formerly placed in *Seicercus*) belongs with *Phylloscopus*.

1099 CLAUDIA'S WARBLER

Phylloscopus claudiae Not illustrated

IDENTIFICATION 11.5–12 cm. Monotypic. **Adult** On current knowledge, not separable in field from Blyth's Leaf-warbler; has similar or slightly less white on outertail-feathers. **VOICE** Not yet documented. **HABITAT & BEHAVIOUR** Broadleaved evergreen forest; c.600–1,500 m, sometimes lower. Behaves in similar way to Blyth's. **RANGE & STATUS** Breeds W,C,E China; some winter to south. **SE Asia** Uncommon to fairly common winter visitor C,S Myanmar, north Tenasserim, NW,NE Thailand. **BREEDING** Undocumented. **NOTE** Split from Blyth's Leaf-warbler, following Olsson *et al.* (2005), but pending data on vocal differences from other similar taxa. Identity of wintering/passage birds in W,C Thailand, C,S Laos, E Tonkin, N,C Annam, previously assigned to Blyth's, now unclear.

1100 BLYTH'S LEAF-WARBLER

Phylloscopus reguloides Plate **106**

IDENTIFICATION 11.5–12 cm. *P.r.assamensis* (W,N Myanmar): **Adult** Probably indistinguishable in field from Claudia's and Hartert's Warblers, which see for differences. Difficult to separate from Davison's Warbler, but shows less yellow on supercilium and underparts and less white on tail (narrower white fringe to inner web of two outermost feathers). See White-tailed Leaf-warbler, note voice, behaviour and range. *P.r.ticehursti* (rest of breeding range) has yellower supercilium, median crown-stripe, wing-bars and underparts and slightly greener upperparts. However, populations in NW Thailand and W Tonkin, at least, are thought to be more similar to *assamensis* morphologically. **Juvenile** Similar to adult but crown much plainer, with paler lateral stripes and almost obsolete median stripe, underparts duller. **Other subspecies in SE Asia** *P.r.reguloides* (visitor Myanmar?). **VOICE** Song is a strident, undulating, usually alternating series of phrases, interspersed with lengthy intervals: ***wit tissu-tissu-tissu wit tewi-tewi-tewi-wit chewi-chewi-chewi...***; ***wit-chuit wit-chuit wit chuit chi-tewsi chi-tewsi chi...*** and ***wit-chi wit-chi wit-chi-wit wit-chi wit-chi wit-chi-wit...*** etc. Calls include a high-pitched ***PIT-chee*** and ***PIT-chee wi'chit*** with stressed first note. **HABITAT & BEHAVIOUR** Broadleaved evergreen forest; up to 2,700 m (breeds above 1,500 m). Has distinctive habit of clinging to tree-trunks and branches, often hanging downwards, almost vertically, in manner of nuthatch. Agitated birds on territory flick one wing at a time, relatively slowly, unlike White-tailed and Emei Leaf-warblers and Davison's Warbler. **RANGE & STATUS** Breeds N Pakistan, NW,N,NE Indian subcontinent, S,SE Tibet, SW China; some winter to south, northern India, Bangladesh. **SE Asia** Locally common resident W,N,E Myanmar, W,NW,SE Thailand, W Tonkin, east S Laos, S Annam; ? SE Thailand. ? scarce winter visitor N Myanmar, north Tenasserim. **BREEDING** February–June. **Nest** Ball-shaped, in recess in bank, sometimes in tree-trunk or stump. **Eggs** 4–5; white; 15.3 x 11.9 mm (av.; *assamensis*). Brood-parasitised by Oriental Cuckoo.

1101 **HARTERT'S WARBLER**
Phylloscopus goodsoni Not illustrated

IDENTIFICATION 11.5–12 cm. *P.g.fokiensis*: **Adult** Shows less white on tail than Claudia's and Blyth's Leaf-warblers (still more than Emei Leaf-warbler), similar to *ticehursti* race of Blyth's but perhaps even yellower below. **VOICE** Apparently undocumented. **HABITAT** Broadleaved evergreen forest; c.500–1,100 m. **RANGE & STATUS** Breeds S,SE China; some winter to south/south-west. **SE Asia** Uncommon to fairly common winter visitor N Laos, probably also parts of Vietnam. **NOTE** Split from Blyth's Leaf-warbler, following Olsson *et al.* (2005), but pending data on vocal differences from other similar taxa.

1102 **ARCTIC WARBLER**
Phylloscopus borealis Plate **106**

IDENTIFICATION 12.5–13 cm. *P.b.borealis*: **Adult** Differs from very similar Greenish and Two-barred Warblers in following subtle features: slightly larger; slightly longer, heavier bill with darker tip to lower mandible; supercilium falls short of bill-base; eyestripe clearly reaches bill-base; ear-coverts more prominently mottled; legs browner or yellower, toes more yellowish; greyer breast-sides and flanks, often appearing slightly streaked. Note voice. Rest of underparts whitish, sometimes with slight yellowish suffusion. Duller, with narrower wing-bars when worn. Overall proportions similar to Eastern Crowned Warbler but easily separated by plain greenish-olive crown, lack of contrasting yellow on undertail-coverts, and paler legs and feet. *P.b.xanthodryas* (recorded Peninsular Malaysia, E Tonkin at least) shows yellower-tinged underparts (particularly when fresh). For differences from similar Large-billed Leaf-warbler see that species. **VOICE** Song is a fast, rather hard whirring trill: ***dyryryryryryryryr...*** or ***dererererererererere...*** (varying somewhat in pitch/speed). Call is a loud sharp ***dzip*** or ***dzrt***. **HABITAT & BEHAVIOUR** Mixed deciduous and broadleaved evergreen forest, secondary growth, gardens, mangroves; mainly lowlands but up to 1,800 m on migration. Movements quite sluggish and methodical compared to Greenish and Two-barred. **RANGE & STATUS** Breeds N Palearctic, NE China, Japan, N America (Alaska); winters Sundas, Philippines, Wallacea, rarely Bismarck Archipelago, NW Australia. **SE Asia** Scarce to common winter visitor E Myanmar, Tenasserim, W,C,SE,S Thailand, Peninsular Malaysia, Singapore, S Laos, S Annam, Cochinchina; commoner in south. Uncommon to common passage migrant Thailand, Peninsular Malaysia, Singapore, Indochina (except C Laos, W Tonkin).

1103 **LARGE-BILLED LEAF-WARBLER**
Phylloscopus magnirostris Plate **106**

IDENTIFICATION 12.5 cm. Monotypic. **Adult** Very like Greenish and Two-barred Warblers but larger, stockier and much larger-billed, upperparts deeper and greener-olive (in fresh plumage), dark eyestripe broader, supercilium longer (extending to rear of ear-coverts) and contrasting more with darker crown; bill mostly dark with restricted pale orangey-flesh to brownish base of lower mandible, ear-coverts mottled greyish, underparts typically rather dirty, often with diffuse streaking on breast and flanks. Also very like Arctic Warbler but upperparts somewhat darker, crown slightly darker/greyer than mantle, has more extensively dark lower mandible and duller legs and feet. Note distinctive voice and range. **Juvenile** Buffier wing bar/s, more heavily marked and greyer breast. **VOICE** Song is a highly distinctive, sweet, high-pitched, descending ***si si-si su-su***. Calls with diagnostic tit-like ***duu-ti*** or ***dir-tee*** (second note higher). **HABITAT** Broadleaved evergreen forest, mainly near mountain streams when breeding; up to 2,745 m. Breeds above 1,800 m in India. **RANGE & STATUS** Breeds N Pakistan, NW,N India, Nepal, Bhutan, S,SE Tibet, SW,W,NC China; winters S India, Sri Lanka. **SE Asia** Uncommon resident or breeding visitor N Myanmar. Scarce to uncommon winter visitor C,S Myanmar. **BREEDING** India: June–August. **Nest** Rough dome, on ground under log or rock etc. or in recess in bank. **Eggs** 3–5; white; 18.2 x 13.2 mm (av.).

1104 **PALE-LEGGED LEAF-WARBLER**
Phylloscopus tenellipes Plate **106**

IDENTIFICATION 12.5–13 cm. Monotypic. **Adult** Resembles Arctic and Greenish Warblers but shows distinctive combination of uniform dark greyish crown (sometimes slightly paler in centre) contrasting with mantle, pale buff, cream or white supercilium, distinctly paler, warmish olive-brown rump and uppertail-coverts, and pale greyish-pink legs and feet. Usually shows two narrow buffish-white wing-bars. Ear-coverts are mottled and often tinged buffish. Note distinctive voice and behaviour. **VOICE** Song is similar to Arctic but faster, thinner and higher-pitched: ***sresresresresresre...*** Call is a distinctive, very high-pitched, thin, metallic ***tih*** or ***tip***. **HABITAT & BEHAVIOUR** Mixed deciduous, broadleaved evergreen and semi-evergreen forest, secondary growth, also mangroves and gardens on migration; up to 1,500 m (mainly winters below 1,000 m). Usually found close to ground, in undergrowth and understorey trees, rather secretive. **RANGE & STATUS** Breeds E Palearctic, Japan, NE China. **SE Asia** Common winter visitor C,S Myanmar, Tenasserim, Thailand, Peninsular Malaysia (rare), Cambodia, Laos, Cochinchina. Uncommon to fairly common passage migrant C Thailand, E Tonkin (occasionally winters), N,C Annam. Vagrant Singapore.

1105 **GREENISH WARBLER**
Phylloscopus trochiloides Plate **106**

IDENTIFICATION 12 cm. *P.t.trochiloides*: **Adult** Difficult to separate from Two-barred Warbler but only shows one narrow wing-bar on greater coverts. Occasionally shows indistinct second bar (on median coverts) in fresh plumage. Note voice. For differences from similar Arctic and Large-billed Leaf-warblers see those species. Note voice. **Other subspecies in SE Asia** *P.t.obscuratus* (widely claimed from wintering range). **VOICE** Song is a simple hurried repetition of call or similar notes: ***chiree-chiree-chiree-chiree-chee-chee witchu-witchu-witchu-witchu*** etc. Calls with a fairly high-pitched, slurred, disyllabic ***chiree*** or ***chir'ee***. **HABITAT** Broadleaved evergreen forest, secondary growth; up to 2,565 m (mainly in mountains). **RANGE & STATUS** Breeds W,C Palearctic, NW Mongolia, N Pakistan, NW,N,NE Indian subcontinent, S,E Tibet, SW,W,NW China; winters Indian subcontinent, SW China. **SE Asia** Fairly common winter visitor Myanmar, W,NW,NE Thailand, N Laos, W,E Tonkin; likely to be resident N Myanmar. breeding India: May–August. **Nest** Ball-shaped with side-entrance, on or close to ground. **Eggs** 3–4; white; 15.5–15.6 x 11.7–12.5 mm (*trochiloides*).

1106 **TWO-BARRED WARBLER**
Phylloscopus plumbeitarsus Plate **106**

IDENTIFICATION 12 cm. Monotypic. **Adult** Very similar to Greenish Warbler but, when fresh, shows two broader yellowish-white wing-bars. Often worn in winter, with narrower wing-bars (may lack shorter upper wing-bar). When wing-bars are broadest, can resemble Yellow-browed Warbler but is larger and longer-billed and lacks whitish fringes to tertials. More similar to Arctic Warbler, differing in following subtle features: slightly smaller; slightly weaker bill with uniform pinkish lower mandible; supercilium reaches bill-base; less prominent eyestripe in front of eye; paler and plainer ear-coverts; duller to dark legs and feet; paler, more uniform breast-sides and flanks. Also (when fresh) shows broader, longer bar on greater coverts. Note voice. See Large-billed Leaf-warbler. Note range. **VOICE** Song is similar to Greenish but faster and more slurred and jumbled. Usual call is a trisyllabic, slurred, rather sparrow-like ***chireewee*** or ***chir'ee'wee*** (occasionally omits last syllable). **HABITAT &**

BEHAVIOUR Mixed deciduous and semi-evergreen forest, bamboo, sometimes broadleaved evergreen forest, parks and gardens, particularly on migration; up to 1,295 m. Active, working through foliage of mid-storey in forest. **RANGE & STATUS** Breeds Siberia, E Palearctic, Mongolia, NE China; winters to south. **SE Asia** Fairly common to locally common winter visitor Myanmar, Thailand (except extreme S), Cambodia, S Laos, C,S Annam, Cochinchina. Uncommon to fairly common passage migrant C Thailand, N Laos, E Tonkin, N Annam. Vagrant Peninsular Malaysia, Singapore.

1107 EMEI LEAF-WARBLER
Phylloscopus emeiensis Plate **106**

IDENTIFICATION 11.5 cm. Monotypic. **Adult** Very similar to Claudia's, Blyth's and Hartert's Leaf-warblers (particularly former) but shows duller crown pattern, with greenish-grey rather than blackish rear part of lateral stripes and less pale median stripe, which is less contrasting at rear; inner webs of two outer pairs of tail-feathers have narrower whitish margins (difficult to see in field), poorly defined or lacking at tip of outermost and lacking (or very nearly so) at tip of penultimate feather. Small amount of white on outertail-feathers rules out Davison's Warbler, and all subspecies of White-tailed Leaf- except *ogilviegranti*, from which it differs by less distinct crown pattern, less greenish upperparts and less yellow median crown-stripe, supercilium, throat, belly-centre and undertail-coverts. **VOICE** Song is a clear, slightly quivering straight trill (usually lasting 3–4 s), recalling Arctic and, to lesser extent, Pale-legged. Call is a distinctive soft ***tu-du*** or ***tu-du-du-du*** etc. **HABITAT & BEHAVIOUR** Broadleaved evergreen forest; recorded at 305 m. Agitated birds (at least on breeding grounds) often quickly flick both wings. **RANGE & STATUS** Breeds China (SC Sichuan); full breeding and winter ranges unknown. **SE Asia** Status uncertain. One winter record east S Myanmar.

1108 ASHY-THROATED WARBLER
Phylloscopus maculipennis Plate **107**

IDENTIFICATION 9–9.5 cm. *P.m.maculipennis*: **Adult** Tiny size, contrasting pale yellow rump and pronounced head-stripes and wing-bars recall Lemon-rumped Warbler; easily separated by greyish throat and upper breast, demarcated from yellow lower breast to vent and largely white outertail-feathers. Also shows greyer crown, with pale greyish median stripe. See Buff-barred Warbler. **Juvenile** Said to have olive-washed crown and throat and brighter buffish-tinged wing-bars. **VOICE** Song recalls White-tailed Leaf-warbler but shorter, with repeated ***sweechoo*** and ***sweeti*** notes. Call is a thin sharp high-pitched ***zip*** or ***zit***. **HABITAT** Broadleaved evergreen forest; 1,525–3,050 m, locally down to 610 m in winter (N Myanmar). Breeds above c.1,800 m. **RANGE & STATUS** Resident N Pakistan, N,NE Indian subcontinent, SW China. **SE Asia** Locally common resident W,N,S(east),E Myanmar, NW Thailand (Doi Inthanon only), N,S Laos, W Tonkin, C,S Annam. **BREEDING** February–April. **Nest** Ball-shaped, suspended from tree-branch; 5 m above ground. Otherwise undescribed.

1109 BUFF-BARRED WARBLER
Phylloscopus pulcher Plate **107**

IDENTIFICATION 11–11.5 cm. *P.p.pulcher*: **Adult** Broad pale orange-yellowish to orange-buff bar on greater coverts, pale yellowish rump (contrasting with rest of upperparts) and extensively white outertail-feathers diagnostic. Rather dark olive above, with darker (but not strongly contrasting) lateral crown-stripes, typically rather dull below. When worn, wing-bars are narrower and buffier. See Ashy-throated and Lemon-rumped Warblers. **VOICE** Song is a high-pitched twitter, either preceded by or ending with a drawn-out trill. Call is a short sharp thin ***swit*** or ***sit***, sharper and more strident than Ashy-throated. **HABITAT** Broadleaved evergreen forest; 1,370–3,655 m (breeds above 2,135 m), locally down to 1,050 m in winter. **RANGE & STATUS** Resident (subject to altitudinal movements) N,NE Indian subcontinent, S,SE Tibet, SW,W China. **SE Asia** Locally common resident W,N Myanmar, N Laos, W Tonkin. Uncommon to locally common winter visitor S Myanmar, north Tenasserim, NW Thailand. Recorded (status uncertain) E Myanmar. **BREEDING** April–July. **Nest** Untidy ball, in outer tree-branches, bush or against tree-trunk; 0.6–4 m above ground. **Eggs** 3–4; white, speckled and blotched reddish-brown to reddish, particularly towards broader end, where may show ring; 14.9 x 11.4 mm (av.; India).

1110 YELLOW-BROWED WARBLER
Phylloscopus inornatus Plate **107**

IDENTIFICATION 11–11.5 cm. Monotypic. **Adult** Generally the commonest and most widespread *Phylloscopus* warbler in the region. Shows distinctive combination of broad yellowish-white supercilium, two broad, yellowish-white wing-bars and plain olive-green upperside with no obvious median crown-stripe (sometimes faintly paler) and no pale rump-patch or white on tail. Underparts off-whitish. Duller above, with narrower wing-bars when worn. Similar to Two-barred Warbler but a little smaller and shorter-tailed, with whitish tips to tertials and broader, more contrasting wing-bars (similar when worn); bill weaker, usually with darker tip to lower mandible. Note voice. See under extremely similar Hume's Warbler for differences. **VOICE** Song is a high-pitched, thin ***tsitsitsui itsui-it seee tsi tsi-u-eee*** etc. Call is a distinctive loud, high-pitched, slightly rising ***tswee-eep***; ***tsweet*** or ***wee-eest***. **HABITAT** Forest, secondary growth, parks, gardens, mangroves, all kinds of wooded areas; up to 2,440 m. **RANGE & STATUS** Breeds C,E Palearctic, NE China; winters N and north-eastern Indian subcontinent, S China. **SE Asia** Common winter visitor throughout; scarce Peninsular Malaysia, Singapore. Also recorded on passage Peninsular Malaysia, Cambodia, E Tonkin.

1111 HUME'S WARBLER
Phylloscopus humei Plate **107**

IDENTIFICATION 11–11.5 cm. *P.h.mandelli*: **Adult** Apart from voice, difficult to separate from Yellow-browed Warbler but shows grey wash on upperparts (particularly crown and mantle) and somewhat darker lower mandible, legs and feet; median covert bar usually slightly less distinct and duller than greater covert bar, may show darker eyestripe (offsetting supercilium) and slightly duller, greyer-tinged throat and breast, has greater tendency to show a faint pale central crown-stripe. **VOICE** Song is a thin falling rasping ***zweeeeeeeeeoooo***, preceded by excited repetition of ***we-soo*** call-notes. Calls with thin ***we-soo*** or ***wi-soo*** (often doubled) and ***tschu'is***; ***tschui*** or ***tschuit***, with stressed first syllable. **HABITAT** Forest, wooded areas, secondary growth; 1,000–2,780 m (lower in S Thailand). **RANGE & STATUS** Breeds C Asia, W Mongolia, N Pakistan, NW,N India, Nepal, Bhutan, SW,E Tibet, SW,W,NW China; winters south S,NE India. **SE Asia** Scarce to locally common winter visitor W,S Myanmar, NW Thailand, N Laos, W,E Tonkin. Vagrant S Thailand.

1112 CHINESE LEAF-WARBLER
Phylloscopus yunnanensis Plate **107**

IDENTIFICATION 10 cm. Monotypic. **Adult** Like Lemon-rumped Warbler but shows slightly paler crown-sides, different median crown-stripe (usually more indistinct on fore-crown, which may appear unmarked when head-on, and sometimes very faint except on hindcrown, which may show contrasting pale spot); dark eyestripe slightly paler and fairly straight-ended without distinct hook-shape at rear, lacks pale spot on rear ear-coverts and has paler neck-sides (may almost appear to merge with rear supercilium). Also lacks distinct darker area at bases of secondaries, greater coverts often show paler centres and less distinct green fringes. In direct comparison with Lemon-rumped, appears slightly larger, more elongated, longer-billed and less round-headed. See Pallas's Leaf-warbler. Note voice. **VOICE** Song is a prolonged monotonous ***tsiridi-tsiridi-tsiridi-tsiridi-tsiridi...*** similar to Striated Prinia. Calls with a rather loud ***tueet***, and (at least on breeding grounds) varied irregular loud clear scolding whistles, ***tueet-tueet-tueet tueet-tueet tueet-tueet-tueet tueet tUEE tuee-tuee-tuee-tuee-tuee...*** etc., and hammering ***tueet tuee-tee-tee-tee-tee-tee-tee...*** **HABITAT** Broadleaved evergreen forest, secondary growth; 450–1,800 m. **RANGE & STATUS** Breeds W,NC and northern E China; at least some populations winter to south. Recorded during non-breeding season S China (Hong Kong). **SE Asia** Scarce to uncommon (perhaps fairly common) winter visitor east C Myanmar, W,NW,NE Thailand, N Laos, W,E Tonkin.

1113 LEMON-RUMPED WARBLER

Phylloscopus chloronotus Plate **107**

IDENTIFICATION 10 cm. *P.c.chloronotus*: **Adult** Told by tiny size, small dark bill, sharply contrasting pale yellow rump, supercilium, median crown-stripe and double wing-bar, and lack of white on outertail-feathers. Very similar to Pallas's Leaf-warbler but head-stripes and wing-bars less bright yellow, mantle slightly duller and browner-tinged, less greenish; supercilium generally very pale yellowish, often appearing off-white to pale buffish) and usually narrower in front of eye. See under very similar Chinese Leaf-warbler for differences; see also Ashy-throated and Buff-barred Warblers. Note distinctive voice and range. **VOICE** Song is of two types: (1) a drawn-out thin rattle followed by rapid, evenly pitched notes: ***tsirrrrrrrrrrr-tsi-tsi-tsi-tsi-tsi-tsi-tsi-tsi-tsi-tsi...***; (2) variable endless stuttering notes, alternating in pitch: ***tsi tsi-tsi tsi-tsi tsu-tsu tsi-tsi tsu-tsu tsi-tsi tsi-tsi tsi-tsi-tsi tsirrp tsi-tsi tsu-tsu...*** Call is a high-pitched, almost disyllabic, sharp metallic sunbird-like ***twit*** or ***tuit***. **HABITAT** Broadleaved evergreen and semi-evergreen forest, secondary growth; 450–2,500 m. Breeds above 2,200 m in India. **RANGE & STATUS** Breeds N Pakistan, N,NE Indian subcontinent, S,E Tibet, SW,W,NC China; some winter to south. **SE Asia** Scarce winter visitor S(east),E Myanmar, north Tenasserim, NW Thailand, W Tonkin. Recorded (status uncertain; perhaps resident) W,N Myanmar. **BREEDING** India: May–July. **Nest** Ball-shaped with side-entrance, in outer tree-branches; 2–15 m above ground. **Eggs** 3–5; white with numerous reddish blotches and dots (mostly at broader end); 14.1 x 10.9 mm (av.; India).

1114 PALLAS'S LEAF-WARBLER

Phylloscopus proregulus Plate **107**

IDENTIFICATION 10 cm. Monotypic. **Adult** Very similar to Lemon-rumped Warbler, but in fresh plumage shows much yellower supercilium and median crown-stripe, often also ear-coverts, chin, throat and wing-bars; mantle greener, supercilium especially bright in front of and above eye and tends to be broader in front of eye. Duller above with narrower wing-bars and less strongly marked with yellow when worn. See Chinese Leaf, Ashy-throated and Buff-barred Warblers. Note distinctive voice and range. **VOICE** Song is very loud, rich and varied, with clear whistles and trills reminiscent of Canary *Serinus canaria*. Call is a subdued, slightly nasal or squeaky ***chuit*** or ***chui***. **HABITAT** Broadleaved evergreen and semi-evergreen forest, secondary growth; 900–1,700 m. **RANGE & STATUS** Breeds E Palearctic, Mongolia, NE China, N Korea; winters southern China. **SE Asia** Scarce to locally common winter visitor NW Thailand, N Laos, W,E Tonkin; commoner in east. Also recorded on passage E Tonkin.

1115 TICKELL'S LEAF-WARBLER

Phylloscopus affinis Plate **107**

IDENTIFICATION 11–11.5 cm. Monotypic. **Adult** Like Buff-throated Warbler but upperparts more greenish-tinged, supercilium, ear-coverts and underparts distinctly washed lemon-yellow, supercilium typically better defined (particularly in front of eye) and often more whitish towards rear, eyestripe slightly more pronounced, contrasting more with paler, yellower ear-coverts; bill shows little or no dark tip to lower mandible. Worn birds are more washed out, less distinctly yellow and have upperparts similar to Buff-throated. Resembles Aberrant Bush Warbler but wings more pointed, tail shorter and more square-cut. Note voice and behaviour. **VOICE** Song is a short ***chip whi-whi-whi-whi***. Call is a short ***chit*** or ***sit***. **HABITAT & BEHAVIOUR** Low vegetation, grass and scrub; up to 2,135 m. Typical leaf-warbler; very active and conspicuous, usually in bushes and small trees. **RANGE & STATUS** Breeds N Pakistan, NW,N India, Nepal, Bhutan, S,E Tibet, W China; winters India, Bangladesh, SW China. **SE Asia** Uncommon to fairly common winter visitor Myanmar (except Tenasserim).

1116 RADDE'S WARBLER

Phylloscopus schwarzi Plate **107**

IDENTIFICATION 13.5–14 cm. Monotypic. **Adult** Rather stocky, and relatively thicker-billed than other *Phylloscopus* warblers. Apart from distinctive call, very difficult to separate from Yellow-streaked Warbler but slightly larger and proportionately larger-headed, with slightly thicker bill and legs; pre-ocular supercilium less defined. Also very like Dusky but larger, with thicker bill and legs, and usually no dark tip to lower mandible; usually has more olive-tinged upperparts, notably rump, uppertail-coverts and wing fringes (but similar when worn); supercilium usually broader and more diffuse in front of eye (normally deeper buffish than above/behind eye), often bordered above by darkish line; usually shows rusty-buff undertail-coverts contrasting with dull pale brownish-buff rest of underparts. Belly may show slight yellow tinge when fresh. See Buff-throated, Brownish-flanked Bush and Aberrant Bush Warblers. Note distinctive voice. **VOICE** Sings with short, loud outbursts of fast variable trilling: ***ty-ty sui-sui-sui-sui-sui-sui......tydydydydy-dydyd ty-tytyrrrrrrrrrrrr ty-ty suisuisuisuisuisuisuisui tuee-tuee-tuee-tuee-tuee...*** etc. Calls with low, rather soft ***tyt***, ***tuc*** or ***tet***, often in irregularly repeated stuttering series: ***tyt tyt tyt tyt-tyt-tyteryt tyt...*** **HABITAT & BEHAVIOUR** Low vegetation in more open forest, clearings, forest edge, road and track verges, scrub and grass in more open areas; up to 2,135 m. Fairly skulking, usually close to ground. **RANGE & STATUS** Breeds E Palearctic, N Mongolia, NE China; winters extreme S China, rarely N,NE Indian subcontinent. **SE Asia** Fairly common to common winter visitor (except SW,N Myanmar, S Thailand, Peninsular Malaysia, Singapore, W Tonkin). Also recorded on passage E Tonkin. Vagrant S Thailand, west Peninsular Malaysia.

1117 YELLOW-STREAKED WARBLER

Phylloscopus armandii Plate **107**

IDENTIFICATION 13–14 cm. Monotypic. **Adult** Difficult to separate from Radde's but at very close range shows fine yellow streaking on breast and belly (notably when fresh); slightly smaller and proportionately smaller-headed, with thinner bill and legs (slightly thicker than Dusky), supercilium slightly more defined in front of eye. See Dusky and Buff-throated Warblers. Note distinctive voice and range. **First winter** Similar to fresh adult but throat also buffish (though paler than rest of underparts) with yellow streaking (hard to see in field). **VOICE** Sings with short rapid husky slurred undulating phrases, introduced by series of ***tzic*** notes. Call is a sharp bunting-like ***tzic*** or ***tick***. **HABITAT & BEHAVIOUR** Low vegetation, scrub and small trees in borders and clearings in deciduous woodland, broadleaved evergreen and mixed broadleaved and pine forest. Up to 2,500 m (mostly below 1,500 m) in winter; breeds 1,220–3,355 m. Often 2–3 m above ground in small trees. **RANGE & STATUS** Breeds W,C,N China; winters SW China. **SE Asia** Local resident N Myanmar. Scarce to locally common winter visitor W,C,S,E Myanmar, NW Thailand, N Laos, E Tonkin. Also recorded on passage E Tonkin. Vagrant NE,SE Thailand. **BREEDING** Undocumented?

1118 DUSKY WARBLER

Phylloscopus fuscatus Plate **107**

IDENTIFICATION 12–12.5 cm. *P.f.fuscatus*: **Adult** Rather nondescript, without yellowish or greenish plumage tones. Upperparts vary from rather dark brown (fresh) to paler greyish-brown (worn), underparts dirty whitish, often with buffish wash on breast-sides, flanks and undertail-coverts (more pronounced in fresh plumage). Like Radde's Warbler but a little smaller/slimmer, with finer, more pointed bill (typically with dark-tipped lower mandible) and thinner legs; long supercilium is narrower, whiter and more sharply defined in front of eye, usually distinctly buffish-tinged behind and rarely with dark line above; usually lacks contrasting rusty-buff undertail-coverts. Note voice. See Yellow-streaked and Pale-footed Bush Warblers. **VOICE** Song is like Radde's but distinctly higher-pitched, slower and less varied (lacking strong rattling trills) and has fewer syllables per phrase, which lack ***ty-ty*** introductory notes. Often begins song with thin ***tsirit***. Call is a rather hard ***tett*** or ***tak***, often repeated when agitated and sometimes extended in a rattling series. **HABITAT** Low vegetation and small trees in open areas, often near water, mangroves; mostly lowlands but up to 1,830 m on migration. More often in wet habitats than similar *Phylloscopus* warblers. **RANGE & STATUS** Breeds Siberia, E Palearctic, E Tibet, W,N,NE China; winters N,NE Indian subcontinent, S China. **SE Asia** Fairly common to common winter visitor (except S Thailand, Peninsular Malaysia, Singapore). Rare winter visitor/vagrant S Thailand, Peninsular Malaysia, Singapore. Also recorded on passage Cambodia, E Tonkin.

1119 **BUFF-THROATED WARBLER**

Phylloscopus subaffinis Plate **107**

IDENTIFICATION 11–11.5 cm. Monotypic. **Adult** Rather uniform yellowish-buff supercilium and underparts (including throat) and lack of crown-stripes or wing-bars distinctive. Very like Tickell's Leaf-warbler but upperparts browner (worn Tickell's Leaf- similar), supercilium, ear-coverts and underparts more yellowish-buff, supercilium typically less defined (particularly in front of eye) and more uniform buffish-yellow (only slightly yellower than underparts), eyestripe slightly less pronounced; bill typically shows extensive dark tip to lower mandible. See Aberrant Bush, Yellow-streaked and Radde's Warblers. Note distinctive voice. **VOICE** Song is slower and weaker than Tickell's and lacks introductory ***chip*** note; may be introduced by short, subdued ***trr*** or ***trr-trr***. Call is a sibilant cricket-like ***trrup*** or ***tripp***. **HABITAT** Scrub and low vegetation in open areas; 1,200–2,565 m, locally down to 200 m in E Tonkin (breeds above c.1,800 m). **RANGE & STATUS** Breeds W,C and southern China; winters S China. Recorded (status uncertain) NE India. **SE Asia** Local resident W Tonkin. Scarce to fairly common winter visitor W,N,E Myanmar, NW Thailand, N Laos, E Tonkin, C Annam. Also recorded on passage E Tonkin. Vagrant NE Thailand. **BREEDING** SW China: May–July. **Nest** Dome with side-entrance, in low vegetation just above ground. **Eggs** 4; white.

1120 **COMMON CHIFFCHAFF**

Phylloscopus collybita Plate **107**

IDENTIFICATION 12 cm. *P.c.tristis*: **Adult** Recalls Dusky and Buff-throated Warblers but grey-brown above, and whitish below with buffy-washed breast and flanks; has olive-green wing-fringing, slight pale wing-bar, and blackish bill, legs and feet. When worn, appears greyer and whiter, with duller wings, and no wing-bar. **VOICE** Song is ***ch-ch-chewy-chewy-chewy-chewy-ch***, followed by pause. Calls with weak ***peu***, ***sie-u***, ***peeep*** or ***vii(e)p***. **HABITAT & BEHAVIOUR** Low vegetation and smaller trees in wooded as well as more open areas; found at c.1,500 m. Forages actively at all levels. **RANGE & STATUS**. Breeds W Palearctic, Siberia (east to Kolyma R), C Asia, NW Mongolia, N Iran; mostly winters Africa (except S), Arabia, SW Asia, northern, Indian subcontinent. **SE Asia** Vagrant W Tonkin.

TIMALIIDAE: Babblers

Worldwide c.432 species. SE Asia 175 species. Highly variable and poorly defined grouping. Characteristics usually stated to include strong legs and feet, soft plumage and relatively short rounded wings. Largely sedentary; generally feed on insects and other invertebrates but some also consume small fruits, seeds and nectar. Can be roughly grouped as follows: ***Sylvia* warblers** Similar to 'true' warblers (Phylloscopidae). Typically inhabit bushes and scrub, where they feed mainly on insects and some berries etc. **Parrotbills** Short deep bills (relatively long in *Conostoma*) with sharp cutting edges, at least partly developed for feeding on bamboo stems, very soft plumage. ***Chrysomma* babblers** Shortish thick bills and longish to long tails, mostly in grassland (also scrub). **Fulvettas** (*Lioparus, Fulvetta*) Small, robust and stout-billed, mostly gregarious and noisy, mainly in lower storey of forest, secondary growth and bamboo. **White-eyes** (*Zosterops*) Small and warbler-like but with rather slender, slightly downcurved bills and brush-tipped tongues. Highly active and gregarious. Feed mainly on nectar, berries, seeds, insects and caterpillars. **Yuhinas** (*Yuhina*) Variable, but all crested, bills narrowish and slightly downcurved (*Yuhina*) or thicker, shorter and straight (*Staphida*); mostly inhabit middle storey of forest, some feed on nectar. **Nun-babblers** (*Alcippe, Schoeniparus*) Small, robust and stout-billed, nondescript, mainly brownish to greyish, most with black lateral crown-stripes, gregarious and noisy, in lower to middle storey of forest and secondary growth (*Alcippe*); similar but with white supercilia and rufous on crowns, inhabiting forest understorey (*Schoeniparus*). ***Stachyris* babblers** Shortish, broad-based straight bills, mostly rather plain with distinctive head patterns; many have colourful neck-skin which may be exposed when singing. **Scimitar-babblers** (*Xiphirhynchus, Pomatorhinus*) Have relatively (but variably) long, downcurved bills; mostly inhabit ground layer to middle storey of forest, many favouring bamboo. **Chevron-breasted Babbler** Broad-based and pointed bill, recalls but larger than most wren-babblers, relatively arboreal. ***Spelaeornis* wren-babblers** Forest dwelling; similar to *Pnoepyga* wren-babblers but with longer tails, mostly in undergrowth but also on ground. ***Pnoepyga* wren-babblers** Very small and almost tail-less, rather narrow-billed, plumage scaly, almost exclusively on ground. ***Stachyridopsis* babblers** Reminiscent of *Stachyris* babblers but smaller, with mostly olive, yellow and buffish plumage. **Chestnut-capped Babbler** and **Rufous-rumped Grass-babbler** Thickset and relatively stout-billed, with rather broad graduated tails; inhabiting grassland. **Tit-babblers** (*Macronus*) Shortish straight bills, rather plain plumage with somewhat more distinctive head patterns, colourful neck-skin may be exposed when singing; inhabiting forest edge and secondary growth. **Jungle babblers** Relatively plain, mostly forest-dwelling; generally short-tailed, inhabiting ground layer and lower storey (*Pellorneum, Ophrydornis, Malacocincla, Trichastoma*) or longer-tailed and mostly in lower to middle storey (*Malacopteron*). **Wren-babblers** Forest dwelling; tails shortish, bills straight and rather stout, plumage scaly, inhabit ground layer and undergrowth (*Kenopia, Turdinus, Gypsophila, Napothera*); tails shortish, bills downcurved, plumage with streaks, found on ground and in undergrowth (*Rimator*). **White-hooded Babblers** (*Gampsorhynchus*) Thick bills, long tails, mostly white head and upper breast; only in bamboo. ***Pseudominla* fulvettas** Small and small-billed, and quite short-tailed, well-marked head-patterns; foliage-gleaning in lower to middle storey of montane forest. ***Turdoides* babblers** Relatively large, slim and long-tailed, bills straightish to slightly downcurved, only occur in flocks, in grassland or open areas. **Cutias** (*Cutia*) Thickish, pointed and slightly downcurved bills, thickset bodies, well-patterned and sexually dimorphic, strongly arboreal, inhabiting montane forest. **Laughingthrushes** (*Dryonastes, Garrulax, Melanocichla, Rhinocichla, Babax, Grammatoptila, Stactocichla, Leucodioptron, Strophocincla, Pterorhinus, Ianthocincla, Trochalopteron*) and **Liocichlas** (*Liocichla*) Medium-sized and stocky, with strong bills, legs and feet and longish tails, plumage very varied; mostly gregarious, inhabiting forest and secondary growth, many species foraging on ground amongst leaf-litter, searching for beetles, flies and other insects, as well as snails, leeches and seeds etc.; many species also feed on fruit and nectar. **Minlas** (*Chrysominla, Minla*) Smallish, fairly long-tailed and boldly patterned (particularly wings); mostly inhabit lower to middle storey of forest. **Silver-eared Mesia** and **Red-billed**

Leiothrix Stout bills, colourfully patterned heads and wings; inhabit thick secondary growth. **Sibias** (*Crocias, Heterophasia, Malacias, Leioptila*) Mostly rather slim and long-tailed with relatively narrow and pointed bills, quite gregarious, mainly in middle to upper storey of forest; many are fond of berries and nectar. **Barwings** (*Actinodura*) Stout bills, longish to long tails, floppy crests, barred wings; inhabit forest and secondary growth. **NOTE** Recent DNA studies have clearly demonstrated that *Sylvia* warblers and white-eyes are babblers (see Collar & Robson 2007).

1121 LESSER WHITETHROAT
Sylvia curruca Plate **107**

IDENTIFICATION 14 cm. *S.c.blythi*: **Adult** Rather dark grey crown, contrastingly darker lores and ear-coverts and rather square-ended tail with prominent white on outer feathers distinctive. Shows greyish-brown remainder of upperparts and whitish underparts, with greyish-brown wash on flanks. **First winter** Like adult but crown sullied brownish, has less blackish lores and ear-coverts, narrow pale supercilium. **VOICE** Song is a rather low-pitched scratchy warble, often followed by a distinctive dry throbbing rattle. Calls with a fairly subdued dry ***tett***, similar to Dusky Warbler. **HABITAT** Scrub and secondary growth in more open areas; up to 1,300 m. **RANGE & STATUS** Breeds W,C Palearctic, Siberia, northern Middle East, Mongolia, NW,NE China; mainly winters NE Africa, Middle East, Indian subcontinent (except NE). **SE Asia** Vagrant NW,C Thailand.

1122 GREAT PARROTBILL
Conostoma oemodium Plate **108**

IDENTIFICATION 27.5–28.5 cm. Monotypic. **Adult** Much larger than other parrotbills, bill relatively long and yellowish, plumage rather uniform brown with greyish-white forehead, blackish-brown lores and orbital area, and greyer underparts and fringes to primaries and tail-feathers. See Brown Parrotbill. **Juvenile** Upperside a little more rufescent. **VOICE** Sings with variable loud clear full ***whip whi-uu***; ***uu-chip uu-chip***; ***eep whu-eep*** or ***ee-uu braah*** etc. Also ***wip-puwuu-wip-pwraaow*** (***wip*** short and quick, ***puwuu*** rising slightly towards end, ***pwraaow*** rather harsh, gravelly and purring), sometimes with 1–2 extra ***ra*** notes added on. Shorter ***chu-pwerr*** (***chu*** very short quick) may be song variant. Calls include nasal wheezes, squeals, cackling and churring. Combinations of rather high nasal ***u-err***, quick high nasal ***ii*** and rolling, purring ***pwaaowa*** calls by groups of two or more birds. **HABITAT** Open broadleaved evergreen forest, bamboo, scrub; 2,775–3,660 m, down to 2,285 m in winter. **RANGE & STATUS** Resident N India, Nepal, Bhutan, W,C China. **SE Asia** Uncommon resident N Myanmar. **BREEDING** India: May–July. **Nest** Broad, quite shallow cup (sometimes domed), in bamboo. **Eggs** 2–3; white, sparsely blotched, streaked and smudged pale yellowish-brown, with pale inky-purple undermarkings at broader end or speckled reddish, mostly at broader end; 27.8–28.2 x 20.3–20.8 mm.

1123 BROWN PARROTBILL
Cholornis unicolor Plate **108**

IDENTIFICATION 21 cm. Monotypic. **Adult** Overall greyish-brown plumage with browner upperparts, narrow blackish supercilium and short, deep-based yellowish bill distinctive. See Great Parrotbill. **Juvenile** Warmer-washed overall than adult, buffier below, lores and supercilium slightly duller, bill noticeably thinner. **VOICE** Sings with a clear, loud, rather high ***II-WUU-IIEW***; ***II WIU'UU*** or ***II-WUU*** (last note clear, louder and rising slightly) repeated after shortish intervals, or quickly repeated ***WHIIIU*** or ***WHIIIUU***. May be introduced by one to several low, short (often barely audible) ***t***; ***it***; ***ik*** or ***ch*** notes, or low rolling ***chrrr***. Also utters high ***wee-ee*** or ***wee-hiu***, repeated after clear intervals, and high-pitched ***wee-ee-ee-ee-ee...*** Calls include shrill whining ***whi-whi-whi***, low ***brrh***, and harsh crackling ***chrrr***, ***churrrh*** and ***churr'rr'rr***. **HABITAT & BEHAVIOUR** Bamboo, open broadleaved evergreen forest; 2,135–3,660 m. Often in small parties. **RANGE & STATUS** Resident Nepal, Bhutan, NE India, SE Tibet, SW,W China. **SE Asia** Uncommon resident N Myanmar. **BREEDING** India: June–July. Otherwise undocumented.

1124 GREY-HEADED PARROTBILL
Psittiparus gularis Plate **108**

IDENTIFICATION 15.5–18.5 cm. *P.g.transfluvialis*: **Adult** Greyish crown, nape and head-sides, white lores and eyering, long black lateral crown-stripes and pale buff underparts with black throat-centre diagnostic. Rest of upperside rufous-brown, bill rather bright orange-yellow, short and deep-based. *P.g.rasus* (W Myanmar [south Chin Hills only]) is somewhat smaller and lacks black on throat; *laotianus* (extreme E Myanmar, NW[east] Thailand, N Indochina) has paler grey on crown and nape, paler ear-coverts and whiter underparts. **Juvenile** Much brighter rufescent upperside than adult, less pure greyish crown, less defined blackish throat-centre, buffier-tinged underparts, narrower bill. **VOICE** Song of two types. (1) Clear loud, quite shrill ***eu'chu'chu*** or ***eu-chu'chu'chu***, sometimes ***eu'chu'chu-cho*** (with lower-pitched ***cho***). (2) Quite high, clear, stressed ***wi-wuu***; occasionally 3–4 note ***wi wuu wuu...*** Both types have been heard simultaneously and may be pair-duet. Calls with short, quite harsh, rather slurred ***jiow*** or ***jieu*** and less harsh ***djer*** notes, sometimes mixed with harsh scolding scimitar-babbler-like ***chit'it'it'it'it'it'it...***, soft ***chip*** notes and very soft, short rattled ***chrrrat***. **HABITAT & BEHAVIOUR** Broadleaved evergreen forest, secondary growth, scrub bordering forest, bamboo; 1,000–1,830 m, locally down to 610 m. Often in small parties; joins mixed-species feeding flocks. **RANGE & STATUS** Resident NE Indian subcontinent, southern China. **SE Asia** Locally common resident W,N,S(east),E Myanmar, NW Thailand, N,C,S(east) Laos, W,E Tonkin, N,C Annam. **BREEDING** February–July. **Nest** Compact cup, in sapling, tree, bush, bamboo or tall grass; 1–9 m above ground. **Eggs** 2–4; whitish, pale sea-green, grey, yellowish or reddish, spotted, blotched, streaked and clouded pale yellow or yellowish-brown to reddish over grey to dark purple undermarkings (may be capped or ringed at broader end); 21.1 x 15.9 mm (av.; *transfluvialis*).

1125 BLACK-CROWNED PARROTBILL
Psittiparus margaritae Plate **108**

IDENTIFICATION 15.5 cm. Monotypic. **Adult** Like Grey-headed Parrotbill, but somewhat smaller and shorter-tailed, with all-black crown, darker and more rufous-chestnut upperside, blackish mottled ear-coverts and whiter underparts (like Grey-headed race *P.g.laotianus*), with no black on throat (though may show some dark speckles). **Juvenile** Undocumented. **VOICE** Song similar to Grey-headed, but type (1) somewhat huskier and more slurred 3–4 note ***jhu'jhu'jhu*** or ***jchew'jchew'jchew***. Variation includes a slightly clearer, slightly descending ***eu'ju'jhu'jhu***. Type (2) is perhaps a little more hurried than that of Grey-headed. Both songs have also been heard simultaneously and may be a duet. Similar harsh alarm call to Grey-headed. **HABITAT & BEHAVIOUR** Broadleaved evergreen forest, secondary growth, scrub bordering forest, bamboo; 850–1,500 m. Often in flocks; joins mixed-species feeding flocks. **RANGE & STATUS** Endemic. Locally fairly common resident extreme east Cambodia (south-east Mondulkiri Province), S Annam. **BREEDING** Undocumented. **NOTE** Split from Grey-headed Parrotbill, following Robson (2007).

1126 GREATER RUFOUS-HEADED PARROTBILL *Psittiparus bakeri* Plate **108**

IDENTIFICATION 19–19.5 cm. *P.b.bakeri*: **Adult** Like Lesser Rufous-headed Parrotbill but larger and rounder-headed with larger, relatively longer bill, more uniform rufous face and cheeks, and no black eyebrow. Also has duller bluish facial skin (mainly restricted to narrower area on lores and patch behind eye) and mostly dark brown upper mandible. See Spot-breasted Parrotbill and juvenile White-hooded Babbler. **Juvenile** Crown and head-sides paler, upperside more rufescent-brown. **Other subspecies in SE Asia** *P.b.magnirostris* (E Tonkin). **VOICE** Clear, jolly, well-phrased and emphasised song, introduced by plodding introductory notes: ***whit chu WI WI'WII WI'WUUUU***; ***whit tu whit u WI'WIII WI'WUUUU*** and ***whit tu wi'WIIII WI'WUUUUU*** etc., and ***whup chuk chip CHIIII-CHUU-WUUU***. Also gives a shorter, slightly breathless ***wi-wi-wi-wu*** (higher first note, lower end-note), interspersed with ***jhaowh*** call-notes. Calls with a loud metallic machine-gun-like spluttering rattle: ***TRRRRT TRRRRRRRRRRRT TRRRRRRRRRRRT... or PRRRRRRT...*** and distinctive twangy ***JHEW*** or ***JHAOWH***. Contact calls are a short, rather sharp ***wic-wic-wic***; ***wic-it***; ***wic-it-chi*** and ***wic chi-chi-chi*** etc. **HABITAT & BEHAVIOUR** Bamboo in or near broadleaved evergreen forest, forest edge; 500–1,850 m. Usually in small groups; often associates with White-hooded and Collared Babbler flocks, and Lesser Rufous-headed Parrotbill where ranges overlap. **RANGE & STATUS** Resident Bhutan, NE India, E Bangladesh, SE Tibet, SW China. **SE Asia** Scarce to uncommon resident N,E,S(east) Myanmar, N Laos, E Tonkin. **BREEDING** India: April–October. **Nest** Very neat deep cup with distinctive yellowish exterior, in bamboo, grass or sapling; 1–2 m above ground. **Eggs** 2–4; white or dull cream-coloured to blue, spotted, blotched and smeared dingy yellowish- or reddish-brown to dark brown over pale grey to lilac undermarkings (often mainly around broader end); 21.5 x 16.7 mm (av.; *bakeri*). **NOTE** Split from extralimital *P. ruficeps* (White-breasted Parrotbill) following King & Robson (2008).

1127 **SPOT-BREASTED PARROTBILL**

Paradoxornis guttaticollis Plate **108**

IDENTIFICATION 18–22 cm. Monotypic. **Adult** Distinguished by relatively large size and long tail, rufous crown and nape, mostly blackish face, broad black patch on rear ear-coverts, and white remainder of head-sides, throat and upper breast with pointed blackish spots/streaks (often faint). Rest of underparts whitish-buff, bill yellow, short and very deep-based. **Juvenile** Crown paler, rest of upperside more rufescent. **VOICE** Song is a loud staccato series of usually 3–7 notes, repeated every 1.5–15 s: ***whit-whit-whit-whit...*** and jollier ***wui-wui-wui-wui***; ***whi-whi-whi-whi...***, ***dri-dri-dri-dri-dri...*** and ***tui-tui-tui-tui-tui...***, or shorter strident ***du-du-du***. When excited, may give longer ***ju-ju-jiu-witwitwitwit witwitwitwitwitwit*** or ***ju-jujujuju witwitwitwitwit...*** etc., starting with hurried rising notes and culminating with thin, higher-pitched, quite metallic notes. Also gives introductory series alone, and makes coarse sounds (harsher than similar territorial calls): ***ee-cho-cho-cho-cho***; ***chow-chow-chow-chow-chow***; ***jieu-jieu-jieu-jieu*** and ***juju-dui-dui-dui*** etc. Contact calls include low ***ruk-ruk***; ***ruk-uk-uk***; ***rut-rut-rut-rut***; ***chi-cho-cho*** and sibilant ***chu-chu*** and ***chut-chut-chut*** etc. **HABITAT & BEHAVIOUR** Tall grass, scrub, secondary growth; 1,050–2,135 m. Often in small flocks. **RANGE & STATUS** Resident E Bangladesh, NE India, W and southern China. **SE Asia** Common resident W,N,C(east),S(east),E Myanmar, NW Thailand, N Laos, W,E Tonkin. **BREEDING** April–July. **Nest** Compact deep cup, in bamboo, tall grass, bush or low vegetation; 0.9–3.7 m above ground (usually low down). **Eggs** 2–4; whitish to pale olive-grey, spotted and streaked (may also be blotched and smudged) pinkish-brown and buffy-red to dark brown over lilac and grey undermarkings; 22.1–23.5 x 15.8–16.5 mm.

1128 **LESSER RUFOUS-HEADED PARROTBILL**

Chleuasicus atrosuperciliaris Plate **108**

IDENTIFICATION 15 cm. *C.a.atrosuperciliaris*: **Adult** Told by bright buffy-rufous head, short black eyebrow, plain warm olive-brown upperparts with more rufescent wings, and rather uniform buffish-white throat and underparts. Has distinctly peaked crown, short deep-based pink bill, pale bluish lores and neat eyering, and somewhat paler cheeks. See Greater Rufous-headed Parrotbill. Note range. **Juvenile** Undescribed, but probably very similar to adult. **VOICE** Possible song is a series of sharp chipping notes, rapidly repeated at varying speeds, after variable but short intervals: ***tik-tik-tik-tik-tik-tik-tik...***; ***tit-tit-tit-tit-tit-tit-tit-tit...***; ***chit-chit-chit-chit-chit-chit...***; ***tsu-tsu-tsu-tsu-tsu-tsu...*** etc. Flocks give subdued but rapid jumbled chattering interspersed with harsher, more metallic ***chut-chut-chut*** or ***chip-chip-chip***. Soft contact calls include ***wik-wik*** and ***tchip-tchip-tchip-tchip***. **HABITAT & BEHAVIOUR** Bamboo in or near broadleaved evergreen forest, forest edge; 550–2,000 m, locally down to 215 m in Myanmar. Usually in small active flocks, often with White-hooded and Collared Babbler flocks, and Greater Rufous-headed Parrotbill where ranges overlap. **RANGE & STATUS** Resident NE India, Bhutan, SW China. **SE Asia** Uncommon resident N,S(east) Myanmar, NW Thailand, N,C Laos, W,E Tonkin. **BREEDING** India: April–July. **Nest** Deep cup with distinctive yellowish-white exterior, in bamboo or grass; up to 2 m above ground. **Eggs** 3; pale blue; 19.5 x 15.2 mm.

1129 **BROWN-WINGED PARROTBILL**

Suthora brunneus Plate **108**

IDENTIFICATION c.12–13 cm. *S.b.brunneus*: **Adult** Similar to Vinous-throated Parrotbill but has darker, more chestnut crown to upper mantle and head-sides (latter also more concolorous with crown), no chestnut on wings, much more vinous throat and upper breast with darker chestnut streaks (well demarcated from belly), more buff-tinged rest of underparts (particularly in centre) and brownish-yellow bill with blackish culmen. **Juvenile** Forehead to uppermost mantle duller/paler and more concolorous with rest of upperside, which is darker and more rufescent, underparts buffier and more uniform, with less obvious streaks. **VOICE** Continuous twittering when feeding. Otherwise undocumented. **HABITAT & BEHAVIOUR** Scrub and grass, thickets; 1,525–2,375 m. Often in small fast-moving flocks. **RANGE & STATUS** Resident SW China. **SE Asia** Fairly common resident N(east),C(east) Myanmar. **BREEDING** April–June. **Nest** Neat deep cup, in grass or low tangled vegetation; up to 60 cm above ground. **Eggs** 2–4; pale to fairly deep blue; 15.2–17.5 x 12.7–13.5 mm.

1130 **VINOUS-THROATED PARROTBILL**

Suthora webbianus Plate **108**

IDENTIFICATION 11–12.5 cm. *S.w.suffusus*: **Adult** Told by combination of small size, long tail, stubby bill, warm dark brown upperside, contrasting rufous-chestnut to chestnut crown to upper mantle and wing-feather fringes, and pale pinkish-rufous head-sides. Throat and upper breast pale pinkish with warm brown streaks, rest of underparts dull buffish. Bill dark greyish to brownish with paler tip. *S.w.elisabethae* (north W Tonkin) has somewhat duller head and upper mantle and somewhat duller throat and upper breast, with less distinct streaks (may just be worn examples of *S.w.suffusus*?). See Brown-winged and Ashy-throated Parrotbills. **Juvenile** Darker and more uniform above than adult, crown less bright and contrasting, slightly warmer remainder of upperparts, plainer and buffier below, with less pink and less obvious streaking. **VOICE** Song in NE China and Korea is ***rit-rit chididi TSSU-TSSU-TSSIU*** or ***ri rit ri ri chididi WII-TSSI-TSSU*** (2–5 soft quick introductory notes, very rapid ***chididi***, then loud high thin stressed notes). Flocks call with subdued rapid chattering, interspersed with occasional thin, high piercing 2–3 note ***tsiu-tsiu*** or ***tiu-tiu***, and chuntering ***chrr'rr'rr*** and ***chur'ir'it*** etc. More scolding ***wutitit***, and ***wutitich'it'it'it'it*** when agitated. **HABITAT & BEHAVIOUR** Scrub, grass, bamboo thickets; up to 1,500 m. Often in small fast-moving flocks. **RANGE & STATUS** Resident China (except NW,N), Taiwan, N,S Korea, S Ussuriland. **SE Asia** Fairly common resident W(north),E Tonkin. **BREEDING** May–July. **Nest** Neat deep cup, in grass, bamboo, undergrowth, bush or small tree; 0.3–3 m above ground. **Eggs** 3–4; slightly glossy greenish-blue, blue, bluish- to greenish-white or white; 15.5–16.7 x 11.7–12.7 mm. **NOTE** Where range overlaps with Ashy-throated Parrotbill, mixed flocks can occur, and occasional hybrids have been reported. However, considering the level of association between the two here, and elsewhere, hard evidence of interbreeding is surprisingly scarce.

1131 **ASHY-THROATED PARROTBILL**
Suthora alphonsianus Plate **108**

IDENTIFICATION 12.5–13 cm. *S.a.yunnanensis*: **Adult** Like Vinous-throated Parrotbill but lores and sides of head and neck rather dark brownish-grey, throat and breast whitish with narrower, indistinct greyish-brown streaks. **Juvenile** More uniformly chestnut-tinged above, somewhat paler/browner head-side, paler/buffier and more uniform below. **VOICE** Sings with loud, high-pitched 3–4 note (usually 4 note) series, delivered with varying rapidity: ***TSSU-TSSU-TSSU-TSSU*** or ***TSSER-TSSER-TSSER-TSSER***; ***TSSU TSSU-TSSU-TSSU*** (with uneven note-spacing), and longer ***TSSU TSSU-TSSU-TSSU-TSSU-TSSU*** when excited in response to tape-playback. Also ***TSSU-TSSU-SWI-TSSU-ssi*** (with weaker end-note). Calls include rather harsh, chuntering, sometimes scolding ***twer-trr'ir'ir'irrit***; ***twi'it'ti***; ***trr'it'it'it***; ***tcher'der'der***; ***chip-ip-ip-ip***; ***titch'it'it*** etc., recalling certain *Alcippe* nun-babblers. Contact calls are very light ***tu***, ***twi***, ***ti*** and ***du*** notes etc., often rapidly combined. Low purring ***prrr'eet*** and ***prrr'ee*** at close range. **HABITAT & BEHAVIOUR** Scrub and grass, thickets; 1,100–1,500 m. Usually in fast-moving flocks, sometimes quite large. **RANGE & STATUS** Resident SW,W China. **SE Asia** Locally fairly common resident north W Tonkin. **BREEDING** China: April–August. **Nest** Neat deep cup, in sapling, bush, bamboo or grass; 0.5–1.5 m above ground. **Eggs** 2–6; white to pale blue; 14.9–19 x 12–13 mm. **NOTE** May best be treated as a race of Vinous-throated, but the two are altitudinally allopatric in parts of their range, and research is ongoing.

1132 **FULVOUS PARROTBILL**
Suthora fulvifrons Plate **108**

IDENTIFICATION 12–12.5 cm. *S.f.albifacies*: **Adult** Recalls Grey-breasted Parrotbill but has rather uniform warm buff head and breast, with dark brownish-grey lateral crown-stripes. Rest of upperparts buffish-grey, belly whitish, undertail-coverts warm buff. **Juvenile** Duller, particularly below. **VOICE** Song in Indian subcontinent is high-pitched ***si-tsiiii CHUU*** or ***si-siii JUU***; ***si-ti'ti'tituuuu-JHIIU*** and ***si-tituuuu-JHIIU*** etc., the first two notes thin (first one lowest, sometimes doubled, tripled or omitted, the second more drawn-out and very thin/piercing) and the end-note harsher; or ***si-si'ssuuu-JUUU*** (with rising ***suuu***). Also, ***si-si'sissu-suu-u*** and ***si-si'sissu-suuu***, with thin end-notes. Short contact calls include combinations of subdued ***twip*** and ***tip*** notes and husky low ***chew-chew-chew*** and ***cher-cher-cher-cher...*** etc. Repeated, subdued, slightly spluttering ***trrrip*** call-notes notes from W China. **HABITAT & BEHAVIOUR** Bamboo, edge of broadleaved evergreen forest; 2,895–3,660 m, sometimes down to 2,440 m in winter. Usually in fast-moving flocks. **RANGE & STATUS** Resident Nepal, Bhutan, NE India, SE Tibet, SW,W China. **SE Asia** Uncommon resident N Myanmar. **BREEDING** Tibet: June–July. Otherwise undocumented.

1133 **GREY-BREASTED PARROTBILL**
Suthora poliotis Plate **108**

IDENTIFICATION 11.5 cm. *S.p.feae*: **Adult** Told by small size, buffy-rufous crown to upper mantle, long black lateral crown-stripes, narrow buffy-rufous supercilium, rather dark grey head-sides, lower throat and breast and contrasting black remainder of throat, whitish lores and cheeks, and broad pointed white submoustachial streak. Rest of upperparts rufescent-brown with buffy-rufous rump and uppertail-coverts, rest of underparts rufous-buff with whitish centre; has rufous-fringed tail-feathers, tertials and secondaries, whitish-fringed outer primaries and contrasting blackish primary coverts; eyering mostly buffy-rufous. *S.p.poliotis* (W[north],N Myanmar) has broader black lateral crown-stripes, mostly white narrow supercilium and white eyering, more black on throat. See Buff-breasted and Black-eared Parrotbills and Black-throated Tit. Note range. **VOICE** Sings with strange, extended, high, wheezy, buzzy, nasal notes: ***dwit-awitawi-DUUUU-DIRRRR*** and ***dwut'utu-DUUUU-DIRR***, with extended, hurried introduction. Soft high ***tu***, ***ti***, ***tit*** and ***tip*** contact notes are jumbled together into constant twittering by flock members. **HABITAT & BEHAVIOUR** Bamboo, broadleaved evergreen forest, forest edge; 980–2,650 m. Usually in fast-moving flocks, sometimes quite large. **RANGE & STATUS** Resident NE India, SE Tibet, SW China. **SE Asia** Uncommon resident W(north),N,E(south),S(east) Myanmar, W,NW Thailand. **BREEDING** February–July. **Nest** Neat compact cup (apparently sometimes bag-shaped), in bamboo or thick bush; up to at least 60 cm above ground. **Eggs** 2–4; pale to deep blue; 15.7 x 11.9 mm (av.; *poliotis*). **NOTE** Split from extralimital *S. nipalensis* (Grey-capped Parrotbill) following Penhallurick & Robson (in prep.).

1134 **BUFF-BREASTED PARROTBILL**
Suthora ripponi Plate **108**

IDENTIFICATION 11.5 cm. *S.r.ripponi*: **Adult** Told from Grey-breasted Parrotbill by rufous-buff breast, broader black lateral crown-stripes, mostly white narrow supercilium and white eyering. See Black-eared Parrotbill and Black-throated Tit. Note range. **VOICE** Song-types so far recorded are shrill, thin high ***SIIR-EE*** or ***SWEE-EE*** or ***SIII-II*** (stressed at start). Flocks keep up subdued chattering, interspersed with more rattled ***chrrrp'rrp*** or ***chrrr'rrt*** calls. **HABITAT & BEHAVIOUR** Bamboo, broadleaved evergreen forest, forest edge; 2,100–2,600 m. Usually in fast-moving flocks, sometimes quite large. **RANGE & STATUS** Resident NE Indian (Mizoram). **SE Asia** Uncommon resident W Myanmar (south Chin Hills). **BREEDING** Undocumented. **NOTE** Split from extralimital *S. nipalensis* (Grey-capped Parrotbill) following Penhallurick & Robson (in prep.).

1135 **BLACK-EARED PARROTBILL**
Suthora beaulieui Plate **108**

IDENTIFICATION 11.5 cm. *S.b.beaulieui*: **Adult** Told from Grey-breasted Parrotbill by browner hindcrown to upper mantle, shorter but broader black lateral crown-stripes, narrow white supercilium which turns grey behind ear-coverts, incomplete white eyering, black ear-coverts, all-black throat fanning slightly onto uppermost breast, broad, less pointed white submoustachial streak, greyish-white then warm buff remainder of breast, and broader white centre to abdomen. *S.b.kamoli* (east S Laos, southern C Annam) has supercilium rufous behind eye, rear ear-coverts with more grey, narrower lateral crown-stripes. See Grey- and Buff-breasted Parrotbills and Black-throated Tit. Note range. **VOICE** Songs so far documented are shorter than Grey-breasted: ***TIRRRR'III*** or ***TWIRRRR-IIII***, rising towards end, and with no introduction. Contact calls similar to Grey-breasted. Also, harsh rattled ***trrrt***; ***trrrrrr***; and ***chrrrr*** etc., and scolding squeaky nasal notes. **HABITAT & BEHAVIOUR** Bamboo, broadleaved evergreen forest, forest edge; 1,200–2,000 m. Usually in fast-moving flocks, sometimes quite large. **RANGE & STATUS** Endemic. Uncommon resident NE Thailand (Phu Luang), N,C,S(east) Laos, N,C(southern) Annam. **BREEDING** Undocumented. **NOTE** Split from extralimital *S. nipalensis* (Grey-capped Parrotbill) following Penhallurick & Robson (in prep.).

1136 **GOLDEN PARROTBILL**
Suthora verreauxi Plate **108**

IDENTIFICATION 11.5 cm. *S.v.craddocki*: **Adult** Similar to Grey-breasted and Black-eared Parrotbills but lacks black lateral crown-stripes, has rufous-buff ear-coverts and no grey on head-sides, neck or breast. Note range. **VOICE** Based on one recording of Chinese nominate, song shorter and weaker than Grey-breasted: ***CHUUR-DII***. Also, a thin high-pitched, wispy

hsu-ssu-ssu-ssi (? a kind of 'second song'). Typical calls are quite harsh, low, slightly spluttering ***trr'it*** or ***trr'eet***; ***trrit***; ***trr-rr-rr*** and ***trr-rrt*** etc., strongly recalling certain calls of Golden-breasted Fulvetta. Foraging flocks also utter a jumble of high ***it***, ***twit***, ***tit*** and ***tip*** notes. Also gives a thin, high, rather hoarse, breathless ***tssuu*** and ***tssuu'tirr***. **HABITAT & BEHAVIOUR** Bamboo, edge of broadleaved evergreen forest; 1,500–3,000 m. Usually in small fast-moving flocks. **RANGE & STATUS** Resident C and southern China, Taiwan. **SE Asia** Uncommon to locally common resident extreme E Myanmar, northern N Laos, W,E(north-west) Tonkin. **BREEDING** Taiwan: July. **Nest** Deep cup or dome, in bamboo; 1.5 m above ground. **Eggs** 3; pale milky-blue; 15–16 x 11.8–12 mm.

1137 SHORT-TAILED PARROTBILL

Neosuthora davidiana Plate **108**

IDENTIFICATION 9.5–10 cm. *N.d.thompsoni* (E Myanmar, NW,NE Thailand, N[west & north] Laos, north W Tonkin): **Adult** Small size, short tail, very deep-based whitish bill and plain chestnut head and upper mantle with mostly black throat diagnostic. Has contrasting brownish-slate rest of upperparts, pale buffish underparts with whiter uppermost breast and centre of abdomen, greyer flanks and buffy-rufous vent. Shape recalls munias. *N.d.tonkinensis* (N[extreme east],C[north-east] Laos, E Tonkin, N Annam) Centre of throat tipped with rufous-buff to buffy greyish-white (giving mottled or streaked appearance), underparts purer grey; possibly paler grey above. See Black-throated and Golden Parrotbills. Note habitat and range. **Juvenile** Undescribed, but probably very similar to adult. **VOICE** Song is very thin, high-pitched, rapid ascending 6–9 note ***ih'ih'ih'ih'ih'ih..*** or ***zu'zu'zu'zu'zu'zu'...*** Also ***tit tiwit tit tit-tew hiuuu-ti-di'di'di*** (soft spaced notes then thin rising ***hiuuu***, followed by short ditty). Foraging flocks utter soft ***tip***, ***tit*** and ***tut*** contact notes, which can jumble together to sound like low twittering. Also, occasional thin ***tssu*** or ***chu***. Harder, shrill stressed ***si'si'si'sit'sit...*** and rather harsh ***tidit t'di-di'dit***, are presumed alarm calls. **HABITAT & BEHAVIOUR** Bamboo in or near broadleaved evergreen forest, grass, scrub; 50–1,200 m. Often in small flocks; joins bird-waves. **RANGE & STATUS** Resident S,SE China. **SE Asia** Scarce to uncommon resident E Myanmar, NW(east),NE(north-west) Thailand, N Laos, E Tonkin, N Annam. **BREEDING** March–April. **Nest** Cup; in bamboo; 60–90 cm above ground. Otherwise undocumented.

1138 JERDON'S BABBLER

Chrysomma altirostre Plate **108**

IDENTIFICATION 16–17 cm. *C.a.altirostre*: **Adult** Similar to Yellow-eyed Babbler but lores and short supercilium grey, throat and breast greyer, bill has pale lower mandible, eyes brown to golden-brown, eyering greenish-yellow. *C.a.griseigularis* (N Myanmar) has upperparts darker and tinged more rufous-chestnut, lores blacker, throat and upper breast greyer and lower breast to vent deeper, richer buff. See Slender-billed Babbler. Note range. **Juvenile** Warmer brown overall, lower mandible more pinkish. **VOICE** Sings with repeated, rather weak ***chi-chi-chi-chew-chew-chew*** or ***ih-ih-ih-ih chew chitit chew i'wwiuu*** etc. Sometimes more slurred and wispy, recalling Oriental White-eye; may start with uneven rapid ***itch***, ***itit*** and ***tchew*** notes. Calls include short ***tic*** notes. **HABITAT & BEHAVIOUR** Tall grass; lowlands. Skulking. **RANGE & STATUS** Resident Pakistan, S Nepal, NE India; formerly Bangladesh. **SE Asia** Scarce local resident N,S Myanmar. **BREEDING** April–July. **Nest** Neat deep cup, fairly low down in grass. **Eggs** 2–3; pale or bright pink, blotched and smeared reddish-brown to light red over greyish undermarkings; 18.1 x 14.6 mm (av.; *griseigularis*).

1139 YELLOW-EYED BABBLER

Chrysomma sinense Plate **108**

IDENTIFICATION 16.5–19.5 cm. *C.s.sinense*: **Adult** Long tail, short black bill, and white lores, eyebrow (short), throat and breast distinctive. Has orange-yellow eyes and orange eyering. See Jerdon's and Slender-billed Babblers and Plain Prinia. **Juvenile** Paler above, shorter-tailed, bill browner. **VOICE** Sings with variable clear high phrases: ***wi-wu-chiu***; ***wi-wu'chrieu***; ***wi-wu-wi'tchu-it***; ***wi-wi-chu*** and ***wi-tchwi-wi-tchiwi*** etc.; repeated after fairly short intervals. Calls include trilling ***chrr-chrr-chrr...*** and ***chr'r'r'r'r***. **HABITAT & BEHAVIOUR** Grassland and scrub, secondary growth; up to 1,830 m. Skulking. **RANGE & STATUS** Resident Indian subcontinent, SW,S China. **SE Asia** Common resident (except S Thailand, Peninsular Malaysia, Singapore, Cambodia, W Tonkin, C Annam). **BREEDING** March–October. **Nest** Compact deep cup, in sapling, bush or low vegetation; 0.5–2.1 m above ground. **Eggs** 3–5; pinkish-white, mottled and streaked deep brick-red to bright blood-red, sometimes over purplish undermarkings; 17.9 x 14.9 mm (av.).

1140 GOLDEN-BREASTED FULVETTA

Lioparus chrysotis Plate **109**

IDENTIFICATION 11 cm. *L.c.forresti*: **Adult** Unmistakable, with blackish head, white median crown-stripe, silvery ear-coverts, blackish throat with grey feather-fringes and orange-yellow underparts. Rest of upperparts bluish-grey, wings and tail black with yellow to orange-yellow fringing, white inner fringes of tertials and white tips to flight-feathers. *L.c.amoena* (W Tonkin) has all-blackish throat and conspicuous yellow eyering; *robsoni* (C Annam) is drab olive-grey above and on ear-coverts, has off-white median crown-stripe, yellow throat and eyering, bright orange wing-fringing. **Juvenile** Throat mostly yellowish. **VOICE** Song is a rather rapid, very thin, high-pitched ***si-si-si-si-suu***, repeated at intervals. Calls with a low rattling ***witrrrit***; ***wit*** and ***wittit*** etc. **HABITAT & BEHAVIOUR** Bamboo, broadleaved evergreen forest, secondary growth; 1,765–2,650 m. Often in small flocks. **RANGE & STATUS** Resident Nepal, NE Indian subcontinent, SW,W,C,S China. **SE Asia** Fairly common to common resident N Myanmar, W Tonkin, southern C Annam. **BREEDING** April–June. **Nest** Cup or dome with side-entrance, in bush or bamboo; 0.4–1 m above ground. **Eggs** 3–5; pinkish to whitish, lightly speckled and/or blotched brown, reddish-brown or grey (often mostly at broader end); 16–17 x 11–11.5 mm (av.; China).

1141 WHITE-BROWED FULVETTA

Fulvetta vinipectus Plate **109**

IDENTIFICATION 11.5–12.5 cm. *F.v. perstriata*: **Adult** Blackish head-sides, long white supercilium, narrow blackish lateral crown-stripes and white throat and upper breast with dark brownish streaks diagnostic. Crown and nape warm-tinged dark brown (slightly greyish), mantle and back somewhat paler and greyer, scapulars, rump and uppertail-coverts rufescent-brown, rest of underparts pale buffish; wings appear largely rufous with contrasting blackish inner primaries and whitish-fringed outer primaries. *F.v.ripponi* (W Myanmar) has head-sides as crown, narrower and less distinct lateral crown-stripes and warm brown streaks on throat and upper breast; *austeni* (west N Myanmar [Mt Saramati]) is like *ripponi* but supercilium starts above eye, ear-coverts more blackish-brown; *valentinae* (W Tonkin) has mainly greyish crown to mantle, lower breast and upper belly, and blacker streaks on throat and upper breast. **VOICE** Song-types in W Myanmar include a thin, high-pitched, quite quickly repeated ***si wi-su***, also ***tsi-si-si-si-su-su-su***, repeated after longer intervals. Usual calls include a low, repeated ***trr*** and ***trr-trr*** and higher ***whit-it-it-it*** etc. **HABITAT & BEHAVIOUR** Bamboo, scrub, open broadleaved evergreen forest and forest edge; 1,830–3,355 m. Often in small fast-moving flocks. **RANGE & STATUS** Resident N,NE Indian subcontinent, S,SE Tibet, SW,W China. **SE Asia** Common resident W,N Myanmar, W Tonkin. **BREEDING** March–May. **Nest** Rather deep compact cup, in bush, bamboo or underbrush 0.9–2 m above ground. **Eggs** 2–3; grey to grey-blue, sparsely but boldly marked dark brown to blackish over pale greyish undermarkings (mostly around broader end). 16.3 x 13 mm (av.; *ripponi*).

1142 **LUDLOW'S FULVETTA**

Fulvetta ludlowi Plate **109**

IDENTIFICATION 11.5–12.5 cm. Monotypic. **Adult** Combination of quite dark upperparts, plain-looking greyish-brown crown and head-sides (without white supercilium), and bold dark brownish streaks on white throat and upper breast diagnostic. **Juvenile** Somewhat paler than adult. **VOICE** Songs include a high, thin sibilant ***see-see-spir'r'r*** (first two notes clear, ending a short, lower, sputtery trill). **HABITAT & BEHAVIOUR.** More open broadleaved evergreen, bamboo; found at 2,900 m. Typically occurs in small flocks, often joining bird-waves. **RANGE & STATUS** Resident E Bhutan, NE India, SE Tibet. Uncommon/local resident north N Myanmar. **BREEDING** Undocumented.

1143 **STREAK-THROATED FULVETTA**

Fulvetta manipurensis Plate **109**

IDENTIFICATION 11.5–12.5 cm. *F.m.manipurensis*: **Adult** Combination of rather pale greyish-brown crown and mantle, plain pale brownish-grey head-sides (supercilium a shade paler), strongly contrasting blackish-brown lateral crown-stripes, white throat with dark brown streaks and pale greyish breast distinctive. Back, rump and vent buffy-rufous, wings like White-browed Fulvetta. See Indochinese Fulvetta. **Other subspecies in SE Asia** *F.m.tonkinensis* (W Tonkin). **VOICE** Song is a rather well-spaced, very high-pitched ***ti ti si-su***. Utters low ***tirrru*** call-notes when on the move. **HABITAT & BEHAVIOUR** Broadleaved evergreen forest, forest edge, bamboo, scrub; 1,525–2,800 m (breeds above 1,765 m). Often in small flocks. **RANGE & STATUS** Resident NE India, SW China. **SE Asia** Common resident W(north Chin Hills),N Myanmar, W Tonkin. **BREEDING** March–April. Otherwise undocumented.

1144 **INDOCHINESE FULVETTA**

Fulvetta danisi Plate **109**

IDENTIFICATION 13 cm. *F.d.danisi*: **Adult** Resembles Streak-throated Fulvetta but has darker crown and mantle, less contrasting lateral crown-stripes and strong vinous-pinkish tinge to head-sides, throat and upper breast, which are uniformly and more boldly streaked; wings plain-looking apart from silvery-white fringing at base of primaries. *F.d.bidoupensis* (C[southern],S Annam; east S Laos?) lacks any visible whitish fringes to primaries and has browner crown. **VOICE** Song is undocumented. Calls with quickly repeated, rapid ***chrrrrit*** and ***chrrt-chrrt-chrrt...*** etc. **HABITAT & BEHAVIOUR** Broadleaved evergreen forest, bamboo, scrub; 1,800–2,440 m (above 2,160 m in S Annam). Often in small flocks. **RANGE & STATUS** Endemic. Locally common resident N(south-east),C,S(east) Laos, C(southern),S Annam. **BREEDING** Undocumented. **NOTE** For split from extralimital Spectacled Fulvetta *F. ruficapilla* see Collar & Robson (2007).

1145 **CHESTNUT-FLANKED WHITE-EYE**

Zosterops erythropleurus Plate **110**

IDENTIFICATION 11–12 cm. Monotypic. **Adult** From other white-eyes by chestnut patch on flanks. Some individuals appear to lack chestnut or show fainter, pinker suffusion. Otherwise similar to Japanese White-eye but colder green above, without yellow on forehead. **VOICE** Similar to Japanese White-eye. **HABITAT & BEHAVIOUR** Open forest, secondary forest; up to 2,590 m, mostly above 800 m. Gregarious, often found in mixed flocks with Japanese. **RANGE & STATUS** Breeds SE Siberia, Ussuriland, NE China, N Korea; winters SW China. **SE Asia** Uncommon to locally common winter visitor W,C,S,E Myanmar, W,NW,NE Thailand, Cambodia, N,C Laos, W,E Tonkin. Also recorded on passage E Tonkin.

1146 **ORIENTAL WHITE-EYE**

Zosterops palpebrosus Plate **110**

IDENTIFICATION 10.5–11 cm. *Z.p.siamensis*: **Adult typical morph** Told from Japanese White-eye by whitish to yellowish-white ventral stripe, slightly yellower-green upperparts and more extensive yellow on forehead. More difficult to separate from Everett's White-eye but upperparts somewhat yellower-green, shows yellow on forehead, grey and yellow of underparts paler. **Adult yellow morph** Underparts completely yellow; apparently only occurs in this subspecies. *Z.p.auriventer* (Tenasserim, coastal west S Thailand, Peninsular Malaysia, Singapore) has clear yellow mid-ventral stripe and less yellow on forehead and lores than other subspecies; *williamsoni* (coastal C,W,S[east] Thailand) has little or no mid-ventral stripe, paler yellow and grey on underparts and duller green on upperparts and wing-feather fringes. **Juvenile** Probably differs as Japanese. **VOICE** Has short thin wispy song made up of slurred call-notes. Usual call is a repeated wispy sibilant ***jeww*** or ***cheuw***. **HABITAT & BEHAVIOUR** Deciduous, broadleaved evergreen and swamp forest, secondary growth, mangroves, wooded and cultivated areas, parks, gardens. Up to 1,525 m, locally 1,830 m; restricted to mangroves and non-forest (mainly coastal) habitats in S Thailand, Peninsular Malaysia and Singapore. Highly gregarious. **RANGE & STATUS** Resident Indian subcontinent, SW China, Greater Sundas, W Lesser Sundas; Afghanistan? **SE Asia** Common resident (except E Tonkin, N Annam, Singapore). Scarce (probably now originating from captivity) Singapore. **BREEDING** December–September. Multi-brooded. **Nest** Delicate cup, slung from twigs of tree-branch; 1.5–18 m above ground. **Eggs** 2–4; pale blue to greenish-blue; 15.2 x 11.4 (av.; *williamsoni*).

1147 **JAPANESE WHITE-EYE**

Zosterops japonicus Plate **110**

IDENTIFICATION 10–11.5 cm. *Z.j.simplex*: **Adult** Only likely to be confused with Oriental White-eye. Differs by darker, less yellowish-green upperparts and lack of ventral stripe; yellow on head restricted to loral band. See Chestnut-flanked and Everett's. Note range. **Juvenile** Like adult but eyering greyer at first. **VOICE** Like Oriental. **HABITAT & BEHAVIOUR** Forest, secondary growth, cultivated areas, parks, gardens; up to 2,590 m (breeds in lowlands). Highly gregarious, often in large restless flocks in winter. **RANGE & STATUS** Breeds C,E and southern China, Taiwan, S Korea, Japan; some northern populations winter to south. Introduced Hawaii. **SE Asia** Common resident E Tonkin. Common winter visitor C,S,E Myanmar, north Tenasserim, NW,NE Thailand, Laos, W,E Tonkin, N,C Annam. Also recorded on passage E Tonkin. Recorded in summer NW Thailand. Fairly common feral resident Singapore (perhaps not yet established?). **BREEDING** April–August. **Nest** Cup slung from twigs of tree, bush or bamboo. **Eggs** 2–4, pale blue to greenish-blue; 15.5 x 12 mm (av.).

1148 **EVERETT'S WHITE-EYE**

Zosterops everetti Plate **110**

IDENTIFICATION 11–11.5 cm. *Z.e.wetmorei*: **Adult** Similar to Oriental White-eye race *auriventer* but upperparts darker green, without yellow on lores/forehead, flanks deeper grey, yellow of throat more greenish-tinged, ventral stripe and undertail-coverts deeper yellow. **Juvenile** Like adult but duller with paler, less pronounced mid-ventral stripe and greener wing-feather fringes. **Other subspecies in SE Asia** *Z.e.tahanensis* (extreme S Thailand, Peninsular Malaysia). **VOICE** Call-notes are thinner and higher-pitched than Oriental: ***tsieu*** or ***tschew***. **HABITAT & BEHAVIOUR** Broadleaved evergreen forest; up to 2,010 m. Highly gregarious. **RANGE & STATUS** Resident Borneo, Philippines, Talaud Is. **SE Asia** Uncommon to locally fairly common resident NE(south),SE,W(south),S Thailand, Peninsular Malaysia. **BREEDING** February–August. **Nest** Delicate cup, slung from twigs of tree-branch, or in bamboo; up to 13 m above ground. Otherwise undocumented.

1149 **BLACK-CHINNED YUHINA**

Yuhina nigrimenta Plate **110**

IDENTIFICATION 10.5–11.5 cm. Monotypic. **Adult** Identified by small size, short tail, grey head, contrasting black face and chin and black bill with red base and lower mandible. Forehead and crown-feathers broadly centred blackish, rest of upperside dull greyish-brown, lower throat buffish-white, rest of underparts mostly pale buff (somewhat richer on vent). **Juvenile** Upperparts browner, crest shorter. **VOICE** Song is a clear high-pitched ringing ***uu ii uui ii uui uu ii uui ii uui uu ii uui ii uui...***, with higher ***ii*** notes. Calls include a nervous harsh ***whit'rr'u***; ***wh'rr'rr*** and ***whrr'rr'ik***. **HABITAT & BEHAVIOUR** Broadleaved evergreen forest, forest edge, secondary growth; 610–2,135 m, locally down to 200 m in N Myanmar. Usually in restless, sometimes quite large flocks. **RANGE & STATUS** Resident N,NE Indian subcontinent, SE Tibet, C and southern China. **SE Asia** Locally common resident N Myanmar, east Cambodia, N,C Laos, W,E Tonkin, N,C,S Annam. **BREEDING** February–July. **Nest** Compact cradle, attached to moss or lichen on underside of tree-branch, amongst roots on bank or in low vegetation; up to 8 m above ground. **Eggs** 3–4; pale sea-green to pale clay-coloured, freckled and/or stippled very pale brown to reddish over reddish-grey to violet undermarkings (often ringed at broader end); 16.2 x 12.3 mm (av.).

1150 WHITE-COLLARED YUHINA

Yuhina diademata Plate **110**

IDENTIFICATION 17.5–18 cm. Monotypic. **Adult** Unmistakable, with mostly brownish-grey plumage, tall crest and conspicuous broad white nuchal patch. Has darker face, pale streaks on ear-coverts, white undertail-coverts and mostly blackish flight-feathers with some whitish fringing on primaries. **Juvenile** Darker parts of body tinged rufescent-brown, crest shorter. **VOICE** Calls include subdued, worrisome ***wi wrrr'i wrrr wrrr'i***. **HABITAT** Edge of broadleaved evergreen forest, secondary growth, scrub and grass; 1,675–2,745 m, sometimes down to 1,250 m in winter. **RANGE & STATUS** Resident SW,W,C China. **SE Asia** Fairly common resident N,E(north) Myanmar, W Tonkin. **BREEDING** April–May. **Nest** Cup, in sapling or low vegetation; 0.2–1.5 m above ground. **Eggs** 2–3; whitish to pale greenish-blue, blotched and speckled umber to light brown and bluish- to blackish-purple (often mostly at broader end); 19–20.3 x 14.7–15.2 mm.

1151 STRIPE-THROATED YUHINA

Yuhina gularis Plate **110**

IDENTIFICATION 14–15.5 cm. *Y.g.gularis*: **Adult** Told from other yuhinas by robust shape, tall forward-curved crest, pale pinkish to buffish throat (centre whiter) with blackish streaks, and prominent pale orange-buff fringes to secondaries. Upperparts rather warm-tinged drab brown, breast greyish-pink, belly and vent deep ochre-buff with browner flanks and centres of undertail-coverts, primaries mostly blackish with some whitish fringing. *Y.g.uthaii* (C Annam) has much broader black throat-streaks, somewhat colder upperside and paler, greyer head-sides. Birds from W Tonkin (syn. *Y.g.sordidior*) appear to have darker, warmer-tinged upperside, deeper, richer ochre-buff to orange-buff on underparts, and darker, more extensive orange-rufous fringes to secondaries. **Juvenile** Slightly darker and warmer brown above, crest shorter. **VOICE** Typically calls with very nasal ***mherr*** or ***wherr*** or ***skyeer*** notes, often in long series and sometimes followed by hurried ***whu'whu'whu'whi'whi'whi*** (possibly song); sometimes alternated with clearer, quick, upslurred ***squick!*** Also, short ***wiht*** notes. **HABITAT & BEHAVIOUR** Broadleaved evergreen forest, mixed coniferous and broadleaved forest, forest edge, secondary growth; 1,830–3,200 m, sometimes down to 1,675 m in winter. Usually in small parties; fond of flowering trees. **RANGE & STATUS** Resident N,NE Indian subcontinent, S,SE Tibet, SW,W China. **SE Asia** Common resident W,N,E Myanmar, N Laos, W Tonkin, C(southern) Annam. **BREEDING** March–June. **Nest** Cradle-shaped or globular, in hole in bank or rock-face or amongst moss on underside of tree-branch; up to 9 m above ground. **Eggs** 4; pale olive, speckled dark reddish-brown to light reddish; 17–17.5 x 12.3–12.8 mm (*gularis*).

1152 RUFOUS-VENTED YUHINA

Yuhina occipitalis Plate **110**

IDENTIFICATION 13 cm. *Y.o.obscurior*. **Adult** Easily told by combination of slim build, rather narrow pointed bill, tall greyish crest, pronounced white eyering, black malar streak, prominent rufous nuchal patch and rufous-buff vent. Ear-coverts and neck greyish, upperparts brown, sides of forehead light rufous-buff, throat and breast plain light vinous-pinkish (sometimes buff-tinged). **Juvenile** Nuchal patch paler rufous, breast duller, crest shorter. **VOICE** Song is a simple, weak, rather high-pitched ***swi'si'si su'su'su swi'si'si si'si'si su'su'su...*** etc. Calls with nasal, buzzy ***bee*** or ***bzzee*** and ***beebee*** notes. **HABITAT & BEHAVIOUR** Broadleaved evergreen forest, forest edge, secondary growth; 1,830–2,500 m, sometimes down to 800 m in winter. Often in flocks; associates with mixed-species feeding flocks; visits flowering trees. **RANGE & STATUS** Resident Nepal, NE Indian subcontinent, S,SE Tibet, SW China. **SE Asia** Common resident N Myanmar. **BREEDING** Indian subcontinent: April–June. **Nest** Cup, embedded in moss or among lichen streamers on (or hanging from) tree-branches; 0.9–4.5 m above ground. **Eggs** 2+; otherwise undocumented.

1153 WHITE-NAPED YUHINA

Yuhina bakeri Plate **110**

IDENTIFICATION 12.5–13 cm. Monotypic. **Adult** Dark rufous head with broad white streaks on ear-coverts, white nape-patch, white upper throat and shortish crest diagnostic. Forehead and crown-feathers have dark brown centres, upperparts drab brown with indistinct pale shaft-streaks and rufescent-brown rump and uppertail-coverts, underparts mostly buffish with pinkish-tinged breast, olive-tinged flanks and rather narrow brown streaks on lower throat and breast. See Striated Yuhina race *castaneiceps*. Note range. **Juvenile** Mantle browner, underpart-streaks paler and less distinct. **VOICE** Possible song is hurried series of high thin notes, repeated every 1–3 s: ***tsu'tsu'tsu*** or ***du'du'du***; ***tsu'tst***; and ***tsu'tsu'tsut*** etc. Very thin, piercing, high-pitched ***tsit***; ***tssu*** and ***tsidit*** notes. **HABITAT & BEHAVIOUR** Broadleaved evergreen forest; 800–1,400 m (should occur c.2,200 m). Often in small flocks. **RANGE & STATUS** Resident Nepal, NE Indian subcontinent, SW China (W Yunnan). **SE Asia** Uncommon resident N Myanmar. **BREEDING** April–July. **Nest** Cup or dome, in hole in bank, amongst moss on tree-trunk or in low bush. **Eggs** 3–4; white, blotched reddish-brown to umber (often mostly at broader end); 19.3 x 14.2 mm.

1154 WHISKERED YUHINA

Yuhina flavicollis Plate **110**

IDENTIFICATION 12.5–13.5 cm. *Y.f.rouxi*: **Adult** Told by combination of prominent erect crest, broad white eyering, blackish-brown face and moustachial stripe, rufous and golden-yellow nuchal collar and dull buffish olive-brown flanks with prominent white streaks. Ear-coverts pale brownish-grey, paler towards moustachial stripe, lower throat and upper breast with narrow blackish-brown shaft-streaks, and undertail-coverts buffish. *Y.f.rogersi* (east NW Thailand) is greyer above, has considerably paler, duller nuchal collar and duller, browner flanks; *constantiae* (south-east N Laos) has uniform rich rufous nuchal collar, browner upperparts and more rufescent upper flanks. See Burmese Yuhina. **Juvenile** Mantle to uppertail-coverts and scapulars somewhat browner, moustachial stripe duller and narrower. **VOICE** Song is a repeated, shrill, high-pitched ***tzii-jhu ziddi***, with stressed first note and slightly undulating end-note. Calls include a thin squeaky ***swii swii-swii*** and sudden harsh nasal ***jhoh***. **HABITAT & BEHAVIOUR** Broadleaved evergreen forest, forest edge, secondary growth; 1,000–2,620 m (mostly above 1,220 m), sometimes descending to 215 m in winter (N Myanmar). Often in small flocks; associates with bird-waves; visits flowering trees. **RANGE & STATUS** Resident N,NE Indian subcontinent, S,SE Tibet, SW China. **SE Asia** Common resident W,N,E Myanmar, NW Thailand (local),

N,C Laos, W,E(north) Tonkin, N Annam. **BREEDING** March–June. **Nest** Cup-shaped to pendant, suspended between bush or tree twigs, amongst moss, roots or grass on bank; up to 4 m above ground. **Eggs** 2–4, white, stippled pink to red-brown; 19.8 x 14.2 mm (av.; India).

1155 **BURMESE YUHINA**

Yuhina humilis Plate **110**

IDENTIFICATION 13 cm. *Y.h.clarki*: **Adult** Like Whiskered Yuhina but has greyish-brown crown and ear-coverts, grey nuchal collar, grey base colour to flanks, greyer streaks on lower throat and breast and pale greyish undertail-coverts. **Other subspecies in SE Asia** *Y.h.humilis* (N Tenasserim, W Thailand). **VOICE** Flock members utter low ***chuck-chuck***, occasionally ***chir-chir-chir-chir***; also ***chit-a-wit***. **HABITAT & BEHAVIOUR** Broadleaved evergreen forest, forest edge, secondary growth; 1,065–2,285 m. Often in small flocks, visits flowering trees. **RANGE & STATUS** Endemic. Uncommon to locally common resident south E Myanmar, north Tenasserim, W(north),NW(south-west) Thailand. **BREEDING** February–April. **Nest** Sling-shaped and sewn along twig, with entrance about half-way down; c.15 m above ground. Otherwise undocumented.

1156 **STRIATED YUHINA**

Staphida castaniceps Plate **110**

IDENTIFICATION 13–14 cm. *S.c.striata*: **Adult** From other yuhinas (apart from Chestnut-collared, which see for differences) by slim build, short crest and relatively long, graduated tail with white-tipped outer feathers (outermost pair also edged white). Has cold brownish upperparts with greyish feather-fringing on crown and nape and prominent whitish shaft-streaks elsewhere, pale dull rufous-brown ear-coverts with paler streaks, narrow whitish eyebrow and buffish-white underparts (throat and breast whiter). *S.c.castaniceps* (SW,W Myanmar) has warmer brown forehead and crown with broad greyish-brown fringes, rufous-chestnut nape, drab greyish-brown upperparts with very faint pale shaft-streaks and slightly stronger creamy-buff tinge to underparts; *plumbeiceps* (N Myanmar) has decidedly greyer upperside (notably crown) with less distinct pale shaft-streaks and plainer, brighter rufous ear-coverts. **Juvenile** Duller with shorter crest. **VOICE** Song is a simple series of high-pitched shrill ***tchu***, ***tchi*** or ***tchi-chi*** notes. Flocks call with continuous loud chattering, interspersed with squeaky high-pitched notes. **HABITAT & BEHAVIOUR** Broadleaved evergreen forest, forest edge, scrub; 610–1,800 m, locally down to 180 m (winter only?). Usually in fairly large restless flocks. **RANGE & STATUS** Resident NE Indian subcontinent. **SE Asia** Common resident SW,W,N,E,S(east) Myanmar, north Tenasserim, W,NW(west) Thailand. **BREEDING** January–June. **Nest** Neat cup, in hole in bank (often by trail) or rock; up to 2 m above ground. **Eggs** 2–4; glossy white (may be bluish- or greenish-tinged), spotted, speckled and blotched brown to reddish-brown, often over greyish to pale purple undermarkings (markings often mainly around broader end); 16.6 x 13.3 mm (av.; India).

1157 **CHESTNUT-COLLARED YUHINA**

Staphida torqueola Plate **110**

IDENTIFICATION 14–15 cm. Monotypic. **Adult** From similar Striated Yuhina, by broad chestnut nuchal collar and ear-coverts with very prominent whitish streaks, greyer base colour to upperparts (particularly crown) and cleaner white throat and upper breast. Note range. **Juvenile** Duller with shorter crest. **VOICE** Song is a double ***di-duit***. Flocks call with continuous loud chattering, interspersed with squeaky high-pitched notes. **HABITAT & BEHAVIOUR** Broadleaved evergreen forest, forest edge, scrub; 350–1,800 m, mainly above 900 m. Usually in fairly large restless flocks. **RANGE & STATUS** Resident southern China. **SE Asia** Common resident east NW Thailand, Laos, W,E Tonkin, N,C Annam. **BREEDING** February–June. **Nest** Neat cup, in hole in bank (often by trail) or rock; up to 2 m above ground. **Eggs** 3–5; glossy white (may be bluish- or greenish-tinged), dotted and speckled shades of brown, dark grey and black, with some light and dark grey spots; 18.5 x 13.7 mm (av. China). **NOTE** Split from Striated Yuhina following Collar & Robson (2007).

1158 **GREY-CHEEKED FULVETTA**

Alcippe fratercula Plate **109**

IDENTIFICATION 13–15 cm. *A.f.fratercula*: **Adult** Told from other nondescript fulvettas ('nun-babblers') by combination of grey crown to upper mantle (very faintly tinged brown) and head-sides, obvious blackish lateral crown-stripes, prominent white eyering (often faintly interrupted above/in front of eye) and distinctly buff underparts. Rest of upperparts warm-tinged mid-brown, centre of throat whiter, eyes dark crimson. *A.f.yunnanensis* (N Myanmar) has somewhat narrower and fainter lateral crown-stripes and usually more distinctly buff underparts. See Nepal Fulvetta, note range. **Juvenile** Body washed rufescent, grey of crown to upper mantle tinged brown, eyes greyish. **VOICE** Song is a repeated series of high-noted phrases: ***it-chi wi-wi***; ***ii chu chi-wi***; ***ii yu yu-wi*** and ***ii yu yu-wi wi-you*** etc.; normally ending with 2–3 distinctive, curious buzzy ***eerh*** sounds. Calls with nervous ***chrr'rr'r*** and ***chrr'rr'rrt*** etc. and harsh ***chitti-tit***. **HABITAT & BEHAVIOUR** Broadleaved evergreen forest, forest edge, secondary growth, scrub, bamboo; 900–2,565 m. Usually in small restless flocks, mostly in lower storey; often associates with mixed-species feeding flocks. **RANGE & STATUS** Resident SW China. **SE Asia** Common resident N,E,C(east),S(east) Myanmar, Tenasserim, W,NW,NE(north-west) Thailand, north-west N Laos. **BREEDING** February–July. **Nest** Compact cup, in bush or low vegetation; 0.2–2 m above ground. **Eggs** 2–4; white to pinkish-white, spotted, mottled and sometimes streaked rusty-red, pale red or reddish-purple over lavender undermarkings (often denser at broader end); 19.6–22.1 x 14.5–14.7 mm (*fratercula*). **NOTE** Split from *A. morrisonia* (Huet's Fulvetta) following Zou *et al.* (2007).

1159 **SCHAEFFER'S FULVETTA**

Alcippe schaefferi Plate **109**

IDENTIFICATION 13–15 cm. *A.s.schaefferi* (W,E Tonkin): **Adult** from similar Grey-cheeked Fulvetta by very faint lateral crown-stripes, usually warmer upperparts, much less buff underparts, greyish-white throat with very faint darker streaks, greyer malar area, pale, slightly vinous-grey wash on breast and whitish centre of abdomen. *A.s.laotiana* (N[south-east],C Laos) has somewhat more pronounced crown-stripes, less warm-tinged upperparts, and slightly buffier underparts. See Black-browed Fulvetta; note range. **Juvenile** Body washed rufescent, grey of crown to upper mantle tinged brown, eyes greyish. **VOICE** Similar to Grey-cheeked. Song phrases include ***it'i u-iwi-u-i-ii*** and ***it'i-u-iwi-u-ii*** etc. **HABITAT & BEHAVIOUR** Broadleaved evergreen forest, forest edge, secondary growth, scrub, bamboo; 600–2,565 m (mostly above 900 m). Behaves as Grey-cheeked. **RANGE & STATUS** Resident W,C,S China. **SE Asia** Common resident N(east),C Laos, W,E Tonkin, N Annam. **BREEDING** February–July. **Nest** Compact cup, in bush or low vegetation. **Eggs** 2–4; white to pinkish-white, spotted, mottled and squiggled dark brown to reddish-brown, over pale purple undermarkings (markings denser at broader end). **NOTE** Split from *A. morrisonia* (Huet's Fulvetta) following Zou *et al.* (2007).

1160 **NEPAL FULVETTA**

Alcippe nipalensis Plate **109**

IDENTIFICATION 13.5–14 cm. *A.n.commoda* (N Myanmar): **Adult** Only likely to be confused with Grey-cheeked Fulvetta race *yunnanensis* in N Myanmar. Differs by brown tinge to crown and nape, more pronounced dull blackish lateral crown-stripes, more rufescent upperparts, even more pronounced white eyering (faintly interrupted by dark grey above/in front of eye), whitish centre to throat and abdomen and less distinctly buff rest of underparts. Note habitat and range. *A.n.stanfordi* (elsewhere Myanmar) is paler and colder-tinged with greyer crown and nape and slightly less distinct lateral crown-stripes. **Juvenile** Upperparts, flanks and vent warmer-tinged. **VOICE** Song is considerably lower, slower and more spaced than Grey-cheeked, with no buzzy end-notes: ***chu-chui-chiwi*** and ***ew-ew-ui-iwi*** etc. Other calls are similar. **HABITAT & BEHAVIOUR** Broadleaved evergreen forest, forest edge, secondary growth, scrub, bamboo; 440–2,400 m. Behaviour as Grey-cheeked. **RANGE & STATUS** Resident Nepal, NE Indian subcontinent, SE Tibet. **SE Asia** Common resident SW,W,N,S(west) Myanmar. **BREEDING** March–July. **Nest** Compact cup, in bush, bamboo or low vegetation; 0.3–1.5 m above ground. **Eggs** 2–5; white to pinkish-white, finely speckled or blotched reddish-brown to purplish- or blackish-red, sometimes over faint purple undermarkings; 17.8–20.1 x 14–15.2 mm (India).

1161 **MOUNTAIN FULVETTA**

Alcippe peracensis Plate **109**

IDENTIFICATION 14–15.5 cm. *A.p.peracensis*: **Adult** Told by relatively small size and long tail, dull slate-grey crown to upper mantle (clearly tinged drab brown), prominent blackish lateral crown-stripes, mid-grey head-sides (without obvious brown), prominent white eyering, whitish centre to throat and abdomen, very light buffish creamy-brown wash on throat-sides and breast and pale greyish olive-brown lower flanks and undertail-coverts (slightly buffier toward belly and thighs). Rest of upperparts faintly rufescent-tinged drab olive-brown. *A.p.annamensis* (Indochina) has paler grey on head, distinctly more olive upperside (less rufescent-tinged) and somewhat paler and duller underparts. See Grey-cheeked and Schaeffer's Fulvettas. **Juvenile** Head and upperparts duller, crown contrasts less with upperparts, lateral crown-stripes less distinct or almost absent, underparts duller. **VOICE** Song recalls Grey-cheeked (with similar buzzy end-notes) but somewhat faster and shorter. In Indochina, ***iti-iwu uwi-u wheer wheer*** and ***it'iti-iwu wi-wui wheer wheer*** etc. In Peninsular Malaysia, similar ***iti iwu-wi-wi*** and ***it iwu-u-wi*** etc., though apparently with no buzzy end-notes. Other calls are similar to Grey-cheeked. **HABITAT & BEHAVIOUR** Broadleaved evergreen forest, forest edge, secondary growth, scrub, bamboo; 900–2,100 m. Behaviour as Grey-cheeked. **RANGE & STATUS** Endemic. Common resident extreme S Thailand, Peninsular Malaysia, north-east Cambodia, C,S Laos, C,S Annam. **BREEDING** January–June. **Nest** Small cup, in bush, sapling or low vegetation; 1.5–1.8 m above ground. **Eggs** 2+; whitish, blotched/spotted brown to dull brownish-red (mostly around broader end); 19 x 14.2 mm (*annamensis*).

1162 **BLACK-BROWED FULVETTA**

Alcippe grotei Plate **109**

IDENTIFICATION 15.5–16.5 cm. *A.g.grotei*: **Adult** Recalls Mountain Fulvetta race *annamensis* but larger and relatively shorter-tailed, slaty-grey crown (only faintly tinged brownish) contrasts sharply with dark rufescent-tinged brown upperside, has brownish-tinged head-sides (particularly behind ear-coverts), indistinct broken whitish eyering, and whitish underparts with faintly buffish-tinged pale brown wash on breast and flanks. From Schaeffer's Fulvetta race *laotiana* by lack of obvious whitish eyering, greyer lores, slightly blacker lateral crown-stripes, slightly more rufescent upperparts, brown wash on head-sides, less obvious buff on underparts, whitish centre of throat and white centre of abdomen. See Brown-cheeked Fulvetta. Note habitat and range. **Juvenile** Upperparts washed a little more rufous-chestnut and somewhat more uniform, grey duller and restricted to crown, lateral crown-stripes browner and shorter (not reaching mantle), ear-coverts paler/browner, breast and flanks slightly warmer. **Other subspecies in SE Asia** *A.g.eremita* (SE Thailand, south-west Cambodia). **VOICE** Song is similar to Brown-cheeked but usually rises less at end: ***yu-chi-chiwi-chuwoo***; ***yu-uwii-ii-uwoo***; ***yu-uwii-uwoo***; ***yii-uu-iwi-iwi-uwoo*** and ***yi-yuii-yui-uwee-uwee*** etc. When alarmed, utters harsh rasping spluttering ***wit-it-itrrrt***; ***wit-tritt***; ***witchititit*** and ***err-rittirrirrrt*** etc. **HABITAT & BEHAVIOUR** Broadleaved evergreen forest, forest edge, secondary growth, scrub, bamboo; up to 1,200 m. Usually in small nervous flocks in middle to lower storey; often with mixed-species feeding flocks. **RANGE & STATUS** Endemic. Common resident SE Thailand, south-west and east Cambodia, C,S Laos, south E Tonkin, N,C,S Annam, Cochinchina. **BREEDING** February–July. **Nest** Cup, in low undergrowth; 1 m above ground. **Eggs** 3; whitish, blotched and spotted warm brown. **NOTE** For split from Mountain Fulvetta see Collar & Robson (2007).

1163 **BROWN-CHEEKED FULVETTA**

Alcippe poioicephala Plate **109**

IDENTIFICATION 15.5–16.5 cm. *A.p.haringtoniae*: **Adult** Told from other nondescript fulvettas ('nun babblers') by greyish-buff head-sides, distinctly buff underparts and lack of white eyering. Has grey crown with long, rather narrow blackish lateral stripes and contrasting warm-tinged olive-brown upperparts. *A.p.phayrei* (SW,W Myanmar) has paler crown and nape with no obvious lateral stripes, less contrasting (much paler and colder-tinged) upperparts, and much paler underparts with whitish centre to throat and abdomen; *fusca* (north W,C,E Myanmar) resembles *phayrei* but is slightly warmer above and has faint suggestion of lateral crown-stripes, underparts a little paler and warmer than *haringtoniae*; *karenni* (E[south],S[east] Myanmar) is roughly intermediate between *haringtoniae* and *phayrei* and has indistinct lateral crown-stripes; *davisoni* (Tenasserim, S Thailand) resembles *karenni* but has upperparts (apart from crown and nape) closer to *haringtoniae* but with less distinct lateral crown-stripes (like *fusca*); *alearis* (NW[east],NE Thailand, N Indochina) has more olive (colder-tinged) upperparts and duller buff underparts. **Juvenile** Upperparts and wing/tail fringing browner, bill more yellowish. **VOICE** Sings with repeated pleasant phrases of spaced, fairly even notes, which usually rise at end: ***chu'uwi-uwi-uwee***; ***i'chiwi-uwi-uwee***; ***yi'chiwi-wi-uwuuee*** etc.; sometimes more slurred ***ch'uwi-u-uu-uwi-uwee*** etc. Calls with harsh buzzy spluttering rattles when agitated: ***wit-i-rrrr***; ***witt-witt***; ***witch-itititit*** and ***whi-chirru*** etc. and higher ***whi-sihihihi***. **HABITAT & BEHAVIOUR** Broadleaved evergreen and mixed deciduous forest, forest edge, secondary growth, bamboo; up to 1,520 m (mainly below 1,200 m). Often in small parties; inhabits middle to lower storey. **RANGE & STATUS** Resident C,NE and southern India, E Bangladesh, SW China. **SE Asia** Common resident Myanmar, W,NW,NE,S(north) Thailand, N,C Laos, W,E(north-west) Tonkin. **BREEDING** January–September. **Nest** Quite compact deep cup, in bush, sapling, bamboo or low herbage; 0.6–1.8 m above ground. **Eggs** 2–3; whitish to salmon-pink, blotched, smudged and sometimes streaked pinkish-red, reddish- to purplish-brown or blackish, often over pale greyish to pinkish-brown undermarkings (often capped or ringed at broader end); 17.3–20.1 x 13.5–15 mm (*davisoni*).

1164 **BROWN FULVETTA**

Alcippe brunneicauda Plate **109**

IDENTIFICATION 14–15.5 cm. *A.b.brunneicauda*: **Adult** Similar to southern subspecies of Brown-cheeked Fulvetta but crown darker brownish-grey, without lateral stripes, upperparts rather dark warm-tinged brown with more rufous-chestnut tinge to rump and uppertail-coverts, head-sides pale brownish-grey, underparts whitish with pale greyish-brown wash on breast and flanks and warmer, browner lower flanks. Note habitat and range. See *Malacopteron* babblers (particularly juveniles) and Mangrove Whistler. **Juvenile** Undescribed, but probably very similar to adult. **VOICE** Sings with rather slow, measured, high-pitched and slightly undulating ***hi-tu-tu ti-tu ti-tu*** and ***hi-tu hi-tu hi-tu*** or descending ***hi ti-tu ti ti-tu ti-tu*** etc., repeated after rather long intervals. Calls with rather stressed ***whit*** notes and short harsh rattles. **HABITAT** Middle to lower storey of broadleaved evergreen forest, often near streams; up to 900 m. **RANGE & STATUS** Resident Sumatra,

Borneo. **SE Asia** Uncommon to fairly common resident S Thailand, Peninsular Malaysia. **BREEDING** April–May. **Nest** Open cup in low vegetation, sapling etc.; up to 3 m above ground. **Eggs** 2+; said to be pure white.

1165 **RUFOUS-THROATED FULVETTA**

Schoeniparus rufogularis Plate **109**

IDENTIFICATION 12–14 cm. *S.r.major*: **Adult** Easily told by rufous-brown crown, long black lateral crown-stripes, broad white supercilium and eyering, dark ear-coverts and whitish underparts with broad rufous-chestnut band extending from neck-side across upper breast/lower throat (often fainter in centre). Rest of upperparts rufescent-tinged cold dark brown, flanks greyish-brown, undertail-coverts warm buffish. *S.r.collaris* (N Myanmar) has distinctly broader breast/throat-band; *khmerensis* (SE Thailand) has darker upperparts and more chestnut crown; *stevensi* (W,E Tonkin, N Annam) has duller upperparts, buff-tinged supercilium, eyering, throat and centre of underparts and more rufous breast/throat-band; *kelleyi* (C Annam) has darker breast/throat-band. **Juvenile** Apparently undescribed. **VOICE** Song consists of quite loud, shrill phrases (usually starting with high note), repeated every 6–14 s: ***wi-chuw-i-chewi-cheeu***; ***wi-chi-wichi-chwi***; ***wi-ti-ti-tuee*** or ***chuu-chu-wichu-chi-chu*** etc. Female may interject with low harsh ***jrrjrrjrr...*** or ***jujuju...*** When alarmed, gives worried low ***wrrr-it wrrr wrrr wrrr-it...*** or ***wrrreet wrrreet...***, undulating ***chrr-chrrr-chrr...***, whining ***whiri-whiri-whiri...***, shrill explosive ***whit whit-whit...*** and ***wiwit-wiwit..wit...*** etc. **HABITAT & BEHAVIOUR** Broadleaved evergreen forest; up to 1,100 m. Skulks in low undergrowth. **RANGE & STATUS** Resident NE Indian subcontinent, SW China. **SE Asia** Scarce to locally common resident N Myanmar, NW,NE,SE Thailand, N,C Laos, W,E Tonkin, N,C Annam. **BREEDING** February–July. **Nest** Dome, semi-dome or cup; on ground or in low vegetation or sapling; up to 1 m above ground. **Eggs** 3–4; pale greyish with dark spots and lines, brownish clouds and purplish spots and blotches; 19.5 x 14.7 mm (av.; *collaris*).

1166 **RUSTY-CAPPED FULVETTA**

Schoeniparus dubius Plate **109**

IDENTIFICATION 13.5–15.5 cm. *S.d.genestieri*: **Adult** Recalls Rufous-throated Fulvetta but has uniform dark brown head-sides, without white eyering, lacks rufous/rufous-chestnut breast/neck-band, has contrasting rufous-buff patch on forehead, browner crown with blackish scales and browner flanks. *S.d.mandellii* (W Myanmar) has blackish ear-coverts and distinctive bold black and buff to whitish streaks on neck-sides to upper mantle; *intermedius* (N,C,E Myanmar) is roughly intermediate; *dubius* (east S Myanmar, north Tenasserim, W Thailand) is warmer above, slightly browner on head-sides and buffier below, particularly throat-sides and breast; *cui* (east S Laos, C Annam) has slightly darker/colder ear-coverts and upperside, rather more contrastingly rufous forehead and crown-sides above greyer supercilium, and distinctly buffish underparts with slightly paler centre to throat and belly. **Juvenile** Undescribed, but probably very similar to adult. **VOICE** Song recalls Rufous-throated but less shrill, with fewer notes per phrase: ***chu-chi-chiu***; ***chu-chi-twi***; ***chu-witee-wee***; ***chu-witchui-chu***; ***wi-chi-chu-chiu*** and ***wi-witu-chu*** etc. Calls with low grumbling rattles when agitated (sometimes ending with high squeaky note): ***chrrr-rr...***; ***chrrr-rrr-ritz*** and ***chrrr-rrr-rititz*** etc. **HABITAT & BEHAVIOUR** Edge of broadleaved evergreen forest, secondary growth, scrub, bamboo; 1,000–2,600 m. Always close to ground; skulking. **RANGE & STATUS** Resident Bhutan, NE India, south-western China. **SE Asia** Fairly common to common resident W,N,C,E,S(east) Myanmar, north Tenasserim, W Thailand, N,C,S(east) Laos, W Tonkin, southern C Annam. **BREEDING** February–June. **Nest** Dome, semi-dome or deep cup, on or near ground, amongst leaves or low vegetation, up to 90 cm. **Eggs** 2–4; whitish to pale buff or pale olive, spotted, smudged and streaked dull brown or light reddish-brown to blackish-brown over grey to inky-purple undermarkings; 20.8 x 15.6 mm (av.; *mandellii*).

1167 **SOOTY BABBLER**

Stachyris herberti Plate **111**

IDENTIFICATION 18 cm. Monotypic. **Adult** Overall sooty dark brown plumage with whiter throat and pale bill and eyering diagnostic. **Juvenile** More uniformly brownish, lacks obvious eyering. **VOICE** Very soft, repeated ***tip*** and ***tu-tip*** contact notes uttered by foraging flocks. Short, hard ***wittitit*** when agitated. **HABITAT & BEHAVIOUR** Broadleaved evergreen forest on limestone, usually close to limestone outcrops and boulders; up to 610 m. Typically in small flocks, fairly low down. **RANGE & STATUS** Endemic. Locally fairly common resident C Laos, C Annam. **BREEDING** March–May. **Nest** In rock-hollow. Otherwise undocumented.

1168 **BLACK-THROATED BABBLER**

Stachyris nigricollis Plate **111**

IDENTIFICATION 15.5–16 cm. Monotypic. **Adult** Rufescent upperparts and wings, black face, throat and upper breast, broad white cheek-patch, white forehead-streaking, short white eyebrow and white scales on lower breast diagnostic. Neck-side and rest of underparts greyish. See White-necked and Spot-necked Babblers. Note range. **Juvenile** Lacks white forehead-streaks, dark areas of head more sooty-brownish, lacks black-and-white breast-scaling and shows much browner, less contrasting underparts. **VOICE** Song is a repeated series of spaced, rather weak-sounding piping notes: ***pu-pu-pu-pu-pu-pu*** or ***too-too-too-too-too...*** and faster ***pupupupupupupupu***. Also more convoluted series of rapidly repeated hollower notes: ***puwut-puwut-puwut-puwut***; ***pwut-pwut-pwut-pwut-pwut*** and ***chu-chuwu-chu-chu-chu-chu*** etc. Other calls include a harsh slow rattled ***tchrrr-rrt*** and ***chrrrt-trrerrt-trrerrt***, harsh ***ti-tu-chu-chu*** and high ***tchi-tchu***. **HABITAT** Broadleaved evergreen forest, freshwater swamp forest; up to 455 m. **RANGE & STATUS** Resident Sumatra, Borneo. **SE Asia** Uncommon to fairly common resident S Thailand, Peninsular Malaysia; formerly Singapore. **BREEDING** May–July. **Nest** Loosely domed structure, in concealed position on ground. **Eggs** 2; white; 20.3–21.3 x 13.7–14.2 mm.

1169 **WHITE-NECKED BABBLER**

Stachyris leucotis Plate **111**

IDENTIFICATION 14–15 cm. *S.l.leucotis*: **Adult** Similar to Black-throated Babbler but has pale buffish lores, lacks white cheek-patch and breast-scaling, ear-coverts greyer, bordered with white spots. Shows slightly paler tips to wing-coverts and tertials. See Grey-throated and Spot-necked Babblers. Note range. **Juvenile** Upperparts rather uniform dull chestnut, lores buffish, neck spots smaller and buffish, ear-coverts browner, throat sooty-brown, lower breast and belly much darker and browner. **VOICE** Song is a repeated, simple whistled ***uu-wi-u-wi***; ***uu-wi-u-wi-u*** or ***uui-wi-oi-wi*** (second note lower-pitched). **HABITAT & BEHAVIOUR** Broadleaved evergreen forest; up to 800 m. Rather shy. **RANGE & STATUS** Resident Sumatra, Borneo. **SE Asia** Rare to scarce resident S Thailand, Peninsular Malaysia. **BREEDING** Not documented in region. Borneo: **Nest** Cup, in low vegetation. **Eggs** 3; white.

1170 **SNOWY-THROATED BABBLER**

Stachyris oglei Plate **111**

IDENTIFICATION 16 cm. Monotypic. **Adult** Similar to Spot-necked Babbler but has more prominent white supercilium (extending to bill-base), broadly black lores and upper ear-coverts, rather clean white lower ear-coverts to throat (without malar streak) and grey breast. **Juvenile** Apparently undescribed. **VOICE** Song is very reminiscent of Spot-necked Babbler: usually a three-note ***tiii tii tuu*** or ***phii tii tuu*** descending note by note, or ***tuuu TII tuu*** or ***tchuu TII tuu*** with middle note highest; sometimes only the last two notes. May be accompanied (presumably by female) with a jaunty ***up uip-uip*** or ***uh uif-uif*** (with ***uip*** or ***uif*** rising), or just ***up uip***. Rapid metallic rattling calls from agitated flocks. **HABITAT & BEHAVIOUR** Broadleaved evergreen forest, secondary growth, bamboo; 450–800 m. Often in small flocks, which can move very quickly. **RANGE & STATUS** Resident NE India. **SE Asia** Uncommon local resident N Myanmar. **BREEDING** NE India: April–June. **Nest** Large globular structure, on ground. **Eggs** 4; white; 22.8 x 17.1 mm.

1171 SPOT-NECKED BABBLER

Stachyris strialata Plate **111**

IDENTIFICATION 16–16.5 cm. *S.s.guttata* (north Tenasserim, W Thailand): **Adult** White throat and upper breast, deep rufous to rufous-chestnut lower breast and belly, blackish lores and malar streak and prominent white-streaked supercilium, neck and mantle-sides diagnostic. Head-sides greyish with prominent white cheek-patch. *S.s.nigrescentior* (S Thailand) has deeper rufous-chestnut lower breast and belly. See Snowy-throated and Grey-throated Babblers. **Juvenile** Undescribed for regional races. **Other subspecies in SE Asia** *S.s.tonkinensis* (eastern N&C Laos, Vietnam), *helenae* (north NW Thailand, western N Laos). **VOICE** In Indochina, song is a rather high-pitched, well-spaced, whistled ***tuh tih tih*** or ***tuh tih tuh***, repeated after longish intervals; sometimes slightly faster, more breathless ***tu-ti-u***. Two-note ***tuh tih*** has been noted in W Thailand and occasionally N Annam. May be accompanied (by female?) with a high rising note and hard rattle: ***whii-ii-titititititi*** etc. Other calls include a scolding ***tirrrrirrirr***, ***tchrrrt-tchrrrt...*** and short high-pitched ***tip*** notes. **HABITAT & BEHAVIOUR** Broadleaved evergreen forest, secondary growth, scrub and grass; 50–1,525 m. Often in small groups; skulking. **RANGE & STATUS** Resident S China, Sumatra. **SE Asia** Uncommon to locally common resident N Myanmar, Tenasserim, W,NW,S Thailand, N,C Laos, W,E Tonkin, N,C Annam. **BREEDING** February–June. **Nest** Dome with large side-entrance or cup, just above ground in rattan or other low vegetation. **Eggs** 3–4; glossy white; 22.3 x 16.5 mm.

1172 GREY-HEADED BABBLER

Stachyris poliocephala Plate **111**

IDENTIFICATION 14–15 cm. Monotypic. **Adult** Deep, dark rufescent-brown upperparts, rufous-chestnut underparts and greyish head with whitish streaks on forehead and throat diagnostic. Eyes contrasting creamy- to buffish-white. **Juvenile** Upperparts dull chestnut, darker on crown, underparts paler than adult with colour extending to throat-centre and rear ear-coverts, eyes browner. **VOICE** Sings with pleasant, clear, quite high-pitched phrases, repeated after longish intervals: ***chit-tiwi-wioo-iwee*** and ***yit-uip-ui-wiee*** and higher, longer-spaced ***chu-chi-chiee*** and ***chi-chu-chiee*** etc. Other calls include a quiet descending ***dji-dji-dji-du*** and more even ***dji-dji-dji-dji-dji...***, harsh scolding ***chrrrrttutut*** and ***chrrrrtut*** when alarmed, and very soft ***tip-tip-tip...*** contact calls. **HABITAT** Broadleaved evergreen forest, secondary growth; up to 760 m. **RANGE & STATUS** Resident Sumatra, Borneo. **SE Asia** Uncommon to fairly common resident S Thailand, Peninsular Malaysia. **BREEDING** March–September. **Nest** Dome or cup, in concealed position on ground. **Eggs** 3; white.

1173 GREY-THROATED BABBLER

Stachyris nigriceps Plate **111**

IDENTIFICATION 12.5–14 cm. *S.n.spadix*: **Adult** Black-and-silvery streaked crown with black lateral bands, narrow whitish supercilium, grey throat with broad white submoustachial patch and warm buffish underparts diagnostic. *S.n.coltarti* (N,E[north] Myanmar) has darker throat, warmer buff ear-coverts and underparts; *yunnanensis* (E Myanmar, east NW Thailand, northern Indochina) is darker overall; *rileyi* (C,S Annam) shows uniform pale greyish throat, paler washed-out underparts, paler crown and lores with black streaks restricted to forecrown and greyish-white streaks restricted to forehead and nape-sides; *dipora* (south Tenasserim, S Thailand), *davisoni* (extreme S Thailand, Peninsular Malaysia) and *tionis* (Pulau Tioman, off east Peninsular Malaysia) have duller dark crown-streaks (including lateral stripe) and narrower whitish streaks. **Juvenile** Mantle to uppertail and wings strongly washed chestnut, head-sides and underparts washed dark rufous-buff, hindcrown unstreaked. **VOICE** Song is a very high-pitched, quavering or undulating, rising ***ti tsuuuuuuuueee*** or ***tsi tuuuu-uuuuuiiii*** etc., resembling highly speeded-up song of Rufous-fronted Babbler. Sometimes gives more spaced ***si-si-siiiiii-u*** or ***ti-ti-ti-tuu***. Alarm calls include scolding ***chrrrt*** and ***chrrrrit*** notes and prolonged ***chrrrrrr-rrr-rrt*** etc. **HABITAT & BEHAVIOUR** Broadleaved evergreen forest, secondary growth; up to 1,830 m. Often in mixed-species feeding flocks. **RANGE & STATUS** Resident Nepal, NE Indian subcontinent, SE Tibet, SW China, Sumatra, Borneo. **SE Asia** Common resident (except C,SE Thailand, Singapore, Cambodia). **BREEDING** January–August. **Nest** Dome or ball with wide side-entrance, sometimes semi-dome or cup, on ground, in hole in bank, or in shrub or sapling up to 1.2 m. **Eggs** 2–5; fairly glossy white; 17.3–21.3 x 14–15.5 mm.

1174 CHESTNUT-WINGED BABBLER

Stachyris erythroptera Plate **111**

IDENTIFICATION 12.5–13.5 cm. *S.e.erythroptera*: **Adult** Drab brown plumage, greyish head-sides, throat and breast and blue orbital skin diagnostic. Eyes dark reddish, belly and vent dull pale buffish, wing-feathers fringed rufous-chestnut. Bluish neck-skin often visible when singing. **Juvenile** Crown and upperparts paler and more rufescent, grey plumage-parts much paler and brown-tinged, belly paler and browner, orbital skin duller. **VOICE** Song is a soft, mellow, quite quick, piping 7–10 note ***hu-hu-hu-hu-hu-hu***, occasionally a faster tremulous ***hu hu'u'u'u'u'u'u*** or slow ***chu hu-hu-hu-hu***; may be accompanied (by female?) with low ***chrrr*** notes. Calls with harsh scolding ***trrrrrt-trrrrrt...*** and soft ***wip*** and ***wit*** contact notes. **HABITAT** Broadleaved evergreen forest, secondary forest; up to 800 m. **RANGE & STATUS** Resident Sumatra, Borneo. **SE Asia** Common resident south Tenasserim, S Thailand, Peninsular Malaysia, Singapore. **BREEDING** All year. **Nest** Ball or oval with side-entrance, in bush, sapling, palm or amongst creepers etc.; 0.3–8 m above ground. **Eggs** 2–3; glossy white, sometimes faintly bluish-tinged (some said to show reddish spots); 16.5 x 14 mm (av.). Brood-parasitised by Drongo Cuckoo.

1175 CHESTNUT-RUMPED BABBLER

Stachyris maculata Plate **111**

IDENTIFICATION 17–18.5 cm. *S.m.maculata*: **Adult** Larger than other *Stachyris* babblers, with black throat, distinctive broad black to dark brown streaks/spots on whitish lower throat to upper belly, and rufous-chestnut lower back to uppertail-coverts. Has cream to yellowish-white eyes, blue orbital skin and blue neck-skin (usually visible when singing). **Juvenile** Said to have pale grey throat and breast, without black. **VOICE** Highly varied, several birds often calling simultaneously with combinations of: loud full ***wup wup wup or wup wup wup wup...***, similar ***wu wup-wuhup-wup-wuhup***; ***wuhup-wuhup*** and ***wuoo-wuoo-wuoo-wuoo...*** and tremulous ***t'u'u'u'u'u'u***, quavering ***ph'u'u'u'u'u*** and ***tik-tik-wrrrrrrrr*** etc. Often also interspersed (by female?) with mix of quite harsh scratchy jumbled notes and quiet conversational notes: ***jriii-jriii***; ***tchup-tchup tchup***; ***jrrt-jrrrt-jrr-jrr-jrr-jrr*** and ***ju-ju-ju-wiiii*** etc. **HABITAT & BEHAVIOUR** Broadleaved evergreen forest; up to 200 m. Usually in small flocks. **RANGE & STATUS** Resident Sumatra, Borneo. **SE Asia** Common resident S Thailand, Peninsular Malaysia; formerly Singapore. **BREEDING** March–September. **Nest** Loose ball or cup, in palm etc.; 0.6–1 m above ground. **Eggs** 3; slightly glossy white; 21–21.3 x 15.5–16.1 mm.

1176 SICKLE-BILLED SCIMITAR-BABBLER

Xiphirhynchus superciliaris Plate **114**

IDENTIFICATION 20 cm. *X.s.forresti* (east N Myanmar): **Adult** Easily told by very long, sharply downturned blackish bill. Head blackish with untidy whitish supercilium and dark-streaked whitish throat; underparts deep rufous. *X.s.rothschildi* (W,E Tonkin) shows paler crown and narrower supercilium (post-ocular only). **Other subspecies in SE Asia** *X.s.intextus* (W,N[west] Myanmar). **VOICE** Has two territorial calls: an extremely rapid, staccato, hollow piping ***wuwuwuwuwuwu*** and slower, clear, even staccato ***put-put-put-put-put-put*** or ***whit'whit'whit'whit'whit'whit***. Both may be accompanied (by female?) with a high-pitched clear ***u-WI*** (second note stressed); sometimes ***ti-WEE***. **HABITAT** Bamboo, broadleaved evergreen forest; 915–2,745 m. **RANGE & STATUS** Resident Nepal, NE Indian subcontinent, SW China (Yunnan). **SE Asia** Uncommon resident W,N Myanmar, W,E Tonkin. **BREEDING** India: April–July. **Nest** Large ball, on or close to ground. **Eggs** 3–5; white; 23.7 x 17.8 mm (av.).

1177 LARGE SCIMITAR-BABBLER

Pomatorhinus hypoleucos Plate **114**

IDENTIFICATION 25.5–28 cm. *P.h.tickelli*: **Adult** Told from other scimitar-babblers by larger size, brown ear-coverts, chestnut neck-patch and dark greyish breast-sides and flanks with broad white streaks. Shows whitish supercilium behind eye (mixed with rufous) spreading into short white streaks on neck-side; bill brown. *P.h.hypoleucos* (SW,W,S[west],N Myanmar) shows fewer white markings on breast-sides, dark rufous rear supercilium and neck-patch and slightly darker upperparts; *wrayi* (extreme S Thailand, Peninsular Malaysia) is much colder and darker above, has dark chestnut neck-patch, blackish-brown base colour of breast-sides, flanks and vent, and less white and no rufous on supercilium. **Juvenile** Markings on breast-sides and flanks more diffuse, crown and upperparts much more rufescent. **Other subspecies in SE Asia** *P.h.brevirostris* (southern Indochina, west to east Cambodia). **VOICE** Song is a variable series of usually three loud hollow piping notes, often from duetting pairs: ***wiu-pu-pu—wup-up-piu***; ***wiu-pu-pu—wo-hu***; ***whiu-pu-pu—whip-up-up*** and ***wiao-pu-pu—hu-pwuhu*** etc. Individual phrase variants also include ***oh-pu***; ***whiu-wao*** and hurried ***wiupupu*** and ***whipuwup*** etc. Usual calls are a loud, harsh, grating ***whit-tchtchtchtch*** and ***hekhekhekhekhekhek*** etc., when agitated, and a short, hard ***puh***. **HABITAT & BEHAVIOUR** Broadleaved evergreen and mixed deciduous forest, bamboo; up to 1,550 m (915–2,135 m in Peninsular Malaysia). Usually on or close to ground. **RANGE & STATUS** Resident NE India, E Bangladesh, SW,S China. **SE Asia** Fairly common to common resident (except C,E Myanmar, C Thailand, Singapore); extreme south of S Thailand only. **BREEDING** November–June. **Nest** Large oval or semi-dome, on ground or in palm, rattan etc. up to 3 m. **Eggs** 2–5; whitish; 30.1 x 21.7 mm (*hypoleucos*) & 29.5 x 22.3 (*wrayi*).

1178 RUSTY-CHEEKED SCIMITAR-BABBLER

Pomatorhinus erythrogenys Plate **114**

IDENTIFICATION 23.5–25 cm. *P.e.celatus*: **Adult** Combination of plain deep orange-rufous head-sides, flanks and vent and white throat, centre of breast and belly diagnostic. See Spot-breasted Scimitar-babbler. **Juvenile** Underparts rufous, with white on centre of throat and narrowly down centre of abdomen. **Other subspecies in SE Asia** *P.e.imberbis* (C[east],E[south],S[east] Myanmar). **VOICE** Song is very similar to Spot-breasted. Typical duets include ***whi-u-ju-whi-u...***; ***iu-chu-ip-iu-chu...*** and ***yu-u-yi-yu-u...*** etc. Also a high clear ***pu*** or ***ju***, and repeated, slightly rolling, well-spaced ***jrr-jrr-jrr-jrr...*** When alarmed, gives a rattling ***whih-whihihihihi*** and harsh ***whit-it*** or ***whoi-whitititititit***, the ***whoi*** note recalling sound of stone dropped in water and sometimes given singly. **HABITAT** Scrub and grass, open broadleaved evergreen forest; 915–2,000 m. **RANGE & STATUS** Resident NE Pakistan, NW,N India, Nepal, Bhutan. **SE Asia** Common resident C(east),E,S(east) Myanmar, NW Thailand. **BREEDING** February–May. **Nest** Loose dome with broad side-entrance, placed on ground in sheltered situation or in bush up to 1.2 m above ground. **Eggs** 2–4; fairly glossy, white; 24.9–30.5 x 18.5–21.6 mm.

1179 SPOT-BREASTED SCIMITAR-BABBLER

Pomatorhinus mcclellandi Plate **114**

IDENTIFICATION 22–23 cm. Monotypic. **Adult** Recalls Black-streaked Scimitar-babbler, but shows drab brown flanks and breast markings (latter more spot-shaped). Note range. **Juvenile** Like adult but breast markings fainter. **VOICE** Commonly sings in pair-duet, very like Rusty-cheeked Scimitar-babbler: a quick, loud, far-carrying fluty quick two-note phrase is answered by a very short sharp note: ***wi-wru-pi-wi-wru...***; ***wip-uip-ju-wip-uip...*** and ***wi-wu-jrr-wi-wu...*** etc. When alarmed, utters a harsh rattle preceded by quick high notes: ***wi-wi-chitit***, and a rapid mechanical rattle preceded by a loud frog-like sound: ***whoip-tutututututut***. **HABITAT** Scrub and grass, open broadleaved evergreen forest; 1,000–1,830 m. **RANGE & STATUS** Resident E Bhutan, NE India, E Bangladesh. **SE Asia** Common resident W Myanmar. **BREEDING** March–June. **Nest** Untidy oval or dome with side-entrance, or bowl, on ground or in sapling or bush, up to 1.5 m. **Eggs** 3–4; white; 26.5 x 19.3 mm (av.). **NOTE** Split from *P. erythrocnemis* (Spot-breasted Scimitar-babbler) following Collar & Robson (2007).

1180 BLACK-STREAKED SCIMITAR-BABBLER *Pomatorhinus gravivox* Plate **114**

IDENTIFICATION 23–25 cm. *P.g.odicus*: **Adult** Similar to Rusty-cheeked Scimitar-babbler but with distinctive blackish spots/streaks on breast. See Spot-breasted Scimitar-babbler, but note range. **Juvenile** Like adult but breast markings fainter. **VOICE** Similar to Spot-breasted. Rapid-fire ***whi'chu*** or ***whi'tu*** from single birds. Duets not yet documented. When alarmed, ***whoi-t't't't't*** and ***whoip-tut'ut'ut'ut'ut'ut*** or ***whup-which'ch'ch'ch'ch'ch***. **HABITAT** Scrub and grass, open broadleaved evergreen forest; 1,220–2,600 m. **RANGE & STATUS** Resident SE Tibet, SW,W,C China. **SE Asia** Common resident N,E(east) Myanmar, N Laos, W,E Tonkin. **BREEDING** China: March–June. **Nest** Untidy oval or dome with side-entrance, on ground or in sapling or bush, up to 1.5 m. **Eggs** 2–6; glossless white; 28.4–31.8 x 21–21.7 mm. **NOTE** Split from *P. erythrocnemis* (Spot-breasted Scimitar-babbler) following Collar & Robson (2007).

1181 WHITE-BROWED SCIMITAR-BABBLER

Pomatorhinus schisticeps Plate **114**

IDENTIFICATION 21–23 cm. *P.s.olivaceus* (Tenasserim [except north], W,S Thailand): **Adult** Combination of yellowish bill, drab olive-brown upperparts and unmarked white throat, breast and belly-centre diagnostic. Shows rufous nuchal collar and brownish flanks and vent. *P.s.salimalii* (north-west N Myanmar), *mearsi* (SW,W,C,S Myanmar west of Irrawaddy R) and *annamensis* (east Cambodia, S Annam, Cochinchina) show more blackish-brown crown, contrasting with mantle, and chestnut flanks with white streaks; *ripponi* (C[east],E[north & east] Myanmar, extreme north NW Thailand, north-west N Laos) and *nuchalis* (C,E,S Myanmar east of Irrawaddy R) show duller, greyer upperparts, broader rufous-chestnut nuchal collar and some rufous-chestnut on flanks (more on *nuchalis*); *klossi* (SE Thailand, south-west Cambodia) is a little darker above and shows chestnut on flanks. See Streak-breasted Scimitar-babbler. **Juvenile** Similar to adult. **Other subspecies in SE Asia** *P.s.difficilis* (S[south-east] Myanmar, north Tenasserim, W,NW[west] Thailand), *humilis* (NW[east],NE Thailand, north Cambodia, N[south],C,S Laos, N,C Annam). **VOICE** Song is a series of usually 3–7 clear, quite quickly delivered hollow piping notes: ***hu-hu-hu-hu-hu***; ***whu-pu-pu-pu-pu-po***; ***whu-wu-wu-pu***; ***whu-wu-wu-wu-wu-wu-wu*** and ***whi-hu-wi*** etc., sometimes with a more abrupt first note or longer pause between first and second notes. May give faster ***whuhuhuhuhuhu*** etc. or, occasionally, shorter ***wu-hup***; ***oo-hu*** etc. (by female?), which may be combined in duets: ***wu-hu-hu-hu-wu-hup-wu-hu-hu-hu...*** etc. When alarmed, utters a harsh mocking ***whihihihihi*** and ***whichitit*** etc.

Excited pairs give a variety of mixed calls: rather nasal, husky ***whor-whor-whor***, very nasal ***wiaaah*** or ***woieee***, jumbled chattering mixed with quite high ***woh*** notes, short clear ***oh*** and ***whu*** notes, and throaty ***wuhu-wuhu*** and ***wuhu wuhu-wip*** etc. **HABITAT** Broadleaved evergreen and deciduous forest, secondary growth, bamboo, scrub and grass; up to 2,135 m, locally 2,600 m. **RANGE & STATUS** Resident N,NE Indian subcontinent, SW China (west Yunnan). **SE Asia** Fairly common to common resident Myanmar, Thailand (except C), Cambodia, N(west & south),C,S Laos, N,C,S Annam, Cochinchina. **BREEDING** November–July. **Nest** Loose globe or dome with side-entrance; on ground or in low vegetation up to 1 m. **Eggs** 2–5; white; 26.6 x 19.2 mm.

1182 CHESTNUT-BACKED SCIMITAR-BABBLER *Pomatorhinus montanus* Plate 114

IDENTIFICATION 19 cm. *P.m.occidentalis*: **Adult** Resembles White-browed Scimitar-babbler but shows diagnostic black crown and dark chestnut mantle to rump and flanks. Note range. **Juvenile** Duller than adult, chestnut of plumage paler and less extensive on flanks, ear-coverts mostly chestnut, crown washed rufous. **VOICE** Song is a clear loud resonant 2–3 note ***whu-whoi***; ***woi-woip*** and ***yu-hu-hu*** etc. Duets include ***whu-whi-whu-woi-whu-whi...*** **HABITAT** Broadleaved evergreen forest; up to 1,370 m. **RANGE & STATUS** Resident Greater Sundas. **SE Asia** Fairly common resident extreme S Thailand, Peninsular Malaysia. **BREEDING** November–September. **Nest** Large ball or sheltered cup, in depression on bank or in low dense vegetation. **Eggs** 2–3; fairly glossy white; 24.5 x 16.2–16.9 mm.

1183 STREAK-BREASTED SCIMITAR-BABBLER *Pomatorhinus ruficollis* Plate 114

IDENTIFICATION 17–19.5 cm. *P.r.reconditus* (W,E Tonkin, N Annam): **Adult** Similar to White-browed Scimitar-babbler but smaller and shorter-billed, with heavy chestnut streaking on lower throat and breast. *P.r.bakeri* (W Myanmar) shows dull warm-tinged brown underpart-streaking and slightly paler upperparts; *similis* (N Myanmar), *albipectus* (northern N Laos) and *beaulieui* (N[west & southern],C Laos) show duller rufous-tinged underpart-streaking, with centre of breast and abdomen whiter on *albipectus* and particularly *beaulieui*. **Juvenile** Mask warm dark brown, supercilium shorter, nuchal patch indistinct, underparts uniform buffish-brown with whiter throat. **VOICE** Song is a loud, clear, rather high piping ***u-hu-hu***; ***wu-wu-wu*** and ***wu-wee-wu*** etc., uttered quite quickly. Sometimes a slower ***wu-wu*** or fuller-sounding ***whu-whi*** or ***u-whi wi***. When alarmed, gives harsh scolding rattles: ***whi-whi-whi whi-whi-whi-whichitit***; ***chrrurururur***; ***whi-wi chutututut***; ***chutitititititit*** and ***whi-wir-irrirrirr*** etc. **HABITAT** Open broadleaved evergreen forest, forest edge, bamboo, scrub and grass; 900–2,750 m, locally down to 50 m in Indochina. **RANGE & STATUS** Resident N,NE Indian subcontinent, W,C and southern China, Taiwan. **SE Asia** Uncommon to common resident W,N Myanmar, N(eastern),C Laos, W,E Tonkin, N Annam. **BREEDING** January–May. **Nest** Crude ball or dome with side-entrance toward top, in concealed position on ground, in hole in bank or in low vegetation up to 1 m. **Eggs** 2–5; fairly glossy white; 23.4 x 17.4 mm (av.; *bakeri*).

1184 ORANGE-BILLED SCIMITAR-BABBLER *Pomatorhinus ochraceiceps* Plate 114

IDENTIFICATION 22–24 cm. *P.o.ochraceiceps*: **Adult** Long narrow downcurved reddish-orange bill, uniform rufescent-brown crown and upperparts and white breast and belly-centre distinctive. *P.o.austeni* (southern N Myanmar) has duller brown crown and upperparts; *stenorhynchus* (northern N Myanmar) shows warm buff breast and belly. See Coral-billed Scimitar-babbler. **Juvenile** Blackish bill with pale flesh tip and base, shorter tail. **Other subspecies in SE Asia** *P.o.alius* (NE Thailand, S Laos, C,S Annam). **VOICE** Song is a hurried hollow piping ***wu-wu-wu***; ***wu-wu-woi***; ***wu-wu-whip*** and ***pu-pu*** etc., sometimes answered (by female?) with nasal ***wyee***. Also, very rapid ***wi-wuwu*** and loud whistled ***u-wip***, repeated after pauses. When alarmed, gives harsh scratchy ***whi-chutututut***; ***whi-trrrrt whi-trrrrt...*** and ***tchrrrtututut tchrrrt...*** etc. Variable sounds (often combined) when excited: rapid purring ***wrrrrp***, clear ***wuhu-wuhu***, nasal ***woiee-woiee***, weak ***whiu-whiu***, high ***wheep*** and high rising ***whi*** etc. **HABITAT & BEHAVIOUR** Broadleaved evergreen forest, bamboo; 230–1,800 m. Often associates with mixed-species feeding flocks and particularly White-hooded and Collared Babblers. **RANGE & STATUS** Resident NE India, SW China (Yunnan). **SE Asia** Uncommon resident N,S(east),E Myanmar, north Tenasserim, W,NW,NE Thailand, Laos, W,E Tonkin, N,C,S Annam. **BREEDING** February–June. **Nest** Oval ball, on ground or amongst low undergrowth. **Eggs** 3–5; white; 25.2 x 18.3 mm (av.; *austeni*).

1185 CORAL-BILLED SCIMITAR-BABBLER *Pomatorhinus ferruginosus* Plate 114

IDENTIFICATION 21–23 cm. *P.f.albogularis*: **Adult** Similar to Orange-billed Scimitar-babbler but has shorter, thicker, deep red bill, black line above supercilium and broadly black head-sides. Crown and upperparts warmish brown, lower throat to belly buff. *P.f.phayrei* (SW,W Myanmar) has duller, more olive-brown upperparts and orange-buff lower throat to belly; *stanfordi* (N Myanmar) and *orientalis* (east N Laos, W,E Tonkin, N Annam) are a little warmer buff below, the latter also deeper, dark warm-tinged brown above; *dickinsoni* (S Laos, C Annam) have buff of underparts restricted to flanks and lack black line above supercilium. Population in C Laos is paler below than *orientalis* and may be undescribed race. **Juvenile** More rufescent above and below. **VOICE** Several birds may unite to produce a wide variety of sounds, including soft, questioning ***whu***; ***whiu*** and ***whoiee***, oriole-like meeowing ***whheeeei***, shrill yelping ***yep-yep-yep...*** and short squeaky notes, along with more typical harsh scolding ***whit whit-tchrrrrt***; ***tchrrrt-tchrrrrt*** and ***whitchitit*** etc. Also a scratchy ***weeitch-oo*** and shrill ***wheep-wheep***. Clear whistled ***ch-wooa*** also recorded. When alarmed, gives harsh dry ***krrrrrt***; ***krururuttt*** and ***krrrrirrrurut*** etc., less scratchy and piercing than Orange-billed. **HABITAT & BEHAVIOUR** Broadleaved evergreen forest, bamboo; 800–2,000 m. Sometimes associates with mixed-species feeding flocks. **RANGE & STATUS** Resident E Nepal, Bhutan, NE India, SE Tibet, SW China (Yunnan). **SE Asia** Uncommon resident Myanmar (except C), W,NW Thailand, Laos, W,E Tonkin, N,C Annam. **BREEDING** March–August. **Nest** Rough oval with entrance at one end, on ground or in bush, sapling etc.; up to 2 m above ground. **Eggs** 3–5; fairly glossy white; 27.1 x 19.3 mm (av.; NE India).

1186 CHEVRON-BREASTED BABBLER *Sphenocichla roberti* Plate 112

IDENTIFICATION 16.5–18 cm. Monotypic. **Adult** Unmistakable large wren-babbler with heavily scaled plumage, dark-barred wings and tail, and conical, pointed bill. Forehead rufous, rest of upperparts warmish brown with black-and-whitish scales; supercilium, throat and centre of abdomen whitish, rest of underparts brown with pronounced black-and-white chevron-shaped scales. **Juvenile** Warmer brown overall, scaling slightly duller and less contrasting. **VOICE** Song is a clear, loud, fluty, melodious ***uu-wii-wu-yu*** (***wii*** highest, ***yu*** lowest), lasting 1.1–1.2 sec. **HABITAT & BEHAVIOUR** Broadleaved evergreen forest, secondary growth, bamboo; 915–1,525 m. Often found in small groups. **RANGE & STATUS** Resident NE India, SW China (NW Yunnan). **SE Asia** Scarce resident N Myanmar. **BREEDING** India: May–June. **Nest** Pad wedged behind loose bark (like treecreeper *Certhia*); 6.5 m above ground. **Eggs** 4; white; 20.7–22.3 x 17–17.4 mm.

1187 **BAR-WINGED WREN-BABBLER**

Spelaeornis troglodytoides Plate **112**

IDENTIFICATION 13 cm. *S.t.souliei*: **Male** Combination of buffish-grey wings and tail with prominent dark bars, rufescent mantle to uppertail-coverts, white throat and rufous flanks and vent diagnostic. Crown appears blackish (tinged rufous) with white spots, has prominent black-tipped white streaks on rest of upperparts and lower flanks. See other *Spelaeornis* wren-babblers and Spotted Wren-babbler. **Female** More rufous-streaked throat than male, more extensive and brighter rufous below. **Juvenile** Crown, head-sides and upperparts indistinctly mottled dark brown and dull rufescent-brown, underparts rather plain deep rufous, wing- and tail-barring a little less distinct. **VOICE** Song is a repeated husky rolling ***chi'whi-whi'whi-whi'whi-whi'whi*** or ***ch-whi-whi-whi-whi*** etc. **HABITAT & BEHAVIOUR** Broadleaved evergreen forest, bamboo; 2,440–2,895 m. Inhabits undergrowth, usually not far from ground. **RANGE & STATUS** Resident Bhutan, NE India, SE Tibet, SW,W,C China. **SE Asia** Scarce resident N Myanmar. **BREEDING** March–June. Otherwise undocumented.

1188 **GREY-BELLIED WREN-BABBLER**

Spelaeornis reptatus Plate **112**

IDENTIFICATION 11–12 cm. Monotypic. **Male** Told from other small wren-babblers by combination of relatively long, unmarked tail and plain wings (no bars or spots). Crown and upperparts rather uniform brown with darker scaling, head-sides grey, throat white with some indistinct dark markings, rest of underparts brownish with pale greyish centre of breast and belly and black and whitish spots/bars. **Female** Throat and breast variably washed rufescent-buff. See Pale-throated and Chin Hills Wren-babblers, but note range. **VOICE** Sings with a repeated, decelerating trill: ***pwwrriii'i'i'i'i*** or ***pwwr-ree'e'e'e'e***, and ***pr'r'r'r'r'ir'ir'dir'dir'déér***. Song is sometimes accompanied (by female?) with a rasping ***pitcherrr-pitcherrr***. Soft ***pt...pt...*** contact call. **HABITAT & BEHAVIOUR** Broadleaved evergreen forest, forest edge, secondary growth, scrub and grass near forest; 1,400–2,800 m. Stays close to ground. **RANGE & STATUS** Resident NE India (SE Arunachal Pradesh), SW China (W Yunnan). **SE Asia** Locally common resident N,E Myanmar, W Thailand (Doi Kajela). **BREEDING** March–June. **Nest** Loose ball, on ground or bank. **Eggs** 3; pinkish-white, unmarked or sparingly streaked with darker pink, or speckled with dark red or reddish-brown; 18.1–20.7 x 14–14.6 mm. **NOTE** Split from extralimital *S. chocolatinus* (Naga Wren-babbler) following Rasmussen & Anderton (2005).

1189 **PALE-THROATED WREN-BABBLER**

Spelaeornis kinneari Plate **112**

IDENTIFICATION 11–12 cm. Monotypic. **Male** Recalls Grey-bellied Wren-babbler, but shows darker moustachial line, darker grey on breast and belly, and more distinct scales/bars on breast. **Female** Compared to Grey-bellied, shows buffier throat, darker and browner base colour of breast and flanks, more distinct dark bars/scales (as well as buff scales) on breast, and darker grey on centre of abdomen. **Juvenile** Undescribed. **VOICE** Sings with two types of trill (repeated after intervals), which slow towards end: (1) loud rapid ***chwi'i'i'i'witchu-wit***; ***chwiwiwi'i'witchu-wit*** and ***chwwwiwiwi-witchu-wit*** etc. (1.5–2 s long), with stressed ending; (2) ***churrrrr'r'r-r-rt-rt-yut-yut-yut-yut*** (2–2.5 s long), stuttering in middle and becoming spaced and fuller-sounding. Song is sometimes accompanied (by female?) with a hoarse, buzzy ***titcher tcher-tcher*** or ***titcher-tcher-tcher-tcher-tcher***. **HABITAT & BEHAVIOUR** Broadleaved evergreen forest, forest edge, secondary growth, scrub and grass near forest; 1,600–2,500 m. Stays close to ground. **RANGE & STATUS** Resident SW (SE Yunnan) China. **SE Asia** Locally fairly common resident W,E(north-west) Tonkin. **BREEDING** Undocumented. **NOTE** Split from extralimital *S. chocolatinus* (Naga Wren-babbler) following Collar & Robson (2007).

1190 **CHIN HILLS WREN-BABBLER**

Spelaeornis oatesi Plate **112**

IDENTIFICATION 11–12 cm. Monotypic. **Adult** Told by grey of head-sides being restricted to lores and around eye, mostly brownish ear-coverts and white throat, breast- and belly-centre with prominent black spots (larger on sides of throat and breast) and brown wash on throat-sides, breast and flanks. Note range. **Juvenile** Warmer above, with little scaling, washed warm brown below, with only throat and mid-belly white; fewer spots. **VOICE** Sings with repeated loud abrupt undulating ***chiwi-chiwi-chiwi-chew*** and ***witchi-witchi-witchi-wu*** etc., sometimes shortened, or hurried to ***wituwituwitu-wu***, or extended to ***witchu-witchu-witchu witchu-witchu-witchu...*** and a hurried ***witchuwiwitchuwi-witchuwi***. Song is sometimes accompanied (by female?) with a hoarse, buzzy ***titcher tcher-tcher*** or ***titcher-tcher-tcher-tcher-tcher***. Calls with a soft ***tuc tuc tuc...*** and ***chit-chit-chit...*** and very quiet ***ik ik ik...*** (audible at close range). **HABITAT & BEHAVIOUR** Broadleaved evergreen forest, forest edge, secondary growth, scrub and grass near forest; 1,400–2,800 m. Stays close to ground. **RANGE & STATUS** Resident NE India (Mizoram). **SE Asia** Locally common resident W Myanmar. **BREEDING** March–June. **Nest** Dome with side-entrance or oval with entrance near top, on ground amongst plant stems or in fern up to 75 cm. **Eggs** 2–4; dull white to pinkish-white, sparingly freckled reddish and light purple; 17.5–18.5 x 15–15.2 mm. **NOTE** Split from extralimital *S. chocolatinus* (Naga Wren-babbler) following Rasmussen & Anderton (2005).

1191 **SPOTTED WREN-BABBLER**

Elachura formosa Plate **112**

IDENTIFICATION 10 cm. Monotypic. **Adult** Brown plumage, peppered with white specks and black-barred rufous wings and tail diagnostic. Greyish-brown above, speckled white, with row of larger white spots across lower nape/upper mantle, paler and warmer below, intricately speckled and vermiculated white and black; plainer on throat. **Juvenile** Much darker than adult, but with stronger white spotting. **VOICE** Song is an extremely high-pitched, drawn-out, tinkling ***ti-ti-ti-i tit-si-ii ti-ti-ti-i tit-si-ii...*** or ***tit-tit-ti-i tit-tsii-ii tit-tit-ti-i tit-tsii-ii***, with ***si-ii*** and ***sii-ii*** parts slurred and rising (hard to hear). **HABITAT & BEHAVIOUR** Broadleaved evergreen forest, scrub and weeds in gullies; 480–1,975 m. On or close to ground, skulking. **RANGE & STATUS** Resident E Nepal, NE Indian subcontinent, southern China. **SE Asia** Uncommon resident W,N Myanmar, NW Thailand (Doi Lang, Doi Pha Hom Pok), N,C Laos, W Tonkin, N Annam. **BREEDING** India (not definitely authenticated): April–May. **Nest** Semi-dome, on ground or bank. **Eggs** 3–4; glossy white, with a few reddish-brown specks; 16.5 x 12.4 mm.

1192 **SCALY-BREASTED WREN-BABBLER**

Pnoepyga albiventer Plate **112**

IDENTIFICATION 8.5–10.5 cm. *P.a.albiventer*: **Adult dark morph** Small size, scaly unstreaked plumage and tail-less appearance distinctive. Underparts mostly dark brown with buff scales (throat and breast plainer). In some populations (i.e. W Myanmar) throat often white. Hard to separate from Pygmy Wren-babbler but slightly larger and more robust, shows pale buff to whitish speckling on ear-coverts and sides of crown and neck. Note voice and range. **Adult pale morph** Buff of underparts replaced by white (except lower flanks). Differs from Pygmy as dark morph. **Juvenile** Rather plain brown; crown, upperparts and head-sides spotless and only slightly scaled, underparts densely mottled dark brown and dark buff with paler centre of throat and abdomen (dark morph), or cold drab brown with paler throat-centre and pale and dark mottling on breast and belly (pale morph). Gradually acquires patches of adult plumage below. See Pygmy; also Winter Wren. **VOICE** Has rapid high-pitched jumbled warbler-like song (1.5–2 s long): roughly ***wisisitititiwi*** or ***wiswisiwitwisititui***, repeated after long intervals. Call is a repeated short, loud, slightly explosive ***tschik*** or ***tchik***. **HABITAT & BEHAVIOUR** Broadleaved evergreen forest; 1,200–3,000 m (breeds above 2,200 m). On or close to ground; often adopts less upright posture than Pygmy. **RANGE & STATUS** Resident N,NE India, Nepal, Bhutan, S,SE Tibet, SW China. **SE Asia** Fairly common resident W,N Myanmar, W Tonkin. **BREEDING** May–July. **Nest** Ball with side-entrance, embedded in moss on bank, boulder or tree-trunk. **Eggs** 2–5; white (may sometimes show a few reddish-brown specks); 19.1 x 14.1 mm (av.).

1193 PYGMY WREN-BABBLER

Pnoepyga pusilla Plate **112**

IDENTIFICATION 7.5–9.5 cm. *P.p.pusilla*: **Adult dark morph** Very like Scaly-breasted Wren-babbler but slightly smaller and slimmer, lacks speckling on crown, neck-sides and ear-coverts. Note voice and range. *P.p.annamensis* (Cambodia, S Annam; S Laos?) and *harterti* (S Thailand, Peninsular Malaysia) tend to show plainer throat and centre of underparts, more rufescent lores and warmer upperparts. **Adult pale morph** Differs from Scaly-breasted in same way as dark morph. May show variable buff wash on breast and flanks. *P.p.annamensis* and *harterti* tend to show more rufescent lores and head-sides, warmer upperparts. **Juvenile** Probably indistinguishable on plumage from Scaly-breasted. **VOICE** Song is an unmistakable, very high-pitched, well-spaced ***ti-ti-tu*** (c.4 s long), repeated every 3–5 s, sometimes with shorter space between last two notes (NW Thailand, W Tonkin); or just ***ti-tu*** (W Myanmar, S Annam). Call is a repeated sharp loud ***tchit*** or ***chit***, very like Scaly-breasted. **HABITAT & BEHAVIOUR** Broadleaved evergreen forest, secondary forest; 500–2,565 m (breeds above 750 m); locally down to 180 m in winter E Tonkin. On or near ground; often adopts more upright posture than Scaly-breasted. **RANGE & STATUS** Resident N,NE Indian subcontinent, SE Tibet, C and southern China, Sumatra, Java, Flores, Timor. **SE Asia** Common resident Myanmar (except SW,C), W,NW,S Thailand, Peninsular Malaysia, south-west Cambodia, Laos, W Tonkin, N,C,S Annam. **BREEDING** November–August. **Nest** Ball with side-entrance or built-in cup, embedded in moss or amongst ferns or creepers on bank, boulder, tree-trunk etc., rarely in isolated sapling; 0.5–2 m above ground (rarely to 6 m). **Eggs** 2–6 (2 in south); white (may sometimes show one or two faint spots); 17.1 x 13.1 mm (av.; *pusilla*) & 19.7–20.5 x 14.1–15.4 mm (*harterti*).

1194 GOLDEN BABBLER

Stachyridopsis chrysaea Plate **111**

IDENTIFICATION 10–12 cm. *S.c.assimilis*: **Adult** Unmistakable, with bright yellow forehead and underparts, black face and dark crown-streaking. Ear-coverts and nape yellowish-olive, upperparts and flanks greyish-olive, eyes dark crimson. *S.c.chrysaea* (N Myanmar) is brighter yellow, has more contrasting black and yellow crown-streaks and greener upperparts; *binghami* (W Myanmar) has grey ear-coverts and more contrasting crown-streaks. **Juvenile** Underparts washed out, yellowest on throat and upper breast, eyes brownish. **Other subspecies in SE Asia** *S.c.aurata* (eastern E Myanmar, north NW Thailand, northern Indochina), *chrysops* (S Thailand southwards). Populations in east NW,NE(north-west) Thailand, S Laos, C Annam not yet assigned to race. **VOICE** Song is like Rufous-fronted Babbler but notes tend to be clearer, often sounding more spaced, and usually with more obvious pause (occasionally no pause) after first note: rapid ***tu tu-tu-tu-tu-tu-tu*** or slower ***ti tu-tu-tu-tu-tu***. Usually with 5–10 notes and lasting 1–1.25 s. Introductory notes sometimes given singly. When alarmed, utters scolding ***chrrrrr-rr-rr***, ***chrirrrrr*** and ***chrrrrrr*** etc. **HABITAT & BEHAVIOUR** Broadleaved evergreen forest; 750–2,600 m, locally down to 450 m N Myanmar. Often in mixed-species flocks. **RANGE & STATUS** Resident Nepal, NE Indian subcontinent, SE Tibet, SW China, Sumatra. **SE Asia** Common resident (except SW,C Myanmar, C,SE Thailand, Singapore, Cambodia, S Annam, Cochinchina). **BREEDING** January–July. **Nest** Dome or ball, with side-entrance towards top, on ground, in hole in bank or in bamboo clump or bush up to 0.3–4 m. **Eggs** 3–4; fairly glossy white to pinkish-white (possibly rarely showing some faint reddish-brown spots); 15.5 x 12.2 mm (av.; *chrysaea*).

1195 RUFOUS-CAPPED BABBLER

Stachyridopsis ruficeps Plate **111**

IDENTIFICATION 12.5 cm. *S.r.pagana* (S Laos, C,S Annam): **Adult** Very similar to Rufous-fronted Babbler but upperparts paler and more olive, has yellowish-washed lores, head-sides and underparts, lacks pale greyish eyebrow, forehead to crown more orange-rufous; tends to have more pinkish lower mandible. Note range. *S.r.rufipectus* (west N Myanmar) shows more rufous-buffish wash on mantle and buffish-yellow head-sides and underparts (whiter on centre of throat and belly), mixed with greyish-olive on flanks; *bhamoensis* (N[east],E Myanmar) and *davidi* (north Indochina) show broadly greyish-olive flanks and darker, more rufescent mantle, *davidi* also with yellower lores, throat and upper breast. See Pin-striped Tit-babbler. **Juvenile** Crown paler, underparts paler and more washed out. **VOICE** Song is possibly inseparable from Rufous-fronted. However, phrases may be generally slower, with a typical duration of 1.25–2 s. Sometimes gives 1–2 introductory notes on their own. When alarmed, utters low harsh scolding ***trrrrt-trrrrt-trrrrt*** and ***trrutut-trrrrt-trrrrt*** etc. **HABITAT & BEHAVIOUR** Broadleaved evergreen forest, secondary forest, bamboo; 950–2,195 m, down to 455 m in winter (N Myanmar). Often in mixed-species feeding flocks. **RANGE & STATUS** Resident Nepal, Bhutan, NE India, C and southern China, Taiwan. **SE Asia** Common resident N,E Myanmar, N,S(east) Laos, Vietnam (except Cochinchina). **BREEDING** April–July. **Nest** Ball, oval or cone with side-entrance or deep cup, in bush, bamboo or low vegetation; up to 3 m above ground, rarely on ground. **Eggs** 3–5; white to brownish-white, speckled brown to reddish, usually over purple to grey underspecks (often mostly towards broader end); 15.5–17.3 x 12.4–13.7 mm (*davidi*).

1196 RUFOUS-FRONTED BABBLER

Stachyridopsis rufifrons Plate **111**

IDENTIFICATION 11.5–12.5 cm. *S.r.rufifrons*: **Adult** Rufous forehead to midcrown and buffish underparts distinctive. Very like Rufous-capped, but upperparts somewhat darker and warmer, shows pale greyish (sometimes whitish or buffy-tinged) lores, eyebrow and eyering, and buffier underparts (shade and extent very variable); forehead to crown duller rufous. Note range. *S.r.poliogaster* (extreme S Thailand, Peninsular Malaysia) has darker, duller crown and upperparts and much paler buff lower breast to vent (contrasts with upper breast); *obscura* (S Thailand) is roughly intermediate. Told from similar Pin-striped Tit-babbler by smaller size, lack of prominent yellow supercilium and lack of yellow on underparts or prominent breast-streaking. See Grey-faced Tit-babbler. **Juvenile** Crown paler than adult and underparts paler, more washed out. **Other subspecies in SE Asia** *S.r.pallescens* (SW,W,C,S[north-west] Myanmar), *ambigua* (north-west N Myanmar), *planicola* (south-east N Myanmar), *adjuncta* (east NW Thailand, northern Indochina), *insuspecta* (S Laos). **VOICE** Song typically consists of 5–7 rather high-pitched monotone piping notes: ***tuh tuh-tuh-tuh-tuh-tuh***, usually with brief pause after first note (sometimes quite well spaced or lacking); generally delivered quite quickly but speed varies. Phrases usually last 1.25–1.75 s and are repeated at varying intervals. Sometimes gives more spaced variants: ***tuh-tuh-tuh-tuh-tuh-tuh***, combined with low slurred ***churr-churr-churr-churr...*** or ***chrerr-chrerr-chrerr...*** etc. Other calls include a short querulous rolling ***wirrrri*** when alarmed, and very fast ***wu-yu-yu-yu-yu-yu-yi*** like subdued version of song. Also, very soft ***wit*** and ***wi*** contact notes. **HABITAT** Forest edge, secondary growth, scrub and grass, bamboo, broadleaved evergreen forest; up to 2,100 m. **RANGE & STATUS** Resident NE,E India, Bhutan,

Bangladesh, SW China (NW Yunnan), Sumatra, Borneo. **SE Asia** Common resident Myanmar, Thailand (except C,SE), Peninsular Malaysia, Laos, W Tonkin, N Annam. **BREEDING** March–May. **Nest** Dome, semi-dome or cup, in bamboo clump, palm etc.; up to 2 m above ground, rarely on ground or bank. **Eggs** 3–5; white, speckled brown to reddish-brown (usually mostly towards broader end); 16.1 x 12.4 mm (India). **NOTE** Includes *S.rodolphei* Deignan's Babbler (NW Thailand [Doi Chiang Dao]), considered here to be synonymous with *S.r.rufifrons*.

1197 PIN-STRIPED TIT-BABBLER

Macronus gularis Plate **111**

IDENTIFICATION 12.5–14 cm. *M.g.sulphureus*: **Adult** Rufous crown, olive-brown upperparts and largely yellowish head-sides and underparts distinctive. From Rufous-fronted Babbler by broad yellow supercilium, dark loral eyestripe, pale yellow throat and centre of underparts, olive-washed flanks and prominent dark narrow streaks on lower throat and breast. *M.g.lutescens* (north-east E Myanmar, NW(north),NE Thailand, Laos, W,E Tonkin, N Annam) has much brighter throat and underparts; *archipelagicus* (Tenasserim [Mergui Archipelago]) is yellower below with broader blacker streaks and more rufous-chestnut wash on crown and upperparts; *kinneari* (C Annam), *versuricola* (east Cambodia, S Annam, Cochinchina), *connectens* (southern Tenasserim east to SE Thailand) and *inveteratus* (islands off SE Thailand and Cambodia) show broader, dark throat- and breast-streaks (darkest on *connectens*), broader, darker olive flanks and much darker upperside contrasting less with crown; *condorensis* (Cochinchina [Con Son I]) is like *connectens* but forecrown more chestnut, upperparts darker; *gularis* (southern Peninsular Malaysia, Singapore) is similar to *condorensis* but upperparts much darker and more chestnut-tinged. **Juvenile** Upperparts more uniformly rufous, yellow of underparts whiter, olive parts browner, supercilium narrower but head-sides and throat similar to adult. **Other subspecies in SE Asia** *M.g.ticehursti* (SW,W Myanmar), *saraburiensis* (south-west NE Thailand, west Cambodia). **VOICE** Song varies somewhat. In northern Vietnam gives repeated loud clear bouncing ***ti chut-chutut-chut*** or ***tit-chutut-chutut-chutut...***; in C Annam and Thailand an even, well-spaced, 4–5 note ***chut-chut-chut-chut-chut...*** or ***chut chut-chut-chut-chut...*** etc. Typical calls include a harsh ***chrrrt-chrr***, ***chrrrt-chrr-chrri*** and ***tititit-chrreeoo***. **HABITAT** Open broadleaved evergreen and deciduous forest, peatswamp forest, mangroves, secondary growth, scrub and grass, bamboo; up to 1,525 m. **RANGE & STATUS** Resident N,NE Indian subcontinent, S,E India, SW China, Philippines (Palawan). **SE Asia** Common resident (except C Thailand). **BREEDING** December–August. **Nest** Rough ball or dome with side-entrance, in bush, palm, bamboo clump or dense undergrowth; 0.3–6 m above ground. **Eggs** 2–5; fairly glossy white to dull white (sometimes faintly pinkish-tinged) or sea-green, liberally speckled various shades of brown, red or purple (often mostly towards broader end); 17.8 x 12.7 mm (av.; *gularis*). Brood parasitised by Drongo Cuckoo.

1198 GREY-FACED TIT-BABBLER

Macronus kelleyi Plate **111**

IDENTIFICATION 14 cm. Monotypic. **Adult** Very similar to Pin-striped Tit-babbler but forehead, crown and upperparts more uniform rufescent-brown, has grey supercilium, greyer head-sides and paler yellow underparts with much finer dark streaks. Note voice and range. **Juvenile** Underparts greyer and less yellow, with diffuse narrow dark throat-streaking. **VOICE** Song is a soft, even, fairly well-spaced 2–20 note (usually c.8–12) ***tuh-tuh-tuh-tuh-tuh-tuh-tuh-tuh...*** Calls include harsh ***chrrrrii-chrrruu-chrrrii-chru***; ***chrrreechrrrer*** and ***chit-chrrerr***, coarse ***wi-ti-ti-chu*** and harsh squeaky ***trrrrrrt trrrrrt...*** **HABITAT** Broadleaved evergreen forest, secondary forest; 50–1,165 m. **RANGE & STATUS** Endemic. Fairly common resident E Cambodia, C,S Laos, N,C,S Annam, Cochinchina. **BREEDING** February–July. **Nest** Untidy ball with side-entrance, in banana, tree or vine; 3–15 m above ground. **Eggs** 4. Otherwise undocumented.

1199 FLUFFY-BACKED TIT-BABBLER

Macronus ptilosus Plate **111**

IDENTIFICATION 16.5 cm. *M.p.ptilosus*: **Adult** Overall darkish brown plumage with rufous crown, black cheeks and throat and blue spectacles diagnostic. Elongated white-shafted plumes on lower back and flanks rarely visible in field; bluish neck-skin often visible when singing. See Chestnut-winged Babbler. **Juvenile** Crown paler, mantle darker, throat less black, breast more rufous. **VOICE** Song is a repeated, rather low ***puh puh-puh-puh*** or ***puh puh puh-puh-puh*** or slower ***wuh wu-hu wu-hu*** and ***wuh-wuh hu-wu hu-wu*** etc. Often accompanied (by female?) with strange low forced husky ***hherrh herr hherr herr*** or ***iwit-cherrhh-iwit-cherrhh*** etc. Also a strange low frog-like creaking or croaking ***aahkeeah-oh*** etc. **HABITAT** Edge of broadleaved evergreen forest, freshwater swamp forest, secondary forest, bamboo; up to 200 m. **RANGE & STATUS** Resident Sumatra, Borneo. **SE Asia** Scarce to fairly common resident S Thailand, Peninsular Malaysia; formerly Singapore. **BREEDING** February–June. **Nest** Loosely built ball with side-entrance or partly buried cup, on ground, in palm or thick undergrowth; up to 1 m above ground. **Eggs** 2; lightly glossed pale grey-green to white, densely speckled and freckled brown (markings coalescing into blotches over broader end); 19.2–22.7 x 14.6–16.5 mm.

1200 CHESTNUT-CAPPED BABBLER

Timalia pileata Plate **108**

IDENTIFICATION 15.5–17 cm. *T.p.smithi* (N,E Myanmar, NW Thailand, northern Indochina) **Adult** Easily told by combination of thick black bill, rufous-chestnut cap, black mask and contrasting white supercilium, cheeks, throat and upper breast. Upperparts warm-tinged olive-brown, sides of neck and breast grey, lower breast to belly warm buff, fine dark streaks on lower throat and breast. *T.p.bengalensis* (western Myanmar) and *patriciae* (C[west],W[south] Thailand) show less rufescent, more greyish-olive upperparts (warm buffish-tinged in latter), greyer flanks and duller, less buff underparts; *intermedia* (C,S Myanmar east of Irrawaddy R, Tenasserim, W Thailand) and *dictator* (NE,SE Thailand, southern Indochina) are roughly intermediate. **Juvenile** Upperparts and outer fringes of wing-feathers warmer brown, cap and supercilium duller, lower mandible with pale base. **VOICE** Sudden husky notes followed by thin metallic notes: ***wher-wher witch-it-it*** and ***wher-er-itch-it-it*** etc.; often by more than one bird at a time. Calls include short metallic ***tzit*** and harsh ***chrrt*** notes and varied low chuntering grumbles. **HABITAT & BEHAVIOUR** Grassland and scrub, secondary growth; up to 1,500 m. Often in small groups; skulking. **RANGE & STATUS** Resident N,NE Indian subcontinent, SW,S China, Java. **SE Asia** Common resident (except S Thailand, Peninsular Malaysia, Singapore). **BREEDING** April–September. **Nest** Rough ball or dome with large side-entrance, sometimes deep cup, in bush or sapling; up to 90 cm above ground. **Eggs** 2–5; fairly glossy white (rarely pinkish), speckled olive-brown to reddish-brown (at broader end) over purple to grey undermarkings; 19 x 14.5 mm (av.; *bengalensis*).

1201 RUFOUS-RUMPED GRASS-BABBLER

Graminicola bengalensis Plate **111**

IDENTIFICATION 16–18 cm. *G.b.striata*: **Adult** Resembles Rusty-rumped Warbler but larger, tail blackish with broad white crescent-shaped tips (prominent from below), undertail-coverts much shorter; has broader blackish mantle-streaking, contrasting with plain rufous rump and uppertail-coverts, distinctive white streaks on neck/nape-side and relatively shorter, thicker bill. See Striated Grassbird and much smaller cisticolas. **Juvenile** Has warmer, more rufous-buff streaks on crown and mantle, duller and browner dark streaking on upperparts (less sharply contrasting), paler rufous wing-feather fringes. **VOICE** Probable song is a subdued high ***er-wi-wi-wi*** or ***bzz-wi-wi-wi you-wuoo yu-wuoo***, followed by a series of usual call-notes or ***er-wit-wit-wit*** and ending with curious subdued strained wheezy sounds. Usual call-notes are a subdued harsh scolding ***err-err-err-errrr*** and ***jjjerrreah*** etc. (sometimes rising at end), reminiscent of distant Eurasian Jay. Also ***wirrruu wirrruu***, interspersed with wheezy notes and (probable) song phrases. **HABITAT & BEHAVIOUR** Tall emergent vegetation in or bordering freshwater marshes and swamps or along banks of rivers; lowlands. Skulking, typically perches on grass- or reed-stems in upright posture. **RANGE & STATUS** Resident N,NE Indian subcontinent, S China. **SE Asia** Former resident (current status unknown) Tenasserim, E Tonkin. Former resident (probably extinct) C Thailand (last recorded 1923). **BREEDING** June–August. **Nest** Deep cup, low down in grass or small bush. **Eggs** 4; white, cream or pinkish-white, boldly spotted or freckled reddish-brown, with a few pale purple markings (usually ringed at broader end); 18 x 15 mm (av.). **NOTE** Genetic studies have demonstrated that this 'warbler', as it was formerly treated, is actually a babbler (Alström *et al.* 2006).

1202 BUFF-BREASTED BABBLER
Pellorneum tickelli Plate **113**

IDENTIFICATION 13.5–15.5 cm. *P.t.fulvum*: **Adult** Very nondescript. Similar to Abbott's but longer-tailed and thinner-billed, upperparts paler and more olive-brown, head-sides tinged buffish, breast extensively buff with faint darker streaks, buff of flanks and vent paler (lacking rufous), legs and feet pinkish. *P.t.assamense* (N Myanmar) has distinctly darker, more rufescent upperside and shows pale shaft-streaks on crown; *annamense* (S Laos, southern Vietnam) and *tickelli* (Tenasserim and W Thailand south to Peninsular Malaysia) are also more rufescent but not much darker. See Ferruginous, Horsfield's and Spot-throated Babblers. **Juvenile** Upperparts strongly rufescent. **Other subspecies in SE Asia** *P.t.grisescens* (SW,S[west] Myanmar). **VOICE** Song is a loud, sharp, quickly repeated ***wi-twee*** or ***wi-choo***, sometimes incessantly repeated, without pause: ***witweewitweewitwee...*** or ***wituitwituitwituit...*** etc. Also a high-pitched jolly laughing ***swi-tit-tit-titchoo*** and variants. Calls include harsh rattling ***prrree*** or ***trrrit*** notes, interspersed with higher ***pieu*** or explosive ***whit*** or ***twit*** notes. **HABITAT** Broadleaved evergreen forest, secondary growth, bamboo, sometimes mixed deciduous forest; up to 1,550 m. **RANGE & STATUS** Resident NE India, E Bangladesh, SW China, Sumatra (Belitung I). **SE Asia** Common resident (except W,C Myanmar, C,SE Thailand, Singapore). **BREEDING** February–July. **Nest** Dome, semi-dome or deep cup, on ground or amongst low vegetation up to 30 cm, sometimes 1.5 m above ground. **Eggs** 3–4; whitish to pale olive-grey, densely freckled and spotted reddish-brown to olive-brown, sometimes with purplish undermarkings; 19.9 x 15.7 mm (av.; *assamense*).

1203 BLACK-CAPPED BABBLER
Pellorneum capistratum Plate **113**

IDENTIFICATION 17–18.5 cm. *P.c.nigrocapitatum*: **Adult** Unmistakable, with warm dark brown upperparts, uniform deep rufous underparts, black crown, nape and moustachial stripe, pale greyish-white supercilium and white throat. **Juvenile** Like adult but only crown black and duller, upperparts more rufescent, throat rufescent, lacks black moustachial stripe. **VOICE** Sings with loud, high-pitched ***teeu*** (first syllable emphasised); repeated every few seconds. Calls include a subdued ***hekhekhekhek...*** and ***yeryeryer...*** and high nasal ***nwit-nwit-nwit...*** **HABITAT & BEHAVIOUR** Broadleaved evergreen forest; up to 760 m. Usually on or close to ground. **RANGE & STATUS** Resident Greater Sundas. **SE Asia** Common resident south Tenasserim, S Thailand, Peninsular Malaysia. **BREEDING** November–September. **Nest** Rather untidy cup, on ground or in sapling, palm etc., up to 60 cm. **Eggs** 2; whitish to creamy-white, freckled purplish-brown to maroon, over lighter purple to maroon undermarkings; 21.5–23 x 14.9–16.5 mm.

1204 PUFF-THROATED BABBLER
Pellorneum ruficeps Plate **113**

IDENTIFICATION 16–18 cm. *P.r.chthonium* (NW Thailand [except Chiang Rai]): **Adult** Easily identified by combination of rufescent crown, buffy-whitish supercilium and whitish underparts with dark streaking on breast and flanks. Dark streaks vague or lacking on centre of upper mantle. Shows marked and extensive subspecific variation. *P.r.minus*, *subochraceum*, *insularum* (Myanmar south of 20°N), *acrum* (W Thailand to Peninsular Malaysia), *euroum*, *smithi*, *ubonense*, *deignani* and *dilloni* (NE,SE Thailand, S Indochina; south of c.17°N) all lack dark streaks on upper mantle; *stageri* (southern N Myanmar) shows the most prominent streaking on upper mantle and behind ear-coverts. Intervening populations (*chthonium* and those listed below) are generally intermediate. Southern forms in Myanmar and W Thailand southwards also show buffier underparts with narrow breast-streaking. **Juvenile** Upperparts (apart from crown) more rufescent, underpart-streaking narrower to very indistinct. **Other subspecies in SE Asia** *P.r.shanense* (E Myanmar), *hilarum* (C Myanmar), *victoriae* (W Myanmar [Chin Hills]), *indistinctum* (NW Thailand [Chiang Rai]), *dusiti* (west NE Thailand [west slope Dong Phraya Fai range]), *elbeli* (elsewhere north-west NE Thailand), *oreum* (N Laos, W Tonkin), *vividum* (E Tonkin, N,C Annam). **VOICE** Song is a repeated loud, shrill, quite high-pitched ***wi-chu*** or ***wi-ti-chu***, sometimes with a more stressed end-note. Also a jolly, rapid, descending sequence: ***tuituitititi-twititi-tititi...*** and variants. Calls with subdued, nasal ***chi*** and ***erh*** notes and rasping ***rrrrrit***. **HABITAT & BEHAVIOUR** Broadleaved evergreen and mixed deciduous forests, secondary growth, scrub, bamboo; up to 1,800 m. Usually on ground, or close to it; hops and walks. **RANGE & STATUS** Resident N,NE Indian subcontinent, C,E and southern India, SW China. **SE Asia** Common resident (except south Peninsular Malaysia, Singapore). **BREEDING** January–August. **Nest** Dome, semi-dome or cup, on ground in sheltered situation. **Eggs** 2–5; whitish (sometimes tinged buffish to greenish), speckled reddish-brown or brown to purplish-brown or grey (usually more at broader end); 20–21.5 x 14.7–16 mm (*acrum*).

1205 SPOT-THROATED BABBLER
Pellorneum albiventre Plate **113**

IDENTIFICATION 13–14.5 cm. *P.a.cinnamomeum*: **Adult** The only small brown babbler with dark-spotted whitish throat (spots chevron-shaped). Resembles Buff-breasted Babbler but smaller, with shorter, more rounded tail, shorter stouter bill and greyish head-sides. Rest of underparts rufous-buff with a little white on centre of belly. *P.a.albiventre* (W Myanmar) shows broadly white centre to abdomen, greyish-brown breast and much duller flanks and undertail-coverts; *pusillum* (W,E Tonkin) has much darker rufous underparts, particularly breast and flanks. **Juvenile** Shows more rufescent outer fringes to primaries and secondaries. **VOICE** Has surprisingly rich and thrush- or chat-like song. Complex and quickly delivered but with much repetition. Females may give antiphonal fast, high ***tchu-tchu-tchu-tchu...*** Calls include a harsh ***chrrr*** and slightly explosive ***tip*** or ***tchip*** notes. **HABITAT & BEHAVIOUR** Open broadleaved evergreen forest, forest edge, overgrown clearings, secondary growth, scrub and grass, bamboo; 500–2,135 m, locally down to 280 m. Very skulking, in low undergrowth. **RANGE & STATUS** Resident NE Indian subcontinent, SW China. **SE Asia** Fairly common to common resident W,N,E,S(east) Myanmar, NW Thailand, Laos, W,E Tonkin, C,S Annam. **BREEDING** April–July. **Nest** Fairly compact dome or semi-dome, in bush, weeds or bamboo clump etc.; 0.3–1.2 m above ground. **Eggs** 2–5; fairly glossy reddish-white to cream-coloured, profusely covered with reddish speckles (often ringed at broader end); 20 x 15.1 mm (av.; *albiventre*).

1206 **MOUSTACHED BABBLER**

Malacopteron magnirostre Plate **113**

IDENTIFICATION 16.5–18 cm. *M.m.magnirostre*: **Adult** Told from other *Malacopteron* babblers by rather uniform olive-brown crown and mantle, greyish head-sides and distinctive dark moustachial stripe. Tail strongly rufescent, underparts whitish with light greyish wash (or vague streaking) across breast. Eyes reddish-brown to brown. See Brown Fulvetta and Short-tailed Babbler. **Juvenile** Moustachial stripe may be less distinct, lower mandible flesh-coloured to yellowish (sometimes with darker tip), eyes greyish or brownish. **VOICE** Song is a series of usually 3–6 well-spaced, clear, sweet whistled notes; ***tii-tu-ti-tu*** or ***ti-tiee-ti-ti-tu*** etc., usually lasting 2–3 s and sometimes descending somewhat towards end. Female may give antiphonal harsh ***tchew*** or ***tchip*** notes. Call is a repeated soft but quite explosive ***whit***, interspersed with buzzing ***bzzii*** notes. **HABITAT & BEHAVIOUR** Broadleaved evergreen forest; up to 900 m. Usually found in pairs or small parties in middle storey. **RANGE & STATUS** Resident Sumatra, Borneo. **SE Asia** Common resident south Tenasserim, S Thailand, Peninsular Malaysia, Singapore (rare). **BREEDING** February–August. **Nest** Open cup, in sapling; 1 m above ground. **Eggs** 2.

1207 **SOOTY-CAPPED BABBLER**

Malacopteron affine Plate **113**

IDENTIFICATION 15–16.5 cm. *M.a.affine*: **Adult** Similar to Moustached Babbler but lacks dark moustachial stripe and shows diagnostic sooty-blackish crown (contrasting with mantle); smaller and more slender-billed. **Juvenile** Crown paler, sometimes barely contrasting with mantle, shows rufescent-tinged wing-feather fringes; lower mandible dull flesh-coloured. **VOICE** Song is a series of usually 6–9 clear, slow, airy, rising and falling whistles, in variable combination; ***phu-phi-phu-phoo-phu-phi-phu*** etc. (usually lasting 4–7 s). Also gives a variable well-phrased subsong: ***whi-whi-whui***, faster ***chut-whi-whi-whi-whu-whi-whu*** etc., sometimes introduced by short scratchy or jumbled notes. Calls include a short, sharp ***which-it*** or ***pit-pwit*** (sometimes used by female to accompany primary song of male) and harsh scolding rattles when alarmed. **HABITAT** Broadleaved evergreen forest, often near water, forest edge, freshwater swamp forest; up to 455 m. **RANGE & STATUS** Resident Sumatra, Borneo. **SE Asia** Rare to locally common resident S Thailand, Peninsular Malaysia. **BREEDING** April–June. **Nest** Loose shallow cup, in tree or creeper etc.; 0.9–7 m above ground. **Eggs** 2; glossy pink, stippled, blotched and smeared reddish-brown, sometimes over lilac undermarkings (often ringed or capped with reddish-brown at broader end); 21.8–22.6 x 15.2–15.9 mm.

1208 **SCALY-CROWNED BABBLER**

Malacopteron cinereum Plate **113**

IDENTIFICATION 14–17 cm. *M.c.cinereum*: **Adult** Shares distinctive rufous forehead and crown and blackish nape with Rufous-crowned Babbler, but differs by smaller size, slenderer bill, lack of greyish streaks on breast, and pinkish legs and feet. Shows brownish wash across breast and flanks and distinctive dark tips to forehead and crown-feathers (rarely visible in field). *M.c.indochinense* (SE,NE Thailand, Indochina) lacks blackish nape, is slightly paler above and slightly buffier-tinged below (except centre of throat and abdomen). **Juvenile** Similar to adult. **VOICE** Song variable, with four main parts which can be combined in various ways. (1) Rapidly repeated, hard, stressed ***dit-dit-dit-dit-dit-dit...*** and ***du-du-dit-dit-dit...*** etc. (2) Rapid, usually gradually descending ***du-du-du-du-du-du...*** (3) More spaced, gradually ascending ***phu-phu-phu-phu*** and ***phu-pu-pi-pee*** etc. (4) Rapid, high-pitched, even ***wiwiwiwiwi-wi-wi-wi-wu*** and ***wi-wi-dudududududu*** etc., slower at end or beginning. When in flocks, several birds may sing simultaneously. Calls with a short subdued ***chit-chit***; ***chreu-chreu...*** or combinations of loud sharp shrill ***chit***; ***whit***; ***tcheu*** and ***titu*** notes. Calls may be interspersed with song-types 3 & 4 (probably by females). **HABITAT & BEHAVIOUR** Broadleaved evergreen forest; up to 800 m. Often found in small parties, in lower to middle storey. **RANGE & STATUS** Resident Greater Sundas. **SE Asia** Common resident NE,SE,S Thailand, Peninsular Malaysia, Cambodia, C,S Laos, south E Tonkin, N,C,S Annam, Cochinchina. **BREEDING** February–October. **Nest** Rather flimsy cup, in sapling or bush; from near ground up to 1.2 m. **Eggs** 2; fairly glossy whitish to creamy-brown, densely spotted rich reddish-brown over pale grey undermarkings; 22.1–23.1 x 15.7–16.7 mm (*cinereum*).

1209 **RUFOUS-CROWNED BABBLER**

Malacopteron magnum Plate **113**

IDENTIFICATION 17.5–19.5 cm. *M.m.magnum*: **Adult** Similar to Scaly-crowned Babbler but larger and bigger-billed, with broad greyish streaks across breast, greyish legs and feet and overall whiter underparts; lacks dark tips to rufous forehead and crown-feathers. Like Moustached Babbler when seen from below. **Juvenile** Similar to adult. **VOICE** Song recalls Scaly-crowned Babbler, consisting of three main parts: (1) series of clear well-spaced notes, louder and more spaced than Scaly-crowned and usually not descending: ***phu-phu-phi-phi*** etc.; (2) series of well-spaced even notes, sometimes slightly descending or hurried towards end: ***chuwee-chuwee-chuwee-chuwu-chuwu***; ***chu-chi-chi-chi-chi-chu-chu-chu-chu-chu*** etc.; (3) well-spaced, loud, very even-pitched ***chut-chut chut-chut-chut-chut-chut-chut*** etc. When singing, males may give type 1 followed, after an interval, by type 2 etc. Series of ***chut*** notes may be given in duet with song-types 1 and 2 (probably by females). **HABITAT** Broadleaved evergreen forest; up to 455 m. **RANGE & STATUS** Resident Sumatra, Borneo. **SE Asia** Uncommon to fairly common resident south Tenasserim, S Thailand, Peninsular Malaysia. **BREEDING** March–July. **Nest** Cup, in sapling or bush; up to 0.6–1 m above ground. **Eggs** 2; pale blue, spotted and blotched rich- to dark red.

1210 **GREY-BREASTED BABBLER**

Ophrydornis albogularis Plate **113**

IDENTIFICATION 13.5–15 cm. *O.a.albogularis*: **Adult** Small and relatively short-tailed with diagnostic slaty-grey head, contrasting white supercilium and throat and grey breast-band. Tail not strongly rufescent. See Sooty-capped Babbler. **Juvenile** Apparently undescribed. **VOICE** Song is a rather long, subdued, discordant series of ascending ***whu-whi***; ***whit-whu*** and ***uu-whi-u*** phrases (and variants), sometimes interspersed with short ***chit*** notes. **HABITAT & BEHAVIOUR** Freshwater swamp forest, broadleaved evergreen forest; lowlands. Normally in lower storey, not gregarious. **RANGE & STATUS** Resident Sumatra, Borneo. **SE Asia** Scarce and local resident Peninsular Malaysia. **BREEDING** Undocumented.

1211 **ABBOTT'S BABBLER**

Malacocincla abbotti Plate **113**

IDENTIFICATION 15–16.5 cm. *M.a.abbotti*: **Adult** Nondescript, stocky, short-tailed and rather large-billed with bright rufous-buff lower flanks and vent. Very similar to Horsfield's Babbler but crown and upperparts concolorous and paler brown, has pale shaft-streaks on crown, slightly less pronounced greyish supercilium and lacks streaking on upper breast; bill a little longer. Shorter-tailed and bigger-billed than Buff-breasted Babbler, lacks strong buffy wash across breast. See White-chested and Short-tailed Babblers. **Juvenile** Crown and upperparts dark rufescent-brown (similar to adult Ferruginous Babbler). **Other subspecies in SE Asia** *M.a.altera* (N[south],C Laos, C Annam), *williamsoni* (NE[south-west] Thailand, north-west Cambodia), *obscurior* (coastal SE Thailand), *olivacea* (extreme S Thailand southwards). **VOICE** Song consists of repeated variable loud jolly phrases: ***chiu-woo-wooi***; ***wiu-wuoo-wiu*** and ***wi-wu-yu-wi*** etc. Usually three well-spaced notes, sometimes up to five. Calls with short, harsh, explosive ***cheu*** notes, interspersed with high nervous ***wer*** notes. **HABITAT & BEHAVIOUR** Broadleaved evergreen forest, forest edge, secondary growth, scrub; up to 1,100 m. Usually close to ground. **RANGE & STATUS** Resident Nepal, NE Indian subcontinent, Sumatra, Borneo. **SE Asia** Fairly common to common resident SW,S Myanmar, Tenasserim, W,NE,SE,S Thailand, Peninsular Malaysia, Singapore, Cambodia, Laos, N,C Annam, Cochinchina. **BREEDING** January–September. **Nest** Fairly neat cup, in rattan, palm or other low undergrowth; 0.45–1.8 m above ground. **Eggs** 2–5; glossy whitish to salmon-pink, sparsely spotted and squiggled red to brown over purple to grey underspots and lines; 21.8 x 16.2 mm (av.; *abbotti*). Brood-parasitised by Moustached Hawk-cuckoo.

1212 **HORSFIELD'S BABBLER**
Malacocincla sepiaria Plate **113**

IDENTIFICATION 14–15.5 cm. *M.s.tardinata*: **Adult** Like Abbott's Babbler but darker above, particularly on crown, which is darker than mantle and lacks paler shaft-streaks; shows vague broad, dull greyish streaks on lower throat/upper breast, more contrasting grey supercilium over eye, slightly shorter, stubbier bill and slightly shorter tail. See White-chested, Buff-breasted and Short-tailed Babblers. **Juvenile** More rufescent above. **VOICE** Usual song is a strident, clearly spaced ***wi-choteuu***, with first note high and sharp, second short, and third high and shrill. Variations occur, particularly when excited, and first note sometimes omitted. Call is a harsh explosive ***whit-whit-whit...***, interspersed with quieter ***wer*** notes. **HABITAT & BEHAVIOUR** Broadleaved evergreen forest, often near water; up to 700 m. Usually fairly close to ground. **RANGE & STATUS** Greater Sundas. **SE Asia** Rare to fairly common resident S Thailand, Peninsular Malaysia. **BREEDING** January–August. **Nest** Broad, rather flimsy cup, situated in low vegetation or sapling; 0.4–1 m above ground. **Eggs** 2; clay-pink, shading to dusky pink over broader end, with long squiggles and a few flecks of rufous-chestnut (rufous and black in some), mostly over broader end; 21–22.5 x 16–17 mm.

1213 **SHORT-TAILED BABBLER**
Malacocincla malaccensis Plate **113**

IDENTIFICATION 13.5–15.5 cm. *M.m.malaccensis*: **Adult** Told from similar Abbott's and Horsfield's Babblers by smaller size, very short tail and diagnostic blackish moustachial line, contrasting with grey ear-coverts and white throat; legs and feet pale pinkish, bill much thinner. **Juvenile** Crown somewhat paler, head-sides brown-tinged, shows rusty outer fringes to primaries. **VOICE** Song is a series of 6–7 loud rich whistled notes, descending in pitch, introduced by a dry trill: ***pi'pi'pi'pi'pi pew pew pew pew pew pew***. Calls with low, harsh, crackling, rattling sounds and a harsh, mechanical ***chutututututut...***, interspersed with soft ***yer*** notes etc. **HABITAT & BEHAVIOUR** Broadleaved evergreen forest, secondary forest; up to 915 m. Usually close to ground. **RANGE & STATUS** Resident Sumatra, Borneo. **SE Asia** Fairly common to common resident S Thailand, Peninsular Malaysia, Singapore. **BREEDING** December–September. **Nest** Cup (sometimes semi-roofed or protected by large leaf), on ground or slightly above it amongst fallen twigs etc. **Eggs** 2; fairly glossy white with rich chestnut dots and flecks, the markings slightly denser at broader end, where mixed with small dull purple-brown flecks.

1214 **WHITE-CHESTED BABBLER**
Trichastoma rostratum Plate **113**

IDENTIFICATION 15–16.5 cm. *T.r.rostratum*: **Adult** Similar to Abbott's and Horsfield's Babblers but has clean white underparts and longish, rather slender bill. Upperparts fairly cold dark brown, breast-sides lightly washed grey. **Juvenile** Similar to adult. **VOICE** Song is a repeated, quite high-pitched, clear ***wi-ti-tiu***; ***chui-chwi-chew*** or ***chwi-chi-cheei*** etc., sometimes introduced with a short trill: ***chr chr ooi-iwee*** etc. Calls with harsh, scolding rattles. **HABITAT & BEHAVIOUR** Riversides and streams in broadleaved evergreen forest, secondary forest, freshwater swamp forest, mangroves; up to 200 m. Usually on or close to ground. **RANGE & STATUS** Resident Sumatra, Borneo. **SE Asia** Uncommon to common resident south Tenasserim, S Thailand, Peninsular Malaysia, Singapore (now rare). **BREEDING** December–July. **Nest** Loose deep cup, in palm or small rattan; 0.4–1.25 m above ground. **Eggs** 2; pale clay-green, blotched and speckled pale lavender-grey, mostly over the broader and, and more finely and generally speckled light chestnut (but mostly towards broader end); 20.5–21 x 15–15.4 mm.

1215 **FERRUGINOUS BABBLER**
Trichastoma bicolor Plate **113**

IDENTIFICATION 16–18.5 cm. Monotypic. **Adult** Combination of bright rufescent upperparts and rather clean creamy- or buffy-whitish underparts diagnostic. Relatively large and long-tailed. See juvenile Abbott's Babbler. **Juvenile** Brighter and more orange-rufous above. **VOICE** Sings with a repeated loud, clear, rather sharp ***u-wit*** or ***u-wee*** (second note higher). Also variable low jolly phrases: ***wit wi-ti-tu-tu*** etc. Calls are low, harsh, dry, rasping sounds and sharp, explosive ***wit*** notes etc. **HABITAT & BEHAVIOUR** Broadleaved evergreen forest, secondary forest; up to 200 m. Usually fairly close to ground. **RANGE & STATUS** Sumatra, Borneo. **SE Asia** Uncommon to fairly common resident south Tenasserim, S Thailand, Peninsular Malaysia. **BREEDING** February–July. **Nest** Open cup, in sapling, rattan or other low vegetation; 0.5–1 m above ground. **Eggs** 2; fairly glossy pale buff, irregularly blotched reddish-brown (sometimes some squiggles) over grey to pinkish-grey undermarkings (markings often mostly towards broader end); 22 x 16.3 mm.

1216 **STRIPED WREN-BABBLER**
Kenopia striata Plate **117**

IDENTIFICATION 15 cm. Monotypic. **Adult** Striking, with bold black-and-white streaking on crown, nape and breast-sides, whitish head-sides, throat and underparts with warm buff lores and lower flanks, and white streaking on mantle, scapulars, wing-coverts and lower flanks. **Juvenile** Crown and dark scaling on breast-sides browner, upperparts and wings a shade paler and warmer, with buffier streaks; has dark mottling across breast, and paler bill. **VOICE** Song is a short clear whistled ***chuuii***, repeated every 1.5–2 s; sometimes ***chiuuu*** or ***chi-uuu***, with very short space between notes. When agitated, may intersperse song with short twangy nasal notes. **HABITAT & BEHAVIOUR** Broadleaved evergreen forest; up to 750 m. Usually close to ground. **RANGE & STATUS** Resident Sumatra, Borneo. **SE Asia** Uncommon resident S Thailand, Peninsular Malaysia; formerly Singapore. **BREEDING** May–October. **Nest** Loose shallow cup, in palm; 50 cm above ground. **Eggs** 2; whitish, irregularly blotched reddish-brown overall.

1217 **MARBLED WREN-BABBLER**
Turdinus marmoratus Plate **112**

IDENTIFICATION 21.5 cm. *T.m.grandior*: **Adult** Larger than other wren-babblers, with relatively long tail and diagnostic rufous ear-coverts and blackish underparts with white bars/scales. Throat white, barred black at sides, upperparts deep warm brown, scaled blackish. **Juvenile** Shows rufous shaft-streaks on head and upperparts. **VOICE** Sings with clear double or single whistle, similar to Large Wren-babbler: ***puuu-chiiii***; ***pyuuu-jhiiii*** or ***puuui-jhiiii***, with higher, somewhat rising first note and lower, buzzier second note. Also ***piuuu-whiiii*** and ***uuuui-jhiii*** or single ***piuuu***. **HABITAT & BEHAVIOUR** Broadleaved evergreen forest; 610–1,220 m. Very shy. **RANGE & STATUS** Resident Sumatra. SE Asia Rare and local resident Peninsular Malaysia. **BREEDING** Undocumented.

1218 **LARGE WREN-BABBLER**
Turdinus macrodactylus Plate **112**

IDENTIFICATION 19–20.5 cm. *T.m.macrodactylus*: **Adult** Larger than other wren-babblers (except Marbled) with distinctive black lores and ear-coverts, white loral supercilium and throat, and diagnostic light blue orbital skin. Rear ear-coverts streaked whitish, underparts pale with indistinct dark scaling and buffish flanks and vent. **Juvenile** Upperparts and wings paler and more rufous, darker on crown and ear-coverts, lacks pronounced dark scaling on upperparts but shows pale buff shaft-streaks; underparts rather uniform, slightly rufescent, pale drab brown, with whiter throat and belly-centre and indistinct darker breast markings. **VOICE** Song is very variable. Usually consists of short loud clear whistled phrases, repeated every few seconds: ***chuu-chreeh*** and ***chu-chiii*** etc., or descending then rising ***phuu-wiii*** or ***u-wiii***; also, single ***chuuu***, sometimes combined with coarser notes: ***uuu-choriii***; ***chuuu weearh-weearh*** etc. May give longer series of rather slow notes: ***wii-tu-tu-tu-tu-tu*** (first note higher) and ***pu-chuu-chuu-chuu-chuu-chuu*** etc. During duets, slow ***pu-yu-yu...*** is answered by ***chuuu*** or ***chuuu-chii***. Longer song variants include ***uuurr-wi-wi-wi-wi-wrriiu*** (mid-section faster) and ***pi-pi-pi-pi-pi-pi-peeoo***, with slower last note. **HABITAT & BEHAVIOUR** Broadleaved evergreen forest; up to 700 m (mostly below 200 m). Usually on or near ground. **RANGE & STATUS** Resident Java, Sumatra. **SE Asia** Uncommon to fairly common resident S Thailand, Peninsular Malaysia; formerly Singapore. **BREEDING** December–September. **Nest** Large cup, in palm or rattan; 0.4–1.3 m above ground. **Eggs** 2; pale pinkish-white to pinkish-red, irregularly marked with reddish blotches, lines or squiggles; 23.4–24.6 x 17.4–17.9 mm.

1219 LIMESTONE WREN-BABBLER

Gypsophila crispifrons Plate **112**

IDENTIFICATION 18–20.5 cm. *G.c.crispifrons*: **Adult** Similar to Streaked Wren-babbler but considerably larger and longer-tailed, lacks whitish spots on tips of greater coverts, shows bolder dark throat-streaking and much colder, darker greyish-brown underparts, streaked white on centre of belly. Has rufous-buff shaft-streaks on upperparts. **Adult white-throated morph** In Tenasserim, may show white supercilium, head-sides and throat and occasionally patches on nape and upperparts. Intermediates with typical birds also occur. *G.c.calcicola* (NE Thailand [Saraburi Province]) is rufescent-brown below (more similar to Streaked); *annamensis* (Indochina) has slightly paler, greyer crown and mantle with whiter shaft-streaks and greyer lower breast and belly. **Juvenile** Apparently undescribed. **VOICE** Song is a loud rapid faltering series of unevenly pitched harsh slurred notes, starting very abruptly: roughly ***chitu-wi-witchuwitchi-witchiwitchiwitchuwitchiu*** or slower ***titu-titu-titu-titu-tit...*** etc. Bouts last 4–30 s and are usually repeated after long intervals. Alarm calls are harsh, scolding rattles: ***chrrr-chrrr-chrrr...*** and ***chrrrow-chrrrow...*** etc. **HABITAT & BEHAVIOUR** Forest on limestone, vicinity of limestone rocks and outcrops; up to 915 m. Usually in small parties, foraging around rocks and tangled vegetation. **RANGE & STATUS** Endemic. Locally common resident south-east S Myanmar, Tenasserim, W,NW,NE Thailand, N Laos, W,E Tonkin, N Annam. **BREEDING** April–May. **Nest** Dome with large entrance, amongst rocks. **Eggs** 5; white to pinkish-white, sparsely speckled and blotched reddish-brown, dark brown, blackish and mauve (mostly at broader end).

1220 STREAKED WREN-BABBLER

Napothera brevicaudata Plate **112**

IDENTIFICATION 14–14.5 cm. *N.b.brevicaudata*: **Adult** Fairly small and short-tailed with blackish-scaled brown crown and upperparts, greyish head-sides, rufescent-brown underparts and distinctive small whitish spots on tips of tertials, secondaries and greater coverts. Centre of throat and upper breast whitish with dark brown streaks. *N.b.striata* (W,N[west],S[west] Myanmar) and *griseigularis* (SE Thailand, south-west Cambodia) show duller, paler-centred belly, the latter with pale grey base colour to throat and upper breast; *stevensi* (northern Indochina) and *proxima* (S Laos, C Annam) are bigger (particularly former), with colder, darker olive-tinged upperparts, colder, darker brown underparts, and less white on throat and upper breast; *leucosticta* (S Thailand southwards) is like *proxima* but shows whitish throat and upper breast, with very broad sooty-brown streaks. See Limestone and Eye-browed Wren-babblers. Note range. **Juvenile** Rather uniform dark brown, with paler chin and throat-centre, slightly paler and warmer underparts and lower upperparts, small dull wing-spots; shows pale shaft-streaks on crown to upper back. **Other subspecies in SE Asia** *N.b.rufiventer* (S Annam). **VOICE** Song consists of very variable, clear, ringing whistles, repeated after intervals: ***chi-oo***; ***peee-oo***; ***pu-ee***; ***chiu-ree; chewee-chui*** and ***pee-wi*** etc.; sometimes single ***pweeee*** etc. When alarmed, utters harsh prolonged scolding rattles: ***trrreeettt*** and ***chrrreerrrrt*** etc., often interspersed with ***wher*** notes. **HABITAT** Broadleaved evergreen forest, forest on limestone, vicinity of limestone rocks and boulders; up to 1,620 m (above 760 m in Peninsular Malaysia [except extreme NW]). **RANGE & STATUS** Resident NE India, SW China. **SE Asia** Common resident (except SW,C Myanmar, C Thailand, Singapore, Cochinchina). **BREEDING** December–July. **Nest** Dome, semi-dome or deep cup, on ground or in hole in bank or rock etc. **Eggs** 2–4; glossy white, speckled and freckled reddish to purplish-brown (often mainly at broader end); 21.6 x 15.7 mm (av.; *leucosticta*).

1221 EYEBROWED WREN-BABBLER

Napothera epilepidota Plate **112**

IDENTIFICATION 10–11.5 cm. *N.e.davisoni*: **Adult** Resembles Streaked Wren-babbler but smaller and shorter-tailed, with diagnostic long, pale buff supercilium, broad dark eyestripe and large whitish spots on tips of greater and median coverts. Dark brown above, scaled blackish, with pale shaft-streaks on mantle; head-sides buffish, throat and belly-centre whiter, breast and flanks dark brown, streaked pale buff. *N.e.roberti* (N Myanmar) has whiter throat with blackish streaks and duller upperparts without shaft-streaks; *laotiana* (N[eastern],C,S Laos) and *amyae* (W,E Tonkin, N,C Annam) are colder brown above, have blacker eyestripe, whiter throat and supercilium, and darker breast and flanks with narrower pale streaks (latter also larger); *clara* (S Annam) has whiter throat, supercilium and underpart-streaking (latter also broader) and slightly colder brown upperparts; *granti* (S Thailand, Peninsular Malaysia) has paler centre to breast and abdomen. See Asian Stubtail. **Juvenile** Rather uniform warm dark brown above with indistinct paler shaft-streaks on nape; has uniform dark rufous underparts with slightly more whitish moustachial area and chin, indistinct rufous supercilium, dark lores but no eyestripe, buff spots on wing-coverts and tertials and pinkish base of lower mandible. **VOICE** Song is a thin falling clear whistle: ***cheeeoo***; ***cheeeeeu*** or ***piiiiiu***, repeated after intervals. When excited, may alternate song phrases with curious, slightly nasal ***chikachik-chikachik-chikachik...*** When alarmed, gives fairly subdued but prolonged rattles: ***prrrt-prrrt-prrrt***; ***wprrrt wprrrt wprrrt*** and ***chrrut-chrrut-chrrut*** etc. **HABITAT & BEHAVIOUR** Broadleaved evergreen forest; 280–2,135 m, locally down to 50 m in N Annam. On or close to ground. **RANGE & STATUS** Resident Bhutan, NE India, S China, Greater Sundas. **SE Asia** Fairly common to common resident N,C,S(east),E Myanmar, Tenasserim, Thailand (except C,SE), Peninsular Malaysia, Laos, Vietnam (except Cochinchina). **BREEDING** January–June. **Nest** Dome or semi-dome, sometimes deep cup, on ground or amongst boulders, rarely in rattan 30 cm above ground. **Eggs** 2–5 (2 in south); dull white to greenish-white, sparsely speckled reddish to dull brown; 19.3 x 14.8 mm (av.; *roberti*).

1222 LONG-BILLED WREN-BABBLER

Rimator malacoptilus Plate **117**

IDENTIFICATION 12 cm. Monotypic **Adult** Small size, heavily streaked plumage, long, narrow, slightly downcurved bill and very short tail diagnostic. Upperparts warm dark brown, streaked pale buffish, throat and underpart-streaking buffish; shows double dark moustachial/malar stripe. See Indochinese Wren-babbler, but note range. **Juvenile** Undescribed. **VOICE** Song is a short (0.4 s) clear whistle: ***chi-iuuh***, smoothly falling in pitch, but gaining volume. When excited, may intersperse song with 1–3 quickly repeated ***chip'wu*** or ***chitt'wu*** phrases. **HABITAT & BEHAVIOUR** Broadleaved evergreen forest, secondary growth, bamboo; 1,220–2,000 m. Skulking, usually close to ground. **RANGE**

& STATUS Resident NE Indian subcontinent, SW China (W Yunnan). **SE Asia** Scarce resident N Myanmar. **BREEDING** India: May–July. **Nest** Rather loose ball with entrance near top, on ground. **Eggs** 4; white, faintly tinged lilac to pink, finely blotched reddish-brown to purple-brown (with some short red-brown lines) over faint reddish-brown to lilac-grey undermarkings (mostly marked towards broader end); 21.2 x 15.5 mm (av.).

1223 WHITE-THROATED WREN-BABBLER
Rimator pasquieri Plate **117**

IDENTIFICATION 12 cm. Monotypic. **Adult** Like Long-billed Wren-babbler, but has brown of plumage darker and colder (including moustachial/malar stripes), throat white, body-streaking whiter. See Indochinese Wren-babbler, but note range. **Juvenile** Undescribed. **VOICE** Song is whistled ***chiiii'uh*** or ***tiiiii'uh***, very similar to Long-billed, but perhaps slightly longer, and repeated at shorter intervals. When excited, intermingles song phrases with shorter ***pi'wip***; ***pit'wip*** or ***pit'rip***. Calls with low ***prrp*** or ***prrt*** notes **HABITAT & BEHAVIOUR** Broadleaved evergreen forest, secondary growth, bamboo; 1,220–2,000 m. Skulking, usually close to ground. **RANGE & STATUS** Endemic. Scarce resident W Tonkin. **BREEDING** Undocumented. **NOTE** Split from Long-billed Wren-babbler following Collar & Robson (2007).

1224 INDOCHINESE WREN-BABBLER
Rimator danjoui Plate **117**

IDENTIFICATION 18–19.5 cm. *R.d.parvirostris*: **Adult** Combination of long, slightly downcurved bill and shortish tail distinctive. Upperparts dark brown with light shaft-streaks on mantle and scapulars; underparts whitish with double blackish moustachial/malar stripe and rufescent breast with smudgy blackish streaks; shows rufous neck-patch. *R.d.danjoui* (S Annam) has browner moustachial/malar stripes and breast-streaking, paler rufous on neck and breast and longer bill; unnamed subspecies (C Annam [near Ngoc Linh]; still not formally described) has brown of plumage colder and darker, rufous of plumage duller and paler, underparts and bill roughly intermediate between *danjoui* and nominate; unnamed subspecies (north-west E Tonkin) has darker moustachial/malar, bold black and white feathering below; *naungmungensis* (N Myanmar) has colder, darker upperside and flanks, blacker moustachial/malar, whiter breast-centre. See Large Scimitar Babbler and Long-billed Wren-babbler; note range. **Juvenile** Similar to adult. **VOICE** Sings with a series of short clear monotone high-pitched whistles (each 0.5–0.75 s long). When agitated, song often interspersed with a subdued, rather nasal, quick ***chuwut-chuwut-chuwut-chuwut...*** Alarm call is a harsh scolding ***chrrr-chrrr-chrrr...*** **HABITAT & BEHAVIOUR** Broadleaved evergreen forest, secondary forest, bamboo. 50–1,300 m; 1,500–2,100 m in C(southern),S Annam, 675–1,650 m in Laos. Skulking and usually close to ground. **RANGE & STATUS** Endemic. Uncommon resident N Myanmar (Naung Mung area, Kachin State), C Laos, E Tonkin, N,C,S Annam. **BREEDING** January–April. **Eggs** <4. Otherwise undocumented. **NOTE** Formerly called Short-tailed Scimitar Babbler *Jabouilleia danjoui*, but see Collar & Robson (2007).

1225 WHITE-HOODED BABBLER
Gampsorhynchus rufulus Plate **118**

IDENTIFICATION 22.5–24 cm. Monotypic. **Adult** Like Collared Babbler, but more olive-tinged above, has paler underparts with whiter breast and white neck/breast-sides; shows distinctive white shoulder-slash (formed by white median and some lesser coverts). **Juvenile** Crown, nape and head-sides rufous, throat buff-tinged, bill browner. Takes some time to lose rufous on head. See Greater Rufous-headed Parrotbill. **VOICE** Song presumably an infrequently heard, short, soft mellow series of low whistles: slightly rising ***fúit!*** and more level ***fwur***. Typical calls recall Collared, but more structured ***u'u'YER-yrrrt*** or ***uh-oh'jó'r'r'r'k*** (***jó*** greatly stressed and highest). Sometimes hollow, slightly accelerating laughter: ***khúrk khúrk khúrk-khurk-khurk-khurk***. Also brief, edgy, subdued, conversational twangy ***gyurt!*** notes. **HABITAT & BEHAVIOUR** Bamboo in or near broadleaved evergreen and semi-evergreen forest; 500–1,525 m. Always in flocks. **RANGE & STATUS** Resident E Nepal, NE Indian subcontinent, SW China (W Yunnan). **SE Asia** Locally common resident SW,W,N,C[west & north],S[west] Myanmar. **BREEDING** May–July. **Nest** Shallow cup, in bush; 2 m above ground. **Eggs** 3–4; pale yellowish or greenish, freckled reddish-brown to dark brown, mainly towards broader end (may have grey undermarkings), sometimes pale dull reddish with reddish-brown markings; 29.9 x 17.6 mm (av.).

1226 COLLARED BABBLER
Gampsorhynchus torquatus Plate **118**

IDENTIFICATION 22.5–24 cm. *G.r.torquatus*: **Adult** White hood, rufescent-brown upperside and warm buff rest of underparts distinctive. Has variable dark markings on neck/breast-sides and pale pinkish bill. See White-hooded Babbler, but note range. *G.r.saturatior* (extreme S Thailand, Peninsular Malaysia) and *luciae* (northern Indochina) are deeper and more extensively orange-buff below, former being colder above with more pronounced blackish collar (indistinct on nape, broken on centre of upper breast), latter having blackish hindcrown, rufescent-brown nape and broad blackish necklace extending from neck-sides across upper breast. **Juvenile** Apparently similar to adult. **VOICE** Harsh hard stuttering rattle or cackle: ***rrrrtchu-rrrrtchu-rrrrtchu***; ***rrrrut-rrrrut*** or ***rrrt-rrrt-rrrt*** etc. Also soft, very quiet ***wit***, ***wet*** and ***wyee*** notes when foraging. **HABITAT** Bamboo in or near broadleaved evergreen and semi-evergreen forest; 500–1,800 m, locally down to 50 m in Vietnam. Always in flocks. **RANGE & STATUS** Resident SW China (S Yunnan). **SE Asia** Locally common resident C(south-east),S(east),E Myanmar, Tenasserim, W,NW,NE, extreme S Thailand, Peninsular Malaysia, Laos, Vietnam. **BREEDING** June–July. **Nest** Situated high on bamboo culm. Otherwise undocumented. **NOTE** Split from White-hooded Babbler, following Rasmussen & Anderton (2005).

1227 YELLOW-THROATED FULVETTA
Pseudominla cinerea Plate **109**

IDENTIFICATION 9.5–10 cm. Monotypic. **Adult** Greyish-olive upperparts, broad orange-tinged yellow supercilium and bright yellow throat, breast and centre of abdomen distinctive. Has long blackish lateral crown-stripes, blackish crown-feather tips and greyish-olive breast-sides and flanks. **Juvenile** Undescribed, but presumably very similar to adult. **VOICE** Song-types include a repeated series of very thin high notes which start slowly and end in complex jumble: ***titz titz-tsi-si-si titituititu*** and ***titz titziziritititit*** etc.; also thin high ***si-si-si-si-si-si'si'si*** and variants, interspersed with sharp metallic ***titz*** notes. When on the move, utters subdued rapid low ***tit*** notes. **HABITAT & BEHAVIOUR** Broadleaved evergreen forest; 900–2,745 m. Often in small fast-moving flocks in low vegetation. **RANGE & STATUS** Resident Bhutan, NE India, SE Tibet, SW China. **SE Asia** Uncommon resident N Myanmar, N Laos, N Annam. **BREEDING** April–July. **Nest** Deep cup, dome or semi-dome, on ground or low down in bush or bamboo. **Eggs** 2–4; pale to warm buff, pale brown or whitish, stippled reddish-brown to dark brown (often ringed around broader end); 18.3 x 14.3 mm (av.).

1228 RUFOUS-WINGED FULVETTA
Pseudominla castaneceps Plate **109**

IDENTIFICATION 10.5–12 cm. *P.c.castaneceps*: **Adult** Told by dull chestnut crown with buff to buffish-white shaft-streaks, long whitish supercilium, blackish head-sides with whitish patches, blackish greater and primary coverts and light rufous and white flight-feather fringing. Rest of upperparts, breast-sides and flanks warm olive-brown, throat and centre of underparts whitish, eyes dark crimson. *P.c.exul* (NW[east],NE

Thailand, Laos, W Tonkin) has darker base colour of crown and darker rufous on flight-feathers; *soror* (extreme S Thailand, Peninsular Malaysia) has darker, more chestnut flight-feather fringes; *stepanyani* (east S Laos, C Annam) has more blackish-brown base colour to crown, slightly darker upperparts, buffier underparts and darker flight-feather fringing. **Juvenile** Slightly duller/paler crown than adult, slightly more olive-ochre below, generally across breast/belly. **VOICE** Song is a relatively rich but high-pitched, undulating and slightly descending 4–8 note ***si tju-tji-tju-tji-tju***. Usual calls include a mixture of subdued harsh ***tcht***, ***tchit*** and ***tchrr*** notes and thin ***tsi-tsi-tsi-trrt*** etc. **HABITAT & BEHAVIOUR** Broadleaved evergreen forest, secondary growth; 1,000–3,505 m, locally down to 760 m N Myanmar (winter only?). Usually in small hyperactive flocks in middle to lower storey; clings to mossy trunks. **RANGE & STATUS** Resident Nepal, NE Indian subcontinent, SE Tibet, SW China. **SE Asia** Common resident Myanmar (except C), W,NW,NE Thailand, Peninsular Malaysia, Laos, W,E Tonkin, N,C(southern) Annam. **BREEDING** January–June. **Nest** Dome or cup, amongst moss or creepers on tree-trunk, sometimes in bush, sapling or hole in bank; up to 3 m above ground. **Eggs** 3–4; white, stippled and/or spotted inky-black to grey, sometimes over pale purplish to grey or brown undermarkings (often ringed or capped at broader end); 17.7 x 13.4 mm (av.; *castaneceps*).

1229 **BLACK-CROWNED FULVETTA**
Pseudominla klossi Plate **109**
IDENTIFICATION 12–12.5 cm. Monotypic. **Adult** Recalls Rufous-winged Fulvetta, but larger, has distinctive blackish crown with broader whitish shaft-streaks, no obvious rufous on primaries, blackish-brown wing-coverts with dull rufescent fringes, somewhat less pronounced supercilium and grey-brown eyes. **Juvenile** Undocumented. **VOICE** Song is long series of short, thin, shrill, high-pitched notes, starting stressed, and trailing-off at end: ***tit it'is'is'is'is'is'su-sr'r'r'r'r*** or shorter ***tit tit'su'su'su'su***, and variants. Also, an agitated, husky, sawing, slightly grating ***hht't't't't'it'it*** mixed with hard ***tid*** and ***tid-rr*** notes. In alarm, ***ss'trrt...ss'trrt...*** **HABITAT & BEHAVIOUR** Broadleaved evergreen forest, secondary growth; 1,510–2,100 m. Behaves like Rufous-winged. **RANGE & STATUS** Endemic. **SE Asia** Uncommon to locally fairly common resident S Annam. **BREEDING** Undocumented. **NOTE** Split from Rufous-winged Fulvetta *P.castaneceps* following Collar & Robson (2007).

1230 **WHITE-THROATED BABBLER**
Turdoides gularis Plate **114**
IDENTIFICATION 25.5 cm. Monotypic. **Adult** Very distinctive. Relatively large and very long-tailed with black lores, pale eyes and white throat and upper breast, contrasting with rufous-buff remainder of underparts. Upperparts greyish olive-brown with dark streaks, more rufescent on crown and head-sides. **Juvenile** Shows slightly less distinct upperpart-streaking; head-sides, crown and upperparts washed buffish. **VOICE** Flocks utter sibilant low ***trrrr trrrr trrrr...*** and louder, constantly repeated ***chr'r'r'r'r'r'r*** or ***whir'r'r'r'r'r'r***. **HABITAT** Scrub and bushes in semi-desert, borders of cultivation; up to 600 m. **RANGE & STATUS** Endemic. Common resident C,S Myanmar. **BREEDING** March–December. **Nest** Fairly compact cup, in bush, hedge or tree-fork; up to 4.5 m above ground. **Eggs** 3–5; glossy pale to deep dark blue; 21.6–24 x 16.8–17.8 mm. Brood-parasitised by Pied Cuckoo.

1231 **STRIATED BABBLER**
Turdoides earlei Plate **114**
IDENTIFICATION 21–22 cm. *T.e.earlei*: **Adult** Told by size, long cross-barred tail, brown crown and upperparts with blackish streaks, and pale brownish underparts with warmer-tinged, thinly dark-streaked throat and breast. Shows short pale submoustachial stripe and largely pale bill with darker culmen and tip. See Striated Grassbird. **Juvenile** Shows less distinct upperpart-streaking. **VOICE** Song is a loud, repeated series of ***tiew-tiew-tiew-tiew*** calls, interspersed with ***quip-quip-quip*** calls from other flock members. **HABITAT & BEHAVIOUR** Grassland and scrub, usually near water; lowlands. Normally in small flocks. **RANGE & STATUS** Resident Pakistan, northern Indian subcontinent. **SE Asia** Uncommon resident SW,N,C,S Myanmar. **BREEDING** May–July. **Nest** Fairly neat cup, in grass, reeds, bush or sapling; 0.3–1.2 m above ground (sometimes to 3 m). **Eggs** 3–4; fairly glossy pale blue; 22.8 x 17.6 mm (av.). Brood-parasitised by Pied Cuckoo.

1232 **SLENDER-BILLED BABBLER**
Turdoides longirostris Plate **114**
IDENTIFICATION 20–21 cm. Monotypic. **Adult** Rather plain, with long, indistinctly cross-barred tail and slightly downcurved black bill. Superficially resembles smaller Yellow-eyed and Jerdon's Babblers but upperside uniformly darker brown, bill much longer, slenderer and downcurved, lores and narrow eyebrow dusky-whitish, eyes whitish to bluish-white; throat and upper breast whitish, grading to warm buff on rest of underparts. **Juvenile** Upperside paler and more rufescent, underparts more rufescent-buff, basal half of lower mandible pale. **VOICE** At least two song-types: (1) rather strident and high-pitched series of shrill notes with short introduction: ***yi chiwiyu chiwiyu'chiwiyu'chiwiyu'chiwiyu...***; (2) rather clear high-pitched ***wiii-wii-jiu-di***; ***wiii-wii-dju-di*** or ***wi-yu-ii*** etc. Also utters even, fairly strident 4–6 note ***chiu-chiu-chiu-chiu...*** (possibly a song-type) and discordant high-pitched ***tiu-tiu-tiu***; ***tit-tit*** and ***tiu-tiu-tit-tit-tu-tu...*** etc. **HABITAT & BEHAVIOUR** Tall grass; lowlands. Usually in small, very skulking groups. **RANGE & STATUS** Resident Nepal, NE India, Bangladesh. **SE Asia** Scarce and local resident SW Myanmar. **BREEDING** May–June. **Nest** Neat cup, in bush. **Eggs** 3–5; pale blue; 21.5 x 16.7 mm (av.).

1233 **HIMALAYAN CUTIA**
Cutia nipalensis Plate **117**
IDENTIFICATION 17–19.5 cm. *C.n.melanchima*: **Male** Resembles White-browed Shrike-babbler but shows diagnostic combination of bluish-slate crown, black head-sides, rufous-chestnut rest of upperparts and whitish underparts with bold black bars on breast-sides and flanks; wings black with tertials and flight-feathers fringed bluish-grey, flanks and vent washed buff. *C.n.cervinicrissa* (Peninsular Malaysia) has darker, more chestnut upperparts and more buff on flanks and vent. **Female** Like male but mantle, back and scapulars more olive-brown with broad blackish streaks, head-sides dark brown. **Juvenile** Duller than respective adults with browner crown and fainter dark bars on underparts. **Other subspecies in SE Asia** *C.n.nipalensis* (W,N(west) Myanmar). **VOICE** Song (*nipalensis*) a series of 6–10 loud, full, rather high-pitched notes, repeated every 4–8 s, sometimes interspersed with faster high ***jiw*** notes (by same bird?): ***yuip-yuip-yuip-yuip-yuip-yuip jiw-jiw-jiw-jiw yuip-yuip-yuip-yuip-yuip-yuip...*** etc. Also, a series of 5–19 rather hard, stressed, relatively high-pitched notes, repeated every 3–9 s (delivered fairly slowly to quickly), sometimes interspersed with harsh grating ***djrrri*** or ***jorrri*** (by same bird?): ***jorrri-jorrri-jorrri-ip-ip-ip-ip-ip-ip...*** etc. Song of *cervinicrissa* consists of 7–12 quickly repeated ***yip*** or ***wip*** notes, or more rapid ***chip'ip'ip'ip'ip...***; also loud, shrill, strident rising ***wee'yu-wee'yu-wee'yu-wee'yu*** and rapidly oscillating versions of short-note repetition with an introduction, e.g. ***wit'chu-wit'chu'it'it'it'it'it...*** Calls include a light ***chick chick chick...***, sharper and louder ***chit***, and harsh low ***jert jert...*** when foraging. Race *cervinicrissa* also utters continuous high, scolding ***yeet-yeet***; ***yeet-yit-yit*** and ***yeet-u yeet-u yeet-yeet yeet-u...*** when alarmed. **HABITAT & BEHAVIOUR** Broadleaved evergreen forest, sometimes mixed evergreen and pine forest; 1,200–2,500 m, locally down to 850 m N Myanmar. Usually in small parties, often associating with mixed-species feeding flocks; quite slow-moving and unobtrusive, often high up in trees. **RANGE & STATUS** Resident N,NE Indian subcontinent, SW China. **SE Asia** Scarce to fairly common resident

W,N,S(east),E Myanmar, NW Thailand, Peninsular Malaysia, N,C Laos, W Tonkin. **BREEDING** March–June. **Nest** Open cup, at junction of tree-trunk and branch; 3–3.6 m above ground. Otherwise undocumented.

1234 VIETNAMESE CUTIA

Cutia legalleni Plate **117**

IDENTIFICATION 17–19.5 cm. *C.l.legalleni*: **Male** Easily told from Himalayan Cutia by entirely black-barred breast to vent. *C.l.hoae* (C Annam; presumably also east S Laos) has much narrower underpart-barring, broken by white centre to abdomen. **Female** Like male but drab dark brown crown and ear-coverts, greyish-brown mantle, back and scapulars with blackish streaks. Female *C.l.hoae* currently undescribed. **Juvenile male** Undescribed. **Juvenile female** Warmer brown upperparts than adult female and fainter, more spot-like markings on underparts. **VOICE** Song is distinctive ***wuYEET wu wi wi wi wi woo*** (initial part of first phrase rising sharply), repeated every 3–6 s; and loud high-pitched strident ***WIII-chiWU-wipwi-weei-weei...*** or ***WIPWI-weEI-weei-weei*** (first note and second part of second note stressed), repeated at similar intervals and interspersed with a 3–4 note ***wii chiwi-chiwi...*** (by same bird?). Also a quite fast, loud, fairly high-pitched ***WEI-wuu-WEI-wuu*** (***WEI*** notes more stressed). Calls are similar to Himalayan. **HABITAT & BEHAVIOUR** Broadleaved evergreen forest, sometimes mixed evergreen and pine forest; 1,200–2,100 m. Behaviour as Himalayan. **RANGE & STATUS** Endemic. Scarce to fairly common resident S(east) Laos, C,S Annam. **BREEDING** Undocumented.

1235 GREY-SIDED LAUGHINGTHRUSH

Dryonastes caerulatus Plate **115**

IDENTIFICATION 26.5–28.5 cm. *D.c.latifrons*: **Adult** Combination of warm brown upperside and all-white underside with broadly grey flanks distinctive. Has black face, black scaling on crown and white patch on ear-coverts. **Juvenile** Lacks black scaling on crown; flanks and undertail-coverts tinged brown. **Other subspecies in SE Asia** *D.c.livingstoni* (south-west N Myanmar), *kaurensis* (south-east N Myanmar). **VOICE** Sings with clear, loud, high, spaced whistled phrases: ***wii'u wii'u wiii-uu whitwitwit witwitwit whi'i whii'u whii'uu whii'it...*** etc. Also an airy, whistled ***kawheeeer*** (rises slightly before tailing off), raptor-like, squealing ***káléép-káléép***, and raspy, mewing, rising then falling ***nyaoowl*** etc. Harsh grating ***grrrh***; ***grriiih*** and ***grrititit*** when agitated. **HABITAT & BEHAVIOUR** Broadleaved evergreen forest, secondary growth, bamboo; 1,525–2,620 m. Occurs in groups of 3–12; shy. **RANGE & STATUS** Resident Nepal, Bhutan, NE India, SW China. **SE Asia** Uncommon resident N,E(north) Myanmar. **BREEDING** April–June. **Nest** Rather large shallow cup, in bush, tree or bamboo; 1–3.7 m above ground. **Eggs** 2–3; pale blue, often tinged greenish; 30.5 x 22.1 mm (av.; India). Brood-parasitised by Chestnut-winged Cuckoo.

1236 BLACK-THROATED LAUGHINGTHRUSH

Dryonastes chinensis Plate **115**

IDENTIFICATION 26.5–30 cm. *D.c.lochmius*: **Adult** Told by olive-brown upperside with contrasting slate-grey crown, black face and centre of throat and upper breast, and sharply contrasting white ear-coverts and throat-sides. Breast greyish, lower flanks and vent olive-brown; has short black eyestripe and small white mark behind bristly black forehead-feathers. See Grey and White-cheeked Laughingthrushes. *D.c.germaini* (S Annam, Cochinchina) has deep rufescent-brown body-plumage; *propinquus* (Tenasserim, W Thailand) is somewhat intermediate. **Adult dark ('*lugens*') morph** Ear-coverts and throat-sides grey to blackish (various intergrades occur). Probably only occurs in *D.c.chinensis* (N[east],C Laos, W,E Tonkin, N,C Annam), with all records from this range (otherwise, typical examples of this subspecies resemble *lochmius*). **Juvenile** More olive-toned overall; crown much duller, throat at least partly darkish grey. **VOICE** Has a repetitive, rather mournful, fluty thrush-like song, incorporating harsh ***wraaah*** notes and squeaky whistles. Calls include low, husky ***how*** notes. **HABITAT & BEHAVIOUR** Broadleaved evergreen and mixed deciduous forest, secondary growth, scrub, bamboo; up to 1,525 m. Usually found singly or in pairs, sometimes in flocks of up to 12. **RANGE & STATUS** Resident SW,S China. **SE Asia** Fairly common to common resident C,E,S Myanmar, Tenasserim, W,NW,NE Thailand, Indochina. **BREEDING** April–August. **Nest** Cup, in bush or bamboo; up to 2.1 m above ground. **Eggs** 3–5; milky-blue; 31.2–31.7 x 21.8–22.9 mm (*propinquus*).

1237 CHESTNUT-BACKED LAUGHINGTHRUSH

Dryonastes nuchalis Plate **115**

IDENTIFICATION 24–27 cm. Monotypic. **Adult** Resembles Black-throated Laughingthrush but smaller, nape and upper mantle rufous to rufous-chestnut, breast much whiter. Note range. **Juvenile** Undescribed. **VOICE** Song is very similar to Black-throated. **HABITAT & BEHAVIOUR** Secondary growth, scrub and grass, bamboo; 305–915 m. Usually found in pairs or small parties; skulking. **RANGE & STATUS** Resident NE India. **SE Asia** Uncommon resident N Myanmar. **BREEDING** April–June. **Nest** Neat cup, in bush within 1 m of ground. **Eggs** 2–3; very pale blue, rarely white; 28.5 x 20.7 mm (av.).

1238 RUFOUS-VENTED LAUGHINGTHRUSH

Dryonastes gularis Plate **115**

IDENTIFICATION 23–25.5 cm. Monotypic. **Adult** Resembles Yellow-throated Laughingthrush but easily separated by large size, heavier build, plain brown tail, yellow chin and bright rufous vent. Orbital skin ochre-yellow or dark slate (significance of difference unknown), legs yellow-ochre to bright reddish-yellow. **Juvenile** Wings more richly coloured, grey of breast-sides mixed with rusty-brown, may show some brown fringing on crown. **VOICE** Possible songs include clear, very sweet, chiming, slightly up-then strongly downslurred whistles: ***fwééúuuu***, and longer ***fwééúúuueéé***. Flocks utter harsh rattling churrs, interspersed with rather nasal, discordant, high-pitched whistled phrases. **HABITAT & BEHAVIOUR** Broadleaved evergreen forest, secondary growth, scrub, sometimes bamboo; 300–1,220 m. Found in flocks of 6–15, sometimes up to 50. Skulking. **RANGE & STATUS** Resident SE Bhutan, NE India. **SE Asia** Local resident N Myanmar, N,C Laos, N Annam. **BREEDING** April–June. **Nest** Fairly bulky shallow cup, in bush or sapling; 0.9–6 m above ground. **Eggs** 2–3; white to pale blue-green; 25.5–31 x 19.2–21.7 mm.

1239 YELLOW-THROATED LAUGHINGTHRUSH

Dryonastes galbanus Plate **115**

IDENTIFICATION 23–24.5 cm. Monotypic. **Adult** Told by combination of warm brown upperparts, rather pale greyish crown and nape, black mask and chin, pale yellow underparts and greyish tail with broad dark terminal band and white-tipped outer feathers. Size, slim build, black chin, yellow vent and tail pattern rule out similar Rufous-vented Laughingthrush. **Juvenile** Upperside (apart from crown), breast and flanks much warmer than adult, rest of underparts whitish, with yellow restricted to throat; paler post-ocular skin, legs and feet. **VOICE** Flocks utter feeble chirping notes. **HABITAT & BEHAVIOUR** Scrub and grass, edge of broadleaved evergreen forest, open forest, secondary growth; 610–1,800 m. Found in flocks of up to 6, sometimes as many as 50–80. Shy. **RANGE & STATUS** Resident NE India, SE Bangladesh. **SE Asia** Rare resident W Myanmar. **BREEDING** April–June. **Nest** Rather untidy shallow cup, in bush or tangled undergrowth; 0.6–3 m above ground. **Eggs** 2–4; white to creamy-white, sometimes very pale blue; 23.6–27.7 x 17.8–20.1 mm.

1240 WHITE-CHEEKED LAUGHINGTHRUSH

Dryonastes vassali Plate **115**

IDENTIFICATION 26.5–28.5 cm. Monotypic. **Adult** Resembles Black-throated Laughingthrush (particularly race *germaini*) but upper ear-coverts black, white of head restricted mostly to throat-sides, has narrower black strip on centre of throat, no black on upper breast, paler underparts with buffish-brown flanks and distinctive brownish-grey tail with broad black subterminal band and white-tipped outer feathers. **Juvenile** Undescribed. **VOICE** Probable song (rarely heard) is a simple ***whii-u whii-u whii-u..*** Foraging flocks can be very noisy, uttering harsh extended rattles and short quick ***whi*** notes. **HABITAT & BEHAVIOUR** Broadleaved evergreen forest, secondary growth, scrub and grass; 650–1,900 m. Usually found in large flocks of 10–35 or more. Shy. **RANGE & STATUS** Endemic. Uncommon to locally common resident east Cambodia, S Laos, C,S Annam. **BREEDING** March–June. **Nest** Cup. Otherwise unknown.

1241 RUFOUS-NECKED LAUGHINGTHRUSH

Dryonastes ruficollis Plate **115**

IDENTIFICATION 22–26.5 cm. Monotypic. **Adult** Told by combination of smallish size, grey crown, blackish face, throat and upper breast, and light rufous-chestnut neck-patch and vent. Rest of body rather cold olive-brown. **Juvenile** Somewhat paler and browner overall. **VOICE** Variable. Song consists of fairly quickly repeated, jolly whistled phrases: ***wiwi'wi-whu whi-yi-ha*** etc. (hurried at beginning); also more prolonged, slurred, scratchy outpourings. Calls with repeated, rather high, shrill ***ch'yaa*** or ***cher*** and harsh ***whit'it*** notes and slow short rattles. **HABITAT & BEHAVIOUR** Broadleaved evergreen forest, secondary growth, scrub and grass, bamboo; 150–1,220 m. Found in pairs or small flocks, sometimes of up to 20 or more. Quite shy and skulking. **RANGE & STATUS** Resident Nepal, NE Indian subcontinent, SE Tibet, SW China (W Yunnan). **SE Asia** Uncommon resident W,N,E(north) Myanmar. **BREEDING** March–August. **Nest** Rather compact deep cup, in bush or thick undergrowth; 1–6 m above ground. **Eggs** 3–4; almost white with bluish-green tinge or pale milky-blue, rarely white; 25.7 x 20 mm (av.).

1242 RUFOUS-CHEEKED LAUGHINGTHRUSH

Garrulax castanotis Plate **115**

IDENTIFICATION 28–30.5 cm. *G.c.varennei*: **Adult** Resembles White-necked Laughingthrush but shows diagnostic large broad orange-rufous patch on ear-coverts and greyish crown; also shows white on rear supercilium and less white on neck-sides. **Juvenile** Undescribed. **VOICE** Very similar to White-necked. **HABITAT & BEHAVIOUR** Broadleaved evergreen forest; 400–1,700 m. Always found in flocks. Shy. **RANGE & STATUS** Resident S China (Hainan). **SE Asia** Locally common resident N(east),C Laos, E Tonkin (Mt Ba Vi only), N,C Annam. **BREEDING** Undocumented. **NOTE** Split from Grey Laughingthrush (see Collar & Robson 2007).

1243 GREY LAUGHINGTHRUSH

Garrulax maesi Plate **115**

IDENTIFICATION 28–30.5 cm. *G.m.maesi*: **Adult** Overall greyish plumage, black face and chin, whitish-grey ear-coverts and white rear supercilium and neck-patch diagnostic. Resembles Black-throated Laughingthrush but lacks black on throat and breast, lacks black eyestripe, has lighter grey (less slaty) crown and is much greyer above. **Juvenile** Upperparts mixed with brown, throat and breast greyish-brown. **VOICE** Very similar to White-necked Laughingthrush. **HABITAT & BEHAVIOUR** Broadleaved evergreen forest; 800–1,700 m, locally down to 380 m. Always found in flocks. Shy. **RANGE & STATUS** Resident SW,S China. **SE Asia** Locally common resident W(north),E Tonkin. **BREEDING** China: April–May. Otherwise undocumented.

1244 BLACK-HOODED LAUGHINGTHRUSH

Garrulax milleti Plate **115**

IDENTIFICATION 28–30 cm. *G.m.milleti*: **Adult** Recalls White-necked Laughingthrush but shows diagnostic uniform brownish-black head and upper breast bordered by whitish band from upper mantle to lower breast. Has distinctive tear-shaped, naked bluish-white post-ocular patch. *G.m.sweeti* (east S Laos, C Annam) has blacker hood, greyer (less brown-tinged) body and darker wings and tail. **Juvenile** Undescribed. **VOICE** Very similar to White-necked. **HABITAT & BEHAVIOUR** Broadleaved evergreen forest; 800–1,650 m. Always in flocks. Shy. **RANGE & STATUS** Endemic. Scarce to locally common resident east S Laos, C,S Annam. **BREEDING** May–June. Otherwise undocumented.

1245 CAMBODIAN LAUGHINGTHRUSH

Garrulax ferrarius Plate **115**

IDENTIFICATION 28–30 cm. Monotypic. **Adult** Like Black-hooded Laughingthrush but hood browner (particularly throat and upper breast), upper mantle and lower breast darker and slatier-grey, white restricted to neck-patch. **Juvenile** Undescribed. **VOICE** Undocumented but likely to be like Black-hooded. **HABITAT** Broadleaved evergreen forest; elevation undocumented. **RANGE & STATUS** Endemic. Scarce to uncommon resident south-west Cambodia. **BREEDING** Undocumented. **NOTE** Split from Black-hooded (see Round & Robson 2001).

1246 WHITE-NECKED LAUGHINGTHRUSH

Garrulax strepitans Plate **115**

IDENTIFICATION 28.5–31.5 cm. Monotypic. **Adult** Combination of brown crown, blackish-brown face, cheeks, throat and upper breast, dark rufous rear ear-coverts and prominent large white neck-patch distinctive. Sometimes has upper breast similar in colour to crown. Told from similar Rufous-cheeked Laughingthrush by brown crown, much browner upperparts, more restricted and duller rufous on ear-coverts and larger white neck-patch; lacks white on supercilium. Note range. **Juvenile** Undescribed. **VOICE** Flock vocalisations are similar to White-crested Laughingthrush but faster, incorporating more prolonged, rapid rattling calls. Outbursts are preceded and interspersed with clicking ***tick*** or ***tekh*** notes which also serve as contact calls. **HABITAT & BEHAVIOUR** Broadleaved evergreen forest; 500–1,800 m. Always found in flocks. Shy. **RANGE & STATUS** Resident SW China (SW Yunnan). **SE Asia** Fairly common resident E Myanmar, Tenasserim, W,NW,NE Thailand, west N Laos. **BREEDING** Undocumented.

1247 WHITE-CRESTED LAUGHINGTHRUSH

Garrulax leucolophus Plate **115**

IDENTIFICATION 26–31 cm. *G.l.diardi*: **Adult** Broad erect whitish crest and black mask diagnostic. Nape mid-grey, merging into rufous-chestnut of upperparts; underparts whitish with grey wash on breast-sides and rufous flanks and undertail-coverts. *G.l.patkaicus* (SW,W,N,E[north],S[west] Myanmar) has chestnut upper mantle, lower breast and centre of abdomen; *belangeri* (C,S,E Myanmar, Tenasserim, W Thailand) is roughly intermediate. **Juvenile** Similar to adult but tends to have shorter crest and more ashy-brown nape; may be more rufescent above (particularly wing-feather fringes). **VOICE** Varied sounds from different flock-members are combined to produce sudden very loud outbursts of extended cackling laughter, typically involving a combination of rapid chattering and repetitive double-note phrases. Outbursts are often introduced by a few subdued ***ow***, ***u'ow*** or ***u'ah*** notes. **HABITAT & BEHAVIOUR** Broadleaved evergreen, semi-evergreen and dry deciduous forest, secondary growth, bamboo; up to 1,600 m, rarely 2,135 m. Always in flocks, usually 6–12, occasionally up to 40. **RANGE & STATUS** Resident N,NE Indian subcontinent, SE Tibet, SW China, Sumatra. **SE Asia** Common resident (except C,S Thailand,

Peninsular Malaysia, Singapore). Rare to locally common feral resident C Thailand (Bangkok environs), Peninsular Malaysia (Penang I, Kuala Lumpur environs), Singapore. **BREEDING** February–September. Co-operative breeder. **Nest** Largish shallow cup, in bush or tree; 1.8–6 m above ground. **Eggs** 2–6; white; 28.7 x 22.4 mm (av.; *belangeri*).

1248 **LESSER NECKLACED LAUGHINGTHRUSH**

Garrulax monileger Plate **115**

IDENTIFICATION 26.5–31.5 cm. *G.m.mouhoti*: **Adult** White supercilium, complete black eyestripe extending behind whitish ear-coverts and across breast in narrow gorget (broken in centre) and whitish remainder of underside distinctive. Upperside brown with contrasting dark rufous nuchal collar and rufous-buff tips to outertail-feathers, lower breast (bordering necklace) and flanks washed rufous; has black line bordering lower ear-coverts. Always separable from very similar Greater Necklaced Laughingthrush by smaller size, black line on lores, orange-yellow to yellowish-brown eyes, dark orbital ring, and paler primary coverts; often lacks black line below cheeks. *G.m.monileger* (W,N,E[north],C,S Myanmar) has white tips to outertail-feathers and paler rufous nuchal collar; *stuarti* (E Myanmar, north Tenasserim, W,NW Thailand) and *fuscatus* (central Tenasserim, south W Thailand) are somewhat intermediate; *pasquieri* (C Annam) is much smaller, with darker upperside, chestnut nuchal collar, black line above supercilium, deep rufous breast with narrow blackish-brown necklace and rustier-buff tail-tips; *tonkinensis* (W,E Tonkin, N Annam) is roughly intermediate between *mouhoti* and *pasquieri*. **Juvenile** Upperside more rufescent, necklace often less distinct, centre of abdomen buffier, pale tail-tips smaller. **Other subspecies in SE Asia** *G.m.schauenseei* (eastern E Myanmar, NW[east],NE[north-west] Thailand, N Laos). **VOICE** Not very vocal. Possible song-types are a mellow, repeated ***u-wi-uu*** and more subdued, quickly repeated ***ui-ee-ee-wu***; ***wiu-wiu-wiu*** and ***ui-ui-ui*** phrases. Presumed songs also described as sequences of tuneful, short, mellow whistles: ***tu'tu'tu'tu-tuwa...tu'tu'tu'tu-tuwa...*** Foraging flocks make a variety of low sounds, becoming harsh and continuous when alarmed, when thin whining sounds may also be given. Subdued, nervous, sweet irregular peeping. **HABITAT & BEHAVIOUR** Broadleaved evergreen, semi-evergreen and dry deciduous forest, secondary growth, bamboo; up to 1,675 m. Always in flocks, often mixed with other medium-sized species. **RANGE & STATUS** Resident Nepal, NE Indian subcontinent, southern China. **SE Asia** Common resident (except C,S Thailand, Peninsular Malaysia, Singapore). **BREEDING** March–August. **Nest** Broad, rather shallow cup, in bamboo or tree; 0.9–4.5 m above ground. **Eggs** 3–5; slightly greenish pale blue (deepness of tone varies); 28.4 x 21.3 mm (av.; *monileger*). Brood-parasitised by Pied and Chestnut-winged Cuckoos.

1249 **GREATER NECKLACED LAUGHINGTHRUSH**

Garrulax pectoralis Plate **115**

IDENTIFICATION 27–34.5 cm. *G.p.subfusus*: **Adult** Difficult to separate from Lesser Necklaced Laughingthrush but larger, with whitish (sometimes dusky) lores, brown to crimson eyes, golden-yellow orbital ring, and blackish-brown primary coverts which contrast with rest of wing; always shows complete black line below ear-coverts and cheeks, typically has mostly buff throat and breast, with broader necklace. Usually shows much white on cheeks and ear-coverts but occasionally all black, has light rufous nuchal collar and light buff tips to outertail-feathers; black necklace often broken in centre. *G.p.pectoralis* (W,N,E,C,S Myanmar) has broader, darker rufous nuchal collar, white-tipped outertail-feathers and usually more complete, solid black necklace; *robini* (N Indochina) is darker above with rufous-chestnut nuchal collar, deeper buff outertail-tips and browner primary coverts; often shows grey admixed with black behind ear-coverts and typically has broken necklace. **Juvenile** Upperside somewhat warmer-tinged, necklace duller and less distinct, often shows paler buff to whitish tips and fringes to primary coverts. **VOICE** Apparent song-types include repeated, clear, ringing, slightly descending and diminishing sequences: ***kléér-éér-éér-éér-éér(...)*** or just ***kléér-éér***, and a sequence of alternating, upslurred mellow whistles, ***tu-twéé-tu-twéétu-twéé***. Mixed series of loud, quavering, nervous ***wee'i'i***; ***wee'u*** and ***wee'ee'u*** phrases. Calls include clear, rapid, nervous ***chit-it(-it-it)***, louder and less mellow than peeping calls of Lesser Necklaced. Also, a variety of short nasal churring calls and low, gruff contact notes. **HABITAT & BEHAVIOUR** Broadleaved evergreen, semi-evergreen and dry deciduous forest, secondary growth; up to 1,830 m. Always in flocks. **RANGE & STATUS** Resident Nepal, NE Indian subcontinent, C and southern China. **SE Asia** Uncommon to common resident Myanmar, W,NW Thailand, N,C Laos, W,E Tonkin, N Annam. **BREEDING** March–August. **Nest** Rather large, shallow cup, in bush, tree or bamboo; usually 1.5–6 m above ground. **Eggs** 3–7; bright or deep blue to pale greenish-blue; 31.4 x 22.7 mm (av.; *melanotis*). Brood-parasitised by Chestnut-winged Cuckoos.

1250 **WHITE-THROATED LAUGHINGTHRUSH**

Garrulax albogularis Plate **115**

IDENTIFICATION 30.5 cm. *G.a.eous*: **Adult** Shows unmistakable combination of brown upperside, greyish-brown breast-band, warm buffish belly and vent and sharply contrasting white throat and upper breast. Has dull rufous patch on forehead. **Juvenile** Upperparts somewhat warmer with little rufous on forehead, less pronounced breast-band, whitish centre of abdomen. **VOICE** Thin shrill wheezy whistles: ***tsu'ueeeee***; ***hiuuuu***; ***huiiii*** and ***hsiii*** etc. Gentle ***chrrrr***, soft ***teh*** notes and subdued chattering when foraging. Harsh forced ***chrrr-chrrr-chrrr...*** when alarmed. **HABITAT & BEHAVIOUR** Broadleaved evergreen forest, secondary growth, bamboo; above 1,600 m, locally down to 600 m in winter. Usually in noisy flocks of 6–12, sometimes up to 100 or more in winter. **RANGE & STATUS** Resident N Pakistan, N,NE India, Nepal, Bhutan, S,SE Tibet, SW,W,C China, Taiwan. **SE Asia** Uncommon resident W Tonkin. **BREEDING** India: March–July. **Nest** Broad cup, in bush or tree; 0.9–4 m above ground. **Eggs** 2–4; deep dull blue to intense greenish-blue; 29 x 21.1 mm (av.; India).

1251 **MASKED LAUGHINGTHRUSH**

Garrulax perspicillatus Plate **115**

IDENTIFICATION 28–31.5 cm. Monotypic. **Adult** Combination of rather uniform greyish-brown plumage, broad blackish-brown mask and orange-buff vent diagnostic. See dark morph Black-throated Laughingthrush. **Juvenile** Mask fainter, rest of head and breast browner, upperparts warmer-tinged. **VOICE** Loud ***jhew*** or ***jhow*** notes (often doubled) and harsh chattering. **HABITAT & BEHAVIOUR** Scrub, bamboo, hedgerows, cultivation (often near villages); lowlands. Usually in flocks of 6–12; rather skulking. **RANGE & STATUS** Resident C and southern China. **SE Asia** Fairly common resident E Tonkin, N,C Annam. **BREEDING** April–July. **Nest** Fairly large, untidy cup, in tree, bush, thick undergrowth or grass; 1–9 m above ground. **Eggs** 2–5; pale blue to greenish-white or greyish-blue; 26.4–30.5 x 19–21.8 mm.

1252 **BLACK LAUGHINGTHRUSH**

Melanocichla lugubris Plate **115**

IDENTIFICATION 25.5–27 cm. Monotypic. **Adult** Unmistakable, with overall blackish plumage, naked bluish-white post-ocular patch and orange-red bill. Superficially resembles Black Bulbul. **Juvenile** Duller and browner, particularly on mantle and wings. **VOICE** Song, typically given by at least two birds at once, consists of amazing loud hollow whooping ***huup-huup-huup...*** and rapid loud ***okh-ohk-okh-okh-okh...***, accompanied by harsh ***awh*** or ***aak*** notes. **HABITAT & BEHAVIOUR** Broadleaved evergreen forest, sometimes secondary growth; 800–1,500 m. Occurs in pairs or small flocks; quite shy. **RANGE & STATUS** Resident Sumatra. **SE Asia** Uncommon resident extreme S Thailand, Peninsular Malaysia. **BREEDING** October–November. **Nest** Bulky cup, in sapling fork; 2.5 m above ground. **Eggs** 2; clear light blue, thinly scattered with fine dots, irregular spots and small smudges of black; c.31 x 23 mm.

1253 SPECTACLED LAUGHINGTHRUSH
Rhinocichla mitrata Plate **116**

IDENTIFICATION 21.5–24 cm. *R.m.major*: **Adult** Largely greyish plumage, chestnut crown, rufous-chestnut undertail-coverts, broad white eyering and orange-yellow bill diagnostic. Primaries edged white, tail tipped blackish, lores blackish, forehead streaked white. See greyer subspecies of Chestnut-crowned Laughingthrush. Note range. **Juvenile** Duller and browner with reduced white streaking on forehead. **VOICE** Song is fairly subdued but clear and quite shrill, consisting of 3–5 note phrases: ***wi wu-wi-wu-wi*** and ***wi wu-wi*** with stressed first note, ***wi-wu-wiu-wu-wi*** with stressed middle note and ***wiu-wu-wui-wi*** with rather sharp last note etc. Calls include a sibilant ***ju-ju-ju-ju-ju*** and ***wi-jujujujuju***, rapid harsh squirrel-like cackling ***wikakakaka***, and other low, harsh sounds. **HABITAT & BEHAVIOUR** Broadleaved evergreen forest, forest edge; above 900 m. Found in pairs or small parties. **RANGE & STATUS** Resident Sumatra. **SE Asia** Common resident extreme S Thailand, Peninsular Malaysia. **BREEDING** February–August. **Nest** Shallow cup, in tree or tangled vegetation; 1–9 m above ground. **Eggs** 2; deep greenish-blue; 29 x 19.5 mm. Brood parasitised by Dark Hawk-cuckoo.

1254 CHINESE BABAX
Babax lanceolatus Plate **114**

IDENTIFICATION 22.5–26 cm. *B.l.lanceolatus* (north & east N Myanmar): **Adult** Large size and pale plumage with bold blackish, brown and chestnut streaks diagnostic. Throat and breast white, mostly unstreaked, has whitish eyes and broad dark brown submoustachial stripe. *B.l.woodi* (north-west N Myanmar) shows black submoustachial stripe, blacker central streaks on feathers of crown, neck and upperparts, and dark shaft-streaks on throat. **Juvenile** Buffier overall with less prominent streaking, only pronounced on nape, mantle and breast; plainer crown, paler ear-coverts, brown (rather than greyish) outer fringes of wing-feathers. **VOICE** Songs of nominate race include clear, quite jolly ***wee-wer-choh WHI*** or ***phi-phu-chu WHI*** (first three notes clearly spaced and descending, fourth higher or omitted). *B.l.woodi* sings with loud full clear whistled ***pu-i*** or ***tchu-wi*** phrases: ***pu-i pu-i pu-i pu-i pu-i pu-i...***, sometimes hurried to ***pui-pui-pui-pui...*** Calls with harsh, thin, grating buzzing sounds, and sibilant tittering, e.g. ***jerrt jerrt jerrt chit'it'it'it'it...*** etc. *B.l.woodi* utters quiet chuntering ***witchawitchawitcha-wit*** and single ***whit*** notes when on the move. **HABITAT** Open broadleaved evergreen forest, forest edge, secondary growth, scrub and grass, bamboo; 1,200–2,800 m. **RANGE & STATUS** Resident NE India (Mizoram), W,C and southern China. **SE Asia** Uncommon resident W,N,C(east),E(north) Myanmar. **BREEDING** March–June. **Nest** Fairly open cup, in bush, brambles or sapling; 0.6–1.2 m above ground. **Eggs** 2–6; deep blue to bluish-green; 30.2–30.5 x 21.1–21.8 mm (*woodi*).

1255 STRIATED LAUGHINGTHRUSH
Grammatoptila striata Plate **115**

IDENTIFICATION 29.5–34 cm. *G.s.cranbrooki*: **Adult** Combination of broad rounded floppy crest, rich dark brown plumage, heavily pale-streaked upper- and underparts and broad blackish supercilium diagnostic. Crown unstreaked. *G.s.brahmaputra* (W Myanmar) has narrower blackish supercilium, obvious whitish shaft-streaks on forehead and above eye; streaking on rest of head whiter and more pronounced. **Juvenile** Upperparts a little lighter and warmer, shaft-streaks on underparts narrower and less pronounced. **VOICE** Song is a repeated loud, vibrant, rolling ***prrrit-you prrit-pri-prii'u***. When agitated, calls with low grumbling ***aawh aawh aawh'o aawh aawh...*** **HABITAT & BEHAVIOUR** Broadleaved evergreen forest, secondary growth, bamboo; 800–2,500 m. Occurs singly, in pairs or small flocks of 5–8; often feeds quite high up in trees. **RANGE & STATUS** Resident N,NE India, Nepal, Bhutan, S,SE Tibet, SW China. **SE Asia** Uncommon to locally common resident N Myanmar. **BREEDING** April–July. **Nest** Broad, rather shallow cup, in tree or among creepers; 1–6 m above ground. **Eggs** 2–3; very pale greenish-blue or pale blue to almost white, occasionally with some tiny dark brownish-red specks; 31.5 x 23.5 mm (av.; *cranbrooki*). Brood-parasitised by Chestnut-winged Cuckoo.

1256 SPOT-BREASTED LAUGHINGTHRUSH
Stactocichla merulinus Plate **116**

IDENTIFICATION 25–26 cm. *S.m.merulinus*: **Adult** The only plain brown laughingthrush with pronounced blackish-brown spots on buffish-white throat and breast. Has narrow buffish-white supercilium behind eye. *S.m.laoensis* (NW Thailand) has blacker and more separated streaks below, buff ground colour paler; *obscura* (east N Laos, W Tonkin) is richer buff below, with much heavier blackish spotting. See Chinese Hwamei. **Juvenile** Upperside and flanks more rufescent. **VOICE** Song is loud, rich, melodious and thrush-like. Recalls Black-throated Laughingthrush but much richer and more varied, lacking harsh notes and squeaky whistles. **HABITAT & BEHAVIOUR** Edge of broadleaved evergreen forest, secondary growth, overgrown clearings, bamboo; 800–2,000 m. Usually in pairs; very skulking. **RANGE & STATUS** Resident NE India, SW China. **SE Asia** Scarce to uncommon resident W,N Myanmar, NW Thailand, N Laos, W,E(north-west) Tonkin, N Annam. **BREEDING** April–July. **Nest** Rather bulky cup, in thick bush or bamboo clump; close to ground. **Eggs** 2–3; pale blue to blue, often tinged greenish; 28.7 x 21.2 mm (av.; *merulinus*).

1257 ORANGE-BREASTED LAUGHINGTHRUSH
Stactocichla annamensis Plate **116**

IDENTIFICATION 24–25 cm. Monotypic. **Adult** Similar to Spot-breasted Laughingthrush but has black throat, heavy black streaks on deep orange-rufous breast and narrow, pale orange-rufous supercilium. **Juvenile** Undescribed. **VOICE** Very similar to Spot-breasted. **HABITAT & BEHAVIOUR** Broadleaved evergreen forest, forest edge, secondary growth and overgrown clearings bordering forest; 915–1,510 m. Usually found in pairs; very skulking. **RANGE & STATUS** Endemic. Uncommon resident S Annam. **BREEDING** Undocumented. **NOTE** For split from Spot-breasted Laughingthrush see Collar & Robson (2007).

1258 CHINESE HWAMEI
Leucodioptron canorum Plate **116**

IDENTIFICATION 21–24 cm. *L.c.canorum*: **Adult** Warm brown body-plumage, dark brown streaks on crown, nape, mantle, throat and breast and pronounced white spectacles diagnostic. See White-browed and Spot-breasted Laughingthrushes. **Juvenile** Less streaking on head and breast. **VOICE** Has rich, varied, quite high-pitched song, including regular repetition. Usually starts slowly, then increases in volume and pitch (may repeat this during single outburst). Faster and higher-pitched than Black-throated Laughingthrush, reminiscent of Spot-throated Babbler. **HABITAT & BEHAVIOUR** Secondary growth, scrub and grass, bamboo, open woodland; up to 1,600 m. Usually found in pairs, sometimes in small groups; rather skulking. **RANGE & STATUS** Resident W,C and southern China; introduced Taiwan. **SE Asia** Fairly common resident N,C,S(north-east) Laos, W,E Tonkin, N,C(north) Annam. Uncommon feral resident Singapore. **BREEDING** March–July. **Nest** Rather large cup, in bush, tree or undergrowth; up to 2 m above ground. **Eggs** 2–5; light blue or deep bluish-green; 24–28.8 x 19.3–25 mm.

1259 STRIPED LAUGHINGTHRUSH

Strophocincla virgata Plate **116**

IDENTIFICATION 23 cm. Monotypic. **Adult** Smallish size, slim build, rufescent body-plumage with narrow whitish streaks and prominent whitish supercilium and submoustachial stripe diagnostic. Breast and belly paler and buffier with indistinct streaking, wings patterned with chestnut and greyish. **Juvenile** Paler and more rufescent above, paler throat, no streaking on lower underparts. **VOICE** Has two territorial calls, often given antiphonally by pairs: (1) (probably by male) is a clear, hurried ***chwi-pieu***; ***pi-pweu*** or ***wiwi-weu***, repeated every few seconds; (2) (probably by female) is a loud staccato rattling trill, usually with a shorter introductory note, ***cho-prrrrrt*** or ***chrrru-prrrrrt***, repeated after slightly longer intervals. Calls with mixed harsh ***chit*** and ***chrrrrr*** notes: ***chit chrrrrr-chrrrrr-chrrrrr chit...*** etc. **HABITAT & BEHAVIOUR** Scrub and grass, secondary growth, forest edge; 1,400–2,400 m. Found singly or in pairs; skulking. **RANGE & STATUS** Resident NE India. **SE Asia** Fairly common to common resident W Myanmar. **BREEDING** April–July. **Nest** Deep cup, in bush or grass; up to 2 m above ground. **Eggs** 2–3; pale blue, sometimes tinged greenish; 24.1–28.4 x 18.3–20.1 mm.

1260 WHITE-BROWED LAUGHINGTHRUSH

Pterorhinus sannio Plate **116**

IDENTIFICATION 22–24 cm. *P.s.comis*: **Adult** Told by nondescript brownish plumage, with contrasting buffish-white supercilium, which loops in front of eye to meet broad buffish-white cheek-patch, and rufescent vent. Has vague pale streaking on mantle. See Chinese Hwamei. *P.s.sannio* (E Tonkin) has whiter facial markings. **Juvenile** Brown parts more rufescent than adult, lacks vague mantle streaks. **VOICE** Repeated loud harsh emphatic ***jhew*** and ***jhew-jhu***. Harsh buzzy ***dzwee*** notes when agitated. **HABITAT & BEHAVIOUR** Scrub and grass, secondary growth, bamboo, cultivation, open woodland; 600–1,830 m, locally down to 215 m. Usually in small noisy parties; not particularly shy. **RANGE & STATUS** Resident NE India, W,C and southern China. **SE Asia** Locally common to common resident N,C,E Myanmar, NW Thailand, N Laos, W,E Tonkin. **BREEDING** February–August. **Nest** Fairly compact cup, in bush, tree or undergrowth; 0.6–6 m above ground. **Eggs** 3–4; pale blue or pale blue-green to pale greenish-white; 26 x 19.4 (av.; India).

1261 MOUSTACHED LAUGHINGTHRUSH

Ianthocincla cineracea Plate **116**

IDENTIFICATION 21–24.5 cm. *I.c.cineracea*: **Adult** Rather small and long-tailed. Sandy-brown above, with sharply contrasting black crown, nape-centre and subterminal tail-band; head-sides buff-tinged greyish-white with black post-ocular stripe and frayed black submoustachial stripe; throat, breast and tail-tips whitish. Wings are prominently patterned black, grey and white. *I.c.strenua* (E Myanmar) is much buffier-brown overall, particularly head-sides and underparts. **Juvenile** Much more rufescent overall, lacks head pattern, apart from post-ocular stripe. **VOICE** Not certainly documented. **HABITAT & BEHAVIOUR** Edge of broadleaved evergreen and mixed broadleaved and coniferous forest, secondary growth, scrub and grass, bamboo; 1,220–2,500 m. Usually found singly or in pairs, sometimes in small groups. Shy and skulking. **RANGE & STATUS** Resident NE India, W,C and southern China. **SE Asia** Uncommon resident W,E(north) Myanmar. **BREEDING** March–June. **Nest** Compact cup, in bush, thick undergrowth or bamboo; 0.9–2 m above ground. **Eggs** 2–4; blue to turquoise-blue; 25.3 x 18.6 mm (av. *cineracea*).

1262 RUFOUS-CHINNED LAUGHINGTHRUSH

Ianthocincla rufogularis Plate **116**

IDENTIFICATION 23–25.5 cm. *I.r.rufiberbis*: **Adult** Very distinctive. Upperside mostly warm olive-brown with extensively black crown, black-scaled mantle and scapulars, black wing-bands and black subterminal tail-band; underparts whitish with pale rufous-chestnut chin and vent, broadly frayed black submoustachial stripe and dark-speckled breast. *I.r.intensior* (W Tonkin) is deeper rufous above, often with less black on crown; shows more solid black cheeks and submoustachial patch, larger blackish breast markings and paler rufous chin. See Spotted Laughingthrush. **Juvenile** Shows reduced dark markings, particularly on crown and upperparts, and less rufous on chin. **VOICE** Song is a repeated loud, measured, clear, slightly husky ***whi-whi-WHU-whi*** or ***whi-whi-whi-WHI***, with slightly higher penultimate note or end-note. Alarm call is an irregular series of harsh low notes which run into short grating rattles. **HABITAT & BEHAVIOUR** Broadleaved evergreen forest, forest edge, secondary growth; 915–1,600 m. Usually in pairs or small parties; skulking. **RANGE & STATUS** Resident N,NE Indian subcontinent, SW China (Yunnan), formerly NE Pakistan. **SE Asia** Uncommon resident N Myanmar, W Tonkin. **BREEDING** May–September. **Nest** Fairly deep cup, in bush or tree; 0.6–6 m above ground. **Eggs** 2–4; white; 26.2 x 19.4 mm (av.; India).

1263 CHESTNUT-EARED LAUGHINGTHRUSH

Ianthocincla konkakinhensis Plate **116**

IDENTIFICATION 24 cm. Monotypic. **Adult** Like Rufous-chinned Laughingthrush (race *intensior*) but ear-coverts chestnut, lacks rufous chin and dark band across flight-feathers, tail-tip buffish-white, forehead and supercilium mostly grey. Note range. **Juvenile** Said to have only little grey on rear supercilium. **VOICE** Song is a sweet *Turdus* thrush-like rambling series of up to c.20 fairly well-spaced and stressed notes, and some mimicry, lasting c.4–6 sec. Calls including low grumbling ***rrreeek rrreeek rrreeek...***, with notes thinning towards end. **HABITAT** Broadleaved evergreen forest, bamboo; 1,600–1,700 m. Usually in pairs or small parties; skulking. **RANGE & STATUS** Endemic. Uncommon local resident east S Laos (presumably this taxon), southern C Annam. **BREEDING** March–April. Otherwise undocumented.

1264 SPOTTED LAUGHINGTHRUSH

Ianthocincla ocellata Plate **116**

IDENTIFICATION 31–33.5 cm. *I.o.maculipectus*: **Adult** Relatively large size, profuse black-and-white spotting on rufous-olive to chestnut upperparts, black crown and lower throat and rufous-buff underparts with short black-and-white bars on breast diagnostic. Face and supercilium mostly warm buffish, has black and grey fringing on flight-feathers and grey-fringed chestnut tail with black subterminal band and white tip. **Juvenile** Crown and throat browner, white spots on upperparts smaller and absent on rump and uppertail-coverts. **VOICE** Sings with repeated rich mellow fluty phrases: ***wu-it wu-u wu-u wi-u wi-u***; ***w'you w'you uu-i w'you uu'i*** and ***w'you uu-wi'ii uwa*** etc. **HABITAT & BEHAVIOUR** Light broadleaved evergreen and mixed evergreen and coniferous forest, rhododendron scrub; 2,135–2,745 m. Found in pairs or small parties of 5–8; secretive. **RANGE & STATUS** Resident N,NE India, Nepal, Bhutan, S,SE Tibet, W,C China. **SE Asia** Fairly common resident N Myanmar. **BREEDING** India: May–June. **Nest** Large loose cup, in bush, tree or undergrowth; within 2 m of ground. **Eggs** 2+; pale blue to deep blue-green, sometimes with some dark brown specks toward broader end; 30–32.7 x 21.2–21.8 mm.

1265 SCALY LAUGHINGTHRUSH

Trochalopteron subunicolor Plate **116**

IDENTIFICATION 23–25.5 cm. *T.s.griseatus*: **Adult** Slim and rather nondescript, with heavily dark-scaled body-plumage. Recalls Blue-winged but lacks black supercilium, has dark eyes, mostly olive to yellowish-olive wings and brown to golden-brown tail with white-tipped outer feathers. Base colour of crown brownish-grey. *T.s.fooksi* (W Tonkin)

Has darker, slatier-grey crown and extensively dark throat. **Juvenile** Upperparts and breast warmer with less distinct scaling, whitish tail-tips smaller. **VOICE** Sings with a repeated, rather shrill high whistled ***whiu-whiiiu*** and ***whi'ii'i whi'ii'i*** etc. Calls with low buzzing ***thriiii*** notes, interspersed with high-pitched squeaks. **HABITAT & BEHAVIOUR** Broadleaved evergreen forest, secondary growth, dwarf rhododendron, scrub, bamboo; 1,600–3,960 m. Found in pairs or parties of up to 10–20. **RANGE & STATUS** Resident Nepal, Bhutan, NE India, S Tibet, SW China. **SE Asia** Uncommon resident N,E(north) Myanmar, W Tonkin. **BREEDING** June–July. **Nest** Cup, in bush or sapling; 60–90 cm above ground. **Eggs** 3–4; blue (sometimes tinged greenish); 29.5–30.3 x 23 mm (India).

1266 BROWN-CAPPED LAUGHINGTHRUSH
Trochalopteron austeni Plate **116**

IDENTIFICATION 24 cm. *T.a.victoriae*: **Adult** Rather plain warm brown upperside with white streaks on neck and brown-and-whitish scaled underparts diagnostic. Shows white tips to greater coverts, tertials and secondaries, whitish-fringed outer primaries and white-tipped tail-feathers. **Juvenile** Slightly paler below than adult, with broader but less contrasting whitish scaling; small pointed white tail-tips. **VOICE** Sings with repeated loud clear jolly phrases, including ***whit-wee-wi-weeoo***; ***whichi-wi-chooee***; ***whi-chi-weeoo*** and ***whiwiwi-weeee-weeoo***; sometimes more rapid ***whi-wi-wi-wi-wi-wi-wi-weee*** etc. Song is often accompanied (presumably by female) with fairly harsh ***jee-jee-jee*** or ***jee-jee...jee-jee...*** etc. Call is a subdued but harsh ***grrrret-grrrret-grrrret...*** **HABITAT & BEHAVIOUR** Oak and rhododendron forest, bamboo, secondary forest, forest edge; 1,975–3,050 m. Usually found singly or in pairs, sometimes small parties; usually shy and skulking. **RANGE & STATUS** Resident NE India. **SE Asia** Common resident W Myanmar. **BREEDING** April–August. **Nest** Cup, in bush; within 2 m of ground. **Eggs** 2–4; white; 26.3 x 19 mm (av.; India).

1267 BLUE-WINGED LAUGHINGTHRUSH
Trochalopteron squamatum Plate **116**

IDENTIFICATION 22–25 cm. Monotypic. **Male** Robust shape, dark-scaled body-plumage and striking whitish eyes distinctive. Might be confused with Scaly Laughingthrush but shows black supercilium, mostly bright rufous to rufous-chestnut wings with black and white fringed primaries and blackish tail with reddish-rufous tip. **Female** More rufescent on face, back and underparts than male, flight-feathers and tail dark brown (blackish on male). **Juvenile** Eyes brown. **VOICE** Song consists of thin, rather high-pitched rising whistles, of two main types: (1) (male?) with stressed end-note, ***pwuuuuu-wit***; ***piuuuuu-witchi*** and ***weeuwiiiii-it*** etc.; (2) (female?) more mournful, less high-pitched ***pwiiiieeu***; ***pwiiiiiu*** or wheezy ***phwiiiiu***. Calls include a fairly short subdued buzzy ***jrrrrr-rrr-rrr...*** etc. and harsh, rather liquid, quite buzzy ***cher-cher-rrru*** or ***jo-jorrrru***. **HABITAT & BEHAVIOUR** Broadleaved evergreen forest, secondary forest, bamboo; 1,220–2,200 m, down to 900 m N Myanmar. Usually occurs singly or in pairs, sometimes small parties; often very secretive. **RANGE & STATUS** Resident Nepal, Bhutan, NE India, SW China. **SE Asia** Uncommon resident W,N,C(east),E Myanmar, W Tonkin. **BREEDING** April–July. **Nest** Cup, in bush or tangled creepers; 1.2–1.8 m above ground. **Eggs** 2–4; fairly glossy, bright pale blue, with slight greenish tinge; 29.5 x 20.8 mm (av.). Brood-parasitised by Chestnut-winged Cuckoo.

1268 BLACK-FACED LAUGHINGTHRUSH
Trochalopteron affine Plate **116**

IDENTIFICATION 24–26 cm. *T.a.oustaleti*: **Adult** Easily identified by mostly blackish-brown head and sharply contrasting whitish submoustachial and neck-patches. Body-plumage rufescent-brown with greyish scaling/mottling, wings and tail patterned grey and golden- to yellowish-olive. **Juvenile** Crown brown, lacks neck-patch, body more uniform brown. **Other subspecies in SE Asia** *T.a.saturatus* (W Tonkin). **VOICE** Song consists of repeated loud, shrill, rather high-pitched, quite quick phrases: ***wiee-chiweeoo***; ***wiee-chweeiu*** and ***wiee-weeoo-wi*** etc., with first note shriller, higher and rising. Calls include a continuous, rather high rattling. **HABITAT & BEHAVIOUR** Broadleaved evergreen, coniferous and mixed forest, high-altitude juniper and rhododendron scrub, bamboo; 1,705–3,660 m. Occurs in pairs or small flocks. **RANGE & STATUS** Resident Nepal, Bhutan, NE India, S,SE Tibet, SW,W China. **SE Asia** Locally common resident N Myanmar, W Tonkin. **BREEDING** India: May–June. **Nest** Cup, in bush; 1–2.5 m above ground. **Eggs** 2–3; blue (sometimes tinged greenish), with a few blackish-brown to blackish-purple spots, blotches or scrolls towards broader end; 28.5 x 21.2 mm (av.; India).

1269 ASSAM LAUGHINGTHRUSH
Trochalopteron chrysopterum Plate **116**

IDENTIFICATION 23–26 cm. *T.c.erythrolaemum*: **Adult** Mostly rufous head-sides, throat and underparts, grey supercilium, heavily dark-spotted mantle and breast and yellowish-olive primary coverts distinctive. *T.c.woodi* (N Myanmar) has silvery-grey forecrown and ear-coverts with black streaks. See Chestnut-crowned and Silver-eared Laughingthrushes, but note range. **Juvenile** Duller hindcrown, warmer upperparts, little or no body-spotting. **VOICE** Songs include an emphatically ended ***(tu)whééeeerwhít!***, slightly more even ***quwhééyerwhúrt!***, an overall rising, worried ***fu'uwhééyer whéét!*** and similar, and even more anxious ***fyuwééyer-whéét!*** Song of *woodi* is a loud, high, far-carrying, slightly rising ***wi-eeoo*** or ***wieeoo***; also transcribed as ***fu-wééééyúr*** and more even ***fuúwééeer***. Calls include low, musical purring ***squar-squar-squar*** and, from *woodi*, a low, grunting, sputtering, churring ***gnrsh gnrsh...***, or similar but drier, flat, rasping ***krssh***, and a higher-intensity metallic, grating, upslurred ***kr'r'r'reep! kr'r'r'reep!***, sometimes mixed with occasional high short squeals in chorus. **HABITAT & BEHAVIOUR** Broadleaved evergreen and mixed broadleaved and coniferous forest, secondary growth, bamboo; 1,065–3,050 m. Usually in pairs or small groups. Skulking but not particularly shy. **RANGE & STATUS** Resident N,NE Indian subcontinent, S,SE Tibet, SW China. **SE Asia** Common resident W,N,S(west) Myanmar. **BREEDING** April–October. **Nest** Deep cup, in bush, tree, creeper or bamboo etc.; 0.9–5.5 m above ground. **Eggs** 2–3; turquoise-blue to bright blue with dark purplish to black blotches, specks and streaks; 29.9 x 20.5 mm (av.; *erythrolaemum*).

1270 SILVER-EARED LAUGHINGTHRUSH
Trochalopteron melanostigma Plate **116**

IDENTIFICATION 26 cm. *T.m.schistaceum* (eastern E Myanmar, NW Thailand [Doi Chiang Dao, Doi Pha Hom Pok etc.]): **Adult** Very variable. Told by combination of rufous-chestnut to chestnut crown, blackish face, rufescent greater coverts and yellowish-olive fringing on wing- and tail-feathers. Has mostly plain olive-greyish body, mostly silvery-grey supercilium and ear-coverts, chestnut upper throat and black primary coverts. *T.m.melanostigma* (E Myanmar, north Tenasserim, W[north],NW Thailand [Doi Inthanon etc.]) has warm-tinged underparts, more chestnut on throat and variable rufescent suffusion on upperparts; *ramsayi* (east S Myanmar, central Tenasserim; probably south W Thailand) is even warmer below (strongly rufescent); *subconnectens* (extreme E Myanmar, east NW Thailand, west N Laos) and *connectens* (N[east],C Laos, northern Vietnam) are similar to *melanostigma* but have prominent pale greyish to whitish breast-scaling (latter also has yellowish-olive primary coverts). See Golden-winged Laughingthrush, but note range. **Juvenile** Apparently undescribed. **VOICE** Song in NW Thailand is quite loud and quite liquid ***WI-wiwioo***; ***WU-weeoo*** or ***TU-tweeoo***, with more emphasis on first note. Also quickly delivered ***u-wip-weeoo***, with faint introductory note; ***weeOO-wiwip*** with rising first note; sometimes a very fast ***wiu-wip*** or ***wipu-wip***. Moaning, slightly husky, clear mewing ***weeaa-ao***; ***waaaow***; ***wayaaa*** or

aoaaaa may be given antiphonally (presumably by female). **HABITAT & BEHAVIOUR** Broadleaved evergreen and mixed broadleaved and coniferous forest, secondary growth, bamboo; 1,065–2,565 m, locally down to 610 m. Usually in pairs or small groups. Skulking but not particularly shy. **RANGE & STATUS** Resident SW China (SE Yunnan). **SE Asia** Common resident S(east),E Myanmar, north Tenasserim, W,NW Thailand, N,C Laos, W,E(north-west) Tonkin, N Annam. **BREEDING** March–August. **Nest** Deep cup, in bush, tree, creeper or bamboo etc. **Eggs** 2 (central Tenasserim); sky blue, with sepia blotches and spots; 30.5 x 20.4 mm (av. *ramsayi*).

1271 **MALAYAN LAUGHINGTHRUSH**

Trochalopteron peninsulae Plate **116**

IDENTIFICATION 25.5–26.5 cm. Monotypic. **Adult** Told by plain rufescent-brown body-plumage, dark grey head-side, with mostly brownish ear-coverts and indistinct whitish eyering. See Spectacled Laughingthrush. **Juvenile** Crown duller than adult, grey of neck-side suffused with brownish. **VOICE** Sings with clear and far-carrying ***wip-wééOO*** or ***wiw-wééOO*** and similar (with rising, stressed second note). Also, a quickly delivered ***wip-wi-eeoo*** (short first note, somewhat rising middle note). **HABITAT & BEHAVIOUR** Broadleaved evergreen and mixed broadleaved and coniferous forest, secondary growth; above 1,050 m. Usually in pairs or small groups. Skulking but not particularly shy. **RANGE & STATUS** Endemic. Common resident S Thailand, Peninsular Malaysia. **BREEDING** January–May and October. **Nest** Fairly bulky open cup, in sapling, tree-fern crown, creeper, ginger-plant etc.; 1.5–3 m above ground. **Eggs** 2–4; blue, with light scattering of black or deep reddish-brown to black spots.

1272 **GOLDEN-WINGED LAUGHINGTHRUSH**

Trochalopteron ngoclinhense Plate **116**

IDENTIFICATION 27 cm. Monotypic. **Adult** Told from similar Silver-eared Laughingthrush race *connectens* by generally darker-toned head and body, brownish-grey forehead and supercilium with blackish streaks, browner ear-coverts, black primary coverts and diagnostic golden to orange fringing on wing- and tail-feathers. Note range. **Juvenile** Undescribed. **VOICE** Song undocumented. Only described call is double-noted, rather cat-like mewing ***rr'raow rr'raow...*** (first note short and descending, second longer, rising and with slight downward inflection at end, and emphasised). **HABITAT & BEHAVIOUR** Broadleaved evergreen forest; 2,000–2,200 m. Usually found singly or in pairs; very shy. **RANGE & STATUS** Endemic. Uncommon local resident southern C Annam (Mt Ngoc Linh). **BREEDING** Undocumented.

1273 **COLLARED LAUGHINGTHRUSH**

Trochalopteron yersini Plate **116**

IDENTIFICATION 26–28 cm. Monotypic. **Adult** Black head with silvery-grey ear-coverts and deep orange-rufous collar and breast diagnostic. Blackish primary coverts contrast with bright golden- to orange-olive wing-feather fringes. **Juvenile** Orange-rufous parts paler and browner, mantle to back more chestnut-tinged. **VOICE** Song is a repeated loud, quite high-pitched, rising whistle: typically ***wueeeeoo*** or ***u-weeeeoo*** (rising in middle), ***uuuu-weeoo*** (rising more at start) or ***wiu-weeeu*** (quick first note, stressed second note) etc. Whistles are often answered antiphonally (by female?) with a low, quite harsh, mewing ***wiaaah***; ***ayaaa*** or ***ohaaaah*** (sometimes given alone). Alarm calls are subdued, harsh, slurred, slightly rising notes: ***grreet-grreet-grreet-grreet-grreet-grreet-grreet-grreet-grrr-rr...*** etc. **HABITAT & BEHAVIOUR** Broadleaved evergreen forest, secondary growth and scrub bordering forest; 1,500–2,440 m. Usually in pairs or small flocks; very shy. **RANGE & STATUS** Endemic. Locally common resident S Annam. **BREEDING** March–April. Otherwise undocumented.

1274 **RED-WINGED LAUGHINGTHRUSH**

Trochalopteron formosum Plate **116**

IDENTIFICATION 27–28 cm. *T.f.greenwayi*: **Adult** Told from similar Red-tailed Laughingthrush by dark-streaked grey forecrown and ear-coverts, plain brown upperparts (including nape), and browner underparts. Note range. **Juvenile** Similar to adult. **VOICE** Song consists of repeated loud, rather thin, clear whistled phrases: ***chu-weewu*** or slightly rising ***chiu-wee*** etc. Possible duets (intra-sex?) include ***chiu-wee-u-weeoo*** (latter rising slightly in middle) and ***u-weeoo-wueeoo*** (latter quickly delivered). Also a louder ***wu-eeoo***. **HABITAT & BEHAVIOUR** Broadleaved evergreen forest, secondary growth and scrub near forest, bamboo; 2,400–2,800 m. Usually found in pairs or small parties; shy and skulking. **RANGE & STATUS** Resident SW,W China. **SE Asia** Uncommon resident W Tonkin. **BREEDING** June–July (China). Otherwise undocumented.

1275 **RED-TAILED LAUGHINGTHRUSH**

Trochalopteron milnei Plate **116**

IDENTIFICATION 26–27 cm. *T.m.sharpei*: **Adult** Bright rufous crown and nape and extensively red wings and tail diagnostic. Face blackish, ear-coverts silvery-grey, body greyish-brown with slightly paler and darker scaling. *T.m.vitryi* (S Laos; Bolovens Plateau only) and undescribed subspecies (C Annam; presumably also east S Laos) have darker and more chestnut crown and nape, greyer lower throat and breast with pronounced blackish scallops; throat/breast paler with more pronounced scallops on latter. **Juvenile** Similar to adult. **VOICE** Sings with clear loud whistled phrases. Either ***uuu-weeoo***; ***eeoo-wee*** or shorter ***uuuwi*** etc. (all rising at start) or ***uuuu-hiu-hiu*** and ***uuuu-hiu-hiu-hiu*** etc., with slightly rising introduction and faster, soft laughter at end. **HABITAT & BEHAVIOUR** Broadleaved evergreen forest, bamboo, secondary growth, scrub and grass; 800–2,500 m. Usually found singly, in pairs or small parties, sometimes quite large flocks; shy and unobtrusive. **RANGE & STATUS** Resident southern China. **SE Asia** Scarce to uncommon resident N,E Myanmar, NW Thailand, Laos, W,E Tonkin, N,C Annam. **BREEDING** April–May. **Nest** Neat cup, in bush or undergrowth; up to 90 cm above ground. **Eggs** 2–3; white, sparingly spotted and blotched dark red, reddish-brown or blackish (mostly towards broader end), sometimes with some purplish undermarkings; 25.4–33 x 20.3–20.9 mm (*sharpei*).

1276 **CRIMSON-FACED LIOCICHLA**

Liocichla phoenicea Plate **116**

IDENTIFICATION 20.5–23.5 cm. *L.p.bakeri*: **Adult** Similar to Scarlet-faced Liocichla, but has darker red on head which does not extend to throat-sides, browner-tinged crown, and deeper, warmer brown body-plumage. Note range. *L.p.phoenicea* (northern N Myanmar) has browner crown, slightly paler and more extensive red on head, darker and more rufous- to chestnut-tinged lower upperparts, with less extensive grey on centre. **Juvenile** Duller, with less red in plumage and less obvious black lateral crown-stripe. **VOICE** Has loud clear cheerful song of repeated 5–8 note phrases (sometimes shortened to 3), the last part usually rising but sometimes falling. Phrases include ***chewi-ter-twi-twitoo***; ***chi-cho-choee-wi-chu-chooee***; ***chi-wee-ee-ee-weeoo*** and ***chiu-too-ee*** etc. Calls with harsh grating rasping notes: ***chrrrt-chrrrt*** and buzzy, upslurred grumbling ***grssh! grssh! grssh!*** **HABITAT & BEHAVIOUR** Secondary growth, scrub and grass, broadleaved evergreen forest; 800–2,500 m. Usually found singly, in pairs or small groups; skulking and unobtrusive. **RANGE & STATUS** Resident NE Indian subcontinent, SW China (NW Yunnan). **SE Asia** Fairly common resident W,N Myanmar. **BREEDING** March–June. **Nest** Rather deep compact cup, in bush or sapling; up to 3 m above ground. **Eggs** 2–4; greenish-blue to pale blue, marked (towards broader end) with dark brown to dark reddish or maroon dashes, scrolls, fine curly lines and/or spots and a few pale purple spots; 26.1 x 18.5 mm (av.).

1277 **SCARLET-FACED LIOCICHLA**

Liocichla ripponi Plate **116**

IDENTIFICATION 20.5–23.5 cm. *L.r.ripponi*: **Adult** Relatively small size, rather plain brown body-plumage and striking, extensively red head and throat-sides diagnostic. Also has extensively red-fringed wings with black and white markings, blackish tail with pale rufous tip and grey crown with narrow black border. See Crimson-faced Liocichla, and Red-winged and Red-tailed Laughingthrushes. Note range. **Juvenile** Duller, with less red in plumage and less obvious black lateral crown-stripe. **Other subspecies in SE Asia** *L.p.wellsi* (N Indochina). **VOICE** Songs recall Crimson-faced, with phrases including: ***tiyúu-fwéé-túu***; ***chu'u-wiu-WUU***; ***pi-WII-wu***, and falling then rising ***chiu-too-ee***. Also, curious mewing ***ji-uuuu*** or ***chi-uuuu*** or steeply and evenly falling, mellow whistled ***tyúúúúu***. **HABITAT & BEHAVIOUR** Secondary growth, scrub and grass, broadleaved evergreen forest; 1,400–2,200 m. Usually found singly, in pairs or small groups; skulking and unobtrusive. **RANGE & STATUS** Resident SW China (southern Yunnan) **SE Asia** Fairly common resident N(south-east),C(east),S(east),E Myanmar, NW Thailand, N Laos, W,E(north-west) Tonkin. **BREEDING** March–June. **Nest** Cup, placed in bush, sapling or bamboo; up to 0.6–1.5 m above ground. **Eggs** 3; pale blue, spotted and streaked with dark red; 25.8 x 18.6 mm (av. *ripponi*). **NOTE** Split from Crimson-faced (formerly Red-faced) Liocichla *L.phoenicea* following Collar & Robson (2007).

1278 **BAR-THROATED MINLA**

Chrysominla strigula Plate **117**

IDENTIFICATION 16–18.5 cm. *C.s.castanicauda*: **Adult** Golden-rufous crown, greyish head-sides, blackish eyebrow (to above eye), and strong submoustachial streak and mostly whitish throat with broad black bars/scales diagnostic. Has olive-greyish upperparts, dull brownish-chestnut and black tail with whitish tip, yellowish underparts, black primary coverts, contrasting orange-yellow flight-feather fringes and broadly white-bordered tertials. *C.s.yunnanensis* (W,N,E[east of Salween R] Myanmar, N Indochina) has warmer upperparts and darker chestnut on tail; *malayana* (Peninsular Malaysia) has duller crown, slatier-grey ear-coverts, duller, greyer nape to uppertail-coverts, broader black throat-bars, duller greyish-olive wash on breast and flanks, dull yellowish-olive undertail-coverts and much narrower white tips to secondaries; *traii* (southern C Annam) shows mostly whitish lores, cheeks and upper ear-coverts, brighter crown and nape, and greyer rest of upperparts. **Juvenile** Hindcrown, nape and upperparts greyer, head-sides somewhat paler, throat-bars narrower and more broken, underparts more washed out. **VOICE** Sings with repeated, high-pitched, slightly quavering ***tui-twi ti-tu***, ***twi ti-u*** or ***twi-twi TWI twi*** with higher penultimate note. **HABITAT** Broadleaved evergreen forest; 1,600–3,000 m. **RANGE & STATUS** Resident N,NE Indian subcontinent, S,SE Tibet, SW China. **SE Asia** Locally common to common resident W,N,S(east),E Myanmar, Tenasserim, W(local),NW Thailand, Peninsular Malaysia, N,C Laos, W Tonkin, southern C Annam. **BREEDING** March–July. **Nest** Cup, in bush or small tree; 1.5–3 m above ground. **Eggs** 2–4; pale to deep blue or blue-green, lightly spotted black, pale red or brown (mostly at broader end); 20.4 x 15.3 mm (av.; India).

1279 **RED-TAILED MINLA**

Minla ignotincta Plate **117**

IDENTIFICATION 13–14.5 cm. *M.i.mariae*: **Male** Easily told by black crown, nape and head-sides, long broad white supercilium, and pale yellow throat and underparts with faint broad greyish streaks on breast and flanks. Rest of upperparts olive-brownish, wings black with white-fringed coverts, white-tipped tertials and red-fringed flight-feathers (becoming orange-yellow distally), tail black with red outer fringes and reddish-pinkish tip. *M.i.ignotincta* (Myanmar) has chestnut-tinged deep brown upperparts. **Female** Upperparts slightly duller, has much paler pinkish-yellow to whitish flight-feather fringes and pinker tail-feather fringes and tip. **Juvenile** Like female, but mantle slightly darker, white plumage parts duller, faint dark scaling below. **VOICE** Song is a fairly high-pitched, loudish, quite quickly repeated ***wi ti wi-wu***, first two notes higher. Calls with subdued harsh ***wih-wih-wih-wih...***, louder, harsher ***yih-yih-yih-yih...***, short ***wit*** and ***wih*** notes and hurried ***witti-wi-wrrh***. **HABITAT** Broadleaved evergreen forest, secondary forest; 1,100–2,800 m, locally down to 470 m N Myanmar (winter only?). **RANGE & STATUS** Resident Nepal, NE Indian subcontinent, SE Tibet, SW,W,S China. **SE Asia** Common resident W,N,S(east),E Myanmar, N,C,S(east) Laos, W,E(north-west) Tonkin, N,C Annam. **BREEDING** April–June. **Nest** Small cup or purse, in fork of bush or tree; 1.2–3 m above ground. **Eggs** 2–4; pale to deep blue, speckled black or reddish-brown (often ringed at broader end); 19.4 x 14.4 mm (av.; *ignotincta*).

1280 **BLUE-WINGED SIVA**

Siva cyanouroptera Plate **117**

IDENTIFICATION 14–15.5 cm. *S.c.sordida*: **Adult** Slim, long-tailed and rather nondescript with distinctive violet-blue fringing on primaries and tail. Has indistinct long whitish supercilium, narrow dark violet-blue lateral crown-streak, variable faint dark violet-blue streaks on forehead (may show some pale greyish streaks), warm brown upperparts with greyer nape, and somewhat paler uppertail-coverts and greyish-white underparts with whiter vent. From below, tail appears mostly whitish with narrow blackish border. *S.c.aglae* (W Myanmar) and *cyanouroptera* (west N Myanmar) have more extensive dark blue and pale greyish streaks on forecrown, bluer forecrown-streaks and lateral crown-streak, whiter (more prominent) supercilium, warmer mantle, darker greyish throat and breast, and bluer fringes to wings and tail; *wingatei* (N[east],E Myanmar, north-east NW Thailand, Laos, W,E Tonkin, N,C Annam) is roughly intermediate between *sordida* and *aglae*/*cyanouroptera*; *sordidior* (S Thailand, Peninsular Malaysia) has browner crown and upperparts (similar to *orientalis*); *rufodorsalis* (SE Thailand, south-west Cambodia) has distinctly chestnut-washed upperparts (brighter on rump and uppertail-coverts), and darker blue on wings and tail; *orientalis* (extreme east Cambodia, S Annam) is larger (15.5–17.5 cm), has brownish crown and nape, concolorous with paler, browner mantle, but paler and more greyish lower rump and uppertail-coverts, duller wings and tail, more whitish fringes to secondaries and whiter underparts. **Juvenile** Crown and nape browner than adult. **VOICE** Song is a rather quickly repeated, very thin, high-pitched ***psii sii-suuu***, with falling end-note or rising and falling ***tsuit-twoo*** etc. Calls include short ***whit*** and ***bwik*** contact notes. **HABITAT & BEHAVIOUR** Broadleaved evergreen forest, secondary forest; 900–2,600 m, locally down to 460 m N Myanmar (winter only?). Usually in small flocks. **RANGE & STATUS** Resident N,NE Indian subcontinent, SW,W,S China. **SE Asia** Common resident (except SW,C Myanmar, C Thailand, Singapore, Cochinchina). **BREEDING** December–June. **Nest** Cup, in bush or tree, sometimes hole in bank; up to 8 m above ground. **Eggs** 2–5; deep blue to bluish-white with a few small black or rusty-brown and violet spots towards larger end; 18.4 x 14.1 mm (av.; India).

1281 **SILVER-EARED MESIA**

Mesia argentauris Plate **117**

IDENTIFICATION 16.5–18 cm. *M.a.galbana*: **Male** Very distinctive, with yellow bill, black head, silver-grey ear-coverts, orange-yellow throat and upper breast, red patch at base of orange-yellow-fringed flight-feathers. Has small yellow forehead-patch and orange-reddish tail-coverts. *M.a.ricketti* (N Indochina) has more orange forehead-patch, redder nape/upper mantle, orange-red throat and upper breast (throat redder) and slightly darker remainder of underparts; *tahanensis* (S Thailand, Peninsular Malaysia) has deeper orange throat-sides and upper breast and slightly paler reddish wing-patch (feathers fringed orange); *cunhaci* (S Laos, C,S Annam) has much larger yellow forehead-patch. **Female** Forehead-patch, throat and breast duller and paler, nape/upper mantle and uppertail-coverts dull golden-olive; has less yellow on centre of underparts, pale yellowish-rufous undertail-coverts, paler, duller reddish wing-patch and paler, duller yellow fringes to flight-feathers. *M.a.ricketti* has more orange-tinged forehead-patch, throat and upper breast, and more golden-rufous tinge to nape/upper mantle and uppertail-coverts; *tahanensis* has deeper orange throat-sides and upper breast, and slightly paler reddish wing-patch. **Juvenile** Similar to respective adults but black of head duller, rest of body somewhat more washed out. **Other subspecies in SE Asia** *M.a.aureigularis* (W Myanmar), *vernayi* (N Myanmar). **VOICE** Song is a repeated, cheerful, loud, clearly spaced, descending ***che tchu-tchu che-rit*** or ***che chu chiwi chwu*** etc. Calls include a flat piping ***pe-pe-pe-pe-pe*** (sometimes used to accompany song; by female?) and harsh chattering notes. **HABITAT & BEHAVIOUR** Edge of broadleaved evergreen forest, secondary growth, scrub; 450–2,000 m, locally down to 175 m E Tonkin (winter only?). Often in small to quite large parties. **RANGE & STATUS** Resident N,NE Indian subcontinent, SE Tibet, SW China, Sumatra. **SE Asia** Common resident Myanmar (except SW,C), W,NW,NE,S Thailand, Peninsular Malaysia, east Cambodia, Laos, Vietnam (except Cochinchina). **BREEDING** November–September. **Nest** Deep cup; in bush or dense vegetation; just above ground or up to 5 m. **Eggs** 2–5; fairly glossy white, lightly spotted rich pinkish-brown (often more towards broader end); 21.5–23.1 x 16.7–17.6 mm (av.; *tahanensis*).

1282 RED-BILLED LEIOTHRIX
Leiothrix lutea Plate **117**

IDENTIFICATION 15.5–16 cm. *L.l.kwangtungensis*: **Male** Easily told by red bill, yellowish lores and orbital area, golden-olive crown and nape, dark submoustachial stripe, deep yellow throat, orange-rufous upper breast and conspicuous wing pattern with yellow to orange-yellow fringing and large chestnut-red patch at base of primaries. Rest of upperparts olive-slate, ear-coverts pale yellowish-grey, belly and vent pale yellow, flanks broadly greyish; has unusually long, pale-tipped uppertail-coverts and shallow tail-fork. *L.l.yunnanensis* (N Myanmar) has duller, more olive-tinged crown, greyer nape and rest of upperparts, deeper, more reddish-rufous upper breast and little or no chestnut-red at base of much more extensively black-fringed primaries; *calipyga* (W Myanmar) is roughly intermediate and has different wing pattern, with chestnut-red and black fringing on inner primaries and less chestnut-red at base of primaries. **Female** Has more greenish-olive crown, greyer nape, paler lores and orbital area, greyer ear-coverts, less distinct submoustachial stripe, paler throat and breast and duller reddish patch at base of flight-feathers. *L.l.calipyga* and *yunnanensis* lack chestnut-red at base of extensively black-fringed primaries. **Juvenile** Recalls female but crown concolorous with mantle, underparts olive-grey with whitish centre of throat and abdomen, bill paler. **VOICE** Sings with rather rapid, fluty, thrush-like warble of up to 15 notes. Utters harsh scolding buzzy rattles when alarmed. **HABITAT** Edge of broadleaved evergreen forest, secondary forest, scrub; 800–2,135 m. **RANGE & STATUS** Resident NW,N,NE Indian subcontinent, SE Tibet, C and southern China. **SE Asia** Common resident W(scarce),N Myanmar, W,E Tonkin. **BREEDING** April–October. **Nest** Cup, in bush or bamboo clump; 0.6–1.5 m above ground (rarely to 4.5 m). **Eggs** 3–5; bluish- or greenish-white to greenish-blue (rarely white), boldly blotched reddish-brown to brown or various shades of reddish-brown and purple over underlying lilac or grey streaks and clouds (often marked mainly at broader end). 21.9 x 16.1 mm (av.; *calipyga*).

1283 GREY-CROWNED CROCIAS
Crocias langbianis Plate **118**

IDENTIFICATION 22 cm. Monotypic. **Adult** Slaty-grey crown and nape, blackish mask, dull dark rufous upperparts with blackish-brown streaks and white underparts with bold blackish streaks along flanks diagnostic. Has faint pale shaft-streaks on crown, nape and mask, mostly slaty-grey, white-tipped uppertail, mostly grey greater coverts and secondaries, and white-fringed blackish primaries. Females are possibly slightly duller above. **Juvenile** Crown browner with broader buffish streaks, head-sides somewhat duller, flank-streaks narrower and shorter, greater coverts and secondaries browner, white tail-tips much narrower. **VOICE** Sings with repeated high, loud, metallic, almost spluttering ***wip'ip'ip'ip'ip'ip...*** and ***wu-tu-ti'ti'ti'ti'ti'ti...***; high, clear, slower and more spaced ***uee'uee'uee'uee'uee'uee...***; rapid oscillating ***wi'wi'wi'wi'wi'wi'wi'wi'...***; and even, regular ***whi-whi-whi-whi-whi-whi-whi-whi...*** etc.; the various types often combined with grumbly, buzzy ***bidu-bidu-bidu-bidu...*** and ***bidi-wi-di-di-di*** etc. Quiet ***tip***, ***pit*** and ***it*** contact notes. **HABITAT & BEHAVIOUR** Broadleaved evergreen forest; 910–1,450 m. Often singly or in pairs; joins mixed-species feeding flocks; forages slowly, mostly in denser middle storey. **RANGE & STATUS** Endemic. Scarce and local resident S Annam. **BREEDING** April–May. Otherwise unknown.

1284 LONG-TAILED SIBIA
Heterophasia picaoides Plate **118**

IDENTIFICATION 28–34.5 cm. *H.p.cana*: **Adult** Unmistakable, with all-greyish plumage, very long tail with greyish-white tips, and long broad white wing-patch. Bill distinctly narrow and slightly downcurved, eyes dark red. *H.p.picaoides* (W,N Myanmar) is slightly browner overall; *wrayi* (Peninsular Malaysia) is noticeably browner overall and has smaller white wing-patch. See Beautiful Sibia. **Juvenile** Like adult but eyes grey. **VOICE** Thin, metallic high-pitched ***tsittsit*** and ***tsic*** notes, interspersed with a dry, even-pitched rattling: ***tsittsit-tsic-tsic-tsic-tsic chrrrrrrrt tsitsitsittsit-tsic-tsic-chrrrrrrrrt...*** etc. **HABITAT & BEHAVIOUR** Broadleaved evergreen forest, forest edge, secondary growth; 900–2,285 m, rarely down to 460 m. Usually in small flocks; regularly visits flowering trees. **RANGE & STATUS** Resident Nepal, NE Indian subcontinent, SW China, Sumatra. **SE Asia** Locally common resident W,N,E,S(east) Myanmar, north Tenasserim, W,NW Thailand, Peninsula Malaysia, N,C,S(east) Laos, W,E(north-west) Tonkin, C Annam. **BREEDING** February–August. **Nest** Very deep cup, suspended from outer tree-branches; 6–20 m above ground. **Eggs** 5; pale grey-green with very small reddish-brown blotches (particularly around broader end); 24.5 x 18.1 mm (av.; *picaoides*).

1285 GREY SIBIA *Malacias gracilis* Plate **118**

IDENTIFICATION 22.5–24.5 cm. *M.g.dorsalis*: **Adult** Resembles Black-headed Sibia but blackish-brown hindcrown and rear ear-coverts merge into brownish-grey mantle, has pale grey rump and uppertail-coverts, mostly pale grey tertials with black fringes, pale grey central uppertail with broad blackish subterminal band and warm buff-tinged vent. Birds in N Myanmar have much slatier-grey nape to back and scapulars (may be undescribed subspecies). **Juvenile** Forecrown/crown duller and brown-tinged, nape to uppertail-coverts and scapulars paler and browner, body paler overall. **VOICE** Song is a repeated, very loud, strident and far-carrying series of well-spaced high-pitched shrill whistled notes (usually descending towards end or from beginning): ***tu-tu-ti-ti-ti-tu***, ***ti-ti-ti-ti-tu***, ***ti-ti-ti-ti-tiu-tu***, ***ti-ti-titi-ti-tu*** and ***tiu-tiu-tiu-tiu-tiu*** etc. Calls with a harsh grating, slightly metallic ***trrit-trrit***; contact calls include quiet hurried, nasal, rather squeaky ***witwit-witarit*** and ***wit-witarit-warao*** and soft sibilant ***ti-tew***. **HABITAT** Broadleaved evergreen forest; 1,400–2,800 m. **RANGE & STATUS** Resident NE India, SW China (W Yunnan). **SE Asia** Common resident W,N Myanmar. **BREEDING** April–July. **Nest** Deep neat cup, in tree; up to 6 m above ground. **Eggs** 2–4; pale greyish to pale bluish or bluish-green, lightly freckled, blotched and sometimes streaked pale reddish, reddish-brown or dark brown to ashy-brown (often ringed at broader end); 22.6–23.9 x 16.5–18 mm.

1286 BLACK-HEADED SIBIA
Malacias desgodinsi Plate **118**

IDENTIFICATION 21.5–24.5 cm. *M.d.desgodinsi*: **Adult** Black crown, nape, head-sides, wings and tail, grey upperparts with slight mauvish-brown wash and white underparts with faint mauvish-grey wash on breast and flanks diagnostic. Tail has grey-tipped central and white-tipped outer feathers. *M.d.engelbachi* (S Laos [Bolovens Plateau]) has mostly dark brown lower mantle, scapulars and back and distinct broken white eyering;

kingi (east S Laos, southern C Annam) has lower mantle, back and scapulars washed drab brown (not nearly as brown as *engelbachi*) and distinct broken white eyering; *robinsoni* (S Annam) has broad broken white eyering and browner ear-coverts with white streaks. All three southern subspecies are smaller, with mostly white malar area. **Juvenile** *M.d.robinsoni* has cap and base colour to ear-coverts duller, eyering narrower, grey of upperparts slightly brown-tinged, darker band across breast, making throat contrast more. **Other subspecies in SE Asia** *M.d.tonkinensis* (W Tonkin). **VOICE** Song recalls Dark-backed Sibia, but slower, more structured and less descending: ***hi wi-wi wi wi*** or ***hi hwi-wi wi-wi*** etc. Contact calls are thin, repeated ***tsrri*** notes. **HABITAT & BEHAVIOUR** Broadleaved evergreen forest; 800–2,290 m. Often in small flocks. **RANGE & STATUS** Resident SW,W,C China. **SE Asia** Common resident east N Myanmar, N(north-east),S Laos, W,E(north-west) Tonkin, C(southern),S Annam. **BREEDING** February–June. **Nest** Cup-shaped, in tree. **Eggs** 1-3 (in captivity); pale blue, stippled (and with some short lines) faint darkish brown to dull mauve, with odd larger blotches and clouds (mainly in ring around broader end); 24.4 x 17.1 mm (*desgodinsi*). **NOTE** For split from Dark-backed Sibia see Collar & Robson (2007).

1287 DARK-BACKED SIBIA

Malacias melanoleucus Plate **118**

IDENTIFICATION 21–23 cm. *M.m.radcliffei*: **Adult** Unmistakable, with all-blackish upperside, all-white underparts and whitish tail-tips. Mantle and scapulars to uppertail-coverts brownish-black (appear black in field). *M.m.melanoleucus* (N Tenasserim; W Thailand?) has distinctly brown mantle and scapulars to uppertail-coverts; *castanoptera* (south-western E Myanmar) has prominent rufous on inner greater coverts and outer webs of tertials (intergrades occur). **Juvenile** Undescribed. **VOICE** Sings with repeated loud high-pitched wavering whistle, dropping in pitch at end: ***hrrrr'rrr'r'i-u***, ***hrrrr'r'r'i-i*** and ***hrrrr'r'i*** etc. Calls include quite harsh ***trr-trr-trr-trr*** contact notes. **HABITAT & BEHAVIOUR** Broadleaved evergreen forest; 1,000–2,565 m. Often in small flocks. **RANGE & STATUS** Endemic. Common resident C,S(east),E Myanmar, north Tenasserim, W,NW Thailand, west N Laos. **BREEDING** February–June. **Nest** Cup, on outer tree-branches; 2.5–7.6 m above ground. **Eggs** 2–3; pale blue to greenish-blue, usually well marked dark red to dull pale red (often capped at broader end); 29.2 x 17 mm (av.).

1288 BEAUTIFUL SIBIA

Malacias pulchellus Plate **118**

IDENTIFICATION 23.5 cm. *M.p.pulchellus*: **Adult** Easily told by mostly bluish-grey plumage, black mask and wing-coverts, blue-grey fringes to flight-feathers and brown tertials and basal two-thirds of tail. Tail also has broad grey tip and black subterminal band. See Long-tailed Sibia. **Juvenile** Body tinged brown but crown, throat and upper breast similar to adult. **VOICE** Song is a repeated loud strident ***ti-ti-titi-tu-ti*** (descending slightly towards end). Recalls Grey Sibia but higher-pitched, shriller and generally less descending, with more hurried beginning. Calls with continuous low rattling ***chrrrrrrrrr*** or ***churrururr***. **HABITAT & BEHAVIOUR** Broadleaved evergreen forest; 1,830–2,745 m, locally down to 900 m (winter only?). Often in small flocks; regularly visits flowering trees. **RANGE & STATUS** Resident NE India, SE Tibet, SW China. **SE Asia** Uncommon resident N Myanmar. **BREEDING** April–July. **Nest** Cup; in outer tree-branches. **Eggs** 1+; pale blue; 23.8 x 17.9 mm.

1289 RUFOUS-BACKED SIBIA

Leioptila annectens Plate **118**

IDENTIFICATION 18.5–20 cm. *L.a.mixta* (north and east NW Thailand, N Indochina): **Adult** Black crown and head-sides, rufous-chestnut back to uppertail-coverts, broadly white-tipped black tail and white underparts with rufous-buff lower flanks and vent diagnostic. Nape and upper mantle black with white streaks, lower mantle and scapulars mixed rufous-chestnut and black (paler rufous on latter), wings mostly black with dull rufous tips to greater coverts and whitish edges to flight-feathers and tertials. *L.a.annectens* (W,N,C[east] Myanmar) has slightly paler rufous-chestnut on upperparts and slightly darker vent; *saturata* (S[east],E Myanmar, west NW Thailand) and *eximia* (S Annam) have more black on mantle and scapulars, fewer white streaks on nape/upper mantle, dark chestnut on rest of upperparts and slightly paler buff lower flanks and vent; *davisoni* (north Tenasserim, W Thailand) has almost completely black mantle, scapulars and upper back with only few white streaks on lower nape/upper mantle, and paler vent than *saturata*. See Dark-backed Sibia. **Juvenile** Very similar to adult. **VOICE** Sings with clear, loud, strident, jolly, slightly descending ***wip'i-iu-iu-ju*** (*annectens*); ***wii-wii-wii-er-yu*** (*eximia*), slower, more even ***it wi-wiu-jui*** (*saturata*). Calls include harsh chattering when alarmed. **HABITAT & BEHAVIOUR** Broadleaved evergreen forest; 1,000–2,300 m, locally down to 215 m in winter (N Myanmar). Usually forages methodically in middle to upper storey; joins mixed-species feeding flocks. **RANGE & STATUS** Resident E Nepal, NE Indian subcontinent, SW China. **SE Asia** Fairly common resident W,N,C(east),S(east),E Myanmar, north Tenasserim, W,NW Thailand, N,S(east) Laos, W Tonkin, N,C(southern),S Annam. **BREEDING** April–June. **Nest** Compact cup, in fork of outer tree-branch; 2–6 m above ground. **Eggs** 2–4; pale greenish-blue to greenish-grey with reddish-brown blotches, smears, spots and streaks over pale lavender or purplish-red to pale brown undermarkings; 22 x 15.5 mm (av.; *annectens*).

1290 STREAK-THROATED BARWING

Actinodura waldeni Plate **118**

IDENTIFICATION 20–22 cm. *A.w.poliotis*: **Adult** Told by combination of bulky, relatively short-tailed appearance, floppy dark brown crown and nape feathers with very narrow buffish edges, dark brown head-sides with pale brownish-grey streaks on ear-coverts, lack of pale eyering and dull rufescent underparts with warm buff streaks (broader on throat and upper breast). Rest of upperparts strongly rufescent with broad blackish-brown subterminal tail-band; wings like other barwings but with some grey fringing on greater coverts. *A.w.waldeni* (west N Myanmar) has paler base colour to crown, nape and head-sides, more defined pale buffish edges to crown-feathers and more deeply rufescent body-plumage with less distinct paler streaks on throat and breast; *saturatior* (east N Myanmar) has broader, whiter margins to crown and nape feathers, broader, rather more silvery streaks on ear-coverts, and darker base colour to underparts with more even and contrasting buffish streaks. **Juvenile** Browner crown, weaker throat streaking, warmer below; tertials browner, with few widely spaced black bars. **VOICE** Song is a repeated loud, strident, slightly wavering and rising phrase, starting with a slight rattle: ***tchrrrr-jo-jwiee*** or ***dddrrt-juee-iwee***; sometimes a shorter ***jorr-dwidu***. Calls include low, rather nasal grumbling ***grrr-ut grrr-ut...*** and ***grr-grr-grr-grr-grr...*** **HABITAT & BEHAVIOUR** Broadleaved evergreen forest; 1,700–3,300 m. Often in small slow-moving parties; joins mixed-species feeding flocks. **RANGE & STATUS** Resident NE India, SE Tibet, SW China. **SE Asia** Fairly common resident W,N Myanmar. **BREEDING** Undocumented.

1291 STREAKED BARWING

Actinodura souliei Plate **118**

IDENTIFICATION 21–23 cm. *A.s.griseinucha*: **Adult** Similar to Streak-throated Barwing but has blackish cheeks and lores, mostly plain silvery ear-coverts, pale brownish-grey hindcrown and nape with indistinct dark streaks, and predominantly buffish remainder of body-plumage with broad blackish streaks. Back to uppertail-coverts rufous-buff with less distinct streaks, vent mostly plain rufous-buff. Note range. **Juvenile** Undescribed. **VOICE** Song is undocumented. Calls recall Streak-throated. **HABITAT & BEHAVIOUR** Broadleaved evergreen forest; 1,950–2,500 m. Often in small, slow-moving parties; joins mixed-species feeding flocks. **RANGE & STATUS** Resident south-western China. **SE Asia** Uncommon resident W Tonkin. **BREEDING** Undocumented.

1292 **RUSTY-FRONTED BARWING**

Actinodura egertoni Plate **118**

IDENTIFICATION 21.5–23.5 cm. *A.e.ripponi*: **Adult** Told by rather slim, long-tailed appearance, brownish-grey hood with chestnut forehead, face and chin, yellow to pinkish-yellow bill with darker culmen and tip, and mostly rufous-buff underparts. Uppertail warmish brown with narrow dark bars, wings like other barwings but greater coverts and base of primaries plain rufous. See Spectacled Barwing. Note range. *A.e.lewisi* (north-west N Myanmar) has darker chestnut frontal parts, slatier-grey on head, darker/warmer and less olive-tinged above, more rufescent and less buffy below. **Juvenile** Crown and nape washed warm brown. **VOICE** Song is a fairly quickly repeated, clear, rather high ***ti-wi-wi-wu ti-wi-wi-wu*** or ***ti-wi-wi-woi ti-wi-wi-woi***. Calls with low harsh ***grrit***, ***grrrrrrit*** and ***gwah*** notes. **HABITAT** Broadleaved evergreen forest, forest edge, secondary growth; 1,220–2,600 m, locally down to 215 m in winter (N Myanmar). **RANGE & STATUS** Resident Nepal, NE Indian subcontinent, SW China. **SE Asia** Common resident SW,W,S(west),N Myanmar. **BREEDING** April–July. **Nest** Rather deep cup, in bush, sapling or bamboo; 0.9–7.5 m above ground. **Eggs** 2–4; bright blue to pale blue-green, blotched, spotted and streaked dark brown to reddish-brown over pale greyish to purplish undermarkings; 20.3–24.6 x 15.2–18 mm.

1293 **SPECTACLED BARWING**

Actinodura ramsayi Plate **118**

IDENTIFICATION 23.5–24.5 cm. *A.r.ramsayi*: **Adult** Combination of greyish-olive ear-coverts and upperside, buffy-rufous forehead, pronounced white eyering, blackish lores and cheeks, and deep buff underparts distinctive. Blackish primary coverts contrast with largely rufescent base colour of flight-feathers; has blackish bars on wings (including greater coverts) and uppertail. *A.r.yunnanensis* (W,E Tonkin) is much more rufescent above and below, has uniform rufous crown and nape and often shows narrow blackish throat-streaks; *radcliffei* (C[east],E[north] Myanmar) is roughly intermediate and also has rich dark brown lores and cheeks; birds from south-east N Laos (probably an undescribed subspecies) are like *yunnanensis* but have greyer upperparts with deep rufous restricted to forehead. **Juvenile** Undescribed. **VOICE** Song is a quite quickly repeated, rather mournful, high-pitched, bouncing, descending ***iee-iee-iee-iuu***; sometimes accompanied (by female?) with high-pitched, even, forced 2–3 note ***ewh ewh ewh***. Calls include low harsh ***baoh*** or ***berrh*** notes. **HABITAT** Broadleaved evergreen forest, forest edge, secondary growth, scrub and grass; 1,000–2,500 m, locally down to 610 m in Myanmar. **RANGE & STATUS** Resident SW China. **SE Asia** Common resident C(east),E,S(east) Myanmar, NW Thailand, N,C Laos, W,E Tonkin. **BREEDING** March–April. Otherwise undocumented.

1294 **BLACK-CROWNED BARWING**

Actinodura sodangorum Plate **118**

IDENTIFICATION 24 cm. Monotypic. **Adult** Like Spectacled Barwing but has black crown, colder olive-brown upperside, rather broad blackish throat-streaks, mostly black flight-feathers and tertials with some buff bars and buff outer greater coverts with some broad black bars. **Juvenile** Undescribed. **VOICE** Similar to Spectacled. **HABITAT & BEHAVIOUR** Broadleaved evergreen forest, forest edge, secondary growth, tall grass and scrub adjacent to pine woodland; 1,100–2,400 m. Occurs singly or in pairs. **RANGE & STATUS** Endemic. Local resident east S Laos, southern C Annam. **BREEDING** Undocumented.

GENUS INCERTAE SEDIS: Fire-tailed Myzornis

Monotypic family ? Rather slender downcurved bill, specially adapted, bristled tongue for nectar-feeding, bright green plumage, sexually dimorphic, often around rhododendrons.

1295 **FIRE-TAILED MYZORNIS**

Myzornis pyrrhoura Plate **110**

IDENTIFICATION 12.5 cm. Monotypic. **Male** Very striking. Largely bright green, with black-centred crown-feathers, narrow black mask, red-fringed and black-tipped tail, reddish patch on lower throat/upper breast, prominent flame-red to yellow fringes to secondaries, and white tips to secondaries and outer primaries. Has variable golden-brown wash on mantle, scapulars and back, light blue wash on centre of abdomen (sometimes on rump and lower flanks) and deep orange-buff vent. **Female** Upperparts often bluer-tinged (particularly scapulars and rump) and less golden-brown; underparts generally bluer-tinged without distinct reddish patch on throat/breast, centre of underparts buffier-grey, vent duller. **Juvenile** Duller than female. **Immature male** Similar to female but underparts greener with more orange-tinged centre of lower throat/upper breast. **VOICE** Calls include extremely thin, high-pitched, often quickly repeated ***si*** notes. **HABITAT & BEHAVIOUR** Edge of broadleaved evergreen forest, secondary growth, scrub; 2,440–3,660 m, down to 1,800 m in winter. Often found feeding on flowering shrubs and trees, including rhododendrons; sometimes joins mixed-species feeding flocks. **RANGE & STATUS** Resident Nepal, Bhutan, NE India, SE Tibet, SW China. **SE Asia** Uncommon resident N Myanmar. **BREEDING** Nepal: May–June. **Nest** Globular structure, embedded in moss on tree-trunk, bank or rock-face. **Eggs** Uncertain, said to be white; 17 x 13 mm.

ACROCEPHALIDAE: *Acrocephalus* warblers & allies

Worldwide 46 species. SE Asia 9 species. Small to relatively large and active with relatively fine, pointed bills. Nondescript brown with paler underparts and usually a pale supercilium. Feed mainly on insects and spiders, frequenting grass, reeds and scrub.

1296 **BLACK-BROWED REED-WARBLER**

Acrocephalus bistrigiceps Plate **104**

IDENTIFICATION 13.5–14 cm. Monotypic. **Adult** Generally the commonest and most widespread small reed-warbler wintering in the region. Shows distinctive long broad buffy-white supercilium, bordered above by prominent blackish lateral crown-stripe. Crown and upperparts warm olive-brown with slightly more rufescent rump and uppertail-coverts; underparts whitish with warm buff breast-sides and flanks. Brown of plumage paler and more greyish-olive when worn. **VOICE** Song is a series of quickly repeated short phrases, interspersed with dry rasping and churring notes. Calls with soft, clucking ***chuc*** notes. **HABITAT** Emergent vegetation and scrub in and around marshes, rice-paddy margins, sometimes drier areas on passage; up to 800 m. **RANGE & STATUS** Breeds SE Siberia, Ussuriland, Sakhalin, C,E,NE China, N Korea, northern Japan; winters NE Indian subcontinent, S China, rarely N Sumatra. **SE Asia** Fairly common to common winter visitor (except SW,W,N,C Myanmar, W Tonkin, S Annam). Also recorded on passage C Thailand, Cambodia. Rarely recorded in summer S Thailand.

1297 **BLYTH'S REED-WARBLER**

Acrocephalus dumetorum Plate **104**

IDENTIFICATION 13.5–15.5 cm. Monotypic. **Adult** Difficult to separate from Blunt-winged Warbler but slightly longer-billed and shorter-tailed, has less contrasting tertial fringes and slightly longer primary projection. In worn plumage, also differs by somewhat greyer-olive, more uniform upperside (rump and uppertail-coverts warmer-tinged on Blunt-winged); in fresh plumage, has noticeably colder, less rufescent upperside and duller flanks. See other small reed-warblers. **VOICE** Song is made up of pleasant, varied, well-spaced phrases, each repeated 2–10 times and typically beginning with 1–2 'tongue-clicking' notes. Phrases include a ***trek-trek CHUEE***; ***chrak-chrak CHU-EE-LOO*** and ***trek-trek see-ee-hue***. Often includes avian mimicry. Calls are a fairly soft ***chek*** or ***teck*** (slightly softer than Paddyfield Warbler), ***chek-tchr*** and harsh, scraping ***cherr*** or ***trrrr*** notes when agitated. **HABITAT** Scrub and rank vegetation in wet and dry areas, cultivation borders; lowlands. **RANGE & STATUS** Breeds north-eastern W Palearctic east to Siberia (Baikal region), N Afghanistan, NW Mongolia; mainly winters Indian subcontinent. **SE Asia** Uncommon winter visitor SW,C,S,E Myanmar.

1298 LARGE-BILLED REED-WARBLER

Acrocephalus orinus Plate **104**

IDENTIFICATION 13–14.5 cm. Monotypic. **Adult** Difficult to separate from Blunt-winged Warbler and Blyth's Reed-warbler, but bill longer, more wedge-shaped when seen from below, and with entirely flesh-pink lower mandible. Supercilium shorter than Blunt-winged; brown of plumage tends to be very slightly darker and more warm-toned than Blyth's. In hand, legs and claws average longer than Blyth's. See other small reed-warblers. **VOICE** Undocumented. **HABITAT & BEHAVIOUR** Freshwater aquatic habitats, including reed-beds, also bamboo clumps, mango trees etc.; lowlands. Sight records are of foraging birds in mid-storey to canopy of trees and bamboo. **RANGE & STATUS** Recorded once NW India (Sutlej Valley, Himachal Pradesh, mid-November 1867), once near Kolkata (Calcutta), early April 2007, and at Kanha NP (NC India) in early April–early May 2008 (latter two being sight records supported by photographs). **SE Asia** Rare (winter visitor and/or passage migrant?; all three records in late March) NW,C Thailand.

1299 MANCHURIAN REED-WARBLER

Acrocephalus tangorum Plate **104**

IDENTIFICATION 13–14.5 cm. Monotypic. **Adult** Similar to Paddyfield Warbler but bill longer, usually with completely pale lower mandible; always shows more pronounced blackish line on crown-side (though less pronounced in fresh plumage), supercilium typically whiter; in worn plumage, shows darker greyish-brown upperparts. In worn plumage also very similar to Black-browed Reed-warbler but shows longer bill, narrower blackish line on crown-side (contrasting less with darker crown-centre), usually whiter supercilium and longer, somewhat narrower tail. **VOICE** Similar to Paddyfield Warbler. **HABITAT** Reedbeds and emergent vegetation in freshwater marshes, lake borders; lowlands (but up to 1,900 m on passage). **RANGE & STATUS** Breeds NE China. **SE Asia** Local winter visitor S(east) Myanmar, SE,C,W(coastal) Thailand, Cambodia, S Laos. Scarce passage migrant N Laos, W,E Tonkin, Cochinchina. Vagrant NW Thailand.

1300 PADDYFIELD WARBLER

Acrocephalus agricola Plate **104**

IDENTIFICATION 13–14.5 cm. Monotypic. **Adult** Similar to Blunt-winged Warbler but has longer, more distinct supercilium behind eye, often bordered above by faint dark line, bill shorter with extensive dark tip to lower mandible, longer primary projection (beyond tertial tips); in worn plumage (mid- to late winter) slightly greyer above with less contrastingly rufescent rump and uppertail-coverts. See Black-browed and Blyth's Reed-warblers. **VOICE** Song is a series of rich warbling phrases, interspersed with squeakier higher-pitched notes, clearly richer, more musical and more varied than Black-browed, and slightly slower and less forced. Calls with a soft ***dzak*** or ***tack***, fairly gentle ***trrrr*** notes and rather harsh, nasal ***cheeer***. **HABITAT** Reedbeds, emergent vegetation in freshwater marshes, lake borders; lowlands. **RANGE & STATUS** Breeds W(east),C Palearctic, NW China; winters Iran, Indian subcontinent. **SE Asia** Rare winter visitor/vagrant SW,N Myanmar, NW Thailand, N Laos.

1301 BLUNT-WINGED WARBLER

Acrocephalus concinens Plate **104**

IDENTIFICATION 14–14.5 cm. *A.c.concinens*: **Adult** Told from other small reed-warblers (except Blyth's) by combination of relatively long bill (lower mandible pale with darker shadow near tip), short whitish supercilium (ending just behind eye), lack of dark line on crown-side and short primary projection (beyond tertial tips). See Paddyfield Warbler. For differences from very similar Blyth's Reed-warbler, see that species. **Other subspecies in SE Asia** *A.c.stevensi* (Myanmar). **VOICE** Song is relatively slow and deep-throated (rhythm and quality may almost recall miniature Oriental Reed-warbler), broken into short repeated phrases and including some fairly deep churring notes. Calls are a short, quiet ***tcheck*** and soft drawn-out ***churrr***. **HABITAT** Tall grass and reeds, usually near water; up to 1,525 m (to 1,900 m on passage). **RANGE & STATUS** Breeds N Afghanistan, N Pakistan, NW,NE India, N,C China, possibly S China. Chinese populations winter to south, S China; winter range of western populations unclear. **SE Asia** Uncommon winter visitor S,E Myanmar, north Tenasserim, W(coastal),NW,NE,C,northern S Thailand, Cambodia, N,C Laos, E Tonkin. Also recorded on passage W,E Tonkin.

1302 INDIAN REED-WARBLER

Acrocephalus brunnescens Plate **104**

IDENTIFICATION 18.5–20.5 cm. *A.b.amyae*: **Adult** Very similar to Oriental Reed-warbler but has narrower bill, less pronounced supercilium (particularly behind eye), rarely shows obvious streaking on lower throat/upper breast, has slightly shorter primary projection and longer tail without whitish tips; also usually buffier underparts. See Thick-billed Warbler and male Manchurian Bush-warbler. **Juvenile** Undescribed. **VOICE** Song is a loud, prolonged, varied series of mainly harsh, guttural, low-pitched clanging notes, most being repeated several times, though irregularly spaced and with breaks of several seconds. In India, note-types include short raucous ***krsh-krsh-krsh...***, buzzy high staccato notes with sharp ending, e.g. ***vurchIK vurchIK...***, rasping nasal ***gurdy-gurdy...***, burry hollow ascending ***krr-kru-vit...***, high metallic ***tzwink-tzwink-tzwink...***, and raspy ***scratch'it-scratch'it...***. Calls include deliberate, hard, loud ***CHAUK*** or ***TSCHAK***. **HABITAT** Emergent vegetation in marshes; up to 915 m. **RANGE & STATUS** Breeds Central Asia, E,S Iran, Afghanistan, Pakistan, India (local), Sri Lanka, S China; some populations winter Persian Gulf, lowlands of Indian subcontinent (except Sri Lanka). **SE Asia** Uncommon resident SW,C,S,E Myanmar. Recorded (status uncertain) NW,C Thailand, N Laos, C Annam. **BREEDING** India: May–August. **Nest** Neat deep cup, attached to reeds or grass; 0.3–1 m above water or ground. **Eggs** 3–6; greyish-white to pale sea-green or pale brown, liberally speckled and spotted blackish-brown over lavender-grey undermarkings; 22.7 x 15.9 mm (av. India).

1303 ORIENTAL REED-WARBLER

Acrocephalus orientalis Plate **104**

IDENTIFICATION 18–20 cm. Monotypic. **Adult** The commonest and most widespread large reed-warbler in the region. Shows prominent whitish supercilium, dark eyestripe, warmish olive-brown upperparts and whitish underparts with warm buffish wash on flanks and vent. In fresh plumage appears somewhat warmer above. Difficult to separate from Indian Reed-warbler but has somewhat stouter bill, more prominent supercilium (particularly behind eye), indistinct greyish streaks on lower throat/upper breast (more obvious in worn plumage), slightly longer primary projection (beyond tertial tips) and shorter tail with whitish tips (mainly outer feathers); also tends to be less buffish below. See Thick-billed Warbler and male Manchurian Bush-warbler. **VOICE** Song consists of deep guttural churring and croaking notes interspersed with repeated warbling phrases: ***kawa-kawa-kawa-gurk-gurk eek eek gurk kawa...*** etc. Calls with loud ***CHACK*** and soft churring notes. **HABITAT**

Emergent vegetation in marshes, rice paddies, borders of cultivation, grass and scrub in less wet areas; up to 950 m. **RANGE & STATUS** Breeds S,SE Siberia, Sakhalin, E Mongolia, N,NE,C,E China, N,S Korea, Japan; winters NE India, S China, Sundas, Philippines, Wallacea, rarely New Guinea, N Australia. **SE Asia** Uncommon winter visitor and passage migrant (except SW,W,N,E Myanmar, W Tonkin, N Annam). Rarely recorded in summer C Thailand.

1304 THICK-BILLED WARBLER

Acrocephalus aedon Plate **104**

IDENTIFICATION 18.5–21 cm. *A.a.stegmanni*: **Adult** Like Oriental and Indian Reed-warblers but easily separated by relatively short stout bill and lack of long pale supercilium and dark eyestripe. Appears distinctly round-headed, plain-faced (paler area on lores) and relatively shorter-winged and long-tailed. In fresh plumage, has strong rufescent tinge to upperparts, wings and flanks. **Other subspecies in SE Asia** *A.a.aedon* (recorded from Indochina). **VOICE** Song is a fast loud stream of avian mimicry, interspersed with twittering and excitable sounds (many calls repeated 2–3 times). Calls include hard clicky ***teck*** notes and a harsh wheezy ***verrrh*** when agitated. **HABITAT** Scrub and grass in relatively dry areas, forest clearings and edge; up to 1,525 m. **RANGE & STATUS** Breeds S,SE Siberia, Ussuriland, E Mongolia, NE China, N Korea; winters India, Nepal, Bangladesh. **SE Asia** Common winter visitor (except southern S Thailand, Peninsular Malaysia, Singapore, W Tonkin). Scarce to fairly common passage migrant C Thailand, N Laos, W Tonkin. Vagrant Peninsular Malaysia.

MEGALURIDAE: Grasshopper warblers, *Bradypterus* bush-warblers, Striated Grassbird & allies

Worldwide c.41 species. SE Asia 18 species. **Grasshopper warblers** (*Locustella*) Small to relatively large, with relatively fine, pointed bills, strongly graduated and rounded tails, and mostly streaked upperparts; very skulking, close to ground, mostly in grass and scrub. ***Bradypterus* bush-warblers** Similar to grasshopper warblers, but unstreaked and brownish with paler underparts, some with throat/breast spotting). **Striated grassbird** Relatively large and long-tailed with heavily streaked plumage; inhabiting grassland and reedbeds, feeding mainly on insects.

1305 PLESKE'S WARBLER

Locustella pleskei Plate **119**

IDENTIFICATION 15–16 cm. Monotypic. **Adult** Larger and bulkier than Rusty-rumped Warbler, with much longer bill. Plain-looking, greyish-brown above (no rufous); whitish below with greyish-brown flanks (and to lesser extent breast). Can show vague spots/mottling on lower throat and upper breast and darker shaft-streaks on creamy-brownish undertail-coverts (only thin pale tail-feather tips). Has relatively indistinct supercilium, but obvious pale cream eyering (broken at front and rear). Resembles Oriental Reed-warbler. **VOICE** Sings like Rusty-rumped, but much thinner and reedier. **HABITAT** Mangrove scrub and other low coastal vegetation. **RANGE & STATUS** Breeds offshore islands off extreme south-east Russia, southern Japan, S Korea, E China; winters S China (Hong Kong). **SE Asia** Rare to scarce winter visitor E Tonkin (Red River Delta).

1306 RUSTY-RUMPED WARBLER

Locustella certhiola Plate **119**

IDENTIFICATION 14–15 cm. *L.c.certhiola*: **Adult** Resembles Lanceolated Warbler but larger, with more rufescent upperparts, darker crown and darker tail with whitish feather-tips (difficult to see in field); shows rather contrasting warm rufescent rump and uppertail-coverts, longer and more prominent pale supercilium; lacks dark streaks on flanks and undertail-coverts. Underparts whitish, washed buffy-brown on breast, flanks and vent. *L.c.rubescens* (recorded C,E Myanmar, Cambodia; ? NW Thailand) is darker, deeper rufescent-brown above, with much narrower blackish markings on crown and somewhat less distinct supercilium. See larger Rufous-rumped Grass-babbler. **Juvenile** Like adult but washed yellowish below, with faint dark streaks/spots on lower throat and upper breast. **VOICE** Song is markedly different from Lanceolated, consisting of rapidly delivered, well-structured musical warbling phrases, including: ***tri-tri-tri-tri***; ***prt-prt***; ***chi-chi-chi-chi***; ***srrrrrt*** and ***sivih-sivih-sivih*** etc. Calls include thin metallic ***pit*** or ***pt*** notes, clicking chick, dry rolling ***trrrrt*** and rattling trilled ***rit-tit-tit-tit...*** **HABITAT & BEHAVIOUR** Tall reeds, grass and other vegetation in marshes or freshwater wetlands, rice-paddy margins; up to 610 m. Slightly less skulking than Lanceolated. **RANGE & STATUS** Breeds C,E Palearctic, Mongolia, northern China; winters NE,S Indian subcontinent, S China, Sumatra, Java, Borneo. **SE Asia** Fairly common to common winter visitor (except SW,W,N Myanmar, Tenasserim, C Laos, W Tonkin, C,S Annam). Also recorded on passage NW,C,S Thailand, Peninsular Malaysia, Cambodia, N Laos. Recorded (status uncertain S Annam).

1307 LANCEOLATED WARBLER

Locustella lanceolata Plate **119**

IDENTIFICATION 12–13.5 cm. Monotypic. **Adult** Small size and heavily streaked appearance distinctive. Resembles larger Rusty-rumped Warbler and cisticolas but shows prominent streaking on rump, lower throat, breast, flanks and undertail-coverts, and lacks pronounced rufescent plumage tones and pale tips to tail-feathers. Upperparts and flanks somewhat warmer in fresh plumage, colder and more greyish-olive when worn. See Spotted and Baikal Bush-warblers. **Juvenile** Somewhat less heavily/boldly streaked, with looser, fluffier plumage. **VOICE** Song is a sustained shuttling trill: ***zizizizizizizizizizi...*** Calls include a rather subdued ***tack***, thin clicking ***chick*** or ***pit*** and trilled ***rit-tit-tit-tit***. **HABITAT & BEHAVIOUR** Grass, weeds and scrub, often in marshy areas, cultivation borders; up to 1,800 m. Very skulking, on or near ground, difficult to flush. **RANGE & STATUS** Breeds northern W(north-east),C,E Palearctic, NE China, N Japan; N India, S China, Sumatra, Java, Borneo, Philippines. **SE Asia** Common winter visitor (except SW,W,N,C Myanmar, W Tonkin). Also recorded on passage C Thailand, Peninsular Malaysia, Cambodia, E Tonkin.

1308 SPOTTED BUSH-WARBLER

Bradypterus thoracicus Plate **119**

IDENTIFICATION 13–14 cm. *B.t.thoracicus*: **Adult** Best separated from very similar Baikal Bush-warbler by more rufescent-tinged upperparts, greyer supercilium and breast-sides, heavier dark spotting on throat and breast, and completely blackish bill. Note voice and range. **Juvenile** Upperparts like adult (perhaps slightly paler and less deeply rufescent), obvious yellowish wash on supercilium, head-sides and underparts, browner-grey breast and flanks, and only vague breast markings. Gradually attains dark speckling below. **VOICE** Sings with a rhythmic ***trr-tri'tri'tree trr-tri'tri'tree trr-tri'tri'tree trr-tri'tri'tree...*** **HABITAT & BEHAVIOUR** Breeds in scrub and low vegetation bordering broadleaved evergreen and coniferous forest, clearings; 3,050–3,655 m. In Indian subcontinent, winters in similar habitat to Baikal. Behaves like Baikal. **RANGE & STATUS** Resident (subject to altitudinal movements) N,NE Indian subcontinent, SE Tibet, SW,W China. **SE Asia** Uncommon resident (subject to altitudinal movements) N Myanmar. **BREEDING** India: May–July. **Nest** Dome or deep cup, in low vegetation; up to 50 cm above ground. **Eggs** 3–4; white, lightly freckled pinkish-red to brick-red (often ringed or capped at broader end); 18.4 x 13.9 mm (av.).

1309 **BAIKAL BUSH-WARBLER**
Bradypterus davidi Plate **119**
IDENTIFICATION 13–14 cm. *B.t.suschkini*: **Adult** Similar to other *Bradypterus* bush-warblers but shows distinctive combination of cold, dark brown upperparts (slightly rufescent), dark speckling on lower throat and upper breast, greyish breast-sides, dark undertail-coverts with contrasting white tips and relatively short tail. Shows whitish supercilium and dark-tipped pale lower mandible. May show much weaker throat/breast-spotting (possibly age-related). Note voice. See Lanceolated Warbler and Spot-throated Babbler. **VOICE** Sings with monotonous ***dzzzzr dzzzzr dzzzzr dzzzzr...*** Calls with harsh low ***tuk*** and ***rrtuk*** notes. **HABITAT & BEHAVIOUR** Tall grass, scrub and weeds in open areas, often near water; up to 1,400 m. Very skulking, on or close to ground. **RANGE & STATUS** Breeds S Siberia, Mongolia, NE China; winters NE India; S China? **SE Asia** Locally common winter visitor C,E Myanmar, W,NW,NE,C Thailand, N Laos, E Tonkin. **NOTE** Split from Spotted Bush-warbler following Alström *et al.* (in press).

1310 **CHINESE BUSH-WARBLER**
Bradypterus tacsanowskius Plate **119**
IDENTIFICATION 14 cm. Monotypic. **Adult** Told from other *Bradypterus* bush-warblers by combination of relatively long-tailed appearance, more greyish-olive upperparts, whiter centre of underparts and rather indistinct darker centres to undertail-coverts; usually lacks obvious dark speckling on lower throat and upper breast, bill has pale lower mandible. **First winter** Like adult but supercilium and underparts washed yellow, often has light speckling on lower throat and upper breast. **VOICE** Song is a monotonous, husky, crackling insect-like ***hhhhhht hhhhhht hhhhhht hhhhhhhht...*** **HABITAT** Tall grass, reeds and scrub, particularly in alluvial plains; up to 1,500 m. **RANGE & STATUS** Breeds S Siberia, Ussuriland, Mongolia, NC,N,NE China; winters NE Indian subcontinent; S China? **SE Asia** Scarce winter visitor S Myanmar, NW Thailand, N Laos, S Annam. Rare passage migrant/vagrant E Tonkin.

1311 **BROWN BUSH-WARBLER**
Bradypterus luteoventris Plate **119**
IDENTIFICATION 14–14.5 cm. Monotypic. **Adult** From other *Bradypterus* bush-warblers by combination of dark rufescent-brown upperparts, buffy-rufous flanks, lack of throat/breast-speckling and lack of obvious pale tips on plain rufescent undertail-coverts. Bill has pale fleshy-brown lower mandible. Note voice. **Juvenile** Upperparts tinged more rufous-chestnut, light areas of underparts washed pale yellow. **VOICE** Song is a prolonged sewing-machine-like repetition of rather quiet rapid short notes: ***tutututututututututut...*** or ***hehehehehehehehehehe...*** Calls include a harsh deep ***thuck thuck thuck...***, sharp high-pitched ***tink tink tink tink...*** and harsh grating ***tchrrrrk tchrrrrk...*** **HABITAT** Low vegetation in clearings and along forest edge, grass and scrub; 1,830–2,590 m, down to 800 m in winter. **RANGE & STATUS** Resident (subject to altitudinal movements) NE Indian subcontinent, C and southern China. **SE Asia** Locally common resident W,N Myanmar, W Tonkin. Scarce winter visitor S Myanmar, NW,NE Thailand. **BREEDING** June–July. **Nest** Deep cup (sometimes domed), in low vegetation; up to 1 m above ground (rarely on ground). **Eggs** 3–5; white to pale pink, freckled (sometimes blotched) reddish-brown (sometimes capped at broader end); 18.2 x 14.3 mm (av.).

1312 **RUSSET BUSH-WARBLER**
Bradypterus mandelli Plate **119**
IDENTIFICATION 13–14 cm. *B.m.mandelli*: **Adult** Similar to Brown Bush-warbler but shows dark brown undertail-coverts with contrasting broad whitish feather-tips and all-blackish bill; usually shows fine dark speckling on lower throat and upper breast. From Spotted and Baikal Bush-warblers by much more rufescent upperparts and flanks, usually less distinct throat/breast-speckling and longer tail. Note voice. **Juvenile** Similar to Spotted but upperparts more rufescent, flanks more broadly dark and mixed with some rufous, breast-band mixed with some rufous; possibly less yellow-tinged below. See Brown. **Other subspecies in SE Asia** *B.m.idoneus* (S Annam). **VOICE** Song is a distinctive monotonous metallic buzzing ***zree-ut zree-ut zree-ut zree-ut...*** Calls are very similar to Brown. **HABITAT** Low vegetation in clearings and along forest edge, grass and scrub; 400–2,500 m (mainly above 1,000 m). **RANGE & STATUS** Resident NE India, Bhutan, C and southern China. **SE Asia** Locally common resident W,N Myanmar, NW Thailand, N Laos, W,E Tonkin, S Annam. **BREEDING** Not reliably documented.

1313 **STRIATED GRASSBIRD**
Megalurus palustris Plate **119**
IDENTIFICATION Male 25–28 cm, female 21.5–24 cm. *M.p.toklao*: **Adult** Largest warbler in region. Told by size, long graduated pointed tail, buffish-brown upperside with heavy dark streaking, prominent white supercilium and largely whitish underparts with fine dark streaks on breast and flanks. Prominence and contrast of upperpart-streaking, general tone of upperparts and extent of underpart-streaking variable. See Striated Babbler. **Juvenile** Supercilium and underparts washed yellow, streaking on breast and flanks fainter, bill paler. **VOICE** Song consists of loud, rich, fluty warbling notes. Call is an explosive ***pwit***. **HABITAT & BEHAVIOUR** Marshlands with clumps of tall grass and reeds, scrub, sometimes in drier areas; up to 1,525 m. Often perches in open. Has high-soaring and parachuting song-flight. **RANGE & STATUS** Resident NE Pakistan, C,N,NE India, S Nepal, Bangladesh, SW,S China, Java, Bali, N Borneo, Philippines. **SE Asia** Locally common resident Myanmar, C,NW,NE(south-west),SE Thailand, Indochina (except C,S Laos, S Annam). **BREEDING** April–August. **Nest** Large dome, in bush or low vegetation. **Eggs** 3–4; whitish to pale dull pink, rather densely speckled dark brown and purple-brown over lilac-grey undermarkings; 22.7 x 16.7 mm (av.). Brood-parasitised by Eurasian Cuckoo.

CISTICOLIDAE: Cisticolas, tailorbirds, prinias & allies

Worldwide at least 114 species. SE Asia 14 species. **Cisticolas** (*Cisticola*) Small, streaky and warbler-like with shortish tails; inhabit marshy and open grassy habitats, feeding mainly on insects. **Tailorbirds** (*Orthotomus*) Small and plump, with rather long bills and tails (some depending on season), adults with rufous to chestnut on head; mostly in wooded habitats. **Prinias** (*Prinia*) Small and warbler-like with long graduated tails, varied streaked and unstreaked plumage patterns; inhabit grasslands and other open habitats, feeding mainly on insects.

1314 **ZITTING CISTICOLA**
Cisticola juncidis Plate **119**
IDENTIFICATION 10.5–12 cm. *C.j.malaya*: **Adult non-breeding** Upperparts buffish-brown, boldly streaked blackish, rump distinctly rufescent, tail with blackish subterminal markings and broad white tips; underparts whitish with variable warm buff wash on breast and flanks. Very like non-breeding male and female Bright-headed Cisticola, but has whiter supercilium, duller nape and whiter tail-tips. Note voice. Resembles Rusty-rumped Warbler but much smaller and finer-billed. **Adult breeding** Crown and mantle rather uniform with broader dark streaks, tail paler, slightly shorter and more rufescent. **Juvenile** Upperparts intermediate between non-breeding and breeding adult, underparts washed light yellow. **Other subspecies in SE Asia** *C.j.cursitans* (SW,W,N,E,C Myanmar), *tinnabulans* (Indochina). **VOICE** Song is a long, monotonous, clicking ***dzip dzip dzip dzip...*** or ***pip pip pip pip...*** Call is a ***chipp*** or ***plit***.

HABITAT & BEHAVIOUR Rice paddies, marshes, grassland (mainly wet); up to 1,220 m. Skulking but has weakly undulating, often wide-circling song-flight. **RANGE & STATUS** Resident (subject to some minor movements) Africa, southern W Palearctic, Middle East, Indian subcontinent, C and southern China, Taiwan, Japan, Philippines, Sumatra, Java, Bali, Lesser Sundas, Philippines, Sulawesi, New Guinea, N Australia. **SE Asia** Common resident throughout. **BREEDING** January–October. **Nest** Oval dome or semi-dome, in low vegetation. **Eggs** 2–6 (2–3 in south); white to pale blue, with irregular fine reddish-brown to brown or pale red speckles (often mainly at broader end); 15.7–17 x 11–12 mm (av.; *malaya*). Brood-parasitised by Eurasian and Plaintive Cuckoos.

1315 BRIGHT-HEADED CISTICOLA

Cisticola exilis Plate **119**

IDENTIFICATION 9.5–11.5 cm. *C.e.equicaudata*: **Male non-breeding** Similar to female non-breeding but tail longer. *C.e.tytleri* shows somewhat heavier dark streaking above. **Male breeding** Unmistakable, with unstreaked golden-rufous crown and rich buff breast; tail much shorter. *C.e.tytleri* (SW,N Myanmar) has creamy-buffish crown and bolder black streaking on mantle. **Female non-breeding** Hard to separate from Zitting but shows distinctive rufescent supercilium, nuchal collar and neck-sides, and duller brownish-white tail-tips. **Female breeding** Like non-breeding but rump and uppertail-coverts plainer. **Juvenile** Like non-breeding female but upperparts and wing-feather fringes somewhat browner, underparts distinctly pale yellow, washed buff on flanks. **VOICE** Song consists of one or two comical jolly doubled notes introduced by a long buzzy wheeze (often repeated on its own): ***bzzzeeee joo-ee***; ***bzzzeeee joo-ee di-di*** and ***bzzzeeee-dji-shiwi joo-ee*** etc.; sometimes shorter ***trrrt joo-ee***. **HABITAT & BEHAVIOUR** Tall vegetation in marshes, grassland with bushes, sometimes dry croplands; up to 1,450 m. Skulking. Has short song-flight but often sings from perch. **RANGE & STATUS** Resident N,NE Indian subcontinent, C,S India, southern China, Taiwan, Sundas, Philippines, Sulawesi, S Moluccas, New Guinea, Bismarck Archipelago, Australia. **SE Asia** Locally common resident SW,W,N,S Myanmar, Thailand (except S), Cambodia, N(south),C,S Laos, C,S Annam, Cochinchina. **BREEDING** April–August. **Nest** Oval dome or half-cup, attached to low vegetation; 50–75 cm above ground or water. **Eggs** 4–5; bright pale blue, spotted and blotched reddish-brown, black or purple, sometimes with grey undermarkings (often marked mainly at broader end); 14.8 x 11.4 mm (av.; *tytleri*). Brood-parasitised by Plaintive Cuckoo.

1316 ASHY TAILORBIRD

Orthotomus ruficeps Plate **120**

IDENTIFICATION 11–11.5 cm. *O.r.cineraceus*: **Male** Rufous forecrown, chin and head-sides and dark grey throat, breast and flanks diagnostic. Tail brownish with whitish feather-tips and dark subterminal markings. **Female** Mostly whitish on centre of underparts. **Juvenile** Upperparts browner, no rufous on head (except for wash on ear-coverts), underparts like female but more washed out, tail with darker subterminal area. From Rufous-tailed by greyer tail-base (no obvious rufous) with whitish tip, rufous wash on ear-coverts and no yellowish tinge below. **VOICE** Song is a repetitive ***chip-WII-chip chip-WII-chip...*** (***chip*** brief, ***WII*** stressed) and ***chu-IIP chu-IIP chu-IIP*** (***IIP*** stressed). Calls (all repeated) include a spluttering trilled ***prrrrt***, rolling ***prii'u*** and harsh ***thieu***. **HABITAT** Mangroves, coastal scrub and peatswamp forest, rarely inland forest; also all available habitats on certain offshore islands. Lowlands. **RANGE & STATUS** Resident Greater Sundas, Philippines (Palawan). **SE Asia** Common resident Tenasserim, S Thailand, Peninsular Malaysia, Singapore, Cochinchina. **BREEDING** January–October. **Nest** Like Common Tailorbird; up to 0.25–1.5 m above ground. **Eggs** 2–3; blue-green, white or pinkish, evenly spotted and flecked reddish-brown and pale purple, sometimes unmarked; 15.5–17 x 10.7–11.2 mm. Brood-parasitised by Plaintive Cuckoo.

1317 RUFOUS-TAILED TAILORBIRD

Orthotomus sericeus Plate **120**

IDENTIFICATION 12–14 cm. *O.s.hesperius*: **Adult** Combination of rufous-chestnut crown, dull rufous tail and mostly whitish underside diagnostic. See Ashy Tailorbird. **Juvenile** Lacks obvious rufous-chestnut on crown, upperparts rather uniform and browner than adult (crown slightly more rufescent), tail with darker subterminal area, underparts tinged yellowish to buffish. **VOICE** Song consists of rapidly repeated loud couplets with stressed first note: ***chop-wir***, ***chik-wir***, ***tu-twik*** and ***prui-chir*** etc. Partner often joins in with monotonous ***u'u'u'u'u...*** Also a high-pitched wheezy ***tzee-tzee-tzee...*** **HABITAT** Forest edge, secondary growth, overgrown clearings, locally edge of cultivation and mangroves, dense gardens; up to 400 m. **RANGE & STATUS** Resident Sumatra, Borneo, Philippines. **SE Asia** Uncommon to fairly common resident Tenasserim, S Thailand, Peninsular Malaysia, Singapore. **BREEDING** January–September. **Nest** Like Common Tailorbird; up to 1 m above ground. **Eggs** 2–3; blue or buffy-pink, spotted reddish-brown and lavender, mainly over broader end.

1318 DARK-NECKED TAILORBIRD

Orthotomus atrogularis Plate **120**

IDENTIFICATION 10.5–12 cm. *O.a.nitidus*: **Male non-breeding** Resembles Common Tailorbird but shows more extensive and brighter rufous on crown, brighter yellowish-green upperparts, bolder grey streaking on throat/breast-sides and yellow undertail-coverts. Population in E Tonkin and N Annam shows darker rufous on crown and darker green upperparts (undescribed subspecies?). **Male breeding** Shows diagnostic solid blackish-grey lower throat and upper breast; tail much shorter than Common. **Female** Like male non-breeding but has less distinct streaking on throat/breast-sides. **Juvenile** Like female but duller green above, rufous on crown initially lacking and only gradually acquired. **Other subspecies in SE Asia** *O.a.atrogularis* (S Thailand southwards), *annambensis* (Pulau Tioman, off south-east Peninsular Malaysia). **VOICE** Song is a staccato nasal high-pitched ***kri'i'i'i'i*** mixed with short ***tew*** notes. Various dry staccato trilling sounds: ***churrrit churrrit churrrit-churrrit***; ***titttrrrt titttrrrt*** etc., or ringing ***prrrp-prrrp***. Pairs often call antiphonally: ***tittrrruit-churrrit...*** etc. **HABITAT & BEHAVIOUR** Broadleaved evergreen, semi-evergreen and mixed deciduous forest, secondary growth, scrub, mangroves, locally in overgrown rural areas, rarely parks and gardens; up to 1,200 m. Frequently cocks tail, quite skulking. **RANGE & STATUS** Resident NE India, E Bangladesh, SW China, Sumatra, Borneo, Philippines. **SE Asia** Common resident (except SW Myanmar). **BREEDING** February–September. **Nest** Like Common Tailorbird; within 0.5–4 m of ground. **Eggs** 3–5 (3–4 in south); white to pale blue, with or without pale rufous and dark brown blotches, spots and scribbles; 15.4 x 11.4 mm (av.; *nitidus*). Brood-parasitised by Plaintive and Drongo Cuckoos.

1319 COMMON TAILORBIRD

Orthotomus sutorius Plate **120**

IDENTIFICATION 10.5–13 cm. *O.s.inexpectatus*: **Adult** Combination of rufescent forecrown, greenish upperparts, pale underparts and long bill and tail distinctive. Often shows indistinct greyish streaks on throat/breast-sides and some grey on nape. Central tail-feathers of breeding male up to 4 cm longer. *O.s.maculicollis* (southern S Thailand southwards) has darker rufous forehead, darker green upperparts and darker grey nape, ear-coverts and streaks on breast-sides. From similar Dark-necked Tailorbird by duller, less extensive rufous on crown, pale supercilium, duller upperparts and lack of yellow on vent and wing-feather fringes. **Juvenile** Shows less obvious rufous on forecrown (initially lacking). **Other subspecies in SE Asia** *O.s.longicauda* (N Indochina; E Myanmar?). **VOICE** Song is a repeated explosive ***chee-yup chee-yup chee-yup...***; ***pitchik-pitchik-pitchik...*** or ***te-chi te-chi te-chi te-chi...*** etc. Calls include a loud, quickly repeated ***pit-pit-pit...*** and quick ***cheep-cheep...*** **HABITAT & BEHAVIOUR** Gardens, scrub, bamboo clumps, cultivation borders, open deciduous woodland, mangroves; up to 1,525 m. Quite skulking, frequently cocks tail. **RANGE & STATUS** Resident Indian subcontinent, southern China, Java. **SE Asia** Common resident throughout. **BREEDING** December–October. Multi-brooded. **Nest** Deep cup, in cylinder formed by 1–3 sewn-together leaves, in bush or tree; 0.5–2 m (rarely to 5 m) above ground. **Eggs** 3–6 (3–4 in south); white, cream, pale pink or pale blue to greenish-blue, with or without sparse reddish-brown, brown, black and purplish-black spots and blotches (often mainly at broader end); 14–16.5 x 10.5–12 mm (av.; *maculicollis*). Brood-parasitised by Plaintive Cuckoo.

1320 **RUFESCENT PRINIA**
Prinia rufescens Plate **120**
IDENTIFICATION 10.5–12.5 cm. *P.r.beavani*: **Adult non-breeding** Small size, plain rufescent mantle, and strongly graduated tail with pale-tipped feathers distinctive. Difficult to separate from Grey-breasted Prinia but has slightly thicker bill, more rufescent upperparts and -tail, more extensive rufous-buff on flanks and vent, and usually more pronounced, longer whitish supercilium. Note distinctive voice. See Plain Prinia. **Adult breeding** Crown, nape and ear-coverts distinctly slaty-grey, tail slightly shorter. *P.r.objurgans* (SE Thailand), *peninsularis* (south Tenasserim, S Thailand), *extrema* (extreme S Thailand, Peninsular Malaysia) and *dalatensis* (S Annam) have slightly darker grey crown and nape and darker upperparts, contrasting more with underparts. **Juvenile** Like adult non-breeding but crown tinged greyish-olive, underparts tinged yellow, outer fringes of wing-feathers paler and more rufescent. **Other subspecies in SE Asia** *P.r.rufescens* (western & northern Myanmar). **VOICE** Song is a rhythmic ***ti'chew-ti'chew-ti'chew-ti'chew...*** or more rapid ***chewp'chewp'chewp'chewp'chewp...*** Calls with buzzing ***peez-eez-eez-eez*** and single ***tchi*** notes. **HABITAT & BEHAVIOUR** Undergrowth in deciduous and open evergreen forest, forest edge, grass and scrub; up to 1,675 m. Often in small, lively parties. **RANGE & STATUS** E,NE India, Bhutan, Bangladesh, SW China. **SE Asia** Common resident (except C Thailand, Singapore). **BREEDING** December–September. **Nest** Rough cup, sewn into 1–3 upright leaves (forming cone); up to 1 m above ground. **Eggs** 3–4; pale blue to whitish, speckled pale reddish to reddish-brown, or unmarked; 16.1 x 11.8 mm (av.; *rufescens*).

1321 **GREY-BREASTED PRINIA**
Prinia hodgsonii Plate **120**
IDENTIFICATION 10–12 cm. *P.h.erro*: **Adult non-breeding** Difficult to separate from Rufescent Prinia but has slightly thinner bill, little (typically pre-ocular only) or no supercilium, less rufescent upperparts and -tail and often a greyish wash on sides of neck and breast. Note distinctive voice. *P.h.rufula* (N Myanmar) tends to be warmer above. **Adult breeding** Unmistakable with dark grey head and broad breast-band, and whitish throat and belly. Tail slightly shorter, may show slight pale loral eyebrow. **Juvenile** Like non-breeding adult but more rufescent above and on wing-feather fringes, bill pale. **Other subspecies in SE Asia** *P.h.hodgsonii* (SW,W,C,S[west] Myanmar), *confusa* (N Indochina). **VOICE** Song is a repeated ***ti swii-swii-swii-swii***, each ***swii*** louder than the last. Also extended bouts of rapid scratchy warbling, interspersed with thin ***tee-tsi*** and ***tir tir tir...*** Calls include a high-pitched ***ti-chu*** and laughing ***hee-hee-hee-hee***. **HABITAT & BEHAVIOUR** Dry grassland, scrub, secondary growth; up to 1,525 m. Usually in small hyperactive flocks. **RANGE & STATUS** Resident N Pakistan, Indian subcontinent, SW China. **SE Asia** Common resident (except Tenasserim, S Thailand, Peninsular Malaysia, Singapore). **BREEDING** April–August. Multi-brooded. **Nest** Delicate cup, sewn into 1–3 upright leaves (forming cone); up to 90 cm above ground. **Eggs** 3–4; variable, blue, white, pinkish-white, grey-green or blue-green, speckled light reddish to reddish-brown (often ringed or capped at broader end), sometimes unmarked; 14.7 x 11.7 mm (av.; *rufula*). Brood-parasitised by Plaintive Cuckoo.

1322 **YELLOW-BELLIED PRINIA**
Prinia flaviventris Plate **120**
IDENTIFICATION 12–14.5 cm. *P.f.delacouri*: **Adult** Shows diagnostic combination of slaty-greyish crown and head-sides, greenish-olive mantle and bright yellow belly and undertail-coverts, contrasting with whitish throat and upper breast. Tail somewhat shorter when worn (summer). *P.f.sonitans* (E Tonkin) is markedly different, with more rufescent nape to uppertail and wing-feather fringes, light buffish throat (whiter in centre) and breast, and deep buff belly and vent (lighter and faintly yellow-tinged in centre). **Juvenile** Differs from adult by uniform rufescent olive-brown upperparts, uniformly pale yellow supercilium and underparts with buffier flanks and rufescent wing-feather fringes. *P.f.sonitans* has whiter throat, breast and centre of abdomen and duller head. **Other subspecies in SE Asia** *P.f.flaviventris* (SW,W,N,C,S[west] Myanmar), *rafflesi* (south Tenasserim and S Thailand southwards). **VOICE** Song is a repetitive, rhythmic, descending, chuckling ***didli-idli-u didli-idli-u didli-idli-u...*** Call is a drawn-out mewing ***pzeeew***. **HABITAT & BEHAVIOUR** Reeds and grass mainly in marshes, coastal scrub and landward edge of mangroves; up to 1,450 m. Skulks; makes curious snapping sound with wings. **RANGE & STATUS** Resident Pakistan, NW,N,NE Indian subcontinent, S China, Taiwan, Greater Sundas. **SE Asia** Common resident throughout. **BREEDING** January–September. **Nest** Oval dome, with large entrance near top, in grass or bush; 0.3–1.5 m above ground. **Eggs** 3–4; glossy brick-red (rarely dark-ringed or -capped at broader end); 15.2 x 11.7 mm (av.; *flaviventris*). Brood-parasitised by Plaintive Cuckoo.

1323 **PLAIN PRINIA** *Prinia inornata* Plate **120**
IDENTIFICATION 13.5–15 cm. *P.i.herberti*: **Adult** Long broad whitish supercilium distinctive. Larger and longer-tailed than Rufescent and Grey-breasted Prinias, smaller and plainer above than Striated and Brown. Shows slightly variable darker streaking on crown, creamy-whitish underparts washed buffish on flanks, sometimes obscure faint darker mantle-streaking. *P.i.extensicauda* (N Indochina) and *blanfordi* (Myanmar [except Tenasserim], NW Thailand) have a non-breeding plumage, with warmer upperparts, tail and wings, mostly rich buff underparts and supercilium, and longer tail; *extensicauda* is deepest buff below. See Brown Prinia. **Juvenile** Like adult but upperparts and tail warmer, wing-feather fringes rufescent, underparts faintly washed yellowish. **VOICE** Song is a monotonous rattling buzzing ***jit-it-it-it-it-it-it*** or ***jirt'jirt'jirt'jirt'jirt'jirt...*** Calls include a clear ***tee-tee-tee*** and nasal ***beep***. **HABITAT & BEHAVIOUR** Grass, reeds and scrub, usually in marshy areas, rice paddies, landward edge of mangroves; up to 1,450 m. Less skulking than most prinias; makes wing-snapping sounds. **RANGE & STATUS** Resident Indian subcontinent, C and southern China, Taiwan, Java. **SE Asia** Common resident (except W,S Thailand, Peninsular Malaysia, Singapore). **BREEDING** March–October. Multi-brooded. **Nest** Globular or long domed structure, with entrance near top, attached to grass or crop stems; 0.9–1.8 m above ground. **Eggs** 3–6; very glossy creamy-white, deep reddish-pink or dusky-blue, boldly marked with deep chocolate and reddish-brown blotches, clouds and hairlines; 15.7 x 11.9 mm (av.; *herberti*). Brood-parasitised by Plaintive Cuckoo.

1324 **STRIATED PRINIA**
Prinia crinigera Plate **120**
IDENTIFICATION 15.5–20 cm. *P.c.catharia*: **Adult non-breeding** Relatively large size, long tail (longer in late winter/spring) and heavily streaked upperparts distinctive. Crown- and mantle-streaking clearly defined, some dark speckling on throat-sides and upper breast, brown bill. From Brown Prinia by heavier-streaked upperparts, prominent streaking on

ear-coverts and neck-sides, and dark speckling on lower throat and upper breast. Note range. Recalls Striated Grassbird but much smaller, without prominent supercilium. Note habitat and range. See Hill and Plain Prinias. **Adult breeding** Darker above, with less defined streaking, breast more mottled, bill black (male only?); when very worn, shows dark blotching across breast and shorter tail. See Brown. **Juvenile** Similar to non-breeding adult but warmer above with fainter dark streaks, less speckled below. See Brown. **VOICE** Sings with a monotonous wheezy scraping ***chi'sireet-chi'sireet-chi'sireet-chi'sireet-chi'sireet-chi'sireet...*** Also has a harsh ***tchak-tchak...*** contact call. **HABITAT & BEHAVIOUR** More open areas with grass and scrub; up to 1,525 m (lower limit unclear). Skulking. **RANGE & STATUS** Resident NE Afghanistan, Pakistan, NW,N,NE Indian subcontinent, C and southern China, Taiwan. **SE Asia** Locally common resident W,N Myanmar. **BREEDING** India: May–October. **Nest** Oval with entrance near top, in low vegetation; up to 1.2 m above ground. **Eggs** 3–7; pale blue to dull green or white to pale pink, usually speckled reddish (often ringed at broader end); 16.7 x 12.8 mm (av.). Brood-parasitised by Eurasian and Plaintive Cuckoos.

1325 **BROWN PRINIA** *Prinia polychroa* Plate **120**
IDENTIFICATION 14.5–18 cm. *P.p.cooki*: **Adult non-breeding** Very like Striated Prinia but upperparts less prominently streaked, possibly less strongly buff on breast and flanks, no obvious streaking on ear-coverts and neck-sides or speckling on lower throat and upper breast. Bill brown. Resembles Plain but larger, with indistinct supercilium/streaks on crown and mantle. **Adult breeding** Upperparts greyer and generally less distinctly streaked than non-breeding; lacks breast-blotching of Striated. Bill black (male only?). *P.p.rocki* (C,S Annam) is buffier on underparts and richer buff on lower flanks in non-breeding plumage; similar in breeding plumage. **Juvenile** Much more uniform above than adult breeding, streaking on upperparts very indistinct, restricted to forecrown or almost absent; upperparts, tail and wings warmer than adult non-breeding. Shows slight yellowish tinge to underparts. See Plain Prinia. **VOICE** Song is similar to Striated: very rapid ***ts'weu-ts'weu-ts'weu-ts'weu-ts'weu...*** or ***tis'iyu-tis'iyu-tis'iyu-tis'iyu...*** Also utters repeated loud clear ***chii*** or ***chiu*** and ***hu'ee*** notes. **HABITAT** Grass and undergrowth in dry dipterocarp woodland and pine forest; up to 1,450 m. **RANGE & STATUS** Resident SW China, Java. **SE Asia** Uncommon to locally common resident S(north),C,E Myanmar, W,NW,NE,C Thailand, Cambodia, C,S Laos, C,S Annam. **BREEDING** May–June. **Nest** Oval dome, with entrance towards top, in low vegetation. **Eggs** 4; fairly glossy, white to pale pink, speckled and blotched reddish-brown, usually in ring or cap at broader end, sometimes overall; 17.8 x 12.7 mm (av.; *cooki*). Brood-parasitised by Plaintive Cuckoo.

1326 **BLACK-THROATED PRINIA**
Prinia atrogularis Plate **120**
IDENTIFICATION 15–20.5 cm. *P.a.khasiana*: **Adult non-breeding** From similar Hill Prinia by strongly rufescent upperparts, tail and wings (brighter on forehead), paler, browner head-sides and almost complete lack of breast-streaking. **Adult breeding** Markedly different to Hill, with black throat merging to scales on breast, white submoustachial stripe and much reduced supercilium. Note range. **Juvenile** Like non-breeding adult but with much shorter tail and variably, faintly dark-mottled throat. **VOICE** Sings with a loud, clear, mechanical, well-spaced, quite jolly, ringing 4–13 note ***tUUH-tUUH-tUUH-tUUH-tUUH-tUUH...*** or ***i'YUU-i'YUU-i'YUU-i'YUU-i'YUU-i'YUU-i'YUU....*** Calls include long series of well-spaced, clear, quite high notes: ***T'WII T'WII T'WII T'WII T'WII...*** or more spaced ***t'WI t'WI t'WI t'WI...*** and ***t'YU t'YU t'YU...*** or ***t'HU t'HU t'HU...*** Also, a series of well but erratically spaced, sparrow-like ***tchi'UP*** calls. **HABITAT & BEHAVIOUR** Grass and scrub, overgrown clearings; 900–2,530 m. Skulking. **RANGE & STATUS** Resident Nepal, NE Indian subcontinent, SE Tibet, southern China, Sumatra. **SE Asia** Common resident W Myanmar. **BREEDING** March–June. **Nest** Globular structure with large side-entrance, in low undergrowth. **Eggs** 3–5; variable, often pale greenish, sometimes white or pink, speckled and blotched reddish (usually ringed at broader end); 16.9 x 12.7 mm (India).

1327 **HILL PRINIA** *Prinia superciliaris* Plate **120**
IDENTIFICATION 15–20.5 cm. *P.s.erythropleura*: **Adult non-breeding** From other prinias by combination of size, very long tail (longest in late winter/spring), unstreaked upperparts, distinct white supercilium, greyish head-sides and dark-streaked breast. **Adult breeding** Breast-streaking sparser and more spot-like; worn birds have shorter tails. *P.s.waterstradti* (Peninsular Malaysia) has darker, colder crown and upperparts, almost lacks supercilium and shows heavier breast-streaking (particularly at sides). **Juvenile** Like non-breeding adult but upperparts, tail and wings warmer, breast-streaking diffuse and smudgy. **Other subspecies in SE Asia** *P.s.superciliaris* (N Indochina), *klossi* (S Indochina). **VOICE** Sings with continuous, sprightly, mechanical 2–7 note ***TRIIH-TRIIH-TRIIH-TRIIH...*** (harsher and more stressed than Black-throated). Also utters a long series of spaced, clear, quite high notes: ***tHU-tHU-tHU-tHU-tHU...***, or even more spaced ***CHUU CHUU CHUU CHUU CHUU...*** or ***T'iUU T'iUU T'iUU T'iUU T'iUU...*** Duets of high, ringing ***t'EE*** and ***tHU*** either from pairs or competing males. **HABITAT & BEHAVIOUR** Grass and scrub, bracken-covered slopes, overgrown clearings; 900–2,565 m. Skulking. **RANGE & STATUS** Resident NE India (SE Arunachal Pradesh to Patkai Hills), southern China, Sumatra. **SE Asia** Common resident N,E,S(east) Myanmar, north Tenasserim, W,NW,NE Thailand, Peninsular Malaysia (Gunung Tahan), Laos, W,E Tonkin, C,S Annam, north Cochinchina. **BREEDING** March–June. **Nest** Globular structure with large side-entrance, in low undergrowth; up to 60 cm above ground. **Eggs** 3–5; variable, often pale to greyish-bluish, speckled and blotched reddish to reddish-brown (usually ringed at broader end). Brood-parasitised by Plaintive Cuckoo. **NOTE** Split from Black-throated Prinia, following Rasmussen & Anderton (2005).

Numbers in **bold** refer to plate numbers. Numbers in roman refer to species numbers, not page numbers.

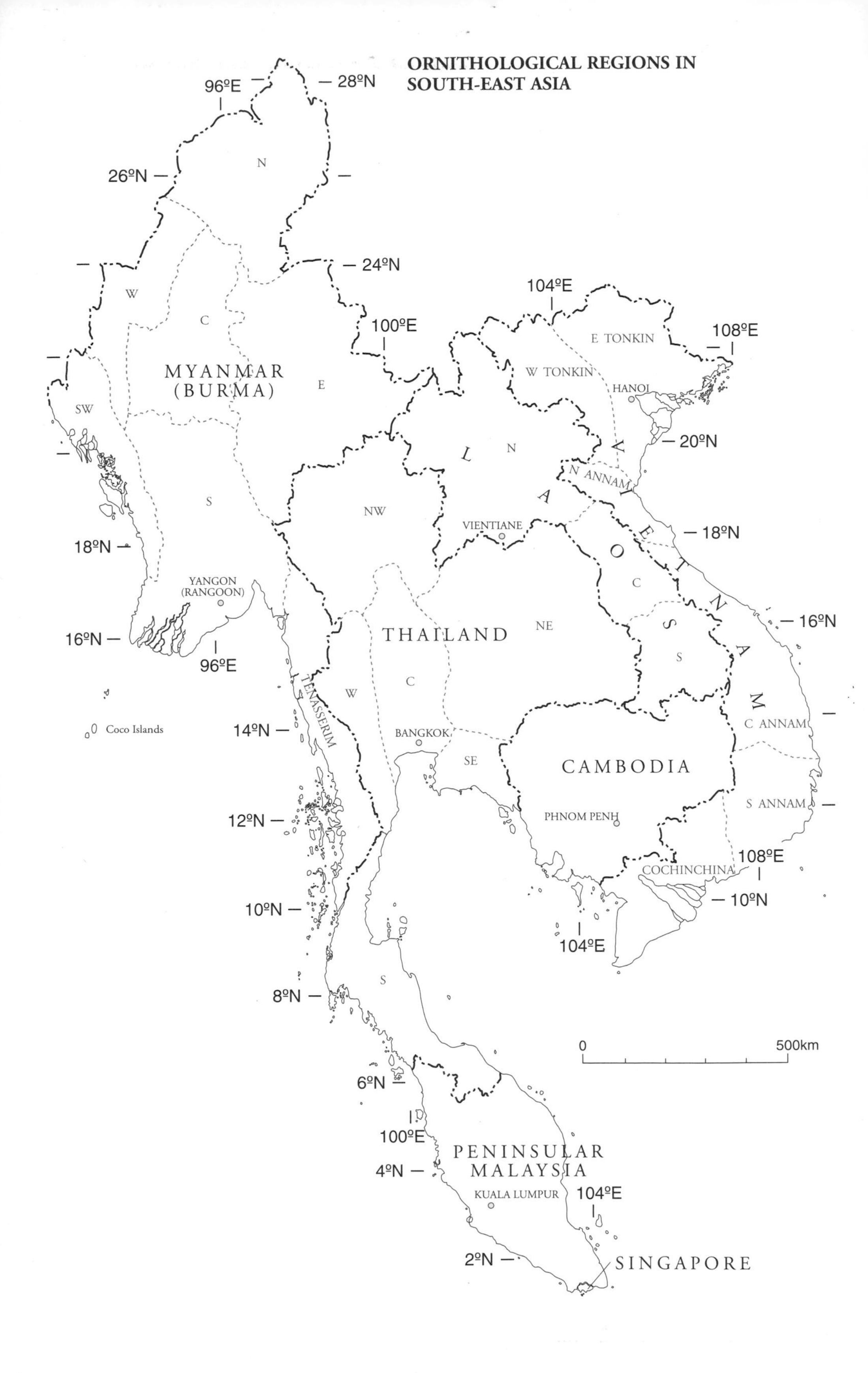

ORNITHOLOGICAL REGIONS IN SOUTH-EAST ASIA
96ºE
28ºN
N
26ºN
24ºN
W
C
E
MYANMAR (BURMA)
SW
S
18ºN
YANGON (RANGOON)
16ºN
96ºE
Coco Islands
104ºE
100ºE
E TONKIN
108ºE
W TONKIN
HANOI
20ºN
N
LAOS
N ANNAM
VIETNAM
NW
VIENTIANE
18ºN
C
16ºN
S
NE
THAILAND
S
TENASSERIM
W
C
14ºN
BANGKOK
C ANNAM
SE
CAMBODIA
12ºN
S ANNAM
PHNOM PENH
108ºE
COCHINCHINA
10ºN
10ºN
104ºE
S
8ºN
0
500km
6ºN
100ºE
PENINSULAR MALAYSIA
4ºN
KUALA LUMPUR
104ºE
2ºN
SINGAPORE